S0-AKY-016

Chemistry of Organic Fluorine Compounds II

Chemistry of Organic Fluorine Compounds II

A Critical Review

Miloš Hudlický, Editor
Virginia Polytechnic Institute and State University

Attila E. Pavlath, Editor
Agricultural Research Service,
U. S. Department of Agriculture

ACS Monograph 187

American Chemical Society
Washington, DC 1995

Library of Congress Cataloging-in-Publication Data

Chemistry of organic fluorine compounds II: a critical review / Miloš Hudlický, editor, Attila E. Pavlath, editor.

p. cm. — (ACS monograph, ISSN 0065–7719; 187)

Includes bibliographical references (p. -) and index.

ISBN 0–8412–2515–X

1. Organofluorine compounds. I. Hudlický, Miloš, 1919– . II. Pavlath, Attila E. III. Series.

QD305.H15C48 1995
547´.02—dc20

95–20195
CIP

This book is printed on acid-free paper.

PRINTED IN THE UNITED STATES OF AMERICA

1995 Advisory Board

FOREWORD

ACS MONOGRAPH SERIES was started by arrangement with the interallied Conference of Pure and Applied Chemistry, which met in London and Brussels in July 1919, when the American Chemical Society undertook the production and publication of Scientific and Technological Monographs on chemical subjects. At the same time it was agreed that the National Research Council, in cooperation with the American Chemical Society and the American Physical Society, should undertake the production and publication of Critical Tables of Chemical and Physical Constants. The American Chemical Society and the National Research Council mutually agreed to care for these two fields of chemical progress.

The Council of the American Chemical Society, acting through its Committee on National Policy, appointed editors and associates to select authors of competent authority in their respective fields and to consider critically the manuscripts submitted. The first Monograph appeared in 1921. Since 1944 the Scientific and Technological Monographs have been combined in the Series.

These Monographs are intended to serve two principal purposes; first, to make available to chemists a thorough treatment of a selected area in a form usable by persons working in more or less unrelated fields so that they may correlate their own work with a larger area of physical science; and second, to stimulate further research in the specific field treated. To implement this purpose the authors of Monographs give extended references to the literature.

ABOUT THE EDITORS

MILOŠ HUDLICKÝ, a native of Czechoslovakia, obtained his Ph.D. from the Technical University in Prague, Czechoslovakia. After spending 1948 at the Ohio State University as a UNESCO postdoctoral fellow, he taught as an assistant professor and later as an associate professor at the Technical University in Prague until 1958. He then worked as a research associate at the Research Institute of Pharmacy and Biochemistry in Prague. After the Russian occupation of Czechoslovakia in 1968, he moved to the United States, where he was offered a professorship at Virginia Polytechnic Institute and State University. He has been Professor Emeritus since 1989. He received the Votoček Medal in Prague in 1992 for his work in chemistry.

His field of interest is organofluorine chemistry. He has written 68 research papers, 19 review papers, 29 patents, and 15 books (9 in English). His chef-d'oeuvres are *Chemistry of Organic Fluorine Compounds, Reductions in Organic Chemistry,* and *Oxidations in Organic Chemistry*, published in 1976, 1984, and 1990, respectively.

ATTILA E. PAVLATH received his diploma as a Chemical Engineer from the Technical University of Budapest and his Ph.D. in chemistry from the Hungarian Academy of Science. He taught at his alma mater before he left Hungary after the 1956 revolution. After spending some time at McGill University in Montréal and Stauffer Chemical in Richmond, California, he joined the U.S. Department of Agriculture at Albany, California, in 1967 where he is a research leader.

He has published more than 100 research papers and holds 25 patents. He has written three books, one of them ACS Monograph 155, *Aromatic Fluorine Compounds,* in 1963. In 1976, he received the ACS California Section Award for Outstanding Contribution to Chemistry for his work of a quarter century on various aspects of fluorine chemistry. In addition to his scientific activities, he is also serving his second term on the ACS Board of Directors.

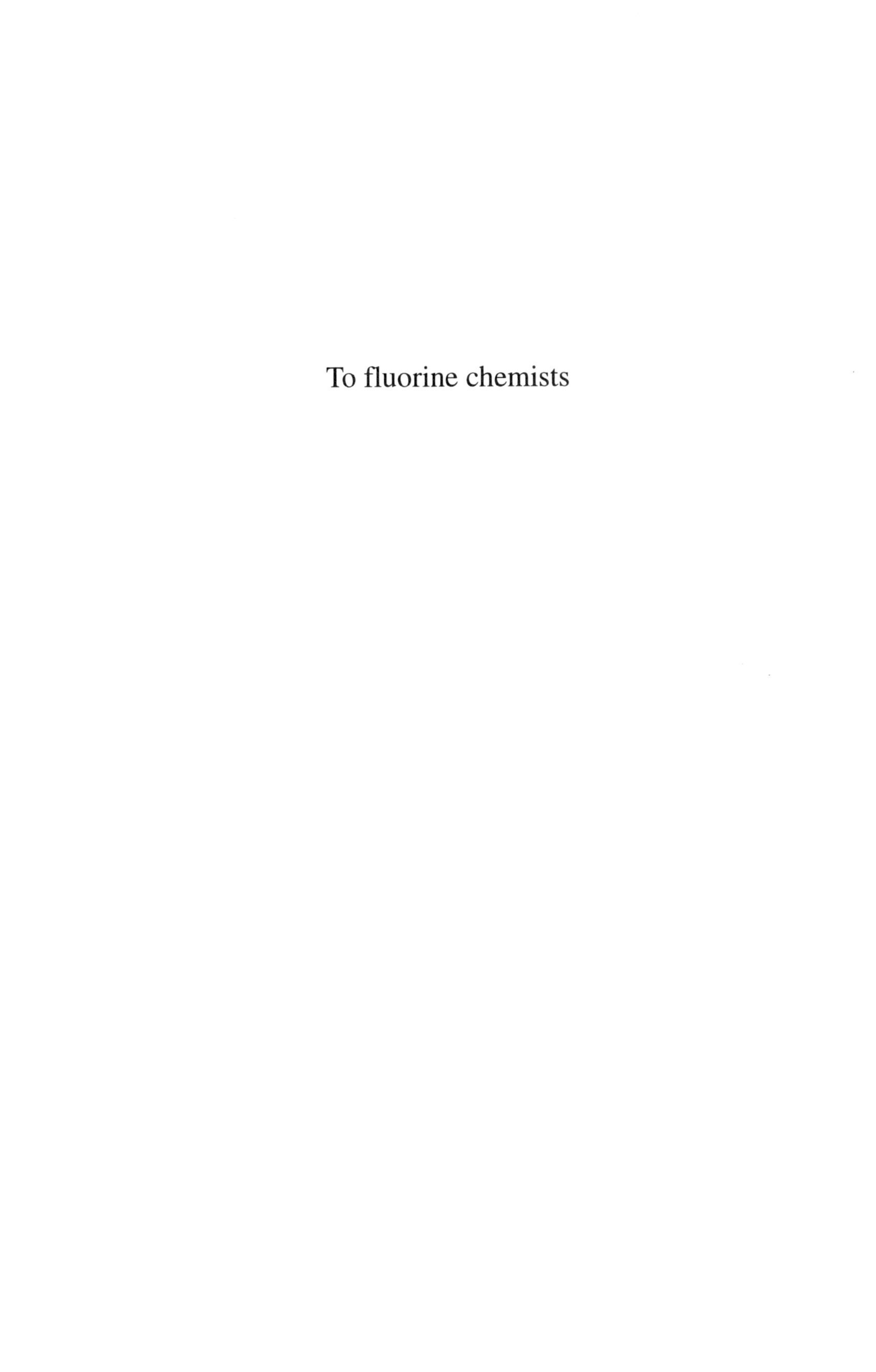

To fluorine chemists

CONTENTS

PREFACE

SINCE THE SECOND EDITION OF THE *Chemistry of Organic Fluorine Compounds,* Miloš Hudlický continued screening *Chemical Abstracts* to keep up with fluorine chemistry. Some 15,000–20,000 clippings accumulated in the file cabinets in his office. After his retirement, he was almost at the verge of discarding this voluminous material when it occurred to him that it could be used as a first-tier literature search for updating the book.

He felt that it would be beyond his capability to work up the enormous amount of data but that the task might be feasible for a team of co-workers. We are now presenting to the fluorine chemistry audience a result of a common effort of 44 contributors who have many years of experience in various fields of fluorine chemistry.

The book is not a new edition of the *Chemistry of Organic Fluorine Compounds* but an update of the 1976 edition with literature coverage from 1972 to 1991. The structure of the original book has been preserved, and it is assumed that its users are familiar with the 1976 edition, which was reprinted by Ellis Horwood in 1992 and to which numerous references are made. The book will be useful especially to organic fluorine chemists engaged in laboratory research.

Acknowledgments
In addition to the contributors, quite a few persons participated in the project. Acknowledgments are due to J. P. Chupp and R. Hakes of Monsanto for critical remarks and to Angie Miller, Wanda Ritter, and Janet LaPalometo for typing the manuscript.

Finally, we would like to express our thanks to the reviewers for their constructive suggestions and to ACS Books Department staff members Catherine Buzzell and Maureen Rouhi, cover designer Eileen Hoff, and indexer Aubrey McClellan for their meticulous work and cooperation.

MILOŠ HUDLICKÝ
Virginia Polytechnic Institute
and State University
Blacksburg, VA 24061–0212

ATTILA E. PAVLATH
Agricultural Research Service
U.S. Department of Agriculture
Albany, CA 94710

February 12, 1992

HOW TO USE THIS BOOK

The book is not a new edition of *Chemistry of Organic Fluorine Compounds*, which was published in 1976. It does not, as a rule, mention topics that were discussed or references up to the end of 1971 that appeared in the second edition of 1976, which has been out of print for a long time. It was reprinted unchanged in 1992 by Ellis Horwood, Chichester, England.

In order to locate items of interest in the book, the subject index lists, in addition to chemical operations, types of compounds rather than specific compounds (with a few exceptions). If, for example, readers do not find what they are looking for under the entry "fluoroolefins", they may try "olefins", "double bonds, additions of", etc.

To locate a citation of a reference in the text, look for the author's name in the author index. The full reference can be found on the page number that appears in **bold** in the first column; the reference number itself appears in *italics* in the second column; and the page(s) of the text where the reference is cited appear in roman type in the third column.

For example, reference *139*, whose full citation appears on page 53, is listed under Chambers, O. R. The number **53** appears in bold, and the number 48, which refers to the page in which reference *139* is cited in the text, is in roman type.

Safety Considerations

Hazardous reactions are pointed out in specific cases. In addition general safety rules for work with fluorine and fluorinated agents are thoroughly discussed on pages 25 and 26 of this book and on pages 13 and 14 in the 1976/1992 editions.

ABBREVIATIONS

CFC	chlorofluorocarbons
CNS	central nervous system
CTFE	chlorotrifluoroethylene
DABCO	1,4-diazabicyclo[2.2.2]octane
DBH	1,3-dibromo-5,5-dimethylhydantoin
DBU	1,8-diazabicyclo[5.4.0]undecane
DG	diethylene glycol
DHP	diheptyl phthalate
DMAC	*N,N*-dimethylacetamide
DMF	dimethylformamide
DMI	1,3-dimethyl-2-imidazolidinone
DMSO	dimethylsulfoxide
DN	donor number
ECF	electrochemical fluorination
ECTFE	chlorotrifluoroethylene-ethylene copolymers
ETFE	tetrafluoroethylene-ethylene copolymers
FEP	fluoroethylenepropylene
FPE	tetrafluoroethylene-hexafluoropropylene copolymers
HCFC	hydrochlorofluorocarbons
HFP	hexafluoropropylene
HFPO	hexafluoropropylene oxide
HMPA	hexamethylphosphoramide
HMPT	hexamethylphosphortriamide
LDA	lithium diisopropylamide
LHMDS	lithium hexamethyldisilazane
LICA	lithium dicyclohexylamide
LOI	limiting oxygen index
LTG	cryoenic zone reactor
MG	monoglyme
MTPA	Mosher's acid
NAFION	perfluororesinsulfonic acid
NMP	*N*-methylpyrrolidone
PCTFE	polychlorotrifluoroethylene
PFA	tetrafluoroethylene-perfluoro(propyl vinyl ether) copolymers
PHFPO	polyhexafluoropropylene oxide
PMVE	perfluoro(methyl vinyl ether)
PTC	phase transfer conditions
PTFE	polytetrafluoroethylene
PVF	poly(vinyl fluoride)
PVDF	poly(vinylidene fluoride)
SET	single electron transfer

TBAF	tetrabutylammonium fluoride
TBAH	tetrabutylammonium hydrogen sulfate
TDA	tris(3,6-dioxaheptyl)amine
TEBA, TEBAC	benzyltriethylammonium chloride
TEBAC	triethylbenzylammonium chloride
TFAP	2-trifluoroacetoxypyridine
TFAT	trifluoroacetyl triflate
TFE	tetrafluoroethylene
THF	tetrahydrofuran
THP	tetrahydropyran
TMBAC	trimethybenzylammonium chloride
TMS	trimethylsilyl, tetramethylsilane
TMSCl	trimethylsilyl chloride
TMSOTF	trimethylsilyl trifluoromethanesulfonate
VDF	vinylidene fluoride
VF	vinyl fluoride

PROCEDURES

CONTRIBUTORS

Adcock, James L., 97
Baucom, K. B., 736
Boudakian, Max M., 271
Brunner, E., 840
Burger, K., 840
Burton, Donald J., 670
Cantrell, G. L., 387, 398, 403, 408
Clark, L. C., 1138
Dmowski, W., 199, 263
Dolbier, W. R., 797
DuBoisson, R. A., 753
Elliott, Arthur J., 1119
Elsheimer, S. R., 364
Everett, T. Stephen, 1037
Feiring, Andrew E., 54, 61, 70
Ferstandig, L. L., 1133
Filler, R., 1011
Gumprecht, W. H., 422
Halpern, Donald F., 172, 1126
Hudlický, Miloš, 3, 120, 888
Kirk, K., 1011
Koroniak, H., 913
Krespan, C. G., 297
Lang, J. F., 387, 398, 403, 408, 525
Lang, R. W., 1143
Meshri, Dayal T., 25, 1023
Paleta, Oldrich, 321
Patrick, Timothy B., 133, 501
Pavlath, Attila E., 3, 888
Powell, R. L., 1089
Purrington, Suzanne T., 41
Robin, Mark L., 1029, 1099
Schierlinger, C., 840
Schmiegel, W. W., 1101
Sellers, S. F., 747
Sewald, N., 840
Smart, B. E., 767, 979
Soulen, R. L., 497, 757
Sprague, L. G., 729, 742
Stang, Peter J., 941
Tamborski, C., 646
Turner-McMullin, Seniz, 545
Vernice, Gerald G., 172
Wakselman, C., 446
Welch, J. T., 545, 615
Yang, Zhen-Yu, 670
Zhdankin, Viktor V., 941

Chapter 1

Survey of the Literature of Organic Fluorine Chemistry

Survey of the Literature of Organic Fluorine Chemistry

by Miloš Hudlický and Attila E. Pavlath

Like the number of fluorinated organic compounds, the number of monographs and reviews keeps increasing at a fast rate. From 1900 until 1960, only 22 monographs were published. During the next 14 years (1960 until 1974), 51 monographs were published. In the next 18 years (1974–1992), this number was 45. The average yearly production of monographs during these three periods are 0.36, 3.6, and 2.5, respectively.

International symposia on fluorine were held in 1976 in Kyoto, Japan; in 1979 in Avignon, France; in 1982 in Vancouver, Canada; in 1985 in Santa Cruz, California; in 1988 in East Berlin, East Germany; in 1991 in Bochum, Germany; and in 1994 in Yokohama, Japan. The 1997 symposium is planned for Vancouver, Canada.

The international fluorine symposia are paralleled by *European symposia:* in 1967 in Leicester, England; in 1968 in Göttingen, Germany; in 1970 in Aix-en-Provence, France; in 1972 in Ljubljana, Yugoslavia; in 1974 in Aviemore, Scotland; in 1977 in Dortmund, Germany; in 1980 in Venice, Italy; in 1983 in Jerusalem, Israel; in 1989 in Leicester, England; and in 1992 in Padua, Italy. The meeting in 1995 will be held in Bled, Slovenia. The series of the *Winter Fluorine Conferences* in Florida was held in 1971, 1974, and 1977 in St. Petersburg; in 1979, 1981, and 1983 in Daytona Beach; in 1985 in Orlando; and in 1987, 1989, 1991, and 1993 again in St. Petersburg. The meeting in 1995 also will be held in St. Petersburg.

Special symposia, outside these three series of symposia, include a special and unique *Moissan symposium,* held in 1986 in Paris at the hundred-year anniversary of the preparation of elemental fluorine; a symposium on *Synthetic Fluorine Chemistry* held in 1990 in Los Angeles; and a symposium on *Fluorinated Monomers and Polymers* held in 1993 in Prague, Czechoslovakia.

A special issue (Volume 6) of the *Bulletin de la Societé Chimique de France* published in 1986 contained papers contributed to the Moissan symposium. Abstracts of the papers presented at this symposium were published as a special volume (Volume 35) of the *Journal of Fluorine Chemistry* in 1987. A separate volume (Volume 54) of the *Journal of Fluorine Chemistry* was also issued in 1991 and contained abstracts of papers presented at the Thirteenth International Fluorine Symposium in Bochum, Germany.

The survey of the organic fluorine literature is subdivided into a section on monographs that is practically exhaustive and a section of review papers published in various journals. The review papers mentioned were selected on the basis of the

0065–7719/95/0187–0003$08.00/1

number of references to the original literature. With a few exceptions, only those reviews listing 40 or more references were included. A list of older reviews is in Sheppard and Sharts's *Organic Fluorine Chemistry* published in 1969 (W. A. Benjamin, New York), pages 487–510.

Following are the contents of the most recent volume of P. Tarrant's *Fluorine Chemistry Reviews.*

Contents of *Fluorine Chemistry Reviews,* Volume 8

- Robert Filler, The Pentafluorophenyl Group: Effects on Reactivity of Organic Compounds
- O. Paleta, Ionic Addition Reactions of Halomethanes with Fluoroolefins
- Friedhelm Aubke and Darryl D. DesMarteau, Halogen Derivatives of Group VIA Oxyacids
- Donald J. Burton and Jerry L. Hahnfeld, The Preparation and Reactions of Fluoromethylenes

Table 1. Fluorine Chemistry Monographs

Year	*Author(s) or Editor(s)*	*Book Title*	*Publisher*	*Place*	*No. of Pages*
1973	Belenkii, G. G.; Vlasov, V. M.; Grebenshchikova, G. F., et al.	*Syntheses of Organofluorine Compounds* (Russ.)	Khimiya	Moscow	312
1976	Hudlický, M.	*Chemistry of Organic Fluorine Compounds: A Laboratory Manual with Comprehensive Literature Coverage*	Ellis Horwood Halsted Press (John Wiley)	Chichester, U.K. New York	903
	Banks, R. E.; Barlow, M. G.	*Fluorocarbons and Related Chemistry,* Vol. 3	The Chemical Society	London	491
	Filler, R.	*Biochemistry Involving Carbon–Fluorine Bonds*	American Chemical Society	Washington, DC	214
1977	Tarrant, P.	*Fluorine Chemistry Reviews,* Vol. 8	Marcel Dekker	New York	206
	Emsley, J. W.; Phillips, L.; Wray, V.	*Fluorine Coupling Constants*	Pergamon Press	Oxford, U.K. New York	674
	Ishikawa, N.; Kobayashi, Y.	*Fluorine Compounds: Their Chemistry and Applications* (Jap.)	Kodasha	Tokyo	237
1978	Shiley, R. H.; Dickerson, D. R.; Finger, G. C.	*Aromatic Fluorine Chemistry at the Illinois State Geological Survey: Research Notes 1934–1976*	Illinois State Geological Survey	Urbana, IL	114
1979	Banks, R. E., Ed.	*Organofluorine Chemicals and Their Industrial Applications*	Ellis Horwood Halsted Press (John Wiley)	Chichester, U.K. New York	255

Continued on next page.

Table 1—Continued

Year	*Author(s) or Editor(s)*	*Book Title*	*Publisher*	*Place*	*No. of Pages*
1980	Webb, G. A.	*Annual Reports on NMR Spectroscopy,* Vol. 10B	Academic Press	New York	511
1981	Ishikawa, N., Ed.	*The Most Advanced Technologies of the Applications of Fluorine Compounds* (Jap.)	CMC	Tokyo	384
1982	Banks, R. E., Ed.	*Preparation, Properties and Industrial Applications of Organofluorine Compounds*	Ellis Horwood Halsted Press (John Wiley)	Chichester, U.K. New York	352
	Eisenberg, A.; Yeager, H. L., Eds.	*Perfluorinated Ionomer Membranes*	American Chemical Society	Washington, DC	500
	Filler, R.; Kobayashi, Y., Eds.	*Biomedical Aspects of Fluorine Chemistry*	Kodasha Elsevier Medical Press	Tokyo Amsterdam New York	246
	Ishikawa, N.; Kobayashi, Y.	*Fluorine Compounds: Their Chemistry and Applications* (Russ.)	MIR	Moscow	276
1983	Webb, G. A.	*Annual Reports on NMR Spectroscopy,* Vol. 14	Academic Press	New York	406
	Yakobson, G. G., Ed.	*Reactivity of Polyfluoro Aromatic Compounds* (Russ.)	Nauka, Sib. Otd.	Novosibirsk, U.S.S.R.	251
1985	Knunyants, I. L.; Yakobson, G. G., Eds.	*Syntheses of Fluoroorganic Compounds*	Springer–Verlag	Berlin	299

1985	Hagemüller, P.	*Inorganic Solid Fluorides: Chemistry and Properties*	Academic Press	New York	628
1986	Banks, R. E.; Sharp, D. W. A.; Tatlow, J. C., Eds.	*Fluorine: The First Hundred Years (1886–1986)*	Elsevier Sequoia	Lausanne, Switzerland New York	399
	Rakhimov, A. I.	*Chemistry and Technology of Organofluorine Compounds* (Russ.)	Khimiya	Moscow	272
1987	German, L. S.; Zemskov, S. V., Eds.	*New Fluorinating Agents in Organic Synthesis* (Russ.)	Nauka, Sib. Otd.	Novosibirsk, U.S.S.R.	257
	Ishikawa, N., Ed.	*Synthesis and Functions of Fluorine Compounds* (Jap.)	CMC	Tokyo	339
1988	Liebman, J. F.; Greenberg, A.; Dolbier, W. R., Jr., Eds.	*Fluorine-Containing Molecules*	VCH Publishers	New York	346
	Furin, G. G.; Zibarev, A. V.; Mazalov, L. N.; Yumatov, V. D., Eds.	*Electronic Structure of Organofluorine Compounds* (Russ.)	Nauka, Sib. Otd.	Novosibirsk, U.S.S.R.	260
	Yagupol'skii, L. M.	*Aromatic and Heterocyclic Compounds with Fluorine-Containing Substituents* (Russ.)	Naukova Dumka	Kiev, U.S.S.R.	320
	Ishikawa, N., Ed.	*The Technologies Coping with CFC/Halon Issue I. CFC Alternatives and Recovery/Recycle Technologies* (Jap.)	CMC	Tokyo	
	Watanabe, N.; Nakajima, T.; Touhara, H.	*Graphite Fluorides* (*Studies in Inorganic Chemistry,* Vol. 8)	Elsevier	Amsterdam New York	

Continued on next page.

Table 1—Continued

Year	*Author(s) or Editor(s)*	*Book Title*	*Publisher*	*Place*	*No. of Pages*
1989	German, L. S.; Zemskov, S., Eds.	*New Fluorinating Agents in Organic Synthesis*	Springer–Verlag	Berlin New York	283
	Ishikawa, N., Ed.	*The Technologies Coping with CFC/Halon Issue II* (Jap.)	CMC	Tokyo	205
1990	Bissell, E. R.	*The Chemistry of Aliphatic Fluoronitrocarbons*	2 5.25" IBM diskettes	E. R. Bissell 101 Via Lucia Alamo, CA 94507	N/A
	Ishikawa, N., Ed.	*Synthesis and Functions of Fluorine Compounds* (Russ.)	MIR	Moscow	405
	Ishikawa, N., Ed.	*90's Fluorine-Containing Bioactive Compounds: Development and Applications* (Jap.)	CMC	Tokyo	264
	Ishikawa, N., Ed.	*Present State and Future of the Development of CFC Alternatives* (Jap.)	The Chemical Daily	Tokyo	211
1991	Welch, J. T., Ed.	*Selective Fluorination in Organic and Bioorganic Chemistry*	American Chemical Society	Washington, DC	216
	Welch, J. T.; Eswarakrishnan, S.	*Fluorine in Bioorganic Chemistry*	John Wiley	New York	261
	Shteingarts, V. D.; Kobrina, L. S.; Bil'kis, I. I.; Starchenko, V. F.	*Chemistry of Polyfluoroarenes* (Russ.)	Nauka, Sib. Otd.	Novosibirsk, U.S.S.R.	272

1991	Kilbourn, M. R.	*Fluorine-18 Labeling Radiopharmaceuticals*	National Academy Press	Washington, DC	149
	Chambers, R. D.	*Fluorine in Organic Chemistry*	Reprinted by Aldrich Chemical Co.		391
1992	Olah, G. A.; Chambers, R. D.; Prakash, G. K. S.	*Synthetic Fluorine Chemistry*	John Wiley	New York	402
	Hudlický, M.	*Chemistry of Organic Fluorine Compounds: A Laboratory Manual with Comprehensive Literature Coverage*	Reprinted by Ellis Horwood	Chichester, U.K.	903

Table 2. Fluorine Chemistry as Part of Monographs and Journals

Year	*Author(s) or Editor(s)*	*Book Title*	*Publisher*	*Place*	*No. of Pages*
		Gmelin's Handbook of Inorganic Chemistry, 8th ed.	Verlag Chemie	Weinheim, Germany	
1973		Vol. 12: *Perfluorohaloorganic Compounds of Main Group Elements*. Part 2: *Compounds of Sulfur (Continuation), Selenium and Tellurium*			247
1975		Supplementary Work, Vol. 24: *Perfluorohaloorganic Compounds of Main Group Elements*. Part 3: *Phosphorus, Arsenic, Antimony and Bismuth Compounds*	Springer–Verlag	Berlin	233
1978		*Perfluorohaloorganic Compounds of Main Group Elements*. Part 5: *Compounds of Nitrogen (Heterocyclic Compounds)*	Springer–Verlag	Berlin	224
1984		System No. 5: *Fluorine Perfluorohaloorganic Compounds of Main Group Elements*, Supplementary, Vol. 1: *Compounds with Elements of Main Groups 1 to 5 (Excluding Nitrogen) and with Sulfur (Partially)*	Springer–Verlag	Berlin	380
1978	Agranat, I.; Selig, H., Eds.	*Isr. J. Chem.*, Special Issue	The Weizmann Science Press	Israel	161

1986	Flahaut, J., Ed.	*Bull. Soc. Chim. Fr.* Numéro Spécial 6	Société Chimique de France	Paris	150
1986	Banks, R. E.; Sharp, D. W. A.; Tatlow, J. C., Eds.	*J. Fluorine Chem.* Celebration Volume to Commemorate the Centenary of the Isolation of Fluorine by Moissan on 26th June 1886	Elsevier Sequoia	Lausanne, Switzerland	399
1987	Banks, R. E.; Sharp, D. W. A.; Tatlow, J. C., Eds.	*J. Fluorine Chem.*, Vol. 35 Abstracts of Papers presented at the Moissan Symposium in Paris, 1986	Elsevier Sequoia	Lausanne, Switzerland	266
1988		*Ullmann's Encyclopedia of Industrial Chemistry*	VCH	Weinheim New York	
		Fluorine Compounds, Organic, Vol. 11a			44
		Fluoropolymers, Organic, Vol. 11a			37
1991	Banks, R. E.; Sharp, D. W. A.; Tatlow, J. C., Eds.	*J. Fluorine Chem.*, Vol. 54 Abstracts of Papers presented at the 13th International Symposium on Fluorine Chemistry in Bochum, Germany, 1991	Elsevier Sequoia	Lausanne, Switzerland	425

Table 3. Selected Fluorine Chemistry Reviews

Year	*Title*	*Author(s)*	*Journal*	*Volume and Pages*	*Number of Citations*
1973	Synthesis of Fluorinated Carbohydrates	Foster, A. B.; Westwood, J. H.	*Pure Appl. Chem.*	*35,* 147–168	77
	Halo Compounds	Brooke, G. M.	*MPT Int. Rev. Sci. Org. Chem. Ser. One*	2, 65–89	182
1974	Nucleophilic Substitution in Polyfluoroaromatic Compounds	Kobrina, L. S.	*Fluorine Chem. Rev.*	*7,* 1–114	189
	Preparation and Reactions of Polyfluorinated Aromatic Heterocyclic Compounds	Yakobson, G. G.; Petrova, T. D.; Kobrina, L. S.	*Fluorine Chem. Rev.*	*7,* 115–223	285
	Fluorination by Sulfur Tetrafluoride	Boswell, G. A., Jr.; Ripka, W. C.; Scribner, R. M.; Tullock, C. W.	*Org. Reactions*	*21,* 1–124	183
	Modern Methods To Prepare Monofluoroaliphatic Compounds	Sharts, C. M.; Sheppard, W. A.	*Org. Reactions*	*21,* 125–146	386
	5-Fluorouracil: Review of the Chemistry and Applications	Anderson, R.	*Am. Lab.*	*6,* 36–46	65
1975	Preparation and Application of Aliphatic Fluorine Compounds. I. Halides, Olefins, Alcohols, and Cyclobutanes (Ger.)	Liebig, H.; Ulm, K.	*Chem. Ztg.*	*99,* 477–485	180
	Advances in the Synthesis and Study of Fluoroorganic Compounds (Russ.)	Knunyants, I. L.; Polishchuk, V. R.	*Usp. Khim.*	*44,* 685–714	277

1976	Production and Uses of Aliphatic Compounds. II. Ether, Epoxide and Polyether, Carboxylic Acids and Their Derivatives, Sulfonic Acids, Toxicological Data of Aliphatic Fluorine Compounds (Ger.)	Liebig, H.; Ulm, K.	*Chem. Ztg.*	*100,* 3–14	270
	Aromatic Compounds with Fluorine-Containing Substituents (Russ.)	Yagupol'skii, L. M.; Kondratenko, N. V.	*Zh. Vses. Khim. O-va*	*21,* 299–306	114
	Methods for Synthesis of Organic Compounds with Nitrogen–Fluorine Bonding (Russ.)	Fokin, A. V.; Studnev, Y. N.; Kuznetsova, L. G.	*Reakts. Metody Issled. Org. Soedin.*	*24,* 7–466	819
	New Data on the Reactions of Organofluorine Compounds (Russ.)	Knunyants, I. L.; Polishchuk, V. R.	*Usp. Khim.*	*45,* 1139–1176	280
	Perfluoro-*t*-butyl Anion in the Synthesis of Organofluorine Compounds (Russ.)	Dyatkin, B. L.; Delyagina, N. I.; Sterlin, S. R.	*Usp. Khim.*	*45,* 1205–1221	70
	Fluorinated Peroxides	Shreeve, J. M.	*Endeavour*	*35,* 79–82	19
	Recent Synthetic Methods for Polyfluoroaromatic Compounds	Yakobson, G. G.; Vlasov, V. M.	*Synthesis*	652–672	220
	The Application of Thermolytic Reactions for the Syntheses of Fluoroorganic Compounds	Platonov, V. E.; Yakobson, G. G.	*Synthesis*	374–384	88
1977	The Pentafluorophenyl Group: Effects on Reactivity of Organic Compounds	Filler, R.	*Fluorine Chem. Rev.*	*8,* 1–37	111

Continued on next page.

Table 3—Continued

Year	Title	Author(s)	Journal	Volume and Pages	Number of Citations
1977	Ionic Addition Reactions of Halomethanes with Fluoroolefins	Paleta, O.	*Fluorine Chem. Rev.*	*8*, 39–71	86
	Halogen Derivatives of Group VI A Oxyacids	Aubke, F.; DesMarteau, D. D.	*Fluorine Chem. Rev.*	*8*, 73–118	165
	The Preparation and Reactions of Fluoromethylenes	Burton, D. J.; Hahnfeld, J. L.	*Fluorine Chem. Rev.*	*8*, 119–188	290
	Fluorinated Isocyanates and Their Derivatives as Intermediates for Biologically Active Substances (Ger.)	Kuehle, E.; Klauke, E.	*Angew. Chem.*	*89*, 797–804	47
	Radical Reactions of Polyfluoroaromatic Compounds (Russ.)	Kobrina, L. S.	*Usp. Khim.*	*46*, 660–684	101
	Organic *N*-Fluoroimides (Russ.)	Fokin, A. V.; Uzun, A. T.; Stolyarov, V. P.	*Usp. Khim.*	*46*, 1995–2026	184
	Nucleophilic Substitution	Brooke, G. M.	*Aromat. Heteroaromat. Chem.*	*5*, 314–338	194
	Aromatic Substitution by Free Radicals, Carbenes, and Nitrenes	Challand, S. R.	*Aromat. Heteroaromat. Chem.*	5, 339–358	78
1978	Vinyl Triflate Chemistry: Unsaturated Cations and Carbenes	Stang, P. J.	*Acc. Chem. Res.*	*11*, 107–114	57
	Aromatic Fluorine Chemistry at the Illinois State Geological Survey: Research Notes, 1934–1976	Shiley, R. H.; Dickerson, D. R.; Finger, G. C.	*Ill. State Geol. Surv. Circ.*	*501*, 1–114	47

1978	Tri- and Tetracoordinate Fluorosulfur(IV) and Pentacoordinate Fluorosulfur (VI) Compounds	Shreeve, J. M.	*Isr. J. Chem.*	*17,* 1–10	107
	Reactions of Electropositive Chlorine Compounds with Fluorocarbons	Schack, C. J.; Christe, K. O.	*Isr. J. Chem.*	*17,* 20–30	99
	Novel Methods for Selective Fluorination of Organic Compounds: Design and Synthesis of Fluorinated Antimetabolites	Kollonitsch, J.	*Isr. J. Chem.*	*17,* 53–59	50
	Application of Fluoroxy Compounds to Organic Synthesis: Electrophilic Fluorination of Unsaturated Molecules	Hesse, R. H.	*Isr. J. Chem.*	*17,* 60–70	43
	Reactions of Organic Compounds with Xenon Fluorides	Filler, R.	*Isr. J. Chem.*	*17,* 71–79	56
	Intramolecular Nucleophilic Displacement of Fluorine	Hudlický, M.	*Isr. J. Chem.*	*17,* 80–91	68
	Photochemistry of Fluorosubstituted Aromatic and Heteroaromatic Molecules	Zupan, M.; Stret, B.	*Isr. J. Chem.*	*17,* 92–99	63
	Pentafluorosulfur Peroxides: Synthesis, Properties, and Vibrational Spectra	DesMarteau, D. D.; Hammaker, R. M.	*Isr. J. Chem.*	*17,* 103–113	74
1980	Polyfluoroaromatic Monocarbonyl Compounds (Russ.)	Gerasimova, T. N.; Fokin, E. P.	*Usp. Khim.*	*49,* 1057–1078	148
1980	Advances in the Chemistry of Fluoroorganic Hypohalites and Related Compounds (Russ.)	Mukhametshin, F. M.	*Usp. Khim.*	*49,* 1260–	1288

Continued on next page.

Table 3—Continued

Year	*Title*	*Author(s)*	*Journal*	*Volume and Pages*	*Number of Citations*
1981	Fluorination Methods in Organic Chemistry (Ger.)	Gerstenberger, M. R. C.; Haas, A.	*Angew. Chem.*	*93,* 659–680	403
	Fluoro-Containing β-Diketones (Russ.)	Pashkevich, K. I.; Saloutin, V. I.; Postovskii, I. Y.	*Usp. Khim.*	50, 325–354	92
	Fluoro-Containing Halomethylene-phosphoranes (Russ.)	Tyuleneva, V. V.; Rokhlin, E. M.; Knunyants, I. L.	*Usp. Khim.*	*50,* 522–713	80
	The Trifluoromethyl Group in Chemistry and Spectroscopy: Carbon–Fluorine Hyperconjugation	Stock, L. M.; Wasiliewski, M. R.	*Prog. Phys. Org. Chem.*	*13,* 253–313	135
	Valence Bond Isomers of Aromatic Compounds Stabilized by Trifluoromethyl Groups	Kobayashi, Y.; Kumadaki, I.	*Acc. Chem. Res.*	*14,* 76–82	46
	Polyfluoroheteroaromatic Compounds	Chambers, R. D.; Sargent, C. R.	*Adv. Heterocycl. Chem.*	*28,* 1–71	381
	Fluoroaldehyde Polymers	Neeld, K.; Vogl, O.	*Macromol. Rev.*	*16,* 1–40	138
1982	Fluorine-Containing α-Dicarbonyl Compounds and Their Derivatives (Russ.)	Saloutin, V. L.; Pashkevich, K. I.; Postovskii, I. Y.	*Usp. Khim.*	*51,* 1287–1304	97
	Organic Synthesis Involving Fluorine-18	Tewson, T. J.	*App. Nucl. Radiochem.*	163–183	60
	Perfluoroalkanesulfonic Esters: Methods of Preparation and Applications in Organic Chemistry	Stang, P. J.; Hannack, M.; Subramanian, L. R.	*Synthesis*	85–126	263

1983	Synthesis and Properties of Aliphatic Fluoronitroso Compounds (Russ.)	Knunyants, I. L.; Sizov, Y. A.; Ukharov, O. V.	*Usp. Khim.*	*52,* 976–1017	304
	Fluorine-Containing Dyes (Russ.)	Yagupol'skii, L. M.; Ilchenko, A. Y.; Gandelsman, L. Z.	*Usp. Khim.*	*52,* 1732	271
	Fluorocarbons	Smart, B. E.	*Chem. Halides, Pseudo-Halides, Azides*	*1,* 603–655	398
	Trifluoromethyl Derivatives of the Transition Metal Elements	Morrison, J. A.	*Adv. Inorg. Chem. Radiochem.*	*27,* 293–316	113
1984	Synthesis and Properties of Aryl Polyfluoromethyl Ethers and Thioethers (Fr.)	Langlois, B.; Desbois, M.	*Ann. Chim. (Paris)*	*9,* 729–741	32
	Present Status of Organofluorosilicate Chemistry (Ger.)	Mueller, R.	*Z. Chem.*	*24,* 41–51	70
	Halogen Fluorides in Organic Syntheses (Russ.)	Boguslavskaya, L. S.	*Usp. Khim.*	*53,* 2024–2055	160
	Chemistry of Perfluoroisobutylene (Russ.)	Zeifman, Y. V.; Ter-Gabryelian, E. G.; Gambaryan, N. P.; Knunyants, I. L.	*Usp. Khim.*	*53,* 431–461	199
	Fluorocarbon Iodides: Versatile Reagents	Tarrant, P.	*J. Fluorine Chem.*	*25,* 69–74	24
	The Synthesis and Biology of Fluorinated Prostacyclins	Barnette, W. E.	*CRC Crit. Rev. Biochem.*	*15,* 201–235	97
1984	Fluorination by Sulfur Tetrafluoride	Wang, C.-L. J.	*Org. Reactions*	*34,* 319–400	110
1985	Preparative Fluorinations with Molecular Fluorine	Vyplel, H.	*Chimia*	*39,* 305–311	36

Continued on next page.

Table 3—Continued

Year	Title	Author(s)	Journal	Volume and Pages	Number of Citations
	Chemistry of Organofluorosilanes (Ger.)	Mueller, R.	*Z. Chem.*	*25,* 421–427	85
	Fluorine-Containing α-Ketoesters (Russ.)	Pashkevich, K. I.; Saloutin, V. I.	*Usp. Khim.*	*54,* 1997–2026	157
	Modern Synthetic Procedures for the Fluorination of Organic Molecules	Haas, A.; Lieb, M.	*Chimia*	*39,* 134–140	128
	α-Fluorocarbonyl Compounds and Related Chemistry	Rozen, S.; Filler, R.	*Tetrahedron*	*41,* 1111–1153	232
	Synthesis of 1,2-Disubstituted Polyfluorobenzenes	Gerasimova, T. N.; Orlova, N. A.	*J. Fluorine Chem.*	*28,* 361–380	105
1986	New Trends in Organofluorine Chemistry (Fr.)	Normant, J. F.; Wakselman, C.	*Bull. Soc. Chim. Fr.*	858–860	89
	Perfluorinated Resinsulfonic Acid (Nafion-H) Catalysis in Synthesis	Olah, G. A.; Iyer, P. S.; Prakash, G. K. S.	*Synthesis*	513–531	181
	Application of Elemental Fluorine in Organic Synthesis	Purrington, S. T.; Kagan, B. S.; Patrick, T. B.	*Chem. Rev.*	*86,* 997–1018	150
	Homolytic Arylation of Aromatic and Polyfluoroaromatic Compounds	Bolton, R.; Williams, G. H.	*Chem. Soc. Rev.*	*15,* 261–289	122
1986	Fluorinated Organic Molecules	Smart, B. E.	*Mol. Struct. Energ.*	*3,* 141–191	241
1987	Perfluoromethylarsines (Ger.)	Kober, F.	*Chem. Ztg.*	*111,* 127–134	112

	Recent Advances in the Chemistry of Haloperfluoroalkanes	Wakselman, C.	*Actual. Chim.*	*5,* 137–141	62
	Advances in the Preparation of Biologically Active Organofluorine Compounds	Welch, J. T.	*Tetrahedron*	*43,* 3123–3197	198
	Catalysis with Nafion	Waller, F. J.; Van Scoyoc, R. W.	*ChemTech*	*17,* 438–441	38
	Fluoro Carbanions	Chambers, R. D.; Bryce, M. R.	*Stud. Org. Chem. (Amsterdam) (Compr. Carbanion Chem. Pt. C)*	*5,* 271–319	128
	Modern Methods for the Introduction of Fluorine into Organic Molecules: An Approach to Compounds with Altered Chemical and Biological Activities	Mann, J.	*Chem. Soc. Rev.*	*16,* 381–436	55
1988	Fluorination with Diethylaminosulfur Trifluoride and Related Aminofluoro-sulfuranes	Hudlický, M.	*Org. Reactions*	*35,* 513–637	173
	Elemental Fluorine as a Legitimate Reagent for Selective Fluorination of Organic Compounds	Rozen, S.	*Acc. Chem. Res.*	*21,* 307–312	74
	Fluoroaromatic Compounds: Synthesis, Reactions, and Commercial Applications	Hewitt, C. D.; Silvester, M. J.	*Aldrichimica Acta*	*21,* 3–10	103
	New Aspects of Carbonylations Catalyzed by Transition Metal Complexes	Ojima, I.	*Chem. Rev.*	*88,* 1011–1030	52
	Polyfluoroaromatics: An Excursion in Carbanion Chemistry	Filler, R.	*J. Fluorine Chem.*	*40,* 387–405	35

Continued on next page.

Table 3—Continued

Year	*Title*	*Author(s)*	*Journal*	*Volume and Pages*	*Number of Citations*
1988	Fluorine-Containing Cyclohexadienones: Synthesis and Properties	Kobrina, L. S.; Shteingarts, V. D.	*J. Fluorine Chem.*	*41*, 111–162	122
	(Fluoroalkenyl)phosphonates	Kadyrov, A. A.; Rokhlin, E. M.	*Usp. Khim.*	*57*, 1488–1509	96
	The Electrocyclic Ring Opening of Fluorinated Cyclobutene Derivatives	Dolbier, W. R., Jr.; Koroniak, H.	*Mol. Struct. Energ.*	*8*, 65–81	51
	Fluorine-Substituted Analogs of Nucleic Acid Components	Bergstrom, D. E.; Swartling, D. J.	*Mol. Struct. Energ.*	*8*, 250–306	133
1989	Fluorine-Containing Organic Derivatives of Polyvalent Halogens	Maletina, I. I.; Orda, V. V.; Yagupol'skii, L. M.	*Usp. Khim.*	*58*, 925–950	162
	Some Peculiarities of Radical Reactions of Polyfluoroaromatic Compounds	Kobrina, L. S.	*J. Fluorine Chem.*	*42*, 301–344	73
	Catalytic Hydrogenolysis of Carbon–Fluorine Bonds: π-Bond Participation Mechanism	Hudlický, M.	*J. Fluorine Chem.*	*44*, 345–359	58
	Fluorine-Substituted Carbocations	Allen, A. D.; Tidwell, T. T.	*Adv. Carbocation Chem.*	*1*, 1–44	173
	Carbocations Destabilized by Electron-Withdrawing Groups: Applications in Organic Chemistry	Charpentier-Marize, M.; Bonnet-Delpon, D.	*Adv. Carbocation Chem.*	*1*, 219–253	54
1990	Vibrational Spectra of Polyfluoroaromatic Compounds	Korobeinicheva, I. K.; Fugaeva, O. M.; Furin, G. G.	*J. Fluorine Chem.*	*46*, 179–209	72

	Kinetics of Nucleophilic Substitution Reactions of Polyfluoroaromatic Compounds	Rodionov, P. P.; Furin, G. G.	*J. Fluorine Chem.*	*47,* 361–434	105
	Methods of Introduction of Fluorine into Aromatic and Nitrogen-Containing Heterocyclic Compounds (Czech.)	Hradil, P.; Radl, S.	*Chem. Listy*	*84,* 952–969	113
	Recent Progress in Perfluoroalkylation by Radical Species with Special Reference to the Use of Bis(perfluoroalkanoyl)peroxides	Yoshida, M.; Kamigata, N.	*J. Fluorine Chem.*	*49,* 1–20	62
	Preparation and Properties of Chiral Fluoroorganic Compounds	Bravo, P.; Resnati, G.	*Tetrahedron: Asymmetry*	*1,* 661–692	300
	Aliphatic Fluoronitro Compounds	Adolph, H. G.; Koppes, W. M.	*Nitro Compd.*	367–605	597
	Kinetics of Nucleophilic Substitution Reactions of Polyfluoro Aromatic Compounds (Russ.)	Rodionov, P. P.; Furin, G. G.	*Izv. Sib. Otd. Akad. Nauk SSSR*	3–26	87
1990	Nucleophilic Substitution in Aromatic Compounds with Fluorinated Substituents (Russ.)	Boiko, V. N.	*Izv. Sib. Otd. Akad. Nauk SSSR*	126–136	53
1991	Synthesis of Ftorafur (Russ.)	Lukevits, E.; Zabolatakya, A.	*Khim. Geterotsikl. Soedin.*	1590–1620	204
	Preparation of Trifluoromethyl Ketones and Related Fluorinated Ketones	Begue, J. P.; Bonnet-Dalpon, D.	*Tetrahedron*	*47,* 3207–3258	245
	Cycloadditions of Fluoroallene and 1,1-Allene	Dolbier, W. R., Jr.	*Acc. Chem. Res.*	*24,* 62–69	36

Continued on next page.

Table 3—Continued

Year	Title	Author(s)	Journal	Volume and Pages	Number of Citations
1991	Perfluoroalkanesulfonic Acids and Their Derivatives (Russ.)	Huang, W; Chen, Q.	*Chem. Sulphonic Acids, Esters, Their Derivatives*	903–946	157
	Fluorine-Containing Aromatic Amino Acids (Russ.)	Kukhar, V. P.; Yagupol'skii, Y. L.; Gerus, I. I.; Kolycheva, M. T.	*Usp. Khim.*	*60*, 2047–2063	100
	Ene Reaction of Trifluoromethyl Carbonyl Compounds (Jap.)	Nagai, T.; Kumadaki, I.	*Yuki Gosei Kagaku Kyokaishi*	*49*, 624–635	50
	Fluorine-Stabilized Sulfur–Carbon Multiple Bonds	Seppelt, K.	*Angew. Chem.*	*103*, 399–413	232
	Recent Advances in Trifluoromethylation (Jap.)	Uneyama, K.	*Yuki Gosei Kagaku Kyokaishi*	*49*, 612–623	169
1992	Fluorinated Organometallics: Perfluoroalkyl and Functionalized Perfluoroalkyl Organometallic Reagents in Organic Synthesis	Burton, D. J.; Yang, Z. Y.	*Tetrahedron*	*48*, 189–275	297
	Trifluoromethylations and Related Reactions in Organic Chemistry	McClinton, D. A.	*Tetrahedron*	*48*, 6555–6666	228
1992	Fluoride Ion as Nucleophile and a Leaving Group in Aromatic Nucleophilic Substitution Reactions	Vlasov, V. M.	*J. Fluorine Chem.*	*61*, 193–216	77
	Recent Advances in the Selective Formation of the Carbon–Fluorine Bond	Wilkinson, J. A.	*Chem. Rev.*	*92*, 505–519	184
	Stereoselective Preparation of Trifluoromethylated Organic Molecules	Yamazaki, T.; Kitazume, T.	*Rev. Heteroat. Chem.*	*7*, 132–148	48

Chapter 2

Survey of Fluorinating Agents

Survey of Fluorinating Agents

by Dayal T. Meshri

As far as handling of fluorinating agents is concerned, no dramatic changes have occurred since the publication of *Chemistry of Organic Fluorine Compounds* in 1976. Reiteration of basic safety regulations, however, is in order.

Safety Advisory

The safety information is included within this book by the American Chemical Society as a precaution to the readers. The book discusses the preparation and use of various fluorinating materials. The extreme caution required when preparing and using these reagents cannot be overemphasized. The hazards generally associated with the preparation, handling, and use of fluorinating agents are explosion, fire, and uncontrolled and violent exothermic reaction that may result in equipment failure and physical injury to the operator. To minimize the potential for any of these occurrences, chemists must prudently exercise all safe practices at their disposal. All new reactions should be conducted on the smallest possible scale and with clean glassware or metalware that has been carefully inspected for defects. Supervision by a chemist experienced in this field is desirable. If possible, all reactions should be conducted in an adequate laboratory fume hood with the sash of laminated glass closed. This measure will provide the operator with a physical barrier against fires, explosions, and chemical splashes, as well as facilitate the removal of potentially toxic, flammable, or offensive vapors. The laboratory should be equipped with the basic safety facilities necessary to handle and use corrosive agents, including approved and periodically inspected and tested eyewash stations, safety showers, and fire suppression equipment. To minimize personal exposure, appropriate and adequate personal protection equipment must be worn when conducting the experiments. Essential safety apparel includes chemical splash goggles; full face shield; chemically impervious neoprene gloves; and laboratory coats or, better still, rubber aprons.

Before following procedures outlined in this book, readers should refer to the original references for any specific handling practice that may be cited. Material safety data sheets also should be reviewed for any additional handling precautions.

A special precaution is imperative in handling elemental fluorine and chlorine trifluoride. Both gases are extremely corrosive and so reactive that they are very seldom used without dilution with inert gases such as nitrogen, helium, or argon. Fluorine is now available in mixtures with nitrogen. *Fluorine* is very toxic in

0065–7719/95/0187–0025$08.00/1

concentrations above ~1 ppm, but because it is perceptible in very small quantities, inhalation of toxic doses can be avoided. It is an irritant to the eyes and mucous membranes. Eyes should be irrigated for at least 15 min with large amounts of gently flowing cold water, and an aftertreatment with 0.5% Pontocaine hydrochloride solution (Winthrop Laboratories, New York, NY; tetracaine hydrochloride) or similar local anesthetic is recommended. No ointments are to be applied unless instructed to do so by a physician. The same treatment is recommended for eyes injured by fumes of *hydrogen fluoride, sulfur tetrafluoride,* and other similar volatile fluorides.

Skin affected by fluorine, hydrogen fluoride, or sulfur tetrafluoride vapors should be immediately and thoroughly washed with water and a solution of iced, 0.21% water solution of Hyamine 1622 (Lonza, Inc., Fairlawn, NJ; benzethonium chloride) or iced, 0.13% water solution of Zephiran (Winthrop Laboratories; benzalkonium chloride) and calcium gluconate should be applied externally afterwards. If a large area of skin is affected, it is even better to apply the calcium gluconate solution subcutaneously. The application of calcium gluconate is especially desirable for burns caused by liquid, anhydrous, or highly concentrated *hydrogen fluoride.* Again, immediate washing with water is imperative. Some fluorides of peroxide nature are potentially explosive and should be handled with special care: *trifluoromethyl hypofluorite, difluoromethylene bis(hypofluorite), acetyl and trifluoroacetyl hypofluorites, and especially perchloryl fluoride.* Commercial production of perchloryl fluoride has been discontinued. When working with perchloryl fluoride, reaction exit gases must not be allowed to condense in a dry ice or liquid nitrogen trap, because perchloryl fluoride may explode on contact with organic material.

All fluorinating agents should be considered toxic in different amounts and, therefore, handled accordingly. Nonvolatile fluorides, however, are not too dangerous in this respect, because it is unlikely that they will be swallowed or that they will penetrate into the blood stream. What is extremely dangerous is inhalation of volatile fluorides, that is, gases, liquids, or solids with considerable vapor pressure. Such fluorides are indicated in the tables in this chapter.

Many fluorides are corrosive to glass and some metals. Even the very corrosive anhydrous fluorides, however, can be handled at room temperature in steel, stainless steel, copper, and Teflon equipment. For reactions at higher temperature, nickel and Monel metal are indispensable. More data on corrosion of materials are given in pages 22 and 23 of *Chemistry of Organic Fluorine Compounds*, published in 1976.

Most of the fluorinating agents in use at present have been described fully in *Chemistry of Organic Fluorine Compounds* (1976). In this chapter, only corrections and changes of applications are listed. There are, however, more recently developed fluorinating agents, especially fluorinated organic compounds, which are usually more selective than the widely used inorganic fluorinating agents. It is especially these fluorinating compounds that are listed in the tables that follow.

Table 1. Inorganic Agents

Fluorinating Agent	*Method of Preparation*	*Yield (%)*	*Properties*	*Ref.*
$PbF_2(OCOCH_3)_2$	$Pb(OCOCH_3)_4$ (25 g), HF^a (4.4 mL), $CHCl_3$ (50 mL), 0 °C	90	White solid, mp, 190–210 °C decomposes	1
$[NF_4]^+[BF_4]^-$	NF_3 (26 mmol), F_2 (9 mmol), BF_3 (27 mmol), –196 °C	62[b]	White hygroscopic solid; mp, >150 °C decomposes	2, 3
$CsSO_4F$	Cs_2SO_4 (8 mL, aq. 2 M), F_2/N_2 (20%, 2 g), –4 °C	50	Detonates at 100 °C	4, 5
$RbSO_4F$	Rb_2SO_4 (8 mL, aq. 1.3 M), F_2/N_2 (20%, 0.76 g), –4 °C	50	mp, decomposes	
$MCoF_4$ (Li, Rb, Cs)	$MCoCl_3$, F_2, 250–500 °C	98		6
$MCoF_4$ (Na, K)	$MCoF_3$, F_2, 250–500 °C	98–99		6
$MCoF_4$ (Na, Cs, K, Rb)	MF, CoF_2, F_2, 250–300 °C	99–100	Decomposes at 350 °C	7[c]
WF_6[d]	W (300 mesh), F_2, 220 °C	98–100	mp, 2.3 °C; bp, 17.5 °C; *d*, 3.441 at 15 °C	7[c]
$[XeF]^+[SbF_6]^-$	XeF_2, SbF_5	~100	mp, 70–75 °C; bp, 220 ± 5 °C	8, 9
$[XeF]^+[Sb_2F_{11}]^-$	XeF_2, $2SbF_5$	~100		10, 11

[a]Anhydrous hydrogen fluoride.
[b]Yield based on fluorine.
[c]The compound is commercially available.
[d]The compound has considerable vapor pressure.

References are listed on pages 36–38.

Table 2. Ammonium Fluorides

Fluorinating Agent	*Method of Preparation*	*Yield (%)*	*Properties*	*Ref.*
$(C_2H_5)_3N \cdot 3HF$	$(C_2H_5)_3N$, HF[a], Et_2O	~100	bp, 78 °C at 1.5 mm Hg; 68 °C at 0.2 mm Hg	14, 15
$(C_2H_5)_3N \cdot 2HF$	$2(C_2H_5)_3N$, 3HF, $(C_2H_5)_3N$, evaporate at 15 mm Hg	~100	Hygroscopic, decomposes at 65 °C	16
$(C_2H_5)_3N \cdot HF$	$2(C_2H_5)_3N$, 3HF, $2(C_2H_5)_3N$	~100	Hygroscopic	16
$(C_4H_9)_4NF \cdot xH_2O$	$(C_4H_9)_4NOH$, HF, 0 °C	70–80	mp, 62–63 °C	7[b]
$(C_4H_9)_4NF \cdot 2HF$	$(C_4H_9)_4NCl$, $NH_4F \cdot HF$, IRA-40		Decomposes at >140 °C	17
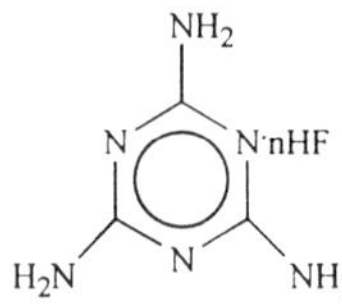 NH_2, N, N·nHF, H_2N, N, NH_2 (triazine structure)	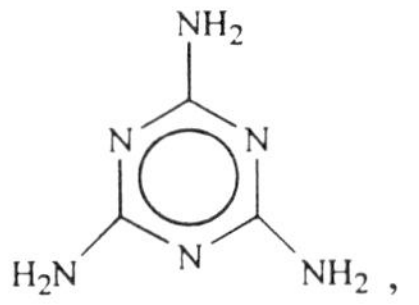 NH_2, N, N, H_2N, N, NH_2 (triazine structure), HF (24–40 mol), Ø °C	~100	Liquid	18

[a]Anhydrous hydrogen fluoride.
[b]The compound is commercially available.

Table 3. Phosphonium Fluorides

Fluorinating Agent	*Method of Preparation*	*Yield (%)*	*Properties*	*Ref.*
$(C_4H_9)_4PF \cdot 2HF$	$(C_4H_9)_4POH$ (40%, aq.), 3HF[a] (47%), 5–20 °C	~100	Colorless	43
$(C_4H_9)_4PF \cdot HF$	$(C_4H_9)_4POH$ (40%, aq.), 2HF (47%), 5–20 °C	~100	Colorless oil mp, 30–34 °C	43
$(C_4H_9)_4PF \cdot H_2O$	$(C_4H_9)_4POH$ (40%, aq.), HF (47%)	~100	Colorless oil	43
$(C_4H_9)_4PF$	$(C_4H_9)_4PF \cdot HF$, C_4H_9Li, C_6H_{14}	~100	Colorless oil	43

[a]Anhydrous hydrogen fluoride.

References are listed on pages 36–38.

Table 4. Fluorinated Amines

Fluorinating Agent	*Method of Preparation*	*Yield (%)*	*Properties*	*Ref.*
$CHClFCF_2N(C_2H_5)_2$	$CClF{=}CF_2$, $HN(C_2H_5)_2$, RT, autoclave	74		12[a]
$CF_3CHFCF_2N(C_2H_5)_2$ (Ishikawa reagent)[b]	$CF_3CF{=}CF_2$ (25 g; 0.17 mol), $HN(C_2H_5)_2$, (11 g; 0.15 mol), dry ether (30 mL)	72	bp, 56–57 °C at 58 mm Hg	13

[a]The compound is commercially available.
[b]The compound has considerable vapor pressure.

Table 5. *N*-Fluoropyridines and Analogues

Fluorinating Agent	*Method of Preparation*	*Yield (%)*	*Properties*	*Ref.*
	F_2/N_2 (5%), $CFCl_3$ -78 °C	45-60		19, 20
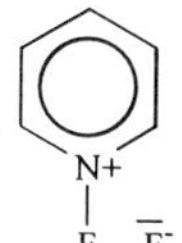	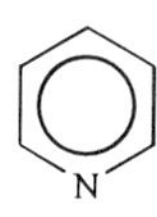 F_2/N_2 (10%), CH_3CN, -40 °C ; F_2/N_2 (10%), $CFCL_3$, -40 °C		Unstable above -2 °C	21, 22

Continued on next page.

Table 5—Continued

Fluorinating Agent	*Method of Preparation*	*Yield (%)*	*Properties*	*Ref.*
N-fluoropyridinium, $^{-}OSO_2CF_3$	pyridine; F_2/N_2 (10%), CF_3SO_3 Na, CH_3CN, -40 °C	67-71	mp, 185-187 °C	23[b]
N-fluoro-2,4,6-trimethylpyridinium (H_3C, CH_3, CH_3), $^{-}OSO_2CF_3$	2,4,6-trimethylpyridine (H_3C, CH_3, CH_3), CF_3SO_3NA; F_2/N_2 (10%), CH_3CN, -40 °C; F_2/N_2 (10%), $CFCL_3$, -40 °C	49	mp, 164–166 °C	21
N-fluoro-3,5-dichloropyridinium (Cl, Cl), $^{-}OSO_2CF_3$	3,5-dichloropyridine (Cl, Cl), CF_3SO_3NA; F_2/N_2 (10%), CH_3CN, -40 °C ; F_2N_2 (10%), $CFCL_3$, -40 °C	41-62	mp, 99–101 °C	23
[pyridine · N-fluoropyridinium] $[B_2F_7]^-$	pyridine	70	mp, 196-200 °C	24[b]

References are listed on pages 36–38.

Table 5—Continued

Fluorinating Agent	*Method of Preparation*	*Yield (%)*	*Properties*	*Ref.*
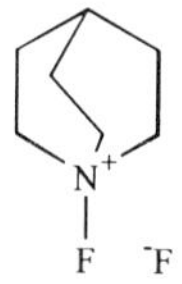 N^+ F $^-$F	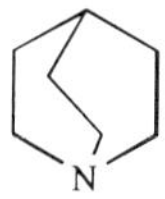 N F_2 (1% in $CFCL_3$)	83–89	mp, 126–128 °C	25
R $\overline{X}$ N^+ N^+ F $\overline{Y}$	R N	83–89	mp, 170 °C decomposes at 195 °C; soluble in water, dilute HCl, CH_3 CHN, dimethylformamide, and acetone	24[b]
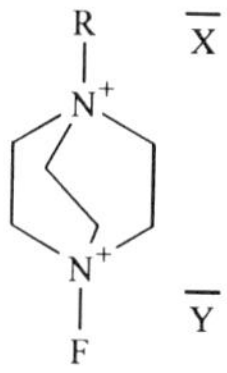 R X, Y CH_3 BF_4 CH_2Cl BF_4 CH_2Cl OSO_2CF_3 CH_2CF_3 OT_3	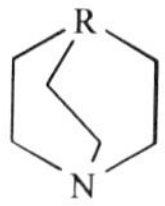 Proprietary information R = CH_2Cl; Y, Y = BF_4			26

[a]Anhydrous hydrogen fluoride.
[b]The compound is commercially available.

References are listed on pages 36–38.

Table 6. *N*-Fluoropyridines and Sulfonimides

Fluorinating Agent	*Method of Preparation*	*Yield (%)*	*Properties*	*Ref.*
$CF_3SO_2NFCH_3$[a]	$CF_3SO_2NHCH_3$, F_2/N_2 (1%), NaF, $CFCl_3$, –75 °C	11	Vapor pressure, 38 mm at 22 °C	27
$(CF_3SO_2)_2NF$	$(CF_3SO_2)_2NH$, F_2/N_2 (10%), –196 to 22 °C	95	mp, –69.8 °C	27
$CF_3SO_2NFSO_2C_4F_9$	$CF_3SO_2NHSO_2C_4F_9$, F_2/N_2 (10%), –196 to 22 °C	96	mp, –56 °C	27
$CF_3SO_2NFSO_2C_6F_{13}$	$CF_3SO_2NHSO_2C_6F_{13}$, F_2/N_2 (10%), –196 to 22 °C	93	mp, –28 °C	27
$C_4F_9SO_2NFSO_2C_6F_{13}$	$C_4F_9SO_2NHSO_2C_6F_{13}$, F_2/N_2 (10%), –196 to 22 °C	88	mp, 60 °C decomposes	27
$(CF_2)_n$<SO_2, SO_2>NF (cyclic)	$(CF_2)_n$<SO_2, SO_2>NH (cyclic)			
	F_2 (10% excess), –196 to 22 °C			
	$n = 2$	77	mp, 59 °C	27
	$n = 3$	61	mp, 97 °C	27
	$n = 4$	86	mp, 54 °C	27
H_3C–C_6H_4–SO_2NFR	H_3C–C_6H_4–SO_2NHR			
	F_2/N_2 (1–5%), $CHCl_3$, $CFCl_3$, –78 °C			
	$R = CH_3$	59		28
	$R = C_5H_{12}$	57		28
	$R = C(CH_3)_3$	14		28

Table 6—Continued

Fluorinating Agent	*Method of Preparation*	*Yield (%)*	*Properties*	*Ref.*
$C_6H_5SO_2NFC(CH_3)_3$	$C_6H_5SO_2NHC(CH_3)_3$ F_2/N_2 (1–5%), $CHCl_3$, $CFCl_3$, –78 °C	71		29, 30
H_3C–C_6H_4–SO_2NF–(EXO-norbornyl)	H_3C–C_6H_4–SO_2NH–(EXO-norbornyl) F_2/N_2 (1–5%), $CHCl_3$, $CFCl_3$, −78 °C, 4–6 h	88		29, 30
N-fluoro camphorsultam (CH_3, CH_3, H_2C, SO_2, R, N–F)	camphorsultam (CH_3, CH_3, H_2C, SO_2, R, NH) F_2/N_2 (10%), NaF, $CHCl_3$, $CFCl_3$, –40 °C			
	R = H	75	mp, 112–114 °C	31, 32
	R = CH_3	80	mp, 151–154 °C	31, 32

Continued on next page.

Table 6—Continued

Fluorinating Agent	*Method of Preparation*	*Yield (%)*	*Properties*	*Ref.*
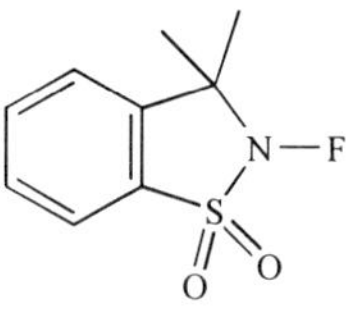 N—F	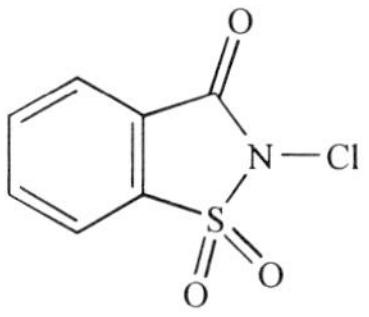 N—Cl; 1. Mg, tetrahydrofuran, Structure O-3; 2. F_2/N_2 (10%), NaF, $CHCl_3:CFCl_3$ (1:1), –40 °C	49–74	mp, 114–116 °C	33
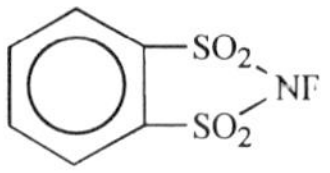 SO_2 NF SO_2	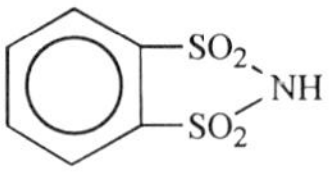 SO_2 NH SO_2; F_2/N_2 (10%), NaF, $CHCl_3:CFCl_3$ (1:1), –40 °C	>90%	mp, 139–140 °C	34
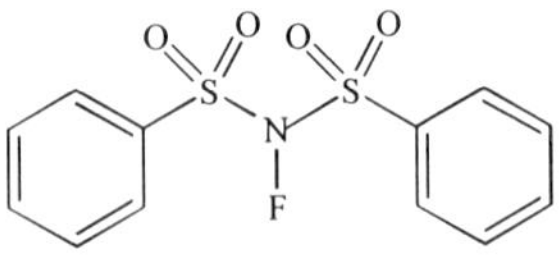 O O O O S N S F	Proprietary information		mp, 114–116 °C	35[b]

[a]The compound has considerable vapor pressure.
[b]The compound is commercially available.

References are listed on pages 36–38.

Table 7. Organic Hypofluorites

Fluorinating Agent	*Method of Preparation*	*Yield (%)*	*Properties*	*Ref.*
$CF_2(OF)_2$[a]	CO_2, F_2/N_2, CsF, –196 °C to RT	~100	Yellow liquid; mp, –184 °C; bp, -65 °C	44, 45
CH_3COOF[a]	$(CH_3CO)_2O$, CH_3COOM (M = Na, K), F_2/N_2 (10%), 7–8 h	80		46, 47
CF_3COOF[a]	CF_3COOH, CH_3COOM (M = Na, K), F_2/N_2 (10%), $CFCl_3$	90		47

[a]The compound has considerable vapor pressure.

Table 8. Fluoroaminosulfuranes

Fluorinating Agent	*Method of Preparation*	*Yield (%)*	*Properties*	*Ref.*
R_2NSF_3[a]	$R_2NSi(CH_3)_3$, SF_4, $CHCl_3$, –65 to –60 °C, R = CH_3	60–80	bp, 117.5 °C; *d*, 1.3648	36, 37
DAST[a]	R = C_2H_5	70–90	Pale yellow liquid; bp, 43–44 °C at 12 mm Hg; 46–47 °C at 10 mm Hg	37–39
	R = $(CH_3)_2CH$	99	Decomposes at 60 °C at 2 mm Hg	37, 40
R_2[a]	$R_2 = (CH_2)_4$	76–83	mp, –18 °C; bp, 54–55 °C at 15 mm Hg	38, 39
R_2[a]	$R_2 = (CH_2)_5$	60–97	bp, 76 °C at 12 mm Hg	41
R_2[a]	$R_2 = O(CH_2CH_3)_3$	55–98	bp, 41–42 °C at 0.5 mm Hg	38, 39, 41
$C_6H_5N(C_2H_5)SF_3$	$C_6H_5N(C_2H_5)Si(CH_3)_3$, SF_4, –65 to –60 °C	60	bp, 50–57 °C at 0.3 mm Hg; decomposes slowly at 20 °C	39

Continued on next page.

Table 8—Continued

Fluorinating Agent	*Method of Preparation*	*Yield (%)*	*Properties*	*Ref.*
$R^1{}_2NSF_2NR^2{}_2$	$R^1{}_2NSi(CH_3)_3$, $R^2{}_2NSF_3$, $CFCl_3$, –78 to 20 °C			
	$R^1, R^2 = CH_3$	60	mp, 64–65.5 °C	37
	$R^1 = CH_3, R^2 = C_2H_5$	92	Liquid, not distilled	37
	$R^1, R^2 = C_2H_5$			37
	$R^1 = CH_3, R^2 = (CH_2)_5$	99	mp, 25–26 °C	37
	$R^1, R^2 = (CH_2)_5$	100	mp, 104–105 °C	38
	$R^1, R^2 = O(CH_2CH_2)_2$	98	mp, 104–105 °C	48
$(R_2N)_3S^+(CH_3)_3Si^-F_2$	$3(R_2N)_2NSi(CH_3)_3$, SF_4, $(C_2H_5)_2O$, –78 °C			
TASF	$R = CH_3$	86–89	mp, 58–62 °C	42[b]
	$R = C_2H_5$	98	mp, 90–95 °C	42
	$R = (CH_2)_5$	89	mp, 87–90 °C	42

[a]The compound has considerable vapor pressure.
[b]The compound is commercially available.

References for Pages 25–36

1. Bornstein, J.; Skatos, L. *J. Am. Chem. Soc.* **1968,** *90,* 5044.
2. Christe, K. O.; Wilson, R. D.; Goldberg, I. B. *Inorg. Chem.* **1979,** *18,* 2578.
3. Christe, K. O.; Wilson, W. W.; Schack, C. J.; Wilson, R. D. *Inorg. Synth.* **1986,** *24,* 42.
4. Appelman, E. H.; Basile, L. J.; Thompson, R. C. *J. Am. Chem. Soc.* **1979,** *101,* 3384.
5. Stavber, S.; Zupan, M. *J. Org. Chem.* **1985,** *50,* 3609.
6. Edwards, A. J.; Plevey, R. G.; Sallomi, I. J.; Tatlow, J. C. *J. Chem. Soc., Chem. Commun.* **1972,** 1028.
7. Meshri, D. T., personal experience.
8. Bardin, V. V.; Furin, G. G.; Yakobson, G. G. *Zh. Org. Khim.* **1982,** *18,* 604 (Engl. Transl. 525).
9. Bardin, V. V.; Furin, G. G.; Yakobson, G. G. *J. Org. Chem.* **1986,** *51,* 1482.
10. Edwards, A. J.; Holloway, J. H.; Peacock, R. D. *Proc. Chem. Soc.* (London) **1963,** 275.

11. Burgess, J.; Fraser, J. W.; McRae, V. M.; Peacock, R. D.; Russell, D. R. *J. Inorg. Chem. Supplement* **1976,** 183.
12. U.S. Patent 3 105 078, 1963; *Chem. Abstr.* **1964,** *60,* 4276.
13. Takaoka, A.; Iwakiri, H.; Ishikawa, N. *Bull. Chem. Soc. Jpn.* **1979,** *52,* 3377.
14. Alvernhe, G.; Laurent, A.; Haufe, G. *J. Fluorine Chem.* **1986,** *34,* 147.
15. Franz, R. *J. Fluorine Chem.* **1980,** *15,* 423.
16. Giudicelli, M. B.; Picq, D.; Veyron, B. *Tetrahedron Lett.* **1990,** *31,* 6527.
17. Bosch, P.; Camps, F.; Chamorro, E.; Gasol, V.; Guerrero, A. *Tetrahedron Lett.* **1987,** *28,* 4733.
18. Fukuhara, T.; Yoneda, N.; Abe, T. *Nippon Kagaku Kaishi* **1985,** *10,* 1951; *Chem. Abstr.* **1985,** *104,* 206662S.
19. Purrington, S. T.; Jones, W. A. *J. Org. Chem.* **1983,** *48,* 761.
20. Purrington, S. T.; Jones, W. A. *J. Fluorine Chem.* **1984,** *26,* 43.
21. Simons, J. H. *Fluorine Chemistry,* Volume 1; Academic Press: New York, 1950; p 420.
22. Meinert, H. *Z. Chem.* **1965,** *5,* 64.
23. Umemoto, T.; Tomita, K. *Tetrahedron Lett.* **1986,** *27,* 3271.
24. Poss, A. J.; Wagner, W. J.; Frenette, R. L. In *Tenth Winter Fluorine Conference, Abstract of Papers;* St. Petersburg, FL; January 28 to February 2, 1991; Abstract 103.
25. Banks, R. E.; Boisson, R.; Tisiliopoulos, E. *J. Fluorine Chem.* **1986,** *32,* 461.
26. *Select Fluor;* Lal, G. S. *J. Org. Chem.* **1993,** *58,* 2791.
27. Singh, S.; DesMarteau, D. D.; Zuberi, S. S.; Witz, M.; Huang, H. N. *J. Am. Chem. Soc.* **1987,** *109,* 7194.
28. Barnette, W. E. *J. Am. Chem. Soc.* **1984,** *106,* 452.
29. U.S. Patent 4 479 901, 1984; *Chem. Abstr.* **1985,** *102,* 113537.
30. Lee, S. H.; Schwartz, J. *J. Am. Chem. Soc.* **1986,** *108,* 2445.
31. Differding, E.; Lang, R. W. *Tetrahedron Lett.* **1988,** *29,* 6087.
32. Auer, K.; Hungerbuhler, E.; Lang, R. W. *Chimia* **1990,** *44,* 120.
33. Differding, E.; Lang, R. W. *Helv. Chim. Acta* **1989,** *72,* 1248.
34. Davis, F. A.; Han, W. *Tetrahedron Lett.* **1991,** *32,* 1631.
35. Allied Signal Chemicals Product Information Bulletin *NFSi;* 1992.
36. Demitras, G. C.; Kent, R. A.; MacDiarmid, A. G.; *Chem. Ind.* (London) **1964,** 1712.
37. Middleton, W. J. *J. Org. Chem.* **1975,** *40,* 574.
38. Markovskii, L. N.; Pashinnik, V. E.; Kirsanova, N. A. *Zh. Org. Khim.* **1976,** *12,* 965 (Engl. Transl. 973).
39. Markovskii, L. N.; Pashinnik, V. E.; Kirsanov, A. V. *Synthesis* **1973,** 787.
40. Gibson, J. A.; Ibbott, D. J.; Janzen, A. F. *Can. J. Chem.* **1973,** *51,* 3203.
41. Middleton, W. J.; Bingham, E. M. *Org. Synth.* **1977,** *57,* 50.
42. Farnham, W. B.; Harlow, R. L. *J. Am. Chem. Soc.* **1981,** *103,* 4608.

43. Yoshika, H.; Seto, H.; Qian, Z. In *Tenth Winter Fluorine Conference, Abstract of Papers;* St. Petersburg, FL; January 28 to February 2, 1991; Abstract 104.
44. Hohorst, F. A.; Shreeve, J. M. *J. Am. Chem. Soc.* **1967,** *87*, 1809.
45. Hohorst, F. A.; Shreeve, J. M. *Inorg. Synth.* **1968,** *11*, 143.
46. Jewett, D. M.; Potocki, J. F.; Ehrenkaufer, R. E. *J. Fluorine Chem.* **1984,** *24,* 477.
47. Rozen, S.; Lerman, O.; Kol, M. *J. Chem. Soc., Chem. Commun.* **1981,** 443.
48. Markovskii, L. N.; Pashinnik, V. E.; Kirsanova, N. A.; *Zh. Org. Khim.* **1975,** *11,* 74 (Engl. Transl. 72).

Chapter 3

Methods of Introducing Fluorine into Organic Molecules

Addition of Fluorine

by Suzanne T. Purrington

Addition of Fluorine to Double and Triple Bonds

Many different reagents are available that allow the addition of fluorine to a double bond as exemplified in equation 1 for the addition to *cis*-stibene [*1, 2, 3, 4*].

$$C_6H_5CH{=}CHC_6H_5 \ (cis) \longrightarrow C_6H_5CHFCHFC_6H_5 \qquad \textbf{1}$$

Ref.	Conditions	%	meso:DL
[*1*]	F_2/N_2, $CHCl_3/CCl_3F$, –78°	60	78:22
[*2*]	XeF_2, HF	90	53:47
[*3*]	CF_3OF, Et_2O	41	76:24
[*4*]	$CsSO_4F$, HF, CH_2Cl_2		49:51

As indicated in equation 2 [*5, 6, 7, 8*], *rearranged products* are sometimes observed during the addition. An extensive study of the fluorination of ***diphenylacetylenes*** with **fluorine** reveals *complex product mixtures* that include rearrangement products [*9*].

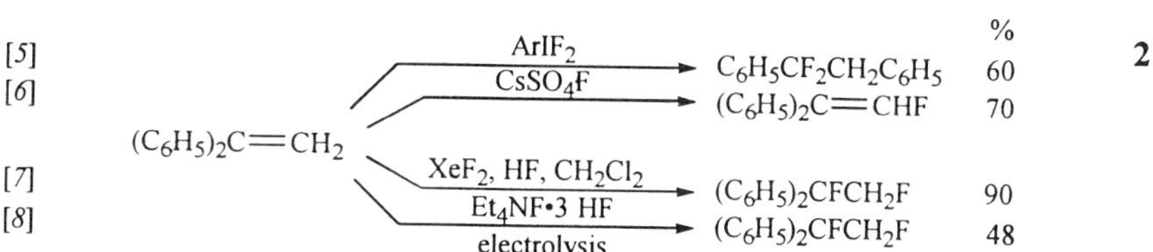

A number of carbonyl derivatives with ***electron-rich double bonds (enols and their derivatives)*** react with **fluorine** to ultimately give *α-fluorocarbonyl compounds,* as exemplified in Table 1 [*10–35*]. Glycals have been fluorinated to give rise to *fluorinated carbohydrates* [*36*], as shown in equation 3 [*37, 38, 39, 40, 41*].

Vanadium pentafluoride and **antimony pentafluoride** add fluorine to polyhaloalkenes in up to 90% yields [*42, 43*]. **Cesium fluoride** [*44, 45*] and ***N,N,N′,N′*-tetramethylformamidinium bifluoride** [*46*] have been used to prepare perfluoro carbanions in 60–90% yields. **Elemental fluorine** adds to perfluoroolefins and perfluorodienes to give addition products accompanied by dimers and oligomers [*47*]. In certain hindered alkenes, long-lived radicals are formed in concentrations as high as 3 M after addition of fluorine [*48, 49*] (equation 4).

0065–7719/95/0187–0041$08.00/1

Table 1. Preparation of α-Fluorocarbonyl Compounds

Precursor	*Reagent*	*α-Fluoro Derivative*	*Yield (%)*	*Ref.*
Silyl enol ether	F_2	Aldehyde or or ketone	52–72	10
	XeF_2		60–77	11, 12
	CF_3OF		70–83	13
	$ArIF_2$		23–38	12
	N-Fluoro-pyridinium triflate		23–87	14
Enol acetate	XeF_2		37–62	15
	Fluoroxy compounds		62–85	16
	Electrochemical fluorination		44–63	17
	$CsSO_4F$		70–80	6
	CF_3OF		—	18
	CH_3CO_2F		45–87	16, 19, 20, 21
Enol ether	CF_3OF		—	18
Enolate	CH_3CO_2F		54–86	22
	$FClO_3$		27–78	23, 24
	N-Fluorosulfonamide		35–81	25
Enolate	*N*-Fluorosulfonamide	β-Dicarbonyl compound	53–81	25
	N-Fluoropyridone		9–39	26
	$FClO_3$		33–78	27, 28
	CH_3CO_2F		75–90	14, 29
	N-Fluoropyridinium triflate		78	14
β-Diketone	XeF_2		53	15
	$C_{19}XeF_6$		40–90	30
	CH_3CO_2F		30	29
	$(CF_3SO_2)_2NF$		80–100	31
Silyl enol ether	F_2		23–42	32
Enol	F_2	Pyruvate	40–70	33, 34

References are listed on pages 49–53.

Table 1—Continued

Precursor	*Reagent*	*α-Fluoro Derivative*	*Yield (%)*	*Ref.*
Silyl ketene acetal derivative	F_2	Carboxylic acid derivative	53–76	35
	CF_3OF		65–90	13
	N-Fluoropyridinium triflate		65	14
Enolate	CH_3CO_2F		67	22

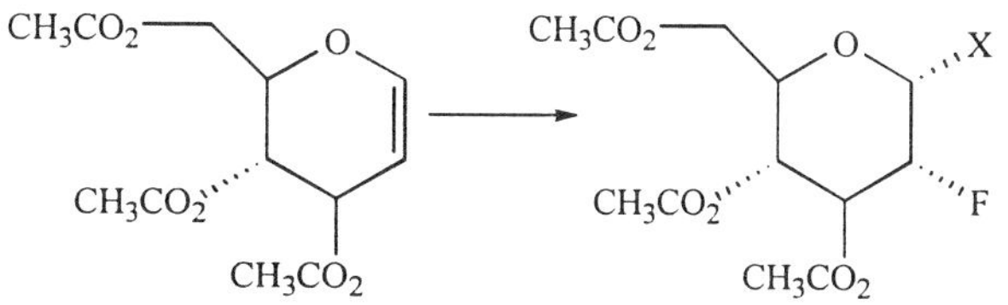

3

		X	%
[*37, 38*]	AcOF, –78°	OAc	78
[*39*]	XeF_2, $BF_3{\cdot}Et_2O$	F	61
[*40*]	F_2/Ar, –78°, CCl_3F	F	40
[*41*]	CF_3OF, –78°, CCl_3F	F, OCF_3	34,26

[*49*] $$[(CF_3)_2CF]_2C{=}CFCF_3 + F_2 \longrightarrow [(CF_3)_2CF]_2\dot{C}CF_2CF_3$$ **4**

Addition of Fluorine to Aromatic Systems

Partial fluorination [*50*] and perfluorination [*51*] of aromatic systems can be accomplished **electrochemically.** A number of other reagents add fluorine to benzene and its derivatives, as elaborated in equation 5 [*52, 53, 54, 55*].

Fluorine has also been added successfully to ***perfluorinated aromatic compounds,*** as illustrated in equation 6 [*57, 58, 59, 60, 61*]. Also **bromine trifluoride** adds fluorine to pentafluorophenol [*56*] and octafluoronaphthalene [*62*].

Fluorination of **graphite** with fluorine gives graphite fluorides that have interesting properties, as recently reviewed [*63*]. **Pyridine** and its derivatives add elemental fluorine to form *unstable N-fluoro adducts* [*14, 26, 64, 65*]. These may decompose to 2-fluoropyridines [*65*] or be stabilized by treatment with triflate salts to form useful electrophilic fluorinating agents [*64*].

X → X, H, F_{10} + X, F_{11} 5

Ref.	X	Conditions	F_{10}	F_{11}
[*52*]	X=H	AgF_2, 300–380°	27%	47%
[*52*]	H	$KAgF_4$, 300–380°	7%	74%
[*53*]	H	K_2NiF_6, 120–350°	6.4%	22%
[*54*]	CF_3	CoF_3, 260–280°		major products
[*55*]	CN	$CsCoF_4$, 300°		55%

F → F_8 + F_{10} 6

Ref.	Conditions	F_8	F_{10}
[*57*]	2 VF_5, –25°, SO_2ClF	61%	13%
[*58*]	$CoF_3{\bullet}CaF_2$, 50°	6.7%	11.4%
[*59*]	$XeF_2{\bullet}BF_3$, CH_2Cl_2, 25°	74%	
[*60*]	$XeF^+SbF_6^-$, SO_2ClF, HF	91%	
[*61*]	28% SbF_5 in BrF_3, –80°, SO_2ClF	77%	

Addition of Fluorine to Carbon–Nitrogen Compounds

Many nitrogen-containing functional groups react with fluorine to give a variety of products. Several reagents add fluorine to nitro anions, as shown in equations 7 and 8 [*66, 67, 68*]. A detailed review of these reactions was recently published [*69*].

$$M^+[(CH_3)_2CNO_2]^- \longrightarrow (CH_3)_2CFNO_2 \qquad 7$$

Ref.	M	Reagent	Yield
[*25*]	M=$(C_4H_9)_4N^+$	CH_3–C_6H_4–$SO_2NF C(CH_3)_3$	83–87%
[*66*]	Na^+	F_2/N_2	62%
[*67*]	Na^+	$FClO_3$	36%

[*68*]

$$RCH{=}CHC(NO_2)_2K \xrightarrow[CH_3CN]{XeF_2} RCH{=}CHC(NO_2)_2F \qquad 8$$

R=CN, $CONH_2$ 63–71%

Diazoketones react with **trifluoromethyl hypofluorite** to give a mixture of α,α-difluoro ketones and α-fluoro-α-trifluoromethoxy ketones [*70, 71*]. With **elemental fluorine,** two fluorine atoms replace the nitrogen in both diazoketones and unactivated diazo compounds [*70, 72, 73*] (equation 9).

$$C_6H_5CO(R)C{=}N_2 \longrightarrow C_6H_5COCF_2R + C_6H_5COCFR(OCF_3) \qquad \mathbf{9}$$

Ref.	R	Conditions	$C_6H_5COCF_2R$	$C_6H_5COCFR(OCF_3)$
[*71*]	R=H	CF_3OF, –78 °, CCl_3F	49%	36%
[*72*]	C_6H_5	F_2/N_2, –70°, CCl_3F	79%	

Aryl ketone hydrazones are oxidized and fluorinated by fluorine to give a mixture of mono- and difluoro hydrocarbons [*74*] (equation 10).

[*74*]

$$(C_6H_5)_2C{=}NNH_2 \xrightarrow[CCl_3F,\ -75^\circ]{F_2/N_2} \underset{11\%}{(C_6H_5)_2CHF} + \underset{69\%}{(C_6H_5)_2CF_2} \qquad \mathbf{10}$$

Isocyanides react with fluorine [*75*] to yield difluoromethylene imines, which tend to dimerize. In an inert solvent, imines accept fluorine from trifluoromethyl hypofluorite [*76, 77*] (equation 11).

[*76*]

$$C_6H_5CH{=}NC_6H_{13} + CF_3OF \xrightarrow{CCl_3F} C_6H_5CF_2NFC_6H_{13}\ \ 52\% \qquad \mathbf{11}$$

However, when the addition is performed in a nucleophilic solvent such as methanol, cleavage of the imine linkage occurs to give *difluoroamino compounds* [*78*] (equation 12). *N,N*-Difluorotrifluoromethylamine can be prepared from ***azides*** or from ***thiocyanates,*** as shown in equation 13 [*79, 80*]. Another way to produce difluoroamino compounds is the addition of fluorine to nitriles by means of AgF_2 [*81*] or CoF_3 [*81*].

[*78*]

$$R^1N{=}CHR^2 + CF_3OF \xrightarrow{CH_3OH} R^1NF_2 + R^2CH(OCH_3)_2 \qquad \mathbf{12}$$

R^1 = alkyl or aryl; $R^2 = C_6H_5$; 55–75%

[*80*]

$$CF_3N_3 \xrightarrow[70^\circ]{F_2} \underset{92\%}{CF_3NF_2} \xleftarrow[-78^\circ]{6\,F_2} KSCN \ (50\%) \qquad \mathbf{13}$$

Fluorination of Oxygen-Containing Compounds

Although fluorination of peroxoanions [*82*] has been examined, the major emphasis in the fluorination of oxygenated material is the *preparation of fluoroxy compounds.* The simplest, **trifluoromethyl hypofluorite,** can be prepared almost quantitatively by the action of fluorine on carbonyl fluoride (fluorophosgene) in the presence of various catalysts [*83, 84*]. Addition of fluorine to trifluoroacetic acid [*85*] or its sodium salt [*86*] gives rise to $F_3CF(OF)_2$. Long-chain fluoroxy compounds can also be formed from their potassium salts [*21, 87*]. **Acyl hypofluorites,** which are useful *electrophilic fluorinating agents,* are also prepared by the addition of fluorine to sodium carboxylates [*16, 19, 88, 89, 90*]. Fluorination of ***cesium or rubidium sulfates*** with fluorine produces **fluoroxysulfates,** MSO_4F (M = Cs or Rb), which are also sources of electrophilic fluorine [*91, 92*].

Oxidative Fluorination of Heteroatoms

Many reagents can be used to oxidatively fluorinate phosphorus(III) compounds, as shown in equation 14 [*93, 94, 95, 96*].

14

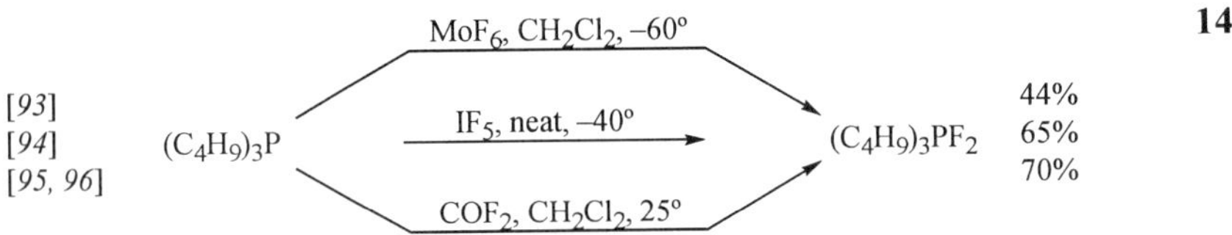

In addition, fluorine [*97, 98*], trifluoromethyl hypofluorite [*99, 100*], bis(trifluoromethyl)peroxide [*100, 101*], trifluoromethyl disulfide [*100*], and xenon difluoride [*102, 103, 104*] react with ***phosphines*** to give the corresponding *difluorophosphoranes* in yields ranging from 25% to near quantitative. ***Phosphites*** are fluorinated by **carbonyl fluoride** [*95, 96*] or **2-hydroperfluoropropyl azide** [*105*] (equation 15).

15

$$(CF_3CH_2O)_3PF_2 \xleftarrow{(CF_3CH_2O)_3P} CF_3CHFCF_2N_3 \xrightarrow{(CH_3O)_3P} (CH_3O)_2P(=O)F$$

[*105*] $(CF_3CH_2O)_3PF_2$ 68% $(CH_3O)_2P(O)F$

Trimethylarsine gives a 98% yield of trimethylarsine difluoride when treated with **xenon difluoride** [*102*] in fluorotrichloromethane, and tris(pentafluorophenyl)arsine gives a 94% yield of tris(pentafluorophenyl)arsine difluoride after reaction with dilute fluorine in fluorotrichloromethane at –40 °C [*106*]. Other trivalent arsenic compounds have also been fluorinated with xenon difluoride [*103*]. In addition, arsines have been oxidatively fluorinated by **iodine pentafluoride** [*107*] or **electrochemically** in 26–34% yield [*108*].

Xenon difluoride has been used to oxidize a number of ***antimony compounds*** [*102, 109*] in yields ranging from 73 to 98%. **Elemental fluorine** oxidized tris(pentafluorophenyl)stibine to tris(pentafluorophenyl)stibine difluoride in 98% yield [*106*]. Oxidative fluorination of ***stibines*** has also been accomplished with **iodine pentafluoride** [*107*].

Trialkylbismuth compounds when treated with **iodine pentafluoride** give unstable trialkylbismuth difluorides or alkyl iodide tetrafluorides, depending on the alkyl group [*107*]. ***Disulfides*** are oxidatively fluorinated by several reagents. Treatment of diphenyl disulfide with **sulfur tetrafluoride** gives a 20% yield of phenyl sulfenyl fluoride, which forms thianthrene [*110*]. Other disulfides and sulfides are oxidatively fluorinated to *fluorinated sulfuranes* as shown in equations 16 [*111, 112*] and 17 [*98, 113, 114*].

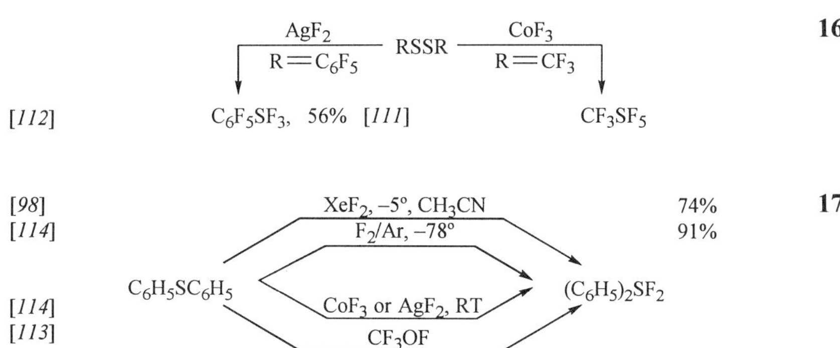

When there is an α-hydrogen in the sulfide, the sulfuranes are unstable and cannot be isolated [*115*]. ***Sulfoxides*** [*116*] and ***sulfinyl fluorides*** [*117*] are oxidatively fluorinated with fluorine to give hexavalent sulfur compounds in yields up to 90%.

Diselenides can also be oxidatively fluorinated, as shown in equation 18 [*118, 119*]. Benzene selenyl fluoride is postulated as an intermediate during the fluoroselenation of alkenes with diphenyl diselenide and xenon difluoride [*73, 120*].

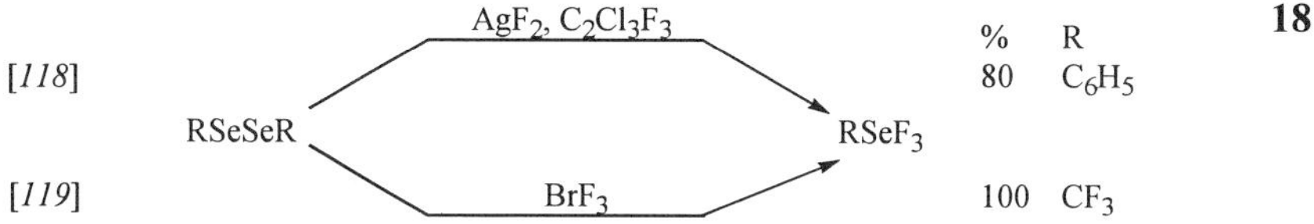

Selenides and tellurides can be fluorinated as shown in equations 19 [*106, 121*] and 20 [*114, 122*].

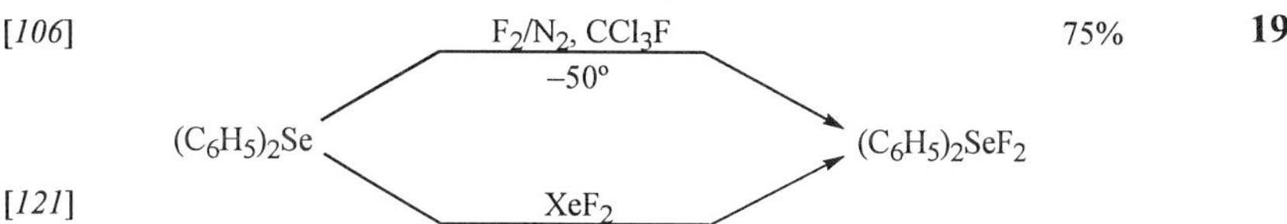

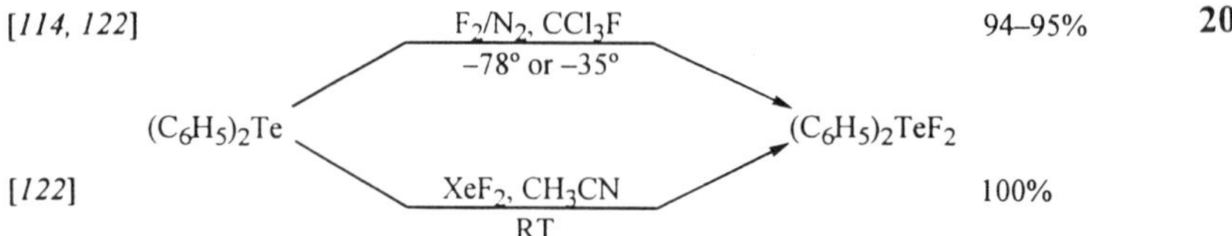

Pentafluorophenylditelluride yields 79–80% pentafluorophenyltellurium trifluoride on treatment with **fluorine or xenon difluoride** [*105*]. The fluorinated tetravalent tellurium can be further oxidized to hexavalent tellurium with either xenon difluoride [*122*] or fluorine [*123*].

Perfluorinated organic bromides can be oxidatively fluorinated with **elemental fluorine** to derivatives containing tri- [*124*] and pentavalent [*125, 126, 127*] bromine in yields up to 42%. **Perfluoroheptylbromine tetrafluoride** has been used to fluorinate double bonds in halogenated alkenes [*127*].

Xenon difluoride is used to prepare **methyliodine difluoride** from methyl iodide [*102, 128*] as well as to convert miscellaneous aryl [*103, 129, 130*], heptafluoropropyl [*129*], and 2,2,2-trifluoroethyl [*103*] iodides to the corresponding organo iodine difluorides in yields ranging from 60 to 100%. **Elemental fluorine** transforms ***aryl iodides*** to their corresponding *aryliodine difluoride compounds* [*131, 132*], which are known to add fluorine to alkenes [*133*] (equation 21).

[*133*] $\xrightarrow{ArIF_2}$ **21**

45–55% + 15–20% + 6–8% + 18–32%

Polymer-bound phenyliodine difluoride, which also has been used as a reagent to add fluorine to alkenes, can be prepared by the addition of xenon difluoride to the polymer [*134, 135, 136*]. Methyl iodide is converted to *trifluoromethyliodine difluoride* by treatment with fluorine at −110 °C [*137*]. Perfluoroalkyliodine tetrafluorides could be synthesized from the perfluoroalkyliodine difluorides and fluorine [*138*] or chlorine trifluoride [*139*]. Perfluoroalkyl [*140*] and perfluoroaryl [*141*] iodides are oxidized to the corresponding iodine difluorides by **chlorine trifluoride.**

References are listed on pages 49–53.

References for Pages 41–48

1. Rozen, S.; Brand, M. *J. Org. Chem.* **1986,** *51,* 3607.
2. Zupan, M.; Pollak, A. *Tetrahedron Lett.* **1974,** 1015.
3. Barton, D. H. R.; Hesse, R. H.; Jackman, G. P.; Ogunkaya, L.; Pechet, M. M. *J. Chem. Soc., Perkin Trans. 1* **1974,** 739.
4. Stavber, S.; Zupan, M. *J. Org. Chem.* **1987,** *52,* 919.
5. Patrick, T. B; Scheibel, J. J.; Hall, W. E.; Lee, Y. H. *J. Org. Chem.* **1980,** *45,* 4492.
6. Stavber, S.; Zupan, M. *J. Chem. Soc., Chem. Commun.* **1981,** 795.
7. Zupan, M.; Pollak, A. *J. Org. Chem.* **1976,** *41,* 4002.
8. Bensadat, A.; Bodennec, G.; Laurent, E.; Tardivel, R. *Tetrahedron Lett.* **1977,** 3799.
9. McEwen, W. E.; Guzikowski, A. P.; Wolf, A. P. *J. Fluorine Chem.* **1984,** *25,* 169.
10. Purrington, S. T.; Lazaridis, N. V.; Bumgardner, C. L. *Tetrahedron Lett.* **1986,** *27,* 2715.
11. Cantrell, G. L.; Filler, R. *J. Fluorine Chem.* **1985,** *27,* 35.
12. Tsushima, T.; Kawada, K.; Tsuji, T. *Tetrahedron Lett.* **1982,** *23,* 1165.
13. Middleton, W. J.; Bingham, E. M. *J. Am. Chem. Soc.* **1980,** *102,* 4845.
14. Umemoto, T.; Kawada, K.; Tomita, K. *Tetrahedron Lett.* **1986,** *27,* 4465.
15. Zajc, B.; Zupan, M. *J. Org. Chem.* **1982,** *47,* 574.
16. Rozen, S.; Menahem, Y. *J. Fluorine Chem.* **1980,** *16,* 19.
17. Laurent, E.; Tardivel, R.; Thiebault, H. *Tetrahedron Lett.* **1983,** *24,* 903.
18. Barton, D. H. R.; Godinho, L. S.; Hesse, R. H.; Pechet, M. M. *J. Chem. Soc., Chem. Commun.* **1968,** 804.
19. Rozen, S.; Menahem, Y. *Tetrahedron Lett.* **1979,** 725.
20. Rozen, S.; Lerman, O. *J. Am. Chem. Soc.* **1979,** *101,* 2782.
21. Barnette, W. E.; Wheland, R. C.; Middleton, W. J.; Rozen, S. *J. Org. Chem.* **1985,** *50,* 3698.
22. Rozen, S.; Brand, M. *Synthesis* **1985,** 665.
23. Barton, D. H. R.; Hesse, R. H.; Pechet, M. M.; Tewson, T. J. *J. Chem. Soc., Perkin Trans. 1* **1973,** 2365.
24. Newman, H.; Fields, T. L. *J. Org. Chem.* **1970,** *35,* 3156.
25. Barnette, W. E. *J. Am. Chem. Soc.* **1984,** *106,* 452.
26. Purrington, S. T.; Jones, W. A. *J. Org. Chem.* **1983,** *48,* 761.
27. Vo Thi, G.; Margaretha, P. *Helv. Chim. Acta* **1976,** *59,* 2236.
28. Shapiro, B. L.; Chrysam, M. M. *J. Org. Chem.* **1973,** *83,* 880.
29. Lerman, O.; Rozen, S. *J. Org. Chem.***1983,** *48,* 724.
30. Yemul, S. S.; Kagan, H. B.; Setton, R. *Tetrahedron Lett.* **1980,** *21,* 277.
31. Xu, Z.-Q.; DesMarteau, D. D.; Gotoh, Y. *J. Chem. Soc., Chem. Commun.* **1991,** 179.

32. Purrington, S. T.; Bumgardner, C. L.; Lazaridis, N. V.; Singh, P. *J. Org. Chem.* **1987,** *52,* 4307.
33. Tsushima, T.; Nishikawa, J.; Sato, T.; Tanida, H.; Tori, K.; Tsuji, T.; Misaki, S.; Suefuji, M. *Tetrahedron Lett.* **1980,** *21,* 3593.
34. Tsushima, T.; Kawada, K.; Tsuji, T.; Misaki, S. *J. Org. Chem.,* **1982,** *47,* 1107.
35. Purrington, S. T.; Woodard, D. L. *J. Org. Chem.,* **1990,** *55,* 3423.
36. Card, P. J. *J. Carbohydr. Chem.* **1985,** *4,* 451.
37. Adam, M. J. *J. Chem. Soc., Chem. Commun.* **1982,** 730.
38. Adam, M. J.; Pate, B. D.; Nesser, J. R.; Hale, L. D. *Carbohydr. Res.* **1982,** *124,* 215.
39. Korytnyk, W.; Valentekovic-Horvat, S. *Tetrahedron Lett.* **1980,** *21,* 1493.
40. Ido, T.; Wan, C.-N.; Fowler, J. S.; Wolf, A. P. *J. Org. Chem.* **1977,** *42,* 2341.
41. Adamson, J.; Foster, A. B.; Hall, L. D.; Hesse, R. H. *J. Chem. Soc., Chem. Commun.* **1969,** 309.
42. Bardin, V. V.; Avramenko, A. A.; Petrov, V. A.; Krasilnikov, V. A.; Kelin, A. I.; Tushin, P. P.; Furin, G. G.; Yakobson, G. G. *Zh. Org. Khim.* **1987,** *23,* 593 (Engl. Transl. 536).
43. Bardin, V. V.; Avramenko, A. A.; Furin, G. G.; Krasilnikov, V. A.; Kelin, A. I.; Tushin, P. P. *J. Fluorine Chem.* **1990,** *49,* 385.
44. Snegirev, V. F.; Delyagina, N. I.; Bakhmutov, V. I. *Izv. Akad. Nauk SSSR, Ser. Khim.* **1986,** 1325 (Engl. Transl. 1201).
45. Bayliff, A. E.; Bryce, M. R.; Chambers, R. D.; Matthews, R. S. *J. Chem. Soc., Chem. Commun.* **1985,** 1018.
46. Delyagina, N. I.; Igumnov, S. M.; Snegirev, V. F.; Knyunants, I. L. *Izv. Akad. Nauk SSSR, Ser. Khim.* **1981,** 2238 (Engl. Transl. 1836).
47. Hotchkiss, I. J.; Stephens, R.; Tatlow, J. C. *J. Fluorine Chem.* **1976,** *8,* 379.
48. Allayarov, S. R.; Sumina, I. V.; Barkalov, I. M.; Bakhmutov, Y. L.; Asamov, M. K. *Izv. Vyssh. Uchebn. Zaved., Khim. Khim. Tekhnol.* **1989,** *32,* 26; *Chem. Abstr.* **1990,** *112,* 197451h.
49. Scherer, K. V.; Ono, T.; Yamanouchi, K.; Fernandez, R.; Henderson, P. *J. Am. Chem. Soc.* **1985,** *107,* 718.
50. Ludman, C. J.; McCron, E. M.; O'Malley, R. F. *J. Electrochem. Soc.* **1972,** *119,* 874.
51. Banks, R. E.; Blow, M. G.; Nickkho-Amiry, M. *J. Fluorine Chem.* **1979,** *14,* 383.
52. Plevey, R. G.; Steward, M. P.; Tatlow, J. C. *J. Fluorine Chem.* **1973,** *3,* 259.
53. Plevey, R. G.; Rendell, R. W.; Steward, M. P. *J. Fluorine Chem.* **1973,** *3,* 267.

54. Alsop, D. J.; Burdon, J.; Carter, P. A.; Patrick, C. R.; Tatlow, J. C. *J. Fluorine Chem.* **1982,** *21,* 305.
55. Phull, G. S.; Plevey, R. G.; Rendell, R. W.; Tatlow, J. C. *J. Chem. Soc., Perkin Trans. 1* **1980,** 1507.
56. Soelch, R. R.; Mauer, G. W.; Lemal, D. M. *J. Org. Chem.* **1985,** *50,* 5845.
57. Bardin, V. V.; Avramenko, A. A.; Furin, G. G.; Yakobson, G. G.; Krasilnikov, V. A.; Tushin, P. P.; Kelin, A. I. *J. Fluorine Chem.* **1985,** *28,* 37.
58. Chambers, R. D.; Clark, D. T.; Holmes, T. F.; Musgrave, W. K. R.; Ritchie, I. *J. Chem. Soc., Perkin Trans. 1* **1974,** 114.
59. Stavber, S.; Zupan, M. *J. Org. Chem.* **1981,** *46,* 300.
60. Bardin, V. V.; Furin, G. G.; Yakobson, G. G. *Zh. Org. Khim.* **1982,** *18,* 604 (Engl. Transl. 525).
61. Bardin, V. V.; Furin, G. G.; Yakobson, G. G. *J. Fluorine Chem.* **1983,** *23,* 69.
62. Bastock, T. W.; Pedler, A. E.; Tatlow, J. C. *J. Fluorine Chem.* **1976,** *8,* 11.
63. Watanabe, N.; Nakajima, T.; Touhara, H. *Graphite Fluorides;* Elsevier: Amsterdam, Netherlands, 1988.
64. Umemoto, T.; Tomita, K.; Kawada, K. *Org. Synth.* **1990,** *69,* 129.
65. VanDerPuy, M. *Tetrahedron Lett.* **1987,** *28,* 255.
66. Fokin, A. V.; Radchenko, V. P.; Galakhov, V. S.; Stolyarov, V. P. *Izv. Akad. Nauk SSSR, Ser. Khim.* **1982,** 1349 (Engl. Transl. 1202).
67. Shechter, H.; Robertson, E. B. *J. Org. Chem.* **1960,** *25,* 175.
68. Tselinskii, I. V.; Mel'nikov, A. A.; Trubitsin, A. E.; Frolova, G. M. *Zh. Org. Khim.* **1990,** *26,* 69.
69. Adolph, H. G.; Koppes, W. M. In *Nitro Compounds: Recent Advances in Synthesis and Chemistry;* Feuer, H.; Nielson, A. T., Eds.; VCH Publishers: New York, 1990.
70. Leroy, J.; Wakselman, C. *J. Chem. Soc., Perkin Trans. 1* **1978,** 1224.
71. Wakselman, C.; Leroy, J. *J. Chem. Soc., Chem. Commun.* **1976,** 611.
72. Patrick, T. B.; Scheibel, J. J.; Cantrell, G. L. *J. Org. Chem.* **1981,** *46,* 3917.
73. Vyplel, H. *Chimia* **1985,** *39,* 304.
74. Patrick, T. B.; Flory, P. A. *J. Fluorine Chem.* **1984,** *25,* 157.
75. Ruppert, I. *Tetrahedron Lett.* **1980,** *21,* 4893.
76. Leroy, J.; Dudragne, F.; Adenis, J. C.; Michaud, C. *Tetrahedron Lett.* **1973,** 2771.
77. Moldavskii, D. D.; Temchenko, V. G.; Antipenko, G. L. *Zh. Org. Khim.* **1971,** *7,* 44.
78. Barton, D. H. R.; Hesse, R. H.; Klose, T. R.; Pechet, M. M. *J. Chem. Soc., Chem. Commun.* **1975,** 97.
79. Schack, C. J. *J. Fluorine Chem.* **1981,** *18,* 583.
80. Ruff, J. K. *J. Org. Chem.* **1967,** *32,* 1675.
81. Phull, G. S.; Plevey, R. G.; Tatlow, J. C. *J. Chem. Soc., Perkin Trans. 1* **1984,** 455.

82. Yu, S.-L.; DesMarteau, D. D. *Inorg. Chem.* **1978,** *17,* 2484.
83. Kennedy, R. C.; Cady, G. H. *J. Fluorine Chem.* **1973,** *3,* 41.
84. Mukhametshin, F. M.; Povroznik, S. V. *Zh. Org. Khim.* **1987,** *23,* 945 (Engl. Transl. 853).
85. Sekiya, A.; DesMarteau, D. D. *Inorg. Nucl. Chem. Lett.* **1979,** *15,* 203.
86. Lerman, O.; Rozen, S. *J. Org. Chem.* **1980,** *45,* 4122.
87. Mulholland, G. K.; Ehrenkaufer, R. E. *J. Org. Chem.* **1986,** *51,* 1482.
88. Rozen, S.; Lerman, O. *J. Am. Chem. Soc.* **1979,** *101,* 2782.
89. Rozen, S.; Lerman, O. *J. Org. Chem.* **1980,** *45,* 672.
90. Rozen, S.; Hebel, D. *J. Org. Chem.* **1990,** *55,* 2621.
91. Ip, D. P.; Arthur, C. D.; Winans, R. E.; Appelman, E. H. *J. Am. Chem. Soc.* **1981,** *103,* 1964.
92. Appelman, E. H.; Basile, L. J.; Hayatsu, R. *Tetrahedron* **1984,** *40,* 189.
93. Mathey, F.; Bensoam, J. *C. R. Acad. Sci., Ser. C* **1972,** *274,* 1095.
94. Frohn, H. J.; Maurer, H. *J. Fluorine Chem.* **1986,** *34,* 73.
95. Gupta, O. D.; Shreeve, J. M. *J. Chem. Soc., Chem. Commun.* **1984,** 416.
96. Williamson, S. M.; Gupta, O. D.; Shreeve, J. M. *Inorg. Synth.* **1986,** *24,* 62.
97. Ruppert, I.; Bastian, V. *Angew. Chem., Int. Ed. Engl.* **1977,** *16,* 718.
98. Marat, R. K.; Janzen, A. F. *Can. J. Chem.* **1977,** *55,* 3031.
99. De'Ath, N. J.; Denney, D. Z.; Denney, D. B.; Hall, C. D. *Phosphorus* **1974,** *3,* 205.
100. De'Ath, N. J.; Denney, D. B.; Denney, D. Z.; Hsu, Y. F. *J. Am. Chem.Soc.* **1976,** *98,* 768.
101. De'Ath, N. J.; Denney, D. Z.; Denney, D. B. *J. Chem. Soc., Chem. Commun.* **1972,** 272.
102. Forster, A. M.; Downs, A. J. *Polyhedron* **1985,** *4,* 1625.
103. Alam, K.; Janzen, A. F. *J. Fluorine Chem.* **1987,** *36,* 179.
104. Gibson, J. A.; Mart, R. K.; Janzen, A. F. *Can. J. Chem.* **1975,** *53,* 3044.
105. Lermontov, S. A.; Popov, A. V.; Sukhozhenko, I. I.; Martynov, I. V. *Izv. Akad. Nauk SSSR, Ser. Khim.* **1989,** 215 (Engl. Transl. 204).
106. Kasemann, R.; Naumann, D. *J. Fluorine Chem.* **1988,** *41,* 321.
107. Frohn, H. J.; Maurer, H. *J. Fluorine Chem.* **1986,** *34,* 129.
108. Nitikin, E. V.; Kazakova, A. A.; Ignatev, Y. A.; Parakin, O. V.; Kargin, Y. M. *Zh. Obshch. Khim.* **1983,** *53,* 230 (Engl. Transl. 201).
109. Yagupol'skii, L. M.; Popov, V. I.; Kondratenko, N. V.; Korsunski, B. L.; Aleinikov, N. N. *Zh. Org. Khim.* **1975,** *11,* 459 (Engl. Transl. 454).
110. Seel, F.; Budenz, R.; Flaccus, R. D.; Staab, R. *J. Fluorine Chem.* **1978,** *12,* 437.
111. Sheppard, W. A.; Foster, S. S. *J. Fluorine Chem.* **1972,** *2,* 53.
112. Tyczkowski, E. A. USNTIS, PB Rep. 1974, No. 233149/4GA; *Chem. Abstr.* **1975,** *82,* 124632v.
113. Denney, D. B.; Denney, D. Z.; Hsu Y. F. *J. Am. Chem. Soc.* **1973,** *95,* 4064.

114. Ruppert, I. *Chem. Ber.* **1979,** *112,* 3023.
115. Forster, A. M.; Downs, A. J. *J. Chem. Soc., Dalton Trans.* **1984,** 2827.
116. Ruppert, I. *Angew. Chem. Int. Ed. Engl.* **1979,** *18,* 880.
117. Ruppert, I. *Chem. Ber.* **1980,** *113,* 1047.
118. Furin, G. G.; Andreevskaya, O. I.; Yakobson, G. G. *Izv. Sib. Otd. Akad. Nauk SSSR, Ser. Khim. Nauk* **1981,** 141; *Chem. Abstr.* **1981,** *95,* 132412g.
119. Lehmann, E. *J. Chem. Res.* **1978,** (S) (1) 42.
120. Uneyama, K.; Kanai, M. *Tetrahedron Lett.* **1990,** *31,* 3583.
121. Yagupol'skii, Y. L.; Savina, T. I. *Zh. Org. Khim.* **1979,** *15,* 438.
122. Alam, K.; Janzen, A. F. *J. Fluorine Chem.* **1985,** *27,* 467.
123. Klein, G.; Naumann, D. *J. Fluorine Chem.* **1985,** *30,* 259.
124. Obaleye, J. A.; Sams, L. C. *J. Inorg. Nucl. Chem.* **1981,** *43,* 2259.
125. Obaleye, J. A.; Sams, L. C. *Inorg. Nucl. Chem. Lett.* **1980,** *16,* 343.
126. Habibi, M. H.; Sams, L. C. *J. Fluorine Chem.* **1981,** *18,* 277.
127. Habibi, M. H.; Sams, L. C. *J. Fluorine Chem.* **1982,** *21,* 287.
128. Gibson, J. A.; Janzen, A. F. *J. Chem. Soc., Chem. Commun.* **1973,** 739.
129. Maletina, I. I.; Orda, V. V.; Aleinikov, N. N.; Korounskii, B. L.; Yagupol'skii, L. M. *Zh. Org. Khim.* **1976,** *12,* 1371 (Engl. Transl. 1364).
130. Gregorcic, A.; Zupan, M. *Bull. Chem. Soc. Jpn.* **1977,** *50,* 517.
131. Ruppert, I. *J. Fluorine Chem.* **1980,** *15,* 173.
132. Naumann, D.; Ruether, G. *J. Fluorine Chem.* **1980,** *15,* 213.
133. Gregorcic, A.; Zupan, M. *J. Chem. Soc., Perkin Trans. 1* **1977,** 1446.
134. Zupan, M. *Collect. Czech. Chem. Commun.* **1977,** *42,* 266.
135. Zupan, M.; Pollak, A. *J. Chem. Soc., Chem. Commun.* **1975,** 715.
136. Sket, B.; Zupan, M.; Zupet, P. *Tetrahedron* **1984,** *40,* 1603.
137. Naumann, D.; Feist, H. R. *J. Fluorine Chem.* **1980,** *15,* 541.
138. Naumann, D.; Schmeisser, M.; Deneken, L. *Inorg. Nucl. Chem., Herbert H. Hyman Memorial Volume,* **1976,** 13.
139. Chambers, O. R.; Oates, G.; Winfield, J. M. *J. Chem. Soc., Chem. Commun.* **1972,** 839.
140. Oates, G.; Winfield, J. M. *J. Fluorine Chem.* **1974,** *4,* 235.
141. Berry, J. A.; Oates, G.; Winfield, J. M. *J. Chem. Soc., Dalton Trans.* **1974,** 509.

Addition of Hydrogen Fluoride

by Andrew E. Feiring

Anhydrous hydrogen fluoride adds across carbon–carbon double and triple bonds and to other unsaturated systems, but wide variations of solvent, catalyst, temperature, and conditions are required with different substrates.

Addition of Hydrogen Fluoride to Alkenes

The addition of hydrogen fluoride to alkenes is a classic method for the preparation of fluorinated alkanes. Monofluoroalkanes are formed from hydrocarbon alkenes with orientation determined by Markovnikov's rule, but cationic polymerization can be a major side reaction because of the strong acidity and relatively low nucleophilicity of fluoride ion in anhydrous hydrogen fluoride. Best results are often obtained by adding the alkene to an excess of hydrogen fluoride with the reaction temperature adjusted according to the reactivity of the olefin [*1*]. With C_3 or larger alkenes, liquid paraffin diluents may be added to minimize polymer formation [*2, 3*]. Ethene is hydrofluorinated at lower temperatures (−20 to 40 °C) by reaction with hydrogen fluoride in the presence of a $SnCl_4$–carboxylic acid catalyst [*4*].

The hydrofluorination of alkenes also occurs in the gas phase, generally at somewhat higher temperatures [*5*]. Fluoroethane is obtained in yields as high as 98% at 100 to 160 °C by reaction in the presence of minor amounts of higher α-olefins [*6*], and 2-fluoropropane is prepared in greater than 90% yield at ≤50 °C from hydrogen fluoride and propene in the presence of activated carbon [*7*].

Hydrogen fluoride adds to more complex molecules, such as ***unsaturated steroids,*** to give fluorinated derivatives [*1, 8*]. Low temperatures and inert diluents, such as tetrahydrofuran or methylene chloride, are generally employed. With bicyclic unsaturated terpenes, rearrangements often accompany addition to the double bond [*1*].

The hydrofluorination of alkenes may be advantageously conducted with mixtures of hydrogen fluoride and tertiary amines. The best known of these complexes is a combination of **70% hydrogen fluoride and 30% pyridine (termed "pyridinium poly[hydrogen fluoride]" or "Olah's Reagent"),** which is stable at 1 atm (101 kPa) to 55 °C (boiling point of anhydrous hydrogen fluoride is 19 °C) [*9*]. Addition of alkenes such as propene, 2-butene, 2-methylpropene, cyclopentene, cyclohexene, and norbornene in tetrahydrofuran to the hydrogen

0065–7719/95/0187–0054$08.00/1

fluoride–amine complex at 0–20 °C affords the corresponding monofluoroalkanes in yields of 35–90% (equation 1).

[*9*]

$$R^1R^2C{=}CR^3R^4 \xrightarrow[\text{THF, 0° - 20°C}]{C_5H_5NH^+ (HF)_xF^-} R^1R^2C(H){-}C(F)R^3R^4 \quad 35 - 90\%$$ **1**

A polymeric form of the reagent from hydrogen fluoride and poly(4-vinylpyridine) is especially easy to handle [*10*]. Other tertiary amines can also be employed [*11*], and a two-phase mixture of hydrogen fluoride–melamine–pentane hydrofluorinates cyclohexene to fluorocyclohexane in 98% yield [*12*].

Additions of **hydrogen fluoride** to ***halogen-containing olefins*** is of special interest because the products are precursors to fluorinated monomers and to chlorofluorocarbon alternatives with decreased ozone-depletion potential [*13*]. Generally, the ease of addition decreases, as does the polymerizability of the olefin, as the number of halogens on the double bond increases. Thus 1,1-dichloroethene will add hydrogen fluoride in the absence of a catalyst at 65 °C, whereas tetrachloroethene is inert at temperatures up to 160 °C [*1*]. With unsymmetrical olefins, the addition product generally has fluorine attached to the most substituted carbon: trichloroethene, for example, gives only 1,1,2-trichloro-1-fluoroethane [*14*] (equation 2).

[*14*]

$$CHCl{=}CCl_2 \xrightarrow[5°C]{HF,\ TaF_5} CH_2ClCCl_2F \quad 89\ \%$$ **2**

Many *catalysts* are employed for gas- and liquid-phase additions of hydrogen fluoride to halogenated olefins; representative examples are shown in Table 1. Typical gas-phase catalysts are alumina or chromium [*15*], manganese, or bismuth salts that have been fluorinated by high-temperature reaction with hydrogen fluoride. Liquid-phase reactions, generally at lower temperatures than their gas-phase counterparts, are catalyzed by boron trifluoride or high-valent metal fluorides that form superacid solutions in hydrogen fluoride. Under these conditions, halogen-exchange reactions may follow addition, and so the products often include more highly fluorinated derivatives (equation 3).

[*14*]

$$Cl_2C{=}CCl_2 \xrightarrow[160°C]{HF} \text{No Reaction}$$

$$Cl_2C{=}CCl_2 \xrightarrow[150°C]{HF,\ TaF_5} CHCl_2CClF_2,\quad 93\ \%$$ **3**

Table 1. Addition of Hydrogen Fluoride to Halogenated Olefins

Olefin	*Catalyst*	*Conditions*	*Products (Yield, %)*	*Ref.*
$CH_2{=}CHCl$	$SnCl_4$	—[a]	CH_3CHF_2	16
	$AlF_3/Cr_2O_3/NiF_2$	—[b]	CH_3CHF_2	17
	Activated carbon/BF_3	100 °C[b]	CH_3CHClF (90)	18
	VCl_3/C	225 °C[b]	CH_3CHF_2 (91) CH_3CHClF (4)	19
$CH_2{=}CCl_2$	$SnCl_4/P(OC_2H5)_3$	60 °C[a]	CH_3CCl_2F (61) CH_3CClF_2 (34)	20
	Fluorinated $Bi(NO_3)_3$ on Al_2O_3	198–210 °C[b]	CH_3CF_3 (99.7) CH_3CClF_2 (0.2)	21
	Fluorinated $CrCl_3$ on Al_2O_3	198 °C[b]	CH_3CF_3 (99)	22
	Fluorinated $Bi(NO_3)_3$–$Mn(NO_3)_2$ on Al_2O_3	250 °C[b]	CH_3CF_3 (100)	23
	AlF_3	74–86 °C[b]	CH_3CClF_2 (90)	24
$CHCl{=}CCl_2$	WF_6	100–120 °C[a]	CH_2ClCCl_2F (81)	25
	TaF_5	5 °C[a]	CH_2ClCCl_2F (89)	14
	BF_3	95 °C[a]	CH_2ClCCl_2F (60) $CH_2ClCClF_2$ (9)	26
	Fluorinated In_2O_3 on Al_2O_3	350 °C[a]	CH_2ClCF_3 (91)	27
	Fluorinated Bi and Mn salts on Al_2O_3	235–250 °C[b]	CH_2ClCF_3 (92) CH_2ClCF_2Cl (3)	28
$CCl_2{=}CCl_2$	TaF_5	150 °C[a]	$CHCl_2CClF_2$ (93)	14
	NbF_5	150 °C[a]	$CHCCl_2F$ (34) $CHCl_2CClF_2$ (4)	14
	$MoCl_5$	150 °C[a]	$CHCl_2CCl_2F$ (53) $CHCl_2CClF_2$ (16)	14
	$TiCl_4$	150 °C[a]	$CHCl_2CCl_2F$ (42) $CHCl_2CClF_2$ (11)	
	$SbCl_5$	150 °C[a]	$CHCl_2CCl_2F$ (21) $CHCl_2CClF_2$ (30)	14
	$TaCl_5$ or $TaBr_5$	119–122 °C[a]	$CHCl_2CCl_2F$ (12) $CHCl_2CClF_2$ (43)	29
	$NbCl_5$ or $NbBr_5$	142–148 °C[a]	$CHCl_2CClF_2$ (85) $CHCl_2CF_3$ (11)	30

Table 1—Continued

Olefin	*Catalyst*	*Conditions*	*Products, (Yield, %)*	*Ref.*
	Fluorinated $NiCl_2$ on Al_2O_3	325 °C[b]	$CHCl_2CF_3$ (70) $CHClFCF_3$ (12)	31
CH_2=CHF	Fluorinated $CrCl_3$ on Al_2O_3	—[b]	CH_3CHF_2	22
CHF=CF_2	Fluorinated Cr_2O_3	350 °C[b]	CH_2FCF_3 (97.8)	32
CH_3CCl=CH_2		14 °C[a]	$CH_3CClFCH_3$ (75)	33
CF_3CF=CF_2	CrO_2F_2	260–270 °C[b]	CF_3CHFCF_3	34

[a]Liquid phase.

[b]Gas phase.

With very ***electrophilic olefins,*** an alternative hydrogen fluoride addition process is often preferred. This process, involving *reaction of the olefin with fluoride ion* in the presence of a proton donor, is applicable to certain perhalogenated alkenes [*1*] and substrates with other electron-attracting groups attached to the double bond [*35, 36*] (equations 4 and 5).

$$[35]\quad \begin{matrix} R^1CH_2 \\ R^2O_2C \end{matrix}\!\!>C{=}CF_2 \xrightarrow[\text{THF, 0°C}]{Bu_4NF,\ CH_3C_6H_4SO_3H} \begin{matrix} R^1CH_2 \\ R^2O_2C \end{matrix}\!\!>CHCF_3 \quad 71\ \% \qquad (4)$$

$$[36]\quad CF_2{=}CFCO_2CH_3 \xrightarrow[\text{t-BuOH}]{KF} CF_3CHFCO_2CH_3 \quad 50\ \% \qquad (5)$$

Fluoride ion sources include ***alkali metal, ammonium, tetraalkylammonium, and silver fluorides.*** With silver fluoride, the polyfluoroalkylsilver intermediates can be isolated [*1, 37*] (equation 6).

$$[37]\quad CF_2{=}CHSF_5 \xrightarrow[CH_3CN]{AgF} CF_3(SF_5)CHAg \xrightarrow{HCl} \underset{69\ \%}{CF_3CH_2SF_5} \qquad (6)$$

Addition of Hydrogen Fluoride to Acetylenes

The addition of **hydrogen fluoride** to ***acetylene*** has been widely investigated because the initial product, vinyl fluoride, is a commercially important monomer. Acetylene reacts with hydrogen fluoride in the liquid phase in the absence of catalyst to give vinyl fluoride and 1,1-difluoroethane in modest yields [*1*], but better results are achieved by conducting the addition with various additives or *catalysts*

(Table 2). Either vinyl fluoride or 1,1-difluoroethane can be obtained as the major product from liquid- or vapor-phase reactions.

Table 2. Addition of Hydrogen Fluoride to Acetylene

Temperature (°C)	*Catalyst*	*Yield $H_2C{=}CHF$ (%)*	*Yield CH_3CHF_2 (%)*	*Ref.*
70[a]	$PhN(CH_3)_2$, Ph_2Hg	97		38
55[a]	KBF_4, FSO_3H		99.7	39
20[a]	FSO_3H, K_2ZrF_6		99.6	40
50[b]	$Hg(NO_3)_2$, $Cd(NO_3)_2$ on C	98	1	41
300[b]	$Cd(BF_4)_2$ on Al_2O_3	91		42
100[b]	$HgCl_2$, $BaCl_2$ on C	82	4	43
350[b]	AlF_3	80	13	44
310[b]	SbF_3 on Al_2O_3	26	70	45
275[b]	AlF_3	19	81	46
230[b]	Al_2O_3, silicic acid	2	98	47
262[b]	Fluorinated H_3BO_3, Fe_2O_3		99	48

[a]Liquid phase.
[b]Vapor phase.

Alkylacetylenes form *geminal difluorides* by reaction with hydrogen fluoride, neat or in ether solution, at –50 to 0 °C [*1*] or by reaction with the pyridinium poly(hydrogen fluoride) reagent in tetrahydrofuran [*9*] (equation 7).

[*9*]

$$R^1C{\equiv}CR^2 \xrightarrow[\text{THF, 0° C}]{C_5H_5NH^+\,(HF)_xF^-} R^1CH_2CF_2R^2 \qquad 7$$

$R^1 = H, R^2 = C_4H_9$ 70 %
$R^1 = R^2 = C_2H_5$ 75 %

Phenylacetylene gives 1-phenyl-1,1-difluoroethane on reaction with a large excess of hydrogen fluoride in ether at 0 °C or, in better yield, in the gas phase over a mercuric oxide catalyst [*1*]. Allene affords 2,2-difluoropropane [*1*].

Electronegatively substituted acetylenes, such as dimethyl acetylenedicarboxylate, do not react under normal conditions but will add the elements of hydrogen fluoride by reaction with fluoride ion (e.g., CsF or tetraalkylammonium dihydrogen trifluoride) and a proton source under phase-transfer conditions [*49, 50*] (equation 8).

[49] $CH_3O_2CC{\equiv}CCO_2CH_3 \xrightarrow[\text{60°C}]{(C_4H_9)_4N^+ H_2F_3^- \,/\, CH_2ClCH_2Cl}$ (F)(CH_3O_2C)C=C(CO_2CH_3)(H) 90 % **8**

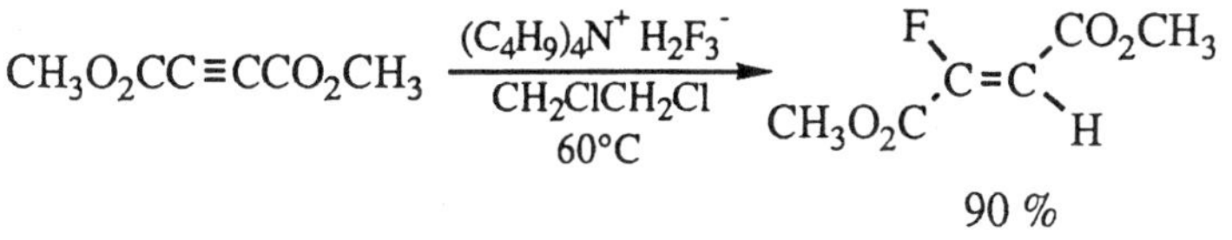

Additions of Hydrogen Fluoride to Other Compounds

Hydrogen fluoride adds to the *carbon–nitrogen multiple bond of imines,* isocyanates, isothiocyanates, and nitriles, although the adducts may be relatively unstable to loss of hydrogen fluoride or hydrolysis. ***Perfluorinated imines*** give secondary amines that can be isolated [*1*]. ***Azirines*** react with hydrogen fluoride by addition and ring opening to give β,β-difluoroamines, α-fluoroketones, or pyrazines, depending on the substitution pattern of the cyclic imine [*51, 52*] (equation 9).

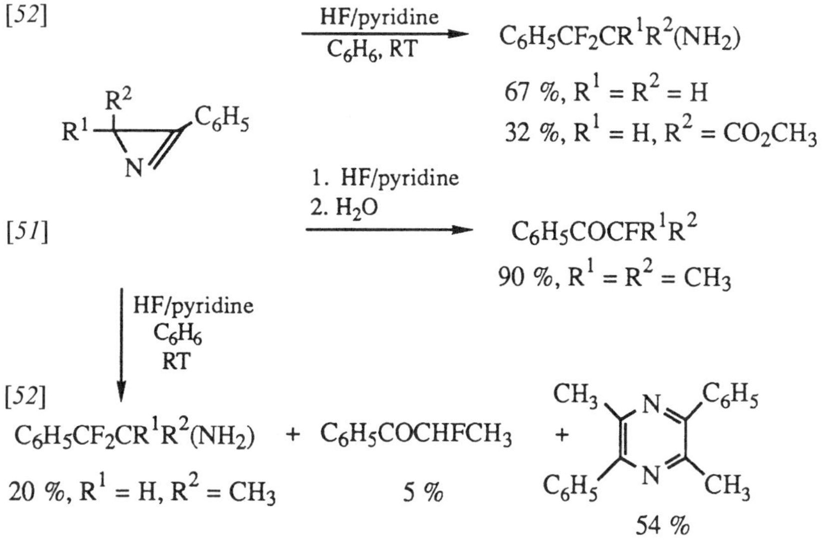

9

Azirine (R^1, R^2, C_6H_5, N):

[52] $\xrightarrow[C_6H_6,\ RT]{HF/pyridine}$ $C_6H_5CF_2CR^1R^2(NH_2)$
67 %, $R^1 = R^2 = H$
32 %, $R^1 = H$, $R^2 = CO_2CH_3$

[51] $\xrightarrow{1.\ HF/pyridine;\ 2.\ H_2O}$ $C_6H_5COCFR^1R^2$
90 %, $R^1 = R^2 = CH_3$

[52] $\xrightarrow{HF/pyridine,\ C_6H_6,\ RT}$ $C_6H_5CF_2CR^1R^2(NH_2)$ + $C_6H_5COCHFCH_3$ + pyrazine (CH_3, C_6H_5, C_6H_5, CH_3)
20 %, $R^1 = H$, $R^2 = CH_3$ — 5 % — 54 %

Alkyl and aryl isocyanates react with anhydrous hydrogen fluoride to give carbamyl fluorides [*1, 53*]; the more conveniently handled pyridinium poly(hydrogen fluoride) reagent can also be used, although the yields tend to be poorer [*9*] (equation 10).

$$RNCO \longrightarrow RNHCOF \quad \mathbf{10}$$

[53]	$R = CH_3$	HF, -80° to 20°C	100 %
[9]	$R = C_6H_5$	$C_5H_5NH^+ (HF)_xF^-$, RT	58 %
	$R = CH_3$	$C_5H_5NH^+ (HF)_xF^-$, RT	40 %

Carbamyl and thiocarbamyl fluorides are obtained from hydrogen fluoride and ***cyanic acid or alkali metal cyanates or thiocyanates*** [*1, 54*]. ***Nitriles*** give imidofluoride salts with hydrogen fluoride [*1*], whereas hydrogen cyanide affords difluoromethylamine, which can be isolated as its hexafluoroarsinate salt [*55*] (equation 11).

[*55*]

$$HCN \xrightarrow[\text{2. } AsF_5]{\text{1. HF}} CHF_2NH_3^+\ AsF_6^- \qquad \mathbf{11}$$

Trifluoromethylisocyanide reacts with hydrogen fluoride to give a mixture of isomeric fluorinated imines [*56*] (equation 12).

[*56*]

$$CF_3N{=}C \xrightarrow[\text{RT}]{\text{HF}} \text{E + Z } CF_3N{=}CHF \quad 55\ \% \qquad \mathbf{12}$$

Diazoketones [*57*] ***and esters*** [*58*] react with hydrogen fluoride in organic solvents to give α-fluoroketones or esters; pyridinium poly(hydrogen fluoride) offers a convenient medium for the reaction (equation 13) [*9*].

[*9*]

$$RCOCHN_2 \xrightarrow[\text{ether, -15° to 25°C}]{C_5H_5NH^+\ (HF)_xF^-} RCOCH_2F \qquad \mathbf{13}$$

$R = C_6H_5$	51 %
$R = c\text{-}C_6H_{11}$	62 %
$R = C_2H_5$	40 %

Hydrogen fluoride will add to ***cyclopropanes*** [*1, 59*] and strained heterocyclic rings. The best-known and most synthetically useful of these ring-opening reactions are with aziridines and epoxides to give β-fluoroamines and alcohols, respectively; these reactions will be covered in another section. ***Diketene*** and other unsaturated lactones react with hydrogen fluoride in the presence of boron trifluoride to give difluorocarboxylic acids [*60*] (equation 14), and a fluorinated oxazetidine gives a fluoroalkylfluoroalkoxyamine [*61*] (equation 15).

[*60*]

$$\begin{matrix} H_2C{=}C{-}O \\ |\quad\ \ | \\ H_2C{-}C{=}O \end{matrix} \xrightarrow[\text{0° to 40°C}]{HF,\ BF_3} CH_3CF_2CH_2CO_2H \quad 94\ \% \qquad \mathbf{14}$$

[*61*]

$$\begin{matrix} CF_3\text{.} \\ N{-}O \\ |\quad\ | \\ F_2C{-}CF_2 \end{matrix} \xrightarrow[\text{23°C}]{HF,\ AsF_5} CF_3NHOCF_2CF_3 \quad 97\ \% \qquad \mathbf{15}$$

References are listed on pages 84–96.

Addition of Halogen Fluorides to Unsaturated Systems

by Andrew E. Feiring

Additions of the elements of the halogen fluorides (chlorine fluoride, bromine fluoride, and iodine fluoride) to olefins are among the most widely investigated methods for introducing fluorine into organic compounds. The products, vicinal halofluoroalkanes, are usually formed with regiochemistry predicted by Markovnikov's rule and with *trans* stereochemistry, although exceptions have been noted, especially in steroid and carbohydrate chemistry. Wagner–Meerwein rearrangements are also observed with some cyclic systems. With nucleophilic olefins, product formation is generally consistent with a polar mechanism involving halonium ion intermediates.

Chlorine fluoride is a stable, albeit highly reactive, compound and can be used for chlorofluorinations of unsaturated compounds [*62*], including ***halogenated olefins, acrylates, dienes, and styrenes*** [*63, 64, 65*]. **Chlorine trifluoride** and its mixtures with chlorine fluoride are also employed [*62*]. **Bromine fluoride** and **iodine fluoride** can be generated in situ by reaction of the elements [*66*] or decomposition of higher halogen fluorides [*67*]. More generally, however, addition of the three halogen fluorides is accomplished by *reaction of the olefin* with a source of positive halogen and a fluoride ion donor.

Reagent combinations for additions of the halogen fluorides to cyclohexene to form *trans*-1-halo-2-fluorocyclohexane (equation 1) are shown in Table 1.

F
X

1

Positive halogen sources include the ***halogens, N-halosuccinimides, 1,3-dibromo-5,5-dimethylhydantoin, hypochlorites, hypobromites, and iodine complexes of pyridine*** and its derivatives. The *fluoride ion donor* is typically hydrogen fluoride or one of its complexes with tertiary amines, although **boron trifluoride, fluoroboric acid, or the fluorinated amine** $CHClFCF_2N(C_2H_5)_2$ are also used. Reactions are generally run at room temperature or below in nonpolar solvents such as ether, methylene chloride, or carbon tetrachloride. The relatively polar tetramethylenesulfone is used with **pyridinium poly(hydrogen fluoride)** [*9*]. Halofluorinations of many other olefins are described in the references in Table 1 (*see also* reference 62). Other reports include the bromo- or iodofluorination of

0065–7719/95/0187–0061$08.00/1

Table 1. Additions of Halogen Fluorides to Cyclohexene

Reagents[a]	*Solvent, Temperature*[b]	*X*	*Yield (%)*	*Ref.*
CH_3OCl, BF_3	CCl_4, RT	Cl	77	68
NCS, $(C_4H_9)_4NHF_2$	CH_2Cl_2, RT	Cl	47	69
NCS, $(C_2H_5)_3N{\cdot}3HF$	Ether, 0–20 °C	Cl	82	70
NCS, $C_5H_5NH^+(HF)_xF^-$	TMS, RT	Cl	85	9
$Pb(O_2CCH_3)_4$, $CuCl_2$, HF	$CClF_2CCl_2F$, RT	Cl	73	71
F_2, Br_2	$CHCl_3/CCl_3F$, –78 °C	Br	61	66
CH_3OBr, BF_3	CCl4, RT	Br	63	68
NBS, $(C_4H_9)_4NHF_2$	CH_2Cl_2, RT	Br	65	69
NBS, $C_5H_5NH^+(HF)_xF^-$	TMS, RT	Br	90	9
NBS, $(C_2H_5)_3N{\cdot}3HF$	Ether, 0–20 °C	Br	88	70
NBS, HF	Ether, –80–0 °C	Br	42	72
NBA, $CHClFCF_2N(C_2H_5)_2$	–15 °C	Br	91	73
I_2, F_2	CCl_3F, –78 °C	I	64	66
$I(C_5H_5N)_2{}^+BF_4{}^-$, HBF_4	CH_2Cl_2, –60 °C	I	89	74
$I(collidine)_2{}^+BF_4{}^-$	CH_2Cl_2, RT	I	58	75
I_2, $C_5H_5NH^+(HF)_xF^-$	TMS, RT	I	60	9
NIS, $(C_4H_9)_4NHF_2$	CH_2Cl_2, RT	I	91	69
NIS, $C_5H_5NH^+(HF)_xF^-$	TMS, RT	I	75	9
NIS, HF	Ether, –80–0 °C	I	72	72
NIS, $(C_2H_5)_3N{\cdot}3HF$	Ether, 0–20 °C	I	75	70
NIS, $CHClFCF_2N(C_2H_5)_2$	–15 °C	I	97	73
NIS, NH_4HF_2, AlF_3	$CH_3O(CH_2)_2OCH_3$, 60 °C	I	75	76
AgF, I_2	CH_3CN, –8 °C	I	60	77

[a]NCS, *N*-chlorosuccinimide; NBS, *N*-bromosuccinimide; NBA, *N*-bromoacetamide; NIS, *N*-iodosuccinimide.

[b]RT, room temperature; TMS, tetramethylenesulfone.

olefins with 1,3-dibromo-5,5-dimethylhydantoin or *N*-iodosuccinimide and **tetrabutylammonium dihydrogen trifluoride** [*78*] or **tetrabutylammonium fluoride** [*79*], bromofluorinations with 1,3-dibromo-5,5-dimethylhydantoin with **silicon tetrafluoride** and water [*80*], bromofluorinations with bromine and **xenon difluoride** in hydrogen fluoride [*81*], iodofluorinations with **methyliodonium difluoride** in hydrogen fluoride [*82, 83*] (equation 2), reactions of ***halogenated olefins*** with bromine in hydrogen fluoride [*84*], and reactions of cinnamates [*85*] or other phenyl-substituted olefins [*86*] with *N*-bromosuccinimide and hydrogen fluoride or hydrogen fluoride–pyridine complex. A polymeric form of the hydrogen fluoride–pyridine complex, prepared by addition of **hydrogen fluoride to a cross-linked styrene–4-vinylpyridine copolymer,** is used with *N*-halosuccinimides to halofluorinate olefins [*87, 88*]. Reaction of this combination with 1,1-diphenylethene (equation 3) illustrates the typical *Markovnikov regiochemistry.*

[*83*] 2

XeF_2 + CH_3I → CH_3IF_2

HF, CH_2Cl_2, RT

F, I

80 %

[*87*] C_6H_5, C_6H_5, H, H 3

Poly-co-styrene-vinylpyridine-HF

N-Bromosuccinimide

CH_2Cl_2 RT

$(C_6H_5)_2CFCH_2Br$ 92 %

The *carbonium ion rearrangements* that can be observed in halofluorinations are illustrated by the reactions of norbornene (Table 2) and norbornadiene (Table 3). Product ratios may vary with the different reagent combinations.

Rearrangements are also observed during *halofluorinations* with ***cyclic medium ring dienes*** [*70, 93*] (equations 4 and 5) and with the monoepoxide of 1,5-cyclooctadiene [*94*] (equation 6) during halofluorinations. Again, there are differences in product mixture with apparently minor variations in reagents (equation 4).

[*93*] 4

F, Br + F, Br + F, Br

Conditions			
NBS, $C_5H_5N{\cdot}10HF$ Ether, RT	0 %	26 %	66 %
NBS, $(C_2H_5)_3N{\cdot}3HF$ CH_2Cl_2, RT	92 %	2 %	6 %

[*70*] 5

NBS, $(C_2H_5)_3N{\cdot}3HF$

Ether, O° to 20°C

F, Br

78 %

Table 2. Products of Addition of Halogen Fluorides to Norbornene

Reagents[a]	*X*						*Ref.*
A	Cl	18%	23%	29%	1.5%	1%	89
B	Br	15%	22%	28%	3%	1.5%	89
C	Br	37%	38%	28%			90

[a]A is *N*-chlorosuccinimide, $C_5H_5NH^+(HF)_xF^-$, ether, 15 °C; B is *N*-bromosuccinimide, $C_5H_5NH^+(HF)_xF^-$, ether, 15 °C; and C is *N*-bromoacetamide, HF, ether, –80 °C.

Table 3. Products of Addition of Halogen Fluorides to Norbornadiene

Reagents[a]	*X*					*Ref.*
A	Cl	40%	20%		3%	91
B	Br	50%	20%			91
C	Br	53%	38%	5%		92

[a]A is *N*-chlorosuccinimide, $C_5H_5NH^+(HF)_xF^-$, ether, 15 °C; B is *N*-bromosuccinimide, $C_5H_5NH^+(HF)_xF^-$, ether, 15 °C; and C is *N*-bromosuccinimide, $(C_2H_5)_3N{\cdot}3HF$, CH_2Cl_2, 15 °C.

[*94*] NXS, $(C_2H_5)_3N{\cdot}3HF$, CH_2Cl_2, RT **6**

X = Cl	71 %	24 %
X = Br	65 %	27 %
X = I	63 %	22 %

Additions of the **halogen fluorides** to ***unsaturated steroids*** [*62, 95, 96, 97, 98, 99*] and ***carbohydrates*** [*62, 75*] are well known. Typical reagent combinations include 1,3-dibromo-5,5-dimethylhydantoin (DBH) or the *N*-halosuccinimides with hydrogen fluoride. *Reversal of the expected regiochemistry* can be observed with certain steroidal olefins [*100, 101*] (equation 7).

References are listed on pages 84–96.

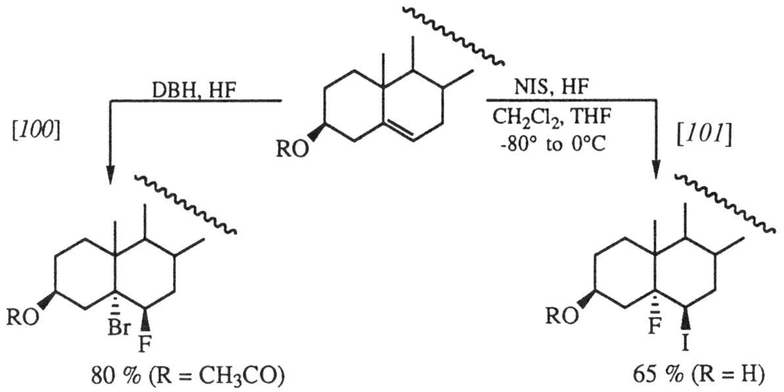

7

Additions of halogen fluorides to the more ***electrophilic perfluorinated olefins*** generally require different conditions. Reactions of **iodine fluoride,** generated in situ from iodine and iodine pentafluoride [*62, 102, 103, 105*] or iodine, hydrogen fluoride, and paraperiodic acid [*104*], with fluorinated olefins (equations 8–10) are especially well studied because the perfluoroalkyl iodide products are useful precursors of surfactants and other fluorochemicals. Somewhat higher temperatures are required compared with reactions with hydrocarbon olefins. Additions of **bromine fluoride,** from bromine and bromine trifluoride, to perfluorinated olefins are also known [*106*].

[*102*] $CF_2{=}CF_2 \xrightarrow[SbF_3\text{ , }CF_3(CF_2)_5I,\ 40-50°C]{IF_5,\ I_2} CF_3CF_2I$ 98 % **8**

[*105*] $CF_3CF{=}CF_2 \xrightarrow[150°C]{IF_5,\ I_2} (CF_3)_2CFI$ 99 % **9**

[*104*] $CF_3CF{=}CF_2 \xrightarrow[150°C]{I_2,\ H_5IO_6,\ HF} (CF_3)_2CFI$ 68 %

[*103*] $SF_5CF{=}CF_2 \xrightarrow[Al,\ AlI_3,\ 50\text{ to }100°C]{IF_5,\ I_2} SF_5CFICF_3$ 22 % **10**

More general procedures for additions of halogen fluorides to highly fluorinated olefins involve reactions with a source of *nucleophilic fluoride ion,* such as an **alkali metal fluoride,** in the presence of a *positive halogen donor* [*62, 107, 108, 109, 110, 111*] (equations 11 and 12). These processes are likely to occur by the generation and capture of *perfluorocarbanionic intermediates.* Tertiary fluorinated carbanions can be isolated as cesium [*112*], silver [*113*], or tris(dimethylamino)sulfonium salts [*114*] and react with sources of positive halogen to give halofluorinated products in high yields (equations 13–15).

[*107*] $CF_3CF{=}CF_2$ —KF (0.1 % K_2CO_3), I_2; CH_3CN, 110°C→ 89 % $(CF_3)_2CFI$ **11**

[*108*] $CF_3CF{=}CF_2$ —KF (calcined), I_2; CH_3CN→ 79 % $(CF_3)_2CFI$

[*109*] $CF_3CF{=}CF_2$ —HF, KF/C, I_2; 290° to 310°C→ 47 % $(CF_3)_2CFI$

[*110*] $CF_3CF{=}CF_2$ —KF, ClCN; CH_3CN, -78° to RT→ $(CF_3)_2CFCl$

[*111*] $CF_2{=}CFCl$ —KF, CCl_3CCl_3; DMF, 100°C→ CF_3CFCl_2 91 % (15 % conversion) **12**

[*112*] $CF_2{=}C(CF_3)SO_2F$ —CsF, CH_3CN, 5 to 10°C→ $(CF_3)_2C(Cs)SO_2F$ **13**

$(CF_3)_2C(Cs)SO_2F$ —SO_2Cl_2→ $(CF_3)_2CClSO_2F$

$(CF_3)_2C(Cs)SO_2F$ —Br_2→ $(CF_3)_2CBrSO_2F$ 93 %

[*113*] $CF_3CF{=}C(CF_3)_2$ —AgF; CH_3CN RT→ $CF_3CF_2C(CF_3)_2Ag$ —I_2→ $CF_3CF_2C(CF_3)_2I$ **14**

$CF_3CF{=}C(CF_3)_2$ —KF, I_2, IF_5; 180 to 200°C→ $CF_3CF_2C(CF_3)_2I$ 73 - 79 %

Brominations and iodinations of tertiary perfluorinated carbanions can be reversible (equations 16 and 17) because of stability of the carbanions and steric crowding in the product [*115*].

[*114*] $CF_3CF_2CF{=}C(CF_3)_2$ **15**

$$CF_3CF_2CF{=}C(CF_3)_2 \xrightarrow[CH_3CN]{[(CH_3)_2N]_3S^+\ (CH_3)_3SiF_2^-} CF_3CF_2CF_2(CF_3)_2C^-\ [(CH_3)_2N]_3S^+$$

$$CF_3CF_2CF_2(CF_3)_2C^-\ [(CH_3)_2N]_3S^+ \xrightarrow[C_6H_5CN,\ 0°C]{Cl_2} CF_3CF_2CF_2CCl(CF_3)_2 \quad 77\ \%$$

$$CF_3CF_2CF_2(CF_3)_2C^-\ [(CH_3)_2N]_3S^+ \xrightarrow[C_6H_5CN,\ 0°C]{Br_2} CF_3CF_2CF_2CBr(CF_3)_2 \quad 99\ \%$$

[*115*] **16**

$$\underset{70\ \%}{C_2F_5CF{=}C(CF_3)_2} \underset{CH_3CN}{\overset{I_2,\ CsF}{\rightleftharpoons}} \underset{30\ \%}{C_3F_7C(CF_3)_2I}$$

[*115*] **17**

$$\text{perfluoro-}\Delta^{9,10}\text{-octalin (F)(F)} \overset{CsF,\ Br_2}{\rightleftharpoons} \text{9-bromoperfluorodecalin (F)(F), Br}$$

Steric effects are presumably also responsible for the surprising formation of a chlorinated product from reaction of a hindered fluorinated olefin with potassium fluoride and iodine chloride [*116*] (equation 18).

[*116*] **18**

$$(CF_3)_2C{=}CFC_2F_5 \xrightarrow[\text{sulfolane, } 90°C]{CsF,\ ICl} (CF_3)_2CClC_3F_7 \quad 59\ \%$$

In contrast to reactions with olefins, there are relatively few reports on *additions of the halogen fluorides to alkynes* [*62*]. Dialkyl and diphenylacetylenes react with the ***N*-halosuccinimides in pyridinium poly(hydrogen fluoride)** to give vicinal halofluoroalkenes [*9*] (equation 19). ***Acetylenes*** also react with **bromine fluoride** and iodine fluoride, generated in situ from the elements, at low temperatures [*117*]. Although the corresponding halofluoroalkenes can be isolated by careful reaction with one equivalent of the reagents, the major products with excess reagent are the *tetrahalides* (equation 20).

[*9*] **19**

$$R^1C{\equiv}CR^2 \xrightarrow[RT]{NXS,\ C_5H_5NH^+(HF)_xF^-} (F)(R^1)C{=}C(R^2)(X)$$

R^1, R^2, X, %: C_2H_5, C_2H_5, Cl, 70; C_6H_5, C_6H_5, Cl, 95; C_2H_5 , C_2H_5, Br, 85; C_6H_5, C_6H_5, Br, 95; C_2H_5, C_2H_5, I, 70; C_6H_5, C_6H_5, I, 90.

[*117*] $R^1\text{-}C\equiv C\text{-}R^2 \xrightarrow[\text{CFCl}_3,\ -75°\text{C}]{\text{XF}} R^1CF_2CX_2R^2$ **20**

R^1, R^2, X, %: C_4H_9, H, Br, 60; C_6H_5, H, Br, 45; C_4H_9, H, I, 80; C_6H_5, H, I, 40; CH_3, CH_3, I, 85.

Diazo ketones and esters react with ***N*****-halosuccinimides in pyridinium poly(hydrogen fluoride)** to give geminal halofluoro derivatives [*9*] (equation 21).

[*9*] $RCOCHN_2 \xrightarrow[0°\text{C}]{\text{NXS, } C_5H_5NH^+(HF)_xF^-} RCOCHXF$ **21**

R, X, %: C_6H_5, Cl, 49; C_6H_5, Br, 63; C_6H_5, I, 62; C_2H_5, Cl, 50; C_2H_5, Br, 32; C_2H_5, I, 80; C_2H_5O, Cl, 30; C_2H_5O, Br, 50; C_2H_5O, I, 50.

Reactions of the halogen fluorides with other unsaturated functional groups are generally reported only with highly fluorinated substrates. ***Hexafluorobenzene*** and derivatives [*118*] and ***octafluoronaphthalene*** [*119*] react with mixtures of ***bromine and bromine trifluoride*** by 1,4 addition of fluorine followed by addition of bromine fluoride across a remaining double bond (equation 22).

[*118*] $C_6F_6 \xrightarrow[0°\text{C}]{\text{BrF}_3,\ \text{Br}_2}$ octafluorocyclohexa-1,4-diene (9 %) + 1-bromononafluorocyclohexene (41 %) **22**

Hypochlorites are obtained by reaction of ***fluorinated carbonyl compounds,*** such as carbonyl fluoride, with an alkali metal fluoride and chlorine fluoride [*62, 120*].

Carbon–nitrogen multiple bonds in fluorinated ***imines*** and ***nitriles*** react with halogen fluoride reagents. Imines provide *N*-chloroamines on reaction with chlorine fluoride [*62, 121, 122, 123*] (equations 23 and 24) or with cesium fluoride and chlorine [*124*] and *N*-bromoamines on reaction with cesium fluoride and bromine (equation 24).

With ***nitriles,*** products from addition of one or two equivalents of halogen fluoride can be obtained [*125, 126, 127, 128*] (equations 25 and 26) on reaction with **chlorine fluoride** or **bromine and an alkali metal fluoride.**

[*122*] $(CF_3)_2NCF{=}NCF_3 \xrightarrow[\text{-78°C to RT}]{\text{ClF}} (CF_3)_2NCF_2N(Cl)CF_3$ 75 % **23**

[*121*] $NCCF{=}NF$: with CsF, Br_2, -78°C → $NCCF_2NBrF$ 20 %; with CsF, ClF, -196° -> RT → $Cl_2NCF_2CF_2NClF$ 58 % + $NCCF_2NClF$ 40 % **24**

CF_3CN: with ClF, F_2, 25°C [*128*] → CF_3CF_2NClF 95 %; with CsF, Br_2, 23°C [*125*] → $CF_3CF{=}NBr$ 79 % **25**

[*127*] $FCN \xrightarrow[\text{22°C}]{\text{KF, Br}_2} CF_2{=}NBr$ 37 % + CF_3NBr_2 37 % **26**

Chlorine fluoride will also oxidize perfluoroalkylsulfenyl chlorides and disulfides to perfluoroalkylsulfur chloride tetrafluorides [*129*].

Addition of Fluorine and Other Elements

by Andrew E. Feiring

Many reagents enable the addition fluorine and another element or group across multiple bonds.

Addition of Fluorine and Oxygen

Fluorine and oxygen can be added by using oxygen difluoride, dioxygen difluoride, and combinations of oxidizing agents (hydrogen peroxide [*130*] or chromium trioxide) and hydrogen fluoride [*131*], but these reagents have received little recent attention. In contrast, ***trifluoromethyl hypofluorite*** [*131*] is of continuing interest for forming vicinal *trifluoromethoxy fluorides and difluorides from olefins,* especially ***unsaturated steroids*** [*132, 133*] and ***carbohydrates*** [*134, 135, 136, 137, 138*]. Additions to ***nucleophilic olefins*** appear to be polar processes that give rise to *cis Markovnikov products* from a reagent polarized in the direction $^{\delta+}F^{\delta-}OCF_3$ (equations 1 and 2). Addition of the reagent to diphenylacetylene gives a trifluoro(trifluoromethoxy) derivative [*132*] (equation 3).

With more ***electrophilic halogenated olefins,*** additions of trifluoromethylhypofluorite appear to go by radical processes [*139*] and give oligomeric products [*140*] or one-to-one adducts [*141, 142*] (equation 4) depending on reaction conditions.

Trifluoromethyl- and perfluoro-*tert*-butylhypofluorites also add to ***hexafluorobenzene*** to give stable adducts [*143*] (equation 5).

Trifluoromethyl hypofluorite will fluorinate *Schiff bases,* giving *N*,α,α-trifluoroamines and α-fluoroimines [*144*] and reacts with **diazoketones** to give adducts in modest yields [*145*] (equation 6). *N*-Substituted ***aziridines*** give ring opened products by 1,3 addition of fluorine on nitrogen and trifluoromethoxy on carbon [*146*] (equation 7).

Perhaps the most generally useful reagent for adding fluorine and oxygen is acetyl hypofluorite. It is produced by bubbling a fluorine–nitrogen mixture into a suspension of sodium acetate in fluorotrichloromethane and acetic acid and is used in situ or as a gas generated by passing fluorine–nitrogen over a solid acetate salt [*147, 148, 149*]. The reagent adds to olefins to afford *vicinal fluoroacetates with Markovnikov regiospecificity* (reagent polarized $^{\delta+}F^{\delta-}O_2CCH_3$) and predominantly or exclusively *syn stereochemistry* [*150*] (equations 8–10). It is widely used to prepare fluorinated carbohydrates [*151, 152, 153*], including fluorine-18 labeled compounds for diagnostic applications [*149, 154, 155, 156*].

0065–7719/95/0187–0070$09.26/1

[132] 1

CF_3OF; $CFCl_3$, CH_2Cl_2; -78°C

F, X, OH

X = OCF_3, 48 %
X = F, 14 %

[134] 2

CH_2OAc, AcO, O, OAc

CF_3OF; CCl_3F, -80°C

CH_2OAc, AcO, O, OAc, OCF_3, F

39 % +

CH_2OAc, AcO, O, OAc, F, F

37 % +

CH_2OAc, AcO, O, OCF_3, OAc, F

3 % +

CH_2OAc, AcO, O, F, OAc, F

2 %

[132] $C_6H_5C{\equiv}CC_6H_5 \xrightarrow[CFCl_3\ \ -78°C]{CF_3OF} C_6H_5CF_2CF(OCF_3)C_6H_5$ 3

75 %

[142] $CHF{=}CF_2 \xrightarrow[-111°\ to\ 20°C]{CF_3OF} CF_3OCHFCF_3 + CF_3OCF_2CHF_2$ 4

50 % 25 %

[143] C_6F_6 5

CF_3OF; -111° to 20°C

CF_3O, F (1-trifluoromethoxy-heptafluorocyclohexa-1,4-diene) + CF_3O, F (1-trifluoromethoxy-heptafluorocyclohexa-1,3-diene)

[*145*] $C_6H_5COCHN_2 \xrightarrow[\text{CFCl}_3,\ \text{CHCl}_3,\ -70°\text{C to RT}]{\text{CF}_3\text{OF}}$ **6**

$C_6H_5COCHFOCF_3$ (10 %) + $C_6H_5COCHF_2$ (14 %)

[*146*] 2,2,3,3-tetramethyl-1-(phenylsulfonyl)aziridine ($SO_2C_6H_5$ on N; CH_3, CH_3 on each ring carbon) $\xrightarrow[\text{CFCl}_3,\ 0°\text{C}]{\text{CF}_3\text{OF}}$ $CH_3C(CH_3)(NFSO_2C_6H_5)C(CH_3)(OCF_3)CH_3$ **7**

[*150*] trans-$C_6H_5CH{=}CHC_6H_5 \xrightarrow[\text{CHCl}_3,\ -78°\text{C}]{\text{CH}_3\text{COOF}}$ $C_6H_5CH(F)CH(O_2CCH_3)C_6H_5$ **8**

threo 50 %, *erythro* 7 %

[*150*] cyclohexene $\xrightarrow[\text{CHCl}_3,\ -78°\text{C}]{\text{CH}_3\text{COOF}}$ cis-1-fluoro-2-(O_2CCH_3)cyclohexane 60 % **9**

[*150*] $C_{10}H_{21}CH{=}CH_2 \xrightarrow[\text{CHCl}_3,\ -78°\text{C}]{\text{CH}_3\text{COOF}} C_{10}H_{21}CH(O_2CCH_3)CH_2F$ **10**

30 %

Trifluoroacetyl hypofluorite is prepared by reacting wet sodium trifluoroacetate with fluorine at low temperatures in an inert solvent. If the dry salt is used, the major product is **pentafluoroethyl hypofluorite** [*157*] (equation 11). Both hypofluorites add to stilbene to give vicinal adducts. Trifluoroacetyl hypofluorite also converts ***enol*** acetates to the corresponding *α-fluoroketones* [*158, 159*], but because of its high reactivity, is less generally useful as a fluorinating agent than acetyl hypofluorite. Acyl hypofluorites can also be prepared from longer chain perfluorocarboxylates [*160*] and from sodium α,α-dichloropropionate (but not from sodium propionate) and used to prepare α-fluoroketones from enol acetates [*161*].

Addition of **cesium fluoroxysulfate** to ***olefins*** gives *vicinal fluoroalkyl sulfates* with low regio- and stereoselectivity [*162, 163*] (equations 12 and 13). Reactions of this reagent with olefins in methanol or acetic acid give *vicinal fluoroalkyl methyl ethers or acetates,* respectively [*164, 165, 166*] (equation 14), with a predominance of *syn* and *Markovnikov selectivity.* ***Acetylenes*** give products from reaction with two equivalents of the reagent [*167*] (equation 15).

[*157*] **11**

$$CF_3CO_2F \xleftarrow[\text{CFCl}_3,\ -78^\circ\text{C}]{F_2} CF_3CO_2Na\ (\text{wet}) \qquad CF_3CO_2Na\ (\text{dry}) \xrightarrow[\text{CFCl}_3,\ -78^\circ\text{C}]{F_2} CF_3CF_2OF$$

cis-$C_6H_5CH{=}CHC_6H_5$

$$C_6H_5CHFCH(O_2CCF_3)C_6H_5 \qquad C_6H_5CHFCH(OCF_2CF_3)C_6H_5$$

erythro 65 %   *erythro* 38%, *threo* 7 %

[*163*] **12**

$$C_6H_5CH{=}CH_2 \xrightarrow[CH_3CN,\ RT]{CsSO_4F} C_6H_5CH(OSO_3Cs)CH_2F + C_6H_5CHFCH_2OSO_3Cs$$

20 %   51 %

[*163*] **13**

$$\text{cyclohexene} \xrightarrow[CH_2Cl_2,\ RT]{CsSO_4F} \text{cis-2-fluorocyclohexyl-}OSO_3Cs + \text{trans-2-fluorocyclohexyl-}OSO_3Cs$$

20 %   20 %

[*164*] **14**

$$(C_6H_5)_2C{=}CH_2 \xrightarrow[ROH,\ RT]{CsSO_4F} (C_6H_5)_2C(OR)CH_2F$$

R = CH_3 37 %
R = $OCCH_3$ 48 %

[*167*] **15**

$$C_6H_5C{\equiv}CH \xrightarrow[CH_3OH,\ RT]{CsSO_4F} C_6H_5C(OCH_3)_2CHF_2 + C_6H_5COCHF_2$$

25 %   26 %

Vicinal fluoroalkyl ethers are also obtained from ***olefins*** and **methyl hypofluorite** at low temperatures [*168*] and from the unstable reagents generated from additions of xenon difluoride to methanol [*169*] or other alcohols [*170*]. Alcohols tend to give complex product mixtures depending on the olefin structure, and best results are often achieved when reactions are run in the presence of boron trifluoride. Additions of xenon difluoride to trifluoromethanesulfonic, fluorosulfonic, or nitric acids give **fluoroxenonium reagents** that are stable to about –10 °C and react with ***olefins*** to give *vicinal adducts* with predominantly *syn stereochemistry* [*171*] (equation 16).

[*171*] cyclohexene $\xrightarrow[\text{CH}_2\text{Cl}_2,\ -78^\circ \text{ to } 10^\circ\text{C}]{\text{FXeOSO}_2\text{CF}_3}$ *cis*-1-fluoro-2-(trifluoromethylsulfonyloxy)cyclohexane (OSO_2CF_3, F) 75 % **16**

The highly reactive fluorinating agent ***N*-fluoropentachloropyridinium triflate** [*172*] reacts with ***olefins*** such as styrene and 2-methyl-1-pentene in acetic acid to give *vicinal fluoroalkyl esters* (equation 17). Use of alcohols or trimethylsilyl ethers as solvents affords the corresponding *vicinal fluoroalkyl ethers.*

[*172*] **17**

$$C_6H_5CH{=}CH_2 \xrightarrow[25^\circ C]{C_5Cl_5NF^{+}\,{}^{-}OSO_2CF_3,\ CH_3COOH} C_6H_5CH(O_2CCH_3)CH_2F \quad 72\ \%$$

Fluorine perchlorate [*173*] (equation 18), **pentafluorotellurium hypofluorite** [*174*] (equation 19), and **trifluoroamine oxide** [*175, 176*] (equation 20) add to certain ***fluorinated olefins.***

[*173*] **18**

$$CF_3CF{=}CF_2 \xrightarrow[-45^\circ C]{FOClO_3} \underset{50\ \%}{CF_3CF_2CF_2OClO_3} + \underset{24\ \%}{CF_3CF(OClO_3)CF_3}$$

[*174*] **19**

$$CF_3CF{=}CF_2 \xrightarrow[-196^\circ \text{ to RT}]{TeF_5OF} TeF_5OC_3F_7 \quad 78\ \%$$

[*175*] **20**

$$CF_2{=}CF_2 \xrightarrow[-196^\circ \text{ to RT}]{NF_3O,\ BF_3} CF_3CF_2ONF_2 \quad 60\ \%$$

Addition of Fluorine and Sulfur and Fluorine and Selenium

Fluorine and sulfur (in the form of a methylthio group) are added to ***nucleophilic olefins*** with *Markovnikov regioselectivity* and *anti stereoselectivity* by **dimethyl(methylthio)sulfonium fluoroborate and triethylamine tris(hydrogen fluoride)** [*177*] (equation 21).

A variety of reagent combinations add *fluorine and sulfur* to ***fluorinated olefins.*** Typically the olefin is reacted with fluoride ion and a source of electrophilic sulfur, such as sulfur tetrafluoride [*131, 178, 179, 180*], alkyl or amino sulfur trifluorides [*131, 181*], sulfenyl halides [*182, 183*] (equation 22), disulfur dichloride [*184*] (equation 23), thionyl fluoride [*178*], dialkyldisulfides [*185, 186*] (equation 24), sulfuryl fluoride [*131, 178*] (equation 25), thiazyl fluoride [*187*], or thiazyl trifluoride [*188*] (equation 26).

[*177*] cyclohexene $\xrightarrow[CH_2Cl_2,\ 0°C\ to\ RT]{(CH_3)_2SSCH_3\ BF_4,\ (C_2H_5)_3N{\cdot}3HF}$ trans-1-fluoro-2-(methylthio)cyclohexane (SCH_3, F) **21**

[*183*] $(CF_3)_2C{=}CF_2 \xrightarrow[monoglyme,\ 0\ to\ 20°C]{1.\ CsF;\ 2.\ C_6H_5SCl} (CF_3)_3CSC_6H_5$ 82 % **22**

[*184*] $(CF_3)_2C{=}CFCF_2CF_3 \xrightarrow[DMF]{CsF,\ S_2Cl_2} [C_3F_7C(CF_3)_2]_2S_3$ 50 % **23**

[*186*] $CF_3CF{=}CF_2 \xrightarrow{CsF,\ CH_3SSCH_3} (CF_3)_2CFSCH_3$ 38 % **24**

[*178*] $CF_3SF_4CF{=}CF_2 \xrightarrow[80°C]{CsF,\ SO_2F_2} [CF_3SF_4CF(CF_3)]_2SO_2$ 80 % **25**

[*188*] $(CF_3)_2C{=}CF_2 \xrightarrow[130°C]{CsF,\ NSF_3} (CF_3)_3CSF_2N$ 43 % **26**

Tetrakis(trifluoromethyl)dithietane is generated by dimerization of hexafluorothioacetone, which is prepared in situ from hexafluoropropylene with potassium fluoride and sulfur [*189*] or with sulfur and antimony pentafluoride [*190*] (equation 27):

[*189*] $CF_3CF{=}CF_2 \xrightarrow{KF,\ S;\ 120°C}$ 70 % tetrakis(trifluoromethyl)-1,3-dithietane ($(CF_3)_2C$, S, S, $C(CF_3)_2$) **27**

[*190*] $CF_3CF{=}CF_2 \xrightarrow{S,\ SbF_5;\ SO_2,\ 40°C}$ 40 %

The KF–S reaction presumably involves attack of a fluorinated carbanion on sulfur, whereas the S–SbF_5 reaction may involve electrophilic attack by a cationic sulfur species on the olefin under the strong Lewis acid conditions. Electrophilic attack on a fluorinated olefin may also account for formation of a perfluorinated sulfide from reaction of bis(pentafluorophenyl)disulfide with hexafluoropropylene under superacid conditions [*185*] (equation 28).

[*185*] $CF_3CF{=}CF_2 \xrightarrow[SbF_5,\ HF,\ 80°C]{C_6F_5SSC_6F_5} C_6F_5SCF(CF_3)_2$ 54 % **28**

A number of reagents, including **perfluoroalkylsulfur trifluorides** [*191*] (equation 29), will add fluorine and sulfur to *carbon–nitrogen double bonds* [*131*].

[*191*] $(CF_3)_2C{=}NH \xrightarrow[70°C]{CsF,\ (CF_3)_2CFSF_3} (CF_3)_2CFN{=}SFCF(CF_3)_2$ 18 % **29**

Fluorine and selenium, in the form of a phenylselenenyl group, add to ***nucleophilic olefins*** with *Markovnikov regioselectivity* and *anti stereoselectivity* on reaction with several reagents that may form phenylselenenyl fluoride in situ [*192, 193, 194*] (equation 30).

30

[*192*] cyclohexene $\xrightarrow[CH_3CN,\ RT]{AgF,\ C_6H_5SeCl}$ 92 %

[*194*] cyclohexene $\xrightarrow[CH_2Cl_2,\ -20°C]{C_6H_5SeSeC_6H_5,\ XeF_2}$ 73 % trans-1-fluoro-2-(SeC_6H_5)cyclohexane

[*193*] cyclohexene $\xrightarrow{AgF,\ C_6H_5SeBr}$ 57 %

Fluorinated tertiary selenoethers are prepared by reaction of branched ***perfluoroolefins*** with an **alkali metal fluoride** and **phenylselenenyl chloride** [*182*] (equation 31).

[*182*] $(CF_3)_2C{=}CFCF_2CF_3 \xrightarrow[DMF,\ RT]{1.\ KF;\ 2.\ C_6H_5SeCl} C_3F_7C(CF_3)_2SeC_6H_5$ 21 % **31**

Addition of Fluorine and Nitrogen

Fluorine and nitrogen may be added to olefins with the nitrogen in different oxidation states. Fluorine and a nitro group are added by reaction of an olefin with **nitryl fluoride** [*131*], nitronium tetrafluoroborate [*195*] (equation 32), or a combination of **nitric acid and hydrogen fluoride** [*131, 196*] (equation 33):

[*195*] cyclopentene $\xrightarrow{NO_2BF_4}$ 1-fluoro-2-(NO_2)cyclopentane 44 % **32**

[*196*] $CF_2{=}CH_2 \xrightarrow[FSO_3H,\ -10° \text{ to } 10°C]{HNO_3,\ HF} CF_3CH_2NO_2$ 93 % **33**

Nitrosyl fluoride reacts with ***steroidal olefins*** to give, ultimately, *α-fluoroketones* [*131*]; with steroidal vinyl fluorides, the reaction provides *α,α-difluoroketones* from *intermediate nitroimines* [*197*] (equation 34):

[*197*] **34**

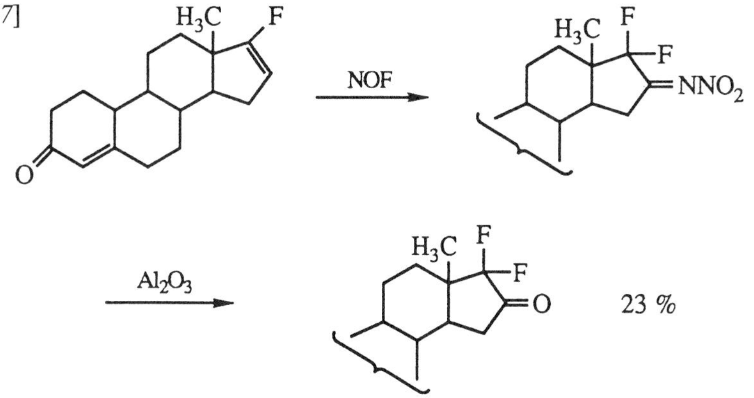

H_3C F — NOF → H_3C F F =NNO_2

O

$\xrightarrow{Al_2O_3}$ H_3C F F =O 23 %

Fluorinated olefins react with nitrosyl fluoride [*131*], dinitrogen tetroxide and fluoride ion [*131*], or nitrosyl chloride and fluoride ion [*198, 199*] (equations 35 and 36) to afford fluoronitroso compounds. Tertiary fluoronitroso compounds are a convenient source of tertiary perfluoroalcohols.

[*198*] $CF_3CF{=}CF_2 \xrightarrow[CH_3CN\ RT]{KF,\ CF_3CO_2Ag} (CF_3)_2CFAg$ **35**

$\xrightarrow{NOCl} (CF_3)_2CFNO$ 85 %

[*199*] $(CF_3)_2C{=}CFCF_2CF_3 \xrightarrow[DMAC,\ 0°C]{KF,\ NOCl} CF_3CF_2CF_2C(CF_3)_2NO$ **36**

$\longrightarrow CF_3CF_2CF_2C(CF_3)_2OH$ 65 %

Reaction of ***fluoroolefins*** with **fluoride ion and benzenediazonium chloride** affords perfluoroalkylazobenzenes [*200*] (equation 37).

Electrochemical oxidation of 1,2-dihydronaphthalene or an indene in acetonitrile containing **triethylamine tris(hydrogen fluoride)** provides a mixture of *stereoisomeric difluorides and vicinal fluoroacetamides* [*201*] (equation 38).

[*200*] $CF_3C(R){=}CF_2 \xrightarrow[CH_3CN]{CsF,\ C_6H_5N_2Cl} (CF_3)_2C(R)N{=}NC_6H_5$ **37**

R = F, CF_3 41 - 53 %

[*201*] **38**

$C(CH_3)_3$ (indene) $\xrightarrow[CH_3CN,\ Electrolysis]{(C_2H_5)_3N\cdot 3HF}$

$C(CH_3)_3$ ···F, F 22 % + $C(CH_3)_3$ ···F, $NHCOCH_3$ 74 %

N-Fluoropyridinium triflates react with cyclic enol ethers to give addition products as mixtures of *cis* and *trans isomers* [*172*] (equation 39).

[*172*] **39**

(dihydropyran, O) $\xrightarrow[CH_2Cl_2,\ reflux]{C_5H_5NF^{+}\ {}^{-}OSO_2CF_3}$ F, O, $\overset{+}{N}C_5H_5$ $^{-}OSO_2CF_3$ 86 %

Addition of Fluorine and Carbon

Addition of fluorine and carbon to nucleophilic olefins is little studied, a rare example being the photochemical addition of benzotrifluoride to electron-rich cyclic olefins [*202*] (equation 40):

[*202*] **40**

CH_3, CH_3, O, O, CH_3, CH_3 + $C_5H_5CF_3$ $\xrightarrow[dioxane]{h\nu}$ F, CH_3, CH_3, O, O, CH_3, CH_3, $CF_2C_6H_5$

60 - 80 %

In contrast, *additions of fluorine and carbon to fluorinated olefins* are widely investigated. The best known processes involve reactions of olefins with fluoride ion to generate carbanionic intermediates [*203*] that are trapped in situ by carbon-based electrophiles.

A classic example of this process is *dimerization* or *oligomerization* of the olefins in the presence of fluoride ion [*131, 204, 205*], in which the electrophile is the fluoroolefin itself. ***Hexafluoropropylene*** can give dimers in yields as high as 92% [*206*]. Dimers of fluorinated dienes [*207*] and cyclic [*208*] (equation 41) and functionalized [*209*] (equation 42) fluoroolefins can be obtained, as can mixed oligomers from two different fluoroolefins [*210, 211*] (equation 43).

[*208*] (cyclic perfluorocyclopentene) $\xrightarrow[\text{tetramethylenesulfone, 80°C}]{\text{CsF}}$ (F)=(F) 86 % **41**

[*209*] $CClF{=}CFCO_2CH_3 \xrightarrow[CH_3CN,\ 80°C]{KF}$ **42**

$CH_3O_2CCF(CF_3)CF{=}CFCO_2CH_3$ 89 %

[*210*] $(CF_3)_2C{=}CF_2 + CF_2{=}CF_2 \xrightarrow[CH_3CN]{CsF}$ **43**

$(CF_3)_3CCF_2CF_2CF{=}C(CF_3)_2$ 60 %

Other well-known reactions are those of ***fluorinated olefins*** with fluoride ion and negatively substituted aromatic compounds leading to the formation of *perfluoroalkylated aromatic compounds*. The reaction may be considered an *anionic version* of a *Friedel–Crafts process* and can result in introduction of one or several perfluoroalkyl substituents [*131*]. Aromatic substrates include substituted and unsubstituted ***perfluorobenzenes*** [*131, 212, 213, 214*], ***fluorinated heterocycles*** [*131, 203, 215, 216, 217, 218, 219, 220, 221, 222, 223*], ***perchlorinated heterocycles*** [*224*] (equation 44), and other ***activated aromatic compounds*** [*225*] (equation 45). The fluorinated olefins can be linear or cyclic [*208*] (equation 46).

[*224*] (trichloro-1,2,4-triazine: Cl, Cl, Cl) $+ CF_3CF{=}CF_2 \xrightarrow[\text{sulfolane, 25°C}]{KF}$ **44**

(tris-substituted triazine: $(CF_3)_2CF$, $(CF_3)_2CF$, $CF(CF_3)_2$) 61 %

[*225*] (4-fluoronitrobenzene: F, NO_2) $+ (CF_3)_2C{=}CF_2 \xrightarrow[\text{sulfolane, 80°C}]{CsF}$ (4-nitrophenyl: $C(CF_3)_3$, NO_2) 75 % **45**

References are listed on pages 84–96.

[*208*] **46**

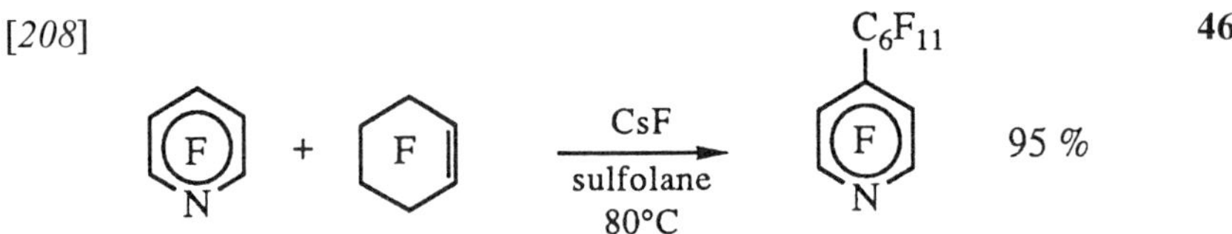

Other electrophiles that react with in situ-generated perfluorocarbanions include epoxides [*226*] (equation 47), carbon dioxide [*227*] (equation 47), acyl halides [*228, 229, 230, 231, 232, 233*] (equation 48), fluoroformates [*234*], carbonyl fluoride [*235, 236, 237*], hexafluorothioacetone (generated from its dimer) [*238*] (equation 48), an α-fluoroalkylamine [*239*] (equation 48), cyanuric fluoride [*240*], and reactive alkyl halides [*241, 242, 243, 244, 245*] (equation 49). Interestingly, an in situ-generated carbanion will also react with dibromodifluoromethane via a mechanism involving *difluorocarbene* [*246*] (equation 50).

47

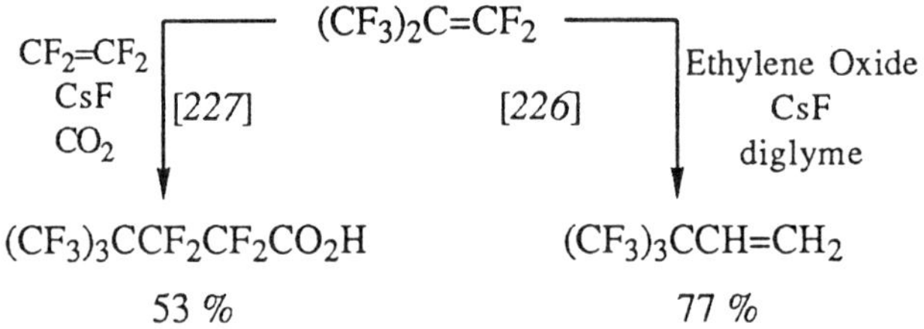

$(CF_3)_2C{=}CF_2$

$CF_2{=}CF_2$, CsF, CO_2 [*227*] → $(CF_3)_3CCF_2CF_2CO_2H$ 53 %

Ethylene Oxide, CsF, diglyme [*226*] → $(CF_3)_3CCH{=}CH_2$ 77 %

48

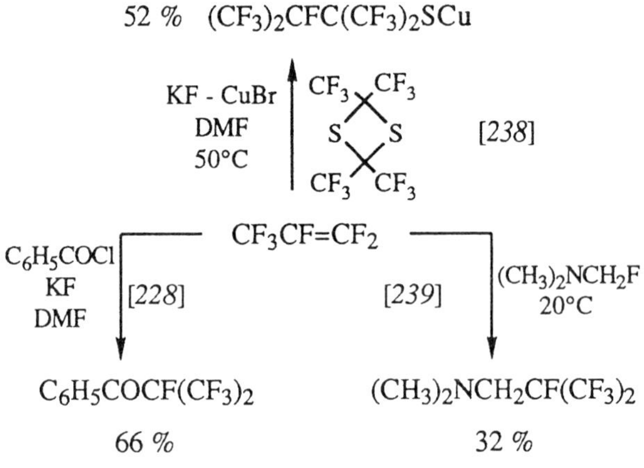

52 % $(CF_3)_2CFC(CF_3)_2SCu$

↑ KF - CuBr, DMF, 50°C; CF_3, CF_3 / S S / CF_3, CF_3 (dimer) [*238*]

$CF_3CF{=}CF_2$

C_6H_5COCl, KF, DMF [*228*] → $C_6H_5COCF(CF_3)_2$ 66 %

$(CH_3)_2NCH_2F$, 20°C [*239*] → $(CH_3)_2NCH_2CF(CF_3)_2$ 32 %

49

$C_2F_5CF{=}C(CF_3)_2$

CH_3I, CsF, DMF, 20° to 60°C [*241*] → $C_3F_7C(CF_3)_2CH_3$ 79 %

$C_6H_5CH_2Br$, CsF, diglyme, 50° - 60°C [*244*] → $C_3F_7C(CF_3)_2CH_2C_6H_5$ 90 %

References are listed on pages 84–96.

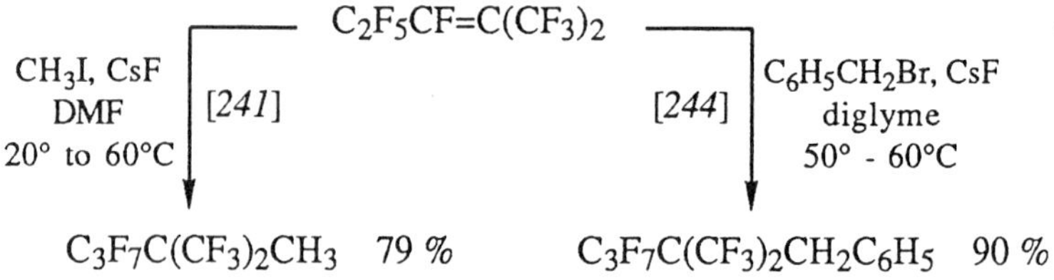

[*246*] $C_2F_5CF{=}C(CF_3)_2 \xrightarrow[\text{diglyme}]{CF_2Br_2,\ CsF} C_3F_7C(CF_3)_2CF_2Br$ 75 % **50**

An allylic perfluorinated carbanion can be generated from ***tetrakis(trifluoromethyl)allene*** and cesium fluoride and can be trapped by reactive alkylating agents [*247*] (equation 51).

[*247*] $(CF_3)_2C{=}C{=}C(CF_3)_2 \xrightarrow[\text{diglyme, 120°C}]{CH_3I,\ CsF} CH_3C(CF_3)_2CF{=}C(CF_3)_2$ 44 % **51**

Hexafluoro-2-butyne readily undergoes anionic oligomerization in the presence of fluoride ion [*248*], but the intermediate vinylic carbanion can be trapped by highly electrophilic fluorinated heterocycles [*249*] (equation 52).

[*249*] $CF_3C{\equiv}CCF_3$ + tetrafluoropyrimidine $\xrightarrow[\text{sulfolane, 20°C}]{CsF}$ trifluoropyrimidinyl–$C(CF_3){=}CF(CF_3)$ **52**

trans 63 %, *cis* 7 %

Addition of carbon and fluorine can also be initiated by *electrophilic attack on a fluorinated olefin* under strongly acidic conditions [*250, 251, 252, 253, 254, 255*]. Best known are *fluoroalkylations* of **tetrafluoroethylene** by tertiary or highly halogenated allylic or benzylic cations in the presence of antimony pentafluoride (equation 53):

53

$CF_2{=}CF_2$ →

[*254*] hexafluorocyclobutene, SbF_5 → 1-pentafluoroethyl-pentafluorocyclobutene (CF_2CF_3) 73 %

[*252*] $CF_3CCl{=}CCl_2$, SbF_5 → $C_3F_7CCl{=}CCl_2$ 60 %

[*255*] $(CH_3)_3CCl$, SbF_5 → $(CH_3)_3CCF_2CF_3$ 60 %

[*250*] perfluoroindene, SbF_5 → 1-C_2F_5-perfluoroindene 33 %

Acyl fluorides [*256*] and formaldehyde [*257*] add to fluorinated olefins, such as difluoroethene, under Lewis or Brønsted acid conditions (equation 54).

$$CF_2{=}CH_2 \xrightarrow[[257]]{CH_2O,\ CH_3OH,\ HF} CF_3CH_2CH_2OCH_3\ \ 56\,\%$$

$$CF_2{=}CH_2 \xrightarrow[[256]]{C_6H_5COF,\ SbF_5,\ SO_2} C_6H_5COCH_2CF_3\ \ 33\,\%$$

54

Additions of fluorine and carbon across the *carbonyl group* of highly fluorinated ketones and acid fluorides are important reactions for the synthesis of fluorinated monomers and intermediates. Generally, the carbonyl compound is treated with an alkali metal fluoride or other source of fluoride ion [*258*] in a polar aprotic solvent to generate an intermediate alkoxide. The alkoxide is alkylated by a variety of reagents, including allyl bromide [*259, 260*] (equation 55), α-haloketones [*261, 262*] and nitriles [*263*](equation 56), alkyl iodides [*131*], and alkyl sulfates [*264, 265*] (equation 57) and can react with the combination of iodine and tetrafluoroethylene to add an iodotetrafluoroethyl group [*266, 267, 268, 269*] (equation 58). *Intramolecular fluoroalkylations* are also observed [*270*] (equation 59).

[*259*]

$$C_3F_7OCF(CF_3)COF \xrightarrow[\text{diglyme},\ 0°\ \text{to}\ 90°C]{CsF,\ CH_2{=}CHCH_2Br} C_3F_7OCF(CF_3)CF_2OCH_2CH{=}CH_2\ \ 84\,\%$$

55

$$CF_3COCF_3 \xrightarrow[[263]]{KF,\ ClCH_2CN,\ \text{diglyme}} (CF_3)_2CFOCH_2CN$$

$$CF_3COCF_3 \xrightarrow[[261]]{CsF,\ ClCH_2COCH_2Cl,\ \text{diglyme}} [(CF_3)_2CFOCH_2]_2CO\ \ 88\%$$

56

[*264*]

$$CF_3COCOCF_3 \xrightarrow[\text{tetraglyme},\ 100°\ C]{KF,\ (CH_3SO_3CH_2)_2} \text{2,3-difluoro-2,3-bis(trifluoromethyl)-1,4-dioxane (O, CF}_3\text{, F, F, CF}_3\text{)}\ \ 46\,\%$$

57

[*267*]

$$CF_2O \xrightarrow[\text{tetraglyme}]{CsF,\ CF_2{=}CF_2\ -\ ICl} CF_3OCF_2CF_2I\ \ 69\,\%$$

58

[*270*] **59**

$$ClCH_2CH_2OCF_2COF \xrightarrow[\text{diglyme, } 110°C]{CsF} \text{2,2,3,3-tetrafluoro-1,4-dioxane} \quad 33\ \%$$

A particularly important process for the *synthesis of fluorinated monomers* is attack of **functionalized perfluoroalkoxides** onto ***hexafluoropropylene oxide*** [*271, 272, 273, 274*] (equation 60) or other fluorinated epoxides [*275, 276*], resulting in ring-opening of the epoxide and formation of a new perfluoroalkoxide or acid fluoride:

[*274*] **60**

$$C_2H_5O_2CCF_2CF_2COF + \text{hexafluoropropylene oxide} \xrightarrow[\text{tetraglyme, } -10° \text{ to } 0°C]{CsF}$$

$$C_2H_5O_2CCF_2CF_2CF_2OCF(CF_3)COF \quad 57\ \%$$

Fluoride ion can also initiate ring opening of hexafluoropropylene oxide to give a heptafluoropropoxide, which can react in turn with the epoxide to give dimers or higher oligomers that are precursors to perfluorinated stable fluids [*277, 278, 279*] (equation 61):

[*278*] **61**

$$\text{hexafluoropropylene oxide} \xrightarrow[\text{18-crown-6, diglyme, RT}]{KF}$$

$$CF_3CF_2CF_2O[CF(CF_3)CF_2O]_nCF(CF_3)COF \quad n = 0\text{: } 41\ \%;\ n = 1\text{: } 18\ \%$$

Fluorinated lactone rings are also opened by fluoride ion to give alkoxides that can be trapped by electrophiles [*280*] (equation 62).

[*280*] **62**

$$\text{tetrafluoro-}\beta\text{-propiolactone} \xrightarrow{KF,\ CF_2{=}CF_2,\ I_2} ICF_2CF_2O(CF_2)_2COF$$

Carbon and fluorine can be added to ***carbon–nitrogen multiple bonds,*** generally in highly fluorinated substrates. Perfluoromethanimine adds fluoride ion to give perfluoromethanamine ion, which will react with acid fluorides or with another molecule of the imine to give linear or cyclic adducts [*281*] (equation 63). Other ***fluorinated linear*** [*282*] (equation 64) and ***cyclic imines*** [*283, 284, 285, 286*] (equation 65) give azanions that can be alkylated. ***Alkyl, aryl, and perfluoroalkyl nitriles*** [*287, 288, 289*] (equation 66) and ***fluorinated imines*** [*290*] (equation 67) react with **carbonyl fluoride** to give *fluorinated isocyanates.* A fluorinated isocyanate reacts further with fluoride ion to give a nitroanion intermediate that can be alkylated [*291*] (equation 68).

[*281*] **63**

$CF_2{=}NF \xrightarrow{KF} CF_3NF^- K^+$

$CF_3NF^- K^+ \xrightarrow[22°C]{RCOF} CF_3NFCOR$ (R = F: 93 %; R = CF_3: 68 %)

$CF_3NF^- K^+ \xrightarrow{CsF,\ CF_2{=}NF}$ $FN{-}CF_2{-}NCF_3$ (three-membered ring, C bearing F_2) 70 %

$CF_3NF^- K^+ \xrightarrow[22°C]{KF,\ CF_2{=}NF} CF_3NFCF{=}NF$ 80 %

[*282*] **64**

$$CF_3N{=}CF_2 \xrightarrow[CH_3CN]{CsF,\ BrCH_2CO_2C_2H_5} (CF_3)_2NCH_2CO_2C_2H_5 \quad 44\ \%$$

[*283*] **65**

F-ring cyclic imine (perfluoro-tetrahydropyridine) $\xrightarrow[CH_3CN,\ RT]{CsF,\ CH_3I}$ F-ring N-CH_3 (perfluoro-N-methylpiperidine) 69 %

[*287*] **66**

$$C_2H_5CN \xrightarrow[RT]{COF_2,\ HF} C_2H_5CF_2NCO \quad 95\ \%$$

[*290*] **67**

$$(CF_3)_2C{=}NH \xrightarrow[CH_3CN,\ -190°C\ to\ RT]{COF_2,\ KF} (CF_3)_2CFNCO$$

[*291*] **68**

$$CF_3NCO \xrightarrow[diglyme,\ 40°-50°C]{CsF,\ BrCH_2CH{=}CH_2} CF_3N(COF)(CH_2CH{=}CH_2) \quad 59\ \%$$

References for Pages 54–84

1. Hudlický, M. *Chemistry of Organic Fluorine Compounds,* 2nd ed.; Ellis Horwood Ltd.: Chichester, United Kingdom; John Wiley: New York, 1976; pp 36–41.

2. Kraus, W. P.; Hutson, T. U.S. Patent 4 049 728, 1977; *Chem. Abstr.* **1977,** *87,* 200767.
3. Sobel, J. E. U.S. Patent 3 928 486, 1975; *Chem. Abstr.* **1976,** *84,* 104957.
4. Pazderskii, Y. A.; Romanyuk, N. P.; Kostyk, G. P. U.S.S.R. Patent 565 908, 1977; *Chem. Abstr.* **1977,** *87,* 167525.
5. Carter, C. O. U.S. Patent 3 906 051, 1975; *Chem. Abstr.* **1976,** *84,* 30391.
6. Hutson, T.; Carter, C. O. U.S. Patent 4 052 469, 1977; *Chem. Abstr.* **1977,** *87,* 200769.
7. Sweeney, R. F.; Woolf, C. U.S. Patent 2 917 559, 1959; *Chem. Abstr.* **1960,** *54,* 10858.
8. Neder, A.; Uskert, A.; Mehesfalvi, Z.; Kuszmann, J. *Acta Chim. Acad. Sci. Hung.* **1980,** *104,* 123.
9. Olah, G. A.; Welch, J. T.; Vankar, Y. D.; Nojima, M.; Kerekes, I.; Olah, J. A. *J. Org. Chem.* **1979,** *44,* 3872.
10. Olah, G. A.; Li, X. Y. *Synlett* **1990,** 267.
11. Tojo, M.; Fukuoka, S. Jpn. Kokai Tokkyo Koho JP 63 088 146 A2, 1988; *Chem. Abstr.* **1988,** *109,* 189978.
12. Yoneda, N.; Nagata, S.; Fukuhara, T.; Suzuki, A. *Chem. Lett.* **1984,** 1241.
13. Manzer, L. *E. Science* (Washington, D.C.) **1990,** *249,* 31.
14. Feiring, A. E. *J. Fluorine Chem.* **1979,** *14,* 7.
15. Von Halasz, S. P. Ger. Offen. DE 3 009 760, 1981; *Chem. Abstr.* **1981,** *95,* 186620.
16. Golubev, A. N.; Gol'dino, A. L.; Panshin, Yu. A.; Kolomenskov, V. I. U.S.S.R. Patent 341 788, 1972; *Chem. Abstr.* **1972,** *78,* 3663.
17. Usmanov, K. U.; Yul'chibaev, A. A.; Sirlibaev, T. S.; Akrambhodzhaev, A. U.S.S.R. Patent 466 202, 1975; *Chem. Abstr.* **1975,** *83,* 58068.
18. Kuroda, T.; Furukawa, Y. Jpn. Kokai Tokkyo Koho 78 116 305, 1978; *Chem. Abstr.* **1979,** *90,* 103374.
19. Martens, G.; Godfroid, M. Ger. Offen. DE 2 215 019, 1972; *Chem. Abstr.* **1973,** *78,* 42840.
20. Franklin, J.; Janssens, F. Eur. Pat. Appl. EP 361 578, 1990; *Chem. Abstr.* **1990,** *113,* 77674.
21. Schultz, N.; Vahlensieck, H. J.; Gebele, R. Ger. Offen. DE 1 900 241, 1970; *Chem. Abstr.* **1971,** *74,* 3310.
22. Stolkin, I.; Koetzsch, H. J. Ger. Offen. DE 1 910 529, 1970; *Chem. Abstr.* **1970,** *73,* 120052.
23. Schultz, N.; Vahlensieck, H. J. Ger. Offen. DE 2 000 200, 1971; *Chem. Abstr.* **1971,** *75,* 129319.
24. Gumprecht, W. H. Eur. Pat. Appl. EP 353 059, 1990; *Chem. Abstr.* **1990,** *113,* 5701.

25. Van der Puy, M. U.S. Patent 4 374 289, 1983; *Chem. Abstr.* **1983,** *98,* 197593.
26. Henne, A. L.; Arnold, R. C. *J. Am. Chem. Soc.* **1948,** *70,* 758.
27. Hirayama, H.; Kobayashi, H.; Oho, H.; Tomota, S.; Takaichi, A. Jpn. Kokai Tokkyo Koho JP 02 095 438, 1990; *Chem. Abstr.* **1990,** *113,* 58474.
28. Schultz, N.; Vahlensieck, H. J. Ger. Offen. DE 2 108 951, 1972; *Chem. Abstr.* **1972,** *77,* 139405.
29. Rao, V. M. N. PCT Int. Appl. WO8 912 614, 1989; *Chem. Abstr.* **1990,** *113,* 5700.
30. Rao, V. M. N. Eur. Pat. Appl. EP 348 190, 1989; *Chem. Abstr.* **1990,** *112,* 216210.
31. Manzer, L. E.; Rao, V. N. M. U.S. Patent 4 766 260, 1988; *Chem. Abstr.* **1989,** *110,* 97550.
32. Darragh, J. I.; Potter, S. E. Ger. Offen. DE 2 837 515, 1979; *Chem. Abstr.* **1979,** *91,* 19875.
33. Webb, J. L.; Corn, J. E. *J. Org. Chem.* **1973,** *38,* 2091.
34. Halasz Von, S. P. Ger. Offen. 2 712 732, 1978; *Chem. Abstr.* **1978,** *89,* 21488 [57].
35. Kitasume, T.; Onogi, T. *Synthesis* **1988,** 614.
36. Paleta, O.; Havlu, V.; Dedek, V. *Collect. Czech. Chem. Commun.* **1980,** *45,* 415.
37. Einer, H. F.; Kirk, R.; Noftle, R. E.; Uhrig, M. *Polyhedron* **1982,** *1,* 723.
38. Maximovich, M. J.; Stevens, H. C.; Trager, F. C. Br. Patent GB 1 170 396, 1969; *Chem. Abstr.* **1970,** *72,* 44361.
39. Meussdoerffer, J. N.; Niederpruem; H. Ger. Offen. DE 2 139 993, 1973; *Chem. Abstr.* **1973,** *78,* 123998.
40. Meussdoerffer, J. N.; Niederpruem; H. Ger. Offen. DE 2 234 305, 1974; *Chem. Abstr.* **1974,** *80,* 107970.
41. Ogura, E.; Hatabu, K.; Nomura, N. U.S. Patent 3 555 102, 1971; *Chem. Abstr.* **1971,** *74,* 64586.
42. Usmanov, Kh. U.; Yul'chibaev, A. A.; Sirlibaev, T. S.; Saparniyazok, K. *Izv. Vyssh. Ucheb. Zared., Khim. Khim. Tekhnol.* **1973,** *16,* 77; *Chem. Abstr.* **1973,** *78,* 123901.
43. Newkirk, A. E. *J. Am. Chem. Soc.* **1946,** *68,* 2467.
44. Gardner, L. E. U.S. Patent 3 607 955, 1971; *Chem. Abstr.* **1971,** *75,* 129320.
45. Usmanov, Kh. U.; Sirlibaev, T. S.; Akrambhodzhaev, A.; Saparniyazov, K.; Yul'chibaev, A. A. *Zh. Prikl. Khim. (Leningrad)* **1978,** *51,* 1839; *Chem. Abstr.* **1978,** *89,* 147253.
46. Daikin Kogyo Co. Fr. Patent 1 570 306, 1969; *Chem. Abstr.* **1970,** *72,* 132015. Fr. Patent 1 601 442, 1970; *Chem. Abstr.* **1971,** *74,* 126389.
47. Shinoda, K.; Watanabe, T.; Mizusawa, S. Ger. Offen. DE 1 941 234, 1970; *Chem. Abstr.* **1970,** *72,* 110762.

48. Paucksch, H.; Massonne, J.; Derleth, H. Ger. Offen. DE 2 105 748, 1972; *Chem. Abstr.* **1972,** *77,* 164009.
49. Cousseau, J.; Albert, P. *Bull. Soc. Chim. Fr.* **1986,** 910.
50. Gorgues, A.; Stephan, D.; Cousseau, J. *J. Chem. Soc., Chem. Commun.* **1989,** 1493.
51. Alvernhe, G.; Kozlowska-Gramaz, E.; Lacombe-Bar, S.; Laurent, A. *Tetrahedron Lett.* **1978,** 5203.
52. Wade, T. N.; Kheribet, R. *J. Org. Chem.* **1980,** *45,* 5333.
53. Nelson, S. J. Ger. Offen. 3 019 590, 1980; *Chem. Abstr.* **1981,** *94,* 191729.
54. Feiring, A. E. *J. Org. Chem.* **1976,** *41,* 148.
55. Gillespie, R. J.; Hulme, R. *J. Chem. Soc., Dalton Trans.* **1973,** 1261.
56. Lentz, D.; Oberhammer, H. *Inorg. Chem.* **1985,** *24,* 4665.
57. Hanack, M.; Dolde, J. *Liebigs Ann. Chem.* **1973,** *9,* 1557.
58. Kent, P. W.; Wood, K. R.; Welch, V. A. *J. Chem. Soc.* **1964,** 2493.
59. Gregorcic, A.; Zupan, M. *Vestn. Slov. Kem. Drus.* **1978,** *25,* 135; *Chem. Abstr.* **1978,** *89,* 129103.
60. Crochemore, M. Fr. Demande FR 2 627 488, 1989; *Chem. Abstr.* **1990,** *112,* 76398.
61. DesMarteau, D. D.; Kotun, S. P.; Malacrida, A. Eur. Pat. Appl. EP 353 721, 1990; *Chem. Abstr.* **1990,** *113,* 58476.
62. Hudlicky', M. *Chemistry of Organic Fluorine Compounds,* 2nd ed.; Ellis Horwood Ltd.: Chichester, United Kingdom; Wiley: New York, 1976; pp 52–56.
63. Gambaretto, G. P.; Napoli, M. *J. Fluorine Chem.* **1976,** *7,* 569.
64. Boguslavskaya, L. S.; Chuvatkin, N. N.; Panteleeva, I. Y. *Zh. Org. Khim.* **1982,** *18,* 2082 (Engl. Transl. 1832).
65. Boguslavskaya, L. S.; Panteleeva, I. Y.; Ternovskoi, L. A.; Krom, E. N. *Zh. Org. Khim.* **1980,** *16,* 2525 (Engl. Transl. 2155).
66. Rozen, S.; Brand, M. *J. Org. Chem.* **1985,** *50,* 3342.
67. Boguslavskaya, L. S.; Chuvatkin, N. N.; Kartashov, A.; Ternovskoi, L. A. *Zh. Org. Khim.* **1987,** *23,* 262 (Engl. Transl. 230).
68. Heasley, V. L.; Gipe, R. K.; Martin, J. L.; Wiese, H. C.; Oakes, M. L.; Shellhamer, D. F.; Heasley, G. E.; Robinson, B. L. *J. Org. Chem.* **1983,** *48,* 3195.
69. Camps, F.; Chamorro, E.; Gasol, V.; Guerrero, A. *J. Org. Chem.* **1989,** *54,* 4294.
70. Alvernhe, G.; Laurent, A.; Haufe, G. *Synthesis* **1987,** 562.
71. Serguchev, Yu. A.; Gutsulyak, R. B. *Zh. Org. Khim.* **1986,** *22,* 668 (Engl. Transl. 597).
72. Bowers, A.; Ibañez, L. C.; Denot, E.; Becerra, R. *J. Am. Chem. Soc.* **1960,** *82,* 4001.
73. Moural, J.; Mícková, R.; Schwarz, V. Czech. Patent 178 287, 1979; *Chem. Abstr.* **1980,** *92,* 42247.

74. Barluenga, J.; Campos, P. J.; Gonzalez, J. M.; Suarez, J. *J. Org. Chem.* **1991,** *56,* 2234.
75. Evans, R. D.; Schauble, J. H. *Synthesis* **1987,** 551.
76. Ichihara, J.; Funabiki, K.; Hanafusa, T. *Tetrahedron Lett.* **1990,** *31,* 3167.
77. Schmidt, H.; Meinert, H. *Angew. Chem.* **1960,** *72,* 493.
78. Kuroboshi, M.; Hiyama, T. *Tetrahedron Lett.* **1991,** *32,* 1215.
79. Maeda, M.; Abe, M.; Kojima, M. *J. Fluorine Chem.* **1987,** *34,* 337.
80. Shimizu, M.; Nakahara, Y.; Yoshioka, H. *J. Chem. Soc., Chem. Commun.* **1989,** 1881.
81. Stavber, S.; Zupan, M. *J. Fluorine Chem.* **1977,** *10,* 271.
82. Zupan, M.; Pollak, A. *Tetrahedron Lett.* **1975,** 3525.
83. Stavber, S.; Zupan, M. *J. Fluorine Chem.* **1978,** *12,* 307.
84. Mel'nikova, N. B.; Krom, E. N.; Kartashov, V. R.; Ternovskii, L. A. *Izv. Vyssh. Uchebn. Zaved., Khim. Khim. Tehknol.* **1981,** *24,* 1070; *Chem. Abstr.* **1982,** *96,* 19523.
85. Hamman, S.; Beguin, C. G. *J. Fluorine Chem.* **1983,** *23,* 515.
86. Zupan, M.; Pollak, A. *J. Chem. Soc., Perkin Trans. 1* **1976,** 971.
87. Gregorcic, A.; Zupan, M. *J. Fluorine Chem.* **1984,** *24,* 291.
88. Gregorcic, A.; Zupan, M. *Bull. Chem. Soc. Jpn.* **1987,** *60,* 3083.
89. Gregorcic, A.; Zupan, M. *Collect. Czech. Chem. Commun.* **1977,** *42,* 3192.
90. Dean, F. H.; Marshall, D. R.; Warnhoff, E. W. Pattison, F. L. M. *Can. J. Chem.* ***1967,*** *45,* 2279.
91. Gregorcic, A.; Zupan, M. *Tetrahedron* **1977,** *33,* 3243.
92. Alvernhe, G.; Anker, D; Laurent, A; Haufe, G.; Beguin, C. *Tetrahedron* **1988,** *44,* 3551.
93. Haufe, G.; Alvernhe, G.; Laurent, A. *Tetrahedron Lett.* **1986,** 4449.
94. Haufe, G.; Alvernhe, G.; Laurent, A. *J. Fluorine Chem.* **1990,** *46,* 83.
95. Grinenko, G. S.; Samsonova, N. V.; Gusarova, T. I. U.S.S.R. Patent 422 242, 1975; *Chem. Abstr.* **1977,** *87,* 168271.
96. Kerb, U.; Wieske, R. Ger. Patent 1 593 499, 1974; *Chem. Abstr.* **1975,** *82,* 17009.
97. Mícková, R.; Moural, J.; Schwarz, V. *Tetrahedron Lett.* **1978,** 1315.
98. Samsonova, N. V.; Grinenko, G. S.; Alekseeva, L. M.; Sheinker, Yu. N. *Khim.-Farm. Zh.* **1976,** *10,* 106; *Chem. Abstr.* **1976,** *85,* 33259.
99. Skibinska, M.; Ksiezny, C. D.; Cieslik, H.; Jaworska, R.; Rzasa, J.; Uszycka Hoirawa, T.; Wajcht, J. Pol. Patent 95 883 (1978); *Chem. Abstr.* **1979,** *91,* 20904.
100. Breazu, D.; Kovendi, A.; Deesy, A. I.; Fey, L.; Moraru, L.; Laurentiu, O. Rom. Patent 58 603, 1975; *Chem. Abstr.* **1978,** *88,* 191249.
101. Bowers, A.; Denot, E.; Becerra, R. *J. Am. Chem. Soc.* **1960,** *82,* 4007.
102. Daikin Kogyo Co., Ltd. Jpn. Kokai Tokkyo Koho JP 60 23 33, 1985; *Chem. Abstr.* **1985,** *103,* 5882.

103. Gard, G. L.; Woolf, C. *J. Fluorine Chem.* **1972,** *1,* 487.
104. Millauer, H. Ger. Offen. 2 100 140, 1972; *Chem. Abstr.* **1972,** *77,* 125944.
105. Chambers, R. D.; Musgrave, W. K. R.; Savory, J. *J. Chem. Soc.* **1961,** 3779.
106. Lo, E. S.; Readio, J. D.; Iserson, H. *J. Org. Chem.* **1970,** *35,* 2051.
107. Nagai, M.; Imazu, S.; Shinkai, H.; Kato, T.; Asaoka, M.; Nakatsu, T. Jpn. Kokai 74 18 806, 1974; *Chem. Abstr.* **1974,** *80,* 132775.
108. Ono, H.; Higaki, H. Jpn. Patent 72 07 529, 1972; *Chem. Abstr.* **1972,** *76,* 139916.
109. Nomura, N.; Ikubo, Y. Jpn. Kokai 77 68 109, 1977; *Chem. Abstr.* **1977,** *87,* 133862.
110. Dear, R. E.; Woolf, C. U.S. Patent 3 770 838, 1973; *Chem. Abstr.* **1974,** *80,* 14547.
111. Krespan, C. G.; Smart, B. E. U.S. Patent 4 922 038, 1990; *Chem. Abstr.* **1990,** *113,* 97028.
112. Eleev, A. F.; Vasil'ev, N. V. Sokol'skii, G. A. *Zh. Org. Khim.* **1985,** *21,* 280 (Engl. Transl. 250).
113. Probst, A.; Raab, K.; Ulm, K.; Werner, v. K. *J. Fluorine Chem.* **1987,** *37,* 223.
114. Smart, B. E.; Middleton, W. J.; Farnham, W. B. *J. Am. Chem. Soc.* **1986,** *108,* 4905.
115. Pletnev, S. I.; Igumnov, S. M.; Zakharova, E. V.; Makarov, K. N. *Izv. Akad. Nauk SSSR, Ser. Khim.* **1990,** 636 (Engl. Transl. 557).
116. Nazarenko, T. I.; Deev, L. E.; Ponomarev, V. G.; Pashkevich, K. I. *Izv. Akad. Nauk SSSR, Ser. Khim.* **1989,** 2869 (Engl. Transl. 2632).
117. Rozen, S.; Brand, M. *J. Org. Chem.* **1986,** *51,* 222.
118. Bastock, T. W.; Harley, M. E.; Pedler, A. E.; Tatlow, J. C. *J. Fluorine Chem.* **1975,** *6,* 331.
119. Bastock, T. W.; Pedler, A. E.; Tatlow, J. C. *J. Fluorine Chem.* **1976,** *8,* 11.
120. Haspel-Hentrich, F.; Shreeve, J. M. *Inorg. Synth.* **1986,** *24,* 58.
121. Mir, Q.-C.; DesMarteau, D. D. *J. Fluorine Chem.* **1990,** *48,* 367.
122. Patel, N. R.; Kirchmeier, R. L.; Shreeve, J. M. *J. Fluorine Chem.* **1990,** *48,* 395.
123. Sarwar, G.; Kirchmeier, R. L.; Shreeve, J. M. *Inorg. Chem.* **1990,** *29,* 571.
124. Bailey, A. R.; Banks, R. E. *J. Fluorine Chem.* **1983,** *23,* 87.
125. O'Brien, B. A.; DesMarteau, D. D. *J. Org. Chem.* **1984,** *49,* 1467.
126. O'Brien, B. A.; DesMarteau, D. D. *Rev. Chim. Miner.* **1986,** *23,* 621.
127. O'Brien, B. A.; Thrasher, J. S.; Bauknight, C. W.; Robin, M. L.; DesMarteau, D. D. *J. Am. Chem. Soc.* **1984,** *106,* 4266.
128. Sekiya, A.; DesMarteau, D. D. *Inorg. Chem.* **1981,** *20,* 1.
129. Abe, T.; Shreeve, J. M. *J. Fluorine Chem.* **1973,** *3,* 187.
130. Berrier, C.; Jacquesy, J. C.; Jouannetaud, M. P.; Vidal, Y. *Tetrahedron* **1990,** *46,* 815.

131. Hudlicky′, M. *Chemistry of Organic Fluorine Compounds,* 2nd ed.; Ellis Horwood Ltd.: Chichester, United Kingdom; Wiley: New York, 1976; pp 56–60.
132. Barton, D. H. R.; Danks, L. J.; Ganguly, A. K.; Hesse. R. H.; Tarzia, G.; Pechet, M. H. *J. Chem. Soc., Perkin Trans. 1* **1976,** 101.
133. Barton, D. H. R.; Danks, L. J.; Hesse, R. H.; Pechet, M. M.; Wilshire, C. *Nouv. J. Chim.* **1977,** *1,* 315.
134. Adamson, J.; Marcus, D. M. *Carbohydr. Res.* **1972,** *22,* 257.
135. Bischofberger, K.; Brink, A. J.; Jordaan, A. *J. Chem. Soc., Perkin Trans. 1* **1975,** 2457.
136. Butchard, C. G.; Kent, P. W. *Tetrahedron* **1979,** *35,* 2439.
137. Butchard, C. G.; Kent, P. W. *Tetrahedron* **1979,** *35,* 2551.
138. Kent, P. W.; Dimitrijevich, S. D. *J. Fluorine Chem.* **1977,** *10,* 455.
139. Dos Santos Afonso, M.; Czarnowski, J. *Z. Phys. Chem. (Munich)* **1988,** *158,* 25.
140. Campbell, D. H.; Fifolt, M. J.; Saran, M. S. Ger. Offen. DE 3 438 934, 1985; *Chem. Abstr.* **1986,** *104,* 5565.
141. Dos Santos Afonso, M.; Schumacher, H. J. *Int. J. Chem. Kinet.* **1984,** *16,* 103.
142. Sekiya, A.; Ueda, K. *Chem. Lett.* **1990,** 609.
143. Toy, M. S.; Stringham, R. S. *J. Fluorine Chem.* **1975,** *5,* 31.
144. Leroy, J.; Dudragne, F.; Adenis, J. C.; Michaud, C. *Tetrahedron Lett.* **1973,** 2771.
145. Leroy, J.; Wakselman, C. *J. Chem. Soc., Perkin Trans. 1* **1978,** 1224.
146. Seguin, M.; Adenis, J. C.; Michaud, C.; Basselier, J. J. *J. Fluorine Chem.* **1980,** *15,* 201.
147. Rozen, S.; Lerman, O.; Kol, M. *J. Chem. Soc., Chem. Commun* **1981,** 443.
148. Lerman, O.; Tor, Y.; Hebel, D.; Rozen, S. *J. Org. Chem.* **1984,** *49,* 806.
149. Haaparanta, M.; Bergman, J.; Solin, O.; Roeda, D. *Nuklearmedizin, Suppl. (Stuttgart)* **1984,** *21,* 823.
150. Rozen, S.; Lerman, O.; Kol, M.; Hebel, D. *J. Org. Chem.* **1985,** *50,* 4753.
151. Dax, K.; Glaenzer, B. I.; Schulz, G.; Vyplel, H. *Carbohydr. Res.* **1987,** *162,* 13.
152. Satyamurthy, N.; Bida, G. T.; Padgett, H. C.; Barrio, J. R. *J. Carbohydr. Chem.* **1985,** *4,* 489.
153. Shiue, C.-Y.; Wolf, A. P. *J. Fluorine Chem.* **1986,** *31,* 255.
154. Ehrenkaufer, R. E.; Potocki, J. F.; Jewett, D. M. *J. Nucl. Med.* **1984,** *25,* 333.
155. Ishiwata, K.; Ido, T.; Nakanishi, H.; Iwata, R. *Appl. Radiat. Isot.* **1987,** *38,* 463.
156. Oberdorfer, F.; Traving, B. C.; Maier-Borst, W.; Hull, W. E. *J. Labelled Compd. Radiopharm.* **1988,** *25,* 465.

157. Rozen, S.; Lerman, O. *J. Am. Chem. Soc.* **1979,** *101,* 2782.
158. Rozen, S.; Menahem, Y. *Tetrahedron Lett.* **1979,** 725; *J. Chem. Soc., Chem. Commun.* **1979,** 479.
159. Rozen, S.; Menahem, Y. *J. Fluorine Chem.* **1980,** *16,* 19.
160. Middleton, W. J.; Rozen, S. U.S. Patent 4 568 478, 1986; *Chem. Abstr.* **1986,** *105,* 42337.
161. Rozen, S.; Hebel, D. *J. Org. Chem.* **1990,** *55,* 2621.
162. Zefirov, N. S.; Zhdankin, V. V.; Gakh, A. A.; Ugrak, B. I.; Romaniko, S. V.; Koz'min, A. S.; Fainzil'berg, A. A. *Izv. Acad. Nauk SSSR, Ser. Khim.* **1987,** 2636 (Engl. Transl. 2451).
163. Zefirov, N. S.; Zhdankin, V. V.; Koz'min, A. S.; Fainzil'berg, A. A.; Gakh, A. A.; Ugrak, B. I.; Romaniko, S. V. *Tetrahedron* **1988,** *44,* 6505.
164. Stavber, S.; Zupan, M. *Tetrahedron* **1986,** *42,* 5035.
165. Stavber, S.; Zupan, M. *J. Org. Chem.* **1987,** *52,* 919.
166. Stavber, S.; Zupan, M. *Tetrahedron* **1990,** *46,* 3093.
167. Stavber, S.; Zupan, M. *J. Org. Chem.* **1987,** *52,* 5022.
168. Kol, M.; Rozen, S.; Appelman, E. *J. Am. Chem. Soc.* **1991,** *113,* 2648.
169. Shellhamer, D. F.; Curtis, C. M.; Dunham, R. H.; Hollingsworth, D. R.; Ragains, M. L.; Richardson, R. E.; Heasley, V. L.; Shackelford, S. A.; Heasley, G. E. *J. Org. Chem.* **1985,** *50,* 2751.
170. Shellhamer, D. F.; Carter, S. L.; Dunham, R. H.; Graham, S. N.; Spitsbergen, M. P.; Heasley, V. L.; Chapman, R. D.; Druelinger, M. L. *J. Chem. Soc., Perkin Trans. 2* **1989,** 159.
171. Zefirov, N. S.; Gakh, A. A.; Zhdankin, V. V.; Stang, P. J. *J. Org. Chem.* **1991,** *56,* 1416.
172. Umemoto, T.; Fukami, S.; Tomizawa, G.; Harasawa, K.; Kawada, K.; Tomita, K. *J. Am. Chem. Soc.* **1990,** *112,* 8563.
173. Schack, C. J.; Christe, K. O. *Inorg. Chem.* **1979,** *18,* 2619.
174. Shack, C. J.; Christe, K. O. *J. Fluorine Chem.* **1984,** *24,* 467.
175. Kinkead, S. A.; Shreeve, J. M. *Inorg. Chem.* **1984,** *23,* 3109.
176. Wilson, R. D.; Maya, W.; Pilipovich, D.; Christe, K. O. *Inorg. Chem.* **1983,** *22,* 1355.
177. Haufe, G.; Alvernhe, G.; Anker, D.; Laurent, A.; Saluzzo, C. *Tetrahedron Lett.* **1988,** *29,* 2311.
178. Gupta, K. D.; Shreeve, J. M. *Inorg. Chem.* **1985,** *24,* 1457.
179. Krügerke, T.; Seppelt, K. *Z. Anorg. Allg. Chem.* **1984,** *517,* 59.
180. Muratov, N. N.; Mohamed, N.-M.; Kunshenko, B. V.; Alekseeva, L. A.; Yagupol'skii, L. M. *Zh. Org. Khim.* **1986,** *22,* 964 (Engl. Transl. 862).
181. Radchenko, O. A.; Il'chenko, A. Y.; Yagupol'skii, L. M. *Zh. Org. Khim.* **1980,** *16,* 863 (Engl. Transl. 758).

182. Suzuki, H.; Satake, H.; Uno, H.; Shimizu, H. *Bull. Chem. Soc. Jpn.* **1987,** *60,* 4471.
183. Zeifman, Y. V.; Lantseva, L. T.; Knunyants, I. L. *Izv. Akad. Nauk SSSR, Ser. Khim.* **1978,** 2640 (Engl. Transl. 2362).
184. Suzuki, H.; Satake, H.; Uno, H.; Shimizu, H. *Bull. Chem. Soc. Jpn.* **1987,** *60,* 1157.
185. Belen'kii, G. G.; German, L. S.; Knunyants, I. L.; Furin, G. G.; Yakobson, G. G. *Zh. Org. Khim.* **1976,** *12,* 1183 (Engl. Transl. 1191).
186. Haszeldine, R. N.; Hewitson, B.; Tipping, A. E. *J. Chem. Soc., Perkin Trans. 1* **1976,** 1178.
187. Bludssus, W.; Mews, R.; Glemser, O.; Alange, G. G. *Isr. J. Chem.* **1978,** *17,* 137.
188. Waterfeld, A.; Bludssus, W.; Mews, R.; Glemser, O. *Z. Anorg. Allg. Chem.* **1980,** *464,* 268.
189. Dyatkin, B. L.; Sterlin, S. R.; Zhuravkova, L. G.; Martynov, B. I.; Mysov, E. I.; Knunyants, I. L. *Tetrahedron* **1973,** *29,* 2759.
190. Belen'kii, G. G.; Kopaevich, Y. L.; German, L. S.; Knunyants, I. L. *Dokl. Akad. Nauk SSSR* **1971,** *201,* 603; *Chem. Abstr.* **1972,** *76,* 99030.
191. Lensch, C.; Glemser, O. *Z. Naturforsch., B: Anorg. Chem., Org. Chem.* **1982,** *37B,* 401.
192. McCarthy, J. R.; Matthews, D. P.; Barney, C. L. *Tetrahedron Lett.* **1990,** *31,* 973.
193. Tomoda, S.; Usuki, Y. *Chem. Lett.* **1989,** 1235.
194. Uneyama, K.; Kanai, M. *Tetrahedron Lett.* **1990,** *31,* 3583.
195. Mursakulov, I. G.; Talybov, A. H.; Guseinov, M. M.; Aslanova, M. R.; Smit, V. A.; Verdieva, S. S. *Dokl. Akad. Nauk SSSR* **1978,** *34,* 35; *Chem. Abstr.* **1979,** *90,* 38561.
196. Baasner, B.; Hagemann, H.; Klauke, E. Ger. Offen. DE 3 305 201, 1984; *Chem. Abstr.* **1985,** *102,* 5684.
197. Ripka, W. C. U.S. Patent 3 634 466, 1971; *Chem. Abstr.* **1972,** *76,* 99924; U.S. Patent 3 629 301, 1971; *Chem. Abstr.* **1972,** *76,* 86012.
198. Banks, R. E.; Dickinson, N.; Morrissey, A. P.; Richards, A. *J. Fluorine Chem.* **1984,** *26,* 87.
199. Scherer, K. V.; Terranova, T. F.; Lawson, D. D. *J. Org. Chem.* **1981,** *46,* 2379.
200. Dyatkin, B. L.; Zhuravkova, L. G.; Martynov, B. I.; Sterlin, S. R.; Knunyants, I. L. *J. Chem. Soc., Chem. Commun.* **1972,** 618.
201. Laurent, E.; Lefranc, H.; Tardivel, R. *Nouv. J. Chim.* **1984,** *8,* 345.
202. Mattay, J.; Runsink, J.; Rumbach, T.; Ly, C.; Gersdorf, J. *J. Am. Chem. Soc.* **1985,** *107,* 2557.
203. Bayliff, A. E.; Chambers, R. D. *J. Chem. Soc., Perkin Trans. 1* **1988,** 201.

204. Dmowski, W.; Flowers, W. T.; Haszeldine, R. N. *J. Fluorine Chem.* **1977,** *9,* 94.
205. Ishikawa, N.; Maruta, M. *Yuki Gosei Kagaku Kyokaishi* **1981,** *39,* 51; *Chem. Abstr.* **1981,** *95,* 5897.
206. Fujiyama, M.; Mizuno, T.; Nakamura, S.; Mikami, T.; Yakura, T. Jpn. Kokai Tokkyo Koho 78 144 508, 1978; *Chem. Abstr.* **1979,** *90,* 203464.
207. Chambers, R. D.; Lindley, A. A.; Fielding, H. C. *J. Chem. Soc., Perkin Trans. 1* **1981,** 939.
208. Chambers, R. D.; Gribble, M. Y.; Marper, E. *J. Chem. Soc., Perkin Trans. 1* **1973,** 1710.
209. Svoboda, J.; Paleta, O.; Dedek, V. *Collect. Czech. Chem. Commun.* **1981,** *46,* 1272.
210. Postovoi, S. A., Mysov, E. I.; Zeifman, Y. V.; Knunyants, I. L. *Izv. Akad. Nauk SSSR, Ser. Khim.* **1982,** 1586 (Engl. Transl. 1409).
211. Drayton, J. V.; Flowers, W. T.; Haszeldine, R. N.; Parry, T. A. *J. Chem. Soc., Chem. Commun.* **1976,** 490.
212. Anderson, R. W.; Frick, H. R. U.S. Patent 3 661 967, 1972; *Chem. Abstr.* **1972,** *77,* 75028.
213. Drayton, C. J.; Flowers, W. T.; Haszeldine, R. N.; Morton, W. D. *J. Chem. Soc., Perkin Trans. 1* **1975,** 1035.
214. Yakhlakova, O. M.; Krokhalev, A. M.; Mokrinskii, V. V.; Bezmaternykh, N. A.; Neifel'd, P. G.; Zabolotskikh, V. F. U.S.S.R. Patent 706 393, 1979; *Chem. Abstr.* **1980,** *92,* 128553.
215. Bell, S. L.; Chambers, R. D.; Gribble, M. Y.; Maslakiewicz, J. R. *J. Chem. Soc., Perkin Trans. 1* **1973,** 1716.
216. Barlow, M. G.; Haszeldine, R. N.; Dingwall, J. G. *J. Chem. Soc., Perkin Trans. 1 1973,* 1542.
217. Banks, R. E.; Mullen, K,; Nicholson, W. J.; Oppenheim, C.; Prakash, A. *J. Chem. Soc., Perkin Trans. 1* **1972,** 1098.
218. Chambers, R. D.; Corbally, R. P.; Holmes, T. F.; Musgrave, W. K. R. *J. Chem. Soc., Perkin Trans. 1* **1974,** 108.
219. Chambers, R. D.; Corbally, R. P.; Musgrave, W. K. R.; Jackson, J. A.; Matthews, R. S. *J. Chem. Soc., Perkin Trans. 1* **1972,** 1286.
220. Chambers, R. D.; Corbally, R. P.; Musgrave, W. K. R. *J. Chem. Soc., Perkin Trans. 1* **1973,** 1281.
221. Chambers, R. D.; Gribble, M. Y. *J. Chem. Soc., Perkin Trans. 1* **1973,** 1405.
222. Chambers, R. D.; Gribble, M. Y. *J. Chem. Soc., Perkin Trans. 1* **1973,** 1411.
223. Chambers, R. D.; Jackson, J. A.; Partington, S.; Philpot, P. D.; Young, A. C. *J. Fluorine Chem.* **1975,** *6,* 5.

224. Chambers, R. D.; Musgrave, W. K. R.; Wood, D. E. *J. Chem. Soc., Perkin Trans. 1* **1979,** 1978.
225. Delyagina, N. I.; Pervova, E. Y.; Dyatkin, B. L.; Knunyants, I. L. *Zh. Org. Khim.* **1972,** *8,* 851 (Engl. Transl. 859).
226. Gervits, L. L.; Makarov, K. N.; Cheburkov, Y. A.; Knunyants, I. L *J. Fluorine Chem.* **1977,** *9,* 45.
227. Zeifman, Y. V.; Postovoi, S. A.; Knunyants, I. L. *Dokl. Acad. Nauk SSSR* **1982,** *265,* 347; *Chem. Abstr.* **1982,** *97,* 215438.
228. Ishikawa, N.; Shin-Ya, S. *Bull. Chem. Soc. Jpn.* **1975,** *48,* 1339.
229. Ishikawa, N.; Iwamoto, K.; Ishiwata, T.; Kitazume, T. *Bull. Chem. Soc. Jpn.* **1982,** *55,* 2956.
230. Kato, S.; Suyama, T. Jpn. Kokai Tokkyo Koho 79 163 521, 1979; *Chem. Abstr.* **1980,** *93,* 71035.
231. Sud'enkov, Y. Y.; Zapevalova, T. B.; Plashkin, V. S.; Kolenko, I. P. *Zh. Org. Khim.* **1975,** *11,* 1626 (Engl. Transl. 1612).
232. Tokuyama Soda Co., Ltd. Jpn. Kokai Tokkyo Koho 81 25 133, 1981; *Chem. Abstr.* **1981,** *95,* 97094.
233. Vilenchik, Y. M.; Lekontseva, G. I.; Semerikova, L. S. *Zh. Vses. Khim O-va.* **1981,** *26,* 210; *Chem. Abstr.* **1981,** *95,* 203256.
234. Katoh, S.; Suyama, T. Jpn. Kokai Tokkyo Koho 80 15 447, 1980; *Chem. Abstr.* **1980,** *93,* 70994.
235. England, D. C. U.S. Patent 3 733 357, 1973; *Chem. Abstr.* **1973,** *79,* 41977.
236. England, D. C. Can. Patent 982 144 (1976); *Chem. Abstr.* **1976,** *84,* 179694.
237. Glazkov, A. A.; Ignatenko, A. V.; Krukovskii, S. P.; Ponomarenko, V. A. *Izv. Akad. Nauk SSSR, Ser. Khim.* **1976,** 918 (Engl. Transl. 896).
238. Kitazume, T.; Ishikawa, N. *Bull. Chem. Soc. Jpn.* **1975,** *48,* 361.
239. Knunyants, I. L.; Delyagina, N. I.; Igumnov, S. M. *Izv. Akad. Nauk SSSR, Ser. Khim.* **1981,** 857 (Engl. Transl. 637).
240. Liang, W.-X.; Chen, Q.-Y. *Hua Hsueh Hsueh Pao* **1980,** *38,* 269; *Chem. Abstr.* **1981,** *94,* 65630.
241. Dmowski, W.; Wozniacki, R. *J. Fluorine Chem.* **1987,** *36,* 385.
242. Ikeda, I.; Tsuji, M.; Okahara, M. *J. Fluorine Chem.* **1987,** *36,* 171.
243. Knunyants, I. L.; Pervova, E. Y.; Delyagina, N. I. U.S.S.R. Patent 340 657, 1972; *Chem. Abstr.* **1972,** *77,* 151510.
244. Makarov, K. N.; Gervits, L. L.; Knunyants, I. L. *J. Fluorine Chem.* **1977,** *10,* 157.
245. Scherer, K. V. U.S. Patent 4 173 654, 1979; *Chem. Abstr.* **1980,** *92,* 58210.
246. Postovoi, S. A.; Lantseva, L. T.; Zeifman, Y. V. *Izv. Akad. Nauk SSSR, Ser. Khim.* **1982,** 210 (Engl. Transl. 199).
247. Mirzabekyants, N. S.; Gervits, L. L.; Cheburkov, Y. A.; Knunyants, I. L. *Izv. Akad. Nauk SSSR, Ser. Khim.* **1977,** 2772 (Engl. Transl. 2563).

248. Chambers, R. D.; Jones, C. G. P. *J. Fluorine Chem.* **1981,** *17,* 581.
249. Chambers, R. D.; Partington, S.; Speight, D. B. *J. Chem. Soc., Perkin Trans. 1* **1974,** 2673.
250. Karpov, V. M.; Mezhenkova, T. V.; Platonov, V. E.; Yakobson, G. G. *Zh. Org. Khim.* **1984,** *20,* 1341 (Engl. Transl. 1220).
251. Petrov, V. A.; Belen'kii, G. G.; German, L. S.; Mysov, E. I. *Izv. Akad. Nauk SSSR, Ser. Khim.* **1981,** 2098 (Engl. Transl. 1723).
252. Petrov, V. A.; Belen'kii, G. G.; German, L. S.; Kurbakova, A. P.; Leites, L. A. *Izv. Akad. Nauk SSSR, Ser. Khim.* **1982,** 170 (Engl. Transl. 160).
253. Petrov, V. A.; Belen'kii, G. G.; German, L. S. *Izv. Akad. Nauk SSSR, Ser. Khim.* **1982,** 1591 (Engl. Transl. 1414).
254. Belen'kii, G. G.; Lur'e, E. P.; German, L. S. *Izv. Akad. Nauk SSSR, Ser. Khim.* **1976,** 2365 (Engl. Transl. 2208).
255. Belen'kii, G. G.; Savicheva, G. I.; German, L. S. *Izv. Akad. Nauk SSSR, Ser. Khim.* **1978,** 1433 (Engl. Transl. 1250).
256. Belen'kii, G. G.; German. L. S. *Izv. Akad. Nauk SSSR, Ser. Khim.* **1974,** 942 (Engl. Transl. 913).
257. Feiring, A. E. *J. Fluorine Chem.* **1978,** *12,* 471.
258. Igumnov, S. M.; Delyagina, N. I.; Zeifman, Y. V.; Knunyants, I. L. *Izv. Akad. Nauk SSSR, Ser. Khim.* **1984,** 827 (Engl. Transl. 762).
259. Izeki, Y.; Nakahara, A.; Nakajima, J. Jpn. Kokai Tokkyo Koho JP 02 25 439, 1990; *Chem. Abstr.* **1990,** *113,* 39939.
260. Nakahara, A.; Izeki, Y. Jpn. Kokai Tokkyo Koho JP 01 172 353, 1989; *Chem. Abstr.* **1990,** *112,* 98021.
261. Makarov, K. N.; Abroskina, T. N.; Cheburkov, Y.A.; Knunyants, I. L. *Izv. Akad. Nauk SSSR, Ser. Khim.* **1976,** 940 (Engl. Transl. 922).
262. Krespan, C. G. *J. Org. Chem.* **1978,** *43,* 637.
263. Ishikawa, N.; Osawa, T.; Edamura, K.; Hayashi, S. *Nippon Kagaku Kaishi* **1977,** 141; *Chem. Abstr.* **1977,** *86,* 139374.
264. Schwertfeger, W.; Siegemund, G. *Angew. Chem. Int. Ed. Engl.* **1980,** *19,* 126.
265. Volkov, N. D.; Nazaretyan, V. P.; Yagupol'skii, L. M. *Zh. Org. Khim.* **1982,** *18,* 519; (Engl. Transl. 454).
266. Izeki, Y.; Nakahara, A.; Nakajima, J. Jpn. Kokai Tokkyo Koho JP 02 169 532, 1990; *Chem. Abstr.* **1990,** *113,* 171498.
267. Okamoto, S.; Abe, M. Jpn. Kokai Tokkyo Koho JP 01 70 427, 1989; *Chem. Abstr.* **1989,** *111,* 114748.
268. Psarras, T. U.S. Patent Appl. 942 571, 1979; *Chem. Abstr.* **1979,** *91,* 56380.
269. Psarras, T. U.S. Patent Appl. 954 943, 1979; *Chem. Abstr.* **1979,** *91,* 56379.
270. Muffler, H.; Siegemund, G.; Schwertfeger, W. Ger. Offen. 2 928 602, 1981; *Chem. Abstr.* **1981,** *94,* 175133.
271. Kawaguchi, T.; Tamura, Y.; Negishi, S. Jpn Kokai 77 10 221, 1977; *Chem. Abstr.* **1977,** *87,* 22430.

272. Kondo, A.; Yanagihara, T. Jpn. Kokai Tokkyo Koho 79 135 722, 1979; *Chem. Abstr.* **1980,** *92,* 163576.
273. Resnick, P. R. U.K. Patent Appl. GB 2 081 267, 1982; *Chem. Abstr.* **1982,** *96,* 217257.
274. Yamabe, M.; Munekata, S.; Samejima, S. Ger. Offen. 2 708 677, 1978; *Chem. Abstr.* **1979,** *90,* 86759.
275. Baucom, K. B. U.S. Patent Appl. 70 473, 1980; *Chem. Abstr.* **1980,** *93,* 132229.
276. Izeki, Y.; Takesue, M.; Takada, K. Jpn. Kokai Tokkyo Koho JP 01 40 443, 1989; *Chem. Abstr.* **1990,** *112,* 7037.
277. Asahi Glass Co. Jpn. Kokai Tokkyo Koho JP 82 45 132, 1982; *Chem. Abstr.* **1982,** *97,* 91756.
278. Daikin Kogyo Co., Ltd. Jpn. Kokai Tokkyo Koho JP 82 64 641, 1982; *Chem. Abstr.* **1982,** *97,* 127479.
279. Shchibrya, T. G.; Ignatenko, A. V.; Krukovskii, S. P.; Ponomarenko, V. A. *Izv. Akad. Nauk SSSR, Ser. Khim.* **1980,** 700; *Chem. Abstr.* **1980,** *93,* 8589.
280. Munekata, S.; Yamabe, M.; Kaneko, I. Fr. Demande FR 2 463 115, 1981; *Chem. Abstr.* **1982,** *96,* 68354.
281. Chang, S. C.; DesMarteau, D. D. *J. Org. Chem.* **1983,** *48,* 771.
282. Gontar, A. G.; Bykovskaya, E. G.; Knunyants, I. L. *Zh. Vses. Khim. O-va.* **1975,** *20,* 232; *Chem. Abstr.* **1975,** *83,* 9063.
283. Bailey, A. R.; Banks, R. E.; Barlow, M. G.; Nickkho-Amiry, M. *J. Fluorine Chem.* **1980,** *15,* 289.
284. Barnes, R. N.; Chambers, R. D.; Hewitt, C. D.; Silvester, M. J.; Klauke, E. *J. Chem. Soc., Perkin Trans. 1* **1985,** 53.
285. Barnes, R. N.; Chambers, R. D.; Matthews, R. S. *J. Fluorine Chem.* **1982,** *20,* 307.
286. Barnes, R. N.; Chambers, R. D.; Silvester, M. J.; Hewitt, C. D.; Klauke, E. *J. Fluorine Chem.* **1984,** *24,* 211.
287. Clifford, A. F.; Thompson, J. W. *Inorg. Nucl. Chem., Herbert H. Hyman Memorial Volume* **1976,** 37.
288. Duncan, L. C.; Rhyne, T. C.; Clifford, A. F.; Shaddix, R. E.; Thompson, J. W. *Inorg. Nucl. Chem., Herbert H. Hyman Memorial Volume* **1976,** 33.
289. Thompson, J. W.; Howell, J. L.; Clifford, A. F. *Isr. J. Chem.* **1978,** *17,* 129.
290. Beyleveld, W. M.; Oxenrider, B. C.; Woolf, C. U.S. Patent 3 795 689, 1974; *Chem. Abstr.* **1974,** *80,* 120248.
291. Gontar, A. G.; Bykovskaya, E. G.; Knunyants, I. L. *Izv. Akad. Nauk SSSR, Ser. Khim.* **1976,** 209 (Engl. Transl. 202).

Replacement of Hydrogen by Fluorine

by James L. Adcock

Fluorination with Elemental Fluorine

For the past 20 years, four main ways of replacement of hydrogen by fluorine have been used: fluorination using elemental fluorine, electrochemical fluorination, fluorination using high-valency metal fluorides, and selective electrophilic fluorination.

Since 1972, significant advances in the chemistry of elemental fluorine have occurred in two main areas: The first area is so-called *"deep" fluorination* in which molecules having few or no fluorines and often containing functional groups are converted to highly fluorinated or perfluorinated molecules with significantly reduced fragmentation. The second area of achievement results from the intermediacy of fluorine-generated active intermediates capable of selective fluorination on substrate molecules. Both of these advances have as their hallmarks significantly reduced side reactions, fragmentation, rearrangement, and oligomerization.

Perfluorination

Successful direct perfluorination requires at least two separate sets of conditions. Initially an organic molecule containing many replaceable hydrogen bonds to carbon must be protected from too rapid attack by elemental fluorine. Most fragmentation occurs at this stage. However as the molecule is more highly fluorinated, reaction slows and more vigorous conditions are necessary for completion of the final substitution steps of the reaction. A direct fluorination sequence for a complex organic molecule therefore requires milder conditions at the outset and more forcing conditions at the finale if an effective substitution rate is to be maintained; the alternative is to conduct the dilute-fluorine, low-temperature conditions over a sufficiently long period allowing for the generally slower hydrogen substitution rates in highly fluorinated molecules [*1*].

Use of a hydrogen fluoride scavenger during the fluorination of especially functionalized organic compounds is advantageous in limiting Lewis acid-induced skeletal rearrangements and rearrangements induced or enhanced in strong protic acids. In fact the presence of sodium or potassium fluoride appears to enhance the degree or rate of fluorination itself. This recent work also suggests that hydrogen fluoride inhibits direct fluorination [2].

0065–7719/95/0187–0097$08.54/1

Several distinct technologies have developed to take advantage of this new concept in kinetic control. They can be divided into two basic types: batch and flow. *Batch* types have proliferated in recent years because of their engineering simplicity. All technologies basically alter conditions over time: processing of a batch proceeds from inception to completion prior to initiation of a second cycle. Typical of these technologies are the so-called *La-Mar* and *low-temperature gradient* (LTG) or *cryogenic zone reactor* methods [*1*]. The La-Mar technique is most readily applicable to *microcrystalline solid*–gaseous fluorine heterogeneous reactions or to *bulk solids* when only *surface fluorination* is desired. Temperatures vary from 200 to 600 K according to the resistance of the material to fluorination and may be increased as the process proceeds. Reaction rate is controlled by the introduction of a fixed or slowly increasing flow of elemental fluorine, diluted initially with a high flow of helium or nitrogen. The concentration of fluorine is initially very low, but it is increased as the reaction progresses principally by reduction of the flow of diluent gas. The absolute rate of fluorine injection is fixed or varied independently of concentration, and thus no run-away reaction is possible because fluorine can be consumed only as fast as it is injected. This condition is a significant safety feature.

Later developments in this technique to handle *volatile organic compounds* came with the realization that the activation energy for the reaction of molecular fluorine with a carbon–hydrogen bond is very small and that fluorination occurs in the dark at temperatures near 185 K. In the LTG or cryogenic zone reactor, volatile organic compounds are condensed as solids in a copper turnings-packed tube divided into four or more distinct zones of temperature control. Again reactions occur in a solid state–gaseous fluorine heterogeneous system. Many deep-fluorination reactions have been carried out by using various modifications of this reactor; many report quite high yields (Table 1), although reaction cycles might extend over several days.

Although most deep fluorinations have resulted in fluorocarbons, a few have not, even under more forcing conditions. The results are explained by invoking steric hindrance induced by faster peripheral fluorination shielding more internal hydrogen atoms from attack [*15*] (equation 1).

[*15*]

$$(CH_3)_3CCH_2C(CH_3)_3 \xrightarrow[-78^\circ\,C]{F_2} \underset{14\%}{(CF_3)_3CCH_2C(CF_3)_3} + \underset{66\%}{(CF_3)_3CCHF(CF_3)_3} + \underset{20\%}{(CF_3)_3CCF_2C(CF_3)_3}$$

The action of elemental fluorine on R_fCH_3, $R_fCH_2CH_3$, and $R_fCH_2CH_2$-$CF(CF_3)_2$ (where R_f is $C_3F_7(CF_3)_2C$–) was investigated, and the products were analyzed. The remarkable inertness of R_fCH_3 toward elemental fluorine was noted. In bicyclic compounds a significant reduction in electron density at bridgehead hydrogen atoms due to fluorine inductive effects seems to be more important than steric effects. The resulting more acidic hydrogens are more resistant to fluorine

Table 1. Direct Fluorination of Organic Compounds Using Batch Reaction Techniques

Reactant	*Product*	*Yield (%)*	*Ref.*
$(CH_3)_3CC(CH_3)_3$	$(CF_3)_3CC(CF_3)_3$	9.3	3
$C(CH_3)_4$	$C(CF_3)_4$	10.4	4
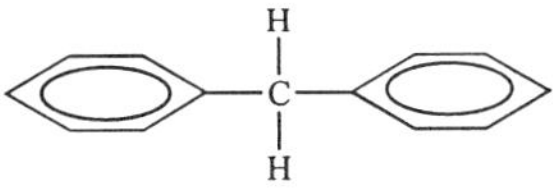	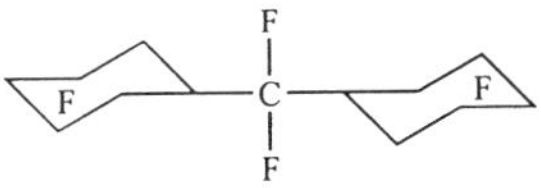	93	5
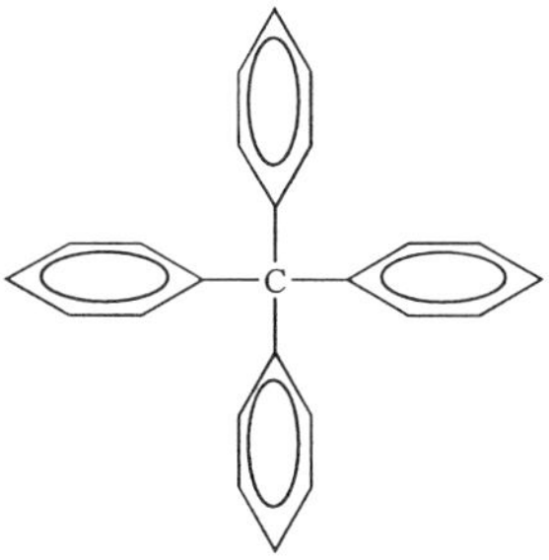	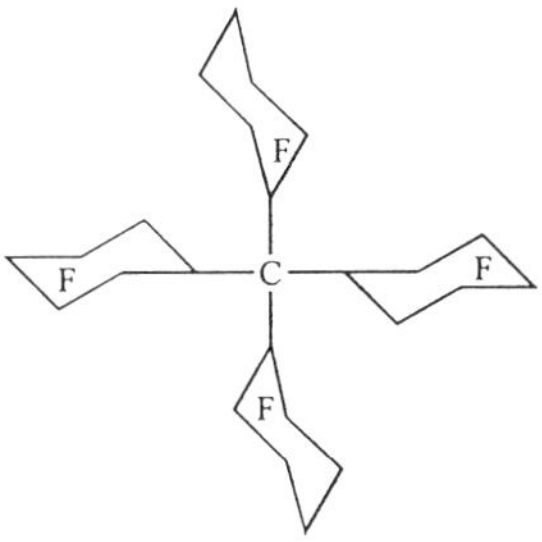	96	5
$CH_3OCH_2CH_2OCH_3$	$CF_3OCF_2CF_2OCF_3$	21	6
	$CF_3OCF_2CF_2OCHF_2$	0–25	6
$CH_3(OCH_2CH_2)_2OCH_3$	$CF_3(OCF_2CF_2)_2OCF_3$	16	6
$O(CH_2CH_2)_2O$	$O(CF_2CF_2)_2O$	40	7
$C_2H_5OCH_2CH_2OC_2H_5$	$C_2F_5OCF_2CF_2OC_2F_5$	18	8
$C(OCH_3)_4$	$C(OCF_3)_4$	49.5	9
$C(OC_2H_5)_4$	$C(OC_2F_5)_4$	56.5	9
$C(OC_3H_7)_4$	$C(OC_3F_7)_4$	18.6	9
$[(CH_3)_2CH]_2O$	$[(CF_3)_2CF]_2O$	49	10
$[(CH_3)_2CHCH_2]_2O$	$(CF_3)_2CFCF_2]_2O$	55.8	10
$[(CH_3)_2CHCH_2CH_2]_2O$	$[(CF_3)_2CFCF_2CF_2]_2O$	52	10
$[(CH_3)_3CCH_2]_2O$	$[(CF_3)_3CCF_2]_2O$	>50	10

Continued on next page.

Table 1—Continued

Reactant	*Product*	*Yield (%)*	*Ref.*
$CH_3CO_2C_2H_5$	$CF_3CO_2C_2F_5$	5	7
	$CF_3CO_2CHFCF_3$	20	
	CF_3COF, CHF_2COF		
$(CH_3)_3CCOF$	$(CF_3)_3CCOF$	52	7
	$(CHF_2)(CF_3)_2CCOF$	20	8
$(CH_3)_2C(COF)_2$	$(CF_3)_2C(COF)_2$	14	8
c-C_8H_{16}	c-C_8F_{16}	18.7	11
	c-C_8HF_{15}	6.2	
$(CH_3)_3CCH_2CH_2C(CH_3)_3$	$(CF_3)_3CCF_2CF_2C(CF_3)_3$	89.2	12
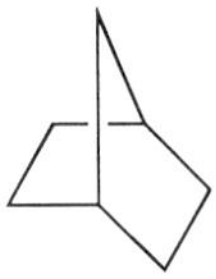	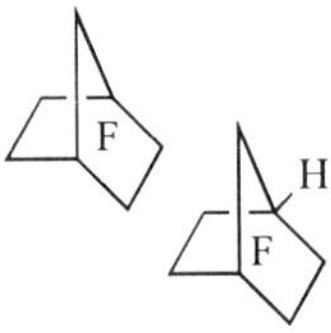 F; F, H	12.22 8.14	11
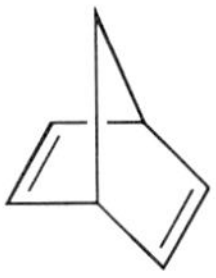	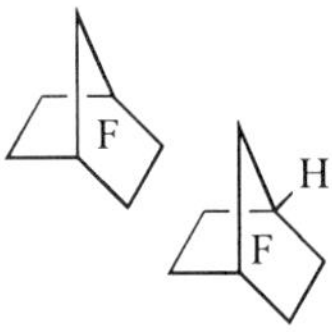 F; F, H	8.22 1.88	11
	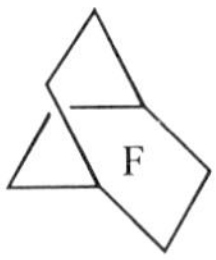 F	3.5	11

References are listed on pages 116–119.

Table 1—Continued

Reactant	*Product*	*Yield (%)*	*Ref.*
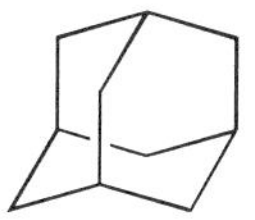 adamantane	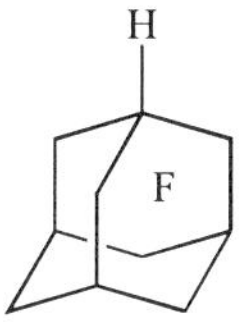 H; F (perfluoroadamantane)	4.4	11
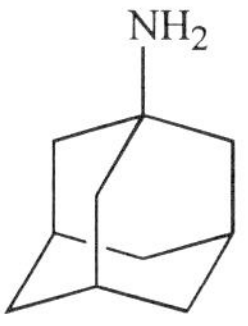 NH_2 (1-aminoadamantane)	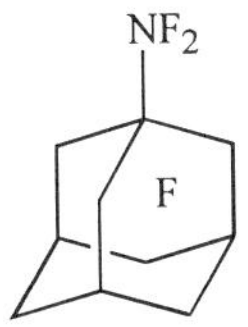 NF_2; F	6.3	13
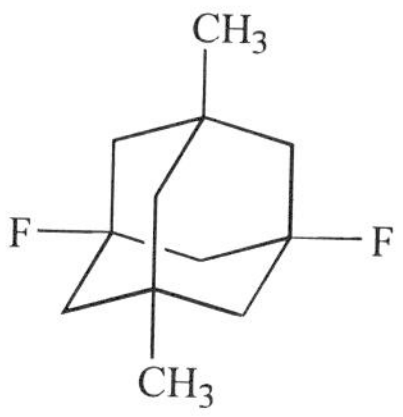 CH_3, F, F, CH_3	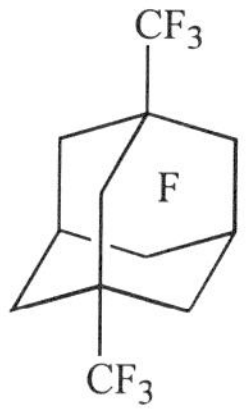 CF_3, F, CF_3	26	13
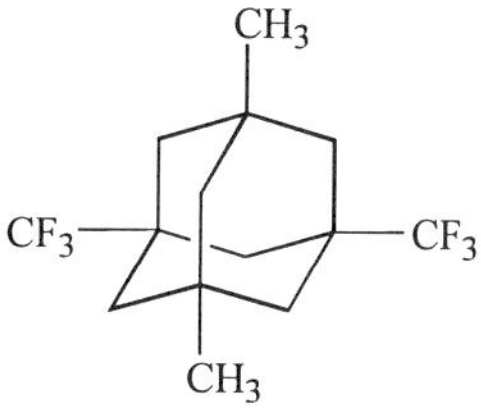 CH_3, CF_3, CF_3, CH_3	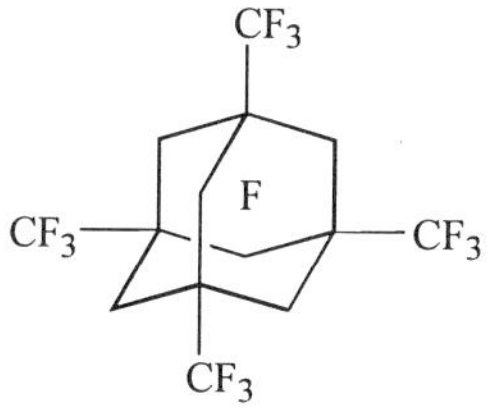 CF_3, F, CF_3, CF_3, CF_3	4.2	13

Continued on next page.

References are listed on pages 116–119.

Table 1—Continued

Reactant	*Product*	*Yield (%)*	*Ref.*
			14
			14
		33.5	14

attack [*16*]; this increased resistance can be seen in Table 1 for the products of deep fluorination of norbornane and, to a lesser extent, of norbornadiene [*11*]. The results reported for adamantane were subsequently shown to be in error; the product was perfluoroadamantane [*17*].

Deep fluorination using the La-Mar technique was carried out on polymers such as polyethylene and polypropylene [*18*], on polyethers [*19, 20, 21*], and on polyesters subsequently treated with sulfur tetrafluoride [*22*]. Deep fluorinations carried out under conditions producing limited fragmentation produced oligomeric perfluoropolyethers from powdered polyethylene oxide [*23*]. Deep fluorinations carried out in the limited presence of molecular oxygen result in the conversion of

References are listed on pages 116–119.

pendant methyl groups on polypropylene to carbonyl fluoride groups (isolated and characterized as carboxylic acid groups) simultaneously with their perfluorination [*24*]. The simultaneous functionalization–fluorination reaction can be carried out on small molecules such as neopentane producing perfluoropivaloyl fluoride, but conversions are low under conditions giving perfluorination [*25*]. Other deep fluorinations of functionalized polymers have been more extensively reviewed [*1*].

Deep fluorination in solvents is widely accepted in industrial applications. Perfluorination of the "glymes" (ethylene glycol dimethyl ethers) and their homologues were conducted in a Fomblin perfluoropolyether solvent in the presence of sodium fluoride as hydrogen fluoride scavenger [*26*] (equations 2–5).

$$CH_3OCH_2CH_2OCH_2CH_2OCH_3 \xrightarrow{F_2} CF_3OCF_2CF_2OCF_2CF_2OCF_3 \quad 56\% \tag{2}$$

$$CH_3O(CH_2CH_2O)_4CH_3 \xrightarrow{F_2} CF_3O(CF_2CF_2O)_4CF_3 \quad 43\% \tag{3}$$

$$\text{HCFClCFCl-(tetrahydrofuran-2-yl)} \xrightarrow{F_2} \text{CF}_2\text{ClCFCl-(perfluorotetrahydrofuran-2-yl)} \quad 67\% \tag{4}$$

$$\text{HCFClCFCl-(1,4-dioxan-2-yl)} \xrightarrow{F_2} \text{CF}_2\text{ClCFCl-(perfluoro-1,4-dioxan-2-yl)} \quad 74\% \tag{5}$$

A variation of the solution technique uses inverse addition of substrate to a well-stirred, fluorine-saturated solvent and ultraviolet light to hasten reaction at lower temperatures [*27*].

The most recent development in direct fluorination technology is the *aerosol direct fluorination process* [*28*]. Unlike previous technologies, it is a *continuous flow process.* Unfluorinated material enters the reactor as an aerosol, with the compounds being adsorbed on airborne sodium fluoride particulates. Fluorine concentration, initially zero, increases over the *length of the reactor*, as does the temperature. The most significant advantage over the LTG or cryogenic zone reactor is reaction time. Residence time in the reactor is of the order of 1–2 min. A significant difference in the two technologies is the reliance on a photochemical finishing stage to efficiently utilize elemental fluorine and produce perfluorination with insignificant levels of residual hydrogen. Significant improvements to the technology were reviewed [*29*].

The most important contribution of this technology to the science of fluorine chemistry has been its ability to probe mechanistic aspects of elemental fluorine attack on organic molecules. As has been long accepted, fluorine attack on nonaromatic organic molecules is free radical in nature. The great oxidizing power of fluorine and its strongly electrophilic nature have tended to be neglected. The adsorption of organic molecules on sodium fluoride particulates keeps those molecules in low-level vibrational and electronic ground states. The particulates quickly absorb excess vibrational energy and remove endogenous hydrogen fluoride, thereby preventing Lewis and protic acid-catalyzed rearrangements, and minimize intermolecular free radical interactions between adsorbed organic molecules. Reaction of organic free radicals is generally limited to reaction with gaseous molecular fluorine. Structural rearrangements of organic molecules must therefore be intramolecular in nature and be fast relative to fluorine attack. These factors significantly reduce the numbers of interfering side reactions seen predominantly in gaseous but also in liquid- or solution-phase fluorinations except at very high dilution.

Deep fluorination of ***alkanes, ethers, acid halides, esters, alkyl chlorides, most ketones, ketals, orthoesters,*** and combinations of these functional groups produces principally the perfluorinated analogues (Table 2). Chlorine substituents (or chloro groups) usually survive fluorination.

The mechanism of aerosol direct fluorination, although generally free radical in nature, provides a unique opportunity to probe the nature of fluorine attack on organic molecules. The *fluorination of neopentyl bromide* is an interesting example. Neopentyl bromide is perfluorinated to perfluoroisopentane, following loss of bromine and structural rearrangement. This reaction was shown with the help of fluorination reactions of lower fluorine–neopentyl bromide molar ratios to be the result of initial attack by fluorine on bromine, oxidizing it to what is presumed to be $(CH_3)_3CCH_2BrF_2$. This oxidation is followed by an internal redox reaction that produces neopentyl cation, which quickly rearranges to the more stable tertiary 2-methyl-2-butyl cation. This product was not intercepted, but its product resulting from proton loss, 2-methyl-2-butene, was isolated together with unreacted neopentyl bromide. Even more significant, fluorination at higher fluorine/neopentyl bromide ratios produces significant quantities of "abnormal" nonstatistically predicted compounds having a 3,3-difluoromethylene moiety. This group of unexpected products is explained by fluorine attack at bromine to produce presumably $(CH_3)_3CCH_2BrF_4$, a species that fluorinates the *alpha* carbon prior to disproportionation and cation formation. The scheme rationalizing all products isolated is described elsewhere [*33*]. The intermediacy of neopentyl cation is inescapable [*39*]. *Neopentyl chloride*, however, is fluorinated "normally" to *perfluoroneopentyl chloride* in very high yields. The neopentyl radical is known not to undergo rearrangement [*40*].

The generally accepted fact that free radicals do not rearrange is perhaps more crucial to the success of direct fluorination in maintaining the structural integrity of molecules undergoing fluorination than any other "truth". However, even here we find exceptions. Freidlina [*41*] showed conclusively in a series of works the *tendency of halogen atoms, fluorine excluded, to migrate by 1,2-shifts to produce more stable free radicals*. The aerosol direct fluorination of a series of alkyl chlorides showed conclusively the prominent role of these migrations in determining the

Table 2. Direct Fluorination of Organic Compounds Using Flow Reaction Techniques

Reactant	*Product*	*Yield (%)*	*Ref.*
$C(CH_3)_4$	$C(CF_3)_4$	38	28
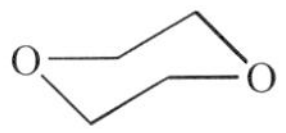	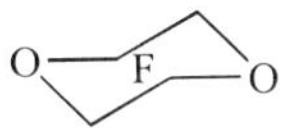	57	28
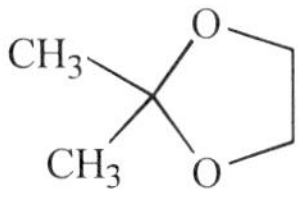	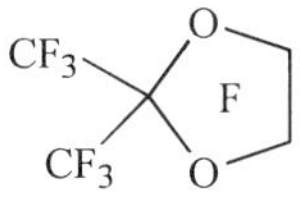	51	28
$C_2H_5COC_2H_5$	$C_2F_5COC_2F_5$	13	30
$C_3H_7COC_3H_7$	$C_3F_7COC_3F_7$	23	30
$C_2H_5COC_4H_9$	$C_2F_5COC_4F_9$	13	30
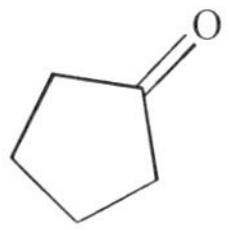	C_4F_9COF	24	30
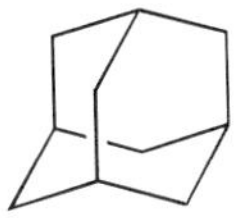	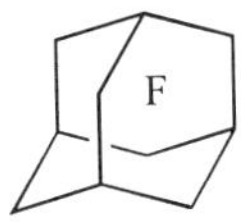	28	31
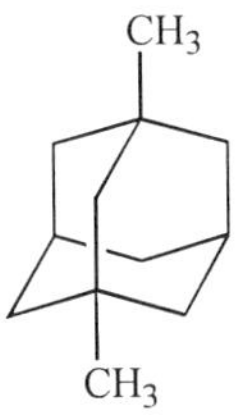	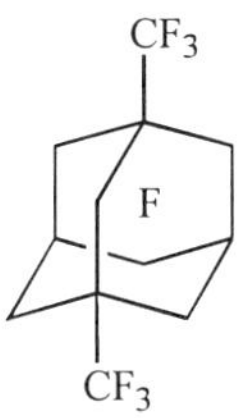	41.9	32
$(CH_3)_3CCH_2Cl$	$(CF_3)_3CCF_2Cl$	74	33
$(CH_3)_3CCH_2Br$	$(CF_3)_2CFC_2F_5$	62.5	33

Continued on next page.

Table 2—Continued

Reactant	*Product*	*Yield (%)*	*Ref.*
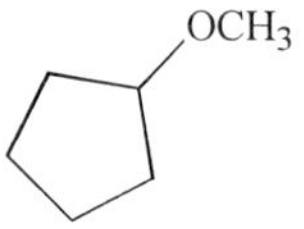 OCH_3	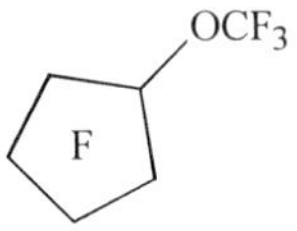 OCF_3 F	22	34
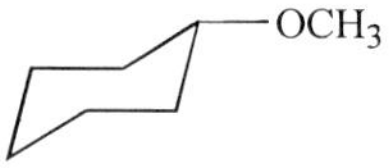 OCH_3	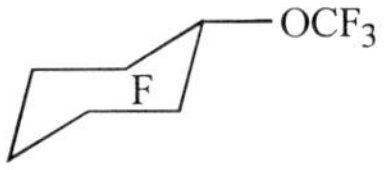 OCF_3 F	32	34
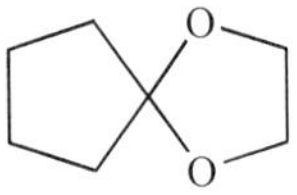 O O	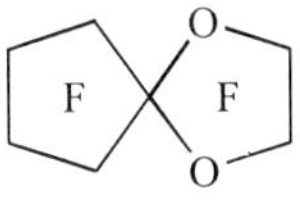 F O F O	14	34
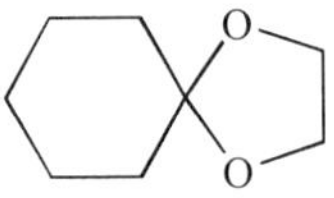 O O	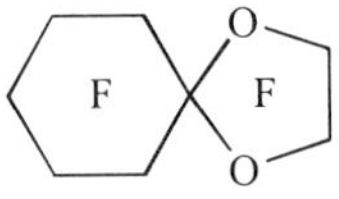 F O F O	12	34
$(CH_3)_3CCOCH_3$	$(CF_3)_3CCOCF_3$	12	35
	$(CF_2H)(CF_3)_2CCOCF_3$	13	
	$(CF_2H)_2(CF_3)CCOCF_3$	8	
$(CH_3)_3CCOC(CH_3)_3$	$(CF_3)_3CCOCF_2CF(CF_3)_2$	9	35
C_3H_7Cl	C_3F_7Cl	63	36
$CH_3CHClCH_3$	$CF_3CFClCF_3$	32	36
	C_3F_7Cl	16	
C_4H_9Cl	C_4F_9Cl	43.8	36
$(CH_3)_2CHCH_2Cl$	$(CF_3)_2CFCF_2Cl$	41.6	36
$C_2H_5CHClCH_3$	$C_2F_5CFClCF_3$	13.7	36
	C_4F_9Cl	19.5	
$(CH_3)_3CCl$	$(CF_3)_2CFCF_2Cl$	46.8	36

References are listed on pages 116–119.

Table 2—Continued

Reactant	*Product*	*Yield (%)*	*Ref.*
$(CH_3)_2CHCH_2CH_2Cl$	$(CF_3)_2CFCF_2CF_2Cl$	32	36
$C_2H_5CH(CH_3)CH_2Cl$	$C_2F_5CF(CF_3)CF_2Cl$	39	36
$C_2H_5CHClC_2H_5$	$C_2F_5CFClC_2F_5$	10	36
	$C_3F_7CFClCF_3$	16	
	$C_5F_{11}Cl$	5	
$(CH_3)_2CClC_2H_5$	$(CF_3)_2CFC_2F_5$	10	36
	$CF_2ClCF(CF_3)C_2F_5$	22	
	$(CF_3)_2CFCF_2CF_2Cl$	8	
	$(CF_3)_2CFCFClCF_3$	2	
c-C_5H_9Cl	c-C_5F_9Cl	40.2	36
$CH_3OCH_2CH_2OCH_3$	$CF_3OCF_2CF_2OCF_3$	36	37
$CH_3O(CH_2CH_2O)_2CH_3$	$CF_3O(CF_2CF_2O)_2CF_3$	22	37
$CH_3O(CH_2CH_2O)_3CH_3$	$CF_3O(CF_2CF_2O)_3CH_3$	30	37
$CH_3O(CH_2CH_2O)_4CH_3$	$CF_3O(CF_2CF_2O)_4CH_3$	15	37
$C(OCH_3)_4$	$C(OCF_3)_4$	8	
	$CF(OCF_3)_3$	20	
$CH_3C(OCH_3)_3$	$CF_3CF(OCF_3)_2$	24	38
[spiro structure: O O / O O]	[spiro structure: O O, F F / O O]	17	38
CH_2Cl_2	CF_2Cl_2	67	2
$CHCl_3$	$CFCl_3$	87.5	2
CH_3CCl_3	$CF_2ClCFCl_2$	98	2
CH_2ClCH_2Cl	CF_2ClCF_2Cl	57	2

product(s) formed as a result of the direct fluorination of chloro and polychloro alkanes. In all cases the direct fluorination of primary alkyl chlorides results in the unrearranged perfluorinated analogues [*33, 36*]. Those molecules having a tertiary chlorine rearrange completely in virtually all cases to molecules resulting from at least one 1,2-chlorine shift [*36*]. In tertiary amyl chloride one-third of the rearranged product is the result of two sequential 1,2-chlorine shifts. The sequential shifts must somehow act in concert because approximately 90% of the precursor to the product that would have been formed by a single 1,2-chlorine shift, perfluoro-2-methyl-3-

chlorobutane, rearranged to form the precursor to perfluoro-2-methyl-4-chlorobutane. Furthermore it could be shown by reactions of low fluorine–substrate molar ratios that the rearrangements occurred very early in the direct fluorination, at the mono- and difluorination stage(s) [*36*].

Direct fluorination of secondary alkyl chlorides produces both rearranged and unrearranged products [*36*]. The relative amounts are apparently determined by the rate of rearrangement relative to the rate of radical capture by molecular fluorine. If the rate of intramolecular rearrangement is slow relative to fluorination, less rearranged product is formed than when intramolecular rearrangement is faster. Any equilibration of a radical intermediate can be ruled out because rearrangement occurs when secondary alkyl chlorides are fluorinated to produce perfluorinated primary and secondary alkyl chlorides, but the analogous primary alkyl chloride produces only the perfluorinated primary chloride.

In radical reactions not involving bromine or chlorine on the substrate, rearrangements are much rarer. One example is the fluorination of di-*tert*-butyl ketone which produces perfluorinated *tert*-butyl isobutyl ketone [*35*]. Although isolated yields are poor only the rearranged ketone could be isolated. This is perhaps only the second example of a 1,2-acyl shift. Low fluorine:substrate ratios show that this rearrangement occurs after monofluorination.

In a few cases extensive cleavage of the substrate molecule occurs. The case documented [*38*] involves *beta* cleavage of tetramethylorthocarbonate and trimethylorthoacetate. The mechanism of this reaction involves the formation of a radical which can form a more stable radical by eliminating a stable molecule by *beta* cleavage. Interestingly tetramethylorthocarbonate seems to suffer less cleavage by the LTG method [*9*]. Examples of beta cleavage during direct fluorination are shown in equations 6–9, where AF is aerosol direct fluorination procedure.

$$(CH_3O)_4C \xrightarrow{-AF^a} \underset{7\%}{(CF_3O)_4C} + \underset{14\%}{(CF_3O)_3CF} + \underset{6\%}{(CF_3O)_2CF_2} + \underset{66\%}{CF_3OCF_3} + COF_2 \tag{6}$$

$$(CH_3O)_3CH \xrightarrow{-AF} \underset{5\%}{(CF_3O)_3CF} + \underset{29\%}{(CF_3O)_2CF_2} + \underset{60\%}{CF_3OCF_3} + COF_2 \tag{7}$$

$$CH_3C(OCH_3)_3 \xrightarrow{-AF} \underset{0\%}{CF_3C(OCF_3)_3} + \underset{24\%}{CF_3CF(OCF_3)_2} + CF_3CF_2OCF_3 + COF_2 \tag{8}$$

$$(CH_3)_3COCH_3 \xrightarrow{-AF} \underset{36\%}{(CF_3)_3COCF_3} + \underset{40\%}{(CF_3)_3CF} \tag{9}$$

[a]Aerosol direct fluorination procedure

The significance of such rudimentary probes of mechanistic pathways rests with their dispelling the myth of fluorine's unpredictability. In fact, fluorine, though highly reactive and with few activation barriers, is highly predictable if the energy density (exothermicity per unit volume) of its reactions can be controlled so as not to disturb the integrity of the substrates with which it reacts. Precise control of reactions will eventually lead to exploitation of what selectivity exists in the energy differences of potential reaction sites. Significant strides have occurred in the selectivity of direct fluorination, principally reactions in which elemental fluorine generates an in situ active agent.

Selective Fluorination

The simplest method for obtaining selective fluorination is to conduct reactions under conditions that invigorate the electrophilicity of fluorine. In practice this method entails the creation of anionic or strongly nucleophilic reactive centers on substrate molecules while suppressing or reducing the tendency toward radical attack. Numerous examples of selective fluorine attack on carbanionic, amido and carboxylato species are documented. Especially abundant is *alpha* fluorination of nitroalkanes in polar solvents [*42, 43, 44, 45, 46*] (equations 10–14).

[*42*] $$NH_4^+\ EtO_2CC^-HNO_2 \xrightarrow[5\text{–}10\ ^\circ C,\ H_2O]{3\%\ F_2/N_2} EtO_2CCFHNO_2 \quad 72\% \qquad \mathbf{10}$$

[*43*] $$O_2NCH_2(CH_2)_2CH_2NO_2 \xrightarrow[-30\ ^\circ C]{MeOH/NaOH} \xrightarrow[-30\ ^\circ C,\ 3\text{–}4\ hr]{7\text{–}10\%\ F_2/N_2} \underset{49\%}{O_2NCHFCH_2CH_2CHFNO_2} \qquad \mathbf{11}$$

[*44*] $$HC(NO_2)_3 \xrightarrow[H_2O,\ 0\ ^\circ C]{50\%\ F_2/N_2} FC(NO_2)_3 \quad 92.3\% \qquad \mathbf{12}$$

[*45*] $$CH_3(CH_2)_2CH_2N^-NO_2 \xrightarrow[H_2O,\ 0\text{–}5\ ^\circ C]{15\%\ F_2/N_2} CH_3(CH_2)_2CH_2NFNO_2 \quad 86\% \qquad \mathbf{13}$$

[*46*] $$NaO_2C(CH_2)_4CO_2Na \xrightarrow{15\%\ F_2/N_2} \xrightarrow[0\text{–}5\ ^\circ C]{H_2SO_4} F(CH_2)_4CO_2H \quad 40\% \qquad \mathbf{14}$$

Selective fluorination in polar solvents has proved commercially successful in the synthesis of 5-fluorouracil and its pyrimidine relatives, an extensive subject that will be discussed in another section. Selective fluorination of enolates [*47*], enols [*48*], and silyl enol ethers [*49*] resulted in preparation of *alpha*-fluoro ketones, *beta*-diketones, *beta*-ketoesters, and aldehydes. The reactions of fluorine with these functionalities is most probably an addition to the *ene* followed by elimination of fluoride ion or hydrogen fluoride rather than a simple substitution. In a similar vein, selective fluorination of pyridines to give 2-fluoropyridines was shown to proceed through pyridine difluorides [*50*].

The high reactivity of N–H bonds has also been exploited to produce N–F derivatives without significant substitution on neighboring C–H bonds, Diethylphosphoramidates of ammonia, alkylamines, and α,ω-diaminoalkanes were fluorinated in polar solvents to produce difluoroamine [*51*], *N,N*-difluoroalkylamines, and α,ω-bis(*N,N*-difluoroamino)alkanes [*52*]. Acetamide undergoes fluorination to give modest yields of *N,N*-difluoroacetamide and acetyl fluoride when fluorinated at –78 °C in the presence of potassium fluoride [*53*]. Good yields, 55%, of *N*-fluoroimine esters were reported in the fluorination of imine esters of perfluorocarboxylic acids [*54*]. A rather surprising example is the reaction of fluorine with glycolic acid amidine hydrochloride to produce the first example of an α,α-bis(difluoroamino)fluoroacetaldehyde [*55*].

The work of Rozen, which is the subject of an excellent account [*56*], has substantially elaborated the use of elemental fluorine as a selective fluorinating agent of complex organic compounds. The suppression of free radicals coupled with enhancement of the electrophilic nature of fluorine by low temperatures, dilution, and most importantly, dramatic solution effects by polar and radical scavenger solvents (chloroform, acetic acid, and nitromethane) make it possible to selectively attack otherwise unactivated, tertiary aliphatic hydrogens on steroids and other complex molecules with excellent regiospecificity and good stereospecificity. The commonly used solvent chloroform (neat or diluted with $CFCl_3$) acts as a radical scavenger, provides a polar medium, and acts as an acceptor for the developing fluoride ion by hydrogen bonding. The active agent is viewed as a polar complex of F–F···$HCCl_3$ whose positive pole attacks electrophilically to form a nonclassical $[H–C–F]^+$ carbocation at the tertiary C–H bond that has the highest p-orbital contribution to its hybridization and is most removed from the effect of electron-withdrawing groups. Such attack would be expected to lead to complete retention of configuration, and this expectation was confirmed [*57*].

In other work Rozen added molecular fluorine to a steroidal ene–one dissolved in ethanol at low temperatures to produce a *vicinal* difluoride in a cleaner, better yield reaction than previously obtainable [*58*]. Although the reaction was not general, the stereoselectivity was very high, and contrary to addition of other halogens, addition was *syn*, characteristic of an electrophilic addition pathway.

Rozen used molecular fluorine to generate, in suspensions of sodium trifluoroacetate [*59*] and sodium acetate [*60*], fluoroxy derivatives as indirect fluorinating agents. The products from fluorination of sodium trifluoroacetate were characterized as mostly pentafluoroethyl hypofluorite under anhydrous conditions and trifluoroacetyl hypofluorite when traces of water or hydrogen fluoride are present. This reagent solution produces α-fluorocarbonyl derivatives when reacted with electron-rich enol acetates [*61*]; when anhydrous, it can add the pentafluoroethoxy group and fluorine to various olefins [*62*], whereas with a trace of water it forms the fluorohydrins [*63*].

The product from fluorination of sodium acetate is acetyl hypofluorite [*64*], which is isolated and characterized [*65*]. The value of this reagent lies in its relative mildness, because it reacts cleanly with most olefins, adding the elements of acetoxyl and fluorine [*66*]. Trifluoroacetyl hypofluorite adds cleanly only to benzylic or electron-rich double bonds.

In related work, fluorine reacts with iodine in fluorotrichloromethane, and the iodine fluoride thus formed adds the elements of iodine and fluorine to olefins at –78 °C with full regio- and stereoselectivity [*67*]. Bromine–fluorine, on the other hand,

References are listed on pages 116–119.

produces an unselective, uncontrolled reaction. Fluorine reacts with water in acetonitrile and apparently forms a relatively stable hypofluorous acid–acetonitrile (HOF–H_3N) complex that smoothly epoxidizes olefins [*68*]. Water is the source of the epoxide oxygen, and so this reagent is an excellent way to incorporate ^{17}O and ^{18}O.

Elemental fluorine is becoming more widely used in organic chemistry. Three excellent reviews [*69, 70, 71*] of this work provide a wealth of detailed information.

Experimental Procedure: Aerosol Fluorination of Dimethoxymethane [29]

For the safety advisory, see pages 25–26 of the book.

Dimethoxymethane (Aldrich) as received is drawn into a 10.0-mL Precision Scientific Pressure-Lok syringe and is placed in a Sage Instruments model 341A syringe pump. The needle of the syringe is attached to a length of 1/16-inch SS tubing leading to the hydrocarbon inlet of the heated (65 °C) evaporator unit (*see* Figure 1 of reference 29). Previously two nickel combustion boats filled with approximately 10.0 g of anhydrous sodium fluoride were placed inside the tube–furnace preaerosol particulate generator. Also, approximately 200 g of 1/8-inch sodium fluoride pellets were placed in the product trap to absorb excess hydrogen fluoride not absorbed by the aerosol particulates. The main helium carrier gas flow through the preaerosol furnace–tube is set to 500 mL/m, and the furnace is heated to a 950 °C thermocouple reading for the furnace effluent. (*Note*: The NaF melt is at least 100 °C hotter than this reading.)

The secondary hydrocarbon carrier (helium) flow entering the top of the reactor is set to 500 mL/m, and the primary hydrocarbon carrier entering alongside the hydrocarbon liquid inlet is set to 55 mL/m. The circulating coolant (methanol at –15 °C) is fed into the bottom of each module until the thermocouple temperature controller at the top of each module shuts off the flow at the preset temperatures of –10 and –8 °C for the top (No. 1) and bottom (No. 2) modules, respectively. Fluorine gas is then introduced into the four inlets in the following amounts diluted by helium (in parentheses): module No. 1 inlet 1 (top), 8 mL/m F_2 (170 mL/m He); module No. 1 inlet 2 (bottom), 50 mL/m F_2 (170 mL/m He); module No. 2 inlet 1, 35 mL/m F_2 (170 mL/m He); and module No. 2 inlet 2, 5 mL/m F_2 (170 mL/m He).

When all flows and temperatures are stabilized, the photochemical stage is energized (1200-W Hanovia medium-pressure mercury arc in a water-cooled quartz immersion well that produces approximately 337 W of ultraviolet energy or 573 W of total UV–visible–IR energy over 10 inches). When the system is stabilized, the Dewar flask surrounding the product trap is filled with liquid nitrogen and the syringe pump is started. Dimethoxymethane (molecular weight, 76.1 amu) is introduced at 1.0 mL/h (0.860 g/h; 11.3 mmol/h). A total of 5.0 mL (4.3 g; 56.5 mmol) is introduced over 5 h. The fluorine–ether stoichiometry is 2.6:1, a 21:1 mole ratio of F_2 to ether.

References are listed on pages 116–119.

When 5.00 mL of ether has been delivered by the syringe pump, the pump is shut off. The reactor is allowed to run an additional 15 min before the fluorine and the mercury arcs are shut off. The preaerosol furnace, the evaporator heater unit, and the coolant pump are shut off. Once the system approaches ambient conditions, all the helium carriers are shut off and the product trap valves are closed. The product trap and its Dewar flask filled with liquid nitrogen are removed to the vacuum line where the trap is evacuated.

The contents of the trap are transferred to the glass vacuum line by vacuum distillation. The contents are condensed into a 150-mL stainless steel cylinder containing an excess of 20% aqueous sodium carbonate and allowed to react with occasional shaking for 24 h. The products are fractionated through –45, –78, –131, and –196 °C traps under active pumping (3 µm of Hg). The product (8.5 g) is collected in the –131 °C trap. The remaining vacuum line traps were virtually empty.

The product assayed by capillary gas chromatography (SP2100, 50 m, inlet temperature of 35 °C; column temperature of 35–50 °C at 1 °C/m; flame ionization detector at 200 °C) proved to be essentially pure (>99%) perfluorinated dimethoxymethane in 68% yield. ^{19}F NMR $\phi CF_2 = -56.55$ ppm (heptet, 2); $\phi CF_3 = -57.70$ ppm (triplet, 6). Elemental analysis, C_3F_8. Calc.: C, 17.65; F, 74.49. Found: C, 17.25; F, 74.05.

Electrochemical Fluorinations
by James L. Adcock

Although occasionally used for selective fluorinations [*72*], *electrochemical fluorination, ECF,* is principally used for deep fluorinations. Despite low to moderate yields of perfluorinated compounds and many side products, it remains competitive because of the relatively low cost of electricity as chemical reactant and hydrogen fluoride, HF, as fluorine source. Fluorinations are conducted in nickel or steel cells equipped with nickel or steel cathodes and nickel anodes. Anode potential versus the **hydrogen reference electrode** must be greater than 3.5 V to achieve appreciable fluorination, but potentials greater than 5.0 V (cell potential, 7.5–8 V) result in extensive breakdown of the carbon skeleton. Conditioning of the anodes so that they are properly coated with nickel fluorides is essential for good results, as is drying of the hydrogen fluoride electrolytically [*73*].

Although mechanistically complex, electrochemical fluorination is generally thought to be a free radical process. However, compelling evidence supports carbocation intermediates [*74, 75, 76*]. The electrochemical fluorination mechanism postulated by Burdon et al. [*74*], the so-called EC_bEC_N mechanism, involves for saturated substrates a two-electron oxidation, proton loss, and fluoride-ion capture. In the first stage, E, the organic molecule adsorbed on the anode surface, undergoes oxidation to a radical cation; a saturated radical cation in the second stage, C_b, eliminates a hydrogen ion to form a radical. In the third stage, E, that radical is oxidized to a cation, which in the fourth stage, C_N, reacts with a fluoride ion to form the C–F bond. Modifications to the mechanism occur for olefinic and aromatic substrates [*77*]. In olefins, initial anodic oxidation results in predominantly *syn* addition to the olefin. In methylene chloride–triethylamine·3HF, methylbenzenes are initially oxidized at the methyl groups to form benzylic cations. Benzylic cations formed from benzylic ketones, esters, or nitriles may react with acetonitrile as well as fluoride ion when acetonitrile is part of the electrolyte [*78*]. Electrochemical generation of carbocations was convincingly demonstrated in a perfluorinated ionomer (Nafion) divided cell [*79*]. Perhaps the most convincing evidence that ease of anodic oxidation is critical, is the electrochemical fluorination of acetonitrile to trifluoroacetonitrile in yields exceeding that for C_2F_6 and $C_2F_5NF_2$ combined when helium sparging is used to reduce electrode contact time [*80*].

Both alkanesulfonyl fluorides and fluoride ions were shown by tensammetric measurements to be adsorbed on the NiF_2-coated nickel anode, and alkanesulfonyl fluorides are oxidized at high potentials. Increasing the fluoride-ion concentration increases adsorption of alkanesulfonyl fluorides, but interestingly, the addition of fluoride ion lowers yields by inhibiting the oxidation of the alkanesulfonyl fluorides [*81*]. This study shows that C_1 to C_8 alkanesulfonyl fluorides are equally well adsorbed on the anode; however, yields decrease as the alkyl chain lengthens. This decrease in yield was attributed to the increasing difficulty in oxidation of the longer chain sulfonyl fluorides. The current density at constant potential decreases as the chain lengthens at a given molar concentration. A similar decrease in current density was observed for α-chloroethanesulfonyl fluoride relative to ethanesulfonyl fluoride [*81*]. Electrochemical fluorination of α-chloroethanesulfonyl chloride yields at best 8.4% $CF_3CFClSO_2F$ plus 9.3% $C_2F_5SO_2F$ with considerable fragmentation, even though carried out at −10 °C with a current density of 0.004 A/cm^2 [*82*]. Electrochemical fluorination of chloromethanesulfonyl chloride occurs in considerably better yield, 57.3%, but that is still considerably less than the methanesulfonyl chloride yield, 87% [*83*]. Electrochemical fluorination of propanesulfonyl fluoride and 1,3-propanedisulfonyl difluoride produces the perfluorinated analogues in yields of 42 and 34%, respectively [*84*].

During electrochemical fluorination retention of important functional groups or atoms in molecules is essential. Acyl fluorides and chlorides, but not carboxylic acids and anhydrides (which decarboxylate), survive perfluorination to the perfluorinated acid fluorides, albeit with some cyclization in longer chain ($>C_4$) species [*73*]. Electrochemical fluorination of acetyl fluoride produces perfluoroacetyl fluoride in 36–45% yields [*85*]. Electrochemical fluorination of octanoyl chloride results in perfluorinated cyclic ethers as well as perfluorinated octanoyl fluoride. Cyclization decreases as initial substrate concentration increases and has been linked to hydrogen-bonded onium polycations [*73*]. Cyclization is a common phenomenon involving longer ($>C_4$) and branched chains. α-Alkyl-substituted carboxylic acid chlorides, fluorides, and methyl esters produce both the perfluorinated cyclic five- and six-membered ring ethers as well as the perfluorinated acid fluorides in good yields [*86*]. Yields of the various cyclic perfluorinated ethers, which are valuable as potential blood substitutes and lubricant fluids, are sufficient to make the side product of greater importance. Abe et al. [*87*] investigated electrochemical fluorination–cyclization as a route to many new perfluorinated ether species. Cycloalkyl-substituted carboxylic acid esters produce perfluorinated bicyclic and monospiro ethers as well as the corresponding acid fluorides [*87*] (equations 15–18).

R
CHCOCH$_3$
O
ECF
R$_f$
F
F
O
15

R = H, CH$_3$, C$_2$H$_5$

R$_f$ = F, CF$_3$, C$_2$F$_5$
10.5%, 18.8%, 14.7%

CH$_2$COCH$_3$
O
ECF
F
F
O
16

11.8%

CH$_2$CH$_2$COCH$_3$
O
ECF
F
F
O
17

12.8%

R$_f$
CH$_2$CH$_2$COCH$_3$
O
ECF
F
F
O
18

R = H, CH$_3$

R$_f$ = F, CF$_3$
16.6%, 13.8%

Electrochemical fluorination of α-cyclohexenyl-substituted carboxylic (acetic, propanoic, butanoic, and pentanoic) acid esters (methyl, ethyl, and propyl) results in a series of both perfluoro-9-alkyl-7-oxabicyclo[4.3.0]nonanes and perfluoro-8-alkoxy-9-alkyl-7-oxabicyclo[4.3.0]nonanes [*88*] (equation 19).

Electrochemical fluorination of primary alcohols and aldehydes having four to eight carbon atoms produces principally the perfluorinated cyclic ethers but also perfluorinated acid fluorides [*89*]. Alkyl-substituted and unsubstituted five-, six-, and seven-membered cyclic ethers themselves give fair to good yields of perfluorinated cyclic ethers; however, significant rearrangement occurs and involves ring contraction, ring expansion, and some ring scission [*90, 91*].

References are listed on pages 116–119.

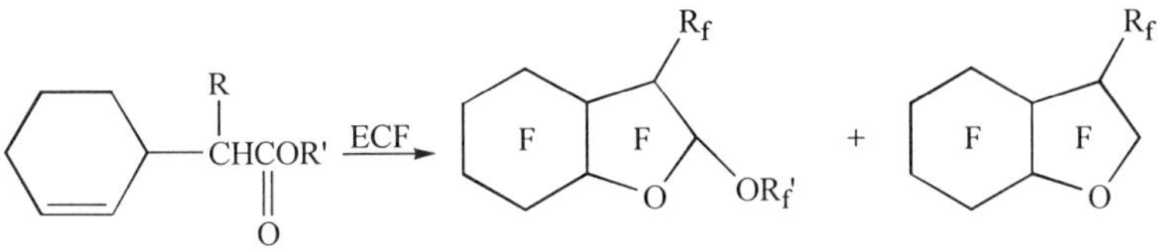

(19)

R = H, CH_3, C_2H_5, C_3H_7; R' = CH_3
R = H; R' = C_2H_5, C_3H_7

Yields (%) of Cyclic Ethers from Equation 19 [a]

R'_f	$R_f = F$	$R_f = CF_3$	$R_f = C_2H_5$	$R_f = C_3H_7$
CF_3	8.8 (19.3)	9.7 (21.9)	9.5 (22.9)	6.9 (17.1)
C_2H_5	5.4 (16.9)			
C_3H_7	2.2 (16.7)			

[a] Values in parentheses are yields of fluorinated cyclic ethers without pendant ether group (OR'_f).

Ethers having an α-chloromethyl group produce the chlorine-containing perfluorinated ethers in only 1–2% yields [*90*]. Earlier investigations of chlorine-containing ethers reveal that chlorine bonded to the α-carbon of an ether linkage is readily removed during electrochemical fluorination, whereas a chlorine bonded to a β-carbon is retained [*92*]. This stability pattern is repeated for electrochemical fluorination of tertiary amines; chloromethyl amines uniformly lose chlorine and are degraded, but 2-chloroalkyl and higher amines are fluorinated in fair to good yields [*93*]. A trend to higher yields is observed as the chlorine becomes more remote from the nitrogen atom, but the yields of chlorine-containing perfluorinated compounds are almost always considerably less than those of the corresponding fully perfluorinated structures. Ring contraction from seven- to six-membered rings in cyclic amines, and presumably ring contraction in cyclic ethers as well, is attributed to a carbocation generated by anodic oxidation of the adsorbed radical [*94*]. Retention of a chlorine atom during electrochemical fluorination is modest at best in virtually all cases; even the excellent yield of chloromethanesulfonyl chloride is considerably less than that of the unchlorinated alkyl analogue [*83*].

Electrochemical fluorination of tertiary amines is perhaps the most effective process for producing perfluorinated tertiary amines. These valuable materials, useful as fluids and in blood-substitute preparations, can be prepared in moderate to good yields. The best yield reported, 52%, is for the fluorination of diethylcyclohexylamine [*95*]. Complex mixtures of products are commonly obtained with the perfluorinated analogue produced in 10–25% yields. Reduction of the electrolyte temperature to –4 °C from 19 °C significantly reduces degradation products but slows the electrolysis somewhat [*96*]. Electrochemical fluorination of partially fluorinated amines does not significantly improve yields of perfluorinated amines [*97*]. Trimethylamine gives the previously unknown mono-, bis-, and tris(difluoromethyl)amines [*98*], as well as perfluorinated ethyl amines [*99*] (equation 20).

$$
\begin{array}{lll}
 & & + \ (CF_3)_2NCF_2H \ (2\%) + CF_3N(CF_2H)_2 \ (0.5\%) \\
(CH_3)_3N & \longrightarrow & (CF_3)_3N \ (11\%) + (CF_2H)_3N \ (0.1\%) + C_2F_5N(CF_3)_2 \ (0.2\%) \\
 & & + \ C_2F_5N(CF_3)(CF_2H) \ (0.02\%) + C_2F_5N(CF_2H)_2 \ (0.005\%)
\end{array}
\qquad \mathbf{20}
$$

Electrochemical fluorination of *N,N*-dialkylamino-substituted carboxylic acids as their methyl esters produces the analogous perfluorinated tertiary amine carboxylic acid derivatives in 18–30% yields as well as cyclic amine ethers [*100*].

References for Pages 97-116

1. Lagow, R. J.; Margrave, J. L. *Progr. Inorg. Chem.* **1979,** *26,* 161.
2. Adcock, J. L.; Kunda, S. A.; Taylor, D. R.; Sievert, A. C.; Nappa, M. J. *Ind. Eng. Chem. Res.* **1989,** *28,* 1547.
3. Maraschin, N. J.; Lagow, R. J. *J. Am. Chem. Soc.* **1972,** *94,* 8601.
4. Maraschin, N. J.; Lagow, R. J. *Inorg. Chem.* **1973,** *12,* 1458.
5. Aikman, R. E.; Lagow, R. J. *J. Org. Chem.* **1982,** *47,* 2789.
6. Adcock, J. L.; Lagow, R. J. *J. Org. Chem.* **1973,** *38,* 3617.
7. Adcock, J. L.; Lagow, R. J. *J. Am. Chem. Soc.* **1974,** *96,* 7588.
8. Adcock, J. L.; Beh, R. A.; Lagow, R. J. *J. Org. Chem.* **1975,** *40,* 3271.
9. Lin, W. H.; Clark, W. D.; Lagow, R. J. *J. Org. Chem.* **1989,** *54,* 1990.
10. Persico, D. F.; Huang, H. N.; Lagow, R. J.; Clark, L. C. *J. Org. Chem.* **1985,** *50,* 5156.
11. Maraschin, N. J.; Catsikis, B. D.; Davis, L. H.; Jarvinen, G.; Lagow, R. J. *J. Am. Chem. Soc.* **1975,** *97,* 513.
12. Liu, E. K. S.; Lagow, R. J. *J. Fluorine Chem.* **1979,** *13,* 71.
13. Robertson, G.; Liu, E. K. S.; Lagow, R. J. *J. Org. Chem.* **1978,** *43,* 4981.
14. Lin, W. H.; Bailey, W. I., Jr.; Lagow, R. J. *J. Chem Soc., Chem. Commun.* **1985,** 1350; *Pure Appl. Chem.* **1988,** *60,* 473.
15. Shimp, L. A.; Lagow, R. J. *J. Org. Chem.* **1977,** *42,* 3437.
16. Dmowski, W. *J. Fluorine Chem.* **1990,** *49,* 281.
17. Adcock, J. L.; Luo, H.; Zuberi, S. S. *J. Org. Chem.* **1992,** *57,* 4749.
18. Margrave, J. L.; Lagow, R. J. *J. Polym. Lett.* **1974,** *12,* 177.
19. Gerhardt, G. E.; Lagow, R. J. *J. Polym. Sci. Polym. Chem. Ed.* **1979,** *18,* 157.
20. Gerhardt, G. E.; Lagow, R. J. *J. Chem. Soc., Perkin Trans. 1* **1981,** 1321.
21. Persico, D. F.; Lagow, R. J. *Macromolecules* **1985,** *18,* 1383.

22. Persico, D. F.; Gerhardt, G. E.; Lagow, R. J. *J. Am. Chem. Soc.* **1985,** *107,* 1197; *Makromol. Chem., Rapid Commun.* **1985,** *6,* 85.
23. Gerhardt, G. E.; Lagow, R. J. *J. Chem. Soc., Chem. Commun.* **1977,** 259; *J. Org. Chem.* **1978,** *43,* 4505.
24. Adcock, J. L.; Inoue, S.; Lagow, R. J. *J. Am. Chem. Soc.* **1978,** *100,* 1948.
25. Adcock, J. L. *J. Fluorine Chem.* **1980,** *16,* 297–300.
26. Moggi, M.; Calini, P.; Gregorio, G.; Moggi, G. *J. Fluorine Chem.* **1988,** *40,* 349.
27. Scherer, K. V.; Yamonouchi, K.; Ono, T. *J. Fluorine Chem.* **1990,** *50,* 47.
28. Adcock, J. L.; Horita, K.; Renk, E. B. *J. Am. Chem. Soc.* **1981,** *103,* 6937.
29. Adcock, J. L.; Cherry, M. L. *Ind. Eng. Chem. Res.* **1987,** *26,* 208.
30. Adcock, J. L.; Robin, M. L. *J. Org. Chem.* **1983,** *48,* 2437.
31. Adcock, J. L.; Robin, M. L. *J. Org. Chem.* **1983,** *48,* 3128.
32. Huang, S.; Klein, D. H.; Adcock, J. L. *Rapid Commun. Mass Spectr.* **1988,** *2,* 204.
33. Adcock, J. L.; Evans, W. D.; Heller-Grossman, L. *J. Org. Chem.* **1983,** *48,* 4953.
34. Adcock, J. L.; Robin, M. L. *J. Org. Chem.* **1984,** *49,* 191.
35. Adcock, J. L.; Robin, M. L. *J. Org. Chem.* **1984,** *49,* 1442.
36. Adcock, J. L.; Evans, W. D. *J. Org. Chem.* **1984,** *49,* 2719.
37. Adcock, J. L.; Cherry, M. L. *J. Fluorine Chem.* **1985,** *30,* 343.
38. Adcock, J. L.; Robin, M. L.; Zuberi, S. *J. Fluorine Chem.* **1987,** *37,* 327.
39. March, J. *Advanced Organic Chemistry: Reactions, Mechanisms, and Structure;* McGraw Hill: New York, 1968; pp 781–795.
40. Whitmore, F. C.; Popkin, A. N.; Bernstein, H. T.; Wilkins, J. P. *J. Am. Chem. Soc.* **1941,** *63,* 124.
41. Freidlina, R. K. *Adv. Free Radical Chem.* **1965,** *1,* 211.
42. Oreshko, G. V.; Eremenko, L. T. *Izv. Akad. Nauk SSSR,* **1971,** 2791; *Chem. Abstr.* **1971,** *76,* 126336b.
43. Eremenko, L. T.; Oreshko, G. V. *Izv. Akad. Nauk SSSR,* **1973,** 1174; *Chem. Abstr.* **1973,** *79,* 52745x.
44. Grakauskas, V.; Baum, K. *J. Org. Chem.* **1968,** *33,* 3080.
45. Grakauskas, V.; Baum, K. *J. Org. Chem.* **1972,** *37,* 334.
46. Grakauskas, V. *J. Org. Chem.* **1969,** *34,* 2446.
47. Misaki, S.; Suefuji, M.; Tsushima, T.; Tanida, H. *J. Org. Chem.* **1982,** *47,* 1107.
48. Purrington, S. T.; Bumgardner, C. L.; Lazaridis, N. V. *J. Org. Chem.* **1987,** *52,* 4307.
49. Purrington, S. T.; Lazaridis, N. V.; Bumgardner, C. L. *Tetrahedron Lett.* **1986,** *27,* 2716.
50. Van der Puy, M. *Tetrahedron Lett.* **1987,** *28,* 255.
51. Mathey, F.; Bensoam, J. *C. R. Acad. Sci., Ser. C* **1972,** *274,* 933.
52. Bensoam, J.; Mathey, F. *C. R. Acad. Sci., Ser. C* **1974,** *278,* 1313.

53. Davydov, A. V.; Stolyarov, V. P. *Zh. Vaes. Khim. O-va.* **1977,** *22,* 107; *Chem. Abstr.* **1977,** *86,* 189156v.
54. Fokin, A. V.; Studnev, Y. N.; Stolyarov, V. P.; Baranov, N. N. *Izv. Akad. Nauk SSSR* **1982,** 937; *Chem. Abstr.* **1982,** *97,* 38475z.
55. Fokin, A. V.; Uzun, A. T.; Stolyarov, V. F. *Izv. Akad. Nauk. SSSR* **1982,** 1438; *Chem. Abstr.* **1982,** *97,* 162294d.
56. Rozen, S. *Acc. Chem. Res.* **1988,** *21,* 307.
57. Rozen, S.; Gal, C. *J. Org. Chem.* **1987,** *52,* 2769.
58. Rozen, S.; Brand, M. *J. Org. Chem.* **1986,** *51,* 3607.
59. Rozen, S.; Lerman, O. *J. Am. Chem. Soc.* **1979,** *101,* 2782.
60. Rozen, S.; Lerman, O.; Kol, M. *J. Chem. Soc., Chem. Commun.* **1981,** 443.
61. Rozen, L.; Menahem, Y. *J. Fluorine Chem.* **1980,** *16,* 19.
62. Mulholland, G. K.; Ehrenkaufer, R. E. *J. Org. Chem.* **1986,** *51,* 1482.
63. Rozen, S.; Lerman, O. *J. Org. Chem.* **1980,** *45,* 672.
64. Hebel, D.; Lerman, O.; Rozen, S. *J. Fluorine Chem.* **1985,** *30,* 141.
65. Appelman, E. H.; Mendelsohn, M. H.; Kim, H. *J. Am. Chem. Soc.* **1985,** *107,* 6515.
66. Rozen, S.; Lerman, O.; Kol, M.; Hebel, D. *J. Org. Chem.* **1985,** *50,* 4753.
67. Rozen, S.; Brand, M. *J. Org. Chem.* **1985,** *50,* 3342.
68. Rozen, S.; Kol, M. *J. Org. Chem.* **1990,** *55,* 5155.
69. Purrington, S. T.; Kagen, B. S.; Patrick, T. B. *Chem. Rev.* **1986,** *86,* 997.
70. Vypel, H. *Chimia* **1985,** *39,* 305.
71. Haas, A.; Lieb, M. *Chimia* **1985,** *39,* 134.
72. O'Malley, R. F.; Mariani, H. A.; Buehler, D. R.; Jerina, D. M. *J. Org. Chem.* **1981,** *46,* 2816–2818.
73. Drakesmith, F. G.; Hughes, D. A. *J. Appl. Electrochem.* **1979,** *9,* 685.
74. Burdon, J.; Parsons, I. W.; Tatlow, J. C. *Tetrahedron,* **1972,** *28,* 43.
75. Rozhkov, I. N. *Russ. Chem. Rev.* **1976,** *45,* 615.
76. Gambaretto, G. P.; Napoli, M.; Conte, L.; Scipioni, A.; Armelli, R. *J. Fluorine Chem.* **1985,** *27,* 149–155.
77. Bensadat, A.; Bodonner, G.; Laurent, E.; Tardivel, R. *J. Fluorine Chem.* **1982,** *20,* 333.
78. Laurent, E.; Marquet, B.; Tardivel, R.; Thiebault, H. *Tetrahedron Lett.* **1987,** *28,* 2359.
79. Huba, F.; Yeager, E. B.; Olah, G. A. *Electrochim. Acta* **1979,** *24,* 489.
80. Haruta, M.; Watanabe, N. *J. Fluorine Chem.* **1976,** *7,* 159.
81. Cauquis, G.; Keita, B.; Pierre, G.; Jaccaud, M. *J. Electroanal. Chem. Interfacial Electrochem.* **1979,** *100,* 205.
82. Sartori, P.; Habel, W. *J. Fluorine Chem.* **1981,** *18,* 131.
83. Sartori, P.; Habel, W. *J. Fluorine Chem.* **1980,** *16,* 265.
84. Hollitzer, E.; Sartori, P. *J. Fluorine Chem.* **1987,** *35,* 329.

85. Wasser, D. J.; Johnson, P. S.; Klink, F. W.; Kucera, F.; Liu, C. C. *J. Fluorine Chem.* **1987,** *35,* 557.
86. Abe, T.; Kodaira, K.; Baba, H.; Nagase, S. *J. Fluorine Chem.* **1978,** *12,* 1.
87. Abe, T.; Baba, H.; Hayashi, E.; Nagase, S. *J. Fluorine Chem.* **1983,** *23,* 123.
88. Abe, T.; Hayashi, E.; Baba, H.; Nagase, S. *J. Fluorine Chem.* **1984,** *25,* 419.
89. Abe, T.; Nagase, S.; Baba, H. *Bull. Chem. Soc. Jpn.* **1976,** *49,* 1888.
90. Abe, T.; Nagase, S. *J. Fluorine Chem.* **1979,** *13,* 519.
91. Abe, T.; Hayashi, E.; Baba, H.; Kodaira, K.; Nagase, S. *J. Fluorine Chem.* **1980,** *15,* 353.
92. Okazaki, K.; Nagase, S.; Baba, H.; Kodaira, K. *J. Fluorine Chem.* **1974,** *4,* 387.
93. Hayashi, E.; Abe, T.; Baba, H.; Nagase, J. *J. Fluorine Chem.* **1983,** *23,* 371; Omori, K.; Nagase, S.; Baba, H.; Kodaira, K. *J. Fluorine Chem.* **1977,** *9,* 279.
94. Hayashi, E.; Abe, T.; Baba, H.; Nagase, S. *J. Fluorine Chem.* **1984,** *26,* 417.
95. Conte, L.; Fraccaro, C.; Napoli, M.; Mistrorigo, M. *J. Fluorine Chem.* **1986,** *34,* 183.
96. Conte, L.; Napoli, M.; Gambaretto, G. P. *J. Fluorine Chem.* **1985,** *30,* 89.
97. Ono, T.; Inoue, Y.; Fukaya, C.; Arakawa, Y.; Naito, Y.; Yokoyama, K.; Yamanouchi, K.; Kobayashi, Y. *J. Fluorine Chem.* **1985,** *27,* 333.
98. Burger, H.; Niepel, H.; Pawelke, G.; Frohn, H. J.; Sartori, P. *J. Fluorine Chem.* **1980,** *15,* 231.
99. Burger, H.; Eujen, R.; Niepel, H.; Pawelke, G. *J. Fluorine Chem.* **1981,** *17,* 65.
100. Abe, T.; Hayashi, E.; Baba, H.; Fukaya, H. *J. Fluorine Chem.* **1990,** *48,* 257.

Replacement by Means of High-Valency Fluorides

by Miloš Hudlický

Some high-valency fluorides applied at moderate temperatures are capable of replacing individual hydrogens in aromatic rings. Thus ***benzene*** affords fluorobenzene on treatment with **silver difluoride** [*1*] and with **chlorine pentafluoride** [*2*] (equations 1 and 2).

[*1*] benzene $\xrightarrow[\text{reflux}]{AgF_2,\ C_6H_{14}}$ fluorobenzene 61% **1**

[*2*] benzene $\xrightarrow[CCl_4, 0°C]{ClF_5/N_2}$ fluorobenzene (54%) + chlorobenzene (37%) **2**

Lead tetrafluoride, generated in situ from lead dioxide and hydrogen fluoride, can replace benzylic hydrogen by fluorine [*3*]. Under similar conditions phenol is simultaneously oxidized to 4,4-difluoro-2,5-cyclohexadienone [*4*] (equations 3 and 4).

[*3*] 4-$NO_2C_6H_4CH_3$ $\xrightarrow[\text{-196°C to RT, 17 h}]{PbO_2,\ HF}$ 4-$NO_2C_6H_4CH_2F$ (21%) + 4-$NO_2C_6H_4CHF_2$ (71%) **3**

[*4*] C_6H_5–OH $\xrightarrow{PbO_2,\ HF/C_5H_5N}$ 4,4-difluoro-2,5-cyclohexadienone 50% **4**

Vanadium pentafluoride replaces benzylic hydrogen by fluorine but also adds fluorine to the aromatic system, giving fluorinated cyclohexadienes and cyclohexenes [*5*] (equation 5).

0065–7719/95/0187–0120$08.00/1

[5] **5**

CH_3 (F) $\xrightarrow[\text{SO}_2\text{ClF}]{2\ \text{VF}_5}$ -30 to -25°C

CH_2F (F) 30% + CHF_2 (F) 1% + CH_3 F 15% + CH_2F F 29%

CHF_2 F 6% + CH_3 F 4% + CH_2F F 6% + CHF_2 F 3%

High-valency fluorides such as **silver difluoride** [*1, 6*], **manganese trifluoride** [*7, 8*] and especially **cobalt trifluoride** [*8, 9, 10, 11, 12, 13, 14, 15, 16, 17, 18, 19, 20, 21, 22, 23*] convert, by combined addition of fluorine and replacement of hydrogen, ***aromatic hydrocarbons and heterocycles*** to poly- and perfluorinated products containing variable numbers of hydrogen atoms and double bonds. Such fluorinations are carried out at elevated temperatures so that elimination of hydrogen fluoride, skeletal rearrangements, and fragmentation often take place.

Complex fluorinating agents such as **potassium tetrafluoroargentate** [*6*], **potassium hexafluoronickelate** [*24*] and especially **potassium tetrafluorocobaltate** [*8, 9, 25*] and **cesium tetrafluorocobaltate** [*26, 27, 28, 29*] affect organic compounds in a similar way as cobalt trifluoride but give more unsaturated fluorinated products [*25*].

The fluorinations using high-valency metal fluorides and their complexes are carried out at elevated temperatures in tubular reactors fitted with rotating shafts with paddles. The technique has been sufficiently described in the literature [*30*].

In the past 20 years, fluorinations with the high-valency fluorides received much competition from electrophilic fluorinations in the field of replacement by fluorine of individual hydrogen atoms and from aerosol fluorinations in the field of poly- and perfluorinations. Research in the area of applications of **cobalt trifluoride and its analogues** has continued, and additional results have been obtained in separation of very complex mixtures of products and in isolation and identification of the components of the mixtures. The merits of recent work are better knowledge of the process, better understanding of the mechanism, and better techniques for isolation and identification of new compounds. As many as 50 products result from such fluorinations; thus, the actual yields of individual compounds are generally low and sometimes difficult to figure out from published results.

Up to 20 products with different numbers of fluorine atoms and double bonds have been isolated from the fluorination of ***benzene*** with **silver difluoride** [*6*], **manganese trifluoride** [*7*], **potassium tetrafluoroargentate** [*6*], and **potassium hexafluoronickelate** [*24*]. The composition of the products depends on the fluorinating agents and on the temperature (Table 1).

Considerable differences are evident in the results of fluorinations of ***toluene*** with **potassium tetrafluorocobaltate** at 340 °C [*25*] and with **cesium tetrafluorocobaltate** at 320 °C [*29*] (Table 2).

Table 1. Comparison of Fluorination of Benzene with Manganese Trifluoride [7], Potassium Hexafluoronickelate [24], Silver Difluoride [6], and Potassium Tetrafluorocobaltate [6]

Reagent	F_{12}	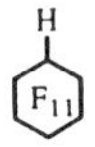 H, F_{11}	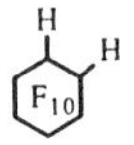 H, H, F_{10}	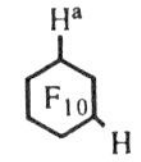 H[a], F_{10}, H	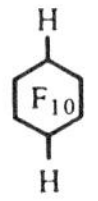 H, F_{10}, H	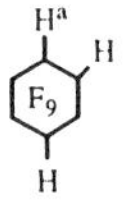 H[a], H, F_9, H	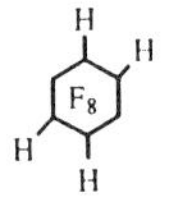 H, H, F_8, H, H
MnF_3 [a]		3.5%	1.1%		6.2%	+	
K_3NiF_6 [c]	+	+	+	+	+ [b]	+	+
AgF_2 [d]	47	3	3	5[e]	2	4	
K_2AgF_4 [f]	18	10	9	8[e]	3	16	
K_2AgF_4 [d]	74	7	–	–	3	–	

Reagent	F_{10}	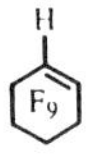 H, F_9	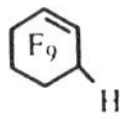 F_9, H	 F_9, H	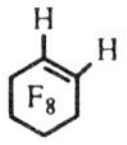 H, F_8, H	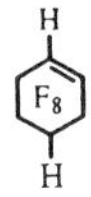 H, F_8, H	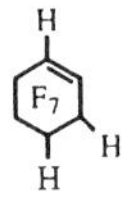 H, F_7, H, H
MnF_3 [a]		3.3%	0.2%	1.6%	0.04%	15.8%	+
K_3NiF_6 [c]		+	+			+	+
AgF_2 [d]							
K_2AgF_4 [f]	2						
K_2AgF_4 [d]	8						

Reagent	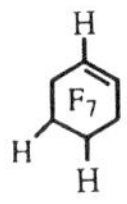 H, F_7, H, H	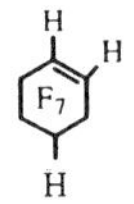 H, F_7, H, H	 H, F_6, H	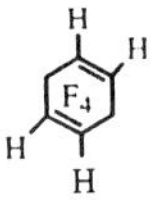 H, H, F_4, H, H	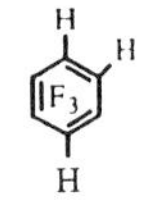 H, F_3, H, H	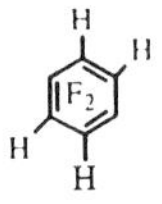 H, H, F_2, H, H	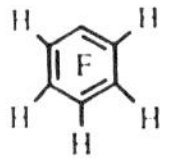 H, H, F, H, H, H
MnF_3 [a]	+	8.0%	4.0%		+	+	3.9%
K_3NiF_6 [c]	+	+		+	+	+	+
AgF_2 [d]							
K_2AgF_4 [f]							
K_2AgF_4 [d]							

[a]300 °C.
[b]Mixture of stereoisomers.
[c]120–200 °C.
[d]300–380 °C; percentage of the product (not yields).
[e]Mixture of 1,3- and 1,4-dihydrodecafluorocyclohexanes.
[f]180–260 °C; percentage of the product (not yields).

Table 2. Fluorination of Toluene with Potassium Tetrafluorocobaltate at 340 °C [*25*] and Cesium Tetrafluorocobaltate at 360 °C [*29*]

Reagent	CHF_2 (H)	CH_2F (F_5)	CHF_2 (F_5)	CHF_2, H, (F_4), H, H	CHF_2, H, (F_6), H, H	CHF_2, H, (F_6), H, H
$KCoF_4$	+			+	+	+
$CsCoF_4$	+	+	+			

Reagent	CHF_2, H, (F_7), H	CHF_2, H, (F_7), H	CHF_2, H, (F_8)	CHF_2 (F_9)	CHF_2 (F_9)	CHF_2 (F_{11})
$KCoF_4$	+	+	+	+	+	+
$CsCoF_4$						+

Reagent	CF_3, H, (F_4), H, H	CF_3, H, (F_6), H, H	CF_3, H, (F_8)	CF_3 (F_9)	CF_3 (F_{11})
$KCoF_4$	+	+	+	+	+
$CsCoF_4$				+	+

At 320 °C, **cesium tetrafluorocobaltate** converts ***benzotrifluoride*** to *m*-fluorobenzotrifluoride, 2*H*-heptafluorotoluene, octafluorotoluene, perfluoro-1-methylcyclohexene, and perfluoromethylcyclohexane [*29*]. ***1,3-Bis(trifluoromethyl)-benzene*** is converted at 420 °C to 4,5,6-trifluoro-1,3-bis(trifluoroethyl)benzene, perfluoro-1,3-dimethylbenzene, and perfluoro-1,3-dimethylcyclohexane [*29*]. ***p-Xylene*** gives at 350 °C 1,4-bis(difluoromethyl)tetrafluorobenzene, 1-difluoromethyl-3-trifluoromethyltetrafluorobenzene, perfluoro-1,3-dimethylbenzene, and perfluoro-1,3-dimethylcyclohexane.

Some derivatives of benzene can be fluorinated over **cobalt trifluoride** at lower temperatures. At 125 °C, ***2H-tetrafluoronitrobenzene*** gives *trans*-2*H*,1-nitrodecafluorocyclohexane, *cis*-2*H*,1-nitrodecafluorocyclohexane, *trans*-5*H*,4-nitrooctafluorocyclohexene, and perfluoronitrocyclohexane [*14*]. ***Pentafluorobenzaldehyde*** at 135 °C affords 9% of perfluoro-3-cyclohexenecarbonyl fluoride and 11% of perfluorocyclohexanecarbonyl fluoride [*14*].

Fluorination of ***naphthalene*** in a stream of dilute fluorine over **cobalt trifluoride** yields a mixture of perfluorodecalins and partially fluorinated naphthalene derivatives that are fully fluorinated to perfluorodecalins by a subsequent treatment with dilute fluorine [*23*]. *Perfluorodecalins* are also final products of fluorination of ***tetralin*** over cobalt trifluoride at 250 °C. A more thorough investigation of this reaction revealed seven additional compounds as products and intermediates [*11*] (equation 6).

[*11*] **6**

CoF_3, 250°C, 4h

9.6% + 4.7% + 2.2%

3.4% + 7.4% + 10.1%

22.4% + 18.2% + 22%

[a]The percentage refers to the composition of the mixture, not the actual yields.

[*23*]

1. CoF_3, F_2/N_2, 260°
2. F_2/N_2, 140°, 27h

~100%

Fluorination of ***perhydroacenaphthene*** gives a 45% yield of perfluoroperhydroacenaphthene. The same compound is also obtained from *cis,trans,trans*-1,5,9-cyclododecatriene in 50% yield [*18*] (equation 7).

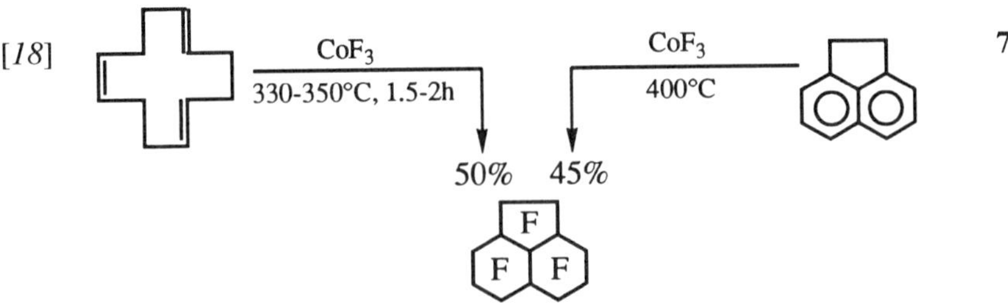

Fluorination of ***N*-methylpyrrole** over **cobalt trifluoride** gives six polyfluorinated *N*-methylpyrrolidines. The same compounds and two others and perfluoro-1-methylpyrrolidine are obtained by fluorination of ***N*-methylpyrrolidine.** The yields of the individual products are low [*13*] (equation 8).

[*13*] **8**

N-methylpyrrole —(CoF_3, 120-140°C, 2.5h)→ F_6 (H, H; N–CH_2F) + F_6 (H, H; N–CHF_2) + F_7 (H; N–CH_2F) + F_7 (H; N–CHF_2) + F_8 (N–CH_2F) + F_8 (N–CHF_2) + F_7 (H; N–CH_2F) + F_7 (H; N–CHF_2) + F_8 (N–CF_3) ←(CoF_3, 90°C, 1.5h)— N-methylpyrrolidine (N–CH_3)

Perfluoro-N-methylpyrrolidine is also the main product of fluorination of ***pyridine*** by **cesium tetrafluorocobaltate** at 310–405 °C, together with *pentafluoropyridine.* Depending on the reaction temperature and time, up to 10 more products were isolated. Yields of the isolated compounds obtained at 310 °C after 3.5 h, or at 395–405 °C after 25–35 min (in parentheses), are as follows: perfluoro-*N*-methylpyrrolidine, 10.6% (7.4%); pentafluoropyridine, 6.8% (14.9%); 2,3,4,5-tetrafluoropyridine, 3% (1.6%); 2,3,4-trifluoropyridine, 1.5%; 2,3,6-trifluoropyridine, 2.0%; 2,5-difluoropyridine, 1.6%; 2,3-difluoropyridine, 1.6%; 2,6-difluoropyridine, 1.2%; tetrafluoro-3-(fluoromethyl)pyridine, 1.3%; undecafluoro-2-aza-2-hexene, 0.4% (0.6%), obtained by the cleavage of the ring; and bis(trifluoromethyl)amine, 4.5% (3.5%), resulting from degradation [*26*].

Similarly complex is the fluorination of the three ***methylpyridines*** (α-, β-, and γ-picolines) with **cesium tetrafluorocobaltate.** 2-Methylpyridine was fluorinated at 270 °C for 180–200 min, 3- and 4-methylpyridines were fluorinated at 330 to 340 °C for 150 min. All of them afforded the respective *polyfluorinated pyridines* and *perfluoro-1,2-, 1,3-dimethyl-, and 1-ethylpyrrolidines.* In addition, perfluoro-2-aza-2-hexene and bis(trifluoromethyl)amine were isolated in variable yields [*27*]. All the isolated products of the fluorination of 3-methylpyridine (3-picoline) are shown in equation 9.

In ***quinoline*** and ***isoquinoline,*** the benzene ring is more receptive to fluorination, its double bonds being saturated and the hydrogen atoms replaced in preference to those in the pyridine ring. As with pyridine and its homologues, ring contraction takes place during fluorination with **cesium tetrafluorocobaltate** at

335–350 °C [*22, 28*] (equation 10). Nitrogen-free perfluoro compounds such as perfluorocyclohexene, perfluorocyclohexane, perfluoromethylcyclohexane, and perfluoroheptane result from fragmentation [*22, 28*].

[*27*] 9

$CsCoF_4$, 330-340°C, 2.5h

CH_3 + CH_2F + CHF_2 + H CHF_2 + CHF_2 H +

F_7 CF_3 CF_3 + CF_3 F_7 CF_3 + F_8 C_2F_5 + $CF_3N{=}CFC_3F_7$ + CF_3NHCF_3

[*28*] 10

$CsCoF_4$, 335-350°C

F_8 N + F_8 N F + F_8 N F

F_8 F N H + H F_8 F N + F_8 F N

[*22*]

CoF_3, 350°C

F_9 F_6 N + F_{11} N F_6

Fluorination of ***isoquinoline*** gave a mixture of partly and fully fluorinated compounds of which perfluoro-3,4,5,6,7,8-hexahydroisoquinoline and perfluoromethylcyclohexane were fully identified [*28*] (equation 11).

[*28*] 11

$CsCoF_4$, 335-340°C

F_8 F_5N + CF_3 F_{11}

References are listed on pages 130–132.

Fluorination of ***aliphatic hydrocarbons*** with **cobalt trifluoride** gives complex mixtures. Isobutane (2-methylpropane) fluorinated at 140–200 °C affords a mixture of 30 products of different degrees of fluorination and of isobutane as well as butane skeletons. The *tertiary hydrogen is replaced preferentially.* Products containing 5–10 atoms of fluorine including a small amount of perfluoroisobutane were isolated [*10*].

Similar preference in replacement by fluorine of tertiary versus secondary and secondary versus primary hydrogens is observed in the fluorination of alkanes with chlorine trifluoride in 1,2-difluorotetrachloroethane at room temperature (Table 3). Skeletal rearrangements accompany the fluorination [*31*].

Table 3. Fluorination of Alkanes with Chlorine Trifluoride [*31*]

Alkane	*Product (Yield, %)*		
$CH_3CH_2CH_3$[a]	$CH_2FCH_2CH_3$ (1)	$CHF_2CH_2CH_3$ (7)	CH_3CHFCH_3 (48)
$CH_3CHCH_2CH_3$[b] | CH_3	$CH_3CFCH_2CH_3$ | CH_3 (33)	$CH_3CHCF_2CH_3$ | CH_3 (5)	$CH_3CFCF_2CH_3$ | CH_3 (2)
CH_3 | $CH_3CCH_2CH_3$[c] | CH_3	$CH_3CF—CHCH_3$ | | CH_3 CH_3 (69)	$CH_3CF—CFCH_3$ | | CH_3 CH_3 (26)	
$CH_3CH—CHCH_3$ | | CH_3 CH_3[d]	$CH_3CF—CHCH_3$ | | CH_3 CH_3 (55)	$CH_3CF—CFCH_3$ | | CH_3 CH_3 (20)	

[a]Propane (42%) was recovered.

[b]2-Methylbutane (28%), 2-methyl-1-butene (8%), and 2-methyl-2-butene (24%) were recovered.

[c]2,2-Dimethylbutane (5%) was recovered.

[d]2,3-Dimethylbutane (25%) was recovered.

Of alicyclic compounds, fluorination of cyclopentane and cyclohexane has been sufficiently investigated in the past. ***Cycloheptane*** on fluorination with **cobalt trifluoride** gives a mixture of highly fluorinated and perfluorinated cycloheptanes and methylcyclohexanes [*12*] (equation 12).

In bicyclic systems, *bridgehead hydrogens are most resistant to replacement by fluorine.* Cobalt trifluoride converts 1*H*-nonafluorobicyclo[3.2.0]hept-6-ene to

[*12*] 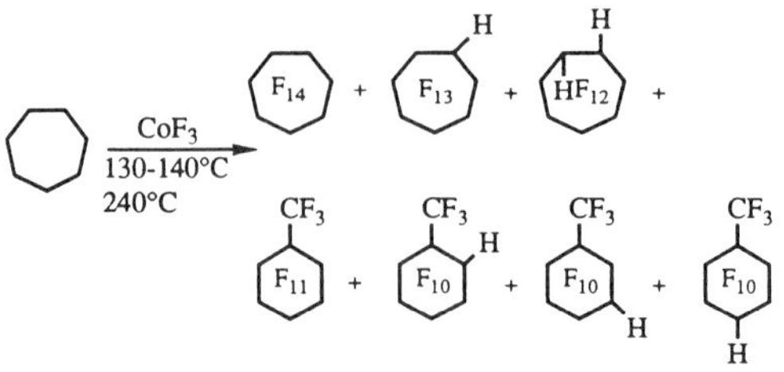 **12**

1*H*-undecafluorobicyclo[3.2.0]heptane at 100 °C, but the replacement of the last hydrogen by fluorine is incomplete even at 250 °C [*15*] (equation 13).

[*15*] 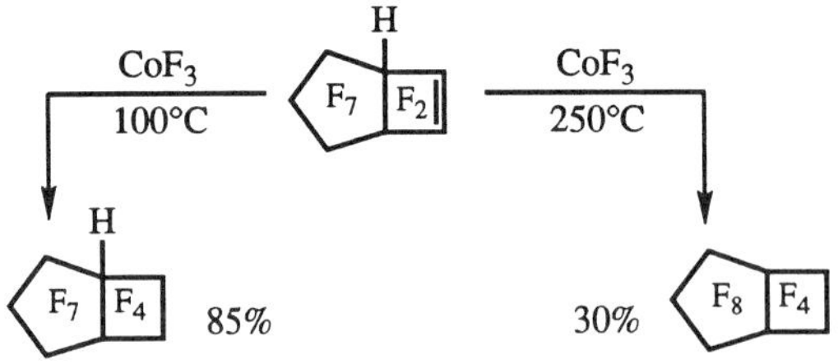 **13**

Fluorination of *cis,trans,trans*-1,5,9-cyclododecatriene with cobalt trifluoride gives a 50% yield of perfluoroperhydroacenaphthene [*18*] (equation 7).

Fluorination over cobalt trifluoride over a temperature gradient of 200 to 250 °C converts *exo*-dicyclopentadiene to a mixture of *endo*- and *exo*-perfluorotetrahydrodicyclopentadiene and perfluorobicyclo[5.3.0]decane [*21*] (equation 14).

[*21*] 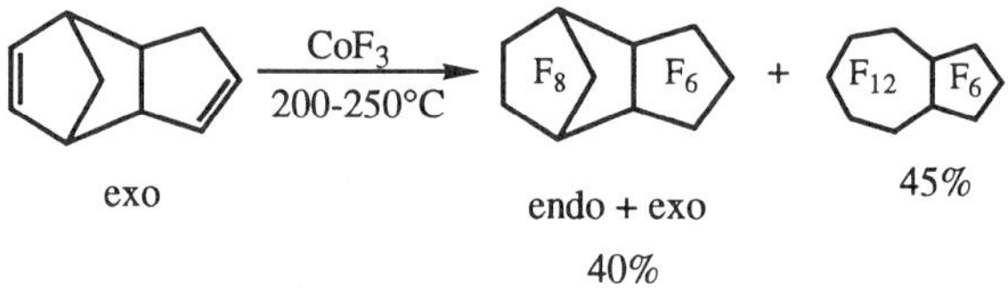 **14**

Perfluorodimethyladamantane is prepared from ***adamantane dicarboxylic acid*** by treatment with sulfur tetrafluoride followed by energetic fluorination with ***cobalt trifluoride*** over two temperature ranges [*8*] (equation 15).

Fluorination of ***aliphatic ethers*** at gentle conditions with **cobalt trifluoride** or **potassium tetrafluorocobaltate** do not give perfluorinated products and cause only negligible cleavage of the ether bond. Complex mixtures are formed from ethyl methyl ether and from diethyl ether [*9*] (equations 16 and 17).

Cyclic ethers having perfluorinated side chains or perfluorinated rings attached in the α-position with respect to oxygen are remarkably stable toward **cobalt trifluoride** and can be successfully fluorinated to perfluoro ethers at 440 °C without being cleaved to a considerable extent at the oxygen [*16, 17*]. Selected examples are shown in Table 4.

[8] 15

$$\text{1,3-adamantanedicarboxylic acid (}CO_2H\text{, }CO_2H\text{)} \xrightarrow[\text{110°C, 24h, sealed}]{SF_4} \text{1,3-bis(}CF_3\text{)adamantane (80\%)} \xrightarrow[\text{250-300°C, 275-380°C}]{CoF_3} \text{F, F-substituted 1,3-bis(}CF_3\text{)adamantane (68\%)}$$

[9] 16

$$CH_3CH_2OCH_3 \xrightarrow[75\text{-}100°C]{CoF_3} CHF_2CHFOCH_3 + CHF_2CF_2OCH_3 + (CH_2F)_2 + CHF_2CHFOCH_2F + CHF_2CHFOCHF_2$$

$$CH_3CH_2OCH_3 \xrightarrow[200°C]{KCoF_4} CHF_2CHFOCH_3$$

[9] 17

$$CH_3CH_2OCH_2CH_3 \xrightarrow[60\text{-}80°C]{CoF_3} CH_2FCHFOCHFCHF_2 \text{ (DL-}erythro + threo\text{ 3-9\%)} + (CHF_2CHF)_2O \text{ (}meso\text{ + DL 70-83\%)} + CHF_2CF_2OCHFCHF_2 \text{ (DL 4-14\%)} + (CH_2F)_2 \text{ (2-5\%)} + (CHF_2)_2 \text{ (1-2\%)}$$

$$CH_3CH_2OCH_2CH_3 \xrightarrow[200°C]{KCoF_4} CHF_2CHFOCH_2CH_3$$

Fluorination of ***trimethylamine*** over **cobalt trifluoride** gives six fluorinated products [*20*] (equation 18).

[20] 18

$$(CH_3)_3N \xrightarrow[100°C]{CoF_3} CF_3N(CHF_2)_2 + (CHF_2)_3N + HCON(CHF_2)_2 + \text{1,3-bis(}CHF_2\text{)-}F_6\text{-imidazolidine} + \text{1-}CH_2F\text{-3-}CHF_2\text{-}F_6\text{-imidazolidine} + \text{1,3-bis(}CHF_2\text{)-H-}F_5\text{-imidazolidine}$$

4-Methylmorpholine on treatment with **cobalt trifluoride** at 100 °C for 3 h yields 10 fluorinated morpholines and a small amount of bis(trifluoromethyl)-fluoromethylamine [*19*] (equation 19).

[19] 19

N-CHF_2, F_6 (2.1%) + N-CHF_2, F_6 (1.6%) + N-CF_3, F_6 (6.1%) + N-CHF_2, F_6 (3.9%) + N-CHF_2, F_7 (38.6%) +

N-CF_3, F_7 (5.6%) + N-CH_2F, F_8 (15.2%) + N-CHF_2, F_8 (16.0%) + N-CF_3, F_8 + N-F, F_8 (2.4%) + $(CF_3)_2NCH_2F$ (8.2%)

Table 4. Selected Examples of Preparation of Perfluoro Ethers [*16, 17*] by Heating with Cobalt Trifluoride and Calcium Fluoride at 440 °C

Partly Fluorinated Ether	*Perfluorinated Ether*	*Yield (%)*	*By-product*	*Yield (%)*	*Ref.*
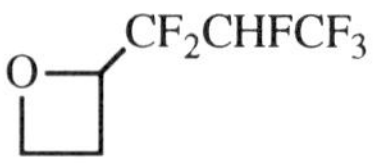 O CF_2CHFCF_3	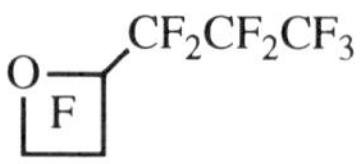 O F $CF_2CF_2CF_3$	8	C_5F_{12} C_6F_{14}	3 10	17
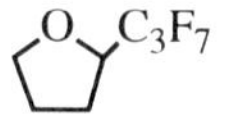 O C_3F_7	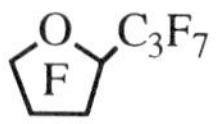 O F C_3F_7	70			16[a]
 O CF_2CHClF	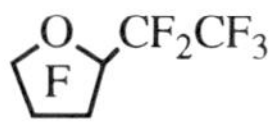 O F CF_2CF_3	12	 O CF_2CF_2Cl	45	17
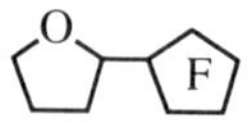 O F	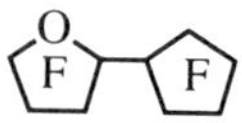 O F F	65	 F $CF_2CF_2CF_3$	11	17
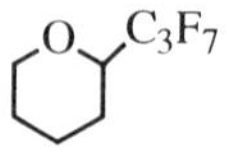 O C_3F_7	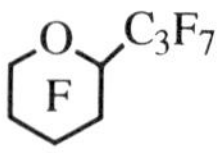 O F C_3F_7	68			16[a]
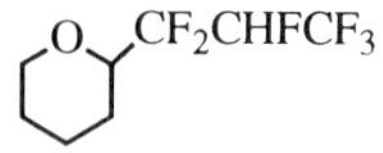 O CF_2CHFCF_3	 O F $CF_2CF_2CF_3$	62			17
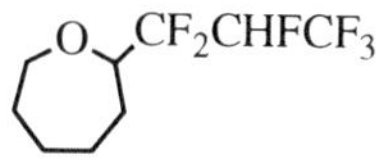 O CF_2CHFCF_3	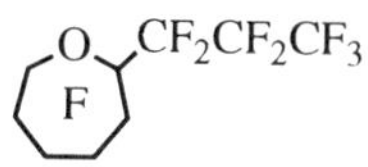 O F $CF_2CF_2CF_3$	45	C_8F_{18}	12	17

[a]Only cobalt trifluoride without calcium fluoride was used.

References for Pages 120–130

1. Fisher, R. G., Jr.; Zweig, A. U.S. Patent 4 394 527, 1983; *Chem. Abstr.* **1983,** *99,* 175351m.
2. Boudakian, M. M.; Hyde, G. A. *J. Fluorine Chem.* **1984,** *25,* 435.

3. Feiring, A. E. U.S. Patent 4 051 168, 1977; *Chem Abstr.* **1977,** *87,* 184189u.
4. Meurs, J. H. H.; Sopher, W.; Eilenberg, W. *Angew. Chem.* **1989,** *101,* 955; *Angew. Chem. Int. Ed. Engl.* **1989,** *28,* 927.
5. Avramenko, A. A.; Bardin, V. V.; Karelin, A. I.; Krasil'nikov, V. A.; Tushin, P. P.; Furin, G. G., Yakobson, G. G. *Zh. Org. Khim.* **1986,** *22,* 2584 (Engl. Transl. 2318).
6. Plevey, R. G.; Steward, M. P.; Tatlow, J. C. *J. Fluorine Chem.* **1973/74,** *3,* 259.
7. Pedler, A. E.; Rimmington, T. W.; Stephens, R.; Uff, A. J. *J. Fluorine Chem.* **1972/73,** *2,* 121.
8. Moore, R. E. Ger. Offen. 2 808 112, 1978; *Chem. Abstr.* **1979,** *90,* 22436g.
9. Brandwood, M.; Coe, P. L.; Ely, C. S.; Tatlow, J. C. *J. Fluorine Chem.* **1975,** *5,* 521.
10. Burdon, J.; Huckerby, T. N.; Stephens, R. *J. Fluorine Chem.* **1977,** *10,* 523.
11. Coe, P. L.; Habib, R. M.; Tatlow, J. C. *J. Fluorine Chem.* **1982,** *20,* 203.
12. Oliver, J. A.; Stephens, R.; Tatlow, J. C. *J. Fluorine Chem.* **1983,** *22,* 21.
13. Coe, P. L.; Holton, A. G.; Sleigh, J. H.; Smith, P. T.; Tatlow, J. C. *J. Fluorine Chem.* **1983,** *22,* 287.
14. Phull, G. S.; Plevey, R. G.; Tatlow, J. C. *J. Fluorine Chem.,* **1984,** *25,* 111.
15. Dodsworth, D. J.; Jenkins, C. M.; Stephens, R.; Tatlow, J. C. *J. Fluorine Chem.* **1984,** *24,* 509.
16. Chambers, R. D.; Grievson, B. *J. Fluorine Chem.* **1984,** *25,* 523.
17. Chambers, R. D.; Grievson, B.; Drakesmith, F. G.; Powell, R. L. *J. Fluorine Chem.* **1985,** *29,* 323.
18. Platonov, V. E.; Prokudin, I. P.; Popkova, N. V.; Rodionov, P. P.; Asovich, V. S.; Mel'nichenko, B. A.; Andreevskaya, O. I.; Yakobson, G. G. *Izv. Akad. Nauk SSSR* **1984,** 2409 (Engl. Transl. 2205).
19. Rendell, R. W.; Wright, B. *Tetrahedron* **1978,** *34,* 197.
20. Rendell, R. W.; Wright, B. *Tetrahedron* **1979,** *35,* 2405.
21. Moore, R. E. U.S. Patent 4 220 606, 1980; *Chem. Abstr.* **1981,** *94,* 156420p.
22. Tatlow, J. C.; Plevey, R. G.; Wotton, D. E. M.; Sargent, C. R. Br. Patent 2 113 223, 1983; *Chem. Abstr.* **1984,** *100,* 34526z.
23. Nishimura, M.; Okada, N.; Tokunaga, S. Jpn. Patent 01 186 828, 1989; *Chem. Abstr.* **1990,** *112,* 20799b.
24. Plevey, R. G.; Rendell, R. W.; Steward, M. P. *J. Fluorine Chem.* **1973/74,** *3,* 267.
25. Bailey, J.; Plevey, R. G.; Tatlow, J. C. *J. Fluorine Chem.* **1988,** *39,* 23.
26. Plevey, R. G.; Rendell, R. W.; Tatlow, J. C. *J. Fluorine Chem.* **1982,** *21,* 159.

27. Plevey, R. G.; Rendell, R. W.; Tatlow, J. C. *J. Fluorine Chem.* **1982,** *21,* 265.
28. Plevey, R. G.; Rendell, R. W.; Tatlow, J. C. *J. Fluorine Chem.* **1982,** *21,* 413.
29. Bailey, J.; Plevey, R. G.; Tatlow, J. C. *J. Fluorine Chem.* **1987,** *37,* 1.
30. Stacey, M.; Tatlow, J. C. In *Advances in Fluorine Chemistry,* Vol. 1; Stacey, M; Tatlow, J. C.; Sharpe, A. G., Eds.; Butterworths: London, 1960; p 166.
31. Brower, K. R. *J. Org. Chem.* **1987,** *52,* 798.

Electrophilic Fluorination of Carbon–Hydrogen Bonds

by Timothy B. Patrick

The synthetic requirements associated with the regioselective fluorination of organic molecules has led to the creation of an intriguing class of reagents known as electrophilic or electropositive fluorinating agents. The reagents have the important synthetic attribute of delivering a fluorine atom to a nucleophilic site, an uncanny reversal of the role usually taken by the most electronegative element known. To achieve the apparent reversal of fluorine character, the fluorine atom must be bound to a group of atoms that is a better leaving group than fluoride ion. In many cases a nucleophilic reagent simply attacks the fluorine atom in an S_N2-type manner, but sometimes single electrons are transferred to the electropositive fluorinating agent followed by the formation of radical intermediates. However, much mechanistic uncertainty exists. In all, we behold an extremely valuable synthetic methodology coupled with exciting mechanistic possibilities.

Systems usually fluorinated by electropositive fluorine reagents include ***activated alkenes (enol ethers, enol acetates, silyl enol ethers, and enamines), activated aromatic systems, certain slightly activated carbon–hydrogen bonds, and selected organometallics.***

Electropositive fluorinating agents are categorized in distinct classes as (1) **fluoroxy reagents,** in which the fluorine is bound to an oxygen atom (for reviews, *see* references 1, 2, 3, 4, 5, and 6); (2) **fluoraza reagents,** in which the fluorine atom is bound to the nitrogen atom of either an amide or ammonium ion structure (for a partial review, *see* reference 6); (3) **xenon difluoride,** in which the fluorine atoms are bound to xenon (for reviews, *see* references 5, 7, 8, and 114); and (4) **perchloryl fluoride,** in which the fluorine atom is bound to the chlorine atom of the perchloryl function (for a review, *see* reference 9).

Many fluorinations by electropositive fluorine reagents produce *α-fluoro carbonyl compounds as the final result. An extensive review exists on the preparation of α-fluorocarbonyl compounds [10]. Also, electropositive reagents are used widely in the preparation of ^{18}F-labeled radioactive materials* required in positron emission tomography for biomedical research. Excellent reviews are available on fluorine-18 labeling [*11*, *12*].

The types of reactions covered in this segment are those in which the overall transformation is the conversion of a carbon–hydrogen bond to a carbon–fluorine bond through the use of electropositive fluorine reagents [*1*, *2*, *3*, *4*, *5*, *6*].

0065–7719/95/0187–0133$11.42/1

Fluoroxy Reagents

Fluoroxytrifluoromethane (trifluoromethyl hypofluorite) and acetyl hypofluorite are the most commonly used neutral hypofluorites; cesium fluoroxysulfate is an inorganic anionic fluoroxy derivative.

Fluoroxytrifluoromethane is prepared in a process that uses cesium fluoride as a catalyst for the reaction between fluorine and carbon monoxide [*13*] (equation 1). Bisfluoroxydifluoromethane is prepared in a similar manner from carbon dioxide [*13*]. **Fluoroxymethane** was prepared recently [*14*].

[*13*] $$C{\equiv}O + 2F_2 \xrightarrow{\text{CsF, light, 2 h}} \underset{95\ \%}{CF_3OF} \qquad 1$$

Fluoroxy reagents became of general synthetic utility after the discovery of a relatively simple preparation of acetyl hypofluorite [*15*] (equation 2).

[*15*] $$CH_3COONa + F_2 \xrightarrow[\text{-78 °C, 30 min}]{CH_3COOH,\ CFCl_3} \underset{80\ \%}{CH_3COOF} + NaF \qquad 2$$

The synthetic procedure has been extended to the preparation of **trifluoroacetyl hypofluorite** [*16, 17*], **long-chain perfluorinated hypofluorites** [*18, 19*], and radiolabeled hypofluorites [*11, 12*].

Cesium fluoroxysulfate [*5, 6*] is prepared by reacting fluorine with cesium sulfate in aqueous solution [*20, 21*] (equation 3).

[*20, 21*] $$Cs_2SO_4 + F_2 \xrightarrow[\text{0 °C}]{H_2O} \underset{65\ \%}{CsSO_4F} + CsF \qquad 3$$

The white cesium fluoroxysulfate precipitates from the reaction medium and may be kept for several months in the cold (0 to –15 °C). Metal surfaces can cause detonation of the reagent. The reaction scope of cesium fluoroxysulfate seems narrower than that of acetyl hypofluorite because of its limited solubility in organic solvents. Cesium fluoroxysulfate has not been prepared with a fluorine-18 label.

Reactions with Aromatic Systems

The fluoroxy reagents react readily with ***activated aromatic systems*** (Table 1) to give moderate yields of fluoroaromatic compounds. The fluorine atom shows a preference for *ortho* orientation because of complexation between the fluoroxy reagent and the ring substituent [*15, 22*]. Nucleophilic attack by the substrate on the fluorine atom of the fluoroxy reagent is a frequently encountered path, but reactions also proceed by a free radical mechanism [*23*, *51*].

4-Fluororesorcinol can be prepared from derivatives of resorcinol and fluoroxytrifluoromethane [*15*, *24*] (equation 4) or by fluorination of resorcinol with cesium fluoroxysulfate [*25*] (equation 5). Reaction of acetyl hypofluorite with *N*-acetyl-*O*-methyltyrosine methyl ester provides a good route to 3-fluorotyrosine derivatives [*26*] (equation 6).

[*15*, *24*] OCH_3 … OCH_3 — 1. CF_3OF, $CFCl_3$, -78 °C; 2. HBr, reflux, 4 h — 60 % → OH, OH, F **4**

OCH_3, $COCH_3$, OCH_3 — 1. CF_3OF, $CFCl_3$, -78 °C; 2. HBr, reflux, 4 h — 74 %

[*25*] OH, OH — $CsSO_4F$, CH_3CN, BF_3; RT, 3 h → OH, F, OH + OH, OH, F **5**

1 mol $CsSO_4F$	2 %	21 %
2 mol $CsSO_4F$	25 %	27 %

[*26*] $COOCH_3$, $NHCOCH_3$, CH_3O —i→ $COOCH_3$, $NHCOCH_3$, CH_3O, F 75 % **6**

i: CH_3COOF, $CHCl_3$, cold

Table 1. Fluorination of Aromatics

Entry	*Substrate*	*Product*
1	C_6H_6	C_6H_5F
2	$C_6H_5OCH_3$	$FC_6H_4OCH_3$
3	C_6H_5OH	FC_6H_4OH
4	$C_6H_5NHCOCH_3$	$FC_6H_4NHCOCH_3$
5	$C_6H_5CH_3$	$FC_6H_4CH_3$
6	$C_6H_5NO_2$	$FC_6H_4NO_2$

[a]Reagents are shown in Table 3.
[b]I, CF_3OF; II, $CsSO_4F$; III, CH_3COOF.
[c]Yield (%).
[d]References.
[e]K: $R_y = Cl_5$; X = OTf.
[f]Yields (%) of *ortho, meta,* and *para,* respectively.

References are listed on pages 167–171.

Table 1—Continued

Entry	*Solvent*[b]	*Reagents*[a] *ROF*	*XeF*$_2$	=*N*$^+$–*F*
1	I	39[c] [*13*][d]	13 [*90*]	C, 88 [*66*]
	II	12 [*33*]	68 [*97*]	B, 50 [*64*]
	III	18 [*51*]		K[e], 48 [*75*]
2	I	9[f], 0[f], 13[f] [*13*]	38, 4, 30 [*96*]	C, 24, 0, 74 [*64*]
	II	49, 0, 16 [*32*]		B, 69, 0, 24 [*65*]
	III	76, 0, 9 [*22*]		K[g], 44, 0, 48 [*75*]
3	I		19, 19, 9 [*97*]	B, 60, 0, 40 [*65*]
	II	56, 0, 11 [*32*]		K[h], 84, 0, 10 [*75*]
	III	45, 0, 30 [*51*]		J, 61, 0, 30 [*73*]
4	I	37, 17, 10 [*13*]	37, 3, 16 [*98*]	K[i], 58, 0, 16 [*75*]
	II	74, 0, 11 [*32*]		F, 24, 0, 16 [*82*]
	III	48, 0, 7 [*22*]		G, 50, 0, 30 [*75*]
5	I	5, 0, 2 [*13*]	16, 3, 14 [*7*]	B, 94, 4, 22 [*64*]
	II	31, 4, 8 [*16*]		F, 13, 1, 6 [*82*]
	III	8, 1, 4 [*51*]		
6	I	0	19, 15, 11 [*7*]	
	II	6, 16, 3 [*6*]		
	III	0		

[g]K: R_y = 2,6-diCO_2CH_3; X = OTf.
[h]K: R_y = 3,5-Cl_2; X = BF_4.
[i]K: R_y = Cl_5; X = OTf.

Continued on next page.

Table 1—Continued

Entry	*Substrate*	*Product*
7	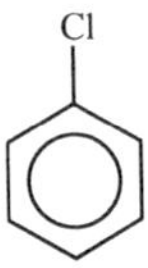 Cl	 Cl, F
8	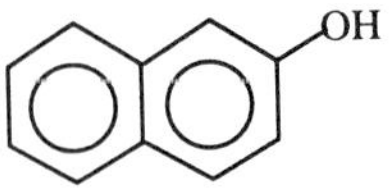 OH	a) 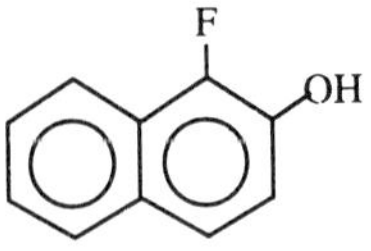 F, OH b) F, F, O 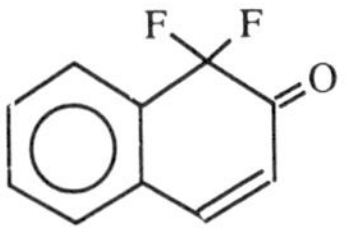
9	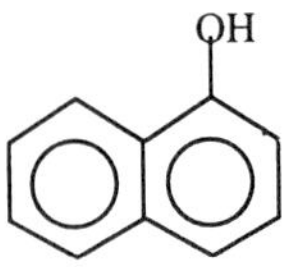 OH	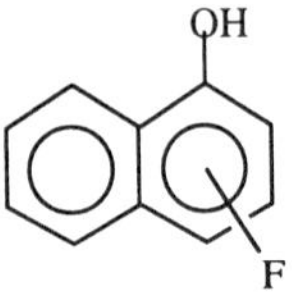 OH, F
10	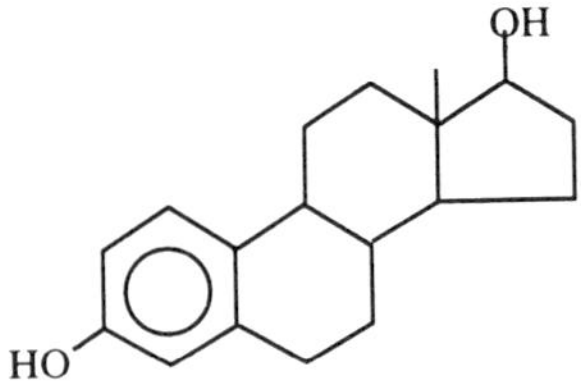 OH, HO	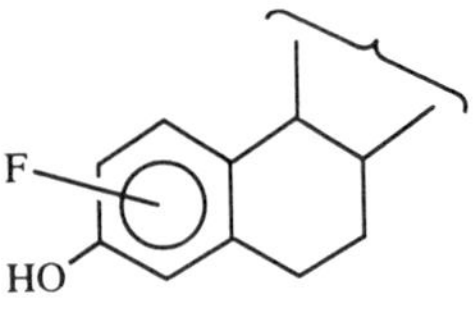 F, HO
11	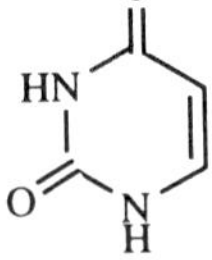 O, HN, O, N H	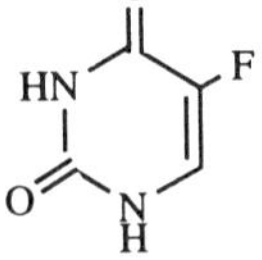 O, F, HN, O, N H

[j]K: R_y = 3,5-Cl_2; X = OTf. [k]K: R_y = H_5; X = OTf.

Table 1—Continued

Entry	Solvent[b]	Reagents[a] ROF	XeF_2	$=N^+–F$
7	I		16, 3, 46 [*7*]	
	II			
	III	5, 0, 5 [*51*]		
8	I	a14, b20 [*27, 28*]	a40, b0 [*97*]	J, a83, b15 [*73*]
	II	a15, b13 [*25, 32*]		K[j], a84, b11 [*75*]
	III	a65 (as OCH_3) [*18*]		
9	I	2F 26, 4F 0 [*28*]		B, 2F 60, 4F 0 [*64*]
	II	2F 78, 4F 10 [*32*]		K[k], 2F 42, 4F 9 [*75*]
	III	2F 10, 4F 16 [*18*] (as OCH_3)		
10	I	6F 65 [*55*]	2F-25, 4F-25 [*97*]	B, 2F 10, 4F 15, 10 F57 [*65*]
	II	2F 20, 4F 20 [*25*]		K[l], 2F 30, 4F 30 [*77*]
	III			
11	I	94 [*35*]		
	II	89 [*59*]	10 [*7*]	
	III			
	F_2	55 [*58*]		
	$FClO_3$	0		

[l]K: R_y = 3,5-Cl_2: X = OTf. [m]K: R = CH_2Cl; X = BF_4.

References are listed on pages 167–171.

Polynuclear aromatics react with fluoroxy reagents to give high yields of *ortho* substitution products accompanied by varying yields of geminal difluoro products. The geminal difluorination occurs presumably by an addition–elimination mechanism [*27, 28, 29, 30, 31, 32*]. Unactivated aromatic systems are fluorinated in lower yield to give monofluorinated products (Table 1, entries 6 and 7). Examples of fluorination of polynuclear systems [*15, 21, 25, 30, 32, 33*] are shown in equations 7–10.

OCH_3 i or ii F OCH_3 + F F O **7**

i. CF_3COOF, AcOH, $CFCl_3$, - 75 °C 65 % 0 % [*15*]

ii. $CsSO_4F$, CH_3CN, BF_3, RT, 3 h 51 % 13 % [*21, 25*]

OCH_3 i or ii OCH_3 F + OCH_3 F **8**

i. CF_3COOF, AcOH, $CFCl_3$, -75 °C 10 % 60 % [*15*]

ii. $CsSO_4F$, CH_3CN, BF_3, RT 14 % 48 % [*21*]

OH i or ii O F F **9**

i. $CsSO_4F$, CH_3CN, BF_3, RT, 3 h 22 % [*25*]

ii. CF_3OF, $CHCl_3$, RT, 1 h 40 % [*31*]

[30, 33] **10**

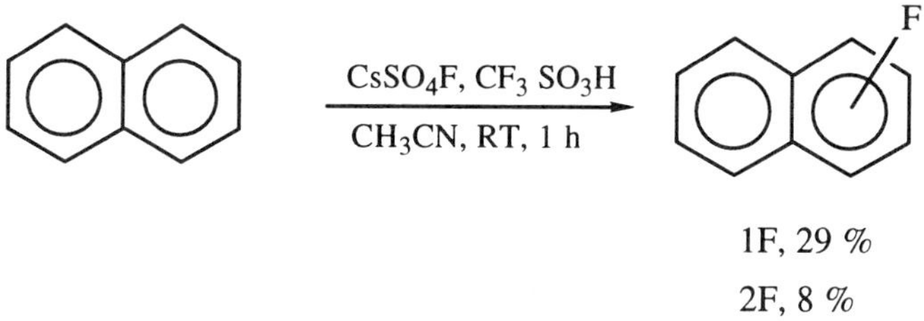

$CsSO_4F$, CF_3SO_3H; CH_3CN, RT, 1 h

1F, 29 %

2F, 8 %

The ***heterocyclic system*** bimane produces a difluorinated product on fluorination with acetyl hypofluorite [*34*] (equation 11).

Cesium fluoroxysulfate reacts with derivatives of porphyrin to give 5-fluoroporphyrins accompanied by polyfluorinated products [*35, 36*].

5-Fluorouracil can be produced by a variety of positive fluorine reagents as well as elemental fluorine (Table 1, entry 11).

[*34*] 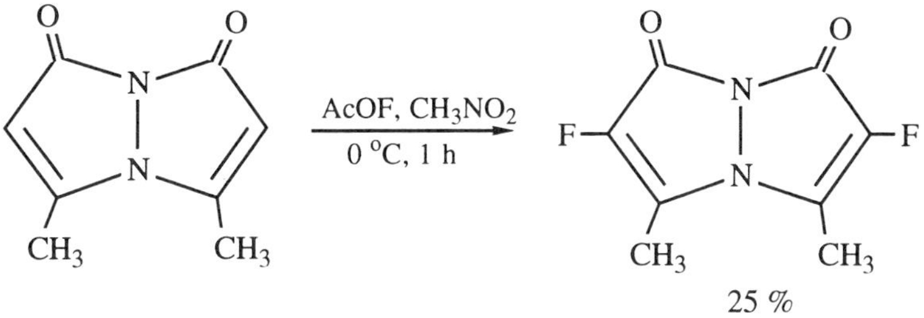 **11**

AcOF, CH_3NO_2; 0 °C, 1 h

O O N N CH_3 CH_3 → F F O O N N CH_3 CH_3

25 %

Reactions with Enols and Enolates

Reactions of **fluoroxytrifluoromethane** with ***enol ethers, enol acetates,*** and ***enamines*** [*1, 2, 3*] are very useful, especially for the preparation of *steroidal* α*-fluoro ketones* (Table 2, entries 1, 3, 5, 6, and 7) [*1*] (equation 12).

[*1*] 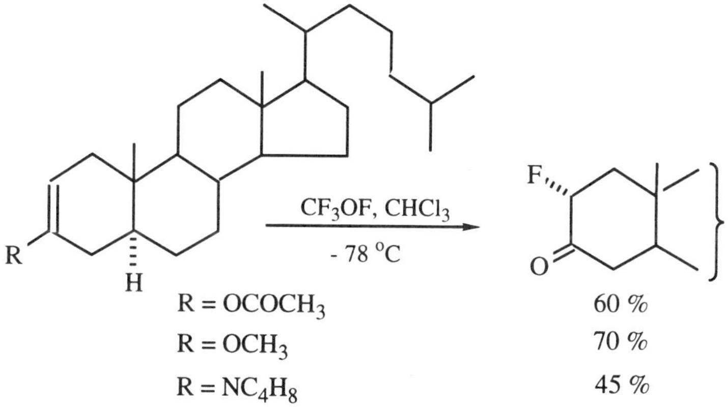 **12**

R, H

CF_3OF, $CHCl_3$; - 78 °C

F, O

R = $OCOCH_3$	60 %
R = OCH_3	70 %
R = NC_4H_8	45 %

Trimethylsilyl enol ethers are effective substrates in fluorination with fluoroxytrifluoromethane for the *preparation of α-fluoro esters, amides, and aldehydes* [*37*] (equations 13–15).

[*37*] **13**

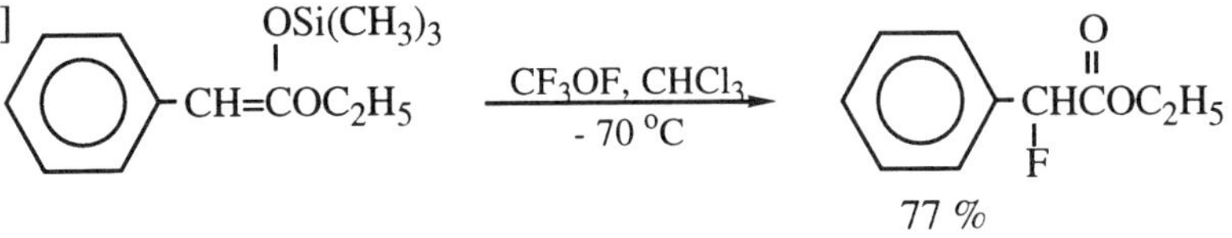

$C_6H_5CH{=}C(OSi(CH_3)_3)OC_2H_5 \xrightarrow[-70\,^{\circ}C]{CF_3OF,\ CHCl_3} C_6H_5CHFCOOC_2H_5$ (77 %)

[*37*] **14**

$C_6H_5CH{=}C[OSi(CH_3)_3]_2 \xrightarrow[-70\,^{\circ}C]{CF_3OF,\ CHCl_3} C_6H_5CHFCOOH$ (90 %)

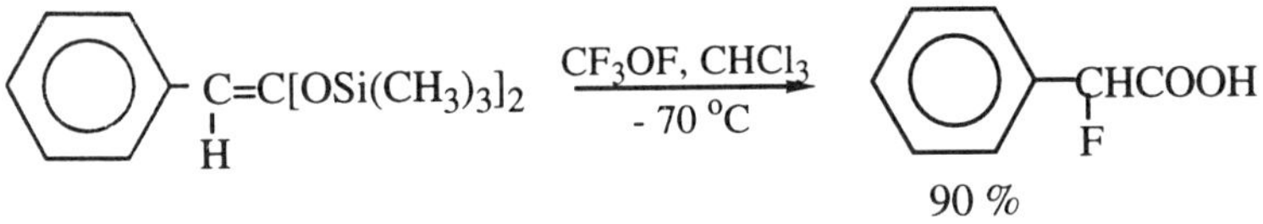

[*37*] **15**

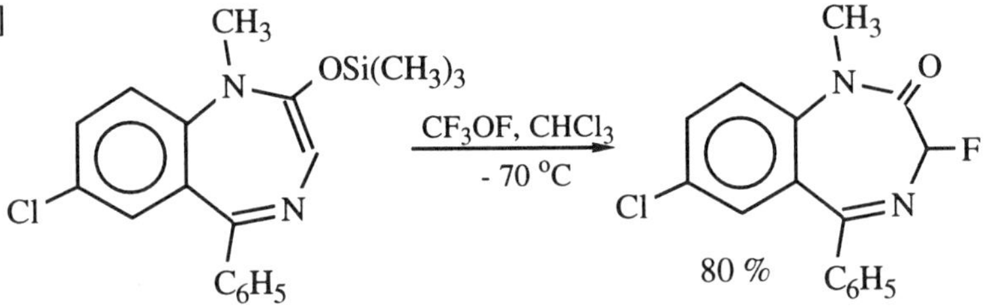

CH3, N, OSi(CH3)3, Cl, N, C6H5 — CF3OF, CHCl3, - 70 °C → CH3, N, O, F, Cl, N, C6H5 (80 %)

These reactions liberate carbonyl fluoride and fluorotrimethylsilane and thus require no hydrolysis. A fluorinated erythromycin derivative is obtained from fluorination of 3-*O*-mycarosyl-8,9-anhydroerythronolide B 6,9-hemiacetal, an enol ether [*38*] (equation 16).

[*38*] **16**

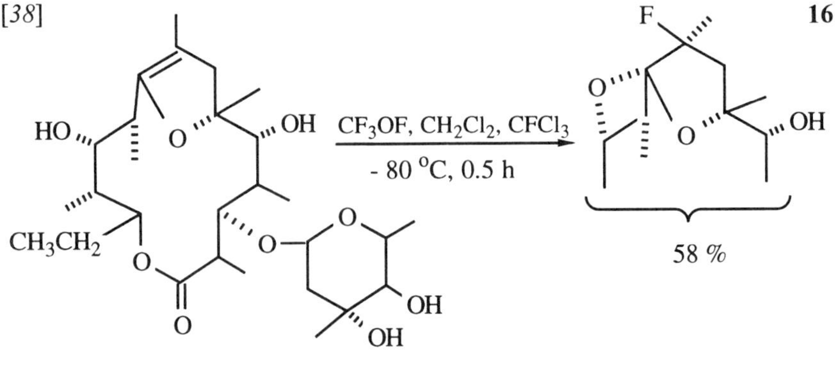

HO, O, OH, CH3CH2, O, O, O, O, OH, OH — CF3OF, CH2Cl2, CFCl3, - 80 °C, 0.5 h → F, O, O, OH (58 %)

Acyl hypofluorites react readily with ***enols, enol acetates, vinyl ethers, and enolates*** [*39, 40, 41*] (Table 2, entries 1–7). Steroidal enol acetates are also fluorinated smoothly [*39*] (equations 17–19).

[*39*] **17**

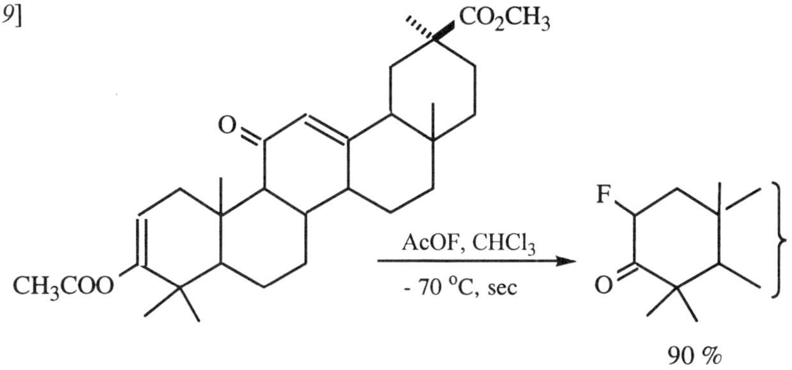

CO_2CH_3 O CH_3COO → AcOF, $CHCl_3$, - 70 °C, sec → F O 90 %

[*41*] **18**

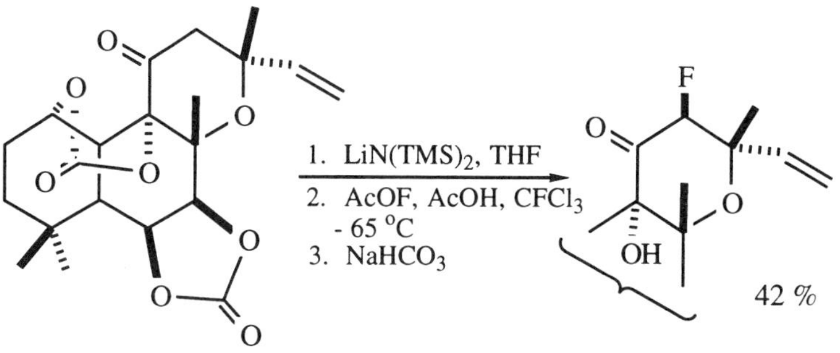

O O O O O O O → 1. $LiN(TMS)_2$, THF 2. AcOF, AcOH, $CFCl_3$ - 65 °C 3. $NaHCO_3$ → F O O OH 42 %

[*40*] **19**

$$R^1CH_2\overset{O}{\overset{\|}{C}}R^2 \xrightarrow[\text{2. AcOF, CFCl}_3\text{, -78 °C}]{\text{1. LDA, THF, -78 °C}} R^1\overset{F}{\overset{|}{C}}H\overset{O}{\overset{\|}{C}}R^2$$

R^1	R^2	%
C_6H_5	H	75
C_4H_9	C_3H_7	54
2-Naphthyl	H	55

Table 2. Fluorination of Enols and Enolates

Entry	*Substrate*	*Product*
1	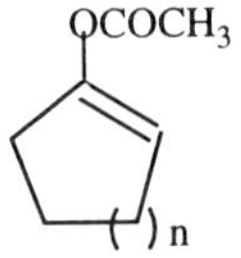 $OCOCH_3$ (cycloalkenyl acetate), n = 1 - 4	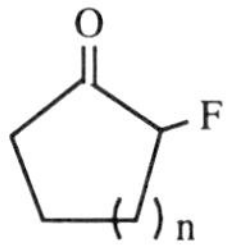 2-fluorocycloalkanone, O, F, n
2	$CH_3COCH_2COCH_3$	a $CH_3COCHFCOCH_3$ b $CH_3COCF_2COCH_3$
3	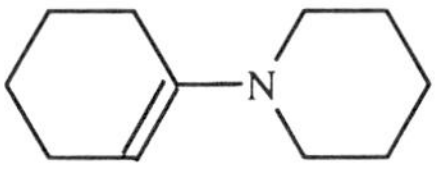 1-(cyclohexenyl)piperidine, N	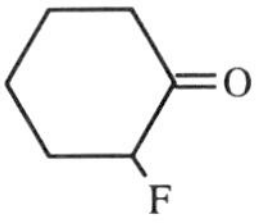 2-fluorocyclohexanone, =O, F
4	$R\bar{C}(CO_2C_2H_5)_2$	$RCF(CO_2C_2H_5)_2$
5	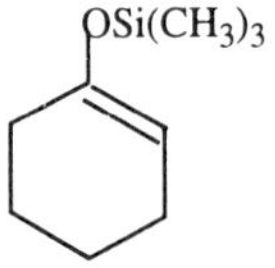 $OSi(CH_3)_3$ (1-trimethylsilyloxycyclohexene)	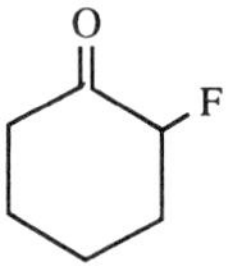 2-fluorocyclohexanone, O, F
6	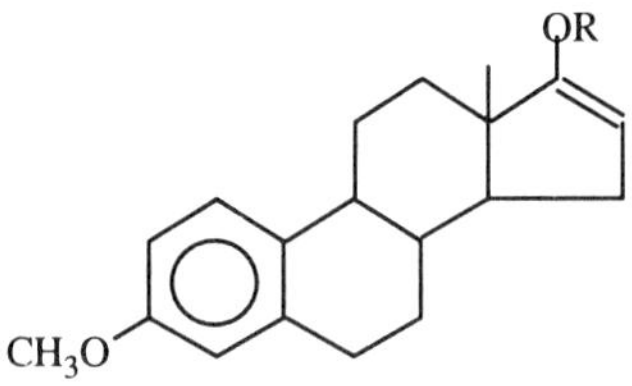 steroid enol, OR, CH_3O	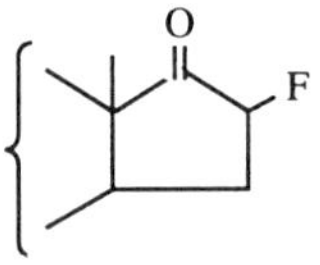 16-fluoro-17-ketone, O, F

Table 2—Continued

	Reagents[a]		
ROF[b]	*ROF*	*XeF_2*	*N–F*
I	Steroids[c], 50–90[d] [*1*][e]	37–62 [*88*]	K[f, j], 73 [*75*]
II	78–80 [*60*]		
III	70 [*16, 39*]		
I			
II	44[a], 19[b] [*59*]	Oa, 43b [*88*]	B, Oa, 54b [*83*]
III			
I	Steroids[d] [1]		H, 43 [*82*]
II			I, 44 [*69*]
III	Steroids[d], 50 [*16*]		K[i], Steroids, 54 [*75*]
I		32 [*89*]	A, R = CH_3, 53; C_6H_5, 81 [*64*]
II			B, R = CH_3, 96 [*65*]
III	CH_3, 77 [*58*][g, h]		C, R = C_6H_5, 93 [66] H, R = CH_3, 17; R = C_6H_5, 39 [*69*] I, R = C_6H_5, 56 [*71*] K[i], R = CH_3, 78; C_6H_5, 83 [*75*] L, R = C_2H_5, 58 [*66*] K[j], 87 [*75*]
I	74 [*37*]	90 [*10*]	
II			
III			
I	R = Ac, 65, α/β = 2/1 [*62*] R = $SiMe_3$, 65, all α	R = Ac, 99, 9/1 [*62*] R = $SiMe_3$, 44, all α	
II	R = Ac, 22, α/β = 20/1 R = $SiMe_3$, 15, α/β = 9/1		
III	0		

Continued on next page.

Table 2—Continued

Entry	*Substrate*	*Product*
7	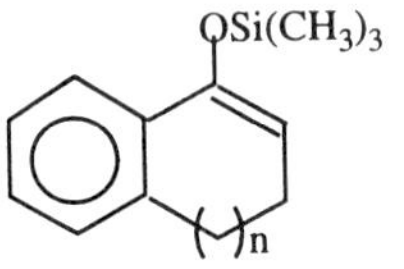OSi(CH3)3 n n = 0, 1	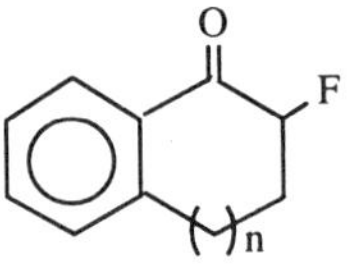O F n

[a]Reagents are shown in Table 3.
[b]I, CF_3OF; II, $CsSO_4F$: III, CH_3COOF.
[c]Steroid substrates.
[d]Yield (%).
[e]References.

Vinyl acetates are also fluorinated by mixtures containing both acyl hypofluorite and perfluoroalkyl hypofluorite, obtained from fluorination of the carboxylic acid salts [*16, 19*] (equations 20 and 21).

Both enol acetates and enols are fluorinated by cesium fluoroxysulfate at room temperature in either acetonitrile or methylene chloride solutions (Table 2, entries 1, 2, and 6).

Fluorination of Tertiary Carbon–Hydrogen Bonds

Fluorine is well-known for its ability to fluorinate selectively *tertiary carbon–hydrogen bonds* that are slightly activated by polar substituents. *Hypofluorite reagents* also react with carbon–hydrogen bonds, both secondary and tertiary, but tertiary bonds react better [*6*].

[*16*]

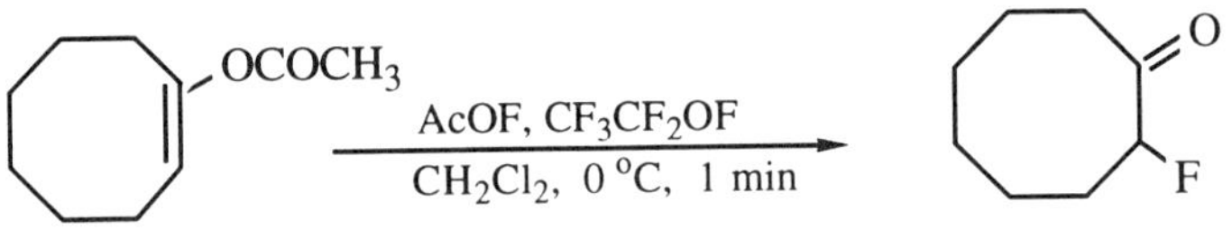
OCOCH3
AcOF, CF3CF2OF
CH2Cl2, 0 °C, 1 min
O
F

20

72 %

[*19*]

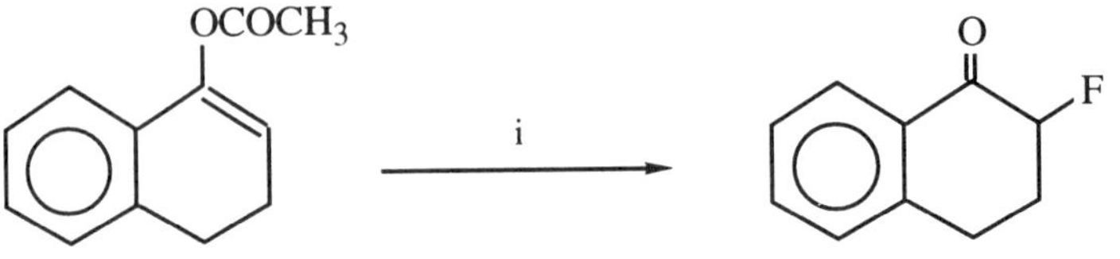
OCOCH3
i
O
F

21

i: $CF_3(CF_2)_6CF(OF)_2$, $CF_3(CF_2)_6COOF$, CH_2Cl_2
- 20 °C, 1 min

85 %

References are listed on pages 167–171.

Table 2—Continued

	Reagents[a]		
ROF[b]	*ROF*	*XeF_2*	*N–F*
I	85, [*19*]	88 [*87*]	K[j], Steroid[c], 78 [*75*]
II			J[k], Steroid[c], 90 [*73*]
III	85 [*19*]		

[f] K: R_y = 2,6-di-CH_2OCH_3; X = OTf.
[g] R = C_2H_5.
[h] R = C_6H_5.
[i] K: R_y = 2,4,6-$(CH_3)_3$; X = OTf.
[j] K: R_y = H_5; X = OTf.
[k] R = CH_2Cl; X = BF_4.

Several reagents are compared in their ability to fluorinate adamantane [*2, 42, 43, 44, 45, 46, 47, 48*] (equation 22). Cyclohexane behaves in a similar fashion but gives lower yields [*3, 42, 49*].

Fluoroxytrifluoromethane effectively fluorinates tertiary carbon–hydrogen bonds in materials with biological applications [*2, 3*] (equations 23 and 24).

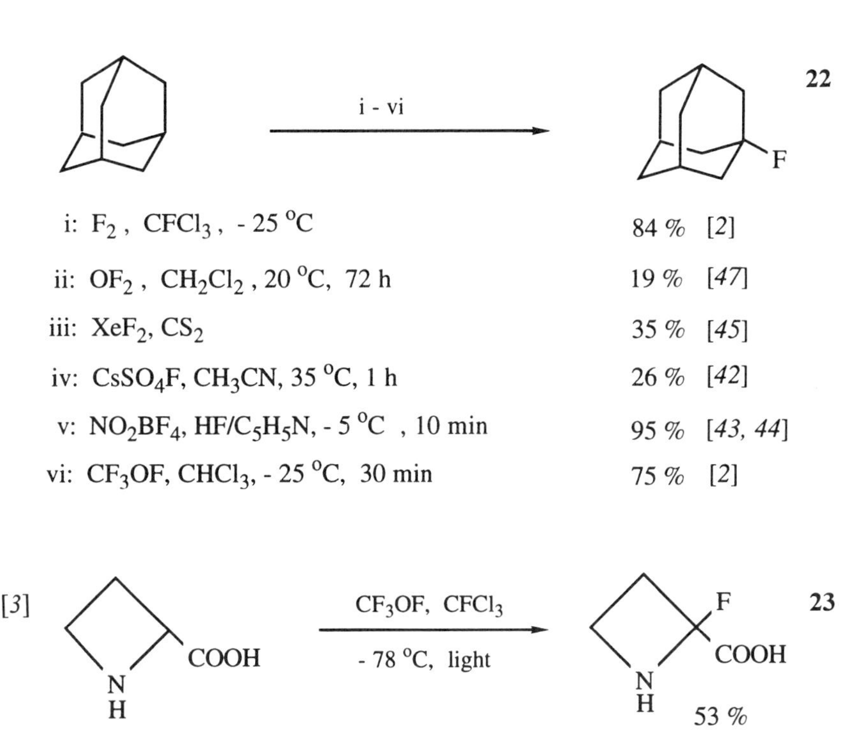

[*2*] **24**

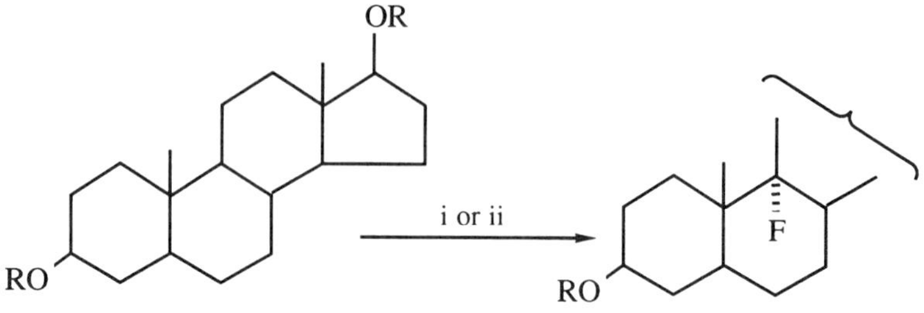

OR

i or ii

RO RO F

$R = CF_3CO$ i: CF_3OF, $CHCl_3$ 34 %

$R = CH_3CO$ ii: F_2, $CFCl_3$ 70 %

Reactions with Organometallics

The preparation of aromatic fluorine compounds may be accomplished by direct fluorination or by fluorination of organometallic intermediates. The ipso fluorination of an aryl organometallic derivative with a positive fluorine reagent allows control over the regioselectivity of the fluorination and offers advantages in the preparation of ^{18}F-labeled materials [*11, 50*].

Acetyl hypofluorite is very effective in the fluorination of the aryl–metal (Hg, Ge, or Si) bond, but yields are frequently low. With aryl silicon compounds some competition exists for replacement of an aromatic hydrogen [*51, 52, 53, 54*] (equations 25–27). **Fluoroxytrifluoromethane** fluorinates *p*-methoxyphenyl mercuric acetate to give *p*-fluoroanisole in 86% yield [*52*].

[*51, 52*] **25**

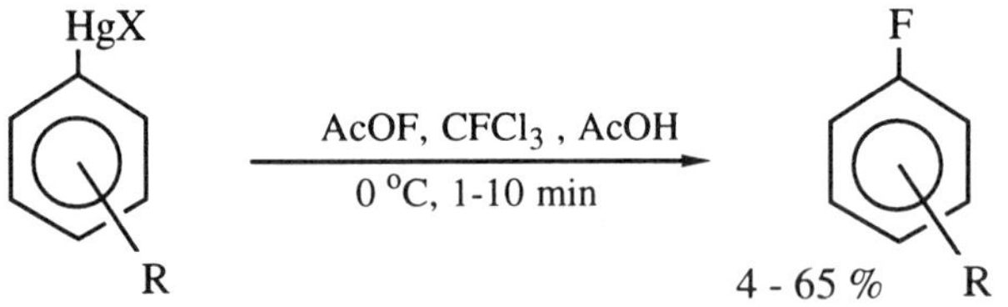

HgX F

AcOF, $CFCl_3$, AcOH

0 °C, 1-10 min

R 4 - 65 % R

$R = OCH_3$, OH, $NHCOCH_3$, CH_3, Cl, H

X = OAc, Cl

[*54*] 26

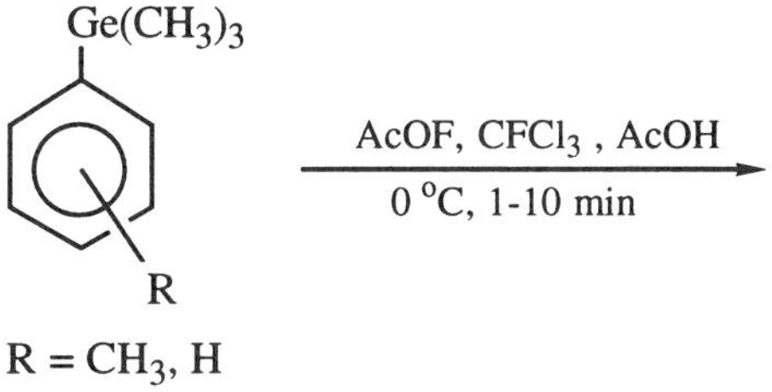

$Ge(CH_3)_3$ arene (R) → AcOF, $CFCl_3$, AcOH; 0 °C, 1-10 min → F arene (R), 8 - 16 %

R = CH_3, H

[*53*] 27

$Si(CH_3)_3$ arene (R) → AcOF, $CFCl_3$, AcOH; 25 °C, 5-10 min → F arene (R), 6 - 16 % + $Si(CH_3)_3$, F arene (R), 1 - 13 %

R = H, CH_3, OCH_3, Cl, Br, CH_3CO

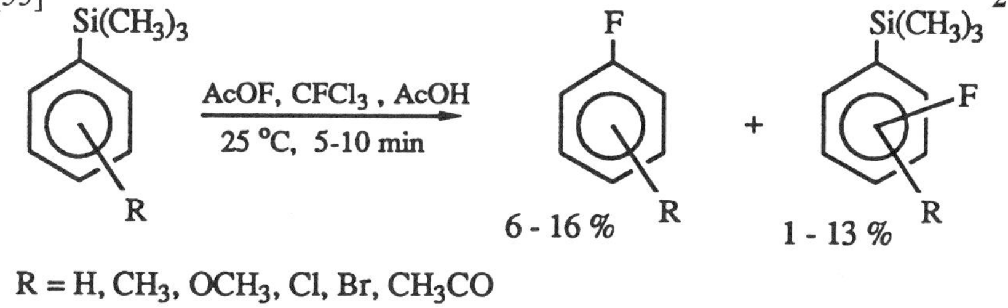

The ***aryl tin compounds*** are better substrates for fluorination because they give high yields of fluorinated aromatics and they may be fluorinated with acetyl hypofluorite, cesium fluoroxysulfate, or fluorine [*52, 54*] (equation 28). ***Aryl boronic esters*** react with **cesium fluoroxysulfate** to produce fluoroaromatics [*55*] (equation 29).

Acetyl hypofluorite gas is used in a reaction with an arylmercuric acetate to produce a derivative of L-DOPA [*56*] (equation 30).

Alkenes may be converted in a two-step procedure into ***vicinal fluoro ethers*** [*57*] (equation 31).

28

$Sn(CH_3)_3$ arene (R) → F arene (R)

[*54*] AcOF, $CFCl_3$, AcOH; 0 °C → 36 - 68 %

[*52*] $CsSO_4F$, CH_3CN; RT, 15 h → 11 - 86 %

[*52*] F_2, $CFCl_3$; - 78 °C, sec → 30 - 67 %

R = H, CH_3, OCH_3, Cl, CF_3

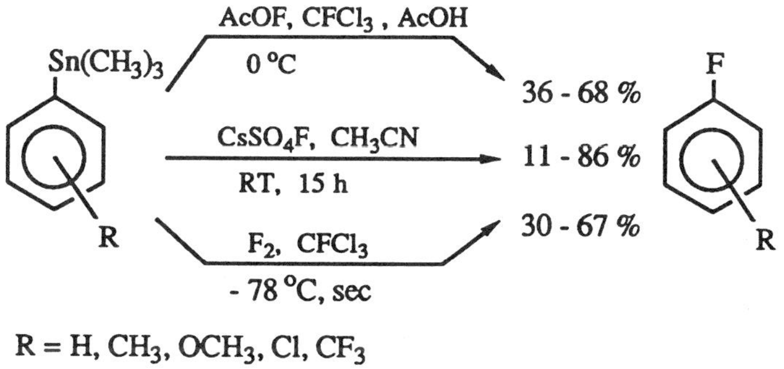

[*55*] **29**

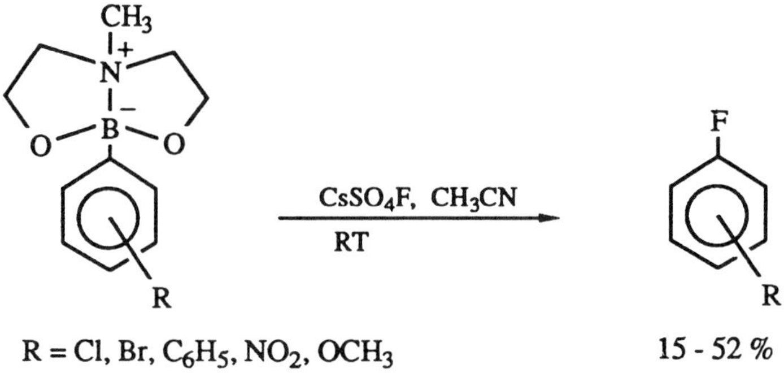

CH_3, N^+, B^-, O, O, R

$CsSO_4F$, CH_3CN, RT

F, R

R = Cl, Br, C_6H_5, NO_2, OCH_3

15 - 52 %

[*56*] **30**

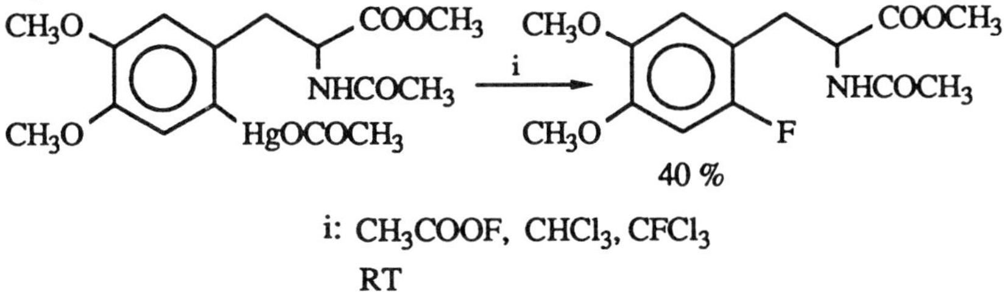

CH_3O, $COOCH_3$, $NHCOCH_3$, CH_3O, $HgOCOCH_3$

i

CH_3O, $COOCH_3$, $NHCOCH_3$, CH_3O, F

40 %

i: CH_3COOF, $CHCl_3$, $CFCl_3$ RT

[*57*] **31**

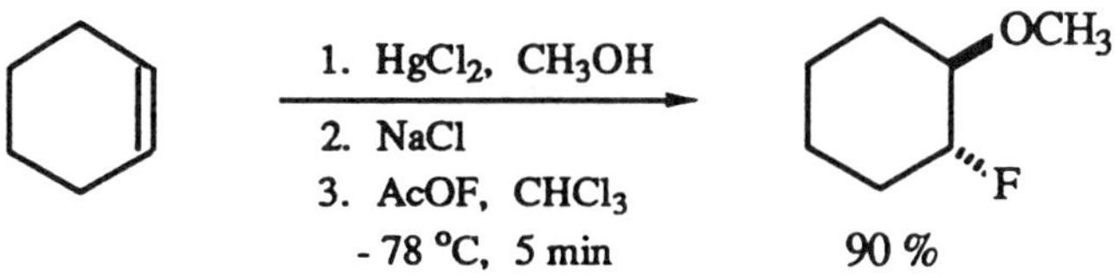

1. $HgCl_2$, CH_3OH
2. NaCl
3. AcOF, $CHCl_3$ - 78 °C, 5 min

OCH_3, F

90 %

Fluoraza Reagents [*6*]

The fluoraza reagents consist of two types of compounds: one in which a fluorine atom is bound to the nitrogen atom of an amide or, more often, a sulfonamide and one in which a fluorine atom is bound to the nitrogen atom of a tertiary amine such as pyridine, quinuclidine, or triethylenediamine 1,4-diazabicyclo[2.2.2]octane. The positive charge on the nitrogen is counterbalanced by a non-nucleophilic anion such as triflate or tetrafluoroborate.

The ***N*-fluoroamide reactants** were initially prepared by fluorination of amides with fluoroxytrifluoromethane [*63*]. Current methods involve the direct fluorination of amide reagents with elemental fluorine under a variety of conditions [*48, 64, 65, 66, 67, 68, 69, 71, 72, 73, 74, 76, 81, 82*] (Tables 3a and 3b).

The ***N*-fluoroamine salt reagents** are also prepared by direct fluorination. Thus, *N*-fluoroquinuclidinium salts [*71, 72*], *N*-fluorotriethylenediammonium salts [*73*], and *N*-fluoropyridinium salts [*74*] are prepared as shown in equations 32–34 (Table 3b, entries I, J, and K).

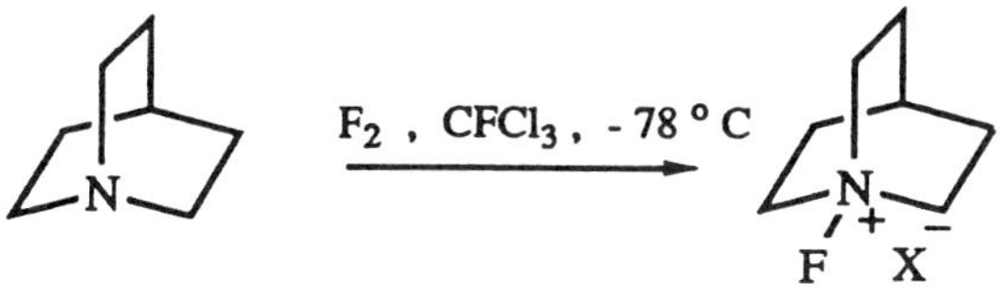

F_2 , $CFCl_3$, - 78 °C 32

[*71, 81*] X = F, 89 %

[*72*] X = OTf, 88 %

[*73*] F_2 , $CHCl_3$ / RX, -78 °C 33

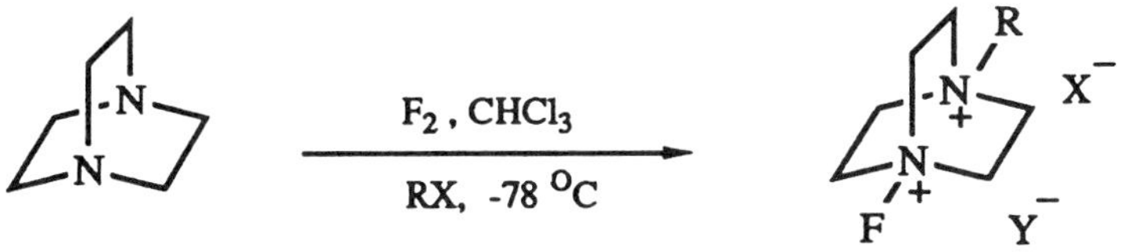

$R = CH_3$, CH_2Cl , CH_2CF_3

X, Y = F, BF_4, OTf, OTs

[*74*] 1. F_2/N_2 , CH_3CN -40 °C / 2. HX or MX or Me_3SiX or Lewis Acid 34

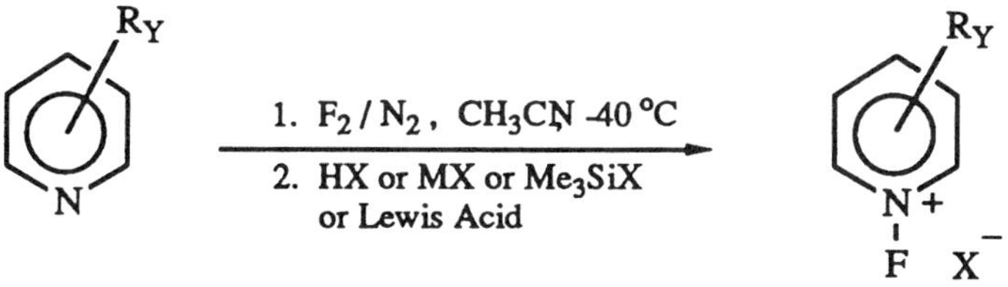

22 - 90 %

$R = CH_3$, OCH_3, $C(CH_3)_3$, CH_2OCH_3, CN, CH_3CO_2, NO_2, Cl, CF_3, $COCH_3$, $COOCH_3$ (R often in 2 position)

X = OTf, BF_4, OSO_2F, SbF_6, ClO_4 , B_2F_7

Y = 1 - 5

The effects of ring substituents and counterions on the preparation and reactivity of the *N*-fluoropyridinium salts have been reported in detail [*74, 75, 76*].

Table 3a. Preparation of *N*-Fluorosulfonamides

$$RSO_2NHR^1 \xrightarrow[\text{below}]{F_2 \text{ Conditions}} RSO_2NFR^1$$

Entry	N-*Fluorosulfonamide*	*Yield (%)*	*Conditions*	*Ref.*
A	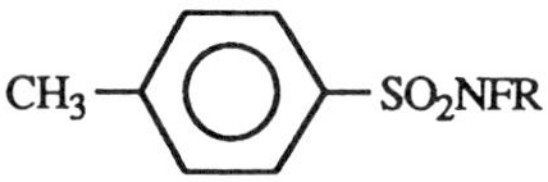 CH_3–C_6H_4–SO_2NFR; R = CH_3, $C(CH_3)_3$, C_6H_{11}, norbornyl, neopentyl	11–59	F_2 (N_2), $CFCl_3$, $CHCl_3$ –78 °C, 4–6 h	64
B	$(CF_3SO_2)_2NF$	95	F_2, neat, –196 °C	65
C	 N(F-ring)–$NFSO_2CF_3$	89	F_2, CH_3CN, –60 °C 6h	66
D	 NF, SO_2 (bicyclic sultam)	75	F_2, $CHCl_3$, $CFCl_3$ – 40 °C, 30 min	67
E	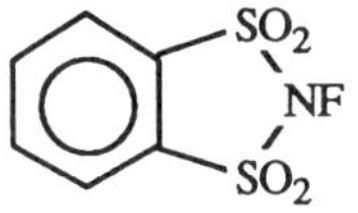 SO_2, NF, SO_2 (benzo-fused)	90	F_2, $CHCl_3$, NaF – 40 °C	48
F	$(C_6H_5SO_2)_2NF$	70	F_2, N_2, CH_3CN, – 40 °C	82
G	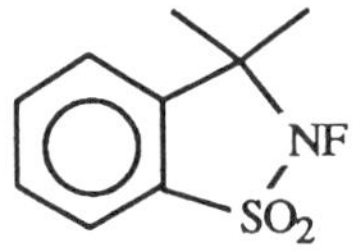 NF, SO_2 (benzo-fused, gem-dimethyl)	90	F_2, $CHCl_3$, – 40°C	68

Reactions with Aromatics

Several fluoraza reagents shown in Tables 3a and 3b (B, C, E, F, J, and K) are reactive enough to fluorinate an aromatic ring (Table 1). The *ortho* isomer predominates in the *o*/*m*/*p* mixture. Reagent K has been used to prepare fluorinated derivatives of tyrosine and estradiol [*77*] (equation 35) (Table 1, entry 10).

The ***N*-fluoropyridinium salts** produce 2-fluoropyridines on treatment with base. A carbene mechanism has been proposed [*78*] (equation 36).

Activated aromatic compounds are fluorinated in the *ortho* and *para* positions with both triethylenediammonium reagents (J) and *N*-fluorosulfonamides (E and F) [*73, 82*] (equation 37).

Table 3b. Preparation of *N*-Fluoroamide and *N*-Fluoroamine Salts

Entry	N-*Fluoroamide or* N-*Fluoroamine Salt*	*Yield (%)*	*Conditions*	*Ref.*
H	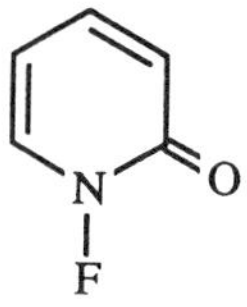 1-fluoro-2-pyridone (N–F)	63	F_2, N_2, 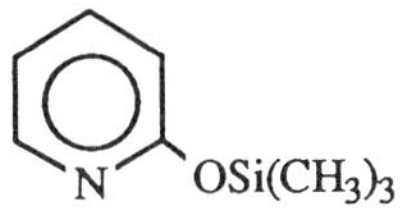 2-(trimethylsilyloxy)pyridine ($OSi(CH_3)_3$)	69
I	 N-fluoroquinuclidinium X^-; X = F, OTf	89	F_2, $CFCl_3$, –78 °C	71, 72, 81
J	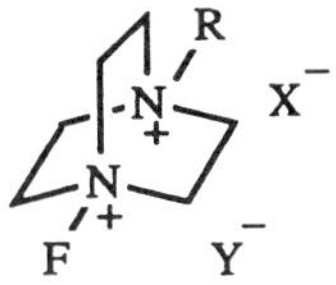 1-R-4-fluoro-1,4-diazoniabicyclo[2.2.2]octane X^- Y^-; R = CH_3, CH_2Cl, CH_2CF_3; X, Y = F, BF_4, OTf, OTs	90	F_2, $CHCl_3$, RX –78 °C	73
K	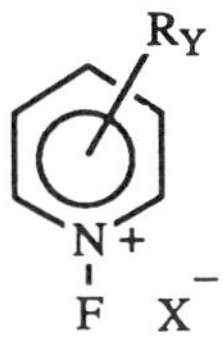 N-fluoropyridinium (R_Y) X^-; R = CH_3, OCH_3, $C(CH_3)_3$, CH_2OCH_3, CN, CH_3CO_2, NO_2, Cl, CF_3, $COCH_3$ (R often in position 2); X = OTf, BF_4, OSO_2F, SbF_6, ClO_4, $C_6H_5NB_2F_7$; Y = 1–5	90	F_2, CH_3CN, –40 °C MX	74, 76
L	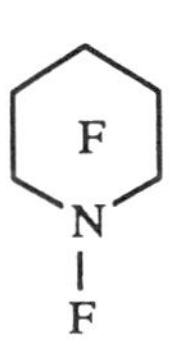 1,4-difluoropiperidine (F, N–F)	8	Electrochemical fluorination	66

[77] 35

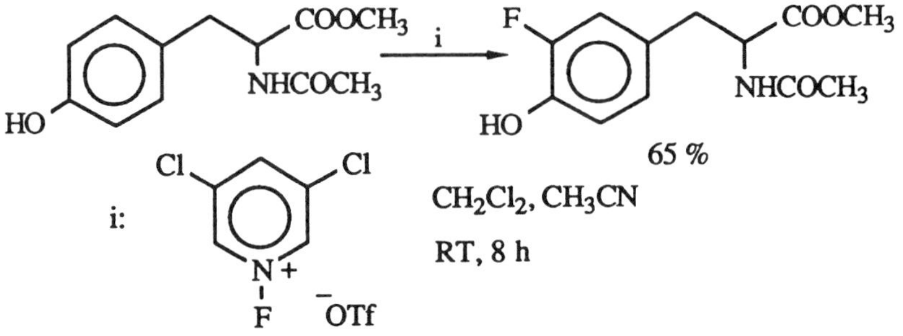

HO–C$_6$H$_4$–CH$_2$CH(NHCOCH$_3$)COOCH$_3$ —i→ F, HO–C$_6$H$_3$–CH$_2$CH(NHCOCH$_3$)COOCH$_3$ 65 %

i: 3,5-Cl$_2$-pyridinium N–F $^-$OTf, CH_2Cl_2, CH_3CN, RT, 8 h

[78] 36

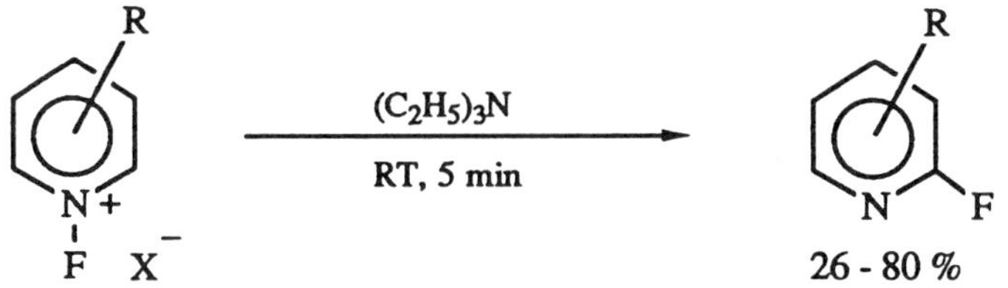

R-pyridinium N+–F X^- —$(C_2H_5)_3N$, RT, 5 min→ R-2-fluoropyridine 26 - 80 %

R = alkyl, polyalkyl, C_6H_5, OCH_3, Cl, CN, NO_2

X = BF_4, SbF_6, PF_6

37

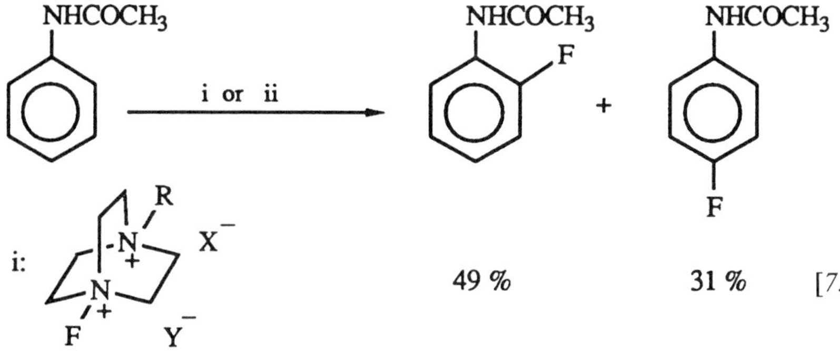

$C_6H_5NHCOCH_3$ —i or ii→ 2-F-$C_6H_4NHCOCH_3$ + 4-F-$C_6H_4NHCOCH_3$

i: R–N+ / N+–F bicyclic salt, X^-, Y^- — 49 % — 31 % [73]

R = CH_3, CH_2Cl, CH_2CF_3

X, Y = BF_4, OTf, OTs

CH_3OH, reflux, <5 min

ii. $(PhSO_2)_2NF$, 100 °C neat, 18 h — 25 % — 15 % [82]

Reactions with Enols and Enolates

Both the *N*-fluorosulfonamides and the *N*-fluoroammonium salts are very effective in the fluorination of ***enol acetates, enamines, silyl enol ethers,*** and ***enolates*** (Table 2). The reactions are thought to proceed through a mechanism which involves S_N2 attack on the fluorine atom, but contributions from electron-transfer pathways also exist [*65, 68, 73, 75, 76, 79, 80, 81, 82*].

The fluorination of β-diketones and β-ketoesters with ***N-fluorobis(trifluoromethanesulfonyl)imide*** (Table 3a, B) can be controlled to give either monofluorination or difluorination. Monofluorination occurs when the strong acid, bis(trifluoromethanesulfonyl)imide, a reaction product, is removed by addition of water, which prevents further enolization and fluorination of the monofluoro adduct [*83*] (equation 38).

[*83*] 38

$$R^1C(=O)CH_2C(=O)R^2 \xrightarrow[22\,^\circ C,\ 3\text{-}24\ h]{(CF_3SO_2)_2NF,\ CH_2Cl_2} R^1C(=O)CF_2C(=O)R^2 \quad 54 - 100\ \%$$

$$R^1C(=O)CH_2C(=O)R^2 \xrightarrow[22\,^\circ C,\ 8\text{-}14\ h]{(CF_3SO_2)_2NF,\ CH_2Cl_2,\ H_2O} R^1C(=O)CHFC(=O)R^2 \quad 86 - 90\ \%$$

$R^1 = CH_3, OC_2H_5, C_6H_5$
$R^2 = CH_3, OC_2H_5, C_6H_5$

N-Fluorobis(trifluoromethanesulfonyl)imide is also effective in the monofluorination of ***ester*** and ***amide enolates,*** and of neutral ***dicarbonyl compounds.*** Excellent stereoselectivity is observed [*48, 80, 81*] (equations 39–41).

[*80*] 39

$$CH_3CH(C_6H_5)COX \xrightarrow[2.\ (CF_3SO_2)_2NF,\ -80\,^\circ C,\ 10\ min]{1.\ LDA,\ THF,\ -80\,^\circ C} CH_3CF(C_6H_5)COX$$

$X = OC_2H_5$ — 83 %
$X = N(i\text{-}C_3H_7)_2$ — 85 %

[*81*] 40

$$(CH_3)_2CHCOCO_2R \xrightarrow[22\,^\circ C,\ 24\ h]{(CF_3SO_2)_2NF,\ CHCl_3} (CH_3)_2CFCOCO_2R \quad 95\ \%$$

$R = H, C_2H_5$

[*48*] **41**

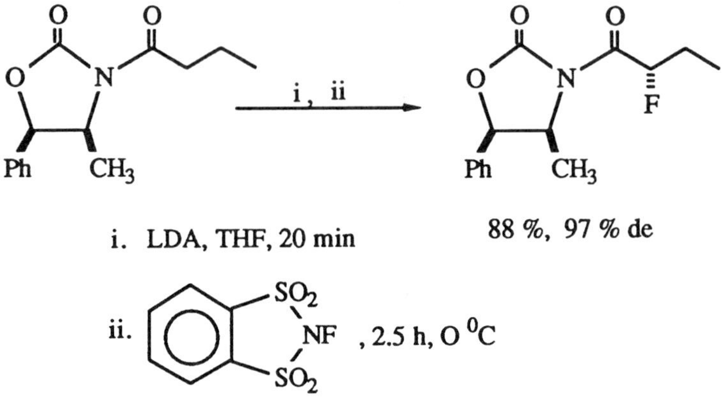

i, ii

i. LDA, THF, 20 min

ii. SO_2 NF SO_2 , 2.5 h, O ^{0}C

Ph CH_3 → Ph CH_3 F

88 %, 97 % de

The extent of mono- versus difluorination of enolates may also be controlled through the use of ***N-fluorosultam*** (Table 3a, G). With a base/*N*-fluoro compound/substrate ratio of 1.2/1.3–1.6/1, the reaction yields monofluoro products, but with a ratio of 2.4–3.6/2.6–3.6/1, the difluorinated compounds are obtained [*68*] (equation 42).

[*68*] **42**

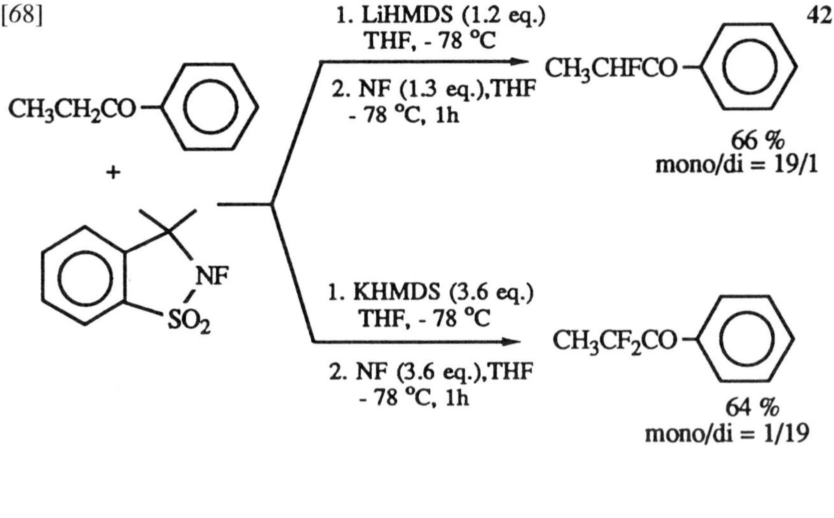

CH_3CH_2CO–Ph + NF SO_2

1. LiHMDS (1.2 eq.) THF, - 78 °C
2. NF (1.3 eq.),THF - 78 °C, 1h

→ CH_3CHFCO–Ph

66 %
mono/di = 19/1

1. KHMDS (3.6 eq.) THF, - 78 °C
2. NF (3.6 eq.),THF - 78 °C, 1h

→ CH_3CF_2CO–Ph

64 %
mono/di = 1/19

Fluorinations of enolates with chiral reagent D (Table 3a) give moderate degrees of enantiomeric excess [*67*] (equation 43).

[*67*] **43**

R = H — 63 %, 70 % ee

R = CH_3 — 10 %, 5 % ee

Reaction conditions: $(C_2H_5)_2O$, 0 °C

When chiral enolates are allowed to react with *N*-fluoro-2,4,6-trimethylpyridinium triflate (K, Table 3b), moderate diastereomeric excesses are achieved [*79*] (equation 44).

[*79*] **44**

i. LiHMDS, THF

ii. N-fluoro-2,4,6-trimethylpyridinium triflate (F, ^{-}OTf), - 78 °C, 15 h

R	%	Diastereo Ratio
CH_3	87	3.8
C_2H_5	96	2
C_3H_7	95	2
$C_6H_5CH_2$	88	1.6

Both *stereoselectivity* and *regioselectivity* occur in the reaction of steroid vinyl esters, ethers, and related compounds with ***N*-fluoropyridinium salts** [*75, 76*] (equation 45).

Reactions with Organometallics

The fluorination of organometallics with *N*-fluoroamide reagents has received only limited attention. ***Grignard reagents,*** both aliphatic and aromatic, are converted to organofluorine compounds. Both the electron transfer and the S_N2 mechanisms have been considered in these processes [*80, 81, 82*]. The reactions are exemplified in equation 46 [*48, 69, 70, 71, 75*]. **Organosilanes** are also fluorinated [*71*] (equation 47).

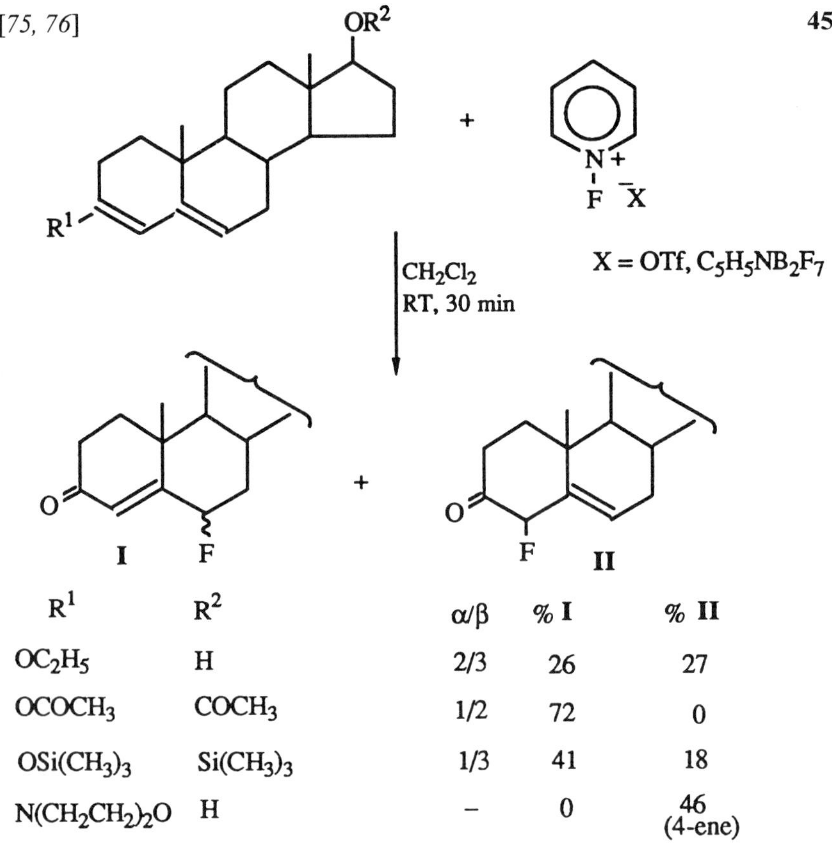

R^1	R^2	α/β	% I	% II
OC_2H_5	H	2/3	26	27
$OCOCH_3$	$COCH_3$	1/2	72	0
$OSi(CH_3)_3$	$Si(CH_3)_3$	1/3	41	18
$N(CH_2CH_2)_2O$	H	–	0	46 (4-ene)

_N_-Fluoro-_N_-_t_-butylbenzenesulfonamide (Table 3a, A) reacts with ***vinyl lithium reagents*** prepared from iodoalkenes to give high yields of fluoroalkenes with high stereoselectivity [*84*] (Table 4).

Xenon Difluoride

Xenon difluoride [*4, 5, 7, 8, 10*] is a white crystalline material obtained through the combination of fluorine and xenon in the presence of light. The reagent is commercially available and possesses a relatively long shelf-life when stored cold (freezer). Xenon difluoride is very effective for small-scale fluorination of alkenes and activated nucleophilic substrates. The reactions are usually conducted between 0 °C and room temperature in chloroform or methylene chloride solutions. Hydrogen fluoride catalysis is sometimes helpful. Xenon difluoride reacts in a manner that usually involves some complexation between the substrate and reagent followed by the formation of radical and radical cation intermediates.

MgBr (phenyl) —(i - v)→ F (phenyl) 46

[70] i. (N-fluoro caprolactam) NF, $(C_2H_5)_2O$, 20 %

[71] ii. (N-fluoroquinuclidinium fluoride) N+ F F−, $(C_2H_5)_2O$, 26 %

[69] iii. (N-fluoro-2-pyridone) N–F, O, $(C_2H_5)_2O$, 0 °C, 15 %

[75] iv. (N-fluoropyridinium triflate) N+ F $\bar{O}$Tf, THF, - 50 °C, 22 %

[48] v. (SO_2 NF SO_2), 0 °C, 80 %

[71] $SiCl_3$ (phenyl) —(N-fluoroquinuclidinium fluoride, N+ F F−; THF, -50 °C)→ F (phenyl) 22 % 47

Table 4. Preparation of Fluoroalkenes from Iodoalkenes [*84*]

[*84*]

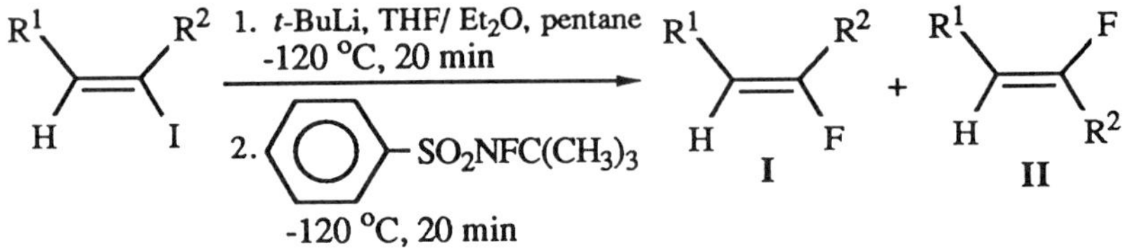

1. *t*-BuLi, THF/ Et_2O, pentane -120 °C, 20 min
2. C_6H_5–$SO_2NFC(CH_3)_3$ -120 °C, 20 min

R^1	R^2	Yield (%) I	Yield (%) II
C_6H_{13}	H	71	15
C_6H_5	H	76	10
C_3H_7	C_3H_7	85	3
C_5H_{11}	CH_3	75	12
CH_3	C_5H_{11}	75	12
C_2H_5	C_4H_9	83	7
C_4H_9	C_2H_5	88	7

Reactions with Aromatics

The results given in both Table 1 and in the reviews [*4, 5, 7, 8, 10*] show that xenon difluoride reacts with a wide variety of aromatic substrates to produce *regioselectively monofluorinated aromatics.* An example is the preparation of 6-fluoro-L-DOPA [85] (equation 48).

[*85*] **48**

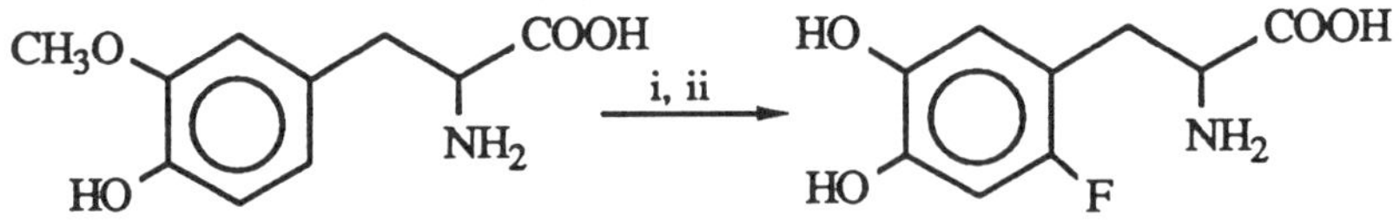

i: XeF_2, CH_2Cl_2, -60 °C

ii: HBr, 140 °C, 15 min

25 %

Xenon difluoride may be used as the pure reagent or as a graphite intercalate for the effective fluorination of polynuclear aromatics [*86, 87*] (equations 49 and 50).

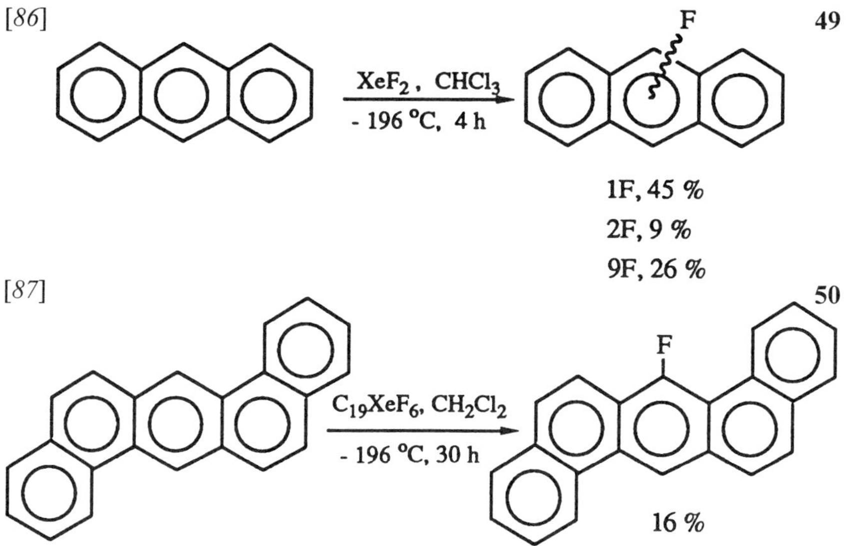

Reactions with Acids, Enols, and Enolates

Xenon difluoride [*88*], xenon difluoride complexed with dialkyl sulfides [*89*], and xenon difluoride intercalated with graphite [*90*] are all effective reagents for the fluorination of ***acids, enolates,*** or ***enols*** (Table 2).

Carboxylic acids react with xenon difluoride to produce unstable xenon esters. The esters decarboxylate to produce free radical intermediates, which undergo fluorination or reaction with the solvent system. Thus aliphatic acids decarboxylate to produce mainly fluoroalkanes or products from abstraction of hydrogen from the solvent. Perfluoro acids decarboxylate in the presence of aromatic substrates to give perfluoroalkyl aromatics. Aromatic and vinylic acids do not decarboxylate [*91*] (equation 51).

[*91*] **51**

$$XeF_2 \xrightarrow[\text{RT, 8h}]{C_6H_5CH_2CH_2COOH,\ CH_2Cl_2} C_6H_5CH_2CH_2F \quad 76\%$$

$$XeF_2 \xrightarrow[\text{RT,28h}]{CF_3COOH,\ C_6H_6,\ CH_2Cl_2} C_6H_5CF_3 \quad 33\%$$

Xenon difluoride reaction with nitro enolates provides a useful entry into a wide variety of fluorinated synthetic intermediates [*91, 92*] (equation 52).

Silyl enol ethers are fluorinated in high yields with xenon difluoride [*62, 93, 94, 95*]. Applications of this reaction to the preparation of fluorinated materials of biomedicinal interest have proven very successful [*62, 94, 95*] (equations 53 and 54).

[*91, 92*] 52

$$R\bar{C}(NO_2)_2 \xrightarrow[20\ ^\circ C,\ 1\ h]{XeF_2,\ CH_3CN} RCF(NO_2)_2$$

R	%
H	27
NO_2	78
CH_3	95
C_6H_5	97
-CH=CHCN	71
-CH=CHCONH$_2$	63

[*95*] 53

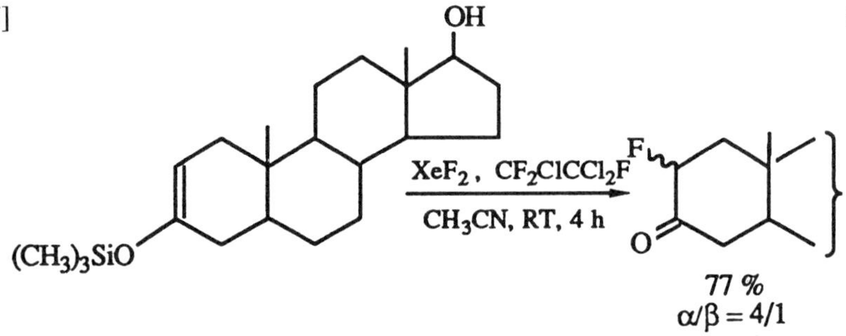

[*94*] 54

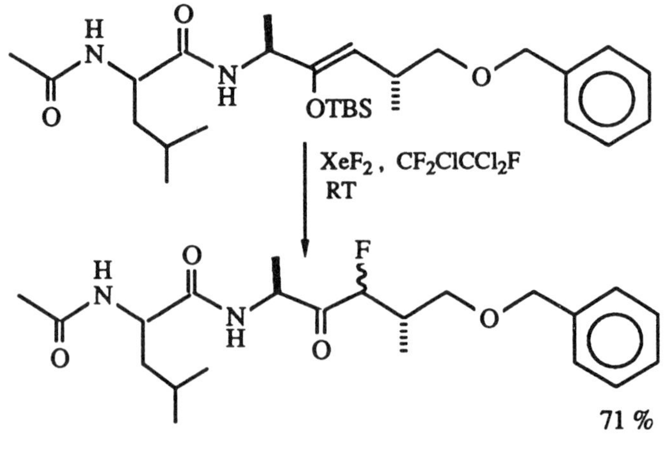

Reactions with Carbon–Hydrogen Bonds

Xenon difluoride fluorinates adamantane in low yield [*45*] (equation 22).

When the carbon–hydrogen bond is activated by an α-sulfur atom, fluorination occurs readily. The reactions involve intermediates that contain sulfur–fluorine bonds. *N*-Fluoropyridinium reagents behave similarly. [*99, 100, 101, 102*] (equations 55–57).

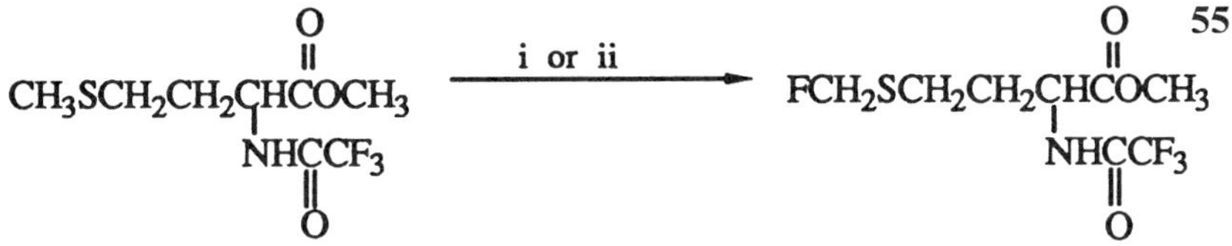

$$CH_3SCH_2CH_2CH(NHC(O)CF_3)C(O)OCH_3 \xrightarrow{\text{i or ii}} FCH_2SCH_2CH_2CH(NHC(O)CF_3)C(O)OCH_3 \quad (55)$$

[*99*] i. XeF_2, CH_3CN, - 20 °C, 30 min 90 %

[*100*] ii. *N*-fluoro-2,4,6-trimethylpyridinium triflate 39 %
CH_2Cl_2, reflux, 2 h

[*102*] 56

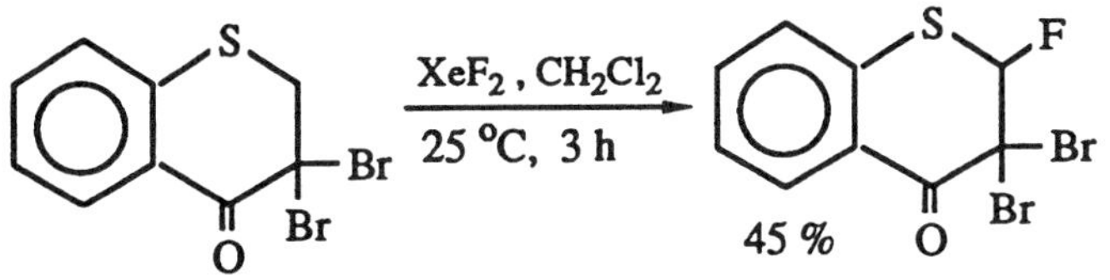

XeF_2, CH_2Cl_2; 25 °C, 3 h; S, F, Br, Br, O; 45 %

[*101*] 57

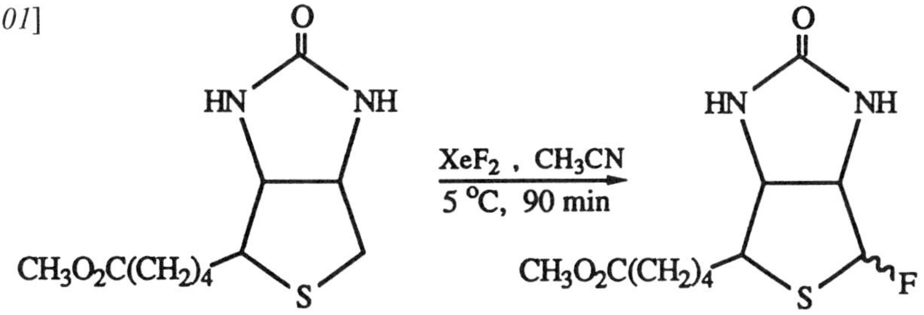

HN, NH, O, $CH_3O_2C(CH_2)_4$, S; XeF_2, CH_3CN; 5 °C, 90 min; F

15 %
α/β = 3/1

***Perchloryl Fluoride* [*9*]**

Perchloryl fluoride, well known for its ability to fluorinate organic substrates, has become a reagent with limited appeal. Its loss of popularity stems from difficult handling, threatening explosions, unwanted chlorinated by-products, and unavailability [*9, 10*]. Perchloryl fluoride, however, still has a useful role in the synthesis of fluorinated compounds, because it reacts very well with highly ***nucleophilic anions*** and ***enamines*** [*103*].

Aromatic substrates react well with **perchloryl fluoride** after conversion into organolithium intermediates [*104, 105*] (equations 58 and 59).

[*105*] $OCH_2CH_2N(CH_3)_2$; C_6H_5, $C=C$, C_6H_5, CH_2CH_3 — 1. *s*-BuLi, THF, - 78 °C, 7 h; 2. $FClO_3$, 30 sec → $OCH_2CH_2N(CH_3)_2$, F; C_6H_5, $C=C$, C_6H_5, CH_2CH_3 **58**

70 %

[*104*] CH_3O, CH_3O, CHO — 1. LiTMEDA, -45 °C, THF; 2. $FClO_3$, 0 °C → F, CH_3O, CHO, CH_3O **59**

63 %

The anions of ***malonaldehyde*** [*106, 107*] and of organophosphonates [*108, 109, 110*] are fluorinated in good yields to provide interesting fluorinated intermediates. The *N*-fluoro compound B in Table 3a is also effective in the fluorination of phosphonate anions [*109*] (equations 60 and 61).

Fluorination of a ***sulfoxide-stabilized carbanion*** provided a route to fluorinated estrones after elimination of the sulfoxide [*111*] (equation 62).

The fluorination of both ***mono and dinitro enolates*** proceeds well with perchloryl fluoride. The mononitro fluorinated intermediates have been used to develop fluorinated materials of general synthetic utility [*112, 113*] (equation 63).

Experimental Procedures

For the safety advisory, see pages 25–26 of the book.

Preparation of *o*- and *p*-Fluoroacetanilide [*30*]

To a solution of cesium fluoroxysulfate (1 mmol) in 1.5 mL of dried acetonitrile is added 1 mmol of boron trifluoride etherate with stirring at room temperature. Acetanilide (1 mmol) dissolved in 0.5 mL of acetonitrile is added, and the mixture is stirred at room temperature for 30 min. Methylene chloride (10 mL) is added, the mixture is filtered, and the filtrate is concentrated. The products are isolated by preparative thin-layer chromatography (silica gel; CH_2Cl_2–CH_3OH, 20:1) to give 74% of 2-fluoroacetanilide and 11% of 4-fluoroacetanilide.

[*106, 107*] **60**

$$R\bar{C}(CHO)_2 \xrightarrow[28\ ^\circ C,\ 4\ h]{FClO_3,\ CH_3OH} RCF[CH(OCH_3)_2]_2$$

R = F 48 %, CH_3 70 %, C_6H_5 75 %

[*109*] **61**

$$(C_2H_5O)_2POCH_2R \xrightarrow{i\ or\ ii} (C_2H_5O)_2POCHFR + (C_2H_5O)_2POCF_2R$$

	$(C_2H_5O)_2POCHFR$	$(C_2H_5O)_2POCF_2R$
i. R = $PO(OC_2H_5)_2$ 1. *t*-BuOK, toluene, 5 °C 2. $FClO_3$, <22 °C, 50 min	47 %	16 %
ii. R = CN 1. *n*-BuLi, THF, - 78 °C 2. $(CF_3SO_2)_2NF$, THF, - 78 °C 1 h	51 %	0 %

[*111*] **62**

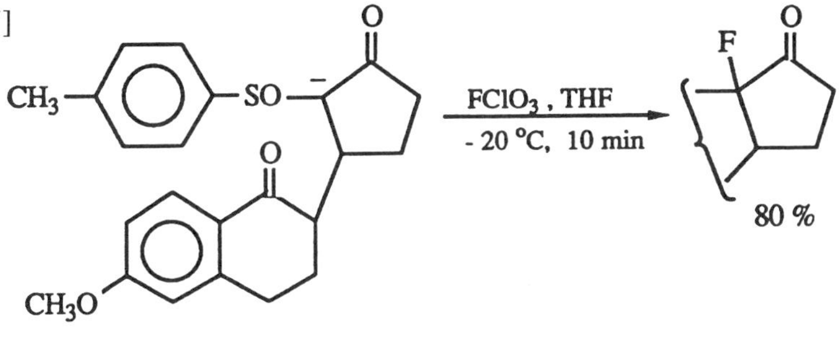

CH_3–C_6H_4–SO– (cyclopentanone carbanion bearing 6-methoxy-1-tetralon-2-yl substituent, CH_3O) $\xrightarrow[-20\ ^\circ C,\ 10\ min]{FClO_3,\ THF}$ 2-fluorocyclopentanone product 80 %

[*112*] **63**

$$O_2NC(H)(R)CO_2C_2H_5 \xrightarrow[\text{2. } FClO_3,\ THF,\ 22\ ^\circ C,\ 1\ h]{\text{1. KF, spray dried}} O_2NC(F)(R)CO_2C_2H_5$$

100 %

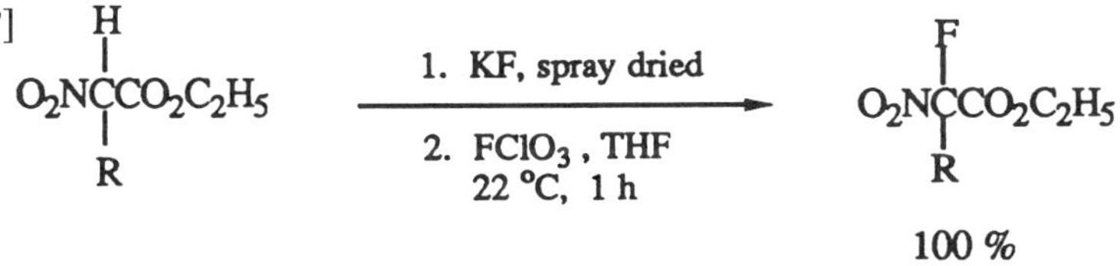

R = $CH_2CO_2C_2H_5$, $CH_2CH_2CO_2C_2H_5$, $CH_2C_6H_5$
$CH_2CH_2COCH_3$, CH_2CH_2CN, $CH(C_6H_5)CH_2COCH_3$

Preparation of Acetyl Hypofluorite and 2-Fluoro-4-Methylacetanilide [*15*]

Into a suspension of 8 g of sodium acetate in 400 mL of a solution of 1 part acetic acid and 10 parts fluorotrichloromethane is passed at –75 °C a stream of fluorine diluted to 10% with nitrogen. The reaction is stirred with a Vibromixer. A solution of 4-methylacetanilide (20 mmol) in a mixture of dichloromethane and fluorotrichloromethane cooled to –75 °C is added to 20 mmol of acetyl hypofluorite as determined by titration with potassium iodide. After 5 min the mixture is poured into water, and the organic layer is washed with sodium bicarbonate solution and dried over anhydrous magnesium sulfate. After concentration and column chromatography over silica gel and elution with chloroform, 2-fluoro-4-methylacetanilide is obtained in 85% yield.

Preparation of 1-Fluoro-1-Octene [*84*]

1-Iodo-1-octene(1 mmol) in 9 mL of tetrahydrofuran–ether–pentane (4:1:1) is treated at –120 °C under argon with 2 mmol of *tert*-butyllithium solution in pentane. After 20 min, 1.5 mmol of *N-tert*-butyl-*N*-fluorobenzenesulfonamide is added at –120 °C. The mixture is allowed to warm slowly to room temperature. 1-Fluoro-1-octene is obtained after preparative gas–liquid chromatography as a mixture of *cis* and *trans* isomers in yields of 15 and 31%, respectively.

Preparation of Diethyl Fluorophenylmalonate [*75*]

N-Fluoro-2,4,6-trimethylpyridinium triflate (1 mmol) is added in several portions at room temperature to a tetrahydrofuran solution of sodium diethyl phenylmalonate, obtained from 1 mmol of diethyl phenyl malonate and sodium hydride at 0 °C in tetrahydrofuran. The reaction mixture is poured into dilute hydrochloric acid and extracted with ether. The ether extract is washed with sodium bicarbonate and water and dried over magnesium sulfate. The oily residue obtained after removal of the ether is chromatographed on silica gel (dichloromethane–hexane, 1:1) to give diethyl fluorophenylmalonate in 83% yield.

Preparation of 2-Fluoro-1-Indanone [93]

The trimethylsilyl enol ether of 1-indanone (3.2 mmol) in 2 mL of methylene chloride is added to a mixture of xenon difluoride (4 mmol) and a catalytic amount of pyridinium poly(hydrogen fluoride) in 5 mL of methylene chloride. The mixture is stirred at 0 °C for 2 h and poured into dilute sodium bicarbonate solution; the organic layer is separated and dried. After concentration and column chromatography (silica gel; hexanes), 2-fluoro-1-indanone (mp, 59 °C) is obtained in 87% yield.

References for pages 133–166

1. Hesse, R. H. *Israel J. Chem.* **1978,** *17*, 60.
2. Barton, D. H. R. *Pure Appl. Chem.* **1977,** *49*, 1241.
3. Kollonitsch, J. *Israel J. Chem.* **1978,** *17*, 53.
4. Rozen, S. *Acc. Chem. Res.* **1988,** *21,* 307.
5. Zupan, M. *Vestn. Slov. Kem. Drug/Suppl.* **1984,** *31,* 151; *Chem. Abstr.* **1984,** *101,* 229451z.
6. Purrington, S. T.; Kagen, B. S.; Patrick, T. B. *Chem. Rev.* **1986,** *86*, 997.
7. Zupan, M. In *The Chemistry of Functional Groups, Supplement D: The Chemistry of Halides, Pseudohalides and Azides,* Parts 1 and 2; Patai, S.; Rappoport, Z., Eds.; Wiley: Chichester, United Kingdom, 1983.
8. Filler, R. *Israel J. Chem.* **1978,** *17,* 71.
9. Sharts, C. M.; Sheppard, W. A. *Org. React.* **1974,** *21,* 125.
10. Rozen, S.; Filler, R. *Tetrahedron* **1985,** *41,* 1111.
11. Kilbourn, M. R. *Fluorine-18 Labeling of Radiopharmoceuticals;* Nuclear Science Series, NAS-NS-3203; National Academy Press: Washington, DC, 1990.
12. Fowler, J. S.; Wolf, A. P. *The Synthesis of Carbon-11, Fluorine-18 and Nitrogen-13 Labeled Radiotracers for Biomedical Applications;* Monograph NAS-NS-3601; Technical Information Center, U.S. Department of Energy.
13. Fifolt, M. J.; Olczak, R. T.; Mundhenke, R. F.; Bieron, J. F. *J. Org. Chem.* **1985,** *50,* 4576.
14. Rozen, S. *J. Am. Chem. Soc.* **1991,** *113,* 2648.
15. Lerman, O.; Tor, Y.; Hebel, D.; Rozen, S. *J. Org. Chem.* **1984,** *49,* 806.
16. Rozen, S.; Menhem, Y. *J. Fluorine Chem.* **1980,** *16,* 19.
17. Mulholland, G. K.; Ehrenkaufer, R. E. *J. Org. Chem.* **1986,** *51,* 1482.
18. Rozen, S.; Hebel, H. *J. Org. Chem.* **1990,** *55,* 2621.
19. Barnette, W. E.; Wheland, R. C.; Middleton, W. J.; Rozen, S. *J. Org. Chem.* **1985,** *50,* 3698.
20. Appelman, E. H.; Basile, L. J.; Thompson, R. C. *J. Am. Chem. Soc.* **1979,** *101,* 3384.
21. Stavber, S.; Zupan, M. *J. Org. Chem.* **1985,** *50,* 3609.
22. Lerman, O.; Tor, Y.; Rozen, S. *J. Org. Chem.* **1981,** *46,* 4629.
23. Levy, J. B., Sterling, D. M. *J. Org. Chem.* **1985,** *50,* 5615.
24. Belanger, P. C.; Lau, C. K.; Williams, H. W. R.; Defresne, C.; Scheigetz, J. *Can. J. Chem.* **1988,** *66,* 1479.
25. Patrick, T. B.; Darling, D. L. *J. Org. Chem.* **1986,** *51,* 3242.
26. Hebel, D.; Lerman, O.; Rozen, S. *Bull. Soc. Chim. Fr.* **1986,** 861.
27. Airey, J.; Barton, D. H. R.; Ganguly, A.; Hesse, R.; Pechet, M. M. *Ann. Quim.* **1974,** *70,* 871.
28. Patrick, T. B.; Hayward, E. C. *J. Org. Chem.* **1974,** *39,* 2120.

29. Patrick, T. B.; Cantrell, G. L.; Chang, C. Y. *J. Am. Chem. Soc.* **1979,** *101*, 7434.
30. Stavber, S.; Zupan, M. *J. Fluorine Chem.* **1981,** *17,* 597.
31. Patrick, T. B.; LeFaivre, M. H., Koertge, T. E. *J. Org. Chem.* **1976,** *41*, 3413.
32. Stavber, S.; Zupan, M. *J. Chem. Soc., Chem. Commun.* **1981,** 148.
33. Appelman, E. H.; Basile, L. J.; Hayatsu, R. *Tetrahedron* **1984,** *40*, 189.
34. Kosower, E. M.; Hebel, D.; Rozen, S.; Radkowski, A. E. *J. Org. Chem.* **1985,** *50,* 4152.
35. Andrews, L. E.; Bonnett, R.; Kozyrev, A. N.; Appelman, E. H. *J. Chem. Soc., Perkin Trans. 1* **1988,** 1735.
36. Michalski, T. J.; Appelman, E. H.; Bowman, M. K.; Hunt, J. E.; Norris, J. R.; Cotton, T. M.; Raser, L. *Tetrahedron Lett.* **1990,** 31, 6847.
37. Middleton, W. J.; Bingham, E. M. *J. Am. Chem. Soc.* **1980,** *102,* 4845.
38. Toscano, L.; Seghetti, E. *Tetrahedron Lett.* **1983,** *24,* 5527.
39. Rozen, S.; Lerman, O., Kol, M.; Hebel, D. *J. Org. Chem.* **1985,** *50,* 4753.
40. Rozen, S.; Brand, M. *Synthesis* **1985,** 665.
41. Shutske, G. M. *J. Chem. Soc., Perkin Trans. 1* **1989,** 1544.
42. Stavber, S.; Zupan, M. *Tetrahedron* **1989,** *45,* 2737.
43. Hashimoto, T.; Prakash, G. K. S.; Shih, J. G.; Olah, G. A. *J. Org. Chem.* **1987,** *52,* 931.
44. Olah, G. A.; Shih, J. G.; Singh, B. P.; Gupta, B. H. B. *J. Org. Chem.* **1983,** *48,* 3356.
45. Podkhalyzin, A. T.; Nazarova, M. P. *Zh. Org. Khim.* **1975,** *11,* 1568 (Engl. Transl. 1547).
46. Alker, D.; Barton, D. H. R.; Hesse, R.; Lister-James, J.; Markwell, R. E.; Pechet, M. M.; Rozen, S.; Takeshita, T.; Toh, H. T. *Nouv. J. Chim.* **1980,** *4,* 239.
47. Bolte, G.; Haas, A. *Chem. Ber.* **1984,** *117,* 1982.
48. Davis, F. A.; Han, W. *Tetrahedron Lett.* **1991,** *32,* 1631. Davis, F. A.; Han, W. *Tetrahedron Lett.,* **1992,** *33,* 1153.
49. Visser, G. W. M.; Bakker, C. N. M.; Halteren, v. B. W.; Herscheid, J. D. M.; Brinkman, G. A.; Hoekstra, A. *Recl. Trav. Chim. Pays-Bas* **1986,** *105,* 214.
50. Adam, M. J.; Ruth, T. J.; Jivan, S.; Pate, B. D. *J. Fluorine Chem.* **1984,** *25,* 329.
51. Visser, G. W. M.; Bakker, C. N. M.; Halteren, v. B. W.; Herscheid, J. D. M.; Brinkman, G. A.; Hoekstra, A. *J. Org. Chem.* **1986,** *51*, 1886.
52. Bryce, M. R.; Chambers, R. D.; Mullins, S. T.; Parkin, A. *Bull. Soc. Chim. Fr.* **1986,** 930.
53. Speranza, M.; Shiue, C.-Y.; Wolf, A. P.; Wilbur, D. S.; Angelini, G. *J. Fluorine Chem.* 1985, 30, 97.
54. Coenen, H. H.; Moerlein, S. M. *J. Fluorine Chem.* **1987,** *36,* 63.

55. Clough, J. M.; Diorazio, L. J.; Widdowson, D. A. *Syn. Lett.* **1990**, 761.
56. Luxen, A.; Barrio, J. R. *Tetrahedron Lett.* **1988,** *29,* 1501.
57. Hebel, D.; Rozen, S. *J. Org. Chem.* **1987,** *52,* 2588.
58. Barton, D. H. R.; Bubb, W. A.; Hesse, R. H.; Pechet, M. M. *J. Chem. Soc., Perkin Trans. 1* **1974,** 2095.
59. Stavber, S.; Zupan, M. *J. Chem. Soc., Chem Commun.* **1983,** 563.
60. Stavber, S.; Zupan, M. *Tetrahedron* **1986,** *42,* 5035.
61. Lerman, O.; Rozen, S. *J. Org. Chem.* **1983,** *48,* 724.
62. Patrick, T. B.; Mortezania, R. *J. Org. Chem.* **1988,** *53,* 5153.
63. Barton, D. H. R.; Hesse, R. H. Ger. Patent 2 332 430, 1974; *Chem. Abstr.* **1974,** *80,* 108209r.
64. Barnette, W. E. *J. Am. Chem. Soc.* **1984,** *106,* 452.
65. Singh, S.; DesMarteau, D. D.; Zuberi, S. S.; Witz, M.; Huang, H. N. *J. Am. Chem. Soc.* 1987, 109, 7194; DesMarteau, D. D.; Witz, M. *J. Fluorine Chem.* 1991, 52, 7; Pennington, W. T.; Resnati, G.; DesMarteau, D. D. *J. Org. Chem.* 1992, 57, 1536.
66. Banks, R. E.; Murtagh, V.; Tsiliopoulos, E. *J. Flourine Chem.* **1991,** *52,* 389; Banks, R. E.; Khazaei, A. *J. Fluorine Chem.* **1990,** *46,* 297.
67. Differding, E.; Lang, R. W. *Tetrahedron Lett.* **1988,** *29,* 6087.
68. Differding, E.; Ruegg, G. M.; Lang, R. W. *Tetrahedron Lett.* **1991,** *32,* 1779.
69. Purrington, S. T.; Jones, W. A. *J. Org. Chem.* **1983,** *48,* 761. *J. Fluorine Chem.* **1984,** *26,* 43.
70. Satyamurthy, N.; Bida, G. T.; Phelps, M. E.; Barrio, J. R. *J. Org. Chem.* **1990,** *55,* 3373.
71. Banks, R. E.; du Boisson, R. A.; Morton, W. D.; Tsiliopoulos, E. *J. Chem. Soc., Perkin Trans. 1* **1988,** 2805.
72. Banks, R. E.; Sharif, I. *J. Fluorine Chem.* **1988,** *41,* 297.
73. Banks, R. E.; Besheesh, M. K.; Khaffaf, S. N.; Sharif, I. *J. Fluorine Chem.* **1991,** *54,* 207; Lal, G. S.; Syvret, R. G. *J. Fluorine Chem.* **1991,** *54,* 208 (Supplemental Issue, Abstracts of the 13th International Symposium on Fluorine Chemistry, September 1–6, 1991, Bochum, Germany).
74. Umemoto, T.; Harasawa, K.; Tomizawa, G.; Kawada, K.; Tomita, K. *Bull. Chem. Soc. Jpn.* **1991,** *64,* 1081; Umemoto, T.; Tomita, K.; Kawada, K. *Org. Synth.* **1991,** *69,* 129.
75. Umemoto, T.; Fukami, S.; Tomizawa, G.; Harasawa, K.; Kawada, K.; Tomita, K. *J. Am. Chem. Soc.* **1990,** *112,* 8563.
76. Poss, A. J.; Van Der Puy, M.; Nalewajek, D.; Shia, G. A.; Wagner, W. J.; Frenette, R. L. *J. Org. Chem.* **1991,** *56,* 5962; Dauben, W. G.; Greenfield, L. D. *J. Org. Chem.* **1992,** *57,* 1597.
77. Hebel, D.; Kirk, K. L. *J. Fluorine Chem.* **1990,** *47,* 179.
78. Umemoto, T.; Tomizawa, G. *J. Org. Chem.* **1989,** *54,* 1726.

79. Ihara, M.; Kai, T.; Taniguchi, N.; Fukumoto, K. *J. Chem. Soc., Perkin Trans. 1* **1990,** 2357.
80. Resnati, G.; DesMarteau, D. D. *J. Org. Chem.* **1991,** *56,* 4925.
81. Differding, E.; Ruegg, G. M. *Tetrahedron Lett.* **1991,** *32,* 3815; Differding, E.; Wehrli, M. *Tetrahedron Lett.* **1991,** *32,* 3819.
82. Differding, E.; Ofner, H. *Synlett* **1981,** 187.
83. DesMarteau, D. D.; Gotoh, Y. *J. Chem. Soc., Chem. Commun.* **1991,** 179.
84. Lee, S. H.; Schwartz, J. *J. Am. Chem. Soc.* **1986,** *108,* 2445.
85. Firnau, G. *Can. J. Chem.* **1980,** *58,* 1449.
86. Anand, S. P.; Quarterman, L. A.; Christian, P. A.; Hyman, H. H.; Filler, R. *J. Org. Chem.* **1975,** *40,* 3796.
87. Agranat, I.; Rabinovitz, M.; Selig, H.; Lin, C.-H. *Synthesis* **1977,** 267.
88. Zajc, B.; Zupan, M. *J. Org. Chem.* **1982,** *47,* 573.
89. Patrick, T. B.; Nadji, S. *J. Fluorine Chem.* **1988,** *39,* 415.
90. Yemul, S. S.; Kagan, H. B.; Setton, R. *Tetrahedron Lett.* **1980,** *21,* 277.
91. Patrick, T. B.; Johri, K. K.; White, D. H.; Bertrand, W. S.; Mokhtar, R.; Kilbourn, M. R.; Welch, M. J. *Can. J. Chem.* **1986,** *64,* 138; Patrick, T. B.; Johri, K. K.; White, D. H. *J. Org. Chem.* **1983,** *48,* 4159; Tanabe, Y.; Matsuo, N.; Ohno, N. *J. Org. Chem.* **1988,** *53,* 4582; Della, E. W.; Head, N. J. *J. Org. Chem.* **1992,** *57,* 2850.
92. Tselinskii, I. V.; Melnikov, A. A.; Trabitsin, A. E.; Frolova, G. M. *Zh. Org. Khim.* **1990,** *26,* 69 (Engl. Transl. 56); Tselinskii, I. V.; Melnikov, A. A.; Trubitsin, A. E. *Zh. Org. Khim.* **1987,** *23,* 1657 (Engl. Transl. 1485).
93. Cantrell, G. L.; Filler, R. *J. Fluorine Chem.* **1985,** *27,* 35.
94. Garrett, G. S.; Emge, T. J.; Lee, S. C.; Fischer, E. M.; Dyehouse, K.; McIver, J. M. *J. Org. Chem.* **1991,** *56,* 4823.
95. Tsushima, T.; Kawada, K.; Tsuji, T. *Tetrahedron Lett.* **1982,** *23,* 1165.
96. Shieh, T. C.; Feit, E. D.; Chernick, C. L.; Yang, N. C. *J. Org. Chem.* **1970,** *35,* 4020.
97. Anand, S. P.; Quarterman, L. A.; Hyman, H. H.; Migliorese, K. G.; Filler, R. *J. Org. Chem.* **1975,** *40,* 807.
98. Anand, S. P.; Filler, R. *J. Fluorine Chem.* **1976,** *7,* 179.
99. Markat, R. K.; Janzen, A. F. *Can. J. Chem.* **1977,** *55,* 3031; Janzen, A. F.; Wang, P. M. C.; Lemire, A. E. *J. Fluorine Chem.* **1983,** *22,* 557.
100. Umemoto, T.; Tomizawa, G. *Bull. Chem. Soc. Jpn.* **1986,** *59,* 3625.
101. Huang, X.; Blackburn, B. J.; Janzen, A. F. *J. Fluorine Chem.* **1990,** *47,* 145.
102. Zupan, M.; Zajc, B. *J. Chem. Soc., Perkin Trans. 1* **1978,** 965.
103. Nakanishi, S.; Jensen, E. V. *Chem. Pharm. Bull.* **1977,** *25,* 3395.
104. Peet, N. P.; McCarthy, J. R.; Sunder, S.; McCowan, J. *Synth. Commun.* **1986,** *16,* 1551.
105. Seitz, D. E.; Blaszczak, L. C. *Synth. Commun.* **1988,** *18,* 2353.
106. Dersch, R.; Reichardt, C. *Liebigs Ann. Chem.* **1982,** 1330.

107. Dersch, R.; Reichardt, C. *Liebigs Ann. Chem.* **1982,** 1348.
108. Blackburn, G. M.; England, D. A.; Kolkmann, F. *J. Chem. Soc., Chem. Commun.* **1981,** 930.
109. McKenna, C. E.; Shen, P. D. *J. Org. Chem.* **1981,** *46,* 4573; Xu, Z.-Q.; DesMarteau, D. D. *J. Chem. Soc. Perkin Trans. 1* **1992,** 313.
110. Hall, C. R.; Inch, T. D.; Williams, N. E. *J. Chem. Soc., Perkin Trans. 1* **1985,** 233.
111. Posner, G. H.; Frye, L. L. *J. Fluorine Chem.* **1985,** *28,* 151.
112. Takeuchi, Y.; Nigata, K.; Koizumi, T. *J. Org. Chem.* **1987,** *52,* 5061; Takeuchi, Y.; Ogura, H.; Kamada, A. *J. Org. Chem.* **1992,** *57,* 2196.
113. Witucki, E. F.; Rowley, G. L.; Warner, M.; Frankel, M. B. *J. Org. Chem.* **1972,** *37,* 152.
114. Tius, M. A.; Kawakami, J. K. *Tetrahedron* **1995,** *51,* 3997.

Replacement of Halogens by Fluorine

by Donald F. Halpern and Gerald G. Vernice

The ease of replacement of halogen by fluorine is governed by the strength of the halogen–substrate bond. The conversion of sulfonyl or acyl halides to their respective fluorides can be accomplished with any number of reagents. Halogen bonded to a carbon adjacent to a trifluoromethyl group of an alkane will resist displacement by most fluorinating reagents. Between these two extremes the number of classes of organic compounds containing halogens that can be replaced by fluorine or fluoride ion is enormous. Only recently has the number of fluorinating reagents grown dramatically. Traditional reagents like mercuric fluoride or silver fluoride are rarely used now, but the popularity of hydrogen fluoride and hydrogen fluoride–antimony pentachloride remains high. The use of potassium fluoride has grown considerably with the development of anhydrous potassium fluoride, spray-dried potassium fluoride, potassium fluoride with protic and/or aprotic solvents, potassium fluoride with phase-transfer catalysts, potassium fluoride with crown ethers, and potassium fluoride supported on calcium fluoride. Examples of other fluorinating agents now being used are the interhalogens, quaternary ammonium and phosphonium fluorides, and even ammonium fluoride. The incorporation of ^{18}F into pharmacologically active molecules has a new important practical application, positron emission tomography (PET), which is reviewed in Chapter 8.

Replacement by Means of Elemental Fluorine

Although **elemental fluorine** can displace halogen, it usually *reacts preferentially with the hydrogen atoms* of a molecule with statistical selectivity. In one example of a successful selective replacement of a halogen by fluorine, 1-bromo- or 1-iodoadamantane and fluorine at –70 °C produces 1-fluoroadamantane in 90% yield. 2-Fluoroadamantane can be generated from either 2-bromo- or 2-iodo-adamantane with yields in the 80% range. The best yields are obtained when either fluorotrichloromethane or fluorotrichloromethane–chloroform is used as the solvent [*1*]. It should be noted that chloroadamantanes are not included in this study. This is not surprising because chloro groups are often used to block or protect a specific site during the direct fluorination of a molecule. As a generalization, *the replacement of halogen by means of elemental fluorine is impractical;* reaction yields are not very good, isolation of a pure compound from the mixture obtained is tedious, and replacement can often be effected more conveniently and certainly more selectively by other reagents that are easier to handle [*2*].

0065–7719/95/0187–0172$09.26/1

Primary alkyl chlorides are fairly stable to fluorine displacement. When fluorinated, 1-chloropropane is converted to 1-chloroheptafluoropropane and 1-chloro-2-methylbutane produces 39% 1-chlorononafluoro-2-methylbutane and 19% perfluoro-2-methylbutane. ***Secondary and tertiary alkyl chlorides*** can undergo 1,2-chlorine shifts to afford perfluorinated primary alkyl chlorides. 2-Chloro-2-methylpropane gives 1-chlorononafluoro-2-methylpropane, and three products are obtained by the fluorination of 3-chloropentane [*3*] (equation 1). Aerosol fluorination of dichloromethane produces dichlorodifluoromethane which is isolated in 98% purity [*4*] (equation 2). If the molecule contains only carbon and halogens, the picture is different. Molecular beam analysis has shown that the reaction of fluorine with carbon tetrachloride, iodotrichloromethane, or bromotrichloromethane proceeds first by abstraction of halogen to form a trichloromethyl radical [*5*].

[*3*] **1**

$$CH_3CH_2CH_2Cl \xrightarrow[\text{-40 to -30°C}]{2.4\%F_2,He} CF_3CF_2CClF_2 \quad 63\%$$

$$(CH_3)_3CCl \xrightarrow[\text{-40 to -30°C}]{4.7\%\ F_2,He} (CF_3)_2CFCClF_2 \quad 47\%$$

$$(CH_3CH_2)_2CHCl \xrightarrow[\text{-45 to -30°C}]{4.5\%F_2,He}$$

$$\underset{15\%}{(CF_3CH_2)_2CClF} + \underset{23\%}{CF_3(CF_2)_3CClF_2} + \underset{7\%}{CF_3(CF_2)_2CClCF_3}$$

[*4*] **2**

$$CH_2Cl_2 \xrightarrow[\text{-40 to -30°C}]{2.2\%\ F_2,\ He} CCl_2F_2 \quad 66\%$$

Replacement by Means of Interhalogens

Warning: *Depending on the ratio of hydrogen to halogen in a given molecule, addition of an interhalogen to that substance in the absence of solvent can result in spontaneous ignition or worse.*

Because of the extreme reactivity of chlorine trifluoride and the relatively mild nature of iodine pentafluoride, bromine trifluoride seems to be the best reactant in this group. The decision to use an interhalogen as an agent to displace a halogen is complicated by the ability of the interhalogen to add across double bonds and replace hydrogen atoms in the molecule with fluorine. Interhalogens can displace halogen very selectively in a wide variety of molecules and functional groups. This potential for selectivity might make this underutilized reagent more attractive to the synthetic chemist of the future. An example of an exchange with a halogen bonded to an element other than carbon is the conversion of ***polyfluoroalkyl***

chlorosulfites to their respective fluorosulfites with **chlorine monofluoride** [*6*] (equation 3).

[*6*] **3**

$$\mathrm{ROS(=O)Cl + ClF \xrightarrow{-196^{\circ}C\ to\ RT} ROS(=O)F} \quad \sim 75\%$$

$$\mathrm{R = CF_3CH_2,\ (CF_3)_2CH,\ CH_3(CF_3)_2C}$$

Conversion of 2-chloro-2-difluoromethoxy-1,1,1-trifluoroethane to 2-difluoromethoxy-1,1,1,2-tetrafluoroethane (98% purity) is accomplished with **bromine trifluoride.** The starting material is the major impurity [7] (equation 4).

[*7*] **4**

$$\mathrm{CF_3CHClOCHF_2 \xrightarrow[20\,-\,40^{\circ}C]{BrF_3} CF_3CHFOCHF_2} \quad 87\%$$

At times an induction period is observed before bromine trifluoride begins to react with the substrate. Adding bromine to the substrate often shortens this induction period. Treatment of 2-methoxy-1,1,1,3,3,3-hexachloropropane or 2-chloromethoxy-1,1,1,3,3,3-hexachloropropane with bromine trifluoride gives a variety of products and presents a very interesting picture of the selectivity of bromine trifluoride with and without internally generated hydrogen fluoride [*8*] (equations 5 and 6).

[*8*] **5**

$$\mathrm{CH_3OCH(CCl_3)_2 \xrightarrow[20\,-50^{\circ}C]{BrF_3}}$$

$$\underset{3\%}{\mathrm{CH_2FOCH(CF_3)_2}} + \underset{59\%}{\mathrm{CH_2FOCH(CF_3)CClF_2}} + \underset{29\%}{\mathrm{CH_2FOCH(CClF_2)_2}}$$

[*8*] **6**

$$\mathrm{CH_2ClOCH(CCl_3)_2 \xrightarrow[<50^{\circ}C]{BrF_3}} \underset{35.5\%}{\mathrm{CH_2FOCH(CF_3)_2}} + \underset{43.7\%}{\mathrm{CH_2FOCH(CF_3)CClF_2}}$$

With bromine trifluoride and 1% tin tetrachloride as a catalyst, 1,3-dichloro-2-fluoropropane gives 1,2,3-trifluoropropane. Without the catalyst, 1-chloro-2,3-difluoropropane provides a four-component product mixture: 1,2,3-trifluoropropane, 3-chloro-1,1,2-trifluoropropane, 1-chloro-2,2,3-trifluoropropane, and 1,1,2,3-tetrafluoropropane [*9*] (equations 7 and 8).

When 1,1,1,3-tetrachloropropane is combined with 1% tin tetrachloride and bromine trifluoride, a mixture of 1-fluoro-1,1,3-trichloropropane, 1,3-dichloro-1,1-difluoropropane, and 1,1-dichloro-1,3-difluoropropane is obtained [*9*] (equation 9).

[9] $(CH_2Cl)_2CHF \xrightarrow[\text{Freon 113, RT}]{BrF_3\text{,1\% }SnCl_4} CH_2FCHFCH_2F$ 43% 7

[9] $CH_2ClCHFCHF_2 \xrightarrow[\text{Freon 113, RT}]{BrF_3} \underset{29\%}{CH_2FCHFCH_2F}$ 8

$+ \underset{35\%}{CHF_2CHFCH_2Cl} + \underset{6\%}{CH_2FCF_2CCH_2Cl} + \underset{29\%}{CHF_2CHFCH_2F}$

[9] $CCl_3CH_2CH_2Cl \xrightarrow[\text{Freon 113, RT}]{BrF_3\text{,1\% }SnCl_4} \underset{51\%}{CCl_2FCH_2CH_2Cl}$ 9

$+ \underset{11\%}{CClF_2CH_2CH_2Cl} + \underset{10\%}{CCl_2FCH_2CH_2F}$

With Freon 112 or 113 as a solvent, fluorination of primary butyl halides with bromine trifluoride can give mixtures of primary and secondary fluorides. When 1,4-dibromobutane is the substrate, 93% 1-bromo-4-fluorobutane and 7% 1-bromo-3-fluorobutane is obtained; with 1,4-dichlorobutane, the product contains 65% 1-chloro-3-fluorobutane and 35% 1-chloro-4-fluorobutane. When 4-bromo- or 4-chlorobutyl trifluoroacetate is used, the ratio of 4-fluorobutyl trifluoroacetate to 3-fluorobutyl trifluoroacetate is 1:4. The effect of solvent is measured in another set of experiments. When the reaction of bromine trifluoride and 1,3-dichloro-2-fluoropropane in either Freon 113 or hydrogen fluoride is allowed to proceed to 40% conversion, the product mixture has the composition shown in Table 1 [*10*].

When 1-chloro-2,3-dibromopropane is combined with one-third of a mole of bromine trifluoride, both 1-bromo-3-chloro-2-fluoropropane and 1-chloro-2,3-difluoropropane are formed [*11*] (equation 10).

[11] 10

$CH_2ClCHBrCH_2Br \xrightarrow[\text{F-113, 20°C}]{{}^1\!/_3 BrF_3} \underset{60\%}{CH_2ClCHFCH_2Br} + \underset{20\text{-}25\%}{CH_2ClCHFCH_2F}$

$CH_2ClCHFCH_2Br \xrightarrow{{}^1\!/_3 BrF_3} CH_2ClCHFCH_2F$

$CH_2ClCHBrCH_2Br \xrightarrow[\text{Freon-113, 20°C}]{{}^2\!/_3 BrF_3} \underset{80\text{-}85\%}{CH_2ClCHFCH_2F}$

The formation of *fluoroalkanes* by the displacement of the halogen by fluorine using **chlorine monofluoride** was studied in depth. Halogens vicinal to the leaving halogen slow the reaction sharply, in contrast to geminal halogens, which have little

Table 1. Effect of Solvent on Displacement of Halogen to 40% Conversion by Bromine Trifluoride [10]

Compound	*Solvent*	*$CH_2ClCHFCH_2F$ (%)*	*$CH_2ClCF_2CH_3$ (%)*
$CH_2ClCHFCH_2Cl$	Freon 113	78.6	2.4
	HF	16.3	31.7[a]
$CH_2ClCHFCH_2Br$	Freon 113	42	2.8
	HF	15.3	29.7[a]

[a] CHF_2CHFCH_2Cl (20–25%)is also isolated.

effect on the rate. Fluorinations of primary alkyl bromides containing vicinal bromo or chloro groups usually occur with rearrangement. Hydride shifts take place if stable tertiary or secondary carbocations can be formed, as illustrated by the fluorination of 1-bromo-2-fluoropropane with chlorine monofluoride to give 2,2-difluoropropane [*12*]. Chlorine monofluoride in hydrogen fluoride selectively substitutes fluorine for chlorine in a variety of chloroalkanes. At –30 to 0 °C, 1,2,3-trichloropropane and one equivalent of chlorine monofluoride gives 1,3-dichloro-2-fluoropropane; a second equivalent forms the rearranged product, 1-chloro-2,2-difluoropropane, and a third equivalent does not react at all with 1-chloro-2,2-difluoropropane (equation 11).

[*13*] **11**

$$CH_2ClCHClCH_2Cl \xrightarrow[-30° - 0°C]{ClF} \underset{88\%}{CH_2ClCHFCH_2Cl} \xrightarrow[-30° - 0°C]{ClF} \underset{81\%}{CH_2ClCF_2CH_3}$$

In contrast, the addition of antimony pentachloride increases the reactivity of chlorine monofluoride toward 1,2,3-trichloropropane but reduces its selectivity [*13*] (equation 12).

[*13*] **12**

$$CH_2ClCHClCH_2Cl \xrightarrow[-60 \text{ to } -20°C]{ClF,\ 3\text{-}5\%\ SbCl_5} \underset{68\%}{CH_2ClCHFCH_2Cl}$$

$$+ \underset{6\%}{CH_2ClCHFCH_2F} + \underset{8\%}{CH_2ClCH_2CHF_2} + \underset{10\%}{CH_2FCH_2CHF_2}$$

With successive equivalents of chlorine monofluoride 1,1,1,2-tetrachloroethane gives 1-fluoro-1,1,2-trichloroethane, 1,2-dichloro-1,1-difluoroethane, 2-chloro-1,1,1-trifluoroethane with some 1-chloro-1,1,2-trifluoroethane, and finally 1,1,1,2-tetrafluoroethane (equation 13).

[*13*] **13**

$$CCl_3CH_2Cl \xrightarrow[-60° \text{ to } -20°C]{ClF} \underset{90\%}{CCl_2FCH_2Cl} \xrightarrow[-60° \text{ to } -20°C]{ClF} \underset{86\%}{CClF_2CH_2Cl}$$

$$\xrightarrow[-60° \text{ to } -20°C]{ClF} \underset{15\text{-}20\%}{CClF_2CH_2F} + \underset{54\%}{CF_3CH_2Cl} \xrightarrow[-60° \text{ to } -20°C]{ClF} \underset{22\%}{CF_3CH_2F}$$

Once again, the addition of antimony pentachloride increases the reactivity of chlorine monofluoride toward alkanes but reduces its selectivity and thus allows more than one product to be generated [*13*] .

The *effect of the substituent group* can be seen in the reactions of the following series of halogenated esters. Methyl 2,3-dichloro-3-fluoro-2-methylpropionate and chlorine monofluoride first form methyl 2-chloro-3,3-difluoro-2-methylpropionate. An additional equivalent of chlorine monofluoride produces methyl 2-methyl-2,3,3-trifluoropropionate (equation 14).

[*13*] **14**

$$CHClFC(Cl)(CH_3)CO_2CH_3 \xrightarrow[\text{Anhyd. HF, } -60° \text{ to } 0°C]{ClF} \underset{83\%}{CHF_2C(Cl)(CH_3)CO_2CH_3} \xrightarrow[\text{Anhyd. HF, } -60° \text{ to } 0°C]{ClF} \underset{76\%}{CHF_2C(F)(CH_3)CO_2CH_3}$$

However, methyl 2,2,3-trichloropropionate and chlorine monofluoride form methyl 2,3-dichloro-2-fluoropropionate. An additional equivalent of chlorine monofluoride selectively generates methyl 3-chloro-2,2-difluoropropionate [*13*] (equation 15).

[*13*] **15**

$$CH_2ClCCl_2COOCH_3 \xrightarrow[\text{Anhyd. HF, } -60° \text{ to } 0°C]{ClF} CH_2ClCClFCOOCH_3 \ (62\%) \xrightarrow[\text{Anhyd. HF, } -60° \text{ to } 0°C]{ClF} \underset{72\%}{CH_2ClCF_2COOCH_3}$$

Chlorine monofluoride can be tamed by reacting it with cyanuric fluoride to form the trimer $(CF_2NCl)_3$. The trimer reacts with 1,1-dichloro-2,2,2-trifluoroethanesulfenyl chloride to form a mixture of 1-chloro-1,2,2,2-tetrafluoroethanesulfenyl chloride and perfluoroethanesulfenyl chloride. In the same way, trifluoroacetyl bromide yields trifluoroacetyl fluoride[14] (equation 16).

[*14*] **16**

$$(FCN)_3 \xrightarrow{\text{excess ClF}} (CF_2NCl)_3$$

$$(CF_2NCl)_3 + CF_3COBr \xrightarrow[72\ h]{20°C} CF_3COBr\ (37\%) + CF_3COF\ (55\%)$$

Replacement by Means of Quaternary Nitrogen and Phosphorus Fluorides

Reactions using ammonium fluoride proceed only with substrates requiring the mildest of conditions. An 80% conversion of trichloromethanesulfenyl chloride to dichlorofluoromethanesulfenyl chloride is obtained with ammonium fluoride in a 1:2 molar ratio at 150 °C, with the isolated product comprising 75% dichlorofluoromethanesulfenyl chloride and 25% chlorodifluoromethanesulfenyl chloride [*15*]. In contrast, with chromium trifluoride in a 1:0.5 molar ratio at 150 °C, the material isolated is 52% dichlorofluoromethanesulfenyl chloride, 9% chlorodifluoromethanesulfenyl chloride, and 39% trifluoromethanesulfenyl chloride [*16*]. The fluorination of carbon tetrachloride can be accomplished with **ammonium hydrogen fluoride** and a variety of metal fluoride catalysts. A 1.7:1 molar ratio of ammonium hydrogen fluoride to carbon tetrachloride, passing over calcium fluoride at 390 to 465 °C, gives 76% fluorotrichloromethane [*17*]. Ammonium hydrogen fluoride and hydrogen fluoride react with *m*-nitrobenzotrichloride at 27 to 46 °C to form the dichlorofluoro and chlorodifluoro products. At 80 to 85 °C all three substitution products are observed [*18*] (equation 17).

[*18*] **17**

$$m\text{-}CCl_3C_6H_4NO_2 \xrightarrow[HF]{NH_4HF_2} \text{products}$$

	27-46°C	~85°C
m-$CCl_2FC_6H_4NO_2$	61%	4%
+ m-$CClF_2C_6H_4NO_2$	36%	91%
+ m-$CF_3C_6H_4NO_2$		4%

Chloromethoxy-1,2,2,2-tetrafluoroethane is converted by **triethylammonium fluoride** or **piperidine hydrogen fluoride** to fluoromethoxy-1,2,2,2-tetrafluoroethane in greater than 60% yield. Piperidine hydrogen fluoride or **butylammonium fluoride** can convert methoxy-1-chloro-2,2,2-trifluoroethane to methoxy-1,2,2,2-tetrafluoroethane in 80% yield [*19*].

Early workers using "anhydrous" **tetrabutylammonium fluoride** (TBAF) may have actually used mixtures of TBAF and tetrabutylammonium bifluoride. It is known that excessive drying at 77 °C in a vacuum below 2 torr (1 torr = 133.322 Pa) produces the bifluoride salt [*20*]. TBAF can be obtained by drying commercially available TBAF trihydrate at 40 °C below 2 torr (Procedure 1, p 192). "Anhydrous" TBAF formed by this procedure still contains between 0.1 to 0.3 equivalent of water and should be used immediately. It is not only a potent source of nucleophilic fluoride but also a strong base. An evaluation of the nucleophilicity of selected fluorinating systems provides the following qualitative order [*21*]: *anhydrous TBAF > CsF·dibenzo-24-crown-8 ≈ CsF·18-crown-6 > KF·18-crown-6.* Allyl bromide is converted to *allyl fluoride* in 85% yield with TBAF. *Benzyl fluoride* is isolated in 66% yield from benzyl bromide and TBAF [*22*]. On the other hand, the reaction of TBAF with 2-bromooctane favors elimination. Hydrolysis of the leaving group is also a significant side reaction due to the moisture present in "anhydrous" TBAF. Bromophenyldiazirine is converted to fluorophenyldiazirine in 4 h with a fourfold excess of "anhydrous" TBAF, in contrast to its preparation from chlorophenyldiazirine, which requires 16 h. Fluorophenoxydiazirine can also be prepared from its chloro precursor [23]. Fluorotrifluoromethyldiazirine is synthesized in 50% yield from 3-bromo-3-trifluoromethyldiazirine and TBAF [*24*]. The preparation of the diazirines is important, because they are easily converted to their respective carbenes and then trapped with alkenes to give cyclopropanes. TBAF is also used as a phase-transfer catalyst with cesium fluoride.

In contrast to the nature of TBAF, **tetrabutylammonium bifluoride** converts benzyl bromide to its fluoride in 100% yield and 4-chloronitrobenzene to 4-fluoronitrobenzene in 70% yield. 1-Bromodecane is transformed by tetrabutylammonium bifluoride to 1-fluorodecane in 88% yield, and 1-chlorododecane forms 1-fluorododecane in 83% yield. In neither case are significant amounts of the elimination products formed [*25*].

Both **methyltriethylphosphonium fluoride** and **methyltributylphosphonium fluoride** have been prepared. The latter generates benzyl fluoride from benzyl chloride in 80% yield and ethyl fluoroacetate from ethyl bromoacetate in 53% yield. Methyltributylphosphonium fluoride converts 1-bromododecane to a 50:50 mixture of 1-fluorododecane and 1-dodecene. Methyltributylphosphonium fluoride also quantitatively forms styrene from 1-bromo-1-phenylethane [*26*]. Methyltributylphosphonium fluoride is the reagent of choice for the conversion of *N,N*-dimethylchloroacetamide to its fluoride, but it is not able to convert chloroacetonitrile to fluoroacetonitrile. Methyltributylphosphonium fluoride changes chloromethyl octyl ether to the crude fluoromethyl ether in 66% yield. The *stereoselectivity* of methyltributylphosphonium fluoride is illustrated by the reactions of the 2-*tert*-butyl-3-chlorooxiranes [*27*] (Table 2).

The reactivity of **tetraphenylphosphonium hydrogen difluoride** is comparable with more conventional chlorine–fluorine exchange reagents. It can be used and regenerated as illustrated in equation 18 and its preparation is outlined in Procedure 3, p 193. The rate of reaction of tetraphenylphosphonium hydrogen difluoride with benzyl bromide using a fluoride concentration of 0.12 M in acetonitrile ($t_{1/2}$ = 25 min at 52 °C) can be compared with that achieved by replacing it with potassium fluoride-18-crown-6 ($t_{1/2}$ = 11.5 h at 82 °C) [*28*]. The

Table 2. Reaction of Tributylmethylphosphonium Fluoride and 2-*tert*-Butyl-3-Chlorooxirane [27]

Oxirane Reactant, Amount (%)	*Solvent*[a]	*Fluorooxirane, Amount (%)*	*Unreacted Chlorooxirane, Amount (%)*
E, 85	Et_2O	*E*, 60	*E*, 28
Z, 15			*Z*, 12
E, 30	Et_2O	*E*, 47	*Z*, 53
Z, 70	THF	*E*, 65	*Z*, 35
	Hexane	*E*, 45	*Z*, 55
Z, 100	THF	*E*, 52	*Z*, 48

[a]THF, tetrahydrofuran.

[29]

$$(C_6H_5)_4P^+ X^- \xrightarrow[H_2O]{HF_2^-} (C_6H_5)_4P^+ HF_2^- (H_2O)_x \qquad \mathbf{18}$$

$$(C_6H_5)_4P^+ HF_2^- (H_2O)_x \xrightarrow{-H_2O} (C_6H_5)_4P^+ HF_2^-$$

$$(C_6H_5)_4P^+ HF_2^- + RX \longrightarrow (C_6H_5)_4P^+ X^- + RF$$

stoichiometry is less than straightforward. Quantitative fluorination of an organic substrate often requires 2 equivalents of tetraphenylphosphonium hydrogen difluoride because of the formation of the unreactive tetraphenylphosphonium dihydrogen trifluoride. With tetraphenylphosphonium hydrogen difluoride, benzyl bromide in acetonitrile is completely converted to benzyl fluoride in 2.5 h at 52 °C. 1-Fluorodecane and a small amount of alkene are obtained at 130 °C from 1-bromodecane and tetraphenylphosphonium dihydrogen trifluoride. 1-Chloro-2,4- dinitrobenzene in acetonitrile with tetraphenylphosphonium hydrogen difluoride at 80 °C results in a 100% yield of 1-fluoro-2,4-dinitrobenzene. On the other hand, tetraphenylphosphonium hydrogen difluoride in dimethyl sulfoxide displaces the nitro group of 2-chloro-6-nitrobenzonitrile to give a 100% yield of 2-chloro-6-fluorobenzonitrile [29] .

Tetrabutylphosphonium fluoride, tetrabutylphosphonium hydrogen difluoride and *tetrabutylphosphonium dihydrogen trifluoride* can be prepared from tetrabutylphosphonium hydroxide and hydrogen fluoride (equation 19).
These quaternary phosphonium fluorides provide a range of activities for the displacement of halogen by fluorine [30] (Table 3).

[*30*] **19**

$$(C_4H_9)_4POH + 2HF \xrightarrow{-H_2O} (C_4H_9)_4PHF_2 \quad 100\%$$

$$(C_4H_9)_4POH + 3HF \xrightarrow{-H_2O} (C_4H_9)_4PH_2F_3 \quad 100\%$$

$$(C_4H_9)_4PHF_2 \xrightarrow{C_4H_9Li} (C_4H_9)_4PF + LiF$$

Table 3. Halogen Displacement Using $(C_4H_9)_4P^+(HF)_nF^-$ (*n* = 0, 1, 2) [*30*]

		Reaction		*Product Yield (%)*	
Reagent	*Substrate*	*Time (h)*	*Temperature (°C)*	*Fluoride*	*Alkene*
$(C_4H_9)_4PF$	$CH_3C_6H_4CH_2Br$	0.25	Ambient	92	
$(C_4H_9)_4PHF_2$	$CH_3C_6H_4CH_2Br$	4	Ambient	99	
	$C_{14}H_{29}Cl$	42	Reflux	72	6
	$C_{14}H_{29}Br$	20	Reflux	74	15
	$C_{12}H_{25}CHBrCH_3$	10	Reflux	23	57
$(C_4H_9)_4PH_2F_3$	$CH_3C_6H_4CH_2Br$	28	60	93	
	2-Cl-5-NO_2-C_5H_3N	2	110	95	
	2-Cl-5-NO_2-C_5H_3N	11	110	40	

Polymer-supported **tetraphenylphosphonium bromide** is a recyclable catalyst for halogen-exchange reactions. The reaction of 1 equivalent of chloro-2,4-dinitrobenzene with 1.5 equivalents of spray-dried potassium fluoride and 0.1 equivalent of this catalyst in acetonitrile at 80 °C for 12 h gives 2,4-dinitrofluorobenzene in 98% yield. An 11% yield is obtained without the catalyst [*31*].

Replacement by Means of Anhydrous Hydrogen Fluoride

Anhydrous hydrogen fluoride, alone or with a catalyst, is unquestionably the primary industrial source of fluoride for the replacement of a number of leaving groups, including halogen. A review of earlier work has shown the ease of replacement of halogen when bonded to silicon, phosphorus, or carbon atoms with reduced electron densities (i.e., in allylic, carbonyl, or benzylic groups) [*2,* p 95ff]. A one-step synthesis of *trifluoromethyl aryl ethers* provides a route to this useful class of compounds from a variety of phenols, carbon tetrachloride, and hydrogen fluoride at 100–150 °C and autogenous pressure. Best results are obtained with phenols containing ring-deactivating groups such as nitro, chloro, or trifluoromethyl groups or protonated amines. The only limitation is that phenols having *ortho* substituents that can hydrogen bond to the phenol group fail to react. If bromotrichloromethane or fluorotrichloromethane is used in place of carbon tetrachloride, the yields are substantially reduced [*32*] (equation 20).

[*32*] (Substituted phenol, R–C_6H_4–OH) $\xrightarrow[150°C,\ 8h]{CCl_4,\ HF}$ (R–C_6H_4–OCF_3) **20**

R =	3-NO_2	4-NO_2	4-Cl	3-CF_3	3-NH_2	4-NH_2	2,4-Cl_2	2-F
% Yield =	69	56	70	60	26	42	73	35

If the aromatic ring does not contain an electron-withdrawing group, a trifluoromethyl group can be added to a variety of aromatic substances by using carbon tetrachloride and hydrogen fluoride. The newly added trifluoromethyl group favors a position *para* to the strongest *ortho–para* director on the ring, with isomer ratios in the order of 1:1 to 3:1 and conversions greater than 50%. Neither polysubstitution nor incomplete fluorination of the trihalomethyl group is observed [*33*] (equation 21).

[*33*] R^1–C_6H_4–R^2 $\xrightarrow[CCl_4]{HF}$ R^1–$C_6H_3(R^2)$–CF_3 + R^2–$C_6H_3(R^1)$–CF_3 **21**

R^1 = H, Br, Cl, F, CH_3, OCF_3,
R^2 = H, Cl, F, CH_3

Side chain *fluorinated aryl isocyanates* are an important group of substances used in the manufacture of biologically active compounds. The synthetic routes to the *meta*- and *para*-trifluoromethylphenyl isocyanates permit the chlorine–fluorine exchange with hydrogen fluoride to proceed under moderate conditions and in a virtually quantitative manner (equation 22).

[*34*] CH_3–C_6H_4–NCO $\xrightarrow{Cl_2}$ CCl_3–C_6H_4–NCO $\xrightarrow{HF}$ CF_3–C_6H_4–NCO **22**

These routes are of no use for the synthesis of *ortho*-trifluoromethylphenyl isocyanate, because the *ortho* trichloro precursor reacts with hydrogen fluoride and then rearranges to give *N*-trifluoromethylanthranilic acid fluoride [*34*]. This *rearrangement* has an aliphatic analogue. Perchlorination of methyl isocyanate gives chlorocarbonyl isocyanide dichloride, which reacts with hydrogen fluoride to displace either one, two, or all three chlorine atoms. Displacement of three chlorine atoms forms trifluoromethyl isocyanate. When sodium fluoride is combined with chlorocarbonyl isocyanide dichloride, fluorocarbonyl isocyanide dichloride is isolated. The yields are listed in the references of the review article. It should be noted that fluorocarbonyl isocyanide dichloride and dichlorofluoromethyl isocyanate are not in a state of equilibrium and each has distinct physical properties [*34*] (equation 23).

[*34*] 23

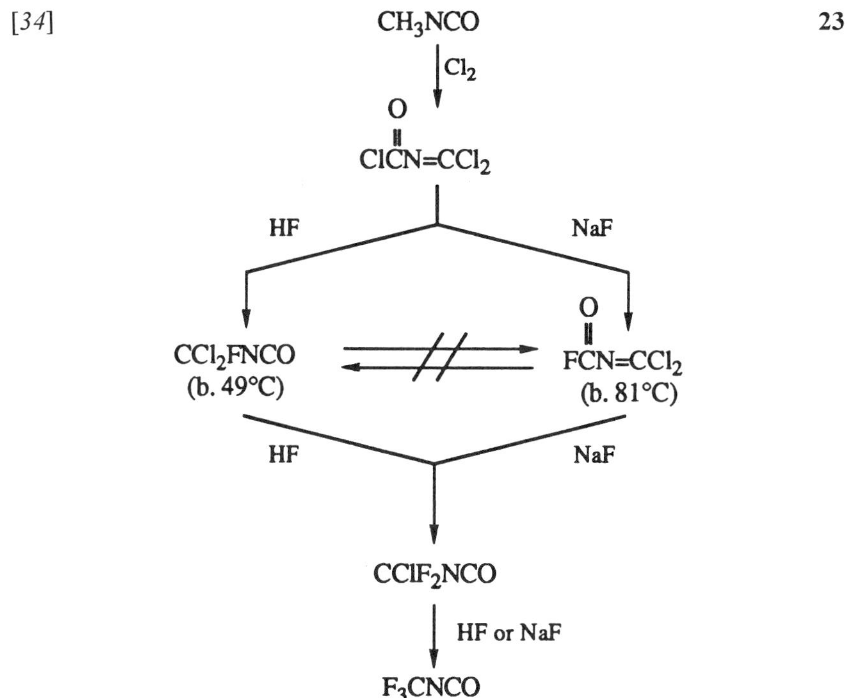

The addition of hydrogen fluoride to an aromatic isocyanide dichloride provides a route to the respective aryl-*N*-trifluoromethyl amine [*34*] (equation 24).

[*34*] 24

$$C_6H_5N{=}CCl_2 \xrightarrow[-15°C\ to\ RT]{HF,\ Et_2O} C_6H_5NHCF_3 \quad (68.5\%)$$

References are listed on pages 196–198.

Another example of a simultaneous chlorine–fluorine exchange and rearrangement is the reaction of *ortho*-methylphenylchloroformate with hydrogen fluoride to give a stable tetrafluorobenzodioxin [*34*] (equation 25).

[*34*] 25

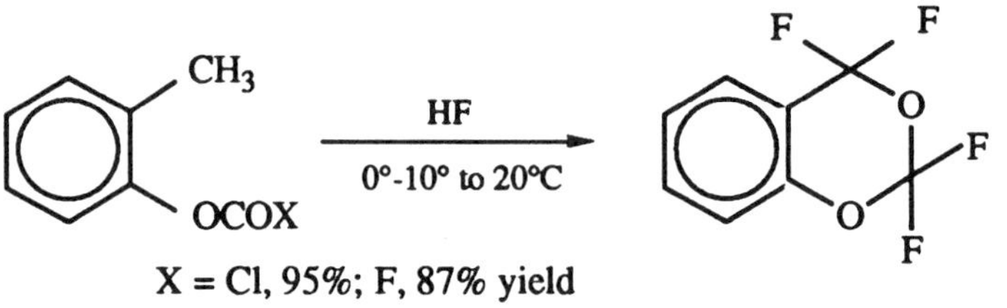

CH_3 / OCOX —HF, 0°-10° to 20°C→ (tetrafluorobenzodioxin: F, F, O, F, O, F)

X = Cl, 95%; F, 87% yield

A related reaction allows 1,1,3,3-tetrafluoro-1,3-dihydroisobenzofuran to be prepared from 2-trichloromethylbenzoyl chloride and anhydrous hydrogen fluoride [*35*] (equation 26).

[*35*] 26

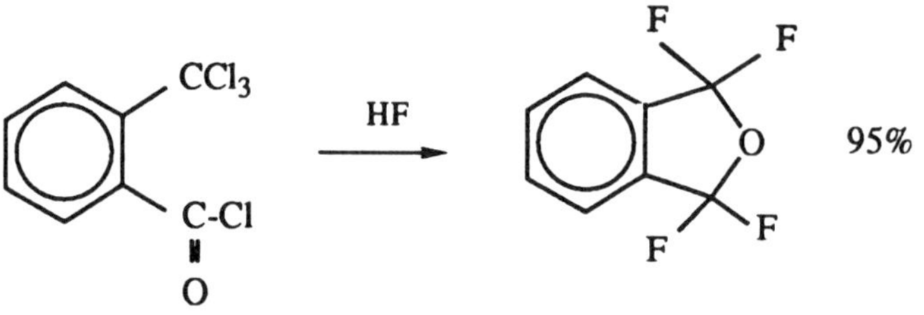

CCl_3 / C-Cl ‖ O —HF→ (F, F, O, F, F) 95%

In contrast, 1,1,3,3-tetrachloro-1,3-dihydroisobenzofuran and anhydrous hydrogen fluoride form 2-trifluoromethylbenzoyl fluoride [*36*] (equation 27).

[*36*] 27

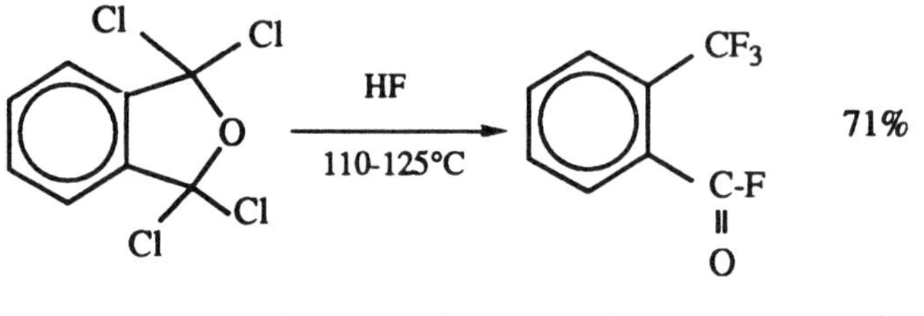

(Cl, Cl, O, Cl, Cl) —HF, 110-125°C→ CF_3 / C-F ‖ O 71%

The combination of a hydrogen fluoride addition and a chlorine–fluorine displacement converts 1,1,2,2-tetrachloroethylene-1,2-di(isocyanide dichloride) to perfluoro-*N,N*-dimethylethylenediamine [*37*] (equation 28).

[*37*] 28

$$(Cl_2C{=}NCCl_2)_2 \xrightarrow[110°C,\ N_2,\ 12\ bar]{HF} CF_3NHCF_2CF_2NHCF_3 \quad 80\%$$

Anhydrous hydrogen fluoride, potassium fluoride, and sodium fluoride have all been used to promote the chlorine–fluorine exchange in *tetrachloropyridazine.* All three fluorine sources replace the chlorine in position 3 first. Both sodium and potassium fluoride produce a mixture of isomers when the second chlorine–fluorine exchange takes place. In contrast, hydrogen fluoride selectively replaces the chlorine at position 6 with fluorine. The third chlorine atom is exchanged by hydrogen fluoride to form exclusively 4-chloro-3,5,6-trifluoropyridazine [*38*] (equation 29; Table 4).

[*38*] **29**

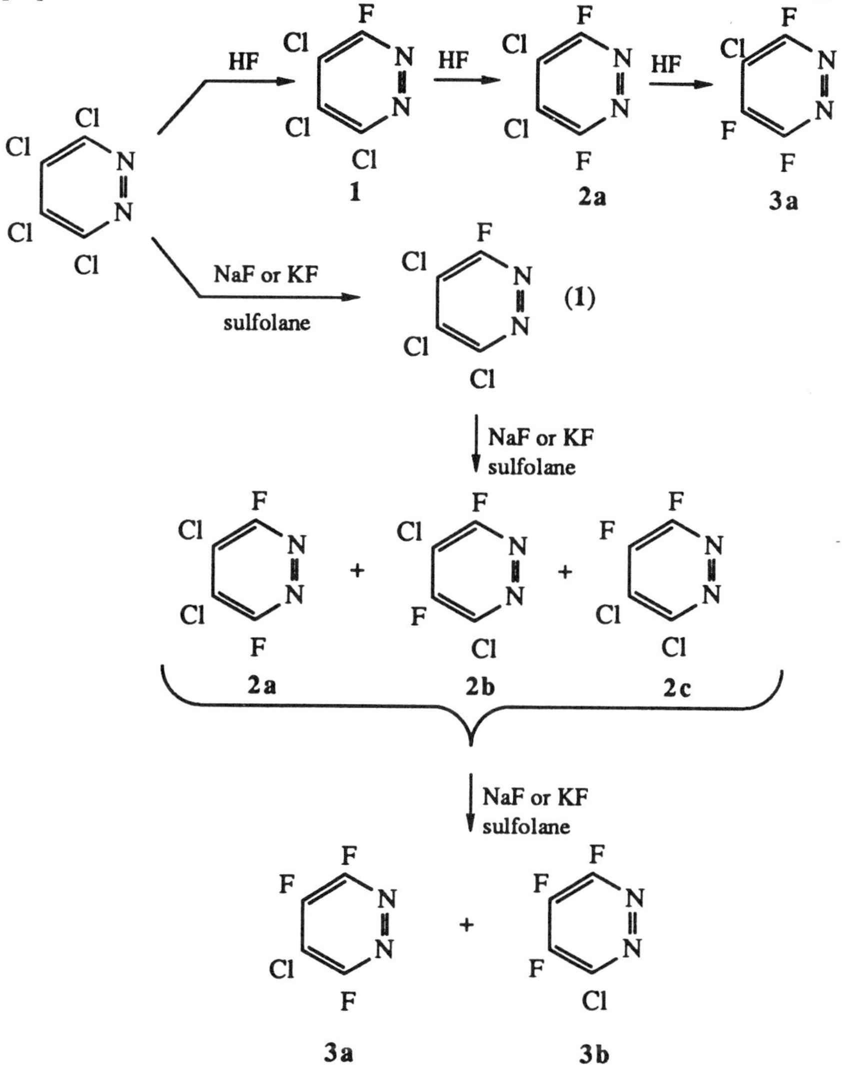

HF → **1** HF → **2a** HF → **3a**

NaF or KF, sulfolane → (1)

NaF or KF, sulfolane → **2a** + **2b** + **2c**

NaF or KF, sulfolane → **3a** + **3b**

A polymer (i.e., Amberlite IRA 900)-supported reagent, $P^+(HF)_nF^-$ $(n=0-2)$, where P^+ is the cationic part of the anion-exchange resin, acts as a source of nucleophilic fluorine and yet exhibits limited basic character, thereby reducing the possibility of elimination. The ability of this series of fluoride sources to exchange bromine for fluorine is described in Table 5 [*39*].

Table 4. Yield of Fluorinated Halopyridazine Based on Fluorinating Agent [*38*]

Agent	No. of Moles	Temp. (°C)	Time (h)	Yield (%)[a] *1*	*2a*	*2b*	*2c*	*3a*	*3b*
HF	10	155	2	71					
	20	155	3		24				
NaF	3	220	5	36	33	4	2	7	1
KF	3	160	5	11	18	18	6	17	2

[a]Refer to equation 29 for structures of halopyridazines.

Table 5. Reactions of $C_6H_5COCBrR^1R^2$ with $P^+(HF)_nF^-$ (n = 0–2) [*39*]

Compound	Reactant	Isolated Yield (%)	Distribution of Products (%)[a] Fluoride	Alkene	Ketol	Chlorideb
$R^1 = R^2 = H$	$n = 0$	Degradation	61		33	6
	$n = 1$	96.1	85.2		9.8	4.9
	$n = 2$	93.7	97			
$R^1 = H, R^2 = CH_3$	$n = 0$	98	79		20	1.5
	$n = 1$	91.8	91.9		4.3	3.7
	$n = 2$	95	98			2
$R^1 = R^2 = CH_3$	$n = 0$	100	32.5	47.5	19.5	Trace
	$n = 1$	87	41.5	43.7	9.4	5.4
	$n = 2$	90	59.7	36.1	2.5	1.7
2-Bromooctane	$n = 0$	20	20	73		
	$n = 1$	20	20	19.5[c]		

[a]Determined by gas chromatography.

[b]The chloride source is the residual chlorine present in commercial P^+Cl^-.

[c]Only 2-octene.

Replacement by Means of Hydrogen Fluoride and Catalysts

The vapor-phase catalytic replacement of chlorine by fluorine with hydrogen fluoride as the fluorine source has been the subject of a number of patents for the synthesis of Freons or Genetrons. This topic has been carefully reviewed in the literature [2, p 97ff]. One advantage of using a catalyst with hydrogen fluoride is to allow some degree of *selectivity* in the displacement of a specific chlorine from a polychlorinated compound. Fluorinated alumina [*40*], antimony halides, fluorinated chromic oxide [*41*], fluorinated cupric oxide [*42*], fluorinated ferric oxide, nickel fluoride, fluorinated molybdenum oxides, and stannic halides are a few of the many metal salts used to catalyze this displacement. The most commonly used catalyst systems are based on the antimony halides. One proposed mechanism uses a transition state to explain the formation of saturated and unsaturated products from the reaction of trichloroethylene with hydrogen fluoride and a catalyst described as antimony fluorochloride [*43*]. Others have shown that a 1:1 mixture of hydrogen fluoride and antimony pentachloride forms a fluorinating agent whose empirical formula is $SbCl_4F$. Kinetic runs using a 1:1 mixture of hydrogen fluoride and antimony pentachloride or antimony tetrachlorofluoride with chloroform demonstrated that the fluorination proceeds in four consecutive steps. The first and second steps are fast and reversible. The major product, chlorodifluoromethane, is generated in the third step. In the fourth step, the formation of trifluoromethane is very slow. If the rate of formation of chlorodifluoromethane is taken as 1.0, the relative reaction rates of the four consecutive steps are 150:7:1.0:0.03 [*44*].

Replacement by Means of Group 1 Fluorides

Reactions of alkali metal fluorides in aprotic solvents to displace halogen in an organic substrate have been reviewed [2, p 112ff]. Halogen exchange with **potassium fluoride** remains popular, and contemporary research is expanding the uses of this versatile reagent. One variable in a reaction of this type is the surface area of the potassium fluoride particles, with the larger surface area providing the more efficient fluoride source. *Spray-dried potassium fluoride* can have a particle size of 10–50 μm and a surface area of 1.3 m^2/g whereas the particle size of calcined potassium fluoride is not smaller than 200–300 μm with a surface area of 0.1 m^2/g. Yields of products fluorinated with spray-dried potassium fluoride are listed in Table 6 [*45*]. Tetrafluoroisophthalonitrile can be prepared in good yield from tetrachloroisophthalonitrile and spray-dried potassium fluoride [*47*] (equation 30).

[*47*] **30**

Tetrachloroisophthalonitrile (CN, Cl, Cl, CN, Cl, Cl) —spray-dried KF, sulfolane, 220°C→ Tetrafluoroisophthalonitrile (CN, F, F, CN, F, F) 80%

A second variable is the amount of residual moisture present in the extremely hygroscopic potassium fluoride. Solvents are often used to remove this moisture by azeotropic distillation. When the reaction mass is anhydrous, the solvent is

Table 6. Fluorination with Spray-Dried Potassium Fluoride in Acetonitrile[a] [*45, 46*]

Compound	Temp. (°C)	Time (h)	Product	Yield (%) Spray-Dried	Yield (%) Calcined
CH_3COCl	RT	3	CH_3COF[b]	83	
C_3H_7COCl	RT	3	C_3H_7COF[b]	83	13
i-C_3H_7COCl	RT	3	*i*-C_3H_7COF[b]	96	
C_4H_9COCl	RT	3	C_4H_9COF[b]	92	
C_6H_5COCl	RT	3	C_6H_5COF[b]	89	19
$C_8H_{17}Br$	Reflux	10	$C_8H_{17}F$[b]	65	
$C_6H_5CH_2Br$	Reflux	10	$C_6H_5CH_2F$[b]	68	0
$ClCH_2CO_2C_2H_5$	Reflux	24	$FCH_2CO_2C_2H_5$[b]	47	
2,4-$(O_2N)_2C_6H_3Cl$	Reflux	10	2,4-$(O_2N)_2C_6H_3F$[b]	58	0
2,4-$Cl_2C_6H_3CHO$	220	12	2,4-$F_2C_6H_3CHO$[c]	66	
3,4-$Cl_2C_6H_3CHO$	220	12	3,4-$ClFC_6H_3CHO$[c]	98	
2,6-$Cl_2C_6H_3CHO$	220	14	2,6-$F_2C_6H_3CHO$[c]	67	
3,4-$Cl_2C_6H_3COCl$	220	12	3,4-$ClFC_6H_3COF$[c]	63	
4-ClC_6H_4COCl	220	12	4-FC_6H_4COF[c]	60	

[a]Yield determined by ^{19}F NMR. RT, room temperature.
[b]Reference 45.
[c]Reference 46.

distilled. If the reaction of pentachlorobenzonitrile and potassium fluoride to form pentafluorobenzonitrile is run solvent free or in benzonitrile, the yield is the same [*48, 49*].

Another approach is to omit the solvent and run the reaction under *supercritical conditions* where potassium fluoride dissolves in the superheated reactant. This approach is illustrated by the conversion of 2-chloromethoxy-1,1,1,3,3,3-hexafluoropropane to 2-fluoromethoxy-1,1,1,3,3,3-hexafluoropropane with either potassium fluoride or sodium fluoride as the fluorine source (equation 31).

[*50*] **31**

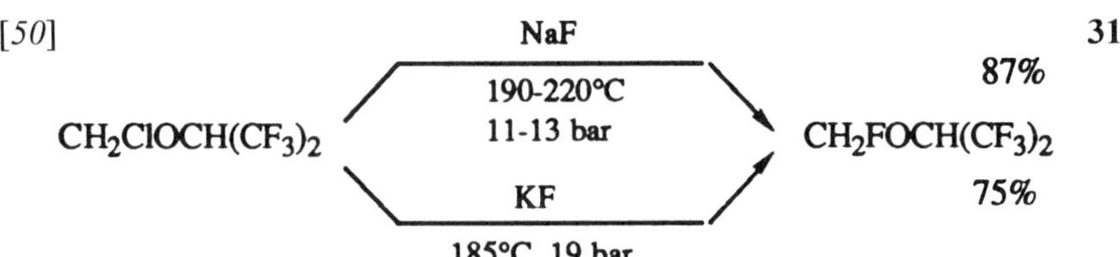

2-Difluoromethoxy-2-chloro-1,1,1-trifluoroethane and potassium fluoride produce 2-difluoromethoxy-1,1,1,2-tetrafluoroethane [*50*]. The yield of the latter reaction is improved by adding a phase-transfer catalyst or crown ether; tetramethylammonium chloride, tetrabutylammonium chloride, or 18-crown-6 with a solvent like sulfolane can be used for this purpose [*51*] (equation 32).

32

$$CHF_2OCHClCF_3 \xrightarrow[\text{278°C, 34 bar}]{\text{KF}} CHF_2OCHFCF_3 \quad 33\% \quad [50]$$

$$CHF_2OCHClCF_3 \xrightarrow[\text{DMF, 220°C}]{\text{KF, 2\% }Me_4NCl} CHF_2OCHFCF_3 \quad 44\% \quad [51]$$

The use of potassium fluoride in combination with a *phase-transfer catalyst* provides a simple efficient method of using metal fluorides in nonpolar and dipolar aprotic reaction media. Displacing a halogen from an electron-deficient aromatic ring is faster in polar aprotic protophilic solvents such as sulfolane or dimethyl sulfoxide than in acetonitrile or dimethoxyethane, but potassium fluoride is not very soluble in these solvents. The addition of catalytic amounts of *18-crown-6* overcomes this problem. As an example, the rate of fluorination of benzyl bromide with potassium fluoride is solvent dependent [*52*]: *18-crown-6·acetonitrile* > *dimethylacetamide* > *1,2-dimethoxyethane* > *sulfolane* > *acetonitrile.* The rates of halogen displacement in simple aliphatic compounds correlate well with the solubility of the alkali metal fluoride in a given solvent. For example, the reaction of 1-bromooctane with potassium fluoride and 18-crown-6 in either acetonitrile or benzene results in a 92% yield of 1-fluorooctane and an 8% yield of 1-octene. Under the same conditions 2-bromooctane generates a 32% yield of 2-fluorooctane, with the remainder being a mixture of 1- and 2-octene [*53*]. Benzyl chloroacetate and potassium fluoride with 18-crown-6 in sulfolane gives benzyl fluoroacetate. The reaction of benzyl trichloroacetate and potassium fluoride under the same conditions produces chloroform and benzyl fluoride, with no halogen-exchange product detected. Benzyl dichloroacetate yields the products of halogen exchange and cleavage [*54*] (equation 33).

[*54*] **33**

$$CH_2ClCO_2CH_2C_6H_5 \xrightarrow[\text{sulfolane, 100°C}]{\text{KF, 18-crown-6}} CH_2FCO_2CH_2C_6H_5 \quad 93\%$$

$$Cl_3CCO_2CH_2C_6H_5 \xrightarrow[\text{sulfolane, 100°C}]{\text{KF, 18-crown-6}} C_6H_5CH_2F + CHCl_3$$

$$CHCl_2CO_2CH_2C_6H_5 \xrightarrow[\text{sulfolane, 100°C}]{\text{KF, 18-crown-6}} CHF_2CO_2CH_2C_6H_5 + CHCl_3 + C_6H_5CH_2F + CH_2Cl_2 \text{ etc.}$$

The conversion of octachloronaphthalene to octafluoronaphthalene with potassium fluoride and either 18-crown-6, dibenzo-18-crown-6, *cis,syn,cis*-dicyclohexano-18-crown-6, *cis,anti,cis*-dicyclohexano-18-crown-6, or *trans,syn,trans*-dicyclohexano-18-crown-6 demonstrates that 18-crown-6 or dibenzo-18-crown-6 increases the yield and selectivity and decreases the reaction time [*55*]. Treatment of 3,4-dichloro-1,2,5-thiadiazole with potassium fluoride in sulfolane with and without 18-crown-6 present shows that less severe conditions can be used with either 18-crown-6 or dibenzo-18-crown-6 to form 3,4-difluoro-1,2,5-thiadiazole (equation 34).

[*56*] 34

3,4-dichloro-1,2,5-thiadiazole → (KF, 18-crown-6, sulfolane, 140°C; or KF, sulfolane, 180°C) → 3-fluoro-4-chloro-1,2,5-thiadiazole (24%) + 3,4-difluoro-1,2,5-thiadiazole (48%)

On the other hand, oxidizing fluorinating agents like silver difluoride, xenon difluoride, or bromine trifluoride replace one chlorine group and then cleave the sulfur–nitrogen bond [*56*].

When potassium fluoride is combined with a variety of **quaternary ammonium salts** its reaction rate is accelerated and the overall yields of a variety of halogen displacements are improved [*57*, p 112ff]. Variables like catalyst type and moisture content of the alkali metal fluoride need to be optimized. In addition, the maximum yield is a function of two parallel reactions: direct fluorination and catalyst decomposition due to its low thermal stability in the presence of fluoride ion [*58, 59, 60*]. One example is trimethylsilyl fluoride, which can be prepared from the chloride by using either 18-crown-6 (Procedure 3, p 192) or *Aliquat 336* in wet chlorobenzene, as illustrated in equation 35 [*61*].

[*61*] 35

$$(CH_3)_3SiCl + KF(H_2O)_x \xrightarrow[90\text{-}100^\circ C]{18\text{-crown-6}} (CH_3)_3SiF + [(CH_3)_3Si]_2O$$

41%

Triethylbenzylammonium chloride and potassium fluoride containing 1% water can provide chlorine–fluorine exchange of activated halides in the absence of a solvent. This mixture also can displace the chlorine of a chlorodifluoromethyl group to form a trifluoromethyl group [*62*] (equation 36).

[*62*] 36

$$CH_3CH_2COCH_2CClF_2 \xrightarrow[THF,\ RT]{KF,\ 1\%\ H_2O,\ Bu_4NF} CH_3CHCOCH_2CF_3$$

100%

Polyethylene glycol (molecular weight, 300–600) can aid in the displacement of activated halogen by fluorine. Propionyl chloride is converted to propionyl fluoride with potassium fluoride and polyethylene glycol in acetonitrile [*63*]. Treatment of benzyl chloride with a mixture of potassium fluoride and potassium iodide for 5 h in acetonitrile containing polyethylene glycol 200 gives benzyl fluoride in 62% yield [*64*].

Alkanoyl chlorides and *chloroformates* can be treated with potassium fluoride and **benzyltriethylammonium chloride** to obtain the respective alkanoyl fluorides or fluoroformates [*62*]. 2-Amino-6-chloro-9-(2,3,5-tri-*O*-acetyl-β-D-ribofuranosyl)purine is converted to the 2-amino-6-fluoro product with potassium fluoride and trimethylamine in dimethylformamide under rigorously anhydrous conditions [*65*] (equation 37).

[*65*] **37**

Cl; H_2N; N; N; N; N; R — KF, $(CH_3)_3N$ / DMF → F; H_2N; N; N; N; N; R — 93%

R = 2,3,5-Tri-*O*-acetyl-β-D-ribofuranosyl

The preparation of 2,6-difluoropyridine in 97% yield from 2,6-dichloropyridine can be accomplished at lower temperatures (150 °C) by using a catalytic amount of tetramethylammonium chloride in dimethyl sulfoxide containing less than 1% water [*66*].

Tetrasubstituted phosphonium halides are just as effective as their ammonium counterparts. A combination of **tetraphenylphosphonium bromide** and either 18-crown-6 or polyethylene glycol dimethyl ether with spray-dried potassium fluoride converts 4-chlorobenzaldehyde to 4-fluorobenzaldehyde in 74% yield [*67*]. In addition, the halogen of a primary alkyl chloride or bromide is easily displaced by fluorine in aqueous saturated potassium fluoride and a catalytic amount of **hexadecyltributylphosphonium bromide** [*68*] (Table 7; Procedure 4, p 194).

The use of **cesium fluoride** is limited because of its cost and its availability as a truly anhydrous reagent. Its use with 18-crown-6 shows a 5 times higher rate for the formation of benzyl fluoride from benzyl bromide when compared with cesium fluoride or potassium fluoride supported on calcium fluoride [*21*]. Either cesium fluoride or potassium fluoride supported on calcium fluoride (Procedures 5a and 5b, p 194) provides about a twofold improvement over either unsupported alkali metal fluoride [*58, 69*]. Cesium fluoride and Aliquat 336 convert benzyl bromide to the fluoride in 94% yield. Using tetrabutylammonium fluoride in place of Aliquat

Table 7. Reactions of Primary Alkyl Halides with Potassium Fluoride and Hexadecyltributylphosphonium Bromide[a] [68]

Compound	KF (mol equiv)	Temp. (°C)	Time (h)	Conversion (%)	Yield (%)[b] Alkene	ROH	RF
$C_6H_{13}Cl$	5	160	7.5	98	8	10	80
$C_8H_{17}Br$	5	160	3.5	92	15	13	64
	10[c]	100	41	87	14	6	67
$C_8H_{17}Cl$	1.5	160	16	94	11	10	73
	5	160	7	95	6	7	82
$C_{12}H_{25}Cl$	5	160	5.5	98	13	8	77
$C_6H_5CH_2Cl$	1.5	120	15	100		10	90
	5	120	7	100		5	95
$C_6H_{13}CHCl_2$	5	160	16	35	100[d]		

[a]Ten mole percent based on alkyl halide.
[b]Determined by gas chromatography.
[c]One change of the aqueous potassium fluoride solution.
[d]1-Chloroheptene.

336 reduces the yield to 76%. Octyl bromide is converted to its fluoride in 74–77% yield by cesium fluoride and either tetrabutylammonium bromide or Aliquat 336, with or without the addition of an equivalent of water. A trace of the ether and about 15–17% octenes are also formed. Using potassium fluoride under the same conditions provides a 60% yield of octyl fluoride [*70*].

Replacement by Means of Other Metal Fluorides

This topic has been reviewed [*2*, pp 94, 100–111, 130–134]. All of the standard approaches to the synthesis of a compound like methyl 2-fluorostearate from methyl 2-bromostearate result in a 1:1 yield of the 2-fluoro ester and the unsaturated esters. Although **silver fluoride** is not a new reagent, its use moist in wet acetonitrile to convert methyl 2-bromostearate to its fluoro ester is a departure from the traditional set of anhydrous conditions (Procedure 6, p 194) [*71*]. In contrast, **silver tetrafluoroborate** converts α-chloroketones to their respective fluoroketones under anhydrous conditions. The displacement of less activated halogen groups by silver tetrafluoroborate to form their respective fluorides is novel. Although silver tetrafluoroborate could not be used to convert an aliphatic terminal dichloromethyl or trichloromethyl group to its corresponding fluoro derivative, it is an effective fluorine source in other situations [*72*] (Table 8).

Although both **cuprous** and **cupric fluorides** have been studied in the past, an active fluorine donor can be formed from cupric oxide and hydrogen fluoride. This donor, in combination with 2,2′-bipyridine, effectively displaces the halogen of primary alkyl halides. Table 9 summarizes this reaction and the recent progress in the exchange of fluorine in alkyl halides [73] .

Table 8. Halogen Exchange with Silver Tetrafluoroborate [72]

Substrate	*Product*	*Yield (%)[a]*
$C_6H_5CHCl_2$	$C_6H_5CHF_2$	40
$C_6H_4CCl_2CH_3$	$C_6H_5CF_2CH_3$	50
$C_3H_7CCl_2C_7H_{15}$	$C_3H_7CF_2C_7H_{15}$	60
1,1-Dichlorocyclohexane	1,1-Difluorocyclohexane	40
$C_6H_5CCl_3$	$C_6H_5CF_3$	48

Experimental Procedures

For the safety advisory, see pages 27–28 of the book.

1. Preparation of Tetrabutylammonium Fluoride (TBAF) [22]

Tetrabutylammonium fluoride trihydrate (Aldrich) is heated in a round-bottom flask with magnetic stirring at 40–45 °C under vacuum (<0.1 mm of Hg). After several hours, the sample liquefies. Heating is continued until the sample loses 20% of its original weight (usually ca. 48 h). The resulting "anhydrous" TBAF (singlet at –99 ppm, ^{19}F NMR) contains 0.1–0.3 molar equivalent of water (by ^1H NMR) and ca. 10% tetrabutylammonium bifluoride (a doublet at –146 ppm, (J = 123 Hz, ^{19}F NMR). **This oil must be used immediately.**

2. Preparation of Tetraphenylphosphonium Hydrogen Difluoride [29]

Tetraphenylphosphonium hydrogen difluoride is prepared by passing an acetonitrile solution of tetraphenylphosphonium bromide through a column of Amberlite IRA 410 resin in its HF_2–form. Resin activation is achieved by passing 0.5 M NH_4HF_2 through the column until no further precipitate is obtained with silver nitrate. The column is washed first with water and then with acetonitrile. The solution of tetraphenylphosphonium hydrogen difluoride in acetonitrile is evaporated to dryness in a rotary evaporator. The resulting solid is dried at ca. 60 °C and 0.1 mm of Hg for 24 h. This substance is stable under anhydrous conditions.

3. Preparation of Trimethylsilyl Fluoride [59]

Dry potassium fluoride (10 g; 0.17 mol), trimethylsilyl chloride (20 g; 0.18 mol), 18-crown-6 (200 mg; 0.76 mol), and water (0.6 mL) are heated in 70 mL of chlorobenzene with vigorous stirring at 90–100 °C in a two-necked flask with a

Table 9. Halogenfluorine Exchange for Fluoride in Haloalkanes [73]

Reagent	*Substrate*	*Temp. (°C)*	*Time (h)*	*Yield of RF (%)*	*Ref.*
Anion-exchange resins	1-Bromooctane	86	20	82	74
	2-Bromooctane	86	20	20	
Aqueous KF with phase-transfer catalyst	1-Bromooctane	160	3.5	64	68
	2-Bromooctane	160	5		
KF with 18-crown-6	1-Bromooctane	83	230	92	53
	2-Bromooctane	83	150	32	
KF in diethylene glycol	1-Bromooctane	23	135	48	75
Spray-dried KF	1-Bromooctane	82	10	65	45
KF-CaF_2	1-Bromooctane	100	48	40	69
KF-CaF_2	1-Bromododecane	160	5	84	58
CuF_2 and 2,2′-bipyridyl	1-Bromooctane	82	82	88	76
CuF and 2,2′-bipyridyl	1-Bromooctane	130	0.75	83	73
	1-Chlorooctane	160	3	64	
	2-Bromooctane	130	0.75	51	
	1,10-Dibromooctane	130	3	66	
CuF_2 and 2,2′-bipyridyl	1-Bromooctane	130	0.75	Trace	73
Cu + CuF_2 in collidine	1-Bromooctane	130	0.75	19	73
Cu + CuF_2 and 2,2′-bipyridyl	1-Bromooctane	130	0.75	75	73
AgF and 2,2′-bipyridyl	1-Bromooctane	130	0.75	99	73
	1-Chlorooctane	130	0.75	86	
	2-Bromooctane	130	0.75	55	
	1,10-Dibromododecane	130	1	90	

References are listed on pages 196–198.

reflux condenser connected to a cold trap cooled to –50 to –60 °C. After 1 h, an additional 10 g of potassium fluoride and 0.6 mL of water are added. This addition is repeated after the second hour. The liquid in the cold trap is purified by trap-to-trap distillation (boiling point, 15–16 °C; yield, 6.9 g, 41%).

4. Preparation of 1-Fluorooctane [68]

1-Chlorooctane (14.9 g; 0.1 mol), potassium fluoride dihydrate (47 g; 0.5 mol), hexadecyltributylphosphonium bromide (5.1 g; 0.01 mol), and water (30 mL) are mixed in an autoclave equipped with a magnetic stirrer and heated to 160 °C (bath temperature) for 7 h. After this time gas–liquid chromatographic analysis (10% Carbowax 20M on Chromosorb) shows a 95% conversion to a mixture of 1-fluorooctane (82%), octenes (6%), and 1-octanol (7%). The organic layer is separated, washed with water, washed with concentrated sulfuric acid, washed once again with water, dried over calcium chloride, and distilled to give 10 g (77%) of 1-fluorooctane.

5a. Preparation of Benzyl Fluoride by Using Potassium Fluoride Supported on Calcium Fluoride (Methanol Procedure) [58]

A slurry of pure calcium fluoride in a solution of potassium fluoride in methanol is slowly evaporated to dryness ($KF:CaF_2$ molar ratio of 1:5) for ca. 1 h at 80 °C under reduced pressure. Fluorination of benzyl bromide for 2 h at 120 °C in sulfolane gives 74% yield (92% conversion), compared with 36% with potassium fluoride alone.

5b. Attrition Procedure [69]

Commercially available potassium fluoride is mixed with calcium fluoride (Wako's guaranteed reagent) at a 1:2 or 1:4 weight ratio. After the mixture is ground, it is dried at 150 °C for several hours. This heterogeneous mixture (2.5 g) and benzyl bromide are heated to reflux with stirring in acetonitrile. The reaction proceeds smoothly to afford the corresponding fluoride in 69, 89, and 95% yield (determined by gas–liquid chromatography) after 6, 10, and 15 h, respectively. After 15 h the solid materials are filtered off and washed with ether. The solvent is evaporated, and benzyl fluoride is isolated in 81% yield after distillation.

6. Preparation of Methyl 2-Fluorostearate by Using Wet Silver Fluoride [71]

Crude methyl 2-bromostearate (33 g; 0.087 mol) is dissolved in 200 mL of acetonitrile containing 0.5 mL of water, and silver(I) fluoride (50 g; 0.393 mol) is added rapidly in one portion. The slurry is stirred vigorously for 20 h in an oil bath at 80 °C. At the end of this time thin-layer chromatographic analysis (petroleum

References are listed on pages 196–198.

ether–methylene chloride, 60:40) shows no remaining bromo ester and a new single spot. A bed of silica gel (200 g in methylene chloride) is prepared in a medium-porosity sintered-glass Buchner funnel, and the reaction mixture is poured through it. About 500 mL of dichloromethane is allowed to percolate slowly through the funnel to elute the 2-fluoro ester. Removal of the solvents gives 23.2 g (84%; melting point, 42–44 °C, recrystallized from methanol).

References for Pages 172–196

1. Rozen, S.; Brand, M. *J. Org. Chem.* **1981,** *46,* 733.
2. Hudlicky', M. *Chemistry of Organic Fluorine Compounds;* Ellis Horwood: Chichester, United Kingdom, 1976.
3. Adcock, J. L.; Evans, W. D. *J. Org. Chem.* **1984,** *49,* 2719.
4. Adcock, J. L.; Sastry, A. K.; Taylor, D. R.; Nappa, M. J.; Sievert, A. C. *I&EC Res.* **1989,** *28,* 1547.
5. Bozzelli, J.; Kaufman, M. *U.S. Nat. Tech. Inform. Serv.* A D Rep 1972, No. 750783; *Chem. Abstr.* **1973,** *78,* 96820.
6. DeMarco, R. A.; Kovacina, T. A.; Fox, W. B. *J. Fluorine Chem.* **1975,** *6,* 93.
7. Robin, M. L.; Halpern, D. F. U.S. Patent 5 015 781, 1991.
8. Huang, C.; Vernice, G. G. U.S. Patent 4 874 902, 1989; *Chem. Abstr.* **1990,** *112,* 157679g.
9. Kartashov, A. V.; Chuvatkin, N. N.; Boguslavskaya, L. S. *Zh. Org. Khim.* **1988,** *24,* 2522; *Chem. Abstr.* **1989,** *111,* 56996g.
10. Kartashov, A. V.; Chuvatkin, N. N.; Boguslavskaya, L. S. *Zh. Org. Khim.* **1988,** *24,* 2525; *Chem. Abstr.* **1989,** *111,* 56722q.
11. Boguslavskaya, L. S.; Chuvatkin, N. N.; Morozova, T. V.; Panteleeva, I. Y.; Kartashov, A. V.; Sineokov, A. P. *Zh. Org. Khim.* **1987,** *23,* 1173 (Engl. Trans. 1060); *Chem. Abstr.* **1988,** *108,* 74804k.
12. Morozova, T. V.; Chuvatkin, N. N.; Panteleeva, I. Y.; Boguslavskaya, L. S. *Zh. Org. Khim.* **1984,** *20,* 1379; *Chem. Abstr.* **1985,** *102,* 5300g.
13. Chuvatkin, N. N.; Panteleeva, I. Y.; Boguslavskaya, L. S. *Zh. Org. Khim.* **1982,** *18,* 946; *Chem. Abstr.* **1982,** *97,* 144272r.
14. Kirchmeier, R. L.; Sprenger, G. H.; Shreeve, J. M. *Inorg. Nucl. Chem. Lett.* **1975,** *11,* 699.
15. Lehms, I.; Kaden, R.; Oese, W.; Mross, D.; Kochmann, W.; Ziegenhagen, D. Ger. (East) Patent DD 274 819, 1990; *Chem. Abstr.* **1990,** *113,* 58491d.
16. Lehms, I.; Kaden, R.; Oese, W.; Mross, D.; Kochmann, W.; Ziegenhagen, D. Ger. (East) Patent DD 274 821, 1990; *Chem. Abstr.* **1990,** *113,* 58492e.
17. Schindel, W. G. U.S. Patent 4 053 530, 1977; Chem. Abstr. 1978, 88, 22127r.
18. Boudakian, M. M. *J. Fluorine Chem.* **1987,** *36,* 283.
19. Muffler, H.; Franz, R. Ger. Offen. 28 23 969, 1979; *Chem. Abstr.* **1980,** *92,* 197322k.
20. Sharma, R. K.; Fry, J. L. *J. Org. Chem.* **1983,** *48,* 2112.

21. Gingras, M.; Harpp, D. N. *Tetrahedron Lett.* **1988,** *29,* 4669.
22. Cox, D. P.; Terpinski, J.; Lawrynowicz, W. *J. Org. Chem.* **1984,** *49,* 3216.
23. Cox, D. P.; Moss, R. A.; Terpinski, J. *J. Am. Chem. Soc.* **1983,** *105,* 6513.
24. Dailey, W. P. *Tetrahedron Lett.* **1987,** *28,* 5801.
25. Bosch, P.; Camps, F.; Chamorro, E.; Gasol, V.; Guerrero, A. *Tetrahedron Lett.* **1987,** *28,* 4733.
26. Bensoam, J.; Leroy, J.; Mathey, F.; Wakselman, C. *Tetrahedron Lett.* **1979,** 353.
27. Leroy, L.; Bensoam, J.; Humiliere, M.; Wakselman, C.; Mathey, F. *Tetrahedron,* **1980,** *36,* 1931.
28. Brown, S. J.; Clark, J. H. *J. Chem. Soc., Chem. Commun.* **1985,** 672.
29. Brown, S. J.; Clark, J. H. *J. Fluorine Chem.* **1985,** *30,* 251.
30. Seto, H.; Qian, Z.; Yoshioka, H.; Uchibori, Y.; Umeno, M. *Chem. Lett.* **1991,** 1185; *Chem. Abstr.* **1992,** *115,* 136232c.
31. Yoshida, Y.; Kimura, Y.; Tomoi, M. *Chem. Lett.* **1990,** 769.
32. Feiring, A. E. *J. Org. Chem.* **1979,** *44,* 2907.
33. Markhold, A.; Klauke, E. *J. Fluorine Chem.* **1981,** *18,* 281.
34. Baasner, B.; Klauke, E. *J. Fluorine Chem.* **1982,** *19,* 553 and references therein.
35. Kodaira, T.; Kohayashi, Y.; Kurono, H. Eur. Pat. Appl. 15 557, 1980; *Chem. Abstr.* **1981,** *94,* 121303a; Alles, H.-U.; Klauke, E.; Laverer, D. *Liebigs Ann. Chem.* **1966,** *730,* 16.
36. Nihon Nohyaku Co. Jpn. Patent 80 129 243, 1980; Chem. Abstr. 1981, 94, 121125u.
37. Lenthe, M.; Doering, F. Ger. Offen. DE 3 324 905, 1985; *Chem. Abstr.* **1985,** *103,* 5884w.
38. Klauke, E.; Oehlmann, L.; Baasner, B. *J. Fluorine Chem.* **1983,** *23,* 301.
39. Cousseau, J.; Albert, P. *J. Org. Chem.* **1989,** *54,* 5380.
40. Vecchio, M.; Grappelli, G.; Tatlow, J. C. *J. Fluorine Chem.* **1974,** *4,* 117.
41. Maragoni, L.; Rasia, G; Gervasutti, C.; Colombo, L. *Chim. Ind. (Milan)* **1982,** *64,* 135.
42. Yoneda, N.; Fukuhara, T.; Nagata, S.; Suzuki, A. *Chem. Lett.* **1985,** *11,* 1693.
43. Scipioni, A.; Gambaretto, G.; Troilo, G. *Atti dell'Instituto Veneto di Scienze Lettre ed Arti* **1966,** *124,* 203; *Chem. Abstr.* **1967,** *67,* 99545z.
44. Santacesaria, E.; Di Serio, M.; Basili, G.; Carra, S. *J. Fluorine Chem.* **1989,** *44,* 87.
45. Ishikawa, N.; Kitazume, T.; Yamazaki, T.; Mochida, Y.; Tatsuno, T. *Chem. Lett.* **1981,** 761.
46. Banks, R. E.; Mothersdale, K. N.; Tipping, A. E.; Tsiliopoulos, E.; Cozens, B. J.; Wotton, D. E. M.; Tatlow, J. C. *J. Fluorine Chem.* **1990,** *46,* 529.
47. Takaoka, A.; Yokokohji, O.; Yamaguchi, Y.; Isono, T.; Motoyoshi, M.; Ishikawa, N. *Nippon Kagaku Kaisi,* **1985,** 2155; *Chem. Abstr.* **1986,** *105,* 152649t.

48. Ishihara,Sangyo Kaisha, Jpn. Patent 60 36 453; *Chem. Abstr.* **1985,** *103,* 53791k.
49. Nippon Shokubia Kagaku Kogyo, Jpn. Patent 59 152 361, 1984; *Chem. Abstr.* **1985,** *102,* 78592p.
50. Halpern, D. F.; Robin, M. L. U.S. Patent 4 874 901, 1989; *Chem. Abstr.* **1990,** *112,* 157680a.
51. Kawai, T. U.K. Patent Appl. GB 2 219 292A, 1989; *Chem. Abstr.* **1990,** *112,* 234804y.
52. Clark, J. H.; MacQuarris, D. *J. Fluorine Chem.* **1987,** *35,* 591.
53. Liotta, C. L.; Harris, H. P. *J. Am. Chem. Soc.* **1974,** *96,* 2250.
54. Yasujima, J.; Fukunishi, K.; Nomura, M.; Yamana, H. *Nippon Kagaku Kaishi,* **1981,** *11,* 1744; *Chem. Abstr.* **1982,** *96,* 68516k.
55. Kusov, S. Z.; Lubenets, E. G.; Kobrina, V. N.; Chmel'nitskii, A.G. *Izv. Sib. Otd. Akad. Nauk SSSR* **1986,** *81*; *Chem. Abstr.* **1986,** *105,* 190622m.
56. Geisel, M.; Mews, R. *Chem. Ber.* **1982,** *115,* 2135.
57. Starks, C.M.; Liotta, C. *Phase Transfer Catalysis, Principles and Techniques;* Academic Press: New York, 1978.
58. Clark, J. H.; Hyde, A. J.; Smith, D.K. *J. Chem. Soc., Chem., Commun.* **1986,** 791.
59. Dermiek, S.; Sasson, Y. *J. Fluorine Chem.* **1983,** *22,* 431.
60. Dermiek, S.; Sasson, Y. *J. Org. Chem.* **1985,** *50,* 87.
61. Dehmlow, E. V.; Fastabend, U.; Kessler, M. *Synthesis,* **1988,** 966.
62. Tordeaux, M.; Wakselman, C. *Synth. Commun.* **1982,** *12,* 513.
63. Kitazume, T.; Ishikawa, N. *Chem. Lett.* **1978,** 283.
64. Mason, T. J.; Lorimer, J. P.; Turner, A. T.; Harris, A. R. *J. Chem. Res.* **1986,** 300.
65. Robins, M. J.; Uznanski, B. *Can. J. Chem.* **1981,** *59,* 2601.
66. Giacobbe, T. J. U.S. Patent 4 031 100, 1977; Chem. Abstr. 1977, 87, 135092y.
67. Yoshida, Y.; Kimura, Y. *Chem. Lett.* **1988,** 1355.
68. Landini, D.; Montanari, F.; Rolla, F. *Synthesis,* **1974,** 428.
69. Ichihara, J.; Toshiya, M.; Terukio, H. *J. Chem. Soc., Chem. Commun.* **1986,** 793.
70. Bram, G.; Loupy, A.; Pigeon, P. *Synth. Commun.* **1988,** *18,* 1661.
71. Pogany, S. A.; Zentner, G. M.; Ringeisen, C. D. *Synthesis,* **1987,** 718.
72. Bloodworth, A. J.; Bowyer, K. J.; Mitchell, J. C. *Tetrahedron Lett.* **1987,** *28,* 5347.
73. Yoneda, N.; Fukukara, T.: Yamagishi, K.; Suzuki, A. *Chem. Lett.* **1987,** 1675.
74. Cainelli, C.; Montanari, F. *Synthesis,* **1976,** 472.
75. Pattison, F. L. M.; Norman, J. J. *J. Am. Chem. Soc.* **1957,** *79,* 2311.
76. Sonoda, H.; Sonoda, T.; Kobayashi, H. *Chem. Lett.* **1985,** 233.
77. Ishikawa, N.; Kitazume, T.; Yamazaki, T.; Mochida, Y.; Tatsuno, T. *Chem. Lett.* **1981,** 761.

Replacement of Oxygen by Fluorine

by W. Dmowski

Cleavage of Ethers

In contrast to hydrogen-type ethers, ***α-haloethers,*** both linear and cyclic, are relatively easily cleaved by **anhydrous hydrogen fluoride.** Bis(1,1-difluoroalkyl) ethers are converted to 1,1,1-trifluoroalkanes and alkanoyl fluorides. The cleavage temperature depends on the substituents present: ethers having no electronegative substituents other than α-fluorines are readily cleaved below 20 °C, 3-halo-1,1-difluoroethers require approximately 70 °C, but 2-halo-1,1-difluoroethers are practically resistant toward hydrogen fluoride [*1*] (equation 1).

[*1*] **1**

$$(XCH_2CHYCF_2)_2O \xrightarrow[\text{autoclave}]{HF} XCH_2CHYCF_3 + XCH_2CHYCOF$$

X	Y	°C	h	%
H	H	20	20	100
Cl	H	70	3	100
H	Cl	120	6	0

A nonconventional synthesis of the known inhalation anaesthetic, 2-bromo-2-chloro-1,1,1-trifluoroethane (Halothane), based on the reaction of ethyl 1,2-dibromo-1,2-dichloroethyl ether with anhydrous hydrogen fluoride and sulfur tetrafluoride, has been patented. The reaction presumably involves cleavage of the ether linkage, followed by fluorination of the intermediate bromochloroacetyl halide with sulfur tetrafluoride; ethyl halides are the by-products [2] (equation 2).

[*2*] **2**

$$CHBrClCBrClOC_2H_5 \xrightarrow[\substack{60-150°C \\ 3-6\ h}]{HF,\ SF_4} CHBrClCF_3 + C_2H_5X$$

51–92 % (X=F,Cl,Br)

Likewise open-chain, ***cyclic five- and six-membered α,α,α′,α′-tetrafluoroethers*** are opened by anhydrous hydrogen fluoride at elevated temperatures to give acyl fluorides terminated with a trifluoromethyl group. A six-membered ether such

0065–7719/95/0187–0199$15.92/1

as 2,2,6,6-tetrafluoropyrane is much more liable to cleave than five-membered tetrafluoroethers. Halogen substituents and the double bond, on the other hand, impart greater stability to these ethers [*3*] (equation 3).

[*3*] **3**

$$\text{cyclic ether } (X, (H_2C)_n, O, HF) \xrightarrow[\text{autoclave, 3 h}]{\text{HF}} CF_3CHX(CH_2)_nCHXCOF$$

n	X	°C	%
0	H	100	20
		170	52
1	H	100	96
0	=	100	0
0	Cl	180	0

Treatment of 1,1,3,3-tetrafluoro-1,3-dihydroisobenzofuran with anhydrous hydrogen fluoride, followed by hydrolysis of the intermediate acid fluoride, gives a high yield of 2-trifluoromethylbenzoic acid [*4*] (equation 4).

[*4*] **4**

$$\xrightarrow[\text{95°C, 3h autoclave}]{\text{HF}} C_6H_4(CF_3)COF \xrightarrow[\text{2. HClaq.}]{\text{1. KOHaq.}} C_6H_4(CF_3)COOH \quad 86\%$$

In the field of nucleosides, an interesting example of a cleavage of a nonfluorinated six-membered cyclic ether containing oxygen and nitrogen atoms to give a fluoro nucleoside has been reported [*5*] (equation 5).

Perfluorinated small-ring heterocycles, such as ***oxetanes*** and ***1,2-oxazetidines,*** are cleaved by hydrogen fluoride–antimony pentafluoride or hydrogen fluoride–arsenic pentafluoride superacid mixtures. The ring opening is regioselective with formal addition of hydrogen fluoride to give, respectively, linear perfluorinated alcohols or amines. Surprisingly, for oxazetidines the normally fragile nitrogen–oxygen bond usually remains intact but the carbon–nitrogen bond is cleaved [*6*] (equations 6 and 7).

[*5*] **5**

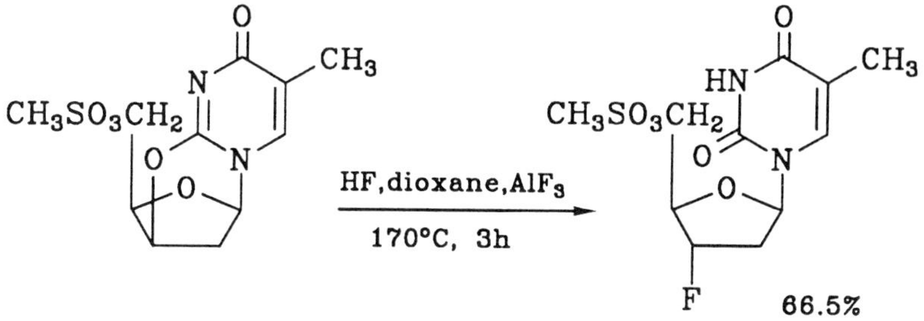

$CH_3SO_3CH_2$ (anhydronucleoside) $\xrightarrow[170°C,\ 3h]{HF,dioxane,AlF_3}$ 3'-fluoro nucleoside ($CH_3SO_3CH_2$, HN, O, CH_3, N, O, F) 66.5%

[*6*] **6**

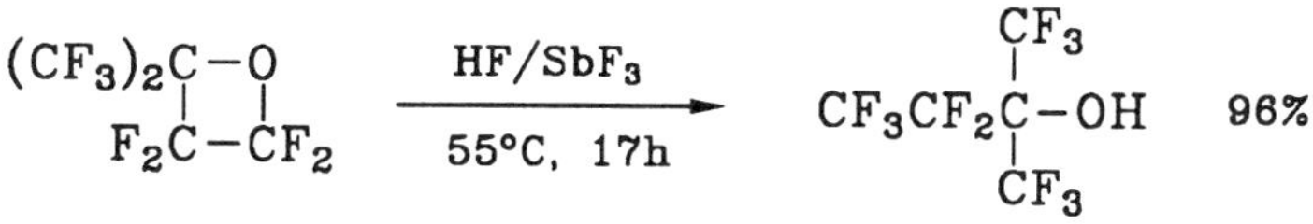

$$(CF_3)_2C{-}O / F_2C{-}CF_2 \xrightarrow[55°C,\ 17h]{HF/SbF_3} CF_3CF_2C(CF_3)_2{-}OH \quad 96\%$$

[*6*] **7**

$$CF_3N{-}O / F_2C{-}CFX \xrightarrow[20°C,\ 12{-}24h]{HF/AsF_5} CF_3NH{-}O{-}CFXCF_3$$

X = H, F 50–97%

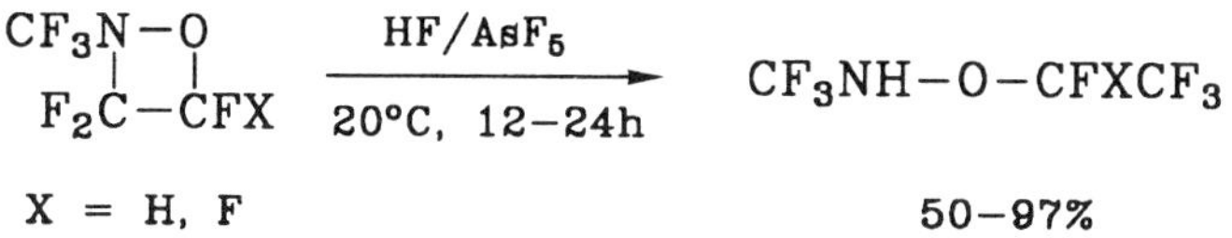

Perfluorinated ***epoxides,*** such as hexafluoro-1,2-epoxypropane [*7*] and epoxides derived from oligomers of tetrafluoroethylene or hexafluorocyclobutene [*8*], are opened by fluoride ion. These reactions, however, lead to isomeric compounds, acyl fluorides or ketones, via addition–elimination processes, rather than to the replacement of oxygen by fluorine. The only exception reported so far is 4a,8a-epoxyperfluorodecalin, which when heated with cesium fluoride in diglyme, undergoes highly stereospecific oxirane ring opening to form trans-perfluorobicyclo[4.4.0]decan-5-ol upon acidification [9] (equation 8).

[*9*] **8**

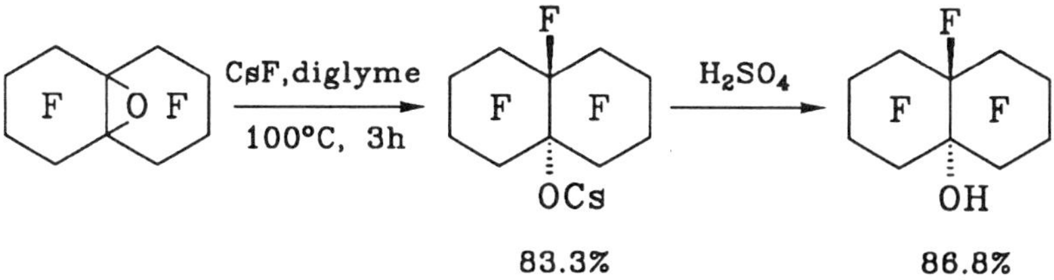

$$\text{F|O F} \xrightarrow[100°C,\ 3h]{CsF,diglyme} \text{F F (F, OCs)} \ 83.3\% \xrightarrow{H_2SO_4} \text{F F (F, OH)} \ 86.8\%$$

There is continuous interest in cleavage of ***epoxides*** as a general method for the preparation of *fluorohydrins.* Epoxides are effectively cleaved under mild conditions by a **70% hydrogen fluoride–pyridine complex (Olah's reagent).** The reactions proceed at room temperature with high regio- and stereoselectivity to give the corresponding fluorohydrins in almost quantitative yields. Numerous examples of cleavage of alkyl phenyl 2,3-epoxycarboxylates, 2,3-epoxyamides, and 2,3-epoxynitriles (glycidic esters, glycidic amides, and glycidic nitriles) to give, respectively, 3-fluoro-3-phenyllactates, amides, and cyanohydrins have been reported; the glycidates of known *cis* and *trans* isomers compositions give the corresponding fluorohydrins containing the *threo* and *erythro* isomers, respectively, at about the same compositions [*10, 11*] (equation 9 and Table 1).

[*10,11*] **9**

$$R^1R^2C\overset{O}{\frown}CR^3Y \xrightarrow[25°C,\ 1-4h]{70\%HF-C_6H_5N,\ CH_2Cl_2} R^1R^2CF-C(OH)R^3Y$$

88–95%

Table 1. Cleavage of Glycidic Esters, Glycidic Amides, and Glycidic Nitriles by Hydrogen Fluoride–Pyridine Complex [*10, 11*]

$R^1R^2C(O)CR^3Y$				$R^1R^2CFC(OH)R^3Y$		
R¹	*R²*	*R³*	*Y*	*cis/trans* (%/%)	*Yield* (%)	*threo/erythro* (%/%)
C_6H_5	H	H	CO_2Et	22/78	95	20/80
C_6H_5	CH_3	H	CO_2Me	41/59	95	46/54
C_6H_5	H	CH_3	CO_2Me	0/100	95	5/95
$-(CH_2)_5-$	H	H	CO_2Et		90	
C_6H_5	H	H	$CONH_2$	0/100	94	8/92
C_6H_5	H	H	CN	59/41	91	57/43
C_6H_5	CH_3	H	CN	56/44	96	55/45
C_6H_5	H	CH_3	CN	49/51	90	50/50
$-(CH_2)_4-$		H	CN		88	

In contrast to pyridine–hydrogen fluoride, which is acidic and acts as a protonating agent, in alkylamine–hydrogen fluoride complexes, fluorine is a nucleophile. The difference of the nature of these two types of reagents has been demonstrated by reactions of 9-oxabicyclo[6.1.0]non-4-ene: the reaction of this epoxide with triethylamine trishydrofluoride gave quantitatively *trans*-2-fluorocyclooct-5-en-1-ol, whereas with 70% HF–pyridine, two isomeric bicyclic fluoroalcohols, *exo*/*endo*-6-fluorobicyclo[3.3.0]octan-2-ol and its C-6 epimer, were obtained as a result of transannular cyclization of the intermediate carbocations [*12*] (equation 10).

[*12*] **10**

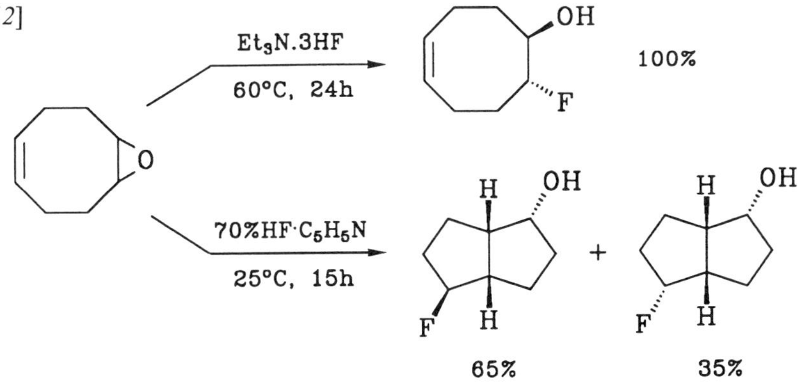

$Et_3N.3HF$, 60°C, 24h → 100%

70%HF·C_5H_5N, 25°C, 15h → 65% + 35%

Regioselectivity in opening of α-functionalized epoxides by treatment with **trimethylamine dihydrofluoride** has been observed. Thus, *cis*-isophorol epoxide gives exclusively 3-fluoro-1,2-diol, whereas from the *trans* isomer, 2-fluoro-1,3-diol is obtained as the main product together with 3-methylenecyclohexane-1,2-diol. This behavior has been discussed in terms of the influence of α-substituents on the transition state conformations [*13*] (equations 11 and 12).

[*13*] **11**

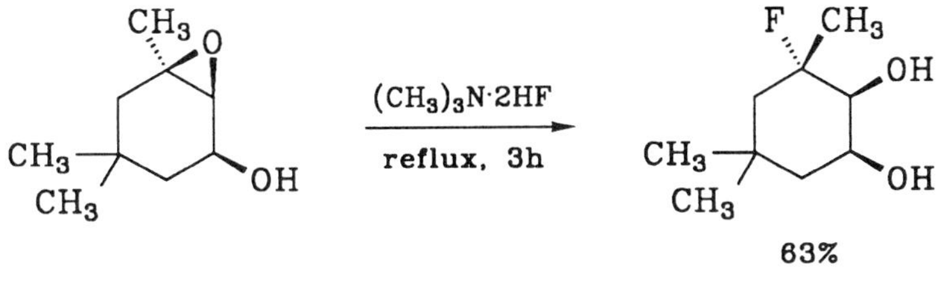

$(CH_3)_3N·2HF$, reflux, 3h → 63%

Recently, new fluorinating agents, **tetrabutylphosphonium fluoride** and its mono- and dihydrofluoride, were used for preparation of fluorohydrins from epoxides [*14*] (equation 13).

[13] 12

$(CH_3)_3N{\cdot}2HF$, reflux, 6h

57% 34%

[14] 13

Reagent, neat

Reagent	Temp.(°C)	Time (h)	Yield (%)	Ratio (**A** : **B**)
n−Bu_4PF	R.T.	20	82	97 : 3
n−Bu_4PHF_2	100	4	94	93 : 7
n−$Bu_4PH_2F_3$	100	10	92	78 : 22

The combination of alkali metal acid fluorides and porous aluminum fluoride is a stable, solid, and efficient substitute for anhydrous hydrogen fluoride for promoting the ring-opening reactions of simple aliphatic oxiranes to give the fluorohydrins under sonication [*15*] (equations 14 and 15).

Diethylaminosulfur trifluoride (DAST) reacts with epoxides to give complex mixtures in which vicinal *cis*-difluorides, geminal difluorides, and bis(2-fluoroalkyl) ethers are the main components [*16*] (equations 16 and 17).

Epoxides are regio- and stereoselectively transformed into fluorohydrins by **silicon tetrafluoride** in the presence of a Lewis base, such as diisopropylethylamine and, in certain instances, water or tetrabutylammonium fluoride. The reactions proceed under very mild conditions (0 to 20 °C in 1,2-dichloroethane or diethyl ether) and are highly chemoselective: alkenes, ethers, long-chain internal oxiranes, and carbon–silicon bonds remain intact. The stereochemical outcome of the epoxide ring opening with silicon tetrafluoride depends on an additive used; without addition of water or a quaternary ammonium fluoride, *cis*-fluorohydrins are formed, whereas in the presence of these additives, only *anti* opening leading to *trans* isomers is observed [*17, 18*] (Table 2).

[*15*] 14

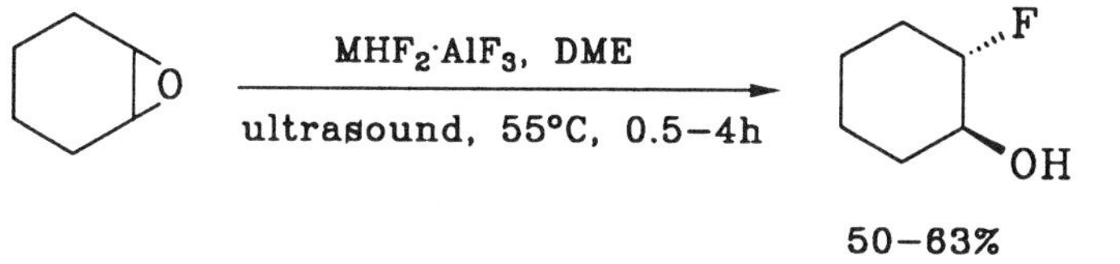

Cyclohexene oxide $\xrightarrow[\text{ultrasound, 55°C, 0.5–4h}]{MHF_2\cdot AlF_3\text{, DME}}$ trans-2-fluorocyclohexanol (F, OH)

50–63%

[*15*] 15

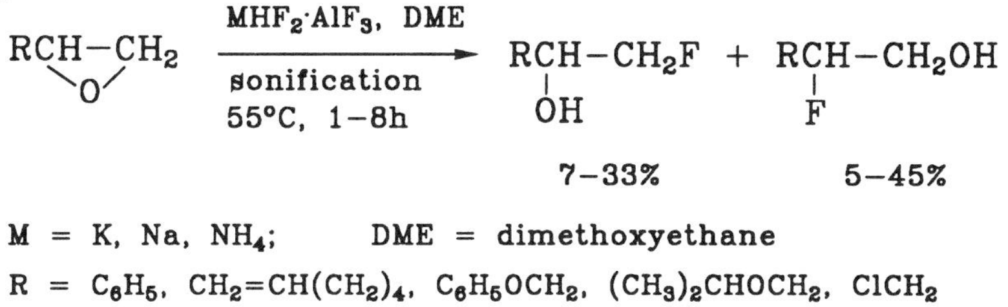

$$RCH{-}CH_2 \text{ (epoxide, O)} \xrightarrow[\text{sonification, 55°C, 1–8h}]{MHF_2\cdot AlF_3\text{, DME}} RCH(OH){-}CH_2F + RCH(F){-}CH_2OH$$

7–33% 5–45%

M = K, Na, NH_4; DME = dimethoxyethane

R = C_6H_5, $CH_2{=}CH(CH_2)_4$, $C_6H_5OCH_2$, $(CH_3)_2CHOCH_2$, $ClCH_2$

[*16*] 16

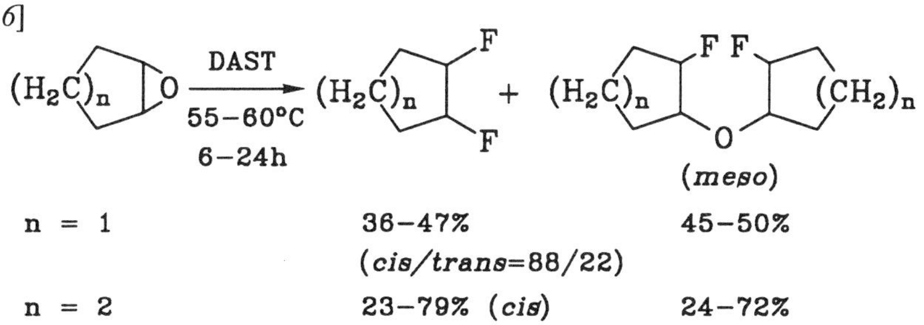

$(H_2C)_n$ (cycloalkene oxide, O) $\xrightarrow[\text{55–60°C, 6–24h}]{\text{DAST}}$ $(H_2C)_n$ (1,2-difluorocycloalkane, F, F) + $(H_2C)_n$ (F F, O-bridged bis(2-fluorocycloalkyl) ether) $(CH_2)_n$ (*meso*)

n = 1 36–47% (*cis/trans*=88/22) 45–50%

n = 2 23–79% (*cis*) 24–72%

[*16*] 17

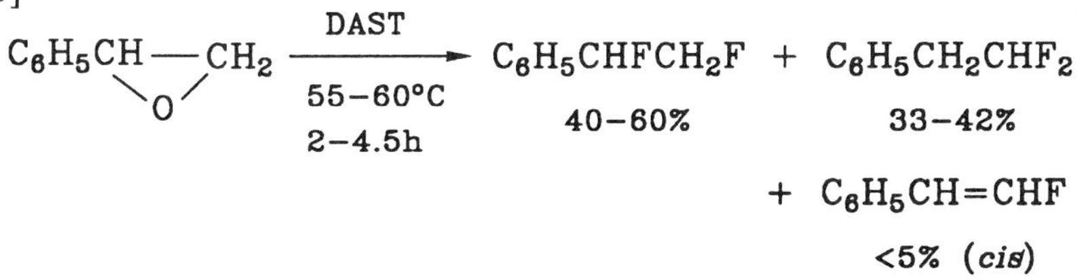

$$C_6H_5CH{-}CH_2 \text{ (epoxide, O)} \xrightarrow[\text{55–60°C, 2–4.5h}]{\text{DAST}} C_6H_5CHFCH_2F + C_6H_5CH_2CHF_2 + C_6H_5CH{=}CHF$$

40–60% 33–42% <5% (*cis*)

Alkyl silyl ethers are cleaved by a variety of reagents. Whether the silicon–oxygen or the carbon–oxygen bond is cleaved depends on the nature of the reagent used. Treatment of alkoxysilanes with electrophilic reagents like antimony trifluoride, 40% hydrofluoric acid, or a boron trifluoride–ether complex results in the cleavage of the silicon–oxygen bond to form mono-, di-, and trifluorosiloxanes or silanes [*19, 20, 21*] (equations 18–20).

Table 2. Cleavage of Epoxides by Silicon Tetrafluoride [*17, 18*][a]

Epoxide	*Solvent Additives*[b]	*Product*	*Yield (%)*
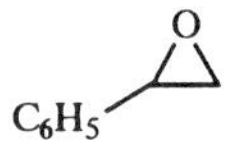 C_6H_5-epoxide (O)	A	$C_6H_5CHFCH_2OH$	71
 C_6H_5, H / O / H, CH_3	A	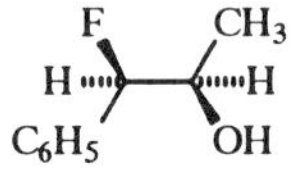 F, CH_3; H, H; C_6H_5, OH	78[c]
 C_6H_5, CH_3 / O / H, H	B	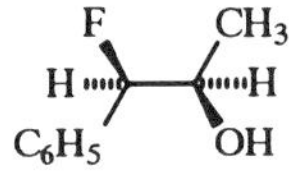 F, CH_3; H, H; C_6H_5, OH	31[c]
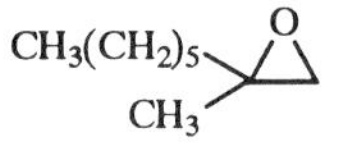 $CH_3(CH_2)_5$, CH_3 / O	C	$CH_3(CH_2)_5CF(CH_3)CH_2OH$	70
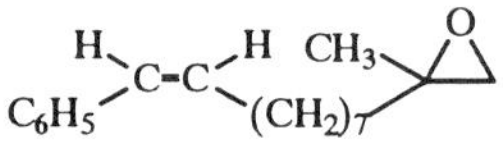 H, C_6H_5 C=C H, $(CH_2)_7$; CH_3 / O	D	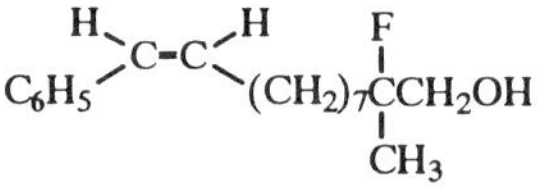 H, C_6H_5 C=C H, $(CH_2)_7CF(CH_3)CH_2OH$	76
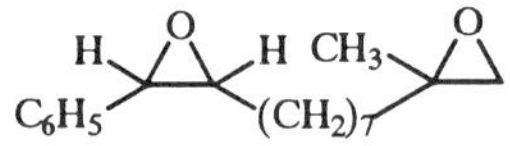 H, C_6H_5 / O / H, $(CH_2)_7$; CH_3 / O	D	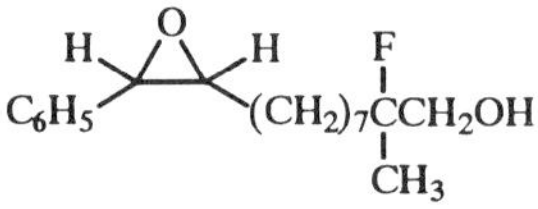 H, C_6H_5 / O / H, $(CH_2)_7CF(CH_3)CH_2OH$	59
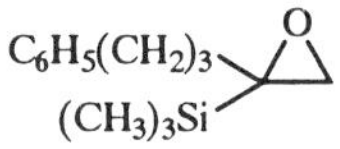 $C_6H_5(CH_2)_3$, $(CH_3)_3Si$ / O	D	$C_6H_5(CH_2)_3CF(Si(CH_3)_3)CH_2OH$	66
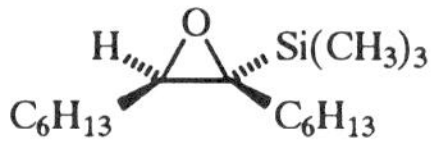 H, C_6H_{13} / O / $Si(CH_3)_3$, C_6H_{13}	D	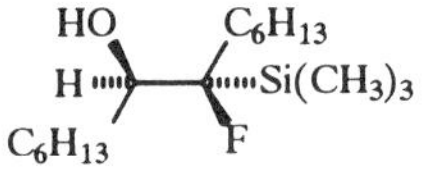 HO, C_6H_{13}; H, $Si(CH_3)_3$; C_6H_{13}, F	70
 H, C_6H_{13} / O / C_6H_{13}, $Si(CH_3)_3$	D	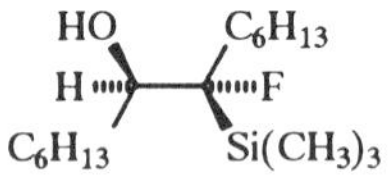 HO, C_6H_{13}; H, F; C_6H_{13}, $Si(CH_3)_3$	61

[a]The reactions were conducted at 0–20 °C for 1–5 h.
[b]A, *i*-Pr_2NEt, $(CH_2Cl)_2$; B, $(CH_2Cl)_2$; C, Bu_4NF, *i*-Pr_2NEt, Et_2O; D, *i*-Pr_2NEt, Et_2O, H_2O.
[c]1-Phenyl-2-propanone was the by-product (11–39%).

[*19*] **18**

$$(CH_3O)_4Si \xrightarrow[80-100°C]{SbF_3} \underset{41\%}{(CH_3O)_3SiF} + \underset{25\%}{(CH_3O)_2SiF_2} + \underset{15\%}{CH_3OSiF_3} + \underset{19\%}{SiF_4}$$

[*20*] **19**

$$(R^1R^2)_{3-n}(R^3O)_nSiC{\equiv}CH \xrightarrow[20°C]{40\%HF} \underset{62-70\%}{(R^1R^2)_{3-n}F_nSiC{\equiv}CH} + n\ R^3OH$$

$n = 1, 2$

$R^1 = CH_3, C_2H_5;\ R^2 = CH_3, C_2H_5, CH_2{=}CH;\ R^3 = CH_3$

[*21*] **20**

$$p\text{-}XC_6H_5OCH_2Si(CH_3)_n(OCH_3)_{3-n} \xrightarrow[\text{heat, 3 h}]{n/3\ BF_3{\cdot}Et_2O} \underset{69-82\%}{p\text{-}XC_6H_5OCH_2Si(CH_3)_nF_{3-n}} + n/3\ B(OCH_3)_3$$

$n = 0, 1;\quad X = H, Br, Cl, F, CH_3, CH_3COO$

The cleavage of the carbon–oxygen bond in alkyl silyl ethers to give alkyl fluorides is readily achieved by using reagents of a rather nucleophilic character, **phenyltetrafluorophosphorane** being the most widely studied. Reactions of phenyltetrafluorophosphorane have been investigated over a large range of silylated alcohols belonging to the various usual structural classes for which widely different reactivities, reaction products, and yields have been found. The yields of the expected or rearranged fluoroalkyl compounds range from 15 to 95% for the primary and secondary alkoxysilanes and are generally higher, sometimes near quantitative, for tertiary alkoxysilanes and when the alkyl group is β-branched or contains electron-attracting groups. The common by-products include alkenes, ethers, and phenylfluorophosphonates. The cleavage reaction involves quantitative immediate formation of an intermediate monoalkoxytrifluorophosphorane, generally under very mild conditions, and appears to be mainly controlled by steric factors. The subsequent decomposition or redistribution reactions giving the final products require heating in some cases and are strongly dependent on the structure of the alkoxy group [*22, 23*] (equation 21 and Table 3).

[*23*] **21**

$$ROSi(CH_3)_3 \xrightarrow[-40 \text{ to } 20°C]{C_6H_5PF_4} C_6H_5PF_3OR + (CH_3)_3SiF$$

$$C_6H_5PF_3OR \xrightarrow{20 \text{ to } 170°C} RF + \text{alkene} + C_6H_5P(O)F_2$$

Side reactions:

$$2\ C_6H_5PF_3OR \longrightarrow C_6H_5PF_2(OR)_2 + C_6H_5PF_4$$

$$C_6H_5PF_2(OR)_2 \longrightarrow ROR + C_6H_5P(O)F_2$$

$$C_6H_5PF_2(OR)_2 \longrightarrow RF + \text{alkene} + C_6H_5PF(O)OR$$

The reactions of phenyltetrafluorophosphorane with numerous silylated secondary or tertiary α- or β-hydroxy esters, ketones, nitriles, ethers, nitro, and trichloromethyl derivatives have been investigated; the corresponding α- or β-fluoro derivatives are obtained in yields varying from reasonable to nearly quantitative [*24, 25, 26, 27*]. The application of phenyltetrafluorophosphorane for fluorination of silyloxy steroids has also been reported [*28*].

Benzyl and alkyl trialkylsilyl ethers undergo clean fluorination to give good yields of benzyl and alkyl fluorides, respectively, when reacted with a combination of a **quaternary ammonium fluoride** and **methanesulfonyl or *p*-toluenesulfonyl fluoride.** The reactions are applicable strictly to a primary carbon–oxygen bond; secondary and tertiary alkyl silyl ethers remain intact or, under forcing conditions, are dehydrated to olefins [*29*] (equation 22).

[*29*] **22**

$$R^1OSi(CH_3)_2R^2 \xrightarrow[24\text{ h, RT or reflux}]{R^3{}_4NF,\ R^4SO_2F,\ THF} R^1F \quad 49\text{–}87\%$$

$R^1 = p\text{-}CH_3OC_6H_5CH_2,\ C_{14}H_{29};\quad R^2 = CH_3,\ (CH_3)_3C$

$R^3 = C_4H_9$ or 1 $C_6H_5CH_2$ and 3 CH_3; $\quad R^4 = CH_3,\ p\text{-}CH_3C_6H_5$

Reactions of aryl or alkyl bis(siloxy)isopropyl ethers with **tetrabutylammonium fluoride–mesyl fluoride reagent** lead to replacement of one siloxy group by fluorine and dehydrosiloxylation, providing an efficient access to fluoroisopropenyl ethers, which are useful as specific building blocks in drug design. The reactions proceed via the intermediate allyl methanesulfonates [*30, 31*] (equation 23).

1-(Trimethylsiloxy)- and 2-(trimethylsiloxy)adamantanes are nearly quantitatively converted to the corresponding *fluoroadamantanes* by treatment with **sulfur tetrafluoride** at 20 °C for 60 h [*32*].

Table 3. Cleavage of Alkoxytrimethylsilanes by Phenyltetrafluorophosphorane [23]

	Yield of Products (%)				
Alkoxy Group	*Alkyl Fluoride*		*Alkene*	*Ether*	*Phos-phonate*
CH_3O–	CH_3F	15	—	45	40
CH_3CH_2O–	CH_3CH_2F	55	—	15	30
$CH_3CH_2CH_2O$–	$CH_3CH_2CH_2F$ CH_3CHFCH_3	35 5	10	10	40
$CH_3CH_2CH_2CH_2O$–	$CH_3CH_2CH_2CH_2F$ $CH_3CH_2CHFCH_3$	35 5	10	10	40
$CH_3(CH_2)_3CH_2O$–	$CH_3(CH_2)_3CH_2F$ $CH_3(CH_2)_2CHFCH_3$	35 5	10	10	40
$CH_3(CH_2)_8CH_2O$–	$CH_3(CH_2)_8CH_2F$	30	30	10	40
$(CH_3)_2CHO$–	$(CH_3)_2CHF$	95	5	—	—
$CH_3CH_2CH(CH_3)O$–	$CH_3CH_2CH(CH_3)F$	50	40	—	10
$(CH_3CH_2)_2CHO$–	$(CH_3CH_2)_2CHF$ $CH_3(CH_2)_2CHFCH_3$	40 10	40	—	10
c-C_5H_9O–	c-C_5H_9F	35	50	—	15
c-$C_6H_{11}O$–	c-$C_6H_{11}F$	30	55	—	15
$CH_3CH_2C(CH_3)_2O$–	$CH_3CH_2C(CH_3)_2F$	100	—	—	—
$CH_3CH_2CH(CH_3)O$–	$(CH_3)_3CF$	60	5	—	35
$CH_3CHCH(CH_3)O$– | CH_3	$CH_3CH_2C(CH_3)_2F$	100	—	—	—
$ClCH_2CH_2O$–	$ClCH_2CH_2F$	70	—	25	5
$ClCH_2CH_2CH_2O$–	$ClCH_2CH_2CH_2F$	65	15	10	10
$ClCH_2CH(CH_3)O$–	$ClCH_2CHFCH_3$	80	20	—	—
$ClCH_2C(CH_3)_2O$–	$ClCH_2CF(CH_3)_2$	60	40	—	—
$ClCH_2CH(C_6H_5)O$–	$ClCH_2CHFC_6H_5$	90	10	—	—

Cleavage of Esters

Few examples of preparation of fluorides by cleavage of ***acetates*** and ***carbonates*** have been reported. Thus, in the field of sugars, 1-*O*-acetylated α-D-glucopyranoses treated with a 50% or **70% hydrogen fluoride–pyridine reagent** at –20 or 0 °C give *glucosyl fluorides* with total retention of configuration (97% α) and 68–89% yields [*33*]. Organotin trifluorides are prepared by treatment of organotin tricarboxylates with hydrogen fluoride [*34*] (equation 24).

[*30,31*] 23

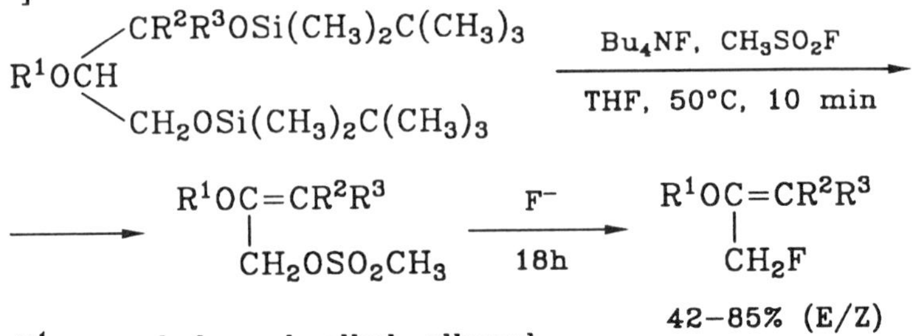

$$R^1OCH(CR^2R^3OSi(CH_3)_2C(CH_3)_3)(CH_2OSi(CH_3)_2C(CH_3)_3) \xrightarrow[\text{THF, 50°C, 10 min}]{Bu_4NF,\ CH_3SO_2F}$$

$$\longrightarrow R^1OC(CH_2OSO_2CH_3)=CR^2R^3 \xrightarrow[\text{18h}]{F^-} R^1OC(CH_2F)=CR^2R^3$$

42–85% (E/Z)

R^1 = aryl, benzyl, alkyl, alkenyl

R^2 = H, phenyl, styryl; R^3 = H, benzyl, naphthyl

[*34*] 24

$$R^1Sn(OC(O)R^2)_3 \xrightarrow[\text{20–25°C, 1.5h}]{\text{42\% aqueous HF},\ C_6H_6} R^1SnF_3 + 3\ R^2COOH$$

85–95%

R^1 = C_2H_5, C_4H_9, C_6H_5, $CH_2{=}CH$; R^2 = C_5H_{11}, C_7H_{15}

On treatment with a potassium fluoride–crown ether complex, alkyl 1,2,2,2-tetrachloroethyl carbonates are cleaved at the carbonyl group–oxygen bond to give high yields of *alkyl fluoroformates* [*35*] (equation 25).

[*35*] 25

$$(CH_3)_3COC(O)OCHClCCl_3 \xrightarrow[\text{30–35°C}]{\text{KF, 18-crown-6}} (CH_3)_3COC(O)F$$

79%

Aryl or ***vinyl fluoroformates,*** when passed in the vapor phase with helium over an alumina or platinum-on-alumina catalyst, eliminate carbon dioxide to give, respectively, *fluoroarenes* or *vinyl fluorides* [*36*] (equation 26).

[*36*] 26

$$C_6H_5OC(O)F \xrightarrow[\text{255°C}]{Pt{-}Al_2O_3,\ He} C_6H_5F + CO_2$$

54%

Esters of penta- and trivalent phosphorus acids and their derivatives readily undergo cleavage of the phosphorus–oxygen bond under extremely mild conditions with formation of the phosphorus–fluorine bond. Phosphates and phosphinates react chemoselectively at low temperatures with sulfuryl chloride fluoride to give *phosphoro- and phosphonofluoridates* in almost quantitative yields [*37, 38*] (equation 27).

[*37,38*] **27**

$$R^1R^2P(X)OR^3 \xrightarrow[\substack{-70 \text{ or } -50°C \\ 0.5 \text{ to } 1h}]{SO_2ClF,\ CH_2Cl_2} R^1R^2P(X)F + R^3Cl + SO_2$$

X = O, S, Se; R^1,R^2 = alkoxyl, alkyl; R^3 = $(CH_3)_3Si$, alkyl

Phosphonofluoridates are also formed in high yields by treatment of vinyl or phenyl phosphates ($R^3 = CH_2{=}CCl_2$, $N{=}CCl_2$, C_6H_5, or p-$NO_2C_6H_5$) with **triethylamine trihydrofluoride** in acetonitrile at 20 °C [*39*]. Anodic fluorination of di- and triamidophosphites in the presence of tetraethylammonium tetrafluoroborate gives diamido phosphorofluoridates, albeit in only 19–25% yields [*40*]. Alkyl amido phosphonofluoridites are obtained by reacting esters of alkylamidophosphonous acids with antimony trifluoride or zinc difluoride [*41*] (equation 28).

[*41*] **28**

$$(CH_3)_2CHP(OR)N(CH_2CH_3)_2 \xrightarrow[40-50°C]{SbF_5\ (ZnF_2)} (CH_3)_2CHP(F)N(CH_2CH_3)_2$$

R = CH_3, C_2H_5 55–60%

In contrast to phosphorus esters, sulfur esters are usually cleaved at the carbon–oxygen bond with carbon–fluorine bond formation. Cleavage of ***esters of methanesulfonic acid, p-toluenesulfonic acid,*** and especially ***trifluoromethanesulfonic acid (triflic acid)*** by fluoride ion is the most widely used method for the *conversion of hydroxy compounds to fluoro derivatives.* **Potassium fluoride, triethylamine trihydrofluoride,** and **tetrabutylammonium fluoride** are common sources of the fluoride ion. For the cleavage of a variety of alkyl mesylates and tosylates with potassium fluoride, polyethylene glycol 400 is a solvent of choice; the yields are limited by solvolysis of the leaving group by the solvent, but this phenomenon is controlled by bulky substituents, either in the sulfonic acid part or in the alcohol part of the ester [*42*] (equation 29).

[*42*] **29**

$$R^1OSO_2R^2 \xrightarrow[50-60°C,\ 50-168h]{KF,\ \text{polyethylene glycol 400}} R^1F \quad 35-88\%$$

R^1 = C_8H_{17}, $(CH_3)_2CCH_2$, $C_6H_5CH_2CH_2$

R^2 = CH_3, p-$CH_3C_6H_4$, naphthyl, mesityl

Displacement of a sulfonyloxy group by fluorine can be achieved by **tetrabutylphosphonium fluoride** and its mono- and dihydrofluoride under mild conditions and in good yields [*14*] (equation 30).

[*14*] **30**

$$R^1OSO_2R^2 \xrightarrow[20-60°C]{Bu_4PHF_2,\ THF} R^1F + \text{alkene}$$

$R^1 = C_{12}H_{25}CH(CH_3)$, $C_{14}H_{29}$, 3α– and 3β–androstanonyl

$R^2 = CH_3$, p–$CH_3C_6H_4$

A one-pot conversion of ***benzyl alcohols*** to *benzyl fluorides* by treatment of the alcohols with a combination of methanesulfonyl fluoride, cesium fluoride, and 18-crown-6 ether in tetrahydrofuran has been reported. The reaction involves mesylation of the alcohols followed by cleavage of the resultant mesyl esters with a fluoride ion. The reaction has been extended also to certain heterocycles bearing the *N*-hydroxymethyl group [*43*] (equation 31).

[*43*] **31**

$$XC_6H_5CH_2OH \xrightarrow[\text{reflux, 5–15h}]{CsF,\ CH_3SO_2F,\ 18\text{-crown-6},\ THF} XC_6H_5CH_2F \quad 23\text{–}80\%$$

X = 2–OCH_3, 3–OCH_3, 4–OCH_3, 4–CH_3, 4–Cl, 4–CF_3
4–$OCH_2C_6H_5$, 2,5–$(OCH_3)_2$

Treatment of some ***mesylated N,N-diallylamino sugars*** with triethylamine trihydrofluoride results in the fluorination and regioselective transfer of the nitrogen atom to give, after platinum-catalyzed reductive deallylation, 2,3-aminofluorodeoxysugars. A mechanism involving aziridinium ions has been proposed [*44*] (equation 32).

[*44*] **32**

(All)$_2$N, OCH_3, OSO_2CH_3 —i→ [$CH_3SO_2O^-$ $\overset{+}{N}$(All)$_2$ OCH_3] —ii→ H_2N, OCH_3, F 71–83%

All = allyl

i: $Et_3N{\cdot}3HF$, CH_3CN, DMF or HMPA, 70°C

ii: 1. $Et_3N{\cdot}3HF$, 2. Pt/C

In contrast to the usual behavior, replacement of the mesyl group in 2-*O*-mesyl-3-diallylaminodeoxy-α-D-altropyranoside by treatment with triethylamine trihydrofluoride leads to, because of neighboring-group participation, the fluorinated product with retention of configuration [*45*] (equation 33).

[*45*] **33**

C_6H_5CH ... OSO_2CH_3 ... OCH_3 ... $N(All)_2$ —($Et_3N{\cdot}3HF$, 70°C, 2h)→ C_6H_5CH ... F ... OCH_3 ... $N(All)_2$

All = allyl 99%

Cleavage of mesyl or tosyl esters with $K^{18}F$ in the presence of 18-crown-6 ether [*46*] or Kryptofix 222 [*47*] provides a reliable method for the preparation of ^{18}F-labeled biologically active compounds.

Cleavage of ***trifluoromethanesulfonates*** with fluoride ion is, because of the unusual ease of displacement of the triflyl group, of particular importance for the preparation of monofluoro derivatives. The leaving ability of the triflyl group is four to five orders of magnitude higher than that of the tosyl group [*48*]. In the preparation of triflate esters, *the triflic group is so reactive that it is displaced by the alcohol to give an ether* [*49*].

Various sources of fluoride ion have been investigated, of which highly nucleophilic tetraalkylammonium fluorides are the most effective. Thus, *fluoroalkyl halides* and *N-(fluoroalkyl)amines* are efficiently synthesized by treatment of the corresponding trifluoromethanesulfonic esters with **tetrabutylammonium fluoride trihydrate** in aprotic solvents [*50*] (equation 34). The displacement reactions proceed quantitatively at room temperature within seconds, but fail with hydrogen fluoride–pyridine and give reasonable yields only with hydrogen fluoride–alkylamine reagents.

$$[50] \quad X(CH_2)_nOSO_2CF_3 \xrightarrow[\text{RT, solvent}]{Bu_4NF\cdot 3H_2O} X(CH_2)_nF \quad \textbf{34}$$

50–100%

n = 2,3; X = Br, I, piperazinyl, piperonyl

The effectiveness of various sources of fluoride ion in the displacement of the trifluoromethanesulfonic group has been demonstrated while introducing a fluorine atom into the five-membered carbocyclic ring of a prostaglandin precursor. Treatment of the corresponding triflyl derivative with potassium fluoride in acetonitrile or with cesium fluoride in refluxing dimethylformamide or hexamethylphosphoric triamide leads to disappointingly low yields (2–16%). The use of tetrabutylammonium fluoride in refluxing tetrahydrofuran affords a 55% reproducible yield of the desired fluoro derivative [*51*] (equation 35).

[*51*] **35**

CF_3SO_2O, $C_6H_5CH_2O$, OR — Bu_4NF, THF; reflux, 15 min → F, $C_6H_5CH_2O$, OR

R = $CH_2OCH_2CH_2OCH_3$ 68–81%

Replacement of the triflate group by fluorine with **tetraalkylammonium fluorides** has been successfully applied in the field of ***carbohydrates.*** The reactions of suitably protected 2-*O*-trifluoromethanesulfonyl-β-D-glucopyranosides [*52, 53*] or 2-*O*-trifluoromethanesulfonyl-β-D-mannopyranosides [*54*] in acetonitrile, tetrahydrofuran, or acetone under mild conditions result in displacement of the triflyl group by fluorine with inversion of configuration to give, respectively, 2-deoxy-2-fluoro-β-D-mannopyranosides or 2-deoxy-2-fluoro-β-D-glucopyranosides in high yields (equation 36).

[*53*] **36**

C_6H_5CH, OR^1, OCH_3, OSO_2CF_3 — $R^2{}_4NF$, CH_3CN; reflux, 30–60 min → C_6H_5CH, OR^1, F, OCH_3 68–81%

R^1 = CH_3, $C_6H_5CH_2$; R^2 = CH_3, C_2H_5, C_4H_9

Fluorination of trifluoromethanesulfonyl carbohydrates with cesium fluoride under forcing conditions (130 °C in dimethylformamide) [*55*] or the use of acetyl-protected substrates [*56*] gives considerably lower yields.

An example of cleavage of the sulfur–oxygen bond in trifluoromethanesulfonic ester has been reported. Trifluoromethyl triflate reacts with neutral or anionic nucleophiles with elimination of carbonyl difluoride and formation of trifluoromethanesulfonyl fluoride [*57*] (equation 37). The mechanism of this reaction involves elimination of fluoride ion, which is a chain carrier in the substitution of fluorine for the trifluoromethoxy group.

[57] **37**

$$CF_3SO_2OCF_3 \xrightarrow[25-55°C,\ 5-15\ min]{Nu,\ THF} CF_3SO_2F + COF_2$$

100%

Nu = C_5H_5N, $(C_2H_5)_3N$, $(C_7H_{15})_4N^+I^-$, CsF

Cyclic sulfates rapidly react with the fluoride ion sources to give monofluoro derivatives. Thus, the 2,3-cyclic sulfate of methyl-4,6-*O*-benzylidene-β-D-mannopyranoside cleanly reacts with **tetramethylammonium fluoride** to give methyl 4,6-*O*-benzylidene-2-deoxy-2-fluoro-β-D-glucopyranoside-3-sulfate. Acid hydrolysis followed by acetylation gives 2-deoxy-2-fluoro-β-D-glucopyranoside triacetate in 84% isolated yield [*58*] (equation 38).

[*58*] **38**

C_6H_5CH O O O OCH_3 O SO_2 —i→ C_6H_5CH O O OSO_3^- OCH_3 O F —ii→ AcO O OAc OCH_3 AcO F 84%

i: $(CH_3)_4NF$, CH_3CN, reflux 10 min; ii: 1. H_2O, 2. Ac_2O

A conversion of 2- and 3-hydroxyalkyltrifluoromethanesulfonamides to 2- and 3-fluoroalkylamines on treatment with tetrabutylammonium fluoride has been described. In these reactions the sulfonyl center undergoes intramolecular nucleophilic attack by the hydroxyl group with the expulsion of trifluoromethyl anion. The resulting cyclic sulfamide–esters then undergo stereospecific nucleophilic ring opening with fluoride ion [*59*] (equation 39). Examples yielding acyclic fluoroalkylamines have also been reported. This chemistry is useful for the incorporation of ^{18}F atoms into biologically active amines [*59*].

Replacement of Hydroxyl Group by Fluorine

One of the most useful ways of introducing fluorine into organic compounds is the *replacement of the hydroxyl group in alcohols, hydroxy compounds, and carboxylic acids.*

Methyl alcohol reacts with anhydrous hydrogen fluoride at 100-500 °C in the presence of aluminum fluoride [*60, 61*], zinc fluoride [*62*], chromium fluoride [*63*], or a mixture of aluminum and chromium fluorides [*64*] to give a 20–78% yield of fluoromethane. Attempted fluorinations of higher alcohols by this method failed [*60*].

[*59*] **39**

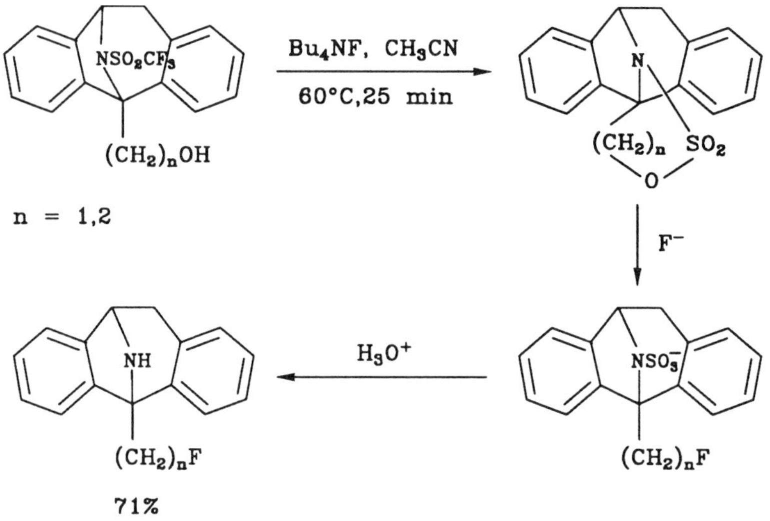

NSO_2CF_3

$(CH_2)_nOH$

n = 1,2

Bu_4NF, CH_3CN

60°C,25 min

N

$(CH_2)_n$ SO_2

O

F^-

NH

$(CH_2)_nF$

71%

H_3O^+

NSO_3^-

$(CH_2)_nF$

Mixtures of anhydrous hydrogen fluoride and tetrahydrofuran are successfully used as fluorinating agents to convert 1,1,2-trifluoro-1-alken-3-ols, easily prepared from bromotrifluoroethene via lithiation followed by the reaction with aldehydes or ketones, to 1,1,1,2-tetrafluoro-2-alkenes. The yields are optimal with a 5:1 ratio of hydrogen fluoride to tetrahydrofuran. The fluorination reaction involves a fluoride ion-induced rearrangement (S_N2' mechanism) of allylic alcohols [*65*] (equation 40).

[*65*] **40**

$$CF_2{=}CFBr \xrightarrow[ii]{i} CF_2{=}CFC(R^1)(R^2)OH \xrightarrow{iii} CF_3CF{=}CR^1R^2$$

i: CH_3Li, Et_2O, −78°C; ii: $R^1R^2C{=}O$, Et_2O, −78°C

iii: 5HF·THF, −78°C to RT, 4–6h

R^1	R^2	%[a]	Z/E
H	C_6H_5	72	100/0
H	C_2H_5	68	100/0
H	$CH_2{=}CH$	72	95/0
CH_3	CH_3	60	
CH_3	C_6H_5	58	50/50

[a] Overall yields from $CF_2{=}CFBr$

A **70:30% hydrogen fluoride–pyridine mixture (Olah's reagent)** can *replace benzylic hydroxyl groups by fluorine.* 1-Chloro-2-hydroxy-2-phenylethanes react at room temperature to give 50–80% isolated yields of 1-chloro-2-fluoro-2-phenylethanes. The conversions of alcohols are almost quantitative, but the effective yields depend on the stability of the resulting fluorides. The method is suitable for fluorination of benzylic alcohols in which the Hammett coefficient for a benzene ring substituent is within the range –0.2 to 0.4; for higher values the reactivity is too low, and for lower values polycondensation of the fluorides occurs [*66*] (equation 41).

[*66*] **41**

$$XC_6H_4CH(OH)CH_2Cl \xrightarrow[\text{RT, 1-48h}]{HF-C_5H_5N} XC_6H_4CHFCH_2Cl \quad (50\text{-}80\%)$$

X = H, 4-CH_3, 4-Br, 4-F, 2,4-Cl_2

A *fluorine–hydrogen rearrangement* occurs during the reaction of α-isopropylbenzyl alcohols in hydrogen fluoride–pyridine to give α- or β-fluorides or their mixture. The product ratio depends on the mole fraction of hydrogen fluoride X_{HF}. The rearrangement is suppressed by electron-withdrawing substituents either on the benzene ring or on the aliphatic carbons [*67*] (equation 42).

[*67*] **42**

$$C_6H_5CH(OH)CH(CH_3)_2 \xrightarrow[\text{RT, 24h}]{HF-C_5H_5N} C_6H_5CHFCH(CH_3)_2 + C_6H_5CH_2CF(CH_3)_2$$

X_{HF}	$C_6H_5CHFCH(CH_3)_2$ %	$C_6H_5CH_2CF(CH_3)_2$ %
0.82	100	0
0.85	50	50
0.87	0	100

Reactions of benzylic α,β-aminoalcohols and α-hydroxyaziridines with Olah's reagent provide a highly efficient way to α,β-fluoroamines, α-fluoroaziridines, and α,γ-difluoroamines [*68*] (equations 43 and 44).

Cyclopropyl methanols when treated with a combination of hydrogen fluoride, pyridine, potassium hydrogen fluoride, and diisopropylamine undergo fluorination and rearrangement to give excellent yields of homoallylic fluorides. Chlorobenzene is a solvent of choice [*69*] (equation 45). A similar treatment of 1-substituted cyclopropyl methanols at low temperatures leads to ring expansion to give

1-fluoro-1-substituted cyclobutanes; the yields dramatically decrease at 0 °C [*70*] (equation 46).

[*68*] **43**

$$\underset{\text{OH}\ \ \text{NHR}^4}{C_6H_5CR^1CR^2R^3} \xrightarrow[\text{20–120°C, 48h}]{\text{10HF–}C_5H_5N} C_6H_5CFR^1\underset{\text{NHR}^4}{CR^2R^3}$$

$R^1,R^2,R^3 = CH_3$; $R^4 = H$	82%
$R^1 = C_2H_5$; $R^2,R^3 = CH_3$; $R^4 = H$	100%
$R^1,R^2 = H$; $R^3,R^4 = CH_3$	95%
$R^1,R^2 = C_2H_5$; $R^3,R^4 = -CH_2-$	82%
$R^1 = H$; $R^2 = C_2H_5$; $R^3,R^4 = -CH_2-$	65%

[*68*] **44**

$$(C_6H_5)_2C(OH)CH\text{—}CHC_6H_5\ (\text{NH aziridine}) \xrightarrow[\text{20°C, 9h}]{\text{10HF–}C_5H_5N} (C_6H_5)_2CFCH(NH_2)CHFC_6H_5$$

65%

[*69*] **45**

$$\text{1-(}CHR^1R^2\text{)-1-OH-2-}R^3\text{-2-}R^4\text{-cyclopropane} \xrightarrow[C_6H_5Cl\text{, RT, 10 min}]{HF\text{–}C_5H_5N\text{, }KHF_2\text{, }i\text{-}Pr_2NH} R^3R^4CFCH_2CH{=}CR^1R^2$$

37–80%

$R^1 = C_6H_5$, $C_6H_5CH{=}CH$, $C_6H_5C{\equiv}C$
$R^2 = H$, CH_3, C_4H_9, $C_6H_5C{\equiv}C$; $R^3 = H$, CH_3; $R^4 = H$, CH_3

[*70*] **46**

$$\text{1-}R^1\text{-1-(}CR^2R^3OH\text{)cyclopropane} \xrightarrow[CH_2Cl_2\text{, −50°C, 10 min}]{HF\text{–}C_5H_5N\text{, }KHF_2\text{, }i\text{-}Pr_2NH} \text{1-}R^1\text{-1-F-2-}R^2\text{-2-}R^3\text{-cyclobutane}$$

55–71%

$R^1 = CH_3$, C_6H_5; $R^2 = H$, CH_3, C_4H_9; $R^3 = C_6H_5$

α-Fluoroamines are versatile reagents for the replacement of hydroxyl groups by fluorine. **2-Chloro-1,1,2-trifluoroethyldiethylamine [fluoroamino reagents (FAR); Yarovenko–Raksha reagent]** has been used extensively to convert alcohols into monofluorides; the earlier literature covering the scope of application, particularly in the field of steroids and hydrocarbons, has been reviewed [*71, 72*].

In contrast to reactions with hydrogen fluoride–pyridine [*68*] (equation 43), the fluorination of benzylic α,β-aminoalcohols with FAR is highly stereospecific and yields the corresponding α,β-fluoroamines with retention of configuration [*73*] (no yields are reported). The fluorination of ethyl 2-hydroxy-2-phenylacetates with FAR also proceeds stereoselectively to give 2-fluoro-2-phenylacetates with a 46% enantiomeric excess [*74*] (no yields are reported), but the reaction of benzylic azidoalcohols is nonstereospecific and gives mixtures of diastereoisomers independently of the configuration of the alcohols [*75*] (equation 47).

[*75*] **47**

$$C_6H_5CH(OH){-}CH(N_3)R \xrightarrow[0°C\text{ to RT, }15h]{CHClFCF_2NEt_2,\ CH_2Cl_2} C_6H_5CHFCH(N_3)R$$

(*erythro* or *threo*) 90%

R = D, CH_3, C_6H_5 (*erythro* and *threo*)

Fluorination of ***bridgehead hydroxyl groups*** with 2-chloro-1,1,2-trifluoroethyldiethylamine leads to the corresponding *bridgehead fluorides.* Thus, esters of gibberellic and tetrahydrogibberellic acids react with FAR in methylene chloride at room temperature to give 7-fluorogibberellins [*76*] (no yields are reported). Conversion of bridgehead bicyclo[2.2.2]octanols to the corresponding fluorobicyclo[2.2.2]octanes requires heating with FAR without a solvent [*77*] (equation 48).

[*77*] **48**

4-R-bicyclo[2.2.2]octan-1-ol (OH, R) $\xrightarrow[60-140°C,\ 30-90\text{ min}]{CHClFCF_2NEt_2}$ 1-fluoro-4-R-bicyclo[2.2.2]octane (F, R)

R = H, CH_3, F, C_6H_5, $O_2NC_6H_5$, CH_3COO 50–64%

Replacement of hydroxyl by fluorine converts ***perfluoroalkanecarboxylic acids*** to *perfluoroalkanoyl fluorides.* The short-chain acids react exothermally with FAR added in a dropwise manner, whereas the longer chain acids require heating [*78*] (equation 49). Attempts to fluorinate pentafluorobenzoic acid with FAR leads to decarboxylation to give pentafluorobenzene [*78*].

[*78*] **49**

$$R_FCOOH \xrightarrow[20-100°C]{CHClFCF_2NEt_2} R_FCOF + CHClFCONEt_2$$

53–88%

R_F = CF_3, CHF_2, O_2NCF_2, CF_3OCF_2, $CClF_2CF_2$
C_3F_7, C_4F_9, $(CF_3)_2CH$, C_6F_{13}

3-Hydroxy-1-adamantanecarboxylic acid reacts exothermally with FAR to give 3-fluoro-1-adamantanoyl fluoride, which, after treatment with ammonium hydroxide, yields 3-fluoro-1-adamantanecarboxamide [*79*] (equation 50).

[*79*] **50**

COOH (adamantane)–OH $\xrightarrow[\text{exothermic reaction}]{CHClFCF_2NEt_2}$ COF (adamantane)–F $\xrightarrow{NH_4OH}$ $CONH_2$ (adamantane)–F

68.8%

2-Chloro-1,1,2-trifluoroethyldiethylamine, despite its popularity, is rarely available commercially owing to its short shelf-life, which demands that it be refrigerated and used within 4 weeks of isolation and preferably straightaway. It is advisable to redistil the reagent if it is to be used after a longer period of storage. To avoid this problem, a stable, transportable form of FAR has been developed by embedding it in a styrene polymer. This polymer-supported fluoroamine reagent converts alcohols into the corresponding fluorides with over 90% yields, retaining its fluorinating capacities for months when stored under nitrogen in polyethylene bottles [*80*] (equation 51).

[*80*] **51**

$$ROH \xrightarrow[20°C]{PFAR,\ CH_2Cl_2} RF \quad >90\%$$

$$PFAR = -(CH_2CH_2)_n- \text{ with } C_6H_4CH_2N(CH_2CH_3)CF_2CHFCl$$

ROH = 1–hexanol, 1–adamantanol, cholesterol

A stable fluorinating agent for the replacement of hydroxyl group by fluorine is **1,1,2,3,3,3-hexafluoropropyldiethylamine (Ishikawa reagent).** This reagent, readily prepared by addition of diethylamine to hexafluoropropene, does not deteriorate and can be used even months after it is prepared [*81*]. 1,1,2,3,3,3-Hexafluoropropyldiethylamine is usually contaminated with a certain amount of 1-diethylaminopentafluoropropene, which itself has no fluorinating ability but, by the addition of hydrogen fluoride formed by the reaction of the former with hydroxylic compounds, is converted to 1,1,2,3,3,3-hexafluoropropyldiethylamine. Therefore, both components of the Ishikawa reagent function ultimately as fluorinating agents. In general, 1,1,2,3,3,3-hexafluoropropyldiethylamine reacts with hydroxylic compounds to give fluorides and the equivalent amounts of *N,N*-diethyl-2,3,3,3-tetrafluoropropionamide and hydrogen fluoride. In many cases, however, considerable amounts of by-products, lowering the yields of the fluorides, are formed (equation 52 and Table 4).

[*81,82,83,84*] **52**

$$ROH + CF_3CHFCF_2N(CH_2CH_3)_2 \xrightarrow[\text{RT, overnight}]{\text{solvent}} RF + CF_3CHFCON(CH_2CH_3)_2 + HF \text{ (+ byproducts)}$$

Lower aliphatic primary alcohols including octanol, halogeno alcohols, and benzylic alcohols yield only alkyl fluorides [*81, 82*]. The reaction of higher primary alcohols gives a mixture of fluorides and alkyl 2,3,3,3-tetrafluoropropionates [*83*] and 2-nitro alcohols; alcohols branched at C-2 [*82, 84*] and unsaturated alcohols [*85*] give 2,3,3,3-tetrafluoropropionates exclusively.

Secondary and tertiary alcohols react with the Ishikawa reagent to give the corresponding fluorides and, usually, considerable amounts of alkenes or ethers [*81*] (Table 5).

1,1,2,3,3,3-Hexafluoropropyldiethylamine is a particularly useful reagent for conversion of secondary benzylic hydroxy esters into the corresponding secondary benzyl fluorides. The reactions proceed with inversion of configuration and a high degree of stereospecificity [*86, 87*] (equation 53).

[*86,87*] **53**

$$R^1C_6H_4CH(OH)R^2 \xrightarrow[\text{RT, overnight}]{CF_3CHFCF_2NEt_2,\ CH_2Cl_2} R^1C_6H_4CHFR^2 \quad 50\text{–}75\%$$

R^1 = H, *o,m,p*-CH_3, *p*-C_2H_5, *p*-$(CH_3)_2CH$, *p*-Cl, *p*-Br *p*-CH_3O, *p*-C_2H_5O, *p*-$(CH_3)_2CHO$

R^2 = $CO_2C_2H_5$, $CHR^3CO_2C_2H_5$; R^3 = H, alkyl

Fluorination of tertiary benzylic hydroxy esters with the Ishikawa reagent gives somewhat lower yields of fluorides because of the formation of dehydrated products, 2-aryl acrylates [*87*] (equation 54).

The reactions of 1,1,2,3,3,3-hexafluoropropyldiethylamine with secondary aliphatic β-hydroxy esters give reasonable to good yields of their corresponding fluorides, whereas aliphatic α-hydroxy esters yield mostly 2,3,3,3-tetrafluoropropionates [*88*] (equations 55 and 56).

Monoesters and monoethers of 1,2-diols react with the Ishikawa reagent to give mixtures of the corresponding fluorides and 2,3,3,3-tetrafluoropropionates in variable ratios [*89*]. The reactions of 2-halogenocyclohexanols, 3-halogenoborneols [*81*], 1,2-dihalogeno-1-propanols, and 1,3-dihalogeno-2-propanols give exclusively the 2,3,3,3-tetrafluoropropionates [*90*]. Unprotected 1,2-diols and 1,3-diols react with 1,1,2,3,3,3-hexafluoropropyldiethylamine to yield cyclic products,

Table 4. Fluorination of Primary Alcohols with 1,1,2,3,3,3-Hexafluoropropyldiethylamine

R of ROH	*Solvent*	*RF (%)*	*CF_3CHFCO_2R (%)*	*Ref.*
$C_2H_5OCH_2CH_2$–	Et_2O	60		81
$C_6H_5CH_2CH_2$–	Et_2O			81
$C_6H_5CH_2CH(CH_3)CH_2$–	CH_2Cl_2		55	82
$C_6H_5CH_2CH(CH_2CH_3)CH_2$–	CH_2Cl_2		70	82
$(CH_3)_2C(NO_2)CH_2$–	CH_2Cl_2		45	84
$CH_3CH_2CH(NO_2)CH_2$–	CH_2Cl_2		51	84
$C_6H_5CH{=}CHCH_2$–	CH_2Cl_2		29	85
$CH_3(CH_2)_7CH{=}CH(CH_2)_8$–	CH_2Cl_2		56	85
$Cl(CH_2)_4$–	Et_2O	64		82
$CH_3CH_2CHBrCH_2$–	Et_2O	50		82
$CH_3(CH_2)_2CHBrCH_2$–	Et_2O	55		82
$Cl(CH_2)_6$–	Et_2O	69		82
$Br(CH_2)_6$–	Et_2O	68		82
$Br(CH_2)_8$–	Et_2O	55		82
$Br(CH_2)_{11}$–	Et_2O	50		82
$CH_3(CH_2)_7$–	Et_2O	87		81
$CH_3(CH_2)_9$–	THF[a]	53	15	83
$CH_3(CH_2)_{11}$–	THF[a]	45	22	83
$CH_3(CH_2)_{13}$–	THF[a]	36	27	83
$CH_3(CH_2)_{15}$–	THF[a]	37	12	83
$CH_3(CH_2)_{17}$–	THF[a]	51	16	83
$C_6H_5CH_2$–	Et_2O	66	—[b]	81
p-$ClC_6H_4CH_2$–	CH_2Cl_2	48		82
m-$ClC_6H_4CH_2$–	CH_2Cl_2	50		82
o-$ClC_6H_4CH_2$–	CH_2Cl_2	49		82
p-$BrC_6H_4CH_2$–	CH_2Cl_2	40		82
m-$BrC_6H_4CH_2$–	CH_2Cl_2	38		82
o-$BrC_6H_4CH_2$–	CH_2Cl_2	49		82

Table 4—Continued

R of ROH	*Solvent*	*RF (%)*	*CF_3CHFCO_2R (%)*	*Ref.*
m-$IC_6H_4CH_2$–	CH_2Cl_2	50		82
p-$NO_2C_6H_4CH_2$–	CH_2Cl_2	60	—[c]	84
m-$NO_2C_6H_4CH_2$–	CH_2Cl_2	66	—[c]	84
o-$NO_2C_6H_4CH_2$–	CH_2Cl_2	17	—[c]	84

[a]The mixture was kept at 40–50 °C for 3 h and then overnight at room temperature.
[b]Dibenzyl ether (25%) was isolated.
[c]Nitrobenzyl fluorides are obtained by adding the alcohol to an excess of the fluorinating agent. When hexafluoropropyldiethylamine is added to the alcohol, dibenzyl ethers are formed exclusively.

Table 5. Fluorination of Secondary and Tertiary Alcohols with 1,1,2,3,3,3-Hexafluoropropyldiethylamine [*81*][a]

		Yield of Products (%)			
ROH	*Solvent*	*RF*	*Alkene*		*R_2O*
$C_6H_{13}CH(CH_3)OH$	Et_2O	62	1- and 2-C_8H_{16}	25	
$C_6H_{13}CH(CH_3)OH$	CH_3CN	18	1- and 2-C_8H_{16}	52	
c-$C_6H_{11}OH$	Et_2O		c-C_6H_{10}	78	
$C_6H_5CH(OH)CH_3$	Et_2O	56			29
$C_6H_5CH(OH)CHCH_3$	Et_2O	65			27
$C_6H_5CH_2OH$	Et_2O	60			25
$C_6H_5CH(OH)COOC_2H_5$	Et_2O	66			
Borneol	Et_2O	58	Camphene	19	
Cholesterol	CH_2Cl_2*b*	83			
$(CH_3)_3COH$	CCl_4	78	$(CH_3)_2C{=}CH_2$	9	
1-Adamantanol	THF[c]	81			

[a]The mixture was kept at room temperature for 20 h.
[b]The mixture was kept at 0–5 °C for 16 h.
[c]The mixture was refluxed for 5 h.

2-diethylamino-2-(1,2,2,2-tetrafluoroethyl)-1,3-dioxolanes and -dioxanes, respectively, but no replacement of hydroxyl groups by fluorine occurs [*90, 91*].

The use of **diethyltrifluoromethylamine** and ***N,N*-dimethyl-α,α-difluorobenzylamine** as fluorinating reagents to replace hydroxyl groups in alcohols and carboxylic acids has been examined. These fluoroamines are fairly stable com-

References are listed on pages 253–262.

[87] **54**

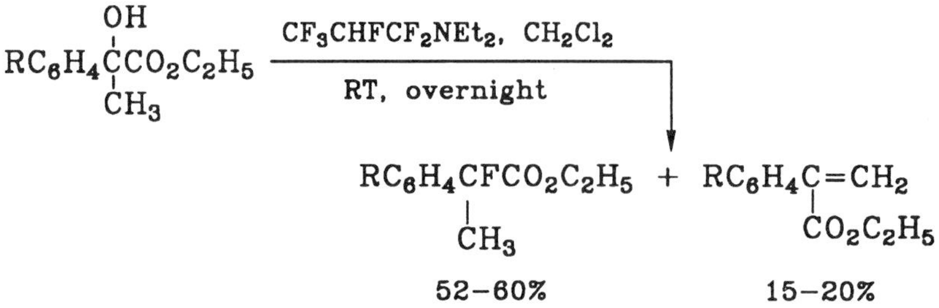

$$RC_6H_4C(OH)(CH_3)CO_2C_2H_5 \xrightarrow[\text{RT, overnight}]{CF_3CHFCF_2NEt_2,\ CH_2Cl_2} RC_6H_4CF(CH_3)CO_2C_2H_5 + RC_6H_4C(=CH_2)CO_2C_2H_5$$

52–60% 15–20%

R = H, p-CH_3, p-C_2H_5, p-i-C_4H_9, p-s-C_4H_9

[88] **55**

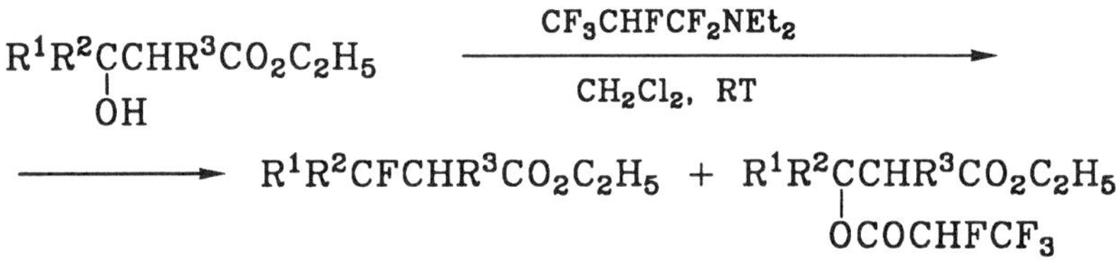

$$R^1R^2C(OH)CHR^3CO_2C_2H_5 \xrightarrow[CH_2Cl_2,\ \text{RT}]{CF_3CHFCF_2NEt_2} R^1R^2CFCHR^3CO_2C_2H_5 + R^1R^2C(OCOCHFCF_3)CHR^3CO_2C_2H_5$$

R^1	R^2	R^3	%	%
CH_3	H	H	38	0
C_2H_5	H	H	50	3
C_5H_{11}	H	H	81	10
$C_{11}H_{23}$	H	H	41	20
$C_{11}H_{23}$	H	CH_3	38	50
C_6H_{13}	CH_3	H	74	0

[88] **56**

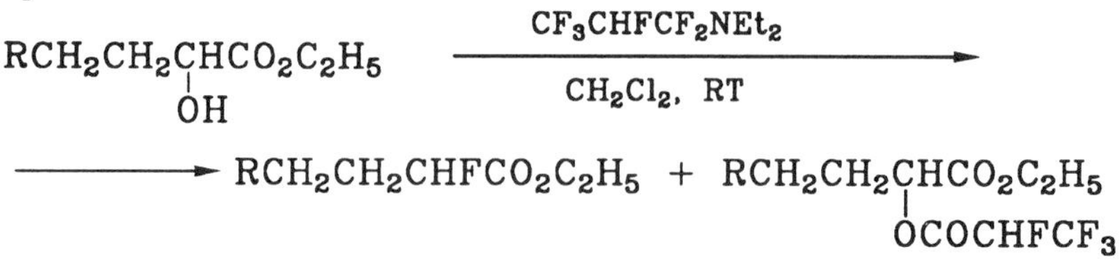

$$RCH_2CH_2CH(OH)CO_2C_2H_5 \xrightarrow[CH_2Cl_2,\ \text{RT}]{CF_3CHFCF_2NEt_2} RCH_2CH_2CHFCO_2C_2H_5 + RCH_2CH_2CH(OCOCHFCF_3)CO_2C_2H_5$$

R	%	%
H	10	31
CH_3	13	42
C_4H_9	3	10
C_6H_{11}	1	3

pounds and can be stored in polyethylene bottles for a long time. Their fluorinating ability, in general, is similar to that of the Yarovenko–Raksha and Ishikawa reagents. Diethyltrifluoromethylamine is particularly useful for the conversion of secondary and tertiary alcohols and low-molecular-weight alkanecarboxylic acids into their respective fluorides, but it is of no use for the fluorination of primary alcohols and higher carboxylic acids. *N,N*-Dimethyl-α,α-difluorobenzylamine gives good results with primary and secondary alcohols and only reasonable yields of *tert*-butyl fluoride from *tert*-butyl alcohol [*92*] (equations 57 and 58).

[*92*] **57**

$$ROH \xrightarrow[\text{reflux, 15–60 min}]{(C_2H_5)_2NCF_3} RF + (C_2H_5)_2NCOF + HF$$

R	CH_3	i-C_3H_7	C_4H_9	t-C_4H_9	CH_3CO	CF_3CO	C_2H_5CO	C_6H_5CO
%	40	75	0	70	82	68	33	16

[*92*] **58**

$$ROH \xrightarrow[\text{reflux, 15–60 min}]{C_6H_5CF_2N(CH_3)_2} RF + C_6H_5CON(CH_3)_2 + HF$$

R	i-C_3H_7	C_4H_9	t-C_4H_9	c-C_6H_{11}	C_2H_5CO
%	74	79	49	trace	49

Dialkylaminotrifluorosulfuranes, despite the potential hazard of explosion if not properly handled [*93*], are very useful and versatile fluorinating agents [*94*]. Among them, commercially available **diethylaminosulfur trifluoride (DAST)** has gained widespread popularity as a fluorinating agent for the *replacement of the hydroxyl group as well as carbonyl oxygen by fluorine.* The principal application of DAST is the conversion of alcohols into monofluorides. Primary, secondary, and tertiary alcohols, both aliphatic and benzylic, are converted into the corresponding fluorides in high yields [*94, 95, 96, 97, 98, 99, 100, 101, 102, 103, 104, 105, 106, 107, 108*]. Fluorinations of alcohols with DAST are most frequently carried out in dichloromethane, fluorotrichloromethane, glyme, or diglyme (Tables 6a–6c). The use of chloroform [*107*], carbon tetrachloride [*109*], tetrahydrofuran [*110, 111, 112, 113, 114*], benzene and toluene [*115, 116, 117, 118*], and isooctane [*95*] as the reaction media has also been reported. Most fluorinations of alcohols with DAST are started at –78 °C and finished by allowing the mixtures to warm to room temperature. However, successful fluorinations have been accomplished also at room temperature [*110*] or at ice-bath temperature [*119*], without solvent, and by reverse addition of the reactant to DAST [*119*]. The fluorination reactions proceed according to the general scheme shown in equation 59.

Table 6a. Fluorination of Primary Alcohols with Diethylaminosulfur Trifluoride (DAST)

Alcohol	*Conditions*	*Product*	*Yield (%)*	*Ref.*
$ClCH_2CH_2OH$	Diglyme, –78 °C to RT	$ClCH_2CH_2F$	69	95
$HOCH_2CH_2OH$	Diglyme, –78 °C to RT	FCH_2CH_2F	70	95
$Br(CH_2)_6OH$	CH_2Cl_2, –78 to 25 °C	$Br(CH_2)_6F$	53	96
$Br(CH_2)_8OH$	CH_2Cl_2, –78 to 25 °C	$Br(CH_2)_8F$	61	96
$C_8H_{17}OH$	CH_2Cl_2, –70 to 25 °C	$C_8H_{17}F$	90	95
$HC{\equiv}C(CH_2)_8OH$	CH_2Cl_2, –78 to 25 °C	$HC{\equiv}C(CH_2)_8F$	82	96
$C_2H_5OCOC{\equiv}CCH_2OH$	CH_2Cl_2, –78 °C 45 min, RT 3 h	$C_2H_5OCOC{\equiv}CCH_2F$	59	97
2-(5-Thiazolinyl)-ethanol	$CHCl_3$, 0 °C to RT, 0.5 h	5-(2-Fluoroethyl)-thiazole	55	107
$C_6H_5CH_2CH_2OH$	CH_2Cl_2, –78 to 40 °C	$C_6H_5CH_2CH_2F$	60	95
3,5-$Br_2C_6H_3CH_2$-CH_2OH	CH_2Cl_2, –78 °C to RT, 1 h	3,5-$Br_2C_6H_3CH_2CH_2F$	60	100
$C_6H_5CH_2OH$	CCl3F, –78 °C to RT	$C_6H_5CH_2F$	75	95
2,6-$Cl_2C_6H_3CH_2OH$	CH_2Cl_2, –70 °C to RT	2,6-$Cl_2C_6H_3CH_2F$	50–80	98
4-$O_2NC_6H_4CH_2OH$	CH_2Cl_2, 45 min to RT	4-$O_2NC_6H_4CH_2F$	90–95	99
2,6-$(CH_3)_2C_6H_3$-CH_2OH	CH_2Cl_2, –70 °C to RT	2,6-$(CH_3)_2C_6H_3CH_2F$	50–80	98

Table 6b. Fluorination of Secondary Alcohols with Diethylaminosulfur Trifluoride (DAST)

Alcohol	*Conditions*	*Product*	*Yield (%)*	*Ref.*
i-C_4H_9OH	–78 °C to RT	*i*-C_4H_9F *t*-C_4H_9F	49 21	95
c-$C_8H_{15}OH$	CCl_3F, –78 °C to RT	*c*-$C_8H_{15}F$ Cyclooctene	70 30	108
1-Hydroxyindane	CH_2Cl_2, –78 °C to RT	1-Fluoroindane	50–80	98
Menthol	CCl_3F, –78 °C to RT	3-Fluoromenthane	50	95
$BrCH_2CH_2CH(OH)$-CH_2Br	CH_2Cl_2, –78 °C to RT	$BrCH_2CH_2CHFCH_2Br$	76	105
$CH_3CH(OH)COO$-C_2H_5	CH_2Cl_2, –78 °C to RT	$CH_3CHFCOOC_2H_5$	78	95
4-$CH_3C_6H_4CH(OH)$-$CF_2SO_2C_6H_5$	CH_2Cl_2, –78 °C, 0.5 h	4-$CH_3C_6H_4CHF$-$CFSO_2C_6H_5$	71	106
$(C_6H_5)_2CHOH$	CH_2Cl_2, –30 °C, 0.5 h	$(C_6H_5)_2CHF$ $[(C_6H_5)_2CH]_2O$	40 44	104
9-Hydroxyfluorene	CH_2Cl_2, –30 °C, 0.5 h	9-Fluorofluorene Difluoroenyl ether	48 13	104

Table 6c. Fluorination of Tertiary Alcohols with Diethylaminosulfur Trifluoride (DAST)

Alcohol	*Conditions*	*Product*	*Yield (%)*	*Ref.*
$C_2H_5C(CH_3)_2OH$	Diglyme, –78 °C	$C_2H_5CF(CH_3)_2$	88	95
$C_6H_5C(CH_3)_2OH$	CH_2Cl_2, –70 °C to RT	$C_6H_5CF(CH_3)_2$	50–80	98
2-$FC_6H_4C(CH_3)_2OH$	Glyme, –78 °C, 60 °C, 6 days	2-$FC_6H_5CF(CH_3)_2$	40	101
$(C_6H_5)_3COH$	CH_2Cl_2, –30 °C, 0.5 h	$(C_6H_5)_3CF$	85	104

59

$$ROH + (C_2H_5)_2NSF_3 \longrightarrow RF + (C_2H_5)_2NSOF + HF$$

Fluorinations with DAST proceed with high chemoselectivity. In general, under very mild reaction conditions usually required for the replacement of hydroxyl groups, other functional groups, including phenolic hydroxyl groups [*112*], remain intact. This provides a method for selective conversion of hydroxy esters [*95, 97*] (Table 6), hydroxy ketones [*120, 121*], hydroxy lactones [*122, 123*], hydroxy lactams [*124*], and hydroxy nitriles [*125*] into fluoro esters, fluoro ketones, fluoro lactones, fluoro lactams, and fluoro nitriles, respectively (equations 60–63).

[*121*] **60**

$$XC_6H_4COCH(OH)CH_3 \xrightarrow[0-25°C,\ 1h]{DAST,\ CH_2Cl_2} XC_6H_4COCHFCH_3$$

80–99%

X = H, 4-*i*-C_4H_9, 4-C_6H_5

[*123*] **61**

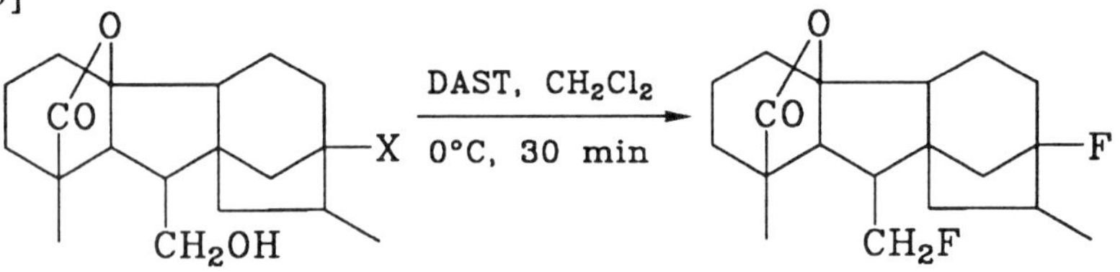

DAST, CH_2Cl_2, 0°C, 30 min

X = F 83%

X = OH 77%

[*124*] **62**

DAST, CH_2Cl_2, −70 to −10°C, 25 min

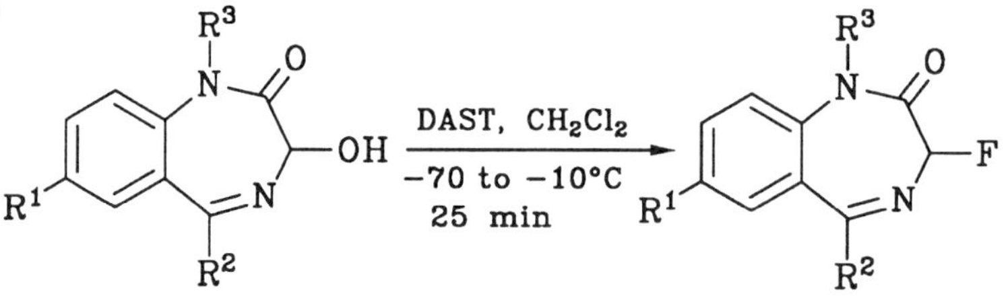

R^1 = H,Br, Cl, NO_2 53–96%

R^2 = C_6H_5, 2-ClC_6H_5, 2-FC_6H_5; R^3 = H, CH_3, C_2H_5

[*125*] **63**

$$R^1_nC_6H_{5-n}CR^2(OH)CN \xrightarrow[5°C\ to\ RT]{DAST,\ CH_2Cl_2} R^1_nC_6H_{5-n}CR^2FCN$$

48–100%

R^1 = Cl, F, NO_2, CH_3O, CH_3COO

R^2 = H, CH_3, C_2H_5; n = 0, 1, 2

α-Amino-β-hydroxy esters in which the amino group is protected as a 4,5-diphenyl-3-oxazoline-2-one moiety undergo fluorination with DAST to give the corresponding β-fluoro compounds as the main products along with the dehydrated compounds [*126*] (equation 64).

[*126*] **64**

DAST, CH_2Cl_2
−78 to 20°C

R	%	%
CH_3	45	24
C_2H_5	48	17
C_3H_7	65	14
$i\text{-}C_4H_7$	64	12

Dehydration to olefins, which sometimes accompanies the reaction of alcohols with DAST [*95, 108*], is seldom as extensive as with α-fluoroamines (FAR and 1,1,2,3,3,3-hexafluoropropyldiethylamine) but occurs in a few cases to the exclusion of fluorination; thus, 9α–fluoro-11-hydroxysteroids give 9α-fluoro-Δ^{11}-steroids [*127, 128*]. Dehydration accompanied by Wagner–Meerwein rearrangement occurs during the fluorination of testosterone [*129*]. Intermolecular dehydration to form ethers in addition to fluorides is observed in the reaction of benzhydryl alcohols [*104*] (Table 6).

Carbocation rearrangements occur in the reactions of some secondary alcohols with DAST; thus isobutyl alcohol gives a mixture of isobutyl fluoride and *tert*-butyl fluoride [*95*] (Table 6), and both borneol and isoborneol rearrange to the same 3-fluoro-2,2,3-trimethylbicyclo[2.2.1]heptane (72–74%) accompanied by camphene [*95*].

Allylic rearrangements are observed in the reactions of allylic alcohols with dialkylaminosulfur trifluorides [*95, 130*]. Both crotyl alcohol and buten-3-ol give mixtures of crotyl fluoride and 3-fluorobutene in ratios that are slightly affected by the polarity of the solvent [*95*] (equations 65 and 66).

[*95*] **65**

$$CH_3CH{=}CHCH_2OH \xrightarrow{\text{DAST, }-78\text{ to }0^\circ C} CH_3CH{=}CHCH_2F + CH_3CHFCH{=}CH_2$$

	$CH_3CH{=}CHCH_2F$	$CH_3CHFCH{=}CH_2$
isooctane	36%	64%
diglyme	28%	72%

[*95*] **66**

$$CH_3CHOHCH{=}CH_2 \xrightarrow{\text{DAST, }-78\text{ to }0^\circ C} CH_3CH{=}CHCH_2F + CH_3CHFCH{=}CH_2$$

	$CH_3CH{=}CHCH_2F$	$CH_3CHFCH{=}CH_2$
isooctane	9%	91%
diglyme	22%	78%

Some ***carbohydrate derivatives,*** in which a free hydroxyl group is adjacent to an acetal oxygen and to an azido group, when treated with DAST undergo specific rearrangements with the participation of the neighboring groups. Either the replacement of the hydroxyl group occurs with retention of configuration or an exchange of the neighboring substituents takes place with inversion of configuration on both chiral centers involved [*116, 117, 131, 132*] (equation 67).

A similar rearrangement takes place with *N,N*-dibenzyl-L-serine benzyl ester [*110*] (equation 68) and with partially protected carbocyclic nucleosides, such as 1-hydroxymethyl-4-(2,4-dinitroanilino)cyclopentane-2,3-diol [*133*].

Skeletal rearrangements involving a ring size change are observed in reactions of DAST with some cyclic alcohols [*131, 134, 135*] (equations 69 and 70).

The replacements of hydroxyl groups by fluorine with DAST usually proceed with *complete inversion of configuration* Thus, fluorination of (+)-(*S*)-2-octanol gives (–)-(*R*)-2-fluorooctane of 97.6% optical purity [*136*], and dimethyl (*S*)-hydroxysuccinate gives dimethyl (*R*)-fluorosuccinate in 85% yield [*137*]. Complete inversions have been reported in the syntheses of fluorinated carbohydrates [*115, 138, 139, 140*] and steroids [*129, 141, 142*]. However, mixtures of stereoisomers resulted from the treatment with DAST of some 4-hydroxyproline derivatives [*143*]. A two-step expedient method for the preparation of diastereomerically and, by extension, even enantiomerically pure vicinal difluoroalkanes involves an oxirane ring opening by addition of hydrogen fluoride and subsequent treatment of the resulting fluorohydrin with DAST [*144*] (equation 71).

[*112*] **67**

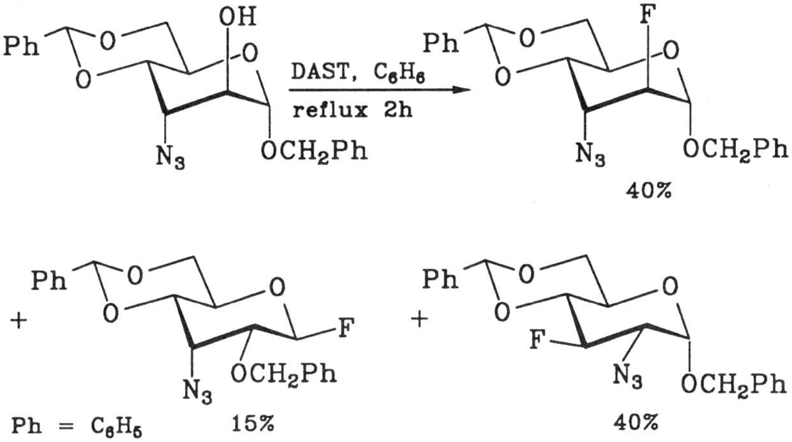

DAST, C_6H_6
reflux 2h

OH, O, Ph, N_3, OCH_2Ph, F

40%

15%

40%

Ph = C_6H_5

[*110*] **68**

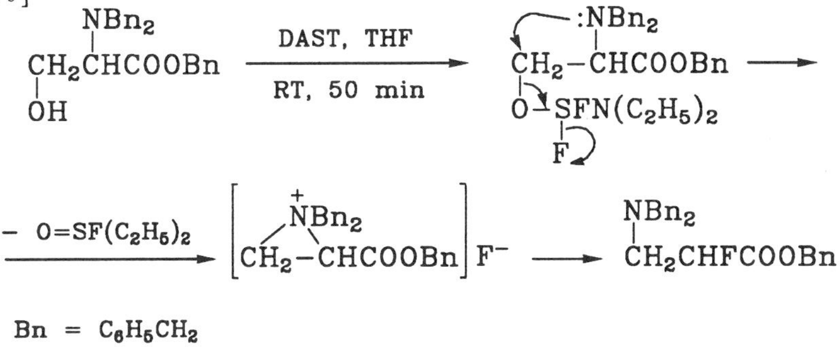

NBn_2
$CH_2CHCOOBn$
OH

DAST, THF
RT, 50 min

$:NBn_2$
$CH_2-CHCOOBn$
$O-SFN(C_2H_5)_2$
F

$- O=SF(C_2H_5)_2$

$[\overset{+}{N}Bn_2\ CH_2-CHCOOBn]\ F^-$

NBn_2
$CH_2CHFCOOBn$

Bn = $C_6H_5CH_2$

[*131*] **69**

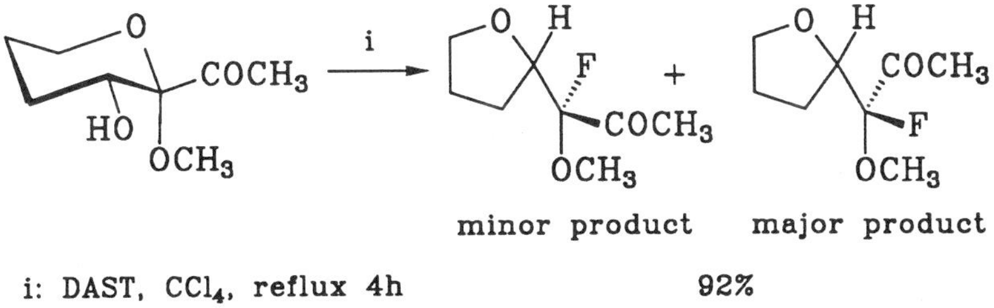

$COCH_3$, HO, OCH_3

i

H, F, $COCH_3$, OCH_3

minor product major product

i: DAST, CCl_4, reflux 4h 92%

[*134*] **70**

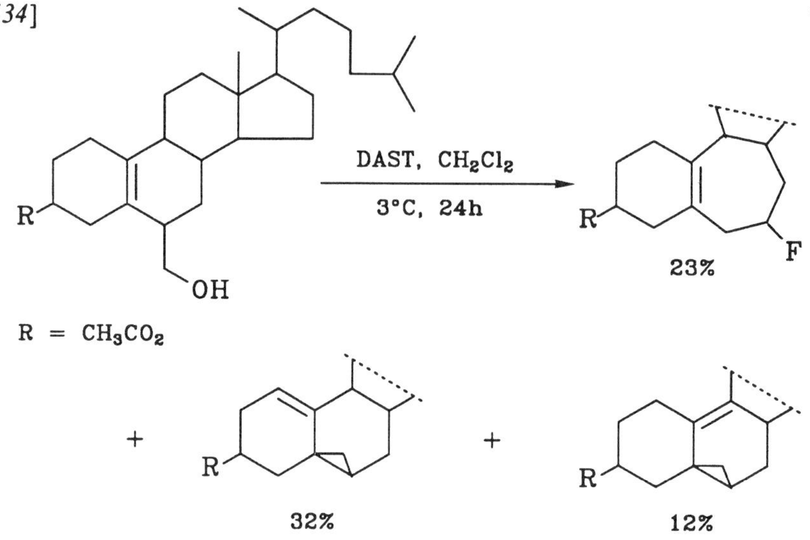

DAST, CH_2Cl_2

3°C, 24h

R = CH_3CO_2

23%

32%

12%

[*144*] **71**

cis *threo* 81% *meso* 61%

trans *erythro* 82% DL 48%

i: $Et_3N{\cdot}3HF$, 150°C, 3h; ii: DAST, CH_2Cl_2, −50°C to RT

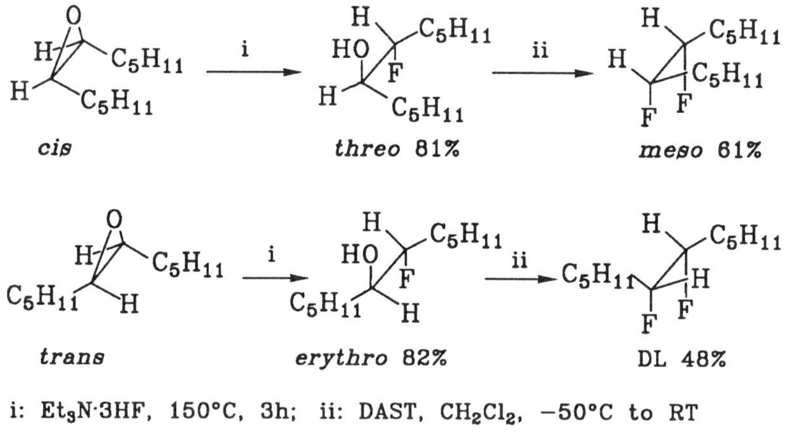

On the other hand, as a result of participation of a neighboring group, *complete or predominant retention of configuration* takes place in many reactions of hydroxylic compounds with DAST. A number of examples have been reported in the field of steroids [*127, 129*], in the conversion of vitamins D into fluoro vitamins D [*145*], and in the fluorination of liquid crystals [*146*] (equation 72).

[*146*] **72**

$$R^1CH(OH)-\underset{\text{O}}{CH-CH}-R^2 \xrightarrow[-78°C,\ 2h]{\text{DAST, } CH_2Cl_2} R^1CHF-\underset{\text{O}}{CH-CH}-R^2 \ (69\%) + R^1CHF\text{-}\underset{\text{O}}{C(H)-CH}-R^2 \ (23\%)$$

$R^1 = C_{10}H_{21}OC_6H_4OCOC_6H_5$; $R^2 = C_6H_{13}$

Reactions of polyhydroxyl compounds such as carbohydrates with DAST lead to replacement of one or two hydroxyl groups by fluorine; more fluorine atoms are not introduced even when a large excess of the reagent is used [*132, 139, 147*].

Although diethylaminosulfur trifluoride (DAST) is the most popular, other dialkylaminosulfuranes, such as diisopropylamino- [*95*], pyrrolidino- [*95, 109, 127*], dimethylamino- [*148*], piperidino- [*148*], and particularly morpholinosulfur trifluoride [*148, 149, 150*], are also used as fluorinating agents to convert alcohols into fluorides.

Morpholinosulfur trifluoride is more thermally stable and therefore safer to handle than DAST and gives a slightly higher yield of the fluoride in the fluorination of cyclohexanol [*93, 150*]. Both solvent and conformational effects are pronounced in the fluorination of cyclohexanols [*150*]. The chiral (*S*)-2-(methoxymethyl)pyrrolidin-1-ylsulfur trifluoride is an effective enantioselective fluorodehydroxylating agent [*151*].

Bis(dialkylamino)sulfur difluorides, $R^1_2NSF_2NR^2_2$, albeit less reactive than DAST, also substitute hydroxyl groups with fluorine [*95, 97, 152*].

Reactions of ***alcohols*** with **sulfur tetrafluoride,** because of decomposition and/or polymerization, usually do not give fluorinated products. However, in the presence of a hydrogen fluoride scavenger like triethylamine or pyridine, even such sensitive substrates as benzylic alcohols [*153*], 2-phenylethanol, and 2-furylmethanol [*154*] can be fluorinated to give the expected fluoro derivatives (equation 73).

[*154*] **73**

$$\text{2-furyl-}CH_2OH + SF_4 + Et_3N \xrightarrow[-50°C,\ 5min]{CH_2Cl_2 \text{ or } C_6H_{12}} \text{2-furyl-}CH_2F \ (20\%) + SOF_2 + Et_3N{\cdot}HF$$

The reaction of ***1,2-diols*** with sulfur tetrafluoride leads to fluorosulfites, which on hydrolysis, give monofluoro alcohols [*155*] (equation 74).

[*155*] **74**

$$\underset{\displaystyle OH}{CH_3\overset{}{C}HCH_2OH} \xrightarrow[20°C]{SF_4} CH_3CHFCH_2OSOF \xrightarrow{H_2O} CH_3CHFCH_2OH$$

α-Hydroxy ketones are selectively fluorinated to *α-fluoro ketones* by treatment with sulfur tetrafluoride in diethyl ether, which probably serves as a hydrogen fluoride scavenger [*156*] (equation 75).

[*156*] **75**

$$\underset{\displaystyle OH}{RCHCOR} \xrightarrow[20°C,\ 70h]{SF_4,\ Et_2O} \underset{70-75\%}{RCHFCOR}$$

$R = C_2H_5,\ C_3H_7$

On the other hand, the **sulfur tetrafluoride–hydrogen fluoride** combination is a selective system for the conversion of hydroxyamines and hydroxyamino acids into fluoroamines and fluoroamino acids [*157, 158*]. In this system, hydrogen fluoride plays various roles: as a protecting reagent for amino groups, as a catalyst, and as a solvent. The displacement of hydroxyl groups by fluorine proceeds with *predominant inversion of configuration* as exemplified by fluorination of L-threonine and L-allothreonine [*158*] (equation 76).

[*158*] **76**

$$CH_3CH(OH)\overset{\displaystyle NH_2}{C}HCOOH \xrightarrow[-78°C\ to\ RT]{SF_4,\ HF} CH_3CHF\overset{\displaystyle NH_2}{C}HCOOH$$

	erythro	*threo*
threo	44.2%	3.8%
erythro (allo)	12.5%	44.5%

Highly fluorinated tertiary alcohols usually give olefins on fluorination with sulfur tetrafluoride [*159*], but in certain cases, replacement of the hydroxyl group with fluorine occurs under mild conditions. Hexafluoro-2-arylpropan-2-ols react with sulfur tetrafluoride at low temperatures to give high yields of heptafluoroisopropylarenes [*160*] (equation 77), and similarly, 3,8-dihydroxy-9,9,9,10,10,10-hexafluoro-*p*-menthane affords 3,8,9,9,9,10,10,10-octafluoromenthane [*160*] (equation 78).

[*160*] **77**

$$\text{X-C}_6\text{H}_4\text{-C(CF}_3)_2\text{-OH} \xrightarrow[\text{-78 to 20 °C, 2 h}]{\text{SF}_4} \text{X-C}_6\text{H}_4\text{-CF(CF}_3)_2$$

X = H, CH_3 75–86%

[*160*] **78**

$$\text{4-CH}_3\text{-2-OH-C}_6\text{H}_9\text{-C(CF}_3)_2\text{-OH} \xrightarrow[\text{-78 to 20 °C, 2 h}]{\text{SF}_4} \text{4-CH}_3\text{-2-F-C}_6\text{H}_9\text{-CF(CF}_3)_2$$

55%

Perfluoropinacol reacts with sulfur tetrafluoride in an unconventional way: instead of replacement of the hydroxyl groups by fluorine, the substitution of four fluorine atoms in the sulfur tetrafluoride molecule with oxygen occurs to give the corresponding spirosulfurane [*161*] (equation 79).

[*161*] **79**

$$(CF_3)_2C(OH)-C(OH)(CF_3)_2 \xrightarrow[\text{20 °C}]{\text{SF}_4} [(CF_3)_2C-O]_2S[O-C(CF_3)_2]_2$$

85%

The reaction of diethyl tartrate with sulfur tetrafluoride at 25 °C results in replacement of one hydroxyl group, whereas at 100 °C, both hydroxyl groups are replaced by fluorine to form α,α′-difluorosuccinate [*162*]. The stereochemical outcome of the fluorination of tartrate esters is retention of configuration at one of the chiral carbon atoms and inversion of configuration at the second chiral center [*163, 164, 165*]. Thus, treatment of dimethyl (+)-L-tartrate with sulfur tetrafluoride gives dimethyl *meso*-α,α′-difluorosuccinate as the final product [*163, 164*], whereas dimethyl *meso*-tartrate is converted into a racemic mixture of D- and L-α,α′-difluorosuccinates [*165*] (equation 80).

80

$$CH_3OCO[CH(OH)]_2COOCH_3 \xrightarrow{SF_4} CH_3OCO[CHF]_2COOCH_3$$

[*163*]	DL	HF,60°C,6h	*meso*	97%
[*164*]	D or L	110°C,6h	*meso*	19.4%
[*165*]	*meso*	110°C,7h	DL	22–23%

Both aliphatic and aromatic carboxylic acids are converted to their fluorides by FAR [*78, 79*], by the Ishikawa reagent [*81*], and by fluoroaminosulfuranes such as DAST [*128, 166*] and its analogues [*128*].

Replacement of Carbonyl Oxygen by Fluorine

Sulfur tetrafluoride is the most common fluorinating agent for replacement of carbonyl oxygen by fluorine. When acted upon by sulfur tetrafluoride, carbonyl groups of aldehydes and ketones are converted into difluoromethylene groups, and carboxylic groups of aliphatic and aromatic carboxylic acids are transformed, via the corresponding acid fluorides, into trifluoromethyl groups. Hundreds of successful applications of sulfur tetrafluoride for selective introduction of fluorine into organic compounds, even as complex as steroids and terpenes, appeared in the literature through 1971 and have been reviewed [*167*]. However, in the past 20 years a number of new and, in many cases, unconventional results involving reactions of sulfur tetrafluoride with carbonyl compounds have been discovered [*168*].

Most of the straight-chain aliphatic ***aldehydes*** react with sulfur tetrafluoride in the usual manner to give the expected 1,1-difluoroalkanes as the only isolable products; reasonable yields are obtained in a narrow temperature range, from –20 to 40 °C. An exception is butyraldehyde, which, when treated with sulfur tetrafluoride below 0 °C, gives, in addition to 1,1-difluorobutane, considerable amounts of bis(1-fluorobutyl) ether [*169*] (equation 81).

[*169*] **81**

$$C_3H_7CHO \xrightarrow[18h]{SF_4} C_3H_7CHF_2 + (C_3H_7CHF)_2O$$

	$C_3H_7CHF_2$	$(C_3H_7CHF)_2O$
−20°C	61%	38%
0°C	85%	15%
20°C	95%	trace

Polyhalodiethyl ethers are also formed in the reaction of sulfur tetrafluoride with perhaloacetaldehydes; in the case of chloral, a chlorine–fluorine rearrangement occurs [*170*] (equations 82 and 83).

[*170*] **82**

$$CF_3CHO \xrightarrow[30°C]{SF_4} \underset{58\%}{CF_3CHF_2} + \underset{16\%}{(CF_3CHF)_2O}$$

[*170*] **83**

$$CCl_3CHO \xrightarrow[30°C]{SF_4} CFCl_2CHFCl\ (57\%) + CCl_3CHF{-}O{-}CCl_2CHFCl\ (12.7\%) + (CCl_3CHF)_2O\ \ 1.3\%$$

A *fluorine–hydrogen migration* is typical for the reactions of aldehydes branched at the carbon atom α to the formyl group. Comparable amounts of 1,1-difluoroalkanes and 1,2-difluoroalkanes together with bis(1-fluoroalkyl) ethers are obtained [*169*] (equation 84).

[*169*] **84**

$$R^1R^2CHCHO \xrightarrow[18h]{SF_4} R^1R^2CHCHF_2 + R^1R^2CFCH_2F + (R^1R^2CHCHF)_2O$$

R^1	R^2	°C	%	%	%
CH_3	CH_3	−10	21	37	35
		0	30	44	22
		40	25	25	12
CH_3	C_2H_5	−17	33	24	43
		20	tar only		
C_2H_5	C_2H_5	0	no reaction		
		20	48	52	0
		40	49	51	0
		60	tar only		
c−C_5H_9CHO		20	41	49	0
c−$C_6H_{11}CHO$		0	79	16	0
		20	80	12	0

Trimethylacetaldehyde reacts with sulfur tetrafluoride with a skeletal rearrangement: 2,3-difluoro-2-methylbutane is formed in high yield as the only fluoroalkane along with bis(1-fluoro-2,2-dimethylpropyl) ether [*169*] (equation 85).

[*169*] **85**

$$(CH_3)_3CCHO \xrightarrow[0-40°C,\ 18h]{SF_4} (CH_3)_2CFCHFCH_3\ (65-81\%) + [(CH_3)_3CCHF]_2O\ (9-26\%)$$

A powerful solvent effect is observed in the reaction of *p*-nitrobenzaldehyde with sulfur tetrafluoride. The reaction conducted in a benzene solution gives the expected *p*-nitrobenzylidene fluoride; without a solvent, bis(*p*-nitro-α-fluorobenzyl) ether is formed as the sole product [*171*] (equation 86).

[*171*] **86**

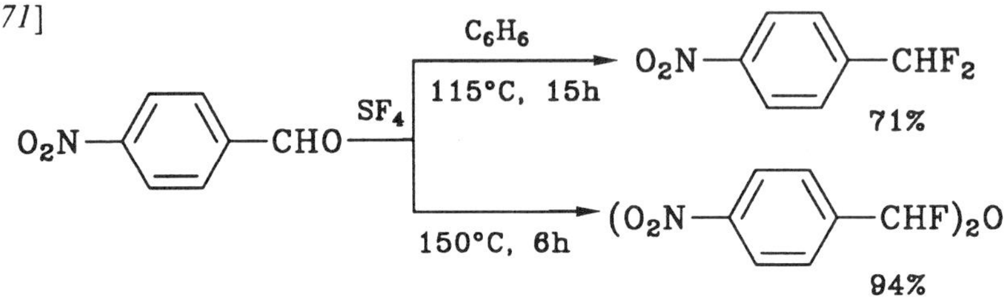

$$O_2N{-}C_6H_4{-}CHO \xrightarrow[115°C,\ 15h]{SF_4,\ C_6H_6} O_2N{-}C_6H_4{-}CHF_2 \quad 71\%$$

$$O_2N{-}C_6H_4{-}CHO \xrightarrow[150°C,\ 6h]{SF_4} (O_2N{-}C_6H_4{-}CHF)_2O \quad 94\%$$

Unsaturated and some aromatic aldehydes react cleanly with sulfur tetrafluoride in the presence of potassium fluoride (a hydrogen fluoride scavenger) to give reasonable to good yields of the corresponding difluoromethyl derivatives [*172*] (equation 87).

[*172*] **87**

$$RCHO + SF_4 \xrightarrow[20°C]{KF} RCHF_2 + SOF_2$$

R	h	%
1-Cyclohexenyl	48	43
trans-Styryl	72	66
2-Furyl	72	17
2-Thienyl	72	73

Fluorination of a ***ketone,*** 4-protoadamantanone, with sulfur tetrafluoride gives a 93% yield of a rearrangement product, 1,2-difluoroadamantane [*173*] (equation 88).

[*173*] **88**

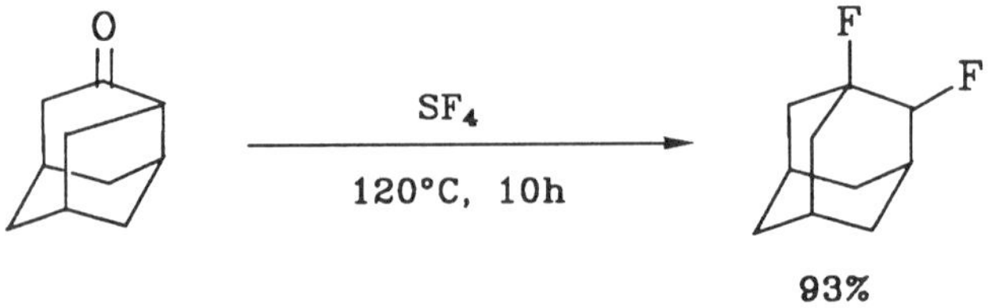

$$\text{4-protoadamantanone} \xrightarrow[120°C,\ 10h]{SF_4} \text{1,2-difluoroadamantane} \quad 93\%$$

Treatment of glyoxal with sulfur tetrafluoride in the presence of sodium fluoride results in the formation of difluoroethylene glycol orthosulfite [*174*] (equation 89). Similarly, perfluorinated 1,2-diketones react with sulfur tetrafluoride to give tetraoxyspirosulfuranes as the only products. Thus, perfluorobiacetyl gives a crystalline product, perfluorobutylene glycol 2,3-orthosulfite [*175*] (equation 89).

$$XCOCOX + SF_4 \xrightarrow[20°C]{} \text{(XFC)}_2\text{O}_2\text{S}\text{O}_2\text{(CFX)}_2 \qquad 89$$

[*174*]	X = H	NaF, 12h	100%
[*175*]	$X = CF_3$	24h	100%

Nonfluorinated α- ***and*** β-***diketones*** give the corresponding tetrafluoroalkanes as the major products together with considerable amounts of alkenes, ethers, and sulfites. Formation of side products is reduced, and yields of tetrafluoroalkanes are greatly improved by conducting the fluorination reactions in an excess of anhydrous hydrogen fluoride [*176, 177*].

1,5-Diphenylperfluoropentan-1,5-dione, in contrast to its shorter chain homologue, which gives the normal fluorination product, on treatment with sulfur tetrafluoride and hydrogen fluoride forms a cyclic ether as the only product [*178*] (equation 90).

[*178*] 90

$$C_6H_5CO(CF_2)_3COC_6H_5 \xrightarrow{SF_4,\ HF}$$ 2,6-difluoro-2,6-diphenyl-3,3,4,4,5,5-hexafluorotetrahydropyran (F_2, F_2, F_2; C_6H_5, F, O, F, C_6H_5)

80%

Results of fluorination of β-***keto esters*** with sulfur tetrafluoride depend on the presence of hydrogen on the α-carbon. Unsubstituted aceto- and propionylacetates give 3,3-difluoroalkanoates, 3-fluoro-2-alkenoates, and 2-alkynoates [*179*]. α-Alkylacetoacetates yield cumulene esters instead of acetyl- enic esters, but fluorination of α,α-dialkylacetoacetates affords exclusively 2,2-dialkyl-3,3-difluorobutanoates [*180*]. However, in the presence of anhydrous hydrogen fluoride, all types of α-keto esters give the corresponding 3,3-difluoro derivatives in high yields [*180, 181*] (equations 91–93). In contrast to the reactions with sulfur tetrafluoride, β-keto esters react with diethylaminosulfur trifluoride quite differently (*see* equation 100).

On treatment with sulfur tetrafluoride at 20 °C, aliphatic α-***hydroxy ketones*** afford mixtures of α,β,β-trifluoroalkanes, difluoroalkylsulfites, and bis(difluoroalkyl)sulfates [*182*].

[*179*] **91**

$$RCOCH_2COOC_2H_5 \xrightarrow[100°C]{SF_4} RCF_2CH_2COOC_2H_5 + RCF{=}CHCOOC_2H_5 + RC{\equiv}CCOOC_2H_5$$

36% 21% 38%

$R = CH_3,\ C_2H_5$

[*180*] **92**

$$CH_3COCHRCOOC_2H_5 \xrightarrow[100°C]{SF_4} CH_3CF_2CHRCOOC_2H_5 + CH_3CF{=}CRCOOC_2H_5 + CH_2{=}C{=}CRCOOC_2H_5$$

22% 19% 33%

$R = C_2H_5,\ i\text{-}C_3H_7$

[*180,181*] **93**

$$CH_3COCR^1R^2COOC_2H_5 \xrightarrow[20°C]{SF_4,\ HF} CH_3CF_2CR^1R^2COOC_2H_5$$

76–85%

$R = H,\ C_2H_5$

Although sulfur tetrafluoride is the most universal reagent for converting a carbonyl group to a difluoromethyl or difluoromethylene group, a few other compounds act in a similar way. **Diethylaminosulfur trifluoride (DAST)** and its analogues [*94*] are used extensively to replace the *carbonyl oxygen in aldehydes and ketones.* Fluorination of carbonyl compounds with DAST requires temperatures ranging from ambient to 80 °C, that is, higher than those applied for the replacement of a hydroxyl group by fluorine. Consequently, hydroxy aldehydes and hydroxy ketones are easily converted to the corresponding fluoro carbonyl compounds (equation 60). ***Aliphatic, cycloaliphatic, and aromatic aldehydes*** are converted into the respective difluoromethylene derivatives in 60–95% yields [*95, 148*] (equations 94 and 95).

Because aldehydes react with aminofluorosulfuranes more readily than ketones, keto aldehydes can be selectively fluorinated at the formyl group [*94, 183*].

Haloacetaldehydes react with DAST to give bis(1-fluorohaloethyl) ethers as the only or main products [*170*] (equation 96).

Most ***aliphatic*** [*95, 184, 185, 186, 187*], ***aromatic*** [*95, 186, 188*], and ***heterocyclic*** [*189, 190*] ***ketones*** react with **DAST** in the usual way to give geminal difluoro derivatives. Fluorination of cycloaliphatic ketones, in particular, is often accompanied by a spontaneous dehydrofluorination to form considerable amounts of fluoroalkenes [*184*] (equation 97). Such dehydrofluorination can be purposely achieved under very gentle conditions by treatment of the geminal difluoride with neutral alumina in hexane [*191*].

[*95*] **94**

$$RCHO \xrightarrow[25°C,\ 0.5-1h]{DAST,\ CFCl_3} RCHF_2 \quad 78-95\%$$

$R = C_2H_5,\ i\text{-}C_3H_7,\ t\text{-}C_4H_9$

[*148*] **95**

$$XC_6H_4CHO \xrightarrow[25-40°C,\ 15\ min]{DAST,\ CH_2Cl_2} XC_6H_4CHF_2 \quad 70-75\%$$

$X = H,\ Br,\ Cl,\ NO_2$

[*170*] **96**

$$CX_nH_{3-n}CHO \xrightarrow[0-60°C,\ 2-6h]{DAST,\ solvent} (CX_nH_{3-n}CHF)_2O \quad 37-67\%$$

$X = Cl,\ F;\quad n = 1-3$

Solvent: CCl_4, diglyme, $C_6H_5CH_3$

[*184*] **97**

2-Methylcyclohexanone → (DAST, solvent; RT, 40h) → 1,1-difluoro-2-methylcyclohexane + 1-fluoro-2-methylcyclohexene + 1-fluoro-6-methylcyclohexene

Solvent			
CH_2Cl_2	60%	30%	10%
glyme	51%	36%	13%

On treatment with DAST, ***4-keto carboxylic acids*** undergo cyclization to form γ-fluorolactones almost quntitatively [*192*] (equations 98 and 99).

Numerous examples of application of DAST for replacement of carbonyl oxygen by fluorine have been reported in the field of ***carbohydrates*** [*193, 194, 195*] and particularly in the field of ***steroids*** [*141, 142, 183, 196, 197, 198, 199*].

[*192*] **98**

$$CH_3CO(CH_2)_2COOH \xrightarrow[0°C,\ 30\ min]{DAST,\ CHCl_3}$$ 5-fluoro-5-methyl-tetrahydrofuran-2-one (F, CH_3, O, =O) 90%

[*192*] **99**

2-acetylbenzoic acid (COOH, $COCH_3$) $\xrightarrow[0°C,\ 30\ min]{DAST,\ CHCl_3}$ 3-fluoro-3-methylphthalide (O, O, F, CH_3) 95%

On treatment with DAST, β-***keto esters*** undergo oxidative fluorination: ethyl acetoacetate and DAST in *N*-methylpyrrolidone give a 48–58% yield of a mixture of equal parts of ethyl cis- and trans-2,3-difluoro-2-butenoate [*200*] (equation 100).

[*200*] **100**

$$CH_3COCH_2COOC_2H_5 \xrightarrow[-70°C,\ RT\ \ 48-64h]{DAST,\ NMP} CH_3CF{=}CFCOOC_2H_5$$

48–58% *cis/trans* 1:1

NMP = *N*–methylpyrrolidone

Phosphorinano-4-ones are fluorinated with **molybdenum hexafluoride** at –25 °C to ambient temperature to give 4,4-difluorophosphorinanes in 36–100% yields [*201*].

Tungsten hexafluoride reacts in the presence of boron trifluoride with aliphatic aldehydes and ketones to form the corresponding difluorides albeit in low yields [*202*].

Carboxylic acids react both with fluoroaminosulfuranes such as DAST and with sulfur tetrafluoride. Whereas DAST converts the acids to *acyl fluorides* only, sulfur tetrafluoride further fluorinates the primarily formed acyl fluorides and *ultimately converts the carboxyl group to a trifluoromethyl group.*

Aliphatic carboxylic acids react with **sulfur tetrafluoride** to give, in addition to 1,1,1-trifluoromethylalkanes, considerable amounts of symmetrical bis(1,1-difluoroalkyl)ethers. Yields of the ethers are related to the nature of the acids and to the reaction conditions. The optimum conditions for the formation of the ethers depend on their stability in highly acidic reaction medium and on the reactivity of the acids toward sulfur tetrafluoride. Simple unsubstituted acids form the ethers only at low temperatures, whereas longer chain and cycloaliphatic acids give the corresponding ethers at somewhat higher temperatures. Halosubstituted acids form the ethers at the relatively high reaction temperatures necessary for these reactions to proceed [*203, 204, 205*] (equation 101).

[*203,204,205*] **101**

$$RCOOH \xrightarrow{SF_4} RCF_3 + RCF_2{-}O{-}CF_2R$$

R	°C	h	%	%
CH_3	−10	48	28	16
	4	48	78	0
CH_2F	60	3	71	11
CH_2Cl	65	3	51	24
CH_2Br	60	3	67	24
CH_3CH_2	−15	48	27	16
	20	20	96	0
CH_2ClCH_2	18	18	29	34
	70	3	45	21
CH_3CHCl	20	20	32	16
	75	6	48	20
$CH_3(CH_2)_4$	15	16	45	7
	22	20	56	0
$c{-}C_6H_{11}$	40	10	22	15
	60	3	74	0

The total yield of products from alkanecarboxylic acids increases, in most cases, by addition of anhydrous hydrogen fluoride. The optimum hydrogen fluoride concentration is much higher than catalytic and is related to the basicity of a carbonyl group. A mechanism for the formation of both 1,1,1-trifluoroalkanes and bis(1,1-difluoroalkyl) ethers has been proposed [*206*] (equation 102).

[*206*] **102**

$$SF_4 + HF \rightleftharpoons SF_3^+ + HF_2^-$$

$$R{-}\underset{F}{\underset{|}{C}}{=}O + SF_3^+ \longrightarrow R{-}\underset{F}{\underset{|}{\overset{+}{C}}}{-}O{-}\overset{F}{\overset{|}{S}}F_2 \xrightarrow{-SOF_2} RCF_2^+ \xrightarrow[-HF]{HF_2^-} RCF_3$$

$$R{-}\underset{F}{\underset{|}{C}}{=}O + RCF_2^+ \longrightarrow R{-}\underset{F}{\underset{|}{\overset{+}{C}}}{-}O{-}CF_2R \xrightarrow[-HF]{HF_2^-} (RCF_2)_2O$$

Participation of fluorocarbocations, derived from carboxylic acids and from halo acetones, in reactions of carbonyl compounds with sulfur tetrafluoride has been directly evidenced by trapping them with aromatic hydrocarbons [*207, 208*].

Treatment of *α- or β-hydroxyacids* with **sulfur tetrafluoride** leads to conversion of the carboxylic group into the trifluoromethyl group, but the hydroxyl group undergoes either fluorination, fluorosulfination, esterification, or dehydration to form esters, ethers, or alkenes. The ratio of the products depends on β-substitution [*209, 210*] (equations 103 and 104).

[*209*] **103**

$$RCH(OH)COOH \xrightarrow[60°C,\ 15h]{SF_4} \underset{5-26\%}{RCHFCF_3} + \underset{40-42\%}{RCH(OSOF)CF_3}$$

$R = H, CH_3$

[*210*] **104**

$$RCH(OH)CH_2COOH \xrightarrow[80°C,\ 8h]{SF_4} \underset{A}{RCHFCH_2CF_3} + \underset{B}{RCH{=}CHCF_3}$$

$$+ \underset{C}{(CF_3CH_2CHR)_2O} + \underset{D}{RCHFCH_2COOCHRCH_2CF_3}$$

R	%A	%B	%C	%D	
H	25	0	35	40	
CH_3	32	35	0	33	Total yield ~60%
CF_3	70	30	0	0	

Reactions of ***alkanedicarboxylic acids*** with **sulfur tetrafluoride** afford, in general, mixtures of bis(trifluoromethyl)alkanes, cyclic α,α,α′,α′-tetrafluoroethers, linear bis(pentafluoroalkyl) ethers, and polyfluoroethers. The cyclic ethers constitute the major products of the reactions with alkane-1,2-dicarboxylic acids; they are also formed in the reactions with alkane-1,3-dicarboxylic acids but not with 1,1- nor 1,4-dicarboxylic acids [*211*] (equation 105).

An illustration of the tendency of alkane-1,2- and alkane-1,3-dicarboxylic acids to a ring closure during the reaction with sulfur tetrafluoride is the reaction of propane-1,2,3-tricarboxylic acid. The corresponding six- and five-membered cyclic ethers are formed in a 1:4 ratio [*211*] (equation 106).

The reaction of cyclohexane-*cis*-1,2-dicarboxylic acid with sulfur tetrafluoride affords the correspoding cyclic ether in a 70% yield with very limited formation of the bis(trifluoromethyl) derivative [*211*] (equation 107).

On treatment with sulfur tetrafluoride followed by hydrolysis, D- and L-hydroxysuccinic acids and D-tartaric acid give complex mixtures of products in which the predominant components are five-membered cyclic ethers, derivatives of 2,2,5,5-tetrafluorotetrahydrofuran [*209, 212*] (equation 108).

Results of the sulfur tetrafluoride fluorination of benzenecarboxylic acids strongly depend on the nature of a benzene ring substituent. Benzoic, toluic, and particularly p-methoxybenzoic acids give poor yields of the respective benzotri-

[211] **105**

$$HOOC(CHX)_nCOOH \xrightarrow{SF_4} CF_3(CHX)_nCF_3 \ (A) + \text{2,2,5,5-tetrafluoro-oxolane ring } F_2C\text{–}(CHX)_n\text{–}CF_2\text{–}O \ (B)$$

$$+ [CF_3(CHX)_nCF_2]_2O \ (C) + \text{poly ethers} \ (D)$$

n	X	°C	h	%A	%B	%C	%D
1	H	130	3	70	0	12	2
2	H	60	3	11	35	12	4
2	Cl	200	6	20	52	0	0
2	Br	180	6	30	37	0	0
3	H	18	20	42	7	18	0
		60	3	78	9	1	0
4	H	5	48	46	0	20	0
		60	3	81	0	0	0

[211] **106**

$$HOOCCH_2CH(COOH)CH_2COOH \xrightarrow[140°C,\ 3h]{SF_4} HOOCCH_2CH(CF_3)CH_2COOH \ (17\%) + \text{six-membered ring } H_2C\text{–}CH(CF_3)\text{–}CH_2\text{–}CF_2\text{–}O\text{–}CF_2 \ (12\%)$$

$$+ \text{five-membered ring } H_2C\text{–}CH(CH_2CF_3)\text{–}CF_2\text{–}O\text{–}CF_2 \ (51\%)$$

[211] **107**

$$\text{cyclohexane-1,2-dicarboxylic acid (H)} \xrightarrow[60°C,\ 3h]{SF_4} \text{1,1,3,3-tetrafluoro-hexahydroisobenzofuran (H)} \ (70\%) + \text{1,2-bis(trifluoromethyl)cyclohexane (H)} \ (3\%)$$

[*212*] **108**

$$\text{HOOC–CHOH–CHOH–COOH} \xrightarrow[\text{2. KOH aq.}]{\text{1. SF}_4\text{, 20°C}}$$

2,2,5,5-tetrafluoro-3,4-dihydroxytetrahydrofuran, 38%, *cis* and *trans* + 2,2,5,5-tetrafluoro-3-fluoro-4-hydroxytetrahydrofuran, 24%, *cis* and *trans* + 2,2,3,4,5,5-hexafluorotetrahydrofuran, 2%, *cis* and *trans*

$$+ \ CF_3CHFCHFCF_3 \ (4\%) + CF_3CHFCH(OH)CF_3 \ (22\%) + CF_3CH(OH)CH(OH)CF_3 \ (10\%)$$

fluorides, but the yields increase with the increase in the electron-withdrawing power of the substituent [*213*] and are substantially increased, even under milder conditions, in the presence of an excess of anhydrous hydrogen fluoride [*214*] (equation 109).

[*213,214*] **109**

$$p\text{-}XC_6H_4COOH \xrightarrow{SF_4} p\text{-}XC_6H_4CF_3$$

	X =	OCH_3	CH_3	H	Cl	F	CF_3	NO_2
160°C, 6h, C_6H_6:	%	8	12	16	24	25	61	66
80–140°C, 10h: excess HF	%	55	80	93	70			92

Hydroxybenzoic acids are easily converted into hydroxybenzotrifluorides by treatment with sulfur tetrafluoride in a hydrogen fluoride solution under mild conditions [*215*]. An exception is salicylic acid, which becomes a resin; however, a 70–75% yield of 2-hydroxybenzotrifluoride can be obtained by carrying out the fluorination in a mixture of three parts of hydrogen fluoride and one part of benzene [*216*] (equation 110).

Fluorination of tetrachlorophthalic acid hemihydrate with an excess of sulfur tetrafluoride, or of tetrachlorophthalic anhydride with a sulfur tetrafluoride–hydrogen fluoride reagent, provides an efficient synthesis of 4,5,6,7- tetrachloro-1,1,3,3-tetrafluoro1,3-dihydroisobenzofuran. 3,4,5,6-Tetrachloro-2-(trifluoromethyl)-benzoyl fluoride is formed as a by-product [*217*] (equation 111).

Dihalo- and dinitropyromellitic acids undergo cyclization when reacted with sulfur tetrafluoride and hydrogen fluoride to give high yields of 4,8-dihalo- and 4,8-dinitro-1,1,3,3,5,5,7,7-octafluoro-1,3,5,7-tetrahydrobenzo[1,2-c:4,5-c′5]difurans, respectively [*218, 219, 220*] (equation 112).

[*215,216*] **110**

$$o,m,p\text{-}HOC_6H_4COOH \xrightarrow{SF_4} o,m,p\text{-}HOC_6H_4CF_3$$

	solvent	°C	%
ortho-OH	HF	25	resin
	HF, C_6H_6	25	72
meta-OH	HF	25	75
para-OH	HF	85–90	80

[*217*] **111**

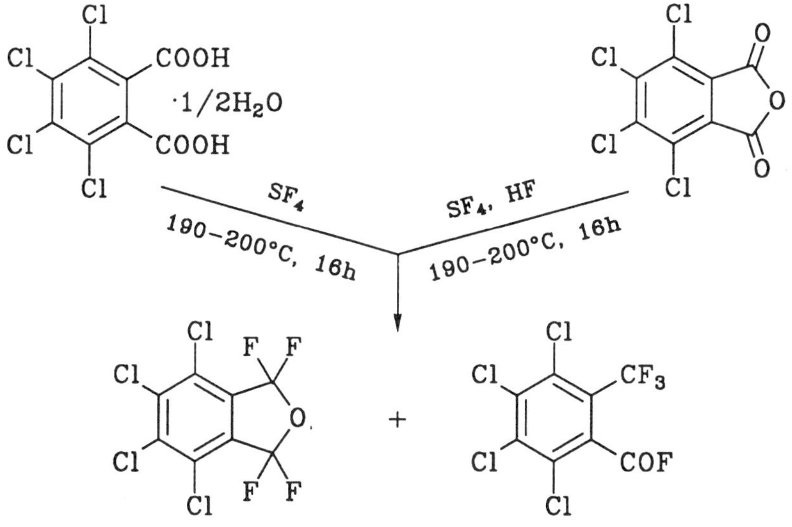

Tetrachlorophthalic acid $\cdot 1/2H_2O$ — SF_4, 190–200°C, 16h; tetrachlorophthalic anhydride — SF_4, HF, 190–200°C, 16h → 4,5,6,7-tetrachloro-1,1,3,3-tetrafluoro-1,3-dihydroisobenzofuran (89%) + 2,3,4,5-tetrachloro-6-(CF_3)benzoyl fluoride (COF) (8%)

112

1,2,4,5-(HOOC)$_4$-3,6-X_2C_6 $\xrightarrow{SF_4,\ HF}$ octafluoro bis(dihydrofuran)benzene (4,8-X)

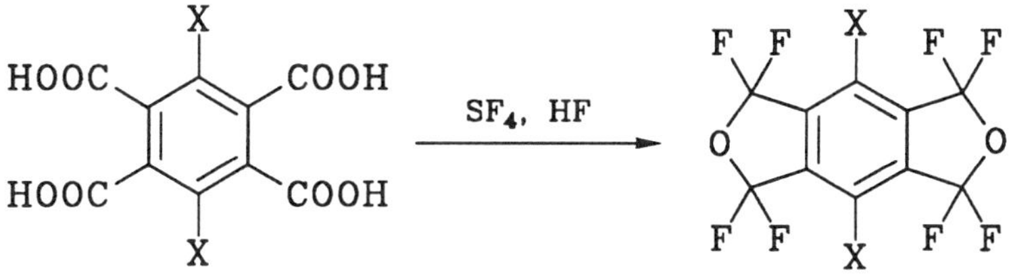

	X		%
[*218,219*]	Br	240°C, 30–40h	84
[*219*]	Cl	210°C, 20h	84
[*220*]	NO_2	160°C, 15h	41

References are listed on pages 253–262.

Naphthalene-1,8-dicarboxylic acid, its nitro-substituted derivatives, and naphthalene-1,4,5,8-tetracarboxylic acid, on treatment with **sulfur tetrafluoride** at 0 °C, undergo dehydration to form quantitatively the corresponding anhydrides [*218, 221*]. Unsubstituted and mononitrated monoanhydrides react further at 200–250 °C to give derivatives of 1,1,3,3-tetrafluoro-1*H*-naphtho[1,8-c,d]pyran. Dinitronaphthalene-1,8-dicarboxylic acid anhydrides and naphthalene-1,4,5,8- tetracarboxylic acid dianhydride give the respective tetra- and octafluoroethers only in the presence of an excess of anhydrous hydrogen fluoride [*221*] (equations 113 and 114).

[*218,221*] **113**

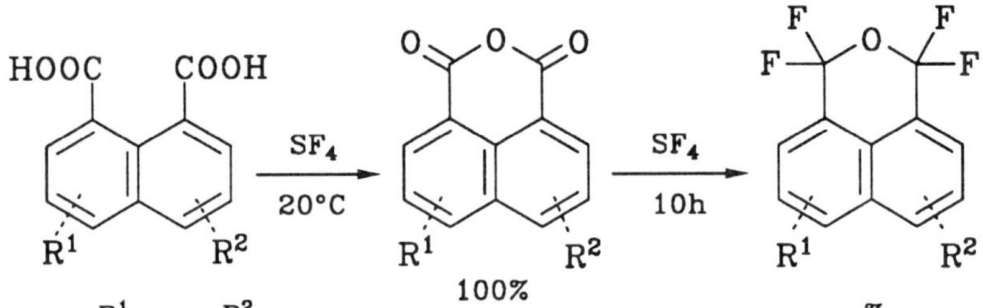

HOOC COOH (R¹, R²) —SF_4, 20°C→ anhydride (R¹, R²) 100% —SF_4, 10h→ tetrafluoro ether (R¹, R²)

R^1	R^2		%
H	H	220°C	63
H	3–NO_2	200°C	91
H	4–NO_2	250°C	85
6–NO_2	3–NO_2	HF, 150°C	100
5–NO_2	4–NO_2	HF, 180°C	80

[*221*] **114**

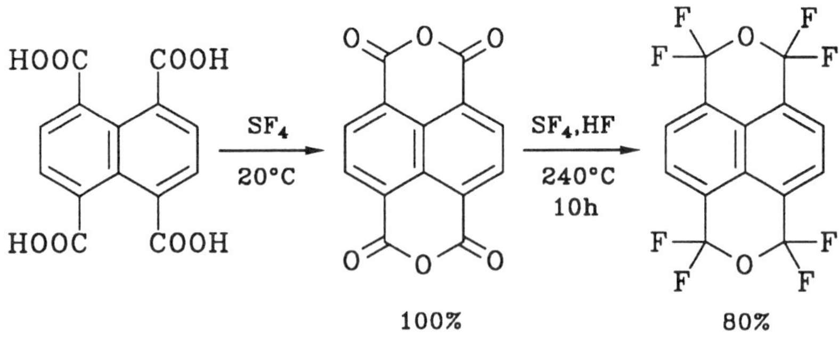

Naphthalene-1,4,5,8-tetracarboxylic acid (HOOC, COOH, HOOC, COOH) —SF_4, 20°C→ dianhydride 100% —SF_4,HF, 240°C, 10h→ octafluoro diether 80%

Furan-2-carboxylic acid reacts with **sulfur tetrafluoride** at 0 °C to give 2-furoyl fluoride, but attempts at further fluorination results in resin formation [222]. A second carboxylic group and/or electron-withdrawing substituents, such as the nitro or trifluoromethyl group, stabilize the furan ring. Thus, furandicarboxylic and tricarboxylic acids [*222, 223*] and their nitro- [*224*] and trifluoromethyl-substituted derivatives [*225, 226*] afford the corresponding polytrifluoromethylfurans in good yields (equation 115).

[*222*] **115**

HOOC–furan–COOH (2,5–),(2,4–),(3,4–) —SF_4, 115–185°C, 12–45h→ F_3C–furan–CF_3 50–70%

Furantetracarboxylic acid reacts with sulfur tetrafluoride to give perfluoro-2,5-dimethyl-2′,5′-dihydro-3′,4′-difuran as the only product [*223*] (equation 116).

[*223*] **116**

HOOC, HOOC, COOH, COOH —SF_4, HF, 190°C, 25h→ F, F, F, F, O, O, CF_3, CF_3 54%

Treatment of 2,5- and 2,4-furandicarboxylic acids with sulfur tetrafluoride in an excess of hydrogen fluoride leads simultaneously to conversion of the carboxylic groups into trifluoromethyl groups and addition of two fluorine atoms to the furan ring to give highly fluorinated diastereoisomers of 2,5-dihydrofuran [*225, 226, 227*] (equations 117 and 118).

[*225*] **117**

HOOC–furan–COOH —SF_4, HF, 200°C, 40h→ F, CF_3, O, F, CF_3 76%

[*227*] **118**

HOOC, COOH, O —SF_4, HF, 200°C, 40h→ CF_3, F, H, O, F, CF_3 64%

The analogous reactions of 5-nitro- [*227*] and 5-bromofuran-2-carboxylic acids [*228*] lead to addition of two fluorine atoms and replacement of the nitro group or bromine to give 2,5,5-trifluoro-2,5-dihydrofuran derivatives. In the latter case, migration of bromine to the side carbon atom occurs [*228*] (equation 119).

[*228*] **119**

5-Bromofuran-2-COOH $\xrightarrow[\text{25°C, 12h}]{SF_4,\ HF}$ 2,2,5-trifluoro-5-(CF_2Br)-2,5-dihydrofuran (35%) + 2,2,5-trifluoro-5-(CF_3)-2,5-dihydrofuran (10%)

Fluorination of **thiophene-2,5-dicarboxylic acid** with a **sulfur tetrafluoride–hydrogen fluoride** mixture provides 2,5-bis(trifluoromethyl)thiophene in a 69% yield; no fluorine addition to the thiophene ring occurs [*229*]. Also, imidazole mono- and dicarboxylic acids yield only the respective trifluoromethylimidazoles [*230*].

Perfluorinated tertiary amides and cyclic imides react at elevated temperatures with a **sulfur tetrafluoride–hydrogen fluoride** mixture in the usual way by replacing the carbonyl oxygen atom by fluorine to give tertiary *perfluoroamines* in high yields [*231*]. ***Nonfluorinated amides,*** when treated with standard-quality sulfur tetrafluoride, which is contaminated with hydrogen fluoride, undergo a nitrogen–carbonyl group bond cleavage to give *acyl fluorides* and then *trifluoromethyl compounds*. However, when the reaction is conducted in the presence of dry potassium fluoride (a hydrogen fluoride scavenger), *N,N*-dialkylbenzamides give high and reproducible yields of *N,N*-dialkyl-α,α-difluorobenzylamines. The yields are evidently lowered by the presence of electron-withdrawing substituents [*232*] (equation 120).

[*232*] **120**

$$XC_6H_5CON(CH_3)_2 \xrightarrow[\text{150°C, 48–72h}]{SF_4,\ KF} XC_6H_5CF_2N(CH_3)_2$$

X:	p-CH_3O	p-CH_3	m-CH_3	H	p-Br	m-Br	p-CF_3	p-NO_2
%:	75	72	90	71	33	31	54	16

The reaction of formamides with sulfur tetrafluoride in the presence of potassium fluoride leads to replacement of both carbonyl oxygen and hydrogen with fluorine. The formyl group is directly converted into the trifluoromethyl group; *N*-(trifluoromethyl)amines are formed in near quantitative yields [*233*] (equation 121).

Results of fluorination of ***lactones*** with **sulfur tetrafluoride** depend on the ring size. γ-Butyrolactone undergoes ring cleavage to give γ-fluorobutyryl fluoride, which is further fluorinated to 1,1,1,4-tetrafluorobutane. The six-membered 1,4-di-

[*233*] **121**

$$R_2NCHO \xrightarrow[150°C,\ 48h]{SF_4,\ KF} R_2NCF_3 \quad 90–94\%$$

$R = CH_3, C_2H_5$; $R_2 = (CH_2)_5, CH_2CH_2–O–CH_2CH_2$

oxane-2,5-diones react without ring opening to give high yields of 2,2,5,5-tetrafluoro-1,4-dioxanes [*234*] (equations 122 and 123).

[*234*] **122**

$$\text{γ-butyrolactone} \xrightarrow[130°C,\ 8h]{SF_4} CH_2FCH_2CH_2COF\ (80\%) \xrightarrow[160°C,\ 12h]{SF_4} F(CH_2)_3CF_3\ (90\%)$$

[*234*] **123**

$$\text{3,6-R,H-1,4-dioxane-2,5-dione} \xrightarrow[130–160°C,\ 9h]{SF_4} \text{2,2,5,5-tetrafluoro-3,6-R,H-1,4-dioxane} \quad 80–85\%$$

Esters of carboxylic acids are resistant toward sulfur tetrafluoride up to 300 °C. However, in the presence of an excess of hydrogen fluoride, they react like the acids to give trifluoromethyl compounds [*187, 235*]. In contrast, esters of highly fluorinated acids or alcohols react with sulfur tetrafluoride in a hydrogen fluoride solution under mild conditions without cleavage of the ester bond and give products of the replacement of carbonyl oxygen by fluorine: α,α-difluoroethers [*235, 236, 237, 238*] (equations 124 and 125).

[*235*] **124**

$$p\text{–}RCOOC_6H_4COOCH_3 \xrightarrow[75–80°C,\ 16h]{SF_4} p\text{–}RCF_2–O–C_6H_4CF_3 \quad 52–70\%$$

$R = CF_3, C_3F_7, C_4F_9, C_6F_{13}, H(CF_2)_4, H(CF_2)_6$

[*238*] **125**

$$FSO_2CF_2COOR \xrightarrow[70–90°C,\ 90h]{SF_4} FSO_2CF_2CF_2–O–R \quad 69–90\%$$

$R = CF_3CH_2, CF_3(CF_2)_nCH_2, (CF_3)_2CH, CF_3(CH_3)CH, C_2H_5$

Cyclic ethylene carbonate and its halogeno derivatives are converted into 2,2-difluoro-1,3-dioxolanes, which are useful as inhalation anaesthetics, by treatment with sulfur tetrafluoride in an anhydrous hydrogen fluoride solution at 100–150 °C [*239*] (equation 126).

[*239*] **126**

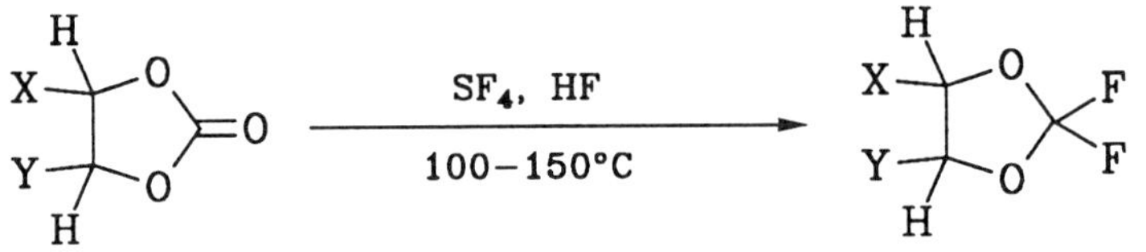

$$\text{XCH–CHY cyclic carbonate} \xrightarrow[100-150°C]{SF_4,\ HF} \text{2,2-difluoro-1,3-dioxolane}$$

X,Y = H,H; H,Cl; Br,Br; Cl,Cl

Besides sulfur tetrafluoride, only a very few reagents can convert a carboxylic group into a trifluoromethyl group.

Molybdenum hexafluoride, in the presence of boron trifluoride, reacts with acetic acid and haloacetic acids at 130–160 °C to give, respectively, 1,1,1-trifluoroethane and 1,1,1-trifluorohaloethanes in 60–89% yields [*240, 241*]. Prolonged treatment of pyridine mono- and dicarboxylic acids with an excess of molybdenum hexafluoride at elevated temperatures provides the respective mono- and bis(trifluoromethyl)pyridines in good yields [*241*] (equation 127).

[*241*] **127**

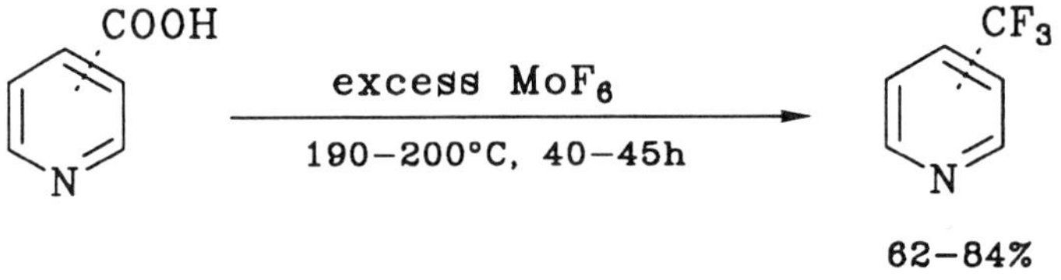

$$\text{C}_5\text{H}_4\text{N–COOH} \xrightarrow[190-200°C,\ 40-45h]{\text{excess } MoF_6} \text{C}_5\text{H}_4\text{N–CF}_3$$

62–84%

Chlorine monofluoride in a hydrogen fluoride solution reacts rapidly at low temperature with haloacetate and halopropanoate esters to give 61–80% yields of α,α-difluoroalkyl ethers [*242*] (equation 128).

[*242*] **128**

$$R^1COOR^2 \xrightarrow[-60\ to\ -40°C,\ 18\ min]{ClF,\ HF} R^1CF_2OR^2 + Cl_2O$$

61–80%

$R^1 = CH_2Cl,\ CH_2FCHCl;\quad R^2 = CH_3,\ C_2H_5$

References are listed on pages 253–262.

Elemental fluorine in the absence of hydrogen fluoride and water produces fluoroxypoly- and fluoroxyperfluoroalkanes from sodium pentafluoropropanoate [*243*] and various poly- and perfluoroalkanoyl fluorides [*244*] (equation 129).

[*244*] **129**

$$CF_3COF \xrightarrow[0.011\ \text{atm},\ -10°C]{F_2,\ CsF,\ Cu\ chips} CF_3CF_2OF \quad 100\%$$

References for Pages 199–253

1. Dmowski, W.; Kolinski, R. A. *Pol. J. Chem.* **1973,** *47,* 1211.
2. Dmowski, W.; Nantka-Namirski, P.; Wozniacki, R. Pol. Patent 96 979, 1978; *Chem. Abstr.* **1979,** *90,* P86709f.
3. Dmowski, W.; Kolinski, R. A. *Pol. J. Chem.* **1978,** *52,* 71.
4. Dmowski, W.; Kolinski, R. A. *Pol. J. Chem.* **1974,** *48,* 1697.
5. Kowollik, G.; Langen, P. *Nucleic Acid Chem.* **1978,** *1,* 199.
6. Kotun, S. P.; DesMarteau, D. *Can. J. Chem.* **1989,** *67,* 174.
7. Millauer, H.; Schwertfeger, W.; Siegemund, G. *Angew. Chemie* **1985,** *97,* 164 and references therein.
8. Bryce, M. R.; Chambers, R. D.; Kirk, J. R. *J. Chem. Soc., Perkin Trans. 1* **1984,** 1391.
9. Makarov, K. N.; Pletnev, S. I.; Gervits, L. L.; Natarov, V. P.; Prokudin, I. P.; Knunyants, I. L. *Izv. Akad. Nauk SSSR* **1986,** 1313; Chem. Abstr. **1987,** *107,* 11568p.
10. Ayi, A. I.; Remli, M.; Condom, R.; Guedj, R. *J. Fluorine Chem.* **1980,** *17,* 565.
11. Ayi, A. I.; Remli, M.; Guedj, R. *Tetrahedron Lett.* **1981,** *22,* 1505.
12. Alverne, G.; Laurent, A.; Haufe, G. *J. Fluorine Chem.* **1986,** *34,* 147.
13. Amri, H.; El Gaied, M. M. *J. Fluorine Chem.* **1990,** *46,* 75.
14. Seto, H.; Qian, Z.-H.; Yoshioka, H.; Uchibori, Y.; Umeno, M. *Chem. Lett.* **1991,** 1185.
15. Ichihara, J.; Hanafusa, T. *J. Chem. Soc., Chem. Commun.* 1989, 1848.
16. Hudlicky, M. *J. Fluorine Chem.* **1987,** *36,* 373.
17. Shimizu, M.; Yoshioka, H. *Tetrahedron Lett.* **1988,** *29,* 4101.
18. Shimizu, M.; Yoshioka, H. *Tetrahedron Lett.* **1989,** *30,* 967.
19. Noskov, V. T.; Kalinina, L. M.; Englin, M. A. *Zh. Obshch. Khim.* **1972,** *42,* 2028; *Chem. Abstr.* **1973,** *78,* 15425b.
20. Yarosh, O. G.; Voronkov, M. G.; Tsetlina, E. O. *Izv. Akad. Nauk SSSR* **1976,** 1633; *Chem. Abstr.* **1976,** *85,* 177529k.

21. Voronkov, M. G.; Trofimova, O. M.; Tchernov, N. F. *Izv. Akad. Nauk SSSR* **1985,** 2148; *Chem. Abstr.* **1986,** *105,* 191180c.
22. Robert, D. U.; Riess, J. R. *Tetrahedron Lett.* **1972,** 847.
23. Robert, D. U.; Flatau, G. N.; Cambon, A.; Riess, J. G. *Tetrahedron,* **1973,** *29,* 1877.
24. Costa, D. J.; Butin, N. E.; Riess, J. G. *Tetrahedron* **1974,** *30,* 3793.
25. Ayi, A. I.; Condom, R.; Maria, P. C.; Wade, T. N.; Guedj, R. *Tetrahedron Lett.* **1978,** 4507.
26. Ayi, A. I.; Condom, R.; Wade, T. N.; Guedj, R. *J. Fluorine Chem.* **1979,** *14,* 43.
27. Ayi, A. I.; Remli, M.; Guedj, R. *J. Fluorine Chem.* **1981,** *17,* 127.
28. Boutin, N. E.; Robert, D. U.; Cambon, A. R. *Bull. Soc. Chim. Fr.* **1974,** 2861.
29. Shimizu, M.; Nakahara, Y.; Yoshioka, H. Tetrahedron Lett. 1985, 26, 4207.
30. Shimizu, M.; Nakahara, Y.; Yoshioka, H. *J. Chem. Soc., Chem. Commun.* **1986,** 867.
31. Shimizu, M.; Nakahara, Y.; Kanemoto, S.; Yoshioka, H. *Tetrahedron Lett.* **1987,** *28,* 1677.
32. Sorochinskii, A. E.; Aleksandrov, A. M.; Gamaleya, V. F.; Kukhar, V. P. *Zh. Org. Khim.* **1981,** *17,* 1642; *Chem. Abstr.* **1981,** *95,* 186706z.
33. Masahiko, H.; Shunichi, H.; Ryoji, N. *Chem. Lett.* **1984,** 1747.
34. Shiryaev, V. I.; Makhalkina, L. V.; Kuz'mina, T. T.; Krylov, V. D.; Osipov, V. G.; Mironov, V. F. *Zh. Obshch. Khim.* **1973,** *43,* 2232; *Chem. Abstr.* **1974,** *80,* 48116w.
35. Piteau, M.; Senet, J. P.; Wolf, P.; Vu A. D.; Olofson, R. A. Fr. Patent 2 571 049, 1986; *Chem. Abstr.* **1987,** *106,* 4532e.
36. Asawon, D. P.; Ryan, T. A.; Brett, B. A. Eur. Patent Appl. EP 118 241, 1984; *Chem. Abstr.* **1984,** *101,* 210709t.
37. Lopusinski, A.; Michalski, J. *J. Am. Chem. Soc.* **1982,** *104,* 290.
38. Dabkowski, W.; Michalski, J. *J. Chem. Soc., Chem. Commun.* **1987,** 755.
39. Zavorin, S. I.; Lermontov, S. A.; Martynov, I. V. *Izv. Akad. Nauk SSSR* **1988,** 1174; *Chem. Abstr.* **1989,** *110,* 95378f.
40. Nikitin, E. V.; Ignat'ev, Y. A.; Romakhin, A. S.; Parakin, O. V.; Kosachev, I. P.; Romanov, G. V.; Kargin, Y. M.; Pudovnik, A. N. *Zh. Obshch. Khim.* **1982,** *52,* 2792; *Chem. Abstr.* **1983,** *98,* 107426u.
41. Krolevets, A. A.; Antipova, V. V.; Popov, A. G. *Zh. Obshch. Khim.* **1989,** *59,* 954; *Chem. Abstr.* **1989,** *111,* 232999w.
42. Badone, D.; Jommi, G.; Pagliani, R.; Tavecchia, P. *Synthesis* **1987,** 920.
43. Makino, K.; Yoshioka, H. *J. Fluorine Chem.* **1987,** *35,* 677.
44. Picq, D.; Anker, D.; Rousset, C.; Laurent, A. *Tetrahedron Lett.* **1983,** *24,* 5619.
45. Picq, D.; Anker, D. *J. Carbohydr. Chem.* **1985,** *4,* 113.

46. Irie, T.; Fukushi, K.; Ido, T.; Nozaki, T.; Kashida, Y. *J. Labelled Compd. Radiopharm.* **1979,** *16,* 17.
47. Hwang, D. R.; Dence, C. S.; Bonasera, T. A.; Welch, M. J. *Appl. Radiat. Isot.* **1989,** *40,* 117.
48. Su, T. M.; Sliwinski, W. T.; Schleyer, P. R. *J. Am. Chem. Soc.* **1969,** *91,* 5386 and references therein.
49. Burdon, J.; McLoughlin, V. C. R. *Tetrahedron* **1965,** *21,* 1.
50. Chi, D. Y.; Kilbourn, M. R.; Katzenellenbogen, J. A.; Welch, M. J. *J. Org. Chem.* **1987,** *52,* 658.
51. Grieco, P. A.; Williams, E.; Sugahara, T. *J. Org. Chem.* **1979,** *44,* 2194.
52. Ogawa, T.; Takashi, Y. *J. Carbohydr. Chem.* **1983,** *2,* 461.
53. Haradahira, T.; Maeda, M.; Omae, H.; Yano, Y.; Kojima, M. *Chem. Pharm. Bull.* **1984,** *32,* 4758.
54. Haradahira, T.; Maeda, M.; Kai, Y.; Kojima, M. *J. Chem. Soc., Chem. Commun.* **1985,** 364.
55. Levy, S.; Livni, E.; Elmaleh, D.; Curatolo, W. *J. Chem. Soc., Chem. Commun.* **1982,** 972.
56. Binkley, R. W.; Ambrose, M. G.; Hehemann, D. G. *J. Carbohydr. Chem.* **1987,** *6,* 203.
57. Taylor, S. L.; Martin, J. C. *J. Org. Chem.* **1987,** *52,* 4147.
58. Tewson, T. J. *J. Org. Chem.* **1983,** *48,* 3507.
59. Lyle, T. A.; Magill, C. A.; Pitzenberger, S. M. *J. Am. Chem. Soc.* **1987,** *109,* 7890.
60. Politanski, S. F.; Ivanyk, G. D.; Sarancha, V. N.; Shevchuk, V. U. *Zh. Org. Khim.* **1974,** *10,* 693; *Chem. Abstr.* **1974,** *81,* 12963b.
61. Showa Denko, K. K. Jpn. Patent 60 115 538, 1985; *Chem. Abstr.* **1986,** *104,* 33755u.
62. Showa Denko, K. K. Jpn. Patent 60 115 536, 1985; *Chem. Abstr.* **1986,** *104,* 33753s.
63. Showa Denko, K. K. Jpn. Patent 60 115 537, 1985; *Chem. Abstr.* **1986,** *104,* 33754t.
64. Showa Denko, K. K. Jpn. Patent 60 116 637, 1985; *Chem. Abstr.* **1986,** *104,* 33756v.
65. Dolbier, W. R., Jr.; Gray, T. A.; Onishi, K. *Synthesis* **1987,** 956.
66. Beguin, C. G.; Charlon, C.; Coulombeau, C.; Luu D. C. *J. Fluorine Chem.* **1976,** *8,* 531.
67. Dahbi, A.; Hamman, S.; Beguin, C. G. *J. Chem. Res. (S)* **1989,** 128.
68. Alvernhe, G.; Lacombe, S.; Laurent, A.; Rousset, C. *J. Chem. Res. (S)* **1983,** 246.
69. Kanemoto, S.; Shimizu, M.; Yoshioka, H. *Tetrahedron Lett.* **1987,** *28,* 663.
70. Kanemoto, S.; Shimizu, M.; Yoshioka, H. *Tetrahedron Lett.* **1987,** *28,* 6313.
71. Liska, F. *Chem. Listy* **1972,** *66,* 189.

72. Sharts, C. M.; Sheppard, W. A. *Org. React.* **1974,** *21,* 158.
73. Hamman, S.; Beguin, C. G. *J. Fluorine Chem.* **1987,** *37,* 343.
74. Hamman, S.; Barelle, M.; Tetaz, F.; Beguin, C. G. *J. Fluorine Chem.* **1987,** *37,* 85.
75. Benaissa, T.; Hamman, S.; Beguin, C. G. *J. Fluorine Chem.* **1988,** *38,* 163.
76. Banks, R. E.; Cross, B. E. *Chem. Ind.* **1975,** 90.
77. Kopecky, J.; Smejkal, J. *Collect. Czech. Chem. Commun.* **1980,** *45,* 2971.
78. Fokin, A. V.; Studnev, Y. N.; Rapkin, A. I.; Sultanbekov, D. A.; Potarina, T. M. *Izv. Akad. Nauk SSSR* **1984,** 411; *Chem. Abstr.* **1984,** *100,* 209142a.
79. Anderson, G. L.; Burks, W. A.; Harruna, I. I. *Synth. Commun.* **1988,** *18,* 1967.
80. Banks, R. E.; Barrage, A. K.; Khoshdel, E. *J. Fluorine Chem.* **1980,** *17,* 93.
81. Takaoka, A.; Hiroshi, I.; Ishikawa, N. *Bull. Chem. Soc. Jpn.* **1979,** *52,* 3377.
82. Watanabe, S.; Fujita, T.; Usui, Y.; Kimura, Y. *J. Fluorine Chem.* **1980,** *31,* 135.
83. Watanabe, S.; Fujita, T.; Suga, K.; Nasuno, I. *J. Am. Oil Chem. Soc.* **1983,** *60,* 1678.
84. Watanabe, S.; Fujita, T.; Sakamoto, M.; Endo, H. *J. Fluorine Chem.* **1988,** *38,* 243.
85. Watanabe, S; Fujita, T.; Sakamoto, M. *J. Fluorine Chem. 1988,* 39, 17.
86. Watanabe, S.; Fujita, T.; Usui, T. *J. Fluorine Chem.* **1986,** *31,* 247.
87. Watanabe, S.; Fujita, T.; Sakamoto, M.; Endo, H. *J. Fluorine Chem.* **1990,** *47,* 187.
88. Watanabe, S.; Fujita, T.; Sakamoto, M.; Arai, T.; Kitazume, T. *J. Am. Oil Chem. Soc.* **1989,** *66,* 131.
89. Watanabe, S.; Fujita, T.; Sakamoto, M.; Kuramochi, T. *J. Fluorine Chem.* **1987,** *36,* 361.
90. Watanabe, S.; Fujita, T.; Nasuno, I.; Suga, K. *J. Am. Oil Chem. Soc.* **1984,** *61,* 1479.
91. Watanabe, S.; Fujita, T.; Suga, K.; Nasuno, I. *Synthesis* **1984,** 31.
92. Dmowski, W.; Kamienski, M. *J. Fluorine Chem.* **1983,** *23,* 219.
93. Messina, P. A.; Mange, K. C.; Middleton, W. J. *J. Fluorine Chem.* **1989,** *42,* 137.
94. Hudlický, M. *Org. React.* **1988,** *34,* 513.
95. Middleton, W. J. *J. Org. Chem.* **1975,** *40,* 574.
96. Carvalho, J. F.; Prestwich, G. D. *J. Org. Chem.* **1984,** *49,* 1251.
97. Poulter, C. D.; Wiggins, P. L.; Plummer, T. L. *J. Org. Chem.* **1981,** *46,* 1532.
98. Adcock, W.; Abeywickrema, A. N. *Aust. J. Chem.* **1980,** *33,* 181.
99. Middleton, W. J.; Bingham, E. M. *Org. Synth.* **1977,** *57,* 72.

100. Schaefer, T.; Kruczyski, L. J.; Krawchuk, B.; Sebastian, R.; Charltin, J. L.; McKinnon, D. M. *Can. J. Chem.* **1980,** *58,* 2452.
101. Rae, I. D.; Burgess, D. A.; Bombaci, S.; Baron, M. L.; Woolcock, M. L. *Aust. J. Chem.* **1984,** *37,* 1437.
102. Olah, G. A.; Singh, B. P.; Liang, G. *J. Org. Chem.* **1984,** *49,* 2922.
103. Olah, G. A.; Singh, B. P. *J. Am. Chem. Soc.* **1984,** *106,* 3265.
104. Johnson, A. L. *J. Org. Chem.* **1982,** *47,* 5220.
105. Saito, K.; Digenis, G. A.; Hawi, A. A.; Chaney, J. *J. Fluorine Chem.* **1987,** *35,* 663.
106. Stahly, G. P. *J. Fluorine Chem.* **1989,** *43,* 53.
107. Lowe, G.; Potter, B. V. L. *J. Chem. Soc., Perkin Trans. 1* **1980,** 2026.
108. Weigert, F. J.; Middleton, W. J. *J. Org. Chem.* **1980,** *45,* 3289.
109. Müller, B.; Peter, H.; Schneider, P.; Bickel, H. *Helv. Chim. Acta* **1975,** *58,* 2469.
110. Somekh, L.; Shanzer, A. *J. Am. Chem. Soc.* **1982,** *104,* 5836.
111. Posner, G. H.; Haines, S. R. *Tetrahedron Lett.* **1985,** *26,* 5.
112. Goswami, R.; Harsy, S. G.; Heiman, D. F.; Katzenellenbogen, J. A. *J. Med. Chem.* **1980,** *23,* 1002.
113. Katzenellenbogen, J. A.; Carslon, K. E.; Heiman, D. F; Goswami. R. *J. Nucl. Med.* **1980,** *21,* 550.
114. Hecht, S. S.; Loy, R.; Mazzarese, R.; Hoffman, D. *J. Med. Chem.* **1978,** *21,* 38.
115. Yang, S. S.; Beattie, T. R.; Shen, T. Y. *Tetrahedron Lett.* **1982,** *23,* 5517.
116. Castillon, S.; Dessignes, A.; Faghih, R.; Lukacs, G.; Olesker, A.; Thang, T. T. *J. Org. Chem.* **1985,** *50,* 4913.
117. Hagesawa, A.; Goto, M.; Kiso, M. *J. Carbohydr. Chem.* **1985,** *4,* 627.
118. Tsuchiya, T.; Takahashi, Y.; Endo, M.; Umezawa, S.; Umezawa, H. *J. Carbohydr. Chem.* **1985,** *4,* 587.
119. Welch, J. T., personal communication.
120. Cantrell, G. L.; Filler, R. *J. Fluorine Chem.* **1985,** *27,* 35.
121. Yamauchi, T.; Hattori, K.; Ikeda, S.; Tamaki, K. *J. Chem. Soc., Perkin Trans. 1* **1990,** 1683.
122. Cross, B. E.; Simpson, I. C. *J. Chem. Res. (S)* **1980,** 118.
123. Boulton, K.; Cross, B. E. *J. Chem. Soc., Perkin Trans. 1* **1981,** 427.
124. Middleton. W. J.; Bingham, E. M.; Smith, D. H. *J. Fluorine Chem.* **1983,** *23,* 557.
125. Le Tourneau, M. E.; McCarthy, J. R. *Tetrahedron Lett.* **1984,** *25,* 5227.
126. Pansare, S. V.; Vederas, J. C. *J. Org. Chem.* **1987,** *52,* 4804.
127. Biollaz, M.; Kalvoda, J. *Helv. Chim. Acta* **1977,** *60,* 2703.
128. Green, M. J.; Shue, H. J.; Tanabe, M.; Yasuda, D. M.; McPhail, A. T.; Onan, K. D. *J. Chem. Soc., Chem. Commun.* **1977,** 611.

129. Rozen, S.; Faust, Y.; Ben-Yakov, H. *Tetrahedron Lett.* **1979,** *20,* 1823.
130. Blackburn, G. M.; Kent, D. E. *J. Chem. Soc., Chem. Commun.* **1981,** 511.
131. Venkian, J.; Cornille, F.; Deshayes, Ch.; Doutheau, A. *J. Fluorine Chem.* **1990,** *49,* 183.
132. Street, I. P.; Withers, S. G. *Can. J. Chem.* **1986,** *64,* 1400.
133. Biggadike, K.; Borthwick, A. D.; Evan, D.; Exal, A. M.; Kirk, B. E.; Roberts, S. M.; Stephenson, L.; Youds, P. *J. Chem. Soc., Perkin Trans. 1* **1988,** 549.
134. Kobayashi, T., et al. *Chem. Pharm. Bull.* **1982,** *30,* 3088.
135. Newman, M. S.; Khanna, J. M.; Kanakarajan, K. *J. Am. Chem. Soc.* **1979,** *101,* 678.
136. Leroy, J.; Hebert, E.; Wakselman, C. *J. Org. Chem.* **1979,** *44,* 3406.
137. Lowe, G.; Potter, B. V. L. *J. Chem. Soc., Perkin Trans. 1* **1980,** 2029.
138. Card, P. J. *J. Org. Chem.* **1983,** *48,* 393.
139. Card, P. J.; Reddy, G. S. *J. Org. Chem.* **1983,** *48,* 4734.
140. Klemm, G. H.; Kaufman, R. J.; Sidhu, R. *Tetrahedron Lett.* **1982,** *23,* 2927.
141. Bird, T. G. C.; Felsky, G.; Fredericks, P. M.; Jones, E. R. H.; Meakins, G. D. *J. Chem. Res. (M)* **1979,** 4728; *(S)* **1979,** 388.
142. Bird, T. G. C.; Fredericks, P. M.; Jones, E. R. H.; Meakins, G. D. *J. Chem. Soc. , Chem. Commun.* **1979,** 65.
143. Hudlický, M.; Merola, J. *Tetrahedron Lett.* **1990,** *31,* 7403.
144. Hamatani, T.; Matsubara, S.; Matsuda, H.; Schlosser, M. *Tetrahedron* **1988,** *44,* 2875.
145. Paaren, H. E.; Fivizzani, M. A.; Schnoes, H. K.; DeLuca, H. F. *Arch. Biochem. Biophys.* **1981,** *209,* 579.
146. Walba, D. M.; Razavi, H. A.; Clark, N. A.; Parmar, D. S. *J. Am. Chem. Soc.* **1988,** *110,* 8686.
147. Somawardhana, C. W.; Brunngraber, E. G. *Carbohydr. Res.* **1983,** *121,* 53.
148. Markovskii, L. N.; Pashinnik, V. E.; Kirsanova, N. A. *Synthesis* **1973,** 787.
149. Kornilov, A. M.; Sorochinskii, A. E.; Yagupolskii, Y. L.; Kukhar, V. P. *Zh. Org. Khim.* **1988,** 24, 1343; *Chem. Abstr.* **1989,** *110,* 57098g.
150. Mange, K. C.; Middleton, W. J. *J. Fluorine Chem.* **1989,** *43,* 405.
151. Hann, G. L.; Sampson, P. *J. Chem. Soc., Chem. Commun.* **1989,** 1650.
152. Markovskii, L. M.; Pashinnik, V. E.; Kirsanova, N. A. *Zh. Org. Khim.* **1975,** *11,* 74; *Chem. Abstr.* **1975,** *82,* 112016j.
153. Schaefer, T.; Rowbotham, J. B.; Parr, W. J. E.; Marat, R. K.; Janzen, A. F. *Can. J. Chem.* **1976,** *54,* 1322.
154. Janzen, A, F.; Marat, R. K. *J. Fluorine Chem.* **1988,** *38,* 205.
155. Burmakov, A. I.; Hassanein, S. M.; Kunshenko, B. V.; Alekseeva, L. A.; Yagupolskii. L. A. *Zh. Org. Khim.* **1986,** *22,* 1273; *Chem. Abstr.* **1987,** *106,* 195830h.

156. Stepanov, I. V.; Burmakov, A. I.; Alekseeva, L. A.; Yagupol'skii, L. M. *Zh. Org. Khim.* **1986,** *22,* 227; *Chem. Abstr.* **1987,** *106,* 17851h.
157. Kollonitsch, J.; Marburg, S.; Perkin, L. M. *J. Org. Chem.* **1975,** *40,* 3808.
158. Kollonitsch, J.; Marburg, S.; Perkin, L. M. *J. Org. Chem.* **1979,** *44,* 771.
159. Gilbert, E. E.; Dear, R. E. A. U.S. Patent 3 655 786, 1972; *Chem. Abstr.* **1972,** *91,* 21109m.
160. Blakitnyi, A. N.; Boiko, V. N.; Konovalov, E. V.; Fialkov, Y. A.; Yagupol'skii, L. M. *Zh. Org. Khim.* **1974,** *10,* 504; *Chem. Abstr.* **1974,** *80,* 146327k.
161. Kryukova, L. Y.; Kryukov, L. N.; Kolomiets, F. A.; Sokol'skii, G. A.; Knunyants, I. L. *Izv. Akad. Nauk SSSR* **1979,** 1913; *Chem. Abstr.* **1980,** *92,* 6515t.
162. Kozlova, A. M.; Sedova, L. N.; Alekseeva, L. A.; Yagupol'skii, L. M. *Zh. Org. Khim.* **1973,** *9,* 1418; *Chem. Abstr.* **1973,** *79,* 91536z.
163. Burmakov, A. I.; Motnyak, L. A.; Kunshenko, B. V.; Alekseeva, L. A.; Yagupol'skii, L. M. *J. Fluorine Chem.* **1981,** *19,* 151.
164. Hudlicky, M. *J. Fluorine Chem.* **1979,** *14,* 189.
165. Bell, H. M.; Hudlický, M. *J. Fluorine Chem.* **1980,** *15,* 191.
166. Mukherjee, J. *J. Fluorine Chem.* **1990,** *49,* 151.
167. Boswell, G. A.; Ripka, W. C.; Scribner, R. M.; Tullock, C. W. *Org. React.* **1971,** *21,* 1.
168. Wang, C.-L. *J. Org. React.* **1985,** *34,* 319.
169. Dmowski W.; Kolinski, R.; Wozniacki, R. *Eleventh International Symposium on Fluorine Chemistry;* Avignon, France; **1979**; Abstract O-4; no full paper has been published.
170. Siegemund, G. *Justus Liebigs Ann. Chem.* **1979,** 1280.
171. Wielgat, J.; Wozniacki, R. *J. Fluorine Chem.* **1984,** *26,* 11.
172. Haas, A.; Plümer, R.; Schiller, A. *Chem. Ber.* **1985,** *118,* 3004.
173. Aleksandrov, A. M.; Sorochinskii, A. E.; Krasnoshchek, A. P. *Zh. Org. Khim.* **1979,** *15,* 336; *Chem. Abstr.* **1979,** *91,* 56443t.
174. Stepanov, I. V.; Burmakov, A. I.; Kunshenko, B. V.; Alekseeva, L. A.; Yagupol'skii, L. M. *Zh. Org. Khim.* **1986,** *22,* 1812; *Chem. Abstr.* **1987,** *107,* 38749s.
175. Hodges, K. C.; Schomburg, D.; Weiss, J.-V.; Schmutzler, R. *J. Am. Chem. Soc.* **1977,** *99,* 6096.
176. Burmakov, A. I.; Stepanov, I. V.; Kunshenko, B. V.; Sedova, L. N.; Alekseeva, L. A.; Yagupol'skii, L. M. *Zh. Org. Khim.* **1982,** *18,* 1163; *Chem. Abstr.* **1982,** *97,* 181672s.
177. Stepanov, I. V.; Burmakov, A. I.; Kunshenko, B. V.; Alekseeva, L. A.; Yagupol'skii, L. M. *Zh. Org. Khim.* **1983,** *19,* 273; *Chem. Abstr.* **1983,** *99,* 21908m.
178. Webster, J. A.; Butler, J. M.; Morrow, T. J. *Adv. Chem. Ser.* **1973,** *129,* 61.

179. Burmakov, A. I.; Bloshchitsa, F. A.; Kunshenko, B. V.; Alekseeva, L. A.; Yagupol'skii, L. M. *Zh. Org. Khim.* **1980,** *16,* 2617; *Chem. Abstr.* **1981,** *94,* 174265w.
180. Bloshchitsa, F. A.; Burmakov, A. I.; Kunshenko, B. V.; Alekseeva, L. A.; Yagupol'skii, L. M. *Zh. Org. Khim.* **1982,** *18,* 728; *Chem. Abstr.* **1982,** *97,* 23271f.
181. Bloshchitsa, F. A.; Burmakov, A. I.; Kunshenko, B. V.; Alekseeva, L. A.; Belferman, A. L.; Pazderskii, Y. A.; Yagupol'skii, L. M. *Zh. Org. Khim.* **1981,** *17,* 1417; *Chem. Abstr.* **1981,** *95,* 186604q.
182. Stepanov, I. V.; Burmakov, A. I. *Zh. Org. Khim.* **1985,** *21,* 45; *Chem. Abstr.* **1985,** *103,* 22128v.
183. Campbell, J. A. U.S. Patent 4 416 822, 1983; *Chem. Abstr.* **1983,** *100,* 139475k; U.S. Patent 4 557 867, 1985; *Chem. Abstr.* **1986,** *104,* 69060c.
184. Boswell, G. A., Jr. U.S. Patent 4 212 815, 1980; *Chem. Abstr.* **1980,** *93,* 239789w.
185. Daub, G. N.; Zuckermann, R. N.; Johnson, W. S. *J. Org. Chem.* **1985,** *50,* 1599.
186. Adcock, W.; Gupta, B. D.; Khor, T.-C. *Aust. J. Chem.* **1976,** *29,* 2571.
187. Sharts, C. M.; McKee, M. E.; Steed, R. F. *J. Fluorine Chem.* **1979,** *14,* 351.
188. Leroy, J. *J. Org. Chem.* **1981,** *46,* 206.
189. Middleton, W. J.; Bingham, E. M. *J. Org. Chem.* **1980,** *45,* 2883.
190. Sufrin, J. R.; Balasubramanian, T. M.; Vora, C. M.; Marshall, G. R. *Int. J. Peptide Protein Res.* **1982,** *20,* 438.
191. Strobach, D. R.; Boswell, G. A. , Jr. *J. Org. Chem.* **1971,** *36,* 818.
192. Patrick, T. B.; Poon, Y.-F.; *Tetrahedron Lett.* **1984,** *25,* 1019.
193. May, J. A.; Sartoralli, A. C. *J. Med. Chem.* **1979,** *22,* 971.
194. Sharma, R. A.; Kavai, I, ; Fu, Y. L.; Bobek, M. *Tetrahedron Lett.* **1977,** 3433.
195. Tsuchiya, T.; Torii, T.; Suzuki, Y.; Umezawa, S. *Carbohydr. Res.* **1983,** *116,* 277.
196. Sialom, B.; Mazur, Y. *J. Org. Chem.* **1980,** *45,* 2201.
197. Marcotte, P. A.; Robinson, C. H. *Biochemistry* **1982,** *21,* 2773.
198. Taguchi, T.; Mitsuhashi, S.; Yamanouchi, A.; Kobayashi, Y.; Sai, H.; Ikekawa, N. *Tetrahedron Lett.* **1984,** *25,* 4933.
199. Yamada, S.; Ohmori, M.; Takayama, H. *Tetrahedron Lett.* **1979,** 1859.
200. Asato, A. E.; Liu, R. S. H. *Tetrahedron Lett.* **1986,** *27,* 3337.
201. Mathey, F.; Muller, G. *C. R. Acad. Sci. (C)* **1973,** *277,* 45.
202. Haas, A.; Maciej, T. *J. Fluorine Chem.* **1982,** *20,* 581.
203. Dmowski, W.; Kolinski, R. A. *J. Fluorine Chem.* **1973,** *2,* 210.
204. Dmowski, W.; Kolinski, R. A. *Pol. J. Chem.* **1973,** *47,* 1211.
205. Dmowski, W.; Kolinski, R. A. *Pol. J. Chem.* **1974,** *48,* 1697.
206. Dmowski, W.; Kolinski, R. A. *Pol. J. Chem.* **1978,** *52,* 547.

207. Wielgat, J.; Domagala, Z. *J. Fluorine Chem.* **1982,** *20,* 785.
208. Wielgat, J.; Domagala, Z.; Kolinski, R. A. *J. Fluorine Chem.* **1987,** *35,* 643.
209. Burmakov, A. I.; Motnyak, L. A.; Kunshenko, B. V.; Alekseeva, L. A.; Yagupol'skii, L. M.; *Zh. Org. Khim.* **1980,** *16,* 1401; *Chem. Abstr.* **1981,** *94,* 15120d.
210. Motnyak, L. A.; Burmakov, A. I.; Kunshenko, B. V.; Neizvestnaya, T. A.; Alekseeva, L. A.; Yagupol'skii, L. M. *Zh. Org. Khim.* **1983,** *19,* 720; *Chem. Abstr.* **1983,** *99,* 104708x.
211. Dmowski, W.; Kolinski, R. A. *Pol. J. Chem.* **1978,** *52,* 71.
212. Motnyak, L. A.; Burmakov, A. I.; Kunshenko, B. V.; Sass, V. P.; Alekseeva, L. A.; Yagupol'skii, L. M. *Zh. Org. Khim.* **1981,** *17,* 728; *Chem. Abstr.* **1981,** *95,* 132589v.
213. Burmakov, A. I.; Alekseeva, L. A.; Yagupol'skii, L. M. *Zh. Org. Khim.* **1972,** *8,* 153; *Chem. Abstr.* **1972,** *76,* 112301z.
214. Kunshenko, B. V.; Burmakov, A. I.; Alekseeva, L. A.; Lukmanov, V. G.; Yagupol'skii, L. M. *Zh. Org. Khim.* **1974,** *10,* 886; *Chem. Abstr.* **1974,** *81,* 25436g.
215. Blakitnyi, A. N.; Zalesskaya, I. M.; Kunshenko, B. V.; Fiyalkov, Y. A.; Yagupol'skii, L. M. *Zh. Org. Khim.* **1977,** *13,* 2149; *Chem. Abstr.* **1978,** *88,* 89251v.
216. Alekseeva, L. A.; Belous, V. M.; Lozinskii, M. O.; Shchendrik, V. P.; Yagupol'skii, L. M. *Ukr. Khim. Zh.* **1983,** *49,* 74; *Chem. Abstr.* **1983,** *98,* 143024v.
217. Dmowski, W.; Wielgat, J. *J. Fluorine Chem.* **1987,** *37,* 429.
218. Yagupol'skii, L. M.; Burmakov, A. I.; Alekseeva, L. A.; Kunshenko, B. V. *Zh. Org. Khim.* **1973,** *9,* 689; *Chem. Abstr.* **1973,** *79,* 18610q.
219. Porwisiak, J.; Dmowski, W. *J. Fluorine Chem.* **1990,** *51,* 131.
220. Lukmanov, V. G.; Alekseeva, L. A.; Yagupols'kii, L. M. *Zh. Org. Khim.* **1977,** *13,* 2129; *Chem. Abstr.* **1978,** *88,* 104804y.
221. Kunshenko, B. V.; Alekseeva, L. A.; Yagupol'skii, L. M. *Zh. Org. Khim.* **1974,** *10,* 1698; *Chem. Abstr.* **1974,** *81,* 151908a.
222. Lyalin, V. V.; Grigorash, R. V.; Alekseeva, L. A.; Yagupol'skii, L. M. *Zh. Org. Khim.* **1975,** *11,* 1086; *Chem. Abstr.* **1975,** *83,* 78992h.
223. Lyalin, V. V.; Grigorash, R. V.; Alekseeva, L. A.; Yagupol'skii, L. M. *Zh. Org. Khim.* **1975,** *11,* 460; *Chem. Abstr.* **1975,** *83,* 9849e.
224. Grigorash, R. V.; Lyalin, V. V.; Alekseeva, L. A.; Yagupol'skii, L. M. *Khim. Geterotsikl. Soedin.* **1977,** 1607; *Chem. Abstr.* **1978,** *88,* 105034j.
225. Grigorash, R. V.; Lyalin, V. V.; Alekseeva, L. A.; Yagupol'skii, L. M. *Zh. Org. Khim.* **1978,** *14,* 2623; *Chem. Abstr.* **1979,** *90,* 137590v.
226. Lyalin, V. V.; Grigorash, R. V.; Alekseeva, L. A.; Yagupol'skii, L. M. *Zh. Org. Khim.* **1984,** *20,* 846; *Chem. Abstr.* **1984,** *101,* 90690f.
227. Grigorash, R. V.; Lyalin, V. V.; Alekseeva, L. A.; Yagupol'skii, L. M. *Zh. Org. Khim.* **1978,** *14,* 844; *Chem. Abstr.* **1978,** *89,* 42951d.

228. Lyalin, V. V.; Grigorash, R. V.; Alekseeva, L. A.; Yagupol'skii, L. M. *Zh. Org. Khim.* **1981,** *17,* 1774; *Chem. Abstr.* **1981,** *95,* 186973j.
229. Nishida, M.; Fuji, S.; Aoki, T.; Hayakawa, Y.; Muramutsu, H.; Morita, T. *J. Fluorine Chem.* **1990,** *46,* 445.
230. Owen, D.; Plevey, R, G, ; Tatlow, J. C. *J. Fluorine Chem.* **1981,** *17,* 179.
231. DePasquale, R. J. *J. Org. Chem.* **1978,** *43,* 179.
232. Dmowski, W.; Kaminski, M. *Pol. J. Chem.* **1982,** *56,* 1369.
233. Dmowski, W.; Kaminski, M. *J. Fluorine Chem.* **1983,** *23,* 207.
234. Muratov, N. N.; Burmakov, A. I.; Kunshenko, B. V.; Alekseeva, L. A.; Yagupol'skii, L. M. *Zh. Org. Khim.* **1982,** *18,* 1403; *Chem. Abstr.* **1982,** *97,* 62910b.
235. Fiyalkov, Y. A.; Moklyachuk, L. I.; Kremlev, M. M.; Yagupol'skii, L. M. *Zh. Org. Khim.* **1980,** *16,* 1476; *Chem. Abstr.* **1981,** *94,* 3820u.
236. DePasquale, R. J. *J. Org. Chem.* **1973,** *38,* 3025.
237. Belous, V. M.; Alekseeva, L. A.; Yagupol'skii, L. M. *Zh. Org. Khim.* **1975,** *11,* 1672; *Chem. Abstr.* **1975,** *83,* 178464f.
238. Huang, T. J.; Dong, Z. X.; Shreeve, J. M. *Inorg. Chem.* **1987,** *26,* 2604.
239. Denson, D. D.; Uyemo, E. T.; Simon, R. L.; Peters, H. M. In *Biochemistry Involving Carbon–Fluorine Bonds*; Filler, R., Ed.; ACS Symposium Series 8; American Chemical Society: Washington, DC, 1976; p 190.
240. Van DerPuy, M. *J. Fluorine Chem.* **1979,** *13,* 375.
241. Shustov, L. D.; Nikolenko, L. N.; Senchenkova, T. M. *Zh. Obshch. Khim.* **1983,** *53,* 103; *Chem. Abstr.* **1983,** *98,* 143326v.
242. Boguslavskaya, L. S.; Panteleeva, I. Yu.; Chuvashkin, N. N. *Zh. Org. Khim.* **1982,** *18,* 222; *Chem. Abstr.* **1982,** *96,* 199015u.
243. Lerman, O.; Rozen, S. *J. Org. Chem.* **1980,** *45,* 4122.
244. Guglielmo, G.; Conte, L. Eur. Patent Appl. EP 194 862, 1986; *Chem. Abstr.* **1986,** *105,* 225774s.

Replacement of Sulfur by Fluorine

by W. Dmowski

The cleavage of a carbon–sulfur bond to form a carbon–fluorine bond occurs under mild conditions and is easily achieved with a number of reagents. Fluorodesulfurization, a method for the conversion of thiols, disulfides, and dithiolates to fluorides, was developed by Kollonitsch and Marburg [*1*]. ***Tertiary 2-aminothiols,*** like D-penicillamine and β-mercaptophenylalanine, on treatment with oxidizing reagents such as **trifluoromethyl hypofluorite, *N*-chlorosuccinimide,** or **chlorine** in an **anhydrous hydrogen fluoride** solution at –78 °C give, respectively, 3-fluoro-D-valine (equation 1) and β-fluorophenylalanine in high yields.

[*1*]

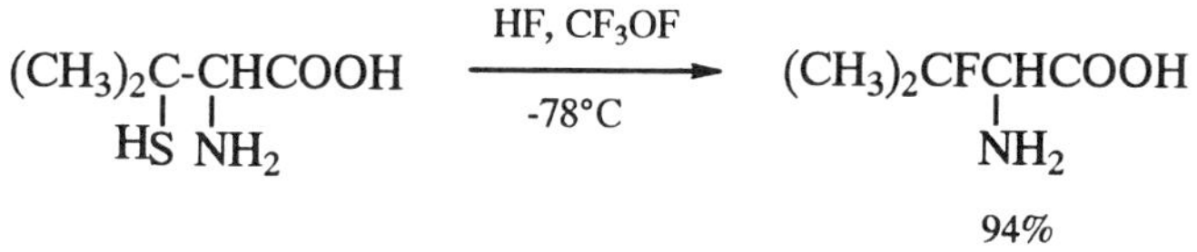

$$(CH_3)_2C(SH)\text{-}CH(NH_2)COOH \xrightarrow[-78°C]{HF,\ CF_3OF} (CH_3)_2CFCH(NH_2)COOH \quad (94\%) \qquad \mathbf{1}$$

Displacement of the sulfhydryl group in ***primary thiols,*** like L-cysteine and 2-diethylaminoethanethiol, requires **elemental fluorine,** the most active oxidant. Elemental sulfur is the major by-product in those reactions [*1*] (equation 2).

[*1*] **2**

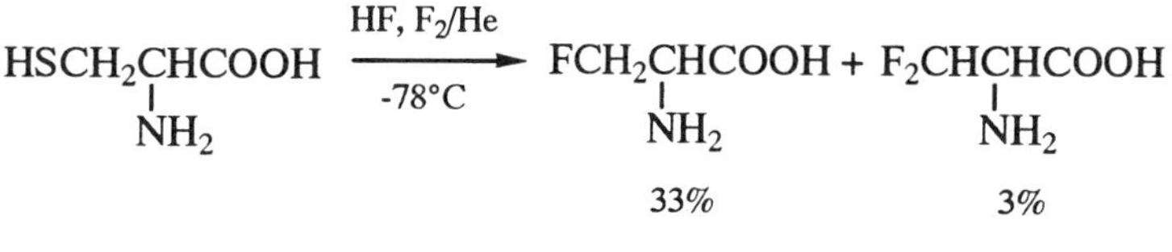

$$HSCH_2CH(NH_2)COOH \xrightarrow[-78°C]{HF,\ F_2/He} FCH_2CH(NH_2)COOH\ (33\%) + F_2CHCH(NH_2)COOH\ (3\%)$$

Cleavage of 2,2-dibutyl-1,3-dithiolane with fluorine in anhydrous hydrogen fluoride results in the formation of 5,5-difluorononane [*1*] (equation 3).

[*1*] **3**

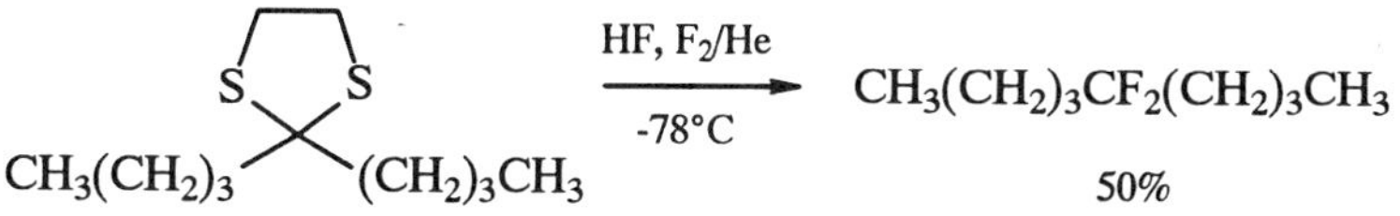

$$\text{2,2-}[CH_3(CH_2)_3]_2\text{-1,3-dithiolane} \xrightarrow[-78°C]{HF,\ F_2/He} CH_3(CH_2)_3CF_2(CH_2)_3CH_3 \quad (50\%)$$

0065–7719/95/0187–0263$08.00/1

The preparation of *gem*-difluoro compounds by the oxidative fluorodesulfurization of ***1,3-dithiolanes*** readily proceeds by treatment with a **pyridinium polyhydrogen fluoride–*N*-halo compound reagent;** the latter serves as a bromonium ion source [*2*]. 1,3-Dibromo-5,5-dimethylhydantoin is the most effective of several *N*-halo oxidants. It is believed that *N*-halo compounds combine with hydrogen fluoride to generate in situ halogen fluorides, the oxidants. Formation of ***gem*-difluorides** from dithiolanes derived from ketones is efficient and rapid, even at –78 °C, whereas the reaction of dithiolanes derived from aldehydes requires higher temperature (0 °C) (equation 4).

[*2*] **4**

2-R^1-2-R^2-1,3-dithiolane $\xrightarrow{C_5H_5N{\cdot}nHF,\ [X^+],\ CH_2Cl_2}$ $R^1CF_2R^2$

R^1	R^2	$[X^+]$	°C	min	%
C_5H_{11}	C_5H_{11}	DBH	-78	10	20
C_6H_5	C_6H_5	DBH	-78	10	52
H	$C_{11}H_{23}$	DBH	0	30	96
H	$4\text{-}NO_2C_6H_5$	DBH	0	30	0
H	mesityl	NIS	-30	15	0

DBH = 1,3-dibromo-5,5-dimethylhydantoin
NIS = N-iodosuccinimide

A drawback of the use of a pyridine–hydrogen fluoride complex is that substrates having an acid-sensitive functionality such as oxirane cannot tolerate the reaction conditions. A mixture of **tetrabutylammonium dihydrogen trifluoride and *N*-halo imide or amides** is a reagent of choice for the oxidative fluorodesulfurization of ***dithioacetals*** and ***dithioketals,*** particularly those containing an acid-sensitive functionality [*3*]. Difluoromethylene compounds are obtained in good yields from 2-aryl-1,3-dithiolanes and 2-aryl-1,2-dithianes derived from the corresponding aromatic aldehydes and ketones. Substrates having an epoxide and a hydroxyl group are converted to the corresponding *gem*-difluoro compounds without any damage to these functionalities (equation 5).

4-Methyliodobenzene difluoride cleaves ***aryl dithioketals.*** The reactions are conducted in dichloromethane solutions at 0 °C to give *gem*-difluoro compounds in 65–90% yields [*4*] (equation 6).

Aromatic thio orthoesters are successfully converted into trifluoromethyl arenes by treatment with a **pyridinium polyhydrogen fluoride–*N*-halo imide reagent.** The reactions are conducted at –30 to –20 °C, and the nature of *N*-halo imide is critical; both 1,3-dibromo-5,5-dimethylhydantoin and *N*-bromosuccinimide give similar yields of trifluoromethyl compounds [*5*] (equation 7).

References are listed on page 269–270.

[3] 5

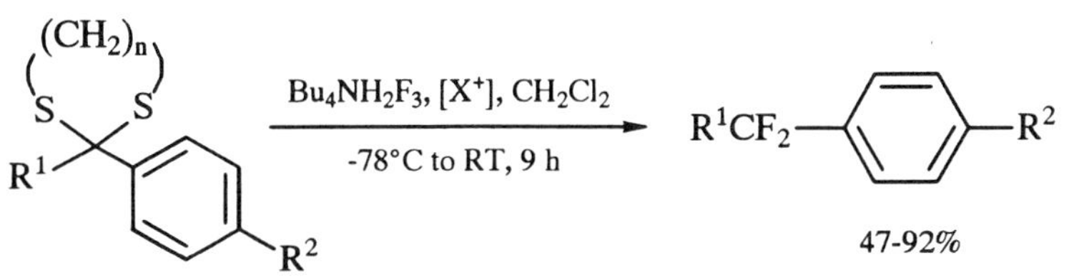

$Bu_4NH_2F_3$, [X^+], CH_2Cl_2, -78°C to RT, 9 h → R^1CF_2–C_6H_4–R^2 47-92%

n = 1 or 2

R^1 = H, C_2H_5, $C_6H_{11}(CH_3)CH$, $CH_3CH(OH)CH_2$, C_6H_5, (oxiranyl)CH_2

R^2 = H, C_3H_7, CH_3O, C_4H_9O, C_6H_5

[X^+] = 1,3-dibromo-5,5-dimethylhydantoin, N-iodosuccinimide, N-bromosuccinimide

[4] 6

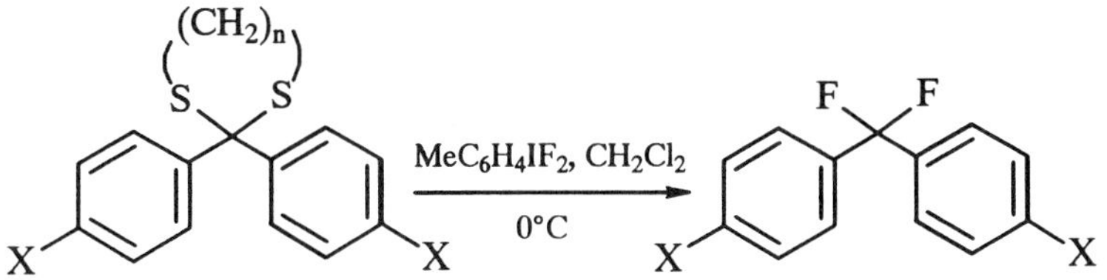

$MeC_6H_4IF_2$, CH_2Cl_2, 0°C → 65-90%

n = 2 or 3; X = H, F, Cl, CH_3

[5] 7

$$ArC(SR)_3 \xrightarrow[-20\text{ to }-30°C]{C_5H_5N\cdot nHF,\ DBH\text{ or }NBS,\ CH_2Cl_2} ArCF_3\ \ 34\text{-}67\%$$

Ar = biphenyl, naphthyl, 3,4-dichlorophenyl, 4-nitrophenyl 3-phenoxyphenyl, 2-benzothiophenyl, 4-phenoxathiinyl

R = CH_3, C_2H_5

DBH = 1,3-dibromo-5,5-dimethylhydantoin; NBS = N-bromosuccinimide

Sulfides are cleaved by treatment with a quaternization reagent and a fluoride ion source to give monofluorinated compounds, generally in good yields. The reactions proceed via formation of sulfonium salts followed by replacement of a sulfide group by fluoride. From among various sulfide groups, sulfonium-generating reagents, and fluoride ion sources tested, a combination of **phenylthio or *p*-chlorophenylthio group, methyl fluorosulfonate, and cesium fluoride** represents the optimal composition [*6*] (equation 8).

When tetrahydrothiophene is treated in a similar manner, ring opening occurs to give 4-methylthiobutyl fluoride. In the benzaldehyde dithioacetal, *gem*-difluorination takes place to form benzylidene fluoride, though in low yield [*6*] (equations 9 and 10).

[*6*] **8**

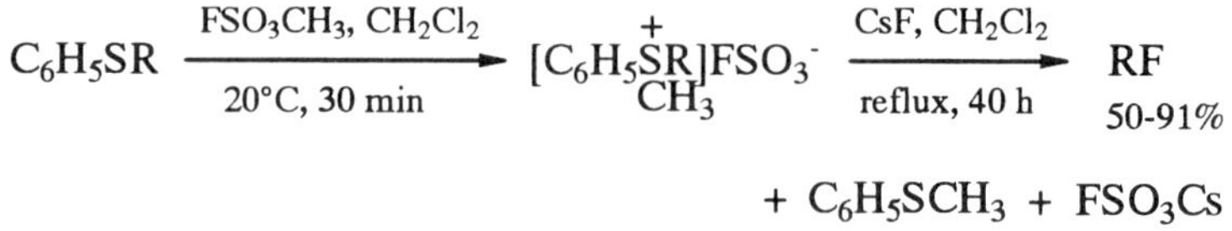

$$C_6H_5SR \xrightarrow[20°C,\ 30\ min]{FSO_3CH_3,\ CH_2Cl_2} [C_6H_5\overset{+}{S}(CH_3)R]FSO_3^- \xrightarrow[reflux,\ 40\ h]{CsF,\ CH_2Cl_2} RF\ (50\text{-}91\%) + C_6H_5SCH_3 + FSO_3Cs$$

R = $C_6H_5CH_2$, *p*-$ClC_6H_5CH_2$, *p*-$C_6H_5OC_6H_5CH_2$, 1-naphthylCH_2, C_8H_{17}

[*6*] **9**

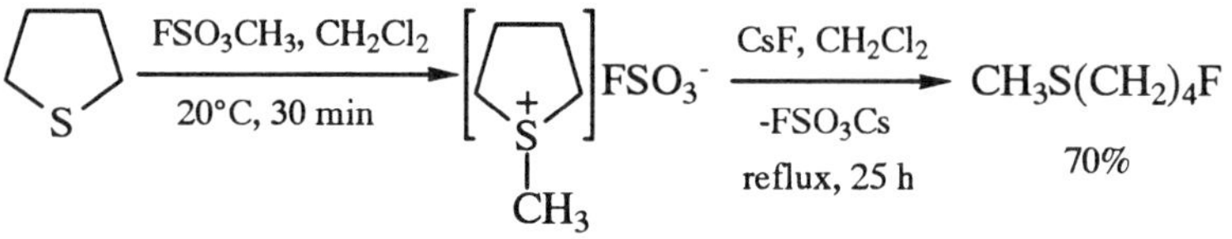

Tetrahydrothiophene $\xrightarrow[20°C,\ 30\ min]{FSO_3CH_3,\ CH_2Cl_2}$ [S-methyltetrahydrothiophenium]FSO_3^- $\xrightarrow[-FSO_3Cs,\ reflux,\ 25\ h]{CsF,\ CH_2Cl_2}$ $CH_3S(CH_2)_4F$ 70%

[*6*] **10**

2-Phenyl-1,3-dithiolane $\xrightarrow[20°C,\ 30\ min]{FSO_3CH_3,\ CH_2Cl_2}$ $\xrightarrow[reflux,\ 26\ h]{CsF,\ CH_2Cl_2}$ $C_6H_5CHF_2$ 25%

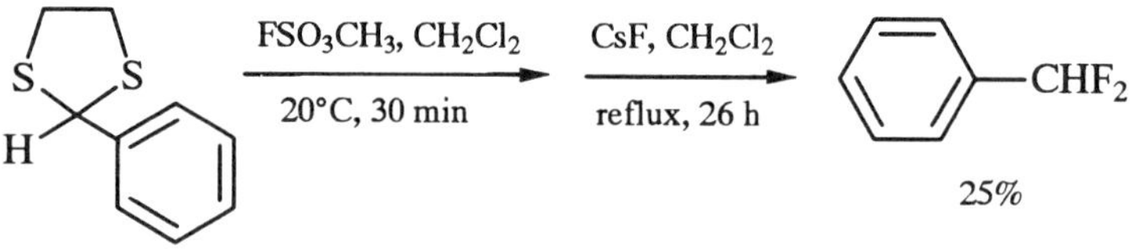

Aryl and alkyl thioglycosides are converted in high yields to the corresponding glycosyl fluorides. The substitution of fluorine for the thiophenyl group proceeds readily at –15 °C with retention of configuration by treatment either with **diethylaminosulfur trifluoride (DAST) and *N*-bromosuccinimide or pyridinium polyhydrogen fluoride and *N*-bromosuccinimide** [*7, 8*] (equation 11).

[*7*] **11**

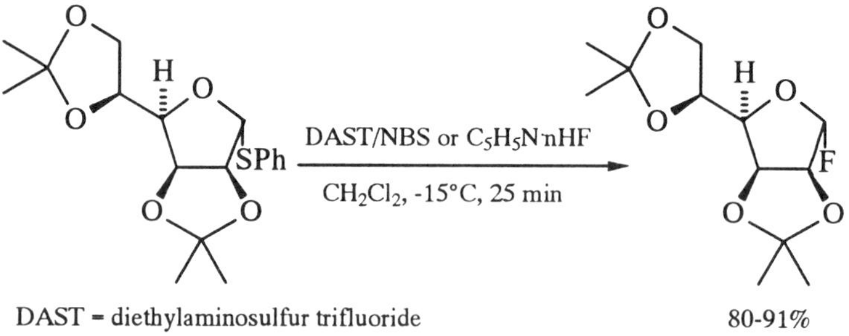

Di-O-isopropylidene furanosyl SPh $\xrightarrow[CH_2Cl_2,\ -15°C,\ 25\ min]{DAST/NBS\ or\ C_5H_5N\cdot nHF}$ Di-O-isopropylidene furanosyl F 80-91%

DAST = diethylaminosulfur trifluoride

NBS = N-bromosuccinimide

In contrast, cleavage of alkyl and also aryl thioglycosides with dimethyl(methylthio)sulfonium tetrafluoroborate takes place with total inversion of configuration [*9*] (equation 12). The latter reagent is commercially available or easily prepared as a crystalline nonhygroscopic solid [*10*]. All the three above-mentioned reagents do not affect glycosidic linkages and most sensitive protecting groups. Dimethyl(methylthio)sulfonium tetrafluoroborate in some instances does not even affect free hydroxyl groups. The reactions are performed at ambient temperature in tetrahydrofuran, but they proceed almost equally well in diethyl ether or dioxane [9].

[*9*] **12**

$$\text{Thioglycoside} \xrightarrow[CH_2Cl_2,\ RT]{[(CH_3)_2SSCH_3]^+BF_4^-} \text{glycosyl fluoride}\quad 75\text{-}100\%$$

R^1 - CH_3COO, 4-$ClC_6H_4CH_2O$, N_3, $C_6H_5CH_2O$

R^2 and R^4 - CH_3CO, $C_6H_4CH_2$, 4-$ClC_6H_5CH_2$

R^3 - H, 4-$ClC_6H_5CH_2$; R^5 - C_2H_5, 4-$CH_3C_6H_5$

Thiocarbonyl compounds can be converted into difluoromethylene compounds usually under milder conditions than the corresponding carbonyl compounds. Ethylene trithiocarbonate reacts smoothly with **sulfur tetrafluoride** at 110 °C in the absence of catalyst to give 2,2-difluoro-1,3-dithiolane in high yield. ***Thiuramsulfides*** under similar conditions are readily converted into *dialkyltrifluoromethylamines* [*11*] (equations 13 and 14).

[*11*] **13**

$$\text{ethylene trithiocarbonate} + 0.5\ SF_4 \xrightarrow{110°C,\ 8\ h} \text{2,2-difluoro-1,3-dithiolane}\ (82\%) + 1.5\ S$$

[*11*] **14**

$$(R^1R^2NC(=S))_2S_2 + 1.5\ SF_4 \longrightarrow 2\ R^1R^2NCF_3 + 5.5\ S$$

R^1 - R^2 - C_2H_5	120°C, 8 h	58%	92%
R^1R^2 - $(CH_2)_5$	100°C, 6 h	70%	88%

These reactions differ from those of sulfur tetrafluoride with carbonyl compounds in that a formal oxidation–reduction of the sulfur atoms in the thiocarbonyl compound and sulfur tetrafluoride molecule occurs, resulting in the formation of free sulfur and the complete utilization of the fluorine atoms in sulfur tetrafluoride.

References are listed on page 269–270.

Carbon disulfide gives an essentially quantitative yield of carbon tetrafluoride and sulfur on reaction with **sulfur tetrafluoride** at 450 °C in the presence of arsenic trifluoride as a catalyst. At lower temperature, *bis(trifluoromethyl) polysulfides* are formed [*11*] (equation 15).

[*11*] **15**

$$CS_2 + SF_4 \xrightarrow[200\text{-}475°C,\ 12.5\ h]{AsF_3} CF_4 + S \quad 100\%$$

$$CS_2 + SF_4 \xrightarrow[150\text{-}180°C,\ 10\ h]{BF_3} CF_3S_2CF_3 + CF_3S_3CF_3 \quad 28\%$$

A number of *aryl trifluoromethyl ethers* are synthesized by fluorination of ***aryl chlorothioformates*** with **molybdenum hexafluoride.** The reagents are progressively heated from –25 to 130 °C, and the products are distilled off. Molybdenum disulfide and carbon disulfide are the by-products [*12*] (equation 16).

[*12*] **16**

$$R\text{-}C_6H_4\text{-}OC(=S)Cl \xrightarrow[-25\ to\ 130°C]{MoF_6} R\text{-}C_6H_4\text{-}OCF_3 \quad 40\text{-}95\%$$

R - H, 2-CH_3, 3-CH_3, 4-CH_3, 4-Br, 4-Cl, 3-F, 4-F, 3-CF_3

The transformation of an ester carbonyl group to a difluoromethylene group, which is usually difficult to perform, can be accomplished by conversion to the ***thioester*** followed by treatment with **diethylaminosulfur trifluoride (DAST).** A variety of ester types react efficiently, although the reaction fails with lactones. Remarkably, ***trimethylsilylmethyl esters*** carry through the procedure with the silyl group intact [*13*] (equation 17).

[*13*] **17**

$$R^1C(=S)OR^2 \xrightarrow[25°C,\ 12\text{-}36\ h]{DAST,\ CH_2Cl_2} R^1CF_2OR^2 \quad 71\text{-}88\%$$

R^1 - C_7H_{15}, c-C_6H_11, adamantyl, C_6H_5, C_6H_5CH-CH

R^2 - CH_3, C_2H_5, $(CH_3)_3SiCH_2$

DAST - diethylaminosulfur trifluoride

Aryl and alkyl xanthates are converted into *difluoro(methylthio)methyl ethers* with **tetrabutylammonium dihydrogen trifluoride and *N*-bromosuccinimide** in

References are listed on page 269–270.

dichloromethane. The latter products, as well as the starting xanthates, give on treatment with **pyridinium polyhydrogen fluoride and 1,3-dibromo-5,5-dimethylhydantoin** under mild conditions *aryl and alkyl trifluoromethyl ethers* in good yields [*14*] (equation 18).

[*14*] **18**

$$\mathrm{ROC(=S)SCH_3} \xrightarrow[\text{0°C to RT, 1 h}]{\mathrm{Bu_4NF{\cdot}2HF,\ NBS,\ CH_2Cl_2}} \underset{15\text{-}58\%}{\mathrm{ROCF_2SCH_3}}$$

$$\mathrm{ROCF_2SCH_3} \xrightarrow[\mathrm{CH_2Cl_2},\ -78\text{ to }0\text{°C, 1 h}]{\mathrm{C_5H_5N{\cdot}9HF,\ DBH}} \mathrm{ROCF_3}$$

$$\mathrm{ROC(=S)SCH_3} \xrightarrow[-78\text{ to }0\text{°C, 1 h}]{\mathrm{C_5H_5N{\cdot}9HF,\ DBH,\ CH_2Cl_2}} \underset{41\text{-}80\%}{\mathrm{ROCF_3}}$$

R = XC_6H_4, $C_6H_5CH_2$, $C_6H_5(CH_2)_3$, $C_{10}H_{21}$

X = 4-alkyl, 4-Br, 3-acetyl, 4-$C_6H_5CH_2O$, 4-(4-acetylphenyl)

NBS = N-bromosuccinimide; DBH = 1,3-dibromo-5,5-dimethylhydantoin

Dithiocarbamates are almost quantitatively converted into *trifluoromethylamines* by reacting at ambient temperature with either **tetrabutylammonium dihydrogen trifluoride, triethylamine trihydrogen trifluoride, or pyridinium polyhydrogen fluoride and an *N*-halo imide reagent** [*15*] (equation 19).

[*15*] **19**

$$\mathrm{R^1R^2NC(=S)SCH_3} \xrightarrow[\text{NBS or NIS or DBH, 0°C to RT, 1 h}]{\mathrm{Bu_4NF{\cdot}2HF\ or\ C_5H_5N{\cdot}9HF\ or\ (C_2H_5)_3N{\cdot}3HF}} \underset{62\text{-}99\%}{\mathrm{R^1R^2NCF_3}}$$

R^1 = XC_6H_4, $C_6H_5CH_2$, C_3H_7

X = 4-Cl, 4-F, 4-CH_3O, 4-CN, 4-NO_2, 3-CH_3O

R^2 = $C_6H_5CH_2$, CH_3, 3-$CH_3C_6H_4$

NBS = N-bromosuccinimide; NIS = N-iodosuccinimide;

DBH = 1,3-dibromo-5,5-dimethylhydantoin

References for Pages 263–269

1. Kollonitsch, J.; Marburg, S.; Perkins, L. M. *J. Org. Chem.* **1976,** *47,* 3107.
2. Sondej, S.; Katzenellenbogen, J. A. *J. Org. Chem.* **1986,** *51,* 3508.
3. Kuroboshi, M.; Hiyama, T. *Synlett* **1991,** 909.

4. Motherwell, W. B.; Wilkinson, J. A. *Synlett* **1991,** 191.
5. Matthews, D. P.; Whitten, J. P.; McCarthy, J. R. *Tetrahedron Lett.* **1986,** *27,* 4861.
6. Ichikawa, J.; Sugimoto, K.; Sonoda, T.; Kobayashi, H. *Chem. Lett.* **1987,** 1985.
7. Nicolaou, K. C.; Dolle, R. E.; Papahatjis, D. P.; Randall, J. L. *J. Am. Chem. Soc.* **1984,** *106,* 4189.
8. Dolle, R. E.; Nicolaou, K. C. *J. Am. Chem. Soc.* **1985,** *107,* 1691.
9. Blomberg, L.; Norberg, T. J. *Carbohydr. Chem.* **1992,** *11,* 751.
10. Meerwein, H.; Zenner, K. F.; Gipp, R. *Liebigs Ann. Chem.* **1965,** *688,* 67.
11. Harder, R. J.; Smith, W. C. *J. Am. Chem. Soc.* **1961,** *83,* 3422.
12. Mathey, F.; Bensoam, J. *Tetrahedron Lett.* **1973,** 2253.
13. Bunnelle, W. H.; McKinnis, B. R.; Narayanan, B. A. *J. Org. Chem.* **1990,** *55,* 768.
14. Kuroboshi, M.; Suzuki, K.; Hiyama, T. *Tetrahedron Lett.* **1992,** *33,* 4173.
15. Kuroboshi, M.; Hiyama, T. *Tetrahedron Lett.* **1992,** *33,* 4177.

Replacement of Nitrogen by Fluorine

by Max M. Boudakian

Primary Amino Compounds

Primary Aliphatic Amines

Until recently, no synthetically useful procedures for replacement of primary aliphatic amines by fluorine had been developed.

A novel fluorination technique features reaction of primary amines with 2,4,6-triphenylpyrlium fluoride to give 2,4,6-triphenylpyridinium fluoride, which on heating forms primary fluorides [*1, 2*] (equation 1 and Table 1).

[*1,2*] **1**

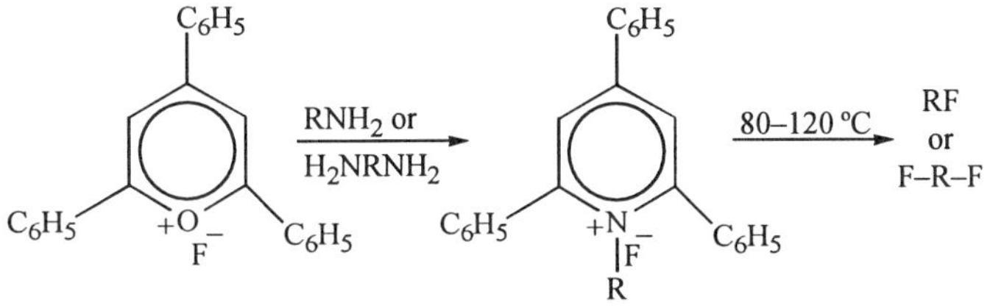

Fluorodediazoniation of 2-fluoro-2-phenylethylamines with sodium nitrite in Olah's reagent (70% anhydrous hydrogen fluoride–30% pyridine) gives high yields of 1,1-difluoro-2-phenylethanes arising from 1,2-migration of phenyl [*3*] (equation 2).

[*3*] **2**

$$C_6H_5CFR^1CH(NH_2)R^2 \xrightarrow[\text{HF/}C_5H_5N\text{, 20–50 °C, 1 h}]{NaNO_2} R^1CF_2CHR^2C_6H_5$$

R^1	R^2	Yield, %
H	H	60
C_2H_5	H	85
C_6H_5	H	80
H	$CO_2C_3H_7$-*i*	95

α-Amino Acids

An example of fluorodediazoniation of α-amino acids is the preparation of β-fluoroaspartic acid in 25% yield from α,β-diaminosuccinic acid and sodium nitrite in anhydrous hydrogen fluoride [*4*].

0065–7719/95/0187–0271$08.54/1

Table 1. Primary Monofluorides and α,ω-Difluorides from 2,4,6-Triphenylpyridinium Fluoride and Primary Amines and Diamines [*1,2*]

RNH_2	*RF*	*Yield of Intermediate (%)*	*Yield of Product (%)*
$C_7H_{15}NH_2$	$C_7H_{15}F$	82	65
$C_8H_{17}NH_2$	$C_8H_{17}F$	78	42
$C_{11}H_{23}NH_2$	$C_{11}H_{23}F$	85	55
$C_6H_5CH_2NH_2$	$C_6H_5CH_2F$	78	62
o-$ClC_6H_4CH_2NH_2$	*o*-$ClC_6H_4CH_2F$	72	62
p-$ClC_6H_4CH_2NH_2$	*p*-$ClC_6H_4CH_2F$	82	65
2,4-$ClC_6H_3CH_2NH_2$	2,4-$ClC_6H_3CH_2F$	86	61
$H_2N(CH_2)_4NH_2$	$F(CH_2)_4F$	83	48
$H_2N(CH_2)_5NH_2$	$F(CH_2)_5F$	85	49
$H_2N(CH_2)_6NH_2$	$F(CH_2)_6F$	86	52

High-yield deaminative fluorination of α-aminocarboxylic acids to α-fluorocarboxylic acids can be effected with sodium nitrite in Olah's reagent [*5, 6*] (equation 3 and Table 2).

[*5,6*] $$RCH(NH_2)CO_2H \xrightarrow[HF/C_5H_5N]{NaNO_2} \left[\begin{array}{c} RCHCO_2H \\ | \\ N_2^+ \end{array}\right] F^- \xrightarrow{-N_2} \underset{28\text{–}88\%}{RCHFCO_2H} \qquad \mathbf{3}$$

These products have biological significance as enzymatic blocking agents. At high hydrogen fluoride/pyridine ratios (70/30), anchimerically assisted rearrangement may occur to give β-fluorocarboxylic acids. Rearrangement can be eliminated and/or suppressed by less acidic reagent (48% hydrogen fluoride/52% pyridine) [*7*]. Similar observations on the effect of hydrogen fluoride/pyridine ratios on regioselectivity (α- versus β-fluorination) in fluorodediazoniation of α-amino esters have also been noted [*8*].

Alkyl Carbamates

Fluorodediazoniation of alkyl carbamates in hydrogen fluoride/pyridine (70/30) represents a novel route to alkyl fluoroformates [*6, 9*] (equation 4).

References are listed on pages 289–293.

Table 2. α-Fluorocarboxylic Acids via Deaminative Fluorination of α-Amino Acids

RCH(NH$_2$)-(CO$_2$H)	*R*	*HF·C$_5$H$_5$N Ratio (w/w)*		*RCH(F)(CO$_2$H)*	
		HF	*C$_5$H$_5$N*	*Yield (%)*	*Ref.*
Glycine	H	48	52	41	7
Alanine	CH_3	48	52	76	7
2-Amino-butanoic acid	C_2H_5	48	52	82	7
Valine	i-C_3H_7	48	52	75	7
Leucine	i-C_4H_9	70	30	88	5, 6
Isoleucine	s-C_4H_9	48	52	71	7
Phenylalanine	$C_6H_5CH_2$	48	52	86	5, 6
Tyrosine	p-$HOC_6H_4CH_2$	48	52	58[a]	7
Serine	$HOCH_2$	70	30	80	5
Threonine	$CH_3CH(OH)$	48	52	43[b]	7
Aspartic acid	HO_2CCH_2	70	30	52	5, 6
Glutamic acid	$HO_2C(CH_2)_2$	70	30	28	5, 6

[a]α-Fluoro, 80%; β-fluoro, 15%.

[b]α-Fluoro, 80%, β-fluoro, 20%.

[*6,9*] $$ROCONH_2 \xrightarrow[HF/C_5H_5N]{NaNO_2} [ROCON_2F] \xrightarrow{-N_2} ROCOF$$ **4**

R	Yield, %	R	Yield, %
CH_3	75	C_4H_9	40
C_2H_5	31	i-C_4H_9	78
C_3H_7	68	s-C_4H_9	75
i-C_3H_7	75	t-C_4H_9	50

Aromatic Amines

Ring-fluorinated aromatics have found wide applications in pharmaceuticals, crop protection chemicals, polymer intermediates, liquid crystals, etc. [*10*]. Routes based on aromatic amines represent one of the major synthetic approaches to these compounds. The scope and the techniques have been sufficiently described in reviews [*11, 12*] and monographs [*13, 14, 15*]. Therefore, only reactions and techniques published after 1971 are discussed.

References are listed on pages 289–293.

Balz–Schiemann Reaction and Modifications

The Balz–Schiemann reaction, thermal decomposition of arenediazonium fluoroborates, is still a favorite approach to the laboratory preparation of fluoroaromatics [*16, 17*]. Caution must be exercised in handling and decomposing nitroarenediazonium fluoroborates and pyridinediazonium fluoroborates because detonations have been reported [*18, 19*].

New diazotization techniques for the Balz–Schiemann reaction feature alternative nitrosating agents in place of aqueous sodium nitrite or substitution of other salts such as arenediazonium hexafluorophosphates for arenediazonium fluoroborates.

High yields of arenediazonium fluoroborates can be obtained from preformed nitrosonium tetrafluoroborate, $NO^+BF_4^-$, prepared from nitrogen dioxide and fluoroboric acid, and aromatic amines in organic solvents [*20, 21*]. In situ diazotization–decomposition (110 °C) of aniline with this reagent in *p*-dichlorobenzene can be accomplished to give fluorobenzene in 75% yield [*22*]. A complementary approach features in situ generation of nitrosonium tetrafluoroborate in solutions of the aromatic amine in diethyl ether [*23*]. This technique gives a 91% yield of the diazonium fluoroborate of 7-aminobenz[*a*]anthracene; pyrolysis results in a 91% yield of 7-fluorobenz[*a*]anthracene.

Nitrite esters are alternative nitrosating agents that can provide high yields of arenediazonium fluoroborates. Treatment of 4-aminoveratrole in methanol and 50% fluoroboric acid with butyl nitrite as the source of nitrous acid gives an 83% yield of veratrole-4-diazonium tetrafluoroborate [*24*]. In situ generation of nitrosyl fluoride from nitrite esters, boron trifluoride, and hydrogen fluoride in anhydrous media provides high yields of *ortho*-substituted diazonium fluoroborates; pyrolysis in paraffin oil gives high yields of aryl fluoride [*25, 26*] (equation 5 and Table 3).

[*26*]

$$ArNH_2 \xrightarrow[i\text{-BuOH, } BF_3\text{, HF}]{i\text{-}C_4H_9ONO} ArN_2BF_4 \xrightarrow[\text{paraffin oil} \; -N_2,\, -BF_3]{\text{heat}} \underset{59\text{–}92\%}{ArF} \qquad \mathbf{5}$$

Arenediazonium hexafluorophosphates represent a promising alternative to diazonium fluoroborates because of decreased water solubility and, in some instances, higher yields of fluoroaromatics [*27, 28*] (equation 6).

[*27,28*]

$$RC_6H_4NH_2 \xrightarrow[2.\ 60\%\ HPF_6]{1.\ NaNO_2,\ HCl} \underset{79\text{–}100\%}{RC_6H_4N_2PF_6} \xrightarrow[\text{mineral oil} \; -N_2,\, -PF_5]{\text{heat}} \underset{60\text{–}78\%}{RC_6H_4F} \qquad \mathbf{6}$$

The relative efficacy of this procedure compared with the standard Balz–Schiemann arenediazonium fluoroborate conditions for difunctional aromatics is shown in Table 66 in reference 14 and in equation 7. The arenediazonium hexafluorophosphate

Table 3. Synthesis of *ortho*-Substituted Fluoroaromatics from Nitrite Esters, Boron Trifluoride, and Hydrogen Fluoride [26]

$ArNH_2$	Yield of ArN_2BF_4 (%)	Yield of ArF (%)
$2\text{-}BrC_6H_4$	97.4	84.1
$2\text{-}ClC_6H_4$	96.7	81.2
$2\text{-}CH_3OC_6H_4$	92.3	58.7
$2\text{-}EtO_2CC_6H_4$	97.7	65.7
$2\text{-}Br\text{-}4\text{-}CH_3C_6H_3$	98.2	92.4
$2\text{-}CF_3\text{-}4\text{-}BrC_6H_3$	93.5	88.5
$2\text{-}Br\text{-}4,6\text{-}(CH_3)_2C_6H_2$	99.0	80.5

$$BrC_6H_4NH_2 \longrightarrow BrC_6H_4N_2^+ \longrightarrow BrC_6H_4F \qquad (7)$$

				Overall yield
o-Br	BF_4	50%	81%	40.5% [*11*]
o-Br	PF_6	97%	77%	74.7% [*27, 28*]
p-Br	BF_4	64%	75%	48.0% [*11*]
p-Br	PF_6	100%	79%	79.0% [*27*]

technique can also be applied to trifunctional aromatics. Thus, 2-methoxy-4-methylaniline is converted in 69% yield to 3-methoxy-4-fluorotoluene, which is important for inhibitor studies on receptor sites in mammalian brains [*29*].

Other less widely used alternative diazonium salts for aryl fluoride synthesis include arenediazonium hexafluoroantimonates, ArN_2SbF_6; arenediazonium hexafluoroarsonates, ArN_2AsF_6; and arenediazonium hexafluorosilicates, $(ArN_2)_2SiF_6$ [*14*].

The Bergmann variation of the Balz–Schiemann reaction is a two-step process featuring copper- or copper halide-catalyzed decomposition of aqueous or acetone solutions of arenediazonium fluoroborates containing alkyl or halogen substituents [*30*]. A recent modification is a one-step technique featuring simultaneous diazotization and decomposition by addition of aqueous sodium nitrite at 25 °C to a mixture of fluoroboric acid, copper powder, and 2-isopropyl-6-methylaniline to give 2-isopropyl-6-methylfluorobenzene in 73% yield [*31*].

Both *photochemical* and *ultrasound variants* of the Balz–Schiemann thermal decomposition step have also been developed. A noteworthy feature of either technique is the use of milder fluorodediazoniation conditions.

Depending on structure, photolysis of films of arenediazonium fluoroborates and hexafluorophosphates at room temperature gives aryl fluorides in 10–75% yield [*32*]. In situ photochemical decomposition of arenediazonium fluoroborates in fluoroboric acid media also can form fluoroaromatics, as shown in the preparation of 3,4-dimethoxy-5-fluorobenzaldehyde [*33, 34*] (equation 8).

[*34*] **8**

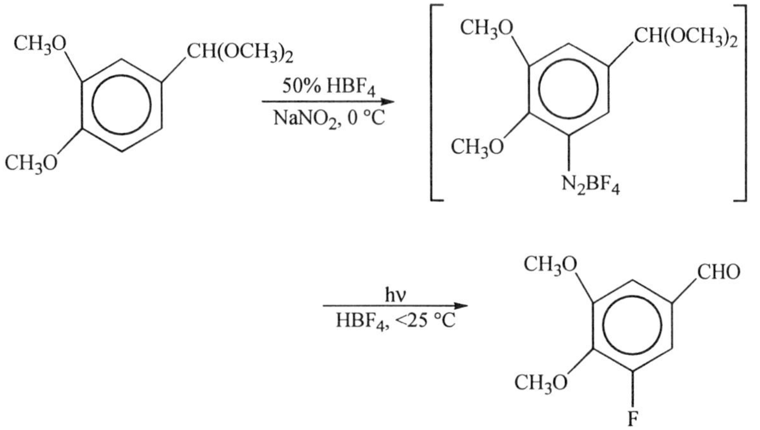

CH_3O, $CH(OCH_3)_2$, CH_3O —(50% HBF_4 / $NaNO_2$, 0 °C)→ [CH_3O, $CH(OCH_3)_2$, CH_3O, N_2BF_4] —(hν / HBF_4, <25 °C)→ CH_3O, CHO, CH_3O, F

This technique has been applied to the synthesis of fluorinated dopamine and other compounds of biological significance. Fluoroheterocycles such as fluoroimidazoles [*35, 36*] and fluoropyrazoles [*37*] can also be prepared by the "photo Balz–Schiemann" technique (equation 9). Photochemically induced in situ fluorodediazoniation can also be applied to arenediazonium fluorides in hydrogen fluoride–pyridine media. Thus, *o*-fluoroanisole is obtained in 73% yield at 20 °C after 18 h [*38*].

[*36*] **9**

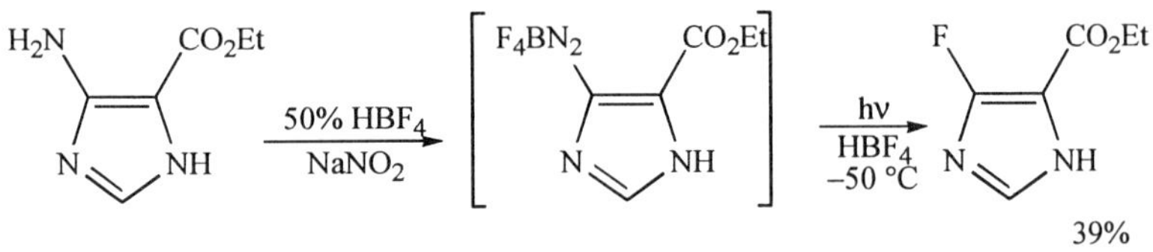

H_2N, CO_2Et, N, NH —(50% HBF_4 / $NaNO_2$)→ [F_4BN_2, CO_2Et, N, NH] —(hν / HBF_4, −50 °C)→ F, CO_2Et, N, NH 39%

Ultrasound-promoted (17 kHz) decomposition of arenediazonium fluoroborates can also be effected in the presence of triethylamine trihydrofluoride in Freon 113 media. For example, fluorobenzene can be obtained in 92–95% yield by this method [*39*] (equation 10).

[*39*] **10**

$$C_6H_5N_2BF_4 \xrightarrow[17\ \text{kHz},\ 40\ ^\circ\text{C},\ 8\ \text{h}]{Et_3N{\cdot}3HF,\ C_2F_3Cl_3} C_6H_5F \quad 92\text{–}95\%$$

The sonification technique does not work when (1) triethylamine trihydrofluoride is eliminated or (2) when triethylamine is used in place of triethylamine trihydrofluoride.

Wallach Aryltriazene Fluorodediazoniation and Modifications

The discovery of the Balz–Schiemann reaction in 1927 replaced the earlier Wallach procedure (1886) based on fluorodediazoniation of arenediazonium piperidides (aryltriazenes) in aqueous hydrogen fluoride [*40, 41*].

Applications in agrochemicals [*42, 43*], pharmaceuticals [*44, 45*], and positron emission tomography (PET) [*46, 47, 48, 49*] have resulted in the resuscitation of the Wallach reaction. The Wallach technique provides high-specific-activity ^{18}F-radiolabeled aromatic fluoride for PET studies, in contrast to the low-specific-activity product by the Balz–Schiemann route.

An illustration of a modified Wallach fluorination is the synthesis of 2,4-dichloro-5-fluorotoluene, an intermediate in the preparation of the fluoroquinolone antibacterial ciprofloxacin. This was prepared in 69% overall yield by heating *N*-(2,4-dichloro-5-methylphenyl)-*N′*,*N′*-dimethyltriazene in anhydrous hydrogen fluoride [*44*] (equation 11).

[*44*] **11**

CH$_3$, Cl, NH$_2$, Cl — 1. NaNO$_2$, HCl; 2. Aq. $(CH_3)_2NH$ → CH$_3$, Cl, N=NN(CH$_3$)$_2$, Cl — 1. HF; 2. 135 °C, 25 bar, 4 h → CH$_3$, Cl, F, Cl

78%

A variety of media have been used for the Wallach fluorination reaction: anhydrous hydrogen fluoride alone or with cosolvents such as methylene chloride, benzene, or tetrahydrofuran and hydrogen fluoride–pyridine alone or with cosolvents such as benzene, glyme, or acetic acid [*42, 43, 46, 50*]. Solutions of cesium fluoride, tetraethylammonium fluoride, or tetrabutylammonium fluoride in strong acids such as methanesulfonic acid or trifluoroacetic acid with numerous cosolvents have also been studied [*48, 49*].

A recently discovered variant of the Wallach technique is the silver ion-catalyzed fluorination of aryl diazo sulfides in hydrogen fluoride–pyridine–toluene solvent [*51*] (equation 12). Electron-withdrawing substituents such as acetyl give higher yields of aryl fluoride (71%) than electron-donating groups (butyl, 39%; methoxy, 2–14%); reductive dediazoniation competes with fluorination.

Fluorodediazoniation in Hydrogen Fluoride

Diazotization of aromatic amines in aqueous 70% hydrogen fluoride at 0 °C followed by in situ decomposition of the arenediazonium fluorides has been replaced by diazotization in anhydrous hydrogen fluoride [*52, 53*] (equation 13).

[*51*] **12**

$$RC_6H_4NH_2 \xrightarrow[C_6H_5SH]{HCl,\ NaNO_2} RC_6H_4N{=}NSC_6H_5 \xrightarrow[HF/C_5H_5N,\ Toluene,\ 90\ ^\circ C]{AgNO_3} RC_6H_4F + RC_6H_5$$

R	Yield, %
i-CH_3CO	71 & 4
p-C_4H_9	39 & 54
p-CH_3O	14 & (–)
o-, *m*-CH_3O	<2 & (–)

Fluoroaromatics are now produced in 75–90% yields on an industrial scale by this method. The nonorganic layer containing water, hydrogen fluoride, and sodium bifluoride is treated with sulfur trioxide, and anhydrous hydrogen fluoride is recycled by distillation [*54*] (equation 13).

Nitrosyl chloride [*55*], nitrosyl fluoride–hydrogen fluoride liquid complexes (NOF·3HF; NOF·6HF) [56], nitrous acid–hydrogen fluoride solutions [*57, 58*], nitrogen trioxide (prepared in situ from nitric oxide and oxygen) [*59*] and *tert*-butyl nitrite–hydrogen fluoride–pyridine [*60*] have been substituted for sodium nitrite in the diazotization step.

[*54*] **13**

$$C_6H_5NH_2 + 15HF + NaNO_2 \xrightarrow{0\ ^\circ C} [C_6H_5N_2F] \xrightarrow[-N_2]{heat}$$

$$\longrightarrow C_6H_5F + 2H_2O + NaF{\cdot}HF + 12HF$$

$$12HF + NaF{\cdot}HF + 2H_2O + 2SO_3 \longrightarrow 14HF + NaHSO_4 + H_2SO_4$$

Thermal stability of arenediazonium fluorides can influence yields of aryl fluorides during decomposition in hydrogen fluoride [*52, 61*]. Use of 70% hydrogen fluoride–30% pyridine (w/w) mixture having a lower pressure of hydrogen fluoride permits higher fluorodediazoniation temperatures and improves yields [*6, 61, 62, 63, 64*]. This technique has also been extended to complexes of hydrogen fluoride with tertiary amines [*65*].

Hydrogen fluoride–ammonium fluoride complexes, $NH_4F{\cdot}(HF)_x$, are stable solvates that also have reduced vapor pressure of hydrogen fluoride. Melting points are 126 °C ($x = 1$), 23 °C ($x = 2$), and –8 °C ($x = 3$) [*66*]. These solvates can serve as diazotization media for aromatic amines. Fluorodediazoniation in $NH_4F{\cdot}(HF)_x$ is achieved at higher decomposition temperatures than in anhydrous hydrogen fluoride and gives less tars and increased yields of fluoroaromatic compound [*59, 67, 68, 69, 70*].

Quaternary Ammonium Salts

Trimethylbenzylammonium hydroxide has been converted to benzyl fluoride in 22–60% yield by treatment with 20% hydrofluoric acid followed by vacuum distillation [*71, 72*].

References are listed on pages 289–293.

Renewed interest in the fluorination of quaternary ammonium salts is prompted by the need for rapid fluorination techniques to incorporate ^{18}F ($t_{1/2}$, 110 min) in positron emission tomography (PET) studies. One promising approach is displacement of trimethylammonium ion, bound directly to an aromatic ring, by fluoride ion. This technique was initially developed with substituted phenyltrimethylammonium perchlorates and unlabeled cesium fluoride in dimethyl sulfoxide or acetonitrile [*73*] (equation 14).

[*73*] $$YC_6H_4\overset{+}{N}(CH_3)_3\ ClO_4^- + CsF \xrightarrow[80\ ^\circ C,\ 20\ min]{DMSO} \underset{11\text{–}91\%}{YC_6H_4F} \qquad \textbf{14}$$

The efficacy of ring fluorination depends on the nature and position of the activating group, Y, in the aromatic ring. The relative extent of F^- for $N^+(CH_3)_3$ displacement decreases in the following order for Y: *p*-NO_2 (71%), *p*-CN (24%), *p*-CH_3CO (15%), *p*-CHO (≤5%) ≈ *m*-NO_2 (≤5%). This fluorodequaternization technique was subsequently adapted to prepare numerous NCA (no-carrier-added) ^{18}F-labeled aryl fluorides [*73, 74*].

The striking nucleofugacity of the trimethylammonium leaving group compared with the nitro group in the reaction with fluoride ion is illustrated by the relative reactivities of *p*-nitrophenyltrimethylammonium perchlorate and *p*-dinitrobenzene [*73*] (Table 4).

Table 4. Comparison of Nucleofugacity of $(CH_3)_3N^+$– and NO_2– Groups with No Carrier Added (NCA) ^{18}F [*73*]

Substrate	*Relative Rate Constant*[a]
p-$NO_2C_6H_4N^+(CH_3)_3ClO_4^-$	30 000
p-$NO_2C_6H_4NO_2$	420

[a]80 °C, 20 min, dimethyl sulfoxide.

Instead of perchlorates, trifluoromethanesulfonates (triflates) can be used. They are readily prepared and claimed to be safer than perchlorates [*75*]. ^{18}F Aryl fluorides have been prepared from *p*-$YC_6H_4N^+(CH_3)_3CF_3SO_3^-$ (Y = COC_6H_5, CN, CHO, NO_2, $COCH_3$, $CO_2C_2H_5$) by this alternative substrate. Fluorodequaternization can also be applied to the preparation of fluorinated heterocyclics such as 4-fluoro-6-phenyl-1,3-pyrimidine [*76*] (equation 15) and fluoropurines [*77*].

[*75*] **15**

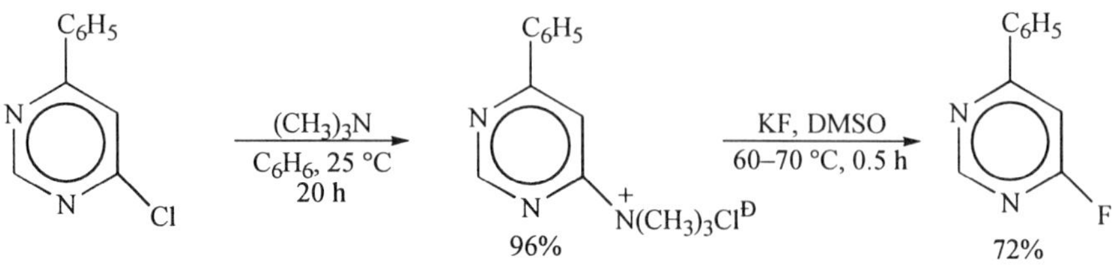

$$\text{4-}C_6H_5\text{-6-Cl-pyrimidine} \xrightarrow[C_6H_6,\ 25\ °C,\ 20\ h]{(CH_3)_3N} \text{4-}C_6H_5\text{-6-}\overset{+}{N}(CH_3)_3Cl^{-}\text{-pyrimidine}\ (96\%) \xrightarrow[60\text{–}70\ °C,\ 0.5\ h]{KF,\ DMSO} \text{4-}C_6H_5\text{-6-F-pyrimidine}\ (72\%)$$

Heteroatom–Amine Systems

Amino groups bound to sulfur can be replaced by fluorine via diazotization. In contrast to carboxylic acid amides, fluorodediazoniation of aromatic sulfonamides is readily accomplished to give sulfonyl fluorides in high yields [*52, 78*] (equation 16). Tetrazotization–fluorination of sulfanilamide can also be effected to give a 38% yield of *p*-fluorobenzenesulfonyl fluoride [*52*].

[*52,78*] **16**

$$RC_6H_4SO_2NH_2 \xrightarrow[HF]{NaNO_2} RC_6H_4SO_2F$$

R	Yields, %	Ref.
H	53	[*78*]
o-CH_3	78	[*78*]
p-CH_3	70	[*52*]
m-NO_2	64	[*78*]

Phosphorus–nitrogen compounds containing alkylated amino groups can be cleaved by fluorinating agents. Phenyldifluorophosphine is formed from the reaction of *N,N,N′,N′*-tetramethylphenylphosphonous amide and benzoyl fluoride [*79*] (equation 17).

[*79*] **17**

$$C_6H_5P[N(CH_3)_2]_2 \xrightarrow[5\ to\ 25\ °C,\ 4.8\ h]{C_6H_5COF} C_6H_5PF[N(CH_3)_2] \xrightarrow{C_6H_5COF} C_6H_5PF_2\ (100\%)$$

Fluorination of hexamethylphosphoramide with ammonium fluoride gives 85–90% yield of the insecticide bis(dimethylamine)fluorophosphine oxide (Dimefox) [*80*] (equation 18).

[*80*] **18**

$$OP[N(CH_3)_2]_3 + NH_4F \xrightarrow[2\ h]{200\ °C} OPF[N(CH_3)_2]_2\ (85\text{–}90\%)$$

The reaction of silylated phosphorus(V) imides with anhydrous hydrogen fluoride gives high yields of alkylenebis(difluorophosphoranes) [*81*] (equation 19).

[*81*] **19**

$$\begin{matrix} & & (CH_2)_n & & \\ & R_2P & & PR_2 & \\ & \| & & \| & \\ & N & & N & \\ & | & & | & \\ & (CH_3)_3Si & & Si(CH_3)_3 & \end{matrix} \xrightarrow[-78\ \text{to}\ 25\ °C]{HF,\ Et_2O} R_2PF_2(CH_2)_nPF_2R_2$$

R	n	Yield, %
C_6H_5	1	51
C_6H_5	2	59
CH_3	3	89

N,N-Acetals (Aminals)

Acid fluorides such as oxalyl fluoride, trichloroacetyl fluoride, propionyl fluoride, as well as Sanger's reagent, 2,4-dinitrofluorobenzene, easily cleave N,N-acetals (aminals) to give high yields of fluoromethyldialkylamines [82] (equation 20) (Table 5).

[*82*] **20**

$$R_2NCH_2NR_2 + CCl_3COF \xrightarrow[-CCl_3CONR_2]{\text{Solvent}} R_2NCH_2F$$

Table 5. Fluoromethyldialkylamines from Fluorination of *N,N*-Acetals, $R_2NCH_2NR_2$ [*82*]

R_2	*Solvent*	*Fluorinating Agent*	*R_2NCH_2F Yield (%)*
$(CH_3)_2$	—[a]	CH_3CH_2COF	91
	$C_6H_5NO_2$	$2,4\text{-}(NO_2)_2C_6H_3F$	91
$(C_2H_5)_2$	Et_2O	CCl_3COF	74
$(C_4H_9)_2$	THF	CCl_3COF	77
$(CH_2)_4$	C_5H_{12}	CCl_3COF	91
$(CH_2)_5$	Et_2O	CCl_3COF	86
$O(CH_2CH_2)_2$	Et_2O	$(COF)_2$	79
	CH_2Cl_2	CCl_3COF	74

[a]No solvent used.

Hydrazine Derivatives

Functionalized hydrazine groups can be replaced by fluorine. The first example is the replacement of trialkylhydrazine groups in bis(1,2,2-trimethylhydrazino)methane, a hydrazine acetal, by trichloroacetyl fluoride to form 1,1,2- trimethyl-2-fluoromethylhydrazine in 87% yield [*83*].

References are listed on pages 289–293.

Substituted aryl hydrazones can be converted to geminal difluorides in satisfactory yields by molecular fluorine [*84*], iodine fluoride [*85*], and *N*-bromo-succinimide–pyridinium polyhydrogen fluoride or *N*-bromosuccinimide–polyvinylpyridinium polyhydrogen fluoride [*86*] (equation 21) (Table 6).

[*84,85,86*] $$R(C_6H_5)C{=}NNH_2 \xrightarrow[\text{agent}]{\text{Fluorinating}} \underset{34\text{–}90\%}{R(C_6H_5)CF_2}$$ **21**

Table 6. Conversion of Aryl Hydrazones to Geminal Difluorides

			Yield of R(C₆H₅)CF₂ (%)			
Reagent	*Solvent*	*Temp. (°C)*	C_6H_5	$C_6H_5CH_2$	CH_3	*Ref.*
F_2	CH_3OH	0 to 10	69	38	34	84
IF	CCl_3F	–78	65	75	45	85
NBS–HF/C_5H_5N	CH_2Cl_2	–78 to 0	90	76	47	86
NBS–HF/PVP	CH_2Cl_2	–45 to 25	85	—[a]	44	86

[a]This compound was not studied.

Iodine fluoride is a more versatile reagent than molecular fluorine in geminal fluorination of other hydrazones and related compounds under milder reaction conditions [*85*]. Substrates fluorinated include hydrazones of simple cyclic or steroidal ketones (e.g., 4-*tert*-butylcyclohexanone, 70%; 3-cholestanone, 70%), *N*-methyl- and *N,N*-dimethylhydrazones [$R_2C{=}NNH(CH_3)$, 70%; $R_2C{=}NNC(CH_3)_2$, 50%], semicarbazones ($R_2C{=}NNHCONH_2$, 25–50%), and 2,4-dinitrophenylhydrazones [$R_2C{=}NNH{-}C_6H_3{-}2,4(NO_2)_2$, 25–50%].

Diazo-Group-Containing Organics

Depending on the fluorinating agent, diazoalkanes, diazoketones, and diazoesters can undergo hydrofluorination, halofluorination, and geminal difluorination reactions.

Diazo Alkanes

Diazoalkanes react with ethyl fluoroformate [*87*] and acyl fluorides [*87, 88*] to give ethyl α-fluorocarboxylates and α-fluoroketones, respectively (equation 22).

Table 7 lists examples of hydrofluorination [*6, 83, 89*] and geminal difluorination [*90*] of aryldiazomethanes.

Diazo Ketones

Diazoketones that are readily prepared from acyl chlorides and diazomethane [*92*] also undergo a variety of fluorination reactions.

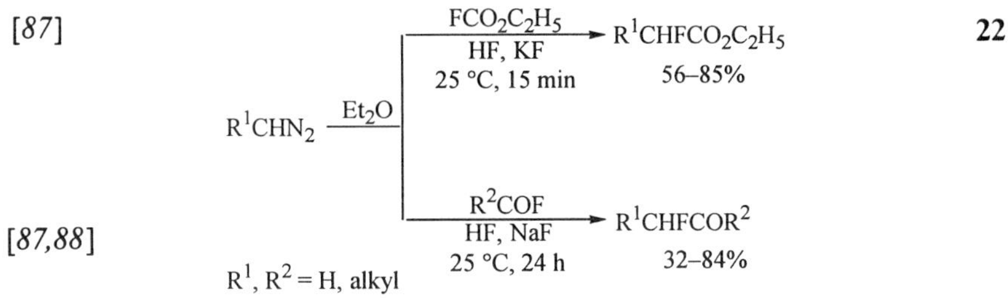

Table 7. Fluorination of Aryldiazomethanes

Aryldiazo-methane	*Reagent*	*Temp. (°C)*	*Time*	*Product*	*Yield (%)*	*Ref.*
$C_6H_5CHN_2$	HF/C_5H_5N	0	—[a]	$C_6H_5CH_2F$	70	6
$C_6H_5CHN_2$	HBF_4, CH_2Cl_2	25	10 min	No reaction		89
p-NO_2-$C_6H_4CHN_2$	HBF_4, CH_2Cl_2	25	10 min	p-NO_2-$C_6H_4CH_2F$	50	89
$(C_6H_5)_2CN_2$	KHF_2, CH_2Cl_2 $(C_4H_9)_4N^+ClO_4^-$	25	48 h	$(C_6H_5)_2CHF$	50	120
$(C_6H_5)_2CN_2$	F_2 (10% in N_2) CCl_3F	–70	15 min	$(C_6H_5)_2CF_2$	71	90

[a]No reaction time cited.

Fluoromethylketones are obtained by dediazoniative hydrofluorination with anhydrous hydrogen fluoride [*91, 92, 93*] or Olah's reagent [*6*] (equation 23).

[*91,92,93*] HF, 0–25 °C, 12 h; 38–74% **23**

$$RCOCHN_2 \xrightarrow{Et_2O} \left[\begin{array}{c} HF,\ 0\text{–}25\ °C,\ 12\ h \\ HF/C_5H_5N,\ 0\text{–}25\ °C,\ 2\ h \end{array}\right] \xrightarrow{-N_2} RCOCH_2F$$

[*6*] HF/C_5H_5N, 0–25 °C, 2 h; 40–50%

This technique can be applied to prepare DL-α-fluoromethylputrescine (5-fluoropentane-1,4-diamine), a potent irreversible inhibitor of *E. coli* ornithine decarboxylase, from 4-phthalimido-1-butyryl chloride, diazomethane, and hydrogen fluoride–pyridine [*94, 95*].

Halofluorination of α-diazoacetophenone by *N*-halosuccinimides in hydrogen fluoride–pyridine provides good yields of α-fluoro-α-haloacetophenones [*6*] (equation 24).

[*6*] $$C_6H_5COCHN_2 \xrightarrow[\text{HF/C}_5\text{H}_5\text{N} \; -15 \text{ to } 25\ ^\circ\text{C, 2 h}]{\text{NBX}} C_6H_5COCHFX$$ **24**

X	Cl	Br	I
Yield, %	49	63	62

High yields (65–94%) of geminal difluoro compounds can be obtained from diazoketones and dilute molecular fluorine [*90*] (equation 25).

[*90*] 2-Diazocyclohexanone $\xrightarrow[\text{CCl}_3\text{F} \; -70\ ^\circ\text{C, 15 min}]{\text{F}_2\text{,N}_2}$ 2,2-difluorocyclohexanone 65% **25**

This method can be adapted for the synthesis of biologically active compounds; the 2,2-difluoro derivative of the antibacterial pleuromutalin, a tricyclic terpenoid, was prepared in 31% yield from 2-diazopleuromutalin and dilute fluorine in chloroform in the presence of potassium fluoride at –50 °C [*96*].

In contrast to molecular fluorine, trifluoromethyl hypofluorite has limited synthetic value for geminal fluorination of diazoketones owing to formation of complex mixtures [*97*].

Diazo Esters

By choice of fluorinating agent, either hydrofluorination [*6, 98, 99*] or halofluorination [*6, 99*] of ethyl diazoacetate is realized (equation 26).

[*6,98,99*] $$N_2CHCO_2C_2H_5 \xrightarrow{\text{Et}_2\text{O}} \xrightarrow[\text{HF/C}_5\text{H}_5\text{N} \; 0\ ^\circ\text{C, 0.5 h}]{\text{HF or}} FCH_2CO_2C_2H_5 \text{ 24–40\%}$$ **26**

[*6,99*] $$\xrightarrow[\text{NBX, HF/C}_5\text{H}_5\text{N} \; 0\ ^\circ\text{C, 0.5 h}]{\text{NBX, HF or}} CHXFCO_2C_2H_5$$

X	Cl	Br	I
Yield, %	30	50–59	50

Geminal difluorination of diethyl diazomalonate can be effected with molecular fluorine (10% in nitrogen) to give diethyl difluoromalonate in 70% yield [*90*] (equation 27).

Aliphatic Azides

Whereas nitrosative decomposition of azidoalkanes such as hexyl, cyclohexyl, and benzyl azides with nitrosonium tetrafluoroborate gives only 0–5% yields of

[*90*] $$N_2C(CO_2C_2H_5)_2 \xrightarrow[\text{CCl}_3\text{F},\ -70\ ^\circ\text{C, 15 min}]{F_2,N_2} \underset{70\%}{CF_2(CO_2C_2H_5)_2} \qquad \textbf{27}$$

fluoroalkane, azidonitriles such as 3-azidopropanenitrile form 3-fluoropropanenitrile in 50% yield under the same conditions [*100, 101*] (equation 28).

[*100,101*] $$NCCH_2CH_2N_3 \xrightarrow[\text{CHCl}_3,\ 25\ ^\circ\text{C}]{NO^+BF_4^-} [NCCH_2CH_2N_4O^+BF_4^-] \xrightarrow{-N_2;\ -N_2O;\ -BF_3} \underset{50\%}{NCCH_2CH_2F} \qquad \textbf{28}$$

Better yields are attributed to intimate association of the basic nitrile group at the surface of the nitrosonium salt causing nitrosative decomposition of the azide to occur in close proximity to the weakly nucleophilic complex fluoride anion. Fluorination yields can be further enhanced to 59–81% by lengthening the azidonitrile chain, but the reaction is accompanied by pronounced secondary fluoronitrile formation arising from rearrangement [*100, 101*] (Table 8).

Table 8. Nitrosative Decomposition of Azidonitriles by $NO^+BF_4^-$ [*100, 101*]

Starting Compound	*Product Distribution (Yield, %)*		
$NC(CH_2)_nN_3$	*$NC(CH_2)_nF$*	*$NC(CH_2)_{n-2}CHFCH_3$*	*$NC(CH_2)_{n-3}CHFCH_2CH_3$*
n = 2	50	0	0
n = 3	42	17	0
n = 4	23	56.5	1.5
n = 6	21	32	8

Nitro Compounds

Replacement of nitro groups by fluoride ion represents a useful route to organic fluorine compounds.

Nitroaliphatics

Fluorodenitration of nitroaliphatics has been primarily restricted to polynitromethanes (Table 9). Side reactions involving potassium nitrite by-product reduce yields of fluoronitromethane. The novel use of the adduct of potassium fluoride with hexafluoroacetone in diglyme as a source of fluoride ion for the fluorodenitration of tetranitromethane significantly increases the yield of fluorotrinitromethane [*102*] (equation 29).

Table 9. Fluorodenitration of Polynitromethanes

Substrate	*Fluorinating Agent*	*Solvent*	*Product*	*Yield (%)*	*Ref.*
$C(NO_2)_4$	KF	DMF	$CF(NO_2)_3$	57	121
	$(CF_3)_2CFO^-K^+$	Diglyme	$CF(NO_2)_3$	91	102
$CCl(NO_2)_3$	CsF	DMF	$CClF(NO_2)_2$	41	121
$CF(NO_2)_3$	CsF	DMF	$CF_2(NO_2)_2$	41	121
	KF	Sulfolane	$CF_2(NO_2)_2$	59	122

[*102*]

$$C(NO_2)_4 + (CF_3)_2CFO^-K^+ \xrightarrow[\substack{0\ °C,\ 1\ h \\ -KNO_2}]{\text{Diglyme}} \underset{91\%}{CF(NO_2)_3} \qquad \mathbf{29}$$

Aliphatic fluorodenitration has also been applied to mononitro compounds, specifically to an α-nitroepoxide. Thus, 1,2-anhydro-3,4:5,6-di-*O*-isopropylidene-1-*C*-nitro-D-mannitol and labeled potassium bifluoride give 2-deoxy-2-fluoro-3,4:5,6-di-*O*-isopropylidene-aldehydo-D-glucose [*103, 104*] (equation 30).

[*103,104*] **30**

CHNO2 / O / CH / OCH / HCO >C(CH3)2 / HCO / H2CO >C(CH3)2 + $KH^{18}F_2$ $\xrightarrow[135\ °C]{HOCH_2CH_2OH}$ CHO / HC—^{18}F / OCH / HCO >C(CH3)2 / HCO / H2CO >C(CH3)2

Nitroaromatics

Aromatic fluorodenitration was first discovered in the reaction of polychloronitrobenzenes with potassium fluoride, when 2,3,5,6-tetrachlorofluorobenzene was prepared in 37% yield from 2,3,5,6-tetrachloronitrobenzene [*105*]. The technique has been adapted to prepare aryl fluorides from other activated nitroaromatics for applications in pharmaceutical and polymer chemistry (equation 31). Fluorodenitration also has been applied to prepare radiolabeled (^{18}F) fluoroaromatics [*74, 106*].

A characteristic feature of aromatic fluorodenitration is modest yield due to side reactions promoted by potassium nitrite and/or its decomposition product, potassium oxide, with the aryl fluoride or starting material.

[*74,106*] $RC_6H_4NO_2 + KF \xrightarrow{\text{Solvent}} RC_6H_4F + KNO_2$ **31**

$R = NO_2; CN; (CO)_2O; CHO; SO_3CH_3; COCl; SO_2Cl; CF_3$

Only 22–45% yields of *m*-fluoronitrobenzene are obtained from the fluorodenitration of *m*-dinitrobenzene by potassium fluoride in *N*-methyl-2-pyrrolidone or hexamethylphosphoramide, along with significant amounts of 3,3′-dinitrodiphenyl ether [*107, 108, 109*] (equation 32).

[*13*] **32**

$m\text{-}C_6H_4(NO_2)_2 \xrightarrow[180\ °C,\ 48\ h]{KF,\ NMP} m\text{-}FC_6H_4NO_2\ (27–45\%) \xrightarrow[-KF]{+KNO_2} [m\text{-}ONOC_6H_4NO_2] \longrightarrow$

$\xrightarrow[-N_2O_3]{+KNO_2} m\text{-}KOC_6H_4NO_2 \xrightarrow[-KF]{m\text{-}FC_6H_4NO_2} (m\text{-}O_2NC_6H_4)_2O\ (\sim 20\%)$

Aryl ethers can also represent the dominant product. For example, 3,3′-bis(trifluoromethyl)-5,5′-dinitrodiphenyl ether is obtained in 54% yield from 3,5-dinitrobenzotrifluoride and potassium fluoride in *N,N*-dimethylformamide with catalytic amounts of water at 160 °C after 24 h [*110*].

The use of phthaloyl dichloride as a scavenger for potassium nitrite and/or potassium oxide [*108, 109, 111*] significantly increases yields of aryl fluorides; *m*-fluoronitrobenzene is obtained in 70% yield with sulfolane as a solvent at 200 °C after 72 h. Addition of tetraphenylphosphonium bromide as a phase-transfer catalyst to the phthaloyl dichloride scavenger system further increases the yield of *m*-fluoronitrobenzene to 89% under less forcing conditions (150–180 °C, 5 h) [*112*]. This optimized technique can be applied to give *m*-fluorobenzonitrile from *m*-nitrobenzonitrile in 86% yield [*112*].

Potassium nitrite by-product can react with nitroaromatic substrate to suppress yields of aryl fluorides. Modest yields (40–60%) of fluorophthalic anhydride are obtained from 3- or 4-nitrophthalic anhydride and potassium fluoride due to formation of by-product dipotassium salt of 3- or 4-nitrophthalic acid [*113, 114, 115*] (equation 33). Higher yields (93%) of 3-fluorophthalic anhydride can be realized by regeneration of 3-nitrophthalic anhydride from the dipotassium salt with thionyl chloride, followed by addition of fresh potassium fluoride [*115*] (equation 33).

[*115*] **33**

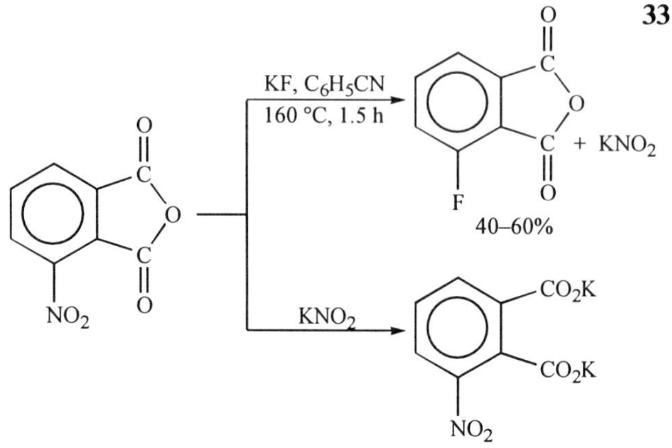

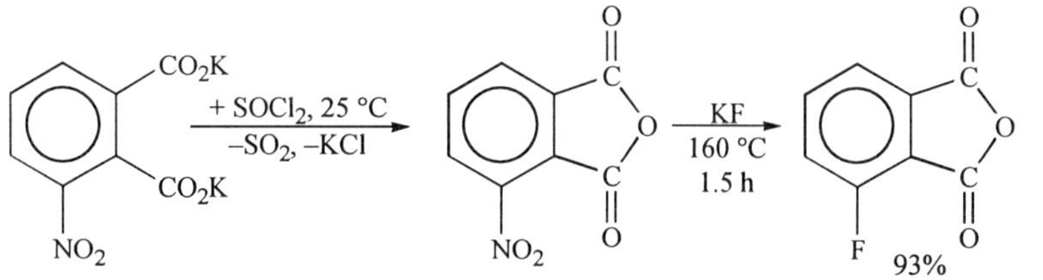

KF, C_6H_5CN, 160 °C, 1.5 h → 3-fluorophthalic anhydride + KNO_2 (40–60%)

KNO_2 → dipotassium 3-nitrophthalate (CO_2K, CO_2K, NO_2)

+ $SOCl_2$, 25 °C, $-SO_2$, –KCl → 3-nitrophthalic anhydride; KF, 160 °C, 1.5 h → 3-fluorophthalic anhydride (93%)

Another fluorodenitration technique features 3-nitrophthaloyl chloride in a dual role as both substrate and potassium nitrite trapping agent to give 3-fluorophthalic anhydride in 82% yield in sulfolane at 130 °C after 1.5 h [*116*].

Although potassium fluoride is the preferred nucleophile, tetrabutylammonium fluoride (TBAF) is successfully used for aromatic fluorodenitration. *o*-Fluoronitrobenzene can be obtained in quantitative yield from *o*-dinitrobenzene in tetra- hydrofuran at 25 °C after 1.5 h [*117*].

Nitroheterocyclics

Fluorodenitration with potassium fluoride has been applied to nitropyridines and nitrothiazoles. 2-Nitropyridine can be converted to 2-fluoropyridine in 55–60% yield [*107*] (equation 34).

[*107*] **34**

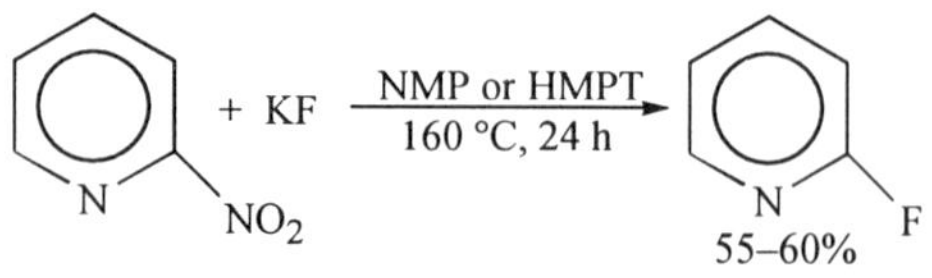

2-nitropyridine (NO_2) + KF → (NMP or HMPT, 160 °C, 24 h) 2-fluoropyridine (F), 55–60%

Anhydrous hydrogen fluoride can also effect fluorodenitration of nitroheterocyclics such as 3(5)-nitro-1,2,4-triazoles and 8-nitropurine to give 3(5)-fluoro-1,2,4-triazoles and 8-fluoropurine, respectively [*118*] (equation 35).

[*118*] **35**

NO_2 → F (HF, 100–150 °C, 20–48 h)

R	Yield, %
H	80
Br	98
CO_2CH_3	70
OH	21

Aqueous hydrofluoric acid is ineffective in fluorodenitration of activated aromatics; hexanitrobenzene in benzene does not react at 25 °C [*119*].

References for Pages 271–289

1. Katritzky, A. R.; Chermprapai, A.; Patel, R. C. *J. Chem. Soc., Perkin Trans. 1* **1980,** 2901.
2. Katritzky, A. R. *Tetrahedron* **1980,** *36,* 679.
3. Wade, T. N. *J. Chem. Res. Synop.* **1980,** 388.
4. Matsumoto, K.; Ozaki, Y.; Iwasaki, T.; Horikawa, H.; Miyoshi, M. *Experientia* **1979,** *35,* 850.
5. Olah, G. A.; Welch, J. T. *Synthesis* **1974,** 652.
6. Olah, G. A.; Welch, J. T.; Vankar, Y. D.; Nojima, M.; Kerekes, I. *J. Org. Chem.* **1979,** *44,* 3872.
7. Olah, G. A.; Prakash, K. B. S.; Chao, Y. L. *Helv. Chim. Acta* **1981,** *64,* 2528.
8. Hamman, S.; Beguin, C. G. *Tetrahedron Lett.* **1983,** *24,* 57.
9. Olah, G. A.; Welch, J. T. *Synthesis* **1974,** 654.
10. Boudakian, M. M. In *Encyclopedia of Chemical Technology,* 3rd ed.; Wiley: New York, 1980; Vol. 10, pp 901–936.
11. Roe, A. *Org. Reactions* **1949,** *5,* 193.
12. Suschitzky, H. *Adv. Fluorine Chem.* **1965,** *4,* 1.
13. Forche, E. *Methoden der Organischen Chemie* (Houben-Weyl); Thieme Verlag: Stuttgart, 1962; Vol. 5, Part 3, G, pp 213–245.
14. Hudlický, M. *Chemistry of Organic Fluorine Compounds;* Ellis Horwood: Chichester, England, 1976; pp 160–169.
15. Pavlath, A. E.; Lefffler, A. J. *Aromatic Fluorine Compounds;* Monograph No. 155; American Chemical Society: Washington, DC, 1962; pp 12–45.
16. Flood, D. T. *Org. Synth.,* Collective Vol. **1943,** *2,* 295.
17. Schiemann, G.; Winkelmueller, W. *Org. Synth.,* Collective Vol. **1943,** *2,* 299.
18. Doak, G. O.; Freedman, L. D. *Chem. Eng. News* **1967,** *45*(53), 8.

19. Johnson, R. P.; Oswald, J. P. *Chem. Eng. News* **1967,** *45* (44), 44.
20. Wannagat, U.; Hohlstein, G. *Chem. Ber.* **1955,** *88,* 1939.
21. Yakobson, G. G.; Dyachenko, A. E.; Belchikova, F. A. *J. Gen. Chem.* (USSR) **1962,** *32,* 849 (Engl. Transl., 842).
22. Milner, D. J. Eur. Pat. Appl. 430 434, 1991; *Chem. Abstr.* **1991,** *115,* 49423x.
23. Newman, M. S.; Lilje, K. C. *J. Org. Chem.* **1979,** *44,* 1347.
24. Furlano, D. C.; Kirk, K. L. *J. Org. Chem.* **1986,** *51,* 4073.
25. Doyle, M. P.; Bryker, W. J. *J. Org. Chem.* **1979,** *44,* 1572.
26. Diehl, H.; Pelster, H.; Habetz, H. U.S. Patent 4 476 320, 1984; *Chem. Abstr.* **1983,** *99,* 70360k. Corresponds to Ger. Offen. 3 141 659, 1983.
27. Rutherford, K. G.; Redmond, W.; Rigamonti, J. *J. Org. Chem.* **1961,** *26,* 5149.
28. Rutherford, K. G.; Redmond, W. *Org. Synth.,* Collective Vol. **1973,** *3,* 133.
29. Honore, T.; Hjeds, H. *Eur. J. Med. Chem.-Chim. Ther.* **1979,** *14,* 285.
30. Bergmann, E. D.; Berkovic, S. *J. Org. Chem.* **1961,** *26,* 919.
31. Wild, J.; Harreus, A.; Goetz, N. Eur. Pat. Appl. 388 870, 1990; *Chem. Abstr.* **1991,** *114,* 206731v.
32. Petterson, R. C.; DiMaggio, A. III; Hebert, A. L.; Haley, T. J.; Mykytka, J. P.; Sarkar, I. M. *J. Org. Chem.* **1971,** *36,* 631.
33. Kirk, K. L. *J. Org. Chem.* **1976,** *41,* 2373.
34. Kirk, K. L.; Cantacuzene, D.; Nimitkitpaisan, Y.; McColloh, D.; Padgett, W. L.; Daly, J. W.; Creveling, C. R. *J. Med. Chem.* **1979,** *22,* 1493.
35. Kirk, K. L.; Cohen, L. A. *J. Am. Chem. Soc.* **1971,** *93,* 3060.
36. Takahashi, K.; Kirk, K. L.; Cohen, L. A. *J. Org. Chem.* **1984,** *49,* 1951.
37. Fabra, F.; Vilarrasa, J.; Coll, J. *J. Heterocyclic Chem.* **1978,** *15,* 1447.
38. Yoneda, N.; Fukuhara, T.; Kikuchi, T.; Suzuki, A. *Synth. Commun.* **1989,** *19,* 865.
39. Mueller, A.; Roth, U.; Siegert, S.; Miethchem, R. *Z. Chem.* **1986,** *26, 1*69.
40. Wallach, O. *Liebigs Ann. Chem.* **1886,** *235,* 255.
41. Wallach, O.; Heusler, F., *Liebigs Ann. Chem.* **1888,** *243,* 219.
42. Furubashi, T.; Hamada, I. Jpn. Kokai Tokkyo Koho JP 59 190 942; *Chem. Abstr.* **1985,** *102,* 131704e.
43. Foerster, H.; Klusacek, H.; Wenz, A. U.S. Patent 4 194 054, 1980; *Chem. Abstr.* **1978,** *89,* 108595z. Corresponds to Ger. Offen. 2 652 810, 1978.
44. Klauke, E.; Grohe, K. Ger. Offen. Patent 3 142 856, 1983; *Chem. Abstr.,* **1983,** *99,* 53378e.
45. Petersen, U.; Grohe, K.; Zeiler, H. J.; Metzger, K. G. Ger. Offen. 3 508 816, 1986; *Chem. Abstr.* **1986,** *105,* 191059v.
46. Ng, J. S.; Katzenellenbogen, J. A.; Kilbourn, M. R. *J. Org. Chem.* **1981,** *46,* 2520.

47. Kilbourn, M. R.; Welch, M. J.; Dence, C. S.; Tewson, T. J.; Saji, H.; Maeda, M. *Int. J. Appl. Radiat. Isot.* **1984,** *35,* 591.
48. Satyamurthy, N.; Barrio, J. R.; Schmidt, D. G.; Kammerer, C.; Bida, G. T.; Phelps, M. E. *J. Org. Chem.,* **1990,** *55,* 4560.
49. Tewson, T. J.; Welch, M. J. *J. Chem. Soc., Chem. Commun.* **1979,** 1149.
50. Rosenfeld, M. N.; Widdowson, D. A. *J. Chem. Soc., Chem. Commun.* **1979,** 914.
51. Haroutounian, S. A.; DiZio, J. P.; Katzenellenbogen, J. A. *J. Org. Chem.* **1991,** *56,* 4993.
52. Ferm, R. L.; Vander Werf, C.A. *J. Am. Chem. Soc.* **1950,** *72,* 4809.
53. Osswald, P.; Scherer, O. Ger. Patent 600 706, 1934; *Chem. Abstr.* **1934,** *28,* 7260.
54. Churchill, J. W. U.S. Patent 2 939 766, 1960; Chem. Abstr. **1960,** *54,* 18906.
55. Schenk, W. J.; Pellon, G. R. U.S. Patent 2 563 796, 1951; *Chem. Abstr.* **1952,** *46,* 9125b.
56. Anello, L. G.; Woolf, C. U.S. Patent 3 160 623, 1964; *Chem. Abstr.* **1962,** *57,* 12380e. Corresponds to Belg. Patent 611 545, 1962.
57. Misaki, S.; Okamoto, M. Jpn. Kokai Patent 74 81 330; *Chem. Abstr.* **1974,** *81,* 169274a.
58. Seel, F. *Angew. Chem. Int. Ed. Engl.* **1965,** *4,* 635.
59. Krackov, M. H.; Rolston, C. H. U.S. Patent 4 912 268, 1990; *Chem. Abstr.* **1990,** *112,* 35424z. Corresponds to Eur. Patent Appl. 330 420, 1989.
60. Robbins, M. J.; Uznanski, B. *Can. J. Chem.* **1981,** *59,* 2608.
61. Fukuhara, T.; Yoneda, N.; Sawada, T.; Suzuki, A. *Synth. Commun.* **1987,** *17,* 685.
62. Fukuhara, T.; Yoneda, N.; Suzuki, A. *J. Fluorine Chem.* **1988,** *38,* 435.
63. Fukuhara, T.; Yoneda, N.; Takumara, K.; Suzuki, A. *J. Fluorine Chem.* **1991,** *51,* 299.
64. Fukuhara, T.; Sasaki, S.; Yoneda, N.; Suzuki, A. *Bull. Chem. Soc. Jpn.* **1990,** *63,* 2058.
65. Boudakian, M. M. U.S. Patent 4 096 196, 1978; *Chem. Abstr.* **1979,** *90,* 103597n.
66. Dove, M. F. A.; Clifford, A. F. *Inorganic Chemistry in Liquid Hydrogen Fluoride;* Pergamon: New York, 1971; p 156.
67. Boudakian, M. M. U.S. Patent 4 075 252, 1978; *Chem. Abstr.* **1978,** *89,* 6086s.
68. Boudakian, M. M. *J. Fluorine Chem.* **1981,** *18,* 497.
69. Boudakian, M. M. *J. Fluorine Chem.,* **1987,** *36,* 283.
70. Boudakian, M. M. U.S. Patent 4 487 969, 1984; *Chem. Abstr.* **1984,** *101,* 130392u. Corresponds to Eur. Patent Appl. 105 641, 1984.
71. Bernstein, J.; Roth, J. S.; Miller, W. T. Jr. *J. Am. Chem. Soc.* **1948,** *70,* 2310.

72. Ingold, C. K.; Ingold, E. H. *J. Chem. Soc.* **1928,** 2249.
73. Angelini, G.; Speranza, M; Wolf, A. P.; Shiue, C.-Y. *J. Fluorine Chem.* **1985,** *27,* 177
74. Shiue, C.-Y.; Fowler, J. S.; Wolf, A. P.; McPherson, D. W.; Arnett, C. D.; Zecca, L. *J. Nucl. Med.* **1986,** *27,* 226.
75. Haka, M. S.; Kilbourn, M. R.; Watkins, G. L.; Toorongian, S. A. *J. Labelled Compd.-Radiopharm.* **1989,** *27,* 823.
76. DeValk, J.; Van der Plas, H. C. *Rec. Trav. Chim. Pays-Bas.* **1972,** *91,* 1414.
77. Kiburis, J.; Lister, J. H. *J. Chem. Soc. C* **1971,** 3942.
78. Halperin, B. I.; Krska, J.; Levy, E.; Vander Werf, C. A. *J. Am. Chem. Soc.* **1951,** *73,* 1857.
79. Brown, C.; Murray, M.; Schmutzler, R. *J. Chem. Soc. C* **1970,** 878.
80. Liberda, H. Br. Patent 1 462 339, 1977; *Chem. Abstr.* **1976,** *84,* 30482k.
81. Appel, R.; Ruppert, I. *Chem. Ber.* **1975,** *108,* 919.
82. Boehme, H.; Hilp, M. *Chem. Ber.* **1955,** *103,* 104.
83. Boehme, H.; Hilp, M. *Chem. Ber.* **1970,** *103,* 3930.
84. Patrick, T. B.; Flory, P. A. *J. Fluorine Chem.* **1984,** *25,* 157.
85. Rozen, S.; Brand, M.; Zamir, D.; Hebel, D. *J. Am. Chem. Soc.* **1987,** *109,* 896.
86. Prakash, G. K. S.; Reddy, V. P.; Li, X.-Y.; Olah, G. A. *Synlett* **1990,** 594.
87. Bergmann, E. D.; Shahak, I. *Israel J. Chem.* **1965,** *3,* 73.
88. Olah, G. A.; Kuhn, S. *Chem. Ber.* **1956,** *89,* 864.
89. Takadate, A.; Tahara, T.; Goya, S. *Synthesis* **1983,** 806.
90. Patrick, T. B.; Scheibel, J. J.; Cantrell, G. L. *J. Org. Chem.* **1981,** *46,* 3917.
91. Bergmann, E. D.; Ikan, R. *Chem. Ind.* (London) **1957,** 394.
92. Fraser, R. R.; Millington, J. E.; Pattison, F. L. M. *J. Am. Chem. Soc.* **1959,** *79,* 1959.
93. Knunyants, I. L.; Kisel, Y. M.; Bykhovskaya, E. G. *Izv. Akad. Nauk SSSR* **1956,** 377; *Chem. Abstr.* **1956,** *50,* 15454c.
94. Danzin, C.; Bey, P.; Schirlin, D.; Claverie, N. *Biochem. Pharmacol.* **1982,** *31,* 3871.
95. Kallio, A.; McCann, P. P.; Bey, P. *Biochem. J.* **1982,** *204,* 771.
96. Vyplel, H. *Chimia* **1985,** *39,* 304.
97. Leroy, J.; Wakselman, C. *J. Chem. Soc., Perkin Trans. 1* **1978,** 1224.
98. Kent, P. W.; Wood, K. R.; Welch, V. A. *J. Chem. Soc.* **1964,** 2493.
99. Machleidt, H.; Wessendorf, R.; Klockow, M. *Liebigs Ann. Chem.* **1963,** *667,* 47.
100. Doyle, M. P.; Whitefleet, J. L.; Zaleta, M. A. *Tetrahedron Lett.* **1975,** 4201.
101. Doyle, M. P.; Whitefleet, J. L.; Bosch, R. J. *J. Org. Chem.* **1979,** *44,* 2923.
102. Grant, L. R. U.S. Patent 4 115 459, 1978; *Chem. Abstr.* **1979**, *90,* 71744q.
103. Beeley, P. A.; Szarek, W. A.; Hay, G. W.; Perlmutter, M. M. *Can. J. Chem.* **1984,** *62,* 2709.

104. Szarek, W. A.; Hay, G. W.; Perlmutter, M. M. *J. Chem. Soc., Chem. Commun.* **1982,** 1253.
105. Finger, G. C.; Kruse, C. W. *J. Am. Chem. Soc.* **1956,** *78,* 6034.
106. Ding, Y.-S.; Shiue, C.-Y.; Fowler, J. S.; Wolf, A. P.; Plenevaux, A. *J. Fluorine Chem.* **1990,** *48,* 189.
107. Bartoli, G.; Latrofa, A.; Naso, F.; Todesco, P. E. *J. Chem. Soc., Perkin Trans. 1,* **1972,** 2671.
108. Effenberger, F.; Streicher, W. U.S. Patent 4 568 781, 1986; *Chem. Abstr.* **1986,** *104,* 148476w. Corresponds to Ger. Offen. 3 400 418, 1985.
109. Effenberger, F.; Streicher, W. *Chem. Ber.* **1991,** *124,* 157.
110. Stults, J. S.; Lin, H. C. U.S. Patent 4 990 670, 1991; *Chem Abstr.* **1991,** *114,* 206762.
111. Kumai, S.; Seki, R.; Furukawa, Y.; Matsuo, M. *Reports Res. Lab. Asahi Glass Co.* **1985,** *35,* 153; *Chem. Abstr.* **1987,** *107,* 6848g.
112. Suzuki, H.; Yazawa, N.; Yoshida, Y.; Furusawa, O.; Kimura, Y. *Bull. Chem. Soc. Jpn.* **1990,** *63,* 2010.
113. Ishikawa, N.; Tanabe, T.; Hayashi, D. *Bull. Chem. Soc. Jpn.* **1975,** *42,* 359.
114. Markezich, R. L.; Zamek, O. S.; Donahue, E. P.; Williams, F. J. *J. Org. Chem.* **1977,** *42,* 3435.
115. Milner, D. J. *Synth. Commun.* **1985,** *15,* 485.
116. Passudetti, M.; Prato, J.; Qaintily, U.; Scorrano, G. *J. Fluorine Chem.* **1990,** *50,* 251.
117. Clark, J. H.; Smith, D. K. *Tetrahedron Lett.* **1985,** 2233.
118. Naik, S. R.; Witkowski, J. T.; Robins, R. K. *J. Org. Chem.* **1973,** *38,* 4353.
119. Nielsen, A. T.; Chafin, A. P.; Christian, S. J. *J. Org. Chem.* **1984,** *49,* 4575.
120. Bethell, D.; McDonald, K.; Rao, K. S. *Tetrahedron Lett.* **1977,** 1447.
121. Khisamutdinov, G. K.; Slovetskii, V. I.; Lvova, M.; Usyshkin, O. G.; Vesprozvanni, M. A.; Fainzilberg, A. A. *Izv. Akad. Nauk SSSR, Ser. Khim.* **1970,** 2553; *Chem. Abstr.* **1971,** *74,* 87266r.
122. Kamlet, M.; Adolph, H. G. *J. Org. Chem.* **1968,** *33,* 3073

Chapter 4

Reactions of Organic Fluorine Compounds

Reduction

by C. G. Krespan

This section covers recent developments in the reduction of functional groups in the following order: carbon–halogen, carbon–oxygen, carbon–carbon, olefin, carbonyl, imine, carboxyl derivatives, nitro and other nitrogen-containing functions, sulfur-containing compounds, and phosphates. Reagents used with each of these functional groups are addressed in the following sequence: hydrogen with catalyst, metal hydrides, metals, photoexcitation, electrochemical cells, and miscellaneous inorganic and organic reducing agents.

Reduction of Carbon–Fluorine Bond

Reductive removal of fluorine from *alkyl fluorides* requires a potent reducing agent and so is not normally encountered. However, hydrogenolysis of an unactivated carbon–fluorine bond in, for example, 3-β-fluorocholestane has been efficiently accomplished in 88% yield with a solution of **potassium** and dicyclohexyl-18-crown-6 in toluene at 25 °C [*1*]. Similarly, **sodium naphthalene** in tetrahydrofuran converts 6-fluorohexene-1 and 1-fluorohexane to hydrocarbons in 50% yield at 25 °C over a 7-h period [*2*].

Highly fluorinated alkanes are also reduced by alkali metals. **Lithium amalgam** converts polytetrafluoroethylene to a carbon polymer composed of monolayer ribbons of six-membered rings with lithium atoms bound to the edges [*3*].

Functional groups nearby can facilitate reduction of tetrahedral carbon–fluorine bonds. Reduction of trifluoromethyl ketones by active metals results in loss of an α-fluorine atom. **Magnesium** and hexafluoroacetone form the enolate salt after loss of fluorine, and the enolate adds to another molecule of fluoroketone to give a ketoalcohol after acidification. The ketoalcohol is itself subject to reduction with formation of monohydroketoalcohol as a major by-product along with the bimolecular reduction product, perfluoropinacol [*4*] (equation 1). **Aluminum** with hexafluoroacetone allows reaction to stop partially at the aluminum enolate, which, on acidification, gives the isolable enol of pentafluoroacetone [*5*].

Solvent plays a key role in the course of reductions by active metals. Changing the solvent of the reaction of hexafluoroacetone with **magnesium** from tetrahydrofuran to dimethylformamide induces bimolecular reduction to the pinacol [*5*] (equation 2).

The carbon–fluorine bond is normally resistant to cleavage by zinc metal so that the carbon–chlorine bond in chlorodifluoromethyl ketones is selectively reduced. The

0065–7719/95/0187–0297$08.72/1

reaction can be carried out in the presence of chlorotrimethylsilane to afford silyl ethers directly [*6*] (equation 3). However, reductive removal of fluorine by zinc is observed with a fluoroimine in dimethylformamide solution [*4*] (equation 4).

1

[*4*] $CF_3COCF_3 \xrightarrow[\text{THF, 20 °C, 1h}]{\text{Mg, HgCl}_2} \xrightarrow{H^+} CF_3COCF_2C(CF_3)_2OH$ *22%* + $CF_3COCHFC(CF_3)_2OH$ *26%*

[*5*] $CF_3COCF_3 \xrightarrow[\text{THF, 20 °C, 1h}]{\text{Al, HgCl}_2} CF_3C(O^-)=CF_2 \xrightarrow{H^+} CF_3C(OH)=CF_2$ *52%*

2

[*5*] $CF_3COCF_3 \xrightarrow[\text{DMF, <20 °C, 1h}]{\text{Mg, HgCl}_2} (CF_3)_2C(OH)-C(OH)(CF_3)_2$ *80%*

3

[*6*] $H(CH_2)_6COCF_2Cl \xrightarrow[\text{CH}_3\text{CN, 60 °C}]{\text{Zn, (CH}_3)_3\text{SiCl}} H(CH_2)_6C(=CF_2)OSi(CH_3)_3$ *68%*

4

[*4*] $(CF_3)_2C=NC_6H_5 \xrightarrow[\text{DMF}]{\text{Zn}} CF_3C(NHC_6H_5)=CF_2$ *80%*

Photochemically induced reduction of alkyl perfluoro carboxylates in hexamethylphosphoramide (HMPA) can replace one, two, or (in the case of trifluoroacetate) three α-fluorine atoms with hydrogen [*7, 8*] (equation 5). Chlorine-containing fluorocarboxylates preferentially lose α-chlorine and may even lose β-chlorine in preference to α-fluorine [*9*] (equation 6). Cyclic α-fluoroketones are also defluorinated photochemically [*10*] (equation 7).

Electrochemical reduction of carbon–fluorine bonds occurs at high pH when a carbonyl group is adjacent. Polarographic reduction of α,α,α-trifluoroacetophenone without loss of fluorine predominates in acidic media to give the alcohol and the corresponding pinacol, whereas reduction of the unprotonated ketone results in hydrogenolysis of the trifluoromethyl group to form acetophenone as product [*11*] (equation 8).

Halogen attached directly to vinylic carbon or in an ***allylic position*** with respect to a double bond is much more liable to undergo reductive removal, so that carbon–fluorine bonds of these types can be reduced fairly easily with the appropriate reagents. The scope of haloolefin reductions with both hydrogen and catalyst and with metal hydrides is described in depth in the previous edition of this book [*12*]. More recently, catalytic hydrogenation of fluorinated substrates has been the subject of reviews incorporating mechanistic interpretations of the varied results [*13, 14*].

[*7,8*] **5**

$$RCF_2CO_2R \xrightarrow[h\nu,\ 10\ ^\circ C]{HMPA} RCHFCO_2R \xrightarrow[h\nu,\ 50\ ^\circ C]{HMPA} RCH_2CO_2R$$

($R = F, CF_3, C_2F_5$) *53-75%* *7-59%*

[*9*] **6**

$$CFYClCHXCO_2R \xrightarrow[h\nu]{(CH_3)_2CHOH}$$

$CFCl_2CH_2CO_2R$ *99%* (X=Y=Cl)

$CFHClCHFCO_2R$ *30%* (X=F, Y=Cl)

$CF_2HCHFCO_2R$ *3%* + $CF_2ClCH_2CO_2R$ *14%* (X=Y=F)

[*10*] **7**

2-fluorocyclohexanone $\xrightarrow[h\nu]{(CH_3)_2CHOH}$ cyclohexanone *60-65%*

[*11*] **8**

$CF_3C(=O)C_6H_5 \xrightarrow{e^-}$

pH < 5: $CF_3CH(OH)C_6H_5$ + $CF_3C(OH)(C_6H_5)$—$C(OH)(C_6H_5)CF_3$

pH > 7: $CH_3C(=O)C_6H_5$

Examples of the use of **lithium aluminum hydride** in stereoselective [*15*] (equation 9), regioselective [*16*] (equation 10), and product-selective [*17*] (equation 11) displacements of fluorine are available.

Stereoselective reduction of vinylic fluorine is also accomplished with **tributylphosphine** [*18*] (equation 12).

Reduction of ***aryl fluorides*** by various reagents to hydro derivatives was also treated extensively in reference 12. A little used method, **electrochemical reduction**, defluorinates simple aryl fluorides [*19, 20*] (equations 13 and 14).

[*15*] **9**

$C_6H_5CF{=}CFCF_3 \xrightarrow[\text{-20 °C, glyme}]{LiAlH_4}$ (E) isomer → (Z) –$C_6H_5CH{=}CFCF_3$ *94%*; (Z) isomer → (E) –$C_6H_5CF{=}CHCF_3$ *84%*

[*16*] **10**

$CF_2{=}CFCO_2Li \xrightarrow[\text{-20 °C, ether-THF}]{LiAlH_4} \xrightarrow{H^+} CHF{=}CFCO_2H$ *50%*

Z:E = 1

[*17*] **11**

$CH_2{=}C(CF_3)CO_2Na \xrightarrow[\text{-78°C, ether-THF}]{LiAlH_4} \xrightarrow{H^+} CF_2{=}C(CH_3)CO_2H$ *54%*

$CH_2{=}C(CF_3)CO_2Na \xrightarrow[\text{-78°C, ether-THF}]{2\ LiAlH_4} \xrightarrow{H^+} CHF{=}C(CH_3)CO_2H$ *46%*

[*18*] **12**

$CF_3CF{=}CF_2 \xrightarrow[\text{-78°C to 25°C, ether}]{(C_4H_9)_3P} \xrightarrow{H_2O}$ (E) –$CF_3CF{=}CHF$ *88%*

[*19*] **13**

1,3-difluorobenzene $\xrightarrow[\text{Hg cathode, diglyme-}H_2O]{e^-}$ fluorobenzene (*15%*) + benzene (*60%*)

[*20*] **14**

$C_6F_6 \xrightarrow[\text{Al cathode, aq. } HCON(CH_3)_2]{e^-} C_6H_6$ *60%*

Reduction of Other Carbon–Halogen Bonds

Hydrodehalogenation, the replacement of a halogen atom (usually chlorine) with hydrogen, as applied to alkanes has been the subject of much recent study, because this process gives *hydrochlorofluorocarbons* (HCFCs) or *hydrofluorocarbons* (HFCs). These classes of fluorocarbons have far shorter half-lives in the troposphere than do chlorofluorocarbons (CFCs) and are therefore sought as alternatives that make little or no contribution to the depletion of the ozone layer or to global warming. **Catalytic hydrogenation** of CFCs having three chlorine atoms on one

References are listed on pages 316–320.

carbon proceeds more readily than reductive removal of lone chlorines, and so considerable selectivity is possible through control of reaction conditions. The following series of examples [*21, 22, 23, 24, 25, 26*] (equations 15–19) is arranged in order of increasing severity of conditions, ending with a thermally induced reduction over carbon without metal catalyst.

[*21*] $CF_3CCl_3 \xrightarrow[120\ ^\circ C]{H_2,\ Ru/C} CF_3CHCl_2$ *95%* **15**

[*22*] $CF_3CCl_3 \xrightarrow[175\ ^\circ C]{H_2,\ Pt/C} CF_3CH_3 + CF_3CH_2Cl + CF_3CHCl_2$ *14%* *2%* *78%*

[*23*] $CF_2ClCF_2CHCl_2 \xrightarrow[200\ ^\circ C]{H_2,\ Pd/C} CF_2ClCF_2CH_2Cl + CF_2ClCF_2CH_3$ *53%* *47%* **16**

[*24*] $CF_3CFCl_2 + ClCF_2CF_2Cl$ (*40 : 60*) $\xrightarrow[250\ ^\circ C]{H_2,\ Rh/C} CF_3CH_2F + CHF_2CHF_2$ *63%* *8%* **17**

[*25*] $CF_3CF_2Cl \xrightarrow[300\ ^\circ C]{H_2,\ Pd/C} CF_3CHF_2$ *100%* **18**

[*26*] $CF_3CHClF \xrightarrow[550\ ^\circ C]{H_2,\ act.\ C} CF_3CH_2F$ *78%* **19**

Replacement of an unactivated lone chlorine with hydrogen is also accomplished in good yield with **lithium aluminum hydride**. Chlorofluoronorcaranes are selectively dechlorinated to give hydro derivatives, largely with retention of configuration. The related monodeuteronorcaranes are prepared with lithium aluminum deuteride [*27*] (equations 20a and 20b).

Chlorofluorocyclopropanes have also been the object of stereoselective dechlorination studies with Group IV hydrides. **Tributyltin hydride** reductions involving radical intermediates proceed with enhanced retention of configuration if electron-donating substituents such as methyl or trimethylsilyl are present in the β position, indicating that these substituents exert a stabilizing effect on the pyramidal configuration of the cyclopropyl radical intermediate. Electron-withdrawing substituents such as fluorine, alkoxy, and alkoxycarbonyl exert a destabilizing influence resulting in lower stereospecificity [*28, 29*] (equations 21 and 22).

[*27*] **20**

a F, Cl — $LiAlH_4$, diglyme, 100 °C, 1.5 hr. → F, H *80%* *(94% stereoselective)*

b Cl, F — $LiAlD_4$, diglyme, 100 °C, 1.5 hr. → D, F *80%* *(90% stereoselective)*

[*28*] **21**

Cl, F, R — $(C_4H_9)_3SnH$, R•, 80-140 °C → H, F, R

77-81%

Selectivity

R = CH_3 > $(CH_3)_3Si$ > H > CO_2CH_3 > CH_3O > F

[*29*] **22**

O, O, F, Cl — $(C_4H_9)_3SnH$, R•, 80 °C → O, O, H, F *60%*

Although **silicon hydrides** are less active reducing agents than tributyltin hydride, they have been used with free radical initiation to debrominate bromofluorocyclopropanes. Stereospecificity is lower than that observed with tributyltin hydride, presumably because of the higher temperatures required and the slowness of hydrogen transfer to the cyclopropyl radical [*30*]. Other indications of greater reactivity in the tin hydride series are monodebromination of dibromofluoromethane by **tributyltin hydride** at only 25–30 °C [*31*] (equation 23) and selective hydrodebromination of bromofluoroandrostenone with triphenyltin hydride and catalyst [*32*] (equation 24).

Reductive cleavages of carbon–chlorine bonds by active metals and with photochemical activation figure in recent studies aimed at HFCs and HCFCs. **Sodium amalgam** [*33*] (equation 25), **zinc** powder [*34*] (equation 26), and **aluminum/tin chloride** [*35*] (equation 26) are all used in conjunction with protic solvents in reactions giving high yields and conversions.

References are listed on pages 316–320.

[*31*] $CBr_2HF \xrightarrow[25\text{-}30\ °C]{(C_4H_9)_3SnH} CBrH_2F$ ***80%*** **23**

[*32*] **24**

$\xrightarrow[\text{R•, 25 °C, THF}]{(C_6H_5)_3SnH}$ ***48%***

[*33*] $CF_3CFCl_2 \xrightarrow[\text{aq. } CH_3OH,\ 25\ °C]{Na/Hg} CF_3CFHCl$ ***93%*** **25**

[*34,35*] $CF_3CCl_3 \xrightarrow[CH_3OH]{Zn\ (80\ °C)\ or\ Al/SnCl_2\ (25\ °C)} CF_3CHCl_2$ ***95-98%*** **26**

Ultraviolet light with a hydrogen source can also be a selective reagent for dehalogenation [*36, 37, 38*] (equations 27–29).

[*36*] $ClCF_2CFCl_2 \xrightarrow[K_2CO_3]{h\nu,\ (CH_3)_2CHOH} ClCF_2CHFCl$ ***94%*** **27**

[*37*] $CFCl_2CF_2OCHCl_2 \xrightarrow[30\ °C,\ 24\ min.]{h\nu,\ (CH_3)_2CHOH} CFHClCF_2OCHCl_2$ ***93%*** **28**

[*38*] $CF_2ClCFClCO_2CH_3 \xrightarrow[20\ °C]{h\nu,\ (CH_3)_2CHOH} CF_2ClCFHCO_2CH_3$ ***89%*** **29**

The responses of ***haloolefins*** and ***haloaromatics*** to various reducing agents were discussed in the previous section; thus only some recent examples are given here, showing the high selectivity attained with aromatic [*39*] (equation 30), olefinic [*40*] (equation 31), and heterocyclic [*41*] (equation 32) substrates with retention of the unsaturated centers.

[*39*] CN, F, F, Cl → CN, F, F; H_2, Pd/C; $N(C_2H_5)_3$, 100 °C; *98%* **30**

[*40*] $(CF_2)_{1-4}$, I, Cl → $(CF_2)_{1-4}$, H, Cl; $NaBH(OCH_3)_3$; THF, 0-25 °C; *100%* **31**

[*41*] F, F, Cl, F, F, N → F, F, F, F, N; hν, C_2H_5OH; *66%* **32**

Reductive Coupling via the Carbon–Fluorine Bond

Coupling reactions and related fluoroalkylations with polyfluoroalkyl halides are induced by various reagents, among them metals such as **copper** and **zinc,** or by an **electrochemical cell.** More recently, examples of carbon–carbon bond formation by coupling of unsaturated fluorides have been reported. Both ***acyclic and cyclic fluoroolefins*** of the type $(R_F)_2C{=}CFR_F$ undergo reductive dimerization on treatment with **phosphines** [*42*] (equation 33). The reaction shown in equation 33 is also accomplished **electrochemically** but less cleanly [*43*].

[*42*] CF_3, F_2, F, F_2, F_2 → CF_3, CF_3, F_2, F_2, F_2 F_2, F_2, F_2; $(C_6H_5)_3P$; CH_3CN, 25 °C; *75%* **33**

Electrochemical reduction of pentafluoronitrobenzene produces an intermediate radical anion that couples at position 4 to form the corresponding biphenyl along with hydroxy derivatives from subsequent nucleophilic substitution *meta* to the nitro groups [*44*] (equation 34). Similar reduction of halopyridines such as pentafluoropyridine leads mainly to 4,4′-bipyridyls [*45*] (equation 35).

Reductive Cleavage of the Carbon–Oxygen Bond

The carbon–oxygen bond is strong; like the normal carbon–fluorine bond, it is difficult to reduce. However, some structural features facilitate its reductive cleavage. Aryl esters of perfluoroalkanesulfonic acids can be cleaved in good yield by

tributylammonium formate and palladium catalyst in an unusual arene synthesis [*46*].

[*44*] **34**

$$\text{C}_6\text{F}_5\text{NO}_2 \xrightarrow[\text{CH}_3\text{CN, R}_4\text{NClO}_4]{e^-} \text{O}_2\text{N–C}_6\text{F}_3\text{X–C}_6\text{F}_3\text{X–NO}_2$$

(X = F, OH) *100%*

[*45*] **35**

$$\text{C}_5\text{F}_5\text{N} \xrightarrow[\text{DMF, R}_4\text{NBF}_4]{e^-} \text{NC}_5\text{F}_4\text{–C}_5\text{F}_4\text{N} \quad 49\%$$

Appropriately constructed ***fluoroallylic alcohols*** are attacked at the double bond by **sodium borohydride** with S_N2' displacement of the hydroxyl group rather than the vinylic fluorine [*47*] (equation 36).

[*47*] **36**

$$\text{CF}_2\text{=CXCH(OH)R} \xrightarrow[\text{diglyme, 70 °C}]{\text{NaBH}_4} \text{CHF}_2\text{CX=CHR} \quad 60\text{-}80\%$$

(X = H, F, Cl;
R = C_6H_5, C_5H_{11})

Perfluorinated epoxides, which are generally susceptible to C–O cleavage during nucleophilic attack, are cleaved by lithium aluminum hydride. Of the available examples, reduction of epoxides from two types of internal olefin to give alcohols is shown [*48*] (equations 37a and 37b).

[*48*] **37**

a $$\text{CF}_3\text{CF–CFCF}_3 \text{ (epoxide, O bridging)} \xrightarrow[\text{ether}]{\text{LiAlH}_4} \text{CF}_3\text{CH(OH)CHFCF}_3 \quad 77\%$$

2 diastereomers

b $$\text{(CF}_3)_2\text{C–CFCF}_2\text{CF}_3 \text{ (epoxide, O bridging)} \xrightarrow[\text{ether}]{\text{LiAlH}_4} \text{(CF}_3)_2\text{CHCH(OH)CF}_2\text{CF}_3 \quad 76\%$$

References are listed on pages 316–320.

Reductive Cleavage of the Carbon–Carbon Bond

Catalytic hydrogenolysis of ***1,1-difluorocyclopropanes*** occurs at the C_2–C_3 bond with partial retention of fluorine in the cases studied [*49*] (equation 38). In contrast, reduction of acetoxy-1,1-difluorocyclopropanes with ***lithium aluminum hydride*** occurs with loss of fluoride to give β-fluoroallyl alcohols, often with high stereoselectivity [*50*] (equation 39). A further variant involving free radical intermediates produces allylidene difluorides after ring opening [*51*] (equation 40).

[*49*] **38**

$$C_6H_5CH{-}CH_2 \; (\text{ring with } CF_2) \xrightarrow[\text{PdO, 25 °C}]{H_2} C_6H_5CH_2CF_2CH_3 + C_6H_5CH_2CHFCH_3$$

44% *40%*

[*50*] **39**

$$RCH{-}CHOCOCH_3 \; (\text{ring with } CF_2) \xrightarrow[\text{ether, 0 °C}]{LiAlH_4} RCH{=}CFCH_2OH$$

71-99%

(R = $C_6H_5CH_2$, C_8H_{11}, etc.)

[*51*] **40**

$$RCH{-}CHCHIR \; (\text{ring with } CF_2) \xrightarrow[\text{R}\bullet\text{, } C_6H_6\text{, 80 °C}]{(C_4H_9)_3SnH} (E){-}RCH_2CF_2CH{=}CHR$$

63-83%

(R = H, CH_3, C_6H_5, $C_6H_5CH_2CH_2$)

Reduction of Olefins

Olefin reduction was thoroughly covered in the previous edition, so that only two aspects need be treated here. Recent work has shown that an acyclic internal ***fluoroolefin*** can be dihydrogenated catalytically with H_2/Pd catalyst, provided that no secondary branching is present at the double bond [*52*] (equation 41). Moreover, for the example shown, the *E* isomer gives the *threo* product exclusively from *cis* addition of hydrogen to the double bond [*53*]. A single branch at the double bond encourages loss of fluoride, and subsequent hydrogenation produces the trihydro and dihydro derivatives (equation 42). Catalytic hydrogenation of a tetrasubstituted fluoroolefin is also accompanied by loss of fluoride after addition of hydrogen (equation 43).

As with both terminal and cyclic ***olefins containing vinylic chlorine or bromine***, acyclic counterparts partly lose the halogen first and then undergo dihydrogenation to the tetrahydro derivative [*54*] (equation 44). When the reaction is carried out in the liquid phase in the presence of base, yields of tetrahydro product are enhanced [*55*] (equation 44).

[*52,53*] **41**

$$(CF_3)_2CFCF{=}CFCF_3 \xrightarrow[\text{Pd/Al}_2\text{O}_3 \text{ or C, } 50\ ^\circ\text{C, vapor}]{H_2} (CF_3)_2CFCHFCHFCF_3 \quad \textit{87\%}$$

[*53*] **42**

$$(CF_3)_2C{=}CFCF_2CF_3 \xrightarrow[\text{Pd/Al}_2\text{O}_3 \text{ or C, } 50\ ^\circ\text{C, vapor}]{H_2} (CF_3)_2CHCH_2CF_2CF_3 \ \textit{11\%} + (CF_3)_2CHCHFCF_2CF_3 \ \textit{72\%}$$

[*52*] **43**

$$CF_3CF_2C(CF_3){=}C(CF_3)CF_2CF_3 \xrightarrow[\text{Pd/Al}_2\text{O}_3\text{, vapor}]{H_2} CF_3CF{=}C(CF_3)\,CH(CF_3)CF_2CF_3$$

44

[*54*]

$$CF_3CH{=}CClCF_3 \xrightarrow[\text{Pd/Al}_2\text{O}_3\text{, vapor}]{H_2} CF_3CH_2CHClCF_3 + CF_3CH_2CH_2CF_3$$

[*55*]

$$CF_3CH{=}CClCF_3 \xrightarrow[\text{diglyme, NaOH, 20-40 °C}]{H_2\text{, Raney Ni}} CF_3CH_2CH_2CF_3 \quad \textit{75\%}$$

A number of publications have appeared revealing that catalytic processes which are unsuccessful with vinylic fluorides can be applied effectively to fluoroalkyl- and fluoroarylethylenes. Hydroformylation [*56*] and related reactions such as amidocarbonylation [*57*] will proceed in high yield and regioselectivity. The choice of the catalyst system, especially of **rhodium and cobalt catalysts,** allows either regioselective reaction to proceed. A few of the large number of synthetic variants are given in equation 45.

[*56*] **45**

$$RCH{=}CH_2 \ (R = CF_3, C_6F_5) \xrightarrow[Rh_6(CO)_{16}\text{, 80 °C}]{H_2\text{, CO}} RCH(CH_3)CHO \quad \textit{94-97\%}$$

$$RCH{=}CH_2 \ (R = CF_3, C_6F_5) \xrightarrow[Co_2(CO)_8\text{, 90-100 °C}]{H_2\text{, CO}} RCH_2CH_2CHO \quad \textit{44-88\%}$$

Reduction of Carbonyl

The well-known reduction of carbonyl groups to alcohols has been refined in recent studies to render the reaction more regioselective and more stereoselective. ***Perfluorodiketones*** are reduced by **lithium aluminum hydride** to the corresponding diols, but the use of **potassium or sodium borohydride** allows isolation of the ketoalcohol. Similarly, a perfluoroketo acid fluoride yields diol with lithium aluminum hydride, but the related hydroxy acid is obtainable with potassium borohydride [*58*] (equations 46 and 47).

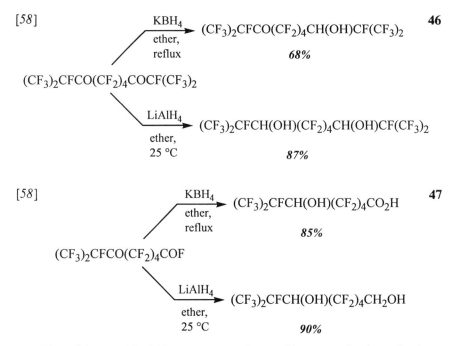

[*58*] **46**

$$(CF_3)_2CFCO(CF_2)_4COCF(CF_3)_2 \xrightarrow[\text{ether, reflux}]{KBH_4} (CF_3)_2CFCO(CF_2)_4CH(OH)CF(CF_3)_2 \quad \textit{68\%}$$

$$(CF_3)_2CFCO(CF_2)_4COCF(CF_3)_2 \xrightarrow[\text{ether, 25 °C}]{LiAlH_4} (CF_3)_2CFCH(OH)(CF_2)_4CH(OH)CF(CF_3)_2 \quad \textit{87\%}$$

[*58*] **47**

$$(CF_3)_2CFCO(CF_2)_4COF \xrightarrow[\text{ether, reflux}]{KBH_4} (CF_3)_2CFCH(OH)(CF_2)_4CO_2H \quad \textit{85\%}$$

$$(CF_3)_2CFCO(CF_2)_4COF \xrightarrow[\text{ether, 25 °C}]{LiAlH_4} (CF_3)_2CFCH(OH)(CF_2)_4CH_2OH \quad \textit{90\%}$$

The milder metal hydride reagents are also used in stereoselective reductions. Inclusion complexes of **amine–borane reagent** with cyclodextrins reduce ***ketones*** to optically active alcohols, sometimes in modest enantiomeric excess [*59*] (equation 48). ***Diisobutylaluminum hydride*** modified by zinc bromide–*N,N,N′,N′*-tetramethylethylenediamine (TMEDA) reduces α,α-difluoro-β-hydroxy ketones to give predominantly erythro-2,2-difluoro-1,3-diols [*60*] (equation 49). The threo isomers are formed on reduction with **aluminum isopropoxide**.

[*59*] **48**

$$C_6H_5COCF_3 \xrightarrow[\beta\text{-cyclodextrin, } H_2O,\ 0\ °C]{C_5H_5N{\cdot}BH_3} (S)\text{–}C_6H_5CH(OH)CF_3 \quad \textit{96\% (13\% ee)}$$

[60] R¹–CH(OH)–CF₂–C(=O)–R² $\xrightarrow[\text{ZnCl}_2\cdot\text{TMEDA, THF, -78 °C}]{(C_4H_9)_2AlH}$ R¹–CH(OH)–CF₂–CH(OH)–R² **49**

91-100% (68-96% erythro)

($R_1 = H(CH_2)_3$, (*E*)–$CH_3CH = CH$, C_6H_5, $(CH_3)_2CH$, $(CH_3)_3C$; $R^2 = C_6H_5$, $H(CH_2)_6$)

Even more highly selective ketone reductions are carried out with **baker's yeast** [*61, 62*] (equations 50 and 51). **Chiral dihydronicotinamides** give carbonyl reductions of high enantioselectivity [*63*] (equation 52), and a crown ether containing a **chiral 1,4-dihydropyridine moiety** is also effective [*64*] (equation 52).

[*61*] *p*–$CH_3C_6H_4SCH_2COCH_2F$ $\xrightarrow[\text{25 °C, 4 h}]{\text{baker's yeast}}$ **50**

(*S*)–(+)–*p*–$CH_3C_6H_4SCH_2CH(OH)CH_2F$

62% (85% ee)

[*62*] $CF_3CH_2COCH_2CO_2C_2H_5$ $\xrightarrow[\text{ClCH}_2\text{CO}_2\text{C}_2\text{H}_5\text{, 25 °C, 4 h}]{\text{baker's yeast}}$ **51**

(*R*)–$CF_3CH_2CH(OH)CH_2CO_2C_2H_5$

60% (84% ee)

Reductive dimerization to form *fluorinated benzopinacols* proceeds in the partly fluorinated case either with **zinc** or by **photolysis** but is not observed with perfluorobenzophenone [*65*] (equation 53). Trifluoroacetophenone is reduced **electrochemically** in dimethylformamide to a stable radical anion, which, in the presence of lithium ion, rapidly dimerizes to pinacol in higher yield than that available by photoreduction [*66*] (equation 54).

The **McMurry reagent** reductively couples trifluoroacetophenones to the corresponding stilbenes [*67*] (equation 55), and cross-coupling of ketones is also reported [*68*] (equation 56).

Reduction of Imines

Fluoroimines are reduced to amines in good yield by **lithium aluminum hydride** [*69, 70*] (equations 57 and 58). Transfer of hydrogen from **2-propanol** to fluoroalkylated benzaldimines is efficiently catalyzed by rhodium [*71*] (equation 59). A new method for the synthesis of optically active fluoroamines is reduction of chiral imines [72] (equation 60).

52

1-propyl-1,4-dihydropyridine-3-CONHCH(CH_3)C_6H_5–(*R*)–(–)

[63] $C_6H_5COCF_3$ $\xrightarrow{CH_3CN,\ [(CH_3)_2C(CN)N=]_2,\ 61\ °C}$ (*S*)–(+)–$C_6H_5CH(OH)CF_3$ ***69% (22% ee)***

[64] $C_6H_5COCF_3$ $\xrightarrow{aq.\ Mg(ClO_4)_2,\ CH_3CN,\ 55\ °C}$ (*S*)–$C_6H_5CH(OH)CF_3$ ***58% (68% ee)***

(macrocyclic N-methyl-1,4-dihydropyridine bis-amide of valine, with (CH$_3$)$_2$CH / NH / HN / CH(CH$_3$)$_2$ groups)

53

[65]

$$C_6F_5COC_6H_5 \xrightarrow[\text{or } h\nu/(CH_3)_2CHOH,\ \text{hexane, } 25\ °C]{Zn/CH_3CO_2H,\ 35\ °C} C_6F_5C(OH)(C_6H_5)-C(OH)(C_6H_5)C_6F_5$$

62-65%

54

[66]

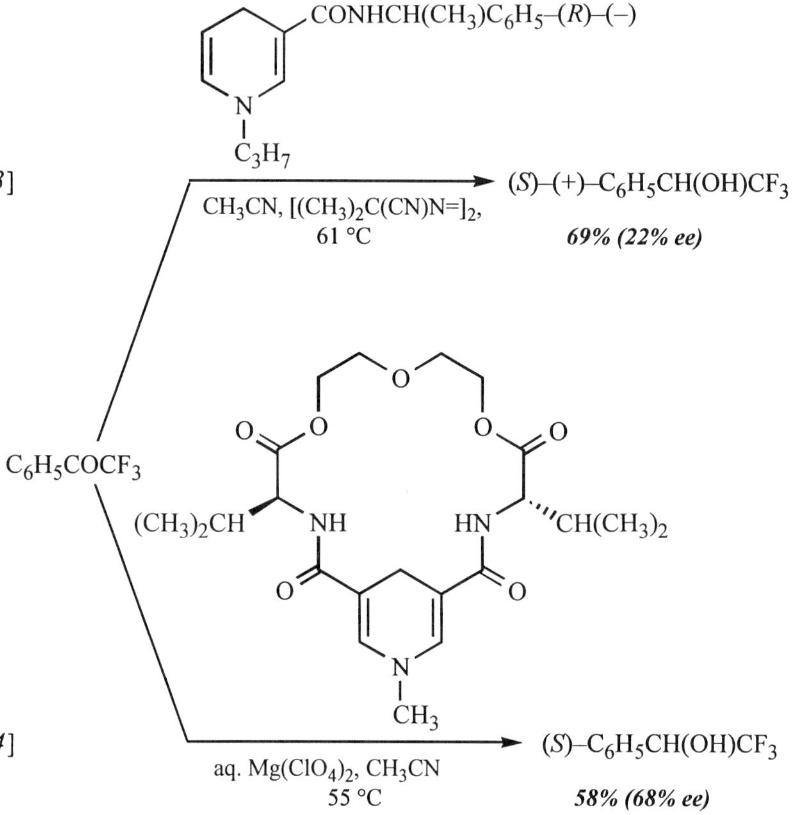

$$C_6H_5COCF_3 \xrightarrow[CH_3CN,\ R_4NClO_4]{e^-} C_6H_5C(=O^{\bullet -})CF_3 \xrightarrow{LiClO_4} C_6H_5C(OH)(CF_3)-C(OH)(CF_3)C_6H_5$$

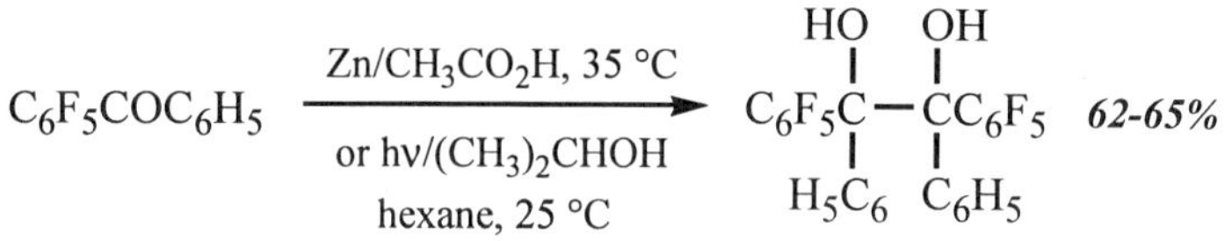

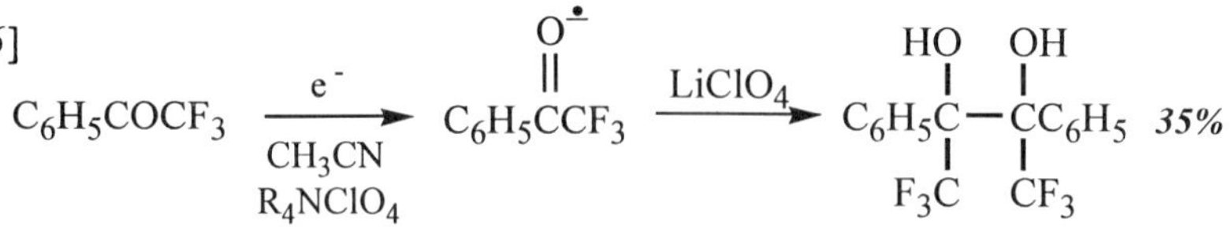

35%

55

[67]

$$CF_3COC_6H_4R \xrightarrow[\text{dioxane, } 100\ °C]{TiCl_4/Zn/C_5H_5N} RC_6H_4C(CF_3)=C(C_6H_4R)-CF_3$$

40-44%

(R = 3–CH_3O, 4–CH_3O)

cis and *trans*

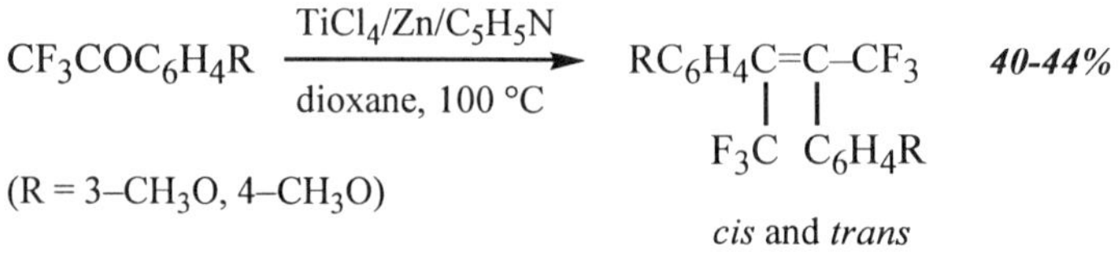

References are listed on pages 316–320.

[*68*] $p\text{-}CF_3C_6F_4COC_6H_5 + C_6H_5COC_2H_5 \xrightarrow[\text{glyme, reflux}]{TiCl_4/Zn} p\text{-}CF_3C_6F_4-C(C_6H_5)=C(C_6H_5)-C_2H_5$ ***39%*** **56**

[*69*] $(CF_3)_2C{=}NH \xrightarrow[\text{diglyme, 25°C}]{LiAlH_4} (CF_3)_2CHNH_2$ ***57%*** **57**

[*70*] octafluorocyclohexanimine (F_2 ring, C=NH) $\xrightarrow[\text{diglyme, 0-25°C}]{LiAlH_4}$ octafluorocyclohexylamine (CH–NH_2) ***28%*** **58**

[*71*] $p\text{-}R^1C_6H_4CH{=}NR^2 \xrightarrow[\text{RhCl(P(C}_6\text{H}_5)_3)_3,\ \text{Na}_2\text{CO}_3,\ \text{reflux}]{(CH_3)_2CHOH} p\text{-}R^1C_6H_4CH_2NHR^2$ ***91-96%*** **59**

(R^1 = H, CF_3; R^2 = $C_6H_5CH_2$, p-$CF_3C_6H_4$, C_6H_5)

[*72*] **60**

$C_6H_5COCF_3$ + (*S*)-(–)-$C_6H_5CH(CH_3)NH_2$ $\xrightarrow{H^+,\ -H_2O}$ $C_6H_5C(CF_3){=}NCH(CH_3)C_6H_5$

$C_6H_5C(CF_3){=}NCH(CH_3)C_6H_5 \xrightarrow[\text{THF, -78 °C}]{(CH_3OCH_2CH_2O)_2AlH}$ chiral amine

chiral amine $\xrightarrow{H_2,\ Pd/C,\ C_2H_5OH,\ 60\ °C}$ (*S*)-(+)-$C_6H_5CH(CF_3)NH_2$ ***95% (80% ee)***

Reduction of Carboxylic Acids and Their Derivatives

Catalytic hydrogenation of trifluoroacetic acid gives trifluoroethanol in high yield [*73*], but higher ***perfluorocarboxylic acids*** and their anhydrides are reduced much more slowly over **rhodium, iridium, platinum, or ruthenium catalysts** [*73, 74*] (equation 61). **Homogeneous catalysis** efficiently produces trifluoroethanol from trifluoroacetate esters [*75*] (equation 61).

Studies of reductions with metal hydrides have concentrated on improvements in selectivity or conditions. Replacement of the usual lithium aluminum hydride–ether combination with **potassium borohydride**–methanol results in high yields of alcohol from ***ester*** [*76*] and less hazard [*77*] (equation 62). Reduction of a fluoroester is effectively limited to the aldehyde stage when sodium **bis(methoxyethoxy)aluminum dihydride** is used as the reducing agent at low temperature [*78*].

[*73,74,75*] **61**

$$CF_3CO_2H \xrightarrow[113\ °C,\ 11\ atm.]{H_2,\ Rh/C} CF_3CH_2OH \xleftarrow[90\ °C,\ 4\ h]{H_2,\ Ru^+\ cat.} CF_3CO_2CH_2CF_3$$

90% *100%*

[*76,77*] **62**

$$CH_3O_2C(CF_2)_4CO_2CH_3 \xrightarrow[CH_3OH,\ 10\ °C]{KBH_4} HOCH_2(CF_2)_4CH_2OH$$

99%

[*78*]

$$\xrightarrow[C_6H_6\ -\ THF,\ -70\ °C]{NaAlH_2(OCH_2CH_2OCH_3)_2} \xrightarrow[H^+]{H_2O} \xrightarrow{P_2O_5} OCH(CF_2)_4CHO$$

65%

Although isolated fluorine atoms can survive reductions with lithium aluminum hydride when they are not in a position α to an ester or nitrile group [*79*] (equations 63a and 63b), those in an α position are reductively cleaved. The milder **borohydride** reagents convert an α-fluorocarboxylate ester to the corresponding alcohol without loss of fluorine [*80*] (equation 64).

[*79*] **63**

a $$R^1R^2CFCH(OH)CO_2CH_3 \xrightarrow[ether,\ 0\ °C]{LiAlH_4} R^1R^2CFCH(OH)CH_2OH$$

$(R^1 = CH_3, C_6H_5; R^2 = CH_3, H)$ *40-45%*

b $$R^1R^2CFCH(OH)CN \xrightarrow[ether,\ -10\ °C\ to\ 0\ °C]{LiAlH_4,\ AlCl_3} R^1R^2CFCH(OH)CH_2NH_2$$

35-40%

[*80*] **64**

$$HOCH_2CHFCO_2C_2H_5 \xrightarrow[THF,\ 25\ °C]{Ca(BH_4)_2} HOCH_2CHFCH_2OH$$

70%

The *N,N*-difluoroamino substituent with no α-hydrogen is resistant to lithium aluminum hydride [*81*] (equation 65), and the selective reduction of the ester functions of ***polychlorofluorocarboxylates*** to alcohols without loss of chlorine is accomplished with sodium borohydride [*82*] (equation 66).

[*81*] $$RC(NF_2)_2CH_2CO_2C_2H_5 \xrightarrow[\text{ether}]{LiAlH_4} RC(NF_2)_2CH_2CH_2OH$$ **65**

$(R = CH_3, F)$ ***43-53%***

[*82*] $$RCO_2CH_3 \xrightarrow[\text{ether, reflux}]{NaBH_4} RCH_2OH$$ ***64-78%*** **66**

$(R = CF_3CFCl, CF_2ClCF_2, CF_2ClCFCl, CF_2Cl)$

The nitro substituent is also preserved during fluoroester reduction with sodium borohydride [*83*] (equation 67). Use of **diborane** itself allows reduction of nitrodifluoroacetanilide to the amine, *N*-nitrodifluoroethylaniline [*84*] (equation 68).

[*83*] $$O_2NCF_2CO_2CH_2C_6H_5 \xrightarrow[\text{aq. THF, 10 °C}]{NaBH_4} O_2NCF_2CH_2OH$$ ***70%*** **67**

[*84*] $$O_2NCF_2C(O)NHC_6H_4R \xrightarrow[\text{THF, 25-65 °C}]{B_2H_6} O_2NCF_2CH_2NHC_6H_5$$ **68**

50-64%

$(R = H, p\text{–}CH_3, m\text{–}OCH_3)$

Reduction of Nitrogen-Containing Functions

Catalytic reduction of fluorinated aliphatic and aromatic ***nitro compounds*** to give oximes and amines was described previously, as was the use of dissolving metals to prepare amines [*85*]. Refinement of these techniques has resulted in optimized yields and, as indicated in equations 69 and 70, in selective reductions [*86, 87*].

[*86*] **69**

$$CF_3CH(CH_3)NO_2 \xrightarrow[\text{Pd/CaCO}_3\text{, CH}_3\text{OH, 30 °C}]{1\text{ eq. }H_2} CF_3C(CH_3){=}NOH$$ ***78%***

$$CF_3CH(CH_3)NO_2 \xrightarrow[\text{95\% aq. CH}_3\text{OH, 50 °C}]{H_2,\ PdCO_3} CF_3CH(CH_3)NHOH$$ ***94%***

[87] **70**

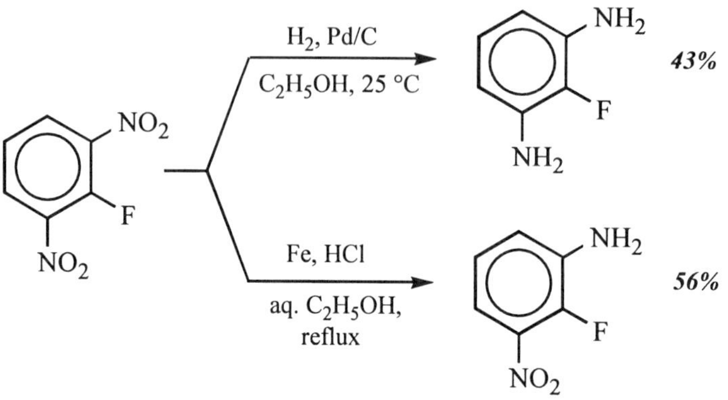

1,3-dinitro-2-fluorobenzene → (H_2, Pd/C; C_2H_5OH, 25 °C) → 1,3-diamino-2-fluorobenzene *43%*

→ (Fe, HCl; aq. C_2H_5OH, reflux) → 3-nitro-2-fluoroaniline *56%*

Controlled reductions are also possible with a variety of other reagents. *m*-Trifluoromethylnitrobenzene is converted to the azoxy derivative with hot **potassium hydroxide–methanol** [*88*] but gives the amine with hot **sodium sulfide**–sodium hydroxide solution [*89*] (equation 71).

71

[*88*] m-$CF_3C_6H_4NO_2$ → (KOH; CH_3OH/toluene, reflux) → m-$CF_3C_6H_4N{=}N(O)C_6H_4CF_3$-m *90%*

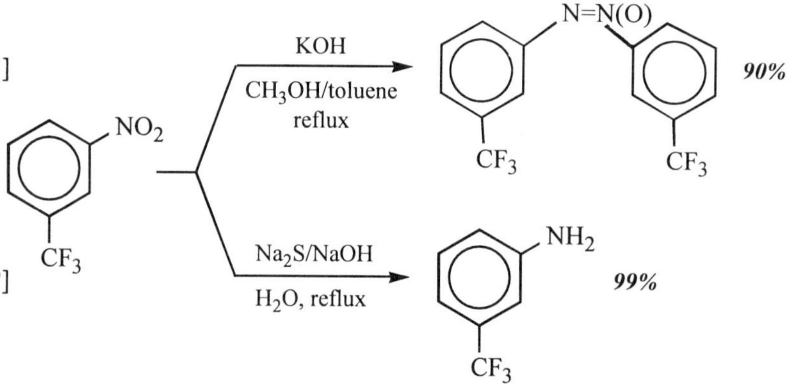

[*89*] → (Na_2S/NaOH; H_2O, reflux) → m-$CF_3C_6H_4NH_2$ *99%*

Potassium bisulfite reduces ***fluoroalkylnitroso compounds*** to the oximes, with the exception that the parent nitrosotrifluoromethane is converted to the corresponding hydroxylamine [*90*] (equation 72). ***Aryl fluoroalkylazo compounds*** are reduced to the hydrazo stage or the fluoroamine by zinc–acid, depending on conditions [*91*] (equation 73). Of the various reagents tested for selective reduction of β-fluoroalkyl ***azides*** to β-fluoroalkylamines, **triphenylphosphine** was the best [*92*] (equation 74).

[*90*] **72**

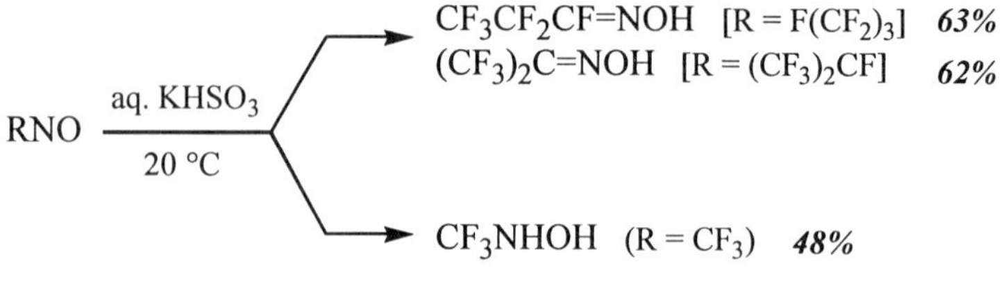

RNO → (aq. $KHSO_3$; 20 °C) →
- $CF_3CF_2CF{=}NOH$ [R = $F(CF_2)_3$] ***63%***
- $(CF_3)_2C{=}NOH$ [R = $(CF_3)_2CF$] ***62%***
- CF_3NHOH (R = CF_3) ***48%***

References are listed on pages 316–320.

[*91*] **73**

$$(CF_3)_2CRN{=}NC_6H_5 \xrightarrow[20\ °C,\ 6\ h.]{ZnCH_3CO_2H} (CF_3)_2CRNHNHC_6H_5 \quad \textit{75\%}$$

$$(CF_3)_2CRN{=}NC_6H_5 \xrightarrow[20\ °C,\ 18\ h.]{ZnCF_3CO_2H} (CF_3)_2CRNH_2 \quad \textit{60-70\%}$$

($R = CF_3, C_3F_7$)

[*92*] **74**

$$C_6H_5CHFCH_2N_3 \xrightarrow[H_2O/THF,\ 25\ °C]{P(C_6H_5)_3} C_6H_5CHFCH_2NH_2 \quad \textit{80\%}$$

Reduction of Sulfur-Containing Functions

Simple ***α-fluorosulfides*** are reduced to the fluoroalkanes by **sodium–ethanol** [*93*] (equation 75). Clean conversion of bis(trifluoromethyl) ***disulfide*** to *trifluoromethyl mercaptan* is accomplished with hydrogen sulfide and ultraviolet irradiation [*94*] (equation 76). ***Perfluoroalkanesulfonyl*** fluorides are converted to the sulfinate salts by **hydrazine** [*95*] (equation 77).

[*93*] **75**

$$C_6H_5CHRCHFSC_6H_5 \xrightarrow{Na,\ C_2H_5OH} C_6H_5CHRCH_2F \quad \textit{44-81\%}$$

($R = H, CH_3, C_6H_5$)

[*94*] **76**

$$CF_3SSCF_3 \xrightarrow[25\ °C]{H_2S,\ h\nu} CF_3SH \quad \textit{90-99\%}$$

[*95*] **77**

$$RSO_2F \xrightarrow[\text{CH}_3\text{OH or ether, reflux}]{N_2H_4} RSO_2H{\bullet}N_2H_4 \xrightarrow{\text{aq. HCl}} RSO_2H \quad \textit{74-82\%}$$

($R = CF_3, C_4F_9, C_8F_{17}$)

Reduction of Enol Phosphates

Polyfluoro-1-alkenyl phosphates are reduced under mild conditions by **diisobutylaluminum hydride** (DIBAL) to the aluminum enolate, which is derivatized by addition to benzaldehyde [*96*] (equation 78).

[*96*] **78**

$$RCF{=}C(C_6H_5)OP(O)(OEt)_2 \xrightarrow[\text{THF, 0 °C}]{\text{DIBAL}} RCF{=}C(OAl(C_4H_9)_2)C_6H_5$$

$(R = CF_3, C_2F_5)$

$$\xrightarrow[\text{0 °C}]{C_6H_5CHO} C_6H_5CH(OH)CFRCOC_6H_5$$

71-78%

Experimental Procedures

Reductive Coupling of 4-Methoxy-α,α,α-Trifluoroacetophenone: cis- and trans-1,1,1,4,4,4-Hexafluoro-2,3-bis(4-methoxyphenyl)-2-butene [*67*]

$TiCl_4$ (2.85 g; 0.015 mol) was added dropwise to 100 mL of cooled, dry dioxane under nitrogen with stirring. A suspension of zinc powder (1.96 g; 0.03 mol) in 20mL of dry dioxane was added in small portions over a period of 10 min. After the addition of 2 mL of dry pyridine, a solution of 4-methoxy-α,α,α-trifluoroacetophenone (2.04 g; 0.01 mol) in 20 mL of dry dioxane was added, and the reaction mixture was heated at reflux. After 4 h at 101 °C, the mixture was cooled, and alkaline hydrolysis was performed with 10% K_2CO_3 solution. After extraction with ether and removal of the solvent, the crude product was purified and separated into the two isomeric compounds by column chromatography (SiO_2; toluene–CCl_4, 4:1) to give 0.43 g (23%) of the *cis* isomer (mp, 105 °C) and 0.32 g (17%) of the *trans* isomer (mp, 174 °C).

F-1-methyl-2-(2-methyl-1-cyclopentenyl)cyclopentene [*42, 97*]

To a mixture of 15.0 g (0.057 mol) of $P(C_6H_5)_3$ and 25 mL of anhydrous acetonitrile stirred at 25 °C was added 22.0 g (0.084 mol) of *F*-1-methylcyclopentene over a period of 0.5h. The mixture was steam distilled, and the lower layer was separated, dried, and fractionated to give 15.2 g (75%) of *F*-1-methyl-2-(2-methyl-1-cyclopentenyl)cyclopentene: bp, 72 °C (100 mm); IR: 1664 (s), 1714 cm^{-1} (m).

References for Pages 297–316

1 Ohsawa, T.; Takagaki, T.; Haneda, A.; Oishi, T. *Tetrahedron Lett.* **1981,** *22,* 2583.
2. Walsh, T. D.; Dabestani, R. *J. Org. Chem.* **1981,** *46,* 1222.

3. Jansta, J.; Dousek, F. P. *Carbon* **1980,** *18,* 433.
4. Zeifman, Y. V.; Vol'pin, I. M.; Postovoi, S. A.; German, L. S. *Izv. Akad. Nauk SSSR* **1987,** *10,* 2396; *Chem. Abstr.* **1988,** *109,* 54292j.
5. Postovoi, S. A.; Vol'pin, I. M.; Mysov, E. J.; Zeifman, Y. V.; German, L. S. *Izv. Akad. Nauk SSSR* **1989,** *5,* 1173; *Chem. Abstr.* **1990,** *112,* 76343r.
6. Yamana, M.; Ishihara, T.; Ando, T. *Tetrahedron Lett.* **1983,** *24,* 507.
7. Portella, C.; Pete, J. P. *Tetrahedron Lett.* **1985,** *26,* 211.
8. Portella, C.; Iznaden, M. *Tetrahedron* **1989,** *45,* 6467.
9. Paleta, O.; Dadak, V.; Dedek, V.; Timpe, H. J. *J. Fluorine Chem.* **1988,** *39,* 397.
10. Reinholdt, K.; Margaretha, P. *Helv. Chim. Acta* **1983,** *66,* 2534.
11. Scott, W. J.; Zuman, P. *Anal. Chim. Acta* **1981,** *126,* 71.
12. Hudlicky, M. *Chemistry of Organic Fluorine Compounds;* Ellis Horwood: Chichester, U.K.; 1976; pp 170–188.
13. Hudlicky, M. *J. Fluorine Chem.* **1989,** *44,* 345.
14. Hudlicky, M. *J. Fluorine Chem.* **1979,** *14,* 189.
15. Dmowski, W. *J. Fluorine Chem.* **1985,** *29,* 273.
16. Sauvetre, R.; Masure, D.; Chiut, C.; Normant, J. F. *C. R. Hebd. Seances Acad. Sci., C* **1979,** *288,* 335.
17. Fuchikami, T.; Shibata, Y.; Suzuki, Y. *Tetrahedron Lett.* **1986,** *27,* 3173.
18. Burton, D. J.; Spawn, T. D.; Heinze, P. L.; Bailey, A. R.; Shinya, S. *J. Fluorine Chem.* 1989, 44, 167.
19. Kariv-Miller, E.; Vajtner, Z. *J. Org. Chem.* **1985,** *50,* 1394.
20. Afanes'ev, V. A.; Efinov, O. N.; Nesterenko, G. N.; Nefedov, O. M.; Pivovarov, A. P.; Rogachev, B. G.; Khidekel, M. L. *Izv. Akad. Nauk SSSR* **1988,** 806; *Chem. Abstr.* **1988,** *109,* 169931j.
21. Morikawa, S.; Yoshitake, M.; Tatematsu, S. Jpn. Pat. 01 258 631, 1989; *Chem. Abstr.* **1990,** *112,* 118250r.
22. Furutaka, Y.; Aoyama, H.; Homoto, Y. Jpn. Pat. 01 149 739, 1989; *Chem. Abstr.* **1989,** *111,* 232067r.
23. Morikawa, S.; Samejima, S.; Yoshitake, M.; Tatematsu, S.; Omori, T. Jpn. Pat. 02 131 437, 1990; *Chem. Abstr.* **1990,** *113,* 131557m.
24. Morikawa, S.; Yoshitake, M., Tatematsu, S. Jpn. Pat. 01 132 538, 1989; *Chem. Abstr.* **1989,** *111,* 173594k.
25. Morikawa, S.; Yoshitake, M.; Tatematsu, S.; Yoneda, S.; Ohira, K. Jpn. Pat. 01 258 632, 1989; *Chem. Abstr.* **1990,** *112,* 118251s.
26. Furutaka, Y.; Aoyama, H.; Homoto, Y. Jpn. Pat. 01 093 549, 1989; *Chem. Abstr.* **1989,** *111,* 114734h.
27. Jefford, C. W.; Burger, U.; Laffer, M. H.; Kabengele, nT. *Tetrahedron Lett.* **1973,** *27,* 2483.
28. Ando, T.; Ishihara, T.; Ohtani, E.; Sawada, H. *J. Org. Chem.* **1981,** *46,* 4446.

29. Yamanaka, H.; Shimamura, T.; Teramura, K.; Ando, T. *Chem. Lett.* **1972,** 921.
30. Ando, T.; Hosaka, H.; Funasaka, W.; Yamanaka, H. *Bull. Chem. Soc. Jpn.* **1973,** *46,* 3513.
31. Robinson, J. M. Ger. Pat. 3 906 273, 1989; *Chem. Abstr.* **1990,** *112,* 54961p.
32. Hofmeister, H.; Laurent, H.; Wiechert, R.; Annen, K.; Steinbeck, H. Ger. Pat. 2 410 442, 1975; *Chem. Abstr.* **1976,** *84,* 59868b.
33. Ballard, D. G. H.; Farrar, J.; Laidler, D. A. Eur. Pat. 164 954, 1985; *Chem. Abstr.* **1986,** *104,* 185970y.
34. Jpn. Pat. 58 222 038, 1983; *Chem. Abstr.* **1984,** *100,* 156218w.
35. Torii, S.; Tanaka, H.; Yamashita, S.; Hotate, M.; Suzuki, A.; Okuma, Y. Jpn. Pat. 02 01 414, 1990; *Chem. Abstr.* **1990,** *112,* 197614p.
36. Furutaka, Y.; Aoyama, H.; Honda, T. Eur. Pat. 308 923, 1989; *Chem. Abstr.* **1989,** *111,* 194094n.
37. Dedek, V.; Liska, F.; Kuzmic, P.; Fikar, J.; Vesely, I.; Salamon, M.; Pavlovska, E.; Simon, J. Czech. Pat. 237 733, 1987; *Chem. Abstr.* **1988,** *108,* 166971r.
38. Paleta, O.; Jezek, R.; Dedek, V. *Collect. Czech. Chem. Commun.* **1983,** *48,* 766.
39. Nishiyama, R.; Fujikawa, K.; Tsujii, Y.; Murai, S.; Jyonishi, H. Eur. Pat. 73 372, 1983; *Chem. Abstr.* **1983,** *99,* 22145x.
40. Natarajan, S.; Soulen, R. L. *J. Fluorine Chem.* **1981,** *17,* 447.
41. Barlow, M. G.; Haszeldine, R. N.; Langridge, J. R. *J. Chem. Soc., Perkin Trans. 1* **1980,** 2520.
42. Stepanov, A. A.; Rozhkov, I. N.; Knunyants, I. L. *Izv. Akad. Nauk SSSR* **1981,** 701; *Chem. Abstr.* **1981,** *95,* 42422f.
43. Stepanov, A. A.; Rozhkov, I. N. *Izv. Akad. Nauk SSSR* **1983,** 890; *Chem. Abstr.* **1983,** *99,* 38105v.
44. Selivanova, G. A.; Starichenko, V. F.; Shteingarts, V. D. *Izv. Akad. Nauk SSSR* **1988,** 1155; *Chem. Abstr.* **1989,** *110,* 57208t.
45. Chambers, R. D.; Musgrave, W. K. R.; Sargent, C. R.; Drakesmith, F. G. *Tetrahedron* **1981,** *37,* 591.
46. Chen, Q. Y.; He, Y. B.; Yang, Z. Y. *J. Chem. Soc., Chem. Commun.* **1986,** 1452.
47. Tellier, F.; Sauvetre, R.; Normant, J. F.; Chuit, C. *Tetrahedron Lett.* **1987,** *28,* 3335.
48. Filyakova, T. I.; Zapevalov, A. Y.; Kolenko, I. P.; Kodess, M. I.; German, L. S. *Izv. Akad. Nauk SSSR* **1980,** 1185; *Chem. Abstr.* **1980,** *93,* 113873a.
49. Isogai, K.; Nishizawa, N.; Saito, T.; Sakai, J. *Bull. Chem. Soc. Jpn.* **1983,** *56,* 1555.
50. Taguchi, T.; Takigawa, T.; Tawara, Y.; Morikawa, T.; Kobayashi, Y. *Tetrahedron Lett.* **1984,** *25,* 5689.

51. Morikawa, T.; Uejima, M.; Kobayashi, Y. *Chem. Lett.* **1988,** 1407.
52. Li, J.; Gong, X.; Zhou, J.; Wu, W.; Huang, Y. *Youji Huaxue* **1984,** *40*(2), 24; *Chem. Abstr.* **1984,** *101,* 37949w.
53. Snegirev, V. F.; Makarov, K. N.; Zabolotskikh, V. F.; Sorokina, M. G.; Knunyants, I. L. *Izv. Akad. Nauk SSSR* **1983,** 2775; *Chem. Abstr.* **1984,** *100,* 208713a.
54. Huang, Y.; Li, J.; Zhou, J.; Zhu, Z. *Youji Huaxue* **1984,** 125; *Chem. Abstr.* **1984,** *101,* 54508u.
55. Bielefeldt, D.; Marhold, A.; Negele, M. Ger. Pat. 3 735 467, 1989; *Chem. Abstr.* **1989,** *111,* 232057n.
56. Fuchikami, T.; Ojima, I. *J. Am. Chem. Soc.* **1982,** *104,* 3527.
57. Ojima, I.; Okabe, M.; Kato, K.; Kwon, H. B.; Horvath, I. T. *J. Am. Chem. Soc.* **1988,** *110,* 150.
58. Saloutina, L. V.; Zapevalov, A. Y.; Kodess, M. I.; Kolenko, J. P. *Zh. Org. Khim.* **1982,** *18,* 788; *Chem. Abstr.* **1982,** *97,* 55274a.
59. Sakuraba, H.; Inomata, N.; Tanaka, Y. *J. Org. Chem.* **1989,** *54,* 3482.
60. Kuroboshi, M.; Ishihara, T.; *Bull. Chem. Soc. Jpn.* **1990,** *63,* 1185.
61. Bucciarelli, M.; Forni, A.; Moretti, I.; Prati, F.; Torre, G.; Resnati, G.; Bravo, P. *Tetrahedron Lett.* **1989,** *45,* 7505.
62. Nakamura, K.; Kawai, Y.; Ohno, A. *Tetrahedron Lett.* **1990,** *31,* 267.
63. Tanner, D. D.; Kharrat, A. *J. Am. Chem. Soc.* **1988,** *110,* 2968.
64. De Vries, J. G.; Kellogg, R. M. *J. Am. Chem. Soc.* **1979,** *101,* 2759.
65. Filler, R.; Kang, H. H. *J. Org. Chem.* **1975,** *40,* 1173.
66. Andrieux, C. P.; Saveant, J. M. *Bull. Soc. Chim. Fr.* **1973,** *6,* 2090.
67. Hartmann, R. W.; Heindl, A.; Schneider, M. R.; Schoenenberger, H. *J. Med. Chem.* **1986,** *29,* 322.
68. Coe, P. L.; Scriven, C. E. *J. Chem. Soc., Perkin Trans. 1* **1986,** *3,* 475.
69. Middleton, W. J.; Krespan, C. G. *J. Org. Chem.* **1965,** *30,* 1398.
70. Coe, P. L.; Sleigh, J. H.; Tatlow, J. C. *J. Fluorine Chem.* **1980,** *15,* 339.
71. Grigg, R.; Mitchell, T. R. B.; Tongpenyai, N. *Synthesis* **1981,** 442.
72. Pirkle, W. H.; Hauske, J. R. *J. Org. Chem.* **1977,** *42,* 2436.
73. Novotny, M. *J. Org. Chem.* **1979,** *44,* 3268.
74. Kolomnikova, G. D.; Kalinkin, M. I.; Tskhurbaeva, Z. T.; Parnes, Z. N.; Kursanov, D. N. Izv. Akad. Nauk SSSR 1978, 1681; Chem. Abstr. 1978, 89, 129011w.
75. Grey, R. A.; Pez, G. P.; Wallo, A.; Corsi, J. *J. Chem. Soc., Chem. Commun.* **1980,** *16,* 783.
76. Mikisheva, L. I.; Krylov, A. I. *Khim. Prom-st., Ser.; Reakt. Osobo Chist. Veshchestva* **1981,** *2,* 28; *Chem. Abstr.* **1981,** *95,* 132219z.
77. Dobina, K. A.; Dolgopol'skii, I. M.; Sinaiskaya, M. I.; Kamysheva, S. A.; Konshin, A. I.; Balashova, L. G. *Zh. Prikl. Khim. (Leningrad)* **1973,** *46,* 687; *Chem. Abstr.* **1973,** *79,* 4911b.
78. Greenwald, R. B.; Evans, D. H. *J. Org. Chem.* **1976,** *41,* 1470.
79. Remli, M.; Guedj, R. *J. Fluorine Chem.* **1983,** *22,* 493.

80. Tolman, V.; Veres, K. *Collect. Czech Chem. Commun.* **1972,** *37,* 2962.
81. Fokin, A. V.; Voronkov, A. N.; Timofeenko, I. A. *Izv. Akad. Nauk SSSR* **1977,** 1453; *Chem. Abstr.* **1977,** *87,* 101878u.
82. Paleta, O.; Danda, A.; Stepan, L.; Kvicala, J.; Dedek, V. *J. Fluorine Chem.* **1989,** *45,* 331.
83. Hill, M. E.; Ross, L. O. U.S. Patent 3 783 144, 1974; *Chem. Abstr.* **1974,** *80,* 70328p.
84. Bissell, E. R.; Swansiger, R. W. *J. Fluorine Chem.* **1981,** *17,* 485.
85. Hudlicky, M. *Chemistry of Organic Fluorine Compounds;* Ellis Horwood, Chichester, U.K., 1976; pp 172–173, 192.
86. Baasner, B.; Ziemann, H.; Klauke, E. Ger. Pat. 3 336 498, 1985; *Chem. Abstr.* ***1985,*** *103,* 214870c.
87. Hudlicky, M.; Bell, H. M. *J. Fluorine Chem.* **1974,** *4,* 19.
88. Prato, M.; Quintily, U.; Scapol, L.; Scorrano, G. *Bull. Soc. Chim. Fr.* **1987,** *1,* 99.
89. Aracs, J.; Benke, B.; Kanakaridisz, S.; Szeszak, W. Hung. Pat. 25 255, 1983; *Chem. Abstr.* **1983,** *99,* 194603m.
90. Banks, R. E.; Dickinson, N. *J. Chem. Soc., Perkin Trans. 1* **1982,** *3,* 685.
91. Zeifman, Y. V.; Lantseva, L. T.; Knunyants, I. L. *Izv. Akad. Nauk SSSR* **1986,** 401; *Chem. Abstr.* **1987,** *106,* 17846k.
92. Hamman, S.; Beguin, C. G. *J. Fluorine Chem.* **1987,** *37,* 191.
93. Purrington, S. T.; Pittman, J. H. *Tetrahedron Lett.* **1988,** *29,* 6851.
94. Zack, N. R.; Shreeve, J. M. *Synth. Commun.* **1974,** *4,* 233.
95. Harzdorf, C.; Meussdoerffer, J. N.; Niederpruem, H.; Wechsberg, M. *Liebigs Ann. Chem.* **1973,** *1,* 33.
96. Ishihara, T.; Yamaguchi, K; Kuroboshi, M. *Chem. Lett.* **1989,** 1191.
97. Stepanov, A. A.; Bekker, G. Y.; Kurbakova, A. P.; Leites, L. A.; Rozhkov, I. N. *Izv. Akad. Nauk SSSR* **1981,** *12,* 2746.

Oxidation

by Oldrich Paleta

Oxidative reactions frequently represent a convenient preparative route to synthetic intermediates and end products. This chapter includes oxidations of alkanes and cycloalkanes, alkenes and cycloalkenes, dienes, aromatic fluorocarbons, alcohols, phenols, ethers, aldehydes and ketones, carboxylic acids, nitrogen compounds, and organophosphorus, -sulfur, -selenium, -iodine, and -boron compounds.

Most frequent are oxidations of alkenes that can be converted to a series of compounds such as epoxides, halohydrins and their esters, ozonides (1,2,4-trioxolanes), α-hydroxyketones, α-hydroxyketone fluorosulfonates, α-diketones, and carboxylic acids and their derivatives.

Oxidation of Alkanes and Cycloalkanes

Interesting, though hardly of practical value, is the oxidation of 1-fluorooctadecane by yeast of genus *Torulopsis gropengiesseri* to give a 9% yield of octadecanedioic acid and a 4.5% yield of hexadecanedioic acid [*1*].

Oxidations of higly fluorinated alkanes and cycloalkanes are rare because of the resistance of these compounds to oxidation agents. Reactive centers include C–H and C–I bonds (oxidations of iodo compounds at iodine atom are described in a special part of this chapter).

The *difluoromethyl group* in ***chlorofluoroalkanes*** can be oxidized into a *carboxylic group* by a mixture of **chlorine and dinitrogen tetroxide** at high temperatures [2] (equation 1).

[2] **1**

$$CClF_2CClF(CF_2)_nCHF_2 \xrightarrow[N_2O_4\ (7.5\ l/h),\ 500\ °C]{Cl_2\ (4.5\ l/h)} CClF_2CClF(CF_2)_nCO_2H$$

n	1	3	5
Yield	75%	84.5%	75.6%

The monofluoromethylene group and difluoromethyl group in 1H-perfluoroalkanes and -cycloalkanes are oxidized at the C–H bond to *perfluoroalkyl and perfluorocycloalkyl fluorosulfates* by **anodic oxidation** in fluorosulfonic acid [*3*, *4*]. Two modifications of the method are used: oxidation by fluorosulfonyl peroxide generated prior to the reaction [*3*] (equation 2A) and direct electrolysis in the acid [*3*, *4*] (equations 2B and 3).

0065–7719/95/0187–0321$12.14/1

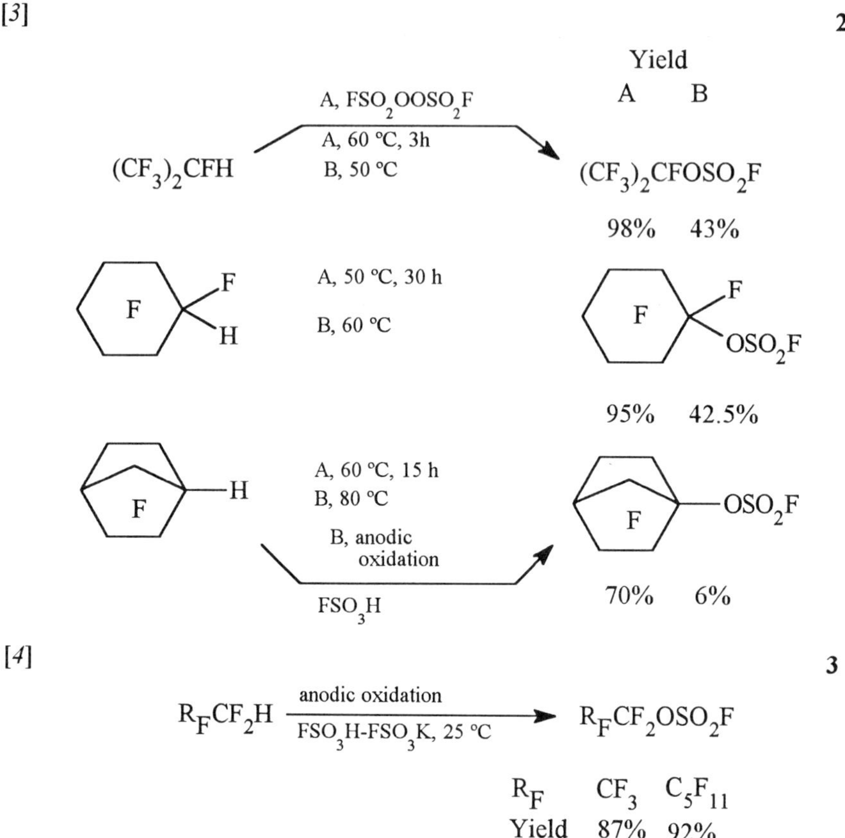

Oxidation of Alkenes and Cycloalkenes

Unsaturated compounds containing both a double bond and a characteristic group such as hydroxyl, carbonyl, or nitrile or an organophosphorus, -selenium -sulfur, -iodine, or -boron group are also included in this section even when only the double bond is oxidized.

Epoxidation

Photochemically induced *epoxidation* of ***tetrafluoroethylene*** by **oxygen** with improved yields (71.1–76.3%; conversion, 21–62.3%) is achieved in the presence of radical generators such as tribromofluoromethane, 1,2-dibromotetrafluoroethane, ethyl nitrite, or 2,2,3,3-tetrafluoropropyl nitrite [*5*]. Trifluoroacrylonitrile can be epoxidized by oxygen under pressure at elevated temperatures [*6*] (equation 4).

Substituted peroxybenzoic acids are used for epoxidation of trifluorovinyl alkenes with attached functional groups [*7*] (equation 5).

[*6*] **4**

$$CF_2{=}CF{-}CN \xrightarrow[CCl_2FCClF_2]{O_2,\ 50\ atm,\ 110\ °C} \underset{\text{48\%}}{CF_2{-}CF{-}CN\ (\text{epoxide, } -O-)}$$

[*7*] **5**

$$CF_2{=}C(CF_3)X \xrightarrow{ArCO_3H,\ Et_2O} CF_2{-}C(CF_3)X\ (\text{epoxide, } -O-)$$

$X = CO_2C_2H_5$ — m-$ClC_6H_4CO_3H$, 0-5 °C, 20 min — 55%

$X = CON(CH_3)_2$ — m-$ClC_6H_4CO_3H$, -10 to 20 °C, 2h — 30%

$X = PO(OCH_3)_2$ — p-$CH_3O_2CC_6H_4CO_3H$, 5-20 °C, 3.5 h — 60%

Hexafluoropropene is converted to its 1,2-epoxide in 55% yield by bubbling through a solution of chromium trioxide in fluorosulfonic acid [*8*]; the analogous reaction with a mixture of chromium trioxide and dichromium trioxide gives pentafluoroacetonyl fluorosulfate [*8*] (equation 6).

[*8*] **6**

$$CF_3CF{=}CF_2 \xrightarrow[40\text{-}45\ °C,\ 3\ h]{CrO_3\text{-}Cr_2O_3,\ FSO_3H} \underset{\text{40\%}}{CF_3C({=}O)CF_2OSO_2F}$$

A convenient agent for the preparation of perfluoroalkene epoxides is sodium hypochlorite in a mixture with aqueous acetonitrile or another aprotic solvent. *cis-* and *trans*-perfluoroalkenes are oxidized with retention of configuration [*9, 10, 11, 12, 13*] (equation 7; Table 1).

Conversion of 3-chloropentafluoropropene to 1-chlorodifluoromethyl-1,2,2-trifluorooxirane in alkaline solution is negatively influenced by the high nucleophilic reactivity of allylic chlorine [*14*]. The reaction is performed at very low temperature to favor the attack of hydroperoxyl anion in a competition with hydroxyl anion. Acceptable yields of 31–38% are obtained in the presence of a phase-transfer catalyst [*14*] (equation 8).

Cyclic perfluoroalkenes are oxidized to the corresponding *epoxides* in high yields by **sodium hypochlorite** in aqueous acetonitrile at 0–20 °C [*13, 15*] (equations 9 and 10). Perfluorocycloheptene gives the epoxide after 1 h in 85% yield [*17*]. Epoxidation of perfluoro-1-methylcyclohexene is accomplished with

Table 1. Oxidation of Perfluoroolefins to Perfluoroepoxides by Sodium Hypochloride

$$R^1R^2C{=}CFR^3 \xrightarrow[\text{conditions}]{NaOCl,\ H_2O} R^1R^2C\overset{O}{—}CFR^3 \qquad \mathbf{7}$$

R¹	*R²*	*R³*	*Conditions*	*Yield (%)*
F	CF_3	F	A	52
F	CF_3	CF_3 *(trans)*	B	*80*
F	C_2F_5	CF_3	B	72
F	C_3F_7	CF_3	B	94
F	$CF(CF_3)_2$	CF_3 *(trans)*	C	72
F	C_2F_5	C_2F_5 *(trans)*	D	84.5
F	CF_3	CF_3	E	90.5
F	C_2F_5	CF_3	E	91.5
F	C_2F_7	CF_3	E	94
F	C_3F_5	C_2F_5	E	91
F	$CF(CF_3)_2$	CF_3	E	89
F	C_5F_{11}	CF_3	E	95
F	$CH(CF_3)_3$	CF_3	E	92
CF_3	CF_3	C_2F_5	F	94
CF_3	C_2F_5	CF_3	G	66
CF_3	$(C_2F_5)_2CCF_3$	CF_3	G	93
$C_2F_5CF(CF_3)$	$(C_2F_5)_2CCF_3$	F	G	89

Conditions:

A: MeCN, 20 °C, 2 h [*9*]; B: diglyme, RT, 1.5 h [*9*];

C: MeCN, 20 °C, several h [*10*]; D: MeCN, 20 °C [*11*];

E: MeCN, 20 °C, 1 h [*12*]; F: MeCN, 20 °C, 1 h [*10*]

G: MeCN, 18-25 °C, 0.5-5 h [*13*].

yield [*18*], and epoxidation of 5,6-dibromooctafluorocyclohept-1-ene takes place at 15 °C in 1 h with 90% yield [*17*]. Perfluorobicyclo[4.4.0]dec-1(6)-ene is also converted easily to the epoxide [*19*] (equation 11).

[*14*] **8**

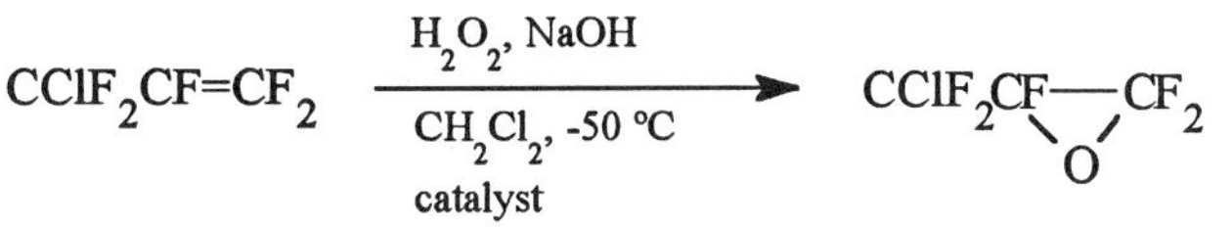

$$CClF_2CF{=}CF_2 \xrightarrow[\text{CH}_2\text{Cl}_2\text{, -50 °C, catalyst}]{\text{H}_2\text{O}_2\text{, NaOH}} CClF_2CF\text{—}CF_2 \text{ (epoxide, O)}$$

	Conversion	Yield
36% H_2O_2, A	17%	8%
45% H_2O_2, A	48%	14%
60% H_2O_2, A	35%	31%
60% H_2O_2, B	72%	38%

A = $(C_4H_9)_4NBr$ B = $C_{16}H_{33}(CH_3)_3NBr$

[*13*] **9**

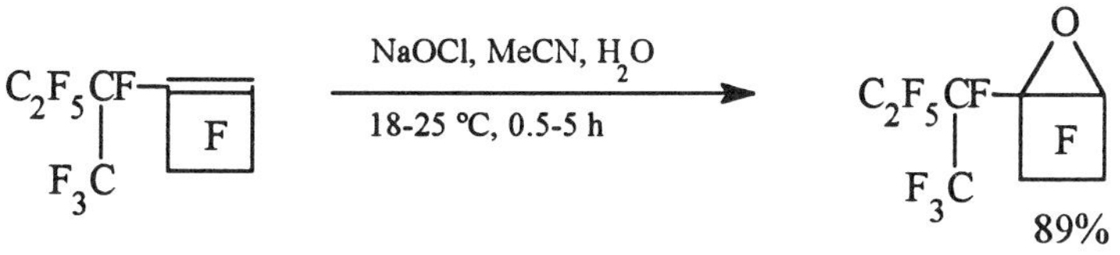

$C_2F_5CF(CF_3)$–perfluorocyclobutene (F) → NaOCl, MeCN, H_2O; 18-25 °C, 0.5-5 h → $C_2F_5CF(CF_3)$–epoxide (O, F) 89%

[*15*] **10**

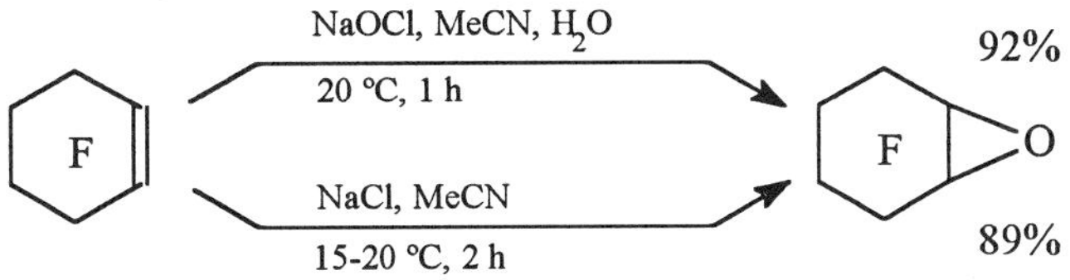

Perfluorocyclohexene (F) → NaOCl, MeCN, H_2O; 20 °C, 1 h → epoxide (F, O) 92%

Perfluorocyclohexene (F) → NaCl, MeCN; 15-20 °C, 2 h → epoxide (F, O) 89%

[*16*]

[*19*] **11**

Perfluorobicyclo[4.4.0]dec-1(6)-ene (F, F) → NaOCl, MeCN, H_2O; -10 °C, 2 h → epoxide (O, F, F) 94%

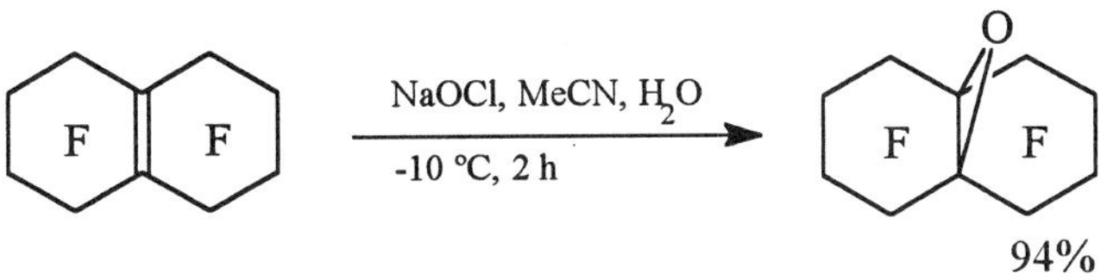

The nonfluorinated double bond in 3-perfluoroalkyl-1-propene is epoxidized with difficulty by ***m*-chloroperoxybenzoic acid** [*20*] (equation 12).

The presence of 9α-fluorine in the molecule of a 11β-hydroxy-δ^4-3-keto steroid causes completely stereospecific alkaline epoxidation with **hydrogen peroxide** in a much slower reaction (4 days vs. 4 h) compared with the nonfluorinated analogue [*21*].

[*20*] **12**

$$\underset{\displaystyle CF_3}{C_3F_7CFCH_2CH{=}CH_2} \xrightarrow[50\ ^\circ C,\ 7\ d]{m\text{-}ClC_6H_4CO_3H,\ CHCl_3} \underset{\displaystyle CF_3}{C_3F_7CFCH_2}\underset{\ \ \ O}{CH{-}CH_2} \quad 50\%$$

A more efficient agent than peroxy compounds for the epoxidation of fluoroolefins with nonfluorinated double bond is the hypofluorous acid–acetonitrile complex [*22*]. Perfluoroalkylethenes react with this agent at room temperature within 2–3 h with moderate yields (equation 13), whereas olefins with strongly electron-deficient double bond or electron-poor, sterically hindered olefins, for example 1,2-bis(perfluorobutyl)ethene and perfluoro-(1-alkylethyl)ethenes, are practically inert [*22*]. Epoxidation of a mixture of 3-perfluoroalkyl-1-propenes at 0 °C is finished after 10 min in 80% yield [*22*]. The trifluorovinyl group in partially fluorinated dienes is not affected by this agent [*22*] (equation 13).

[*22*] **13**

$$R_FCH{=}CH_2 \xrightarrow[CH_2Cl_2]{HOF\ -\ MeCN} R_F\underset{\ \ O}{CH{-}CH_2}$$

R_F		%
C_4F_9	0-20 °C, 3h	63
C_6F_{13}	0-20 °C, 2 h	48
C_6F_5	0 °C, 5 min	85
$CF_2{=}CF(CH_2)_2$	0 °C, 2 min	60
$CF_2{=}CF(CH_2)_4$	-40 °C, 2 min	50
$CF_2{=}CF(CH_2)_6$	-10 °C, 10 min	55

A terminal 1,13-diene with interstitial deka(difluoromethylene) chain is converted to the corresponding diepoxide by repeated reaction with a very large excess of hypofluorous acid–acetonitrile complex [*22*](equation 14).

Ozonization

Ozonides (1,2,4-trioxolanes) are generally obtained by the reaction of ***fluoroalkenes*** with ozone. Thus, vinyl fluoride is oxidized to monofluoroozonide and formyl fluoride [*23*] (equation 15). The same ozonide is formed by ozonolysis of a mixture of *cis*-1,2-difluoroethylene with ethylene [*24*].

[22] **14**

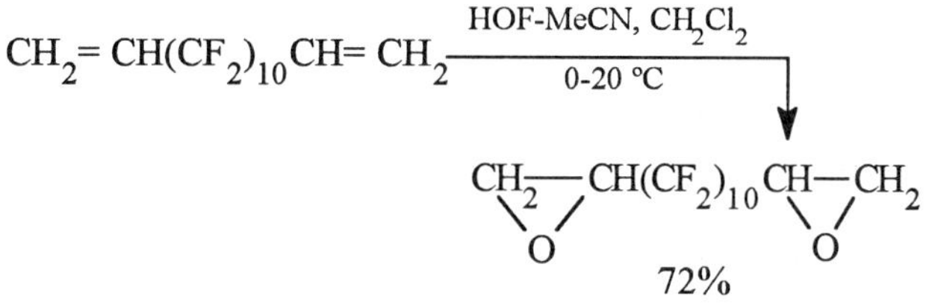

$$CH_2{=}CH(CF_2)_{10}CH{=}CH_2 \xrightarrow[0\text{-}20\ ^\circ C]{HOF\text{-}MeCN,\ CH_2Cl_2} \underset{\text{O}}{CH_2{-}}CH(CF_2)_{10}\underset{\text{O}}{CH{-}}CH_2$$

72%

[23] **15**

CHF=CH$_2$ → (O_3, CH_3Cl, -95 °C) 40%; (O_3, neat, -126 °C) 5–10%

CHF=CHF

CH_2+CH_2 → (O_3, -95 °C, $CH(CH_3)_3$ or $CHCl_3$)

→ F-ozonide + HCOF

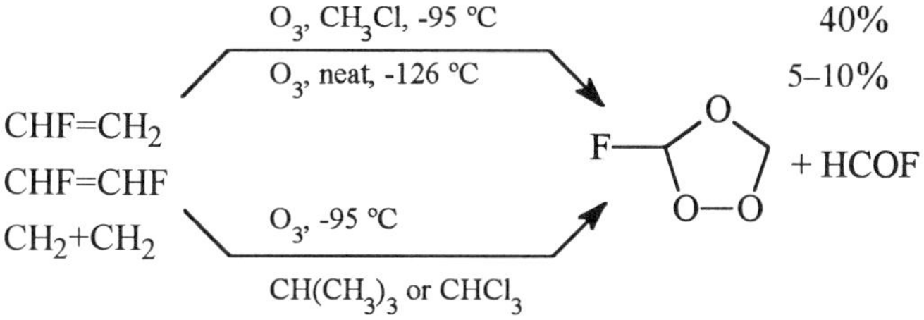

[24]

cis- And *trans*-1,2-difluoroethylene are oxidized by ozone stereoselectively to a mixture of the corresponding epoxides and ozonides with formyl fluoride. The composition of the mixture depends on the solvent used [25] (equation 16).

[25] **16**

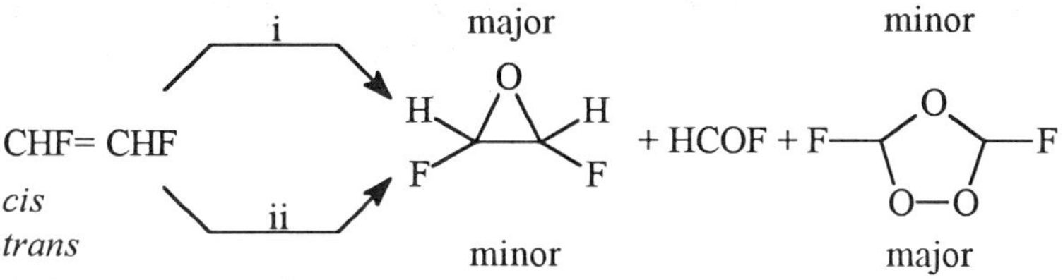

CHF= CHF (*cis*, *trans*) → (i, ii) epoxide (H, F, H, F) (major / minor) + HCOF + F-ozonide-F (minor / major)

i: O_3, CH_3Cl, -78 °C, conversion is 66 %

ii: O_3, $CH(CH_3)_3$, -78 °C

Unusually stable ozonides are prepared by treating mono- and bis(fluroalkyl)ethylenes with ozone [26] (equation 17).

Bubbling ozone into a pentane solution of hexakis(trifluoromethyl)benzvalene gives a pale yellow ozonide, which is stable at room temperatures [27] (equation 18).

Hydroxylation

Hydroxylation, the addition of two hydroxyl groups across a double bond, converts fluorinated alkenes to different products depending on the presence or absence of a fluorine atom at the hydroxylated carbon.

[*26*] **17**

$R^1HC{=}CHR^2 \xrightarrow[\text{i, ii, iii}]{O_3}$ ozonide (R^1, H / H, R^2) + ozonide (H, R^1 / H, R^2)

$R^1 = C_8F_{17}$, $R^2 = H$	i	100%	
$R^1 = R^2 = C_4F_9$	ii	70%	30%
$R^1 = R^2 = C_6F_{13}$	iii	75%	25%

i: 60 °C, 1 h; ii: 20 °C, 6 h; iii: 100 °C, 10h.

[27] **18**

$\xrightarrow[\text{-20 to -10 °C}]{O_3,\ C_5H_{12}}$

R = CF_3 ~ 100%

Hydroxylation of ***terminal ω-fluoroalkenes*** is accomplished by a mixture of **hydrogen peroxide and formic acid.** The first step obviously is the formation of epoxide followed by ring opening [*28*] (equation 19)

[*28*] **19**

$$F(CH_2)_nCH{=}CH_2 \xrightarrow[\text{40 °C, 21 h}]{30\%\ H_2O_2,\ HCO_2H} F(CH_2)_nCH(OH)CH_2OH$$

n = 3 30%
n = 8 75%

Perfluoroalkenes are converted to *vicinal diols* when no fluorine atom is present at the double bond. Configurational isomers of perfluoroalkenes [*29*] (equations 20 and 21) are oxidized stereospecifically. Perfluorbicyclo[4.3.0]non-1(6)-ene gives the corresponding 1,6-diol in a 24% yield upon oxidation with **potassium permanganate** at 18 °C for 1 h [*29*].

Fluorinated alkenes with one fluorine atom attached to the double bond are converted to *α-hydroxyketones* by **potassium permanganate** [*30*] (equation 22).

α-Diketones are formed by permanganate hydroxylation of double bonds flanked by fluorine atoms [*31*] (equation 23).

[29] 20

$(CF_3)(C_2F_5)C{=}C(CF_3)(C_2F_5)$ $\xrightarrow[23\ °C,\ 1.5\ h]{KMnO_4,\ Me_2CO}$ $CF_3C(OH)(C_2F_5)\text{–}C(OH)(C_2F_5)CF_3$

57%

[29] 21

Perfluoro-1,2-bis(trifluoromethyl)cyclohexene (F, CF_3, CF_3) $\xrightarrow[25\ °C,\ 1\ h]{KMnO_4,\ Me_2CO}$ F-cyclohexane-1,2-diol with 1,2-bis(CF_3) (OH, OH, CF_3, CF_3)

79%

[*30*] 22

$R^1(F)C{=}CR^2R^3$ $\xrightarrow[Me_2CO,\ 1.5\text{-}\ 2\ h]{KMnO_4,\ H_2O}$ $R^1COCR^2R^3$ (with OH on C)

R^1	R^2	R^3	°C	%
CF_3	$CF(CF_3)_2$	$CF(CF_3)_2$	20	55
$CF(CF_3)_2$	CF_3	C_3F_7	-15	50
$CF(CF_3)_2$	CF_3	CF_3	-15	45
$CF(CF_3)_2$ with $CF_2C(CF_3)_3$ on C	CF_3	CF_3	20	50
$CH(CF_3)_2$	$CO_2C_2H_5$	$CO_2C_2H_5$	0	51

[*31*] 23

$R^1CF{=}\ CFR^2$ $\xrightarrow[Me_2CO]{KMnO_4,\ H_2O}$ R^1COCOR^2

R^1	R^2	°C	%
C_4H_9	C_4H_9	< 20 °C, 2 h	58
CF_3	C_6H_5	-20 °C, 0.5 h	50
CF_3	C_6H_5	-20 °C, 0.5 h	43
CF_3	$C(CO_2C_2H_5)_2$ with CH_3 on C	-5 to 0 °C, 1 h	60

Hydroxylation of ***1-alkoxy-1-fluoroalkenes*** by **potassium permanganate** followed by spontaneous dehydrofluorination of the primary product gives alkyl-2-hydroxycarboxylates [*32*] (equation 24).

[*32*] **24**

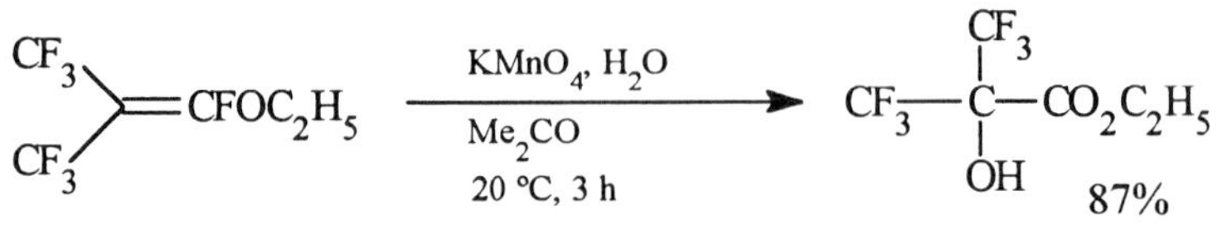

$$(CF_3)_2C{=}CFOC_2H_5 \xrightarrow[\text{Me}_2\text{CO, 20 °C, 3 h}]{\text{KMnO}_4,\ \text{H}_2\text{O}} CF_3{-}C(CF_3)(OH){-}CO_2C_2H_5 \quad 87\%$$

Formation of Halohydrins

Halohydrins are useful intermediates especially in the synthesis of epoxides. The main reaction is usually accompanied by the formation of a dihalide.

Chlorohydrins and 1,2-dichloro derivatives are obtained by *oxidation* of ***alkenes*** with ***tert*-butyl hypochlorite;** when the reaction is performed in acetic acid instead of water, chlorohydrin acetate is formed [*33*] (equation 25).

[*33*] **25**

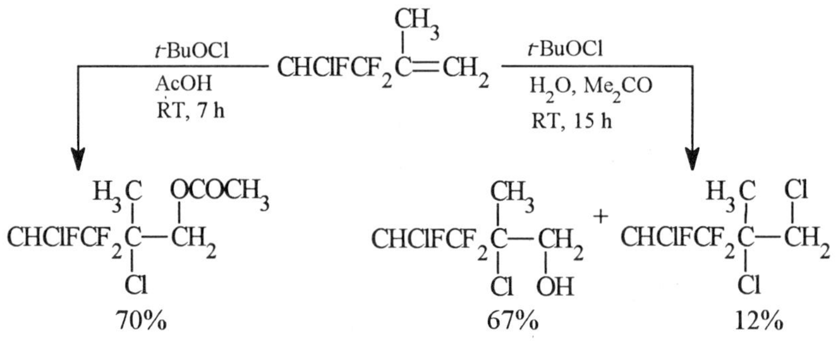

$$CHClFCF_2C(CH_3){=}CH_2 \xrightarrow[\text{AcOH, RT, 7 h}]{t\text{-BuOCl}} CHClFCF_2C(CH_3)(Cl){-}CH_2OCOCH_3 \quad 70\%$$

$$CHClFCF_2C(CH_3){=}CH_2 \xrightarrow[\text{H}_2\text{O, Me}_2\text{CO, RT, 15 h}]{t\text{-BuOCl}} CHClFCF_2C(CH_3)(Cl){-}CH_2OH \ (67\%) + CHClFCF_2C(CH_3)(Cl){-}CH_2Cl \ (12\%)$$

Bromohydrin acetates are formed by the oxidation of the vinyl group in ***perfluoroalkylethylenes*** with **bromine** in acetic acid [*34*] (equation 26).

[*34*] **26**

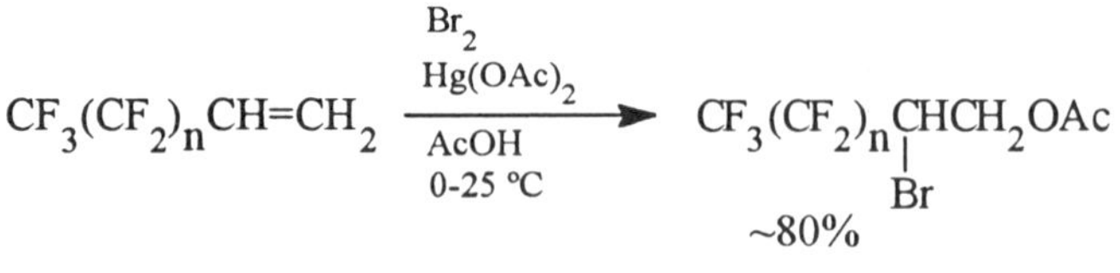

$$CF_3(CF_2)_nCH{=}CH_2 \xrightarrow[\text{AcOH, 0-25 °C}]{\text{Br}_2,\ \text{Hg(OAc)}_2} CF_3(CF_2)_nCHBrCH_2OAc \quad \sim 80\%$$

Similar oxidation with bromine of alkenes having branched perfluoroalkyls in the allylic position gives bromohydrin regioisomers and a dibromo derivative [*35*] (equation 27).

[35] **27**

$$C_3F_7C(CF_3)_2CH_2CH{=}CH_2 \xrightarrow[\text{Hg(OAc)}_2,\ 15\ ^\circ\text{C}]{\text{Br}_2,\ \text{AcOH}} C_3F_7C(CF_3)_2CH_2CHBrCH_2OCOCH_3 + C_3F_7C(CF_3)_2CH_2CH(OCOCH_3)CH_2Br + C_3F_7C(CF_3)_2CH_2CHBrCH_2Br$$

A, B, C

Br_2 addition: A 37-76%, B 7-15%, C 0-56%
Alkene addition, 5 °C: A+B 25%, C 46%

Oxidative Cleavage of the Double Bond

The products formed by oxidative cleavage of an alkene depend on the alkene structure and the agent used. Acid fluorides are formed by ozonization of perfluoroalkenes in trifluoroacetic acid [*36*] (equation 28).

[36] **28**

$$R^1CF{=}CFR^2 \xrightarrow[0\text{-}18\ ^\circ\text{C, 20 min}]{1\text{-}6\%\ O_3,\ CF_3CO_2H} R^1COF + R^2COF$$

$R^1 = F, C_6F_{13}$

$R^2 = CF_3, C_5F_{11}$

conversions 80-100%

The ***trifluorovinyloxy group*** is cleaved by **potassium persulfate** [*37*]. The primarily formed dihydroxy compound undergoes spontaneous hydrolysis with the formation of carboxylic acid (equation 29).

[37] **29**

$$CF_2{=}CFO(CF_2)_3CO_2H \xrightarrow[H_2O,\ 6.5\ h]{K_2S_2O_8} HO_2C(CF_2)_2CO_2H$$

14.2%

Potassium permanganate oxidation of a ***vinyl group*** to a carboxyl group can be used to prepare fluorinated carboxylic acids with aliphatic [*38*] (equation 30) or alicyclic [*39*] (equation 31) skeletons.

[*38*] **30**

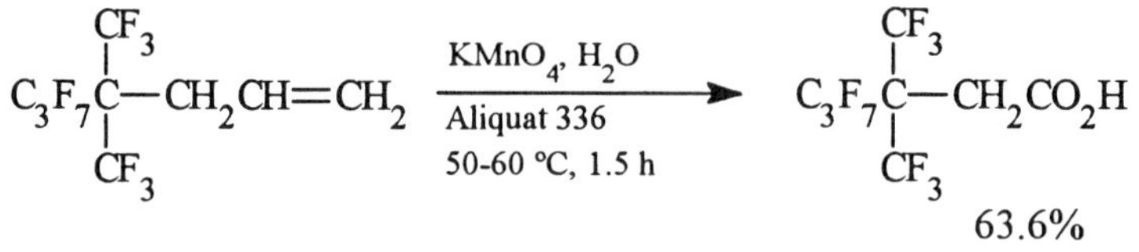

$$C_3F_7C(CF_3)_2{-}CH_2CH{=}CH_2 \xrightarrow[\text{Aliquat 336, 50-60 °C, 1.5 h}]{KMnO_4,\ H_2O} C_3F_7C(CF_3)_2{-}CH_2CO_2H \quad 63.6\%$$

[*39*] **31**

$$\text{1-F-1-X-2-R-2-(CH=CH}_2\text{)cyclopropane} \xrightarrow[\text{RT, 36 h}]{KMnO_4,\ H_2O} \text{1-F-1-X-2-R-cyclopropane-2-}CO_2H$$

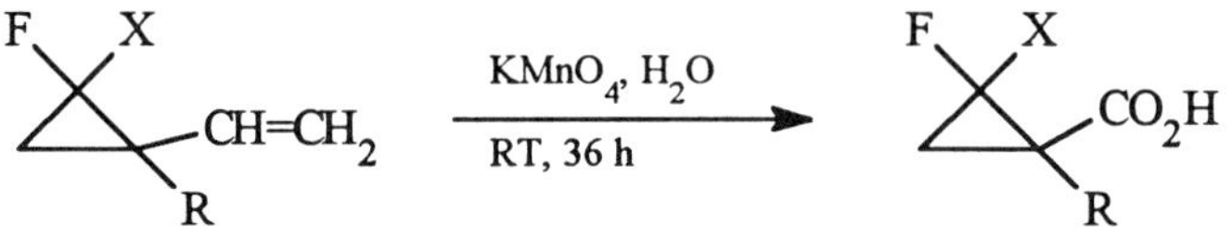

R = H, CH_3 X = H, Br, Cl, F 70-82%

Oxidation of a mixture of ***perfluorononene*** isomers to a mixture of perfluorocarboxylic acids is accomplished with two agents: **potassium permanganate** and ruthenium tetroxide. Oxidation with potassium permanganate is slower and gives lower yields than oxidation with **ruthenium tetroxide** [*40*] (equation 32).

[*40*] **32**

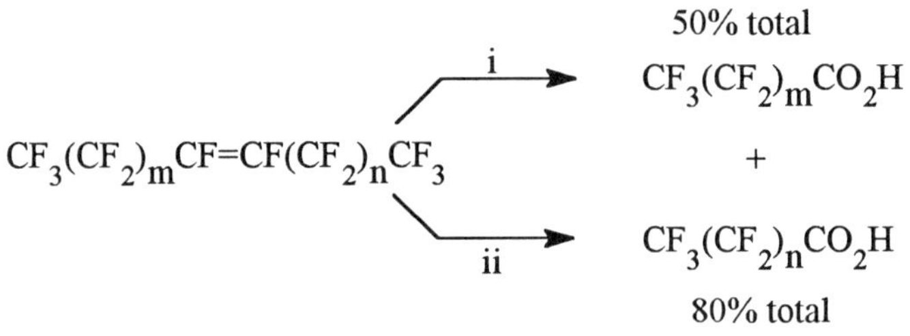

$$CF_3(CF_2)_mCF{=}CF(CF_2)_nCF_3 \xrightarrow{\text{i}} CF_3(CF_2)_mCO_2H + CF_3(CF_2)_nCO_2H$$

i: 50% total; ii: 80% total

i: 1. $KMnO_4$, KOH, H_2O, reflux, 16 h; 2. H_3O^+.

ii: RuO_2, NaOCl, $C_2Cl_3F_3$, RT, 1 h.

A high yield of the resulting perfluorononanoic acid is obtained by the oxidation of (perfluorooctyl)ethylene with a small amount of ruthenium dioxide and an oxidant [*41*] (equation 33).

[*41*] **33**

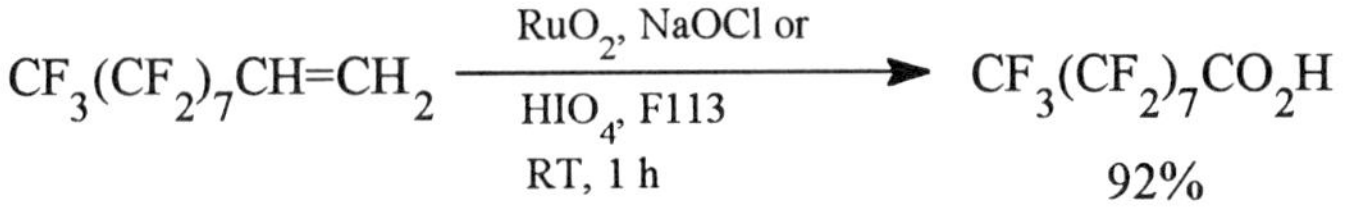

$$CF_3(CF_2)_7CH{=}CH_2 \xrightarrow[\text{HIO}_4\text{, F113, RT, 1 h}]{RuO_2,\ NaOCl\ \text{or}} CF_3(CF_2)_7CO_2H \quad 92\%$$

Oxidation at the Allylic Position

3-(2,2,2-Trifluoro-1-hydroxy-1-methylethyl)cyclohexene is monooxidized at both ***allylic positions*** in the ring with different agents. The major product is the rearranged product C when oxidation is accomplished with **chromium trioxide** in methylene chloride [*42*] (equation 34).

[*42*] **34**

A B C

SeO_2, C_2H_5OH, reflux, 24 h — A 24%

CrO_3, C_5H_5N, CH_2Cl_2, RT, 17 h — B 5%, C 11%

CrO_3, CH_3CO_2H, RT, 15 h — B 13%, C 6%

$(CH_3)_3COOH$, RT, 14 h — B 45%, C 22%

Oxidation of Dienes

Oxidative reactions of ***dienes*** are accomplished under similar conditions as those of alkenes. A bicyclic diene synthesized from hexafluorobenzene and 1,2-dichloroethylene is monoepoxidized by **trifluoroperoxyacetic acid** [*43*] (equation 35).

[*43*] **35**

CF_3CO_2H, CH_2Cl_2

~20%

Nonconjugated perfluorocyclohepta-1,4-diene is oxidized to the corresponding diepoxide by **sodium hypobromite** [*17*] (equation 36), whereas the conjugated-1,3-diene gives a mixture of 1,2-monoepoxide and bridged 2,3:1,4-diepoxide [*17*] (equation 36).

In tetrakis(trifluoromethyl)***allene,*** **potassium permanganate** hydroxylates one double bond [*44*]; the resulting enol tautomerizes to an *α-hydroxyketone* (equation 37).

[*17*] **36**

F (cycloheptadiene) → NaOBr, MeCN, H_2O; 0 °C, 1 h → F (diepoxide) 70%

F (cycloheptadiene) → NaOBr, MeCN, H_2O; 0 °C, 2 h → F (monoepoxide) + F O (bicyclic epoxide)

31% 15%

[*44*] **37**

$$(CF_3)_2C{=}C{=}C(CF_3)_2 \xrightarrow{i} (CF_3)_2C(OH)C(O)CH(CF_3)_2$$

i: $KMnO_4$, MeCN, -15 to 20 °C, 40 min 60%

Oxidation of Aromatic Compounds

Oxidation reactions include electron abstraction, introduction of hydroxyl and oxo groups, and destruction of the carbon skeleton.

Hexafluorobenzene, perfluoromethylbenzene, and octafluoronaphthalene are oxidized to the *corresponding cation salts* by **dioxygenyl hexafluoroarsenate** [45] (equation 38). Salts of monocyclic perfluoroaromatics are unstable above 15 °C, whereas that of octafluoronaphthalene is indefinitely stable at room temperatures.

[*45*] **38**

$$C_6F_6 \xrightarrow[WF_6 \text{ or } SO_2ClF]{O_2^+ AsF_6^-,\ 77\ K} (C_6F_6)^+ AsF_6^-$$

$$C_6F_5CF_3 \rightarrow (C_6F_5CF_3)^+ AsF_6^-$$

$$C_{10}F_8 \rightarrow (C_{10}F_8)^+ AsF_6^-$$

Oxidation of ***hexafluorobenzene*** to *pentafluorophenol* is accomplished by heating with concentrated **hydrogen peroxide** [*46*] (equation 39). Under these conditions, pentafluorobiphenyl is monohydroxylated on the nonfluorinated ring [*46*] (equation 40).

The reaction of ***octafluoronaphthalene*** with 90% **hydrogen peroxide** at 100 °C gives a complex mixture of products that includes naphthoquinone, naphthol derivatives, and benzene derivatives as products of ring degradation [*47*].

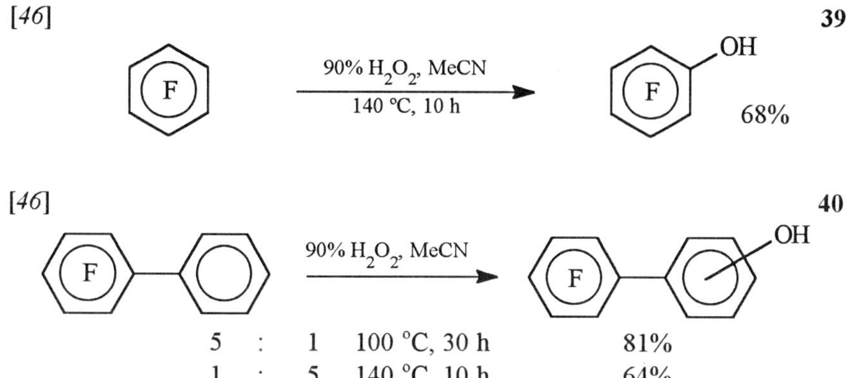

Oxidation of Alcohols

Oxidation of Primary Alcohols

The ***hydroxymethyl group*** in 3-(perfluoro-1,1-dimethylbutyl)propan-1-ol is oxidized to a *carboxylic group* by **potassium permanganate** in the presence of a phase-transfer catalyst [*38*] (equation 41).

[*38*] **41**

$$C_3F_7C(CF_3)_2(CH_2)_2CH_2OH \xrightarrow[\text{Bu}_4\text{NHSO}_4,\ 65\text{-}70\ ^\circ\text{C},\ 24\ \text{h}]{\text{KMnO}_4,\ \text{H}_2\text{O}} C_3F_7C(CF_3)_2(CH_2)_2CO_2H \quad 74\%$$

Anodic oxidation in fluorosulfonic acid converts ***perfluoroalkyl methanols*** into *perfluoroalkyl fluorosulfates* [*4*] (equation 42); ω-hydrododecafluorohexyl methanol can be converted to perfluorohexane-1,6-bisfluorosulfate when a higher ratio of Faraday charge per mole is applied [*4*] (equation 43).

[*4*] **42**

$$C_7F_{15}CH_2OH \xrightarrow[25\ ^\circ\text{C}]{\text{anode, }FSO_3H\text{-}FSO_3K} C_7F_{15}OSO_2F \quad 85\%$$

[*4*] **43**

$$H(CF_2)_6CH_2OH \xrightarrow[FSO_3H\text{-}FSO_3K]{\text{anode, 25 °C}} \begin{cases} \xrightarrow{\text{i}} H(CF_2)_6OSO_2F \quad 40\% \\ \xrightarrow{\text{ii}} FSO_2O(CF_2)_6OSO_2F \quad 82\% \end{cases}$$

i: 5 Faraday per mole
ii: 8.5 Faraday per mole

Oxidation of Secondary Alcohols

Fluorinated ***secondary alcohols*** are easily oxidized to the corresponding *ketones*. This reaction is carried out with a variety of agents.

A hydroxyl group in the neighborhood of fluoroalkyl is oxidized by **sodium dichromate** to yield a ketone [*48*] (equation 44).

[*48*] **44**

$$CF_3CF_2CH(OH)CHBrCH_2Br \xrightarrow[H_2SO_4,\ 90\ ^\circ C,\ 8\ h]{NaCr_2O_7} CF_3CF_2C(=O)CHBrCH_2Br \quad 78.7\%$$

The synthesis of *fluoroalkyl and chloroalkyl fluoromethyl ketones* is achieved by oxidation of the corresponding alcohols by sodium dichromate and sulfuric acid in methylene chloride in the presence of a phase-transfer catalyst [*49*] (equation 45).

[*49*] **45**

$$R(CH_2)_nCH(OH)CH_2F \xrightarrow[CH_2Cl_2,\ Bu_4NBr]{Na_2Cr_2O_7,\ H_2SO_4} R(CH_2)_nC(=O)CH_2F$$

R	n	°C	h	%
C_4F_9	0	50	48	65
C_6F_{13}	0	50	48	65
C_8F_{17}	0	50	24	70
C_4F_9	1	25	10	66
C_6F_{13}	1	25	5	75
CCl_3	0	40	30	67
CCl_3	1	40	27	50
CH_2Cl	0	25	24	55

Oxidation of *trans*-4-(2,2,2-trifluoro-1-hydroxy-1-trifluoromethylethyl)cyclohexanol with **pyridinium chlorochromate** results in the corresponding *cyclic ketone,* whereas oxidation with nitric acid in the presence of a catalyst causes *ring cleavage* [*50*] (equation 46).

Secondary bicyclic alcohols are quantitatively oxidized by **Jones reagent;** however, rearranged products are obtained [*51*] (equation 47).

A variety of secondary alcohols with ***terminal trifluoromethyl group*** are oxidized by the **Dess–Martin "periodinane"** reagent [*52, 53*] (equation 48).

Conversion of 1,6-anhydro-4-*O*-benzyl-2-deoxy-2-fluoro-β-D-glucopyranose to the corresponding oxo derivative is carried out by **ruthenium tetroxide** generated in situ from ruthenium dioxide [*54*] (equation 49).

References are listed on pages 358–363.

[50] 46

$C_6H_5N\text{-}HClCrO_3$, CH_2Cl_2, 20 °C; 60% HNO_3, NH_4VO_3, Cu, 60 °C, 3 h; 70 °C, 1 h

$C(CF_3)_2$ OH — 87.5% ; 58%

HO_2C CO_2H

[51] 47

H_2CrO_4, H_2SO_4, 20 °C, 24 h

~100%

Asymmetric epoxidation of racemic unsaturated fluoro alcohols by the ***chiral Sharpless reagent*** can be exploited for kinetic resolution of enantiomers. The recovered stereoisomer has 14–98% enantiomeric excess [*55*] (equation 50).

Oxidation of ***unsaturated fluoro alcohols*** to *dicarboxylic acids* by **potassium permanganate** is carried out in acetone or in acetic acid [*56*] (equation 51).

The result of oxidation of 8-hydroxy-2,3-tetrafluorobenzobicyclo[3.2.1]octa-2,6-diene depends on the configuration of the hydroxyl group. In the *syn* isomer, the double bond is not epoxidized by the **Jones reagent** [*51*] (equation 52).

Oxidation of Phenols

Fluorinated phenols are usually oxidized to *cyclohexadienone* and *benzoquinone* derivatives and to products of ring rearrangement and ring degradation. Phenols with a trifluoromethyl group at the ring are hydroxylated by **potassium persulfate** [*57*], whereas **chloric acid** causes oxidation to quinone, r*ing contraction,* and incorporation of chlorine in the product [*58*] (equation 53).

48

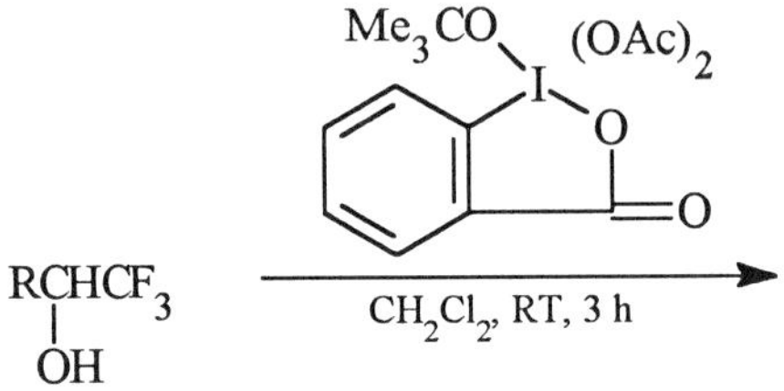

Reagent: Me_3CO–I$(OAc)_2$ (benziodoxolone), CH_2Cl_2, RT, 3 h

$RCH(OH)CF_3 \rightarrow RC(=O)CF_3$

R	%
$C_6H_5CH_2CH_2$	95% [52]
C_9H_{19}	93% [52], [53]
 cyclohexyl	75% [52]
C_6H_5	76% [52], [53]
$C_6H_5C(=CH_2)C(CH_3)_3$	85% [52], [53]
$C_6H_5C{\equiv}C$	90% [52], [53]
$C_8H_{17}S$	0% [53]

[54] 49

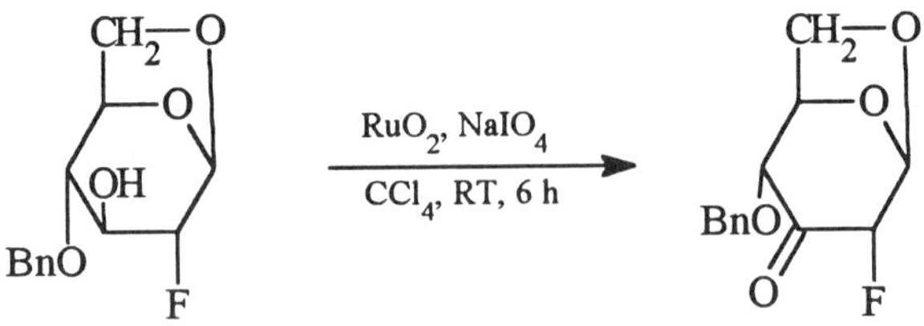

RuO_2, $NaIO_4$, CCl_4, RT, 6 h

85%

[*55*] **50**

$$\text{R-CH=CH-CH(OH)-R}_F \;\; \text{(R)-, (S)-} \xrightarrow[\text{(+)-(L)-[CH(OH)CO}_2\text{-i-Pr]}_2,\ \text{CH}_2\text{Cl}_2,\ -20\ ^\circ\text{C},\ 1\text{-}5\ \text{d}]{\text{Ti(O-i-Pr)}_4,\ t\text{-BuOOH}} \text{R-(epoxide)-CH(OH)-R}_F + \text{R-CH=CH-CH(OH)-R}_F$$

R	R_F	Enantiomeric Yield	(S)-ee
C_6H_{13}	CF_3	60%	46%
C_5H_{11}	CHF_2	97%	31%
C_5H_{11}	CH_2F	98%	43%
C_5H_{11}	$CF_2C_2H_5$	14%	

[*56*] **51**

$$\text{HOCH}_2\text{(CF}_2\text{)}_n\text{CF= CFC}_4\text{H}_9 \xrightarrow{\text{i or ii}} \text{HO}_2\text{C(CF}_2\text{)}_n\text{CO}_2\text{H}$$

i: n = 6 70%; n = 8 60%

ii: n = 4 51%; n = 8 60%

i: $KMnO_4$, Me_2CO, 20-30 °C, 20 h
ii: $KMnO_4$, AcOH, reflux, 2 h

A detailed study of the oxidation of *ortho-*, *meta-*, and *para*-trifluoromethyl phenols showed that oxidation with chloric acid at 5 to 10 °C for 0.5 h gives a complex mixture of products [59] with isolable 2-chloro-1-hydroxy-2-trifluoromethylcyclopent-4-en-3-one-1-carboxylic acid.

The oxidation of ***p-aminophenols*** containing pentafluoroethoxy groups to the corresponding *p-benzoquinones* is carried out with **sodium dichromate** and sulfuric acid [*60*]; the second portion of the oxidizing agent is added after letting the mixture stand overnight [*60*] (equation 54).

Pentafluorophenol is oxidized to different products depending on oxidation agents and reaction conditions. The intermediate is usually pentafluorophenoxy radical, which attacks the aromatic ring to give dimeric and trimeric products. **Electrochemical oxidation** of pentafluorophenol in hydrogen fluoride solvent and in the presence of a strong Lewis base or acid leads to a different ratio of products [*61*] (equation 55).

The structure of the product of the oxidation of pentafluorophenol by hydrogen peroxide is strongly affected by the ratio of substrate to reactant: a higher concentration of **hydrogen peroxide** causes destruction of the aromatic ring [*62*] (equation 56).

[51] **52**

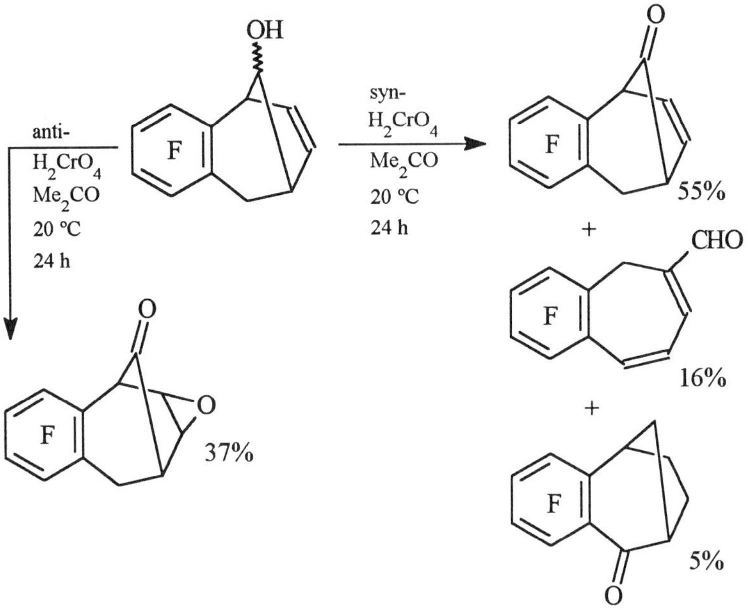

syn-
H_2CrO_4
Me_2CO
20 °C
24 h

55%

16%

5%

anti-
H_2CrO_4
Me_2CO
20 °C
24 h

37%

[58] **53**

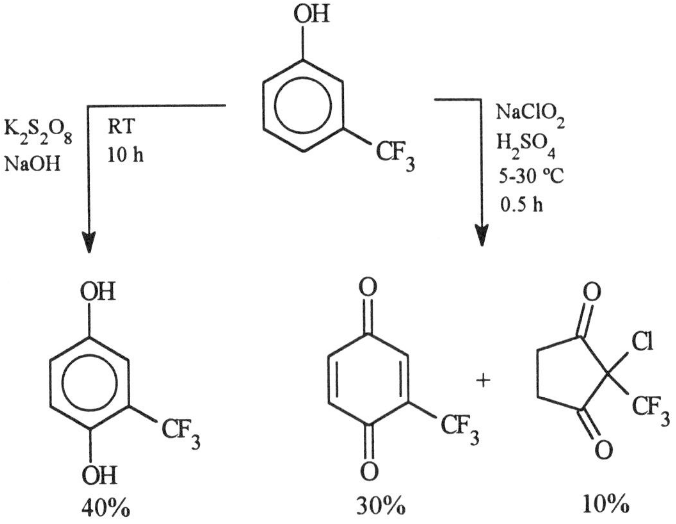

$K_2S_2O_8$
NaOH
RT
10 h

$NaClO_2$
H_2SO_4
5-30 °C
0.5 h

[57] 40% 30% 10%

References are listed on pages 358–363.

[60] **54**

OH, C_2F_5O, OC_2F_5, OC_2F_5, NH_2 —i→ O, C_2F_5O, OC_2F_5, OC_2F_5, O

33.8%

i: $Na_2Cr_2O_7$, H_2SO_4, 0-10 °C, overnight, RT, 6 h

[61] **55**

OH, F —Pt electrode, HF, 0.6-1.55 V→ O, F; O, F, F, O; F, O, F, O

KF	56%	20%	24%
$SbCl_5$	44%	–	56%

Great differences in product structures and distributions are obtained during oxidation with **lead dioxide or tetraacetate** in different solvents and media [*63, 64, 65*]. Oxidation of pentafluorophenol with lead tetraacetate gives perfluoro-2,5-cyclohexadien-1-one in good yield [*63*] (equation 57).

Oxidation of Ethers

Electrochemical oxidation of alkyl aryl ethers results in oxidative dealkylation and coupling of the intermediate radicals. Electrooxidation in the presence of hydrogen fluoride salt leads to fluorinated dienones [*66*] (equation 58).

In the electrochemical oxidation of alkyl tetrahalogenophenyl ethers with hydrogen atoms at para positions, coupled products are obtained [*67*] (equation 59). Under the same conditions, the 2,5-dihydrogen analogue gives no identifiable product [*67*].

[*62*] **56**

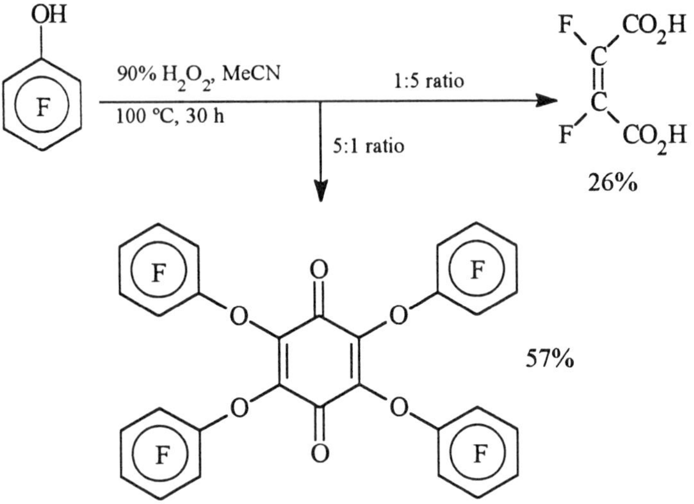

OH — 90% H_2O_2, MeCN; 100 °C, 30 h — 1:5 ratio → F, CO_2H / C=C / F, CO_2H 26%

5:1 ratio → 57%

[*63*] **57**

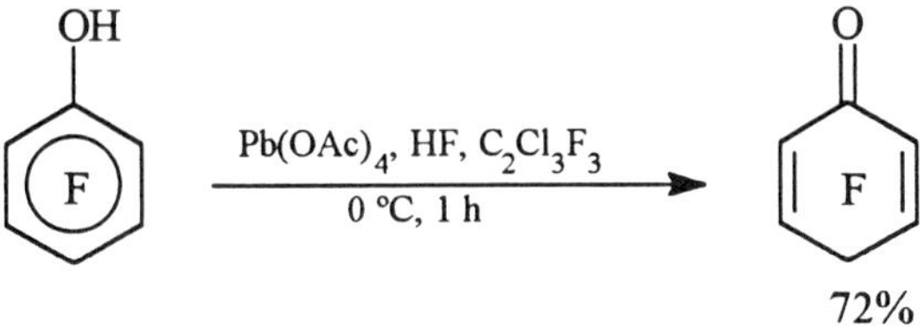

OH — $Pb(OAc)_4$, HF, $C_2Cl_3F_3$; 0 °C, 1 h → O 72%

[*66*] **58**

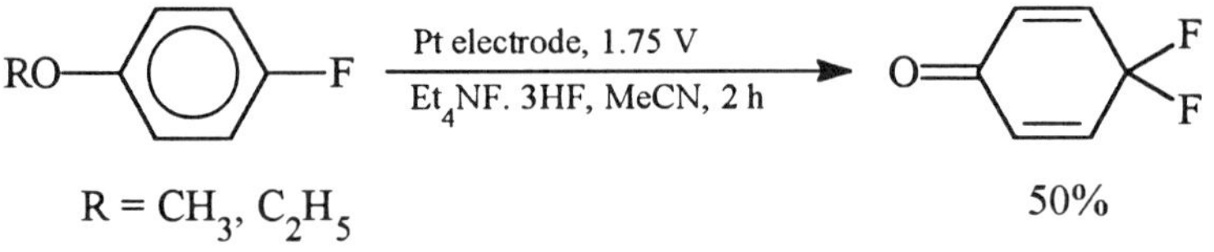

RO—⟨⟩—F — Pt electrode, 1.75 V; $Et_4NF.\ 3HF$, MeCN, 2 h → O=⟨⟩<F F 50%

R = CH_3, C_2H_5

[*67*] **59**

CH_3O—⟨F⟩—H (X, Y) — Pt electrode 2.1-2.4 V → CH_3—⟨F⟩—⟨F⟩—OCH_3 (X, Y, Y, X)

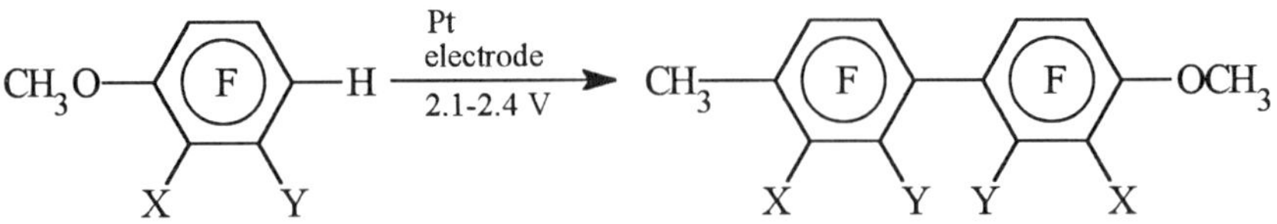

X	Y	% Yield
F	F	65
H	F	56
F	H	26

References are listed on pages 358–363.

The ***furan ring*** in (tetrafluorobenzo)furans is cleaved by **chromium trioxide** with the formation of *ortho-hydroxyketone* derivatives [*68*] (equation 60).

[*68*] **60**

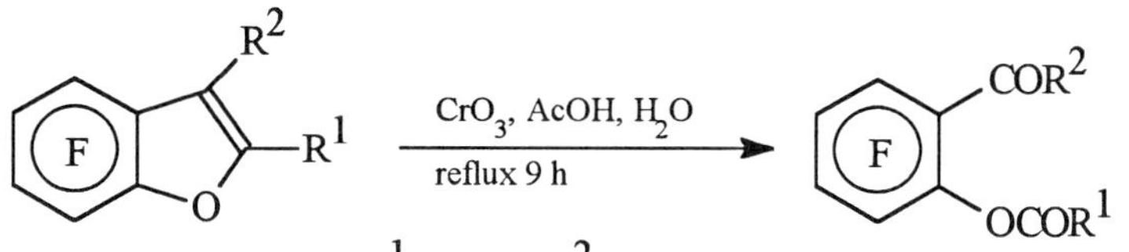

CrO_3, AcOH, H_2O, reflux 9 h

R^1	R^2	% Yield
C_6H_5	C_6H_5	61
C_6H_5	CH_3	48
CH_3	C_6H_5	81

Oxidation of Aldehydes, Ketones, and Ketenes

Aromatic aldehydes and cyclic perfluoroketones are oxidized to *α-hydroxy hydroperoxides* or *bis(α-hydroxy) peroxides,* ***aliphatic ketones*** are converted to *esters,* and ***ketenes*** are converted to *α-lactones.*

Fluorinated benzaldehydes are easily oxidized to peroxy compounds with 30% hydrogen peroxide; oxidation with concentrated **hydrogen peroxide** leads to geminal hydroxy hydroperoxides [*69*] (equation 61).

[*69*] **61**

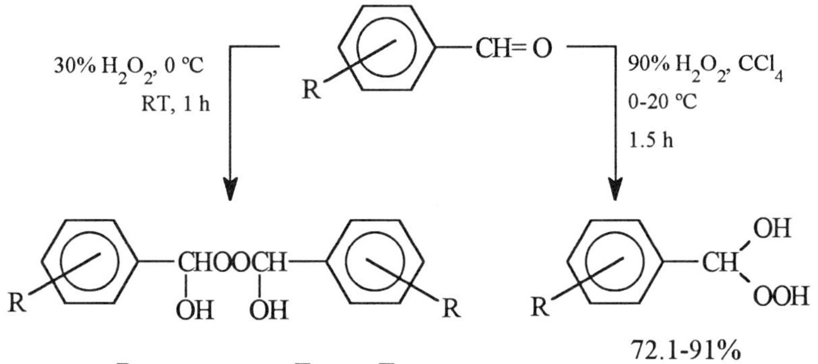

30% H_2O_2, 0 °C, RT, 1 h

90% H_2O_2, CCl_4, 0-20 °C, 1.5 h

R	*m*-F	*p*-F
% Yield	91	100

72.1-91%
R = *m*-F, *p*-F, *o*-CF_3, *m*-CF_3, *p*-CF_3

Pentafluorobenzaldehyde is analogously oxidized to hydroperoxy and peroxy derivatives with yields of 82.6 and 96%, respectively [*69*].

The α-hydroxy hydroperoxides obtained by the above reaction (equation 61) can oxidize highly fluorinated aliphatic aldehydes [*70*] (equation 62).

[*70*] **62**

$$R_FCH{=}O \xrightarrow[Et_2O,\ 20\text{-}40\ ^\circ C,\ 3\ h]{ArCH(OH)OOH} R_FCH(OH)OOH$$

Ar = 4-$CF_3C_6H_4$

R_F	$CF_3(CF_2)_5$,	$H(CF_2)_6$
Yield	78.5%	79.7%

Baeyer–Villiger oxidation by **trifluoroperoxyacetic** acid converts a ***chlorofluoroalkyl ketone*** into an *ester* [*71*] (equation 63).

[*71*] **63**

$$CHClFCH_2COCH(CH_3)_2 \xrightarrow{i} CHClFCH_2COOCH(CH_3)_2\ \ 55\%$$

i: CF_3CO_3H, CH_2Cl_2, 0 °C

Alicyclic perfluoroketones undergo similar oxidation reactions as aromatic fluoroaldehydes (equation 61): a lower concentration of **hydrogen peroxide** oxidizes the ketones to *α-hydroxy hydroperoxides* [72], whereas concentrated hydrogen peroxide converts them to *1,1′-dihydroxydi(perfluorocycloalkyl) peroxides* [*16, 73*] (equation 64).

64

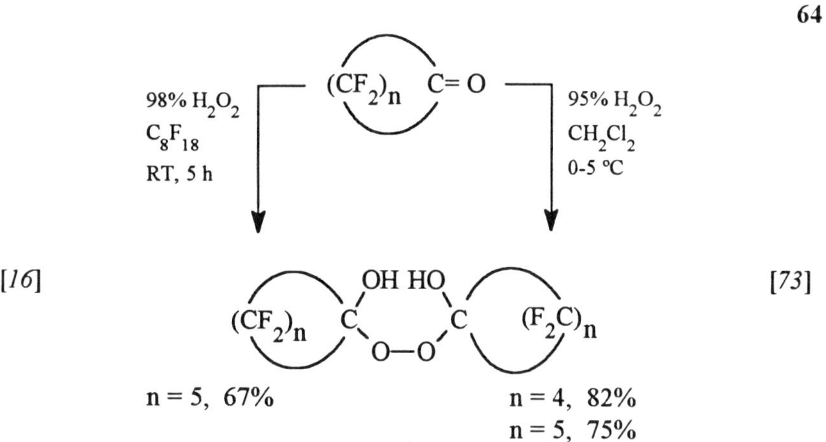

Oxidation of ***perfluoro-1,4-benzoquinone*** by **peroxyacetic acid** gives comparatively good yields of *2,3-difluoromaleic acid* [*74*] (equation 65).

Oxidation of ***pentafluorophenol*** under the conditions described in equation 65 leads to the formation of 2,3-difluoromaleic acid in a yield of 54% [*74*].

Perfluoro(1-ethyl-1-methylpropyl)(1-methylpropyl) ***ketene*** is epoxidized by **sodium hypochlorite** to a stable *α-lactone* [*75*] (equation 66).

[74] **65**

$$\text{perfluoro-}p\text{-benzoquinone} \xrightarrow[75\ ^\circ C,\ 5\ h]{40\%\ MeCO_3H} FC(CO_2H)=C(F)CO_2H \quad 59\%$$

[75] **66**

$$(C_2F_5)(CF_3)FC\text{—}C(=C=O)\text{—}C(CF_3)(C_2F_5)_2 \xrightarrow[20\ ^\circ C]{NaOCl,\ MeCN} (C_2F_5)(CF_3)FC,\ (C_2F_5)_2(CF_3)C\text{—}\overset{O}{C\text{—}C}{=}O \quad 85\%$$

Oxidation of Carboxylic Acids and Their Functional Derivatives

Electrolysis of carboxylic acid salts in solution causes decaboxylative coupling similar to *Kolbe reaction.* Thus, electrolysis of 3,3,3-trifluoro-2-trifluoromethylpropanoic acid in the presence of some of its potassium salt gives the corresponding *fluoroalkane* in a satisfactory yield [76] (equation 67).

[76] **67**

$$(CF_3)_2CHCO_2H \xrightarrow[\text{Pt electrode, 9-30 V, 4 h}]{K\ salt,\ H_2O} CF_3CH(CF_3)\text{—}CH(CF_3)CF_3 \quad 67\%$$

Electrochemical oxidation of a mixture of ***carboxylic acids*** can also give acceptable yields of one product. For example, 1,1,1-trifluoroethane-d_3 and pentafluoropropane-d_3 are prepared in good yields in this way [77] (equation 68).

[77] **68**

$$R_FCO_2H + CD_3CO_2H \xrightarrow[\text{electrode, } N_2,\ 45\ ^\circ C,\ 7\ h]{\text{Pt-foil}} R_FCD_3$$

R_F	CF_3	C_2F_5
Yield %	68	75

References are listed on pages 358–363.

Mixed Kolbe electrosynthesis of two perfluoroacids leads to a mixture of three products, as is generally expected [*78*] (equation 69).

[*78*] **69**

$$RCO_2H + HO_2C(CF_2)_nCO_2CH_3 \xrightarrow{\text{Pt anode, MeCN, MeOH, MeONa, 20 °C}} R(CF_2)_nCO_2CH_3 + RR + CH_3O_2C(CF_2)_{2n}CO_2CH_3$$

24-34% 20-35% 20-35%

$R = CF_3OCF_2CF_2,\ C_2F_5OCF_2,\ C_3F_7OCF(CF_3)$

Aliphatic α, ω-dienes, activated with ester groups at both terminal positions, undergo *trifluoromethylation* with cyclization during **electrochemical oxidation** with sodium trifluoroacetate [*79*] (equation 70).

[*79*] **70**

$$CF_3CO_2Na + (CH_2)_n(CH{=}CHCO_2C_2H_5)_2 \xrightarrow{\text{Pt anode, MeCN, 12 °C, 3 h}}$$

$(H_2C)_n$ ring with CF_3, $CO_2C_2H_5$, $CHCO_2C_2H_5(CF_3)$ + $(H_2C)_n$ ring with CF_3, $CO_2C_2H_5$, F_3C, $CO_2C_2H_5$

n		
n = 0	4%	33%
n = 1	–	37%
n = 2	42%	–
n = 3	26%	–

Electrochemical oxidation of ω-hydrogenperfluoro- and perfluorocarboxylic acids in fluorosulfonic acid gives *fluoroalkyl fluorosulfates* [*3*] (equation 71).

Oxidation of ***5-fluoro-1,3-dimethyluracil*** by *m-chloroperoxybenzoic acid* leads to 4-hydroxy-1,3-dimethylimidazoledione *m*-chlorobenzoate [*80*] (equation 72).

[3] **71**

$$R_FCO_2H \xrightarrow[FSO_3H\text{-}FSO_3F,\ 25\ °C]{\text{anodic oxidation}} R_FOSO_2F$$

R_F	C_3F_7	$H(CF_2)_{10}$
Yield %	85	80

[80] **72**

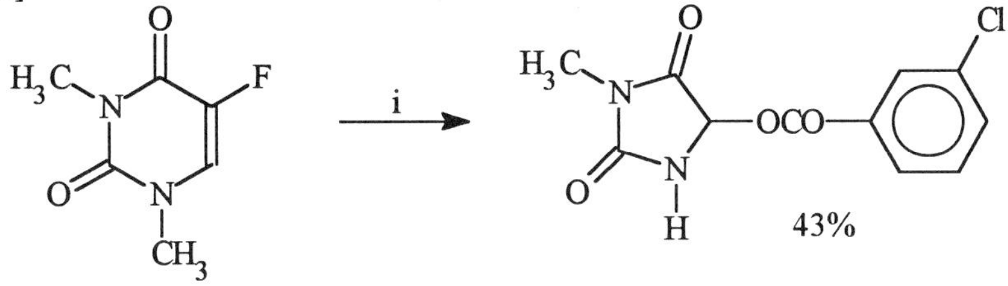

i: m-$ClC_6H_4CO_3H$, CH_2Cl_2, reflux, 72 h

Oxidation of Nitrogen Compounds

Oxidations of nitrogen compounds include oxidation at nitrogen, when it is in a lower oxidation state, or at a carbon atom in the nitrogen compound.

Nitro compounds are *oxidized at carbon* bonded to nitro group: 2-Fluoronitro compounds and silyl nitronates are converted to 2-fluorocarbonyl compounds by **ceric ammonium** salt [*71*] (equation 73).

[*81*] **73**

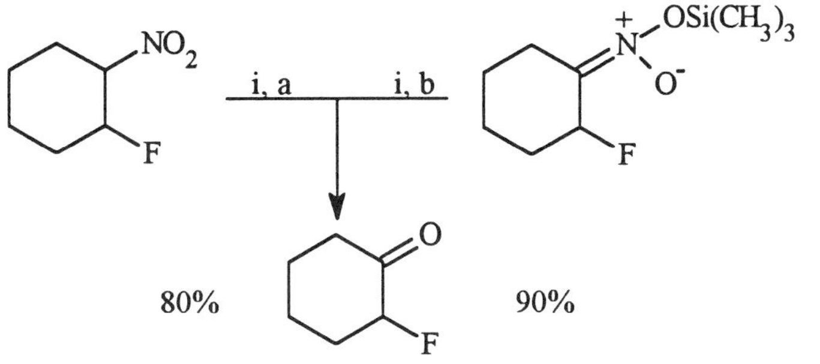

i: $Ce(NH_4)_2(NO_3)_6$, $N(C_2H_5)_3$
a: 50-60 °C, 2.5 h b: 25 °C, 5 min

Nitroso compounds are oxidized to *nitro compounds* by **hydrogen peroxide** and **dinitrogen tetroxide** [*82*] (equation 74).

Various oxidizing agents effectively oxidize *N,N*-bis(trifluoromethyl)hydroxylamine to the corresponding *nitroxyl* [*83*] (equation 75).

N′-[2,4,6-tris(trifluoromethylsulfonyl)phenyl]-*N,N*-diphenylhydrazine is oxidized to the stable *hydrazyl radical* by **lead dioxide** [*84*] (equation 76).

[*82*] **74**

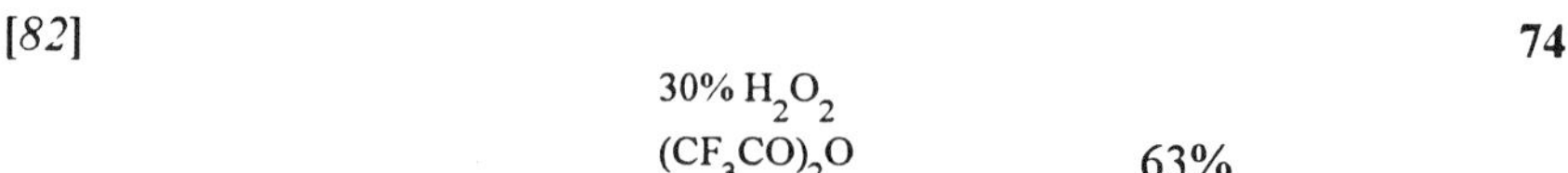

$$ON(CF_2)_nCO_2H \xrightarrow[20\ ^\circ C,\ 7\ d]{30\%\ H_2O_2,\ (CF_3CO)_2O} O_2N(CF_2)_nCO_2H\ \ 63\%$$

$$ON(CF_2)_nCO_2H \xrightarrow[60\ ^\circ C,\ 6\ h]{N_2O_4,\ sealed} O_2N(CF_2)_nCO_2H\ \ 45\%$$

n = 2, 3

[*83*] **75**

$$(CF_3)_2NOH\ (aq) \xrightarrow[RT]{oxidant,\ H_2O} (CF_3)_2NO\ (g)$$

	h	%
$Ce(SO_4)_4$, 10% H_2SO_4	2	90
PbO_2, H_2SO_4	5	90
$K_2S_2O_8$	4	61
NaOCl, KBr	0.5	50
$Ca(OCl)_2$	24	33
Br_2, NaOH, pH 8-9	1	66
Br_2, $Ca(OCl)_2$	0.5	50
Br_2, HgO, H^+	1	35
$KBrO_3$, 2M H_2SO_4	1	98

[*84*] **76**

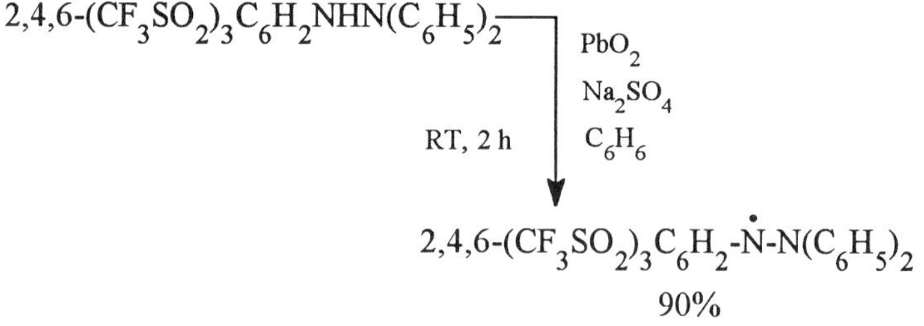

$$2,4,6\text{-}(CF_3SO_2)_3C_6H_2NHN(C_6H_5)_2 \xrightarrow[RT,\ 2\ h]{PbO_2,\ Na_2SO_4,\ C_6H_6} 2,4,6\text{-}(CF_3SO_2)_3C_6H_2\text{-}\dot{N}\text{-}N(C_6H_5)_2$$

90%

2-Fluoro-2,2-dinitroethylamine is oxidized by ***m*-chloroperoxybenzoic acid** to the corresponding *hydroxylamine,* which, on further oxidation by a second equivalent of the **peroxyacid** or **bromine,** gives an *oxime* [*85*] (equation 77).

Oxidation of ***pentafluoroaniline*** by **sodium hypochlorite** under phase-transfer conditions leads to *perfluoroazobenzene* as the main product [*86*] (equation 78).

[*85*] **77**

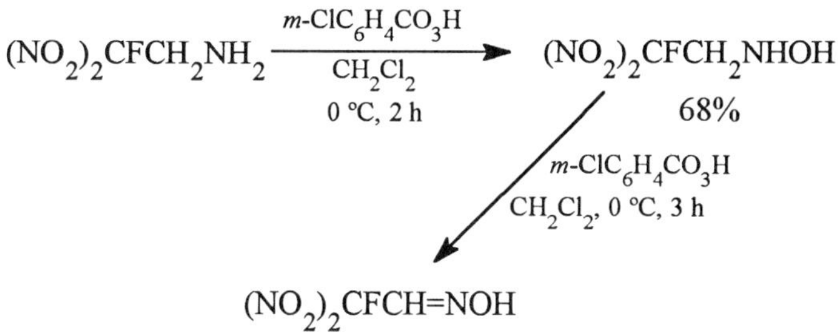

$$(NO_2)_2CFCH_2NH_2 \xrightarrow[\text{CH}_2\text{Cl}_2,\ 0\ ^\circ\text{C, 2 h}]{m\text{-ClC}_6\text{H}_4\text{CO}_3\text{H}} (NO_2)_2CFCH_2NHOH\ (68\%) \xrightarrow[\text{CH}_2\text{Cl}_2,\ 0\ ^\circ\text{C, 3 h}]{m\text{-ClC}_6\text{H}_4\text{CO}_3\text{H}} (NO_2)_2CFCH{=}NOH$$

[*86*] **78**

$$C_6F_5NH_2 \xrightarrow[\text{Bu}_4\text{NHSO}_4,\ 25\ ^\circ\text{C, 3 h}]{\text{NaOCl, CH}_2\text{Cl}_2,\ \text{H}_2\text{O}} C_6F_5N{=}NC_6F_5\ (39\%)$$

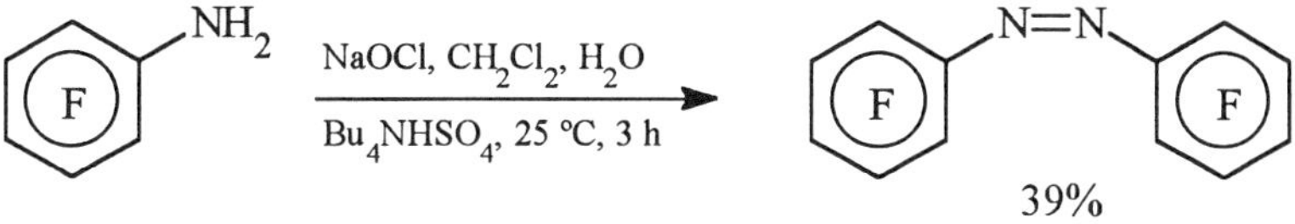

Triaryl-substituted ***imines*** are converted to the corresponding *nitrones* by **peroxyacetic acid** [*87*] (equation 79); substitution of one pentafluorophenyl group by a phenyl group at the carbon atom reduces the yield to ca. 20% [*87*].

[*87*] **79**

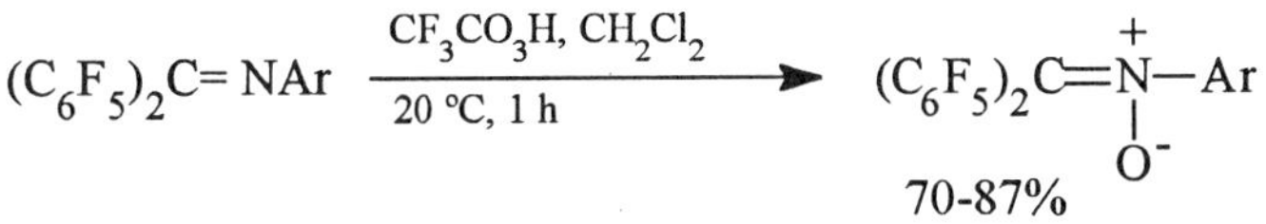

$$(C_6F_5)_2C{=}NAr \xrightarrow[20\ ^\circ\text{C, 1 h}]{\text{CF}_3\text{CO}_3\text{H, CH}_2\text{Cl}_2} (C_6F_5)_2C{=}\overset{+}{N}(O^-){-}Ar\ \ (70\text{-}87\%)$$

$Ar = C_6H_5$, o-$CH_3C_6H_4$, p-$CH_3C_6H_4$, o-FC_6H_4, p-$CH_3OC_6H_4$, C_6F_5

Conjugated ***1,3-diazadienes*** are oxidized by *m-chloroperoxybenzoic acid* to *1-oxa-2,4-diazoles* [*88*] (equation 80).

[*88*] **80**

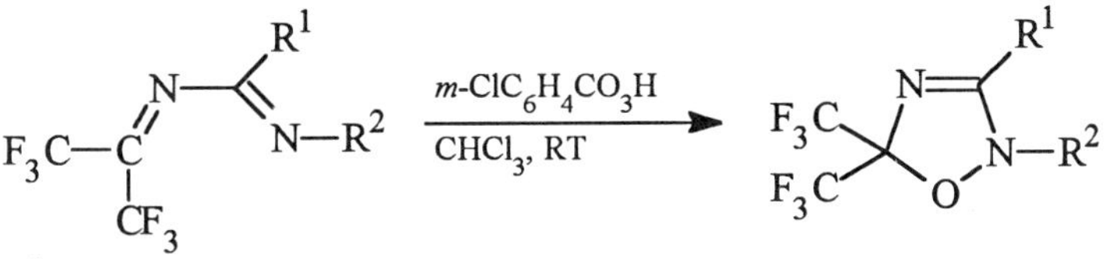

$$(F_3C)_2C{=}N{-}C(R^1){=}N{-}R^2 \xrightarrow[\text{CHCl}_3,\ \text{RT}]{m\text{-ClC}_6\text{H}_4\text{CO}_3\text{H}} \text{5,5-bis(}F_3C\text{)-3-}R^1\text{-2-}R^2\text{-1-oxa-2,4-diazole}\ (35\text{-}56\%)$$

$R^1 = C_6H_5$, 4-$CH_3C_6H_4$, 2-ClC_6H_4
$R^2 = C_8H_9$, C_9H_{11}, $C_{10}H_{13}$

Anodic oxidation of ***N-alkyl-N-fluoroalkylanilines*** in methanol introduces a methoxy group into fluorinated alkyls [*89*] (equation 81).

[*89*] **81**

$$C_6H_5N(R)CH_2CF_3 \xrightarrow[\text{MeOH, KOH}]{\text{anodic cell}} C_6H_5N(R)CH(OCH_3)CF_3 \quad 71\text{-}85\%$$

$R = CH_3, C_2H_5, C_6H_5$

Tertiary *N*-alkyl-*N,N*-bis(2-fluoro-2,2-dinitroethyl)amines containing *N*-methylene or *N*-methyl groups are oxidized by **chromium trioxide** in acetic acid to *N,N*-bis(2-fluoro-2,2-dinitroethyl)formamides [*85*] (equation 82).

[*85*] **82**

$$R_2NCH_3 \xrightarrow{CrO_3,\ AcOH,\ 3\ d} R_2NCHO \quad 97\%$$

$$R_2NCH_2OCH_2NR_2 \xrightarrow{CrO_3,\ AcOH,\ 3\ d} R_2NCHO \quad 95\%$$

$$R_2NCH_2OR \xrightarrow{CrO_3,\ AcOH,\ 20\ h} R_2NCHO \quad 95\%$$

$R = (O_2N)_2CFCH_2$

The ***methyl group*** in 3-methyl-4,5,6,7-tetrafluoroindoles is oxidized to an *aldehydic or a hydroxymethyl group* with high selectivity by **selenium dioxide** [*90*] (equation 83).

Peroxyacid oxidation of bridged 5,6,7,8-tetrafluoro-1,4-dihydronaphthalene-1,4-imines gives aromatic fluorohydrocarbons by elimination of the imine bridge [*91*] (equation 84). Almost the same yields are achieved by oxidation with 30% **hydrogen peroxide** in refluxing methanol [*91*].

Destruction of the aromatic ring is the main reaction in the oxidation of tetrafluoro-*o*-phenylenediamine with **lead tetraacetate:** by-products are tetrafluorobenzotriazole and tetrafluorochinoxaline derivatives [*92*] (equation 85).

Polyfluorinated benzylideneanilines are oxidized by **peroxyacids** to different products depending on reaction contitions: at room temperature the benzylidene carbon is oxidized with the formation of peroxy bonds [*93, 94*] (equation 86), whereas in refluxing agent, the azomethine bond is cleaved [*93*] (equation 86). Pentafluorobenzylideneaniline is oxidized by peroxyacetic acid in dichloromethane at room temperature to perfluorobenzoic acid in a 77% yield [*93*].

Oxidation of more lipophilic ***pentafluorophenyl sulfamimine*** to the corresponding *sulfamyloxaziridine* by ***m*-chloroperoxybenzoic acid** gives better yields in a shorter time compared with the nonfluorinated analogue [*95*] (equation 87).

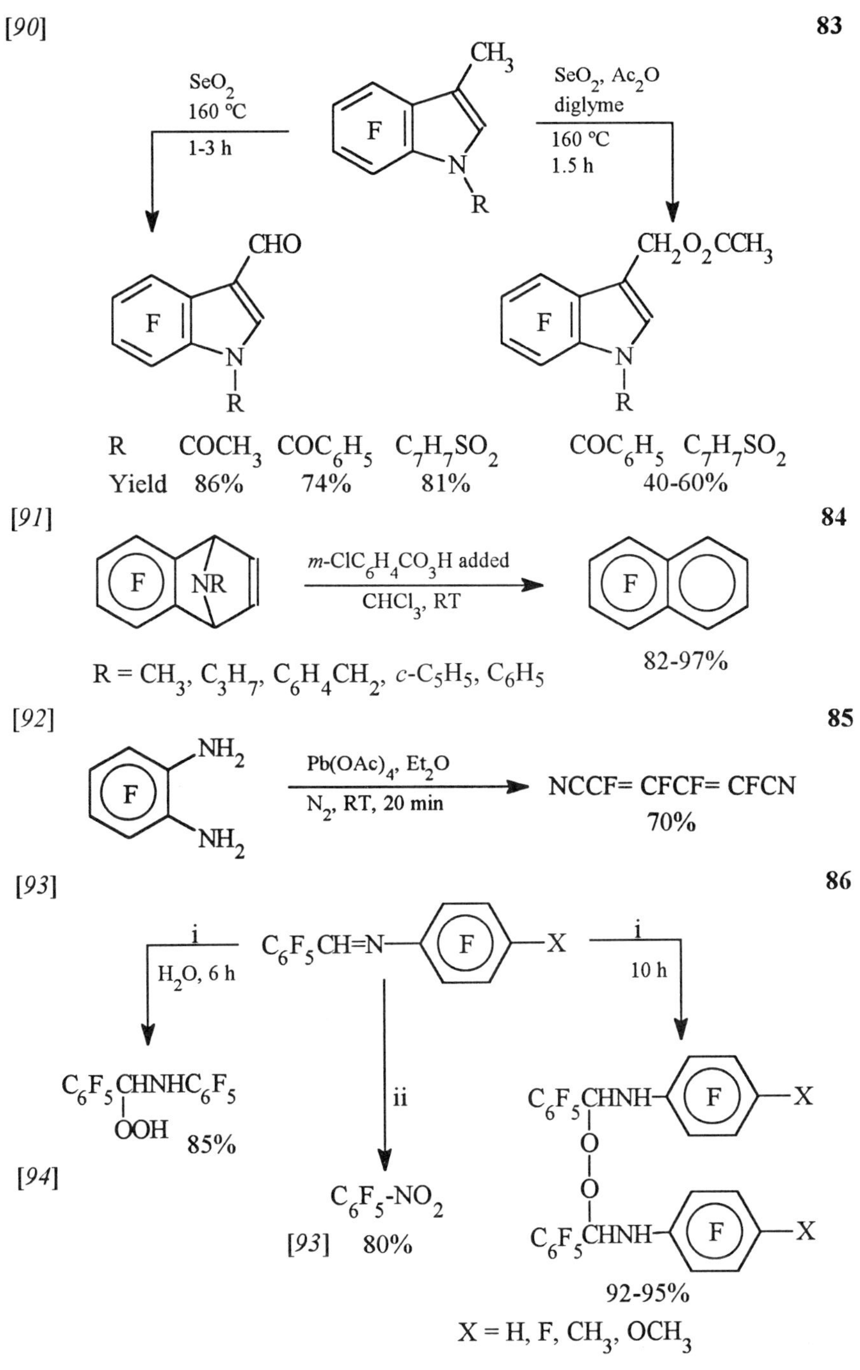
[90]
83
SeO_2
160 °C
1-3 h
CH_3
F
N
R
SeO_2, Ac_2O
diglyme
160 °C
1.5 h
CHO
$CH_2O_2CCH_3$
R COCH$_3$ COC$_6$H$_5$ C$_7$H$_7$SO$_2$ COC$_6$H$_5$ C$_7$H$_7$SO$_2$
Yield 86% 74% 81% 40-60%
[91]
84
NR
m-$ClC_6H_4CO_3H$ added
$CHCl_3$, RT
82-97%
R = CH_3, C_3H_7, $C_6H_4CH_2$, c-C_5H_5, C_6H_5
[92]
85
NH_2
$Pb(OAc)_4$, Et_2O
N_2, RT, 20 min
NCCF= CFCF= CFCN
70%
[93]
86
$C_6F_5CH{=}N$–X
i
H_2O, 6 h
10 h
ii
$C_6F_5CHNHC_6F_5$
OOH 85%
[94]
C_6F_5-NO_2
[93] 80%
C_6F_5CHNH
O
O
92-95%
X = H, F, CH_3, OCH_3
i: CH_3CO_3H, CH_2Cl_2, RT; ii: CF_3CO_3H, reflux, 10 h

[*95*] **87**

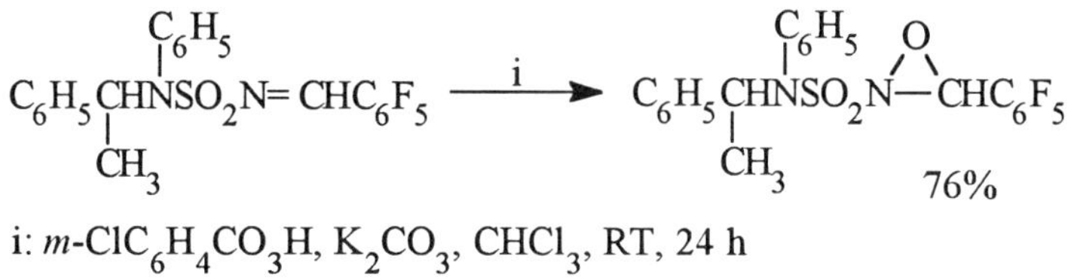

$C_6H_5CH(CH_3)N(C_6H_5)SO_2N{=}CHC_6F_5 \xrightarrow{i} C_6H_5CH(CH_3)N(C_6H_5)SO_2N(O)CHC_6F_5$ 76%

i: *m*-$ClC_6H_4CO_3H$, K_2CO_3, $CHCl_3$, RT, 24 h

1-Perfluoroalkylisoquinolines are obtained by **catalytic dehydrogenation** of 3,4-dihydroisoquinolines [*96*] (equation 88).

[*96*] **88**

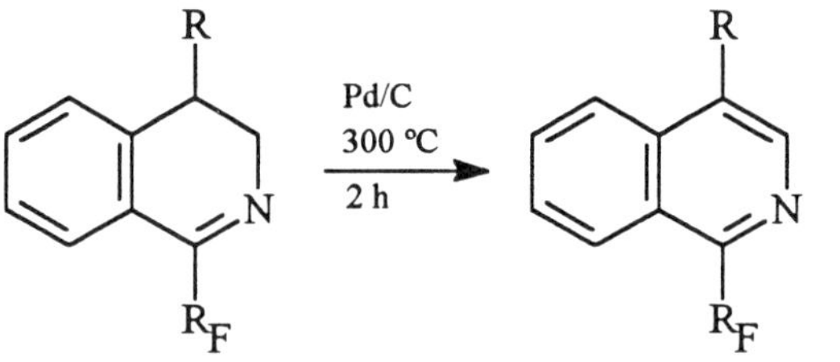

Pd/C, 300 °C, 2 h

R	Yield(%)
H	43-70.5
CH_3	66.6-80

$R_F = CF_3, C_3F_7, C_5F_{11}, C_7F_{15}$

Tetrafluoronitroanilines and 2,5,6-trifluoro-4-nitro-1,3-phenylenediamine react with nitrous acid to give trifluoronitrodiazooxides and 5-fluoro-6-nitro-bis-1,2,3:3,4-diazooxide, respectively [*97*] (equation 89).

[*97*] **89**

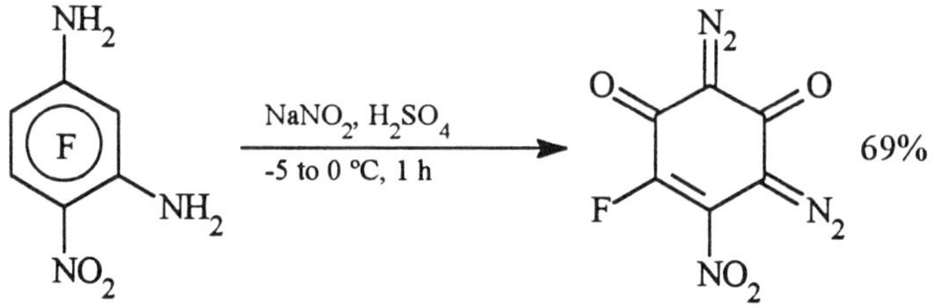

$NaNO_2$, H_2SO_4, -5 to 0 °C, 1 h; 69%

Oxidation of ***perfluoroquinoline*** by **fuming nitric acid** yields pentafluoro-5,8-dioxo-5,8-dihydroquinoline or hexafluoro-2-oxo-1,2-dihydroquinoline in low yields [*98*] (equation 90).

Unlike perfluoroquinoline, the oxidation of ***perfluoroisoquinoline*** gives good yields of dione and trione products [*99*] (equation 91).

[*98*] **90**

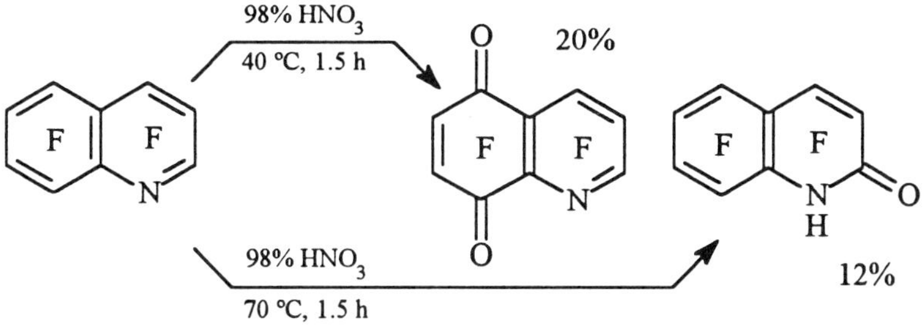

[*99*] **91**

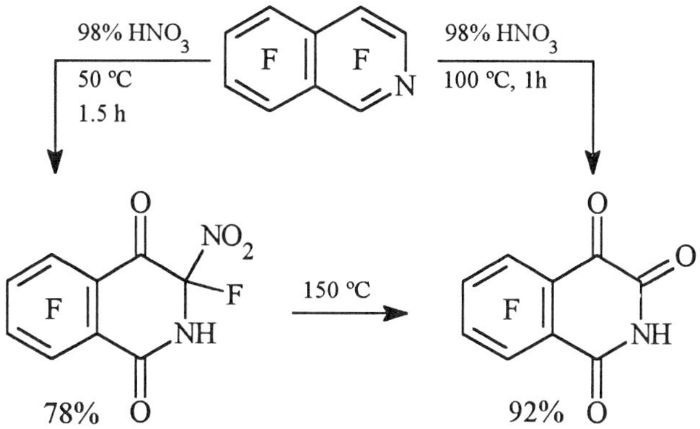

Oxidation of Phosphorus, Sulfur, and Selenium Compounds

Compounds of ***trivalent phosphorus*** are generally oxidized to *pentavalent phosphorus* derivatives: bis(fluoroalkyl) and tris(fluoroalkyl) ***phosphites*** are oxidized to the corresponding *phosphates* [*100, 101*] (equation 92).

[*100*] **92**

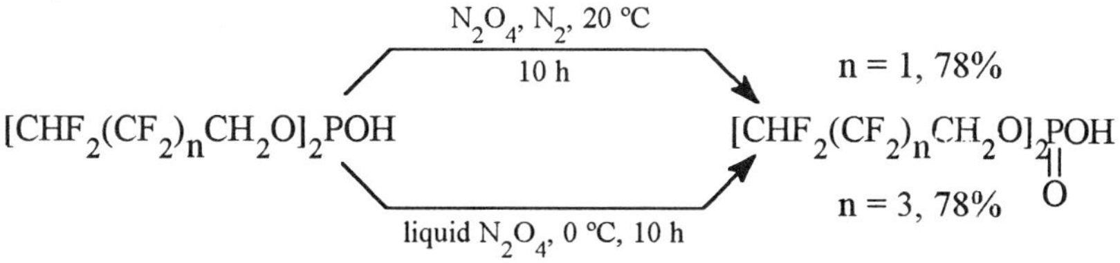

$$[CHF_2(CF_2)_nCH_2O]_2POH \xrightarrow[\text{liquid } N_2O_4,\ 0\ °C,\ 10\ h]{N_2O_4,\ N_2,\ 20\ °C,\ 10\ h} [CHF_2(CF_2)_nCH_2O]_2P(O)OH$$

n = 1, 78%; n = 3, 78%

In the oxidation of tris(2,2,3,3-tetrafluoropropyl) phosphite by oxygen, bromal exhibits a strong catalytic effect, whereas chloral is less effective [*101*] (equation 93).

Gentle oxidation of ***thiols*** leads generally to *disulfides*. When tetrafluoro-1,4-benzenedithiol is added to **dimethyl sulfoxide,** a macrocyclic tetradisulfide is formed [*102*] (equation 94).

p-Thiocresol with a fluorinated 2-hydroxyalkyl group in the *ortho* position is oxidized by **bromine** to a cyclic *sulfenate,* which, by futher oxidation by bromine in the presence of potassium pentafluorophenoxide, gives a bis(aryloxy)sulfurane derivative [*103*] (equation 95).

[*101*] **93**

$(CHF_2CF_2CH_2O)_3P$ —(N_2O_4, O_2, RT; 1-1.5 MPa; sealed)→ $(CHF_2CF_2CH_2O)_3PO$ 88.5%

$(CHF_2CF_2CH_2O)_3P$ —(O_2, 20-60 °C; CX_3CHO, 2-3 h)→ $(CHF_2CF_2CH_2O)_3PO$ X = Cl, 60%; X = Br, 96%

[*102*] **94**

HS–(F)–SH —(DMSO, 25 °C)→ cyclic tetramer with S–S linkages, 95%

[*103*] **95**

SH, OH, H_3C, F_3C, CF_3 —(Br_2, C_5H_5N, CCl_4, -5 °C)→ S, O 85%, CH_3, F_3C, CF_3

—(Br_2, C_6F_5OK)→ C_6F_5O, OC_6F_5, S, O, H_3C, F_3C, CF_3 43%

Conversion of ***trifluoromethyl sulfides*** to the corresponding *sulfoxides* is accomplished by **hydrogen peroxide** [*104*] (equation 96).

N,N*-Dichlorotrifluoromethanesulfonamide** converts ***aryl(trifluoromethyl) sulfides to *sulfimides* [*105*] (equation 97).

[*104*] **96**

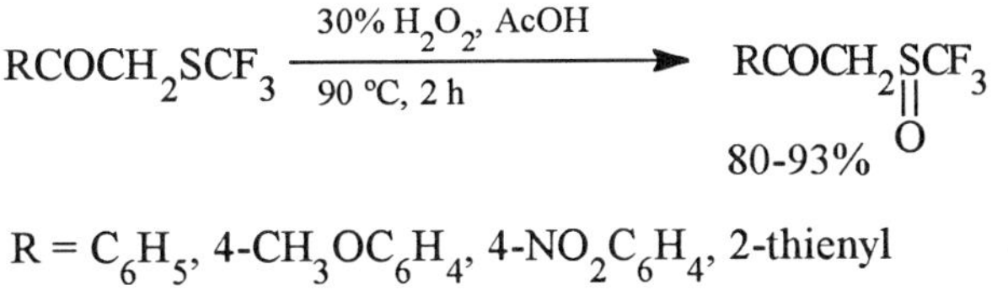

$$RCOCH_2SCF_3 \xrightarrow[90\ °C,\ 2\ h]{30\%\ H_2O_2,\ AcOH} RCOCH_2S(O)CF_3$$

80-93%

R = C_6H_5, 4-$CH_3OC_6H_4$, 4-$NO_2C_6H_4$, 2-thienyl

[*105*] **97**

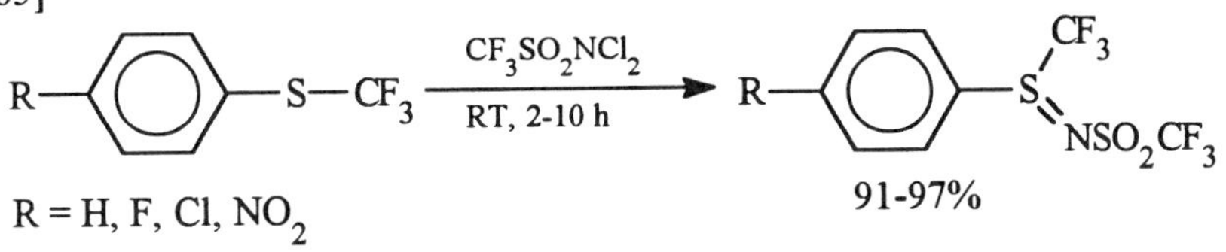

$$R{-}C_6H_4{-}S{-}CF_3 \xrightarrow[RT,\ 2\text{-}10\ h]{CF_3SO_2NCl_2} R{-}C_6H_4{-}S(CF_3){=}NSO_2CF_3$$

R = H, F, Cl, NO_2 — 91-97%

The conversion of ***cyclic sulfides*** to *sulfones* is accomplished by more energetic oxidations. Perhalogenated thiolanes [*106*] and 1,3-dithietanes [*107*] are oxidized to sulfones and disulfones, respectively, by a mixture of **chromium trioxide and nitric acid** (equation 98). The same reagent converts 2,4-dichloro-2,4-bis(trifluoromethyl)-1,3-***dithietanes*** to disulfone derivatives [*107*], whereas **trifluoromethaneperoxysulfonic acid** converts the starting compound to a *sulfone–sulfoxide* derivative [*108*] (equation 99).

[*106*] **98**

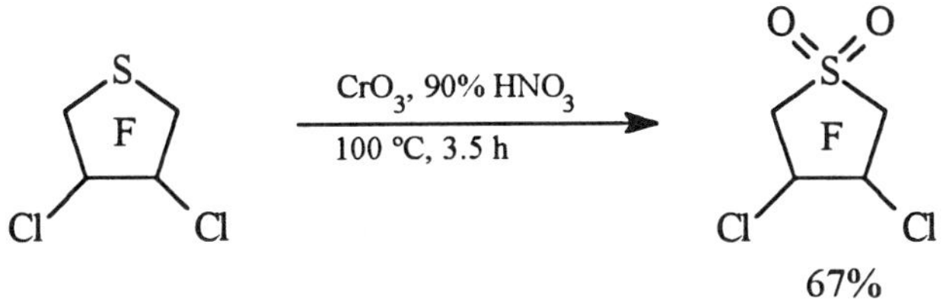

3,4-Dichloro(F)thiolane $\xrightarrow[100\ °C,\ 3.5\ h]{CrO_3,\ 90\%\ HNO_3}$ 3,4-dichloro(F)thiolane 1,1-dioxide, 67%

[*107*] [*108*] **99**

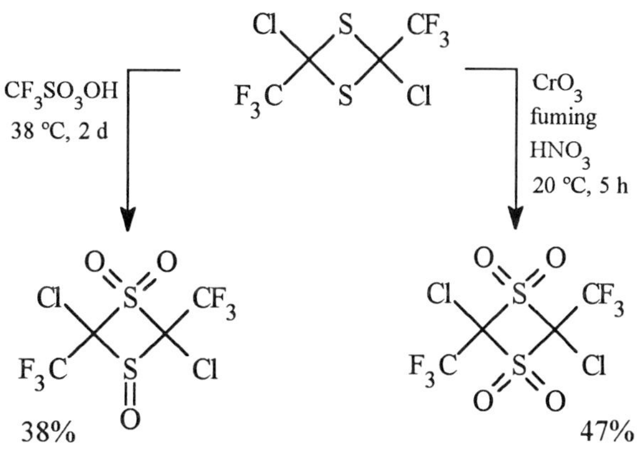

2,4-Dichloro-2,4-bis(trifluoromethyl)-1,3-dithietane:

CF_3SO_3OH, 38 °C, 2 d → sulfone–sulfoxide, 38%

CrO_3, fuming HNO_3, 20 °C, 5 h → disulfone, 47%

Fluoromethyl sulfides are converted to *fluoromethanesulfonyl chlorides* by reaction with **chlorine in waters; at low temperatures, intermediate sulfoxides can be isolated** [*109*] (equation 100).

[*109*] **100**

$$\underset{X=F}{\overset{X=H}{CHFXSCH_2-C_6H_5}} \xrightarrow[-10\text{ to }0\ ^\circ C]{Cl_2,\ H_2O} \underset{X=F,\ 70\%}{\overset{X=H,\ 65\%}{CHFXSO_2Cl}}$$

$$CHFXSCH_2C_6H_5\ (X=F) \xrightarrow[\text{cooling}]{Cl_2,\ H_2O,\ \text{dry-ice}} CHF_2S(=O)CH_2C_6H_5 \xrightarrow{0\ ^\circ C} CHF_2SO_2Cl$$

Trifluoromethanesulfinyl chloride is prepared by oxidation of ***trifluoromethanesulfenyl chloride*** with *m*-chloroperoxybenzoic acid in high yield [*110*] (equation 101).

[*110*] **101**

$$CF_3SCl \xrightarrow[-20\text{ to }25\ ^\circ C,\ 12\text{ h}]{m\text{-}ClC_6H_4CO_3H} CF_3S(=O)Cl \quad 95\%$$

Sulfenamides are usually oxidized to *sulfonamides:* tris(trifluoromethanesulfenyl)- and bis(trifluoromethanesulfenyl)amine are converted to the corresponding sulfonamides by **sodium hypochlorite** at 20 °C for 3 h in 61 and 92% yield, respectively [*111*]. Oxidation of pentafluorobenzenesulfenamide by **manganese dioxide** yields a sulfinamide intermediate that can be trapped [*112*] (equation 102).

[*112*] **102**

$$C_6F_5SNH_2 \xrightarrow[70\ ^\circ C,\ 16\text{ h}]{MnO_2,\ C_6H_6} C_6F_5SO_2NH_2 \quad 83\%$$

$$C_6F_5SNH_2 \xrightarrow[20\ ^\circ C,\ 40\text{ h}]{MnO_2,\ C_6H_6} \underset{70.5\%}{C_6F_5SONH_2} \xrightarrow[C_6H_6,\ 70\ ^\circ C,\ 16\text{ h}]{MnO_2} C_6F_5SO_2NH_2 \quad 96\%$$

Perfluoro(tetramethylene) sulfilimine is easily oxidized to a *sulfoximine* by ***m*-chloroperoxybenzoic acid** [*113*] (equation 103).

References are listed on pages 358–363.

[*113*] **103**

$$\text{(F-cyclic)}\,S{=}NH \xrightarrow{m\text{-}ClC_6H_4CO_3H} \text{(F-cyclic)}\,S({=}O)({=}NH) \quad 98\%$$

Both a carbon–sulfur single bond and a carbon–carbon double bond are oxidized in ***trifluorovinylsulfurpentafluoride*** by **ozone or air** under pressure [*114*] (equation 104).

[*114*] **104**

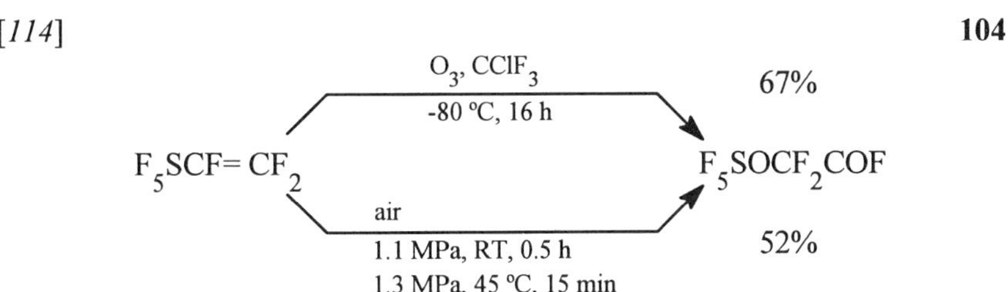

Fluoroalkyl selenides are oxidized to higher oxidation states by ***tert*-butyl hypochlorite** under the mild conditions [*115*] (equation 105).

[*115*] **105**

$$R^1SeR^2 \xrightarrow[10\text{-}15\ ^\circ C,\ 22\text{-}24\ h]{t\text{-}BuOCl,\ CCl_4} R^1Se(Cl)(OC(CH_3)_3)R^2$$

R^1	CF_3	C_3F_7	CF_3
R^2	C_6H_5	C_6H_5	$4\text{-}CH_3C_6H_4$
% Yield	98	90	99

Oxidation of Iodine and Boron Compounds

Fluoroalkyl iodides are oxidized at the iodine atom by **trifluoroperoxyacetic acid** [*119, 120*] or by a mixture of trifluoroacetic anhydride and hydrogen peroxide [*118*]; the products are *bis(trifluoroacetoxy)iodo fluoroalkanes* (equation 106).

Fluoroalkane α,ω-diiodides are oxidized at both iodine atoms under the same conditions, and the corresponding α,ω-bis[di(trifluoroacetoxy)iodo]alkanes are formed in almost quantitative yields [*116, 117, 119*].

β,β-*Difluoroalkenyl boranes,* which are prepared in situ, are converted to *difluoromethyl ketones* by oxidation with **hydrogen peroxide** in alkaline media [*120*] (equation 107).

References are listed on pages 358–363.

106

$$R_FI \longrightarrow R_FI(O_2CCF_3)_2$$

[*116*]	CF_3CO_3H, 0-RT, 24 h	94 -98%
	$R_F = H(CF_2CF_2)_nCH_2$ n = 1-3	
[*117*]	CF_3CO_3H, 0-RT, 24 h	90-95%
	$R = C_2F_5 - C_{10}F_{21}$, $H(CF_2)_{10}$, BrC_2F_4	
[*118*]	$(CF_3CO)_2O$, H_2O, 0 °C, 12 h, RT, 24 h	89%
	$R = CF_3CH_2$	

[*120*] **107**

$$CF_3CH_2OSO_2C_6H_4\text{-}4\text{-}CH_3 \xrightarrow{i} CF_2{=}C(R)BR_2 \xrightarrow{ii} CHF_2C(=O)R$$

R	% Yield
$C_6H_5(CH_2)_4$	81
$C_6H_5CH(CH_3)CH_2$	77
$c\text{-}C_8H_{15}$	63
(bicyclic structure)	67

i: 1. $[(CH_3)_2CH]_2NLi$; 2. BR_3, -78 °C, 30 min, RT, 12 h

ii: 1. CH_3ONa; 2. 30 % H_2O_2, NaOH, 0 °C, 1 h, RT, 1 h

References for Pages 321–358

1. Jones, D. F., Howe, R. *J. Chem. Soc.* **1968**, 2816.
2. Plashkin, V. S.; Zapevalova, T. B.; Zapevalov, A. Y.; Selishchev, B.N. *Zh. Org. Khim.* 1980, *16,* 540; *Chem. Abstr.* **1980,** *93,* 70945.
3. Brunel, D.; Germain, A.; Moreau, P.; Burdon, J.; Coe, P. L.; Plevey, R. G. *J. Chem. Soc., Perkin Trans. 1* **1989,** 2283.
4. Germain, A.; Brunel, D.; Moreau, P. *Bull. Soc. Chim. Fr.* **1986**, 895.
5. Vilenchik, Y. M.; Mitrofanova, L. N.; Senichev, Y. N. *Zh. Org. Khim.* **1978,** *14,* 1587; *Chem. Abstr.* **1978,** *89*, 197230.
6. Kartsov, S. V.; Sokolov, L. F.; Sass, V. P.; Sokolov, S. V. *Zh. Vses. Khim. O-va.* **1979,** *24,* 90; *Chem. Abstr.* **1979,** *90*, 151892.

7. Kadyrov, A. A.; Rokhlin, E. M.; Knunyants, I. L. *Izv. Akad. Nauk SSSR* **1982,** 2344; *Chem. Abstr.* **1983,** *98,* 125776.
8. Kolenko, I. P.; Filyakova, T. I.; Zapevalov, A. Y.; Mochalina, E. P.; German, L. S.; Polishchuk, V. R. *Izv. Akad. Nauk SSSR* **1979,** 667; *Chem. Abstr.* **1979,** *91,* 38866.
9. Kolenko, I. P.; Filyakova T. I.; Zapevalov, A. Y. *Izv. Akad. Nauk SSSR* **1979,** 2509; *Chem. Abstr.* **1980,** *92,* 94148.
10. Zapevalov, A. Y.; Filyakova, T. A.; Kolenko, I. P. *Izv. Akad. Nauk SSSR* **1979,** 2812; *Chem. Abstr.* **1980,** *92,* 128624.
11. Filyakova, T. I.; Zapevalov, A. Y.; Peschanskii, N. V.; Kodess, M. I.; Kolenko, I. P. *Izv. Akad. Nauk SSSR* **1981,** 2612; *Chem. Abstr.* **1982,** *96,* 122142.
12. Filyakova, T. I.; Peschanskii, N. V.; Kodess, M. I.; Zapevalov, A. Y.; Kolenko, I. P. *Zh. Org. Khim.* **1988,** *24,* 371; *Chem. Abstr.* **1988,** *109,* 230652.
13. Coe, P. L.; Sellars, A.; Tatlow, J. C. *J. Fluorine Chem.* **1983,** *23,* 103.
14. Kvíčala, J.; Paleta, O. *J. Fluorine Chem.* **1991,** *54,* 69.
15. Zapevalov, A. Y.; Filyakova, T. I.; Peschanskii, N. V.; Kolenko, I. P.; Kodess, M. I. *Zh. Org. Khim.* **1986,** *22,* 2088; *Chem. Abstr.* **1987,** *107,* 236099.
16. Aleksandrov, A. V.; Kosnikov, A. Y.; Antonovskii, V. L.; Lindeman, S. V.; Struchkov, Y. T.; Gushchin, V. V.; Starostin, E. K.; Nikishin, G. I. *Izv. Akad. Nauk SSSR* **1989,** 918; *Chem. Abstr.* **1989,** *111,* 232134.
17. Coe, P. L.; Mott, A. W.; Tatlow, J. C. *J. Fluorine Chem.* **1985,** *30,* 297.
18. Coe, P. L.; Mott, A. W.; Tatlow, J. C. *J. Fluorine Chem.* **1990,** *49,* 21.
19. Makarov, K. N.; Pletnev, S. I.; Gervits, L. L.; Natarov, V. P.; Prokudin, I. P.; Knunyants, I. L. *Izv. Akad. Nauk SSSR* **1986,** 1313; *Chem. Abstr.* **1987,** *107,* 115268.
20. Dmowski, W.; Plenkiewicz, H.; Porwisiak, J. *J. Fluorine Chem.* **1988,** *41,* 191.
21. Chang, H. -L.; Cimarusti, C. M.; Diassi, P. A.; Grabowich, P. *J. Org. Chem.* **1977,** *42,* 358.
22. Hung, M. H.; Smart, B. E.; Feiring, A. E.; Rozen, S. *J. Org. Chem.* **1991,** *56,* 3187.
23. Mazur, U.; Lattimer, R. P.; Lopata, A.; Kuczkowski, R. L. *J. Org. Chem.* **1979,** *44,* 3181.
24. Lattimer, R. P.; Mazur, U.; Kuczkowski, R. L. *J. Am. Chem. Soc.* **1976,** *98,* 4012.
25. Gilles, C. W. *J. Am. Chem. Soc.* **1975,** *97,* 1276.
26. Caminade, A. M.; Le Blanc, M.; Khatib, F.; Koenig, M. *Tetrahedron Lett.* **1985,** *26,* 2889.
27. Kobayashi, T.; Kumadaki, I.; Oshawa, A.; Hanzawa, Y.; Honda, M.; Iitaka, Y. *Tetrahedron Lett.* **1975,** 3001.

28. Wilshire, J. F. K.; Pattison F. L. M. *J. Am. Chem. Soc.* **1956,** *78,* 4996.
29. Husain, S. Z.; Plevey, R. G.; Tatlow, J. C. *Bull. Soc. Chim. Fr.* **1986,** 891.
30. Postovoi, S. A.; Zeifman, Y. V. *Izv. Akad. Nauk SSSR* **1988,** 892; *Chem. Abstr.* **1989,** *110*, 23296.
31. Zeifman, Y. V.; Postovoi, S. A.; Delyagina, N. I. *Izv. Akad. Nauk SSSR* **1989,** 738; *Chem. Abstr.* **1989,** *111*, 194224.
32. Utebaev, U.; Abduganiev, E. G.; Rokhlin, E. M.; Knunyants, I. L. *Izv. Akad. Nauk SSSR* **1974,** 387; *Chem. Abstr.* **1974,** *81*, 24978.
33. Dědek, V.; Liška, F.; Fikar, J.; Chvátal, Z.; Pohořelsky, L. *Collect. Czech. Chem. Commun.* **1975,** *40,* 1008.
34. Coudures, C.; Pastor, R.; Cambon, A. *J. Fluorine Chem.* **1984,** *24,* 93.
35. *See* reference 20.
36. Moiseeva, N. I.; Gekhman, A. E.; Rumyantsev, E. S.; Klimanov, V. I.; Moiseev, I. I. *Izv. Akad. Nauk SSSR* **1989,** 1706; *Chem. Abstr.* **1990,** *112*, 76130.
37. Chekmarev, P. M.; Makeeva, N. M.; Maksimov, V. L.; Popova, L. A.; Dreiman, N. A. *Zh. Org. Khim.* **1989,** *25,* 2080; *Chem. Abstr.* **1990,** *112*, 178038.
38. Dmowski, W.; Plenkiewicz, H.; Piasecka-Maciejewska, K.; Prescher, D.; Schulze, J.; Endler, I. *J. Fluorine Chem.* **1990,** *48,* 77.
39. Gassen, K. R.; Baasner, B. *J. Fluorine Chem.* **1990,** *49,* 127.
40. Battais, A.; Boutevin, B.; Pitrasanta, Y.; Sierra, P. *J. Fluorine Chem.* **1981,** *19,* 35.
41. Guizard, C.; Cheradame, H.; Brunel, Y.; Beguin, C. G. *J. Fluorine Chem.* **1979,** *13,* 175.
42. Nagal, T.; Morita, M.; Koyama, M.; Ando, A.; Miki, T.; Kumadaki, I. *Chem. Pharm. Bull.* **1989,** *37,* 1751.
43. Takenaka, N. E.; Hamlin, R.; Lemal, D. M. *J. Am. Chem. Soc.* **1990,** *112,* 6715.
44. *See* reference 30.
45. Richardson, T. J.; Tanzella, F. L.; Bartlett, N. *J. Am. Chem. Soc.* **1986,** *108,* 4937.
46. Bogachev, A. A.; Kobrina, L. S.; Yakobson, G. G. *Zh. Org. Khim.* **1986,** *22,* 2578; *Chem. Abstr.* **1987,** *107*, 197634.
47. Bogachev, A. A.; Kobrina, L. S.; Yakobson, G. G. *Zh. Org. Khim.* **1986,** *22,* 2571; *Chem. Abstr.* **1987,** *107*, 217214.
48. Werner, K., von; Gisser, A. *J. Fluorine Chem.* **1977,** *10,* 387.
49. Chaabouni, M. M.; Baklouti, A. *J. Fluorine Chem.* **1990,** *47,* 227.
50. Zalesskaya, I. M.; Fialkov, Y. A.; Parakhnenko, A. I.; Yagupol,skii, L. M. *Zh. Org. Khim.* **1987,** *23,* 2140; *Chem. Abstr.* **1988,** *109*, 129315.
51. Slyńko, N. M.; Mironova, M. K.; Barkhash, V. A. *Zh. Org. Khim.* **1976,** *12,* 1922; *Chem. Abstr.* **1977,** *86,* 43442.
52. Linderman, R. J.; Graves, D. M. *J. Org. Chem.* **1989,** *54,* 661.

53. Linderman, R. J.; Graves, D. M. *Tetrahedron Lett.* **1987,** *28,* 4259.
54. Pacák, J.; Braunová, M.; Stropová, D.; Černy, M. *Collect. Czech. Chem. Commun.* **1977,** *42,* 120.
55. Hanzawa, Y.; Kawagoe, K.; Ito, M.; Kobayashi, Y. *Chem. Pharm. Bull.* **1987,** *35,* 1633.
56. Nguyen, T.; Rubinstein, M.; Wakselman, C. *Synth. Commun.* **1983,** *13,* 81.
57. Feiring, A. E.; Sheppard, W. A. *J. Org. Chem.* **1975,** *40,* 2543.
58. Blazejewski, J. C.; Dorme, R.; Wakselman, C. *Synthesis* **1985,** 1120.
59. Blazejewski, J. C.; Dorme, R.; Wakselman, C. *J. Chem. Soc., Perkin Trans. 1* **1987,** 1861.
60. Belous, V. M.; Litvinova, K. D.; Alekseeva, L. A.; Yagupol'skii, L. M. *Zh. Org. Khim.* **1976,** *12,* 1798; *Chem. Abstr.* **1976,** *85,* 159593.
61. Devynck, J.; Hadid, A. B.; Virelizier, H. *J. Fluorine Chem.* **1979,** *14,* 363.
62. *See* reference 46.
63. Kovtonyuk, V. N.; Kobrina, L. S.; Yakobson, G. G. *J. Fluorine Chem.* **1985,** *28,* 89.
64. Denivelle, L.; Huynh, A. H. *Bull. Soc. Chim. Fr.* **1974,** 487.
65. Kobrina, L. S.; Kovtonyuk, V. N.; Yakobson, G. G. *Zh. Org. Khim.* **1977,** *13,* 1447; *Chem Abstr.* **1977,** *87,* 151814.
66. Aliev, I. Y.; Rozhkov, L. N.; Knunyants, I. L. *Izv. Akad. Nauk SSSR* **1973,** 1430; *Chem. Abstr.* **1973,** *79,* 91647.
67. Chambers, R. D.; Sargent, C. R.; Silvester, M. J.; Drakesmith, F. G. *J. Fluorine Chem.* **1980,** *15,* 257.
68. Inukai, Y.; Sonoda, T.; Kobayashi, H. *Bull. Chem. Soc. Jpn.* **1982,** *55,* 337.
69. Rakhimov, A. I.; Chapurkin, V. V.; Yagupol'skii, L. M.; Kondratenko, N. V. *Zh. Org. Khim.* **1980,** *16,* 1479; *Chem. Abstr.* **1981,** *94,* 3826.
70. Rakhimov, A. I.; Chapurkin, V. V. *Zh. Org. Khim.* **1981,** *17,* 1546; *Chem. Abstr.* **1981,** *95,* 186580.
71. Molines, H.; Wakselman, C. *J. Fluorine Chem.* **1984,** *25,* 447.
72. Rakhimov, A. I.; Volynskaya, E. M.; Chapurkin, V. V.; Alekseenko, A. N.; Il'chenko, A. Y.; Yagupol'skii, L. M. *Zh. Org. Khim.* **1985,** *21,* 656; *Chem. Abstr.* **1985,** *103,* 87529.
73. Rakhimov, A. I.; Volynskaya, E. M.; Chapurkin, V. V. *Zh. Org. Khim.* **1986,** *22,* 879; *Chem. Abstr.* **1987,** *106,* 84023.
74. Kobrina, L. S.; Akulenko, N. V.; Yakobson, G. G. *Zh. Org. Khim.* **1972,** *8,* 2165; *Chem. Abstr.* **1973,** *78,* 42948.
75. Coe, P. L.; Sellars, A.; Tatlow, J. C.; Whittaker, G.; Fielding, H. C. *J. Chem. Soc., Chem. Commun.* **1982,** 362.
76. Krasnikova, G. S.; German, L. S.; Knunyants, I. L. *Izv. Akad. Nauk SSSR* **1973,** 459; *Chem. Abstr.* **1973,** *78,* 158821.
77. Renaud, R. N.; Sullivan, D. E. *Can. J. Chem.* **1972,** *50,* 3084.

78. Berenblit, V. V.; Zapevalov, A. Y.; Panitkova, E. S.; Plashkin, V. S.; Bondarev, D. S.; Sass, V. P.; Sokolov, S. V. *Zh. Org. Khim.* 1979, *15,* 1471; *Chem. Abstr.* **1971,** *91,* 192779.
79. Renaud, R. N.; Stephens, C. J.; Berube, D. *Can. J. Chem.* **1982,** *60,* 1687.
80. Harayama, T.; Kotoji, K.; Yanada, R.; Yoneda, F.; Taga, T.; Osaki, K.; Nagamatsu, T. *Chem. Pharm. Bull.* **1986,** *34,* 2354.
81. Olah, G. A.; Gupta, B. G. B. *Synthesis* **1980,** 44.
82. Sankina, L. V.; Kostikin, L. I.; Ginsburg, V. A *Zh. Org. Khim.* **1972,** *8,* 1362; *Chem. Abstr.* **1972,** *77,* 125879.
83. Booth, B. L.; Kosinski, E. D.; Varley, J. S. *J. Fluorine Chem.* **1987,** *37,* 419.
84. Markovskii, L. N.; Kolomeitsev, A. A.; Polumbrik, O. M.; Yagupol'skii, L. M. *Zh. Org. Khim.* **1982,** *18,* 2217; *Chem. Abstr.* **1983,** *98,* 88902.
85. Adolph, H. G. *J. Org. Chem.* **1975,** *40,* 2626.
86. Deadman, J. J.; Jarman, M.; McCague, R.; McKenna, R.; Neidle, S. J. *Chem. Soc., Perkin Trans. 2* ***1989,*** *971.*
87. Petrenko, N. I.; Gerasimova, T. N.; Fokin, E. P. *Izv. Akad. Nauk SSSR* **1984,** 1378; *Chem. Abstr.* **1982,** *102,* 5791.
88. Burger, K.; Kahl, T. *J. Fluorine Chem.* 1987, *37,* 53.
89. Fuchigami, T.; Nakagawa, Y.; Nonaka, T. *J. Org. Chem.* 1987, *52,* 5489.
90. Fujita, M.; Ojima, I. *Tetrahedron Lett.* **1983,** *24,* 4573.
91. Gribble, G. W.; Allen, R. W.; Anderson, P. S.; Christy, M. E.; Colton, C. D. *Tetrahedron Lett.* **1976,** 3673.
92. Kobrina, L. S.; Akulenko, N. V.; Yakobson, G. G. *Zh. Org. Khim.* **1972,** *8,* 2375; *Chem. Abstr.* **1973,** *78,* 58326.
93. Furin, G. G.; Miller, A. O.; Gatilov, Y. V.; Bagryanskaya, I. Y.; Yakobson, G. G. *J. Fluorine Chem.* **1985,** *28,* 23.
94. Petrenko, N. I.; Gerasimova, T. N. *Izv. Akad. Nauk SSSR* **1985,** 665**;** *Chem. Abstr.* **1985,** *103,* 70998.
95. Davis, F. A.; Chattopadhyay, S.; Towson, J. C.; Lal, S.; Reddy, T. *J. Org. Chem.* **1988,** *53,* 2087.
96. Pastor, R.; Cambon, A. *J. Fluorine Chem.* **1979,** *13,* 279.
97. Hudlický, M.; Bell, H. M. *J. Fluorine Chem.* **1974,** *4,* 149.
98. Sartori, P.; Ahlers, K.; Frohn, H. J. *J. Fluorine Chem.* **1976,** *7,* 363.
99. Sartori, P.; Ahlers, K.; Frohn, H. J. *J. Fluorine Chem.* **1976,** *8,* 457.
100. Fokin, A. V.; Studnev, Y. N.; Rapkin, A. I.; Pasevina, K. I.; Kolomiets, A. F. *Izv. Akad. Nauk SSSR* **1981,** 1641; *Chem. Abstr.* **1981,** *95,* 186571.
101. Sinyashina, T. N.; Mironov, V. F.; Ofitserov, E. N.; Konovalova, I. V.; Pudovik, A. N. *Izv. Akad. Nauk SSSR* **1988,** 1451; *Chem. Abstr.* **1989,** *95,* 186571.
102. Raasch, M. S. *J. Org. Chem.* **1979,** *44,* 2629.
103. Astrologes, G. W.; Martin, J. C. *J. Am. Chem. Soc.* **1975,** *97,* 6909.

104. Yagupol'skii, L. M.; Smirnova, O. D. *Zh. Org. Khim.* **1972,** *8*, 1990; *Chem. Abstr.* **1973,** *78*, 15699.
105. Kondratenko, N. V.; Gavrilova, R. Y.; Yagupol'skii, L. M. *Zh. Org. Khim.* **1988,** *24*, 456; *Chem. Abstr.* **1989,** *110*, 7767.
106. Raasch, M. S. *J. Org. Chem.* **1980,** *45*, 2151.
107. Seelinger, R.; Sundermeyer, W. *Angew. Chem.* **1980,** *92*, 223.
108. Rall, K.; Sundermeyer, W. *J. Fluorine Chem.* **1990,** *47*, 121.
109. Moore, G. G. I. *J. Org. Chem.* **1979,** *44*, 1708.
110. Burton, C. A.; Shreeve, J. M. *Inorg. Chem.* **1977,** *16*, 1039.
111. Haas, A.; Klare, C. *Chimia* **1986,** *40*, 100.
112. Glander, I.; Golloch, A. *J. Fluorine Chem.* **1975,** *5*, 83.
113. Abe, T.; Shreeve, J. M. *J. Chem. Soc., Chem. Commun.* **1981,** 242.
114. Marcellis, A. W.; Eibeck, R. E. *J. Fluorine Chem.* **1975,** *5*, 71.
115. Derkach, N. Y.; Tishchenko, N. P.; Voloshchuk, V. G. *Zh. Org. Khim.* **1978,** *14*, 958; *Chem. Abstr.* **1979,** *91*, 74295.
116. Lyalin, V. V.; Orda V. V.; Alekseeva, L. A.; Yagupol'skii, L. M. *Zh. Org. Khim.* **1972,** *8*, 1019; *Chem. Abstr.* **1972,** *77*, 125808.
117. Umemoto, T.; Kuriu, Y.; Shuyama, H.; Miyano, O.; Nakayama, S. *J. Fluorine Chem.* ***1986,*** *31*, 37.
118. Mironova, A. A.; Soloshonok, I. V.; Maletina, I. I.; Orda, V. V.; Yagupol'skii, L. M. *Zh. Org. Khim.* **1988**, *24*, 593; *Chem. Abstr.* **1988,** *109*, 210610.
119. Mironova, A. A.; Maletina, I. I.; Orda, V. V.; Yagupol'skii, L. M. *Zh. Org. Khim.* **1983**, *19*, 1213; *Chem. Abstr.* **1983,** *99*, 175274.
120. Ishikawa, J.; Sonoda, T.; Kobayashi, H. *Tetrahedron Lett.* **1989**, *30*, 5437.

Halogenation

by S. R. Elsheimer

This section includes not only reactions involving elemental halogens but also those preparations of halogenated fluoroorganics that employ hydrogen halides, nonmetal halides, or metal halides. For simplicity, the term "halogen" is used here to refer to chlorine, bromine, or iodine. Those transformations involving fluorine, hydrogen fluoride, or other fluorinating agents are covered elsewhere in this book and are specifically excluded from consideration here. Emphasis will be placed on results reported since 1971. For a survey of this topic prior to that time, the reader should refer to the previous edition of this text [*1*].

Reactions with Halogens

Addition Reactions

Additions of **elemental halogens** to unsaturated compounds are among the most common preparations of halogenated fluoroorganics. The transformations are usually fairly clean and proceed in good yields. Besides the numerous examples of halogen addition to ***fluoroalkenes*** and fluoroalkyl-substituted alkenes, additions to perfluoropropyl vinyl ether [*2*] and fluorinated styrenes [*3, 4*] have been reported. Both ionic and free-radical processes occur (equations 1 and 2).

[*2*] **1**

$$CF_3CF_2CF_2OCF{=}CF_2 \xrightarrow{Br_2} CF_3CF_2CF_2OCFBrCF_2Br \quad 90\%$$

[*3*] **2**

$$C_6H_5CY{=}CF_2 \xrightarrow{X_2} C_6H_5CYXCF_2X$$

[*3*] Y = H,	X = Cl	81%
	Br	91%
[*4*] Y = F,	X = Cl	95%
	Br	96%

Additions of chlorine or bromine to 2-(4-biphenylyl)pentafluoropropene take place only under free-radical conditions [*5, 6*] (Table 1).

Generally, additions of halogens to fluoroalkenes are less stereoselective than the analogous reactions with nonfluorinated systems. The stereochemical mode of addition can be either *anti* or *syn*. Partitioning between these paths is determined

0065–7719/95/0187–0364$08.54/1

Table 1. Addition of Halogens to 2-(4-Biphenylyl)pentafluoropropene [*5, 6*][a]

$RC(CF_3){=}CF_2 \rightarrow RCX(CF_3)CF_2X$

X_2	*Conditions*	*Solvent*	*% Yield (Conversion)*
Cl_2	hν, 1 day	CF_2ClCCl_2F	99 (98)
Cl_2	Dark, 5 days	CH_3CO_2H	Trace
Br_2	hν, 16 h	CH_3CO_2H	100 (75)
Br_2	Dark, 5 days	CH_3CO_2H	0
Br_2	hν, 1 day	CF_2ClCCl_2F	100 (95)

[a]R = 4-C_6H_5-C_6H_4

Table 2. Stereochemistry of Ionic Addition of Bromine [*7*]

$4\text{-}RC_6H_4CF{=}CFX + Br_2 \rightarrow 4\text{-}RC_6H_4CFBrCFBrX$

X	*R*	*Isomer*	*Time (days)*	*Yield (%)*	*anti/syn Ratio*
Cl	H	*E*	6	55	1.9
		Z	5	60	3.8
Cl	Br	*E*	18	42	3.2
		Z	18	87	3.5
Cl	CH_3	*E*	3	41	1.4
		Z	3	95	1.2
Cl	OCH_3	*E*	1	33	1
		Z	1	95	1
Cl	COOH	*E*	4	90	2.6
		Z	4	80	6.1
CF_3	H	*E*	13	86	1.1
		Z	13	81	0.56
CF_3	Br	*E*	8	76	1.5
		Z	8	36	0.50
CF_3	CH_3	*E*	3	90	0.48
		Z	15	79	0.16
CF_3	OCH_3	*E*	3	100	0.83
		Z	3	94	1.2
CF_3	COOH	*E*	4	47	1.4
		Z	3	27	0.83

by steric and electronic factors. Results from the *ionic bromination* of some fluorostyrenes are shown in Table 2 [*7*].

Analogous reactions under *free-radical conditions* show a preference for the *erythro* isomer when X = Cl; however, when X = trifluoromethyl, the reaction shows essentially no stereoselectivity [*8*] (Table 3).

Table 3. Stereochemistry of Free-Radical Bromine Addition [*8*][a]

$4\text{-}HO_2CC_6H_4CF{=}CFX \rightarrow 4\text{-}HO_2CC_6H_4CFBrCFXBr$

X	*Isomer*	*erythro/threo Ratio*
Cl	*E*	76:24
Cl	*Z*	74:26
CF_3	*E*	53:47
CF_3	*Z*	51:49

[a]Yields are all ≥65%.

Mixtures of 1-phenyl-3,3,3-trifluoropropene which are 97–98% *E* and 2–3% *Z* react quickly with elemental chlorine without irradiation or heating. The reaction is nonstereospecific and produces mixtures of diastereomers in ratios between 1:1.5 and 1:3.5 [*9*] (equation 3).

[*9*]

$$CF_3CH{=}CHC_6H_5 \xrightarrow[30\text{-}35°]{Cl_2,\ CCl_4} CF_3CHClCHClC_6H_5 \quad 93.4\% \qquad \mathbf{3}$$

The reactions of bromine with *E*- or *Z*-1-fluoropropene under ionic conditions result in stereospecific *anti* additions to yield the 1*S*,2*S* and 1*R*,2*S* products, respectively [*10*] (equation 4).

[*10*] **4**

Conjugated, cross-conjugated, and homoconjugated fluoroalkenes react with halogens to yield predominantly *1,4-adducts*. Results from the reactions of a series of conjugated fluoroalkenes with elemental halogens are summarized in Table 4. In nearly all cases, the *trans*-1,4-addition products are formed exclusively [*11*].

A cross-conjugated perfluorotriene is unreactive toward bromine; however, photochemical chlorination yields the 1,4-addition product [*12*] (equation 5).

Table 4. Addition of Halogens to Conjugated Fluoroalkenes [*11*]

Fluorodiene	*Conditions*[a]	*Product*
CH_2=CH-CF=CF_2	Cl_2, CH_2Cl_2 Br_2, $CHCl_3$ I_2, CH_2Cl_2	XCH_2CH=$CFCF_2X$ X = Cl, Br, I
CH_3CH=CH-CF=CF_2	Br_2, $CHCl_3$	CH3CHBrCH=$CFCF_2Br$
CH_3CH=CH-CF=CF_2	Br_2, hν[b] CF_2,ClCF_2Cl	CF_3CHBrCH=$CFCF_2Br$
CHF=CH-CF=CF_2	Cl_2, CH_2Cl_2 Br_2, $CHCl_3$	CHFXCH=CFCF2X X = Cl, Br
CF_2=CH-CF=CF_2	Br2, $CHCl_3$	CF2BrCH=$CFCF_2Br$
CF_2=CF-CF=CF_2	Cl2, CH_2Cl_2 Br_2, $CHCl_3$	XCF2CF=$CFCF_2Br$ X = Cl, Br

[a]All reactions were carried out under argon at –5 °C.

[b]An *E/Z* mixture of 1,4-adducts was obtained under these conditions.

[*12*] **5**

$$\underset{\underset{CF_3CCF_3}{\|}}{(CF_3)_2C{=}CFCCF{=}C(CF_3)_2} \xrightarrow[h\nu,3\ h]{Cl_2} \underset{\underset{CF_3CClCF_3}{|}}{(CF_3)_2C{=}CFC{=}CFCCl(CF_3)_2}$$

Bromination of the homoconjugated 5-(difluoromethylene)-6,6-difluoro-2-norbornene yields different product mixtures under ionic and free-radical conditions [*13*] (equation 6).

[*13*] **6**

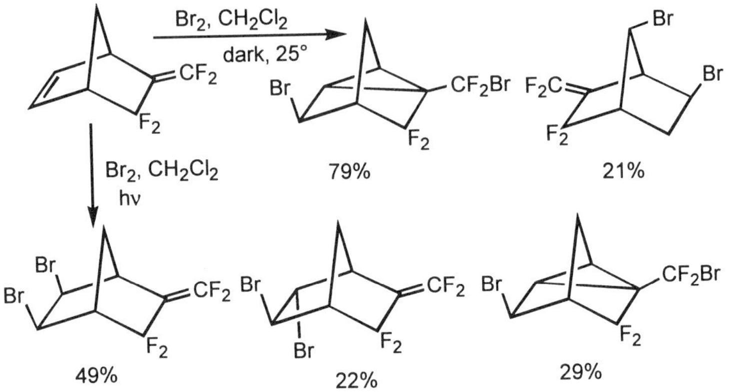

Br₂, CH₂Cl₂ dark, 25° → 79%, 21%; Br₂, CH₂Cl₂ hν → 49%, 22%, 29%

Interhalogen compounds such as **iodine monochloride** have been added to ***fluoroalkyl-substituted alkenes***. The observed unidirectional regiochemistry can be explained by the polarity of the double bond [*14*] (equation 7).

[*14*] **7**

$$CF_3CH{=}CH_2 \xrightarrow{ICl} CF_3CHICH_2Cl \quad 84\%$$

Iodine monochloride adds unidirectionally to perfluoropropene-2-ol to form a chlorohydrin, which easily dehydrochlorinates to yield iodopentafluoroacetone [*15*] (equation 8).

[*15*] **8**

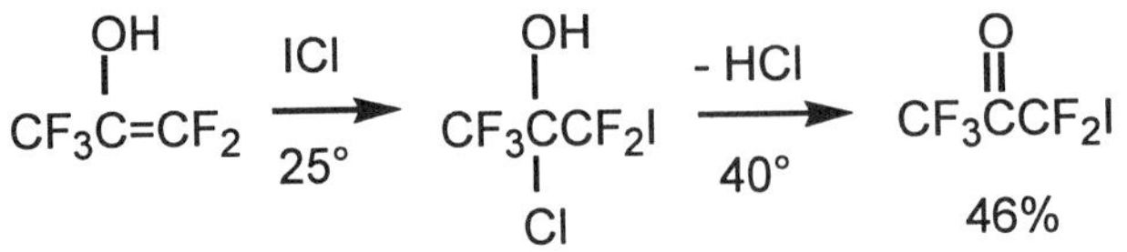

$$CF_3C(OH){=}CF_2 \xrightarrow[25^\circ]{ICl} CF_3C(OH)(Cl)CF_2I \xrightarrow[40^\circ]{-HCl} CF_3C(O)CF_2I \quad 46\%$$

Like simple elemental halogens, iodine monochloride reacts with conjugated fluorodienes to yield mostly 1,4-addition products. These bidirectional reactions lead to mixtures of regioisomers, as shown in Table 5 [*11*].

Table 5. Reactions of Iodine Chloride with Fluorodienes in Methylene Chloride at –5 °C under Argon [11]

Reactant	*Product(s)*	*Yield (%)*
CH_2=CH-CF=CF_2	CH_2ICH=CFCF_2Cl	66
	CH_2ICHClCF=CF_2	33
CH_3CH=CH-CF=CF_2	CH3CHICH=CFCF_2Cl	35
	CH_3CHClCH=CFCF_2I	26
CH_3CH=CH-CF=CF_2	CF3CHICH=CFCF_2Cl	100
CHF=CH-CF=CF_2	CHFICH=CFCF_2Cl	66
	CHFClCH=CFCF_2I	22
CF_2=CH-CF=CF_2	CF2ICH=CFCF_2Cl	47
	CF_2ClCH=CFCF_2I	23
CH_2=CF-CF=CF_2	CF2ClCF=CFCF_2I	70

Terminal (perfluoroalkyl)alkynes react with iodine or iodine chloride to yield *syn* addition products bearing iodine on the terminal carbon [16] (equation 9).

[16] **9**

$$CF_3(CF_2)_5C{\equiv}CH \xrightarrow[65\,°,\ 24\ h]{I_2,\ MeCN} CF_3(CF_2)_5CI{=}CHI \quad Z \quad 87\%$$

$$CF_3(CF_2)_5C{\equiv}CH \xrightarrow[55\,°,\ 3\ h]{ICl,\ AcOH} CF_3(CF_2)_5CCl{=}CHI \quad Z \quad 75\%$$

Photochemical brominations of ***fluorobenzene*** or difluorobenzenes produce hexabromocyclohexane derivatives in low yield. Tri- and tetrafluorobenzenes are unreactive under these conditions [17] (Table 6). Analogous free-radical chlorination of benzotrifluoride produces a 9% yield of the hexachloro adduct [18].

Table 6. Photochemical Bromination of Fluorobenzenes [*17*][a]

Fluoroarene	*Hexabromo Adduct*	*Yield (%)*
C_6H_5F	[a]$C_6H_5FBr_6$	13
o-$C_6H_4F_2$	$C_6H_4F_2Br_6$	4
m-$C_6H_4F_2$	$C_6H_4F_2Br_6$	11
p-$C_6H_4F_2$	$C_6H_4F_2Br_6$	3
1,3,5-$C_6H_3F_3$	$C_6H_3F_3Br_6$	Trace
1,2,3,4-$C_6H_2F_4$	$C_6H_2F_4Br_6$	0
1,2,3,5-$C_6H_2F_4$	$C_6H_2F_4Br_6$	0

[a]Conditions: Br_2, $CFCl_3$, hν, 2 days.

Replacement of Hydrogen with Halogen

The replacement of hydrogen with halogen takes place under thermal or photochemical conditions [*19, 20, 21*] (equations 10–12).

[*19*] **10**

$$CH_3CHF_2 \xrightarrow[18°,\ h\nu]{Cl_2} CH_3CF_2Cl \quad \text{conversion 88.8\%, yield 45\%}$$

[*20*] **11**

$$CF_3CH_2Cl \xrightarrow[450\ °]{Cl_2,\ AlF_3} CF_3CHCl_2 \quad \text{conversion 79\%, yield 45\%}$$

[*21*] **12**

$$CH_3CF_2CH_3 \xrightarrow[AIBN,\ 60\ °,\ 6\ h]{Cl_2,\ CCl_4}$$

$CH_3CF_2CH_2Cl$ 8%	$CH_2ClCF_2CCl_3$ 2%
$CH_3CF_2CHCl_2$ 11%	$CHCl_2CF_2CCl_3$ 1%
$CH_3CF_2CCl_3$ 77%	$CCl_3CF_2CCl_3$ 1%

Chlorine-36-labeled 1,1-dichlorotetrafluoroethane is produced photochemically from chlorine-36 and 1-chloro-1,2,2,2-tetrafluoroethane in better than 90% yield [*22*].

Fluoroform is **brominated** to yield *bromotrifluoromethane*. The reaction rate and yield are both enhanced by various catalysts, as summarized in Table 7 [*23, 24, 25*].

The reaction of 2H-nonafluoro-2-methylpropane with **iodine monochloride** in sulfolane in the presence of potassium fluoride produces 2-chlorononafluoro-2-methylpropane in 92% yield at 50% conversion [*26*].

Table 7. Conversion of Fluoroform to Bromotrifluoromethane

Conditions	*Yield (%)*	*Ref.*
Br_2, quartz, 600 °C, 6 s	72	23
Br_2, $FeCl_3$, KBr, 450 °C, 12 s	71	23
Br_2, Cl_2, 600 °C	67.2	24
Br_2, 600 °C	8.2	24
Br_2, SO_3, 650 °C	84	25
Br_2, 650 °C	77.5	25

Replacement of hydrogen with halogen in ***fluoroarenes*** takes place by an ionic mechanism and is subject to the normal directing effects [*27*, *28*, *29*] (equations 13–15).

[*27*] 13

$$C_6H_5F \xrightarrow[65\text{-}70\ °]{I_2,\ HNO_3} C_6H_4IF \quad \begin{matrix} p & 92\% \\ m & 7.6\% \\ o & 0.5\% \end{matrix}$$

[*28*] 14

2-Fluoroanisole (OCH$_3$, F) $\xrightarrow{Br_2,\ CHCl_3}$ 4-bromo-2-fluoroanisole (OCH$_3$, F, Br) 92%

[*29*] 15

4-F-C_6H_4-NHO_2SCF_3 $\xrightarrow[AcOH,\ 100°,\ 8\ h]{Cl_2,\ AlCl_3}$ 2,6-Cl$_2$-4-F-C_6H_2-NHO_2SCF_3 ~100%

Halogenations of highly fluorinated aromatics also occur [*30, 31*] (equations 16 and 17).

The hydrogens of ***benzotrifluoride*** can be replaced with chlorine, bromine, or iodine. Although *meta* substitution is favored, multiple substitutions are frequently observed [*32*, *33*, *34*, *35*] (Table 8).

Chlorination of ***3-(trifluoromethyl)pyridine*** takes place under fairly harsh conditions [*36*] (equation 18).

Several ***3-(perfluoroalkyl)pyrazoles*** can be brominated at position 4 under mild conditions; however, 5-(perfluoroalkyl)pyrazoles are unreactive [*37*] (equation 19).

[*30*] 16

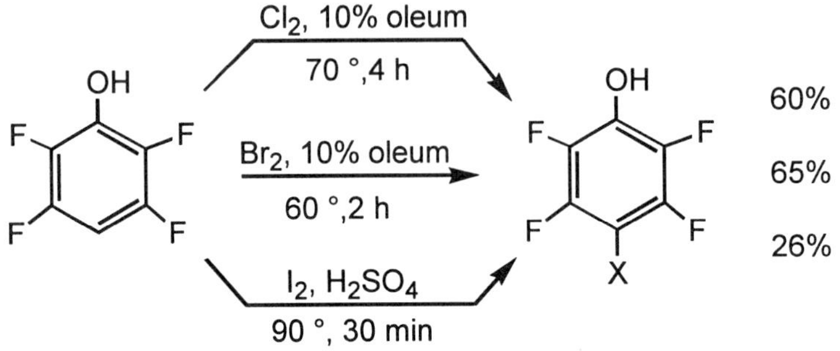

Cl_2, 10% oleum, 70 °, 4 h — 60%

Br_2, 10% oleum, 60 °, 2 h — 65%

I_2, H_2SO_4, 90 °, 30 min — 26%

[*31*] 17

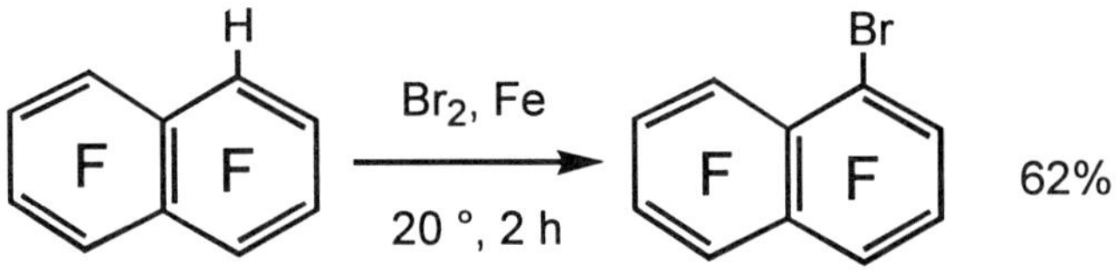

Br_2, Fe, 20 °, 2 h — 62%

[*36*] 18

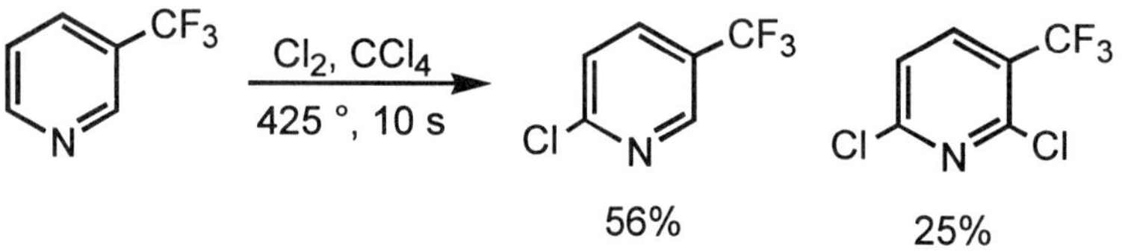

Cl_2, CCl_4, 425 °, 10 s — 56%, 25%

[*37*] 19

Br_2, NaOAc, RT

R^1 = H, CH_3, C_6H_5, CH_2CH_2CN
R^3 = C_3F_7, C_5F_{11}, C_7F_{15} 80-94%
R^5 = CH_3, $C(CH_3)_3$, C_6H_5

R^3 = CH_3, $C(CH_3)_3$, C_6H_5, C_3F_7, C_5F_{11}, C_7F_{15}
R^5 = C_3F_7, C_5F_{11}, C_7F_{15} no reaction

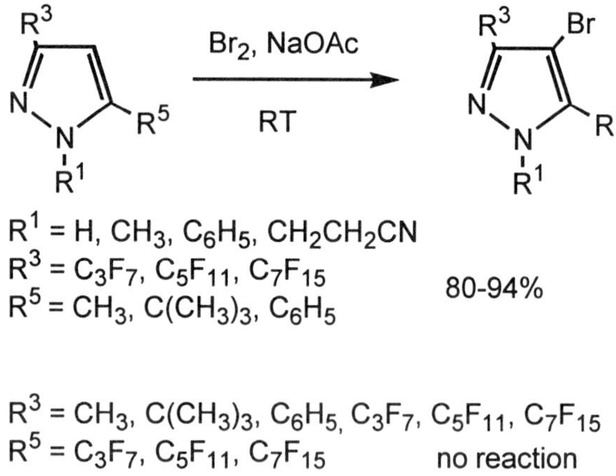

Replacement of hydrogen with halogen can be carried out in the *alpha* position of ***fluorinated ethers, amines, aldehydes, or nitriles***. In 2,2,3,4,4,4-hexafluorobutyl methyl ether, chlorination occurs predominantly at the methyl; however, bromination occurs mostly at the internal position of the fluorobutyl group [*38*] (equation 20).

Table 8. Replacement of Hydrogen with Halogen in Benzotrifluoride

Conditions	*Product(s)*	*Yield (%)*	*Ref.*
Cl_2, $FeCl_3$	*o*-, *m*-, *p*-$ClC_6H_4CF_3$	2.0, 57.7, 3.2	32
20 °C, 3 h	$Cl_2C_6H_3CF_3$[a]	94	
Cl_2, $FeCl_3$	2,3,5-$Cl_3C_6H_2CF_3$	32	33
130 °C, 2 h	2,3,6-$Cl_3C_6H_2CF_3$	63	
	3,4,5-$Cl_3C_6H_2CF_3$	88	
Br_2, $FeCl_3$	*m*-$BrC_6H_4CF_3$	25	34
silica gel			
40–60 °C, 20 h			
I_2, HIO_4, H_2SO_4	2,3,4,5-$I_4C_6HCF_3$	67	35
100 °C, 1 day			

[a]Mixture of isomers.

[*38*] **20**

$$CF_3CFHCF_2CH_2OCH_3 \xrightarrow[h\nu]{Br_2,\ CS_2} CF_3CFHCF_2CHBrOCH_3$$

$$CF_3CFHCF_2CH_2OCH_3 \xrightarrow[h\nu]{Cl_2,\ CS_2} CF_3CFHCF_2CHClOCH_3\ 16\% + CF_3CFHCF_2CH_2OCH_2Cl\ 76\%$$

$CF_3CFHCF_2CH_2OCHCl_2$
$CF_3CFHCF_2CH_2OCCl_3$
7%

Replacement of hydrogen with chlorine adjacent to the nitrogen in polyfluoroalkylamines occurs in excellent yield [*39, 40*] (equations 21 and 22).

[*39*] **21**

$$(CF_3)_2NCHX_2 \xrightarrow[h\nu,\ 3\ days]{Cl_2} (CF_3)_2NCX_2Cl \qquad X = F, Cl \qquad \sim100\%$$

[*40*] **22**

N-methyl perfluoropyrrolidine (ring F) $\xrightarrow[h\nu,\ 7\ days]{Cl_2}$ N-CCl_3 perfluoropyrrolidine (ring F) 95%

Replacement of hydrogen with bromine *alpha* to a carbonyl or a nitrile group takes place under mild conditions and in good yield [*41, 42*] (equations 23 and 24).

[*41*] **23**

$$CF_3CH_2CHO \xrightarrow[40\ °,\ 1\ h]{Br_2,\ AcOH} CF_3CHBrCHO \quad 71.6\%$$

[*42*] **24**

$$(CF_3)CHCN \xrightarrow[15\text{-}50\ °]{Br_2,\ C_5H_5N} (CF_3)CBrCN \quad 71\%$$

Interhalogen compounds react with ***perfluoroalkyl hydrides*** to yield mixtures of halosubstituted products [*43*] (equations 25 and 26).

[*43*] **25**

$$(CF_3)_3CCF_2H \xrightarrow[h\nu,\ 254\ h]{BrCl} (CF_3)_3CCF_2X \quad \text{conversion 97\%} \quad X = Br,\ 52\%;\ X = Cl,\ 45\%$$

$$(CF_3)_3CCF_2H \xrightarrow[h\nu,\ 108\ h]{BrF} (CF_3)_3CCF_2X \quad \text{conversion 97\%} \quad X = Br,\ 78.6\%;\ X = F,\ 6.7\%$$

[*43*] **26**

$$\text{2-}CF_3\text{-2-}CF_2H\text{-perfluoro-1,3-dioxolane } (O\text{–}CF_2\text{–}CF_2\text{–}O) \xrightarrow[h\nu,\ 192\ h]{BrCl} \text{2-}CF_3\text{-2-}CF_2X\text{-perfluoro-1,3-dioxolane} \quad X = Br,\ 59.5\%;\ X = Cl,\ 37.5\%;\ \text{conversion 98\%}$$

Halogenolysis of Fluorinated Compounds

The *Hunsdieckers' method for replacing a carboxyl group by a halogen* has been applied to the preparation of simple fluoroalkyl iodides [*44*], fluoroiodoalkyl ethers [*45*], and fluorovinyl iodides [*46*] (equations 27–29).

[*44*] **27**

$$CF_3CO_2K \xrightarrow[173\text{-}180\ °,\ 4\ h]{I_2,\ sulfolane} CF_3I \quad 80\%$$

[*45*] **28**

$$CF_3CF_2OCF(CF_3)CO_2K \xrightarrow[55\text{-}100\ °,\ 150\ min]{I_2,\ AcNMe_2} CF_3CF_2OCFICF_3 \quad 86\%$$

[46] $R-C_6H_4-CF=CFCO_2M \longrightarrow R-C_6H_4-CF=CFI$ 29

R	M	Conditions	% Yield
H	Ag	I_2, hexane, 60-70 °	64.3
Me_2N	Ag	I_2, CCl_4, 20 °	51
O_2N	Na	I_2, DMF, 150 °	75.5

Halogenolysis of the nitro substituent in fluoroarenes or fluoroalkylarenes provides a route to the corresponding chloro compounds [47, 48] (equations 30 and 31).

[47] 1-NO_2-3,4-$F_2C_6H_3$ $\xrightarrow[400\ °]{Cl_2}$ 1-Cl-3,4-$F_2C_6H_3$ 82% 30

[48] 2-NO_2-4,5-$F_2C_6H_2CF_3$ $\xrightarrow[180\ °]{Cl_2}$ 2-Cl-4,5-$F_2C_6H_2CF_3$ 95% 31

Iodine cleavage of 2,2-difluoroalkenylboranes provides a general route to 1,1-difluoro-2-iodoalkenes [49] (equation 32).

[49] $$CF_2{=}CRBR_2 \xrightarrow[NaOH,\ -10\ °,\ 1h]{I_2,\ THF,\ H_2O} CF_2{=}CRI \quad 51\text{-}76\%$$ 32

R = $(CH_2)_4Ph$, $CH_2CH(CH_3)Ph$, Cyclooctyl

The final step in a preparation of *Z*-1-iodopentafluoropropene involves *iodinolysis* of a fluoroalkenylphosphonium salt [50] (equation 33).

Reactions with Hydrogen Halides

Addition Reactions

Addition of hydrogen halide across fluoroalkenes and fluoroalkylalkenes is an important route to halogen-containing fluoroorganics. Both *ionic and free radical additions* favor the attachment of the halogen to the carbon bearing the greater number of fluorines and/or the fewest number of fluoroalkyl groups [10, 51, 52] (equations 34–36).

[*50*] **33**

$$CF_3CF{=}CF_2 \xrightarrow[-78°\text{-RT}]{Bu_3P, Et_2O} \xrightarrow{BF_3 \cdot Et_2O} (CF_3CF{=}CFPBu_3)BF_4 \xrightarrow{I_2,\ DMF,\ Na_2CO_3} CF_3CF{=}CFI$$

61% overall

[*51*] **34**

$$CH_2{=}CF_2 \xrightarrow[-20\text{-}45\ °]{HI} CH_3CF_2I \quad 96\%$$

[*10*] **35**

$$CH_3CH{=}CHF \xrightarrow[7\ \text{days}]{HBr,\ \text{dark}} CH_3CH_2CHFBr$$

conversion 97%
yield 70%

$$CH_3CH{=}CHF \xrightarrow[20\ h]{HBr,\ h\nu} CH_3CH_2CHFBr\ (36\%) + CH_3CHBrCH_2F\ (4\%) + CH_3CHBrCHFBr\ (58\%)$$

[*52*] **36**

$$CF_3CH{=}CH_2 \xrightarrow[\gamma\text{-irradn., 15-45 atm}]{HCl} CF_3CH_2CH_2Cl$$

Addition of anhydrous **hydrogen chloride** to ***vinylidene fluoride*** is reported to be accompanied by a hazardous (explosive) side reaction [*53*].

Substitution Reactions

Hydrogen halides can, in some cases, be used to replace an atom or group by halogen. Fluoropentanitrobenzene reacts with hydrogen chloride to yield 3-chloro-2,4,5,6-tetranitrofluorobenzene, but the other halopentanitrobenzenes are much less reactive [*54*] (equation 37).

Perfluoroalkylsulfonyl chlorides are efficiently converted to the corresponding perfluoroalkyl bromides with hydrogen bromide and a catalyst [*55*] (equation 38).

[*54*] **37**

$$1\text{-F-}2,3,4,5,6\text{-(NO}_2)_5C_6 \xrightarrow[\text{RT, 2 h}]{\text{HCl, }C_6H_6} 1\text{-F-}2,3,4,5\text{-(NO}_2)_4\text{-}6\text{-Cl-}C_6 \quad 88\%$$

[*55*] **38**

$$CF_3(CF_2)_7SO_2Cl \xrightarrow[\text{Bu}_4\text{NBr, 125 °}]{\text{HBr}} CF_3(CF_2)_7Br \quad 94\%$$

Hydrogen bromide converts ethyl chlorofluoroacetate to bromofluoroacetic acid. [*56*] (equation 39).

[*56*] **39**

$$ClCHFCO_2C_2H_5 \xrightarrow[\text{90 °, 13.5 h}]{\text{HBr}} BrCHFCO_2H \quad 71\%$$

The 2-fluoro substituent of 2-fluoropyridines is replaced by chlorine when hydrogen chloride is used [*57*] (equation 40).

[*57*] **40**

$$\text{3-Cl-2-F-5-CF}_3\text{-pyridine} \xrightarrow[\text{110 °, 20 h}]{\text{HCl}} \text{2,3-Cl}_2\text{-5-CF}_3\text{-pyridine} \quad 98\%$$

Reactions with Nonmetal Halides

Addition Reactions

Polyfluoroalkyl hypochlorites add to ***tetrafluoroethylene*** at room temperature [*58*] (equation 41).

[*58*] **41**

$$CF_2{=}CF_2 \xrightarrow[\text{20 °}]{H(CF_2)_6CH_2OCl} H(CF_2)_6CH_2OCF_2CF_2Cl \quad 91\%$$

$$CF_2{=}CF_2 \xrightarrow[\text{20 °}]{O_2NCF_2CH_2OCl} O_2NCF_2CH_2OCF_2CF_2Cl \quad 72.5\%$$

Fluorinated acyl hypochlorites add unidirectionally to ***vinylidene fluoride*** and 1,1-dichloro-2,2-difluoroethene [*59*] (equation 42).

[*59*] **42**

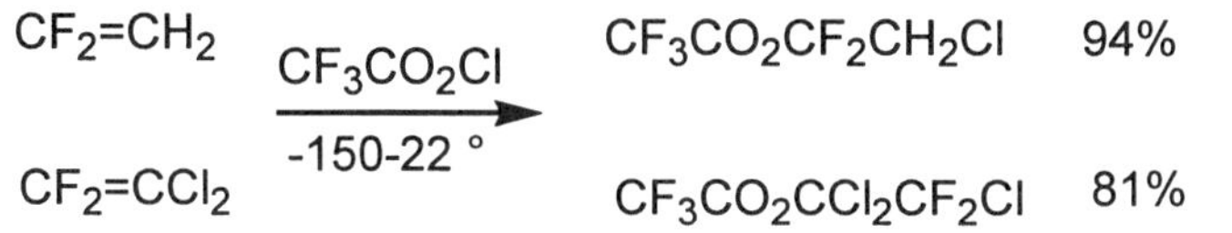

$$CF_2{=}CH_2 \xrightarrow[-150\text{-}22\,°]{CF_3CO_2Cl} CF_3CO_2CF_2CH_2Cl \quad 94\%$$

$$CF_2{=}CCl_2 \xrightarrow[-150\text{-}22\,°]{CF_3CO_2Cl} CF_3CO_2CCl_2CF_2Cl \quad 81\%$$

Perfluoroalkyl alkynes undergo unidirectional *syn* addition of **iodine cyanide** [*16*] (equation 43).

[*16*] **43**

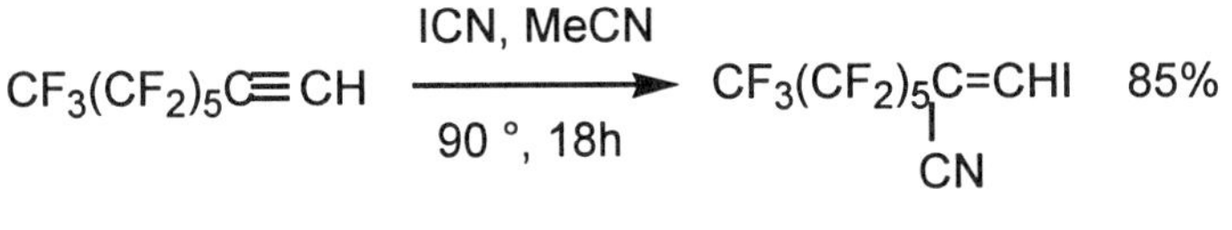

$$CF_3(CF_2)_5C{\equiv}CH \xrightarrow[90\,°,\ 18h]{ICN,\ MeCN} CF_3(CF_2)_5C(CN){=}CHI \quad 85\%$$

Although cyanogen iodide or cyanogen bromide adds to pentafluoropropene-2-ol to produce the halocyanohydrins, cyanogen chloride is unreactive [*15*] (equation 44).

[*15*] **44**

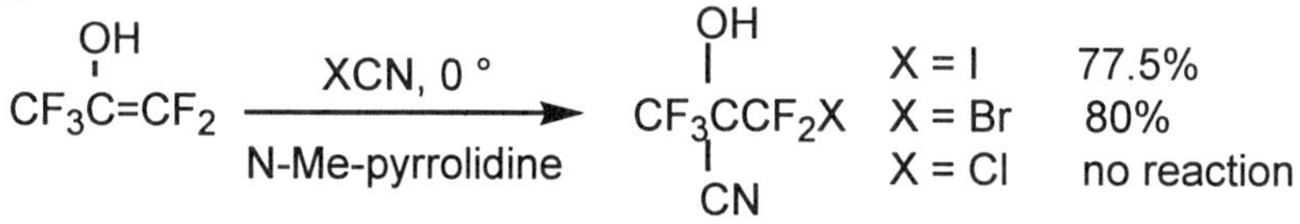

$$CF_3C(OH){=}CF_2 \xrightarrow[\text{N-Me-pyrrolidine}]{XCN,\ 0\,°} CF_3C(OH)(CN)CF_2X$$

X = I 77.5%
X = Br 80%
X = Cl no reaction

Replacement of Hydrogen and Other Elements by Halogen

Replacement of a hydrogen with bromine in the ***polyfluoroalkyl group*** of a ketone or acyl fluoride can be carried out with **phosphorus pentabromide** [*60*] (equation 45).

[*60*] **45**

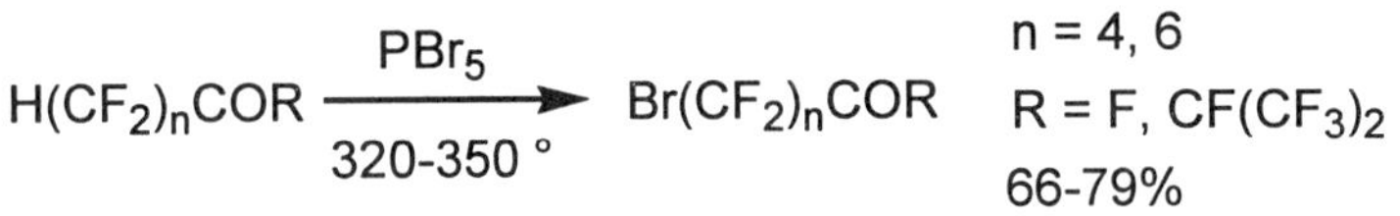

$$H(CF_2)_nCOR \xrightarrow[320\text{-}350\,°]{PBr_5} Br(CF_2)_nCOR$$

n = 4, 6
R = F, $CF(CF_3)_2$
66-79%

Analogous chlorinations with **phosphorus pentachloride** proceed in higher yield and at lower temperature [*61*] (equation 46).

[*61*] **46**

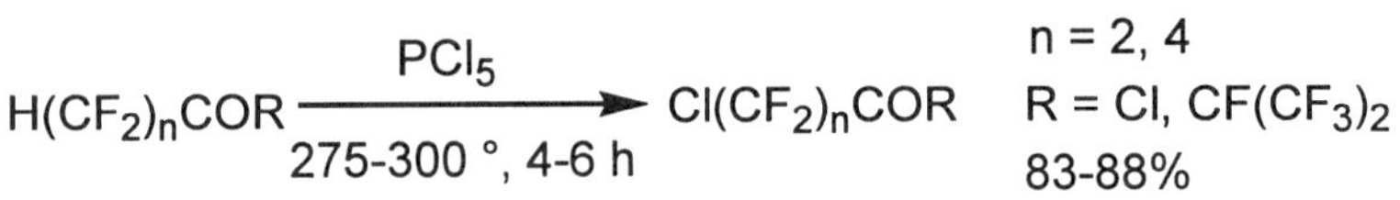

$$H(CF_2)_nCOR \xrightarrow[275\text{-}300\,°,\ 4\text{-}6\ h]{PCl_5} Cl(CF_2)_nCOR$$

n = 2, 4
R = Cl, $CF(CF_3)_2$
83-88%

Sulfuryl chloride in *N,N*-dimethylaniline replaces the *alpha* hydrogen of a polyfluoronitrile [*42*] (equation 47).

[*42*] $$(CF_3)_2CHCN \xrightarrow[0\text{-}5\ °,\ 1\ h]{SO_2Cl_2,\ C_6H_5NMe_2} (CF_3)_2CClCN\quad 73\%$$ **47**

Pentafluorophenol reacts with ***tert*-butyl hypobromite** to give 4-bromo-pentafluorocyclohexadienone [*62*] (equation 48).

[*62*] **48**

OH, F (pentafluorophenol) $\xrightarrow[CCl_4,\ 0\ °]{Me_3COBr}$ O, F, F, F, F, F, Br (4-bromopentafluorocyclohexa-2,5-dienone) 60%

Fluoroalkyl anilines can be halogenated by ***N*-chlorosuccinimide** [*63*] or 2,4,4,6-tetrabromocyclohexadienone [*64*] (equations 49 and 50).

[*63*] **49**

NH_2, CF_3 $\xrightarrow[\text{reflux, MeCN, 3 h}]{\text{N-chlorosuccinimide}}$ NH_2, Cl, CF_3 69%

[*64*] **50**

$N(CH_3)_2$, CF_3 $\xrightarrow[CH_2Cl_2,\ -10\text{-}0\ °]{\text{Br, Br, Br, Br, =O (2,4,4,6-tetrabromocyclohexadienone)}}$ $N(CH_3)_2$, CF_3, Br 82-90%

The amine hydrogens of polyfluorinated anilines can be replaced with chlorine by using ***tert*-butyl hypochlorite** [*65,66*].

Conversion of a carbonyl to a gem-dichloro group is carried out with **phosphorus pentachloride** [*67, 68*] (equations 51 and 52). Under these conditions, 1,4-dicarbonyl compounds give cyclic ethers [*67*] (equation 53).

The carbonyl carbon of ***perfluoro acyl halides*** and ***carboxylic acids*** can be converted to *a trichloromethyl group* [*67*] (equation 54).

[*67*] **51**

$$CF_3CO(CF_2)_7CF_3 \xrightarrow[\text{25 °, 2 days; 260-280 psi}]{PCl_5} CF_3CCl_2(CF_2)_7CF_3 \quad 84\%$$

[*68*] **52**

$$CF_3C(=O)\text{-}C(=O)OCH_3 \xrightarrow[\text{reflux, 16 h}]{PCl_5,\ CCl_4} CF_3CCl_2CCl_2OCH_3 \quad 55\%$$

[*67*] **53**

$$C_2F_5C(=O)CF_2CF_2C(=O)C_2F_5 \xrightarrow[\text{275 °, 7 days}]{PCl_5} \text{2,5-bis}(C_2F_5)\text{-2,5-dichloro-}CF_2\text{-}CF_2\text{ oxolane} \quad 55\%$$

[*67*] **54**

$$R_FCOX \xrightarrow[\text{300 psi, 3-7 days}]{PCl_5,\ 250\text{-}275\ °} R_FCCl_3$$

X = F, Cl, OH

$R_F = C_3F_7,\ C_7F_{15}$

85-93%

A ***perfluorocarboxylic acid*** or its salt reacts with **fluorosulfonylhypohalites** to produce the corresponding perfluoroalkyl halide [*69*] (equations 55 and 56).

[*69*] **55**

$$CF_3CO_2H \xrightarrow[\text{50 °, 1 h}]{ClOSO_2F} CF_3Cl \quad 90\%$$

[*69*] **56**

$$CF_3CF_2CO_2Na \xrightarrow[\text{50 °}]{BrOSO_2F} CF_3CF_2Br \quad 83\%$$

Boron tribromide replaces fluorine with bromine on tertiary or secondary carbons; however, trifluoromethyl groups are inert in this reaction [*70*] (equation 57).

[*70*] **57**

$$\text{1-}CF_3\text{-3,5,7-trifluoroadamantane} \xrightarrow[\text{RT-reflux}]{BBr_3,\ 3\ h} \text{1-}CF_3\text{-3,5,7-tribromoadamantane} \quad 97.4\%$$

The fluorines of a difluoromethylamine group are exchanged for chlorine with **boron trichloride** [*39*] (equation 58).

Simple alkyl fluorides can be converted to alkyl iodides with **iodotrimethylsilane** [*71*].

[*39*] **58**

$$(CF_3)_2NCHF_2 \xrightarrow[\text{RT, 3 days}]{BCl_3} (CF_3)_2NCHCl_2 \quad \sim 100\%$$

The *vinylic fluorine* of perfluoroisobutylene can be replaced with chlorine by using **phosphorus oxychloride** or **benzoyl chloride**; however, somewhat different mixtures result from these two chlorinating agents [*72*] (equation 59).

[*72*] **59**

$$(CF_3)_2C{=}CF_2 \xrightarrow[\text{[Et}_3\text{NCH}_2\text{C}_6\text{H}_5\text{]Cl, 160-170 °, 25 h}]{POCl_3} (CF_3)_2C{=}CFCl\ 34\% \quad + \quad CF_3C(CF_2Cl){=}CFCl\ 9\%$$

$$(CF_3)_2C{=}CF_2 \xrightarrow[\text{C}_5\text{H}_5\text{N, 160-170 °, 25 h}]{C_6H_5COCl} (CF_3)_2C{=}CFCl\ 49\% \quad + \quad (CF_3)_2C{=}CCl_2\ 21\%$$

The ethoxycarbonylimine of perfluoroacetone reacts with phosphorus pentachloride under mild condtions. This reaction is a convenient synthesis of α-chlorohexafluoropropyl isocyanate [*73*] (equation 60).

[*73*] **60**

$$(CF_3)_2C{=}NCO_2C_2H_5 \xrightarrow[\text{165 °, 20 h}]{PCl_5} (CF_3)_2CClNCO \quad 70\%$$

Reactions with Metal Halides

Many reactions of fluorinated organics with metal halides result in the replacement of fluorine with halogen. A general route to 1,1,1-trichloro- or tribromofluoroalkanes involves treating primary ***fluoroalkyl iodides*** with **aluminum trichloride** or **aluminum tribromide** [*74*]. *Benzylic* [*75, 76*] *or vinylic* [*72*] *fluorine* can be exchanged for chlorine when treated with aluminum trichloride.

Vinylic fluorines of ***fluoralkenes*** are replaced with chlorine or bromine when treated with **lithium halide** salts in methoxyethanol, dimethylformamide, and pyridine [*77*].

In ***polyfluoroethers***, fluorines *alpha* to oxygen are replaced with chlorine upon treatment with **aluminum trichloride** [*78, 79, 80*] (equation 61).

[*78*] **61**

(perfluoro spiro ether, F rings) $\xrightarrow[\text{165 °, 20 h}]{AlCl_3}$ (perfluoro spiro ether with CCl_2 alpha to O) 84%

Refluxing perfluorotoluene neat for 4 h with aluminum tribromide produces *p*-bromoperfluorotoluene in 54% yield [*81*]. Fluoropyridines can be transhalogenated at position 2 and/or 4 by heating in the presence of calcium halide in a nonhydroxylic solvent [*82*] (equation 62).

[*82*] 3-chloro-2,4,5,6-tetrafluoropyridine $\xrightarrow[\text{210 °, 2 h, sulfolane}]{CaCl_2}$ 3,4-dichloro-2,5,6-trifluoropyridine 60% **62**

Metal halides can, in some cases, be used to replace other atoms or groups besides fluorine with halogen. Polyfluoroacyl fluorides and chlorides can be converted to fluoroalkyl iodides by simply heating the reactant in the presence of an alkali metal iodide [*83, 84*] (equations 63 and 64).

[*83*]
$$(CF_3)_2NCF_2CF_2COF \xrightarrow[\text{180 °, 6.5 h}]{LiI} (CF_3)_2NCF_2CF_2I \quad 90\%$$ **63**

[*84*]
$$RCO_2(CF_2)_nCOCl \xrightarrow{MI} RCO_2(CF_2)_nI$$ **64**

72--78%

$R = CH_3, C_2H_5$; $n = 3, 4$; $M = Na, K$

Perfluoroallyl fluorosulfonate is converted to perfluoroallyl bromide or iodide in 56 or 75% yield, respectively, by reaction with the potassium halide in monoglyme at room temperature [*85*].

References for Pages 364–382

1. Hudlický, M. *Chemistry of Organic Fluorine Compounds,* 2nd (revised) ed.; Ellis Horwood, Chichester, England: 1976; pp 213–237.
2. Kim, A. Ch.; Glazkov, A. A.; Ignatenko, A. V.; Krukovskii, S. P.; Ponomarenko, V. A. *Izv. Akad. Nauk SSSR* **1984,** 2319; *Chem. Abstr.* **1985,** *102*, 131499.
3. Burton, D. J.; Andreson, A. L.; Takei, R.; Koch, H. F.; Shih, T. L. *J. Fluorine Chem.* **1980**, *16*, 229.
4. Stepanov, M. V.; Panov, E. M.; Kocheshkov, K. A. *Izv. Akad. Nauk SSSR* **1975,** *24*, 2544 (Engl. Transl. 2430).
5. Naae, D. G. *Tetrahedron Lett.* **1976,** 2761.
6. Naae, D. G. *J. Org. Chem.* **1977,** *42*, 1780.
7. Naae, D. G. *J. Org. Chem.* **1980,** *45*, 1394.

8. Naae, D. G. *J. Org. Chem.* **1979,** *44*, 336.
9. Dmowski, W. *J. Fluorine Chem.* **1985,** *29*, 287.
10. Haszeldine, R. N.; Mir, I. D.; Tipping, A. E. *J. Chem. Soc., Perkin Trans. 1* **1976,** 2349.
11. Rondarev, D. S.; Sass, V. P.; Sokolov, S. V. *Zh. Org. Khim.* **1975,** *11*, 937 (Engl. Transl. 927).
12. Ter-Gabrielyan, E. G.; Gambaryan, N. P.; Lur'e, E. P.; Petrovskii, P. V. *Izv. Akad. Nauk SSSR* **1979,** *28,* 1061 (Engl. Transl. 992).
13. Smart, B. E. *J. Org. Chem.* **1974,** *39*, 831.
14. Takahashi, M.; Shuyama, H.; Tsutsumi, Y. Jpn. Pat. 62 178 529, 1987; *Chem. Abstr.* **1988,** *109*, 92267n.
15. Bekker, R. A.; Melikyan, G. G.; Dyatkin, B. L.; Knunyants, I. L. *Zh. Org. Khim.* **1975,** *11*, 1604 (Engl. Transl. 1588).
16. Moreau, P.; Commeyras, A. *J. Chem. Soc., Chem. Commun.* **1985,** 817.
17. Bolton, R.; Owen, E. S. E. *J. Fluorine Chem.* **1990,** *46*, 393.
18. Ushakov, A. A.; Motsarev, G. V.; Rozenberg, V. R.; Kolbasov, V. I.; Belova, L. V.; Chuvaeva, I. N. *Zh. Org. Khim.* **1974,** *10*, 2183 (Engl. Transl. 2196).
19. Semmler, H. J.; Feser, M. Ger. Offen. 2 815 032, 1979; *Chem. Abstr.* **1980,** *92*, 41322.
20. Furutaka, Y.; Homoto, Y.; Honda, T. Jpn. Pat. 89 290 638, 1989; *Chem. Abstr.*, **1990,** *112*, 178061.
21. Morikawa, S.; Samejima, S.; Yoshitake, M.; Tatematsu, S.; Tanuma, T. Jpn. Pat. 90 70 133, 1990; *Chem. Abstr.* **1990,** *113*, 5703.
22. Rowley, L.; Webb, G.; Winfield, J. M. *J. Fluorine Chem.* **1988,** *38*, 115.
23. Bock, H.; Mintzer, J.; Wittmann, J.; Russow, J. *Angew. Chem.* **1980,** *19*, 147.
24. Ikubo, Y.; Kunihiro, K. Jpn. Pat. 77 62 208, 1977; *Chem. Abstr.* **1977,** *87*, 133859.
25. Kouketsu, N.; Inoue, F.; Komatsu, T.; Matsuoka, K. Jpn. Pat. 79 76 504, 1979; *Chem. Abstr.* **1980,** *92*, 22022.
26. Ponomarev, V. G.; Podsevalov, P. V.; Nazarenko, T. I.; Deev, L. E.; Pashkevich, K. I. *Zh. Org. Khim.* **1990,** *26*, 1365 (Engl. Transl. 1179).
27. Fukuoka, S.; Tojo, M. Jpn. Pat. 63 14 743, 1988; *Chem. Abstr.* **1988,** *109*, 109992.
28. Kelly, S. M. *Helv. Chim. Acta* **1984,** *67*, 1572.
29. Trepka, R. D.; McConville, J. W. *J. Org. Chem.* **1975,** *40*, 428.
30. Budnik, A. G.; Kalinichenko, N. V.; Shteingarts, V. D. *Zh. Org. Khim.* **1974,** *10*, 1923 (Engl. Transl. 1934).
31. Osina, O. I.; Shteingarts, V. D. *Zh. Org. Khim.* **1974,** *10*, 329 (Engl. Transl. 329).
32. Inove, F.; Katsuhara, Y.; Okazaki, K. Ger. Offen. DE 3 321 855, 1983; *Chem. Abstr.* **1984,** *100*, 191550.

33. Ushakov, A. A.; Motsarev, G. V.; Kolbasov, V. I.; Survorov, B. A.; Chuvaeva, I. N. *Zh. Org. Khim.* **1976,** *12*, 2204 (Engl. Transl. 2140).
34. Misaki, S.; Furutaka, Y.; Shimoike, T. Jpn. Pat. 75 76 029, 1975; *Chem. Abstr.* **1975,** *83*, 178535.
35. Mattern, D. L. *J. Org. Chem.* **1984,** *49*, 3051.
36. Roberts, N. L.; Whitaker, G. U.S. Pat. 4 393 214, 1983; *Chem. Abstr.* **1983,** *99*, 212421.
37. Peglion, J. L.; Pastor, R.; Greiner, J.; Cambon, A. *Bull. Soc. Chim. Fr.* **1982,** 89.
38. Chambers, R. D.; Grievson, B. *J. Fluorine Chem.* **1985,** *30*, 227.
39. Pawelke, G.; Heyder, F.; Buerger, H. *J. Fluorine Chem.* **1982,** *20*, 53.
40. Coe, P. L.; Holton, A. G.; Sleigh, J. H. *J. Fluorine Chem.* **1983,** *22*, 521.
41. Maruta, T.; Murata K. Jpn. Pat. 02 45 441, 1990; *Chem. Abstr.* **1990,** *113*, 39944.
42. Akataev, N. P.; Butin, K. P.; Sokol'skii, G. A.; Knunyants, I. L. *Izv. Akad. Nauk SSSR* **1974,** *23*, 636 (Engl. Transl. 600).
43. Adcock, J. L.; Evans, W. D. *J. Org. Chem.* **1983,** *48*, 4122.
44. Xu, H.; Zhai, H. *Huaxue Shiji* **1989,** *11*, 123; *Chem. Abstr.* **1989,** *111*, 194056.
45. Werner, K., von; Eur. Pat. EP 295 582, 1988; *Chem. Abstr.* **1989,** *111*, 114746.
46. Sevast'yan, A. P.; Fialkov, Y. A.; Khranovskii, V. A.; Yagupol'skii, L. M. *Zh. Org. Khim.* **1978,** *14*, 204 (Engl. Transl. 191).
47. Kumai, S.; Wada, A.; Morikawa, S. Eur. Pat. EP 355 719, 1990; *Chem. Abstr.* **1990,** *113*, 97175.
48. Kumai, S.; Seki, T.; Matsuo, H. Jpn. Pat. 62 289 534, 1987; *Chem. Abstr.* **1988,** *109*, 210662.
49. Ichikawa, J.; Sonada, T.; Kobayashi, H. *Tetrahedron Lett.* **1989,** *30*, 6379.
50. Heinze, P. L.; Spawn, T. D.; Burton, D. J.; Shin-Ya, S. *J. Fluorine Chem.* **1988,** *38*, 131.
51. Rondestvedt, C. S., Jr. *J. Org. Chem.* **1977,** *42*, 1985.
52. Dobrov, I. V.; Zamyslov, R. A.; Shvedchikov, A. P.; Mungalov, V. E. U.S.S.R. Pat. 642 285, 1979; from *Otkrytiya, Izobret., Prom. Obraztsy, Tovarnye Znaki* **1979,** 94; *Chem. Abstr.* **1979,** *90*, 186341.
53. Jensen, J. H. *Chem. Eng. News* **1981,** *59*, 3.
54. Christian, S. L. *J. Org. Chem.* **1984,** *49*, 4575.
55. Drivon, G.; Durval, P.; Gurtner, B.; Lantz, A. Eur. Pat. EP 298 870, 1989; *Chem. Abstr.* **1989,** *111*, 173590.
56. Drivon, G.; Gurtner, B. Fr. Pat. FR 2 587 334, 1987; *Chem. Abstr.* **1988,** *108*, 221289.
57. Werner, J. A. U.S. Pat. 4 493 932, 1985; *Chem. Abstr.* **1985,** *102*, 131929.

58. Fokin, A. V.; Studnev, Y. N.; Rapkin, A. I.; Pasevina, K. I.; Potarina, T. M.; Verenikin, O. V. *Izv. Akad. Nauk SSSR* **1980,** 2369 (Engl. Transl. 1684).
59. Tari, I.; DesMarteau, D. D. *J. Org. Chem.* **1980,** *45*, 1214.
60. Zapevalov, A. V.; Filyakova, T. I.; Peschanskii, N. V.; Kodess, M. I.; Kolenko, I. P. *Zh. Org. Khim.* **1990,** *26*, 265 (Engl. Transl. 222).
61. Kolenko, I. P.; Plashkin, V.S. *Izv. Akad. Nauk SSSR* **1977,** 1648 (Engl. Transl. 1518).
62. Denivelle, L.; Huynh, A. H. *Bull. Soc. Chim. Fr.* **1974,** 2171.
63. Nickson, T. E.; Roche-Dolson, C. A. *Synthesis* **1985,** 669.
64. Fox, G. J.; Hallas, G.; Hepworth, J. D.; Paskins, K. N. *Org. Synth.* **1973,** *53*, 156; Collective Vol. **1988**, *6*, 181.
65. Banks, R. E.; Noakes, T. J. *J. Chem. Soc., Perkin Trans. 1* **1976,** 143.
66. Banks, R. E.; Barlow, M. G.; Hornby, J. C.; Noakes, T. J. *J. Fluorine Chem.* **1989,** *42*, 179.
67. Chen, L. S.; Chen, G. J. *J. Fluorine Chem.* **1989,** *42*, 371.
68. Saloutin, V. I.; Bobrov, M. B.; Pashkevich, K. I. *Zh. Org. Khim.* **1987,** *23*, 892 (Engl. Transl. 807).
69. Schack, C. J.; Christe, K. O. U.S. Pat. 4 222 968, 1980; *Chem. Abstr.* **1981,** *94*, 65096.
70. Sorochinskii, A. E.; Aleksandrov, A. M.; Kukhar, V. P.; Krasnoshchek, A. P. *Zh. Org. Khim.* **1979,** *15*, 445 (Engl. Transl. 395).
71. Olah, G. A.; Narang, S. C.; Field, L. D. *J. Org. Chem.* **1981,** *46*, 3727.
72. Tyuleneva, V. V.; Rozov, L. A.; Zeifman, Y. V.; Knunyants, I. L. *Izv. Akad. Nauk SSSR* **1975,** 1136 (Engl. Transl. 1042).
73. Sokolov, V. B.; Korenchenko, O. V.; Aksinenko, A. Y.; Martynov, I. V. *Izv. Akad. Nauk SSSR* **1988,** 2189; *Chem. Abstr.* **1989**, *110*, 231092.
74. Eapen, K. C.; Eisentraut, K. J.; Ryan, M. T.; Tamborski, C. *J. Fluorine Chem.* **1986,** *31*, 405.
75. Klauke, E.; Buettner, G.; Scholl, H. J.; Schwarz, H. Ger. Pat. 2 546 533, 1977; *Chem. Abstr.* **1977,** *87,* 134443.
76. Karpov, V. M.; Platonov, V. E.; Yakobson, G. G. *Izv. Akad. Nauk SSSR* **1979**, 2082 (Engl. Transl. 1919).
77. Igumnov, S. M.; Chaplina, I. V. *Izv. Akad. Nauk SSSR* **1988,** 2649; *Chem. Abstr.* **1989,** *110*, 212055.
78. Abe, T.; Hayashi, E.; Baba, H.; Nagase, S. *J. Fluorine Chem.* **1984,** *26*, 295.
79. Abe, T.; Nagase, S. *J. Fluorine Chem.* **1978,** *12*, 359.
80. Barlow, M. G.; Coles, B.; Haszeldine, R. N. *J.Fluorine Chem.* **1980,** *15*, 397.
81. Cherstkov, V. F.; Sterlin, S. R.; German, L. S. *Izv. Akad. Nauk SSSR* **1985,** 2647; *Chem. Abstr.* **1986,** *105*, 133424.

82. Bowden, R. D.; Slater, R. Brit. Pat. 1 367 383, 1974; *Chem. Abstr.* **1975,** *82*, 31266.
83. Fukaya, H.; Abe, T.; Hayashi, E. *Chem. Lett.* **1990,** 813.
84. Ankudinov, A. K.; Ryazanova, R. M.; Sokolov, S. V. *Zh. Vses. Khim. O-va.* **1977,** *22*, 459; *Chem. Abstr.* **1977,** *87*, 200743.
85. Banks, R. E.; Birchall, J. M.; Haszeldine, R. N.; Nicholson, W. J. *J. Fluorine Chem.* **1982,** *20*, 133.

Nitration

by G. L. Cantrell and J. F. Lang

Nitrated fluoro compounds are synthesized by electrophilic (NO_2+), radical (NO_2·), or nucleophilic (NO_2–) methods. Indirect nitration routes can suppress the side reactions associated with severe reaction conditions and some nitration reagents. Novel fluoronitro compounds, unobtainable by direct nitration, can also be prepared. For example, the nitration of (2-fluoro-2,2-dinitroethoxy)acetaldoxime followed by oxidation of the nitroso intermediate with hydrogen peroxide yields 2-fluoro-2,2-dinitroethyl 2,2-dinitroethyl ether [*1*] (equation 1).

[*1*] **1**

$$FC(NO_2)_2CH_2OCH_2CH{=}NOH \xrightarrow[\text{0.5° C, 20 min}]{\text{1. 90\% }HNO_3\text{; 2. 30\% }H_2O_2} FC(NO_2)_2CH_2OCH_2CH(NO_2)_2 \quad 65\%$$

The mild nitrating agents **thionyl chloride nitrate** (equation 2a) and **thionyl nitrate** (equation 2b) react with ***alcohols*** and ***phenols*** to form stable nitrates. The trinitrate of 2,6-di(hydroxymethyl)-4-fluorophenol is prepared by either agent [*2*] (equation 2).

[*2*] **2**

$$\text{a. } SOCl_2 + AgNO_3 \xrightarrow{\text{THF}} SOCl(NO_3) + AgCl$$

$$\text{b. } SOCl_2 + 2AgNO_3 \xrightarrow{\text{THF}} SO(NO_3)_2 + 2AgCl$$

2,6-di(hydroxymethyl)-4-fluorophenol (HO, OH, OH; F) → [3 eq a. or 3 eq b.; THF, 25° C, 1 h] → trinitrate (O_2NO, ONO_2, ONO_2; F)   a. 85% b. 76%

Polyfunctional **fluoronitro alcohols** are provided by the SRN1 reaction of a ***perfluoroalkyl iodide*** or alkylene diiodides with the anhydrous **lithium salt of 2-nitropropane-1,3-diol acetonide.** Hydrolysis of the resulting perfluoroalkyl-

0065–7719/95/0187–0387$08.00/1

substituted nitro acetonide by transketalization gives the corresponding fluoronitro diol or dinitro tetrol. The decyl derivative is insoluble and is not hydrolyzed [*3*] (equation 3).

[*3*] **3**

$$CF_3(CF_2)_nI \xrightarrow[\text{DMF, } N_2\text{; RT, 24 h}]{\text{2,2-dimethyl-1,3-dioxan-5-ylidene nitronate } NO_2^- Li^+} \text{5-}CF_3(CF_2)_n\text{-5-nitro-2,2-dimethyl-1,3-dioxane}$$

$$\xrightarrow[\text{85° C, 1 h}]{BF_3{\cdot}Et_2O,\ (CH_2OH)_2} CF_3(CF_2)_nC(NO_2)(CH_2OH)_2$$

n = 6 58%; 7 83%; 9 no reaction

Nucleophilic displacement of iodide by the nitrite ion in 1-iodo-1*H*,1*H*,2*H*,2*H*-perfluoroalkanes affords the 1-nitro analogue (equation 4). Oxidative nitration of the 1-nitro-1*H*,1*H*,2*H*,2*H*-perfluoroalkane with tetranitromethane yields the *gem*-dinitro compound [*4*].

[*4*] **4**

$$R_fCH_2CH_2I \xrightarrow[\text{25° C, 20 h}]{\text{phloroglucinol}\cdot 2\,H_2O;\ NaNO_2,\ DMF} R_fCH_2CH_2NO_2 + R_fCH_2CH_2OH$$

R_f =	$R_fCH_2CH_2NO_2$	$R_fCH_2CH_2OH$
$CF_3(CF_2)_5$	27%	
$CF_3(CF_2)_9$	51%	40%

$$R_fCH_2CH_2NO_2 \xrightarrow[\text{RT, 30 min}]{C(NO_2)_4,\ K_2CO_3} [R_fCH_2C(NO_2)_2]^- K^+ \xrightarrow{10\%\ HCl} R_fCH_2CH(NO_2)_2$$

R_f =	
$CF_3(CF_2)_5$	36%
$CF_3(CF_2)_9$	62%

Nitrite ion, an ambident nucleophile in aprotic solvents, favors nitrogen atom attack on the double bond of various fluoro(halo)olefins. 2-Monohydroperfluoronitroalkanes can thus be produced [*5*] (equation 5).

[*5*] **5**

$$CF_2{=}CFX \xrightarrow[\substack{H_2O,\ DMF \\ 20^{\circ}\ C,\ day(s)}]{NaNO_2} [O_2NCF_2CFX]^- \longrightarrow O_2NCF_2CFHX$$

X =	
F	50%
Br	30%
I	23.6%
CF_3	10%
OCF_3	50%

Addition of **nitrogen dioxide** to 1′,2′,2′-trifluorostyrene gives a nitro–nitrite or nitro–nitrate adduct, which affords 1′-nitro-1′,1′-difluoroacetophenone [*6*] (equation 6).

[*6*] **6**

$$C_6H_5CF{=}CF_2 \xrightarrow{N_2O_4} C_6H_5\underset{\displaystyle ONO_2}{\underset{|}{C}}FCF_2NO_2 \xrightarrow{H_2O\ \text{or dil.}\ H_2SO_4} C_6H_5COCF_2NO_2$$

Nitronium fluorosulfate in fluorosulfonic acid adds electrophilically across the double bond of ***fluoroolefins*** in a nonspecific manner. Trifluorochloroethylene reacts accordingly with nitronium fluorosulfate to give a 2:1 mixture of regioisomers [*7*] (equation 7). Under these reaction conditions perfluoropropylene is unreactive even after extended heating at 80 °C. 2-Nitroperfluoropropyl fluorosulfate is obtained on treatment of the perfluoropropylene with nitronium fluorosulfate in antimony pentafluoride [*8*] (equation 8).

Radical and ionic nitrations are often competitive pathways in strong nitrating acid mixtures. The predominant reaction pathway is determined by the composition of the nitrating medium. Oxides of nitrogen in the nitrating medium add to fluoroethylenes via *radical nitration. Ionic nitration* of fluoroethylenes with a mixture of 100% sulfuric acid and 99% nitric acid forms *halonitroacetic acids* [*9*] (equation 9), presumably through the corresponding acetyl fluoride. The nitroacetyl fluoride is the major product when sulfur trioxide is present [*9, 10*] (equation 10). If the nitrating mixture contains additional oleum and the substrate is trifluoroethylene, *ionic nitration* affords *nitroethyl fluorosulfates* [*9*] (equation 11).

[*7*] **7**

$$CF_2{=}CFX \xrightarrow[\text{Freon 113, }20^\circ\text{ to }40^\circ\text{ C}]{HSO_3F,\ O_2NSO_3F} O_2NCF_2CFXOSO_2F + O_2NCFXCF_2OSO_2F$$

X =	$O_2NCF_2CFXOSO_2F$	$O_2NCFXCF_2OSO_2F$
H		78.4%
F	71%	
Cl	51.9%	25.9%
CF_3	no reaction	
OCF_3	63.5%	

[*8*] **8**

$$CF_3CF{=}CF_2 \xrightarrow[60^\circ\text{ C, 2 days}]{O_2NSO_3F,\ SbF_5} O_2NCF(CF_3)CF_2OSO_2F$$

78%

[*9*] **9**

$$CFCl{=}CFCl \xrightarrow[8\text{ to }20^\circ\text{ C}]{1\ HNO_3,\ 1.25\ H_2SO_4\ (v/v)} O_2NCFClCO_2H$$

63%

[*9,10*] **10**

$$CRX{=}CFCl \xrightarrow[5\text{ to }15^\circ\text{ C}]{HNO_3,\ H_2SO_4,\ 60\%\text{ oleum }(1.3:1.4:1.0\text{ w/w})} O_2NCRXCOF$$

R =	X =	Yield
H	Cl	76%
H	Br	58%
F	Cl	40%

[*9*] **11**

$$CF_2{=}CFR \xrightarrow[40^\circ\text{ C, 4 h}]{HNO_3,\ H_2SO_4,\ SO_3\ (1:3.5:2\text{ w/w})} O_2NCFRCF_2OSO_2F$$

R = H or F 35%

Unsubstituted positions in ***polyfluorinated aromatics*** are nitrated under vigorous conditions. 1,3,5-Trifluorobenzene reacts with **oleum and a metal nitrate** at elevated temperature to provide 1,3,5-trifluoro-2,4,6-trinitrobenzene [*11*] (equation 12).

Substituted nitrobenzotrifluorides, useful as intermediates for agrochemicals, pharmaceuticals, and liquid crystals, are prepared by direct nitration of the respective benzotrifluorides. The *meta*-directing influence of the trifluoromethyl group is overridden by the *para*-directing effect of fluorine. As a result, nitration of 3,4-difluorobenzotrifluoride generates a 98% yield of 2-nitro-4,5-difluorobenzotrifluoride [*12*] (equation 13). Nitration of 2-chlorobenzotrifluoride in strong acid media gives the expected *meta*-dinitration [*13*] (equation 14), as does 4-chlorobenzotrifluoride [*14, 15, 16, 17, 18*] (equation 15).

[*11*] **12**

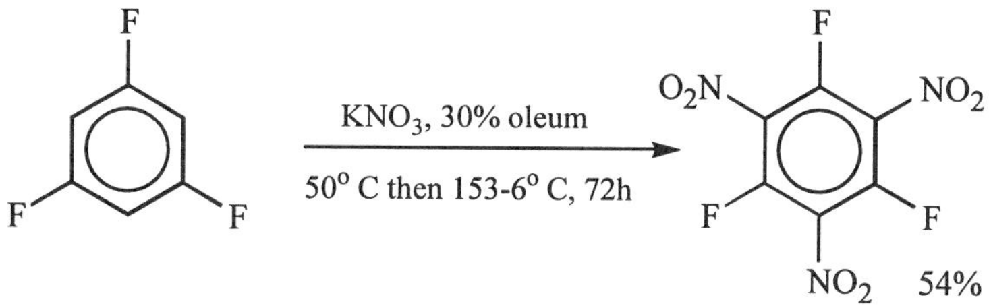

F, F, F (1,3,5-trifluorobenzene) → [KNO_3, 30% oleum; 50° C then 153-6° C, 72h] → F, O_2N, NO_2, F, F, NO_2 54%

[*12*] **13**

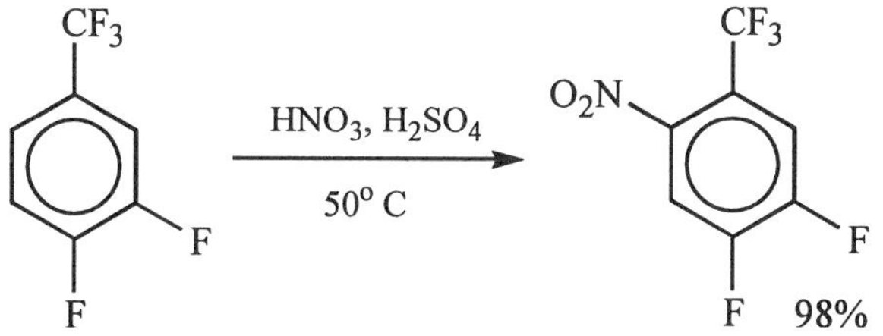

CF_3, F, F → [HNO_3, H_2SO_4; 50° C] → CF_3, O_2N, F, F 98%

[*13*] **14**

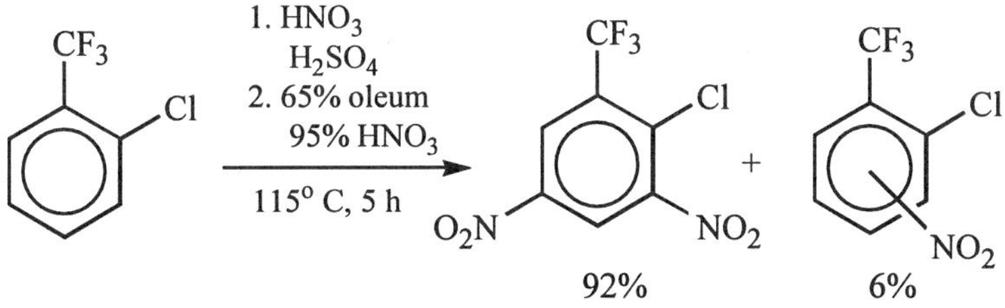

CF_3, Cl → [1. HNO_3 H_2SO_4 2. 65% oleum 95% HNO_3; 115° C, 5 h] → CF_3, Cl, O_2N, NO_2 92% + CF_3, Cl, NO_2 6%

[*14,15,16,17,18*] **15**

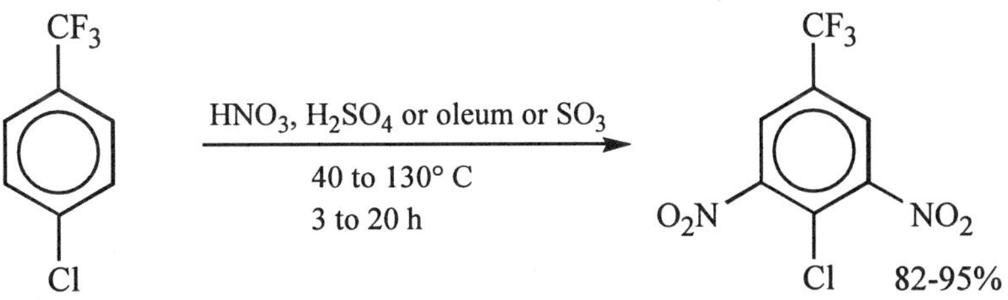

CF_3, Cl → [HNO_3, H_2SO_4 or oleum or SO_3; 40 to 130° C 3 to 20 h] → CF_3, O_2N, NO_2, Cl 82-95%

References are listed on pages 396–397.

Mononitration of a mixture of *3- and 4-chlorobenzotrifluorides,* followed by nucleophilic substitution by hydroxide, ammonia, or a primary or secondary amine in dimethylformamide, leads to 5-chloro-2-nitrobenzotrifluoride. The 4-chloro-3-nitro isomer selectively reacts and can be removed as a water-soluble phenoxide [*19*] (equation 16).

[*19*] **16**

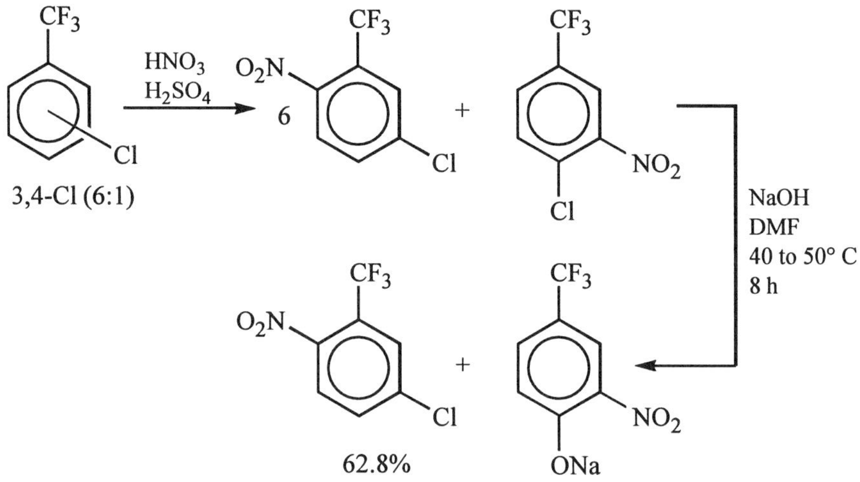

CF_3 ; Cl ; 3,4-Cl (6:1) ; HNO_3, H_2SO_4 → 6 ; O_2N, CF_3, Cl + CF_3, NO_2, Cl ; NaOH, DMF, 40 to 50° C, 8 h → O_2N, CF_3, Cl (62.8%) + CF_3, NO_2, ONa

Upon nitration of 4-trifluoromethylbenzotrichloride or -benzoyl chloride, a 76.3% yield of a mixture of 4-trifluoromethyl-2- and 3-nitrobenzoic acids is obtained [*20*].

Activation of benzotrifluoride by a 3-methyl substitution gives nonspecific nitration at the *ortho* and *para* sites [*21*] (equation 17).

[*21*] **17**

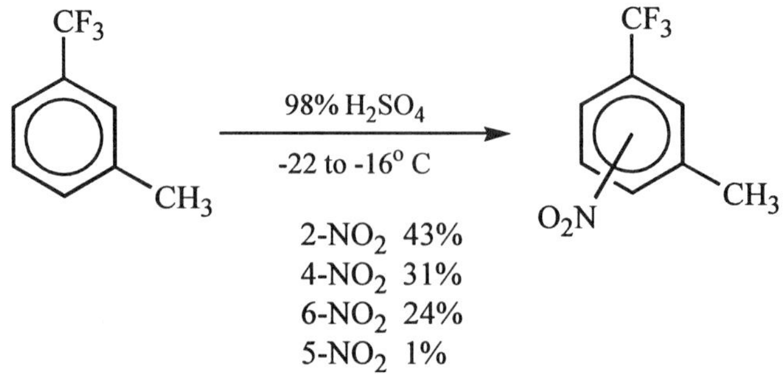

CF_3, CH_3 ; 98% H_2SO_4, -22 to -16° C → CF_3, CH_3, O_2N

2-NO_2 43%
4-NO_2 31%
6-NO_2 24%
5-NO_2 1%

If the benzotrifluoride contains an amino group, *N-nitration* in acetic acid and acetic anhydride furnishes the trifluoromethyl nitroaminobenzene [*22*]. C-nitration occurs when the amino group is N-acylated. 3-Trifluoromethylisobutyranilide therefore yields 3′-trifluoromethyl-4′-nitroisobutyranilide [*23*] (equation 18).

[*23*] **18**

$NHCOCH(CH_3)_2$; CF_3 —[1. oleum; 2. 90% HNO_3; 5° C, 2 h]→ $NHCOCH(CH_3)_2$; CF_3 ; NO_2

Mildly activated aromatics such as ***4-fluorotoluene*** undergo *ipso substitution* at the site of activation. Diastereomeric (*E/Z*) 1,4-adducts (81%) and 1,2-adducts (8%) of nitronium acetate are formed. The *E* and *Z* 1,4-adducts are isolated in 4 and 32% yields, respectively [*24*] (equation 19).

[*24*] **19**

CH_3 ; F —[HNO_3, Ac_2O, $(CF_3CO)_2O$; -40 to -78° C, 1.5 h]→ CH_3 NO_2 ; F OAc (I) + CH_3 NO_2 OAc (II) + O ; CH_3 NO_2 (III)

I: II: III, 81: 8: 11
I E 4%, Z 32%

An *ipso attack* on the fluorine carbon position of **4-fluorophenol** at –40 °C affords 4-fluoro-4-nitrocyclohexa-2,5-dienone in addtion to 2-nitrophenol. The cyclodienone slowly isomerizes to the 2-nitrophenol. Although *ipso* nitration on 4-fluorophenyl acetate furnishes the same cyclodienone, the major by-product is 4-fluoro-2,6-dinitrophenol [*25*]. Under similar conditions, 4-fluoroanisole primarily yields the 2-nitro isomer and 6% of the cyclodienone. The isolated 2-nitro isomer is postulated to form by attack of the nitronium ion *ipso* to the fluorine with concomitant capture of the incipient carbocation by acetic acid. Loss of the elements of methyl acetate follows. The nitrodienone, being the keto tautomer of the nitrophenol, aromatizes to the isolated product [*26*] (equation 20). Intramolecular capture of the intermediate carbocation occurs in nitration of 2-(4-fluorophenoxy)-2-methylpropanoic acid at low temperature to give the spiro products: 3,3-dimethyl-8-fluoro-8-nitro-1,4-dioxaspiro[4.5]deca-6,9-dien-2-one and the 10-nitro isomer [*26*] (equation 21).

References are listed on pages 396–397.

[*25,26*] **20**

R = H

0° C

OR, OR, O, OH

HNO_3 and/ or Ac_2O

-40° C

30 min

NO_2 + O_2N + NO_2

F, F, F NO_2, F

R =					
	H	81%	19%		[*25*]
	AcO		56%	44%	[*25*]
	OCH_3	94%	6%		[*26*]

[*26*] **21**

0° C

$OC(CH_3)_2CO_2H$

HNO_3

Ac_2O

-30° C

10 min

$OC(CH_3)_2CO_2H$ NO_2 + O O O H NO_2 + O O O

F, F, F, F NO_2

53% 26% 21%

An *ipso nitration* on ***polyfluoroaromatics*** such as tetrafluoroxylene, toluenes, or benzenes leads to stable polyfluoro-1,4-cyclohexadienes and polyfluoronitro-1,4-cyclohexadienes. The nitrocyclohexadiene is easily converted to the poly-fluorocyclohexadiene by fluorodenitration in hydrogen fluoride. In examples with an unsubstituted site, the expected tetrafluoronitrotoluene, tetrafluoronitrobenzene, or pentafluoronitrobenzene is produced. With nitronium salts, nitration at the unsubstituted site predominates [*27, 28*] (equation 22). Fully substituted poly-fluoro-aromatics, for example pentafluorotoluene, is exclusively *ipso* nitrated to give 1-methyl-3-nitro-2,3,4,5,6,6-hexafluoro-1,4-cyclohexadiene. Additional hydro-gen fluoride at higher temperature converts the nitrocyclodienone to 1-methyl-2,3,3,4,5,6,6-heptafluoro-1,4-cyclohexadiene [*29*] (equation 22).

Similarly, ***2-methylheptafluoronaphthalene*** when treated with fuming nitric acid in hydrogen fluoride furnishes 1-nitro-3-methyl-1,2,4,4,5,6,7,8-octafluoro-

1,4-dihydronaphthalene in 45% yield. The nitronium ion attacks the *ipso* position meta with respect to the methyl group. Further treatment with hydrogen fluoride at ambient temperature for 20 min provides 2-methyl-1,1,3,4,4,5,6,7,8-nonafluoro-1,4-dihydronaphthalene in 50% yield [*29*].

[*27,28,29*] **22**

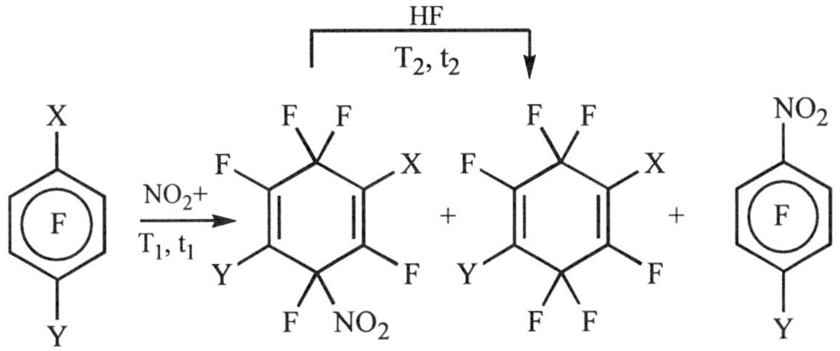

I II III

Nitration Reagents	X =	Y =	T_1 °C	T_2 °C	Time t_1, t_2 (h)	Yield I	Yield II	Yield III	Ref.
HNO_3/HF	CH_3	CH_3	0		1/3	38			[*27*]
"	CH_3	CH_3	0	20	1/2, 1/3		64		"
"	CH_3	Cl	20		5	74			"
"	CH_3	Br	20		5	72			"
"	CH_3	H	20		3	24	13	16	"
"	H	Cl	40		6	35	14		"
"	H	Br	40		5	17	15	14	"
"	H	H	20		3	63			[*28*]
"	H	H	50		12		56		"
NO_2BF_4 + sulfolane	H	H	50		12	60			"
"	H	F	50		12			77	"
HNO_3/HF + SbF_5	H	F	-20 to 59		12			63	"
HNO_3/HF	CH_3	F	20		4	77			[*29*]
"	CH_3	F	20	55	4, 3		66		"

References are listed on pages 396–397.

References for Pages 387–395

1. Grakauskas, V. *J. Org. Chem.* **1973,** *38,* 2999.
2. Haklimelahi, G. H.; Sharghi, H.; Zarrinmayeh, H.; Khalafi-Nezhad, A. *Helv. Chim. Acta* **1984,** *67,* 906.
3. Archibald, T. G.; Taran, C.; Baum, K. *J. Fluorine Chem.* **1989,** *43,* 243.
4. Malik, A. A.; Archibald, T. G.; Tzeng, D.; Garver, L. C.; Baum, K. *J. Fluorine Chem.* **1989,** *43,* 291.
5. Krzhizhevskii, A. M.; Cheburkov, Y. A.; Knunyants, I. L. *Izv. Akad. Nauk SSSR* **1974,** 2144 (Engl. Transl. 2065).
6. Fokin, A. V.; Komarov, V. A.; Davydova, S. M. U.S.S.R. Pat. 351 841, 1972; *Chem. Abstr.* **1973,** *78,* 29442s.
7. Fokin, A. V.; Studnev, Y. N.; Rapkin, A. I.; Chilikin, V. G.; Verenikin, O. V. *Izv. Akad. Nauk SSSR* **1983,** 1437 (Engl. Transl. 1306).
8. Fokin, A. V.; Studnev, Y. N.; Rapkin, A. I.; Chilikin, V. G. *Izv. Akad. Nauk SSSR* **1984,** 473 (Engl. Transl. 439).
9. Martynov, I. V.; Uvarov, V. I.; Brel', V. K.; Anufriev, V. I.; Yarkov, A. V. Izv. *Akad. Nauk SSSR* **1989,** 2732 (Engl. Transl. 2500).
10. Martynov, I. V.; Anufriev, V. I.; Uvarov, V. I.; Brel', V. K. *Izv. Akad. Nauk SSSR* **1987,** 700 (Engl. Transl. 640).
11. Koppes, W. M. et. al. U.S. Pat. Appl. 937 281, 1979; *Chem. Abstr.* **1979,** *91,* 39112s.
12. Kumai, S.; Seki, T.; Matsuo, H. Jpn. Pat. 62 289 534, 1987; *Chem. Abstr.* **1988,** *109,* 210662n.
13. Buettner, G.; Klauke, E.; Wolfrum, G. Ger. Offen. 2 635 695, 1978; *Chem. Abstr.* **1978,** *88,* 136288z.
14. Csanda, E.; Eifert, G.; Hegedus, I.; Kuronya, I.; Schmidt. P.; Sovegjarto, O.; Vilagi, E. *Hung. Teljes* **1973,** 5650; *Chem. Abstr.* **1973,** *79,* 78364x.
15. Milligan, B. U.S. Pat. 3 984 488, 1976; *Chem. Abstr.* **1976,** *85,* 192338r.
16. Macko, J.; Rupcik, M.; Vrzgula, D.; Mrva, B.; Vyboh, P. Czech. Pat. 170 963, 1978; *Chem. Abstr.* **1978,** *88,* 90123q.
17. Bornengo, M. Ger. Offen. 2 746 787, 1978; *Chem. Abstr.* **1978,** *89,* 59768x.
18. Schneider, L.; Graham, D. E. U.S. Pat. 4 096 195, 1978; *Chem. Abstr.* **1978,** *89,* 215053t.
19. Jpn. Pat. 82 102 847, 1982; *Chem. Abstr.* **1982,** *97,* 162562q.
20. Nishimura, Y.; Gotoh, Y.; Kawai, T. Ger. Offen. 3 743 606, 1988; *Chem. Abstr.* **1988,** *109,* 190029f.
21. Chupp, J. P.; Alt. G. H. U.S. Pat. 4 467 125, 1984; *Chem. Abstr.* **1984,** *101,* 191317v.
22. Ayad, K.; Long, F. Br. Pat. 1 329 012, 1973; *Chem. Abstr.* **1974,** *80,* 3238w.
23. Peer, L.; Mayer, J. U.S. Pat. 4 302 599, 1981; *Chem. Abstr.* **1982,** *96,* 103817h.

24. Fisher, A.; Fyles, D. L.; Henderson, G. N.; Mahasay, S. R. *Can. J. Chem.* **1986,** *64,* 1764.
25. Clewley, R. G.; Cross, G. G.; Fischer, A.; Henderson, G. N. *Tetrahedron* **1989,** *45,* 1299.
26. Clewley, R. G.; Fischer, A.; Henderson, G. N. *Can. J. Chem.* **1989,** *67,* 1472.
27. Shtark, A. A.; Shteingarts, V. D. *Izv. Sib. Otd. Akad. Nauk SSSR* **1976,** 123; *Chem. Abstr.* **1976,** *85,* 176911s.
28. Shtark, A. A.; Shteingarts, V. D. *Zh. Org. Khim.* **1986,** *22,* 831 (Engl. Transl. 742).
29. Shtark, A. A.; Shteingarts, V. D. Maidanyuk, A. G. *Izv. Sib. Otd. Akad. Nauk SSSR* **1974,** 117; *Chem. Abstr.* **1975,** *82,* 86237k.

Nitrosation

by G. L. Cantrell and J. F. Lang

Nitrosated fluoro compounds are frequently prepared by electrophilic nitrosation on an electron-rich center of oxygen, nitrogen, or carbon. The preparation of fluorochloronitronitrosomethane from the decarboxylation of fluorochloronitroacetic acid in nitric acid is unique [*1*] (equation 1).

[*1*] **1**

$$\underset{\displaystyle NO_2}{\overset{|}{CClFCO_2H}} \xrightarrow[80^\circ C]{\text{fuming } HNO_3} \underset{\displaystyle NO_2}{\overset{|}{CClFNO}} \quad 52\%$$

Electrophilic nitrosation of the carbanion generated from the reaction of an organic base with a strong organic acid, such as α-hydrohexafluoroisobutyronitrile [*2*]; α-hydrohexafluoroisobutyric acid or its acid chloride [*3*]; or α-hydrotetrafluoroethanesulfonyl fluoride [*4*], yields the corresponding α-nitroso compound as the major product (equations 2 and 3). The α-hydrohexafluoroisobutyric acid or acid chloride reacts with excess trifluoroacetyl nitrite in dimethylformamide to afford the O-substituted oxime [*3*] (equation 4).

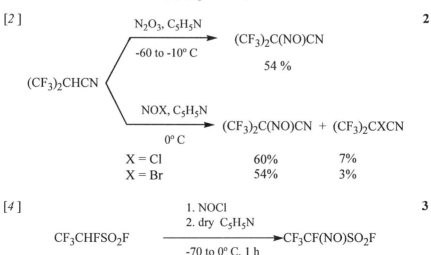

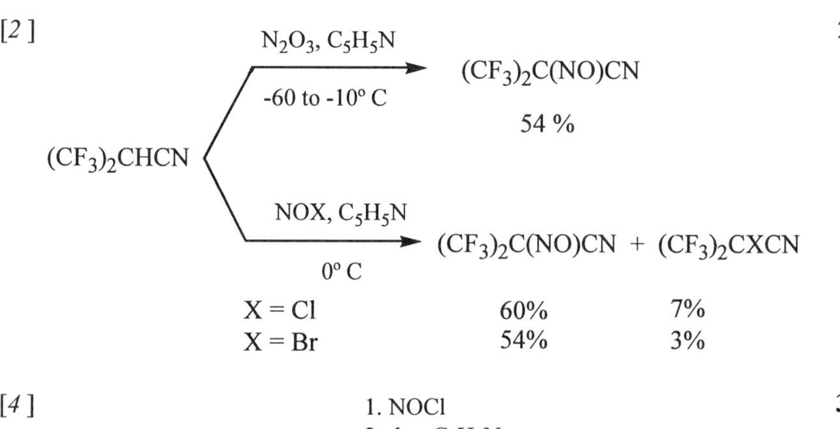

0065–7719/95/0187–0398$08.00/1

[*3*] **4**

$$(CF_3)_2CHX \xrightarrow[]{\text{DMF} \atop CF_3CO_2NO} (CF_3)_2C{=}NOY$$

$X = CO_2H$	$Y = NO$	40%
$X = COCl$	$Y = CF_3CO$	53%

In the presence of potassium fluoride in dimethylformamide, trifluoroacetyl nitrite converts perfluoroisobutylene to tris(trifluoromethyl)nitrosomethane via the carbanion generated by the nucleophilic attack of fluoride ion on perfluoroisobutylene [*3*] (equation 5).

[*3*] **5**

$$(CF_3)_2C{=}CF_2 \xrightarrow{\text{DMF, KF}} [(CF_3)_3C]^- \xrightarrow{CF_3CO_2NO} (CF_3)_3CNO + CF_3CO_2K$$

47%

Sodium nitrite in dimethylformamide acts as a nucleophile and reacts with perfluoropropene to generate a perfluoroalkyl nitrite anion. The intermediate carbanion undergoes intramolecular nitrosation with loss of carbonyl difluoride to give trifluoroacetic acid upon hydrolysis [*5*] (equation 6).

[*5*] **6**

$$CF_3CF{=}CF_2 \xrightarrow[\text{DMF, -78° C}]{NaNO_2} [CF_3CFCF_2ONO]^- \xrightarrow[+H_2O]{-CF_2O} CF_3CO_2H$$

38%

Although nitrosation of β-dicarbonyl compounds becomes increasingly more facile upon successive replacement of the α-alkyl groups with perfluoroalkyl groups because of the increased ionization of the perfluorinated enolate (equation 7), the stability of the nitrosodiketone tautomers decreases. Thus, 1,1,1-trifluoropentane-2,4-dione and 1,1,1,5,5,5-hexafluoropentane-2,4-dione nitrosate much faster than penta-2,4-dione but yield ketoximes, which decompose upon workup [*6*].

[*6*] **7**

$$CF_3COCH_2COCF_3 \rightleftharpoons CF_3C(OH){=}CHCOCF_3 \xrightarrow[H_2SO_4,\ 25°\ C]{BuONO} CF_3COCH(NO)COCF_3 \rightleftharpoons CF_3COC({=}NOH)COCF_3$$

Fluoroalkylamines react with nitrous acid to produce the corresponding unstable fluoroaliphatic diazonium ions. Placement of the trifluoromethyl group at a carbon position α, β, or γ to a diazonium ion was used to probe the inductive effect on the chemistry of the transient carbocation resulting from dediazoniation [*7*]. If the fluoroalkyl group is bound to the same carbon as the amino group, conversion to the more stable diazo compound occurs. For example, 4-diazo-1,1,1,2,2-pentafluoro-3-pentafluoroethyl-3-trifluoromethylbutane is obtained from the reaction of the polyfluoroalkylamine salt with sodium nitrite [*8, 9*] (equation 8).

[*8,9*] **8**

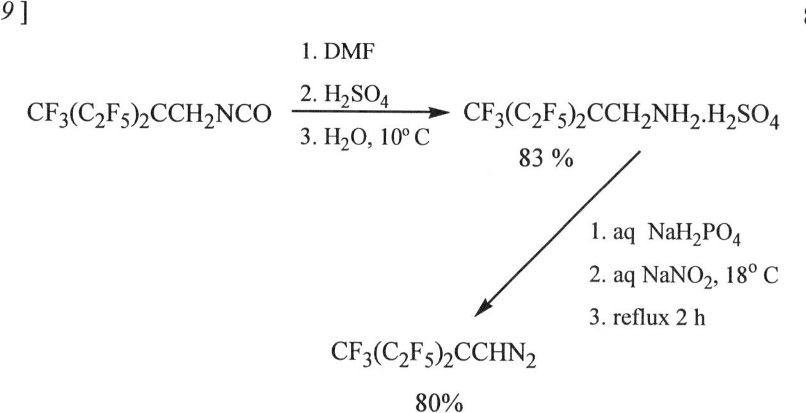

Heavily fluorinated aminobenzenes, pyridines, and pyrimidines are diazotized in strong-acid media. Solid sodium nitrite added directly to the fluorinated amine dissolved in 80% hydrofluoric acid, anhydrous hydrogen fluoride, or (1:1 wt/wt) 98% sulfuric acid in (86:14 wt/wt) acetic and propionic acids affords the electrophilic fluoroarenediazonium ion. Addition of an electron-rich aromatic to the resultant diazonium solution gives the fluoroareneazo compound [*10, 11*] (equations 9 and 10).

[*10*] **9**

4-NH_2-(F)-pyridine $\xrightarrow[\text{AcOH, EtCO}_2\text{H, 3 h, < 30° C}]{\text{NaNO}_2\text{, H}_2\text{SO}_4}$ 4-N_2HSO_4-(F)-pyridine $\xrightarrow[\text{0° C, 3 h}]{\text{ArH}}$ 4-(N=NAr)-(F)-pyridine

Ar	
= 2,4,6- $(CH_3)_3C_6H_2$	77%
4-$HOC_6H_3CH_3$-2	66%
4-$HOC_6H_2(CH_3)_2$-2,6	89%
2-$HOC_{10}H_6$	69%

[*11*] **10**

$$\mathrm{Ar_fNH_2} \xrightarrow[\text{0° C, 3 h}]{\text{NaNO}_2\text{ in 50\% H}_2\text{SO}_4\text{/43\% AcOH/7\% EtCO}_2\text{H}} \mathrm{Ar_fN_2^+} \xrightarrow[\text{RT, 3 h}]{\text{mesitylene}} \text{2,4,6-(CH}_3)_3\text{C}_6\text{H}_2\text{N=NAr}_f$$

Ar_f =	Yield
C_6F_5	33 %
$4\text{-}C_5F_4N$	77 %
$4\text{-}CF_3C_6F_4$	93 %
$4\text{-}C_5F_3NCl\text{-}3$	97 %
$4\text{-}C_5F_2NCl_2\text{-}3,5$	81 %
$2\text{-}C_5F_3NCF(CF_3)_2\text{-}4$	9.4 %
$4\text{-}C_4F_3N_2\text{-}1,3$	77 %

Reaction of 2- or 3-aminopyridine with trifluoronitrosomethane forms the trifluoromethaneazo derivative directly. 4-Aminopyridine fails to give the azo product [*12*] (equation 11).

[*12*] **11**

$$\text{H}_2\text{N-C}_5\text{H}_4\text{N} \xrightarrow[\text{-78° C, 3 h}]{\text{CF}_3\text{NO, CH}_3\text{OH}} \text{CF}_3\text{N=N-C}_5\text{H}_4\text{N}$$

$2\text{-}NH_2$	$2\text{-}CF_3N_2$	38%
$3\text{-}NH_2$	$3\text{-}CF_3N_2$	53%
$4\text{-}NH_2$	$4\text{-}CF_3N_2$	0%

References for Pages 398–401

1. Martynov, I. V.; Brel', V. K.; Uvarova, L. V. *Izv. Akad. Nauk SSSR* **1986,** 952 (Engl. Transl. 870).
2. Aktaev, N. P.; Butin, K. P.; Sokol'skii, G. A.; Knunyants, I. L. *Izv. Akad. Nauk SSSR* **1974,** 636 (Engl. Transl. 600).
3. Isaev, V. L.; Mal'kevich, L. Y.; Truskanova, T. D.; Sterlin, R. N.; Knunyants, I. L. *Zh. Vses. Khim. O-va.* **1975,** *20,* 233; *Chem. Abstr.* **1975,** *83,* 9068z.
4. Sokol'skii, G. A.; Dubov, S. S.; Medvedev, A. N.; Ragulin, L. I.; Chelobov, F. N.; Shalaginov, Y. M.; Knunyants, I. L. *Izv. Akad. Nauk SSSR* **1972,** 129 (Engl. Transl. 116).
5. Krzhizhevskii, A. M.; Cheburkov, Y. A.; Knunyants, I. L. *Izv. Akad. Nauk SSSR* **1974,** 2144 (Engl. Transl. 2065).

6. Crookes, M. J.; Roy, P.; Williams, D. L. H. *J. Chem. Soc., Perkin Trans. 2* **1989,** 1015.
7. Gassen, K. R.; Kirmse, W. *Chem. Ber.* **1986,** *119,* 2233.
8. Coe, P. L.; Sellers, S. F.; Tatlow, J. C.; Fielding, H. C.; Whittaker, G. *J. Chem. Soc., Perkin Trans. 1* **1983,** 1957.
9. Coe, P. L.; Cook, M. I.; Goodchild, N. J.; Edwards, P. N. *J. Fluorine Chem.* **1986,** *34,* 191.
10. Alty, A. C.; Banks, R. E.; Thompson, A. R.; Fishwick, B. R. *J. Fluorine Chem.* **1984,** *26,* 263.
11. Alty, A. C.; Banks, R. E.; Thompson, A. R.; Velis, H. S.; Fishwick, B. R. *J. Fluorine Chem.* **1988,** *40,* 147.
12. Ginsburg, V. A.; Vasil'eva, M. N. *Zh. Org. Khim.* **1973,** *9,* 1080 (Engl. Transl. 1108).

Sulfonation

by G. L. Cantrell and J. F. Lang

Most of the material presented in this section are reactions of **sulfur trioxide.** This compound is ambivalent and frequently forms a carbon–sulfur bond (true sulfonation), but it can form a carbon–oxygen bond as well. Examples of both types of bonding are included.

Upon heating with **sulfur trioxide,** ***1-perfluoroalkynes*** provide perfluoroalkylfluorosulfonylketenes [*1*] (equation 1).

[*1*] **1**

$$R_fC{\equiv}CF \xrightarrow[80^\circ\text{ C, 1 h}]{SO_3} \left[\begin{array}{c} R_fC{=}CF \\ | \quad\; | \\ O_2S{-}O \end{array}\right] \longrightarrow \begin{array}{c} R_fC{=}CO \\ | \\ SO_2F \end{array}$$

R_f	Yield
$R_f = (CF_3)_2CF$	85.9%
$(CF_3)_3C$	90.0%

Perfluoroallyl fluorosulfate is prepared by the treatment of ***perfluoropropene*** with sulfur trioxide in the presence of boron catalysts [*2, 3, 4, 5, 6, 7*] (equation 2). Perfluoroisopropyl allyl ether reacts similarly to give 58% polyfluoroallyl fluorosulfate in a *cis/trans* ratio of 6:4 [*8*]. β-***Sultones*** are the exclusive products without catalyst. ***Polyfluoroolefins*** such as 2-hydropentafluoropropylene [*9*], (2,3-dichloropropyl)trifluoroethylene [*10*], perfluoropropene [*2, 3*], perfluoroisopropyl alkenyl ethers [*8*], and acyclic polyfluoroallyl ethers [*11*] undergo sulfur trioxidation to regioselectively produce the corresponding β-sultones in high yield.

[*2,3,4,5,6,7*] **2**

$$CF_3CF{=}CF_2 \xrightarrow{i} \underset{16\text{-}60\%}{CF_2{=}CFCF_2OSO_2F} + \underset{5\text{-}42\%}{\begin{array}{c} F \\ CF_3C{-}CF_2 \\ | \quad\; | \\ O_2S{-}O \end{array}} + \underset{0\text{-}21\%}{\begin{array}{c} F \\ FC{-}CFCF_2OSO_2F \\ | \quad\; | \\ O{-}SO_2 \end{array}}$$

i = SO_3, 0.5-2 wt % BF_3 or $B(OMe_3)_3$ or B_2O_3
20°- 60° C, 4 - 60 h

On strong heating with sulfur trioxide, perfluoropropylene oxide reacts to furnish two major products, 1,2-perfluoropropylene sulfate and 2-oxoperfluoro-

0065–7719/95/0187–0403$08.00/1

propyl fluorosulfate, and a minor product, 2-oxoperfluoropropyl fluoropyrosulfate. The fluorosulfate hydrates on treatment with water to give the less volatile 2,2-dihydroxyperfluoropropyl fluorosulfate, thus providing a means for separation [*12*] (equation 3).

[*12*] **3**

$$CF_3CF\text{-}CF_2\ (\text{epoxide}) \xrightarrow[150°\ C,\ 10\ h]{SO_3} CF_3CF\text{-}CF_2\ (\text{cyclic } O\text{-}SO_2\text{-}O)\ 17.4\% + CF_3COCF_2OSO_2F\ 40.6\% + CF_3COCF_2OSO_2OSO_2F\ 5\%$$

Octafluoroisobutylene, whose double bond has reduced electron density and limited accessibility, reacts with sulfur trioxide under vigorous conditions. The reaction mixture contains various components including bis-α-trifluoromethyldifluoroethane-β-sultone, bis(α–trifluoromethyldifluoroethane)-β-pyrosultone, the heptafluoroisobutenyl ester of fluorosulfonic acid, and the heptafluoroisobutenyl ester of fluoropyrosulfonic acid [*13*] (equation 4).

[*13*] **4**

$$(CF_3)_2C{=}CF_2 \xrightarrow[170°\ \text{to}\ 190°\ C]{SO_3}$$

$$(CF_3)_2C\text{—}CF_2\ (O_2S\text{—}O)\ 19\% + (CF_3)_2C{=}CFOSO_2F\ 15\%$$

RT, 12 h or -78° C, 1 h → 80 or 84%

$$(CF_3)_2C\text{—}CF_2\ (O_2S\text{—}O\text{—}SO_2\text{—}O)\ 39\% + (CF_3)_2C{=}CFOSO_2OSO_2F\ 24\%$$

-78° C, 4 days → 73%

Polyfluorocyclobutenes produce 1:1, 2:1, and 3:1 adducts on reaction with sulfur trioxide [*14*] (equation 5).

Sulfur trioxide adds to 2,2-difluoroethylenesulfonyl fluoride to afford the β-sultone and its rearrangement product, bis(fluorosulfonyl)acetyl fluoride. Potassium fluoride acts as a base and reacts with the acetyl fluoride to eliminate the elements of hydrogen fluoride and produce bis(fluorosulfonyl)ketene [*15*] (equation 6).

References are listed on pages 406–407.

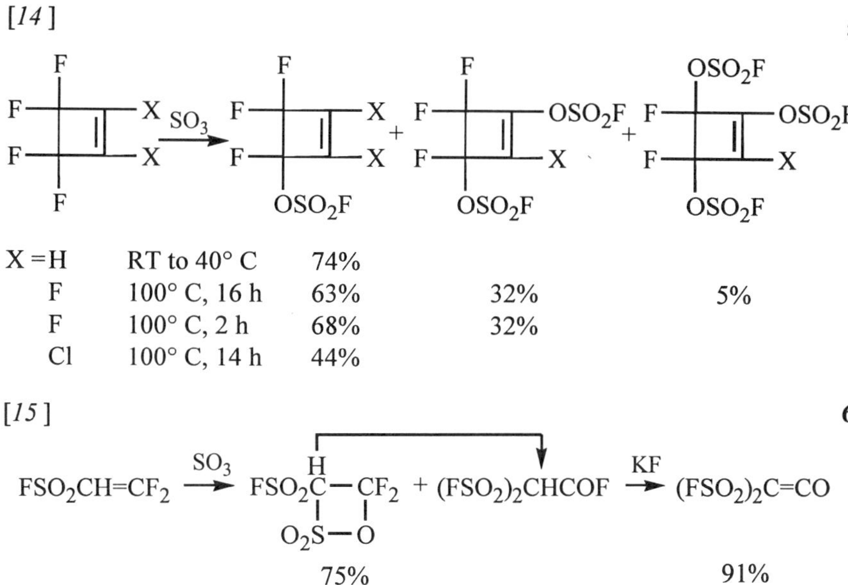

Polyfluorovinyl ethers form unstable β-sultones with sulfur trioxide. The β-sultones isomerize at 25 °C to give β-carbonylsulfonate esters and acids [*16, 17*]. The reaction of sulfur trioxide with ethyl pentafluoroisopropenyl ether to furnish 2-ketopentafluoropropanesulfonic acid and ethyl 2-ketopentafluoropropanesulfonate is an example (equation 7).

[*16,17*] **7**

$$CF_3C(OC_2H_5){=}CF_2 \xrightarrow[-15^\circ \text{ to } 50^\circ\text{ C}]{SO_3} CF_3COCF_2SO_3C_2H_5 + CF_3COCF_2SO_3H$$

($CF_3COCF_2SO_3C_2H_5 \rightarrow CF_3COCF_2SO_3H$: CF_3CO_2H, RT, overnight, $-\,CF_3CO_2C_2H_5$)

67%

Insertion of **sulfur trioxide** into ***perfluorotoluenes*** occurs on extended heating with sulfur trioxide to provide perfluorobenzyl fluorosulfates [*18, 19*] (equation 8).

[*18,19*] **8**

$$C_6F_5CF_3 \xrightarrow[100^\circ\text{ C, 20 h}]{SO_3} C_6F_5CF_2OSO_2F$$

55%

References are listed on pages 406–407.

Sulfonation of the aromatic ring of ***1′,2′,2′-trifluorostyrene*** below 0 °C does not give satisfactory yields with chlorosulfonic acid or a sulfur trioxide–dioxane complex. Tar forms on heating. In contrast, under similar conditions *ipso* substitution is facile at the position of a trialkylsilyl or -stannyl group. Thus, 4-trimethylsilyl-1′,2′,2′D-trifluorostyrene affords the corresponding trimethylsilyl sulfonate [*20*] (equation 9).

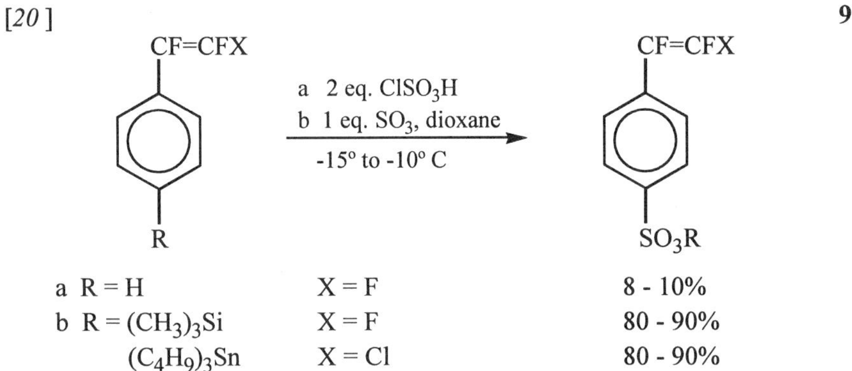

Pentafluorobenzene undergoes ring sulfonation with sulfur trioxide to give pentafluorobenzenesulfonic acid, from which 2,3,5,6-tetrafluorobenzenedisulfonic acid can be obtained indirectly in excellent overall yield [*21*] (equation 10).

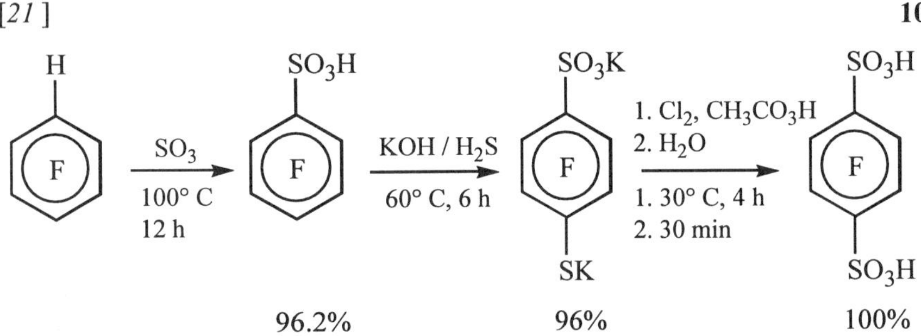

References for Pages 403–406

1. Galakhov, M. V.; Cherstkov, V. F.; Sterlin, S. R.; German, L. S. *Izv. Akad. Nauk SSSR* **1987,** 958 (Engl. Transl. 886).
2. Krespan, C. G. U.S. Pat. 4 235 804, 1980; *Chem. Abstr.* **1981,** *94,* 156305e.
3. Banks, R. E.; Birchall, J. M.; Haszeldine, R. N.; Nicholson, W. J. *J. Fluorine Chem.* **1982,** *20,* 133.
4. England, D.C. U.S. Pat. 4 206 138, 1980; *Chem. Abstr.* **1980,** *93,* 72579p.

5. Krespan, C. G.; England, D.C. Belg. Pat. 878 131, 1980; *Chem. Abstr.* **1980,** *93,* 72579p.
6. England, D. C.; Krespan, C. G. Br. Pat. Appl. 2 027 709, 1980; *Chem. Abstr.* **1981,** *95,* 80150y.
7. Krespan, C. G.; England, D.C. *J. Am. Chem. Soc.* **1981,** *103,* 5598.
8. Cherstkov, V. F.; Sterlin, S. R.; German, L. S.; Knunyants, I. L. *Izv. Akad. Nauk SSSR* **1982,** 2796 (Engl. Transl. 2472).
9. Aktaev, N. P.; Sokol'skii, G. A.; Knunyants, I. L. Izv. *Akad. Nauk SSSR* **1975,** 2530 (Engl. Transl. 2416).
10. Mohtasham, J.; Gard, G. L.; Yang, Z.; Burton, D. J. *J. Fluorine Chem.* **1990,** *50,* 31.
11. Mohtasham, J.; Brennen, M.; Yu, Z.; Adcock, J. L.; Gard, G. L. *J. Fluorine Chem.* **1989,** *43,* 349.
12. Knunyants, I. L.; Shokina, V. V.; Mysov, E. I. *Izv. Akad. Nauk SSSR* **1973,** 2725 (Engl. Transl. 2659).
13. Belaventsev, M. A.; Mikheev, L. L.; Pavlov, V. M.; Sokol'skii, G. A.; Knunyants, I. L. *Izv. Akad. Nauk SSSR* **1972,** 2510 (Engl. Transl. 2441).
14. Smart, B. E. *J. Org. Chem.* **1976,** *41,* 2353.
15. Eleev, A. F.; Sokol'skii, G. A.; Knunyants, I. L. *Izv. Akad. Nauk SSSR* **1980,** 892; *Chem. Abstr.* **1980,** *93,* 132018s.
16. Krespan, C. G.; Smart, B. E.; Howard, E. G. *J. Am. Chem. Soc.* **1977,** *99,* 1214.
17. Krespan, C. G. *J. Org. Chem.* **1979,** *44,* 4924.
18. Cherstkov, V. F.; Sterlin, S. R.; German, L. S.; Knunyants, I. L. *Izv. Akad. Nauk SSSR* **1982,** 2791 (Engl. Transl. 2468).
19. German, L. S.; Knunyants, I. L.; Sterlin, S. R.; Cherstkov, F. *Izv. Akad. Nauk SSSR* **1981,** 1933; *Chem. Abstr.* **1981,** *95,* 203475t.
20. Timofeyuk, G. V.; Sorokina, R. S.; Rybakova, L. F.; Panov, E. M.; Kuznetsova, L. G.; Kocheshkov, K. A. *Dokl. Akad. Nauk SSSR* **1973,** *209,* 367 (Engl.Transl. 225).
21. Sartori, P.; Bauer, G. *J. Fluorine Chem.* **1978,** *12,* 203.

Acid-Catalyzed Syntheses

by G. L. Cantrell and J. F. Lang

Acid-Catalyzed Additions across Double Bonds

Halogenated alkanes *add to* **olefins** *with regioselectivity determined by bond polarity and steric factors* although generalization is difficult. Addition of fluorotrichloromethane to chlorofluoroethenes appears to follow this principle [*1*]. In the case of trifluoroethene, where both the bond polarity and steric effects reinforce each other, the regioselectivity is greater than 99% [*1*] (equation 1). The attacking species is CCl_3+ as well as $CFCl_2+$. In comparison, chlorotrifluoroethene, where polar and steric effects oppose each other, is attacked exclusively at the CF_2 group by 1,2-dichloroethane [*2*] (equation 2).

[*1*] **1**

$$CF_2{=}CHF + CFCl_3 \xrightarrow[10^\circ\,C,\ 10\,h]{AlCl_3} \underset{46\%}{CF_3CHFCCl_3} + \underset{19\%}{CF_2ClCHFCFCl_2}$$

[*2*] **2**

$$CF_2{=}CFCl + ClCH_2CH_2Cl \xrightarrow[20^\circ\,C,\ 8\,h]{HF,\ SbF_5} [CH_3\overset{+}{C}HCl] \longrightarrow \underset{55\%}{CH_3CHClCF_2CF_2Cl}$$

Acylhalogenation of ***haloolefins*** is most often carried out with ***aluminum chloride*** as the catalyst. The yields are variable because of side reactions including halogen exchange. Halogen exchange is avoided and yields are higher when **ferric chloride** is substituted for aluminum chloride in the reaction of fluoroethene with acid chlorides [*3*] (equation 3).

Bis(dimethylamino)carbonium bifluoride catalyzes the addition of acyl fluorides to perfluoroalkenes by the generation of readily acylated carbanions [*4*] (equation 4).

Friedel–Crafts Syntheses

Many ***fluorinated aromatic compounds*** are alkylated with **alkyl halides** under *Friedel–Crafts* conditions. For example, the intramolecular alkylation of 3-fluoro-*N*-(chloroacetyl)aniline with aluminum chloride gives 6-fluorooxindole [*5*] (equation 5). Similarly, 3′-chloro-4-fluoropropiophenone affords 5-fluoroindanone [*6*] (equation 6).

0065–7719/95/0187–0408$08.00/1

[3] 3

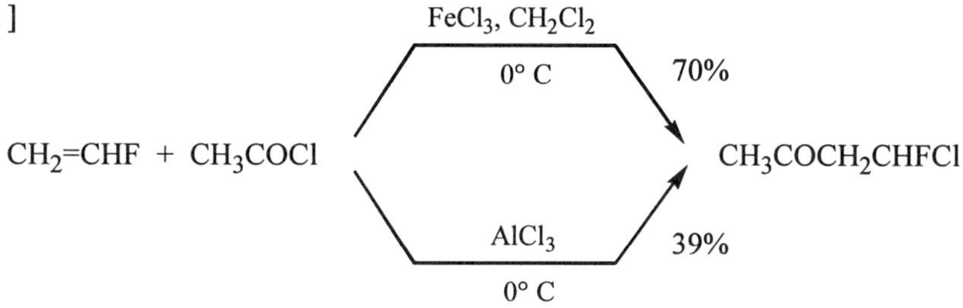

$CH_2{=}CHF + CH_3COCl \xrightarrow[0° C]{FeCl_3, CH_2Cl_2} CH_3COCH_2CHFCl$ 70%

$CH_2{=}CHF + CH_3COCl \xrightarrow[0° C]{AlCl_3} CH_3COCH_2CHFCl$ 39%

[4] 4

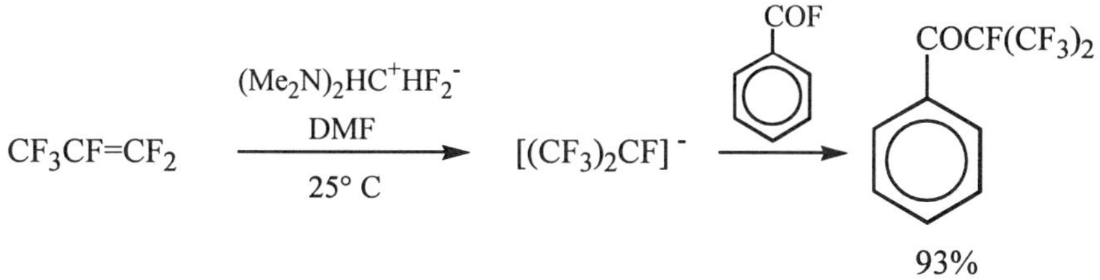

$CF_3CF{=}CF_2 \xrightarrow[25° C]{(Me_2N)_2HC^+HF_2^-,\ DMF} [(CF_3)_2CF]^-$ → (with C_6H_5COF) $C_6H_5COCF(CF_3)_2$

93%

[5] 5

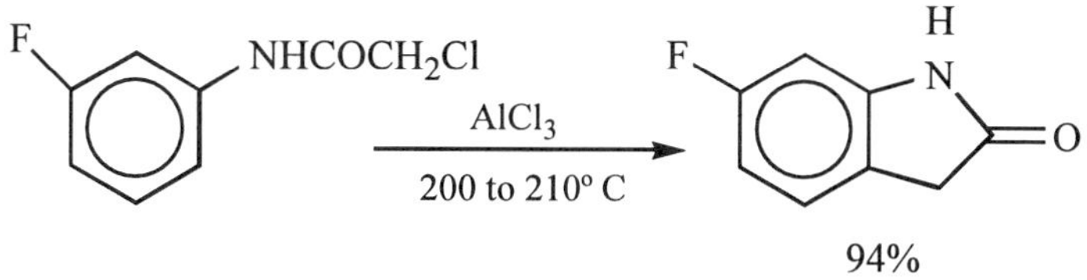

3-fluoro-$C_6H_4NHCOCH_2Cl \xrightarrow[200\ to\ 210° C]{AlCl_3}$ 6-fluorooxindole (F, H, N, O)

94%

[6] 6

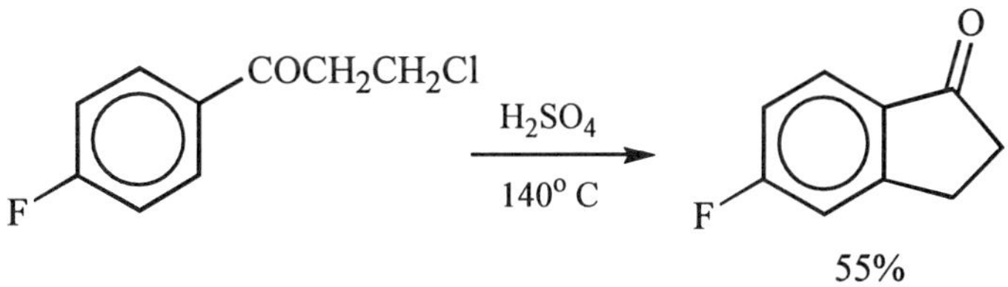

4-fluoro-$C_6H_4COCH_2CH_2Cl \xrightarrow[140° C]{H_2SO_4}$ 5-fluoro-1-indanone (F, O)

55%

In analogy to alkylation with chloroalkyl ethers, methyl α-methoxyperfluoropropionate reacts with 2,6-dimethylphenol to give the bisphenol [*7*] (equation 7).

[*7*] **7**

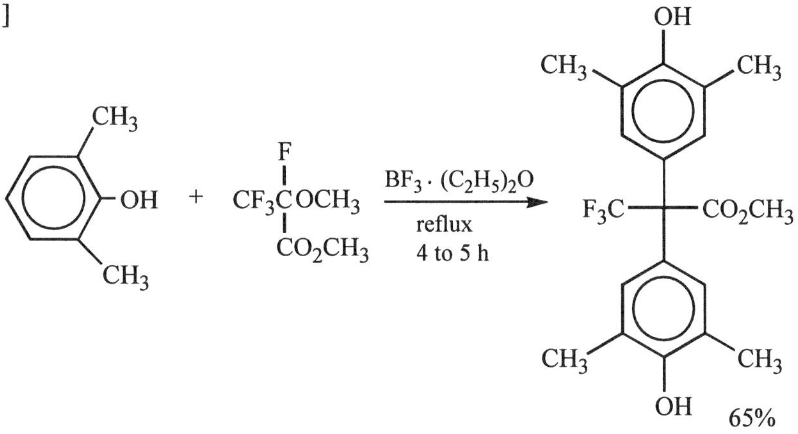

$$2,6\text{-}(CH_3)_2C_6H_3OH + CF_3CF(OCH_3)CO_2CH_3 \xrightarrow[\text{reflux, 4 to 5 h}]{BF_3\cdot(C_2H_5)_2O} F_3C\text{–}C(CO_2CH_3)[C_6H_2(CH_3)_2OH]_2 \quad 65\%$$

Alkyl fluorides are more reactive than other **alkyl halides** under Friedel–Crafts conditions, whereas trifluoromethyl groups are less reactive than other trihalomethyl groups. Thus, a bromoindane is prepared from 1-bromo-1-fluoro-2,2,3,3-tetramethylcyclopropane and benzene [*8*] (equation 8), whereas 3-trifluoromethylphenyldiphenylchloromethane is obtained from 3-trifluoromethylbenzotrichloride and benzene [*9*] (equation 9).

[*8*] **8**

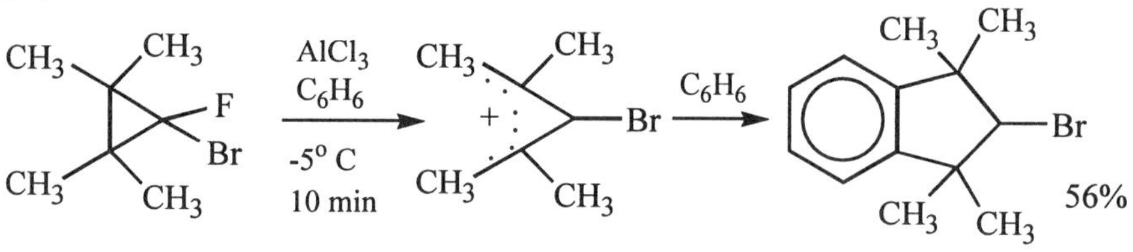

$$\text{1-bromo-1-fluoro-2,2,3,3-tetramethylcyclopropane} \xrightarrow[-5^\circ C,\ 10\ \text{min}]{AlCl_3,\ C_6H_6} [\text{tetramethyl-bromoallyl cation}]^+ \xrightarrow{C_6H_6} \text{2-bromo-1,1,3,3-tetramethylindane} \quad 56\%$$

[*9*] **9**

$$C_6H_6 + 3\text{-}CF_3C_6H_4CCl_3 \xrightarrow[25\ \text{to}\ 30^\circ C]{FeCl_3} (C_6H_5)_2CCl(C_6H_4CF_3\text{-}3) \quad 93\%$$

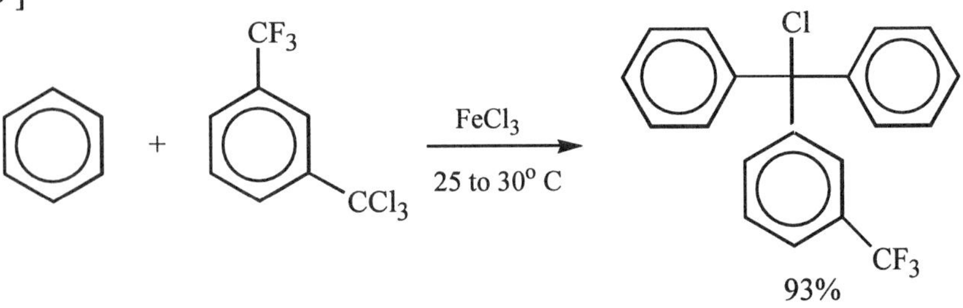

Vinyl and phenyl trifluoromethyl groups are reactive in the presence of aluminum chloride [*10*]. *Replacement of fluorine by chlorine often occurs.* Polyfluorinated trifluoromethylbenzenes form reactive α,α-difluorobenzyl cations in **antimony pentafluoride** [*11*]. 1-Phenylperfluoropropene cyclizes in aluminum chloride to afford 1,1,3-trichloro-2-fluoroindene [*10*] (equation 10). The reaction is hypothesized to proceed via an allylic carbocation, whose fluoride atoms undergo halogen exchange.

[*10*] **10**

$AlCl_3$, CH_3COCl, -15° C, then 60° C; 59%

A similar mechanism operates in the reaction of **3,3,3-trifluoropropene** with ***benzene*** and **aluminum chloride** [*12, 13*]. Perfluorophenylpropene undergoes intramolecular electrophilic attack in a rare example of ring closure at a C_{ar}–F bond [*14*] (equation 11).

[*14*] **11**

$CF_2CF{=}CF_2$; $CF{=}CFCF_3$; $AlCl_3$, 100° C; 54%

Fluorobenzene is readily alkylated with **alkenes** in the presence of protic acids; however, the isomeric purity of the product is poor, and polysubstitution can result. Thus, **propene** and **sulfuric acid** alkylate ***fluorobenzene*** at 20 °C to yield a 45:55 *ortho/para* ratio of the monoalkyl product in addition to di- and triisopropylfluorobenzene [*15*]. The reaction of ***benzene*** and **trifluoropropene** at 25 °C in HF–BF_3 gives a mixture of mono-, bis-, and tris(3,3,3-trifluoropropyl)benzene [*12, 13*] (equation 12).

[*12,13*] **12**

$$C_6H_6 + CF_3CH{=}CH_2 \xrightarrow[25^\circ\ C,\ 4.5\ h]{HF/BF_3} C_6H_5CH_2CH_2CF_3 \text{ + bis and tris products}$$

59%

Fluoral hemiacetal, $CF_3CH(OH)OC_2H_5$, a new trifluoromethylalkylating agent, is reported to produce products controlled by the type of catalyst and solvent used [*16*]. Products including alcohols, ethers, chlorides, and diphenyls can be made with good selectivity. Fluoral hemiacetal, benzene, and sulfuric acid give 1,1-diphenyl-2,2,2-trifluoroethane, whereas substitution of excess aluminum chloride for sulfuric acid leads to the formation of 1-phenyl-1-chloro-2,2,2-trifluoroethane. In a related synthesis, 1,1,1-trifluoroacetone, benzene, and aluminum chloride afford the chloroalkylation product 2-chloro-2-phenyl-1,1,1-trifluoropropane in 50% yield [*17, 18*] (equation 13).

[*17*] **13**

$$CF_3COCH_3 + C_6H_6 \xrightarrow[0^\circ\ C,\ 1\ h]{AlCl_3,\ CH_2Cl_2} C_6H_5CCl(CH_3)CF_3$$

50%

Ketones and alcohols are frequently used as *alkylating agents.* Ketones often condense with two molecules of an aromatic compound through an alcohol intermediate [*19, 20, 21*]. 1′,1′,1′-Trifluoroacetophenone and aniline afford 1,1-bis(4-aminophenyl)-1-phenyl-2,2,2-trifluoroethane [*21*] (equation 14).

Previous efforts to prepare indoles by the cyclization of the corresponding 2-anilino acetals have been largely unsuccessful. However, *N*-trifluoroacetyl-2-anilino diethyl acetal and some derivatives can be converted to indoles in good yield [*22*] (equation 15).

1,2-Epoxy-5-(4′-fluorophenyl)pentane cyclizes in the presence of tin tetrachloride to give 7-fluoro-1,2,3,4-tetrahydro-1-naphthalenemethanol in an unisolated yield of 89% [*23*] (equation 16).

[*21*] 14

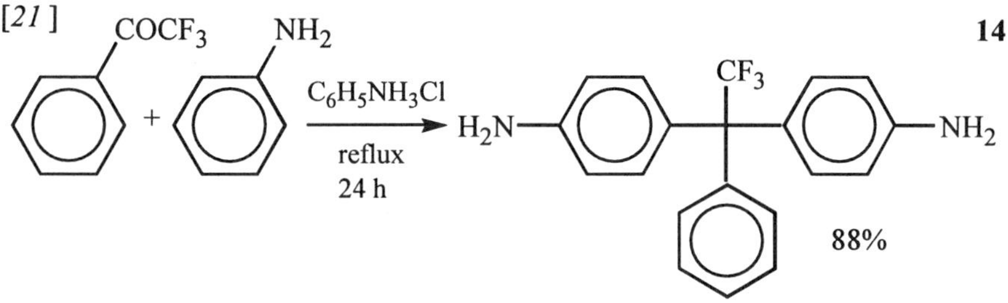

$C_6H_5COCF_3$ + $C_6H_5NH_2$ $\xrightarrow[\text{reflux, 24 h}]{C_6H_5NH_3Cl}$ $H_2N-C_6H_4-C(CF_3)(C_6H_5)-C_6H_4-NH_2$ 88%

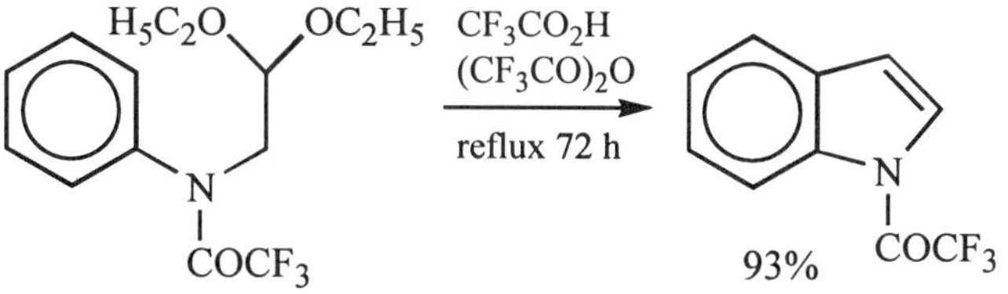

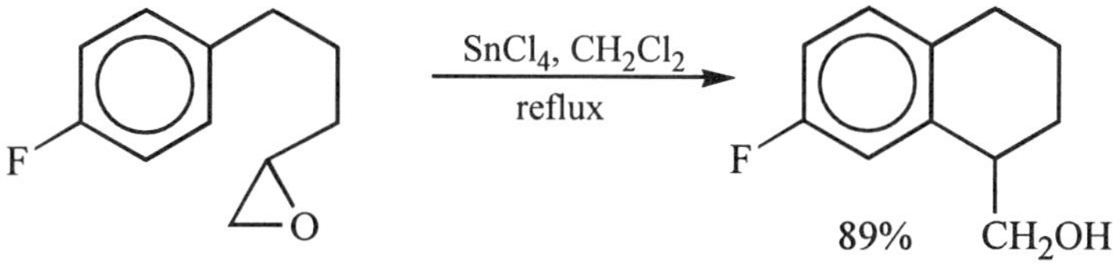

[*22*] 15

$C_6H_5N(COCF_3)CH_2CH(OC_2H_5)_2$ $\xrightarrow[\text{reflux 72 h}]{CF_3CO_2H,\ (CF_3CO)_2O}$ 1-(trifluoroacetyl)indole (N–$COCF_3$) 93%

[*23*] 16

$\xrightarrow[\text{reflux}]{SnCl_4,\ CH_2Cl_2}$ 89% CH_2OH

Vinyl triflates permit alkylation with vinyl cations [*24, 25*]. Fluorobenzene reacts with 2-methyl-1-phenyl-1-propenyl triflate to form a diaryl alkene [*24*] (equation 17).

[*24*] 17

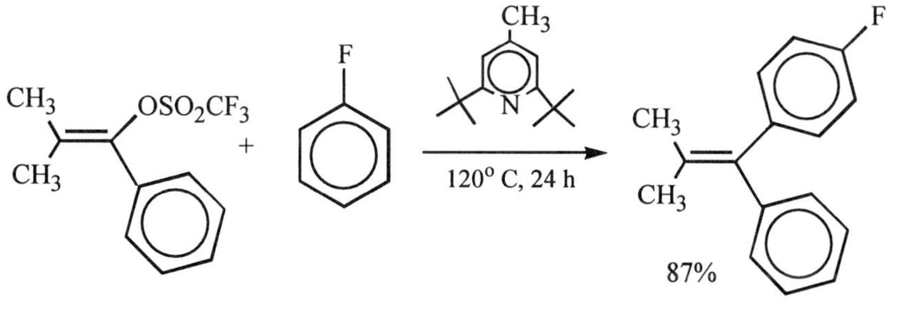

$(CH_3)_2C{=}C(C_6H_5)OSO_2CF_3$ + C_6H_5F $\xrightarrow[120^{\circ}\,C,\ 24\ h]{\text{2,6-di-}tert\text{-butyl-4-methylpyridine}}$ $(CH_3)_2C{=}C(C_6H_5)(C_6H_4F)$ 87%

Fluorides with fluorine–phosphorus bonds also react with Lewis acids. *tert*-Butylpentafluorocyclotriphosphazenes are arylated in the presence of aluminum chloride [*26*] (equation 18).

References are listed on pages 419–421.

[*26*] **18**

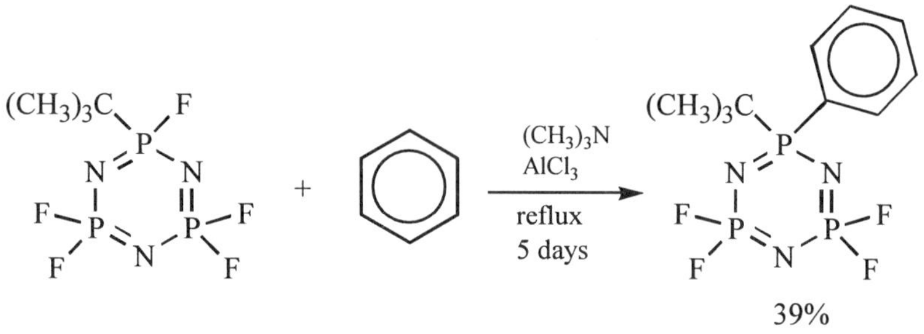

$(CH_3)_3N$, $AlCl_3$, reflux 5 days

39%

The strong *para*-directing influence of fluorine in Friedel–Crafts *acylation* is illustrated by the reaction of ***fluorotoluenes*** with *acetyl chloride.* Treatment of 2-fluorotoluene, 3-fluorotoluene, and 2-fluoro-3-methyltoluene with acetyl chloride and aluminum chloride gives, respectively, 91, 82, and 80% substitution *para* to fluorine [*27*].

2,4-Dihalo-5-fluorobenzoic acids, obtained by oxidation of the corresponding acetophenones, have received considerable attention as intermediates for antibacterial fluoroquinolones [*28, 29*]. In this regard, 1,2,4-trifluorobenzene reacts with acetyl chloride to give 2,4,5-trifluoroacetophenone [*29*] (equation 19). 1,3,5-Trifluorobenzene and 1,2-difluorobenzene yield 2,4,6-trifluoroacetophenone and 3,4-difluoroacetophenone, respectively [*30*] (equations 20 and 21).

[*29*] **19**

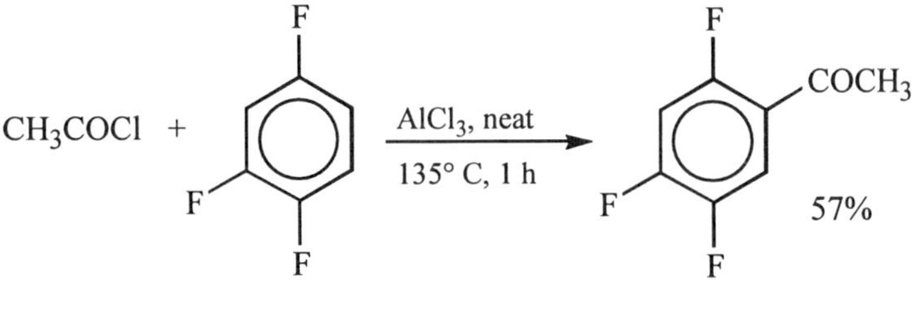

CH_3COCl + ; $AlCl_3$, neat, 135° C, 1 h; $COCH_3$

57%

[*30*] **20**

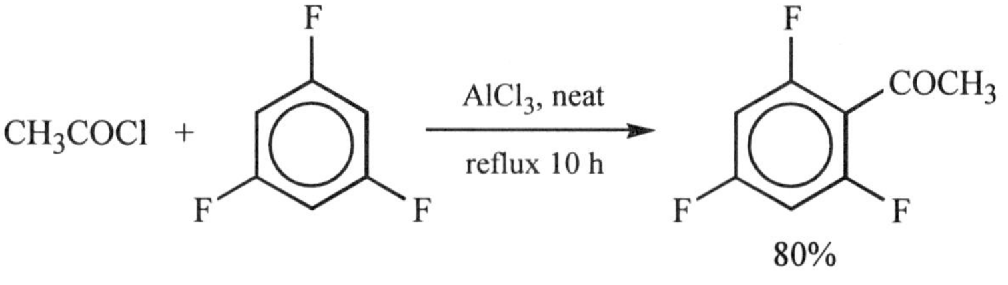

CH_3COCl + ; $AlCl_3$, neat, reflux 10 h; $COCH_3$

80%

1′,1′,1′-Trifluoromethoxybenzene is converted to the *para*-substituted **ketone** with acyl chlorides in HF–BF_3 [*31*]. Benzotrichlorides and benzotrifluorides are converted to acyl benzotrifluorides during acylation in this medium [*32*].

References are listed on pages 419–421.

[*30*] **21**

CH_3COCl + 1,2-difluorobenzene $\xrightarrow[\text{reflux 10 h}]{AlCl_3\text{, neat}}$ 3,4-difluoroacetophenone ($COCH_3$) 80%

Depending on the reagent ratio, **oxalyl chloride** reacts with fluorobenzene in the presence of aluminum chloride to afford either 4-fluorobenzoyl chloride or 4,4′-difluorobenzophenone [*33*] (equation 22). Phosgene, detected by infrared spectroscopy, is an intermediate.

[*33*] **22**

Fluorobenzene $\xrightarrow[25°\ C]{1\ C_6H_5F,\ 2\ ClCOCOCl,\ 1\ AlCl_3}$ 4-fluorobenzoyl chloride ($COCl$) 84%

Fluorobenzene $\xrightarrow[25°\ C]{2\ C_6H_5F,\ 1\ ClCOCOCl,\ 2\ AlCl_3}$ $F–C_6H_4–C(=O)–C_6H_4–F$

Aromatic amino ketones and derivatives, which are of biological interest, can be prepared by the acylation of various benzenes with *N*-(trifluoroacetyl)amino acid chlorides [*34*]. The strongly electron-withdrawing trifluoroacetyl group is superior to other N-protecting groups. The alkaloid halostachine, for example, was prepared by the reaction of benzene with *N*-trifluoroacetylglycyl chloride, followed by N-methylation and reduction [*34*] (equation 23).

Carboxylic acids and their anhydrides acylate a variety of ***benzene*** derivatives, fused ring systems, and heterocyclic compounds. An improved procedure for the preparation of 1,4-difluoroanthracene-9,10-dione involves reacting phthalic anhydride and 1,4-difluorobenzene to prepare an intermediate carboxylic acid [*35*]. *Intramolecular acylation* in polyphosphoric acid completes the synthesis (equation 24).

Desilylation–electrophilic substitution of trimethylsilylbenzenes provides a route to a number of *meta*-substituted fluorobenzene derivatives such as 3-fluoroacetophenone, which is unobtainable under normal Friedel–Crafts conditions [*36*] (equation 25).

Perfluoroalkanoyl chlorides and anhydrides are also acylating agents. Trifluoroacetic anhydride acylates a number of **pyrroles, thiophenes, and furans** without a catalyst [*37, 38, 39*]. ***Azulene*** can be diacylated without a catalyst in 12 h [*40*] (equation 26).

References are listed on pages 419–421.

[*34*] **23**

+ $ClCOCH_2NHCOCF_3$ $\xrightarrow[\text{-15° C, 12 h}]{AlCl_3,\ CH_2Cl_2}$ $COCH_2NHCOCF_3$ 77%

CH_3I, K_2CO_3, Me_2CO, reflux 48 h

$CH(OH)CH_2NHCH_3$ 97% $\xleftarrow[\text{25° C, 12 h}]{NaBH_4,\ EtOH}$ $COCH_2N(CH_3)COCF_3$ 57%

[*35*] **24**

O, O, O + F, F $\xrightarrow[\text{reflux, 48 h}]{AlCl_3}$ O, F, HO_2C, F 75%

polyphosphoric acid, 140° C, 2 h

O, F, O, F 65%

1,2,3-Trialkylindoles undergo Friedel–Crafts reactions at position 6; however, in trifluoroacetic anhydride the α-methyl group of 1,2,3-trimethylindole is acylated through an intermediate enamine [*41, 42*] (equation 27). Similarly, trifluoroacetic anhydride acylates the double bond of the α-methylene compound shown [*42*] (equation 28).

Azupyrene, a nonbenzenoid 4π π-electron system, undergoes electrophilic substitution in trifluoroacetic anhydride suggesting *aromatic character,* which is corroborated by spectral evidence [*43*] (equation 29).

Ferrocene is acylated with a number of poly- and perfluorinated acid anhydrides [*44*].

[*36*] **25**

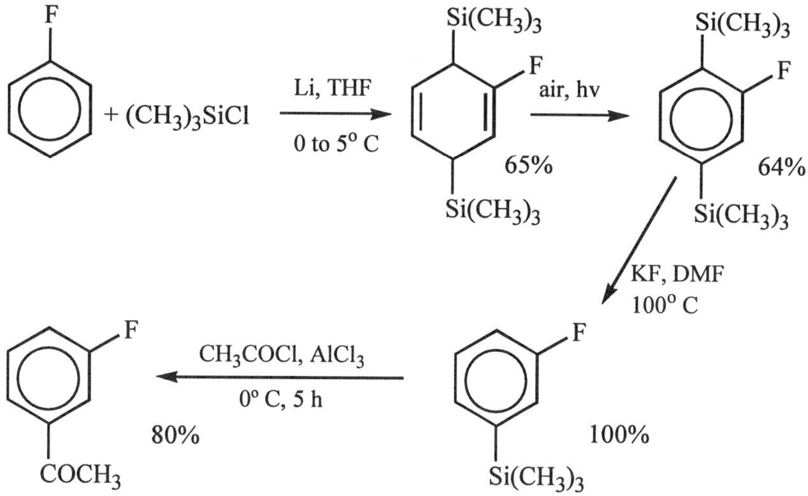

F + $(CH_3)_3SiCl$ —Li, THF / 0 to 5° C→ $Si(CH_3)_3$, F, $Si(CH_3)_3$ 65% —air, hν→ $Si(CH_3)_3$, F, $Si(CH_3)_3$ 64%

—KF, DMF / 100° C→ F, $Si(CH_3)_3$ 100% —CH_3COCl, $AlCl_3$ / 0° C, 5 h→ F, $COCH_3$ 80%

[*40*] **26**

+ $(CF_3CO)_2O$ —CCl_4 / 25° C, 10 min→ $COCF_3$ 100%

—25° C, 12 h→ $COCF_3$, $COCF_3$ 70%

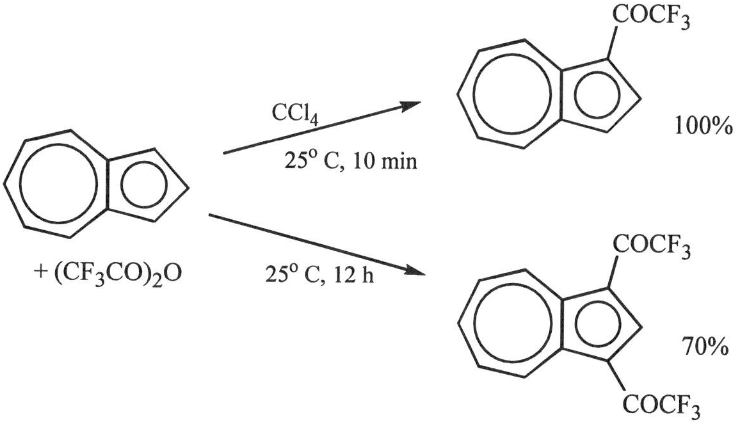

The new reagent **2-trifluoroacetoxypyridine** (TFAP) trifluoroacetylates ***benzene, benzofuran, and dibenzofuran*** at –10 to 0 °C in yields ranging from 53 to 93% [*45*]. TFAP converts isopropylbenzene to *para*-isopropyltrifluoroacetophenone (equation 30).

Mixed anhydrides of a carboxylic acid and trifluoroacetic or triflic acids are especially effective acylating agents because of the high degree of polarization [*46*]. The *nonfluorinated acyl group is the electrophile.* On the basis of the reactivity of 2-methylthiophene, the relative rate of acylation is CF_3SO_2OCOR, 7100; $CF_3COOCOR$, 34; $CCl_3COOCOR$, 1; $(CH_3CO)_2O$, 10^{-3}. Derivatives of ***thiophene and furan*** are acylated by the in situ generation of the mixed anhydride in a solution of trifluoroacetic anhydride, a carboxylic acid, and a phosphoric acid

[*41,42*] **27**

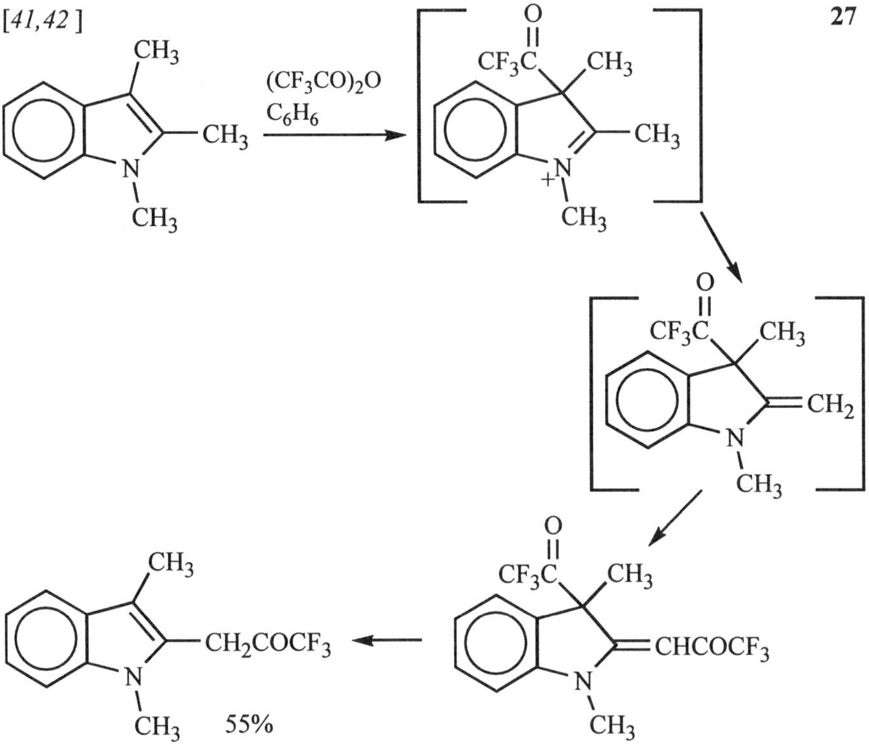

$(CF_3CO)_2O$, C_6H_6

CH_3, CF_3C, CH_2, $CHCOCF_3$, CH_2COCF_3

55%

[*42*] **28**

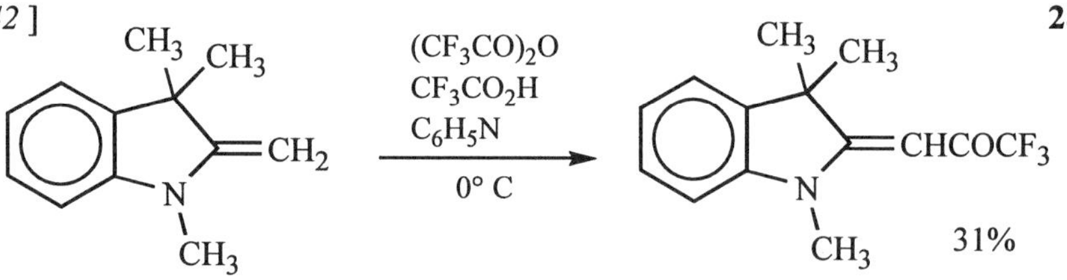

$(CF_3CO)_2O$
CF_3CO_2H
C_6H_5N
0° C

$CHCOCF_3$ 31%

[*43*] **29**

+ $(CF_3CO)_2O$ → 25° C, 3 to 4 days

$COCF_3$ 73%

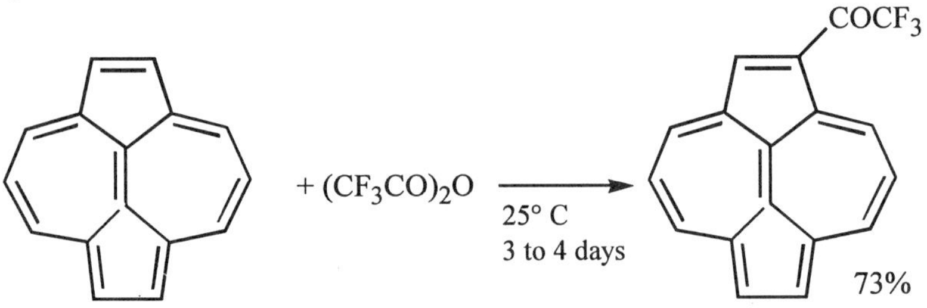

References are listed on pages 419–421.

catalyst [*46, 47, 48*]. This procedure enhances the yield and selectivity of the reaction for position 2 in furan [*48*]. The mixed anhydride **trifluoroacetyl triflate** acylates ***anthracene*** to 9-trifluoroacetylanthracene [*49*] (equations 31 and 32).

[*45*] **30**

$COCF_3$

+ N $OCOCF_3$ $\xrightarrow[0^\circ\ C,\ 4\ h]{AlCl_3,\ CH_2Cl_2}$ + N H O

78%

[*49*] **31**

$$CF_3CO_2H + CF_3SO_3H \xrightarrow[\text{reflux 2 h}]{P_2O_5} CF_3CO_2SO_2CF_3$$

52%

[*49*] **32**

$COCF_3$

+ $CF_3CO_2SO_2CF_3$ $\xrightarrow[80^\circ\ C,\ 44\ h]{C_6H_6}$

81%

References for Pages 408–419

1. Posta, A.; Paleta, O.; Voves, J.; Trska, P. *Collect. Czech. Chem. Commun.* **1974,** *39,* 1330.
2. Belen'kii, G. G.; Petrov, V. A.; German, L. S. *Izv. Akad. Nauk SSSR* **1980**, 1099 (Engl. Transl. 806).
3. Ishikawa, N.; Iwakiri, H.; Edamura, K; Kubota, S. *Bull. Chem. Soc. Jpn.* **1981,** *54,* 832.
4. Igumnov, S. M.; Delyagina, N. I.; Knunyants, I. L. *Izv. Akad. Nauk SSSR* **1981,** 2339 (Engl. Transl. 1924).
5. Sindelar, K.; Protiva, M.; Sedivy, Z. Czech. Pat. 191 777, 1982; *Chem. Abstr.* **1982,** *97,* 91949h.
6. Olivier, M.; Marechal, E. *Bull. Soc. Chim. Fr.* **1973,** 3092.
7. Dyachenko, V. I.; Kolomiets, A. F.; Fokin, A. V. *Izv. Akad. Nauk SSSR* **1989,** 1446 (Engl. Transl. 1327).

8. Mueller, C.; Weyerstahl, P. *Tetrahedron* **1975,** *31,* 1787.
9. Buechel, K. H.; Singer, R. J. Ger. Pat. 2 553 301, 1977; *Chem. Abstr.* **1977,** *87,* 8467a.
10. Fialkov, Y. A.; Sevast'yan, A. P.; Yagupol'skii, L. M. *Zh. Org. Khim.* **1979,** *15,* 1256 (Engl. Transl. 1121).
11. Pozdnyakovich, Y. V.; Shteingarts, V. D. *J. Fluorine Chem.* **1974,** *4,* 297.
12. Kobayashi, Y.; Nagai, T.; Kumadaki, I.; Takahashi, M.; Yamauchi, T. *Chem. Pharm. Bull.* **1984,** *32,* 4382.
13. Kobayashi, Y.; Kumadaki, I.; Takahashi, M.; Yamauchi, T. Fr. Pat. 2 476 639, 1989; *Chem. Abstr.* **1982,** *96,* 34808s.
14. Platonov, V. E.; Dvornikova, K. V. *Izv. Akad. Nauk SSSR* **1989,** 2854 (Engl. Transl. 2617).
15. Plotkina, N. I.; Kachalkova, M. I.; Zhykova, S. M.; Shevchenko, N. A. *Zh. Vses. Khim. O-va.* **1975,** *20,* 598 (Engl. Transl. 39).
16. Guy, A.; Lobgeois, A.; Lemaire, M. *J. Fluorine Chem.* **1986,** *32,* 361.
17. Bonnet-Delpon, D.; Charpentier-Morize, M. *Bull. Soc. Chim. Fr.* **1986,** 933.
18. Bonnet-Delpon, D.; Charpentier-Morize, M.; Jacquot, R. *J. Org. Chem.* **1988,** *53,* 759.
19. Niizeki, S.; Yoshida, M.; Sasaki, M. Jpn. Pat. 90 138 155, 1990; *Chem. Abstr.* **1990,** *113,* 151995f.
20. Dyachenko, V. I.; Galakhov, M. V.; Kolomiets, A. F.; Fokin, A. V. *Izv. Akad. Nauk SSSR* **1989,** 2786 (Engl. Transl. 2550).
21. Kray, W.; Rosser, R. W. *J. Org. Chem.* **1977,** *42,* 1186.
22. Nordlander, J. E.; Catalane, D. B.; Kotian, K. D.; Stevens, R. M.; Haky, J. E. *J. Org. Chem.* **1981,** *46,* 778.
23. Taylor, S. K.; Davisson, M. E.; Hissom, B. R., Jr.; Brown, S. L.; Pristach, H. A.; Schramm, S. B.; Harvey, S. M. *J. Org. Chem.* **1987,** *52,* 425.
24. Stang, P. J.; Anderson, A. G. *Tetrahedron Lett.* **1977,** 1485.
25. Garcia, M. A.; Martinez, A. R.; Garcia, F. A.; Hanack, M.; Subramanian, L. R. *Chem. Ber.* **1987,** *120,* 1255.
26. Allen, C. W.; Bedell, S.; Pennington, W. T.; Cordes, A. W. *Inorg. Chem.* **1985,** *24,* 1653.
27. Valkanas, G. *Chem. Chron.* **1972,** *1,* 255.
28. Mohler, F. et al. Chinese Pat. 85 107 015, 1987; *Chem. Abstr.* **1988,** *108,* 186324k.
29. Baumann, K.; Kuegler, R. Ger. Pat. 3 840 375, 1990; *Chem. Abstr.* **1990,** *113,* 152030z.
30. Joshi, K. C.; Pathak, V. N.; Grover, V. *J. Fluorine Chem.* **1980,** *15,* 245.
31. Desbois, M. *Bull. Soc. Chim. Fr.* **1986,** 885.
32. Desbois, M. Eur. Pat. Appl. 84 742, 1983; *Chem. Abstr.* **1984,** *100,* 6035m.
33. Neubert, M. E.; Fishel, D. L. *Mol. Cryst. Liq. Cryst.* **1979,** *53,*101.

34. Nordlander, J. E.; Payne, M. J.; Njoroge, F. G.; Balk, M. A.; Laikos, G. D.; Vishwanath, V. M. *J. Org. Chem.* **1984,** *49,* 4107.
35. Krapcho, A. P.; Getahun, Z. *Synth. Commun.* **1985,** *15,* 907.
36. Bennetau, B.; Krempp, M.; Dunogues, J. *Tetrahedron* **1990,** *24,* 8131.
37. Clementi, S. I.; Genel, F.; Marino, G. *Chem. Commun.* **1967,** 498.
38. Trofimov, B. A.; Mikhaleva, A. I.; Kalabin, G. A.; Vasil'ev, A. N.; Sigalov, M. V. *Izv. Akad. Nauk SSSR* **1977,** 2639 (Engl. Transl. 2446).
39. Glukhovtsev, V. G.; Brezhnev, L. Y.; Il'in, Y. V.; Karzhavina, N. P.; Nikishin, G. I. Russ. Pat. 1 235 865, 1986; *Chem. Abstr.* **1986,** *105,* 190899g.
40. Mathias, L. J.; Overberger, C. G. *J. Org. Chem.* **1980,** *45,* 1701.
41. Bailey, A. S.; Peach, J. M.; Vandrevala, M. H. *J. Chem. Soc., Chem. Commun.* **1978,** 845.
42. Bailey, A. S.; Haxby, J. B.; Hilton, A. N.; Peach, J. M.; Vandrevala, M. H. *J. Chem. Soc., Perkin Trans. 1* **1981,** 382.
43. Anderson, A. G., Jr.; Masada, G. M.; Kao, G. L. *J. Org. Chem.* **1980,** *45,* 1312.
44. Sokolova, E. G.; Chalykh, G. P.; Malikova, T. A.; Sevost'yanova, L. B.; Nemchinova, O. A. *Zh. Obshch. Khim.* **1973,** *43,* 1333; *Chem. Abstr.* **1973,** *79,* 66530k.
45. Keumi, T.; Shimada, M.; Takahashi, M.; Kitajima, H. *Chem. Lett.* **1990,** 783.
46. Galli, C. *J. Chem. Res. (S.)* **1984,** 272.
47. Gaset, A.; Delmas, M. Fr. Pat. 2 518 999, 1983; *Chem. Abstr.* **1984,** *100,* 6313j.
48. Fayed, S.; Delmas, M.; Gaset, A. *Synth. Commun.* **1982,** *12,* 1121.
49. Forbus, T. R., Jr.; Martin, J. C. *J. Org. Chem.* **1979,** *44,* 313.

Hydrolysis

by W. H. Gumprecht

The primary emphasis of this section is hydrolytic cleavage of carbon–fluorine bonds. The reagents used for this reaction range from highly alkaline to superacidic. In addition, hydrolysis of other groups that are affected by the unusual electronic effects of fluorine will be reviewed, from both the standpoints of preparative and physical chemistries. Other topics include ring-opening reactions, the important hydrolysis of chlorofluorocarbons (CFCs), hydrolysis not involving carbon–fluorine bonds, and the haloform reaction.

Hydrolysis of Monofluorides

Hydrolytic cleavage of single carbon–fluorine bonds generally requires activation by a neighboring group such as a carbonyl, sulfonyloxy, or olefinic bond or a negatively substituted aromatic group.

Alcoholysis of 1-chloro-2-acyl-1,1,2-trifluoroethane, available from trifluoroethylene, an acyl chloride, and aluminum chloride, leads to 1-fluoro-1-acylacetates [*1*] (equation 1). It is surprising that the remaining carbon–fluorine bond resists hydrolysis.

[*1*] **1**

$$\mathrm{RCOCl + CF_2{=}CHF \xrightarrow[RT]{AlCl_3,\ CH_2Cl_2} RCOCF_2CHFCl \longrightarrow}$$

$R = CH_3$ to C_9H_{19} ... 64-82%

$$\mathrm{\xrightarrow[RT]{EtO^-,\ EtOH} RCOCHFCO_2C_2H_5}$$

88-97%

A mixture of either bromine or iodine and sulfur trioxide reacts with either tetrafluoroethylene or chlorotrifluoroethylene to produce an intermediate, which, when treated with fuming sulfuric acid, yields a difluorohaloacetyl fluoride [*2*] (equation 2). 1,1,2-Trichlorotrifluoroethane can be used as a solvent.

β-Perfluoroalkylphenylazoolefins are hydrolyzed in acid medium to monophenylhydrazones of α-ketones [*3*] (equation 3). When the starting material has two perfluoroalkyls attached to the olefin portion, conjugate addition of by-product

0065–7719/95/0187–0422$08.72/1

hydrogen fluoride produces a phenylhydrazone of a perfluoro ketone as well. This side reaction can be suppressed by using sodium bicarbonate buffer.

[2] **2**

$$CF_2{=}CFY + X_2 + SO_3 \xrightarrow{30^\circ C} [XCF_2CFYOSO_2\text{-}] \longrightarrow$$

Y = F or Cl X = Br or I

$$\xrightarrow[\text{or KF, sulfolane, } 50^\circ C]{H_2SO_4 \cdot SO_3\ (30\%),\ 110^\circ C} XCF_2COF$$

40-50%

[3] **3**

$$R_FCF{=}C(R)N{=}NC_6H_5 \xrightarrow[\text{boiling}]{HCl,\ H_2O,\ EtOH} R_FCF_2C(R){=}NNHC_6H_5 + R_FC(=O){-}C(R){=}NNHC_6H_5$$

(R = CF_3) by-product HF

$R_F = CHF_2$ or C_2F_5

R = CH_3, CF_3 or C_6H_5

76 - 91%

Acidic hydrolysis of 1-(*o*-methoxyphenyl)pentafluoropropene gives *o*-hydroxy-2,3,3,3-tetrafluoropropiophenone by hydrolysis of vinylic fluorine and cleavage of the ether to the phenol [*4*] (equation 4).

[*4*] **4**

$$o\text{-}CH_3OC_6H_4CF{=}CFCF_3 \xrightarrow[\text{AcOH, } 140^\circ C,\ 30\ h]{40\%\ HBr,\ H_2O} o\text{-}HOC_6H_4COCHFCF_3$$

86%

Methyl polyfluoro-1-cyclobutenyl ethers react with sulfur trioxide to form fluoride and polyfluorocyclobutenones, but coproducts are their fluorosulfonyloxy derivatives resulting from sulfur trioxide insertion [*5*] (equation 5). The nonsulfonyloxylated polyfluorocyclopentenones are favored when starting with methyl polyfluoro-1-cyclopentenyl ethers (equation 6). Ring insertion of sulfur trioxide occurs during the reaction of 1-methoxytrifluorocyclopropene (equation 7).

[*5*] **5**

$$\text{(F-cyclobutene)}(OCH_3)(X) \xrightarrow[80\text{ - }120\,^\circ C]{2\ SO_3} \text{(F-cyclobutenone)}(X) + FSO_2O\text{-(F-cyclobutenone)}(X)$$

X = F or Cl 22-27% 30%

References are listed on pages 443–445.

[*5*] **6**

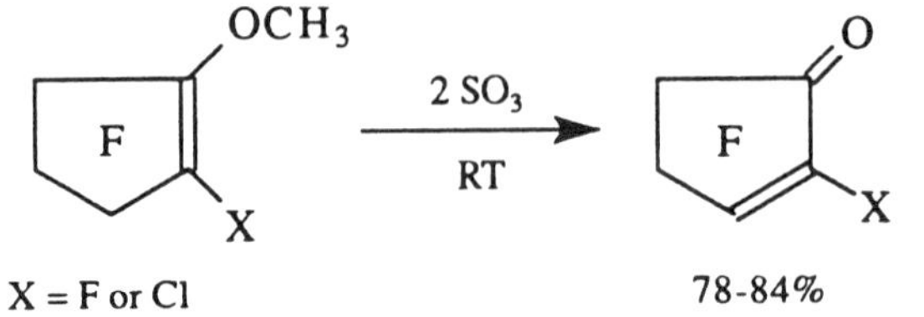

X = F or Cl; $2\,SO_3$, RT; 78-84%

[*5*] **7**

$2\,SO_3$, $CF_2ClCFCl_2$, -50°C; 66%

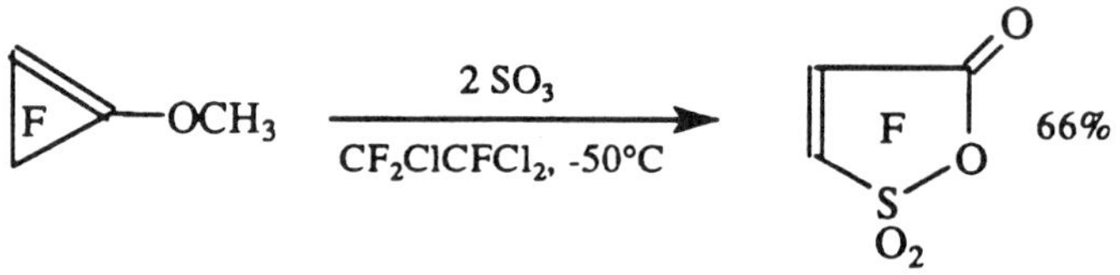

Hexafluorotropone is readily hydrolyzed in aqueous sodium hydroxide to give the three monohydroxy isomers [*6*] (equation 8). The major product is 3-hydroxypentafluorotropone; the other two isomers are formed in about equal quantities. Separation of the 2- and 3-hydroxy isomers from the mixture is based on the ability of the 2-hydroxy isomer to form a cupric chelate, and of the 3-hydroxy isomer to precipitate as an *S*-benzylthiouronium salt.

[*6*] **8**

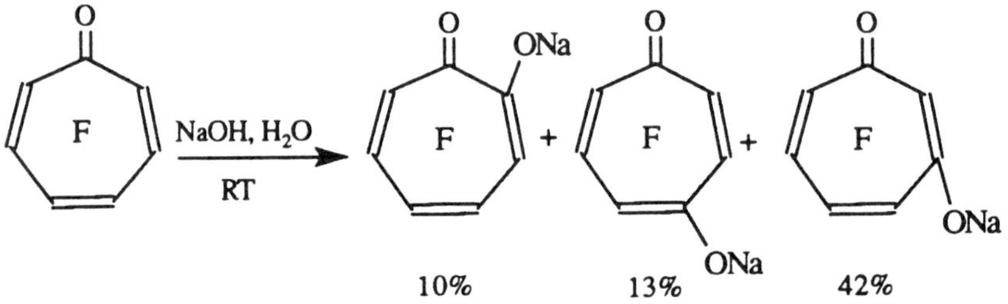

NaOH, H_2O, RT; 10%, 13%, 42%

Under more forcing conditions than had been reported earlier [*7*], pentafluorophenol, when heated in a strong solution of aqueous potassium hydroxide, gives a respectable yield of tetrafluororesorcinol [*8*] (equation 9). The process is much less effective with 2,3,5,6-tetrafluorophenol.

[*8*] **9**

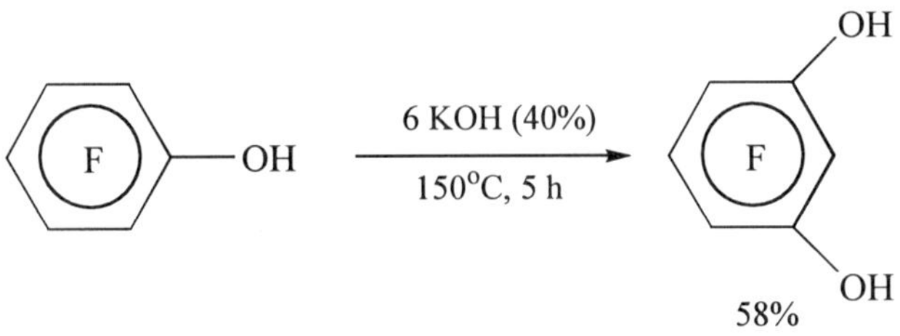

6 KOH (40%), 150°C, 5 h; 58%

Alkaline hydrolysis of pentafluorobenzoic acid, the probable intermediate in the alkaline hydrolysis of perfluorotoluene, gives a high yield of *p*-hydroxytetrafluorobenzoic acid [*9, 10*] (equation 10).

[*9*] **10**

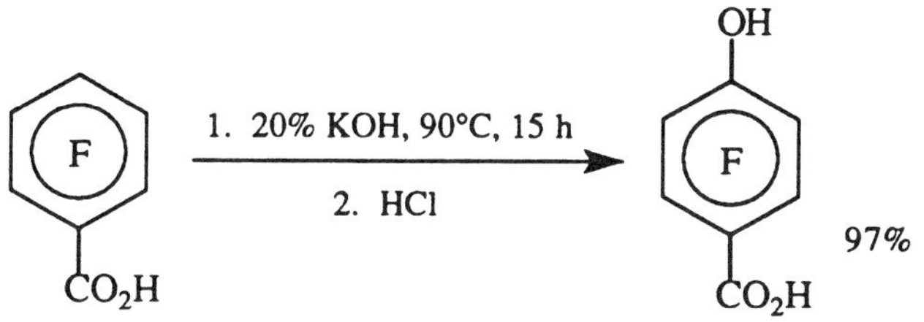

$$C_6F_5CO_2H \xrightarrow[\text{2. HCl}]{\text{1. 20\% KOH, 90°C, 15 h}} p\text{-}HOC_6F_4CO_2H \quad 97\%$$

A nitro group in the *ortho* or *para* position to fluorine is known to enhance its replacement by hydroxyl [*11, 12*]. Bromine and iodine are much less prone to hydrolysis under similar conditions. The effect is much less pronounced with the *meta*-nitro derivative (equation 11). With *o*-nitro-*p*-fluoroaniline, it is the amino group *ortho* to the nitro group, rather than *meta*-fluorine, that is replaced by hydroxyl (equation 12).

[*11*] **11**

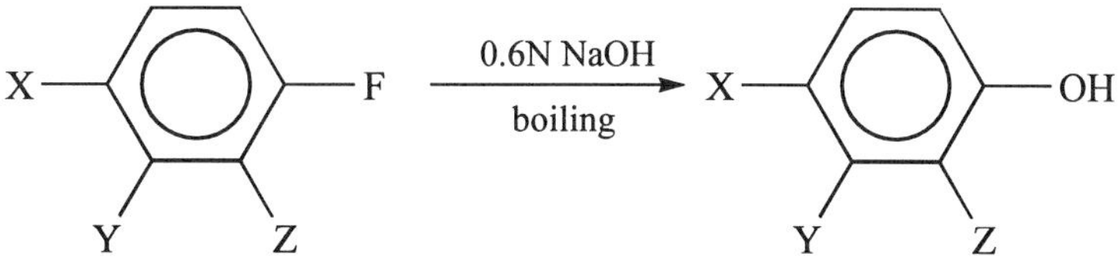

$$X\text{-}C_6H_2(Y)(Z)\text{-}F \xrightarrow[\text{boiling}]{\text{0.6N NaOH}} X\text{-}C_6H_2(Y)(Z)\text{-}OH$$

X	Y	Z	Time, h	Yield, %
NH_2	H	NO_2	2	55
NO_2	H	NH_2	72	30
NO_2	H	NO_2	1	85
H	H	NO_2	72	60
NO_2	H	CH_3	24	65
H	NO_2	H	72	<10

[*11*] **12**

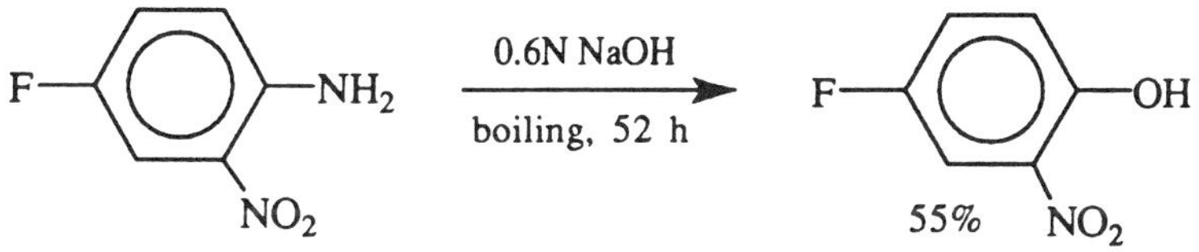

$$F\text{-}C_6H_3(NO_2)\text{-}NH_2 \xrightarrow[\text{boiling, 52 h}]{\text{0.6N NaOH}} F\text{-}C_6H_3(NO_2)\text{-}OH \quad 55\%$$

In an unusual example of displacement of fluorine by hydroxyl, hydroxyl radicals attack fluorinated benzenes. Hexafluorobenzene is the least reactive. The hydroxyl radical generates the pentafluorocyclohexadienonyl radical from it [*13*] (equation 13). These unstable species are detected spectroscopically. Their disappearance is speculated to occur by either dimerization or disproportionation because the spectroscopic decay curves can best be described by bimolecular expressions.

[*13*] **13**

$$C_6F_6 + \cdot OH \xrightarrow{H_2O} \text{F–}C_6F_4(\cdot)\text{=O} + H_3O^+ + F^-$$

Hydrolysis of Geminal Fluorides

Hydrolysis of geminal fluorides requires activation by per- or polyfluoroalkyl or perfluoroaryl groups and by double bonds linked to the same or adjacent carbon. The activation is the result of reducing the electron density at the fluorinated carbon, facilitating attack by, primarily, nucleophilic reagents.

A clue to the relatively easy hydrolysis of some difluoromethylene (and also some trifluoromethyl) groups is addition of a nucleophile to a double bond having two fluorine atoms attached to one of the carbons. Such a double bond may be generated prior to hydrolysis by elimination of hydrogen fluoride or hydrogen halide, provided there is a hydrogen on the vicinal carbon. Regiospecific hydration of such a double bond gives an intermediate that is readily convertible to a carbonyl compound and ultimately a carboxylic acid (equation 14).

14

$$R^1R^2CHCF_2X \xrightarrow[\text{-HX}]{OH^-} R^1R^2C{=}CF_2 \xrightarrow{H_2O} R^1R^2CHCF_2OH$$

$$\xrightarrow[\text{-HF}]{} R^1R^2CHCOF \xrightarrow[\text{-HF}]{H_2O} R^1R^2CHCO_2H$$

Heptyl perfluoroalkylfluoroacetate reacts with secondary amines such as piperidine. Acid hydrolysis of the product yields hydrates of heptyl perfluoroalkanoylfluoroacetate [*14*] (equation 15).

[*14*] **15**

$$R_FCF_2CHFCO_2C_7H_{15} \xrightarrow[\text{boiling; 2. HCl - }H_2O]{\text{1. piperidine, }CH_2Cl_2} R_FC(OH)_2CHFCO_2C_7H_{15}$$

$R_F = C_4F_9$	67%
$R_F = C_5F_{11}$	76%
$R_F = C_6F_{13}$	85%

Reduction of perfluoroalkanesulfonyl halides leads to sulfinic acids, which are unstable in aqueous acid and give perfluoroalkanoic acids [*15*] (equation 16). A variety of reducing agents can be used.

[*15*] **16**

$$R_FCF_2SO_2Y + \text{reducing agents} \xrightarrow{HCl,\ H_2O} R_FCO_2H$$

$R_F = C_7F_{15}$, C_9F_{19} or $(CF_3)_2CFCF_2$ $RF = C_7F_{15}$, 50%

Y = F or Cl

Reducing agents: 57% HI, $LiBH_4$ in THF or Na_2SO_3-H_2O

In the presence of sodium bicarbonate, sodium formaldehyde sulfoxylate (Rongalite) converts perfluoroalkyl iodides and bromides to the corresponding carboxylates. Less decarboxylation to the monohydroperfluoroalkanes occurs when the bromides are used [*16*] (equation 17).

[*16*] **17**

$$CF_3(CF_2)_nX \xrightarrow[DMF,\ H_2O,\ 60\text{-}90^oC]{HOCH_2SO_2Na,\ NaHCO_3} CF_3(CF_2)_{n-1}CO_2^-$$

n = 2 - 11 51-86%

X = Br or I

The methyl ether obtained from detoxifying perfluoroisobutene is dehydrofluorinated with alkali. Hydrolysis of the resulting vinyl ether yields hexafluoroisobutanoic acid [*17*] (equation 18).

[*17*] **18**

$$(CF_3)_2C{=}CF_2 \xrightarrow[RT]{MeONa,\ MeOH} (CF_3)_2CHCF_2OCH_3$$

$$\xrightarrow[Me_4NCl]{48\%\ KOH} (CF_3)_2C{=}CFOCH_3 \xrightarrow[boiling]{37\%\ HCl,\ acetone} (CF_3)_2CHCO_2H$$

95%

Perfluoro-γ-butyrolactone can be prepared from 1,4-diiodoperfluorobutane by reaction with fuming sulfuric acid (oleum) [*18*] (equation 19). The yield depends on the concentration of sulfur trioxide. One of the by-products, 4-iodoperfluorobutyryl fluoride, can be recycled to increase the overall yield of the lactone. Pure sulfur trioxide generates only perfluorotetrahydrofuran, the iodo acyl fluoride, and perfluorosuccinyl fluoride.

Acid hydrolysis of β-perfluoroalkylvinylamines, prepared from secondary amines and perfluoroalkylacetylenes, yields β-aminovinyl perfluoroalkyl ketones as the major products [*19*] (equation 20).

Enamines of an entirely different reactivity can be obtained by reacting compounds having ω-hydroperfluoroalkyl chains with lithium diethylamide. These enamines, with fluorines attached to the double bond, are hydrolyzed in acid to produce substituted diethyl fluoroacetamides [*20*] (equation 21).

[*18*] **19**

$I(CF_2)_4I$ —($H_2SO_4 \cdot SO_3$, 90°C)→ [perfluoro-γ-butyrolactone] + [perfluorotetrahydrofuran] +

$H_2SO_4 . SO_3$ 30% (↑)

9-49%

$I(CF_2)_3COF$ + $FOC(CF_2)_2COF$

$H_2SO_4.SO_3$	Yield of Lactone	Yield of $I(CF_2)_3COF$
30%	53%	3%
60%	13%	2%
100% SO_3	0	3%

[*19*] **20**

$R_FCF_2CH{=}CHNR_2$ —(20%HCl, 35-40°C)→ $R_FCOCH{=}CHNR_2$ 15-46% + $R_FCF{=}CHCHO$ <15%

$R_F = C_5F_{11}$ or C_7F_{15}
$R = C_2H_5$, *i*-C_4H_9 or $C_6H_5CH_2$
$R_2 = (CH_2)_4$, $(CH_2)_5$ or $(CH_2CH_2)_2O$

[*20*] **21**

RCF_2CF_2H —(Et_2NLi, Et_2O, -10°C)→ $RCF{=}CFN(C_2H_5)_2$ —(HCl, H_2O)→ $RCHFCON(C_2H_5)_2$

R =	
$F(CF_2)_4$	60%
$H(CF_2)_4$	60%
$(C_2H_5)_2NCO(CF_2)_4$	40%
$C_6H_5CH_2OCH_2(CF_2)_2$	60%
CH_2O, CH_2O (cyclic acetal) $CH(CF_2)_4$	60%

Neutral hydrolysis of a family of highly reactive perfluoro*bis*azomethines yields *N,N′*-perfluoroalkylureas. In the simplest case (n = 1), the product is hexafluoro-N,N′-dimethylurea [*21*]. Perfluoroethylene*bis*azomethine (n = 2) gives the cyclic *N,N'*-ethyleneurea and *N,N′*-*bis*(trifluoromethyl)oxamide (equation 22).

Decomposition with water of difluorobenzyl cations [*22*], generated from octafluorotoluene (equation 23) or its ring-substituted derivatives (equation 24) and antimony pentafluoride, yields carbonyl compounds or regenerates the starting materials.

[21] **22**

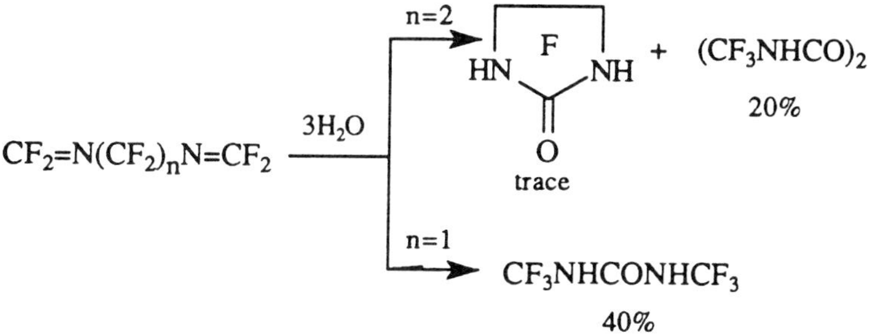

$$CF_2{=}N(CF_2)_nN{=}CF_2 \xrightarrow{3H_2O} \begin{cases} n=2: & \text{cyclic F-urea (HN, NH, C=O), trace} + (CF_3NHCO)_2 \text{ 20\%} \\ n=1: & CF_3NHCONHCF_3 \text{ 40\%} \end{cases}$$

[22] **23**

$$R^2\text{-}C_6F_3(R^1)\text{-}CF_2^+\ SbF_6^- \xrightarrow{H_2O} R^2\text{-}C_6F_3(R^1)\text{-}CO_2H \quad 52\text{-}66\%$$

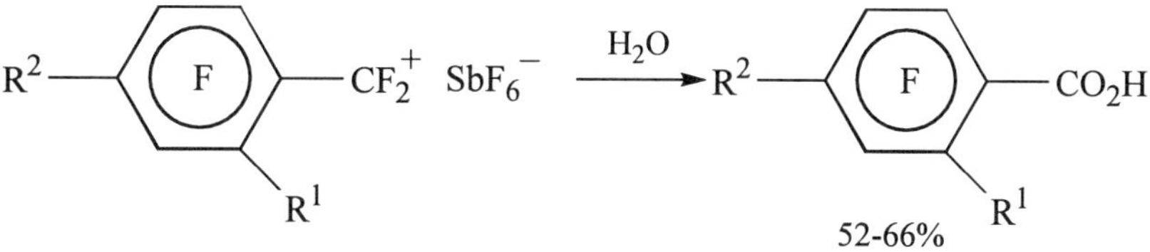

R^1 = F or H

R^2 = F, Cl, Br, CH_3 or H

[22] **24**

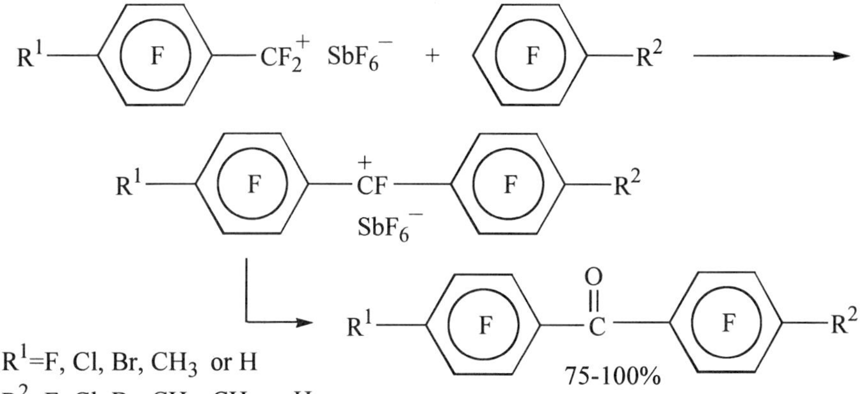

$$R^1\text{-}C_6F_4\text{-}CF_2^+\ SbF_6^- + C_6F_5\text{-}R^2 \longrightarrow R^1\text{-}C_6F_4\text{-}\overset{+}{C}F(SbF_6^-)\text{-}C_6F_4\text{-}R^2 \longrightarrow R^1\text{-}C_6F_4\text{-}C(=O)\text{-}C_6F_4\text{-}R^2 \quad 75\text{-}100\%$$

R^1=F, Cl, Br, CH_3 or H

R^2=F, Cl, Br, CH_3, CH_3 or H

An unusual displacement of both chloride and fluoride occurs when 2,2′-dichloro(perfluorodicyclobuten-1-yl) reacts with pyridine and water, producing a pyridinium betaine of a conjugated dienone [*23*] (equation 25). Apparently, the onium group activates both halogens through the conjugated system.

Terminally unsaturated fluorinated alkenoic acids can be obtained from polyfluorocycloalkenes by reaction with potassium hydroxide in *tert*-butyl alcohol [*24*] (equation 26). The use of a tertiary alcohol is critical because primary and secondary alcohols lead to ethers of the cycloalkenes. The use of a polar aprotic solvent such as diglyme generates enols of diketones [*26*] (equation 27). The compound where n = 2 can also be obtained by hydrolysis of 1-hydroxyperfluoro-1-cyclopentene, available from perfluorocyclopentene by formation of the 1-benzyl ether followed by hydrolysis [*25*] (equation 28).

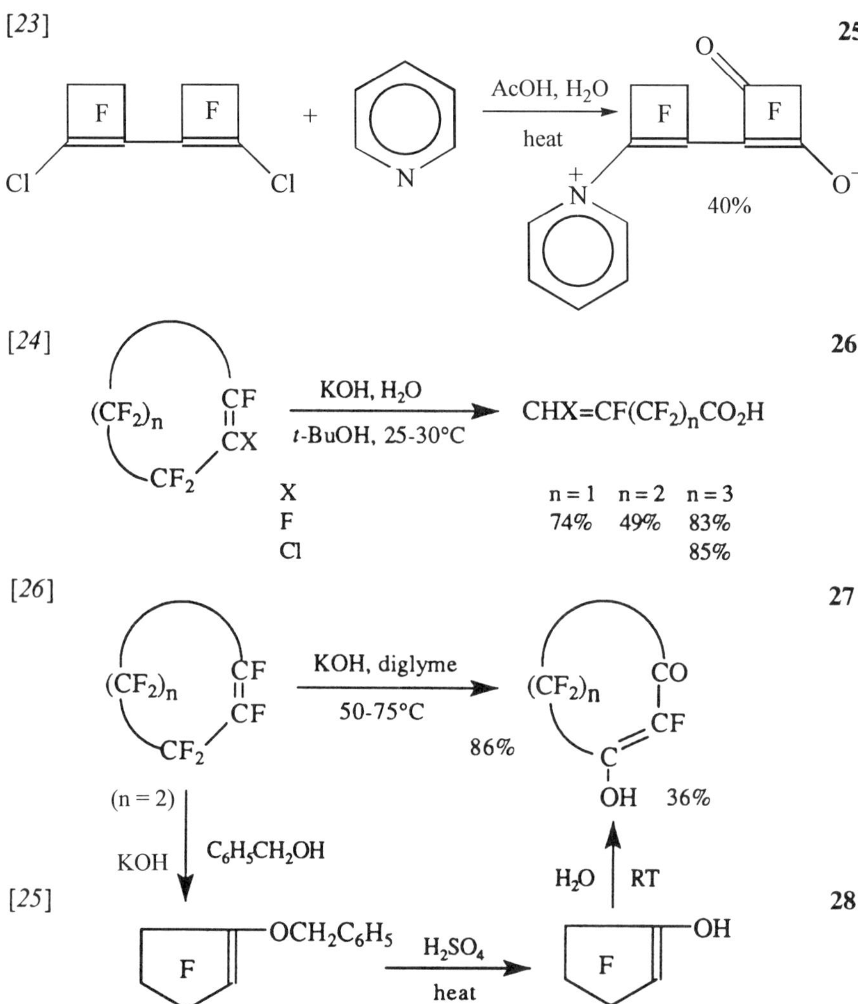

Perfluoroindene and 3-chloroperfluoroindene both react with fuming sulfuric acid [*27*]. However, whereas the former gives only heptafluoro-1-indanone-2-sulfonic acid (equation 29), the latter yields both this product and 3-chloropentafluoroindanone (equation 30).

When treated with fuming sulfuric acid and then with water, decafluorocyclohepta-1,4-diene gives hexafluoro-1,5-dihydroxy-8-oxabicyclo(3.2.1)octan-3-one rather than octafluorocyclohepta-2,6-dienone, which is a probable intermediate [*28*] (equation 31).

References are listed on pages 443–445.

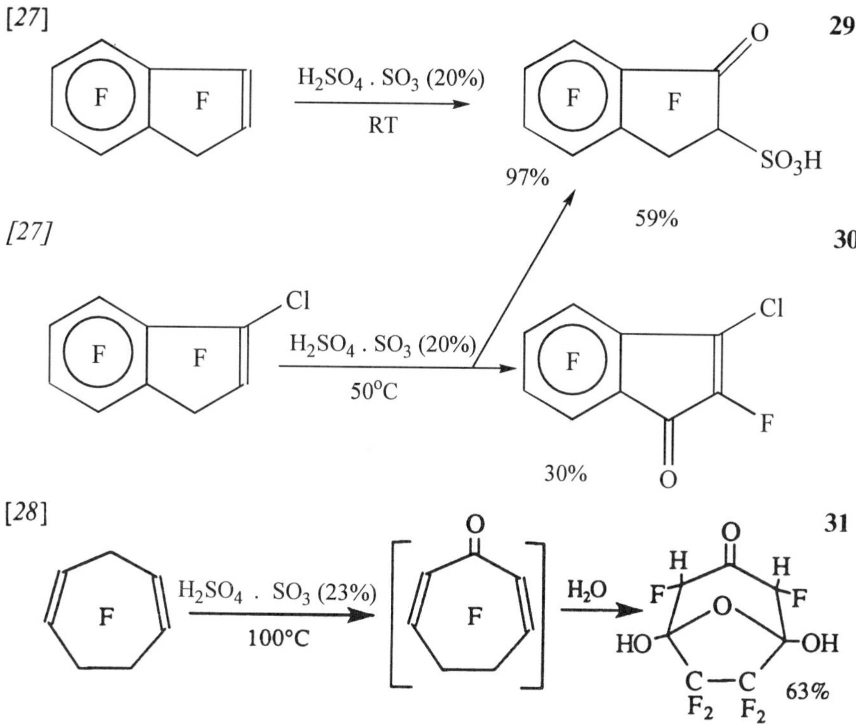

Methyl 2-methoxy-2-polyfluoroalkyl-2-fluoroacetates, generated from polyfluoroalkyltrifluoroethylene oxides and methanol, give, on heating with concentrated or fuming sulfuric acid, methyl polyfluoroalkylglyoxylates [*29*] (equation 32).

[*29*] **32**

$$X(CF_2)_nCF\overset{O}{\frown}CF_2 \xrightarrow[\text{RT}]{CH_3OH} X(CF_2)_nCF(OCH_3)CO_2CH_3$$

X = H or F
n = 1, 2 or 4

$$X(CF_2)_nCF(OCH_3)CO_2CH_3 \xrightarrow{H_2SO_4,\ 100°C \text{ or } H_2SO_4.SO_3\ (20\%),\ 100°C} X(CF_2)_nCOCO_2CH_3$$

Fuming sulfuric acid containing 10–60% sulfur trioxide hydrolyzes perfluoro-*N*-alkylcyclic amines to perfluoro-*N*-alkyl lactams. Mercuric sulfate acts as a catalyst [*30, 31*] (equation 33). The lactams are highly reactive and can be used to prepare polymeric films and surfactants.

Some small-ring compounds containing geminal fluorines suffer hydrolysis of fluorine only after initial opening of the ring. For example, alkaline hydrolysis of *gem*-difluorocyclopropyl alkyl ketones, generated by the addition of difluorocarbene to alkyl vinyl ketones, yields γ-ketoesters. The yields are quite variable [*32*] (equation 34). In contrast, alkaline treatment of methyl *gem*-difluorocyclopropanes with electron-withdrawing substituents on the methyl group causes elimination of hydrogen fluoride and yields fluorodienes [*33*] (equation 35).

[*30,31*] **33**

A F $N(CF_2)_nX$ —($H_2SO_4 . SO_3$ (30%), $HgSO_4$)→ A F $N(CF_2)_nX$ (C=O)

A	n	X	Temp., °C	Time, h	Yield, %
CF_2CF_2	2	F	170	23	72
$CF_2CF_2CF_2$	3	F	170	25	46
CF_2OCF_2	2	Cl	145	24	59
CF_2CF_2	2	N⟨F⟩A	170	24	48[a] (28)
CF_2CF_2	2	$N(C_2F_5)_2$	170	48	53
CF_2OCF_2	3	SO_2F	170	24	62
CF_2CF_2	5	F	170	24	74
CF_2OCF_2	—⟨F⟩		170	24	80

[a] The product contains 48% of bis-pyrrolidone and 28% of the mono-pyrrolidone

[*32*] **34**

R^1CO—(cyclopropane, R^2, F F) —(KOH, CH_3OH, THF, boiling)→ $R^1C(=O)CH_2CH(R^2)CO_2CH_3$

$R^1 = CH_3$, $R^2 = C_6H_5$ 85%
$R^1 = C_4H_9$, $R^2 = H$ 33%
$R^1 = H$, $R^2 = C_6H_5$ 13%

[*33*] **35**

R^1, R^2 (cyclopropane, F F) CH_2Z —(LDA, THF, -78°C or KOH, THF RT to 60°C)→ $R^1R^2C{=}CF{-}CH{=}CH{-}Z$ 28-100%

R^1, $R^2 = CH_3, CH_3$; H, C_6H_{13}; H, C_7H_{15}; H, C_6H_5
$Z = CO_2CH_3$, CN or $SO_2C_6H_5$

Hydrolysis of Trifluoromethyl Groups

Trifluoromethyl groups are very resistant to hydrolysis, unless they are allylic or benzylic, or vicinal to a carbon linked to hydrogen. In the last case, elimination of hydrogen fluoride leads to the formation of a difluoromethylene group which is key to additional reactions.

Alkaline alcoholysis of hexafluoropropylene leads to alkyl 2,3,3,3-tetrafluoropropionates. Subsequent treatment of these with acids and then with alkoxides gives dialkyl fluoromalonates [*34, 35*] (equation 36).

[*34,35*] **36**

$$CF_3CF{=}CF_2 \xrightarrow[\text{2. } H_2SO_4,\ <30°C]{\text{1. } RO^-,\ ROH,\ RT} CF_3CHFCO_2R \xrightarrow[\text{2. HCl, } H_2O]{\text{1. } RO^-,\ ROH} RO_2CCHFCO_2R$$

$R = CH_3$ 71%
$R = C_2H_5$ 63%

The stabilized fluorinated allylic cation, generated from *cis*- or *trans*-1-(*p*-methoxyphenyl)pentafluoropropene and antimony pentafluoride in sulfur dioxide, is solvolyzed by methanol to methyl 2-(*p*-methoxyphenyl)difluoroacrylate [*36*] (equation 37).

[*36*] **37**

$$p\text{-}CH_3OC_6H_4CF{=}CFCF_3 \xrightarrow[-30°C]{SbF_5,\ SO_2} p\text{-}CH_3OC_6H_4\text{-}[CF\overset{+}{\cdots}C(F)\cdots CF_2] \xrightarrow[-78°C]{CH_3OH} p\text{-}CH_3OC_6H_4CF{=}CFCO_2CH_3$$

62%

Perfluorohexamethylbenzene is converted to perfluoropentamethylbenzoic acid by sequential treatment with sodium methoxide, then concentrated sulfuric acid, and finally fuming sulfuric acid [*37*] (equation 38). The intermediate methyl *ortho* ester and methyl ester of the acid can be isolated.

As expected, the rate of alkaline hydrolysis of *p*-hydroxybenzotrifluoride is considerably higher than that of the unsubstituted benzotrifluoride [*38*] (equation 39).

It comes, therefore, as a surprise that with *p*-hydroxy-*p*′-trifluoromethylazobenzene, *p*-hydroxy-*p*′-trifluoromethylstilbene, and *p*-hydroxy-*p*′-trifluoromethyltolane the alkaline hydrolysis is slower than with the nonhydroxylated parent compounds [*39*] (equation 40).

[*37*] **38**

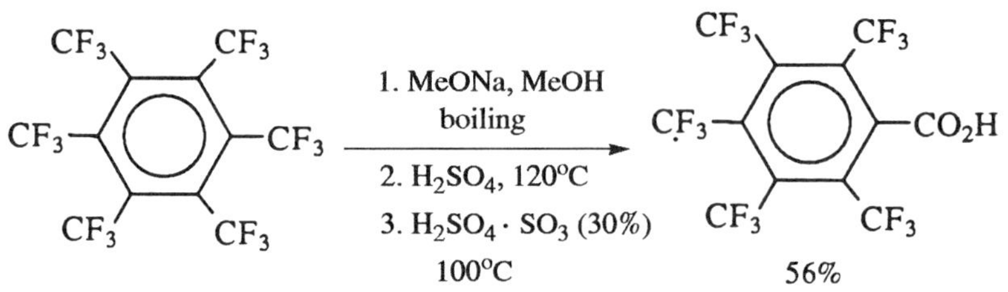

CF_3, CF_3, CF_3, CF_3, CF_3, CF_3

1. MeONa, MeOH boiling
2. H_2SO_4, 120°C
3. $H_2SO_4 \cdot SO_3$ (30%) 100°C

CF_3, CF_3, CF_3, CO_2H, CF_3, CF_3 56%

[*38*] **39**

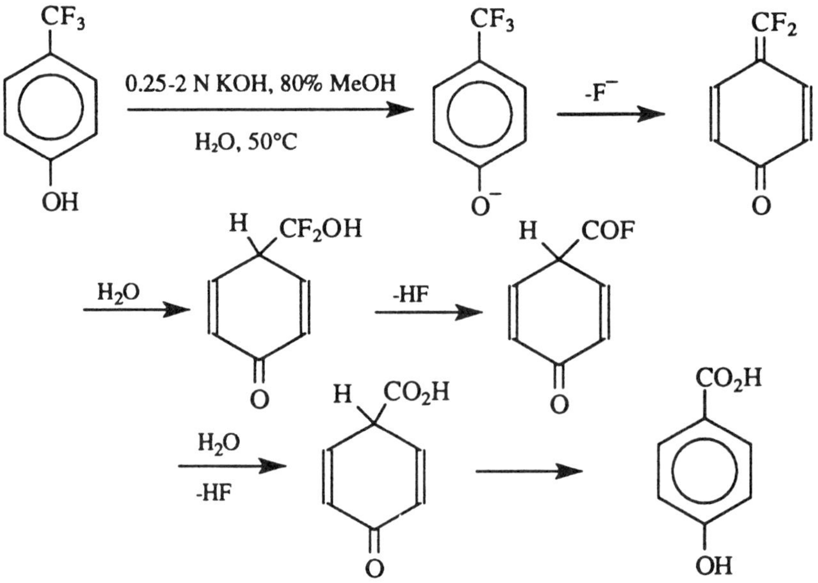

CF_3 / OH

0.25-2 N KOH, 80% MeOH
H_2O, 50°C

CF_3 / O^- —$-F^-$→ CF_2 / O

—H_2O→ H CF_2OH / O —-HF→ H COF / O

—H_2O, -HF→ H CO_2H / O ——→ CO_2H / OH

Hydrolysis of the trifluoromethyl group of 2-trifluoromethylimidazoles is promoted by the formation of the anion, which readily eliminates fluoride. The resultant difluorodiazafulvene then easily adds water. The remaining steps in the hydrolysis are predictable. When aqueous ammonia is used, 2-cyanoimidazoles result [*40*] (equation 41).

The benz analogue of 2-trifluoromethylimidazole is inert to aqueous alkali, presumably because formation of the intermediate diazafulvene would disrupt the benzene ring. However, two quinoline analogues do undergo hydrolysis with subsequent decarboxylation [*41*] (equations 42 and 43). The [4,5-h] isomer is less reactive (equation 42) than the [4,5-f] isomer (equation 43).

[*39*] **40**

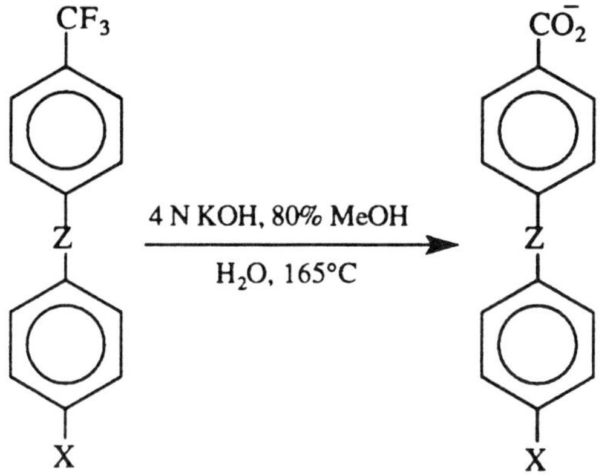

4 N KOH, 80% MeOH

H_2O, 165°C

K_{hyd} for Z = N=N, X = H >> Z = CH=CH, X = H > Z = C≡C, X = H

K_{hyd} for Z = N=N, X = H > Z = N=N, X = OH >> Z = CH=CH, X = OH > Z = C≡C, X = OH

[*40*] **41**

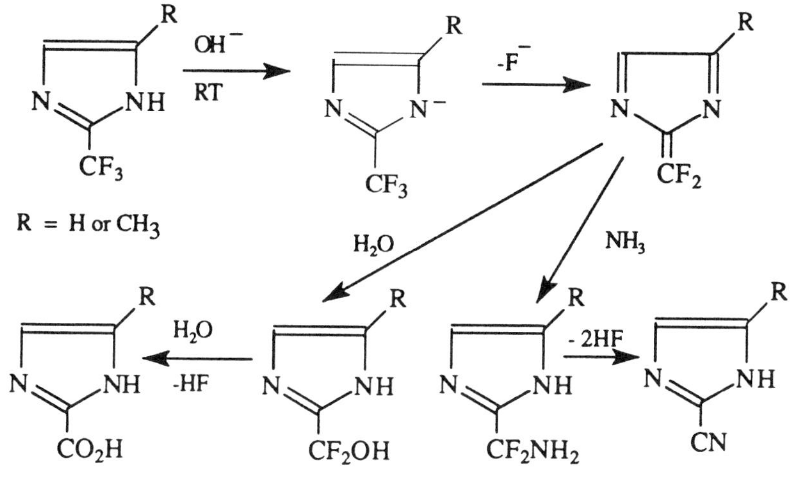

R = H or CH_3

OH^- RT; $-F^-$; H_2O; NH_3; H_2O -HF; - 2HF

63-96% 87-92%

[*41*] **42**

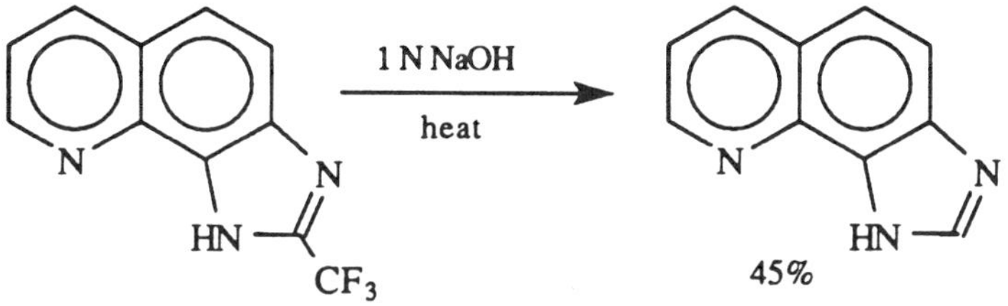

1 N NaOH

heat

45%

[*41*] **43**

1 N NaOH, RT

43%

Hydrolysis of Chlorofluorocarbons (CFCs)

The question of the fate of chlorofluorocarbons upon their release into the atmosphere is of great interest at present because of the potential damage to the earth's protective ozone layer caused by the reaction of ozone with photochemically generated chlorine atoms.

Unfortunately, the thermodynamically favored reactions of trichlorofluoromethane (CFC-11) and dichlorodifluoromethane (CFC-12) with water do not proceed to a significant degree below 300 °C and at least 200 atm (1 atm = 101.325 kPa) or greater [*42*] (equation 44). Even at 4000 atm randomization rather than complete hydrolysis occurs, leaving another chlorofluorocarbon, chlorotrifluoromethane (CFC-13), which is also potentially harmful to the earth's ozone layer.

[*42*] **44**

$$2\ CFCl_3 + 2\ H_2O \longrightarrow CO_2 + 4\ HCl + CF_2Cl_2$$

$$3\ CF_2Cl_2 + 2\ H_2O \longrightarrow CO_2 + 4\ HCl + 2\ CF_3Cl$$

Ferric oxide (3% or more on activated carbon) catalyzes the hydrolytic decomposition of CFC-12 and 1,1,2-trichlorotrifluoroethane (CFC-113) at 450 °C or above [*43*]. The products are carbon monoxide, carbon dioxide, hydrogen chloride, and hydrogen fluoride; with CFC-113, some perhaloacetyl halides are also produced.

Apart from these synthetically impractical examples of hydrolysis of chlorofluorocarbons, there are useful applications converting some chlorofluorocarbons to fluorinated carboxylic acids. As an alternative to the use of the highly corrosive fuming sulfuric acid, normally used in batch processes, a continuous hydrolytic process for converting 1,1,1-trichlorotrifluoroethane (CFC-113a), available by isomerization of CFC-113 [*44*], to trifluoroacetic acid has been developed [*45*] (equation 45). It uses metal chloride catalysts deposited on high-surface-area supports. Unreacted CFC-113a can be recycled.

Hydrolysis Not Involving the Carbon–Fluorine Bond

The versatility of triphenylphosphine in synthesis is well-known. Hydrolysis of its fluorine-containing derivatives does not involve the carbon–fluorine bond but rather the *carbon–phosphorus bond.*

[*45*] **45**

$$CF_3CCl_3 + H_2O \xrightarrow[250\text{-}350°C]{MCl_2,\ support} CF_3CO_2H + HCl \quad (66\text{-}91\%)$$

M = Fe, Mn, Co, Ni, Cu or Zn
support = activated C, alumina or silica

Hydrolysis of bromodifluoromethyl triphenylphosphonium bromide, yielding bromodifluoromethane and triphenylphosphine oxide, proceeds via difluorocarbene rather than by the bromodifluoromethyl carbanion [*46*] (equation 46). Bromodifluoromethane is a candidate for the replacement of Halon 1301 (CF_3Br), a fire extinguishant presumed to cause damage to the stratospheric ozone layer.

[*46*] **46**

$$[(C_6H_5)_3\overset{+}{P}CF_2Br]Br^- \xrightarrow[RT]{H_2O,\ CH_2Cl_2} [CF_2:] + HBr + (C_6H_5)_3PO$$

$$[CF_2:] \longrightarrow CHF_2Br \quad \text{(high yield)}$$

The adducts of perfluoroalkyl cyanides and ethoxycarbonylmethylenetriphenylphosphorane undergo acid hydrolysis to yield ethyl β-perfluoroalkyl-β-ketocarboxylates [*47*] (equation 47).

[*47*] **47**

$$(C_6H_5)_3P{=}CHCO_2C_2H_5 + N{\equiv}CR_F \longrightarrow (C_6H_5)_3PCH(CO_2C_2H_5)\text{–}\bar{N}{=}CR_F$$

$$\xrightarrow[CH_3OH]{HCl,\ H_2O} R_FCOCH_2CO_2C_2H_5 + (C_6H_5)_3PO \quad (>90\%)$$

$R_F = C_5F_{11}, C_7F_{15}$

A one-pot synthesis of alkyl perfluoroalkyl ketones has been developed. Phosphoranes, generated in situ, are acylated with a perfluoroacyl anhydride, and the resulting phosphonium salts are hydrolyzed with alkali [*48*] (equation 48).

Hydrolysis of a carbon–sulfur bond in 2-chloro-2,4,4-trifluoro-1,3-dithietane-*S*-trioxide, which can be obtained from 2,2,4,4-tetrachloro-1,3-dithietane by fluorination with antimony trifluoride followed by selective oxidations, opens the ring to produce 2-chloro-1,1,2-trifluorodimethyl sulfone [*49*] (equation 49).

2-Perfluoroalkylethanols are widely used to prepare useful materials such as oil and water repellants, surfactants, and fire-fighting foams. These alcohols are usually derived from the corresponding iodides by hydrolytic cleavage of a *carbon–iodine bond.*

[*48*] **48**

$$(C_6H_5)_3P{=}CR^1R^2 \xrightarrow[-78°C]{(R_FCO)_2O,\ Et_2O} \left[(C_6H_5)_3\overset{+}{P}C(R^1R^2)(C{=}O)R_F\right]\ {}^{-}O_2CR_F$$

$$\xrightarrow[RT]{OH^-,\ H_2O} \underset{37\text{-}78\%}{R^1R^2CHCOR_F} + R_FCO_2H + (C_6H_5)_3PO$$

$R^1 = CH_3, C_2H_5, C_3H_7, C_4H_9, i\text{-}C_4H_9, i\text{-}C_7H_{15}, C_6H_5$ or $C_6H_5CH{=}CHCH_2$
$R^2 = CH_3$ or $C_6H_5CH_2$
$R_F = CF_3, C_2F_5$ or C_3F_7

[*49*] **49**

$$\text{cyclo-}[S(O_2)\text{–}CFCl\text{–}S(O)\text{–}CF_2] \xrightarrow[\text{boiling}]{CH_3OH,\ H_2O} \underset{60\%}{CHFClSO_2CHF_2} + SO_2$$

One new method for converting 2-perfluoroalkylethyl iodides to the alcohols involves treatment with amides that minimize dehydroiodination. Wet dimethylformamide with a water content of 1.2% favors formation of the alcohol [*50*] (equation 50).

[*50*] **50**

$$C_6F_{13}CH_2CH_2I \xrightarrow[\text{boiling}]{DMF,\ H_2O} C_6F_{13}CH_2CH_2OH$$

Molar ratio DMF/H_2O is 21/1

Hydrolysis of 2-perfluoroalkylethyl iodides in the presence of nitrites also gives 2-perfluoroalkylethanols [*51*] (equation 51). A variety of solvents can be used, but acetonitrile appears the most effective. Solvents can be avoided by using a betaine surfactant as a phase-transfer catalyst.

[*51*] **51**

$$R_FCH_2CH_2I \xrightarrow[150°C]{NaNO_2,\ H_2O} R_FCH_2CH_2OH$$

$R_F = C_8F_{17}$	CH_3CN	92%
	or	
	$C_{12}H_{25}(CH_3)_2\overset{+}{N}CH_2CO_2^-$	93%
$R_F = ICH_2CH_2(CF_2)_6$	CH_3CN	90% (diol)

Another improvement (less by-products) has been made over the fuming sulfuric acid process operated commercially for the hydrolysis of 2-perfluoroalkylethyl iodides. It entails the use of sulfur trioxide in liquid sulfur dioxide [*52, 53*] (equation 52).

[*52*] **52**

$$R_FCH_2CH(R)I \xrightarrow[2.\ H_2O,\ 90°C]{1.\ SO_3,\ SO_2\ (liquid),\ -10°C} R_FCH_2CH(R)OH$$

$R_F = C_8F_{17}$	$R = CH_3$	76%
$R_F = (CF_3)_2CFOCF_2CF_2$	$R = H$	95%
	$R = CH_3$	75%

Miscellaneous Hydrolytic Reactions

One of these reactions is the well-known haloform reaction. Another leads to some unusual trifluoromethylamines.

The *haloform reaction* of unsymmetrical perfluoroalkyl and ω-hydroperfluoroalkyl trifluoromethyl ketones gives the alkane corresponding to the longer alkyl chain [*54*] (equation 53). If the methyl group contains chlorine, the reaction can take different pathways, leading to loss of chlorine (equation 54), because of the variable stability of the chlorine-substituted methyl carbanions in alkali.

[*54*] **53**

$$X(CF_2)_nCOCF_3 \xrightarrow[boiling]{NaOH,\ H_2O} X(CF_2)_nH + CF_3CO_2Na$$

X = F or H
n = 1, 2 or 4

75-80%

[*54*] **54**

$$C_2F_5COCF_xCl_{3-x} \xrightarrow[boiling]{NaOH,\ H_2O}$$

x = 2: $C_2F_5CO_2Na + HCO_2Na + NaF + NaCl$ (83%)

x = 1: $C_2F_5CO_2Na + CHCl_2F + HCO_2Na + NaF + NaCl$ (85%, 34%)

x = 0: $C_2F_5CO_2Na + CHCl_3$ (89%)

Hydrolysis of *N*-substituted trifluoromethyl fluoroformamides leads to *N*-substituted trifluoromethylamines by loss of hydrogen fluoride and carbon dioxide [*55*] (equation 55).

[55] **55**

$$CF_3N(X)COF \xrightarrow[-110°C\ to\ 22°C]{NaF,\ H_2O} CF_3NHX + CO_2 + NaHF_2 \quad (41\text{-}74\%)$$

X = F, CF_3O, $(CF_3)_2CFO$, CH_3O, C_2H_5O, $(CH_3)_2CHO$, $(CH_3)_3CO$ or CH_3CO_2

Physicochemical Impact of Fluorine Substitution on Hydrolytic Processes

Increasing the length of the fluorocarbon chain in linear methyl perfluorocarboxylates decreases the rate of neutral hydrolysis. Thus, the rate coefficients ($k \times 10^6$) in 40 vol % acetonitrile–water at 25 °C for trifluoroacetate, perfluorobutyrate, and perfluorooctanoate are, respectively, 490 ± 0.5, 18.49 ± 0.03, and 8.60 ± 0.005 per s [*56*].

The easy liberation of trifluoroethanol from 2,2,2-trifluoroethyl hydrogen 3,6-dimethylphthalate can be accounted for by ring closure to form 3,6-dimethylphthalic anhydride [*57*] (equation 56). The carboxylate anion is the intermediate needed for ring closure because the rate increases as the pH, while still acidic, increases.

[57] **56**

3,6-dimethyl-2-($CO_2CH_2CF_3$)benzoic acid (CO_2H) $\xrightarrow[39.5°C]{NaOAc,\ HCl,\ H_2O,\ buffer}$ 3,6-dimethylphthalic anhydride + CF_3CH_2OH

1-Trifluoromethyl-1-phenylethyl tosylate has been used to differentiate, as shown in Table 1, the solvolytic power of three fluorinated solvents and to compare these with formic and acetic acids. The three fluorinated solvents are trifluoroacetic acid, trifluoroethanol, and 1,1,1,3,3,3-hexafluoroisopropyl alcohol [*58*].

The electron-withdrawing power of the trifluoromethyl group in the 2,2,2-trifluoroethyl ether of 9-hydroxy-9-(dinitromethyl)fluorene considerably retards its hydrolysis compared with simple alkyl ethers [*59*] (equation 57).

Two distinct mechanisms are operating during the hydrolysis of *p*-nitrotrifluoroacetanilide, depending on the pH of the medium. At pH values greater than 9, breakdown of the *di*anion occurs, whereas at pH 7–9, the *mono*anion decomposes to the products [*60*] (equation 58). The strong electron-withdrawing character of trifluoromethyl and nitro groups profoundly influence these mechanisms. Thus, the trifluoromethyl group makes the carbonyl group more susceptible to attack by hydroxide, and the nitro group stabilizes the amide anion in the mechanism at pH > 9.

Table 1. Solvolysis of $C_6H_5C(CF_3)(CH_3)OTos$ at 25 °C [a]

Solvent	$k_1\ (s^{-1})$
CF_3CO_2H	1.72×10^{-2}
97% $(CF_3)_2CHOH$	2.95×10^{-3}
HCO_2H	3.90×10^{-4}
97% CF_3CH_2OH	5.87×10^{-5}
CH_3CO_2H	3.00×10^{-8}

[a]Calculated from data at higher temperatures.

[*59*] **57**

$CH(NO_2)_2$; OR ; dioxane, H_2O ; pH = 3-12, 25°C ; =O ; + $CH_2(NO_2)_2$ + ROH

K_{hyd} for R = C_2H_5 is about 150 times greater than that for R = CF_3CH_2.

[*60*] **58**

CF_3CONH ; NO_2 ; pH > 9 ; pH = 7-9

O^- ; CF_3CNH ; NO_2 ; O^- ; O^- ; CF_3CNH ; NO_2 ; OH ; H^+

$H\bar{N}$; NO_2 ; H_2N ; NO_2

+ $CF_3CO_2^-$; + CF_3CO_2H

The buffered, imidazole-catalyzed hydrolysis of trifluoroacetanilides appears to be a result of the complexation of the carbonyl group by the imidazole nucleus (however, *see* reference 61). Evidence for such a complex was obtained spectroscopically. There is no proton transfer during complexation because *N*-methylimidazole is also an active catalyst [*62*] (equation 59).

[*62*] **59**

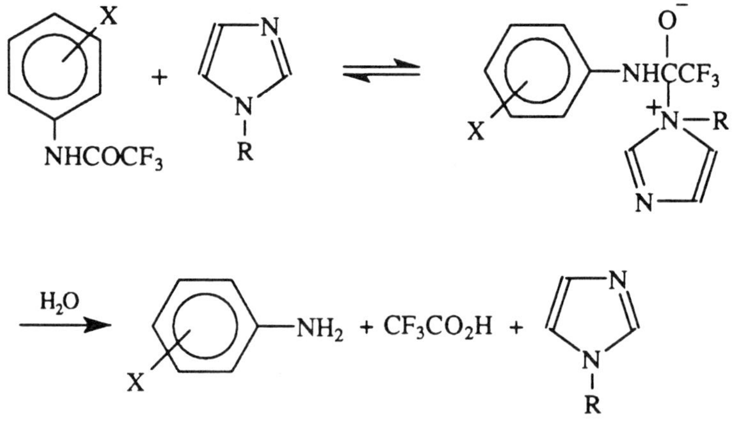

X = H, *p*-F, *m*-Cl, *p*-Cl, *m*-CH_3, *p*-CH_3, *m*-CH_3O or *p*-CH_3O
R = H or CH_3

By using imidazole catalysis, it is possible to get a better understanding of the active forms that water takes in enzymatic processes. Thus, at low concentrations in the presence of an enzyme, the water may not be fully hydrogen bonded and therefore more reactive [*61*]. The rate of hydrolysis of *p*-nitrotrifluoroacetanilide in acetonitrile shows a strong dependence on water concentration at low levels in the presence of imidazole. The imidazolium complex is the approximate transition state (equation 60).

[*61*] **60**

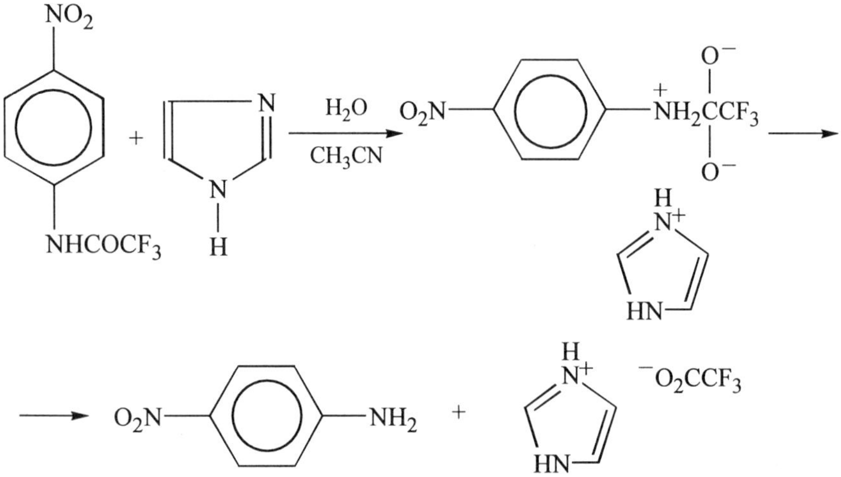

References are listed on pages 443–445.

References for Pages 422–442

1. Ishikawa, N. Jpn. Pat. 82-85 340, 1982; *Chem. Abstr.* **1982,** *97,* 197856r.
2. Yamabe, M.; Munekata, S.; Samejima, S. Jpn. Pat. 82-40 435, 1982; equivalent to U.S. Pat. 4 362 672, 1982; *Chem. Abstr.* **1982,** *97,* 55318t.
3. Pletnev, S. I.; Bargamova, M. D.; Knunyants, I. L. *Izv. Akad. Nauk SSSR* **1982,** 408 (Engl. Transl. 369).
4. Dmowski, W. *J. Fluorine Chem.* **1982,** *20,* 589.
5. Smart, B. E.; Krespan, C. G. *J. Am. Chem. Soc.* **1977,** *99,* 1218.
6. Allen, M. E.; Stephens, R.; Tatlow, J. C. *J. Fluorine Chem.* **1984,** *25,* 309.
7. Birchall, J. M.; Haszeldine, R. N. *J. Chem. Soc.* **1959,** 13.
8. Mobbs, R. H. *J. Fluorine Chem.* **1971/72,** *1,* 365.
9. Kaieda, O.; Okidaka, I.; Nakamura, T. Jpn. Pat. 60-204 742, 1985; *Chem. Abstr.* **1986,** *104,* 68605x.
10. Mazalov, S. A.; Krylova, E. P.; Tataurov, G. P. U.S.S.R. Pat. 270 718, 1970; *Chem. Abstr.* **1970,** *73,* 98622.
11. Anderson, J. S.; Brown, J. C. *Synth. Commun.* **1983,** *13,* 233.
12. Ostaszynski, A.; Tuszko, W. *Rocz. Chem.* **1961,** *35,* 1243; *Chem. Abstr.* **1962,** *57,* 7135e.
13. Koster, R.; Asmus, K-D. *J. Phys. Chem.* **1973,** *77,* 749.
14. Iznaden, M.; Portella, C. *J. Fluorine Chem.* **1989,** *43,* 105.
15. Yamamoto, F. Jpn. Pat. 77-53 812, 1977; *Chem. Abstr.* **1977,** *87,* 117584f.
16. Huang, B-N.; Haas, A.; Lieb, M. *J. Fluorine Chem.* **1987,** *36,* 49.
17. Yokoi, K. Jpn. Pat. 63-35 539, 1988; *Chem. Abstr.* **1988,** *109,* 189842j.
18. Kumai, S.; Samejima, S.; Yamabe, M. *Rep. Res. Lab., Asahi Glass* **1979,** *29,* 105; *Chem. Abstr.* **1980,** *93,* 94779f.
19. Le Blanc, M.; Santini, G.; Gallucci, J.; Riess, J. G. *Tetrahedron* **1977,** *33,* 1453.
20. Wakselman, C.; Nguyen, T. *J. Org. Chem.* **1977,** *42,* 565.
21. Ogden, P. H. U.S. Pat. 3 745 169, 1973; *Chem. Abstr.* **1973,** *79,* 91626d.
22. Pozdnyakovich, Y. V.; Shteingarts, V. D. *J. Fluorine Chem.* **1974,** *4,* 317.
23. Cullen, W. R.; Soulen, R. L. *J. Fluorine Chem.* **1981,** *17,* 453.
24. Bekker, R. A.; Popkova, V. Y.; Knunyants, I. L. *Dokl. Akad. Nauk SSSR* **1978,** *239,* 330 (Engl. Transl. 108).
25. Bekker, R. A.; Popkova, V. Y.; Knunyants, I. L. *Izv. Akad. Nauk SSSR* **1978,** 493 (Engl. Transl. 430).
26. Stockel, R. F.; Beacham, M. T.; Megson, F. H. *J. Org. Chem.* **1965,** *30,* 1629.
27. Karpov, V. M.; Platonov, V. E.; Yakobson, G. G. *Izv. Akad. Nauk SSSR* **1975,** 1593 (Engl. Transl. 1478).
28. Hamor, M. J.; Hamor, T. A.; Jenkins, C. M.; Stephens, R.; Tatlow, J. C. *J. Fluorine Chem.* **1977,** *10,* 605.
29. Saloutin, V. I.; Piterskikh, I. A.; Pashkevich, K. I.; Postovskii, I. Y.; Kodess, M. I. *Izv. Akad. Nauk SSSR* **1980,** 439 (Engl. Transl. 439).

30. Abe, T.; Hayashi, E.; Baba, H.; Nagase, S. Jpn. Pat. 59-164 772, 1984; *Chem. Abstr.* **1985,** *102,* 78718j.
31. Hayashi, E.; Abe, T.; Nagase, S. *J. Chem. Soc. Jpn. Chem. Lett.* **1985,** 375.
32. Kobayashi, Y.; Taguchi, T.; Morikawa, T.; Takase, T.; Takanashi, H. *Tetrahedron Lett.* **1980,** *21,* 1047.
33. Kobayashi, Y.; Morikawa, T.; Yoshizawa, A.; Taguchi, T. *Tetrahedron Lett.* **1981,** *22,* 5297.
34. Ishikawa, N.; Takaoka, A. *J. Chem. Soc. Jpn. Chem. Lett.* **1981,** 107.
35. Minnesota Mining & Manufacturing Co. Br. Pat. 737 164, 1955; *Chem. Abstr.* **1956,** *50,* 13987d.
36. Chambers, R. D.; Matthews, R. S.; Parkin, A. *J. Chem. Soc.,Chem. Commun.* **1973,** 509.
37. Yagupol'skii, L. M.; Lukmanov, V. G.; Alekseeva, L. A. *Zh. Org. Khim.* **1976,** *12,* 470 (Engl. Transl. 467).
38. Kozachuk, D. N.; Serguchev, Y. A.; Failkov, Y. A.; Yagupol'skii, L. M. *Zh. Org. Khim.* **1973,** *9,* 1918 (Engl. Transl. 1936).
39. Fialkov, Y. A.; Kozachuk, D. N.; Yagupol'skii, L. M. *Zh. Org. Khim.* **1973,** *9,* 138 (Engl. Transl. 137).
40. Kimoto, H.; Cohen, L. A. *J. Org. Chem.* **1979,** *44,* 2902.
41. Moores, I. G.; Smalley, R. K.; Suschitzky, H. *J. Fluorine Chem.* **1982,** *20,* 573.
42. Hagen, A. P.; Elphingstone, E. A. *J. Inorg. Nucl. Chem.* **1974,** *36,* 509.
43. Okazaki, S.; Kurosaki, A. *J. Chem. Soc. Jpn. Chem. Lett.* **1989,** 1901.
44. Miller, W. T., Jr.; Fager, E. W.; Griswold, P. H. *J. Am. Chem. Soc.* **1950,** *72,* 705.
45. Kondo, T.; Maruta, M.; Oshio, H. Ger. Offen. Pat. DE 3 509 911, 1985; equivalent to U.S. Pat. 4 647 695, 1987; *Chem. Abstr.* **1986,** *104,* 109034f.
46. Flynn, R. M.; Manning, R. G.; Kessler, R. M.; Burton, D. J.; Hansen, S. W. *J. Fluorine Chem.* **1981,** *18,* 525.
47. Trabelsi, H.; Rouvier, E.; Cambon, A. *J. Fluorine Chem.* **1986,** *31,* 351.
48. Qiu, W.; Shen, Y. *J. Fluorine Chem.* **1988,** *38,* 249.
49. Henn, R.; Sundermeyer, W. *J. Fluorine Chem.* **1988,** *39,* 329.
50. Matsuo, M.; Hayashi, T. *Rep. Res. Lab. Asahi Glass* **1976,** *26,* 55; *Chem. Abstr.* **1977,** *86,* 170772w.
51. Yoshida, C.; Tanaka, K.; Chiba, Y. Jpn. Pat. 2-157 238, 1990; *Chem. Abstr.* **1990,** *113,* 171487v.
52. Mares, F.; Oxenrider, B. C. *J. Fluorine Chem.* **1976,** *8,* 373.
53. Day, R. I. U.S. Pat. 3 283 012, 1966; *Chem. Abstr.* **1967,** *67,* 1773.
54. Saloutina, L. V.; Zapevalov, A. Y.; Kodess, M. I.; Kolenko, I. P.; German, L. S. *Izv. Akad. Nauk SSSR* **1983,** 1114 (Engl. Transl. 1023).
55. Sekiya, A.; DesMarteau, D. D. *J. Fluorine Chem.* **1980,** *15,* 183.
56. Cleve, N. J.; Euranto, E. K. *Finn. Chem. Lett.* **1974,** 82; *Chem. Abstr.* **1974,** *81,* 77795e.

57. Hawkins, M. D. *J. Chem. Soc., Perkin Trans. 2* **1975,** 285.
58. Koshy, K. M. *J. Am. Chem. Soc.* **1980,** *102,* 1216.
59. Hoz, S.; Perach, S. S. *J. Org. Chem.* **1982,** *47,* 4056.
60. Pollack, P. M. *J. Am. Chem. Soc.* **1973,** *95,* 4463.
61. Henderson, J. W.; Haake, P. *J. Org. Chem.* **1977,** *42,* 3989.
62. Stauffer, C. E. *J. Am. Chem. Soc.* **1974,** *96,* 2489.

Alkylation

by C. Wakselman

The reactivities of the substrate and the nucleophilic reagent change when fluorine atoms are introduced into their structures. This perturbation becomes more important when the number of atoms of this element increases. A striking example is the reactivity of alkyl halides. S_N1 and S_N2 mechanisms operate when few fluorine atoms are incorporated in the aliphatic chain, but perfluoroalkyl halides are usually resistant to these classical processes. However, formal substitution at carbon can arise from other mechanisms. For example nucleophilic attack at chlorine, bromine, or iodine (halogenophilic reaction, occurring either by a direct electron-pair transfer or by two successive one-electron transfers) gives carbanions. These intermediates can then decompose to carbenes or olefins, which react further (*see* equations 15 and 47). Single-electron transfer (SET) from the nucleophile to the halide can produce intermediate radicals that react by an $S_{RN}1$ process (*see* equation 57). When these chain mechanisms can occur, they allow reactions that were previously unknown. Perfluoroalkylation, which used to be very rare, can now be accomplished by new methods (*see* for example equations 48–56, 65–70, 79, 107–108, 110, 113–135, 138–141, and 145–146).

Alkylation at Oxygen

Difluoronitroethanol in aqueous alkaline solutions reacts readily with methyl bromoacetate [*1*] and with epoxides [2] (equation 1).

1

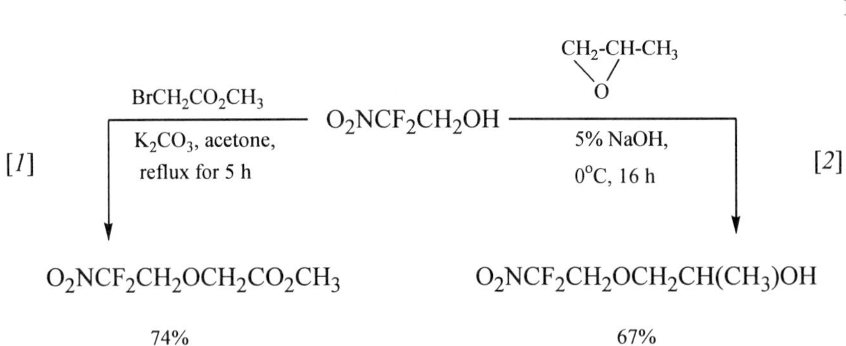

Fluorinated allylic ethers are prepared under *phase-transfer catalysis* (PTC) in the presence of tetrabutylammonium hydrogen sulfate (TBAH) [*3*] (equation 2).

0065–7719/95/0187–0446$13.04/1

[*3*] **2**

$$C_6F_{13}CH_2CH_2OH \xrightarrow[\text{TBAH, 50\% NaOH, 40 °C, 6 h}]{CH_2{=}CHCH_2Cl} C_6F_{13}CH_2CH_2OCH_2CH{=}CH_2$$

95%

Haloforms react with fluorinated alkoxides [*4*] or phenoxides [*5*] by carbene processes (equations 3 and 4).

[*4*] **3**

$$CF_3CH_2OH \xrightarrow[\text{NMP, 25 °C, 3 h}]{CHClF_2\text{, aqueous NaOH}} CF_3CH_2OCHF_2$$

54%

[*5*] **4**

$$C_6F_5OH \xrightarrow[\text{aqueous dioxane, 70 °C, 20 min}]{CHClF_2\text{, NaOH}} (C_6F_5O)_2CHF$$

90%

Nitroalcohols undergo addition to isopropyl 2,2-difluorovinyl ketone and isopropyl 3,3-difluoroacrylate in the presence of amines [*6*] (equation 5).

[*6*] **5**

$$CH_3CH(NO_2)CH_2OH + CF_2{=}CHCOCH(CH_3)_2$$

$$\xrightarrow[\text{dichloromethane, 25 °C, 2 h}]{\text{collidine}} CH_3CH(NO_2)CH_2OCF_2CH_2COCH(CH_3)_2$$

92%

Condensation of sodium phenoxide with 2,2,2-trifluoroethyl iodide gives a product of direct substitution in a low yield; several other ethers are formed by *elimination–addition reactions* [*7*]. Use of mesylate as a leaving group and hexamethyl phosphoramide (HMPA) as a solvent increases the yield of the substitution [*8*]. Even chlorine can be replaced when the condensation is performed with potassium fluoride and acetic acid at a high temperature [*9*] (equations 6–8).

[*7*] **6**

$$CF_3CH_2I \xrightarrow[\text{80 °C, 20 h}]{C_6H_5ONa\text{, DMF}} CF_3CH_2OC_6H_5 + C_6H_5OCF_2CH_2I$$

37% 7%

$$+ C_6H_5OCF{=}CHI + (C_6H_5O)_2C{=}CHI$$

11% 13%

[*8*] **7**

$$p\text{-}O_2NC_6H_4ONa \xrightarrow[\text{HMPA, 140 °C, 20 h}]{CF_3CH_2OSO_2CH_3} p\text{-}O_2NC_6H_4OCH_2CF_3$$

86%

[*9*] **8**

$$CH_3CO_2H \xrightarrow[\text{210 °C, 2 h}]{CF_3CH_2Cl,\ KF} CF_3CH_2OCOCH_3$$

93%

Cyclization of 2-bromo-2-perfluoroalkylethyl acetate under PTC conditions provides perfluoroalkyl oxiranes [*10*] (equation 9).

[*10*] **9**

$$C_6F_{13}CH(Br)CH_2OCOCH_3 \xrightarrow[\text{2. TEBA, ether-dichloromethane, 0 °C, then RT, 1.5 h}]{\text{1. NaOH RT, 1 h then 40 °C, 0.5 h}} C_6F_{13}\underset{\diagdown O \diagup}{CH\text{-}CH_2}$$

80%

Pentafluorobenzyl bromide [*11*] and a partially fluorinated alkyl iodide [*12*] react with potassium carboxylates. An interesting replacement of iodine in a fluorinated alkyl iodide by an acetate group takes place with peroxyacetic acid [*13*] (equations 10–12).

[*11*] **10**

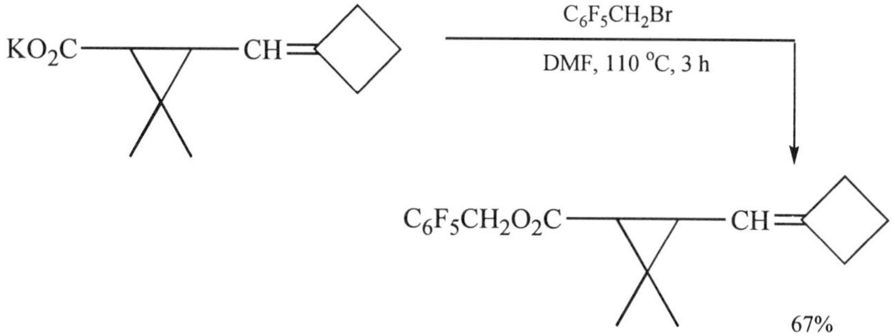

KO_2C–(2,2-dimethylcyclopropyl)–CH=cyclobutylidene $\xrightarrow[\text{DMF, 110 °C, 3 h}]{C_6F_5CH_2Br}$ $C_6F_5CH_2O_2C$–(2,2-dimethylcyclopropyl)–CH=cyclobutylidene

67%

[*12*] **11**

$$(CF_3)_2CF(CF_2)_4CH_2CH_2I \xrightarrow[\text{hydroquinone, }(CH_3)_2CHOH,\ 120\ °C,\ 5\ h]{KO_2CCH{=}CH_2,\ C_6H_{13}NEt_3I} (CF_3)_2CF(CF_2)_4CH_2CH_2O_2CCH{=}CH_2$$

71%

[*13*] **12**

$$C_6F_{13}CH_2CH_2I \xrightarrow[\substack{H_2SO_4 - H_2O \\ 30\ ^\circ C,\ 6\ h}]{CH_3CO_3H} \underset{64\%}{C_6F_{13}CH_2CH_2OCOCH_3} + \underset{23\%}{C_6F_{13}CH_2CH_2OH}$$

Diacetato(1,10-phenanthroline)palladium(II)-catalyzed *transetherification* of polyfluorinated alcohols with enol ethers leads to the corresponding fluorinated vinylic ethers [*14*] (equation 13).

[*14*] **13**

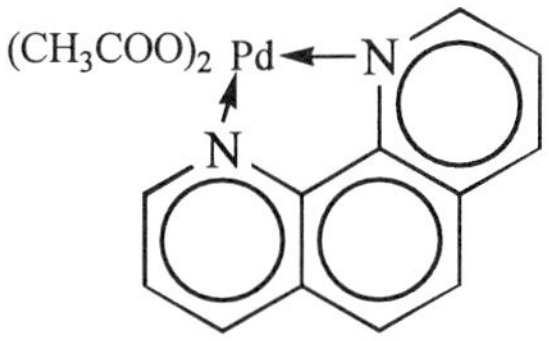

$$C_6F_{13}CH_2CH_2OH + C_2H_5OCH{=}CH_2 \xrightarrow[\text{RT, 12 h}]{(CH_3COO)_2Pd(\text{1,10-phenanthroline})} \underset{56\%}{C_6F_{13}CH_2CH_2OCH{=}CH_2}$$

Condensation of dibromodifluoromethane or bromochlorodifluoromethane with potassium phenoxides usually needs an initiation by a thiol and occurs by an *ionic chain mechanism* [*15, 16*] (equations 14 and 15).

[*15*] **14**

$$p\text{-}O_2NC_6H_4OK \xrightarrow[\substack{C_3H_7SH \\ DMF,\ 4\ h}]{CBr_2F_2} \underset{25\%}{p\text{-}O_2NC_6H_4OCHF_2} + \underset{25\%}{p\text{-}O_2NC_6H_4OCBrF_2}$$

[*16*] **15**

$$BrCF_2X \xrightarrow{Nu^-} {}^-CF_2X \xrightarrow{-X^-} :CF_2 \xrightarrow{Nu^-} NuCF_2^- \xrightarrow{H^+} NuCF_2H$$

$$NuCF_2^- \xrightarrow{BrCF_2X} NuCF_2Br \quad (\text{regenerating } {}^-CF_2X)$$

X = Br, Cl

Reaction of 1,4-dibromohexafluoro-2-butene with sodium alkoxides gives products of allylic rearrangement by an S_N2' process [*17*] (equation 16).

[*17*] **16**

$$BrCF_2CF{=}CFCF_2Br \xrightarrow[\text{- 10 to - 15 °C, then RT, 0.5 h}]{C_2H_5ONa,\ C_2H_5OH} BrCF_2CF(OC_2H_5)CF{=}CF_2 \quad 56\%$$

A peculiar reaction of perfluoroalkyl halides with chlorine fluorosulfate yields perfluoroalkyl fluorosulfates [*18*] (equation 17).

[*18*] **17**

$$C_7F_{15}I \xrightarrow[\text{25 °C, 7 days}]{ClOSO_2F} C_7F_{15}OSO_2F \quad 85\%$$

In fluorinated isocyanide dihalides and imidoyl halides, the halogens are replaced by alkoxyls [*19, 20, 21*] (equation 18).

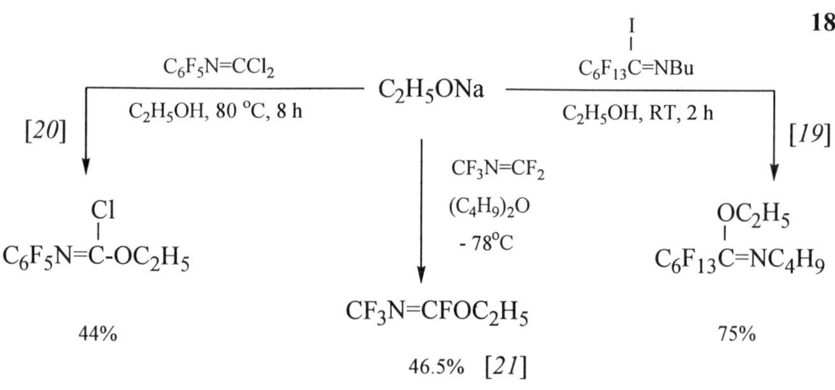

Perfluorocyclopropene reacts readily with sodium methoxide at low temperature [*22*] (equation 19). Slow addition of sodium methoxide to 1,2 dichloro-3,3-difluorocyclopropane yields initially 1-chloro-2-methoxy-3,3-difluorocyclopropane. Further addition of methanol produces probably a ketal. Opening of the cyclopropane ring gives an ortho ester that undergoes facile hydrolysis during the workup to form (*Z*)-methyl 2-fluoro-3-chloroacrylate [*23*] (equation 20). A perfluorocyclobutene dimer is also very reactive and undergoes an easy S_N2' displacement on treatment with ethanol [*24*] (equation 21).

The enol ether formed by a reaction of benzyl alcohol with perfluorocyclopentene is transformed on heating in the presence of concentrated sulfuric acid into a fluorinated enol [*25*] (equation 22).

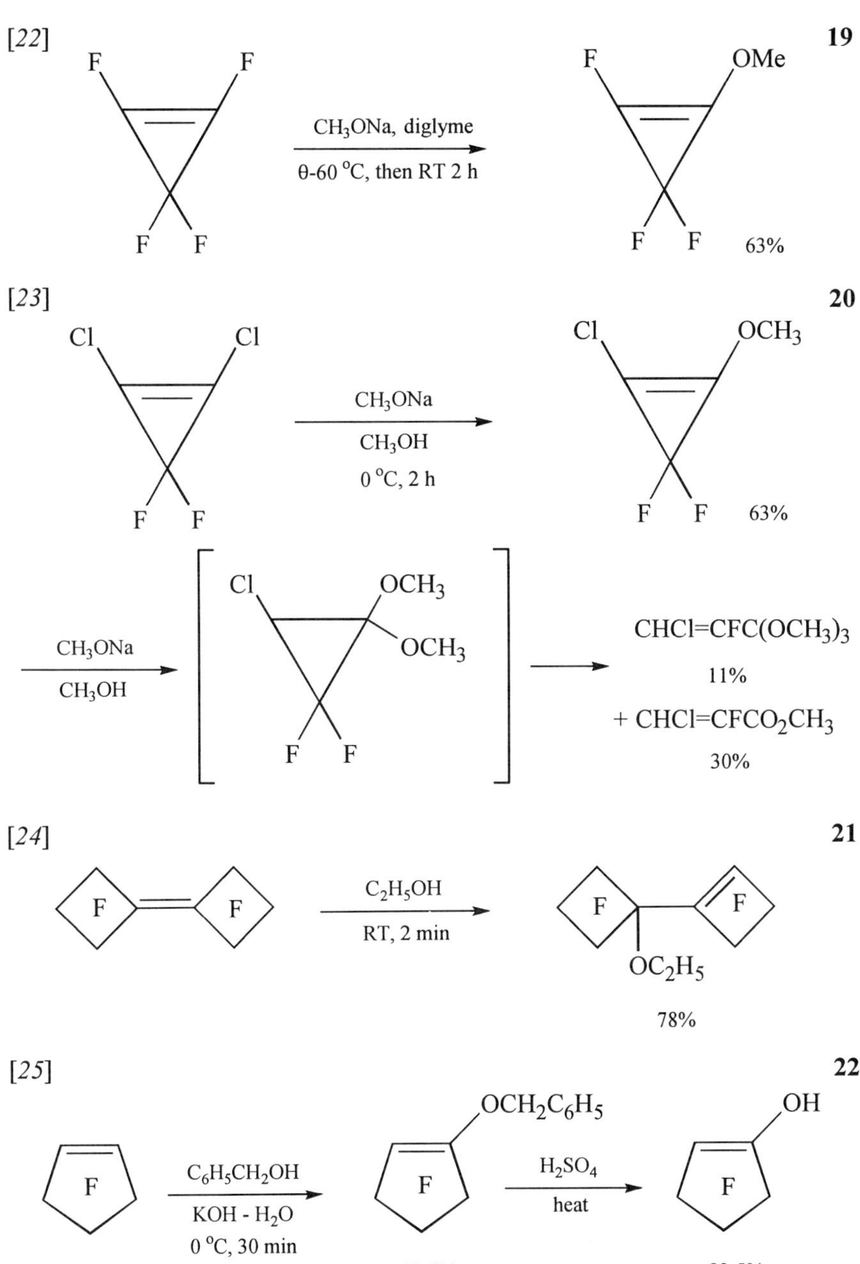

Perfluoro-2-(1-ethyl-1-methylpropyl)-3-methyl-1-pentene, the major hexamer of tetrafluoroethylene, reacts with sodium methoxide to yield an ester, whereas a stable crowded ketene is formed by reaction with sodium hydroxide [*26*] (equation 23).

[*26*] **23**

$$R_FC(=CF_2)CF(CF_3)C_2F_5 \xrightarrow[\text{reflux 5h}]{CH_3ONa,\ CH_3OH} R_FC(CO_2CH_3)=C(CF_3)C_2F_5 \quad 78\%$$

$$R_FC(=CF_2)CF(CF_3)C_2F_5 \xrightarrow[\text{diglyme - }H_2O,\ 80-90\ ^{\circ}C,\ 18\ h]{NaOH} RFC(=C=O)CF(CF_3)C_2F_5 \quad 82\%$$

$$R_F = CF_3C(C_2F_5)_2$$

When the reaction of perfluoro-3,4-dimethyl-3-hexene with methanol is performed in the presence of pyridine, a perfluorinated dihydrofuran is formed, probably by a process involving generation of an anionic oxygen atom by nucleophilic cleavage of a supposed intermediate ether [*27*] (equation 24).

[*27*] **24**

$$C_2F_5C(CF_3)=C(CF_3)C_2F_5 \xrightarrow[\text{pyridine, reflux for 4 h}]{CH_3OH,\ \text{tetraglyme}} C_2F_5C(CF_3)=C(CF_3)CF(OCH_3)CF_3 \longrightarrow$$

$$\text{2,5-difluoro-2,3,4,5-tetrakis(trifluoromethyl)-2,5-dihydrofuran } (F_3C)(F)C{-}C(CF_3)=C(CF_3){-}C(CF_3)(F){-}O \quad 56\%$$

Enolate anions of β-keto esters react with some fluoroolefins, initially by replacement of a vinylic fluorine atom, to give ultimately heterocyclic products [*28, 29*] (equation 25).

A fluorinated enol ether formed by the reaction of sodium ethoxide with chlorotrifluoroethylene is much less reactive than the starting fluoroolefin. To replace the second fluorine atom, it is necessary to reflux the reaction mixture. The nucleophilic substitution proceeds by the addition–elimination mechanism [*30*] (equation 26).

[*28*] **25**

$$R_FC(CF_3)=CFCF_3 \xrightarrow[\text{tetraglyme, RT, 16 h}]{CH_3COCH(Na)CO_2C_2H_5}$$ 4H-pyran ring: C4 bearing F_3C and F; C3 bearing R_F; C5 bearing $CO_2C_2H_5$; C2 bearing F; C6 bearing CH_3; ring O

$R_F = C(CF_3)(CF_2CF_3)_2$ 39%

[*30*] **26**

CF_2=CClF

↓ 3 C_2H_5ONa, THF - 20 °C, 15 min, then reflux for 24 h

$[C_2H_5OCF{=}CClF] \longrightarrow (C_2H_5O)_2C{=}CClF + (C_2H_5O)_3CCHClF$

30% 20%

Trifluorodiazoethane is a convenient reagent for esterification of sulfonic acids [*31*]. A bulky perfluorodiazoalkane reacts with trifluoroacetic acid to form a trifluoroacetate [*32*] (equations 27 and 28).

[*31*] **27**

$$HO_2CCH_2CH_2SO_3H \xrightarrow[\text{RT, 2 h}]{CF_3CHN_2 \text{ acetone-ether}} HO_2CCH_2CH_2SO_3CH_2CF_3$$

96%

[*32*] **28**

$$(C_2F_5)_2(CF_3)CCHN_2 \xrightarrow[Cu(ClO_4)_2,\ 18\ ^\circ C,\ 5\ \text{min}]{CF_3CO_2H} (C_2F_5)_2(CF_3)CCH_2OCOCF_3$$

88%

Polyfluorohalogenated oxiranes are usually opened by nucleophilic reagents, for example alkoxides, at the central carbon atom. However, an increase in the volume of the attacking agent, as in the case of diethylamine, directs the nucleophile toward the terminal carbon atom [*33*] (equation 29).

Oxiranes with a single perfluoroalkyl chain are regioselectively opened by nucleophilic reagents at the more accessible carbon [*34*] (equation 30).

Fluorinated alkoxides, obtained by addition of the fluoride anion to a carboxylic halide group, are weak nucleophiles, which can be alkylated [*35, 36*] (equations 31 and 32).

[*33*] **29**

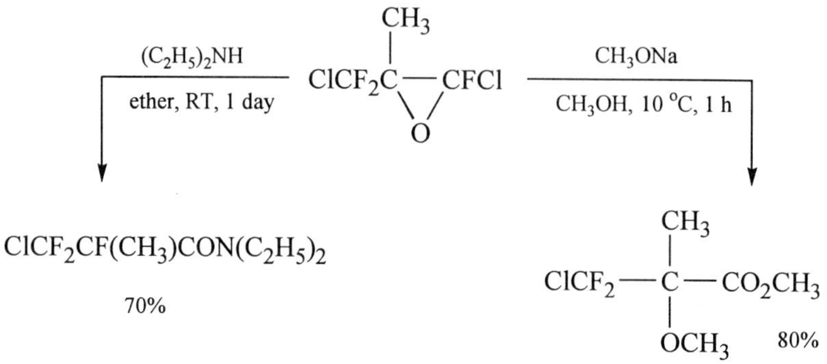

$$ClCF_2C(CH_3)\text{—}CFCl \text{ (epoxide, O bridging)}$$

$$\xrightarrow[\text{ether, RT, 1 day}]{(C_2H_5)_2NH} ClCF_2CF(CH_3)CON(C_2H_5)_2 \quad 70\%$$

$$\xrightarrow[\text{CH}_3\text{OH, 10 °C, 1 h}]{CH_3ONa} ClCF_2\text{—}C(CH_3)(OCH_3)\text{—}CO_2CH_3 \quad 80\%$$

[*34*] **30**

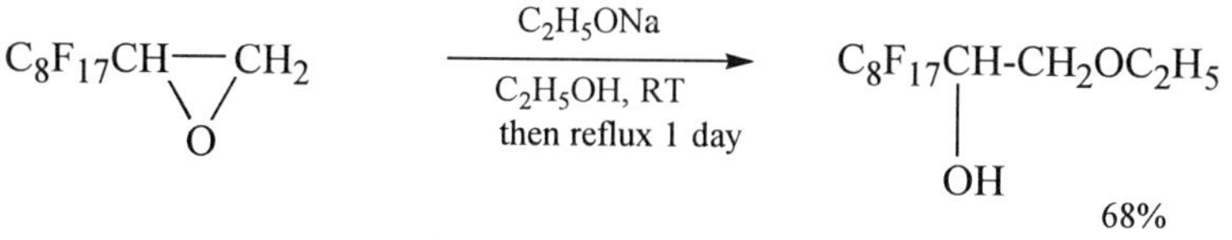

$$C_8F_{17}CH\text{—}CH_2 \text{ (epoxide)} \xrightarrow[\text{C}_2\text{H}_5\text{OH, RT then reflux 1 day}]{C_2H_5ONa} C_8F_{17}CH(OH)\text{-}CH_2OC_2H_5 \quad 68\%$$

[*35*] **31**

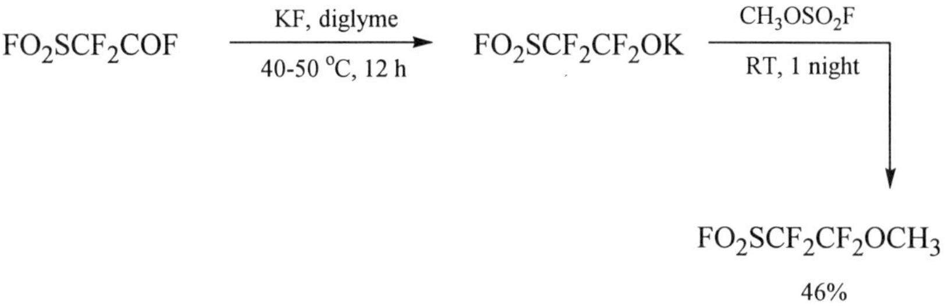

$$FO_2SCF_2COF \xrightarrow[\text{40-50 °C, 12 h}]{\text{KF, diglyme}} FO_2SCF_2CF_2OK \xrightarrow[\text{RT, 1 night}]{CH_3OSO_2F} FO_2SCF_2CF_2OCH_3 \quad 46\%$$

[*36*] **32**

$$ClCH_2CH_2OCF_2COF \xrightarrow[\text{120 °C, 4 h}]{\text{CsF, diglyme}} \text{1,4-dioxane ring: } H_2C\text{—}O\text{—}CF_2\text{—}CF_2\text{—}O\text{—}CH_2 \quad 40\%$$

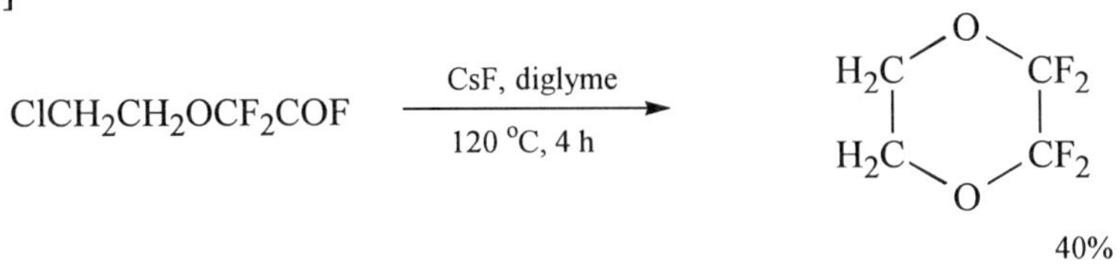

The alkoxide formed by attack of the carbonyl group of perfluorobutyrolactone opens the oxirane ring of hexafluoropropene oxide at the central carbon atom [*37*] (equation 33). A fluorinated sultone reacts with halogenoalkanes in the presence of metal fluoride [*38*] (equation 34).

The sodium derivative of ethyl 4,4,4-trifluoroacetoacetate is usually alkylated at the oxygen atom [*39*]. However, in the presence of HMPA, the ratio of O alkylation to C alkylation depends on reaction time. This observation is explained by a reversible O-alkylation process [*40*] (equations 35 and 36).

[*37*] **33**

$$\text{cyclo-}[CF_2CF_2CF_2C(=O)O] \xrightarrow[\text{-35 °C, then RT, 3 h}]{CF_3-CF\overset{O}{-}CF_2,\ \text{CsF, monoglyme}} FC(=O)(CF_2)_3OCF(CF_3)C(=O)F \quad 87\%$$

[*37*]

[*38*] **34**

$$\text{cyclo-}[CF_2CF_2OSO_2] + CH_2{=}CHCH_2Br \xrightarrow[\text{75 °C, 24 h; then 90-95 °C, 48 h}]{\text{KF, diglyme}} CH_2{=}CHCH_2OCF_2CF_2SO_2F \quad 53\%$$

[*39*] **35**

$$CF_3COCH(Na)CO_2C_2H_5 \xrightarrow[\text{130 °C, 48 h}]{ClCH_2CH_2OH} \text{2-}CF_3\text{-2-}(CH_2CO_2C_2H_5)\text{-1,3-dioxolane} \quad 49\%$$

[*40*] **36**

$$CF_3COCH_2CO_2C_2H_5 \xrightarrow[\text{THF, HMPA, 50 °C, 4 h}]{C_6H_5CH_2Br,\ (C_4H_9)_4NF} CF_3C(OCH_2C_6H_5){=}CHCO_2C_2H_5 \ (13\%) + CF_3COCH(CH_2C_6H_5)CO_2C_2H_5 \ (57\%) + CF_3COC(CH_2C_6H_5)_2CO_2C_2H_5 \ (10\%)$$

References are listed on pages 490–496.

Alkylation of fluorinated carboxylate [*41*] and sulfinate anions [*42*] occurs readily (equations 37 and 38).

[*41*] **37**

$$FO_2SCF_2COOAg \xrightarrow[-196\ ^{\circ}C,\ \text{then RT 24 h}]{CH_3I,\ \text{in the dark}} FO_2SCF_2COOCH_3$$

63%

[*42*] **38**

$$(C_8F_{17}SO_2)_2Zn \xrightarrow[DMSO,\ 70\ ^{\circ}C,\ 6\ h]{C_6H_5CH_2Br} C_8F_{17}SO_2CH_2C_6H_5$$

83%

Alkylation at Sulfur and Selenium

The action of potassium thiocyanate on tosylates derived from fluorohydrins gives the corresponding fluorinated alkyl thiocyanates [*43*] (equation 39).

[*43*] **39**

$$FCH_2CH_2OSO_2C_6H_4CH_3\text{-}p \xrightarrow[C_2H_5OH,\ 80\ ^{\circ}C,\ 24\ h]{KSCN} FCH_2CH_2SCN$$

73%

Replacement of bromine in ethyl bromofluoroacetate by thiolate anions occurs easily [*44*] (equation 40).

[*44*] **40**

$$C_6H_5SH \xrightarrow[THF,\ \text{reflux 1 h}]{BrCHFCO_2C_2H_5,\ (C_2H_5)_3N} C_6H_5SCHFCO_2C_2H_5$$

95%

2-Fluoro- or 2,2-difluoro-1-bromoalkanes readily undergo nucleophilic substitution in the presence of sodium thiophenoxide [*45*] (equation 41).

[*45*] **41**

$$C_8H_{17}CF_2CH_2Br + C_6H_5SNa \xrightarrow[25\ ^{\circ}C,\ 2\ h]{DMF} C_8H_{17}CF_2CH_2SC_6H_5$$

94%

Replacement of iodine in (perfluoroalkyl)ethyl iodides predominates over the usual conversion to olefins when the reagent is very nucleophilic and weakly basic. Soft nucleophiles like sodium thiocyanate and sodium thiolates react well in displacements [*46, 47*] (equation 42).

References are listed on pages 490–496.

42

$C_6F_{13}CH_2CH_2SNa$, THF, 0 °C, then RT, 1 night ← $C_6F_{13}CH_2CH_2I$ → NaSCN, $(CH_3)_3COH$-DMSO, 90 °C, 5 h

$C_6F_{13}CH_2CH_2SCH_2CH_2C_6F_{13}$ [*47*] 74%

$C_6F_{13}CH_2CH_2SCN$ 87% [*46*]

A fluorinated dithiaparacyclophane is prepared from lithium sulfide and 1,4-bis(bromomethyl)tetrafluorobenzene [*48*] (equation 43).

[*48*] **43**

$BrH_2C-C_6F_4-CH_2Br$ → (Li_2S, CH_3OH, RT, 1 night) → cyclophane (H_2C, S, H_2C, CH_2, S, CH_2, two F rings) 54%

A solid–liquid *phase-transfer technique* is used to synthesize aryl *difluoromethyl sulfides and selenides*: thiophenols dissolved in an aromatic solvent are treated with solid sodium hydroxide in the presence of a catalytic amount of tris(3,6-dioxaheptyl)amine (TDA 1) [*49*]. This condensation proceeds by a carbene mechanism (equation 44).

44

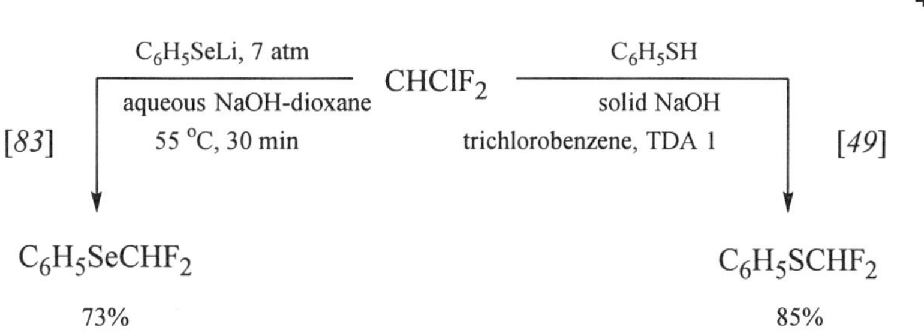

Sodium difluoromethanesulfonate is prepared from chlorodifluoromethane and sodium sulfite. The yield of this carbene reaction is more reproducible in the presence of sodium hydroxide [*50*] (equation 45).

[*50*] **45**

$$CHF_2Cl + Na_2SO_3 \xrightarrow[150\ ^\circ C,\ 20\ h]{NaOH - H_2O} CHF_2SO_3Na \quad 51\%$$

Condensation of thiophenoxide anions with various fluorinated polychloro- or polybromoethanes gives fluoroalkyl phenyl sulfides [*51, 52, 53*]. These formal substitutions involve fluorinated olefins as intermediates [*52, 53*]. In the case of perhalogenated ethanes, the mechanism shows a similarity with that of dihalogenodifluoromethane [*52*] (equations 46 and 47).

46

C_6H_5SM:

- [*52*] $BrCF_2CF_2Br$ (M = Na), aqueous NaOH (50%), TEBA, benzene, 30 °C, 3 h → 25% $C_6H_5SCF_2CF_2Br$ + 25% $C_6H_5SCF_2CF_2H$
- [*53*] (M = K) $ClCF_2CFCl_2$, DMF, RT, 4 h → $C_6H_5SCF_2CFCl_2$ 55%
- [*51*] (M = Na) $ClCF_2CHCl_2$, DMF, 60 °C, 30 min → $C_6H_5SCF_2CHCl_2$ 68%

[*52*] **47**

$$BrCF_2CF_2Br \xrightarrow{Nu^-} {}^-CF_2CF_2Br \xrightarrow{-\ Br^-} CF_2{=}CF_2 \xrightarrow{Nu^-} NuCF_2CF_2^- \xrightarrow{H^+} NuCF_2CF_2H$$

$$NuCF_2CF_2^- + BrCF_2CF_2{-}Br \rightarrow NuCF_2CF_2Br + {}^-CF_2CF_2Br$$

Thiolates react with dibromodifluoromethane by initial displacement of one of the larger halogens [*54, 55, 56*] (equation 48) (*see* equation 15).

Alkylation of thiolate anions by bromochlorodifluoromethane generally follows this mechanism [*55, 57*]. However, depending on the nature of the nucleophile and reaction conditions, disubstitution can arise by a SET process [*57*] (equation 49).

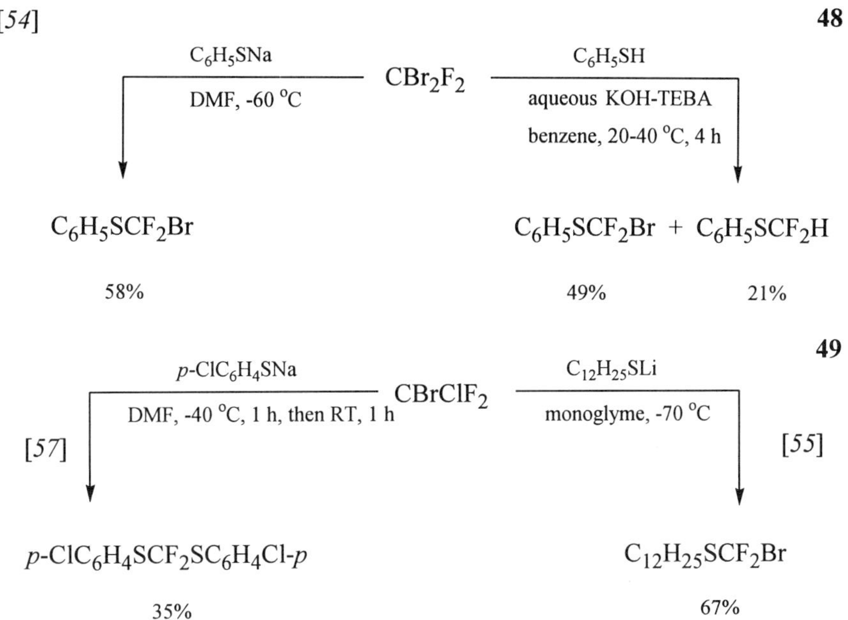

A *SET mechanism* is involved also in the condensation with dichlorodifluoromethane that occurs under moderate pressure [*53, 58*] (equation 50).

[*53*] **50**

$$C_6H_5SK \xrightarrow[\text{DMF, RT, 2.7atm, 3 h}]{CCl_2F_2} \underset{62\%}{C_6H_5SCF_2Cl} + \underset{8\%}{C_6H_5SCF_2H} + \underset{7\%}{C_6H_5SCF_2SC_6H_5}$$

Perfluoroalkylation of thiols by perfluoroalkyl iodides can be performed in liquid ammonia under UV irradiation [*59, 60*]. This photochemical reaction can also occur with thiolates in acetonitrile or under phase-transfer conditions [*60, 61*] (equations 51 and 52).

In fact, **perfluoroalkyl iodides** can react with thiolates even in the absence of UV irradiation [*53, 62, 63*]. However, the *photochemical reaction* is particularly useful for the transformation of aliphatic thiols [*59*], because the spontaneous condensation gives a considerable quantity of disulfide in that case (equations 53–55).

Even the poorly reactive **trifluoromethyl bromide** can react under slight pressure [*53, 58*] or UV irradiation [*64*] (equation 56).

The *perfluoroalkylations* of thiolate anions are interpreted by an $S_{RN}1$ *mechanism* [*58, 63, 64*] (equation 57).

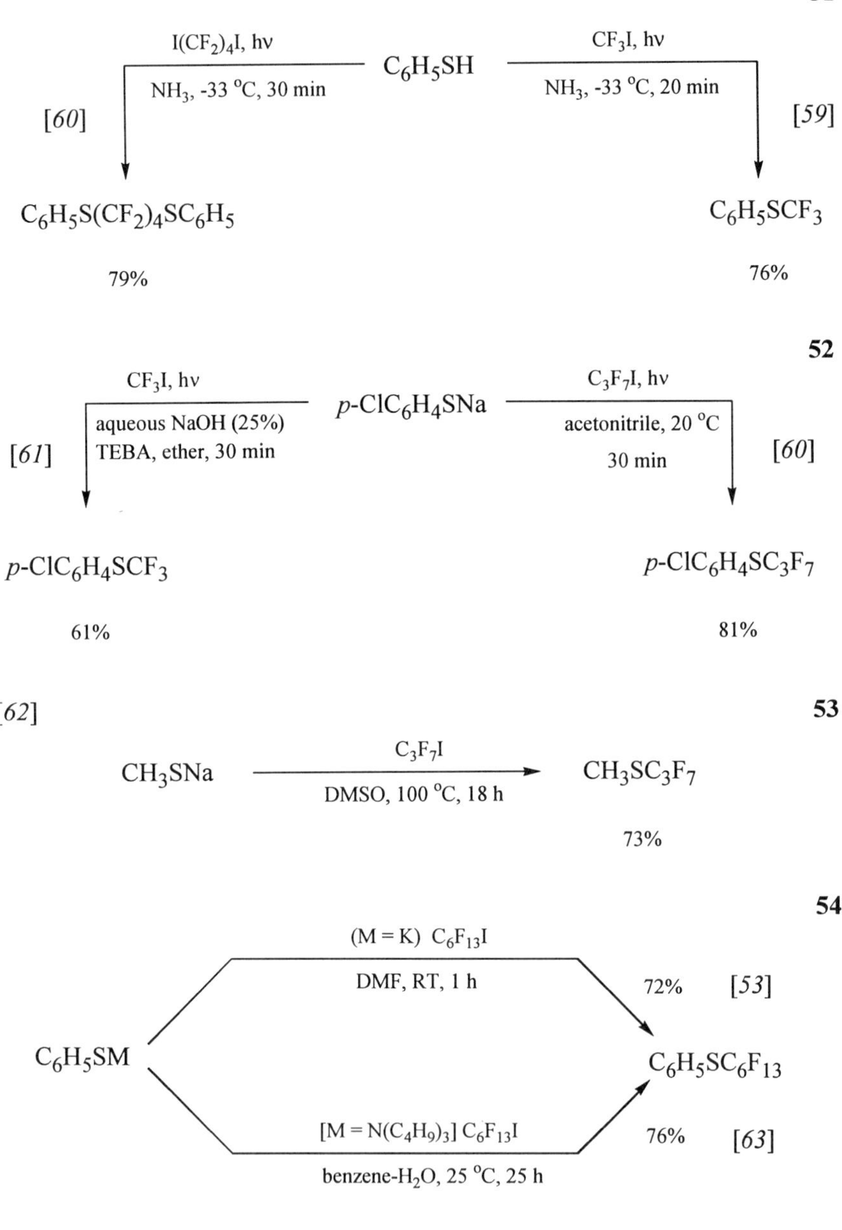

References are listed on pages 490–496.

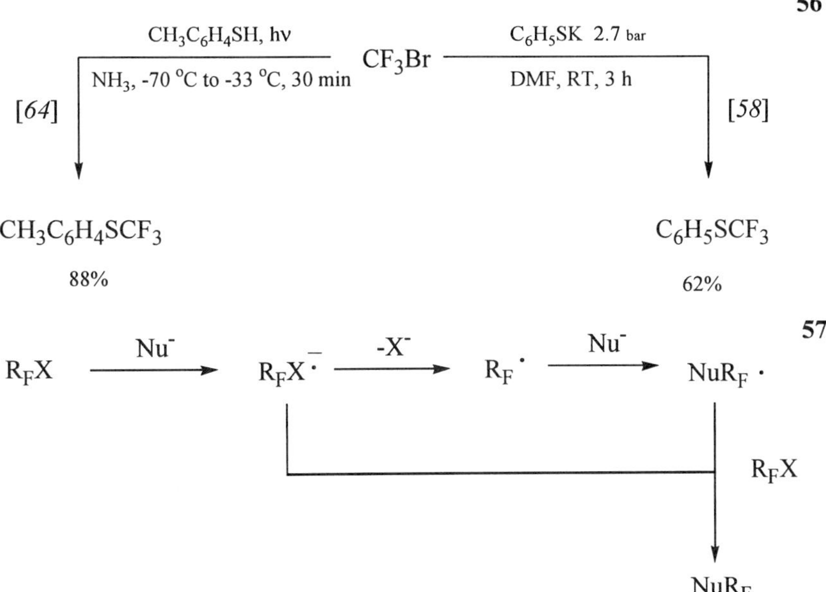

Perfluorodecalin is transformed to octakis(phenylthio)naphthalene by phenyl thiolate in dipolar aprotic solvents. The crucial first step of this reaction may involve an elimination of the tertiary fluorine atoms, which would lead to a symmetric perfluorobicyclic olefin as an intermediate [*65*] (equation 58).

[*65*] **58**

Perfluorodecalin $\xrightarrow[\text{60-70 °C, 10 days}]{C_6H_5SNa,\ \text{DMF}}$ octakis(phenylthio)naphthalene ($C_{10}(SC_6H_5)_8$), 55%

Alkylation of thiolates with **perfluoroalkyliodonium salts** occurs under mild conditions [*66, 67*]. Perfluorocarboxylic peroxides can also be used for this transformation [*68*] (equations 59–61).

Another way to prepare fluorinated sulfides is the photochemical alkylation of sulfides or disulfides by **perfluoroalkyl iodides** [*69, 70, 71*] (equations 62–64). Reaction of trifluoromethyl bromide with alkyl or aryl disulfides in the presence of a sulfur dioxide radical anion precursor, such as sodium hydroxymethanesulfinate, affords trifluoromethyl sulfides [72] (equation 65).

[*66*] **59**

$$p\text{-}CH_3C_6H_4\overset{+}{I}C_3F_7\overset{-}{Cl} \xrightarrow[\text{-30 °C, then 20 °C, 5 h}]{C_6H_5SNa,\ DMF} C_6H_5SC_3F_7 \quad 63\%$$

[*67*] **60**

$$C_6H_5\overset{+}{I}C_8F_{17}\ \ {}^{-}OSO_2CF_3 \xrightarrow[\text{dichloromethane, RT, 15 min}]{C_{12}H_{25}SH,\ \text{pyridine}} C_{12}H_{25}SC_8F_{17} \quad 82\%$$

[*68*] **61**

$$(C_3F_7COO)_2 \xrightarrow[\text{40 °C, 5 h}]{C_6H_5SK,\ C_2Cl_3F_3} C_6H_5SC_3F_7 \quad 49\%$$

[*69*] **62**

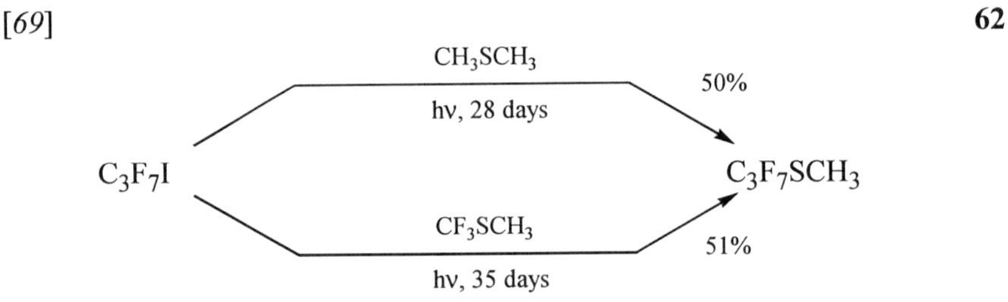

63

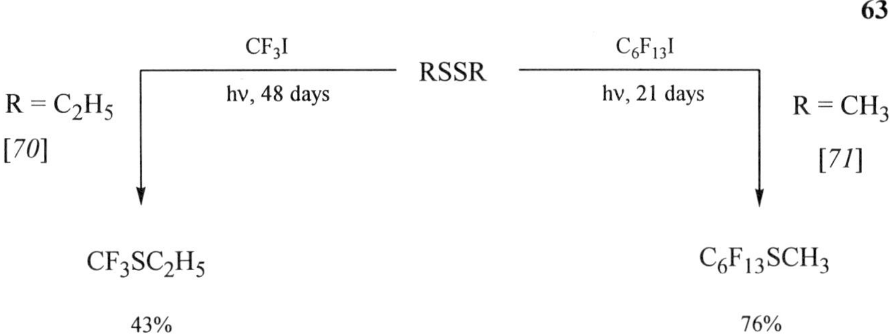

[*70*] **64**

$$CF_3SSCF_3 \xrightarrow[\text{24 days}]{C_6F_5I,\ h\nu} C_6F_5SCF_3 + C_6F_5SSCF_3 + CF_3I$$

$$42\% \qquad 13\% \qquad 44\%$$

References are listed on pages 490–496.

[*72*] **65**

$$CF_3Br + C_6H_5SSC_6H_5 \xrightarrow[\text{NaH}_2\text{PO}_4,\ \text{DMF, H}_2\text{O},\ 20^\circ\text{C, 6 h},\ 2\text{ - }5\text{ bar}]{\text{NaO}_2\text{SCH}_2\text{OH}} CF_3SC_6H_5 \quad 93\%$$

Trifluoromethylation of disulfides occurs also with ***N*-trifluoromethyl-*N*-nitrosobenzene sulfonamide** [*73*] (equation 66).

[*73*] **66**

$$HO_2CCH_2SSCH_2CO_2H \xrightarrow[\text{acetone, 14 h}]{CF_3N(NO)SO_2C_6H_5} CF_3SCH_2CO_2H \quad 63\%$$

Fluorinated sulfinates are prepared from sodium dithionite and liquid perfluoroalkyl halides [*74*] (equation 67). For the transformation of the gaseous and poorly reactive trifluoromethyl bromide, it is necessary to use moderate pressure [*75*] (equation 68). These reactions are interpreted by a SET between the intermediate sulfur dioxide radical anion and the halide. The sodium trifluoromethanesulfinate thus obtained is an intermediate for a chemical synthesis of triflic acid.

[*74*] **67**

$$H(CF_2)_8I + Na_2S_2O_4 + NaHCO_3 \xrightarrow[85\ ^\circ\text{C, 7h}]{\text{acetonitrile-H}_2\text{O}} H(CF_2)_8SO_2Na \quad 94\%$$

[*75*] **68**

$$CF_3Br + Na_2S_2O_4 + Na_2HPO_4 \xrightarrow[65\ ^\circ\text{C, 70 min}]{13\text{ bar, DMF-H}_2\text{O}} CF_3SO_2Na \quad 77\%$$

Under acidic conditions, perfluoroalkyl iodides (but not bromides) react with sodium hydrogen sulfite and cerium(IV) ions to form the corresponding sulfinates [*76*] (equation 69).

[*76*] **69**

$$C_6F_{13}I + NaHSO_3 \xrightarrow[\text{pH 3-4, 70}^\circ\text{C, 6h}]{(NH_4)_2Ce(NO_3)_6\text{, acetonitrile}} C_6F_{13}SO_2Na \quad 76\%$$

Arylsulfinic acids are perfluoroalkylated under the $S_{RN}1$ conditions, but the yield is low [*77*] (equation 70). The isolated product is an aryl perfluoroalkyl sulfone. The possible formation of a perfluoroalkyl aryl sulfinate intermediate, arising from an O-alkylation reaction, has not been discussed.

[*77*] **70**

$$C_6H_5SO_2H \xrightarrow[NH_3,\ -33^\circ C,\ 2h]{CF_3I,\ h\nu} C_6H_5S(=O)_2CF_3 \quad 30\%$$

A disubstitution reaction occurs in the condensation of 1,2-dichloroperfluorocycloalkenes with potassium thiocyanate [*78*] or potassium *p*-nitrothiophenoxide [*79*] (equation 71).

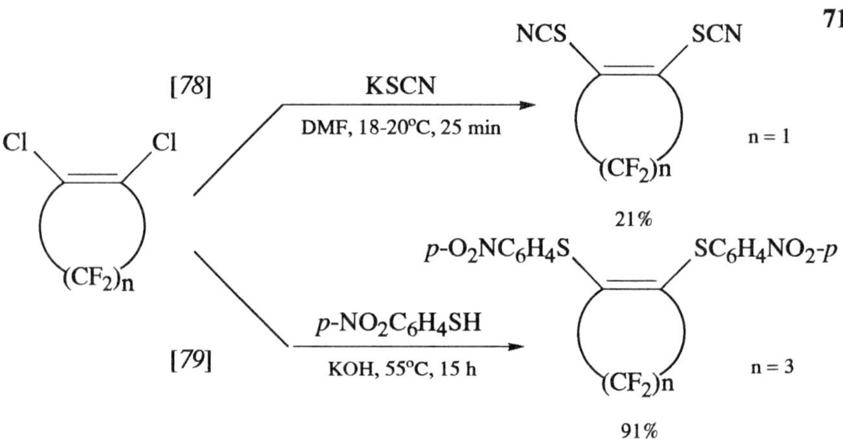

Trifluoromethylmercaptides can be used as nucleophilic reagents in alkylation reactions [*80, 81*] (equation 72).

Trifluoromethyl thiirane is formed by the action of tris(diethylamino)-phosphine on 1-chloromethyl-2,2,2-trifluoroethyldisulfide [*82*] (equation 73).

Difluoromethyl phenyl selenide is prepared by treatment of lithium phenylselenide with **chlorodifluoromethane** via a carbene mechanism [*83*] (equation 44). Bis(2,2,2-trifluoroethyl)diselenide is formed in the reaction of 2,2,2-trifluoroethyl mesylate with lithium diselenide [*84*] (equation 74).

Alkylation at Nitrogen

Replacement of one tosyloxy group in 2-fluoro-2-nitro-1,3-propanediol ditosylate by azide ion occurs easily [*85*] (equation 75).

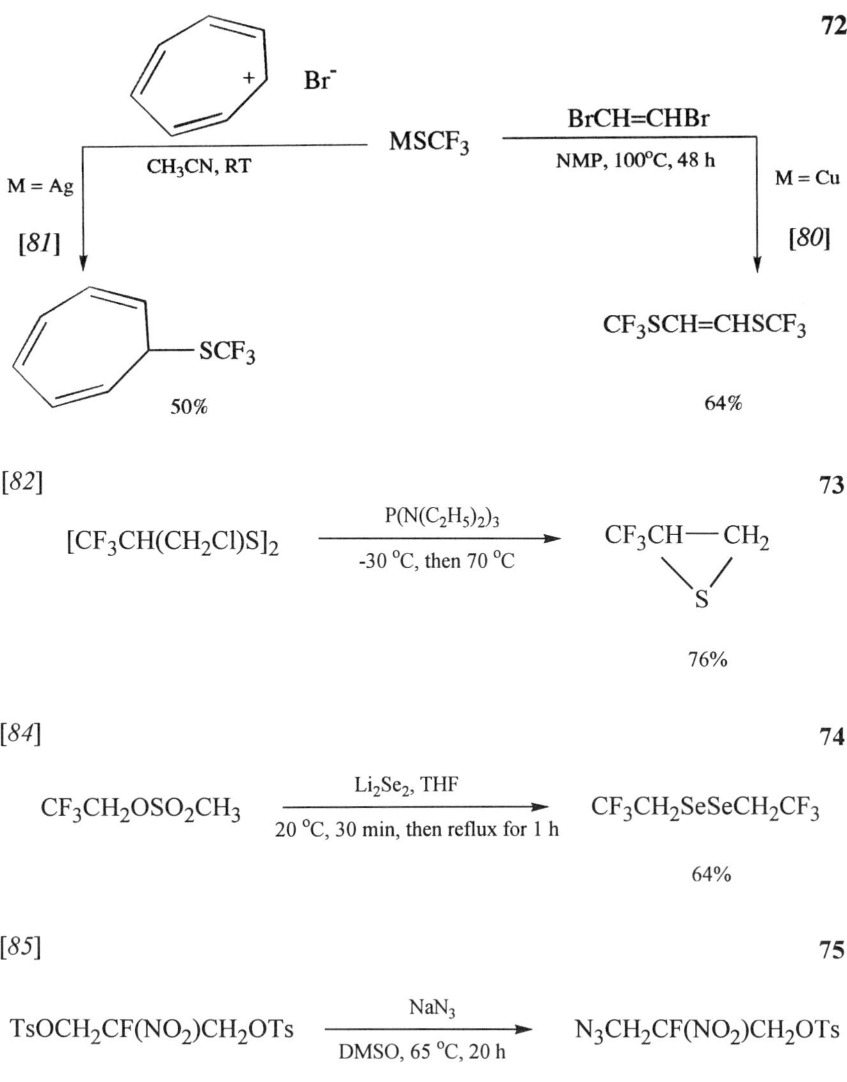

On the contrary, trifluoromethyl groups severely impede S_N1 or S_N2 reactions when located α to the reaction site. However, aminolysis of 1-phenyl-2,2,2,-trifluoroethyl tosylate [*86*] and trifluoroethyl chloride [*87*] and its corresponding nonaflate [*88*] can be performed under severe conditions (equation 76).

In HMPA, *highly fluorinated azides* are prepared. Catalytic hydrogenation of these compounds leads to the corresponding amines [*89*] (equation 77).

Polyfluoroalkyl fluorosulfates react with substituted hydrazines to give the corresponding polyfluorocarboxylic hydrazides [*90*] (equation 78).

76

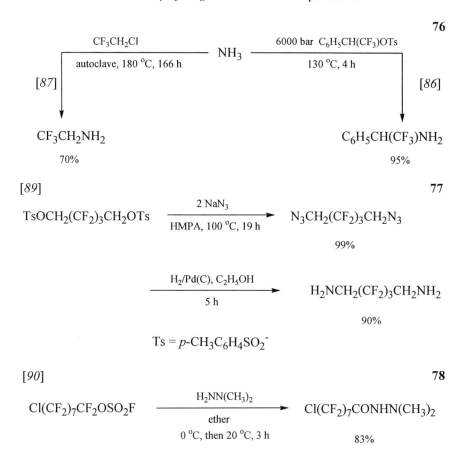

$$NH_3 \xrightarrow[\text{autoclave, 180 °C, 166 h}]{CF_3CH_2Cl} CF_3CH_2NH_2 \; (70\%) \; [87]$$

$$NH_3 \xrightarrow[\text{130 °C, 4 h}]{\text{6000 bar } C_6H_5CH(CF_3)OTs} C_6H_5CH(CF_3)NH_2 \; (95\%) \; [86]$$

[*89*] **77**

$$TsOCH_2(CF_2)_3CH_2OTs \xrightarrow[\text{HMPA, 100 °C, 19 h}]{2\ NaN_3} N_3CH_2(CF_2)_3CH_2N_3 \; (99\%)$$

$$\xrightarrow[\text{5 h}]{H_2/Pd(C),\ C_2H_5OH} H_2NCH_2(CF_2)_3CH_2NH_2 \; (90\%)$$

$$Ts = p\text{-}CH_3C_6H_4SO_2^-$$

[*90*] **78**

$$Cl(CF_2)_7CF_2OSO_2F \xrightarrow[\text{ether, 0 °C, then 20 °C, 3 h}]{H_2NN(CH_3)_2} Cl(CF_2)_7CONHN(CH_3)_2 \; (83\%)$$

Formal replacement of one of the bulky halogens in **polyhalogenofluoroethanes** by nitrogen nucleophiles occurs through the intermediate formation of fluorinated olefins [*91, 92*] (*see* equation 47) (equation 79).

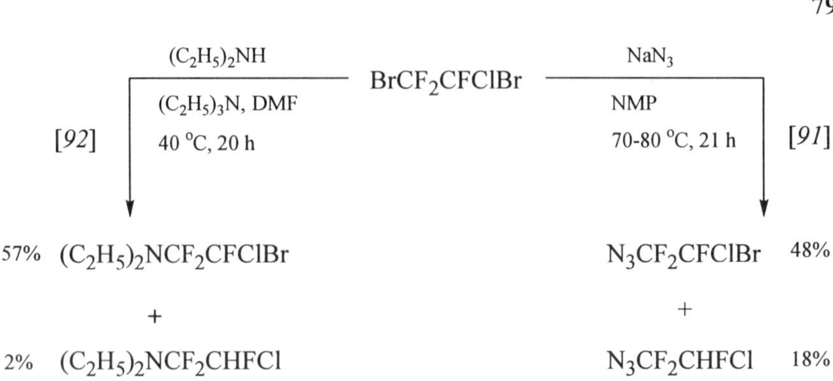

79

$$BrCF_2CFClBr \xrightarrow[(C_2H_5)_3N,\ \text{DMF, 40 °C, 20 h}]{(C_2H_5)_2NH} (C_2H_5)_2NCF_2CFClBr\ (57\%) + (C_2H_5)_2NCF_2CHFCl\ (2\%) \; [92]$$

$$BrCF_2CFClBr \xrightarrow[\text{NMP, 70-80 °C, 21 h}]{NaN_3} N_3CF_2CFClBr\ (48\%) + N_3CF_2CHFCl\ (18\%) \; [91]$$

A nonbasic *enamine* is formed from decafluorocyclohexene and pyrrolidine [*93*], whereas condensation of perfluorocyclobutene with excess hydrazine produces the tetrakishydrazone of cyclobutanetetrone [*94*] (equation 80).

[*93*] **80**

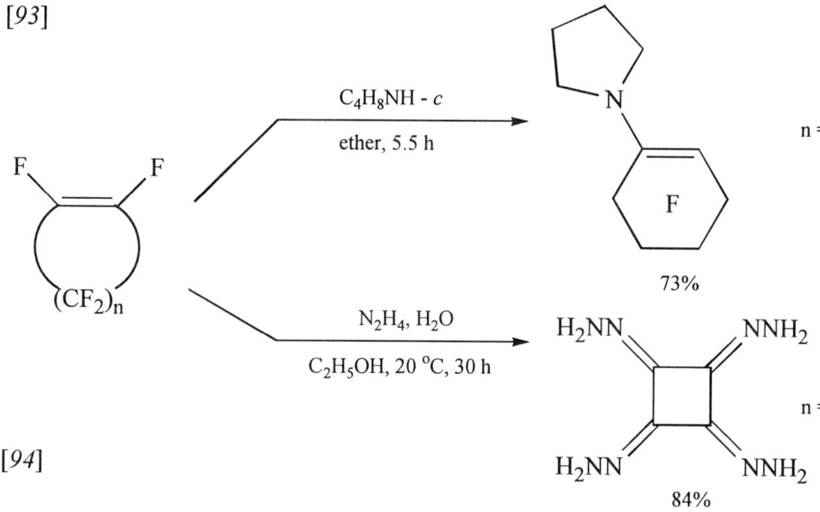

C_4H_8NH - *c*, ether, 5.5 h; n = 4; 73%

N_2H_4, H_2O, C_2H_5OH, 20 °C, 30 h; n = 2; 84%

F, F, $(CF_2)_n$, N, F, H_2NN, NNH_2, H_2NN, NNH_2

[*94*]

Tetrafluoroethene hexamer reacts with ammonia to give a nitrile. The reaction occurs probably by initial formation of a very reactive cyanoolefin [*95*] (equation 81).

[*95*] **81**

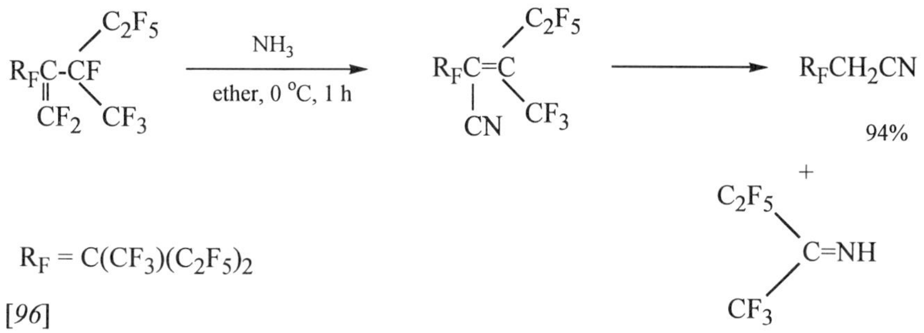

$R_FC(=CF_2)-CF(C_2F_5)(CF_3)$ —(NH_3, ether, 0 °C, 1 h)→ $R_FC(CN)=C(C_2F_5)(CF_3)$ ⟶ R_FCH_2CN 94% + $(C_2F_5)(CF_3)C$=NH

$R_F = C(CF_3)(C_2F_5)_2$

[*96*]

Trifluoroacetylketene *O,O*- and *S,S*-acetals readily react with various amines to give the corresponding *O,N*-, *S,N*- and *N,N*-acetals [*96*] (equation 82).

The addition of a primary amine to methyl perfluoromethacrylate leads to a ketenimine derivative [*97*] (equation 83).

A *carbodiimide* is obtained from perfluoro-2-azapropene and aniline [*98*] (equation 84).

[*96*] **82**

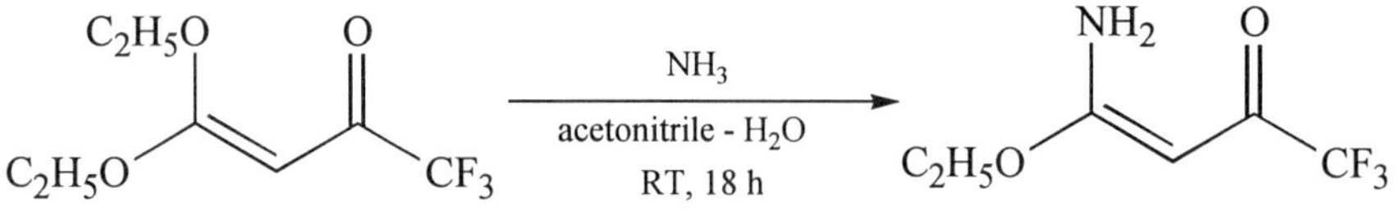

$$(C_2H_5O)_2C{=}CH{-}C(O){-}CF_3 \xrightarrow[\text{acetonitrile - }H_2O\text{, RT, 18 h}]{NH_3} C_2H_5O{-}C(NH_2){=}CH{-}C(O){-}CF_3$$

quantitative

[*97*] **83**

$$CF_2{=}C(CF_3){-}CO_2CH_3 \xrightarrow[\text{ether, - 78 °C}]{C_6H_5NH_2,\ (C_2H_5)_3N} C_6H_5N{=}C{=}C(CF_3){-}CO_2CH_3$$

65%

[*98*] **84**

$$CF_3N{=}CF_2 \xrightarrow[\text{ether, -35 °C, then RT 5 h}]{C_6H_5NH_2,\ KF,\ (C_2H_5)_3N} CF_3N{=}C{=}NC_6H_5$$

80%

Ammonolysis of tetrakis(1,2,3,5-trifluoromethyl)benzene with liquid ammonia in a sealed tube gives a tricyano derivative [*99*] (equation 85).

[*99*] **85**

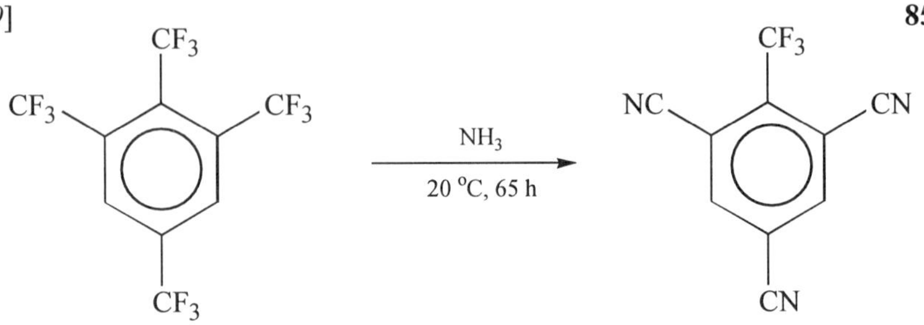

$$1,2,3,5\text{-}(CF_3)_4C_6H_2 \xrightarrow[\text{20 °C, 65 h}]{NH_3} 2\text{-}CF_3\text{-}1,3,5\text{-}(CN)_3C_6H_2$$

78%

N-Acylhexafluoroacetone imines are prepared from hexafluoroacetone and carboxylic amides in the presence of *p*-toluenesulfonic acid (TsOH) [*100*] (equation 86).

Acetoacetic esters react with **difluoroamine** to give alkyl 3,3-bis(difluoroamino)butyrates [*101*] (equation 87).

Normally, phenylhydrazine reacts with the enol form of 1,1,1-trifluoromethylpentane-2,4-dione to give 5-methyl-1-phenyl-3-trifluoromethylpyrazole as the major product. However, the use of pyrrolidine as a transient carbonyl-blocking group can completely reverse the regiochemistry of the addition and leads to 3-methyl-1-phenyl-5-trifluoromethylpyrazole [*102*] (equation 88).

[*100*] **86**

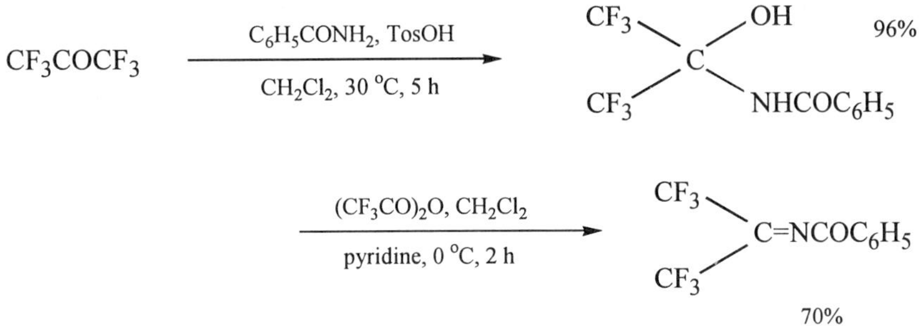

CF_3COCF_3 —($C_6H_5CONH_2$, TosOH; CH_2Cl_2, 30 °C, 5 h)→ $(CF_3)_2C(OH)NHCOC_6H_5$ 96%

—($(CF_3CO)_2O$, CH_2Cl_2; pyridine, 0 °C, 2 h)→ $(CF_3)_2C{=}NCOC_6H_5$ 70%

[*101*] **87**

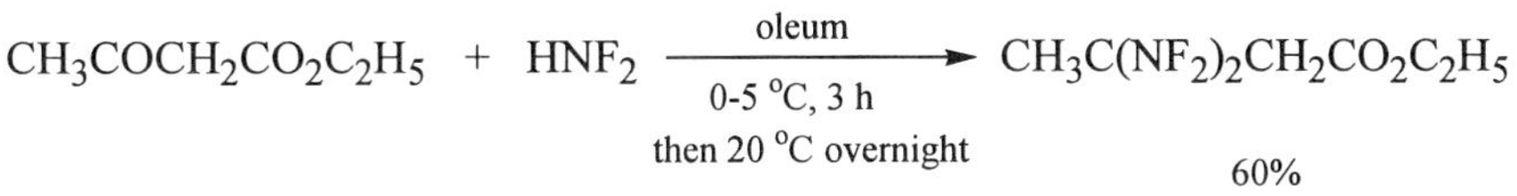

$CH_3COCH_2CO_2C_2H_5$ + HNF_2 —(oleum; 0-5 °C, 3 h, then 20 °C overnight)→ $CH_3C(NF_2)_2CH_2CO_2C_2H_5$ 60%

[*102*] **88**

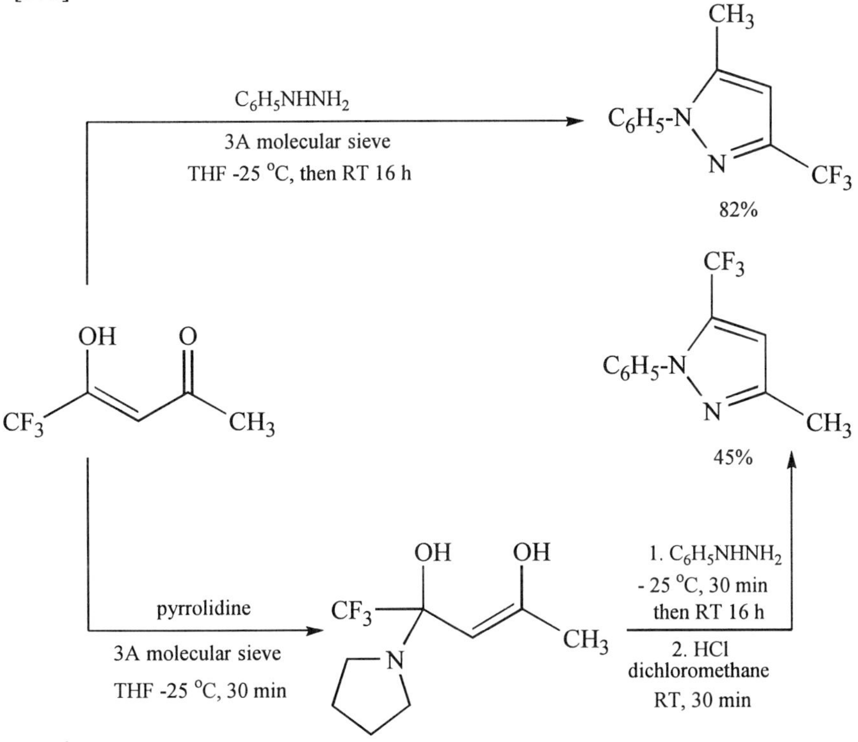

$CF_3C(OH){=}CHCOCH_3$ —($C_6H_5NHNH_2$; 3A molecular sieve; THF -25 °C, then RT 16 h)→ 1-phenyl-5-methyl-3-trifluoromethylpyrazole (C_6H_5-N, CH_3, CF_3) 82%

$CF_3C(OH){=}CHCOCH_3$ —(pyrrolidine; 3A molecular sieve; THF -25 °C, 30 min)→ $CF_3C(OH)(NC_4H_8)CH{=}C(OH)CH_3$ —(1. $C_6H_5NHNH_2$, - 25 °C, 30 min then RT 16 h; 2. HCl dichloromethane RT, 30 min)→ 1-phenyl-5-trifluoromethyl-3-methylpyrazole (C_6H_5-N, CF_3, CH_3) 45%

References are listed on pages 490–496.

The reaction of aliphatic primary amines with alkyl α-hydrogenoperfluorocarboxylates leads to the corresponding β-alkyl iminoesters as the major or the sole tautomers, depending on the length of the perfluoroalkyl chain [*103*] (equation 89).

[*103*] **89**

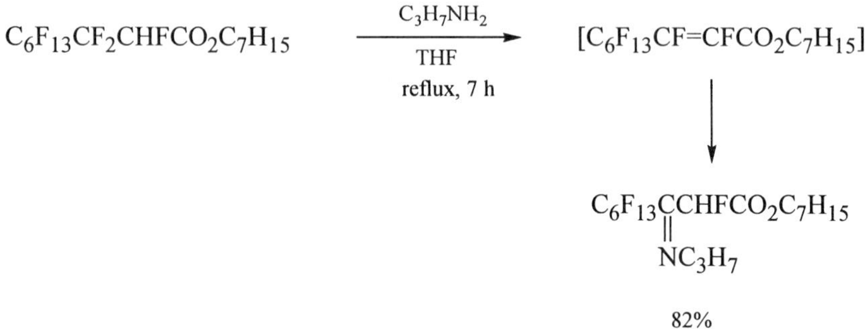

$$C_6F_{13}CF_2CHFCO_2C_7H_{15} \xrightarrow[\text{THF, reflux, 7 h}]{C_3H_7NH_2} [C_6F_{13}CF{=}CFCO_2C_7H_{15}] \longrightarrow C_6F_{13}C({=}NC_3H_7)CHFCO_2C_7H_{15}$$

82%

The reaction of **polyfluoroaldehydes** with phenylhydrazine initially leads to phenylhydrazones; the latter are converted by excess phenylhydrazine to bis-(phenylhydrazones) [*104*] (equation 90).

[*104*] **90**

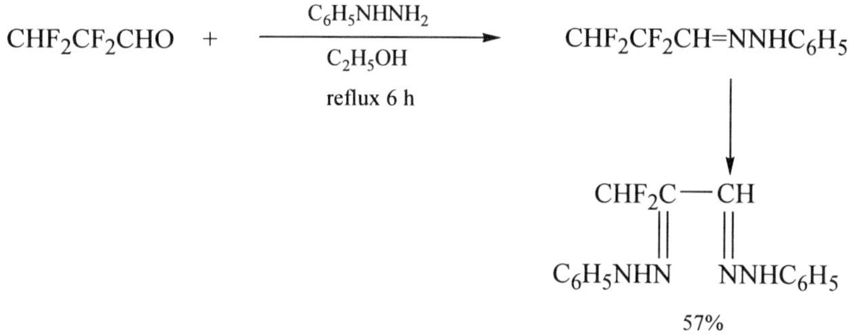

$$CHF_2CF_2CHO + \xrightarrow[\text{C}_2\text{H}_5\text{OH, reflux 6 h}]{C_6H_5NHNH_2} CHF_2CF_2CH{=}NNHC_6H_5 \longrightarrow CHF_2C({=}NNHC_6H_5){-}CH{=}NNHC_6H_5$$

57%

Tetrafluoro-2-azapropenylamine, formed from perfluoro-2-azapropene and ammonia, is cyclized to a triazine under moderate heating [*105*] (equation 91).

A guanidine obtained from a pentafluorophenyl isocyanide dichloride and aniline is cyclized to a tetrafluorobenzimidazole [*106*] (equation 92).

Trifluoropyruvic acid hydrate reacts with *N*-benzylurea to give 3-benzyl-5-hydroxy-5-trifluoromethyl-2,4 imidazoledione [*107*] (equation 93).

Benzylamine adds smoothly to 3,3,3-trifluoro-2-bromopropene. The bromoalkylamine thus formed cyclizes to 1-benzyl-2-trifluoromethylaziridine [*108*] (equation 94).

5-Azido-2-phenyl-4-trifluoromethyl-1,3-azole is prepared from 5-fluoro-4-trifluoromethyl-1,3-azole [*109*] (equation 95).

[*105*] **91**

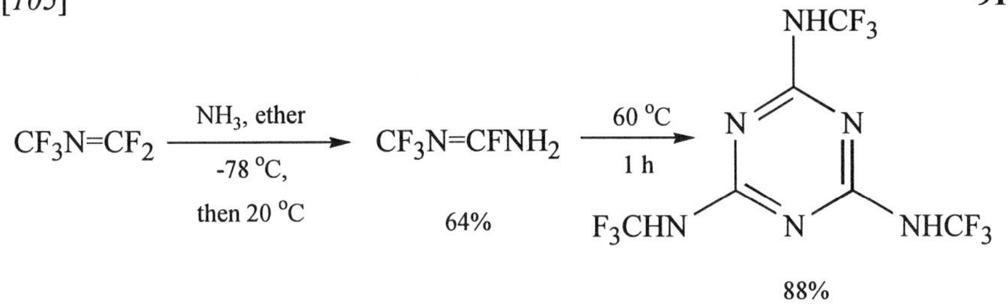

$$CF_3N{=}CF_2 \xrightarrow[\text{-78 °C, then 20 °C}]{NH_3,\ \text{ether}} CF_3N{=}CFNH_2 \ (64\%) \xrightarrow[\text{1 h}]{\text{60 °C}}$$

NHCF$_3$ / N N / F$_3$CHN N NHCF$_3$

88%

[*106*] **92**

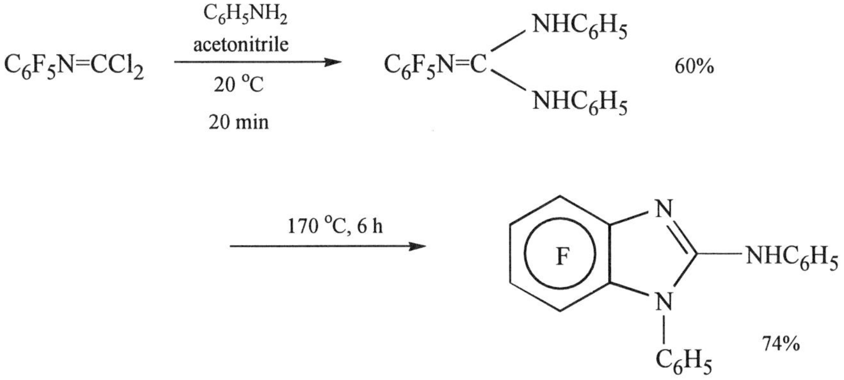

$$C_6F_5N{=}CCl_2 \xrightarrow[\text{20 °C, 20 min}]{C_6H_5NH_2,\ \text{acetonitrile}} C_6F_5N{=}C(NHC_6H_5)_2 \quad 60\%$$

$$\xrightarrow{\text{170 °C, 6 h}}$$

F N NHC$_6$H$_5$ N C$_6$H$_5$

74%

[*107*] **93**

$$CF_3COCO_2H \cdot H_2O \xrightarrow[\text{dioxane, reflux 1.5 h}]{H_2NCONHCH_2C_6H_5,\ H_2SO_4}$$

OH CF$_3$ NH O O N CH$_2$C$_6$H$_5$

88%

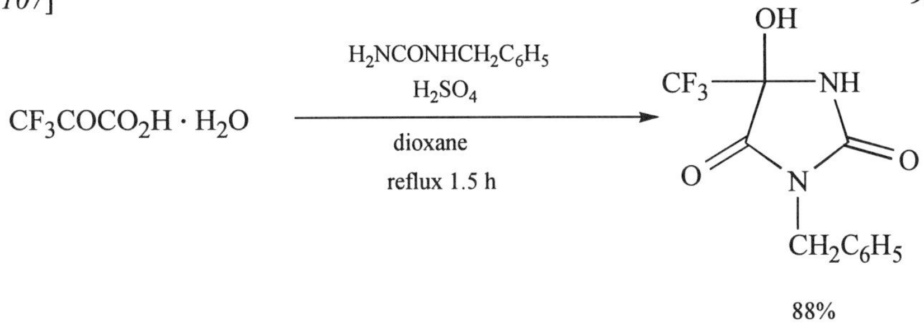

An alternative to Gabriel synthesis is based on the alkylation of trifluoroacetamide [*110, 111*] or *N*-benzyl triflamide [*112*] in place of the classical phtalimide (equations 96 and 97).

Alkylation at Carbon

Deprotonation of *N*-cyclohexylfluoroacetone imines by lithium hexamethyldisilazide or tertiary butyl lithium at very low temperature allows a regioselective alkylation at the carbon carrying fluorine [*113*] (equation 98).

[*108*] **94**

$CF_3CBr{=}CH_2$ $\xrightarrow[\text{20 °C, 10 days}]{C_6H_5CH_2NH_2}$ $CF_3CHBrCH_2NHCH_2C_6H_5$ 50%

$\xrightarrow[\text{153 °C, 3 h}]{(C_2H_5)_3N,\ DMF}$ CF_3-CH—CH_2 (aziridine, N—$CH_2C_6H_5$) 90%

[*109*] **95**

F_3C, F, N, S, C_6H_5 (thiazole) $\xrightarrow[\text{rt, 1 h}]{LiN_3,\ \text{N,N-dimethylpropylene urea}}$ F_3C, N_3, N, S, C_6H_5 (thiazole) 74%

96

CF_3CONH_2 $\xrightarrow[\text{NaH, DMF, 80 °C, 18 h}]{C_8H_{17}I}$ $CF_3CONHC_8H_{17}$ 79% [*110*]

CF_3CONH_2 $\xrightarrow[\text{K}_2\text{CO}_3,\ (\text{C}_4\text{H}_5)_4\text{N Br, DMF, 80 °C, 2 h}]{C_8H_{17}Br}$ $CF_3CONHC_8H_{17}$ 90% [*111*]

[*112*] **97**

$C_6H_5CH_2NHSO_2CF_3$ $\xrightarrow[\text{acetone, RT, 1 night}]{C_7H_{15}Br,\ K_2CO_3}$ $C_7H_{15}N(CH_2C_6H_5)SO_2CF_3$ quantitative

[*113*] **98**

FCH_2—C(=N—C_6H_{11}-*c*)—CH_3 $\xrightarrow[\text{THF, -80 °C, 2 h, then } H_3O^+]{C_4H_9I,\ (CH_3)_3CLi}$ $C_4H_9CHFCOCH_3$ 80%

A fluorinated oxazine, prepared from fluoroacetonitrile and 2-methyl-1,3-pentanediol, is alkylated at low temperature. The resulting products furnish α-fluoroaldehydes after borohydride reduction and hydrolysis [*114*] (equation 99).

[*114*] **99**

1. C_4H_9Li, THF, -78 °C
2. $C_6H_5CH_2Br$

F F $CH_2C_6H_5$ O N O N

76%

1. $NaBH_4$, THF-EtOH, -35 °C
2. HO_2CCO_2H reflux

O H F $CH_2C_6H_5$

51%

Condensation of 2-fluoroacetoacetate with geranyl bromide leads, after basic hydrolysis, to 3-fluorogeranylacetone. The latter is an intermediate for the synthesis of 4-fluorofarnesol [*115*] (equation 100).

Annelation of enamines or enolates with fluorinated methyl vinyl ketones gives the corresponding cyclohexenones [*116, 117*] (equation 101).

Alkylation of ethyl trifluoroacetoacetate enolate occurs at the carbon atom provided that the carbonyl function is masked as a dioxolane or an *N,N*-dimethylhydrazone [*118*] (equation 102).

The enolate of dimethyl trifluoromethylmalonate, formed by the action of cesium fluoride [*119*] or of an electrolytically generated pyrrolidone anion [*120*], can be alkylated with methyl iodide (equation 103).

The condensation of 6-methoxy-1-vinyl-1-tetralol with 2-trifluoromethyl-1,3-cyclopentanedione in the presence of a small amount of triethylamine produces a *secodione* (equation 104). This dione is an intermediate in the total synthesis of 13-trifluoromethyl estrogens, such as 18,18,18-trifluoro-17β-estradiol [*121*].

Regioselective alkylation of a fluorinated β-disulfone occurs at low temperature [*122*] (equation 105).

Claisen rearrangement of the allyl enol ether of trifluoroacetylacetone gives a C-allylated derivative [*123*] (equation 106).

[*115*] **100**

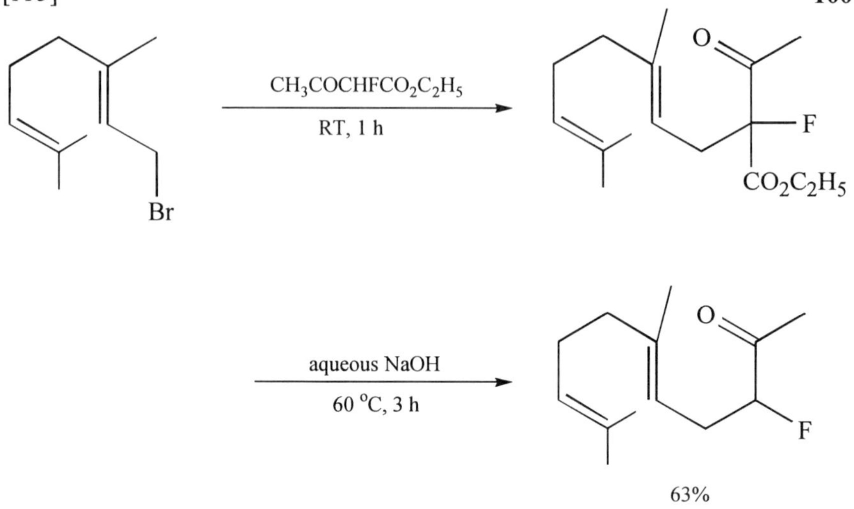

$CH_3COCHFCO_2C_2H_5$, RT, 1 h

F

$CO_2C_2H_5$

Br

aqueous NaOH, 60 °C, 3 h

F

63%

101

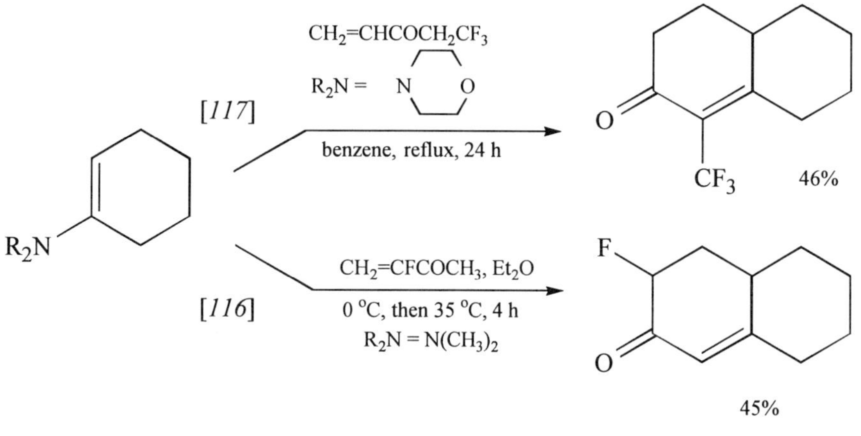

[*117*] $CH_2{=}CHCOCH_2CF_3$, R_2N = morpholino, benzene, reflux, 24 h

CF_3 46%

R_2N

[*116*] $CH_2{=}CFCOCH_3$, Et_2O, 0 °C, then 35 °C, 4 h, $R_2N = N(CH_3)_2$

F

45%

[*118*] **102**

$CF_3C(=N{-}N(CH_3)_2)CH(Li)CO_2C_2H_5$

$CH_3CH_2CH_2I$, THF - HMPA, -78 °C, 2 h, then RT, 22 h

$CF_3C(=N{-}N(CH_3)_2)CH(CH_2CH_2CH_3)CO_2C_2H_5$

85%

References are listed on pages 490–496.

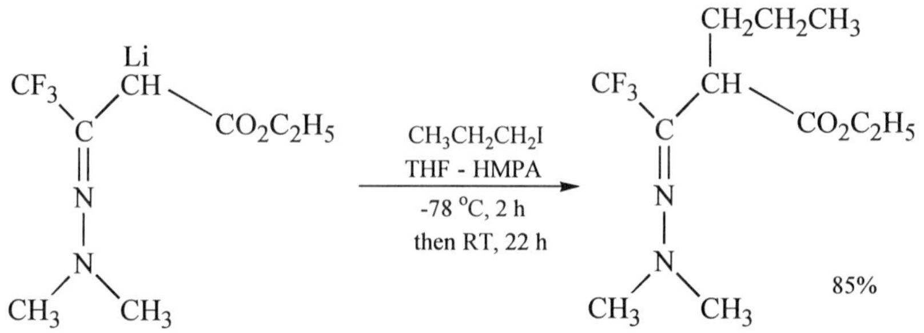

103

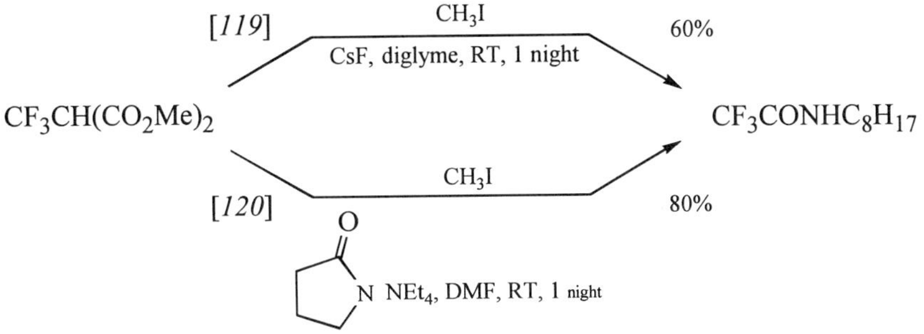

[*119*] $CF_3CH(CO_2Me)_2$ → (CH_3I; CsF, diglyme, RT, 1 night) → $CF_3CONHC_8H_{17}$ 60%

[*120*] $CF_3CH(CO_2Me)_2$ → (CH_3I; 2-oxopyrrolidin-1-yl NEt_4, DMF, RT, 1 night) → $CF_3CONHC_8H_{17}$ 80%

[*121*]

104

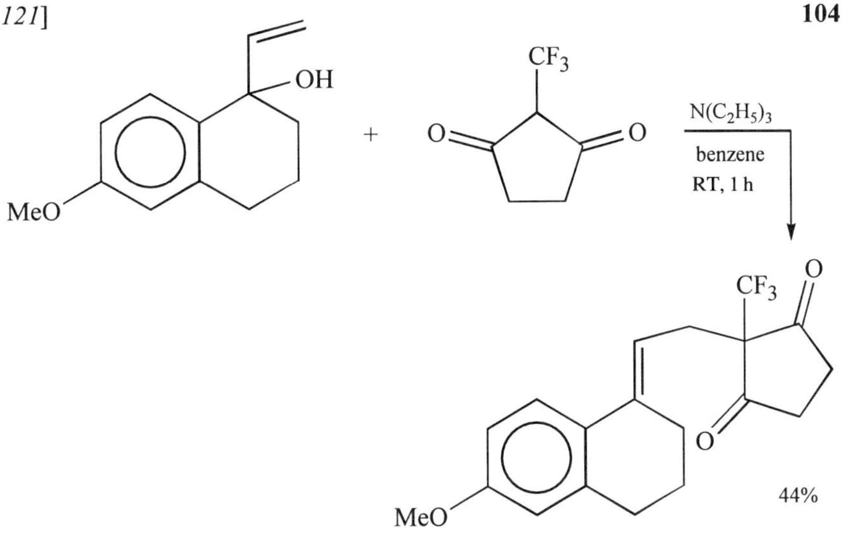

6-methoxy-1-vinyl-1,2,3,4-tetrahydronaphthalen-1-ol (MeO, OH) + 2-(CF_3)-cyclopentane-1,3-dione → ($N(C_2H_5)_3$; benzene; RT, 1 h) → 2-(CF_3)-2-[2-(6-MeO-3,4-dihydronaphthalen-1(2H)-ylidene)ethyl]cyclopentane-1,3-dione

44%

[*122*]

105

$CF_3SO_2CH_2SO_2CH_3$ → (CH_3I; C_4H_9Li, THF-hexane; -70 °C to RT, 1 night) → $CF_3SO_2CH(CH_3)SO_2CH_3$

95%

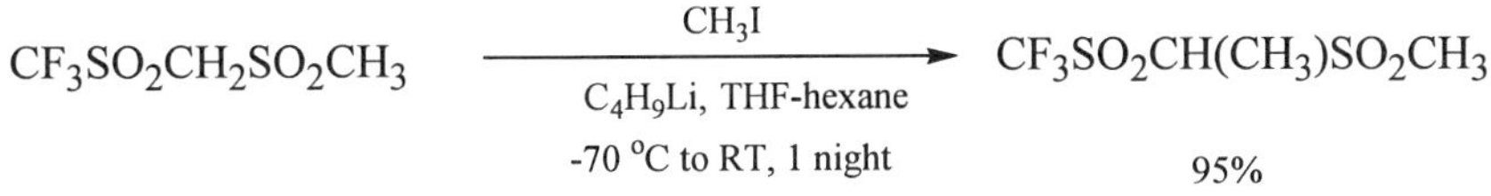

[*123*]

106

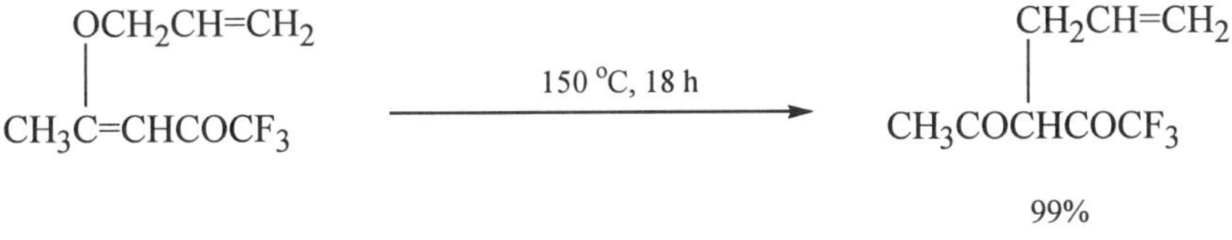

$CH_3C(OCH_2CH{=}CH_2){=}CHCOCF_3$ → (150 °C, 18 h) → $CH_3COCH(CH_2CH{=}CH_2)COCF_3$

99%

Condensation between lithium acetylides and **dibromodifluoromethane** [*124*] or **dichlorofluoromethane** [*125*] leads to fluorohaloacetylenes (equation 107). Sodium alkyl malonates are also alkylated by dihalogenodifluoromethanes [*124*] (equation 108). These reactions involve difluorocarbene as intermediate (for the mechanism of the CF_2Br_2 condensation, *see* equation 15).

107

$C_6H_5C{\equiv}CLi$

[*125*] $CHCl_2F$, THF, -100 ° to -70 ° → $C_6H_5C{\equiv}CCHFCl$ 57%

[*124*] CBr_2F_2, THF, 0 °C then RT for 4 h → $C_6H_5C{\equiv}CCF_2Br$ 35%

[*124*] **108**

$$C_2H_5{-}C(CO_2C_2H_5)_2Na + CBr_2F_2 \xrightarrow[0\,^{\circ}C,\ \text{then RT 2 h}]{THF} C_2H_5{-}C(CO_2C_2H_5)_2{-}CF_2Br\ (45\%) + C_2H_5{-}C(CO_2C_2H_5)_2{-}CF_2H\ (5\%)$$

Intermediate *carbenes* are also involved in the alkylation by haloforms of carbanions derived from Schiff bases of α-amino esters [*126*] or aminomalonates [*127*]. After hydrolysis, fluorinated derivatives of alanine are obtained in both cases because decarboxylation of the substituted malonic acid occurs (equation 109).

The *tertiary carbanion* generated by the addition of fluoride anion to perfluoroisobutylene reacts with dibromodifluoromethane to give perfluoroneopentyl bromide [*128*]. This reaction follows an ionic chain mechanism (*see* equation 15) (equation 110).

The reaction of **perfluoroisobutylene** with malonic esters affords fluorinated allenes [*129*] (equation 111).

Ester enolates react with 3,3,3-trifluoropropene by an S_N2'-type process to give 5,5-difluoro-4-pentenoic acid esters [*130*] (equation 112).

Initial *C-perfluoroalkylation* of β-diketones occurs during their UV irradiation in the presence of **perfluoroalkyl iodides** in liquid ammonia. Fluorinated enaminoketones are obtained by subsequent ammonolysis of a difluoromethylene group and removal of the acetyl group [*131*]. C-alkylation of dimethyl malonate takes place when dimethyl sodiomalonate is treated with polyfluoroalkyl iodides [*132, 133*]. Clean perfluoroalkylation of the 2-nitropropyl anion occurs photochemically [*134*] (equations 113–115). All these condensations are interpreted by a SET process (*see* equation 57).

109

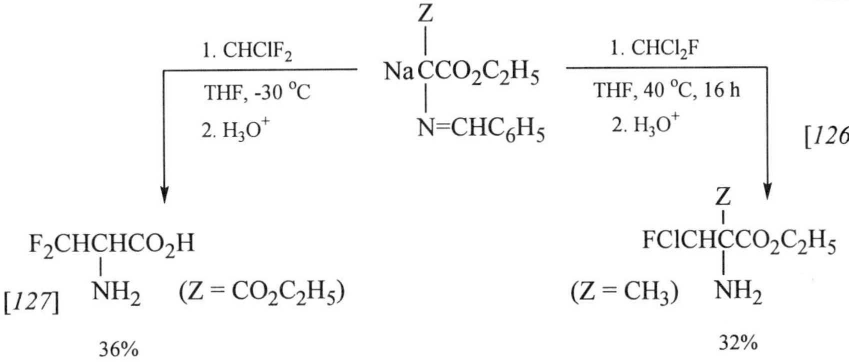

$Na\,C(Z)(CO_2C_2H_5)(N{=}CHC_6H_5)$

Left: 1. $CHClF_2$, THF, -30 °C; 2. H_3O^+ → $F_2CHCH(NH_2)CO_2H$ ($Z = CO_2C_2H_5$) 36% [*127*]

Right: 1. $CHCl_2F$, THF, 40 °C, 16 h; 2. H_3O^+ → $FClCHC(Z)(NH_2)CO_2C_2H_5$ ($Z = CH_3$) 32% [*126*]

[*128*] **110**

$$(CF_3)_2C{=}CF_2 \xrightarrow[\text{diglyme, 0 °C, 6.5 h}]{CsF,\ CBr_2F_2} (CF_3)_3CCF_2Br \quad 50\%$$

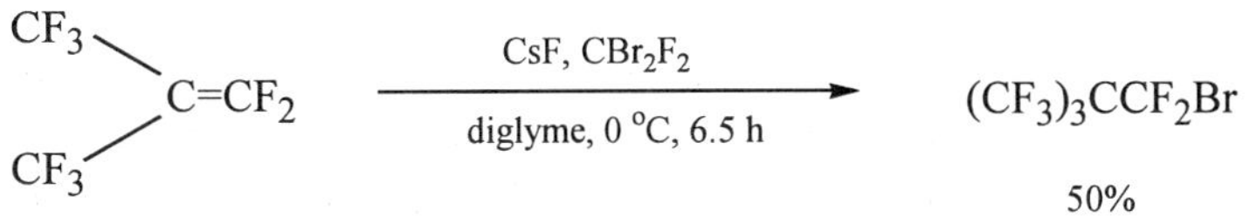

[*129*] **111**

$$(CF_3)_2C{=}CF_2 + H_2C(CO_2CH_3)_2 \xrightarrow[\text{ether, 20 °C, 5 days}]{BF_3\text{-}N(C_2H_5)_3} (CF_3)_2C{=}C{=}C(CO_2CH_3)_2 \quad 45\%$$

[*130*] **112**

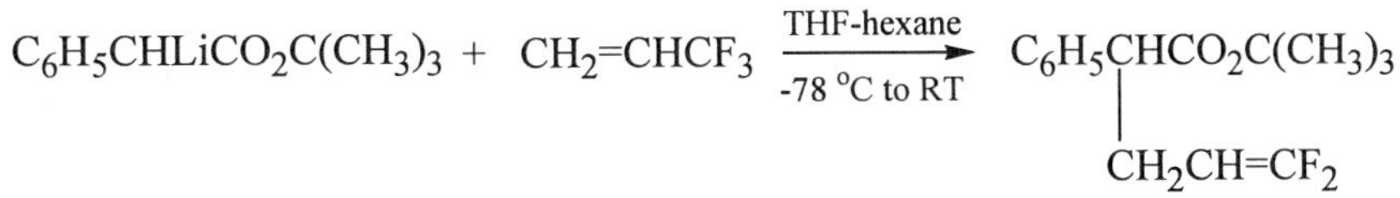

$$C_6H_5CHLiCO_2C(CH_3)_3 + CH_2{=}CHCF_3 \xrightarrow[\text{-78 °C to RT}]{\text{THF-hexane}} C_6H_5CH(CH_2CH{=}CF_2)CO_2C(CH_3)_3$$

[*131*] **113**

$$CH_3COCH_2COCH_3 \xrightarrow[NH_3,\ h\nu,\ \text{-33 °C, 2 h}]{CF_3CF_2I} CH_3CO\overset{-}{C}(C(=NH)CF_3)COCH_3 \longrightarrow CH_3COCH{=}C(NH_2)CF_3 \quad 61\%$$

References are listed on pages 490–496.

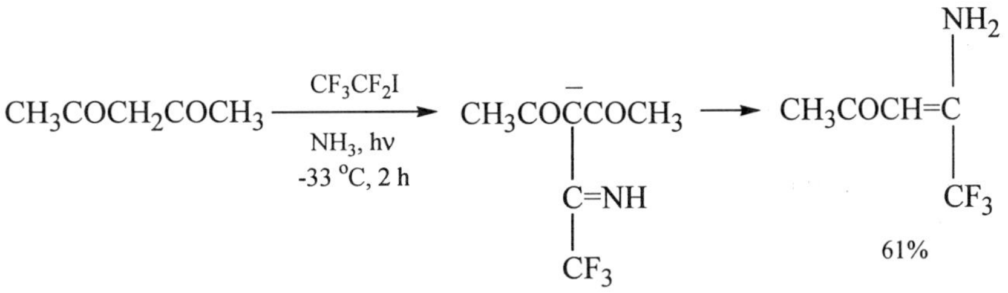

[*132*] **114**

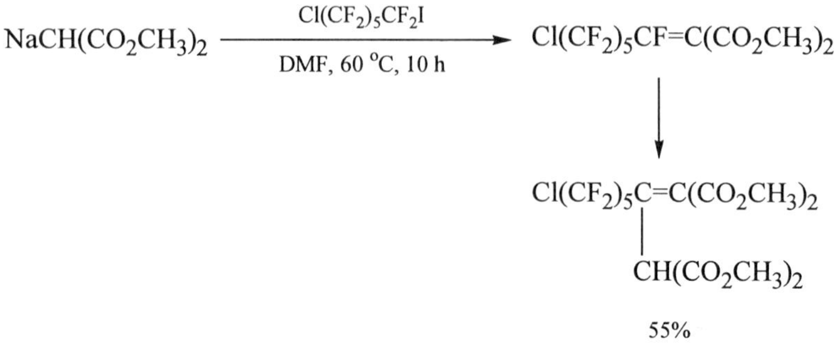

$$NaCH(CO_2CH_3)_2 \xrightarrow[\text{DMF, 60 °C, 10 h}]{Cl(CF_2)_5CF_2I} Cl(CF_2)_5CF{=}C(CO_2CH_3)_2 \longrightarrow Cl(CF_2)_5C(CH(CO_2CH_3)_2){=}C(CO_2CH_3)_2$$

55%

[*134*] **115**

$$(CH_3)_2C{=}NO_2^-\ Li^+ \xrightarrow[\text{DMF, h}\nu\text{, 4 h}]{C_6F_{13}I} C_6F_{13}C(CH_3)_2NO_2$$

79%

Lithium silylamides react smoothly with trifluoronitrosomethane to give diazenes. Traces of water initiate the decomposition of the latter with liberation of a trifluoromethyl carbanion, which is trapped by carbonyl compounds [*135*] (equation 116). Desilylation of trialkyl(trifluoromethyl)silanes by fluoride ion produces also a trifluoromethyl carbanion, which adds to carbonyl carbon atoms [*136, 137*] (equations 117 and 118).

[*135*] **116**

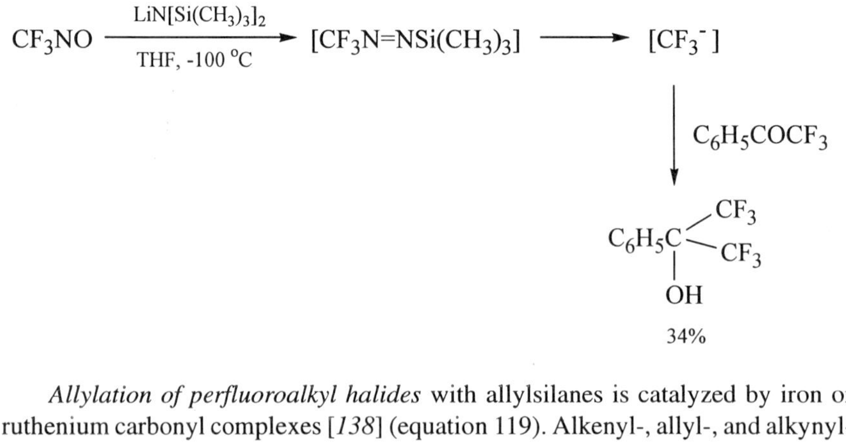

$$CF_3NO \xrightarrow[\text{THF, -100 °C}]{LiN[Si(CH_3)_3]_2} [CF_3N{=}NSi(CH_3)_3] \longrightarrow [CF_3^-] \xrightarrow{C_6H_5COCF_3} C_6H_5C(CF_3)_2OH$$

34%

Allylation of perfluoroalkyl halides with allylsilanes is catalyzed by iron or ruthenium carbonyl complexes [*138*] (equation 119). Alkenyl-, allyl-, and alkynylstannanes react with perfluoroalkyl iodides in the presence of a palladium complex to give alkenes and alkynes bearing perfluoroalkyl groups [*139*] (equation 120).

[*136*] **117**

$$\text{1,4-benzoquinone} \xrightarrow[\text{acetonitrile, RT, 1 h}]{CF_3Si(C_2H_5)_3,\ KF} \text{4-}CF_3\text{-4-}OSi(C_2H_5)_3\text{-cyclohexa-2,5-dienone}$$

72%

[*137*] **118**

$$\text{cyclohexanone} + CF_3Si(CH_3)_3 \xrightarrow[\text{THF, RT, 1 h}]{C_4H_9NF} \text{1-}CF_3\text{-cyclohexan-1-}OH$$

77%

[*138*] **119**

$$CH_2{=}CHCH_2Si(CH_3)_3 \xrightarrow[60\ ^{\circ}C,\ 12\ h]{C_3F_7I,\ Fe_3(CO)_{12}} C_3F_7CH_2CH{=}CH_2$$

80%

[*139*] **120**

$$C_6H_5CH{=}CHSn(C_4H_9)_3 \xrightarrow[\text{hexane, }70\ ^{\circ}C,\ 4\ h]{C_4F_9I,\ Pd\ [P(C_6H_5)_3]_4} C_6H_5CH{=}CHC_4F_9$$

70%

UV irradiation of steroidal dienones in the presence of trifluoromethyl iodide allows the selective introduction of a trifluoromethyl group α to the carbonyl function [*140*] (equation 121).

A rather rare perfluoroalkyl iodide addition to divalent carbon atom has been performed in the case of **isocyanides** under thermal or copper-catalyzed conditions [*141*] (equation 122).

Trifluoromethylation of diethyl maleate or fumarate is performed via the electrochemical oxidation of sodium trifluoroacetate [*142*] (equation 123)

Spontaneous *perfluoroalkylation of enamines* occurs in the presence of fluorinated perhalogenoalkanes [*143, 144*]. This condensation is interpreted by a SET process (mechanism analogous to equation 57 with a neutral nucleophile in place of a charged nucleophilic reagent). Formation of chlorodifluoromethylcyclohexanone and not of its bromo analogue from bromochlorodifluoromethane is in agreement with this radical mechanism (equation 124).

Trifluoromethylation of double bonds or of aromatic nuclei is performed by **bis(trifluoromethyl) tellurium** [*145*] (equation 125).

Cationic fluoroalkylation of alkenes proceeds by action of fluoroalkylphenyliodonium trifluoromethanesulfonates [*146, 147*] (equations 126 and 127).

[*140*] **121**

OAc OAc

CF_3I, hν
pyridine
RT, 8.5 days

O O

CF_3 34%

[*141*] **122**

$$C_6F_{13}I + C{=}NC_4H_9 \xrightarrow[\text{RT, 2 h}]{\text{Cu}} C_6F_{13}(I)C{=}NC_4H_9 \quad 90\%$$

[*142*] **123**

$$C_2H_5O_2CCH{=}CHCO_2C_2H_5 \xrightarrow[\text{acetonitrile-}H_2O,\ 22\ ^\circ C,\ 7\ h]{CF_3CO_2Na,\ e^-} EtO_2CCH(CF_3)CH(CF_3)CO_2C_2H_5$$

(*cis* or *trans*) (*meso* + DL) 47%

$$+ \quad C_2H_5O_2CCH(CF_3)CHCO_2C_2H_5 - C_2H_5O_2CCH(CF_3)CHCO_2C_2H_5 \quad 26\%$$

[*143*] **124**

(1-pyrrolidinylcyclohexene) $\xrightarrow[\text{pentane, RT, 3 h}]{CF_3I}$ 2-(trifluoromethyl)cyclohexanone, 45%

(1-pyrrolidinylcyclohexene) $\xrightarrow[65\%]{BrCF_2Cl}$ 2-(CF_2Cl)cyclohexanone $\xrightarrow[\text{THF, RT, 4 h}]{(C_4H_9)_4NF}$ 2-(trifluoromethyl)cyclohexanone, 90%

[*145*] **125**

cyclohexene $\xrightarrow[155\ ^{\circ}C,\ 72\ h]{(CF_3)_2Te}$ 1,2-bis(CF_3)cyclohexane

85%

[*146*] **126**

$$CH_2{=}CHC_6H_5 \xrightarrow[\substack{C_5H_5N,\ CH_2Cl_2 \\ \text{reflux, 30 min}}]{C_8F_{17}I(C_6H_5)OSO_2CF_3} C_8F_{17}CH{=}CHC_6H_5$$

73%

[*147*] **127**

$$C_6H_5C(OSi(CH_3)_3){=}CH_2 \xrightarrow[\substack{\text{KF, dichloromethane} \\ \text{RT, 1.5 h}}]{CF_3CH_2I(C_6H_5)OSO_2CF_3} C_6H_5COCH_2CH_2CF_3$$

87%

Perfluoroalkylation of substituted benzenes and heterocyclic substrates has been accomplished through *thermolysis* of perfluoroalkyl iodides in the presence of the appropriate aromatic compound. Isomeric mixtures are often obtained. *N*-Methylpyrrole [*143*] and furan [*148*] yield only the α-substituted products (equation 128). Imidazoles are perfluoroalkylated under UV irradiation [*149*] (equation 129). 4-Perfluoroalkylimidazoles are obtained regioselectively by SET reactions of an imidazole anion with fluoroalkyl iodides or bromides under mild conditions [*150*] (equation 130) (for the SET mechanism, *see e*quation 57).

The *photochemical introduction of a trifluoromethyl group* into aromatic and heteroaromatic rings, such as uracil, can be performed also with trifluoromethyl bromide [*151*] (equation 131).

[*148*] **128**

$C_{10}F_{21}I$, 190 °C, 24 h; $C_{10}F_{21}$; 58%

[*149*] **129**

C_2F_5I, CH_3OH, hν, RT, 7 days; C_2F_5; 25% + C_2F_5; 46%

[*150*] **130**

$C_6F_{13}I$, DMF, 10 °C, 6 h; C_6F_{13}; 68%

[*151*] **131**

CF_3Br, acetonitrile, hν, 6 h; CF_3; 6%

Bis(perfluoroacyl) peroxides allow also the alkylation of aromatic compounds [*152, 153*] (equations 132 and 133).

Radical perfluoroalkylation of anilines occurs in the presence of a sufur dioxide radical anion precursor, such as Zn–SO_2 or sodium dithionite [*154, 155*], or of a nickel complex [*156*] (equation 134).

Perfluoroalkylation of pyridines by perfluoroalkyl bromides or iodides occurs in the presence of sulfur dioxide radical anion precursors, such as sodium hydroxymethanesulfinate [*155, 157, 158*] (equation 135).

[*152*] **132**

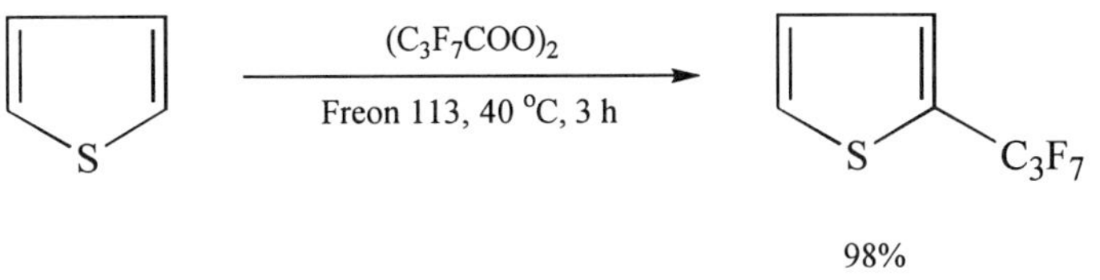

Thiophene $\xrightarrow[\text{Freon 113, 40 °C, 3 h}]{(C_3F_7COO)_2}$ 2-C_3F_7-thiophene

98%

[*153*] **133**

$$C_6H_6 + CF_3C(O)OOC(O)CF_3 \xrightarrow[\text{ether}]{Cl_2CFCClF_2} C_6H_5CF_3$$

70 °C, 4 h

98%

134

NH_2 (aniline)

[*156*] $Cl(CF_2)_6I$, $Ni[P(C_6H_5)_3]_4$, dioxane, 80 °C, 6 h

[*154*] CF_3Br, 3-5 bar, Zn, SO_2, DMF, RT, 2 h

NH_2, $(CF_2)_6Cl$ (para) 45% + NH_2, $(CF_2)_6Cl$ (ortho) 40%

NH_2, CF_3 (para) 20% + NH_2, CF_3 (ortho) 36%

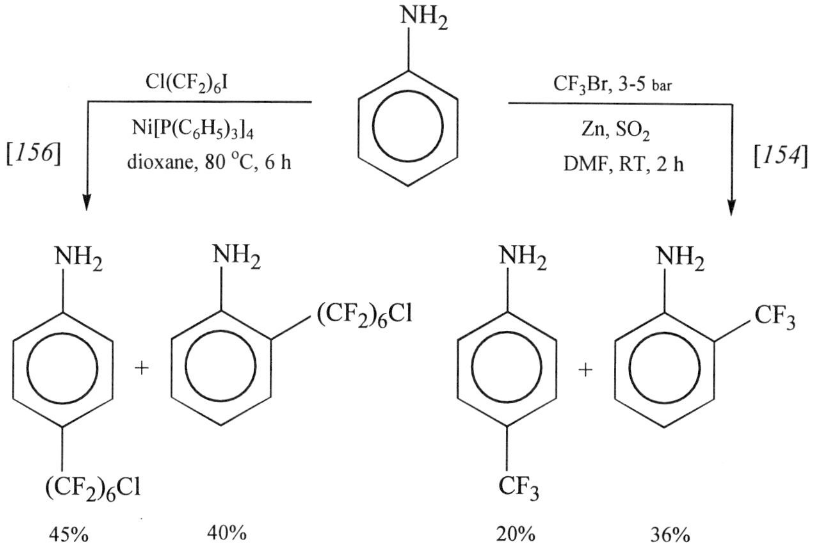

[*157*] **135**

Pyridine (N) + $C_6F_{13}I$ $\xrightarrow[\text{acetonitrile-}H_2O]{NaO_2SCH_2OH,\ NaHCO_3}$ C_6F_{13}-pyridine (N)

70-75 °C, 6 h

(*2,3,4*-isomers)

57%

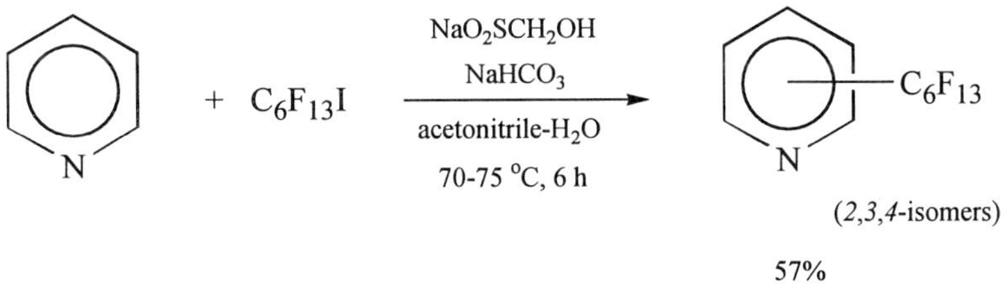

References are listed on pages 490–496.

The main product of the alkylation of pentafluorobenzene with 1,1,2-trichlorotrifluoroethane in the presence of antimony pentafluoride is β-chlorononafluoroethylbenzene [*159*] (equation 136).

[*159*] **136**

$$C_6HF_5 + CFCl_2CF_2Cl \xrightarrow[\text{-10 to -5 °C, 6 h; then 20 °C, 2 h}]{SbF_5} C_6F_5CF_2CF_2Cl \quad 51\%$$

Trichloromethylation of pentafluorobenzene with carbon tetrachloride is performed in the presence of aluminum trichloride [*160*] (equation 137).

[*160*] **137**

$$C_6HF_5 \xrightarrow[\text{reflux 7 h}]{CCl_4,\ AlCl_3} C_6F_5CCl_3\ (18\%) + 4\text{-}ClC_6F_4CCl_3\ (46\%)$$

Cationic alkylation of phenol takes place by the action of perfluoroalkylphenyliodonium trifluoromethanesulfonates [*161*] (equation 138).

[*161*] **138**

$$C_6H_5OH \xrightarrow[\text{pyridine, acetonitrile, RT, 30 min}]{C_2F_5I(C_6H_5)OSO_2CF_3} 4\text{-}C_2F_5C_6H_4OH\ (25\%) + 2\text{-}C_2F_5C_6H_4OH\ (28\%)$$

Reaction of **perfluoroalkanesulfonyl chlorides** with aromatic compounds in the presence of dichloro bis(triphenylphosphine)ruthenium (II) gives perfluoroalkylated products [*162*] (equation 139).

[*162*] **139**

$$C_6H_6 + C_6F_{13}SO_2Cl \xrightarrow[\text{120 °C, 19 h}]{RuCl_2[P(C_6H_5)_3]_2} C_6H_5C_6F_{13} \quad 44\%$$

References are listed on pages 490–496.

***N*-trifluoromethyl-*N*-nitrosotrifluoromethanesulfonamide** is an effective *trifluoromethylating agent* for aromatic compounds under thermal or photochemical conditions [*163*] (equation 140).

[*163*] **140**

4-tert-butylphenol (OH, $C(CH_3)_3$) → ($CF_3N(NO)SO_2CF_3$; hν, biacetyl; acetonitrile, 20 °C, 5 h) → 2-CF_3-4-$C(CH_3)_3$-phenol (65%) + 2,6-bis(CF_3)-4-$C(CH_3)_3$-phenol (17%)

Recently developed trifluoromethylating agents capable of transferring the trifluoromethyl group as a cation to strongly nucleophilic compounds such as carbanions and sulfur and phosphorus nucleophiles are prepared from *o*-biphenyl trifluoromethyl sulfoxide [*164*] and are shown in equation 141.

Direct *perfluoroalkylation* of electron-poor aromatic and heterocyclic systems with perfluorocarboxylic acids is mediated by xenon difluoride [*165*] (equation 142).

Copper-mediated coupling of the aryl iodide derived from 1,3-bis(2-hydroxyhexafluoroisopropyl)benzene with perfluorooctyl iodide gives the desired compound as a dimethyl sulfoxide (DMSO) complex [*166*] (equation 143). Even bromoarenes can be coupled [*167*] (equation 144).

Ullman-type coupling occurs between aryl halides and *trifluoromethyl copper* species generated by the action of copper iodide on sodium trifluoroacetate [*168, 169*] or on methyl fluorosulfonyldifluoroacetate [*170*] (equation 145). Similarly, the pentafluoroethyl group can be introduced from potassium pentafluoropropionate [*171*] (equation 146).

C-alkylation of secondary and tertiary aromatic amines by **hexafluoroacetone or methyl trifluoropyruvate** is performed under mild conditions [*172*] (equation 147). The reaction of phenylhydrazine with hexafluoroacetone leads selectively to the product of the C-hydroxyalkylation at the *ortho* position of the aromatic ring. The change from the *para* orientation characteristic for anilines is apparently a consequence of a cyclic transition state arising from the initial N-hydroxyalkylation at the primary amino group [*173*] (equation 148).

Fluorinated pyrroles are obtained from 2,5-dimethylpyrrole by reductive alkylation with perfluoroaldehyde hydrates [*174*] (equation 149).

Polyfluoropropenes alkylate fluorinated ethylenes in the presence of antimony pentafluoride. This condensation proceeds by initial formation of an allyl cation [*175*] (equation 150).

[*164*] **141**

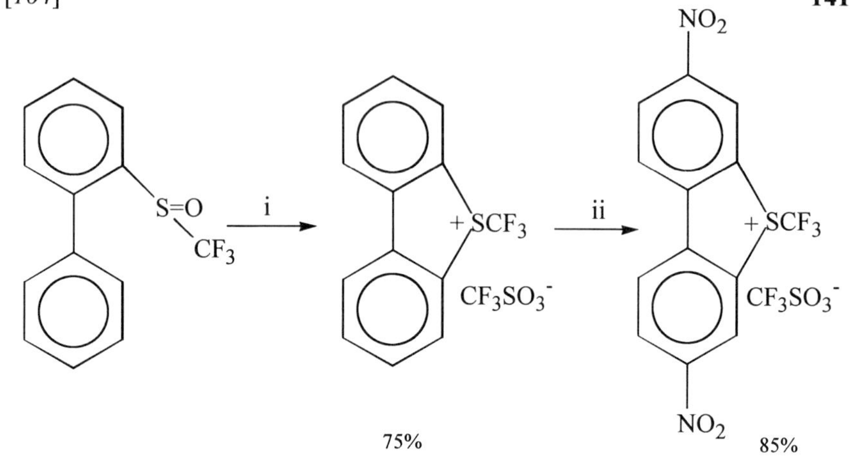

S=O, CF_3 —i→ +SCF$_3$ $CF_3SO_3^-$ —ii→ NO_2, +SCF$_3$, $CF_3SO_3^-$, NO_2

75% 85%

i : $(CF_3CO)_2O$, $C_2Cl_3F_3$, RT, 2d;

ii : 94% HNO_3, $(CF_3SO_2)_2O$, CH_3NO_2, RT

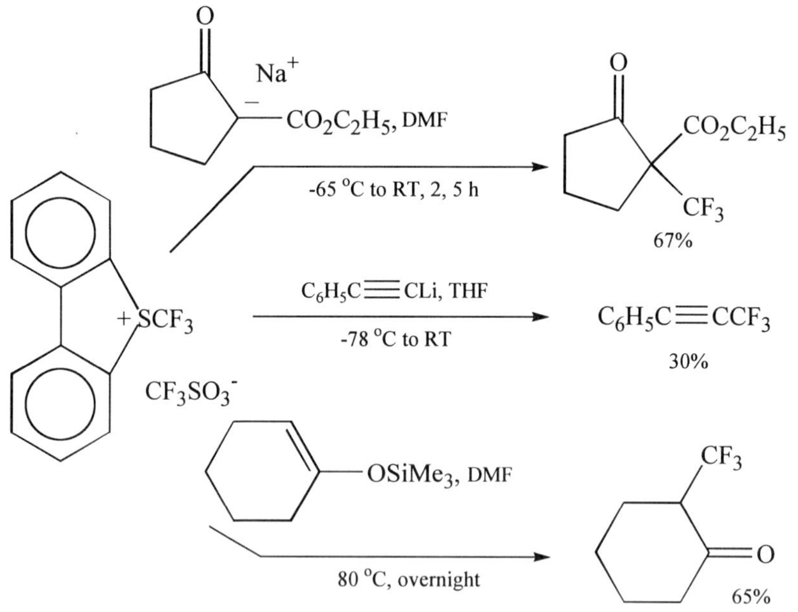

O, Na^+, –, $CO_2C_2H_5$, DMF

-65 °C to RT, 2, 5 h

O, $CO_2C_2H_5$, CF_3

67%

+SCF$_3$

$CF_3SO_3^-$

$C_6H_5C{\equiv}CLi$, THF

-78 °C to RT

$C_6H_5C{\equiv}CCF_3$

30%

$OSiMe_3$, DMF

80 °C, overnight

CF_3, O

65%

[*165*] **142**

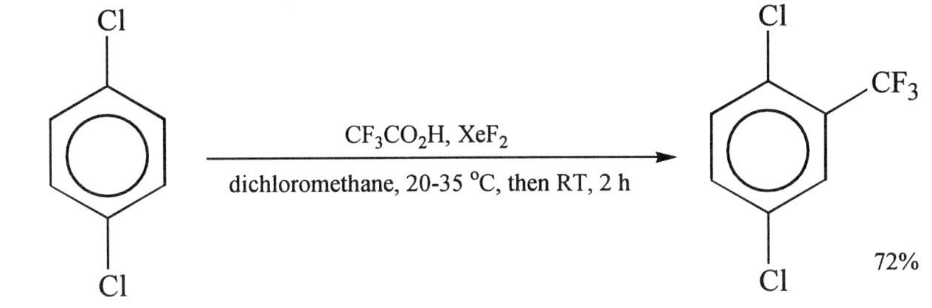

Cl, Cl (1,4-dichlorobenzene) → Cl, CF_3, Cl

CF_3CO_2H, XeF_2

dichloromethane, 20-35 °C, then RT, 2 h

72%

[*166*] **143**

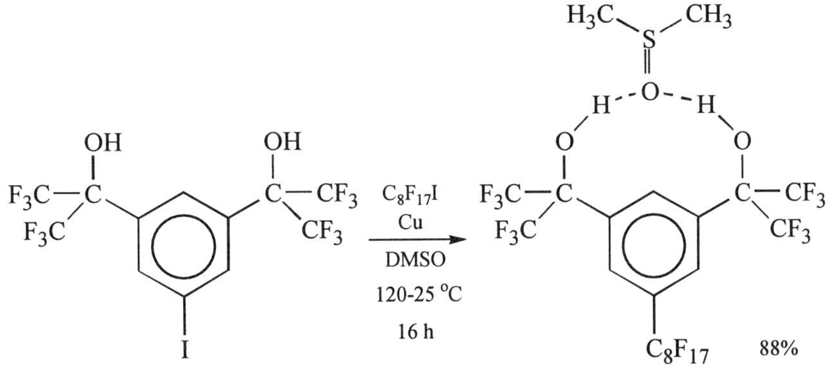

OH, OH, F_3C—C, F_3C, C—CF_3, CF_3, I

$C_8F_{17}I$

Cu

DMSO

120-25 °C

16 h

H_3C, CH_3, S, O, H, H, O, O, F_3C—C, F_3C, C—CF_3, CF_3, C_8F_{17}

88%

[*167*] **144**

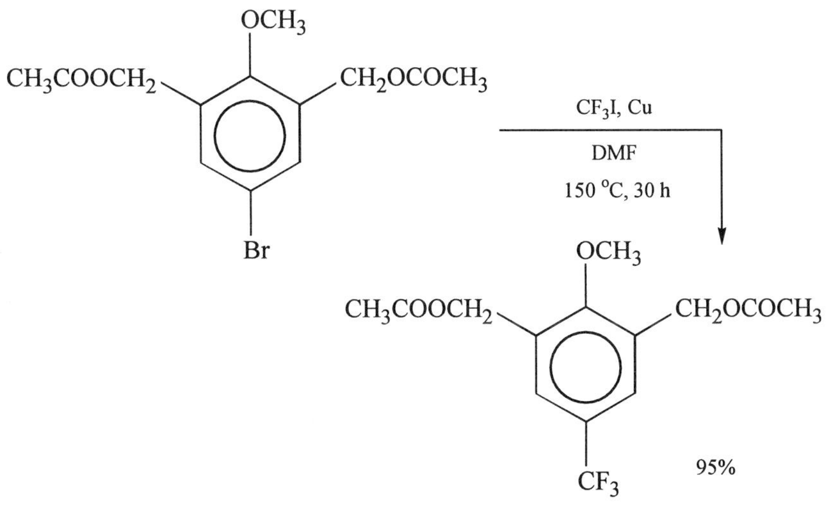

OCH_3, CH_3COOCH_2, CH_2OCOCH_3, Br

CF_3I, Cu

DMF

150 °C, 30 h

OCH_3, CH_3COOCH_2, CH_2OCOCH_3, CF_3

95%

References are listed on pages 490–496.

145

[*168*] C_6H_5I → (CF_3CO_2Na, CuI; NMP, reflux 4h) → $C_6H_5CF_3$ 72%

[*169*] C_6H_5I → ($FO_2SCF_2CO_2CH_3$, CuI; DMF, 70 °C, 2.5 h) → $C_6H_5CF_3$ 72%

[*171*] **146**

CH_3O–C_6H_4–I → ($CF_3CF_2CO_2K$, CuI; DMF - toluene; 120 °C, then 155 °C, 2 h) → CH_3O–C_6H_4–CF_2CF_3

72%

[*172*] **147**

C_6H_5–NHC_2H_5 → (CF_3COCF_3; chloroform, -60 °C; then, 20 °C, 1 h) → HO–C(CF_3)(CF_3)–C_6H_4–NHC_2H_5

91%

[*173*] **148**

C_6H_5–NH–NH_2 + $(CF_3)_2C{=}O$ → (chloroform; -60 °C, then 20 °C, 24 h) → 2-[C(CF_3)$_2$OH]–C_6H_4–NH–NH_2

77%

[*174*] **149**

2,5-dimethylpyrrole (H_3C, N–H, CH_3) → ($C_3F_7CH(OH)_2$; HI, H_3PO_2; CH_3CO_2H; 100 °C, 3.5 h) → 3,4-bis($CH_2C_3F_7$)-2,5-dimethylpyrrole ($C_3F_7H_2C$, $CH_2C_3F_7$, H_3C, N–H, CH_3)

64%

References are listed on pages 490–496.

[*175*] **150**

$$CF_2{=}CFCF_3 \xrightarrow[\text{SbF}_5,\ 40\text{-}50\ ^\circ\text{C, 6 h}]{CF_2{=}CF_2} CF_2{=}CFCF_2CF_2CF_3 \longrightarrow \underset{60\%}{CF_3CF{=}CFCF_2CF_3}$$

Experimental Procedures

For the safety precautions, see page 25–26 of the book.

Preparation of Trifluoromethylthiobenzene from Bromotrifluoromethane [*53*]

A pressure safe glass bottle containing 7.4 g (0.05 mol) of potassium thiophenoxide and 50 mL of dimethylformamide is placed under vacuum. The bottle is charged with 2.7 bars of bromotrifluoromethane and shaken for 3 h. The reaction is slightly exothermic. The mixture is poured in 100 mL of 17% hydrochloric acid. The aqueous phase is extracted with hexane. The organic layer is washed with water and dried over potassium carbonate. The solvent is evaporated, and the residue is distilled to give 5.5 g (62%) of trifluoromethylthiobenzene (bp, 77–78 °C at 754 mm of Hg).

Preparation of 2-(Perfluorohexyl)cyclohexanone from Perfluorohexyl Iodide [*143*]

N-Cyclohex-1-enylpyrrolidine (9 g; 0.06 mol) was dissolved in pentane with *N*-ethyldiisopropylamine (7.8 g; 0.06 mol). Perfluorohexyl iodide (13.4 g; 0.03 mol) is added to the solution. A precipitate of *N*-ethyldiisopropylamine hydroiodide is formed instantly. After 3 h, the precipitate is filtered off, and the solution is evaporated. The crude liquid is hydrolyzed with 6 mL of 40% sulfuric acid. The mixture is stirred for 3 h and extracted with ether. The ether layer is neutralized with aqueous sodium hydrogen carbonate, washed with water, and dried over magnesium sulfate. The solvent is evaporated, and the residue is distilled. A second distillation with a spinning-band column yields 7.9 g (63%) of pure 2-(perfluorohexyl)cyclohexanone (bp, 71–73 °C at 0.4 mm of Hg).

Preparation of N-Butyl Perfluorohexylimidoyl Iodide from Perfluorohexyl Iodide [*141*]

Perfluorohexyl iodide (4.5 g; 0.010 mol) is mixed with *N*-butyl isocyanide (1 g; 0.013 mol). Copper powder (0.635 g; 0.01 mol) is added, and the mixture shaken for 10 s. After a few minutes, an exothermic reaction occurs. The copper is filtered off 2 h later, and the filtrate is distilled to give 4.7 g (90%) of *N*-butyl perfluorohexylimidoyl iodide (bp, 60 °C at 0.5 mm of Hg).

References for Pages 446–489

1. Fokin, A. V.; Voronkov, A. N. *Izv. Akad. Nauk SSSR* **1979,** 2620; *Chem. Abstr.* **1980,** *92,* 128371p.
2. Gervits, L. L.; Makarov, K. N.; Komarova, L. F.; Konovalova, N. V.; Zabolotskikh, V. F.; Cheburkov, Y. A. *Izv. Akad. Nauk SSSR* **1974,** 2256; *Chem. Abstr.* **1975,** *82,* 111517m.
3. Boutevin, B.; Youssef, B.; Boileau, S.; Garnault, A. M. *J. Fluorine Chem.* **1987,** *35,* 399.
4. Terrel, R. C.; Szur, A. J. Ger. Pat. 2 221 043, 1972; *Chem. Abstr.* **1973,** *78,* 29228b.
5. Platonov, V. E.; Malyuta, N. G.; Yakobson, G. G. Izv. *Akad. Nauk SSSR* **1972,** 2819; *Chem. Abstr.* **1973,** *78,* 83949a.
6. Archibald, T. G.; Baum, K. *J. Org. Chem.* **1990,** *55,* 3562.
7. Nakai, T.; Tanaba, K.; Ishikawa, N. *J. Fluorine Chem.* **1977,** *9,* 89.
8. Camps, F.; Coll, J.; Messeguer, A.; Pericas, M. A. *Synthesis* **1980,** 727.
9. Shibuta, D.; Watanabe, M.; Sato, Y. Jap. Pat. 61 69 742, 1986; *Chem. Abstr*. **1986,** *105,* 97019k.
10. Chaabouni, M.; Baklouti, A.; Szonyi , S.; Cambon, A. *J. Fluorine Chem.* **1990,** *46,* 307.
11. Fuchs, R.; Naumann, K.; Behrenz, W.; Hammann, I.; Homeyer, B.; Stendel, W. Eur. Pat. 4022, 1979; *Chem. Abstr.* **1980,** *92,* 94047v.
12. Hayashi, T.; Yamaguchi, H. Jap. Pat. 73 32 611, 1969; *Chem. Abstr.* **1974,** *80,* 26771d.
13. Werner, K., von. Ger. Pat. 3 035 641, 1982; *Chem. Abstr*. **1982,** *97,* 91725g.
14. Boutevin, B.; Youssef, B. *J. Fluorine Chem.* **1989,** *44,* 395.
15. Rico, I.; Wakselman, C. *Tetrahedron* **1981,** *37,* 4209.
16. Rico, I.; Wakselman, C. *J. Fluorine Chem.* **1982,** *20,* 765.
17. Hemer, I.; Moravcova, V.; Dedek, V. *Collect. Czech. Chem. Commun.* **1988,** *53,* 619.
18. Schack, C. J.; Christe, K. O. *J. Fluorine Chem.* **1980,** *16,* 63.
19. Tordeux, M.; Wakselman, C. *Tetrahedron* **1981,** *37,* 315.
20. Petrova, T. D.; Kolesnikova, I. V.; Savchenko, T. I.; Platonov, V. E. *Zh. Org. Khim.* **1984,** *20,* 1197; *Chem. Abstr.* **1985,** *102,* 24240d.
21. Gontar, A. F.; Bykhovskaya, E. G.; Sizov, Y. A.; Knunyants, I. L. *Izv. Akad. Nauk SSSR* **1976,** 2381; *Chem. Abstr.* **1977,** *86,* 54921v.
22. Smart, B. E. *J. Org. Chem.* **1976,** *41,* 2377.
23. Soulen, R. L.; Paul, D. W. *J. Fluorine Chem.* **1977,** *10,* 261.
24. Chambers, R. D.; Taylor, G.; Powell, R. L. *J. Fluorine Chem.* **1980,** *16,* 161.
25. Bekker, R. A.; Popkova, V. Y.; Knunyants, I. L. *Izv. Akad. Nauk SSSR* **1978,** 493; *Chem. Abstr.* **1978,** *88,* 190151n.
26. Coe, P. L.; Sellars, A.; Tatlow, J. C.; Fielding, H. C.; Whittaker, G. *J. Fluorine Chem.* **1986,** *32,* 135.

27. Chambers, R. D.; Lindley, A. E.; Philpot, P. D.; Fielding, H. C.; Hutchinson, J.; Whitaker, G. *J. Chem. Soc., Perkin Trans. 1* **1979,** 214.
28. Chambers, R. D.; Kirk, J. R.; Powell, R. L. *J. Chem. Soc., Perkin Trans. 1* **1983,** 1239.
29. Bryce, M. R.; Chambers, R. D.; Lindley, A. A.; Fielding, H. C. *J. Chem. Soc., Perkin Trans. 1* **1983,** 2451.
30. Sauvetre, R.; Normant, J.; Villieras, J. *Tetrahedron* **1975,** *31,* 897.
31. Meese, C. O. *Synthesis* **1984,** 1041.
32. Coe, P. L.; Cook, M. I.; Goodchild, N. J.; Edwards, P. N. *J. Chem. Soc., Perkin Trans. 1* **1988,** 555.
33. Bekker, R. A.; Asratyan, G. V.; Dyatkin, B. L.; Knunyants, I. L. *Tetrahedron* **1974,** *30,* 3539.
34. Coudures, C.; Pastor, R.; Szonyi, S.; Cambon, A. *J.Fluorine Chem.* **1984,** *24,* 105.
35. Krespan, C. G. *J. Fluorine Chem.* **1980,** *16,* 385.
36. Siegemund, G.; Schwertfeger, W. *J. Fluorine Chem.* **1982,** *21,* 133.
37. Yamabe, M.; Kumai, S. U.S. Pat. 4 151 200, 1979; *Chem. Abstr.* **1979,** *91,* 38940y.
38. Chen, L. F.; Mohtasham, J.; Gard, G. L. *J. Fluorine Chem.* **1990,** *46,* 21.
39. Kondratenko, N. V.; Vechirco, E. P.; Yagupol'skii, L. M. *Zh. Org. Khim.* **1979,** *15,* 704; *Chem. Abstr.* **1979,** *91,* 74506v.
40. Begue, J. P.; Charpentier-Morize, M.; Nee, G. *J. Chem. Soc., Chem. Commun.* **1989,** 83.
41. Terjeson, R. J.; Mohtasham, J.; Petyon, D. H.; Gard, G. L. *J. Fluorine Chem.* **1989,** *42,* 187.
42. Sodoyer, R.; Abad, E.; Rouvier, E.; Cambon, A. *J. Fluorine Chem.* **1983,** *22,* 401.
43. Ellouze, W.; Chaabouni, M. M.; Baklouti, A. *J. Fluorine Chem.* **1987,** *37,* 61.
44. Takeuchi, Y.; Asahina, M.; Hori, K.; Koizumi, T. *J. Chem. Soc., Perkin Trans. 1* **1988,** 1149.
45. Suga, H.; Schlosser, M. *Tetrahedron* **1990,** *46,* 4261.
46. Rondestvedt, C. S.; Thayer, G. L. *J. Org. Chem.* **1977,** *42,* 2680.
47. Dieng, S. Y.; Bertaina, B.; Cambon, A. *J. Fluorine Chem.* **1985,** *28,* 341.
48. Filler, R.; Cantrell, G.; Wolanin, D.; Nakvi, S. M. *J. Fluorine Chem.* **1986,** *30,* 399.
49. Langlois, B. R. *J. Fluorine Chem.* **1988,** *41,* 247.
50. Langlois, B. R. *J. Fluorine Chem.* **1990,** *46,* 407.
51. Korin'ko, V. A.; Serguchev, Y. A.; Yagupol'skii, L. M. *Zh. Org. Chem.* **1975,** *11,* 1268; *Chem. Abstr.* **1975,** *83,* 78783r.
52. Rico, I.; Wakselman, C. *J. Fluorine Chem.* **1982,** *20,* 759.
53. Wakselman, C.; Tordeux, M. *J. Org. Chem.* **1985,** *50,* 4047.
54. Rico, I.; Wakselman, C. *Tetrahedron Lett.* **1981,** *22,* 323.

55. Suda, M.; Hino, C. *Tetrahedron Lett.* **1981,** *22,* 1997.
56. Burton, D. J.; Wiemers, D. M. *J. Fluorine Chem.* **1981,** *18,* 573.
57. Rico, I.; Cantacuzène, D.; Wakselman, C. *J. Org. Chem.* **1983,** *48,* 1979.
58. Wakselman, C.; Tordeux, M. *J. Chem. Soc., Chem. Commun.* **1984,** 793.
59. Boiko, V. N.; Shchupak, G. M.; Yagupol'skii, L. M. *Zh. Org. Khim.* **1977,** *13,* 1057; *Chem. Abstr.* **1977,** *87,* 134226h.
60. Popov, V. I.; Boiko, V. N.; Kondratenko, N. V.; Sambur, V. P.; Yagupol'skii, L. M. *Zh. Org. Khim.* **1977,** *13,* 2135; *Chem. Abstr.* **1978,** *88,* 104823d.
61. Popov, V. I.; Boiko, V. N.; Yagupol'skii, L. M. *J. Fluorine Chem.* **1982,** *21,* 365.
62. Haley, B.; Haszeldine, R. N.; Hewitson, B.; Tipping, A. E. *J. Chem. Soc., Perkin Trans. 1* **1976,** 525.
63. Feiring, A. E. *J. Fluorine Chem.* **1984,** *24,* 191.
64. Ignat'ev, N. V.; Boiko, V. N.; Yagupol'skii, L. M. *Zh. Org. Khim.* **1985,** *21,* 653; *Chem. Abstr.* **1985,** *103,* 141549t.
65. MacNicol, D. D.; Robertson, C. D. *Nature (London)* **1988,** *332*(6159), 59.
66. Yagupol'skii, L. M.; Maletina, I. I.; Kondratenko, N. V.; Orda, V. V. *Synthesis* **1978,** 835.
67. Umemoto, T.; Kuriu, Y. *Chem. Lett.* **1982,** 65.
68. Yoshida, M.; Shimokoshi, K.; Kobayashi, M. *Chem. Lett.* **1987,** 433.
69. Haszeldine, R. N.; Rigby, R. B.; Tipping, A. E. *J. Chem. Soc., Perkin Trans. 1* **1972,** 1506.
70. Haszeldine, R. N.; Rigby, R. B.; Tipping, A. E. *J. Chem. Soc., Perkin Trans. 1* **1972,** 2180.
71. Haszeldine, R. N.; Hewitson, B.; Tipping, A. E. *J. Chem Soc., Perkin Trans. 1* **1974** , 1706.
72. Clavel, J. L.; Langlois, B.; Nantermet, R.; Tordeux, M.; Wakselman, C. Eur. Pat. 374 061, 1990; *Chem. Abstr.* **1991,** *114,* 5483s.
73. Umemoto, T.; Miyano, O. *Tetrahedron Lett.* **1982,** *23,* 3929.
74. Huang, W.; Huang, B.; Wang, W. *Huaxue Xuebao* **1985,** *43,* 663; *Chem. Abstr.* **1986,** *104,* 206683z.
75. Tordeux, M.; Langlois, B.; Wakselman, C. *J. Org. Chem.* **1989,** *54,* 2452.
76. Huang, W.; Chen, J. *Huaxue Xuebao* **1986,** *44,* 484; *Chem. Abstr.* **1986,** *105,* 208425m.
77. Kondratenko, N. V.; Popov, V. I.; Boiko, V. N.; Yagupol'skii, L. M. *J. Org. Chem. USSR* **1977,** *13,* 2086; *Chem. Abstr.* **1978,** *88,* 89273d.
78. Sepiol, J.; Soulen, R. L.; Sepiol, J. *J. Fluorine Chem.* **1983,** *23,* 163.
79. Stockel, R. F. *Can. J. Chem.* **1975,** *53,* 2302.
80. Haas, A.; Kraechter, H. U. *Chem. Ber.* **1988,** *121,* 1833.
81. Hanack, M.; Kuehnle, A. *Tetrahedron Lett.* **1981,** *22,* 3047.
82. Popkova, V. Y.; Galakhov, M. V.; Knunyants, I. L. *Izv. Akad. Nauk SSSR* **1989,** 116; *Chem. Abstr.* **1989,** *111,* 77762t.
83. Suzuki, H.; Yoshinaga, M.; Takaoka, A.; Hiroi, Y. *Synthesis* **1985,** 497.

84. Syper, L.; Mlochowski, J. *Tetrahedron* **1988,** *44,* 6119.
85. Berkowitz, P. T.; Baum, K. *J. Org. Chem.* **1981,** *46,* 3818.
86. Pirkle, W. H.; Hauske, J. R.; Eckert, C. A.; Scott, B. A. *J. Org. Chem.* **1977,** *42,* 3101.
87. Elliot, A. J.; Astrologes, G. W. U.S. Pat. 4 618 718, 1986; *Chem. Abstr.* **1989,** *110,* 192243r.
88. Buerger, H.; Krumm, B.; Pawelke, G. *J. Fluorine Chem.* **1989,** *44,* 147.
89. Greenwald, R. B. *J. Org. Chem.* **1976,** *41,* 1469.
90. Fokin, A. V.; Rapkin, A. I.; Tatarinov, A. S.; Titov, V. A.; Studnev, Y. N. *Izv. Akad. Nauk SSSR* **1986,** 469; *Chem. Abstr.* **1987,** *106,* 4493t.
91. Postovoi, S. A.; Zeifman, Y. V.; Knunyants, I. L. *Izv. Akad. Nauk SSSR* **1986,** 1306; *Chem. Abstr.* **1987,** *107,* 22926m.
92. Li, X.; Pan, H.; Jiang, X. *Tetrahedron Lett.* **1987,** *28,* 3699.
93. Powers, G. A.; Stephens, R.; Tatlow, J. C. *J. Fluorine Chem.* **1982,** *20,* 555.
94. Knunyants, I. L.; Struchkov, Y. T.; Bargamova, M. D.; Espenbetov, A. A. *Izv. Akad. Nauk SSSR* **1985,** 1097; *Chem. Abstr.* **1986,** *104,* 19362g.
95. Coe, P. L.; Sellars, A.; Tatlow, J. C. *J. Chem. Soc., Perkin Trans. 1* **1985,** 2185.
96. Hojo, M.; Masuda, R.; Okada, E.; Yamamoto, H.; Morimoto, K.; Okada, K. *Synthesis* **1990,** 195.
97. Ter-Gabrielyan, E. G.; Lur'e, E. P.; Zeifman, Y. V.; Gambaryan, N. P. *Izv. Akad. Nauk SSSR* **1975,** 1380; *Chem. Abstr.* **1975,** *83,* 96342a.
98. Knunyants, I. L.; Gontar, A. F.; Tikunova, N. A.; Vinogradov, A. S.; Bykhovskaya, E. G. *J. Fluorine Chem.* **1980,** *15,* 169.
99. Yagupol'skii, L. M.; Lukmanov, V. G.; Boiko, V. N.; Alekseeva, L. A. *Zh. Org. Khim.* **1977,** *13,* 2388; *Chem. Abstr.* **1978,** *88,* 62120q.
100. Steglich, W.; Burger, K. *Chem. Ber.* **1974,** *107,* 1488.
101. Fokin, A. V.; Voronkov, A. N.; Timofeenko, I. A.; Kosyrev, Y. M. *Izv. Akad. Nauk SSSR* **1975,** 1445; *Chem. Abstr.* **1975,** *83,* 96374m.
102. Lyga, J. W.; Patera, R. M. *J. Heterocycl. Chem.* **1990,** *27,* 919.
103. Iznaden, M.; Portella, C. *Tetrahedron Lett.* **1988,** *29,* 3683.
104. Knunyants, I. L.; Bargamova, M. D. *Izv. Akad. Nauk SSSR* **1977,** 1812; *Chem. Abstr.* **1977,** *87,* 167938v.
105. Knunyants, I. L.; Gontar, A. F.; Vinogradov, A. S. *Izv. Akad. Nauk SSSR* **1983,** 694; *Chem. Abstr.* **1983,** *98,* 198169b.
106. Kolesnikova, I. V.; Petrova, T. D.; Platonov, V. E.; Mikhailov, V. A.; Popov, A. A.; Savelova, V. A. *J. Fluorine Chem.* **1988,** *40,* 217.
107. Mustafa, M. E. S.; Takaoka, A.; Ishikawa, N. *J. Fluorine Chem.* **1986,** *30,* 463.
108. Ignatova, Y. L.; Karimova, N. M.; Kil'disheva, O. V.; Knunyants, I. L. *Izv. Akad. Nauk SSSR* **1986,** 732; *Chem. Abstr.* **1987,** *106,* 49899w.
109. Burger, K.; Geith, C.; Hoess, E. *Synthesis* **1990,** 357.

110. Harland, P. A.; Hodge, P.; Maughan, W.; Wildsmith, E. *Synthesis* **1984,** 941.
111. Landini, D.; Penso, M. *Synth. Commun.* **1988,** *18,* 791.
112. Hendrickson, J. B.; Bergeron, R. *Tetrahedron Lett.* **1973,** 3839.
113. Welch, J. T.; Seper, K. W. *J. Org. Chem.* **1986,** *51,* 119.
114. Patrick, T. B.; Hosseini, S.; Bains, S. *Tetrahedron Lett.* **1990,** *31,* 179.
115. Ortiz de Montellano, P. R.; Vinson, W. A. *J. Org. Chem.* **1977,** *42,* 2013.
116. Molines, H.; Wakselman, C. *Tetrahedron* **1976,** *32,* 2099.
117. Molines, H.; Wakselman, C. *J. Chem. Soc., Perkin Trans. 1* **1980,** 1114.
118. Aubert, C.; Begue, J. P.; Charpentier-Morize, M.; Née, G.; Langlois, B. *J. Fluorine Chem.* **1989,** *44,* 377.
119. Ishikawa, N.; Yokozawa, T. *Bull. Chem. Soc. Jpn.* **1983,** *56,* 724
120. Fuchigami, T.; Nakagawa, Y. *J. Org. Chem.* **1987,** *52,* 5276.
121. Blazejewski, J. C.; Dorme, R.; Wakselman, C. *J.Chem. Soc., Perkin Trans. 1* **1986,** 337.
122. Hendrickson, J. B.; Boudreaux, G. J.; Palumbo, P. S. *J. Am. Chem. Soc.* **1986,** *108,* 2358.
123. Kamitori, Y.; Hojo, M.; Masuda, R.; Fujitani, T.; Kobuchi, T.; Nishigaki, T. *Synthesis* **1986,** 340.
124. Rico, I.; Cantacuzène, D.; Wakselman, C. *J. Chem. Soc., Perkin Trans. 1* **1982,** 1063.
125. Castelhano, A. L.; Krantz, A. *J. Am. Chem. Soc.* **1987,** *109,* 3491.
126. Bey, P.; Ducep, J. B.; Schirlin, D. *Tetrahedron Lett.* **1984,** *25,* 5657.
127. Tsushima, T.; Kawada, K. *Terahedron Lett.* **1985,** *26,* 2445.
128. Postovoi, S. A.; Lantseva, L. T.; Zeifman, Y. V. *Izv. Akad. Nauk SSSR* **1982,** 210; *Chem. Abstr.* **1982,** *96,* 162065j.
129. Rozov, L. A.; Mirzabekyants, N. S.; Zeifman, Y. F.; Cheburkov, Y. A.; Knunyants, I. L. *Izv. Akad. Nauk SSSR* **1974,** 1355; *Chem. Abstr.* **1974,** *81,* 119875b.
130. Kendrick, D. A.; Kolb, M. *J. Fluorine Chem.* **1989,** *45,* 265.
131. Yagupol'skii, L. M.; Matyushecheva, G. I.; Pavlenko, N. V.; Boiko, V. N. *Zh. Org. Khim.* **1982,** *18,* 14; *Chem. Abstr.* **1982,** *96,* 199023v.
132. Chen, Q.; Qiu, Z. *J. Fluorine Chem.* **1986,** *31,* 301.
133. Chen, Q.; Qiu, Z. *J. Fluorine Chem.* **1987,** *35,* 343.
134. Feiring, A. E. *J. Org. Chem.* **1983,** *48,* 347.
135. Hartkopf, U.; De Meijere, A. *Angew. Chem.* **1982,** *94,* 444.
136. Stahly, G. P.; Bell, D. R. *J. Org. Chem.* **1989,** *54,* 2873.
137. Prakash, G. K. S.; Krishnamurti, R.; Olah, G. A. *J. Am. Chem. Soc.* **1989,** *111,* 393.
138. Fuchikami, T.; Ojima, I. *Tetrahedron Lett.* **1984,** *25,* 307.
139. Matsubara, S.; Mitani, M.; Utimoto, K. *Tetrahedron Lett.* **1987,** *28,* 5857.
140. Rasmusson, G. H.; Brown, R. D.; Arth, G. E. *J. Org. Chem.* **1975,** *40,* 672.
141. Tordeux, M.; Wakselman, C. *Tetrahedron* **1981,** *37,* 315.

142. Renaud, R. N.; Champagne, P. J.; Savard, M. *Can. J. Chem.* **1979,** *57,* 2617.
143. Cantacuzene, D.; Wakselman, C.; Dorme, R. *J. Chem. Soc., Perkin Trans. 1* **1977,** 1365.
144. Rico, I.; Cantacuzene, D.; Wakselman, C. *Tetrahedron Lett.* **1981,** *22,* 3405.
145. Naumann, D.; Wikes, B.; Kischkewitz, J. *J. Fluorine Chem.* **1985,** *30,* 73.
146. Umemoto, T.; Kuriu, Y.; Nakayama, S. *Tetrahedron Lett.* **1982,** *23,* 1169.
147. Umemoto, T.; Goto, Y. *Bull. Chem. Soc. Jpn.* **1987,** *60,* 3823.
148. Cowell, A. B.; Tamborski, C. *J. Fluorine Chem.* **1981,** *17,* 345.
149. Kimoto, H.; Fujii, S.; Cohen, L.A. *J. Org. Chem.* **1982,** *47,* 2867.
150. Chen, Q.; Qiu, Z. *J. Chem. Soc., Chem. Commun.* **1987,** 1240.
151. Akiyama, T.; Kato, K.; Kajitani, M.; Sakaguchi, Y.; Nakamura, J.; Hayashi, H.; Sugimori, A. *Bull. Chem. Soc. Jpn.* **1988,** *61,* 3531.
152. Sawada, H.; Yoshida, M.; Hagii, H.; Aoshima, K.; Kobayashi, M. *Bull. Chem. Soc. Jpn.* **1986,** *59,* 215.
153. Sawada, H.; Nakayama, M.; Akusawa, K. Jap. Pat. 02 62 832, 1990; *Chem. Abstr.* **1990,** *113,* 77894d.
154. Wakselman, C.; Tordeux, M. *J. Chem. Soc., Chem. Commun.* **1987,** 1701.
155. Tordeux, M.; Wakselman, C.; Langlois, B. Eur. Pat. 298 803, 1989; *Chem. Abstr,* **1989,** *111,* 173730b.
156. Zhou, Q.; Huang, Y. *J. Fluorine Chem.* **1988,** *39,* 87.
157. Huang, B.; Liu J. *Tetrahedron Lett.* **1990,** *31,* 2711.
158. Tordeux, M.; Langlois, B.; Wakselman, C. *J. Chem. Soc., Perkin Trans. 1* **1990,** 2293.
159. Brovko, V.; Sokolenko, V. A.; Yakobson G. G. *Zh. Org. Khim.* **1974,** *10,* 300; *Chem. Abstr.* **1974,** *80,* 120417q.
160. Dvonikova, K. V.; Platonov, V. E. *Izv. Akad. Nauk SSSR* **1990,** 468; *Chem. Abstr.* **1990,** *113,* 40029z.
161. Umemoto, T.; Miyano, O. *Bull. Chem. Soc. Jpn.* **1984,** *57,* 3361.
162. Kamigata, N.; Fukushima, T.; Yoshida, M. *Chem. Lett.* **1990,** 649.
163. Umemoto, T.; Ando, A. *Bull. Chem. Soc. Jpn.* **1986,** *59,* 447.
164. Umemoto, T.; Ishihara, S. *Tetrahedron Lett.* **1990,** *31,* 3579; *J. Fluorine Chem.* **1991,** *54,* 203 (Supplement, Abstracts of papers of the 13th International Symposium on Fluorine Chemistry, Sept. 1–6, 1991, Bochum, Germany).
165. Tanabe, Y.; Matsuo, N.; Ohno, N. *J. Org. Chem.* **1988,** *53,* 4582.
166. Sepiol, J.; Soulen, R. L. *J. Fluorine Chem.* **1984,** *24,* 61.
167. Leroy, J.; Wakselman, C.; Lacroix, P.; Kahn, O. *J. Fluorine Chem.* **1988,** *40,* 23.
168. Matsui, K.; Tobita, E.; Ando, M.; Kondo, K. *Chem. Lett.* **1981,** 1719.
169. Carr, G. E.; Chambers, R. D.; Holmes, T. F.; Parker, D. G. *J. Chem. Soc., Perkin Trans. 1* **1988,** 921.

170. Chen, Q.; Wu, S. *J. Chem. Soc., Chem. Commun.* **1989,** 705.
171. Freskos, J. N. *Synth. Commun.* **1988,** *18,* 965.
172. Zelenin, A. E.; Chkanikov, N. D.; Galakhov, M. V.; Kolomiets, A. F.; Fokin, A. V. *Izv. Akad. Nauk SSSR* **1985,** 931; *Chem. Abstr.* **1985,** *103,* 160162c.
173. Sviridov, V. D.; Chkanikov, N. D.; Galkhov, M. V.; Kolomiets, A. M.; Fokin, A. V. *Izv. Akad. Nauk SSSR* **1990,** 948; *Chem. Abstr.* **1990,** *113,* 77779v.
174. Kaesler, R. W.; Legoff, E. *J. Org. Chem.* **1982,** *47,* 5243.
175. Belen'kii, G. G.; Lur'e, E. P.; German, L. S. *Izv. Akad. Nauk SSSR* **1975,** 2728; *Chem. Abstr.* **1976,** *84,* 73567j.

Fluorocarbene Insertions

by R. L. Soulen

Chlorofluorocarbene generated by the thermal decomposition of dichlorofluoromethylphenylmercury reacts with 2,3-dimethylindole to give 3-fluoro-2,4-dimethylquinoline, 3-chloro-2,4-dimethylquinoline and 3-(chlorofluoromethyl)-2,3-dimethylindole [*1*] (equation 1). Similar results are obtained when the chlorofluorocarbene is generated from dichlorofluoromethane by base-catalyzed dehydrohalogenation using a phase-transfer catalyst in an aqueous–organic solvent system [2] (equation 1).

1

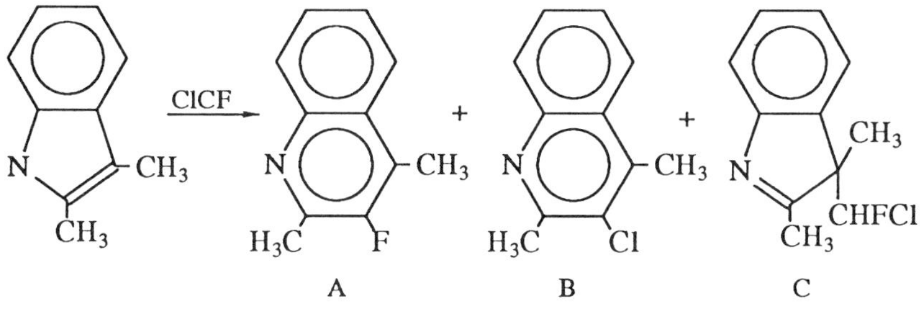

ClCF

A B C

Reaction conditions	°C	%A	%B	%C	Ref.
$C_6H_5HgCCl_2F$, DME, NaI	20	20.5	4.0	3	[*1*]
$C_6H_5HgCCl_2F$, benzene	80	30.2	3.4	trace	[*1*]
$C_6H_5HgCCl_2F$,DME	80	50.2	6.3	trace	[*1*]
$CHCl_2F$, NaOH, H_2O, CH_2Cl_2	0	32	1	18	[2]

Copyrolysis of 1,1-dichloroperfluoroindane and chlorodifluoromethane or tetrafluoroethylene gives 1-perfluoromethyleneindane as the major product and three minor products [*3*] (equation 2). Insertion of difluorocarbene into the benzylic carbon–chlorine bond and subsequent loss of a chlorine molecule is observed in the copyrolysis of chlorodifluoromethane and pentafluorobenzotrichloride to give α-chloroperfluorostyrene as the major product. Aromatic carbon–chlorine bonds are unreactive to the difluorocarbene in this reaction [*4*] (equation 3).

Perfluorocarbenes do not add to the C=N bond to form aziridines. Bis(trifluoromethyl)carbene and hexafluoroisopropylideniminosulfur(II) chloride react by insertion of the carbene into the sulfur–chlorine bond [*5*] (equation 4). Trifluoromethylcarbene

0065–7719/95/0187–0497$08.00/1

and $(NSCl)_3$ give 2,5-bis(trifluoromethyl)-1,3,4-thiadiazole and 2,2,2-trifluoro-1-chloroethyl-(2,2,2-trifluoroethylidenimino)sulfur(II) [*5*] (equation 5).

Similarly, bis(trifluoromethyl)carbene and hexafluoro-2-propanimine or difluorocarbene and perfluoro-2-azapropene also do not form the expected aziridines. Instead, hexafluoro-2-propanimine gives a product by insertion of the carbene into the nitrogen–hydrogen bond (equation 6), and perfluoro-2-azapropene gives products arising from the intermediate bis(trifluoromethyl)amide anion [*6*] (equation 7).

Pentafluorophenylnitrene generated by the photolysis of pentafluorophenylazide in toluene at 25 °C gives pentafluoroaniline and benzyl-substituted anilines. At low temperatures (–78 to –196 °C), perfluoroazobenzene also is formed [*7*] (equation 8).

[*3*] **2**

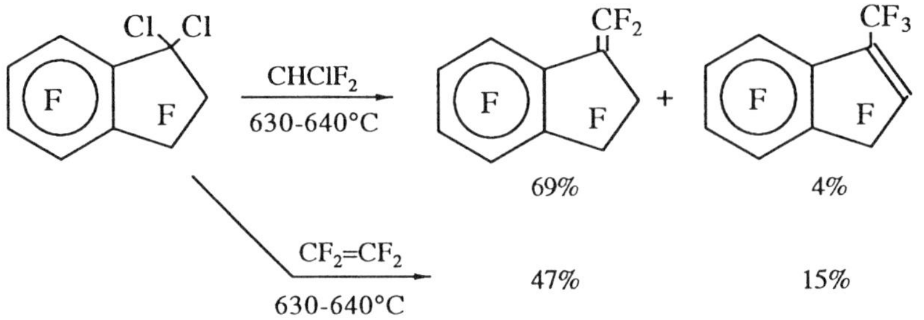

Cl, Cl (F) F —$CHClF_2$, 630-640°C→ CF_2 (F) F (69%) + CF_3 (F) F (4%)

—$CF_2{=}CF_2$, 630-640°C→ 47% 15%

[*4*] **3**

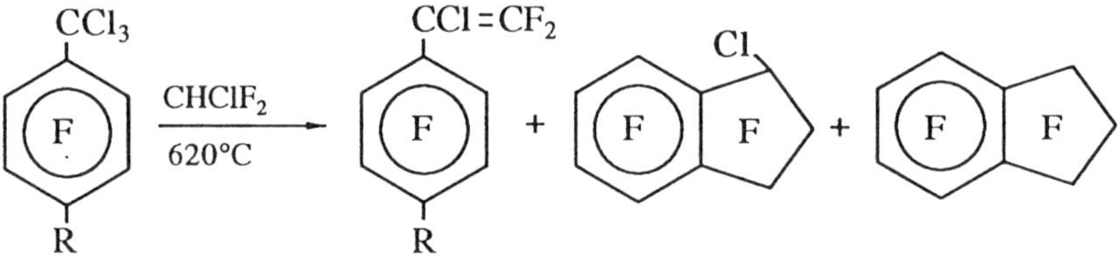

CCl_3 (F) R —$CHClF_2$, 620°C→ $CCl{=}CF_2$ (F) R + Cl (F) F + (F) F

R = F	77%	3%	2%
R = Cl	65%	-	-

[*5*] **4**

$$(CF_3)_2C{=}N_2 + (CF_3)_2C{=}NSCl \xrightarrow[70\ h]{UV} (CF_3)_2C{=}NSC(CF_3)_2Cl$$

68%

[*5*] **5**

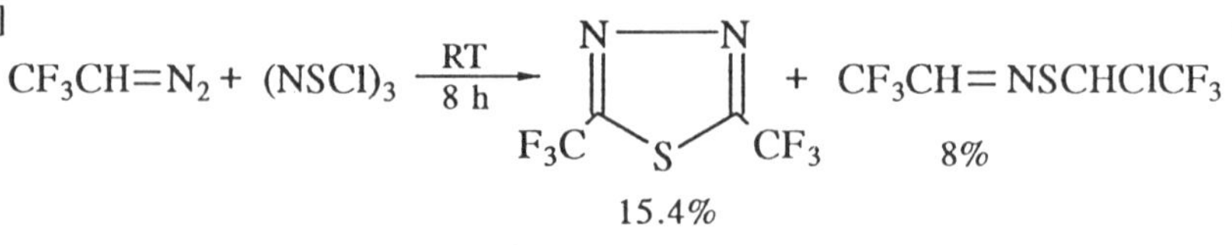

$$CF_3CH{=}N_2 + (NSCl)_3 \xrightarrow[8\ h]{RT} \text{N—N, } F_3C\text{, S, }CF_3 + CF_3CH{=}NSCHClCF_3$$

15.4% 8%

References are listed on page 500.

[6] **6**

$$(CF_3)_2C{=}NH \; + \; (CF_3)_2C{=}N_2 \xrightarrow[5\ days]{60°C} (CF_3)_2C{=}NCH(CF_3)_2 \quad 72\%$$

[6] **7**

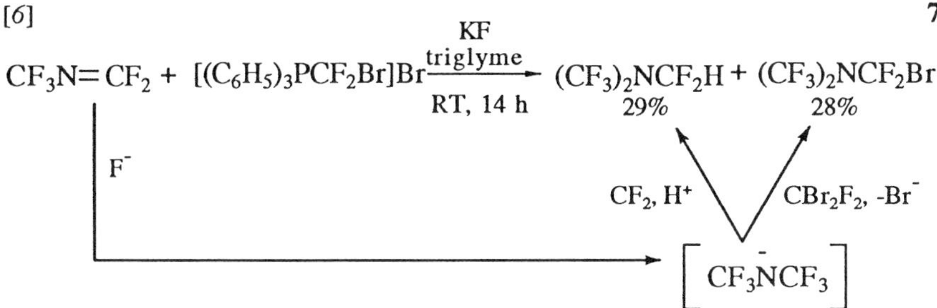

[7] **8**

$$C_6F_5N_3 \xrightarrow[toluene,\ 4\ h]{UV,\ 350\ nm} C_6F_5NHCH_2C_6H_5 + C_6F_5NH_2 + C_6F_5N{=}NC_6F_5$$

temperature	%[a]	%	%
25°C	52	12	0
-78°C	33	9.8	17
-196°C	77	5.9	7.9

a. Yields include ring substitution isomers.

[8] **9**

$$(CH_3)_3SiCF_3 \; + \; (CH_3)_2\underset{F}{Si}\underset{F}{Si}(CH_3)_2 \xrightarrow[3\ days]{160°C,\ sealed\ tube} (CH_3)_2\underset{F}{Si}CF_2\underset{F}{Si}(CH_3)_2 \quad 76\%$$

Difluorocarbene generated by the thermolysis of trimethyltrifluoromethylsilane reacts with disilanes by insertion into the silicon–silicon bond [*8*] (equation 9).

Thermolysis of pentafluoroethyltrifluorosilane at 200 °C gives tetrafluoroethylidene carbene, which reacts with phosphorus trifluoride to give trifluorovinyltetrafluorophosphorane [*9*] (equation 10) and with perfluorotrimethylphosphine to give perfluorodimethylisopropylphosphine and perfluoro-2-butene [*9*] (equation 10).

The relative reactivity of 1,2,2-trifluoroethylidenecarbene insertion into carbon–hydrogen bonds was studied for 11 substrates [*10*] (equation 11).

References are listed on page 500.

[*9*] **10**

$$C_2F_5SiF_3 \xrightarrow[0.5\ \text{atm, 1 h}]{PF_3\ 200°C} CF_2{=}CFPF_4\ (54\%)$$

$$C_2F_5SiF_3 \xrightarrow[0.5\ \text{atm, 1 h}]{P(CF_3)_3\ 200°C} (CF_3)_2PCF(CF_3)_2\ (44\%) + CF_3CF{=}CFCF_3\ (14\%)$$

[*10*] **11**

$$CHF_2CF_2SiF_3 + RH \xrightarrow[150°C, 18\ h]{\text{sealed tube}} RCHFCHF_2$$

Compound	RH bond	%
ethane	primary	16
propane	primary	5
propane	secondary	32
2-methylpropane	primary	1.4
2-methylpropane	tertiary	59

References for pages 497–500

1. Botta, M.; De Angelis, F.; Gambacorta, A. *Gazz. Chim. Ital.* **1983,** *113,* 129.
2. Dehmlow, E. V.; Franke, K. *Liebigs Ann. Chem.* **1979,** *10,* 1456.
3. Karpov, V. M.; Platonov, V. E.; Chuikov, I. P.; Yakobson, G. G. *J. Fluorine Chem.* **1983,** *22,* 459.
4. Dvornikova, K. V.; Platonov, V. E.; Yakobson, G. G. *J. Fluorine Chem.* **1985,** *28,* 99.
5. Steinbeisser, H.; Mews, R. *J. Fluorine Chem.* **1980,** *16,* 145.
6. Sohn, D.; Sundermeyer, W. *Chem. Ber.* **1982,** *115,* 3334.
7. Leyva, E.; Young, M. J. T.; Platz, M. S. *J. Am. Chem. Soc.* **1986,** *108,* 8307.
8. Fritz, G.; Bauer, H. *Angew. Chem.* **1983,** *95,* 740.
9. Sharp, K. G.; Schwager, I. *Inorg. Chem.* **1976,** *15,* 1697.
10. Haszeldine, R. N.; Rowland, R.; Speight, J. G.; Tipping, A. E. *J. Chem. Soc., Perkin Trans. 1* **1979,** 1943.

Arylation

by Timothy B. Patrick

The replacement of a substituent on an aromatic ring by a nucleophile is termed arylation. This chapter considers the replacement by nucleophilic oxygen, sulfur, nitrogen, and carbon of aromatic fluorine atoms, which are often activated by electron-withdrawing groups.

Arylation at Oxygen

The fluorine atom of ***monofluoroaromatic systems*** activated toward the S_NAr reaction by electron-withdrawing groups is readily replaced by alkoxide ion [*1, 2, 3, 4, 5*] (equations 1 and 2).

[*4*] 1

F → OCH_2CF_3

CF_3CH_2ONa HMPA, 250 °C, 18 - 20 h

R = 4-CN (58%), 3-NO_2 (78%), 4-$SO_2C_6H_4F$ (91%)
R = 4-CHO (56%), 4-CH_3CO (86%), H (0%)

[*1*] 2

F, CF_3SO_2, SO_2CF_3 → OCH_3, CF_3SO_2, SO_2CF_3

CH_3ONa, CH_3OH, 1 h

92 %

Activated fluorine is replaced in preference to activated chlorine in the S_NAr reaction (equation 3).

Fluorine replacement by alkoxyl may also be achieved with free alcohol in the presence of a rhodium(III) catalyst (equations 4 and 5) [*6, 7*] or a chromium(VI) complex [*8, 9*] (equation 5).

Oxygen nucleophiles (hydroxyl or nitrite) readily replace fluorine in ***perfluorinated systems*** [*10, 11*] (equations 6 and 7).

0065–7719/95/0187–0501$08.72/1

[5] F, Cl, Cl, Cl, Cl, Cl $\xrightarrow[\text{18-crown-6, } CH_3CN]{p\text{-}CH_3OC_6H_4OK}$ $OC_6H_4OCH_3$, Cl, Cl, Cl, Cl, Cl 88 % **3**

[6] F $\xrightarrow[CH_3OH,\ 80\,°C]{Rh(III),\ CH_3NO_2}$ OCH_3 87 % **4**

$(CH_2)_3OH$, F $\xrightarrow[80\,°C,\ 24\ h]{Rh(III),\ MeNO_2/Me_2CO\ (4/1)}$ 88 % [7] O **5**

$(CO)_3Cr$, $(CH_2)_3OH$, F $\xrightarrow[2.\ I_2]{1.\ (CH_3)_3COK,\ DMSO}$ 75 % [9]

[10] F $\xrightarrow[120\,°C,\ 6\ h]{p\text{-}CH_3OC_6H_4OH,\ K_2CO_3,\ DMF}$ $OC_6H_5OCH_3$, F, $OC_6H_5OCH_3$ 98 % **6**

A useful process for phenol protection and analysis is exemplified by the reaction of estrone with perfluorotoluene [*12*] (equation 8). The hydroxyl function replaces the 4-fluorine in high yield, and the hydroxyl may be deprotected in 87% yield on treatment with hydroxide.

N-Perfluorophenylbenzamide readily undergoes *intramolecular cyclization* to produce the benzoxazole system shown in equation 9 [*13*].

[11] 7

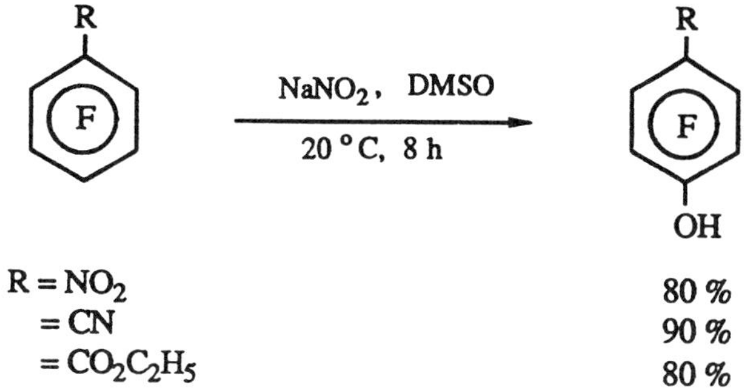

R = NO_2 — $NaNO_2$, DMSO, 20 °C, 8 h — 80 %
= CN — 90 %
= $CO_2C_2H_5$ — 80 %

[12] 8

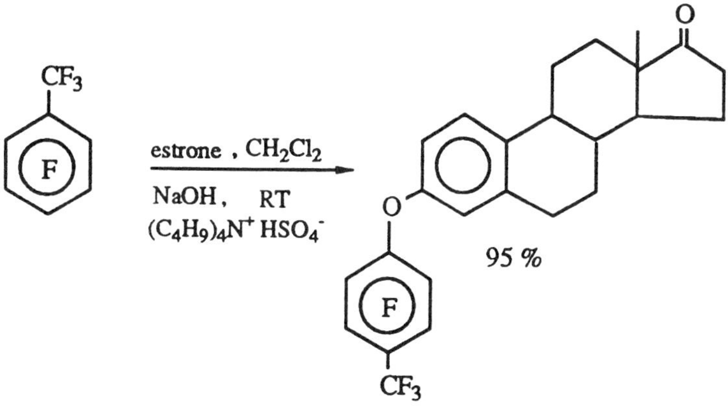

CF_3-C$_6$F$_5$ — estrone, CH_2Cl_2; NaOH, RT; $(C_4H_9)_4N^+ HSO_4^-$ — 95 %

[13] 9

NaH, DMF; heat, 2 h — C_6H_5 — 92 %

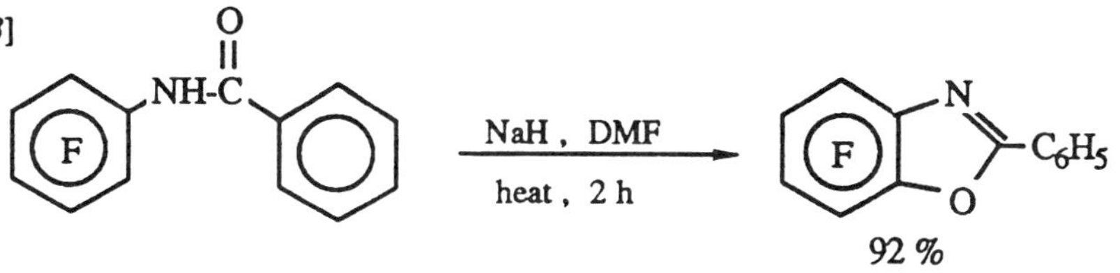

Fluorine replacement in ***condensed polyfluoroaromatics*** can be a complicated process. In perfluoroanthracene, the fluorine reactivity toward replacement by methoxide follows the order 2-F > 9-F [*14*], whereas in perfluoropyrene, the reactivity decreases in the series 1-F > 3-F > 6-F > 8-F [*15*]. In decafluoro-1,4-dihydronaphthalene, the 6-fluorine is most easily replaced [*15*] (equation 10). All fluorine atoms in perfluoronaphthalene are replaced by alkoxide or aryloxide with N^1,N^3-dimethylimidazolidin-2-one (DMI) as the solvent [*16*] (equation 11).

Fluorine atoms in ***aromatic nitrogen heterocycles*** are readily replaced by oxygen nucleophiles [*17*]. Bistrifluoromethyl hydroxylamine anion is an interesting nucleophile for the introduction of oxygen into perfluoropyridine. Rearrangement of the product occurs at 125 °C [*18*] (equation 12).

References are listed on pages 520–524.

[*15*] 10

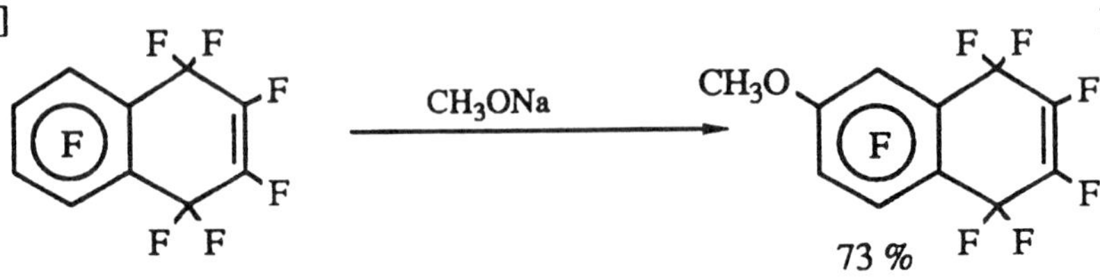

CH_3ONa

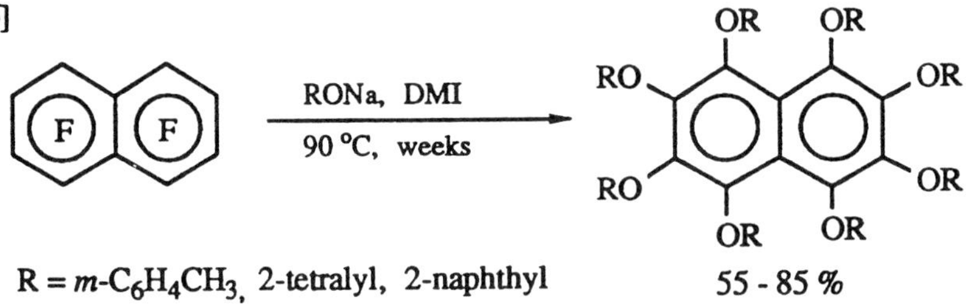

73 %

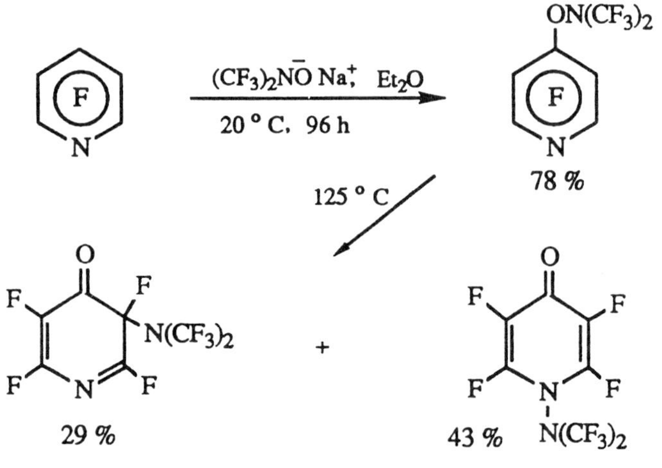

[*16*] 11

RONa, DMI

90 °C, weeks

R = *m*-$C_6H_4CH_3$, 2-tetralyl, 2-naphthyl

55 - 85 %

[*18*] 12

$(CF_3)_2N\bar{O}\ Na^+$; Et_2O

20 ° C, 96 h

78 %

125 ° C

29 %

43 %

Arylation at Sulfur

The S_NAr reaction between **thiolates** and ***monofluorobenzene*** and its derivatives requires high temperatures and polar aprotic solvents [*19, 20, 21*]. Polyfluoroaromatics show very little selectivity for fluorine replacement [*22, 23*] (equation 13).

Chromium complexes of fluoroaromatics undergo fluorine replacement more readily and in high yield [*24*] (equation 14).

The fluorine atoms of both 1- and 2-fluoronaphthalene are readily replaced by butyl and *tert*-butylmercapto groups. *tert*-Butylmercaptide gives lower yields [*25*] (equation 15).

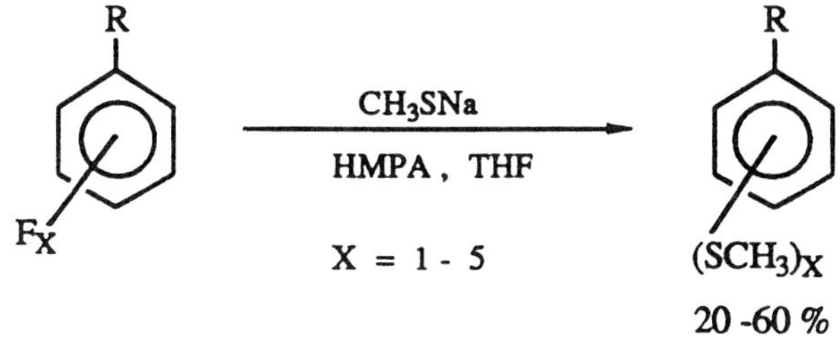

13

R, F_X —[CH_3SNa; HMPA, THF; X = 1 - 5]→ R, $(SCH_3)_X$ 20 -60 %

R = NH_2, $NHNH_2$, F, Br, Cl [*19,20*]
R = NO_2, CH_3, CHO [*21, 22, 23*]

[*24*] 14

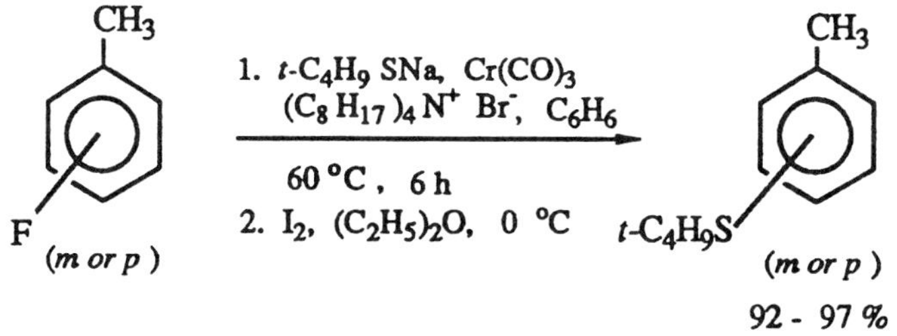

CH_3, F (*m or p*) —[1. *t*-C_4H_9 SNa, $Cr(CO)_3$, $(C_8H_{17})_4N^+$ Br^-, C_6H_6; 60 °C, 6 h; 2. I_2, $(C_2H_5)_2O$, 0 °C]→ CH_3, *t*-C_4H_9S (*m or p*) 92 - 97 %

[*25*] 15

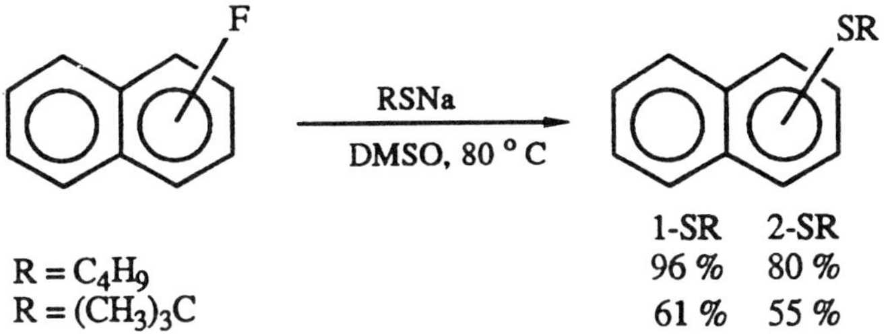

F —[RSNa; DMSO, 80 ° C]→ SR

	1-SR	2-SR
R = C_4H_9	96 %	80 %
R = $(CH_3)_3C$	61 %	55 %

The reaction of 1,2,4,5-tetrafluorobenzene in N^1,N^3-dimethylimidazolidin-2-one (DMI) with sodium or alkaline methyl mercaptide gives 1,2,4,5-tetrakis(methylthio)benzene [*26*] (equation 16).

[*26*] 16

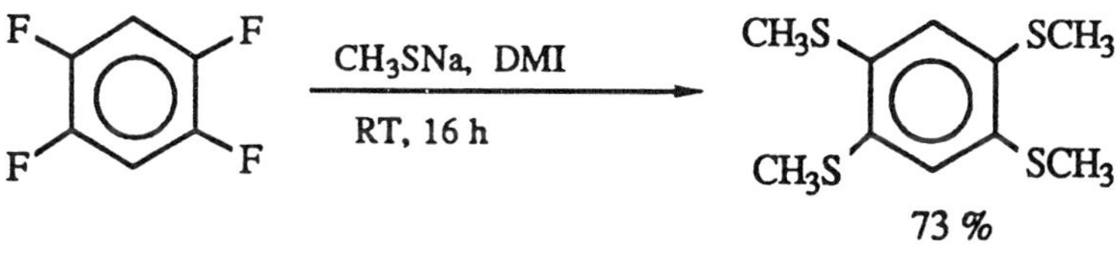

F, F, F, F —[CH_3SNa, DMI; RT, 16 h]→ CH_3S, SCH_3, CH_3S, SCH_3 73 %

Alkylthio groups replace the fluorine atoms of ***pentafluorobenzene derivatives*** with relative ease, but selectivity is difficult to control [*27*] (equation 17).

[27] **17**

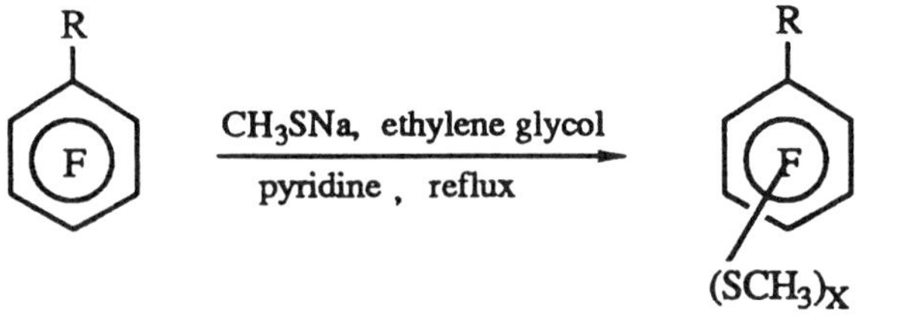

CH_3SNa, ethylene glycol / pyridine, reflux

$(SCH_3)_X$

R = F, Cl, Br, I, NO_2, NH_2, COOH, OH 22 - 80 %
X = 1 - 4

When selectivity is not a problem, as in *intramolecular cyclization,* useful synthetic routes to sulfur-containing heterocycles can be used [*28*] (equation 18).

[28] **18**

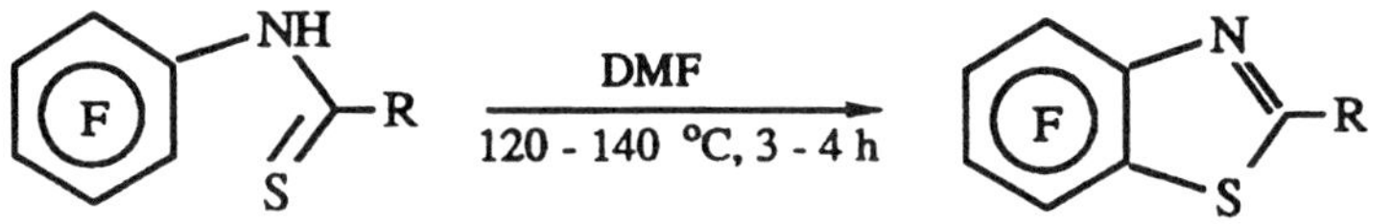

DMF / 120 - 140 °C, 3 - 4 h

R = NH_2, OH, CF_3, CH_3 27 - 89 %

Fluorine replacement in perfluorobenzene can be achieved in good yields and high selectivity when heavy metal thiolates are used as nucleophiles [*29, 30, 31, 32*] (equations 19–21).

[*30, 31*] **19**

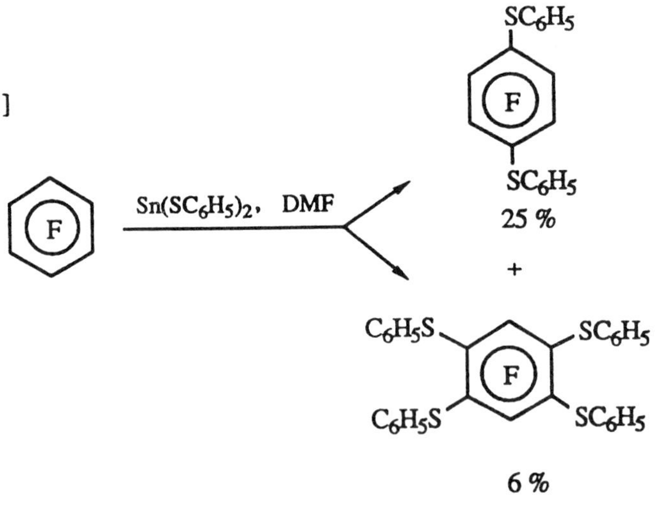

$Sn(SC_6H_5)_2$, DMF

SC_6H_5 / SC_6H_5 25 %

\+

C_6H_5S, SC_6H_5, C_6H_5S, SC_6H_5

6 %

Perfluorobiphenyl is arylthiolated with a potassium mercaptide at elevated temperature [*33*] (equation 22).

Perfluoronaphthalene undergoes an intermolecular substitution followed by intramolecular cyclization on reaction with allyl bromide and mercaptide ion to furnish a dihydrothiophene derivative in high yield [*34*] (equation 23).

[29] 20

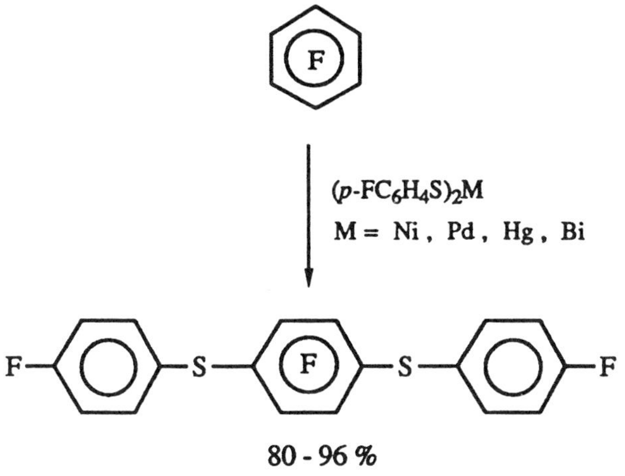

(*p*-$FC_6H_4S)_2M$
M = Ni, Pd, Hg, Bi

F—⟨⟩—S—⟨F⟩—S—⟨⟩—F

80 - 96 %

[32] 21

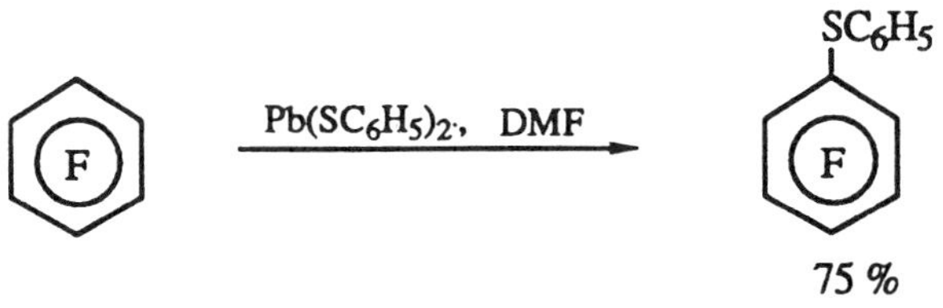

$Pb(SC_6H_5)_2$, DMF

SC_6H_5

75 %

[33] 22

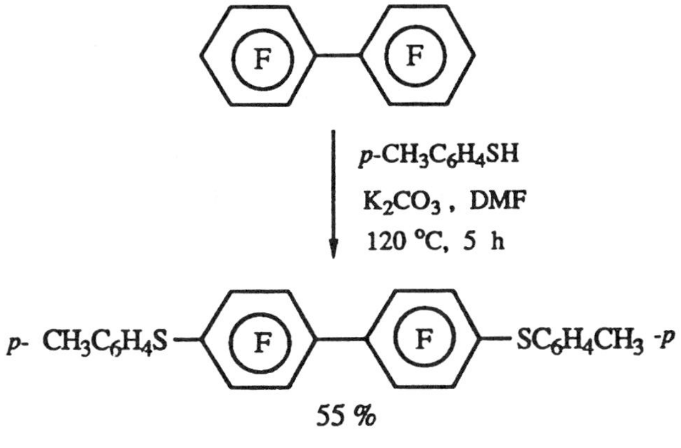

p-$CH_3C_6H_4SH$
K_2CO_3, DMF
120 °C, 5 h

p- $CH_3C_6H_4S$—⟨F⟩—⟨F⟩—$SC_6H_4CH_3$ -*p*

55 %

[34] 23

1. NaSH, CH_2=$CHCH_2Br$
ethylene glycol, DMF
2. $PhNMe_2$, 100 °C, 10 h

CH_3
S

83 %

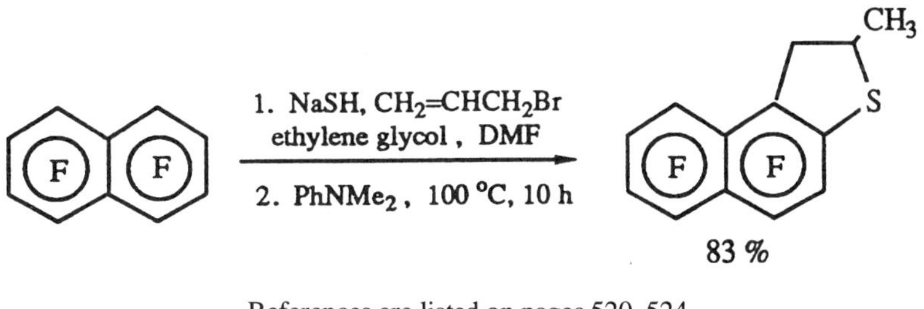

References are listed on pages 520–524.

Perfluoropyridine gives the usual replacement of the 4-fluorine on reactions with either aromatic or aliphatic mercaptides [*35, 36*]. The reaction of perfluoropyridine with cesium trifluoromethyl mercaptide, generated from thiocarbonyl fluoride and cesium fluoride, shows temperature dependence of selectivity in fluorine displacement [*35, 36*] (equation 24).

[*36*] $CF_2{=}S$, CsF, - 15 °C → 4-SCF_3 perfluoropyridine, 84 % **24**

[*35*] $CF_2{=}S$, CsF, 100 °C → 3,5-bis(SCF_3) perfluoropyridine (CF_3S, SCF_3), 75 %

Arylation at Nitrogen and Phosphorus

The replacement of reactive *aromatic fluorine by nitrogen nucleophiles* is a well-known process for the preparation of *aromatic amines*. The aromatic fluorine is activated by the presence of electron-withdrawing substituents on the aromatic ring, especially in *ortho* and *para* positions. [*37, 38, 39*] (equations 25–27).

[*38*] **25**

F, NO_2, SO_2CF_3 benzene —($(C_4H_9)_2NH$, DMF; O_2)→ $N(C_4H_9)_2$, NO_2, SO_2CF_3 benzene 71 %

Reaction times can be shortened and yields improved through the use of high pressure [*40*] (equation 28). Reactions may also be conducted in aqueous medium under ultraviolet irradiation [*41*] (equation 29).

Studies of the S_NAr mechanism have involved the use of ***2,4-dinitrofluorobenzene*** as a substrate. The S_NAr reaction of fluoroaromatics is catalyzed by both aromatic and alphatic amines [*42, 43, 44, 45, 46, 47, 48, 49*]. Diamines [*50*] and β-cyclodextrin [*51*] also catalyze the process. The catalytic effects of amine salts are observed especially with hydrochlorides [*52, 53*]. Further, the S_NAr process is accelerated by the use of polar solvents [*54, 55*].

[*37*] 26

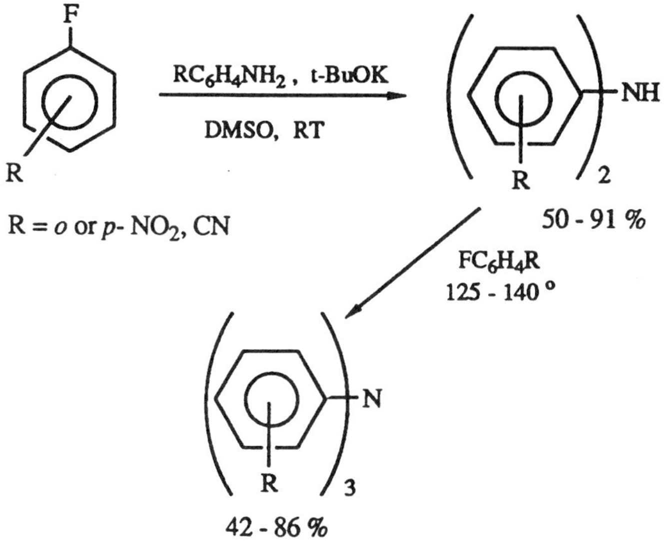

F

R

$RC_6H_4NH_2$, t-BuOK

DMSO, RT

NH

R 2

$R = o$ or p- NO_2, CN

50 - 91 %

FC_6H_4R

125 - 140 °

N

R 3

42 - 86 %

[*39*] 27

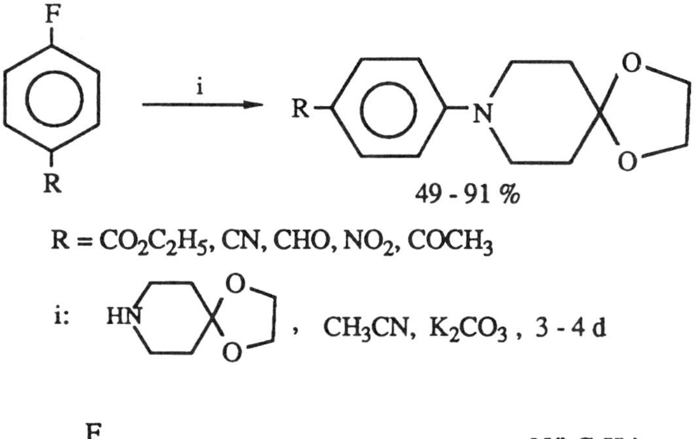

F

R

i

R

N

O

O

49 - 91 %

$R = CO_2C_2H_5$, CN, CHO, NO_2, $COCH_3$

i: HN O O , CH_3CN, K_2CO_3, 3 - 4 d

[*40*] 28

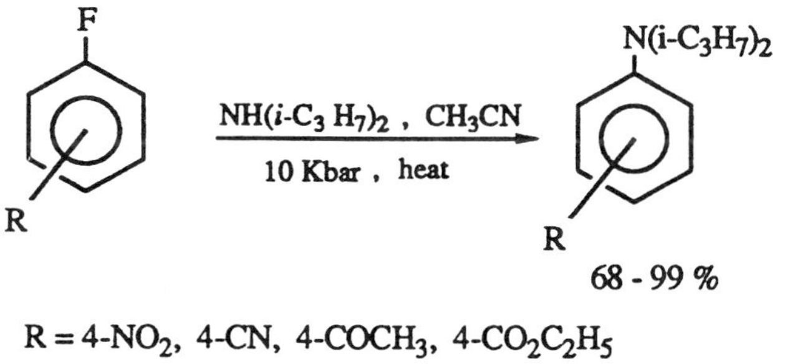

F

R

$NH(i\text{-}C_3H_7)_2$, CH_3CN

10 Kbar , heat

$N(i\text{-}C_3H_7)_2$

R

68 - 99 %

$R = 4\text{-}NO_2$, 4-CN, $4\text{-}COCH_3$, $4\text{-}CO_2C_2H_5$

References are listed on pages 520–524.

[41] 29

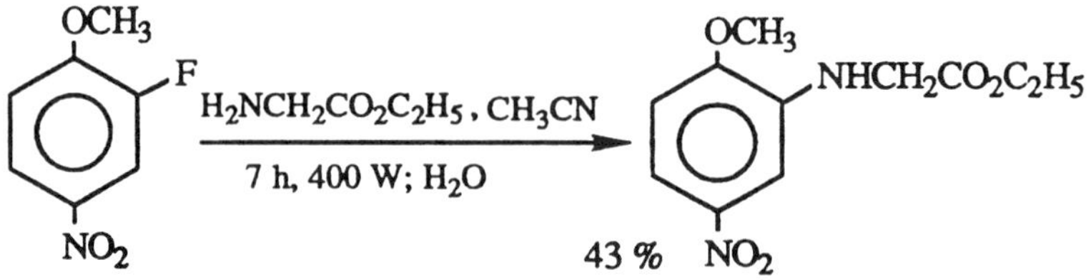

2,4-Dinitrofluorobenzene also serves as an arylation agent for a wide variety of biologically useful amines including aromatic amines [56], amino acids [57], and aminocarbohydrates [58, 59]. Weak nucleophilic amines such as benzimidazole [60] and fluoroamines [61] can also be arylated (equation 30).

30

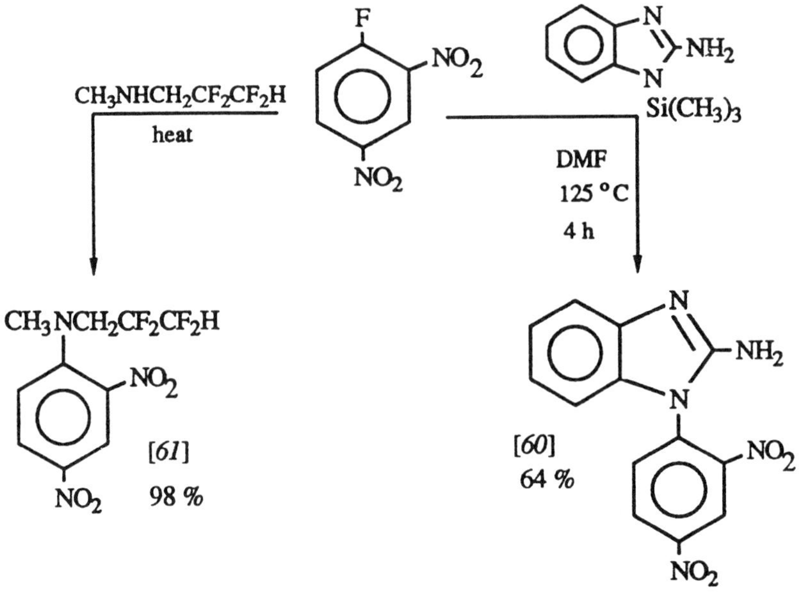

The weakly nucleophilic ring nitrogen of unsubstituted 2-aminothiazole replaces the fluorine atom of 2,4-dinitrofluorobenzene, whereas in 5-methyl-2-aminothiazole, the 2-amino function acts as the nucleophile [62] (equation 31).

Aromatic rings containing electron-withdrawing groups and more than one fluorine atom undergo fluorine replacement with moderate ease and usually with high regioselectivity. Thus the reaction between 2,4,6-trifluoronitrobenzene and ammonia results in the replacement of the 2-fluorine [63] but the reaction with amide ion results in replacement of the 4-fluorine [64] (equation 32).

The reaction of a wide variety of ***pentafluorobenzene*** derivatives reacts with imidazole in tetrahydrofuran or dimethyl sulfoxide results in the replacement of the 4-fluorine [65] (equation 33).

[62] 31

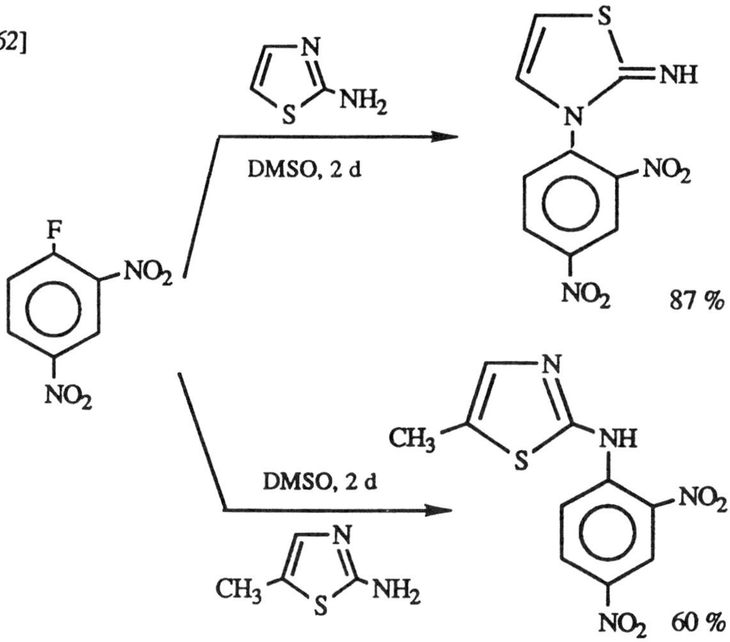

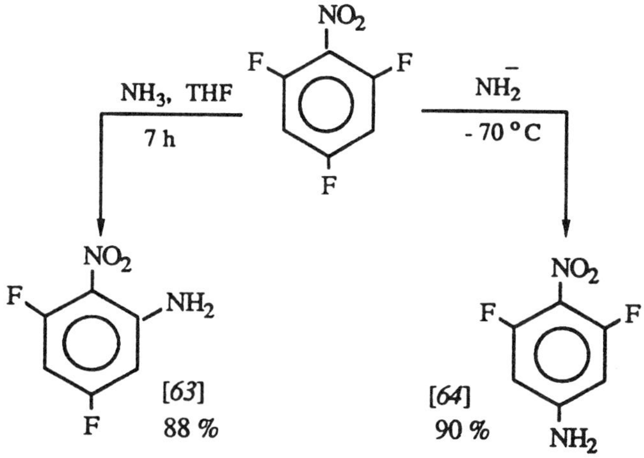

DMSO, 2 d

87 %

DMSO, 2 d

60 %

32

NH_3, THF

7 h

NH_2^-

- 70 °C

[63] 88 %

[64] 90 %

Tris(pentafluorophenyl)phosphine is also replaced when the 4-fluorine reacts with hydrazine [*55, 66*] (equation 34).

When pentafluoronitrobenzene reacts with *ortho-* or *para*-aminophenol in dimethylformamide solution containing sodium hydroxide, the 4-fluorine is replaced. Either the amino or the hydroxyl function of the aminophenol can act as the nucleophile depending on the reaction conditions [*67*] (equation 35).

[*65*] 33

$R = CN, NO_2, CHO, CO_2C_2H_5$ [THF]
$R = I, Br, Cl, H$ [DMSO]

THF or DMSO, 100 °C, 3 - 96 h; 40 - 83 %

[*66*] 34

N_2H_4, C_2H_5OH, reflux, 16 h; 79 %

[*67*] 35

o-$HOC_6H_4NH_2$, KOH; 8 - 66 %; 34 - 92 %

Perfluorobenzene undergoes S_NAr reactions readily with dimethylamine, aniline, *N*-methylaniline [*68*], piperidine, *N*-trimethylsilyliminotrimethylphosphorane [*69*], and lithium anilide [*70*], but the reactions are often accompanied by significant amounts of di- or polysubstitution. However, with the lithium salt of *N*-trimethylsilylaniline, only one fluorine atom in perfluorobenzene is replaced [*71*] (equation 36).

Intramolecular cyclization in perfluoroaromatic systems proves useful for the synthesis of heterocyclic compounds [*72*]. For example, the Fischer indole synthesis, which normally requires the presence of an *ortho* proton, occurs satisfactorily with an *ortho* fluorine in the ***perfluoronaphthalene*** series [*73*] (equation 37).

[71] (F) $\xrightarrow[\text{THF}]{(CH_3)_3Si\bar{N}C_6H_5\ Li^+}$ $(CH_3)_3SiNC_6H_5$–(F) 38 % **36**

[73] (F)(F)–NH–N=C(CH_3)(C_6H_5) $\xrightarrow[\text{reflux, 20 h}]{\text{tetralin}}$ (F)(F) indole, NH, C_6H_5 28 % **37**

The nitrogen atom of ***perfluoropyridine*** activates position 4 toward S_NAr reactions. Position 4 of perfluoropyridine is nearly 10 times as reactive as position 4 of perfluorotoluene [*74*] (equation 38).

[*74*] **38**

Perfluoropyridine: 4: 2×10^5; 3: 8×10^2; 2: 6×10^4

Perfluorotoluene (CF_3): 2: 2×10^3; 3; 4: 5×10^3

$C_6F_6 = 1$

Reaction at position 4 of perfluoropyridine thus occurs readily with nitrogen nucleophiles, as exemplified by its reactions with sodium azide and with hydroxylamine [*75, 76*] (equation 39).

39

NHOH–(F)N 95 % $\xleftarrow{NH_2OH}$ [*76*] (F)N $\xrightarrow[\text{RT, 24 h}]{NaN_3\ ,\ CH_3CN}$ [*75*] N_3–(F)N 69 %

In the presence of 18-crown-6, phthalimide ion replaces the 2-, 4-, and 6-fluorines in perfluoropyridine, whereas only the 4-fluorine is replaced in the absence of the crown ether [*77*] (equation 40).

[77] **40**

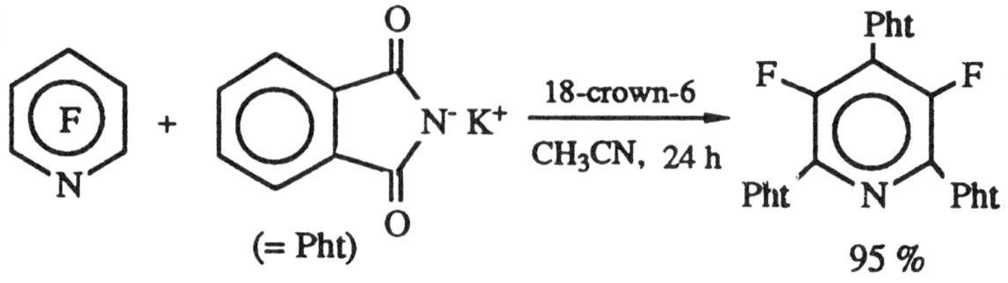

$$\text{C}_5\text{F}_5\text{N} + \text{PhtN}^-\text{K}^+ \xrightarrow[\text{CH}_3\text{CN, 24 h}]{\text{18-crown-6}} \text{2,4,6-(Pht)}_3\text{-3,5-F}_2\text{C}_5\text{N} \quad 95\,\%$$

(= Pht)

When position 4 of perfluoropyridine is blocked with a poor leaving group, ammonia replaces the fluorine in position 2 in good yield. Oxidation of the products obtained with hypochlorite, followed by iodine-catalyzed rearrangement, yields interesting fluorodienes [*78*] (equation 41). Ultraviolet irradiation can be used to assist reactions in which substitution is difficult [*79*].

[78] **41**

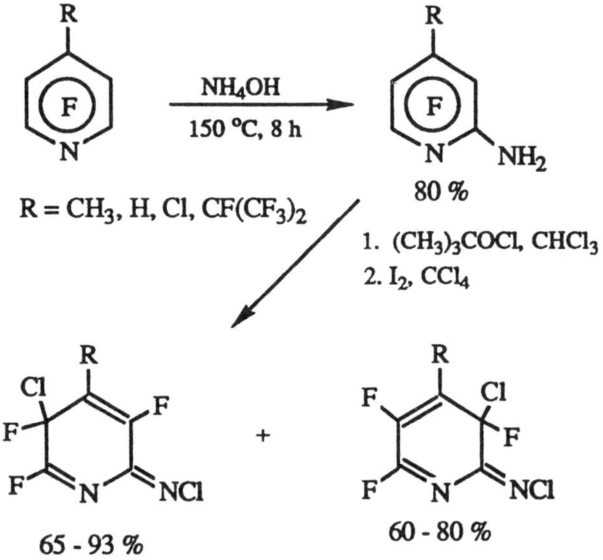

$$\text{4-R-C}_5\text{F}_4\text{N} \xrightarrow[\text{150 °C, 8 h}]{\text{NH}_4\text{OH}} \text{4-R-2-NH}_2\text{-C}_5\text{F}_3\text{N} \quad 80\,\%$$

$R = CH_3, H, Cl, CF(CF_3)_2$

1. $(CH_3)_3COCl$, $CHCl_3$
2. I_2, CCl_4

65 - 93 % + 60 - 80 %

Phosphorus nucleophiles have received little attention compared with nitrogen nucleophiles in reactions with fluorinated systems. Yields with phosphorus nucleophiles are sometimes low, but interesting materials are obtained [*80, 81*] (equations 42 and 43).

Arylation at Carbon

Replacement of an aromatic fluorine atom by a carbon nucleophile is facilitated by the presence of electron-withdrawing groups [*82*] (equation 44). Replacement of an activated aryl hydrogen can occur in preference to a nonactivated aryl fluorine [*83*] (equation 45) in reactions known as "vicarious" substitutions.

[81] F, F; $(C_6H_5)_2PNa$; NH_3, THF; $P(C_6H_5)_2$, $P(C_6H_5)_2$; 35 % **42**

[80] F, N; $(C_2H_5O)_3P$; 80 °C, 24 h; $O{=}P(OC_2H_5)_2$, F, N; 50 % **43**

[82] F, NO_2, NO_2, NH_2, NH_2, NO_2; i; CN, NO_2, NO_2, NH_2, NH_2, NO_2; 82 % **44**

i: $(CH_3)_3SiCN$, CH_3NO_2, reflux, 2 h

[83] Cl; i; F, NO_2; 82 % **45**

i: F, NO_2, $(CH_3)_3COK$, DMF, -20 °C

An *unactivated aryl fluorine* may be activated by complexation with chromium(VI). Replacement of the fluorine in the complexed system occurs readily, and the uncomplexed aromatic product can be generated by treatment with iodine [*84*] (equation 46).

An example of *intramolecular replacement of fluorine* is found in the thermal rearrangement of *syn*-8,16-difluoro[2,2]metacyclophane. In the *cis*-difluoro intermediate, one fluorine atom rearranges while the other is lost as HF in an interesting path to 1-fluoropyrene [*85*] (equation 47).

[*84*] 46

1. $R^1CH_2R^2$, KOH, DMSO; 2. I_2

$Cr(CO)_3$; R^1CHR^2

R^1	R^2	
C_6H_5	CN	80 %
CN	$CO_2C_2H_5$	68 %
$CO_2C_2H_5$	$CO_2C_2H_5$	66 %

[*85*] 47

120 °C, THF, 60 h

99 %

The trifluoromethyl group activates the fluorine in position 4 of perfluorotoluene toward reaction with carbon nucleophiles. Examples on the use of perfluorotoluene as an arylation agent abound, and in all cases, the 4-fluorine atom is replaced predominantly or exclusively [*86, 87, 88, 89, 90*] (equation 48). In ***perfluoromesitylene,*** the aromatic fluorine atoms are activated toward S_NAr reaction, and a reaction with one equivalent of methyllithium causes smooth replacement of one fluorine atom [*91*] (equation 49).

The arylation of **enamines** [*92, 93*] and **enolates** [*3, 94, 95*] (equation 50) with **perfluoroaromatics** proceeds in good yields and provides a variety of synthetically useful intermediates.

Arylations have been used to prepare pentafluorophenylbenzene, but the overall reliability of these methods is questionable. In particular, if phenyl radicals are generated from benzoyl peroxide in the presence of perfluorobenzene, both pure pentafluorophenylbenzene and mixtures of products are reported [*96, 97, 98, 99, 100*] (equation 51).

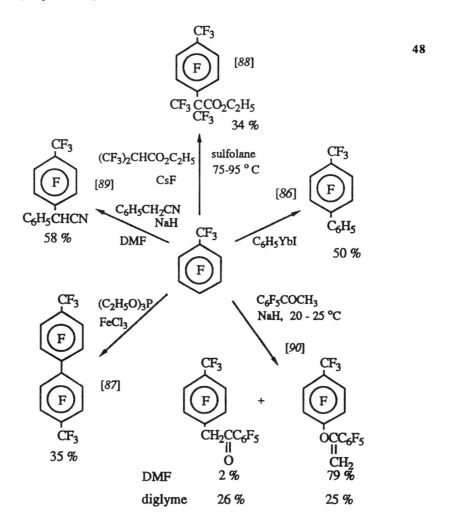

48

$CF_3C_6F_4C(CF_3)_2CO_2C_2H_5$ [*88*] 34 %; reagents: $(CF_3)_2CHCO_2C_2H_5$, CsF, sulfolane, 75-95 °C

$CF_3C_6F_4CH(C_6H_5)CN$ 58 % [*89*]; reagents: $C_6H_5CH_2CN$, NaH, DMF

$CF_3C_6F_4C_6H_5$ 50 % [*86*]; reagent: C_6H_5YbI

$CF_3C_6F_4$–$C_6F_4CF_3$ 35 % [*87*]; reagents: $(C_2H_5O)_3P$, $FeCl_3$

$CF_3C_6F_4CH_2C(=O)C_6F_5$ + $CF_3C_6F_4OC(=CH_2)C_6F_5$ [*90*]; reagents: $C_6F_5COCH_3$, NaH, 20 - 25 °C

	$CH_2CC_6F_5$ (=O)	OCC_6F_5 (=CH_2)
DMF	2 %	79 %
diglyme	26 %	25 %

[*91*] 49

1,3,5-tris(trifluoromethyl)-2,4,6-trifluorobenzene $\xrightarrow{CH_3Li,\ ether}$ 1-methyl-2,4,6-tris(trifluoromethyl)-3,5-difluorobenzene 34 %

50

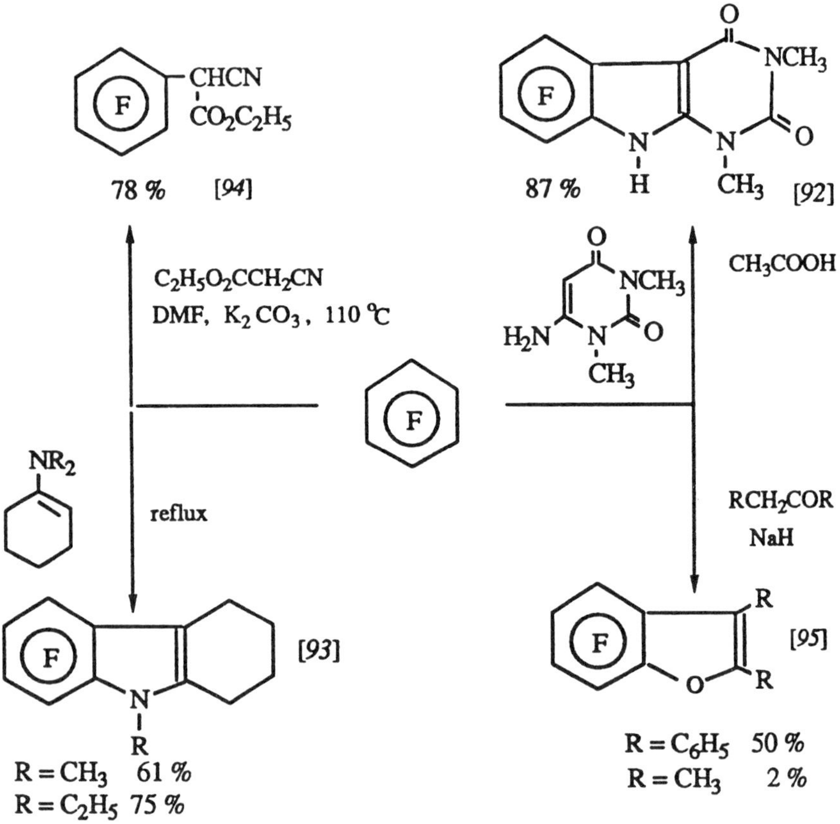

$CHCN$ / $CO_2C_2H_5$

78 % [94]

O, NCH_3, N, O, N, H, CH_3

87 % [92]

$C_2H_5O_2CCH_2CN$

DMF, K_2CO_3, 110 °C

O, NCH_3, H_2N, N, O, CH_3

CH_3COOH

F

NR_2

reflux

RCH_2COR

NaH

[93]

R, R, O

[95]

R

$R = CH_3$ 61 %

$R = C_2H_5$ 75 %

$R = C_6H_5$ 50 %

$R = CH_3$ 2 %

51

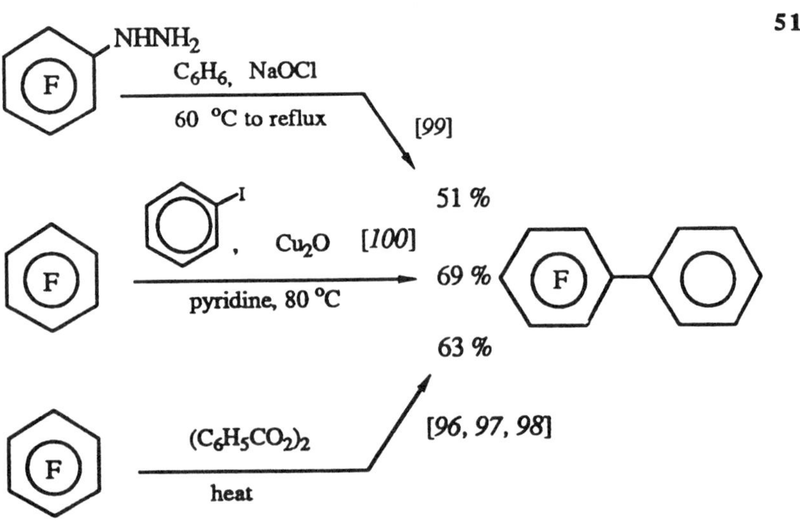

References are listed on pages 520–524.

Perfluoropyridine is often subjected to synthetic studies in comparison with perfluorotoluene. Both substrates react very well with replacement of the fluorine in position 4, but yields from perfluoropyridine are generally higher. Carbanions, enamines, and silanes react readily with ***perfluoropyridine*** [*101, 102, 103*] (equation 52).

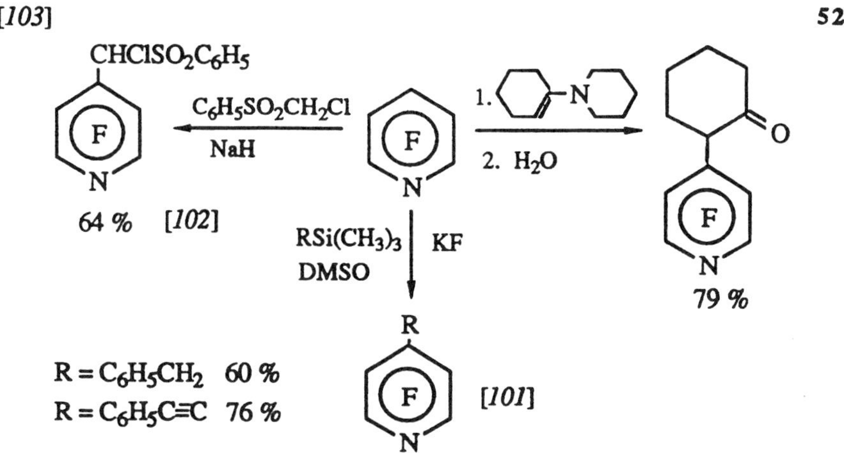

Free radicals, generated in a sensitized photochemical process, also react at position 4 of perfluoropyridine [*104, 105*] (equation 53).

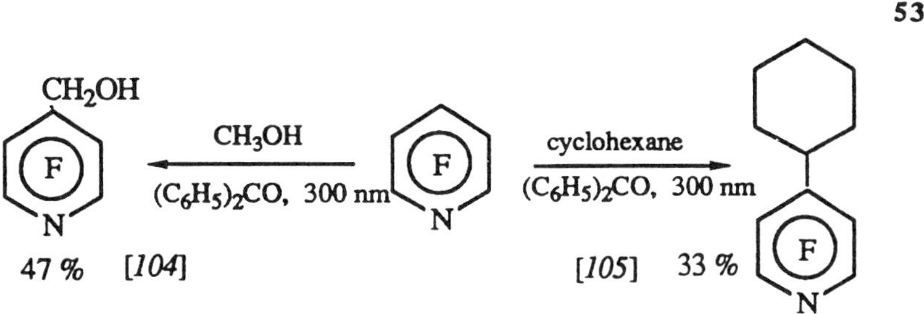

Allyl ethers of perfluoroaromatic phenols have been observed to alkylate the aromatic nucleus [*106*] or to undergo *Claisen rearrangement* [*107]* (equation 54).

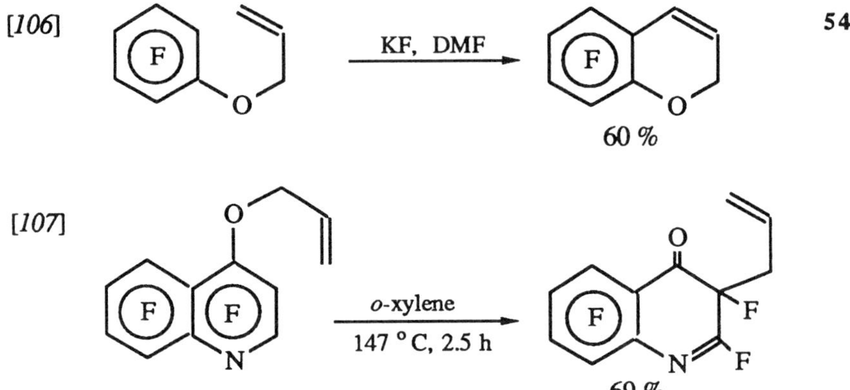

References for Pages 501–520

1. Boiko, V. N.; Shchupak, G. M.; Orlova, R. K.; Yagupol'skii, L. M. *Zh. Org. Khim.* **1985,** *21,* 1477 (Engl. Transl. 1346).
2. Cella, J. A.; Bacon, S. W. *J. Org. Chem.* **1984,** *49,* 1122.
3. Filler, R.; Fiebig, A. E.; Pelister, M. Y. *J. Org. Chem.* **1980,** *45,* 1290.
4. Idoux, J. P.; Madenwald, M. L.; Garcia, B. S.; Chu, D. L.; Gupton, J. T. *J. Org. Chem.* **1985,** *50,* 1876.
5. Kolonko, K. J.; Deinzer, M. L.; Miller, T. L. *Synthesis* **1981,** 133.
6. Goryunov, L. I.; Litvak, V. V.; Shteingarts, V. D. *Zh. Org. Khim.* **1988,** *24,* 401 (Engl. Transl. 354).
7. Houghton, R. P.; Voyle, M.; Price, R. *J. Chem. Soc., Chem. Commun.* **1980,** 884.
8. Aksenov, V. V.; Vlasov, V. M.; Yakobson, G. G. *J. Fluorine Chem.* **1982,** *20,* 439.
9. Houghton, R. P.; Voyle, M.; Price, R. *J. Organomet. Chem.* **1983,** *259,* 183.
10. Kameneva, T. M.; Malichenko, B. F.; Sheludko, E. U.; Pogorelyi, V. K. *Zh. Org. Khim.* **1989,** *25,* 576 (Engl. Transl. 518).
11. Miller, A. O.; Furin, G. G. *J.Org. Chem.* **1989,** *25,* 317.
12. Jarman, M.; McCague, R. *J. Chem. Res., Synop.* **1985,** 114.
13. Inukai, Y.; Oono, Y.; Sonoda, T.; Kobayashi, H. *Bull. Chem. Soc. Jpn.* **1979,** *52,* 516.
14. Burdon, J.; Childs, A. C.; Parsons, I. W.; Tatlow, J. C. *J. Chem. Soc., Chem. Commun.* **1982,** 534.
15. Chuikova, V. D.; Shteingarts, V. D. *Izv. Sib. Otd. Akad. Nauk SSSR* **1973,** 83; *Chem. Abstr.* **1973,** *79,* 104987z.

16. Freer, A.; MacNicol, D. D.; Mallinson, P. R.; Robertson, C. D. *Tetrahedron Lett.* **1989,** *30,* 5787.
17. Estel, L.; Marsais, F.; Queguiner, G. *J. Org. Chem.* **1988,** *53,* 2740.
18. Banks, R. E.; Falou, M. S.; Fields, R.; Olawore, N. O.; Tipping, A. E. *J. Fluorine Chem.* **1988,** *38,* 217.
19. Frazee, W. J.; Peach, M. E.; Sweet, J. R. *J. Fluorine Chem.* **1977,** *9,* 377.
20. Peach, M. E.; Rayner, E. S. *J. Fluorine Chem.* **1979,** *13,* 447.
21. Ulman, A.; Urankar, E. *J. Org. Chem.* **989,** *54,* 4691.
22. Le Blanc, M.; Peach, M. E.; Winter, H. M. *J. Fluorine Chem.* **1981,** *17,* 233.
23. MacDougall, C. T.; Peach, M. E. *Sulfur Lett.* **1987,** *7,* 15.
24. Alemagna, A.; Del Buttero, P.; Gorini, C.; Landini, D. *J. Org. Chem.* **1983,** *48,* 605.
25. Bradshaw, J. S.; South, J. A.; Hales, R. H. *J. Org. Chem.* **1972,** *37,* 2381.
26. Dirk, C. W.; Cox, S. D.; Wellman, D. E.; Wudl, F. *J. Org. Chem.* **1985,** *50,* 2395.
27. Musial, B. C.; Peach, M. E. *J. Fluorine Chem.* **1976,** *7,* 459.
28. Herkes, F. E. *J. Fluorine Chem.* **1979,** *13,* 1.
29. Hergett, S. C.; Peach, M. E. *J. Fluorine Chem.* **1988,** *38,* 367.
30. Hynes, R.; Peach, M. E. *J. Fluorine Chem.* **1986,** *31,* 129.
31. Jesudason, J. J.; Peach, M. E. *J. Fluorine Chem.* **1988,** *41,* 357.
32. Peach, M. E.; Smith, K. C. *J. Fluorine Chem.* **1985,** *27,* 105.
33. Robota, L. P.; Malichenko, B. F. *Zh. Org. Khim.* **1976,** *12,* 236; *Chem. Abstr.* **1976,** *84,* 105136m
34. Brooke, G. M.; Plews, G. *J. Fluorine Chem.* **1986,** *34,* 117.
35. Dmowski, W.; Haas, A. *Chimia* **1985,** *39,* 185.
36. Dmowski, W.; Haas, A. *J. Chem. Soc., Perkin Trans. 1* **1987,** 2119.
37. Gorvin, J. H. *J. Chem Soc., Perkin Trans. 1* **1988,** 1331.
38. Ivanova, T. M.; Shein, S. M. *Zh. Org. Khim.* **1980,** *16,* 1014 (Engl. Transl. 1014).
39. Taylor, E. C.; Skotnicki, J. S. *Synthesis* **1981,** 606.
40. Kotsuki, H.; Kobayashi, S.; Suenaya, H.; Nishizawa, H. *Synthesis* **1990,** *12,* 1147.
41. Figueredo, M.; Marquet, J.; Moreno-Manas, M.; Cantos, A. *Tetrahedron* **1989,** *45,* 7817.
42. Barnkole, T.; Hirst, J.; Onyido, I. *J. Chem. Soc., Perkin Trans. 2* **1981,** 1201.
43. Barnkole, T.; Hirst, J. Onyido, I. *J. Chem. Soc., Perkin Trans. 2* **1979,** 1317.
44. Forlani, L.; Sintoni, M. *J. Chem. Soc., Perkin Trans. 2* **1988,** 1959.
45. Forlani, L.; Torlelli, V. *J. Chem. Res., Synop.* **1982,** 62.
46. Forlani, L.; Tortelli, V. *J. Chem. Res., Synop.* **1982,** 258.

47. Nudelman, N. S.; Palleros, D. *J. Org. Chem.* **1983,** *48,* 1613.
48. Nudelman, N. S.; Cerdeira, S. *J. Chem Soc., Perkin Trans. 2* **1986,** 695.
49. Semmelhack, M. F., Seufert, W., Keller, L. *J. Am. Chem. Soc.* **1980,** *102,* 6584.
50. Guanti, G.; Petrillo, G.; Thea, S.; Pero, F. *J. Chem. Res., Synop.* **1982,** 282.
51. Barra, M.; DeRossi, R. H.; DeVargas, E. B. *J. Org. Chem.* **1987,** *52,* 5004.
52. Barnkole, T. O.; Hirst, J.; Hussain, G. *J. Chem. Soc., Perkin Trans. 2.* **1984,** 681.
53. Hirst, J.; Onyido, I. *J. Chem. Soc., Perkin Trans. 2* **1984,** 711.
54. Forlani, L. *J. Chem. Res., Synop.* **1984,** 260.
55. Naae, D. G.; Lin, T. *J. Fluorine Chem.* **1979,** *13,* 473.
56. Wong, M. P.; Connors, K. A. *J. Pharm. Sci.* **1983,** *72,* 146.
57. Zhong, Z.; Cheng, S.; Te, H. *J. Chem. Soc., Perkin Trans. 2* **1985,** 929.
58. Barends, D. M.; Blauw, J. S.; Mijnsbergen, C. W. *J. Chromatogr.* **1985,** *322,* 321.
59. Talieri, M. J.; Thompson, J. S. *Carbohydr. Res.* **1980,** *86,* 1.
60. Rudyk, V. I.; Troitsakaya, V. I.; Yagupol'skii, L. M. *Zh. Org. Khim.* **1980,** *16,* 2624; *Chem. Abstr.* **1980,** *94,* 208768e.
61. Yagupol'skii, L. M.; Gandelsman, L. Z.; Khomenko, L. A. *Zh. Org. Khim.* **1981,** *17,* 197; *Chem. Abstr.* **1981,** *95,* 24407a.
62. Forlani, L.; DeMaria, P.; Foresti, E.; Pradella, G. *J. Org. Chem.* **1981,** *46,* 3178.
63. Sitzman, M. E. *J. Org. Chem.* **1978,** *43,* 1241.
64. Chuikova, G. A.; Shtark, A. A.; Shteingarts, V. D. *Zh. Org. Khim.* **1988,** *24,* 2513 (Engl. Transl. 2267).
65. Fujii, S.; Maki, Y.; Kimoto, H. *J. Fluorine Chem.* **1989,** *43,* 131.
66. Hanna, H. R.; Miller, J. M. *Can. J. Chem.* **1979**, *57,* 1011.
67. Gerasimova, T. N.; Kolchina, E. F.; Kargapolova, I. *Izv. Akad. Nauk SSSR* **1987,** 2814; *Chem. Abstr.* **1987,** *111,* 96741b.
68. Koppang, R. *Acta Chem. Scand.* **1972,** *25,* 3872.
69. Dahmann, D.; Rose, H. *Chem. Ztg.* **1977,** *101,* 401.
70. Koppang, R. *J. Organometal. Chem.* **1972,** *46,* 193.
71. Koppang, R. *J. Fluorine Chem.* **1977,** *9,* 449.
72. Ames, D. E.; Leung, O. T.; Singh, A. G. *Synthesis* **1983,** 51.
73. Brooke, G. M. *J. Fluorine Chem.* **1988,** *40,* 51.
74. Chambers, R. D.; Martin, P. A.; Waterhouse, J. S.; Williams, D. L. H.; Anderson, B. *J. Fluorine Chem.* **1982,** *20,* 507.
75. Banks, R. E.; Sparkes, G. R. *J. Chem. Soc., Perkin Trans. 1* **1972,** 2964.
76. Miller, A. O.; Furin, G. G. *Izv. Sib. Otd. Akad. Nauk SSSR* **1985,** 123; *Chem. Abstr.* **1985,** *106,* 4588c.
77. Rasshofer, W.; Oepen, G.; Boegtle, F. *Isr. J. Chem.* **1980,** *18,* 249.

78. Banks, R. E.; Barlow, M. G.; Hornby, J. C.; Mamaghani, M. *J. Chem. Soc., Perkin Trans. 1* **1980,** 817.
79. Gilbert, A.; Krestonosich, S. *J. Chem Soc., Perkin Trans. 1* **1980,** 2531.
80. Boenigk, W.; Haegele, G. *Chem. Ber.* **1983,** *116,* 2418.
81. McFarlane, H. C. E.; McFarlane, W. *Polyhedron* **1983,** *2,* 303.
82. Chaykovsky, M.; Adolph, H. G. *Synth. Commun.* **1986,** *16,* 205.
83. Wojciechowski, K. *Bull. Pol. Acad. Sci. Chem.* **1988,** *36,* 235.
84. Baldoli, C.; Del Buttero, P.; Lieandro, E.; Maiorana, S. *Gazz. Chim. Ital.* **1988,** *188,* 409.
85. Boekelheide, V.; Anderson, P. H. *J. Org. Chem.* **1973,** *38,* 3928.
86. Davydov, D. V.; Belotskaya, I. P. *Metalloorg Khim.* **1988,** *1,* 899; *Chem. Abstr.* **1988,** *110,* 231187x.
87. Markovskii, L. N.; Furin, G. G.; Shermolovich, Y. G.; Yakobson, G. G. *Zh. Org. Khim.* **1979,** *15,* 531 (Engl. Transl. 471).
88. Vlasov, V. M.; Aksenov, V. V.; Yakobson, G. G. *Zh. Org. Khim.* **1979,** *15,* 2156 (Engl. Transl. 1953).
89. Vlasov, V. M.; Yakobson, G. G. *Zh. Org. Khim.* **1973,** *9,* 1024 (Engl. Transl. 1051).
90. Zakharova, O. V.; Vlasov, V. M.; Yakobson, G. G. *Zh. Org. Khim.* **1979,** *15,* 2169 (Engl. Transl. 1964).
91. Karpov, V. M.; Ermolenko, N. V.; Platonov, V. E.; Yakobson, G. G. *Zh. Org. Khim.* **1975,** *11,* 1052 (Engl. Transl. 1040).
92. Jiang, J. B.; Roberts, J. *J. Heterocycl. Chem.* **1985,** *22,* 159.
93. Wakselman, C.; Blazejewski, J. C. *J. Chem. Soc., Chem. Commun.* **1977,** 341.
94. Filler, R.; Woods, S. *Org. Synth.* **1977,** *57,* 80; Collective Volume **1988,** *6,* 873.
95. Inukai, Y.; Sonoda, T.; Kobayashi, H. *Bull. Chem. Soc. Jpn.* **1979,** *52,* 2657.
96. Bolton, R.; Moss, W. K.; Sandall, J. P. B.; Williams, G. H. *J. Fluorine Chem.* **1976,** *7,* 597.
97. Bolton, R.; Sandall, J. P. B.; Williams, G. H. *J. Fluorine Chem.* **1974,** *4,* 347.
98. Kobrina, L. S.; Salenko, V. L.; Yakobson, G. G. *J. Fluorine Chem.* **1976,** *8,* 193.
99. Birchall, J. M.; Haszeldine, R. N.; Wilkinson, M., *J. Chem. Soc., Perkin Trans. 1* **1974,** 1740.
100. Ljusberg, H.; Wahren, R. *Acta Chem. Scand.* **1973,** *27,* 2717.
101. Artamkina, G. A.; Kovalenko, S. V.; Beletskaya, I. P.; Reutov, O. A. *Zh. Org. Khim.* **1990,** *26,* 225 (Engl. Transl. 187).
102. Chambers, R. D.; Todd, M. *J. Fluorine Chem.* **1985,** *27,* 237.
103. Suschitzky, H.; Wakefield, B. J.; Whitten, J. P. *J. Chem. Soc., Perkin Trans. 1* **1980,** 2709.

104. Sket, B.; Zupan, M. *J. Heterocycl. Chem.* **1978,** *15,* 527.
105. Sket, B.; Zupan, M. *Synthesis* **1978,** 760.
106. Brooke, G. M. *J. Fluorine Chem.* **1983,** *22,* 483.
107. Brooke, G. M.; Eggleston, I. M.; Hale, F. A. *J. Fluorine Chem.* **1988,** *38,* 421.

Acylation

by R. W. Lang

The discussion of acylation reactions in this chapter is focused on fluorinated carboxylic acid derivatives and their use to build up new fluorine-containing molecules of a general preparative interest. Fifteen years ago, fluorinated carboxylic acids and their derivatives were used mainly for technical applications [*1*]. Since then, an ever growing interest for selectively fluorinated molecules for biological applications [*2, 3, 4, 5*] has challenged many chemists to use bulk chemicals such as trifluoroacetic acid and chlorodifluoroacetic acid as starting materials for the solution of the inherent synthetic problems [*6, 7, 8, 9*].

Acylation at Oxygen

Acylation of various oxygen functions by use of common and commercially available fluorinated carboxylic acid derivatives such as **trifluoroacetic anhydride** or the corresponding **acyl halides** have already been discussed sufficiently in the first edition [*10*]. Therefore only exceptional observations will be described in this section. In the past 15 years, many derivatizations of various nonfluorinated oxygen compounds by fluoroacylation were made for analytical purposes. Thus **Mosher's acid chlorides** for example became ready-to-use reagents for the determination of the *enantiomeric purity* of alcohols and amines by ^{19}F NMR or gas–liquid chromatographic (GLC) techniques [*11*] (equation 1).

[*11*] **1**

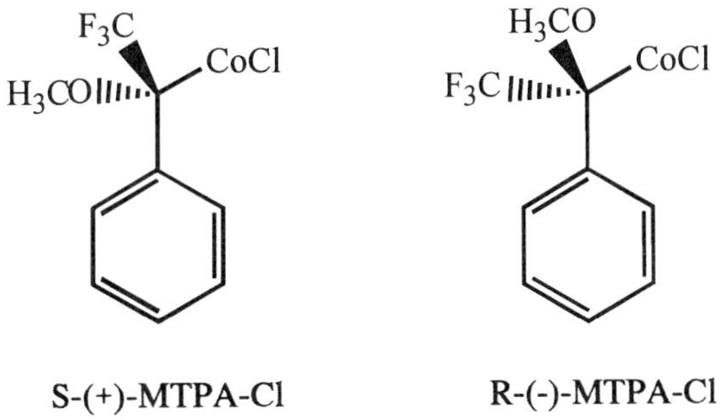

F_3C H_3CO CoCl

S-(+)-MTPA-Cl

H_3CO F_3C CoCl

R-(-)-MTPA-Cl

In sugar chemistry, the *trifluoroacetyl group* became an important tool for different synthetic purposes [*12, 13*]. Many sophisticated and effective acylating

0065–7719/95/0187–0525$08.00/1

agents such as **trifluoroacetyl triflate** (TFAT) [*14, 15*] and **2-trifluoroacetoxy pyridine** (TFAP) [*16*] were developed (equations 2 and 3). The reaction of TFAT with anhydrous ***alcohols*** and ***phenols,*** for example, gives the *trifluoroacetate esters* in high yields [*15*] (equation 4). Unlike other trifluoroacetylating agents such as trifluoroacetic anhydride, TFAT reacts readily with ***alkyl ethers*** (equations 5 and 6). Phenylalkyl ethers are slowly cleaved by TFAT (equation 7), and ketones are trifluoroacetylated by TFAT at oxygen to yield the corresponding enol trifluoroacetates (equation 8).

[*14*] **2**

$$CF_3COOH + CF_3SO_3H \xrightarrow[-(CF_3CO)_2O]{\text{excess } P_2O_5} CF_3\overset{O}{\overset{\|}{C}}OSO_2CF_3$$

TFAT (52%)
bp 62.5°C

[*16*] **3**

2 (2-hydroxypyridine) + $(CF_3CO)_2O$ —THF→ 2-pyridyl $OCOCF_3$ − [2-hydroxypyridinium $^-OCOCF_3$]

TFAP (87%)
bp 81-82°C/21 mm Hg

[*15*] **4**

ROH + TFAT —(2,6-di-t-Bu-4-methylpyridine; CCl_4, 0-25°C)→ $R\text{–}OCOCF_3$

R = $(CH_3)_2CH$	100%
R = $CH_3(CH_2)_6CH_2$	100%
R = C_6H_5	100%
R = p-$NO_2C_6H_4$	100%
R = 2,6-di-t BuC_6H_3	20%

References are listed on pages 541–544.

[*15*] **5**

$\xrightarrow[0°C]{TFAT}$ $OCOCF_3$ + OSO_2CF_3

[*15*] **6**

$\xrightarrow[0°C]{TFAT}$ CF_3COO OSO_2CF_3

[*15*] **7**

OCH_3 OCH_3 $\xrightarrow[65°C, 62 h]{TFAT}$ $OCOCF_3$ $OCOCF_3$

58%

[*15*] **8**

O $\xrightarrow[CCl_4, 25°C, 6 h]{TFAT}$ $OCOCF_3$ + $OCOCF_3$

For most cases, common fluoroacyl derivatives are sufficiently reactive and selective. Thus conversion of **perfluoroglutaric dichloride** to a monomethyl ester by methanol proceeds smoothly under the appropriate reaction conditions [*17*] (equation 9). Perfluorosuccinic acid monoester fluoride, on the other hand, is prepared most conveniently from perfluorobutyrolacetone [*18*] (equation 10).

Owing to the strong acidity of α-fluorinated carboxylic acids, *Fischer esterification* with most aliphatic alcohols proceeds autocatalytically [*19, 20*].

Treatment of *N*-hydroxysuccinimide with trifluoroacetic anhydride gives *N*-trifluoroacetoxysuccinimide quantitatively [*21*]. Some otherwise hardly accessible trifluoroacetylated tertiary alcohols are readily prepared, though in poor yields, by reacting the appropriate anhydride with an excess of an organometallic reagent [*22*] (equation 11).

References are listed on pages 541–544.

[*17*] **9**

$$CH_3OH + ClCO(CF_2)_3COCl \xrightarrow[-5 - 0^\circ C]{\text{tetraglyme}} CH_3OCO(CF_2)_3COCl \quad 70\%$$

[*18*] **10**

$$\text{(hexafluoro-γ-butyrolactone)} \xrightarrow[\text{diglyme}]{\text{ROH, NaF}} ROCO(CF_2)_2COF \quad 40\text{-}71\%$$

$R = CH_3, C_2H_5$

[*22*] **11**

$$(CF_3CO)_2O \xrightarrow{\text{excess RLi}} CF_3COOCR_2CF_3$$

$R = C_6H_5, p\text{-}CH_3C_6H_4$

[*23*] **12**

$$\text{cyclohexanone} \xrightarrow[25^\circ C,\ 73\ h]{(CF_3CO)_2O} \text{1,1-bis}(F_3CCOO)\text{cyclohexane} \quad 83\%$$

A rather interesting trifluoroacetylated acetal is formed in good yield when cyclohexanone is allowed to react with trifluoroacetic anhydride at room temperature [*23*] (equation 12).

Epoxides are easily attacked by **trifluoroacetic anhydride.** The reactions lead to diesters of vicinal diols and monoesters of unsaturated allylic alcohols in ratios depending on the reaction conditions [*24*] (equation 13).

A synthetically valuable reaction sequence is the **chlorodifluoroacetylation** of various substituted ***allylic alcohols*** and the subsequent *Reformatskii–Claisen rearrangement* of the ester thus formed to interesting 2,2-difluoropentenoic acid derivatives [*25*] (equation 14). Comparable sequences have been reported for allyl monofluoroacetates [*26*] and allyl 3,3,3-trifluoropropanoates [*27*] (equations 15 and 16).

[24] **13**

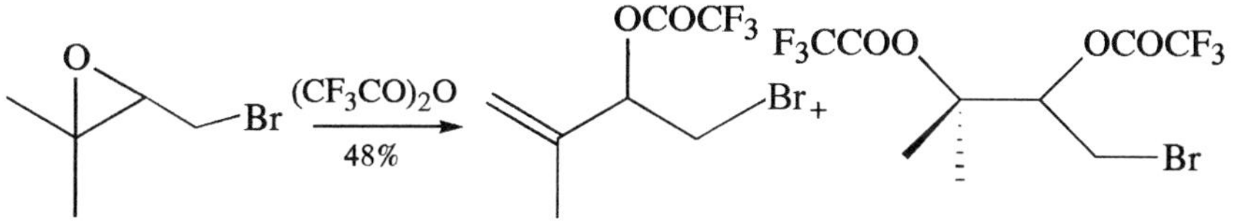

20-40°C	2	3
120-140°C	3	7

[25] **14**

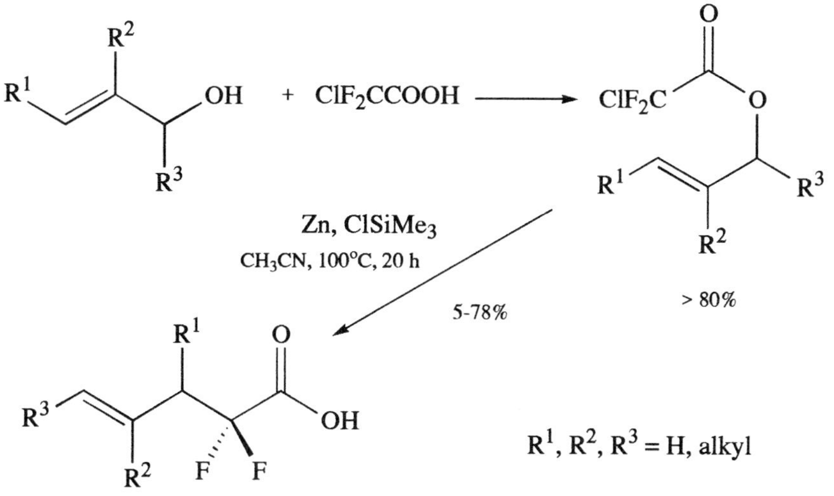

[26] **15**

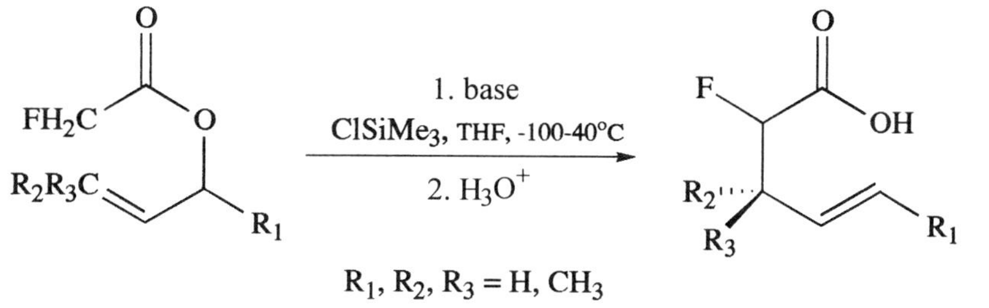

The *Reformatskii reaction* of α-***halogenated carboxylic esters*** with silylated cyanohydrins combined with an intramolecular acylation reaction gives fluorinated derivatives of tetronic acid [*28*] (equation 17). It is noteworthy to mention that this particular reaction sequence only proceeds with *ultrasonic irradiation.* A very similar reaction is used to enter the highly attractive field of *fluorinated sugars* [*29*] (equation 18).

[*27*] **16**

R_1, R_2, R_3 = H, CH_3

CF_3CH_2COOH; Et$_3$N, CH_2Cl_2, 20-25°C; 79-94%

1. $CF_3SO_2OSiMe_3$, Et$_3$N, CH_2Cl_2, 20-25°C; 2. H_3O^+; 15-100%

[*28*] **17**

1. Zn, THF, ultrasound, RT; 2. H_3O^+; 48-69%

$R_1 = CH_3, C_6H_5, p\text{-}CH_3OC_6H_4$ R_2 = H, CH_3 X = F, CF_3

Acylation at Sulfur

Fluoroacetylation of hydrogen sulfide, mercaptans, and thiocyanates have already been discussed in the first edition [*10*]. A useful extension of this particular chemistry is the fluoroacylation of ***trifluoromethyldisulfane*** [*30*] (equation 19).

Acylation at Nitrogen

Fluorinated carboxylic anhydrides and acyl halides as common acylating reagents to convert amines to amides and to acylate suitable heterocyclic nitrogen atoms have already been described in the first edition [*10*]. Like in the acylation at oxygen, much synthetic activity was concentrated in the past few years on the derivatization of biomolecules by fluoroacylation reactions, that is, *trifluoroacetylation of amino sugars,*

[*29*] **18**

2,2-dimethyl-1,3-dioxolane-4-CHO + $BrF_2CCOOEt$ $\xrightarrow[\text{THF, ether}]{\text{Zn}}$ dioxolanyl–CH(OH)–CF_2–COOEt (3 : 1)

Dowex 50 (after separation of the diastereoisomers)

→ HO–CH_2–lactone (OH, F, F) 65%

[*30*] **19**

Reagent	Product	Yield
CF_3SSH + CF_3COF	$CF_3SSCOCF_3$	60%
CF_3SSH + $FCO(CF_2)_3COF$	$CF_3SSCO(CF_2)_3COF$	10%
CF_3SSH + FCOF	CF_3SSCOF	55%
CF_3SSH + FCOCOF	$CF_3SSCOCOF$	30%
CF_3SSH + CF_3COSCl	$CF_3SSSCOCF_3$	5%

amino acids, and peptides for analytical and preparative purposes. The same reagents that were described for acylation at oxygen such as **trifluoroacetyl triflate** (TFAT) and **2-trifluoroacetoxypyridine** (TFAP) are also effective for the acylation at nitrogen [*14, 15*]. For ***amines*** and ***amino acids*** in particular, **polymer-bound trifluoroacetylating reagents** were developed with the advantage to minimize racemization [31]. The use of **bis(trifluoroacetamide)** and ***N*-methyl-bis-(trifluoroacetamide)** as trifluoroacetylating reagents is very attractive for GLC technique [*32*]. *Amide transacetylation* became a very useful trifluoroacetylation technique also for synthetic applications [*33, 34*]. Another widely used reagent—especially for trifluoroacetylation of amino acids and peptides—is **ethyl trifluoroacetate** in the presence of an appropriate base [35]. Selective *monoacylation of diamines* can also be achieved by the use of **alkyl fluorocarboxylates** [*36, 37, 38*]. The reactivity of esters of perfluorocarboxylic acid with aliphatic amines depends on the steric hindrance of the amines [*39*]. A valuable extension of this type of fluoroacylation is the reaction of ethyl (*Z*)-4-bromo-3-trifluoromethyl-2-butenoate with primary amines [*40*] (equation 20).

[*40*] **20**

R		
$R = p\text{-}CH_3OC_6H_4$	59%	-
$R = C_6H_5$	49%	-
$R = p\text{-}ClC_6H_4$	57%	-
$R = n\text{-}C_4H_9$	23%	27%
$R = c\text{-}C_6H_{11}$	23%	23%

Transamination of trifluoroacetylated vinyl ethers followed by monoacylation by diethyl malonate gives an intermediate that is cyclized to a trifluoromethylpyridine derivative [*9, 41*] (equation 21).

Selectivity and reactivity of **mixed anhydrides** toward primary and secondary aliphatic and aromatic amines has been studied in details [*42, 43*] (equation 22).

Fluorocarboxylic acids and their derivatives are used as building blocks in condensation reactions with dinucleophilic species for the synthesis of fluoroalkyl-substituted heteroaromatic systems [*8, 9, 40, 44, 45, 46, 47, 48*] (equations 20, 21, and 23).

An elegant approach to fluorinated β-lactams starts with a *Reformatskii reaction* of alkyl halodifluoroacetates on imines, followed by an *intramolecular acylation* of the amino function formed in situ [*49*] (equation 24).

In the series of highly advanced enzyme inhibitors, fluorinated substrates are playing an important role [*5, 6*]. Many such substrates are synthesized by using fluoroacetic acid derivatives as building blocks [*50, 51*] (equation 25).

Finally, the synthesis of trifluoroacetyl isocyanate has been redesigned such that multigram quantities are produced easily by treatment of trifluoroacetyl chloride with trimethylsilyl isocyanate [*52*] (equation 26).

[*9*] **21**

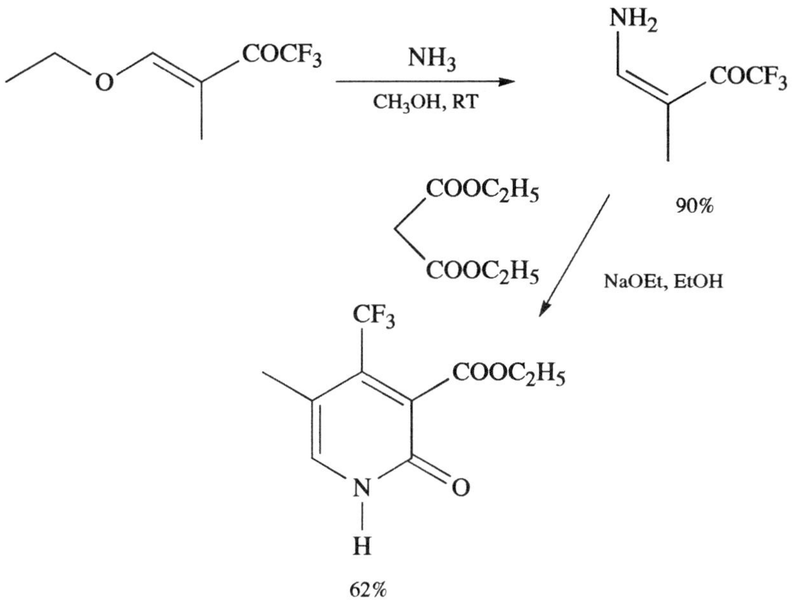

$CH_3CH_2O-CH=C(CH_3)COCF_3$ $\xrightarrow[CH_3OH,\ RT]{NH_3}$ $H_2N-CH=C(CH_3)COCF_3$ 90%

$CH_2(COOC_2H_5)_2$, NaOEt, EtOH

CF₃ / COOC₂H₅ / N–H / O (3-ethoxycarbonyl-4-trifluoromethyl-5-methyl-2-pyridone) 62%

[*42*] **22**

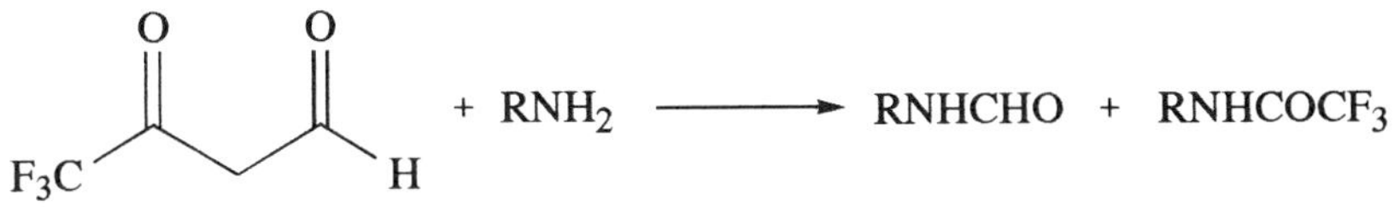

$F_3C-CO-CH_2-CHO$ + RNH_2 ⟶ RNHCHO + $RNHCOCF_3$

R	RNHCHO	:	$RNHCOCF_3$
R = *m*-FC_6H_4	38	:	62
R = $C_6H_5CH_2$	3	:	97
R = C_6H_{13}	11	:	89

[*44*] **23**

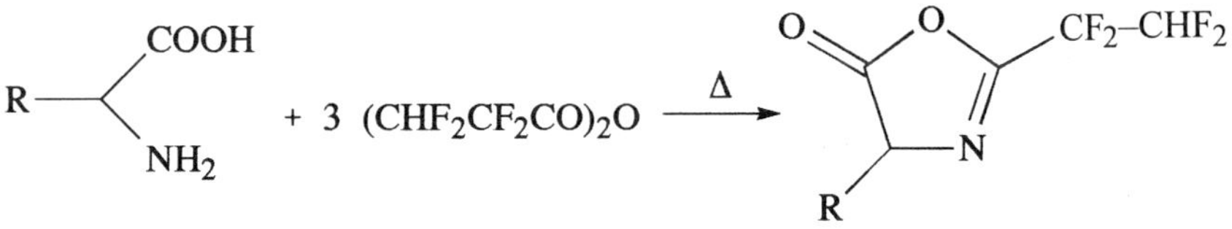

$R-CH(NH_2)COOH$ + 3 $(CHF_2CF_2CO)_2O$ $\xrightarrow{\Delta}$ 4-R-2-(CF_2–CHF_2)-oxazol-5(4H)-one

R = CH_3, C_2H_5, C_3H_7, C_4H_9, *i*-C_4H_9, C_6H_5–CH_2 59-96%

References are listed on pages 541–544.

[*49*] **24**

$$XZnCF_2COOR + A\text{-}CH{=}N\text{-}B \xrightarrow{THF}$$ (β-lactam with F, F, A, N–B, O)

X, R	Imine	Yield
X = I, R = CH_3	(dioxolane imine) NBn	71% (syn/anti = 4.3/1)
X = Br, R = C_2H_5	(dioxolane imine) NBn	67% (syn/anti = 4.7/1)
X = Br, R = C_2H_5	Ph–CH=N–Ph	87%
X = I, R = CH_3	(furyl)–CH=N–CH(CH₃)Ph	66%
X = Br, R = C_2H_5	Ph–CH=N–CH(CH₃)Ph	79%

Acylation at Carbon

Reference should be made to what has already been described in the first edition [*10*] for the well-established *Friedel–Crafts syntheses* and their potential to prepare in particular fluorinated acetophenones. For reactive aromatic, carbocyclic, and heterocyclic nuclei, **trifluoroacetic anhydride** is sufficiently reactive to cause *Friedel–Crafts acylation* without the aid of an external catalyst [*53*]. Less nucleophilic fluorinated benzenes, however, do need a Lewis acid catalyst, even in the case of highly reactive perfluorinated carboxylic acid derivatives [*54*]. More powerful and advantageous, easy-to-handle acylating agents such as **trifluoroacetyl triflate** (TFAT) [*14, 15*] and **2-trifluoroacetoxypyridine** (TFAP) [*16*] have been designed (equations 2 and 3). They can be used for *Friedel–Crafts*-type reactions [*14*] but more likely for acylations at oxygen, sulfur, and nitrogen. For similar synthetic applications, other useful reagents have recently been prepared from trifluoroacetic acid derivatives: the trifluoroacetimidoyl chlorides [*55*] and 2-trifluoromethyl-3-trimethylsilyl-1-propene [*56*] (equations 27 and 28).

[*51*] **25**

i. $BrZnCF_2COOEt$, THF

R_1 = H, alkyl, benzyl

30-61%

NH_3, ether

70-99%

ENZYME INHIBITOR

[*52*] **26**

$$CF_3COCl + (CH_3)_3SiNCO \xrightarrow[180°C,\ 130\ h]{[SnCl_4]} CF_3CONCO + ClSi(CH_3)_3$$

64%

The *regioselectivity of Friedel–Crafts*-type acylations on heteroaromatic compounds has been studied intensively [*57, 58, 59*]. In the case of pyrroles, the orientation of the entering acyl group strongly depends on the bulkiness of the group at the nitrogen atom (equation 29).

A special application of heteroaromatic acylation via a modified *Dakin–West reaction* leads to α-fluoro ketone derivatives [*50*] (equation 30). Such fluoro ketones have been successfully used as enzyme inhibitors in modern bioorganic chemistry [*5, 6*].

A more general way of preparing fluoro ketones proceeds via *acylation of organometallic species* such as organolithium and *Grignard* reagents [*60, 61, 62, 63*] (see the section on Organometallic Syntheses, pages 646, 670).

Trifluoroacetylation of Wittig-type ylides leads to different trifluoromethyl group-containing products, depending on the reaction partners and conditions [*64, 65, 66, 67*] (equations 31–33).

Another well-established process to generate fluoro ketones proceeds via *acylation of enolates* [*68, 69*] or activated methylene compounds [*70, 71*] as well as by *Claisen*-type condensation reactions [*72*]. Because of the electrophilic power of the acylating agents, there is usually no need for a catalyst [*68*].

[55] **27**

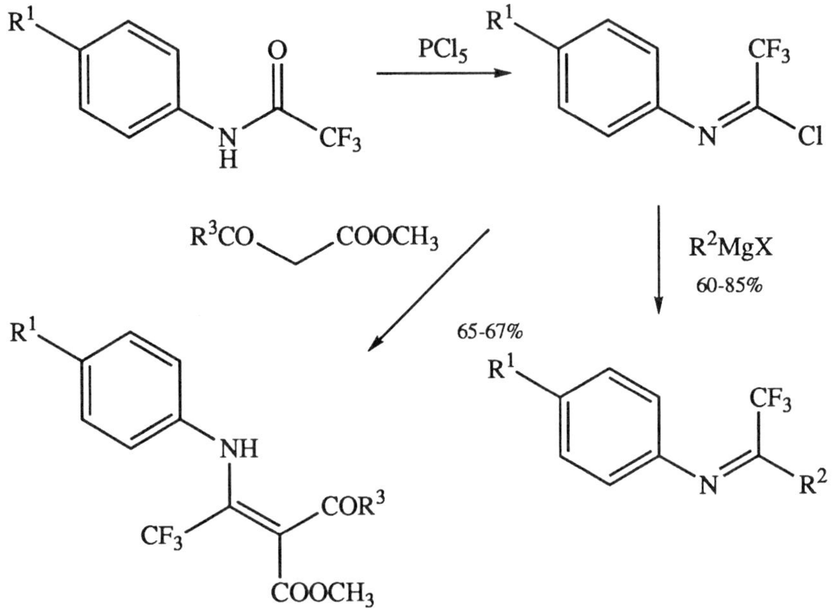

R^1-C$_6$H$_4$-NH-CO-CF$_3$ —(PCl_5)→ R^1-C$_6$H$_4$-N=C(CF$_3$)Cl

R^1-C$_6$H$_4$-N=C(CF$_3$)Cl —(R^2MgX, 60-85%)→ R^1-C$_6$H$_4$-N=C(CF$_3$)R^2

R^1-C$_6$H$_4$-N=C(CF$_3$)Cl —(R^3CO-CH$_2$-COOCH$_3$, 65-67%)→ R^1-C$_6$H$_4$-NH-C(CF$_3$)=C(COR^3)COOCH$_3$

$R^1 = R^3$ = MeO, Me
R^2 = alkyl, vinyl, allyl, benzyl
X = Cl, Br

[56] **28**

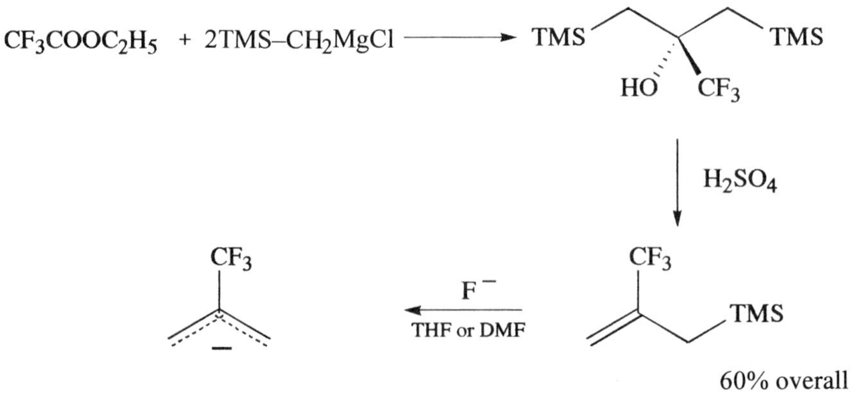

$CF_3COOC_2H_5$ + 2TMS–CH_2MgCl → TMS-CH$_2$-C(OH)(CF$_3$)-CH$_2$-TMS

—(H_2SO_4)→ CH$_2$=C(CF$_3$)-CH$_2$-TMS (60% overall)

—(F^-, THF or DMF)→ [CH$_2$=C(CF$_3$)-CH$_2$]$^-$

[57] **29**

$$\text{N-R pyrrole} \xrightarrow{(CF_3CO)_2O} \text{2-}COCF_3\text{ N-R pyrrole} + \text{3-}COCF_3\text{ N-R pyrrole}$$

R	Yield	2-isomer		3-isomer
R = H	86%	100	:	0
R = CH_3	87%	100	:	0
R = C_3H_7	78%	95	:	5
R = i-C_3H_7	80%	85	:	15
R = t-C_4H_9	79%	10	:	90
R = 1-adamantyl	98%	0	:	100
R = C_6H_5	70%	100	:	0

From the synthetic viewpoint, a particularly interesting trifluoroacetylation reaction of simple ***vinyl ethers*** was reported first by Hojo et al. in 1976 [*73*]. The scope and limitation of this particular reaction were elaborated intensively; the reaction proved to be of general applicability, with practically no restrictions on substituents of the vinyl ether moiety [*9*] (equation 34). This general validity is particularly beneficial because a trifluoroacetylated vinyl ether is the synthetic equivalent of a specifically protected trifluoromethyl-substituted 1,3-dicarbonyl compound [*9*]; thus the reaction provides access to a broad spectrum of variously substituted synthetic building blocks with selective reactivities on each carbon acceptor (a) and donor (d) center (equation 35). Obviously, such building blocks can react as heterodiene systems in cycloaddition reactions [*8, 74*] or can be treated with a wide variety of 1,2- or 1,3-dinucleophilic species to give any desired trifluoromethyl-substituted carbocyclic or heterocyclic system [*8, 75*]. Treatment of simple ***vinyl ethers*** with an excess of **trifluoroacetic anhydride** at elevated temperature leads to doubly acylated products [*76*]. Comparable acylation reactions occur with ***vinyl thioethers*** [*73*], and the mesoionic 1,3-oxathiol-4-ones show, at least in a formal sense, similar behavior [*77*] (equation 36).

[*50*] **30**

Ph
O
N
O
Ph
$(CH_2{=}CHCH_2CF_2CO)_2O$
60°C, 20 h
$CF_2CH_2CH{=}CH_2$
O
O
Ph
N
O
Ph
$(COOH)_2$
O
Ph
$CF_2CH_2CH{=}CH_2$
HN
O
Ph
60% overall

[*66*] **31**

PPh_3
R
$(CF_3CO)_2O$
F_3C
O
$\overset{+}{P}Ph_3$
R
17-71%
Δ
CF_3
R
40-85%
R = H, CH_3, CH_3O, Cl, NO_2

[*65*] **32**

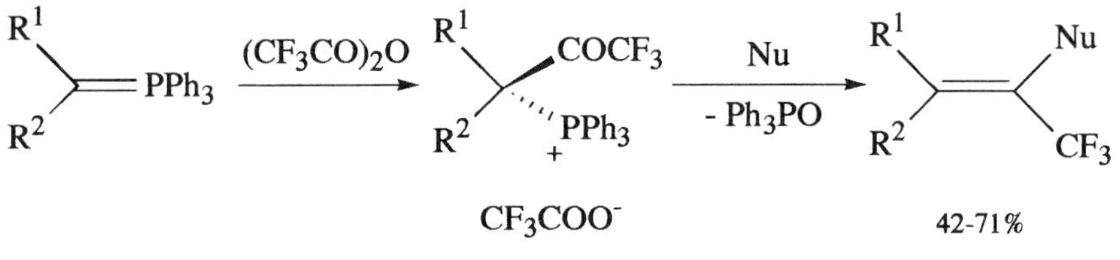

$R^1R^2C{=}PPh_3$ $\xrightarrow{(CF_3CO)_2O}$ $R^1R^2C(COCF_3)\overset{+}{P}Ph_3$ CF_3COO^- $\xrightarrow[-\ Ph_3PO]{Nu}$ $R^1R^2C{=}C(Nu)CF_3$

42-71%

R^1, R^2 = alkyl, allyl, benzyl; Nu = alkyl, phenyl

[*64*] **33**

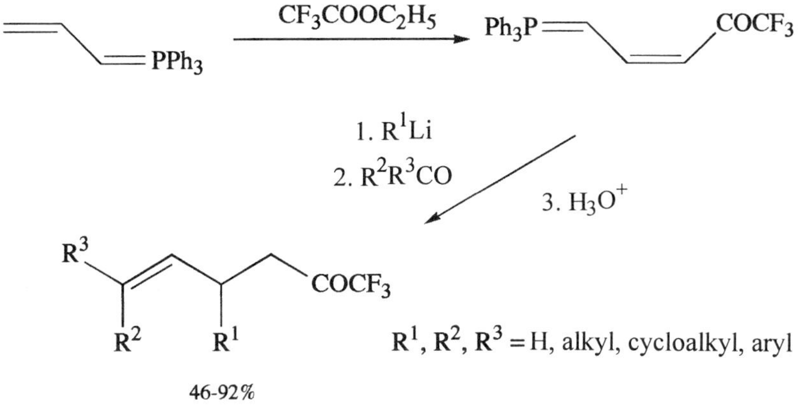

$CH_2{=}CH{-}CH{=}PPh_3$ $\xrightarrow{CF_3COOC_2H_5}$ $Ph_3P{=}CH{-}CH{=}CH{-}COCF_3$

1. R^1Li
2. R^2R^3CO
3. H_3O^+

$R^3R^2C{=}CH{-}CH(R^1){-}CH_2{-}COCF_3$

46-92%

R^1, R^2, R^3 = H, alkyl, cycloalkyl, aryl

[*9*] **34**

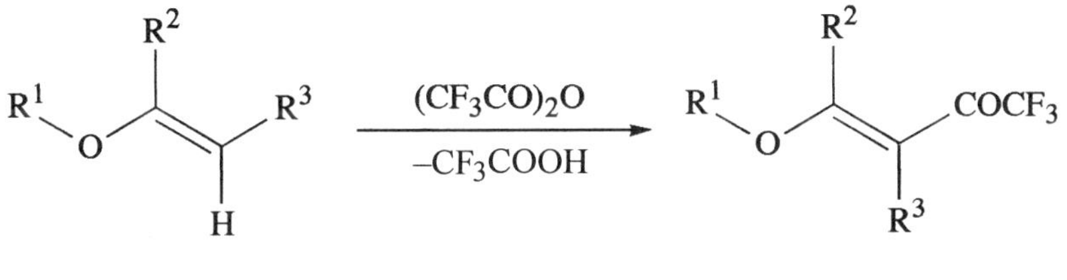

$R^1O{-}C(R^2){=}CH{-}R^3$ $\xrightarrow[-CF_3COOH]{(CF_3CO)_2O}$ $R^1O{-}C(R^2){=}C(R^3){-}COCF_3$

R^1, R^2, R^3 = alkyl, aryl, heterocyclyl

50-95%

References are listed on pages 541–544.

[*9*] **35**

Synthetic Equivalent

RO–C(CH$_3$)=CH–CO–CF_3 ≅ CH_3–CO–CH_2–CO–CF_3 (d a d a)

[*77*] **36**

HOOC–CH_2–O–C(=S)–SR $\xrightarrow{(CF_3CO)_2O}$ [mesoionic 1,3-oxathiolium-4-olate, 2-SR] → 5-$COCF_3$ derivative (F_3CCO, SR)

$R = CH_3, C_2H_5, i\text{-}C_3H_7, t\text{-}C_4H_9, CH_2CH{=}CH_2$

Fluorocylation of enamines and enamides has been intensively studied by different groups [*78, 79, 80, 81*]. The effectiveness of this particular electrophilic substitution reaction becomes obvious when the nitrogen atom of the enamine moiety is engaged in an aromatic system [*82, 83*] or when the olefinic system is part of an aromatic nucleus [*84*] (equations 37 and 38). A further extension of this reaction is demonstrated by the trifluoracetylation of aldehyde dialkyl hydrazones [*85, 86*] (equation 39).

[*83*] **37**

N-(CH=CHR)carbazole $\xrightarrow[CH_2Cl_2,\ -10^\circ C]{(CF_3CO)_2O,\ Py}$ N-(CH=C(R)$COCF_3$)carbazole

$R = H, CH_3, C_2H_5$ 51-96%

[*84*] **38**

$$\text{4-R-C}_6\text{H}_4\text{-N(cyclic X)} \xrightarrow[\text{THF, RT}]{(CF_3CO)_2O} \text{5-R-2-(N-cyclic X)-C}_6\text{H}_3\text{COCF}_3$$

>90%

R = CH_3, OCH_3 X = $(CH_2)_2$, $(CH_2)_3$, CH=CH

[*85*] **39**

$$(CH_3)_2N\text{–}N\text{=CH–Ar} \xrightarrow[\text{2,6-lutidine}]{(CF_3CO)_2O} (CH_3)_2N\text{–}N\text{=C(COCF}_3\text{)–Ar}$$

39-71%

Ar = C_6H_5, *p*-$CH_3C_6H_4$, *p*-$CH_3OC_6H_4$, *p*-ClC_6H_4, *p*-$NO_2C_6H_4$

References for Pages 525–541

1. Liebig, H.; Ulm K. *Chem. Ztg.,* **1976,** *100,* 3.
2. Filler, R.; Kobayashi, Y., Eds.; *Biomedicinal Aspects of Fluorine Chemistry*; Elsevier Biomedical Press: Amsterdam, 1982.
3. Walsh, C. In *Advances in Enzymology*; Meister, A., Ed.; Wiley: New York, 1983; Vol. 55, p. 197.
4. Ojima, I. *L'Actualite Chimique* **1987,** 171.
5. Imperiali, B.; Abeles, R. H. *Biochemistry* **1986,** *25,* 3760.
6. Welch, J. T. *Tetrahedron* **1987,** *43,* 3123.
7. Fuchikami, T. *J. Synth. Org. Chem. Jpn.* **1984,** *42,* 775.
8. Differding, E.; Frick, W.; Lang, R. W.; Martin, P.; Schmidt, C.; Veenstra, S.; Greuter, H. *Bull. Soc. Chim. Belg.* **1990,** *99,* 647.
9. Lang, R. W. *Drug News & Perspectives* **1991,** *4,* 13.
10. Hudlicky, M. *Chemistry of Organic Fluorine Compounds,* Ellis Horwood: Chichester, U.K., 1976, p. 325.
11. Yamaguchi, S. In *Asymmetric Synthesis;* Morrison, J. D., Ed.; Academic: New York, 1983; Vol. 1, p. 128.
12. Kalyanam, N.; Lightner, D. A. *Tetrahedron Lett.* **1979,** 415.
13. Anisimova, N. A.; Belavin, I. Y.; Baukov, Y. I. *Zh. Obshch. Khim.* **1989,** *50,* 2388; *Chem. Abitr.* **1981,** *94,* 103711c.

14. Forbus, T. R., Jr.; Martin, J. C. *J. Org. Chem.* **1979,** *44,* 313.
15. Forbus, T. R., Jr.; Taylor S. L.; Martin J. C. *J. Org. Chem.* **1987,** *52,* 4156.
16. Keumi, T.; Shimada, M.; Morita, T.; Kitajima, H. *Bull. Chem. Soc. Jpn.* **1990,** *63,* 2252.
17. Kawahara, T.; Masahiro, T. Jpn. Kokai Tokkyo Koho JP 62 153 257, 1987; *Chem. Abstr.* **1987,** *108,* 13106n.
18. Yamabe, M.; Munekata, S.; Sugaya, Y.; Jitsugiri, Y. Ger. Offen. 2 651 531, 1977; *Chem. Abstr.* **1977,** *87,* 52808k.
19. Hagen, A. P.; Miller, T. S.; Bynn, R. L.; Kapila, V. P. *J. Org. Chem.* **1982,** *47,* 1345.
20. Johnston, B. H.; Knipe, A. C.; Watts, W. E. *Tetrahedron Lett.* **1979,** 4235.
21. Andreev, S. M.; Pavlova, L. A.; Davidovich, Y. A.; Rogozhin, S. V. *Izv. Akad. Nauk SSSR* **1980,** 1078; *Chem. Abstr.* **1980,** *93,* 167580w.
22. Bassett, N. J.; Ismail, G. H.; Piotis, P.; Tittle, B. *J. Fluorine Chem.* 1976, *8, 89.*
23. Tojo, M.; Fukouka, S. Jpn. Kokai Tokkyo Koho JP 63 41 443, 1988; *Chem. Abstr.* **1988,** *109,* 230380x.
24. Kazarayan, P. I;, Avakyan, S. V.; Simonyan, E. S.; Gevorkyan A. A. *Kim. Geterotsikl. Soedin.* **1990,** 174; *Chem. Abstr.* **1990,** *113,* 39630g.
25. Greuter, H.; Lang, R. W.; Romann, A. J. *Tetrahedron Lett.* **1988,** *29,* 3291.
26. Welch, J. T.; Samartino, J. S. *J. Org. Chem.* **1985,** *50,* 3663.
27. Yokozawa, T.; Nakai, T.; Ishikawa, N. *Tetrahedron Lett.* **1984,** *25,* 3991.
28. Kitazume, T. *Synthesis* **1986,** 855.
29. Hertel, L. W.; Kronin, J. S.; Misner, J. W.; Tustin, J. M. *J. Org. Chem.* **1988,** *53,* 2406.
30. Burton, C. A.; Shreeve, J. M. *J. Am. Chem. Soc.* **1976,** *98,* 6545.
31. Svirskaya, P. I.; Leznoff, C. C.; Steinman, M. *J. Org. Chem.* **1987,** *52,* 1362.
32. Donike, M. *J. Chromatogr.* **1973,** *78,* 273.
33. Barrett, A. G. M. *J. Chem. Soc., Perkin Trans. 1* **1979,** 1629.
34. Medvedeva, E. N.; Kalikhman, I. D.; Chipanina, N. N.; Yushmanova, T. I.; Lopyrev, V. A. *Izv. Akad. Nauk SSSR* **1982,** 1440; *Chem. Abstr.* **1982,** *97,* 126999d.
35. Curphey, T. J. *J. Org. Chem.* **1979,** *44,* 2805.
36. Fokin, A. V.; Pospelov, M. V.; Ovchinnikova, N. S.; Gus'kova, O. V.; Bocharov, B. V.; Galakhov, M. V.; Glazunov, M. P.; Zhuravlev, L. T. *Izv. Akad. Nauk SSSR* **1981,** 863; *Chem. Abstr.* **1981,** *95,* 800091e.
37. Arshady, R. *Chem. Ind. (London)* **1982,** 370.
38. Kita, M.; Yamada, Y.; Tatematsu, R. *Yukagaku* **1983,** *32,* 175; *Chem. Abstr.* **1983,** *99,* 53086h.
39. Kita, M.; Yamada, Y.; Tatematsu, R.; Ozaki, S. *Yukagaku* **1980,** *29,* 514; *Chem. Abstr.* **1980,** *93,* 238805.
40. Taguchi, I.; Saito, S.; Kanai, T.; Kawada, K.; Kobayashi, Y.; Okada, M.; Ohta, K. *Chem. Pharm. Bull.* **1985,** *33,* 4026.

41. Pazenkov, S. V.; Gerus, I. I.; Gorbunova, M. G.; Chaika, E. A. *Zh. Org. Khim.* **1989,** *25,* 1560; *Chem. Abstr.* **1990,** *112,* 157629r.
42. Giumanini, A. G.; Verardo, G. *Zh. Org. Khim.* **1989,** *25,* 650; *Chem. Abstr.* **1989,** *111,* 153313q.
43. Tolmacheva, G. M.; Krikovskii, S. P.; Ignatenko, A. V.; Ponomarenko, V. A. *Izv. Akad. Nauk SSSR* **1979,** 580; *Chem. Abstr.* **1979,** *91,* 4768h.
44. Kayahara, H.; Tomida, I. *Shinshu Daigaku Nogakubu Kiyo* **1974,** *11,* 87; *Chem. Abstr.* **1976,** *81,* 1697930.
45. Eapen, K. C.; Tamborski, C. *J. Fluorine Chem.* **1978,** *12,* 271.
46. Joshi, C.; Chand, P. *Heterocycles* **1981,** *16,* 43.
47. Eapen, K. C.; Tamborski, C. *J. Fluorine Chem.* **1981,** *18,* 243.
48. Kamitori, Y.; Hojo, M.; Masuda, R.; Yoshida, T.; Ohara, S.; Yamada, K.; Yoshikawa, N. *J. Org. Chem.* **1988,** *53,* 519.
49. Taguchi, T.; Kitagawa, O.; Suda, Y.; Ohkawa, S.; Hashimoto, A.; Litaka, Y.; Kobayashi, Y. *Tetrahedron Lett.* **1988,** *29,* 5291.
50. Kolb, M.; Neises, B. *Tetrahedron Lett.* **1986,** *27,* 4437.
51. Schirlin, D.; Baltzer, S.; Altenburger, J. M. *Tetrahedron Lett.* **1988,** *29,* 3687.
52. Kiemstedt, W.; Sundermeyer, W. *Chem. Ber.* **1982,** *115,* 919.
53. Mackie, R. K.; Mhatre, S.; Tedder, J. M. *J. Fluorine Chem.* **1977,** *10,* 437.
54. Furin, G. G.; Yakobson, G. G. *Izv. Sib. Otd. Akad. Nauk SSSR* **1974,** 78; *Chem. Abstr.* **1974,** *80,* 120475c.
55. Uneyama, K.; Morimoto, O.; Yamashita, F. *Tetrahedron Lett.* **1989,** *30,* 4821.
56. Yamazaki, T.; Ishikawa, N. *Chem. Lett.* **1984,** 521.
57. Chadwick, D. J.; Meakins, G. D.; Rhodes, C. A. *J. Chem. Res., Synop.* **1980,** 42.
58. Glukhovtsev, V. G.; Il'in, Y. V.; Ignatenko, A. V.; Brezhnev, L. Y. *Izv. Akad. Nauk SSSR* **1988,** 2361; *Chem. Abstr.* **1988,** *109,* 128733c.
59. Kost, A. N.; Budylin, V. A.; Romanova, N. N.; Matveeva, E. D. *Khim. Geterotsikl. Soedin.* **1981,** 1233; *Chem. Abstr.* **1981,** *95,* 219939h.
60. Rozen, S.; Filler, R. *Tetrahedron* **1985,** *41,* 1111.
61. Creary, X. *J. Org. Chem.* **1987,** *52,* 5026.
62. Shaw, D. A.; Tuominen, T. C. *Synth. Commun.* **1987,** *15,* 1291.
63. Schaub, B. Eur. Pat. Appl. 289 478, 1989; *Chem. Abstr.* **1989,** *111,* 39005r.
64. Shen, Y.; Wang, T. *Tetrahedron Lett.* **1990,** *31,* 3161.
65. Shen, Y.; Qiu, W. *Tetrahedron Lett.* **1987,** *28,* 449.
66. Kobayashi, Y.; Yamashita, T.; Takahashi, K.; Kuroda, H.; Kumadaki, I. *Tetrahedron Lett.* **1982,** *23,* 343.
67. Kobayashi, Y.; Yamashita, T.; Takahashi, K.; Kuroda, H.; Kumadaki, I. *Chem. Pharm. Bull.* **1984,** *32,* 4402.
68. Fokin, A. V.; Uzun, A. T. *Izv. Akad. Nauk SSSR* **1973,** 2294; *Chem. Abstr.* **1974,** *80,* 36665m.
69. Ermolov, A. F.; Totskii, S. A. *Zh. Org. Khim.* **1986,** *22,* 441; *Chem. Abstr.* **1986,** *105,* 78499a.

70. Pashkevich, K. I.; Saloutin, V .I.; Krokhalev, V. M. *Zh. Org. Khim.* **1988,** *24,* 1405; *Chem. Abstr.* **1989,** *110,* 172665b.
71. Ermolov, A. F.; Eleev, A. F.; Benda, A. F.; Sokol'skii, G. A. *Zh. Org. Khim.* **1987,** *23,* 93; *Chem. Abstr.* **1987,** *107,* 197496y.
72. Bizunov, Y. M.; Emel'yanov, V. I.; Knunyants, I. L. *Izv. Akad. Nauk SSSR* **1986,** 1688; *Chem. Abstr.* **1987,** *106,* 138197h.
73. Hojo, M.; Masuda, R.; Kokuryo, Y.; Shioda, H.; Matsuo, S. *Chem. Lett.* **1976,** 499.
74. Hojo, M.; Masuda, R.; Okada, E. *Synthesis* **1989,** 215.
75. Lang, R. W.; Wenk, P. F. *Helv. Chim. Acta* **1988,** *71,* 596.
76. Hojo, M.; Masuda, R.; Okada, E. *Synthesis* **1990,** 347.
77. Gotthardt, H.; Feist, U.; Schoy-Tribbenseem, S. *Chem. Ber.* **1985,** *118,* 774.
78. Bailey, A. S.; Vandrevala, M. H. *Tetrahedron Lett.* **1979,** 4407.
79. Bailey, A. S.; Haxby, J. B.; Hilton, A. N.; Peach, J. M.; Vandrevala, M. H. *J. Chem. Soc., Perkin Trans. 1* **1981,** 382.
80. Hojo, M.; Masuda, R.; Yoshinaga, K.; Munehira, S. *Synthesis* **1982,** 312.
81. Verboom, W.; Reinhoudt, D. N. *J. Org. Chem.* **1982,** *47,* 3339.
82. Moskalev, N. V. *Zh. Org. Khim.* **1989,** *25,* 437; *Chem. Abstr.* **1989,** *111,* 194194v.
83. Moskalev, N. V.; Filimonov, V. D.; Sirotkina, E. E. *Khim. Geterotsikl. Soedin.* **1988,** 1066; *Chem. Abstr.* **1989,** *110,* 192588f.
84. Verboom, W.; van Dijk, B. G.; Reinhoudt, D. N. *Tetrahedron Lett.* **1983,** *24,* 3923.
85. Kamitori, Y.; Hojo, M.; Masuda, R.; Kawamura, Y.; Kawasaki, K.; Ida, J. *Tetrahedron Lett.* **1990,** *31,* 1183.
86. Kamitori, Y.; Hojo, M.; Masuda, R.; Kawamura, Y.; Numai, T. *Synthesis* **1990,** 491.

Fluorinated Sulfur, Phosphorus, Boron, and Silicon Compounds

by J. T. Welch and Seniz Turner-McMullin

Fluorinated Sulfur-Containing Compounds

Sulfides

Preparation of fluorine-containing *sulfides* has been achieved mainly by the reaction of **sulfenyl chlorides** (RSCl), which may react with a variety of nucleophiles. The method is based on the electrophilicity of sulfur and on the leaving-group ability of chlorine.

Trifluoromethanesulfenyl chloride reacts with the indolizine ring to introduce trifluoromethylsulfenyl groups at both positions 1 and 3. An acetyl group at position 3 can be replaced by a trifluoromethanesulfenyl group, whereas the 3-benzoyl and 3-nitroso substituents are unaffected [*1*] (equation 1) (Table 1).

[1] (3-acetyl-2-methylindolizine: CH_3, $COCH_3$) $\xrightarrow{F_3CSCl}$ (1,3-bis(trifluoromethylthio)-2-methylindolizine: SCF_3, CH_3, SCF_3) + HCl + CH_3COCl **1**

One of the methods of introducing the RS group into organic molecules is the reaction of **sulfenyl chlorides** with ***compounds containing active methylene and methine groups.*** The extensive studies reported on the α-sulfenylation of various enolates rely upon the reaction of the electrophilic sulfenyl chlorides with carbon nucleophiles such as enols, enolates, or enamines [*2, 3, 4, 5, 6, 7*]. The acidic α-hydrogen of α,α-bis(trifluoromethyl)methyl pentafluoroethyl ketone reacts with sulfenyl chlorides in the presence of triethylamine to give the corresponding sulfide derivative [*2, 3*] (equation 2).

The reactions of β-*keto acid derivatives* with trifluoromethylsulfenyl chloride give the α-trifluoromethanesulfenyl substitution products [*4*]. The products can be treated with a dimethyl sulfoxide–water solution to form trifluoromethylthioketones or with potassium hydroxide solution to give trifluoromethylthioacetic acid (equation 3) (Table 2).

0065–7719/95/0187–0545$17.00/1

Table 1. Syntheses of Sulfides

Reactants	*Reaction Conditions*	*Product(s)*	*Yield (%)*	*Ref.*
N, R, X	RT, $(C_2H_5)_2O$	SCF_3, N, R, SCF_3		1
CF_3SCl				
R = H, X = H			~100	1
R = CH_3, X = H			~100	1
R = C_6H_5, X= H			~100	1
R = CH_3, X = $COCH_3$				1
R = C_6H_5, X = $COCH_3$			~100	1
R = C_6H_5, X = COC_6H_5			NR	1
R = C_6H_5, X = NO			NR	1
R = CH_3, X = $CH_2C_6H_5$		SCR_3, N, CH_3, $CH_2C_6H_4SCR_3$		1
$(CF_3)_2CHCOX$ RSCl	Et_3N	$(CF_3)_2C(COX)(SR)$		
		X = F, R = C_2H_5	67	3
		X = OCH_3, R = C_6H_5	90	3
		X = OCH_3, R = $N(C_2H_5)_2$	82	3
		X = C_2F_5, R = C_6H_5	85	3
$(CF_3)_2CHCN$ C_6H_5SCl		$(CF_3)_2C(SC_6H_5)CN$	73	3
$CF_3CH(COOCH_3)_2$ C_2H_5SCl		$CF_3C(SC_2H_5)(COOCH_3)_2$	72	3

References are listed on pages 610–614.

Table 1—*Continued*

Reactants	*Reaction Conditions*	*Product(s)*	*Yield (%)*	*Ref.*
$(CF_3)_2C{=}C{=}NC_6H_5$ C_6H_5SCl	C_5H_5N	$(CF_3)_2C(Cl)(SC_6H_5)$-$C{=}NC_6H_5$ (Cl on C-1 of the chain, SC_6H_5 on C-2)	83	3
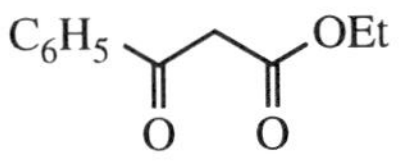 $C_6H_5C(O)CH_2C(O)OEt$ CF_3SCl	RT	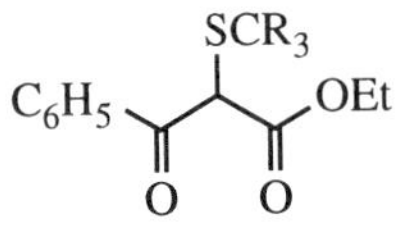 $C_6H_5C(O)CH(SCR_3)C(O)OEt$	59	4
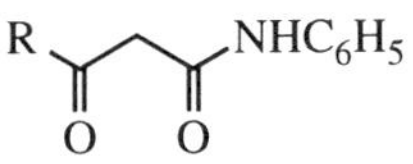 $RC(O)CH_2C(O)NHC_6H_5$ CF_3SCl	$CHCl_3$ RT	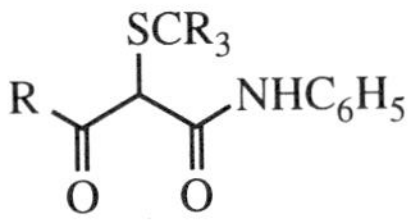 $RC(O)CH(SCR_3)C(O)NHC_6H_5$		
		R = CH_3	75	4
		R = C_6H_5	72	4
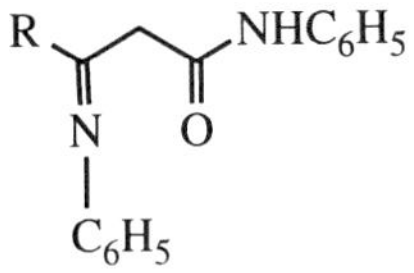 $RC(=NC_6H_5)CH_2C(O)NHC_6H_5$ CF_3SCl	$CHCl_3$ RT	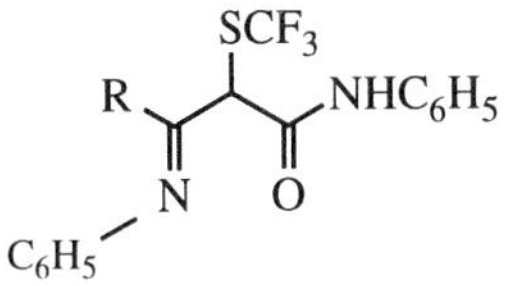 $RC(=NC_6H_5)CH(SCF_3)C(O)NHC_6H_5$		
		R = CH_3	62	4
		R = C_6H_5	83	4
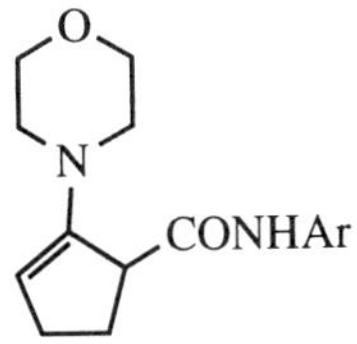 1-morpholino-5-(CONHAr)cyclopentene CF_3SCl	Toluene Pyridine 0 °C	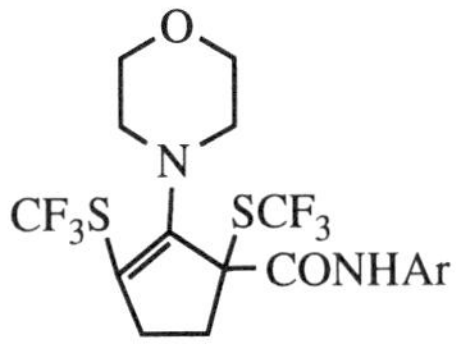 1-morpholino-2-SCF_3-5-SCF_3-5-CONHAr-cyclopentene (CF_3S, SCF_3, CONHAr)		
		Ar = C_6H_5	26	5
		Ar = p-ClC_6H_4	28	5

Continued on next page.

Table 1—*Continued*

Reactants	*Reaction Conditions*	*Product(s)*	*Yield (%)*	*Ref.*
2-morpholino-cyclohexyl-CONHAr	Toluene Pyridine 0 °C	CF_3S, morpholino-N, CONHAr cyclohexane		
CF_3SCl		Ar = C_6H_5	21	5
		Ar = p-ClC_6H_4	26	5
2-morpholino-cycloheptyl-CONHAr	Toluene Pyridine 0 °C	CF_3S, O, CONHAr cycloheptanone		
CF_3SCl		Ar = C_6H_5	45	5
		Ar = p-ClC_6H_4	27	5
C_6H_5(morpholino-N)C=CHCONHAr	Toluene Pyridine 0 °C	C_6H_5(morpholino-N)C=C(SCF_3)CONHAr		
CF_3SCl		Ar = C_6H_5	50	5
		Ar = p-ClC_6H_4	54	5
acetone (O)		F, F, F, F, S, S, O (2-acetyl-4,4,5,5-tetrafluoro-1,3-dithiolane)	95	6
$ClSCF_2CF_2SCl$				

Table 1—*Continued*

Reactants	*Reaction Conditions*	*Product(s)*	*Yield (%)*	*Ref.*
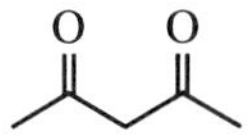 $ClSCF_2CF_2SCl$		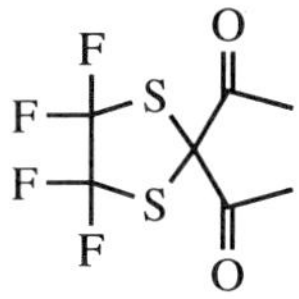	90	6
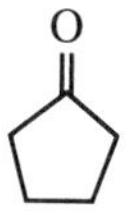 $ClSCF_2CF_2SCl$		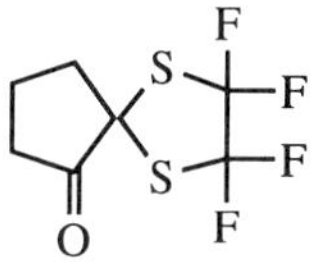	95	6
$(CH_3)_3CCOCH_3$	$ClSCF_2CF_2SCl$	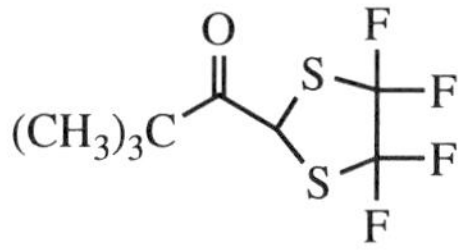 $(CH_3)_3C$	85	7
	$ClSCF_2CF_2SCl$	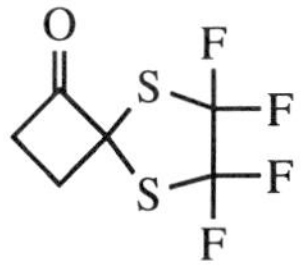	80	7
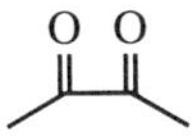	$ClSCF_2CF_2SCl$	 SCF_2CF_2SCl	55	7
 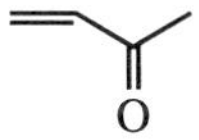	$ClSCF_2CF_2SCl$	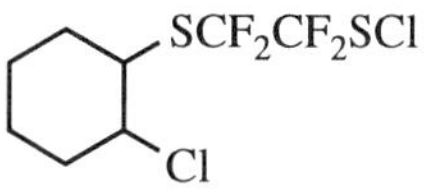 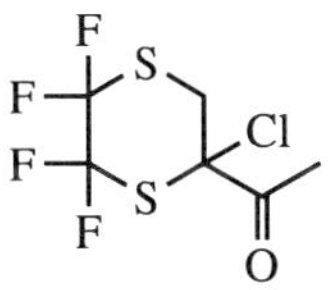 SCF_2CF_2SCl, Cl; Cl	75	7

Continued on next page.

Table 1—*Continued*

Reactants	*Reaction Conditions*	*Product(s)*	*Yield (%)*	*Ref.*
$(C_2H_5)_3N$ CCl_2FSCl	Benzene 20 °C	$(CCl_2FS)_2C{=}CH{-}N(C_2H_5)_2$	26	9
$CH_2{=}CHCH_2OH$ Cl_2FSCl	CH_2Cl_2 C_5H_5N –78 °C	$Cl_2FSO{-}CH_2CH{=}CH_2$ ⇌ $Cl_2FS(O){-}CH_2CH{=}CH_2$	79	11
$CH_2{=}CHCH_2OH$ ClF_2SCl	CH_2Cl_2 C_5H_5N –5 °C	$ClF_2SO{-}CH_2CH{=}CH_2$ ⇌ $ClF_2S(O){-}CH_2CH{=}CH_2$	48	11
$CH_2{=}CHCH_2OH$ $(CF_3)_2CClSCl$	CH_2Cl_2 C_5H_5N –78 °C	$(CF_3)_2CClS(O){-}CH_2CH{=}CH_2$ ⇌ $(CF_3)_2CClSO{-}CH_2CH{=}CH_2$	77	11
$(CF_3)_2CClS(O){-}CH_2CH{=}CH_2$	$KHSO_5$ CH_3OH H_2O 0 °C	$(CF_3)_2CClSO_2{-}CH_2CH{=}CH_2$	31	11

References are listed on pages 610–614.

Table 1—*Continued*

Reactants	*Reaction Conditions*	*Product(s)*	*Yield (%)*	*Ref.*
$(CF_3)_2CClS(O)$–$CH_2CH=CH_2$ 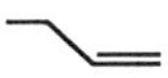	400 °C 1.0 torr	$(CF_3)_2C=SO$		11
$(CF_3)_2CH\,SO_2$–$CH_2CH=CH_2$ 	570 °C $-SO_2$ 0.01 torr	$(CF_3)_2CHCH_2CH=CH_2$	40	11
	SiO_2 CH_2Cl_2	$CF_3CH_2SO_2CH_2CH=CH_2$	53	11
$(CF_3)_2CCl\,SO_2$–$CH_2CH=CH_2$ 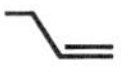	550 °C $-SO_2$ 0.01 torr	$(CF_3)_2CCl$–$CH_2CH=CH_2$ 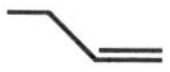	81	11

[2] **2**

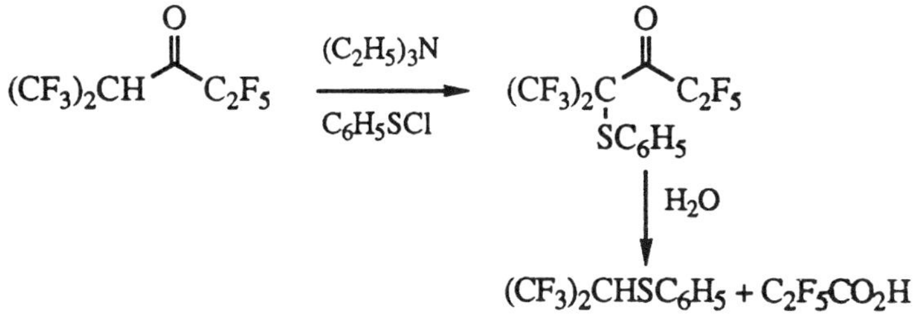

$(CF_3)_2CHC(O)C_2F_5$ $\xrightarrow[C_6H_5SCl]{(C_2H_5)_3N}$ $(CF_3)_2C(SC_6H_5)C(O)C_2F_5$ $\xrightarrow{H_2O}$ $(CF_3)_2CHSC_6H_5 + C_2F_5CO_2H$

[4] **3**

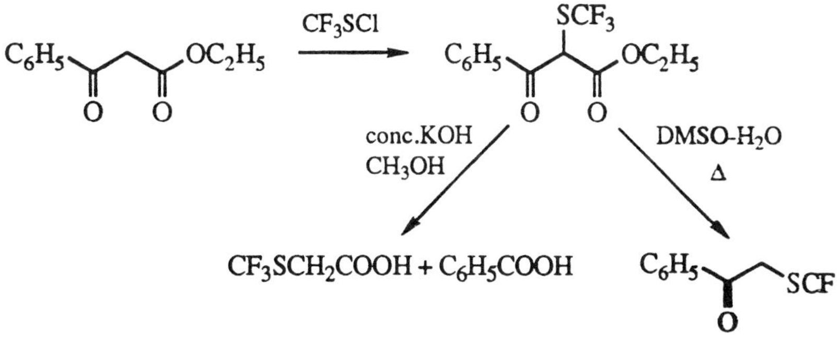

$C_6H_5C(O)CH_2C(O)OC_2H_5$ $\xrightarrow{CF_3SCl}$ $C_6H_5C(O)CH(SCF_3)C(O)OC_2H_5$

conc.KOH, CH_3OH → $CF_3SCH_2COOH + C_6H_5COOH$

DMSO-H_2O, Δ → $C_6H_5C(O)CH_2SCF$

References are listed on pages 610–614.

Table 2. Reactions of Sulfides

Reactants	Reaction Conditions	Product(s)	Yield (%)	Ref.
$(CF_3)_2C(SR)$-COX	H_2O, CO_2	$(CF_3)_2CHSR$	66	3
		R = C_6H_5 X = C_2H_5		
$(CF_3)_2C(SC_2H_5)C(=O)C_6H_5$	F^- CH_3CN	$CF_2=C(SC_2H_5)CF_3$ + C_6H_5COF	58	3
$(CF_3)_2=C(SC_2H_5)CF_3$	CsF CH_3CN	$(CF_3)_2C=C=C(SC_2H_5)CF_3$	3	3
		+ $(CF_3)_2C(SC_2H_5)_2$		
		+ $(CF_3)_2C(SC_2H_5)$-CF=$C(SC_2H_5)$-CF_3		
$RC(=O)CH(SCR_3)C(=O)OC_2H_5$	DMSO H_2O, Δ	$RC(=O)CH_2CH_2SCR_3$		
		R = CH_3	44	4
		R = C_6H_5	66	4
$RC(=O)CH(SCF_3)C(=O)OC_2H_5$	Conc. KOH CH_3OH RT	CF_3SCH_2COOH		
		R = C_6H_5	60	4
		R = CH_3	89	4
1-morpholino-2-(CF_3S)-6-(CONHAr)cyclohexane	Conc. HCl	2-(CF_3S)-7-(CONHAr)cycloheptanone		
		Ar = C_6H_5	74	5
		Ar = *p*-ClC_6H_4	82	5

Table 2—*Continued*

Reactants	*Reaction Conditions*	*Product(s)*	*Yield (%)*	*Ref.*
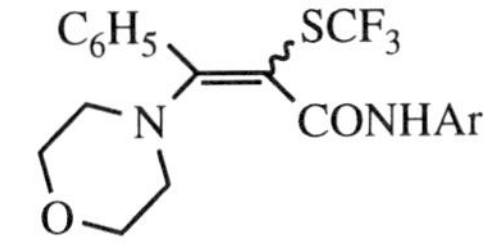 C_6H_5(morpholino)C=C(SCF_3)CONHAr	Conc. HCl, Δ	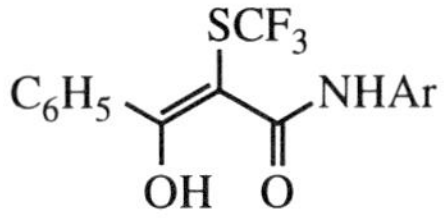 C_6H_5C(OH)=C(SCF_3)C(O)NHAr		
		Ar = C_6H_5	86	5
		Ar = p-ClC_6H_5	81	5
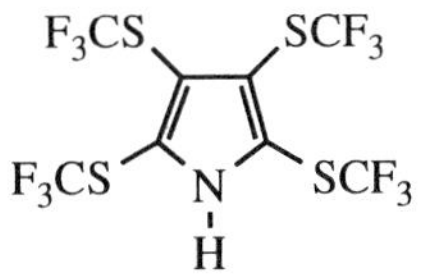 2,3,4,5-tetrakis(SCF_3)pyrrole (N–H)		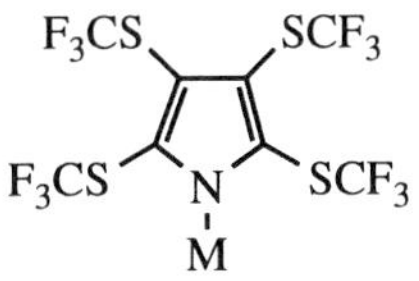 2,3,4,5-tetrakis(SCF_3)pyrrole (N–M)		11
	KOH	M = K		
	NaOH/Δ	M = Na	88	
	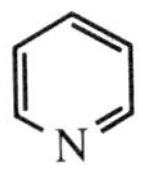 pyridine	M = C_5H_5N	82	
	$(CH_3)_3N$	M = $(CH_3)_3NH$	92	
	NH_3	M = NH_4		
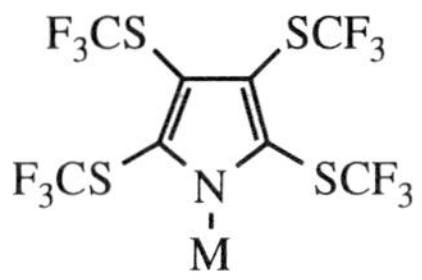 2,3,4,5-tetrakis(SCF_3)pyrrole (N–M)	20–25 °C	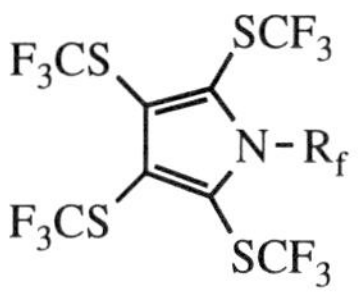 2,3,4,5-tetrakis(SCF_3)pyrrole (N–R_f)		11
R_fCl		R_f = CF_3S, M = Ag	85	
		R_f = CF_2ClS, M = Ag	66	
		R_f = $CFCl_2S$, M = Ag	85	
		R_f = CCl_3S, M = Ag	87	
		R_f = $(CF_3S)_2CClS$, M = K	73	
		R_f = $ClC(O)S$, M = Ag	40	
		R_f = CF_3Se, M = Ag	90	

Continued on next page.

Table 2—*Continued*

Reactants	*Reaction Conditions*	*Product(s)*	*Yield (%)*	*Ref.*
2,3,4,5-tetrakis(SCF_3)-1-M-pyrrole	S_xCl_2	[2,3,4,5-(F_3CS)$_4$pyrrol-1-yl]–S_x–[pyrrol-1-yl-2,3,4,5-(SCF_3)$_4$]		
M = Ag, x = 2	C_5H_{10}		33	11
M = K, x = 1	$(C_2H_5)_2O$ 20 °C		15	11
$Cl_2HCS(O)CH_2CH=CH_2$	$KHSO_5$ CH_3OH H_2O 0 °C	$Cl_2FCSO_2CH_2CH=CH_2$	54	12
$ClCl_2CS(O)CH_2CH=CH_2$	$KHSO_5$ CH_3OH H_2O 0 °C	$ClF_2CSO_2CH_2CH=CH_2$	52	12
$CH_2=CHCH_2OH$ + CF_3SCl	CH_2Cl_2 C_5H_5N –5 °C	$CF_3SOCH_2CH=CH_2$ ⇅ $CF_3S(O)CH_2CH=CH_2$	37	12
$(CF_3)_2CHSCH_2CH=CH_2$	$KHSO_5$ CH_3OH H_2O, 0 °C (3 equiv.)	$(CF_3)_2CHS(O)CH_2CH=CH_2$ ⇅ $(CF_3)_2CHSOCH_2CH=CH_2$	52	12
$(CF_3)_2CClSCH_2CH=CH_2$	$KHSO_5$ CH_3OH H_2O, 0 °C (3 equiv.)	$(CF_3)_2CHSO_2CH_2CH=CH_2$	55	12

Similar sulfenylation reactions of the 2-substituted *cyclic enamines of* β-***keto carboxylic acid anilides*** are also possible. The trifluoromethanesulfenyl substitution takes place according to ring size: Sulfenylation occurs at positions 2 and 5 with five-membered rings, at position 6 with six-membered rings, and at position 7 with seven-membered rings [*5*] (equation 4) (Table 1). Acid hydrolysis of the enamines proceeds readily to form the corresponding keto compounds.

Tetrafluoro-1,2-ethanedisulfenyl dichloride reacts with ***ketones*** to form a heterocyclic ring and with alkenes to give addition products [*6, 7*] (equation 5) (Table 1).

[5] **4**

CF_3SCl; Toluene, Pyridine, 0 °C

$CONHC_6H_5$ → F_3CS, SCF_3, $CONHC_6H_5$

[7] **5**

$ClSCF_2CF_2SCl$

[6]

SCF_2CF_2SCl, Cl

The free-radical chemistry of fluoroalkanesulfenyl chlorides with ***hydrocarbons*** was also investigated [*8, 9*]. Depending upon the structures of the sulfenyl chloride and the hydrocarbon, these reactions yield as major products up to three of the following four types of organic compounds: *thiols, disulfides, sulfides, and chlorohydrocarbons* (equation 6). Perfluoroisobutanesulfenyl chloride is unique in that the only major products detected are the thiol and chlorohydrocarbon [*8*] (equation 6) (Table 3).

In the reaction of $CF_nCl_{3-n}SCl$ with triethylamine, the yield of the product $(CF_nCl_{3-n}S)_2C{=}CHN(C_2H_5)_2$ decreases with decreasing *n* [*10*] (equation 7) (Table 2). 2,3,4,5-Tetrakis(trifluoromethylthio)pyrrole salts react with sulfenyl chloride or S_xCl_2 ($x = 1{,}2$) to give N-sulfenylated pyrroles as well as dipyrrolylsulfane and -disulfene. **Pentakis(trifluoromethylthio)pyrrole** is a mild sulfenylating

[8] **6a**

$$R^1SCl + R^2H \xrightarrow{h\nu} R^1SH + R^1SSR^1 + R^1SR^2 + R^2Cl + HCl$$

6b

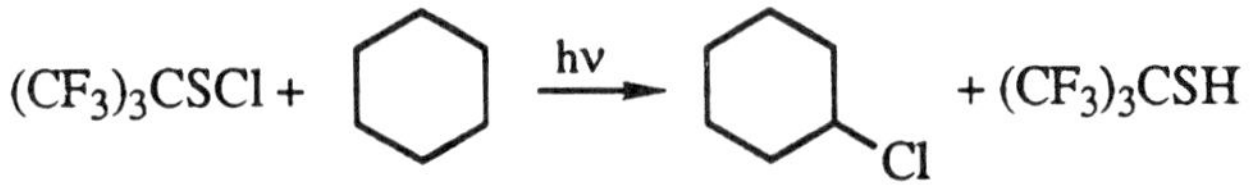

$$(CF_3)_3CSCl + \text{cyclohexane} \xrightarrow{h\nu} \text{chlorocyclohexane (Cl)} + (CF_3)_3CSH$$

Table 3. Chlorination by Sulfenyl Chlorides [*8*]

Reactants	*Reaction Condition*	*Product(s)*	*Yield (%)*
$(CF_3)_3CSCl$ + cyclo-C_6H_{12}	hν	cyclo-$C_6H_{11}Cl$ + $(CF_3)_3CSH$	100
Cl_2CFSCl + cyclo-C_6H_{12}	hν	cyclo-$C_6H_{11}Cl$	64
		$(Cl_2CF)_2S_2$	97
		Cl_2CFS-(cyclo-C_6H_{11})	3
$(CF_3)_3CSCl + CH_3C_6H_5$	hν	$CH_2ClC_6H_5$	100
		$(CF_3)_3CSH$	100
$(CF_3)_2CFSCl + CH_3C_6H_5$	hν	$CH_2ClC_6H_5$	100
		$(CF_3)_2CFSH + [(CF_3)_2CF]_2S_2$	96
$HCF_2CF_2SCl + CH_3C_6H_5$	hν	$(HCF_2CF_2)_2S_2$	13
		$HCF_2CF_2SCH_2C_6H_5$	87
		$CH_2ClC_6H_5$	19
$(CF_3)_3CSCl + C_4H_{10}$	hν	$(CF_3)_3CSH + CH_3CHClCH_2CH_3$	100

[10] **7**

$$(C_2H_5)_3N + CF_3SCl \longrightarrow (F_3CS)_2C{=}CH{-}N(C_2H_5)_2$$

agent; it exchanges the trifluoromethanesulfenyl group for acidic hydrogen atoms in alcohols, thioalcohols, and amines [*11*] (equation 8).

Sulfenyl chlorides react with ***allyl alcohols*** to yield allyl sulfenates, which are in equilibrium with the allyl sulfoxides [*12*] (equation 9a). These products can be oxidized to the corresponding sulfones (equation 9b). Pyrolysis of the sulfoxides gives sulfines or evidence for the presence of sulfines. Pyrolysis of sulfones leads to unsaturated compounds by extrusion of sulfur dioxide [*12*] (equation 9c).

[11] **8**

2,3,4,5-tetrakis(SCF_3)-1-Ag-pyrrole + CF_3SCl ⟶ 1,2,3,4,5-pentakis(SCF_3)pyrrole

$\downarrow C_2H_5OH$

$C_2H_5OSCF_3$ + 2,3,4,5-tetrakis(SCF_3)-1H-pyrrole

[12] **9a**

$CH_2{=}CHCH_2OH + CF_3SCl \longrightarrow F_3CSOCH_2CH{=}CH_2 \rightleftharpoons F_3CS(O)CH_2CH{=}CH_2$

9b

$(CF_3)_2CClS(O)CH_2CH{=}CH_2 \xrightarrow{\Delta} (CF_3)_2C{=}SO$

$\downarrow KHSO_5$

9c

$(CF_3)_2CClSO_2CH_2CH{=}CH_2 \xrightarrow{\Delta} (CF_3)_2CClCH_2CH{=}CH_2$

The reaction of **trifluoromethanesulfenyl chloride** with ***hydrogen sulfide*** yields *trifluoromethyldisulfane* [*13*]. Analogously, chloro(trifluoromethyl)disulfane yields **trifluoromethyltrisulfane** (equation 10) (Table 4). Trifluoromethanesulfanes have acidic protons that enable them to undergo substitution reactions with compounds having labile halogens. Trisulfanes are prepared by nucleophilic attack on the sulfane of chloro(trifluoromethane)disulfane by thiols and thio acids. The similar substitution reaction takes place with several nitrogen nucleophiles as well. The addition reactions of chloro(trifluoromethyl)disulfane with polyhalogenoolefins provide nonsymmetrical disulfanes in 10–60% yields where only tetrafluoroethylene gives a higher telomer [*14*] (equation 11) (Table 4).

[13] $CF_3S_xCl + H_2S \longrightarrow CF_3S_xSH + HCl$ **10**

$x = 1, 2$

[14] CF_3SSCl + $ClFC{=}CFCl$ $\longrightarrow$ $CF_3SSCFClCFCl_2$ **11**

The introduction of a *trifluoromethanethio group* into an aromatic ring has a synthetic importance. The reaction of trifluoromethanethio copper with aryl bromides and iodides provides a convenient route to the synthesis of aryltrifluoromethane sulfides. The reaction is not sensitive to the type of substituents or the aromatic nucleus. Selectivity can be achieved according to the type of halogen or the aromatic ring, because iodides react at lower temperatures than bromides, whereas chlorides do not react [*15*] (equation 12) (Table 5).

[15] **12**

$Hg(SCF_3)_2 + Cu + I{-}C_6H_4{-}CO_2C_2H_5 \longrightarrow F_3CS{-}C_6H_4{-}CO_2C_2H_5$

Perfluorotetramethylene sulfimides are synthesized by the reaction of lithium amide and primary amines with (perfluorotetramethylene)sulfur difluoride [*16*] (equation 13) (Table 6). The products can be oxidized with *m*-chloroperoxybenzoic acid to the corresponding sulfoximides (equation 13) or can be treated with chlorine or bromine to yield N-halo derivatives [*16*]. The reaction of $CF_3SF_3{=}NCF_3$ with nucleophiles takes place by attack of the nucleophile at the positive sulfur center [*17*] (equation 14).

[16] **13**

$(CF_2)_4SF_2$ + $LiNH_2$ $\xrightarrow[-78\,°\rightarrow 0\,°C]{NH_3}$ $(CF_2)_4S{=}NH$ $\xrightarrow{m\text{-}ClC_6H_4CO_3H}$ $(CF_2)_4S({=}O){=}NH$

Table 4. Preparation of Disulfides

Reactants	*Reaction Conditions*	*Product(s)*	*Yield (%)*	*Ref.*
H_2S, CF_3S_xCl $x = 1, 2$		CF_3S_xSH $x = 1, 2$	50–70	13
RSH, CF_3SSCl	20 °C	$RSSSCF_3$ R= CH_3, C_2H_5, CF_3, CH_3CO, CF_3CO		14
$(CH_3)_2NH$, CF_3SSCl		$(CH_3)_2NSSCF_3$		14
AgNCO, CF_3SSCl		CF_3SSNCO		14
$(CF_3)_2C{=}NLi$, CF_3SSCl	–196 to 20 °C	$CF_3SSN{=}C(CF_3)_2$		14
C_4H_9, CF_3SSCl		$CF_3SSC_4H_9$	68	14
$F_2C{=}CF_2$, CF_3SSCl	hν	$CF_3SSCF_2CF_2Cl$		14
$F(CF_3)C{=}CF_2$, CF_3SSCl	hν	$CF_3SSCF(CF_3)CF_2Cl$ (major) + $CF_3SSCF_2CFClCF_3$ (minor)	10–60	14
hexafluorocyclobutene (ring: F F / F, F / F, F), CF_3SSCl	hν	cyclobutane ring: F F / F, F / F_3CSS, Cl / F F	10–60	14
$ClFC{=}CF_2$, CF_3SSCl	hν	$CF_3SSCFClCF_2Cl$ (major) + CF_3SSCF_2CFCl (minor)	10–60	14
$ClC{=}CCl$ (with F on each C) CF_3SSCl	hν	$CF_3SSCFClCFCl_2$	10–60	14

Table 5. Reaction of Trifluoromethylthiocopper[a] with Aryl Halides [15]

Reactant	Reaction Conditions	Product(s)	Yield (%)
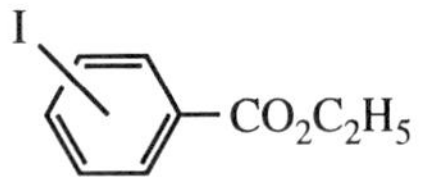 $I-C_6H_4-CO_2C_2H_5$	DMF 110–120 °C	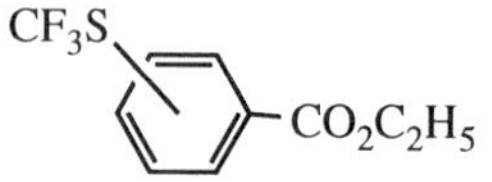 $CF_3S-C_6H_4-CO_2C_2H_5$	*o*, 98 *m*, 90 *p*, 98
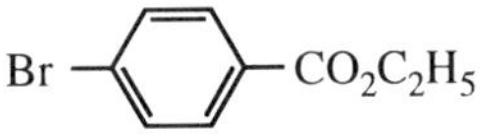 $Br-C_6H_4-CO_2C_2H_5$ (para)	Quinoline 150–190 °C	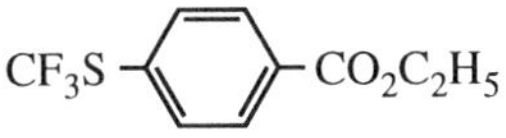 $CF_3S-C_6H_4-CO_2C_2H_5$	87
 $I-C_6H_4-I$ (para)	DMF 110–120 °C	 $CF_3S-C_6H_4-SCF_3$	62
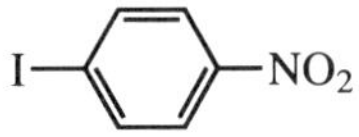 $I-C_6H_4-NO_2$ (para)	DMF 110–120 °C	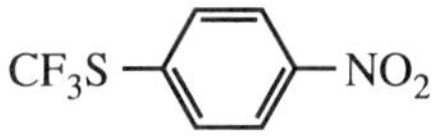 $CF_3S-C_6H_4-NO_2$	89
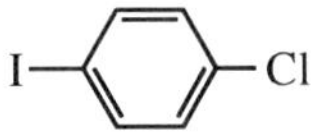 $I-C_6H_4-Cl$ (para)	DMF 110–120 °C	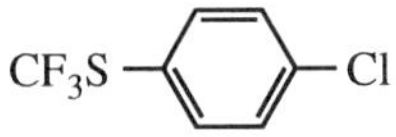 $CF_3S-C_6H_4-Cl$	78
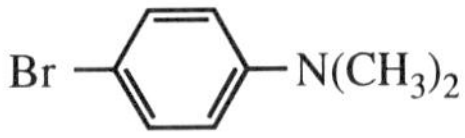 $Br-C_6H_4-N(CH_3)_2$ (para)	HMPA 160–175 °C	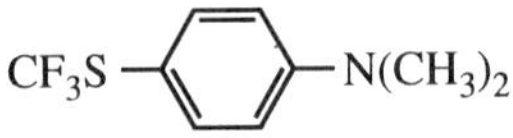 $CF_3S-C_6H_4-N(CH_3)_2$	89
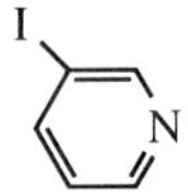 3-iodopyridine	DMF 110–120 °C	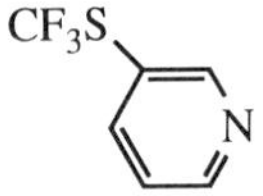 3-(CF_3S)pyridine	95
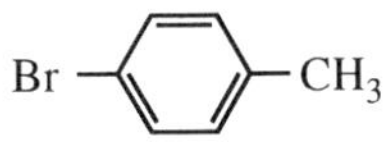 $Br-C_6H_4-CH_3$ (para)	HMPA 160–180 °C	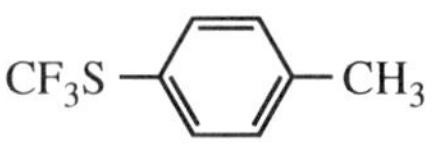 $CF_3S-C_6H_4-CH_3$	79

[a] CF_3SCu is prepared in situ: $Hg(SCF_3)_2 + Cu$.

[17] $$CF_3SF_3{=}NCF_3 + CH_3N[Si(CH_3)_3]_2 \longrightarrow CF_3-S(F)(=NCH_3)=NCF_3 \quad \mathbf{14}$$

Table 6. Synthesis and Reactions of Fluorosulfur Compounds

Reactant	Reaction Conditions	Product(s)	Yield (%)	Ref.
$\overline{(CF_2)_4}SF_2$	$LiNH_2$, NH_3 −78 to 0 °C	$\overline{(CF_2)_4}S{=}NH$	78	16
	RNH_2, NH_3 −78 to 0 °C	$\overline{(CF_2)_4}S{=}NR$		
	0 °C, 25 °C	R= CH_3	70	16
	−78 to 25 °C	R= C_2H_5	37	16
$\overline{(CF_2)_4}SF_2$	$LiN{=}C(CF_3)_2$ 25 °C	$\overline{(CF_2)_4}S{=}N{-}CF(CF_3)_2$	39	16
$\overline{(CF_2)_4}S{=}NCF(CF_3)_2$	$LiN{=}C(CF_3)_2$ 25 °C	$\overline{(CF_2)_4}S{=}NC(CF_3)_2N{=}C(CF_3)_2$	64	16

Continued on next page.

Table 6—*Continued*

Reactant	*Reaction Conditions*	*Product(s)*	*Yield (%)*	*Ref.*
Octafluorotetrahydrothiophene S=NR (ring: $(CF_2)_4$, S=NR)	MCPBA	Octafluorotetrahydrothiophene S(=NR)(=O)		
	0 °C	R = H	97	16
	–78 to 0 °C	R = CH_3	79	16
		R = C_2H_5	84	16
Octafluorotetrahydrothiophene S=NH (ring: $(CF_2)_4$, S=NH)	Cl_2, CsF, 25 °C	Octafluorotetrahydrothiophene S=NCl	87	16
	Br_2, KF, 25 °C	Octafluorotetrahydrothiophene S=NR	23	15
$RC_6H_4N_2^+$ $^-C(SO_2F)_3$	80–120 °C, $-N_2$	$RC_6H_4O–S(=O)(F)=C(SO_2F)_2$		
		R = H	58	26
		R = *p*-F	65	26
		R = *m*-F	43	26

Table 6—*Continued*

Reactant	*Reaction Conditions*	*Product(s)*	*Yield (%)*	*Ref.*
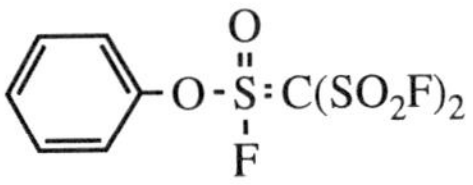 $C_6H_5O-S(=O)(F)=C(SO_2F)_2$	$(CH_3)_3SiNEt_2$, acetonitrile, Δ	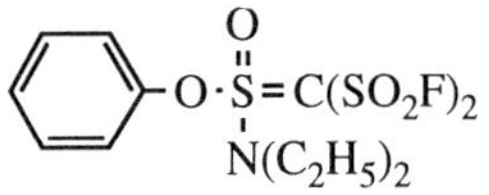 $C_6H_5O-S(=O)(N(C_2H_5)_2)=C(SO_2F)_2$	73	26
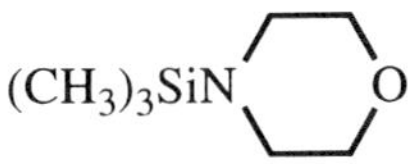 $(CH_3)_3SiN(CH_2CH_2)_2O$ (morpholine)		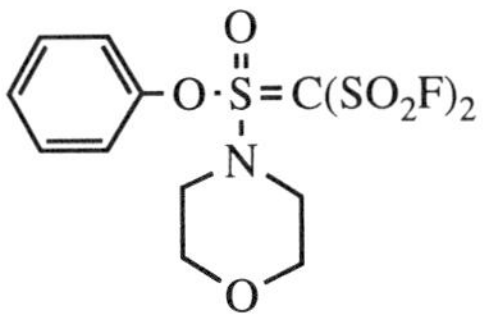 $C_6H_5O-S(=O)(N(CH_2CH_2)_2O)=C(SO_2F)_2$	79	26
	$(CH_3)_3SiOC_6H_5$, CsF	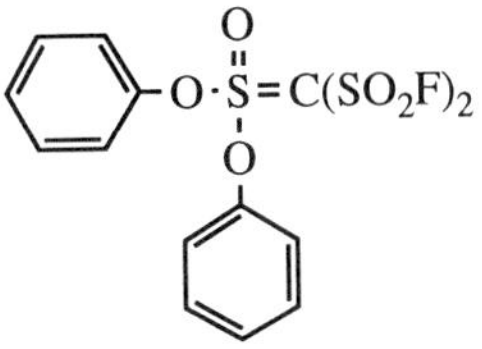 $C_6H_5O-S(=O)(OC_6H_5)=C(SO_2F)_2$	81	26
	C_6H_5Li, −20 °C, Et_2O	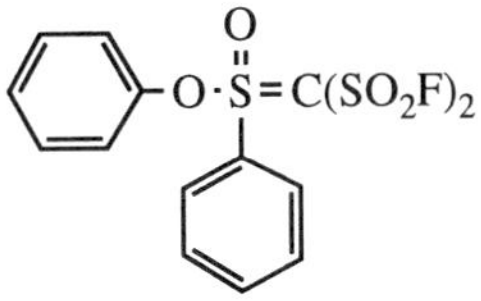 $C_6H_5O-S(=O)(C_6H_5)=C(SO_2F)_2$	10	26
	MF	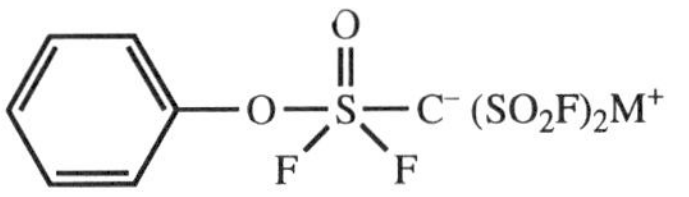 $C_6H_5O-S(=O)(F)_2-C^-(SO_2F)_2M^+$ M = Cs, Ag		26

Continued on next page.

Table 6—*Continued*

Reactant	*Reaction Conditions*	*Product(s)*	*Yield (%)*	*Ref.*
$CF_3SF_4NClCF_3$	Hg, -HgClF or hν or Δ	$CF_3SF_3{=}NCF_3$	80	17
$CF_3SF_3{=}NCF_3$	H_2O, 25 °C or NaOH	$CF_3SF(O)NCF_3$	100	17
$CF_3SF_3{=}NCF_3$	CF_3CH_2OH 25 °C	$CF_3SF(O)NCF_3$ + $CF_3SF_4NHCF_3$		17
$CF_3SF_3{=}NCF_3$ or $CH_3N[Si(CH_3)_3]_2$	CH_3NH_2 –78 to 28 °C	$CF_3SF({=}NCH_3)_2$		17
$CF_3SF_4NHCF_3$	CsF	$CF_3SF_3{=}NCF_3$	100	17
$CF_3SF_4NHR_f$	AgF_2, Δ	$CF_3SF_4NFR_f$		
		$R_f = CF_3$	67	17
		$R_f = C_2F_5$	60	17
$CF_3SF_3{=}NCF_3$	F_2, CsF	$CF_3SF_4NFCF_3$	85	17
$CF_3SF_4N{=}CFCF_3$	F_2, CsF	$CF_3SF_4NFCF_2CF_3$	80	17

Sulfones and Sulfoxides

The trifluoromethanesulfonyl (triflyl group) is one of the strongest electron-withdrawing groups and therefore is a versatile functionality for organic synthesis. *Trifluoromethyl sulfones* have been prepared by using an electrophilic triflyl source such as **triflic anhydride** or by displacing primary halides with a triflate salt.

The preparation of a triflate salt may include the decomposition of triflyl azide by azide ion. Triflyl azide can be prepared by the reaction of the azide ion with trifluoromethanesulfonyl fluoride or trifluoromethanesulfonic anhydride [*18*] (equation 15). Another one-step procedure uses a quaternary ammonium counterion [*19*] (equation 15). This triflate can react with primary halides to form trifluoromethyl sulfones [*19*] (equation 16) (Table 7).

[18] $CF_3SO_2F + NaN_3 \xrightarrow{CH_3OH} CF_3SO_2N_3 + NaF$ (86%) **15**

$CF_3SO_2N_3 + NaN_3 \xrightarrow[66\ °C]{CH_3OH} CF_3SO_2Na + 3N_2$ (93%)

[19] $(C_4H_9)_4NOH + NaN_3 \longrightarrow (C_4H_9)_4N^+N^-_3$ **16**

$(C_4H_9)_4N^+N^-_3 + (F_3CSO_2)_2O \xrightarrow[-78\ °C]{CH_2Cl_2} (C_4H_9)_4N^+CF_3SO^-_3 + CF_3SO_2N_3$

$CF_3SO_2N_3 \xrightarrow{C_6H_5CH_2CH_2Br,\ CH_2Cl_2,\ 50\ °C} C_6H_5CH_2CH_2SO_2CF_3$

The reactions of electrophilic triflyl sources with nucleophiles were investigated. The reaction of triflic anhydride with an organolithium reagent is not synthetically promising because of ditriflylation and other side reactions [*20*]. When phenyllithium reacts with triflic anhydride, dimerization products and acetylenic Michael diadducts are observed [*20*] (equation 17); but using the sodium salt of the alkynes instead of the lithium salt provides the alkynyl trifluoromethyl sulfones [*21*] (equation 18) (Table 7). Alkynyl trifluoromethyl sulfones are of synthetic interest, because they show a pronounced reactivity toward nucleophiles in addition reactions and cyclopentadiene in Diels–Alder reactions [*21*] (Table 7).

[20] **17**

$C_6H_5C{\equiv}CLi + (CF_3SO_2)_2O \xrightarrow[-75°C]{hexane} [C_6H_5C{\equiv}CSO_2CF_3]$

$\xrightarrow{PhC{\equiv}CLi} C_6H_5{-}C{\equiv}C{-}C{\equiv}C{-}C_6H_5$ (35%) $+$ $(C_6H_5C{\equiv}C)(C_6H_5)C{=}C(SO_2CF_3)(C{\equiv}CC_6H_5)$ (23%)

Another triflyl electrophile explored was $C_6H_5N(SO_2CF_3)_2$, which gave better results with organolithium reagents compared with triflic anhydride [*20*] (equation 19) (Table 5). However, some reagents failed to give any triflones, trifluoromethyl sulfones (methyllithium, *tert*-butyllithium, pentafluorophenyllithium, furyllithium, thienyllithium, and ethyl and isopropyl magnesium bromides). The syntheses of aryl triflones are carried out with triflic anhydride–aluminum trichloride

Table 7. Synthesis and Reactions of Trifluoromethyl Sulfones

Reactants	*Reaction Conditions*	*Product(s)*	*Yield (%)*	*Ref.*
$C_6H_5CH_2CH_2Br$ $Bu_4N^+CFS_3O_2^-$	CH_2Cl_2, 50 °C	$C_6H_5CH_2CH_2SO_2CF_3$	74	19
$H_2C{=}CHCH_2Br$ $Bu_4N^+CF_3SO_2^-$	50 °C	$H_2C{=}CHCH_2SO_2CF_3$	98	19
$H_2C{=}CHCH_2SO_2CF_3$	DABCO (cat.), $CHCl_3$	$H_3CCH{=}CHSO_2CF_3$	~100	19
$RC{\equiv}CH$	1. Na/Et_2O	$RC{\equiv}CSO_2CF_3$		21
	2. $(CF_3SO_3)_2O$	$R = C_6H_5$	74	21
	Et_2O	$R = 4\text{-}FC_6H_4$	17	21
	–78 °C to RT	$R = C_4H_9$	60	21
		$R = C_5H_{11}$	50	21
		$R = C_6H_{13}$	50	21
		$R = cyclo\text{-}C_6H_{11}$	72	21
		$R = 4\text{-}CH_3OC_4H_4$	NR	21
$RC{\equiv}CSO_2CF_3$ H_2O	Acetone, RT	$RC(O)CH_2SO_2CF_3$		21
		$R = C_6H_{13}$	92	21
		$R = C_6H_5$	90	21
$C_4H_9C{\equiv}CSO_2F_3$ cyclo-C_4H_8NH	Toluene, 0 °C	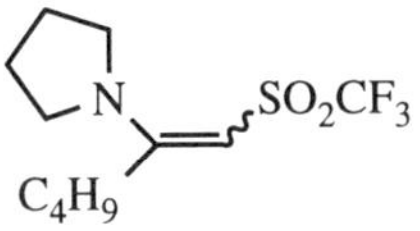 N SO_2CF_3 C_4H_9	93	21
$C_4H_9C{\equiv}CSO_2F_3$ Et_2NH	Toluene, 0 °C	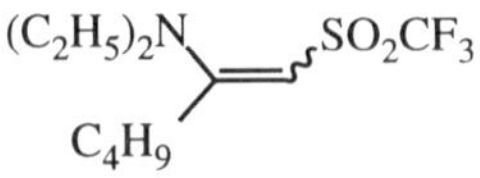 $(C_2H_5)_2N$ SO_2CF_3 C_4H_9	90	21

Table 7—*Continued*

Reactants	*Reaction Conditions*	*Product(s)*	*Yield (%)*	*Ref.*
$C_6H_5C{\equiv}CSO_2CF_3$	Toluene, 0 °C *cyclo*-C_4H_8NH	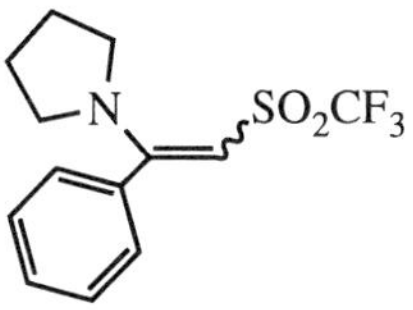 N, SO_2CF_3	90	21
$C_4H_9C{\equiv}CSO_2CF_3$ EtOH	RT	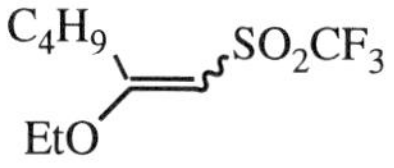 C_4H_9, SO_2CF_3, EtO	86	21
$C_4H_9C{\equiv}CSO_2CF_3$ Cyclopentadiene	Toluene, RT	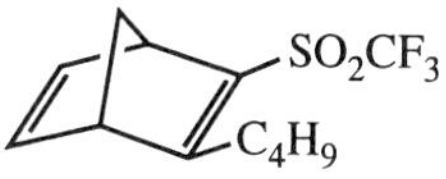 SO_2CF_3, C_4H_9	74	21
$C_6H_5N(SO_2CF_3)_2$ RLi	$(C_2H_5)_2O$ –78 °C	RSO_2CF_3		
		+ $C_6H_5NHSO_2CF_3$		20
		R = C_4H_9	74	20
		R = C_4H_9	67	20
		R = C_2H_5	89	20
C_6H_6, $(CF_3SO_2)_2O$ $AlCl_3$	RT	$C_6H_5SO_2CF_3$	61	20
$C_6H_5CH_3$, $(CF_3SO_2)_2O$ $AlCl_3$	RT	*o*- and *p*-$CH_3CH_6H_4SO_2CF_3$ (1:2)	69	20

Continued on next page.

References are listed on pages 610–614.

Table 7—*Continued*

Reactants	*Reaction Conditions*	*Product(s)*	*Yield (%)*	*Ref.*
p-$(CH_3)_2C_6H_4$, $(CF_3SO_2)_2O$ $AlCl_3$	RT	1,4-$(CH_3)_2$-2-$(SO_2CF_3)C_6H_3$	73	20
C_6H_5Cl, $(CF_3SO_2)_2O$ $AlCl_3$	RT	*o*- and *p*-$ClC_6H_4SO_2CF_3$	10	20
C_6H_5X (X = NO_2, OCH_3) $(CF_3SO_2)_2O$, $AlCl_3$	RT	NR		20

[21] **18**

$$C_6H_5-C{\equiv}C-H \xrightarrow[\text{2. } (CF_3SO_2)_2O,\ (C_2H_5)_2O,\ -78\ °C \to RT]{\text{1. } Na/(C_2H_5)_2O,\ RT} C_6H_5-C{\equiv}C-SO_2CF_3\ \ 74\%$$

pyrrolidine (N–H), toluene, 0 °C → $C_6H_5C(NC_4H_8)=CHSO_2CF_3$

[20] **19**

$$C_2H_5Li + C_6H_5N(SO_2CF_3)_2 \xrightarrow[-78\ °C]{(C_2H_5)_2O} C_2H_5SO_2CF_3 + C_6H_5NHSO_2CF_3$$

89%

as an acylating complex. Only mildly activated aromatic substrates are suitable in this triflylation reaction [*20*] (Table 7).

The trifluoromethanesulfonyl group is a potent electron-withdrawing group. The 2,4,6-tris(trifluoromethanesulfonyl)phenyl group was observed by dynamic ^1H NMR to be rapidly migrating between the nitrogen atoms of the benzamidine system [*22*] (equation 20). This rapid migration is presumably due to the exceptionally high electrophilicity of the 2,4,6-tris(trifluoromethanesulfonyl)phenyl

[22] **20**

$$\text{4-}CH_3C_6H_4N{=}C(C_6H_5)NHC_6H_4CH_3\text{-4} + 1\text{-Cl-}2,4,6\text{-}(SO_2CF_3)_3C_6H_2 \xrightarrow[75\%]{\text{benzene}} \text{4-}CH_3C_6H_4N{=}C(C_6H_5)N(C_6H_4CH_3\text{-4})C_6H_2(SO_2CF_3)_3\text{-}2,4,6$$

$$\rightleftharpoons \left[\text{spiro intermediate: } \text{4-}CH_3C_6H_4N^{+}\text{, }C_6H_5\text{, }NC_6H_4CH_3\text{-4, } (F_3CO_2S)_2\text{, }SO_2CF_3\text{ (anionic ring)}\right] \rightleftharpoons \text{4-}CH_3C_6H_4N(C_6H_2(SO_2CF_3)_3)C(C_6H_5){=}NC_6H_4CH_3\text{-4}$$

group. The mild and convenient synthesis of methylene sulfones [*23*] (equation 21) occurs on treatment of paraformaldehyde with triflone reagent. The methylene group is introduced into the molecule, and the triflate anion is the leaving group.

The sulfone and sulfoxide groups stabilize the α-carbanions, which can react with various electrophiles. Condensation reactions of perfluoroalkyl benzyl sulfones with formaldehyde such as the reaction of perfluoroalkyl methyl sulfones with aromatic aldehydes leads to cyclopropyl perfluoroalkyl sulfones (equation 22). On reaction of the lithium derivatives of alkyl *p*-tolyl sulfoxides with mono- and perfluorinated carboxylic esters, prochiral perfluoro ketones and α-monofluoro ketones are obtained [*24*] (Table 8). Through the use of enantiomerically pure sulfoxides, three-carbon fluorinated chiral carbonyl synthons (the portion of a

[23] $CF_3SO_2CH(R^1)SO_2CH_2R^2$ $\xrightarrow[K_2CO_3]{(CH_2O)_x}$ $R^1C(=CH_2)SO_2CH_2R^2$ **21**

R^1	R^2
CH_3	H
CH_3	CH_3
C_5H_{11}	H
C_5H_{11}	C_5H_{11}
$(CH_2)_4$	
$C_6H_5CH_2$	H

[24] $2\ C_8F_{17}SO_2CH_2CH_2C_6H_5 + (HCOH)_x \xrightarrow[NaOH]{C_2H_5OH}$ 1-$C_8F_{17}SO_2$-1-C_6H_5-cyclopropane **22**

molecule that is recognizably related to a simpler molecule) can be obtained. (For additional information about such reactions, *see* Base-Catalyzed Condensations on page 615.

Convenient syntheses of *vinyl fluorides* are of synthetic interest. The conjugate base of fluoromethyl phenyl sulfone reacts with carbonyl compounds to provide β-fluoro alcohols, which are used to prepare terminal vinyl fluorides [*25*] (equation 23) (Table 9). This reaction offers an alternative to the Wittig reaction, which may be very sensitive to reaction conditions.

Methanetrisulfonyl fluoride, $HC(SO_2F)_3$, is a strong hydrocarbonic acid. When aqueous solutions are mixed with arenediazonium chlorides, the corresponding salts are formed. These salts decompose between 80 and 120 °C with a release of nitrogen and formation of aryloxyfluorooxosulfoniobis(fluorosulfonyl)meth-

[25] **23**

$C_6H_5S(O)CH_3 \xrightarrow[CHCl_3]{Et_2NSF_3} C_6H_5SCH_2F \xrightarrow[CHCl_3]{MCPBA} C_6H_5SO_2CH_2F$

C_6H_5CHO $\xrightarrow[\text{2. } CH_3SO_2Cl,\ (C_2H_5)_3N,\ CH_2Cl_2]{\text{1. } C_6H_5SO_2CHLiF,\ THF,\ -78\ °C}$ $C_6H_5CH{=}C(F)SO_2C_6H_5$ $\xrightarrow[THF\text{-}H_2O,\ \Delta]{\text{Al-amalgam}}$ $C_6H_5CH{=}CHF$

Table 8. Reactions of Sulfoxides

Reactants	Reaction Conditions	Product(s)		Yield (%)	Ref.
4-$CH_3C_6H_4$-S(O)-CH_2R	LDA, THF, –78 °C	4-$CH_3C_6H_4$-S(O)-CH(R)-C(OH)(OH)R_f + 4-$CH_3C_6H_4$-S(O)-CH(R')-C(=O)R_f			24
		R	R_f		
		H	C_3F_7	92	24
		H	C_7F_{15}	84	24
		C_6H_5	CF_3	68	24
		C_6H_5	C_3H_7	64	24
		CH_3	CH_2F	94	24
		$CH_2CH{=}CH_2$	CH_2F	76	24
		$(CH_2)_2CH{=}CH_2$	CH_2F	90	24
		$CH_2CH{=}C(CH_3)_2$	CH_2F	74	24
		C_6H_5	CH_2F	82	24
4-$CH_3C_6H_4$-S(O)-CH_2C(=O)CH_2F	NaH, DMF, 0 °C	4-$CH_3C_6H_4$-S(O)-CH(R)-C(=O)CH_2F			24
				66	24
RX, THF				74	24
		R = CH_3			
		R = $(CH_2)_2CH{=}CH_2$			
4-$CH_3C_6H_4$-S(O)-CH(CH_3)C(=O)CH_2F	NaH, DMF, 0 °C	4-$CH_3C_6H_4$-S(O)-C(CH_3)(CH_3)-C(=O)CH_2F		62	24
CH_3I					

Continued on next page.

Table 8—*Continued*

Reactants	*Reaction Conditions*	*Product(s)*	*Yield (%)*	*Ref.*
4-$CH_3C_6H_4$-S(O)-CH_3; $CH_3CHFCO_2CH_3$	LDA, THF, –78 °C	4-$CH_3C_6H_4$S(O)$CH_2COCHFCH_3$	88	24
4-$CH_3C_6H_4$S(O)CH_2COCH_2F; CH_3I	2 mol LDA, THF, –78 °C	4-$CH_3C_6H_4$S(O)$CH_2COCHFCH_3$ (two diastereomers)	78	24
4-$CH_3C_6H_4$-S(O)-CH_3; $C_2H_5CHFCO_2CH_3$	LDA, THF, –78 °C	4-$CH_3C_6H_4$S(O)$CH_2COCHFC_2H_5$	90	24
4-$CH_3C_6H_4$S(O)CH_2COCH_2F; $BrCH_2CH=C(CH_3)_2$	2 mol LDA, THF –78 °C	4-$CH_3C_6H_4$S(O)$CH_2COCHFCH_2CH=C(CH_3)_2$ + 4-$CH_3C_6H_4$S(O)$CH_2COCHFCH_2CH=C(CH_3)_2$ (two diastereomers)	62	24
4-$CH_3C_6H_4$-S(O)-CH_3; $CH_3OCOCF_2CH_2C_6H_5$	LDA, THF, –78 °C	4-$CH_3C_6H_4$S(O)$CH_2COCF=CHC_6H_5$		

Table 9. Reactions of Sulfones

Reactants	Reaction Conditions	Product(s)		Yield (%)	Ref
(R^1-substituted aryl ketone, $ArCOR^2$) + $C_6H_5SO_2CHFLi$	THF, −78 °C	$ArC(OH)(R^2)CHFSO_2C_6H_5$			25
$ArC(OH)(R^2)CHFSO_2C_6H_5$		$ArC(R^2)=CFSO_2C_6H_5$			
	CH_3SO_2Cl, $(C_2H_5)_3N$, CH_2Cl_2	R^1	R^2		
		H	H	67	25
		4-Cl	H	80	25
		3, 4-$(OCH_3)_2$	H	71	25
	85%, phosphoric acid, Δ	2, 3, 4, 5, 6-F_5	H	78	25
		H	C_6H_5	92	25
$C_6H_5C(OH)(CH_3)CHFSO_2C_6H_5$	CH_3SO_2Cl, $(C_2H_5)_3N$, CH_2Cl_2	$C_6H_5C(=CH_2)CFHSO_2C_6H_5$		82	25
$ArC(R^2)=CFSO_2C_6H_5$	Al amalgam, THF-H_2O, Δ	$ArC(R^2)=CHF$			25
		R^1	R^2		
		H	H	90	25
		4-Cl	H	90	25
		H	C_6H_5	91	25

Continued on next page.

Table 9—*Continued*

Reactants	*Reaction Conditions*	*Product(s)*	*Yield (%)*	*Ref.*
$HC(SO_2F)_3$, $RC_6H_4N_2{+}Cl^-$	5 °C, H_2O	$RC_6H_4N_2{+}C^-(SO_2F)_3$		
		R = H	90	26
		4-F	89	26
		3-F	90	26
		4-NO_2	82	26
$C_6H_5CH_2NH_2$	C_2H_5OH		100	27
$CH_3CH(NH_2)COOC_2H_5$	C_2H_5OH	5:2 diastereoisomeric mix.	73	27
$CH_3CH(NH_2)CH_2OH$	C_2H_5OH	6:5 diastereoisomeric mix.	71	27
F_3C, $SO_2C_6H_5$, $(C_2H_5O_2C)_2CH$	1. 2 equiv. LDA 2. C_6H_5CHO	CF_3, OH, $(C_2H_5O_2C)_2CH$, C_6H_5, $SO_2C_6H_5$	71	27
F_3C, NH, $SO_2C_6H_5$, C_6H_5	1. 2 equiv. LDA 2. C_6H_5CHO	CF_3, OH, $C_6H_5CH_2NH$, C_6H_5, $SO_2C_6H_5$	49	27
F_3C, NH, $SO_2C_6H_5$, C_6H_5	Na(Hg)	CF_3, $C_6H_5CH_2NH$	55	27
F_3C, O, $SO_2C_6H_5$, C_6H_5	Na(Hg)	OH, CF_3, C_6H_5	70	27
CF_3, OH, $(C_2H_5O_2C)_2CH$, C_6H_5, $SO_2C_6H_5$	Na(Hg), THF, CH_3OH	CF_3, $(C_2H_5O_2C)_2C$, C_6H_5	38	27

Table 9—*Continued*

Reactants	*Reaction Conditions*	*Product(s)*	*Yield (%)*	*Ref.*
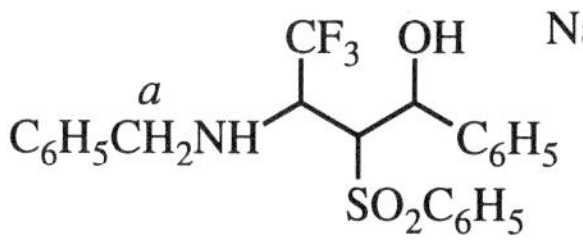 *a* $C_6H_5CH_2NH$–CH(CF_3)–CH($SO_2C_6H_5$)–CH(OH)–C_6H_5	Na(Hg), THF CH_3OH	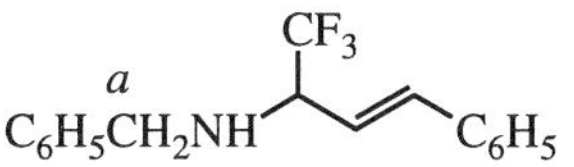 *a* $C_6H_5CH_2NH$–CH(CF_3)–CH=CH–C_6H_5	54	27
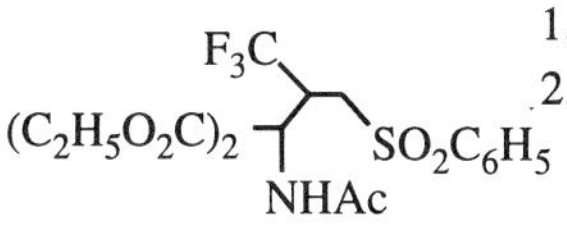 $(C_2H_5O_2C)_2$C(NHAc)–CH(CF_3)–$CH_2SO_2C_6H_5$	1. Na(Hg) 2. H^+, Δ	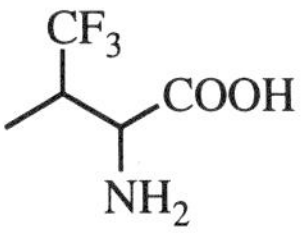 CH_3CH(CF_3)CH(NH_2)COOH	52	27

anides. In the product, the ylide fluorine undergoes substitution reactions with nucleophiles [*26*] (equation 24) (Table 9).

The bioactive properties of the trifluoromethyl group may be important in drug design. One method of introducing the trifluoromethyl group is the use of functionalized molecules containing a trifluoromethyl group as building blocks. Typical of these are vinyl sulfones [*27*] or sulfoxides [*28*], which are good Michael acceptors (equations 25 and 26) (Table 10). The use of vinyl sulfoxides suggests the possibility of the asymmetric synthesis of trifluoromethylated chiral carbons [*28*] (equation 26) (Table 10).

[26] **24**

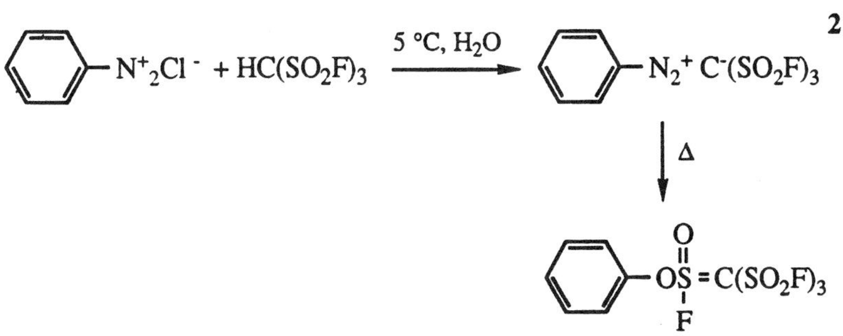

$$C_6H_5N^+_2Cl^- + HC(SO_2F)_3 \xrightarrow{5\ °C,\ H_2O} C_6H_5N_2^+\ C^-(SO_2F)_3 \xrightarrow{\Delta} C_6H_5OS(=O)(F)=C(SO_2F)_3$$

Sulfonates and Sulfinates

The preparation of *trifluoromethyl trifluoromethanesulfonate* was reported by Olah and Ohyama [*29*]. In this safe and practical method, trifluoromethanesulfonic acid and fluorosulfonic acid are used [*29*] (equation 27).

[27] **25**

$F_3C\text{–}CH{=}CH\text{–}SO_2C_6H_5 + AcNHCH(CO_2C_2H_5)_2 \xrightarrow{NaH/THF} (C_2H_5O_2C)_2C(NHAc)\text{–}CH(CF_3)\text{–}CH_2SO_2C_6H_5$

1. Na(Hg)
2. H^+, Δ

$\rightarrow CF_3CH(CH_3)\text{–}CH(NH_2)COOH$

[28] **26**

$4\text{-}CH_3C_6H_4S(O)\text{–}CH{=}CH\text{–}CF_3 + C_6H_5COCH_3 \xrightarrow{LDA, THF} C_6H_5COCH_2CH(CF_3)CH_2S(O)C_6H_4CH_3\text{-}4$

94%

[29] $2\ CF_3SO_3H + FSO_3H \xrightarrow{\Delta} CF_3SO_2OCF_3 + SO_2 + HF + H_2SO_4$ **27**

19%

Table 10. Reactions of Vinyl Sulfones [*27*]

Reactants	*Nucleophile*	*Reaction Conditions*	*Product*	*Yield (%)*
$F_3C\text{–}CH{=}CH\text{–}SO_2C_6H_5$	$CH_2(COOC_2H_5)_2$	NaH, THF	$F_3C\text{–}CH(Nu)\text{–}CH_2\text{–}SO_2C_6H_5$	100
	$CH_3COC_6H_5$	NaH, THF $t\text{-}C_4H_9OH$		60
	$CH_3CH_2COC_6H_5$	NaH, THF		95
	$(CH_3)_2CHCOC_6H_5$	KH $(C_2H_5)_2O$		68
	$C_6H_5C(=NCH(CH_3)C_6H_5)CH_3$	LDA, THF		85
	$AcNHCH(COOC_2H_5)_2$	NaH, THF		100

Other sulfonate derivatives are obtained by the use of **trifluoromethanesulfonyl hypochlorite and hypobromite** (CF_3SO_2OCl and CF_3SO_2OBr) in reactions with ***perfluoroalkyl halides*** and their derivatives [*30*]. These reactions lead to the corresponding trifluoromethanesulfonate derivatives of alkanes (equation 28) (Table 11). The reaction proceeds with complete retention of stereochemistry at the carbon center [*30*].

$$[30] \quad CF_3SO_3Cl + C_3F_7Br \xrightarrow{CF_2Cl_2} CF_3SO_3C_3F_7 \quad \mathbf{28}$$

Table 11. Reactions of Trifluoromethanesulfonyl Hypochlorite with Fluorohalo Compounds [*30*]

Reactants		*Reaction Conditions*	*Products*	*Yield (%)*
CF_3SO_3Cl	$CBrF_3$	CF_2Cl_2 −111 °C	$CF_3SO_3CF_3$	95
	C_2F_5Br	−40--->−5 °C	$CF_3SO_3C_2F_5$	58
	C_3F_7Br	−40--->22 °C	$CF_3SO_3C_3F_7$	65
	CCl_2F_2	−78--->22 °C	$CF_3SO_3CF_2Cl$	91
	CBr_2F_2	−160 °C	$CF_3SO_3CF_2Br$	61
	CCl_3F	−78--->22 °C	$CF_3SO_3CFCl_2$	88
	$CF_3SO_3CF_2Br$	−111-22 °C	$(CF_3SO_3)_2CF_2$	30
	erythro-$CF_3CO_2CHFCHFCl$	−111-22 °C	erythro-$CF_3SO_3CHFCHFO_2CCF_3$	
	threo-$CF_3CO_2CHFCHFCl$	−111-22 °C	threo-$CF_3SO_3CHFCHFO_2CCF_3$	
	erythro-$CF_3SO_3CHFCHFBr$	−111-22 °C	$(CF_3SO_3CHF)_2$	

The nucleophilic displacement of halogens by pentafluorophenoxide ion resulted in the formation of the corresponding esters [*31*] (equation 29) (Table 12). Reactions of trifluoromethanesulfinyl fluoride with fluoro alcohols in the presence of sodium fluoride or cesium fluoride are used to prepare sulfinates [*32*] (equation 30) (Table 12).

$$[31] \quad SO_2F_2 + KOC_6F_5 \longrightarrow FSO_2OC_6F_5 \quad \mathbf{29}$$

$$[32] \quad CF_3S(O)F + CF_3(CH_3)_2COH \xrightarrow[NaF]{25\ °C} CF_3S(O)OC(CH_3)_2CF_3 \quad \mathbf{30}$$

Sulfonates react with a variety of nucleophiles. Synthesis of *N,N*-bis(trifluoromethyl)aminotrifluoromethanesulfonate and its reactions with nucleophiles were investigated [*33*] (equation 31) (Table 13). Nucleophilic attack occurs at either nitrogen or sulfur; amines give complex mixtures [*33*]. **Polyfluoroalkyl fluorosulfates** react with ***amines, alcohols, or alkoxides*** to yield polyfluoroalkyl sulfamates and dialkyl sulfates, respectively [*34*] (equation 32) (Table 13). In these reactions,

Table 12. Preparations and Reactions of Sulfinate Esters [*30*]

Reactants		*Reactions Conditions*	*Products*	*Yield (%)*
$CF_3S(O)F$	ROH	NaF(CsF)	$CF_3S(O)OR$	90
	$CF_3(CH_3)CHOH$	25 °C		87
	$CF_3(CH_3)_2COH$			96
	$(CH_3)_2CHOH$			82
	$(CH_3)_2C(CH_3)OH$			95
	$CF_3(CF_2Cl)CHOH$			70
$CF_3(CH_3)CHOS(O)CF_3$		ClF, –78 °C	$CF_3(CH_3)C(F)OCl$	
$(CF_3)_2C(CH_3)OS(O)CF_3$		ClF, 25 °C	$(CF_3)_2C(CH_3)OCl$	

hard nucleophiles leave the sulfur–oxygen bond intact while the sulfur–fluorine bond is cleaved [*34*]. *N*-(Fluoroalkyl)anilines are prepared by the reaction of aniline with fluoroalkyl sulfonates, where fluoroalkyl *o*-nitrobenzenesulfonate proved to be a highly effective agent [*35*] (equation 33) (Table 13). In these reactions, the nucleophilic attack occurs at the α-carbon.

Amides and ureas may react also with **trifluoromethanesulfonic anhydride** to give resonance-stabilized dicarbonium salts [*36*] (equation 34).

[33] $$(CF_3)_2NONa + CF_3SO_2F \xrightarrow{CFCl_3} (CF_3)_2NOSO_2CF_3 + NaF \quad \mathbf{31}$$

[34] $$CF_3CH_2OSO_2F + NH_2CH_3 \longrightarrow CF_3CH_2OSO_2NHCH_3 \quad \mathbf{32}$$

[35] $$C_6H_5NH_2 + o\text{-}NO_2C_6H_4SO_3CH_2CF_3 \longrightarrow C_6H_5NHCH_2CF_3 \quad \mathbf{33}$$

Table 13. Reactions of Sulfonate Esters with Nucleophiles

Reactants	*Nucleophile*	*Reaction Conditions*	*Product(s)*	*Yield (%)*	*Ref.*
$(CF_3)_2NOSO_2CF_3$	$C_6H_5ONa^+$ $C_6H_5SNa^+$ NH_3	CH_3OH or THF	CF_3SO_2Nu + $(CF_3)_2NOH$	50–100	33
$(CF_3)_2NOSO_2CF_3$	CsF	$(C_2H_5)_2O$ or glyme	CF_3SO_2F + CF_3SO_3Cs + $(CF_3)_2NON(CF_3)_2$		33
$(CF_3)_2NOSOCF_3$	LiI		I_2 + CF_3SO_3Li + $CF_3N{=}CF_2$		33
$CF_3CH_2OSO_2F$	Nu	25 °C	$CF_3CH_2OSO_3Nu$		34
			Nu = $HN(CH_3)_2$	96	34
			Nu = NH_2CH_3	58	34
			Nu = NH_3	50	34
			Nu = $HOCH_2CF_3$	87	34
			Nu = $HOCH(CF_3)_2$	94	34
			Nu = OCH_3	56	34
$CF_3CH_2OSO_2F$	CH_3SH	$(C_2H_5)_3N$	$CF_3CH_2SCH_3$	31	34
o-$NO_2C_6H_4SO_3R$	$C_6H_5NH_2$		C_6H_5NHR		
	R = $CHF_2CF_2CH_2$-			97	35
	$H(CF_2CF_2)_2CH_2$		$H(CF_2)_4CH_2$	78	35
	$H(CF_2CF_2)_3CH_2$		$H(CF_2)_6CH_2$	65	35
	CF_3CH_2			91	35
	$C_6H_5NHCH_2CH_2OH$		$C_6H_5N(R)CH_2CH_2OH$		35

[36] **34**

$$R^1C(=O)R^2 + (CF_3SO_2)_2O \xrightarrow[0\,°C]{CH_2Cl_2} R^1R^2C^+\text{-}O\text{-}C^+R^1R^2 \cdot 2CF_3SO_3^-$$

R^1	R^2
NH_2	NH_2
$N(CH_3)_2$	$N(CH_3)_2$
CH_3	$N(CH_3)_2$
C_6H_5	NH_2
C_6H_5	$N(CH_3)_2$
C_6H_5	$N(C_2H_5)_2$
C_6H_5	$C_5H_{10}N$
C_6H_5	$N[CH(CH_3)_2]_2$

Fluorinated Organophosphorus Compounds

Fluorine-containing organophosphorus compounds have very broad and general synthetic utility. (The reaction of organophosphorus compounds with fluorinated carbonyl-containing species may be found in the section on the aldol reaction.) *Fluoro(organyl)phosphanes,* trivalent phosphorus-containing species where phosphorus directly bears fluorine, have been prepared by exchange reactions with the triethylamine–hydrogen fluoride reagent [*37*] (equation 35).

The reactivity of fluoro(organyl)phosphanes depends on the nucleophilicity of phosphorus. For instance the phosphanes may add to the carbonyl carbon to form *phosphinito phosphoranes* (equation 36) or may react with acid fluorides to form *acyldifluorophosphoranes* [*37*] (equation 37).

[37] $$R_{3\text{-}x}PCl_x + X\,(C_2H_5)_3NHF \xrightarrow{Et_3N} R_{3\text{-}x}PF_x + X\,(C_2H_5)_3NHCl \quad \mathbf{35}$$
$X = 1, 2$ or 3

$$R_2^1PF + R^2CHO \longrightarrow R_2^1PF_2CHR^2OPR^1_2 \quad \mathbf{36}$$
$R^1 = C_4H_9, C_6H_5$
$R^2 = C_6H_5, C_2H_5, (CH_3)_2CH, CH_2{=}CH, C_6H_5CH{=}CH$

[37] $$R^1_2PF + R^2COF \longrightarrow R^1_2F_2PCOR^2 \quad \mathbf{37}$$
$R^1 = C_6H_5, C_6H_5CH_2$
$R^2 = CH_3, C_6H_5$

Trivalent phosphorus species containing organofluorine groups have been prepared by pyrolysis at 330 °C and 0.01 torr of pentafluoroethyltrifluoromethyltrimethylstannylphosphanes [*38*] (equation 38).

These ***phosphaalkenes*** are extremely reactive; they undergo facile [2 + 4] *cycloaddition reactions* (equation 39) or reactions with protic acids.

[38] **38**

$$(CH_3)_3SnP(CF_3)(CF_2CF_3) \xrightarrow[0.01\ Torr]{330\ °C} CF_3P{=}C(CF_3)(F) + CF_3CF_2P{=}CF_2$$

[38] **39**

$$CF_3P{=}CFCF_3 + CF_3CF_2P{=}CF_2 \xrightarrow{\text{2,3-dimethylbutadiene}} \text{cycloadduct (P–}CF_3\text{, }CF(CF_3)\text{)} + \text{cycloadduct (P–}CF_2CF_3\text{, }CF_2\text{)}$$

Much better known are the *fluorinated phosphoranes,* which have been widely used in the *Wittig reaction* for the preparation of fluoroolefins. *Difluoromethylenation reactions* have been effected by using a variety of conditions. Treatment of dibromodifluoromethane with two equivalents of tris(dimethylamino)phosphine in carefully dried triglyme yields a solution of bromodifluoromethylphosphonium broomide, which very effectively converts *ketones* to *difluoromethylene derivatives.* A more sensitive reagent is prepared by the addition of two equivalents of the phosphine to the reaction mixture of fluorohalomethane and a carbonyl compound [*39, 40*] (equation 40) (Table 14).

However, these methods suffer from their sensitivity to the purity of the solvent. Addition of one equivalent of zinc dust and use of dimethylacetamide make the reaction much more reproducible [*41*] (equation 41) (Table 15).

Bromodifluoromethylphosphonium bromide has also been used in chain extension reactions to form dienes [*42*] (equation 42) or in bisdifluoromethylenation of diones [*43*] (equation 43) or halofluoromethanes [*44*], which may themselves be used in the preparation of new phosphonium salts (Table 16).

[39,40] **40**

$$CF_2Br_2 + 2\,[(CH_3)_2N]_3P \xrightarrow[\text{triglyme}]{R^1\,CO\,R^2} R^1\,R_2\,C = CF_2$$

$R^1 = C_6H_5$, $R^2 = CH_3$ 81%

$R^1 = C_6H_5$, $R^2 = CF_3$ 85%

[41] **41**

$$(C_6H_5)_3P + CF_2Br_2 \xrightarrow{\text{2 triglyme}} (C_6H_5)_3P{=}CF_2 \xrightarrow{R\,CHO} RCH{=}CF_2$$

$R = C_6H_{13}, C_{10}H_{21}, C_6H_5CH{=}CH$ 56 - 72%

References are listed on pages 610–614.

Table 14. Synthetic Utility of Fluorinated Phosphonium Salts Prepared From Dibromodifluoromethane [*40*]

>C=O	*R_3P*	*Temp.*	*>C=CF$_2$*	*Yield (%)*
C_6H_5CHO	Ph_3P	70 °C	$C_6H_5CH{=}CF_2$	65
C_6H_5CHO			$C_6H_5CH{=}CF_2$	20
$C_6H_5COCF_3$			$C_6H_5C(CF_3){=}CF_2$	85
$C_6H_{11}COCF_3$			$C_6H_{11}C(CF_3){=}CF_2$	90
$C_6H_5COC_2F_5$			$C_6H_5C(C_2F_5){=}CF_2$	82
$CH_3(CH_2)_5CHO$			$CH_3(CH_2)_5CH{=}CF_2$	72
m-$BrC_6H_4COCF_3$			*m*-$BrC_6H_4C(CF_3){=}CF_2$	83
$C_6H_5COCH_3$			C6$H_5C(CH_3){=}CF_2$	2
$C_6H_5COCH_3$	$(Me_2N)_3P$	RT	$C_6H_5C(CH_3){=}CF_2$	81
$C_6H_5COC_2H_5$	$(Me_2N)_3P$		$C_6H_5C(C_2H_5){=}CF_2$	82
$(C_2H_5)_2CO$	$(Me_2N)_3P$		$(C_2H_5)_2{=}CF_2$	75
=O	$(Me_2N)_3P$		=CF_2	71
$(CH_3)_2CO$	$(Me_2N)_3P$		$(C_2H_5)_2{=}CF_2$	60
$C_6H_5COCF_3$	$(Me_2N)_3P$		$C_6H_5C(CF_3){=}CF_2$	25

$$[(C_6H_5)_3P^+CF_2Br]Br^- + CF_2{=}C(C_6H_5)CF_2CF_3 \xrightarrow[2.\ H_2O]{1.\ Hg\ /\ CH_3CN} \underset{62\%}{CF_2{=}CFC(C_6H_5){=}CFCF_3} \qquad \textbf{42}$$

[*42*]

$$R\overset{O}{\overset{\|}{C}}(CF_2)_n\overset{O}{\overset{\|}{C}}R + \text{excess } (C_6H_5)_3P + \text{excess } CF_2Br \xrightarrow{70\ ^\circ C} \underset{21 - 60\%}{F_2C{=}C(R)(CF_2)_nC(R){=}CF_2} \qquad \textbf{43}$$

[*43*]

n = 2, 3, 4

Difluoromethylenation reactions may be effected also by the addition of chlorodifluoromethane to a solution of nonstabilized phosphorus ylide [*45, 46*] (equation 44) (Table 17).

Dichlorodifluoromethane and triphenylphosphine were used to prepare 3-deoxy-3-*C*-difluoromethylene ribose analogues in up to 65% yield [*47*] (equation 45).

An older procedure based upon the thermally induced decarboxylation of sodium chlorodifluoroacetate in the presence of triphenylphosphine was used to introduce the difluoromethylene group into a substituted benzo[*b*]fluoranthene [*48*] (equation 46).

Table 15. Preparation of *gem*-Difluoroolefins [*41*]

$R^1R^2C=O$	*Product **1**[a]*	*Method[b]*	*^{19}F NMR Yield (Isolated Yield, %)*
$C_6H_{13}CHO$	$C_6H_{13}CH{=}CF_2$	A	72
		A	9
		B	~0
		C	80 (60)
		D	77
		E	
$C_{10}H_{21}CHO$	$C_{10}H_{21}CH{=}CF_2$	C	76 (64)
C_6H_5CHO	$C_6H_5CH{=}CF_2$	A	65
		C	76 (56)
		E	
$C_6H_5CH{=}CHCHO$	$C_6H_5CH{=}CHCH{=}CH_2$	A	(56)[c]
		C	33 (30)
$C_6H_5COCH_3$	$C_6H_5(CH_3)C{=}CF_2$	B	~0
		C	<5
		D	43[d]
		E	

[a]All products exhibited spectral (IR, ^{19}F NMR, and/or MS) data in accord with the assigned structures and/or the reported literature values.

[b]A: CBr_2F_2-Ph_3P (2 equiv.), triglyme; B: CBr_2F_2-$(Me_2N)_3P$ (2 equiv.), triglyme; C: CBr_2F_2-Ph_3P (1 equiv.), Zn dust (1 equiv.), DMAA; D: CBr_2F_2-$(Me_2N)_3P$ (1 equiv.), Zn dust (1 equiv.), DMAA; E: CHF_2I-Ph_3P (1 equiv.), Zn (Cu), triglyme.

[c]GLC yield.

[d]36% Yield in TFH.

Table 16. Preparation of Halofluoromethanes from Halofluoromethyl Phosphonium Salts [44]

$$[R_3P{+}PCFXY]Z^- + \text{halogen} + KF \xrightarrow{\text{solvent}} \text{halofluoromethanes}$$

Phosphonium salt	*Halogen*[a]	*Solvent*	*Product (Yield, %)*[b]
$[Ph_3P^+CF_2Br]Br^-$	I_2	TG[c]	CF_2BrI (41), CF_2I_2 (9)
$[Ph_3P^+CF_2Br]Br^-$	I_2	TG	$CFBr_2I$ (31), $CFBr_3$ (12), $CHFBr_2$ (10)
$[Ph_3P^+CF_2Br]Br^-$	I_2	DMF[d]	$CFBr_2I$ (52), $CFBr_3$ (20), $CFBrI_2$ (15)
$[Ph_3P^+CF_2Br]Br^-$	I_2	Tetr. G[e]	$CFBr_2I$ (57), $CFBr_3$ (15), $CFBrI_2$ (13), $CHFBr_2$ (10)
$[Ph_3P^+CF_2Br]Br^-$	IBr	Tetr. G	$CFBr_2I$ (32), $CFBr_3$ (31), $CHFBr_2$ (9)
$[(Me_2N)_3P^+CF_2Cl]Cl^-$	I_2	TG	CF_2ClI (53; 34)
$[(Me_2N)_3P^+CF_2Cl]Cl^-$	ICl	Tetr. G	CF_2ClI (18), CF_2Cl_2 (28)
$[(Me_2N)_3P^+CF_2Cl]Cl^-$	Br_2	TG	CF_2BrCl (5), CF_2Br_2 (3)
$[(Me_2N)_3P^+CF_2Cl]Cl^-$	I_2	TG	$CFCl_2I$ (63; 31), $CFCl_3$ (27)
$[(Me_2N)_3P^+CF_2Cl]Cl^-$	ICl	Tetr. G	$CFCl_2I$ (43), $CFCl_3$ (24)
$[(Me_2N)_3P^+CF_2Cl]Cl^-$	Br_2	TG	$CFCl_2Br$ (42), $CFCl_3$ (15), $CFClBr_2$ (5), $CFHCl_2$ (25)

[a]Unless otherwise noted, a slight excess of halogen and three-fold excess of KF was used.
[b]^{19}F NMR yield versus $PhCF_3$. When two values are given, the second value is the isolated yield. Products were identified either by spiking with an authentic sample or by combination of IR and mass spectra data of the isolated products.
[c]TG, triethylene glycol.
[d]Dimethylformamide.
[e]Tetr. G, tetraethylene glycol.

$$2\ (C_6H_5)_3P^+\text{-}C^-R^1R^2 + HCF_2Cl \qquad \mathbf{44}$$

$$[45] \longrightarrow (C_6H_5)_3P + [(C_6H_5)_3P^+CHR^1R^2]X^- + F_2C{=}CR^1R^2 \quad 35-93\%$$

$R^1 = H, CH_3, C_6H_5$
R^2 = alkyl, aryl

Monofluoroalkenes have been prepared by the addition of **fluoromethylenetriphenylphosphorane** generated in situ by treatment of fluoroiodomethyltriphenylphosphonium iodide with zinc–copper couple in dimethylformamide [*49*] (equation 47) (Table 18).

Table 17. Reactions of Non-Stabilized Phosphorus Ylides 2 $Ph_3P^+C^-R^1R^2$ with $CHClF_2$ [46]

R^1	R^2	*Solvent*	*Yield of Olefin[a] (%)*
H	C_3H_7	Triglyme	88 (58)[b]
H	C_6H_{13}	Triglyme	93 (60)
H	$C_{11}H_{23}$	Et_2O	90 (68)
H	C_6H_5	Triglyme	92 (60)
H	p-$O_2NC_6H_4$	THF	Trace
H	C_6H_5	Triglyme	Trace
H	$CH{=}CHC_6H_5$	Triglyme	20 (15)
H	$CH{=}CH_2$	Triglyme	Trace
CH_3	CH_3	Triglyme	83 (70)
CH_3	C_2H_5	Triglyme	81 (52)
CH_3	C_6H_5	Triglyme	82 (60)
C_6H_5	C_6H_5	Et_2O	62 (57)
	C_5H_{10}	Et_2O	~100 (80)
	C_4H_8	Triglyme	90 (72)
H	CH_3O	Triglyme	65 (40)
H	$(CH_3)_2CHCH_2O$	Triglyme	35 (12)
H	p-$F_2C{=}CHC_6H_4$	THF	35 (12)

[a]GLC yield is based on thermal conductivity corrections relative to an appropriate internal standard. Product identity was confirmed by comparison of ^{19}F NMR, ^1H NMR, IR, and/or mass spectra with those of authentic samples prepared by alternative routes when possible. Yields are based on ylide. Isolated yields are given in parentheses.

[47] **45**

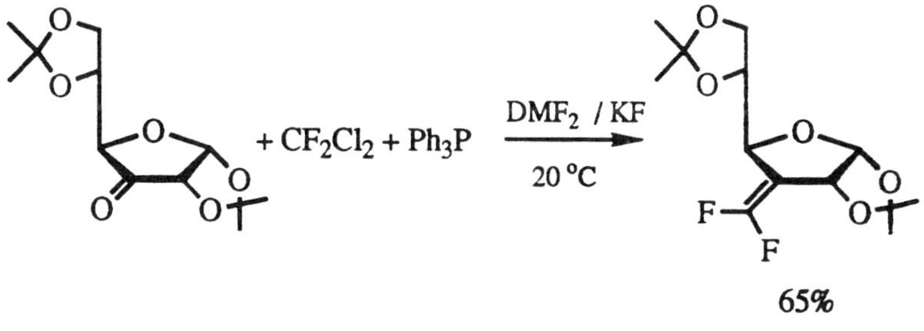

$+ CF_2Cl_2 + Ph_3P \xrightarrow[20\ ^\circ C]{DMF_2\ /\ KF}$

65%

[48] (fluoranthene-type aldehyde, CHO) $\xrightarrow[\text{diglyme, 160 °C}]{ClF_2CCO_2Na / Ph_3P}$ (difluorovinyl product, F, F) **46**

41%

[49] $[(C_6H_5)_3P^+CHFI]I^-$ + Zn/Cu + R CO R^1 $\xrightarrow{\text{DMF}}$ R R^1 C=CHF **47**

12 - 80%

Table 18. Synthesis of Fluoroolefins via Reaction of Fluoromethylene-triphenylphosphorane with Carbonyl Compounds [49]

Product Olefin	*Method*[a]	*GLC Yield*[b] *(%)*	*cis:trans Ratio*
$C_6H_5C(CF_3)$=CHF	A	28.3[c]	39:61
	A	50.4[d]	46:54
$C_6H_5C(CH_3)$=CHF	A	48.8	49:51
$(C_6H_5)_2C$=CHF	A	46.8	
c-C_6H_{10}=CHF	A	69.3	
C_6H_5CH=CHF	A	43.5	44:56[e]
$C_6H_{13}CH$=CHF	A	25.7	48:52
$C_6H_5C(CF_3)$=CHF	A	48.0	49:51
$C_6H_5C(CF_3)$=CHF	B	79.9	52:48
C_6H_5CH=CHF	B	52.4	41:59
C_6H_5CH=CHF	B	65.0	54:46
$C_6H_{13}CH$=CHF	B	54.4	43:57
$C_6H_5C(CF_3)$=CHF	B	12.0	57:43

[a]A: Ylide was pregenerated by reaction with $LiN[CH(CH_3)_2]_2$;
B: ylide was generated in situ.
[b]GLC yields are based on thermal conductivity corrections relative to an appropriate standard.
[c]Betaine decomposes within 42 h at 28 °C.
[d]Betaine decomposes within 2 h at 0 °C
[e]Isomers were not completely separable by GLC; the isomer ratio was determined by using an integrated ^{1}H NMR spectrum of the isomer mixture.

Halofluoroalkenes may be prepared by using fluorodihalomethanes or fluorohalomethanes in olefination procedures similar to those described above. Fluorotrichloromethane treated with tris(dimethylamino)phosphine forms the corresponding phosphonium salt, which can then be used in the *Wittig procedure.* The reaction depends on the nature of the solvent; in tetrahydrofuran, little olefination if any occurs; however, when benzonitrile is added to the mixture, ylide formation is promoted [*50*] (equation 48) (Table 19).

[50] **48**

$$CFCl_3 + [(CH_3)_2N]_3P \longrightarrow \left[[(CH_3)_2N]_3P^+ CFCl_2 \right] Cl^-$$

$$\left[[(CH_3)_2N]_3P^+ CFCl_2 \right] Cl^- + Ph_3P \xrightarrow{R^1COR^2} R^1R^2C{=}CClF \quad 40-83\%$$

Table 19. Preparation of Chlorofluoromethylene Olefins: Carbonyl, Solvent, and Phosphine Survey [*50*]

$$[((CH_3)_2N)_3P^+CFCl_2]Cl^- + R_3P + R^1R^2C{=}O \xrightarrow{60\ °C} \textit{cis \& trans}\ R^1R^2C{=}CFCl$$

R	*R¹*	*R²*	*Solvent*	*Reaction time[a] (h)*	*Olefin Yield[b] (%)*	*cis:trans Ratio*
C_6H_5	CF_3	C_6H_5	THF	2	0	—
C_6H_5	CH_3	C_6H_5	THF	10	3	53:47
C_6H_5	CH_3	C_6H_5	C_6H_5CN	1.5	83	42:58
C_6H_5[c]	H	C_6H_5	C_6H_5CN	3	60	56:44
C_6H_5	CF_3	C_6H_5	C_6H_5CN	11	3	50:50
C_6H_5	CF_3	*i*-C_3H_7O	C_6H_5CN	144	40	10[d]
$(CH_3)_2N$	CH_3	C_6H_5	C_6H_5CN	3	56	52:48
$(CH_3)_2N$	CF_3	*i*-C_3H_7O	C_6H_5CN	2	67	10[d]

[a]Time when maximum olefin yield was initially realized.
[b]GLC yield is based on starting carbonyl compound.
[c]Reaction at 100 °C
[d]Determined by ^{19}F NMR with α,α,α-trifluorotoluene as internal standard.
[e]Reaction at room temperature.

Fluorotrihalomethanes react with two equivalents of triphenylphosphine or a trialkylphosphine to form a fluorinated phosphoranium salt [*51*] (equation 49). The formation of the phosphoranium salts depends upon the solvent system. These salts may in turn be used for *Wittig reactions* [*52, 53*] (equation 50) (Table 20).

$$3\,[C_4H_9]_3P \;+\; CF X_3 \longrightarrow \qquad \textbf{49}$$

[51]

$$\left[[C_4H_9]_3P^+C^-FP^+\,[C_4H_9]_3\right]X^- \;+\; [C_4H_9]_3P\,X_2$$

[52,53]

$$\left[(C_4H_9)_3P^+\text{-}C^-F\text{-}P^+(C_4H_9)_3\right]X^- \xrightarrow[-(C_4H_9)_3PO]{RCHO} CHF{=}CHR \quad 75 - 96\% \qquad \textbf{50}$$

$$\left[(C_4H_9)_3P^+\text{-}C^-F\text{-}P^+(C_4H_9)_3\right]X^- \xrightarrow[-(C_4H_9)_3PO]{R_fCOF} (H)(F)C{=}C(F)(R_f) \quad 20 - 62\%$$

Whereas access to fluoromethylene ylides is easy, higher homologues are more difficult to prepare because of the facile β-elimination of fluorine. However **hexafluoroisopropylidene ylides** may be prepared in situ from tetrakis(trifluoromethyl)-1,3-dithietane and triphenylphosphine [*54*] (equation 51) or 2,2-dichlorohexafluoropropane and triphenylphosphine [*55, 56, 57*] (equation 52) (Table 21).

Perfluoro-2-butene reacts with tributylphosphine to cleanly form a solution of a fluorinated ylide [*58*] (equation 53).

51

[54]

$$4\,Ph_3P + (CF_3)_2C\underset{S}{\overset{S}{\diamond}}C(CF_3)_2 \longrightarrow [2[Ph_3P{=}C\,(CF_3)_2]] \;+\; 2\,Ph_3PS$$

$$\downarrow 2\,R\,CHO$$

$$2\,RCH{=}C(CF_3)_2 \;+\; 2\,Ph_3PO \quad 26 - 64\%$$

[55,56,57]

$$CF_3CCl_2CF_3 \;+\; 2\,Ph_3P \;+\; R\,CHO \qquad \textbf{52}$$

$$\downarrow CH_2Cl_2,\; < -50\,^{\circ}C$$

$$R\,CH{=}C(CF_3)_2 \;+\; Cl_2PPh_3 \;+\; Ph_3PO \quad 4 - 88\%$$

Table 20. Fluorine-Containing Phosphoranium Salts in Fluoroolefin Syntheses [*52, 53*]

Aldehyde [52] *or Acyl Fluoride* [53]	*Fluoroolefin Yield*[a] *(%)*	*Z/E*[b]
C_6H_5CHO	(76) 61	13/87
p-$CH_3C_6H_4CHO$	(74) 54	12/88
p-$CH_3OC_6H_4CHO$	(78) 51	17/83
p-ClC_6H_4CHO	(81) 60	25/75
p-$NO_2C_6H_4CHO$	(73) 57	25/75
m-$CF_3C_6H_4CHO$	50	0/100
o-$CH_3C_6H_4CHO$	54	20/80
o-$CH_3OC_6H_4CHO$	(64)	5/95
$CH_3(CH_2)_5CHO$	(71) 51	100/0
$CH_3(CH_2)_6CHO$	(73) 57	
$C_6H_{11}CHO$	(77) 50	
CF_3COF	45	
CF_3CF_2COF	62	
$CF_3CF_2CF_2COF$	52	
CF_2ClCOF	50	
$CH_3O_2CCF_2COF$	20	
$C_3F_7OCF(CF_3)COF$	49	

[a]Parentheses indicate ^{19}F NMR yield versus benzotrifluoride.
[b] *Z/E* ratios were calculated by means of ^{19}F NMR.
[c]^{19}F NMR yield.
[d]Isolated yield of pure olefin.
[e]The NMR, MS, and IR data were fully consistent with the assigned structures.

[58] **53**

$$CF_3CF{=}CFCF_3 + (C_4H_9)_3P \xrightarrow[-70^{\circ} \text{ to RT}]{Et_2O} \left[CF_3CF{=}\underset{\displaystyle CF_3}{\underset{|}{C}}\text{-}P^{+}(C_4H_9)_3 \right] F^{-}$$

$$\longrightarrow CF_3CF_2\underset{\displaystyle CF_3}{\underset{|}{C^{-}}}\text{-}P^{+}(C_4H_9)_3$$

Table 21. 1,1-Bis(trifluoromethyl)alkenes Prepared from Hexafluoroisopropylidenetriphenylphosphorane [*56*]

Aldehyde	*Yield of $RCH{=}C(CF_3)_2$[a] (%)*
CH_3CH_2CHO	31
$CH_3CH_2CH_2CHO$	22
p-ClC_6H_4CHO	88
p-$CH_3OC_6H_4CHO$	27
(furan-2-yl) O CHO	37
p-$(CHO)C_6H_4CHO$	50
1,2-$(CHO)_2C_6H_4$	10
$Cl_2C{=}CHCHO$	83
$(CH_3)_2CHCHClCHO$	62
C_6H_5CHO	54
C_6H_5CHO	99 (50)[b,c]
p-$CH_3C_6H_4CHO$	65 (42)
p-$CH_3OC_6H_4CHO$	52 (40)
m-$CF_3C_6H_4CHO$	89 (64)
$O_2NC_6H_4CHO$	70 (48)
C_6H_5CHO	100 (62)
$CH_3(CH_2)_5CHO$	100 (38)
$C_6H_5CH{=}CHCHO$	60 (56)
(furan-2-yl) O CHO	69 (31)
(pyridin-2-yl) N CHO	55 (42)

[a]Yield of pure isolated product.
[b]Yields were determined by ^{19}F NMR vs. benzotrifluoride; isolated yields are given in parentheses.
[c]All products exhibited spectral data in accord with the assigned structure and gave satisfactory elemental analyses.

Fluorinated ylides have also been prepared in such a way that fluorine is incorporated at the carbon β to the carbanionic carbon. Various fluoroalkyl iodides were heated with triphenylphosphine in the absence of solvent to form the necessary phosphonium salts. Direct deprotonation with butyllithium or lithium diisopropylamide did not lead to ylide formation; rather, deprotonation was accompanied by loss of fluoride ion. However deprotonation with hydrated potassium carbonate in dioxane was successful and resulted in fluoroolefin yields of 45–80% [*59*] (equation 54).

β-Fluorinated ylides may also be prepared by the reaction of an isopropylidenetriphenylphosphine ylide with a perfluoroalkanoyl anhydride. The intermediate acyl phosphonium salt can undergo further reaction with methylene triphenylphosphorane and phenyllithium to form a new ylide, which can then be used in a Wittig olefination procedure [*60*] (equation 55) or can react with a nucleophile [*61*] such as an acetylide to form a fluorinated enyne [*62*] (equation 56).

$$Ph_3P + ICH_2CH_2R_f \xrightarrow{95\,^oC} [Ph_3P^+CH_2CH_2R_f]\,I^- \quad \textbf{54}$$

$R = C_6H_5$, p-OHC-C_6H_5

[59]

$$\xrightarrow[2.\ RCHO]{1.\ K_2CO_3,\ H_2O} R\,CH{=}CH\,CH_2R_f$$

45 - 80%

$$Ph_3P{=}C(CH_3)_2 \xrightarrow[-78\,^oC]{(R_f CO)_2O} \left[Ph_3P^+C(CH_3)_2C(=O)R_f \right] R_fCO_2^- \quad \textbf{55}$$

[60]

$$\xrightarrow[PhLi,\ -25\,^oC]{Ph_3P=CH_2} Ph_3P{=}CH{-}C(R_f){=}C(CH_3)_2 \xrightarrow{R\,CHO} RCH{=}CH{-}C(R_f){=}C(CH_3)_2$$

20 - 58%

[62]

$$Ph_3P^+C^-R^1R^2C(=O)R_f \xrightarrow[R^3C\equiv CLi]{-\ Ph_3PO} R^1R^2C{=}C(R_f){-}C{\equiv}C{-}R^3 \quad \textbf{56}$$

43 - 80%

$R^1{=}R^2{=}CH_3$, $(CH_2)_4$

$R^1{=}CH_3$, $R^2{=}C_6H_5CH_2$, C_3H_7

$R^3{=}C_4H_9$, C_6H_5

Fluorinated alkynes are readily prepared by treatment of the intermediate formed by acylation with a second equivalent of ylide followed by pyrolysis [*63, 64, 65*] (equation 57).

Hexafluorobenzene may also add to methylene triphenylphosphorane to form a new pentafluorophenyl-bearing ylide. Treatment of this ylide with an acid fluoride or acid anhydride followed by pyrolysis (shown in equation 58) forms the corresponding *pentafluorophenylacetylene* [*66*] (equation 58).

(1-Fluoroacetyl)methylenetriphenylphosphorane has been prepared from the corresponding phosphonium salt. Such *Wittig reagents* react well with aldehydes to form alkenyl fluoromethyl ketones [*67*] (equation 59).

It has also been possible to prepare extremely stable **triphenylphosphonium (trifluoromethanesulfonyl)methylide** from chloromethyl trifluoromethyl sulfone and triphenylphosphine [*68*] (equation 60).

57

$$R_fC(O)Cl + 2\,Ph_3P{=}CHCHO \longrightarrow Ph_3P{=}C(CHO)(COR_f)$$

$R_f = CF_3, C_2F_5, C_3F_7$

[63,64,65]

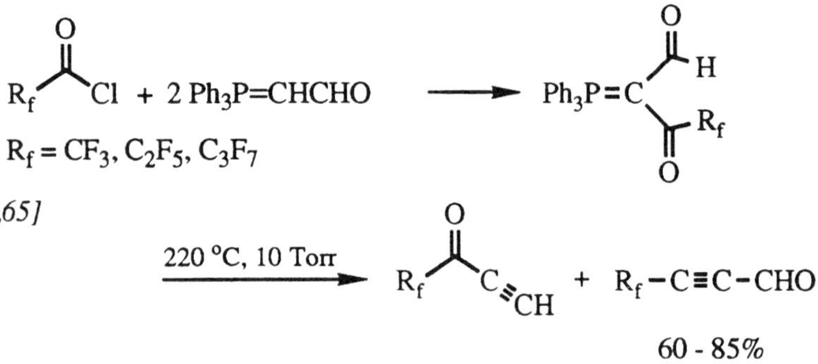

$$\xrightarrow{220\ °C,\ 10\ Torr} R_fC(O)C{\equiv}CH + R_f{-}C{\equiv}C{-}CHO$$

60 - 85%

$$[66]\quad Ph_3P{=}CH_2 \xrightarrow[-30\ °C\ to\ RT]{C_6F_6} Ph_3P{=}CHC_6F_5 \xrightarrow[or\ R\,COF]{(R\,CO)_2O} Ph_3P^{+}C(C_6F_5){=}C(O^{-})R \quad \mathbf{58}$$

51 - 83%

$$\downarrow 230\ °C\quad 2\ Torr$$

$$C_6F_5C{\equiv}C{-}R$$

85 - 96%

$$[67]\quad Ph_3P{=}CHC(O)CH_2F + R\,CHO \longrightarrow R\,CH{=}CH{-}C(O){-}CH_2F \quad \mathbf{59}$$

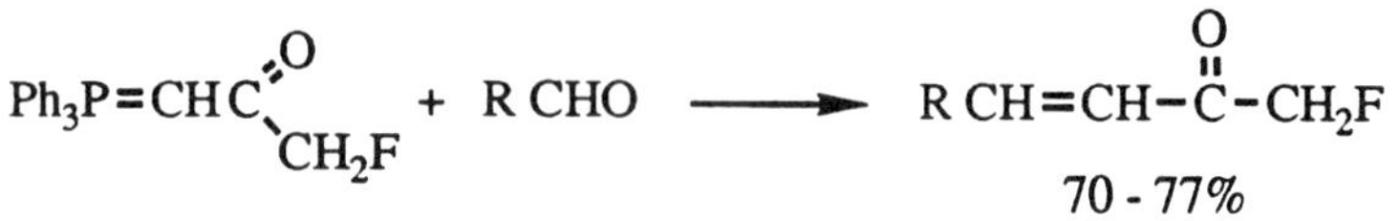

70 - 77%

$$[68]\quad ClCH_2SO_2CF_3 \xrightarrow[12\ h]{Ph_3P\ /\ (CH_3CH_2)_3N} Ph_3P{=}CHSO_2CF_3 \quad \mathbf{60}$$

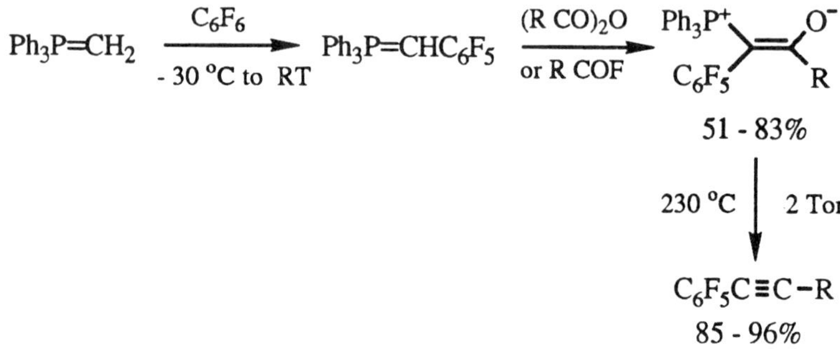

50%

References are listed on pages 610–614.

Alkyl diethylphosphonofluoroacetates have been used extensively in *olefination procedures* [*69*], principally forming the (*E*)-α-fluoro-α,β-unsaturated esters with very high stereoselectivity [*70*] (equation 61) (Table 22). Preparation of the ethyl diethylphosphonofluoroacetate from ethyl fluoroacetate has obviated the necessity to prepare ethyl bromofluoroacetate from bromine fluoride and ethyl diazoacetate [*71*].

[70] **61**

$$(C_2H_5O)_2\overset{O}{\overset{\|}{P}}CHFCO_2CH_3 + RCHO \xrightarrow[-78\ ^\circ C]{BuLi} \underset{H}{\overset{R}{}}\!\!>\!\!=\!\!<\!\!\underset{F}{\overset{CO_2CH_3}{}}$$

55 - 90%

Table 22. Methyl Diethylphosphonofluoroacetate in the Horner–Emmons Reaction [*70*]

Aldehyde	*Temp. (°C)*	*Yield (%)*	*E/Z*[a]
$(CH_3)_2CHCHO$	–78	90	≥ 98/2
$C_6H_5CH(CH_3)CHO$	–78	90	≥ 98/2
	(0)	90	75/25
$CH_3(CH_2)_4CHO$	–78	55	≥ 98/2
$CH_3CH{=}CHCHO$	–78	76	≥ 98/2
$C_6H_5CH{=}CHCHO$	–78	75	≥ 98/2
CHO (perillaldehyde structure)	20	75	≥ 85/15
CHO (myrtenal structure)	20	50	≥ 98/2
C_6H_5CHO	–78	95	≥ 98/2
O (β-ionone structure)	20	90	75/25
	20	90	70/30

[a]The *E/Z* ratio was determined by ^{19}F NMR.

The direct reduction of esters with diisobutylaluminum hydride in the presence of the *Horner–Emmons reagent* prepared from ethyl diethylphosphonofluoroacetate avoids the necessity to work with sensitive aldehydes in the olefination procedure [*72, 73*] (equation 62) (Table 23).

Trapping of the same Horner–Emmons reagent with an acid chloride leads to the formation of the *α-fluoro-β-keto esters* in good yields [*74*] (equation 63) (Table 24).

Both diethylphosphonofluoroacetic acid [*75*] and diethylphosphonofluoroacetonitrile [*76*] have been used in olefination procedures (Table 25).

Fluoroalkylphosphonates may also be deprotonated for use in olefination reactions with aldehydes [*77*] (equation 64).

$$[72,73]\quad (C_2H_5O)_2\overset{O}{\overset{\|}{P}}CHFCO_2C_2H_5 \xrightarrow[\text{THF, -78 °C}]{\text{1. BuLi; 2. }R^1CO_2R^2\text{; 3. DIBAL}} RCH{=}CFCO_2C_2H_5 \quad 45-67\% \quad E/Z\ 20:1-3:1 \qquad \mathbf{62}$$

Table 23. Preparation of Ethyl Alkylidene Fluoroacetates RCH=CFCOOEt from Esters Reduced in situ with Diisobutylaluminum Hydride [*73*]

R^1	R^2	*Method*[a]	*Isolated Yield*[b]	*E/Z*[c]	*bp (°C/mm of Hg)*
CF_3	C_2H_5	A	63	83/17	60–63/144
C_2F_5	C_2H_5	A	44	80/20	50–55/140
C_3F_7	C_2H_5	B	66	77/23	61–65/101
CF_2Cl	C_2H_5	A	64	89/11	64–67/98
CF_2Br	C_2H_5	A	64	88/12	55–60/67
CHFBr	C_2H_5	B	66	88/12	52–54/12
C_5H_{11}	C_2H_5	B	57	93/7	67–70/3
$(CH_3)_2CH$	CF_3	A	54	95/5	52–58/80
C_6H_5	C_4H_9	A	56	100/0	73–77/0.7
(*E*)-$CH_3CH{=}CH$	CH_3	A	55	5/95[d]	56–59/52
-$(CH_2)_3CO_2$		A	76	95/5	75–80/0.7
(*E*)-$C_6H_5CH{=}CF$	C_2H_5	B	45	100/0[d]	80–86/0.3

[a]A: Ester treatment first with DIBAL and then with $[(C_2H_5O)_2P(O)^-CFCO_2C_2H_5]$; B: ester added to a solution of $[(C_2H_5O)_2P(O)^-CFCO_2C_2H_5]$ and then reduced by DIBAL.
[b]Isolated yields are based on R^1COOR^2.
[c]*E/Z* ratios were determined by integration of the vinylic fluorine signals in ^{19}F NMR spectra.
[d]*EE/EZ* ratio.

$$[74]\quad (C_2H_5O)_2\overset{O}{\overset{\|}{P}}CHFCO_2C_2H_5 \xrightarrow[\substack{2.\ R_f\,COCl \\ 3.\ HCO_3^-}]{1.\ BuLi} R_f\overset{O}{\overset{\|}{C}}CHF\overset{O}{\overset{\|}{C}}OC_2H_5 \quad 57-67\% \qquad \mathbf{63}$$

Table 24. Preparation of Ethyl Alkylidenefluoroacetates from Esters Reduced in situ with Diisobutylaluminum Hydride [*74*]

R	*Method* [a]	*Isolated Yield* [b] *(%)*
CH_3	A	60
C_2H_5	A	50
$(CH_3)_2CH$	A	58
$(CH_3)_3C$	A	56
cyclo-C_6H_{11}	A	68
C_2H_5O	A	50
C_2H_5S	B	57
C_3H_7	B	77
CF_3	B	60
CF_2Cl	B	67
C_6H_5	A	70
$CH_3OCOCH_2CH_2$	A	38

[a] A: $Bu_3P{=}CFCO_2Et$ + RCOCl; B: $(EtO)_2P(O)^-CFCO_2Et$ + RCOCl.
[b] Isolated yields are based on acid chloride.

$$[77]\quad \left[(CH_3)_2CHO\right]_2\overset{O}{\overset{\|}{P}}CH_2F \xrightarrow[R\,CHO]{LDA} R\,CH{=}CF\overset{O}{\overset{\|}{P}}\left[OCH(CH_3)_2\right]_2 \qquad \mathbf{64}$$

R= $C_6H_5CH{=}CH$-, (2,2-dimethyl-1,3-dioxolan-4-yl)

Difluoroalkylphosphonates, prepared from difluoroallylic alkoxides and diethyl phosphorochloridate, are remarkably stable [78] (equation 65).

Fluoroalkenol phosphates are not only stable but also sufficiently reactive to undergo olefination reactions with ylides themselves. These enol phosphates are not only precursors to enolates or ketones but also can be used directly as electrophilic reagents [*79*] (equation 66) (Table 26).

Table 25. Olefination Reactions with Diethyl Phosphonofluoroacetonitrile $(EtO)_2P(O)CHFCN$

Carbonyl Compound RCHO or R^1R^2CO	*Yield (%) RCH=CFCN or R^1R^2C=CFCN*	*E/Z* [a]	*Ref.*
3,4-$(OCH_3)_2C_6H_3CHO$	46[b]	0.5/1	76
3,4-methylenedioxybenzaldehyde (structure: O–CH₂–O ring fused to benzene, CHO)	45[b]	4/1	76
4-$OCH_3C_6H_4CHO$	52[b]	1/1	76
2-$NO_2C_6H_4CHO$	62[b]	1.6/1	76
4-$C_6H_5C_6H_4CHO$	82[b]	3/1	76
4-$CH_3C_6H_5CHO$	51[b]	3/1	76
4-$NO_2C_6H_4CHO$	45[b]	2/1	76
i-C_3H_7CHO	76[a]	60/40	75
sec-C_4H_9CHO	74[c]	60/40	75
$C_6H_{13}CHO$	86[c]	55/45	75
CH_3, $(C_2H_5)CO$	91[c]	44/56	75
C_6H_5CHO	91[d]	0/100	75
p-$CH_3OC_6H_4CHO$	90[d]	0/100	75
CH_3, $(C_6H_5)CO$	87[d]	35/65	75

[a] *E/Z* ratios were determined from relative signal intensities in the ^{19}F NMR spectrum.
[b] Constants for the *Z* isomers are greater (33–36 Hz) than those for the *E* isomer (21–22 Hz).
[c] Yield of crude product based on the starting carbonyl compound.
[d] Yield of recrystallized product.

$$\text{[78]}\quad R\text{–C(CH}_3\text{)(OH)–CH=CF}_2 \xrightarrow[\text{DMAP, 48 h}]{(PhO)_2P(O)Cl,\ Et_3N} R(CH_3)C{=}CH\text{∼}CF_2OP(O)(OC_6H_5)_2 \quad 80\% \qquad \mathbf{65}$$

$$\text{[79]}\quad R_fCF_2CF{=}C(R)OP(O)(OC_2H_5)_2 + Ph_3P{=}CR^1R^2 \longrightarrow R_fCF{=}CF\text{–}C(R){=}CR^1R^2 \quad 42\text{ - }90\% \qquad \mathbf{66}$$

Table 26. Reaction between Perfluoro-1-alkenyl Phosphates $R_fCF_2CF{=}CROP(O)(OC_2H_5)_2$ and Phosphonium Ylides $Ph_3P{=}CR^1R^2$ To Give $R_fCF{=}CFCR{=}CR^1R^2$ [79]

R_f	R	R^1	R^2	Solvent	Temp. (°C)	Yield[a] (%)
CF_3	C_6H_{13}	C_5H_{11}	H	THF	67	88
		C_5H_{11}	H	C_6H_{14}	70	70
		C_6H_5	H	THF	RT	42
		Cl	Cl	C_6H_{14}	RT	40
CF_3	C_6H_5	C_5H_{11}	H	THF	RT	64
		C_5H_{11}	H	C_6H_{14}	RT	66
		C_6H_5	H	THF	RT	40
		Cl	Cl	C_6H_{14}	RT	62[b]
n-C_7F_{15}	n-C_3H_7	C_5H_{11}	H	THF	67	90
		C_6H_5	H	THF	RT	46
		Cl	Cl	C_6H_{14}	RT	47

[a]The yields are of pure isolated compounds, mixtures of *E* and *Z* stereoisomers.
[b]*E* stereoisomer only.

Fluorinated Organosilicon Compounds

Fluorinated organosilicon compounds include both those materials where the organic fragment bound to silicon is fluorinated as well as those substances where fluorine is bound directly to silicon. These compounds are useful as building blocks for new materials by serving as monomers [*80*], end groups for polymers [*81*], or surface-conditioning agents [*82*]. Fluorine-containing organosilicon reagents have been widely used as protecting groups and as reagents for the preparation of other fluorinated materials [*83*].

Fluorinated organosilicon compounds may be prepared from readily available ***fluorohalocarbons by using phosphorous amides*** [*84, 85, 86*] or ***tetrakis(dimethylamino)ethylene*** [*87*] for the generation of the fluoroorganic nucleophiles (equations 67–69).

Organometallic reagents, prepared from fluorohalocarbons, including Grignard reagents [*82*] or organolithium reagents prepared by transmetallation [*80*], are silylated with readily available alkylchlorosilanes (equations 70 and 71).

$$[84] \quad BrCF_2Cl + (CH_3)_3SiCl \xrightarrow{P[N(CH_2CH_3)_2]_3} \underset{75\text{-}80\%}{(CH_3)_3SiCF_2Cl} \qquad 67$$

$$[85]\quad CF_3Br + ClSiH[N(CH_3)_2]_3 \xrightarrow{P[N(CH_2CH_3)_2]_3} CF_3SiH[N(CH_3)_2]_2 \quad (86\%) \qquad 68$$

$$[86]\quad CF_3Br + ClSi(CH_2CH_3)_3 \xrightarrow{P[N(CH_2CH_3)_2]_3} CF_3Si(CH_2CH_3)_3 \quad (69\%)$$

$$[87]\quad CF_3I + (CH_3)_3SiCl \xrightarrow[0°C\text{ - }20°C]{(Me_2N)_2C{=}C(NMe_2)_2} CF_3Si(CH_3)_3 \quad (94\%) \qquad 69$$

[82] 70

$$CF_2{=}CCl_2 \xrightarrow[2.\ TMSCl]{1.\ BuLi} CF_2{=}CClSi(CH_3)_3 \xrightarrow[-100°C\text{ - }50°C]{RLi} RCF{=}CClSi(CH_3)_3 \quad (84\text{-}85\%)$$

[80] 71

$$CF_3CH_2CH_2Cl \xrightarrow[Et_2O]{Mg} CF_3CH_2CH_2MgCl \xrightarrow[Et_2O,\ 20°C]{Cl_2SiHCH_3} (CF_3CH_2CH_2)_2SiHCH_3 \quad (95\%)$$

Fluorinated organosilanes are also formed by the addition of **trimethylsilylated ketene acetals** to ***hexafluoroacetone*** [*88*], a process that appears to be driven by the enhanced electrophilicity of the carbonyl carbon, which results from the presence of two adjacent trifluoromethyl groups (equation 72).

Fluorinated organosilanes are used as reagents for the construction of carbon–carbon bonds and for the selective synthesis of phosphate esters.

[88] 72

$$(CH_3)_3SiCH{=}C(OCH_2CH_3)_2 + CF_3COCF_3 \longrightarrow (CF_3)_2C(OSi(CH_3)_3)CH{=}C(OCH_2CH_3)_2 \quad (96\%)$$

Trimethylpentafluorophenylsilane reacts with ***ketones*** in the presence of cyanide ion to form the *silyl ether* derived from the alkoxide created by the addition of the pentafluorophenyl anion to the carbonyl carbon [*89*]. The nucleofugal character of the pentafluorophenyl group is significantly greater than that of the added cyanide ion so that the addition proceeds under nearly neutral conditions (equation 73).

In a different application, the selective alcoholysis of methylphosphonic difluoride is promoted by the capture of fluoride ion by silicon [*83*] (equation 74).

[89] $(CH_3)_3SiC_6F_5 + CN^- \rightleftharpoons (CH_3)_3SiCN + C_6F_5^-$ **73**

$$C_6F_5^- + C_6H_5COCF_3 \rightarrow C_6H_5-C(O^-)(C_6F_5)-CF_3 \xrightarrow[CH_3CN]{(CH_3)_3SiC_6F_5} C_6H_5-C(OSi(CH_3)_3)(C_6F_5)-CF_3$$

85%

[83] $$CH_3P(O)F_2 + Si(OR)_4 \xrightarrow[15\ min]{RT} CH_3P(O)(F)OR + CH_3P(O)(OR)_2$$ **74**

$R = CH_3CH_2CH_2C(CH_3)_2$ — 79% — 19%

The presence of a trialkylsilyl group in a fluorinated organic compound may be useful to direct further transformations of that material. Yet in some instances it is the fluorinated substituent that controls the reactions of the trialkylsilyl group. Contrary to predictions, treatment of *tert*-butyl 3-trifluoromethyl-6-trimethylsilylphenyl carbamate with *tert*-butyllithium results in metallation of one of the methyl groups attached to silicon rather than that of the aromatic ring [*90*] (equation 75).

[90] **75**

$NHCO_2C(CH_3)_3$, $Si(CH_3)_3$, CF_3 (aryl) $\xrightarrow[2.\ PhCHO]{1.\ t\text{-}BuLi,\ 40°C}$ $NHCO_2C(CH_3)_3$, $Si(CH_3)_2CH_2CH(OH)C_6H_5$, CF_3 (aryl)

Fluorinated organosilicon compounds are very useful for introduction of trialkylsilyl protecting groups in synthesis. **Trialkylsilyl trifluoroacetamides** are perhaps the best known of these agents, for they can be used to prepare *silyl ethers or silyl esters* [*91, 92*] (equation 76).

The utility of **trialkylsilyl trifluoromethanesulfonates (trialkylsilyl triflates)** is also well-known [*93, 94, 95*]. The simplest of these reagents, *trimethylsilyl triflate,* may be prepared in high yield by the treatment of hexamethyldisiloxane with triflic anhydride [96] (equation 77).

Handling of this volatile and toxic material may be avoided by the clever use of the **trimethylsilylated perfluorinated resinsulfonic acid** [97]. This solid reagent is prepared by treatment of the acid form of NAFION 511 with chlorotrimethylsilane. This reagent exhibits significant stability in air (equation 78).

[91,92] **76**

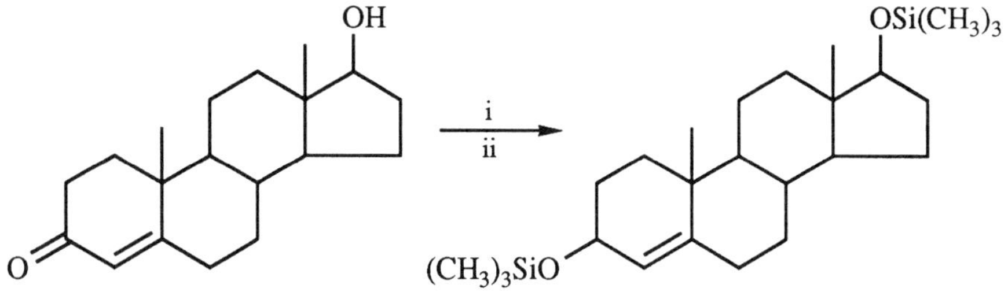

OH → (i, ii) → $OSi(CH_3)_3$; $(CH_3)_3SiO$

i: $CF_3CON(CH_3)Si(CH_3)_3$
ii: 60°C, 20 min

100%

[96] $(CH_3)_3SiOSi(CH_3)_3$ $\xrightarrow[70\text{-}75°C]{(CF_3SO_2)_2O}$ $(CH_3)_3SiOSO_2CF_3$ 88% **77**

[97] NAFION[TM] 511 + $\xrightarrow[80°C]{TMSCl}$ $-[(CF_2CF_2)_mCFCF_2]^-_n$ **78**

$-O-CF_2-CFCF_3-O-CF_2-CF_2-SO_3TMS$

The previously mentioned reagent, **trimethyl(pentafluorophenyl)silane** [*89*], may be used to prepare *silylated enols* [*98*]. The superior ability of the pentafluorophenyl group to distribute negative charge is responsible for the effectiveness of this reagent in transferring the trimethylsilyl group (equation 79).

Compounds where fluorine is bound directly to silicon may be prepared directly from silicon tetrafluoride, obtained by treatment of silicon tetrachloride with antimony trifluoride and an alkyllithium or lithium amide base [*99, 100*] (equation 80).

[98] $(CH_3\overset{O}{\overset{\|}{C}})_2NH$ $\xrightarrow{C_6F_5Si(CH_3)_3}$ $CH_3\overset{OSi(CH_3)_3}{\overset{|}{C}}=N\overset{O}{\overset{\|}{C}}CH_3$ 83% **79**

[99] **80**

$$3\ SiF_4 + 2\ LiNC(CH_3)_3Si(CH_3)_3 \xrightarrow[-70°C]{} 2\ SiF_3NC(CH_3)_3Si(CH_3)_3$$

When vinyl silicon trifluoride is treated with two equivalents of potassium fluoride, a new reagent, a **dipotassium organopentafluorosilicate,** is formed [*101*]. This intermediate has found application as a component in an efficient stereoselective copper chloride-promoted coupling reaction (equation 81).

[101] **81**

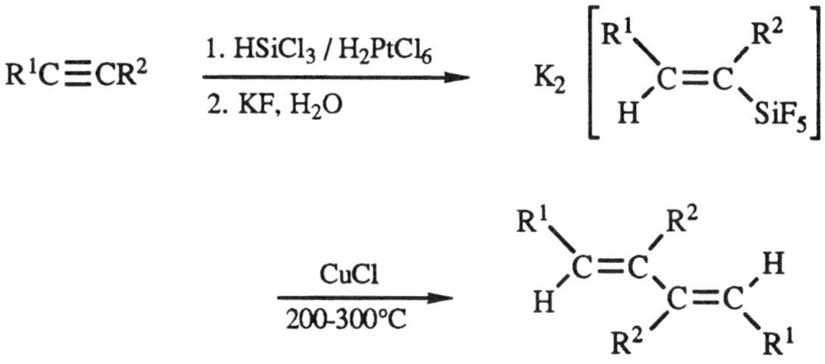

$$R^1C{\equiv}CR^2 \xrightarrow[2.\ KF,\ H_2O]{1.\ HSiCl_3\ /\ H_2PtCl_6} K_2\left[(R^1)(H)C{=}C(R^2)(SiF_5)\right] \xrightarrow[200\text{-}300°C]{CuCl} (R^1)(H)C{=}C(R^2){-}C(R^2){=}C(H)(R^1)$$

35-64%

E, E 92-99%

Alkyltrifluorosilanes and disubstituted difluorosilanes are themselves quite reactive with nucleophiles such as lithium amide bases [*102, 103, 104*], alkyllithium reagents [*105*], Grignard reagents [*105*], or alkoxides [*105*] (equations 82 and 83).

[102] **82**

$$R_{4\text{-}n}SiF_n + [(CH_3)_3Si]_2NH \longrightarrow R_{4\text{-}n}SiF_{n\text{-}1}\ NHSi(CH_3)_3 + FSi(CH_3)_3$$

60-90%

Interestingly it is also possible to form a stable silicon–fluorine bond by treatment of a methoxysilole with boron–trifluoride etherate or of a silole with trityl tetrafluoroborate [*106*]. The resultant fluorosilane is also a building block for further transformations of the silole (equation 84).

Fluorinated Organoboron Compounds

The bond between the trifluoromethyl group and boron is difficult to form. Preparation of *trifluoromethyl aminoboranes* was successfully achieved by transfer of trifluoromethyl group from reagent ***1***, which was prepared in situ [*107*] (equation 85).

[105] **83**

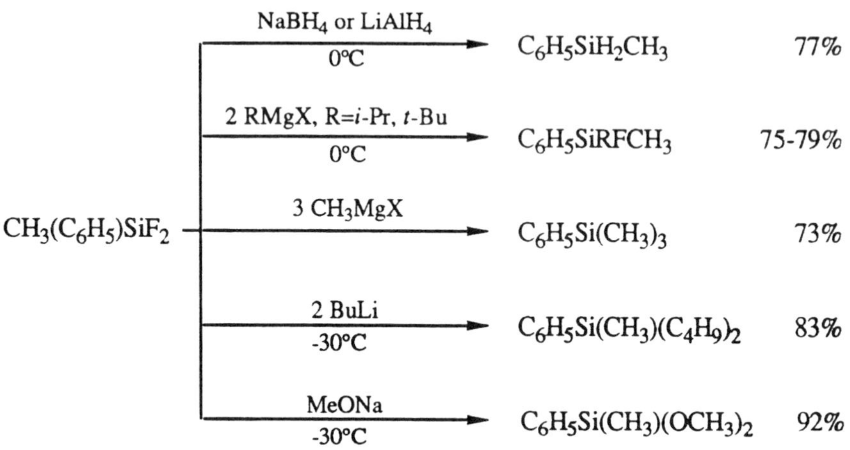

[106] **84**

CH3 CH3 — Si — CH3 OCH3

$\xrightarrow[\text{RT}]{BF_3 \cdot Et_2O}$ $\xleftarrow[0^\circ C]{Ph_3CBF_4}$

CH3 CH3 — Si — H CH3

CH3 CH3 — Si — CH3 F

[107] **85**

$$[(C_2H_5)_2N]_3P \xrightarrow{CF_3Br} [(C_2H_5)_2N]_3\overset{+}{P}Br\overset{-}{C}F_3$$

1

$\downarrow$ $BrB[N(CH_3)_2]_2$

$$CF_3B[N(CH_3)_2]_2 + [(C_2H_5)_2N]_3PBr{+}Br{-}$$

References are listed on pages 610–614.

These products are thermally stable at room temperature but decompose at elevated temperatures by elimination of CF_2. The reaction with hydrogen fluoride is different from that with hydrogen chloride and hydrogen bromide. Presumably this difference is derived from the strength of the boron–fluorine bond [*108*] (equation 86).

[108] **86**

$$CF_3B(N(CH_3)_2)_2 \xrightarrow[X = Br, Cl]{2HX} \left[(F_3C)(H)B(NH(CH_3)_2)_2 \right]^+ X^-$$

$$CF_3B(N(CH_3)_2)_2 \xrightarrow{3HF} CF_3BF_2{\bullet}NH(CH_3)_2 + [H_2N(CH_3)_2{}^+]F^-$$

Pentamethylcyclopentadienyl-substituted boron complexes are obtained by the reaction of the pentamethylcyclopentadienyl anion with boron trifluoride [*109*] (Table 27). Similarly, the Grignard reagent prepared from 3,5-bis(trifluoromethyl)iodobenzene reacts with sodium tetrafluoroborate to form the phase-transfer catalyst **2** under anhydrous conditions [*110*] (equation 87).

[110] **87**

$$3,5\text{-}(F_3C)_2C_6H_3I \xrightarrow[2.\ NaBF_4]{1.\ Mg,\ Et_2O} Na^+ \left[B(C_6H_3\text{-}3,5\text{-}(CF_3)_2)_4 \right]^-$$

2

Boron trifluoride also reacts with hydroxy acridones to form difluoro boron complexes [*111*] (Table 28). Another route to fluoroboranes involves transhalogenation by fluorination of the corresponding chloro- and bromoboranes with lithium or potassium fluoride under mild conditions [*112*] (Table 29).

Boron–nitrogen heterocycles [4,5-diethyl-2,2,3-trimethyl-1-(*O*-trifluoromethyl)phenyl-2,5-dihydro-1(–)-1,2,5-azasilaboroles and -azastannaboroles] are formed in good yields. The products demonstrate atropoisomerism due to the hindered rotation about the N–aryl bond [*113*] (equation 88).

The reaction of symmetrically substituted *N,N′*-diaryloxamides with bromodimethylborane yielded 1,1- or 1,2-addition products according to the ratio of the reactants [*114*] (Table 30).

References are listed on pages 610–614.

Table 27. Alkylation of Boron Halides

Reactants	Reaction Conditions	Product	Yield (%)	Ref.
$P[N(C_2H_5)_2]_3 + CF_3Br + BrB[N(CH_3)_2]_2$	CH_2Cl_2, –70 °C	$CF_3B[N(CH_3)_2]_2$	20	107
$P[N(C_2H_5)_2]_3$ + BrB(N-methyl diazaborolidine) + CF_3Br	CH_2Cl_2, –70 °C	CF_3B(N-methyl diazaborolidine)	25	107
$P[N(C_2H_5)_2]_3 + Br_2BN(CH_3)_2$ + 2 CF_3Br	CH_2Cl_2, –100 °C	$(CF_3)_2BN(CH_3)_2$	8	107
$(CH_3)_5C_5{}^aLi + BF_3{\cdot}Et_2O$	Et_2O, –40 °C	$(CH_3)_5C_5{}^aBF_2$	76	109
2 $(CH_3)_5C_5{}^aLi + BF_3{\cdot}Et_2O$	Et_2O, –40 °C	$[(CH_3)_5C_5{}^a]_2BF$	76	109

[a] $(CH_3)_5C_5$, pentamethylcyclopentadiene.

Table 28. Fluoroboration of Phenols [*111*]

Reactants	Reaction Conditions	Product
$BF_3{\cdot}Et_2O$ + 1-hydroxy-3,5-dimethoxyacridin-9(10H)-one (structure: O, H, O, N H, CH_3O, OCH_3)	Dioxane, RT	BF_2 chelate (structure: F, B, F, ^+O, O, N H, CH_3O, OCH_3)
$BF_3{\cdot}Et_2O$ + 1,3-dihydroxy-5-methoxyacridin-9(10H)-one (structure: O, H, O, N H, CH_3O, OH)	Dioxane, RT	BF_2 chelate (structure: F, B, F, ^+O, O, N H, CH_3O, OH)
1-hydroxy-3,5-dimethoxyacridin-9(10H)-one (structure: O, H, O, N H, CH_3O, OCH_3) + 2-methylbut-3-en-2-ol (OH) + $BF_3{\bullet}Et_2O$	Dioxane, reflux	prenylated BF_2 chelate (structure: F, B, F, ^+O, O, N H, CH_3O, OCH_3)

References are listed on pages 610–614.

Table 29. Preparation of Alkyl and Arylboron Fluorides [*112*]

Reactants	R^1	R^2	*Product*	*Yield (%)*
R^1R^2BCl + LiF			R^1R^2BF	
	C_6H_5	Cl	$C_6H_5BF_2$	65
	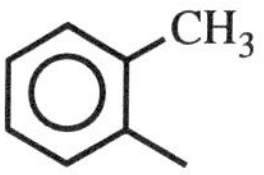 CH_3	Cl	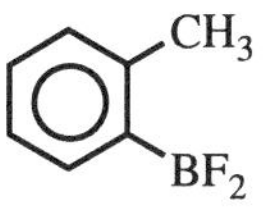 CH_3, BF_2	67
	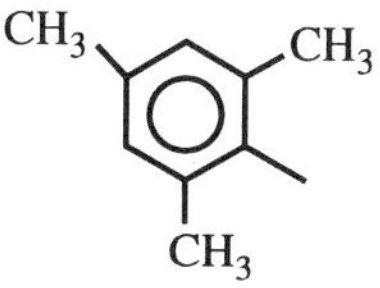 CH_3, CH_3, CH_3	Cl	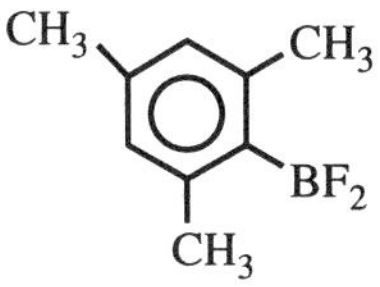 CH_3, CH_3, BF_2, CH_3	72
	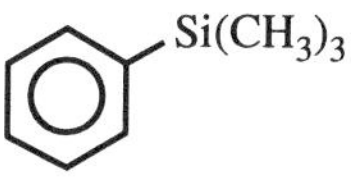 $Si(CH_3)_3$	Cl	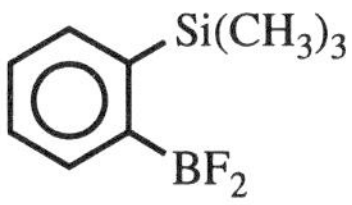 $Si(CH_3)_3$, BF_2	63
	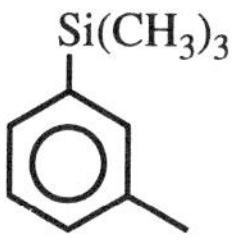 $Si(CH_3)_3$	Cl	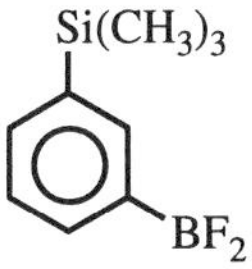 $Si(CH_3)_3$, BF_2	68
	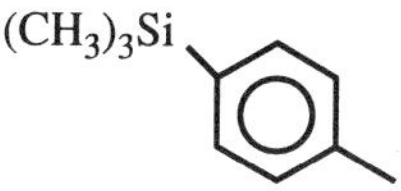 $(CH_3)_3Si$	Cl	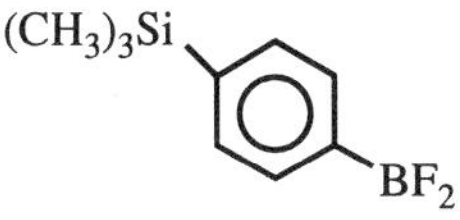 $(CH_3)_3Si$, BF_2	73
	C_8H_{17}	Cl + LiF	$C_8H_{17}BF_2$	73
	R^1, R^2 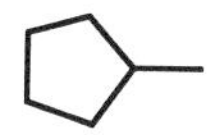 =		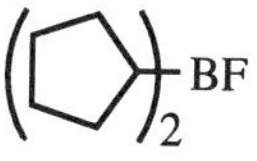 $(\ )_2BF$	76
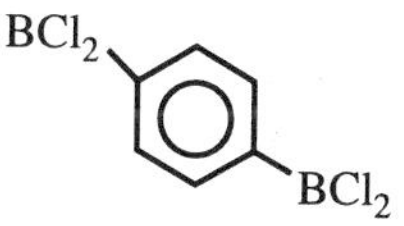 BCl_2, BCl_2	+ LiF		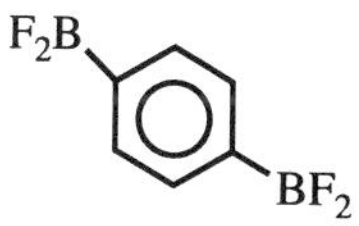 F_2B, BF_2	58

Continued on next page.

Table 29–*Continued*

Reactants	R^1	R^2	*Product*	*Yield (%)*
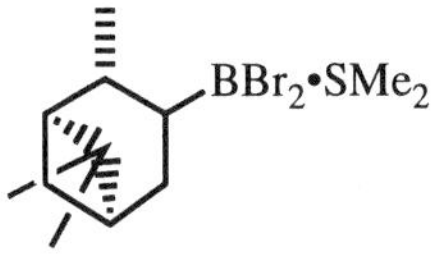 $BBr_2{\bullet}SMe_2$ (pinanyl) + 2 KF			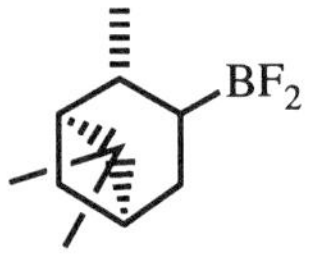 BF_2 (pinanyl)	78
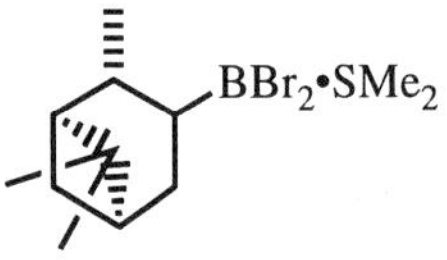 $BBr_2{\bullet}SMe_2$ (pinanyl) + 3 KF			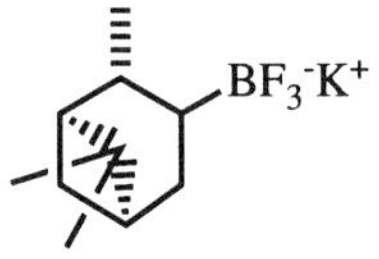 $BF_3^-K^+$ (pinanyl)	89
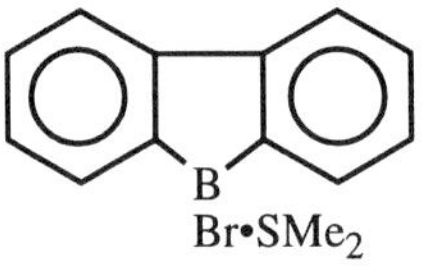 9-borafluorene B–$Br{\bullet}SMe_2$ + 2 KF			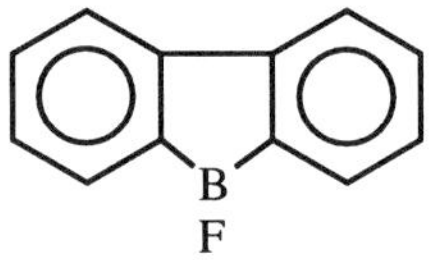 9-borafluorene B–F	73

[113] **88**

$$CH_3CH_2B(NR)Si(CH_3)_2 \text{ ring } (CH_3CH_2, CH_3) + \text{2-}CF_3C_6H_4NH_2 \xrightarrow[\text{160 °C, 100 h; } -CH_3NH_2\text{; toluene, reflux 24 h; } -C_2H_5NH_2]{} CH_3CH_2B(N\text{-}C_6H_4\text{-}2\text{-}CF_3)Si(CH_3)_2 \text{ ring } (CH_3CH_2, CH_3)$$

$R = CH_3$
$R = C_2H_5$

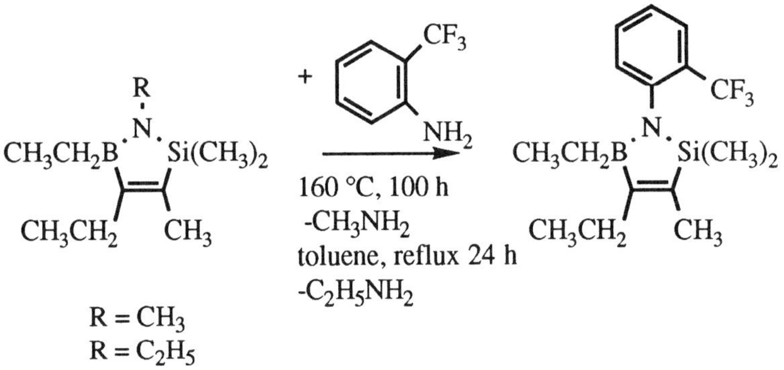

Boranes ***1*** and ***2*** in equation 89 react with 2-fluoro-1,3-dioxa-2-phospholanes to give ring expansion products as the corresponding antimony analogues yield the products of insertion of nitrogen into the antimony–fluorine bond [*115*] (equation 89) (Table 31).

The cyclobutadiene analogue diazadiboretidine reacts with hexafluoroacetone to give a ring expansion product [*116*] (equation 90). 6-(3-Fluoroaryl)decaborane is formed by alkylation of decaboranyl anion and separation of the two isomers (5- and 6-benzyl) formed by reaction with dimethyl sulfide [*117*] (equation 91).

Table 30. Alkylboration of Oxamides [*114*]

Reactants	*Product*	*Yield (%)*
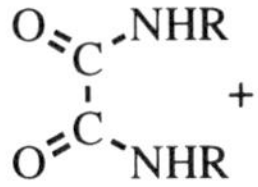 $O{=}C(NHR){-}C(NHR){=}O$ + 2 $(CH_3)_2BBr$ [a]; R		
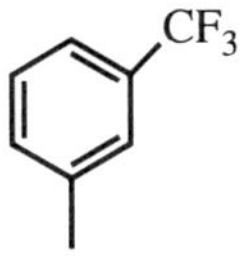 CF_3	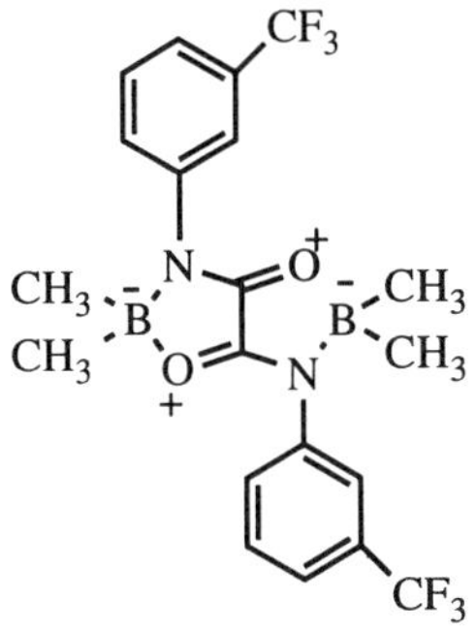 CF_3, CH_3, CH_3, B, N, O, CF_3	64
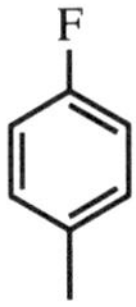 F	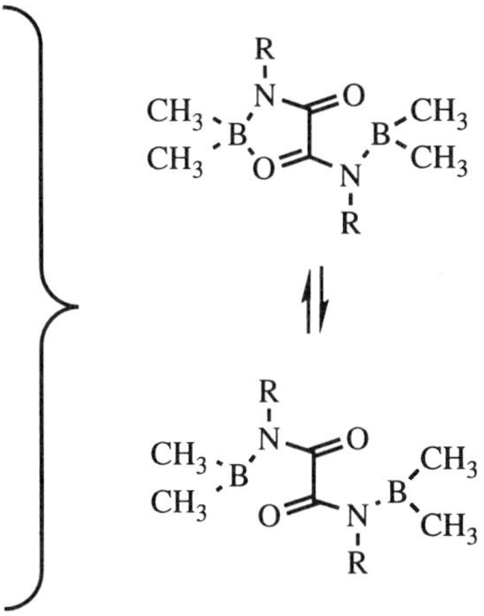 R, N, O, CH_3, B	72
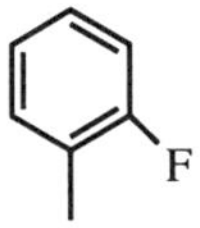 F	⇅	78
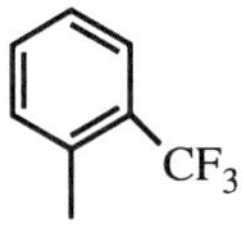 CF_3	R, N, O, CH_3, B	71
 F	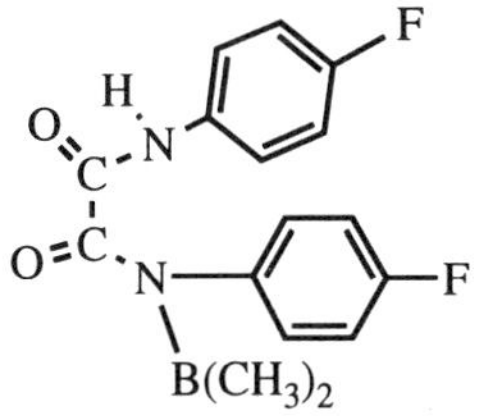 H, N, O, C, F, $B(CH_3)_2$	69
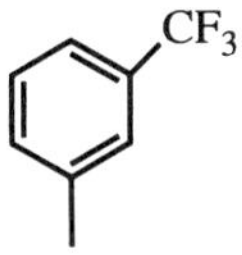 CF_3	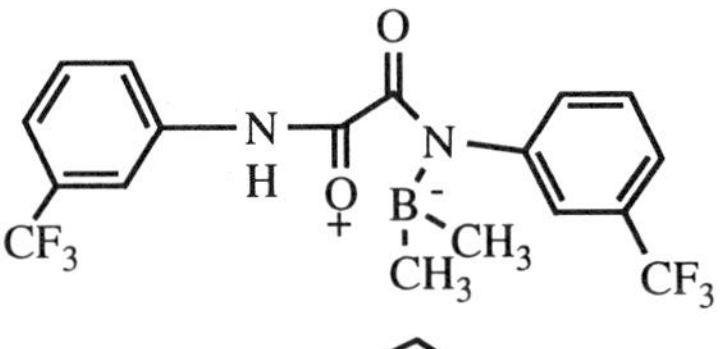 N, H, O, B, CH_3, CF_3	
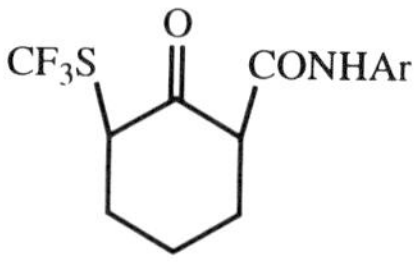 CF_3S, O, CONHAr	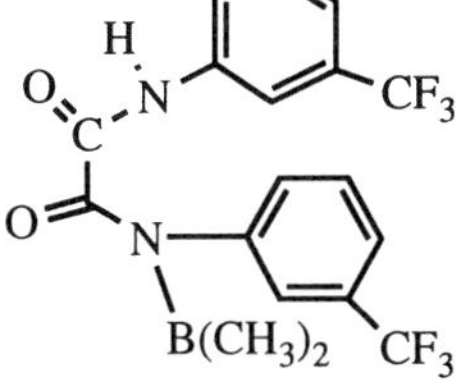 H, N, O, C, CF_3, $B(CH_3)_2$	30

[a] CCl_4; reflux for 24 h.

[115] **89**

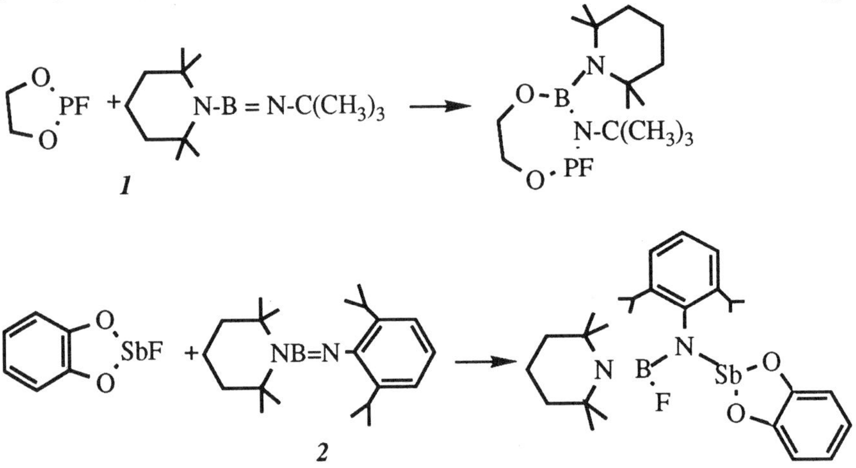

1

2

[116] **90**

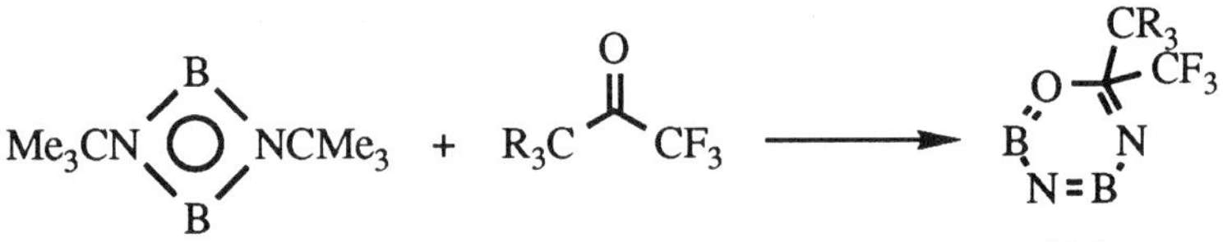

66%

[117]

$B_{10}H_{13}Na$ + (3-fluorobenzyl chloride) $\xrightarrow{Et_2O}$ 5 and 6-(3-fluorobenzyl) decaborane **91**

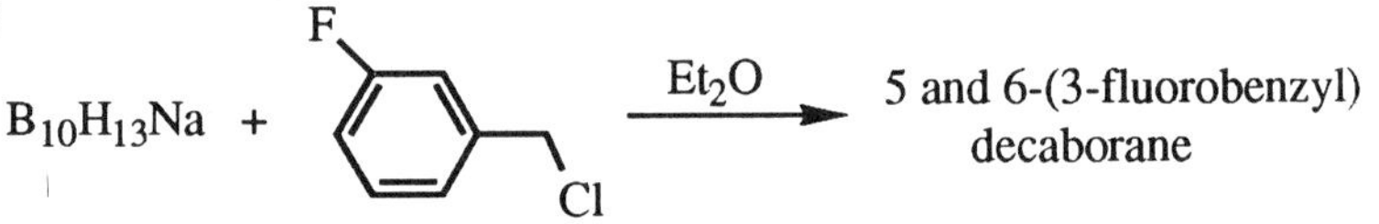

96% (mixture)

Rate constants have been measured for the reactions of boron compounds with a series of bromomethanes and bromofluoromethanes. Previously it was shown that the reactivity of the chlorine in chlorofluoromethane is substantially reduced by increasing fluorine substitution. The corresponding decrease in the reactivity of bromofluoromethane was not observed [*118*].

The reaction of (fluoroalkenyl)carboranes with potassium permanganate in acetone leads to formation of α-diketones [*119*]. These compounds react by photochemical reaction in which the radical formation at boron is followed by addition to the double bond [*120*] (equation 92).

Thermal cleavage of boron tris(trifluoromethane)sulfonate and boron tris(pentafluoroethane)sulfonate was studied. Careful identification of the products has led to the proposal of a mechanism and an account for this process [*121*] (equation 93).

Table 31. Insertion Reactions of Iminoborones [*115*]

Reactants	*Reaction Conditions*	*Product*	*Yield (%)*
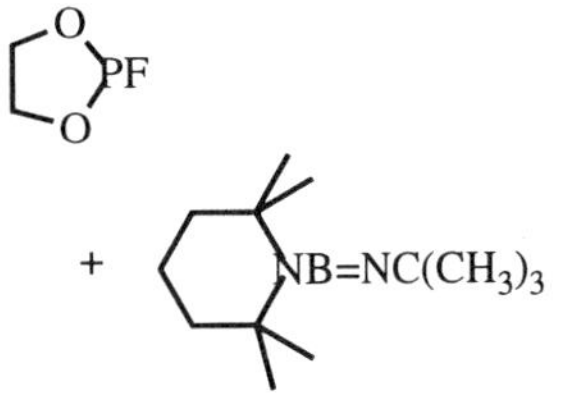 O, PF, O + NB=NC(CH_3)$_3$	Hexane	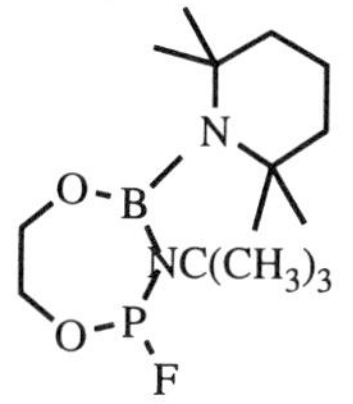 N, O–B, NC(CH_3)$_3$, O–P, F	62
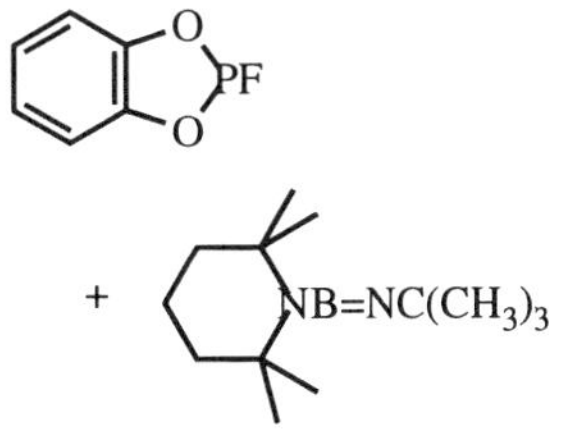 O, PF, O + NB=NC(CH_3)$_3$	Hexane	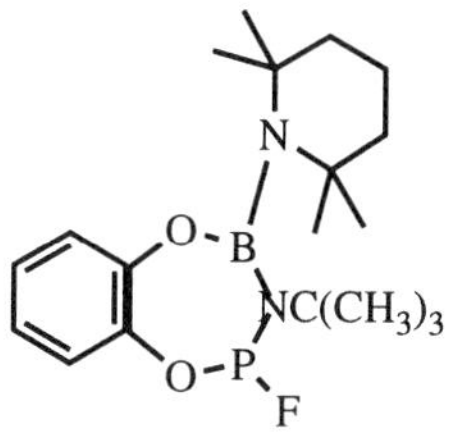 N, O–B, NC(CH_3)$_3$, O–P, F	49
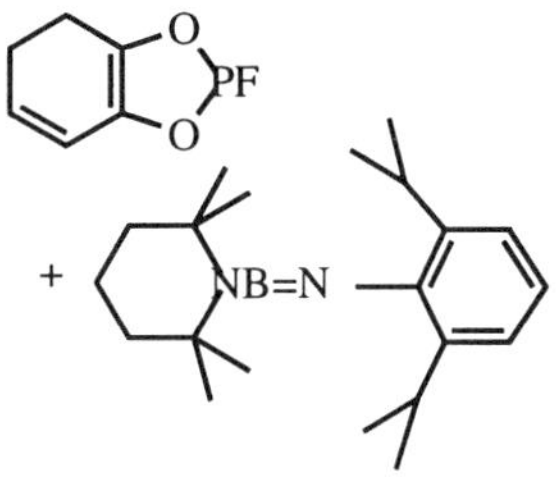 O, PF, O + NB=N	Hexane	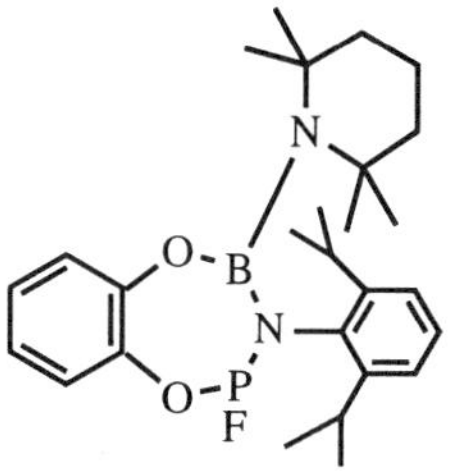 N, O–B, N, O–P, F	62
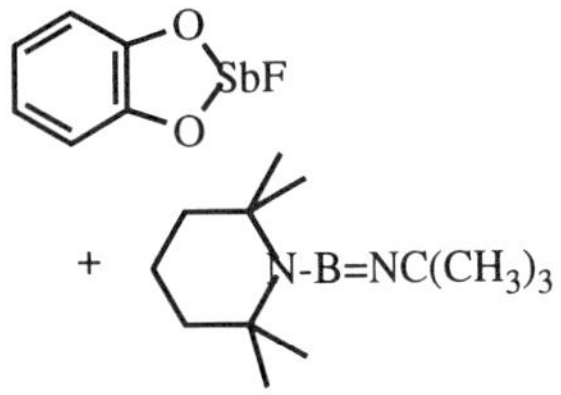 O, SbF, O + N-B=NC(CH_3)$_3$	Hexane	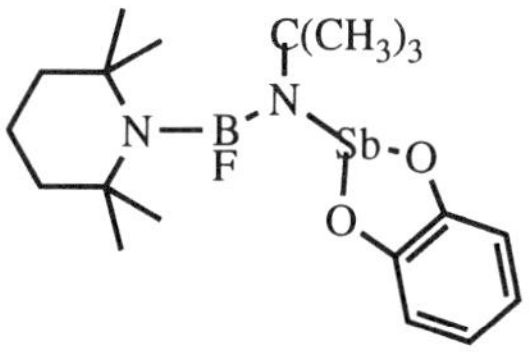 C(CH_3)$_3$, N–B, F, N, Sb–O, O	85
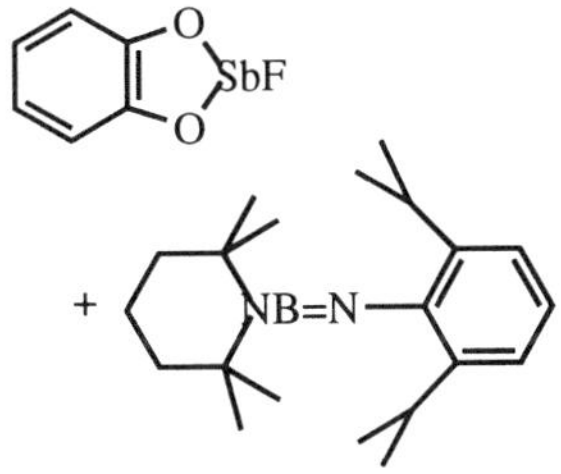 O, SbF, O + NB=N	Hexane	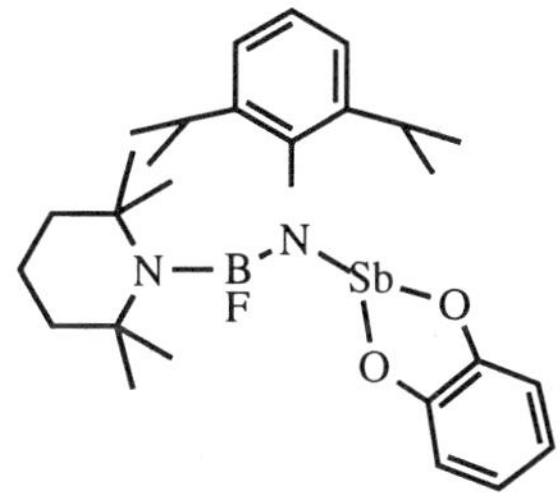 N–B, F, N, Sb–O, O	65

92

[119]

$$RCF=CFCF_3 \xrightarrow[\text{acetone}]{KMnO_4} RC(=O)-C(=O)CF_3$$

$$R = CH_3C\text{-}C\text{-}(B_{10}H_{10}),\ CH_3CB_{10}H_{10}C\text{-}$$

[120]

$$RC\text{-}C(B_{10}H_{10})CF=CFCF_3 \xrightarrow[\text{acetone}]{h\nu} RC\text{-}C(\cdot B_{10}H_{10})CF=CFCF_3 \xrightarrow{RC\text{-}C(B_{10}H_9)CF=CFCF_3} RC\text{-}C(B_{10}H_{10})CHFCF(CF_3)$$

[121]

93

$$B(OSO_2CF_3)_3 \xrightarrow{200\ ^\circ C} SO_2 + BF_3 + COF_2 + CF_3SO_2OSO_2CF_3 + CF_3SO_2OCF_3 + B_2O_3$$

References for Pages 545–610

1. Mirek, J.; Haas, A. *J. Fluorine Chem.* **1981,** *19,* 67.
2. Zeifman, Y. V.; Lantseva, L. T. *Izv. Akad. Nauk SSSR, Ser. Khim.* **1980,** 1102; *Chem. Abstr.* **1980,** *93,* 113882c.
3. Zeifman, Y. V.; Lantseva, L. T.; Knunyants, I. L. *Izv. Akad. Nauk SSSR* **1978,** 1229; *Chem. Abstr.* **1978,** *89,* 146381j.
4. Kolasa, A. *J. Fluorine Chem.* **1987,** *36,* 29.
5. Bogdanowicz-Szwed, K.; Kawalek, B.; Lieb, M. *J. Fluorine Chem.* **1987,** *35,* 317.
6. Roesky, H. W.; Benmohamed, N. *Chem. Ztg.* **1986,** *110,* 417.
7. Roesky, H. W.; Benmohamed, N. *Z. Anorg. Allg. Chem.* **1987,** *545,* 143.
8. Harris, J. F., Jr. *J.Org Chem.* **1979,** *44,* 563.
9. Harris, J. F., Jr. *J. Org. Chem.* **1981,** *46,* 268.
10. Gerstenberger, M. R. C.; Haas, A. *J. Fluorine Chem.* **1983,** *22,* 81.
11. Ceacareanu, D. M.; Gerstenberger, M. R. C.; Haas, A. *Chem. Ber.* **1983,** *116,* 3325.
12. Holoch, J.; Sundermeyer, W. *Chem. Ber.* **1986,** *119,* 269.

13. Burton, C. A.; Shreeve, J. M. *Inorg. Nucl. Chem. Lett.* **1976,** *12,* 373.
14. Zack, N. R.; Shreeve, J. M. *J. Chem. Soc., Perkin Trans. 1* **1975,** 614.
15. Remy, D. C.; Rittle, K. E.; Hunt, C. A.; Freedman, M. B. *J. Org. Chem.* **1976,** *41,* 1644.
16. Abe, T.; Shreeve, J. M. *Inorg. Chem.* **1981,** *20,* 2894.
17. Yu, S.; Shreeve, J. M. *J. Fluorine Chem.* **1976,** *7,* 85.
18. Nazaretyan, V. P.; Yagupol'skii, L. M. *Zh. Org. Khim.* **1978,** *14,* 206; *Chem. Abstr.* **1978,** *88,* 151976r.
19. Hendrickson, J. B.; Judelson, D. A.; Chancellor, T. *Synthesis* **1984,** 320.
20. Hendrickson, J. B.; Bair, K. W. *J. Org. Chem.* **1977,** *42,* 3875.
21. Hanack, M.; Wilhelm, B.; Subramanian, L. R. *Synthesis* **1988,** 592.
22. Mikhailov, I. E.; Minkin, V. I.; Olekhnovich, L. P.; Boiko, V. N.; Ignat'ev, N. V.; Yagupol'skii, L. M. *Zh. Org. Khim.* **1984,** *20,* 454; *Chem. Abstr.* **1984,** *101,* 71979r.
23. Holak, T. A.; Haas, A. *Chem. Scripta* **1980,** *15,* 67.
24. Bravo, P.; Piovosi, E.; Resnati, G.; Demunari, S. *Gazz. Chim. Ital.* **1988,** *118,* 115.
25. Inbasekaran, M.; Peet, N. R.; McCarthy, J. R.; LeTourneau, M. E. *J. Chem. Soc., Chem. Commun.* **1985,** 678.
26. Yagupol'skii, Y. L.; Savina, T. I. *Zh. Org. Khim.* **1985,** *21,* 2048; *Chem. Abstr.* **1986,** *105,* 190571u.
27. Taguchi, T.; Tomizawa, G.; Nakajima, M.; Kobayashi, Y. *Chem. Pharm. Bull.* **1985,** *33,* 678.
28. Yamazaki, T.; Ishikawa, N.; Iwatsuba, H.; Kitazume, T. *J. Chem. Soc., Chem. Commun.* **1987,** 1340.
29. Olah, G. A.; Ohyama, T. *Synthesis* **1976,** 319.
30. Katsuhara, Y.; DesMarteau, D. D. *J. Am. Chem. Soc.* **1980,** *102,* 2681.
31. Falardeau, E. R.; DesMarteau, D. D. *J. Chem. Eng. Data* **1976,** *21,* 386.
32. Majid, A.; Shreeve, J. M. *Inorg. Chem.* **1974,** *13,* 2710.
33. Haszeldine, R. N.; Tewson, T. J.; Tipping, A. E. *J. Fluorine Chem.* **1976,** *8,* 101.
34. Kinkead, S. A.; Kumar, R. C.; Shreeve, J. M. *J. Am. Chem. Soc.* **1984,** *106,* 7496.
35. Yamanaka, H.; Kuwabara, M.; Komori, M.; Otani, M.; Kase, K.; Fukunishi, K.; Nomura, M. *Nippon Kagaku Kaish.* **1983,** 112; *Chem. Abstr.* **1983,** *98,* 178838r.
36. Gramstad, T.; Husebye, S.; Saeboe, J. *Tetrahedron Lett.* **1983,** *24,* 3919.
37. Riesel, L.; Haenel, J. *Phosphorus, Sulfur, Silicon Relat. El.* **1990,** *49/50,* 215.
38. Grobe, J.; Hegemann, M.; LeDuc, V. *Z. Naturforsch. B. Chem. Sci.* **1990,** *45,* 148.
39. Naae, D. G.; Burton, D. J. *Synth. Commun.* **1973,** *3,* 197.
40. Burton, D. J. *J. Fluorine Chem.* **1983,** *23,* 339.

41. Hayashi, S.; Nakai, T.; Ishikawa, N.; Burton, D. J.; Naae, D. G.; Kesling, H. S. *Chem. Lett.* **1979,** 983.
42. Burton, D. J.; Inouye, Y.; Headley, J. A. *J. Am. Chem. Soc.* **1980,** *102,* 3980.
43. Burton, D. J.; Tsao, H. W. *J. Fluorine Chem.* **1988,** *40,* 183.
44. Burton, D. J.; Shunya, S.; Keshung, H. S. *J. Fluorine Chem.* **1982,** *20,* 89.
45. Wheaton, G. A.; Burton, D. J. *J. Org. Chem.* **1983,** *48,* 917.
46. Wheaton, G. A.; Burton, D. J. *Tetrahedron Lett.* **1976,** 895.
47. Tronchet, J. M. J.; Schwarzenbach, D.; Barbalat-Rey, F. *Carbohydr. Res.* **1976,** *46,* 9.
48. Sepiol, J. *J. Fluorine Chem.* **1984,** *25,* 363.
49. Burton, D. J.; Greenlimb, P. E. *J. Org. Chem.* **1975,** *40,* 2796.
50. Van Hamme, M. J.; Burton, D. J. *J. Fluorine Chem.* **1979,** *13,* 407.
51. Cox, D. G.; Burton, D. J. *J. Org. Chem.* **1988,** *53,* 366.
52. Cox, D. G.; Gurusamy, N.; Burton, D. J. *J. Am. Chem. Soc.* **1985,** *107,* 2811.
53. Burton, D. J.; Cox, D. G. *J. Am. Chem. Soc.* **1983,** *105,* 650.
54. Burton, D. J.; Inouye, Y. *Tetrahedron Lett.* **1979,** 3397.
55. Korhummel, C.; Hanack, M. *Chem. Ber.* **1989,** *122,* 2187.
56. Hanack, M.; Korhummel, C. *Synthesis* **1987,** 9447.
57. Naegele, U. M.; Hanack, M. *Liebigs Ann. Chem.* **1989,** 847.
58. Burton, D. J.; Shinya, S.; Howells, R. D. *J. Am. Chem. Soc.* **1978,** *101,* 3689.
59. Escoula, B.; Rico, I.; Laval, J. P.; Lattes, A. *Synth. Commun.* **1985,** *15,* 35.
60. Shen, Y.; Qiu, W. *Tetrahedron Lett.* **1987,** *28,* 4283.
61. Shen, Y.; Qiu, W. *Tetrahedron Lett.* **1987,** *28,* 449.
62. Shen, Y.; Qiu, W. *J. Chem. Soc., Chem. Commun.* **1987,** 703.
63. Huang, Y.; Shen, Y.; Ding, W.; Zheng, J. *Tetrahedron Lett.* **1981,** *22,* 5283.
64. Shen, Y.; Cen, W.; Huang, Y. *Synthesis* **1985,** 159.
65. Shen, Y.; Xin, Y.; Cen, W.; Huang, Y. *Synthesis* **1984,** 35.
66. Shen, Y.; Qiu, W. *Synthesis* **1987,** 42.
67. Leroy, J.; Wakselman, C. *Synthesis* **1982,** 496.
68. Kondo, K.; Cottens, S.; Schlosser, M. *Chem. Lett.* **1984,** 2149.
69. Elkik, E.; Francese, C. *Bull. Soc. Chim. Fr.* **1985,** 783.
70. Etemad-Moghadam, G.; Seyden-Penne, J. *Bull. Soc. Chim. Fr.* **1985,** 448.
71. Elkik, E.; Imbeaux, M. *Synthesis* **1989,** 861.
72. Thenappan, A.; Burton, D. J. *J. Fluorine Chem.* **1990,** *48,* 153.
73. Thenappan, A.; Burton, D. J. *Tetrahedron Lett.* **1989,** *30,* 5571.
74. Thenappan, A.; Burton, D. J. *Tetrahedron Lett.* **1989,** *30,* 6113.
75. Coutrot, P.; Grison, C.; Sauvetre, R. *J. Organomet. Chem.* **1987,** *332,* 1.
76. Patrick, T. B.; Nadji, S. *J. Fluorine Chem.* **1990,** *49,* 147.
77. Blackburn, G. M.; Parratt, M. J. *J. Chem. Soc., Perkin Trans. 1* **1986,** 1425.

78. Ortiz de Montellano, P. R.; Vinson, W. A. *J. Am. Chem. Soc.* **1979,** *101,* 2222.
79. Okada, Y.; Kuroboshi, M.; Ishihara, T. *J. Fluorine Chem.* **1988,** *41,* 435.
80. Martin, S.; Sauvetre, R.; Normant, J. F. *J. Organomet. Chem.* **1986,** *303,* 317.
81. Boutevin, B.; Pietrasanta, Y.; Youssef, B. *J. Fluorine Chem.* **1988,** *39,* 61.
82. Szabo, K.; Ngo Le Ha; Schneider, P.; Zeltner, P.; Kovats, E. S. *Helv. Chim. Acta* **1984,** *67,* 2128.
83. Muller, A. J. *J. Org. Chem.* **1988,** *53,* 3364.
84. Beckers, H.; Buerger, H.; Eujen, R. *Z. Anorg. Allg. Chem.* **1988,** *563,* 38.
85. Broicher, V.; Geffken, I. *J. Organomet. Chem.* **1990,** *381,* 315.
86. Stahly, G. P. U.S. Patent 4 804 772, 1989; *Chem. Abstr.* **1989,** *111,* 174380z.
87. Pawelke, G. *J. Fluorine Chem.* **1989,** *42,* 429.
88. Livantsova, L. I.; Zaitseva, G. S.; Kisin, A. V.; Baukov, Y. I. *Zh. Obshch. Khim.* **1987,** *57,* 708 (Engl. Transl. 708).
89. Gostevskii, B. A.; Vyazankina, O. A.; Vyazankin, N. S. *Zh. Obshch. Khim.* **1984,** *54,* 2613 (Engl. Transl. 2334).
90. MacDonald, J. E.; Poindexter, G. S. *Tetrahedron Lett.* **1987,** *28,* 1851.
91. Samples, M.; Yoder, C. H. *J. Organomet. Chem.* **1987,** *332,* 69.
92. Donike, M.; Zimmermann, J. *J. Chromatogr.* **1980,** *202,* 483.
93. Emde, H.; Domsch, D.; Feger, H.; Frick, U.; Goetz, A.; Hergott, H. H.; Hofmann, K.; Kober, W.; Kraegeloh, K.; Oesterle, T.; Steppan, W.; West, W.; Simchen, G. *Synthesis* **1982,** 1.
94. Oesterle, T.; Simchen, G. *Liebigs Ann. Chem.* **1987,** 687.
95. Frick, U.; Simchen, G. *Liebigs Ann. Chem.* **1987,** 839.
96. Aizpurua, J. M.; Palomo, C. *Synthesis* **1985,** 206.
97. Murata, S.; Noyori, R. *Tetrahedron Lett.* **1980,** *21,* 767.
98. Gostevskii, B. A.; Vyazankina, O. A.; Kalikhman, I. D.; Bannikova, O. B.; Vyazankin, N. S. *Zh. Obshch. Khim.* **1983,** *53,* 229 (Engl. Transl. 200).
99. Klingebiel, U.; Fischer, D.; Meller, A. *Monatsh. Chem.* **1975,** *106,* 459.
100. Klingebiel, U.; Enterling, D.; Meller, A. *J. Organomet. Chem.* **1975,** *101,* 45.
101. Yoshida, J.; Tamao, K.; Kakui, T.; Kumada, M. *Tetrahedron Lett.* **1979,** 1141.
102. Voronkov, M. G.; Basenko, S. V.; Gebel, I. A.; Vitkovskii, V. Y.; Mirskov, R. G. *Dokl. Akad. Nauk SSSR* **1987,** *293,* 362 (Engl. Transl. 120).
103. Stalke, D.; Klingebiel, U.; Sheldrick, G. M. *Chem. Ber.* **1988,** *121,* 1457.
104. Stalke, D.; Keweloh, N.; Klingebiel, U.; Noltemeyer, M.; Sheldrick, G. M. *Z. Naturforsch. B Chem. Sci.* **1987,** 42, 1237.
105. Corriu, R. J. P.; Guerin, C.; Henner, B. J. L.; Man, W. W. C. *Organometallics* **1988,** *7,* 237.
106. Beteille, J. P.; Laporterie, A.; Dubac, J. *Organometallics* **1989,** *8,* 1799.

107. Buerger, H.; Grunwald, M.; Pawelke, G. *J. Fluorine Chem.* **1986,** *31,* 89.
108. Brauer, D. J.; Buerger, H.; Pawelke, G.; Weuter, W.; Wilke, J. *J. Organomet. Chem.* **1987,** *329,* 293.
109. Jutzi, P.; Krato, B.; Hursthouse, M.; Howes, A. *J. Chem. Ber.* **1987,** *120,* 565.
110. Gol'dberg, Y. S.; Abele, E.; Liepins, E. S. M. V. *Zh. Org. Khim.* **1989,** *25,* 1099; *Chem. Abstr.* **1990,** *112,* 77273e.
111. Bahar, M. H.; Sabata, B. K. *Indian J. Chem. Sect. B* **1987,** *26,* 863.
112. Bir, G.; Schacht, W.; Kaufmann, D. *J. Organomet. Chem.* **1988,** *340,* 267.
113. Koester, R.; Seidel, G.; Kerschl, S.; Wrackmeyer, B. *Z. Naturforsch. B Chem. Sci.* **1987,** *42,* 191.
114. Maringgele, W.; Sheldrick, G. M.; Meller, A.; Noltemeyer, M. *Chem. Ber.* **1984,** *117,* 2112.
115. Brandl, A.; Noeth, H. *Chem. Ber.* **1990,** *123,* 53.
116. Schreyer, P.; Paetzold, P.; Boese, R. *Chem. Ber.* **1988,** *121,* 195.
117. Plzàk, Z.; Stibor, B.; Plesek, J.; Heřmánek, S. *Collect. Czech. Chem. Commun.* **1975,** *40,* 3602.
118. McKenzie, S. M.; Stanton, C. T.; Tabacco, M. B.; Sardella, D. J.; Davidovits, P. *J. Phys. Chem.* **1987,** *91,* 6563.
119. Zeifman, Y. V.; Postovoi, S. A.; Lebedev, V. N.; Zakharkin, L. I. *Zh. Obshch. Khim.* **1988,** *58,* 936; *Chem. Abstr.* **1984,** *110,* 95305e.
120. Tumanskii, B. L.; Lebedev, V. N.; Solodovnikov, S. P.; Bubnov, N. M.; Zakharkin, L. I. *Izv. Akad. Nauk SSSR* **1986,** 2824; *Chem Abstr.* **1987,** *107,* 217687s.
121. Olah, G. A.; Weber, T.; Farooq, O. *J. Fluorine Chem.* **1989,** *43,* 235.

Base-Catalyzed Condensations

by J. T. Welch

The condensation of fluorinated carbonyl compounds is a versatile approach to the stereo- and regioselective construction of specifically fluorinated materials. It is possible to use fluorinated reactants in various oxidation states and to utilize them either as the nucleophilic or electrophilic reaction partners. Stereocontrol in those processes using the fluorinated carbonyl compounds in a nucleophilic fashion has lagged behind the achievement of higher yields.

Aldol-Type Condensation

Fluoroalkyl ketones may be used as the electrophilic partners in condensation reactions with other carbonyl compounds. The highly electrophilic ***hexafluoroacetone*** has been used in selective hexafluoroisopropylidenation reactions with **enol silyl ethers** and dienolsilyl ethers [*1*] (equation 1).

[*1*] $F_3C-CO-CF_3$ + $R^1R^2C=C(OSi(CH_3)_3)R^3$ $\xrightarrow[\text{-30 -35 °C}]{SnCl_4}$ $(CF_3)_2C(OH)-CR^1R^2-CO-R^3$ **1**

67-74%

Trifluoromethyl alkyl ketones also undergo directed aldol condensations under thermodynamic conditions in the presence of piperidine and acetic acid [*2, 3*]. Under these reaction conditions, the product suffers a facile dehydration to form the unsaturated trifluoromethyl ketones (equations 2 and 3).

Two equivalents of ethyl trifluoroacetylacetate reacts with one equivalent of an aldehyde and ammonia to give 2,6-bis(trifluoromethyl)-1,4-dihydropyridines in good to fair yields [*4*] (equation 4).

The addition of imidazole to the ethyl hemiacetal of trifluoroacetaldehyde provides 1-(1′-hydroxy-2′,2′,2′-trifluoroethyl)imidazoles in yields depending upon the electronic nature of the substituents [5] (equation 5) (Table 1).

3-Benzyloxy-2-fluoro-2-methylpropionaldehyde was prepared in optically active form from (*S*)-monoethyl 2-fluoro-2-methylmalonate, which had itself been prepared by enzymatic hydrolysis. A number of enol silyl ethers or enolates were added to the aldehyde in processes that occur with fair to good diastereoselectivity [*6*] (equation 6) (Table 2).

0065–7719/95/0187–0615$09.98/1

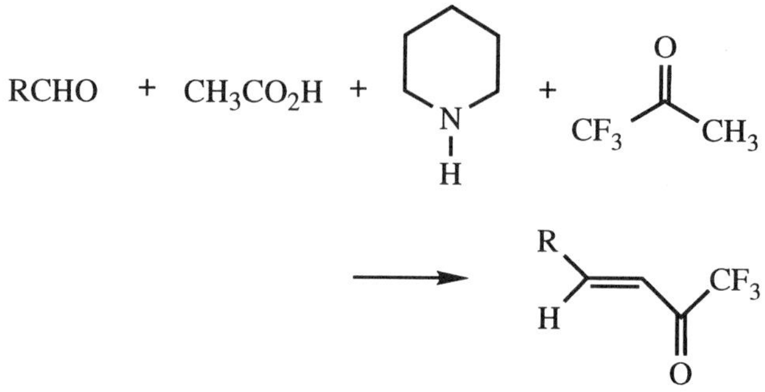

RCHO + CH_3CO_2H + piperidine + $CF_3C(O)CH_3$ **2**

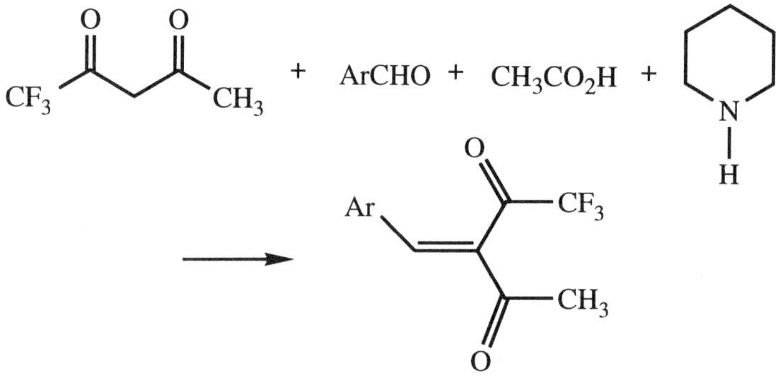

[2] → R(H)C=CH–C(O)CF_3

$CF_3C(O)CH_2C(O)CH_3$ + ArCHO + CH_3CO_2H + piperidine **3**

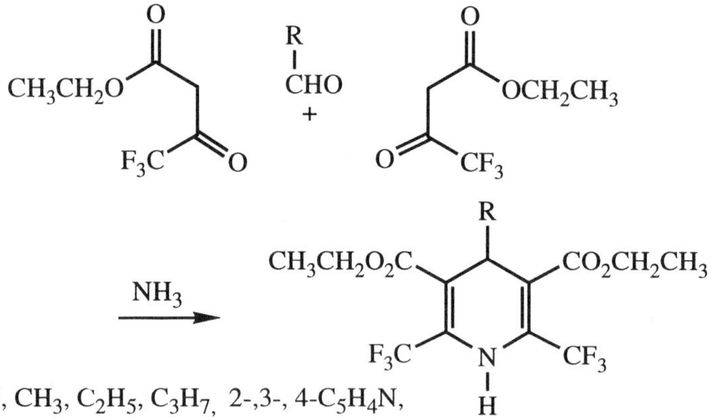

[3] → ArCH=C(COCF_3)(COCH_3)

20-85%

[4] $CH_3CH_2O_2CCH_2C(O)CF_3$ + RCHO + $CH_3CH_2O_2CCH_2C(O)CF_3$ **4**

$\xrightarrow{NH_3}$ 4-R-3,5-bis($CO_2CH_2CH_3$)-2,6-bis(CF_3)-1,4-dihydropyridine (N–H) 52-71%

R = H, CH_3, C_2H_5, C_3H_7, 2-,3-, 4-C_5H_4N, 2, 4-$O_2NC_6H_4$

[5] imidazole (R^1 on N, R^2, R^3, R^4) + $CF_3CH(OH)OCH_2CH_3$ $\xrightarrow{heat}$ imidazole (R^1, R^3, R^4, 2-CF_3CHOH) **5**

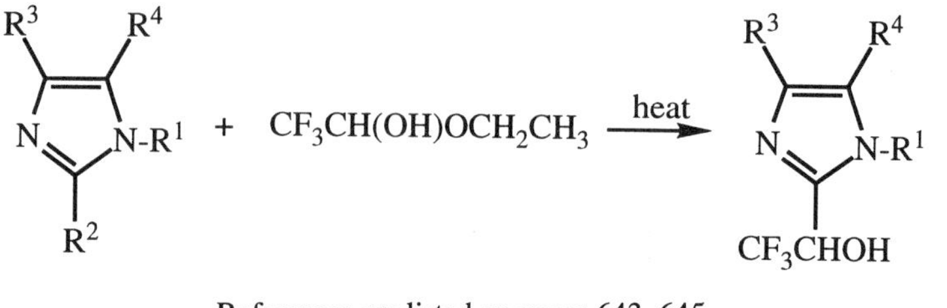

References are listed on pages 643–645.

Table 1. Thermal Condensation of Substituted Imidazoles with Trifluoroacetaldehyde Ethyl Hemiacetal [5]

Imidazole Structure				*Reaction Conditions*			*Yield* [a] *(%)*		
R[1] *(N-1)*	*R*[2] *(C-2)*	*R*[3] *(C-4)*	*R*[5] *(C-5)*	*Amt. (mmol)*	*Method* [b]	*Time (h)*	*I*	*II*	*III*
CH_3	H	H	H	20	B	24	12.5	0	0
C_2H_5	H	H	H	20	B	24	16.5	0	0
$PhCH_2$	H	H	H	20	B	24	23.8	0	0
H	CH_3	H	H	20	A	6	–	62.2	9.7
H	C_2H_5	H	H	20	A	6	–	76.5	7.2
H	Ph	H	H	20	A	6	–	71.2	4.4
H	H	CH_3	H	20	A	3	11.1	73.3	15.5
H	H	Ph	H	•17.3	A	8	5.2	61.1	13.3
H	H	Cl	H	20	A	10	4.0	24.9	11.2
H	H	CH_2OH	H	20	A	10	6.1	20.1	13.9
H	CH_3	CH_3	H	20	A	8		91.7	–
H	C_2H_5	CH_3	H	20	A	2		76.0	–
H	Ph	CH_3	H	20	A	0.5		72.4	–
H	H	Cl	Cl	20	A	6	66.2	–	–

[a] I, 2-adduct; II, 4-adduct; III, bis-adduct.
[b] A, heated at reflux with equimolar amount of trifluoroacetaldehyde ethyl hemiacetal; B, heated in a sealed tube (oil bath, 150–155 °C) with equimolar amount of trifluoroacetaldehyde ethyl hemiacetal.

The Lewis acid-catalyzed addition of ***silyl ketene acetals*** occurred in high yield, and when the ketene acetal bore a substituent, the reactions occurred with modest diastereofacial selectivity [*6*] (equation 7) (Table 3).

In contrast, ***fluorinated ketones*** have been used as both *nucleophilic and electrophilic reaction constituents.* The (*Z*)-lithium enolate of 1-fluoro-3,3-dimethylbutanone can be selectively prepared and undergoes highly diastereoselective aldol condensations with aldehydes [*7*] (equation 8) (Table 4).

The key step to this first reported case of the highly diastereoselective addition of a fluorinated enolate in an aldol process is the selective formation of the enolate. α,α-Difluorinated enolates prepared by a metallation process employing either a zinc–copper couple [*8*] or reduced titanium species [*9*] undergo aldol condensation smoothly (equation 9) (Table 5).

Hexafluoroacetone and nitropentafluoroacetone are sufficiently electrophilic that they condense readily with **phenolates** in excellent yield [*10*] (equation 10).

[*6*] $C_2H_5O(C{=}O)C(CH_3)(F)(C{=}O)OH \xrightarrow[NaBH_4]{(COCl)_2} C_2H_5O(C{=}O)C(CH_3)(F)CH_2OH$ **6**

$BrCH_2C_6H_5$, $LiAlH_4$, Swern → $H(C{=}O)C(CH_3)(F)CH_2OCH_2C_6H_5$ 59% (*S*)

DHP, $LiAlH_4$, $BrCH_2C_6H_5$, Swern → $C_6H_5CH_2OCH_2C(CH_3)(F)(C{=}O)H$ 54% (*R*)

$BrCH_2C_6H_5$, $LiAlH_4$, $t\text{-}C_4H_9(CH_3)_2SiCl$, H_2 / Pd, Swern → $(CH_3)_3CSi(CH_3)_2OCH_2C(CH_3)(F)(C{=}O)H$ 32% (*R*)

Table 2. Reaction of (*S*)-3-Benzyloxy-2-fluoro-2-methyl-propionaldehyde with Various Metal Enolates [*6*]

M	*Additive*	*Yield (%)*	*threo:erythro*[a]
Li		93	33:67
MgBr		54	36:64
ZnCl		33	43:57
$AlEt_2$		45	29:71
TMS	TBAF[b]	0	
TMS	TMSOTF[c]	0	
TMS	$TiCl_4$	85	91:9
TMS	$EtAlCl_2$	55	22:78
TMS	$BF_3{\cdot}Et_2O$	62	24:76

[a]This ratio was determined by HPLC.
[b]TBAF, tetrabutylammonium fluoride.
[c]TMSOTF, trimethylsilyl trifluoromethanesulfonate.

[6] H CH_3 F $OCH_2C_6H_5$ O

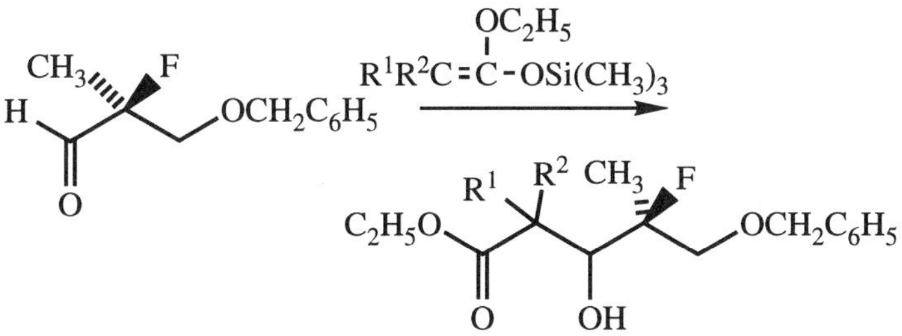

OC_2H_5
$R^1R^2C=C-OSi(CH_3)_3$ → 7

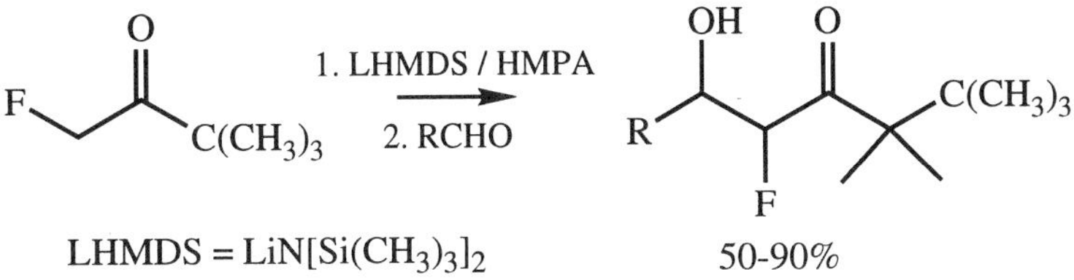

R^1 R^2 CH_3 F
C_2H_5O $OCH_2C_6H_5$
O OH

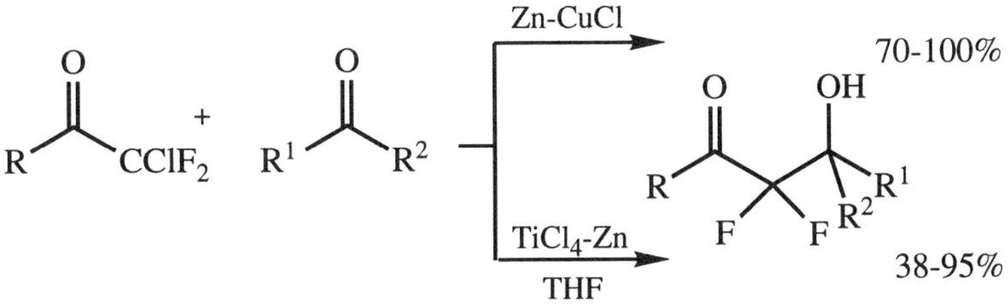

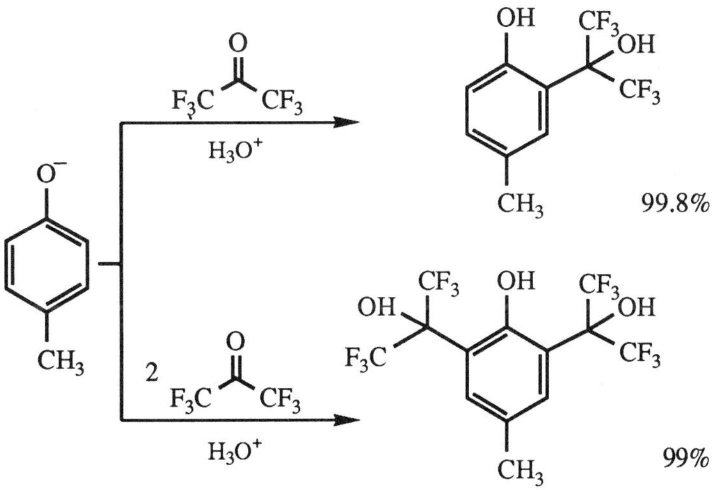

[7] F O $C(CH_3)_3$ 1. LHMDS / HMPA 2. RCHO → OH O R $C(CH_3)_3$ F 8

LHMDS = $LiN[Si(CH_3)_3]_2$ 50-90%

[8] Zn-CuCl 70-100% 9

O + O
R $CClF_2$ R^1 R^2

O OH
R R^1
F F R^2

[9] $TiCl_4$-Zn THF 38-95%

[10] 10

O
F_3C CF_3
H_3O^+

OH CF_3 OH CF_3 CH_3 99.8%

O^- CH_3

2 O F_3C CF_3 H_3O^+

OH CF_3 F_3C OH CF_3 OH CF_3 CH_3 99%

References are listed on pages 643–645.

Table 3. Reaction of (*S*)-3-Benzyloxy-2-fluoro-2-methylpropionaldehyde with Silyl Enol Ethers and Silyl Ketene Acetals [*6*]

R	*R*[1]	*Lewis Acid*	*Yield (%)*	*Diastereoselectivity threo:erythro*[a]
i-Bu	H	$TiCl_4$	83	78:22
		$EtAlCl_2$	51	9:91
t-Bu	H	$TiCl_4$	83	95:5
		$EtAlCl_2$	51	29:71
t-Bu	H	$TiCl_4$[b]	75	9:91
		$EtAlCl_2$[b]	68	12:88
OEt	H[c]	$TiCl_4$	58	20:80
		$EtAlCl_2$	24	14:86
		$BF_3{\cdot}OEt_2$	51	15:85
	$(CH_2)_4$[d]	$TiCl_4$	75	52:48
		$EtAlCl_2$	60	77:23
Et[e]	CH_3	$TiCl_4$[d]	86	30:8:12:50[f]
		$EtAlCl_2$[d]	66	5:33:4:58[f]
Et	CH_3	$TiCl_4$[b]	81	34:7:13:46[f]
		$EtAlCl_2$[b]	62	5:34:4:57[f]
OEt[g]	CH_3	–	90	18:42:11:29[f]
		$TiCl_4$	85	2:7:20:71[f]
		$EtAlCl_2$	79	4:24:10:62[f]
		$BF_3{\cdot}OEt_2$	82	5:37:6:52[f]
SBu-*t*	CH_3	$TiCl_4$	94	78:12:2:8[f]
		$BF_3{\cdot}OEt_2$	81	6:12:81:<1[f]

[a]Isomeric ratios were determined by HPLC,
[b]*E:Z*, 14:86.
[c]The *t*-$BuMe_2Si$ ketene acetal was used.
[d]The *E:Z* ratio of silyl enol ether was 77:23.
[e](*S*)-2 was used instead of the corresponding *R* isomer.
[f]et:ee:te:tt. Isomeric ratios were determined by HPLC, and abbreviations e and t indicate *erythro* and *threo* configurations between C_2–C_3 and C_3–C_4 stereocenters and are arranged in this way.
[g]The corresponding lithium enolate was used.

References are listed on pages 643–645.

Table 4. Directed Aldol Reaction of Lithium Enolate of 1-Fluoro-3,3-dimethylbutanone[a] [7]

RCHO	Yield[b] (%)	Diastereoselectivity[c]
CH_3CH_2	63	16:1
$CH_3CH_2CH_2$	74	19:1
$(CH_3)_2CH$	50	24:1
$(CH_3)_3C$	62	49:1
C_6H_5	70	7:1
3,3-Dimethyl-2,4-dioxol-1-yl	90	32:1

[a]To a solution of 0.01 mol of lithium hexamethyldisilazide and 0.01 mol of HMPA dissolved in 50 mL of anhydrous THF at 78 °C was added 0.5 g (0.004 mol) of 1-fluoro-3,3-dimethylbutanone in THF over 1 min. To the solution of the enolate was then rapidly added 0.003 mol of the aldehyde in THF. After stirring an additional 2 min, the reaction was quenched by rapid addition of a saturated ammonium chloride solution. Extractive workup with hexanes yielded on evaporation the product as a clear colorless oil.
[b]Isolated yield.
[c]Diastereoselectivity was determined by ^{13}C NMR spectroscopy and by gas chromatographic analysis (50 m × 0.025 mm OV-101 open tubular column).

The enhanced reactivity of fluoroalkyl ketones is also manifested in the failure to stop the reaction with **hydrogen cyanide** at the stage of cyanohydrins. Instead, oxazolidinones or dioxolanones are formed (equation 11). If, however, the reaction is conducted under basic conditions with sodium bisulfite and sodium cyanide, the desired cyanohydrin can be prepared [*11*].

[*11*] **11**

$C_7F_{15}-C(=O)-CH_3$ —HCN (DMF or THF or CH_3CN)→ oxazolidinone (C_7F_{15}, CH_3; NH; O; CH_3, C_7F_{15}) 3-25% + dioxolanone (C_7F_{15}, CH_3; O; O; CH_3, C_7F_{15}) 47-60%

Cyclizations such as that which underlies the *Knorr synthesis* have also been successfully used in *Robinson annelation* sequences [*12*] (equation 12).

Lastly, **α-trimethylsilyl** enolates have been added to ***trifluoromethyl ketones*** to effect *Peterson olefination* of the trifluoromethyl ketones [*13*] (equation 13).

Partial control of enolate geometry occurs also when the **enol phosphate,** prepared by treatment of fluoroalkyl ketones with sodium diethyl phosphite, is

Table 5. Aldol Synthesis from Chlorodifluoromethyl Ketones [9]

Chlorodifluoromethyl Ketone	*Metal* [a]	*Carbonyl Compound*	*Yield of Aldol (%)*
$CH_3(CH_2)_5COCF_2Cl$	A	$CH_3(CH_2)_2CHO$	95
$CH_3(CH_2)_5COCF_2Cl$	A	$CH_3(CH_2)_5CHO$	63
$CH_3(CH_2)_5COCF_2Cl$	A	$(CH_3)_2CHCHO$	75
$CH_3(CH_2)_5COCF_2Cl$	A	c-$C_6H_{11}CHO$	38
$CH_3(CH_2)_5COCF_2Cl$	A	$C_2H_5COC_2H_5$	55
$CH_3(CH_2)_5COCF_2Cl$	A	Cyclohexanone	58
c-$C_6H_{11}COCF_2Cl$	A	$CH_3(CH_2)_2CHO$	60
c-$C_6H_{11}COCF_2Cl$	A	$CH_3(CH_2)_5CHO$	61
c-$C_6H_{11}COCF_2Cl$	A	$(CH_3)_2CHCHO$	88
c-$C_6H_{11}COCF_2Cl$	A	c-$C_6H_{11}CHO$	47
c-$C_6H_{11}COCF_2Cl$	A	$C_2H_5COC_2H_5$	50
c-$C_6H_{11}COCF_2Cl$	A	Cyclohexanone	52
CH_3COCF_2Cl	A	$CH_3(CH_2)_2CHO$	62
$C_6H_5CH_2COCF_2Cl$	A	$CH_3(CH_2)_2CHO$	50
$CH_3(CH_2)_5COCF_2Cl$	B	$CH_3(CH_2)_2CHO$	100
$CH_3(CH_2)_5COCF_2Cl$	B	$(CH_3)_2CHCHO$	81
$CH_3(CH_2)_5COCF_2Cl$	B	$(CH_3)_3CCHO$	60
$CH_3(CH_2)_5COCF_2Cl$	B	(*E*)-$CH_3CH{=}CHCHO$	85
$CH_3(CH_2)_5COCF_2Cl$	B	(*E*)-$CH_3CH{=}C(CH_3)CHO$	100
$CH_3(CH_2)_5COCF_2Cl$	B	C_6H_5CHO	90
$CH_3(CH_2)_5COCF_2Cl$	B	$C_2H_5COC_2H_5$ [b]	64
$CH_3(CH_2)_5COCF_2Cl$	B	Cyclohexanone [b]	76
$CH_3(CH_2)_5COCF_2Cl$	B	$C_6H_5COCH_3$ [b]	84
$CH_3(CH_2)_5COCF_2Cl$	B	$CH_2{=}CH(CH_2)_2COCH_3$ [b]	77
c-$C_6H_{11}COCF_2Cl$	B	$CH_3(CH_2)_2CHO$	81
c-$C_6H_{11}COCF_2Cl$	B	$(CH_3)_2CHCHO$	86
c-$C_6H_{11}COCF_2Cl$	B	(*E*)-$CH_3CH{=}CHCHO$	77
c-$C_6H_{11}COCF_2Cl$	B	(*E*)-$CH_3CH{=}C(CH_3)CHO$	93
c-$C_6H_{11}COCF_2Cl$	B	C_6H_5CHO	93
$C_6H_5CH_2COCF_2Cl$	B	$CH_3(CH_2)_2CHO$	70
$C_6H_5CH_2COCF_2Cl$	B	C_6H_5CHO	90
$C_6H_5CH_2COCF_2Cl$	B	$C_2H_5COC_2H_5$ [b]	73
$C_6H_5COCF_2Cl$ [c]	B	$CH_3(CH_2)_2CHO$	70

Table 5—*Continued*

Chlorodifluoromethyl Ketone	*Metal* [a]	*Carbonyl Compound*	*Yield of Aldol (%)*
$C_6H_5COCF_2Cl$ [c]	B	(*E*)-$CH_3CH{=}CHCHO$	100
$C_6H_5COCF_2Cl$ [c]	B	C_6H_5CHO	88
$C_6H_5COCF_2Cl$ [c]	B	$CH_2{=}CH(CH_2)_2COCH_3$ [b,d]	66 [e]

[a] A, $TiCl_4$-Zn [*9*]; B, Zn–CuCl [*8*].
[b] Silver(I) acetate was used as an activator in place of copper(I) chloride, and 1.1 equiv. of diethylaluminum chloride was added. Three equivalents of ketone was used.
[c] The reaction was performed in a mixed solvent of THF–diethyl ether (1:4) under reflux.
[d] No diethylaluminum chloride was added.
[e] Determined by ^{19}F NMR.

[*12*] **12**

F, O, R + O, O, $(CH_2)_n$ → F, O, O, R, $(CH_2)_n$ 65-98% → F, O, O, R, OH, $(CH_2)_n$ 80-85%

[*13*] **13**

O, CF_3; $TMSCH_2CO_2CH_2CH_3$ / LiCA / THF → CF_3, $CO_2C_2H_5$; 98%; *E* : *Z* 1 : 7

treated with a lithium aluminum hydride–copper(II) bromide reagent [*14*]. These enolates react with modest diastereoselectivity (equation 14) (Table 6).

Knorr condensation of ***ethyl trifluoroacetylacetate*** with a variety of partners serves as a facile route to the preparation of a variety of trifluoromethylated pyrroles [*15, 16*] (equation 15).

$$R_fCF{=}C(OP(O)(OCH_2CH_3)_2)(R) \xrightarrow[\text{THF, -30 °C}]{CuBr_2\text{-}LiAlH_4} R_fCF{=}C(O^-M^+){-}R \quad \mathbf{14}$$

$$\xrightarrow{R^1CHO} R^1CH(OH){-}CF(R_f){-}C(O)R$$

[*14*]

38-84%

Table 6. Aldol Reaction of Fluorinated Enol Phosphates with Aldehydes [*14*]

R_f	*R*	*Aldehyde*	*Yield*[a] *(%)*	*Isomer ratio*[b]
CF_3	$CF_3(CF_2)_5$	CH_3CH_2CHO	70	0.7:1
CF_3	$CF_3(CF_2)_5$	$CH_3(CH_2)_2CHO$	70	0.8:1
CF_3CF_2	$CF_3(CF_2)_5$	CH_3CH_2CHO	58	0.7:1
CF_3CF_2	$CF_3(CF_2)_5$	$CH_3(CH_2)_2CHO$	49	1.8:1
CF_3CF_2	$CF_3(CF_2)_5$	$(CH_3)_2CHCHO$	51	0.4:1
CF_3CF_2	$CF_3(CF_2)_5$	$CH_3(CH_2)_5CHO$	51[c]	1.1:1
CF_3CF_2	$CF_3(CF_2)_5$	(*E*)-$CH_3CH{=}CHCHO$	66	1.3:1[d]
CF_3CF_2	$CF_3(CF_2)_5$	C_6H_5CHO	51[c]	1.5:1[d]
CF_3CF_2	*c*-C_6H_{11}	$CH_3(CH_2)_2CHO$	72	0.7:1
CF_3CF_2	*c*-C_6H_{11}	(*E*)-$CH_3CH{=}C(CH_3)CHO$	38	0.8:1[e]
CF_3CF_2	C_6H_5	$CH_3(CH_2)_2CHO$	49	0.7:1
CF_3CF_2	C_6H_5	$(CH_3)_2CHCHO$	37	3.7:1
CF_3CF_2	C_6H_5	(*E*)-$CH_3CH{=}CHCHO$	34	0.7:1
$CF_3(CF_2)_5$	$CF_3(CF_2)_2$	CH_3CH_2CHO	84	0.8:1[e]

[a]The yields refer to pure isolated products, unless otherwise cited.
[b]The values were measured by ^{19}F NMR and represent the ratios of the lower field peaks to the higher field peaks due to the α methine fluorine in an isomeric mixture of the product.
[c]Determined by ^{19}F NMR.
[d]*erythro:threo* ratio.
[e]*threo:erythro* ratio.

[*15,16*]

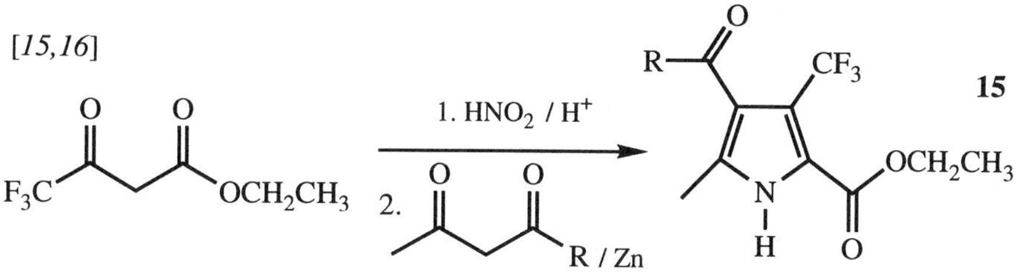

$$CF_3COCH_2COOCH_2CH_3 \xrightarrow[2.\ CH_3COCH_2COR\ /\ Zn]{1.\ HNO_2\ /\ H^+} \textbf{15}$$

35-54%

It is possible to use the enhanced electrophilicity in condensation reactions with ***aldehydes*** in the presence of amines to form imidazoles [*17*] (equation 16).

[*17*] **16**

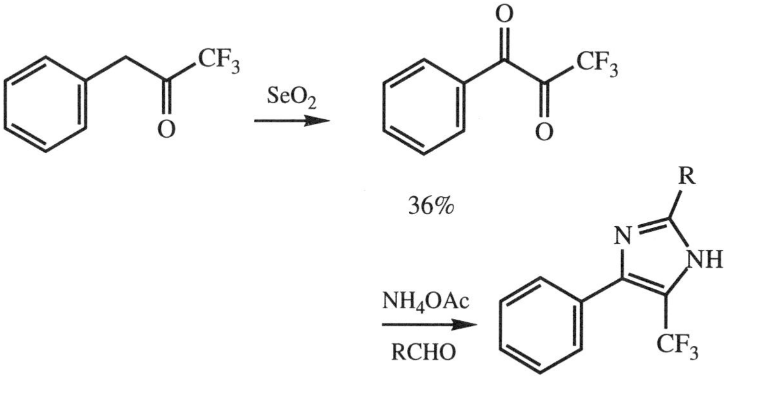

$$C_6H_5CH_2COCF_3 \xrightarrow{SeO_2} C_6H_5COCOCF_3 \ (36\%) \xrightarrow[RCHO]{NH_4OAc}$$

15-47%

For synthetic purposes, *aldol-type condensations of aldehydes with esters or amides* are potentially of great utility because the carbonyl group is easily transformed either by further additions or by oxidation or reduction. Deprotonation of an ester [*18, 19, 20*] or amide of fluoroacetic acid [*9, 21*] has led to aldol condensations in high yields (equation 17) (Table 7).

[*18,19,20*]

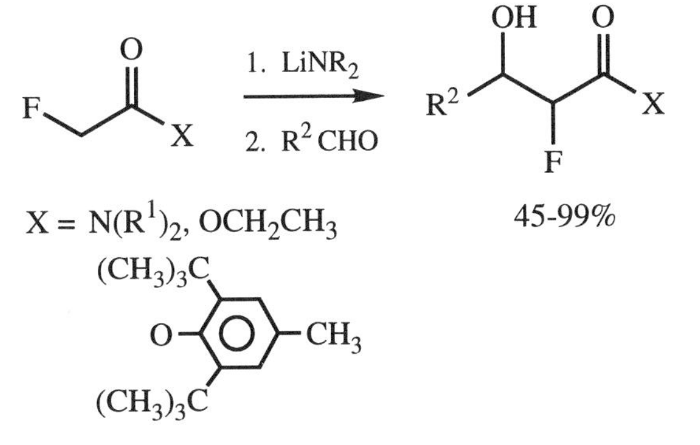

$$FCH_2COX \xrightarrow[2.\ R^2CHO]{1.\ LiNR_2} R^2CH(OH)CHFCOX \quad \textbf{17}$$

$X = N(R^1)_2, OCH_2CH_3$, 2,6-di-$(CH_3)_3C$-4-CH_3-phenoxy (O–)

45-99%

Table 7. Products of the Directed Aldol Reaction of Lithium Enolate of Ethyl Fluoroacetate [*18, 19*]

$$LiCHFCO_2R + R^1R^2CO \rightarrow R^1R^2C(OH)CHFCO_2C_2H_5$$

R	*R¹*	*R²*	*Yield (%)*	*Diastereo-selectivity*
C_2H_5[a]	CH_3	$C(CH_3)_3$	95	1:3:8[b]
C_2H_5[a]	CH_3	C_2H_5	82	1:1
C_2H_5[a]	CH_3	C_6H_5	96	1:1.6[b]
C_2H_5[a]	CH_3	C_5H_{11}	93	1:1.1[b]
C_2H_5[a]	C_6H_5	C_6H_5	70	
C_2H_5[a]	2-Adamantyl		75	
C_2H_5[a]	2-Norbornyl		91	
C_2H_5[a]	H	$C(CH_3)_3$	85	1:3
C_2H_5[a]	H	C_6H_{13}	20	1:2
C_2H_5[a]	H	C_6H_5	93	1:2
C_2H_5[a]	H	3,3-Dimethyl-2,4-dioxol-1-yl	55	1:1.2
BHT[c,d]	H	C_2H_5	70	1:1.3[e]
BHT[d]	H	C_2H_5[f]	65.6	1:2.5[e]
BHT[d]	H	C_2H_5[g]	91.8	1:2.8[e]
BHT[d]	H	C_6H_5	88	1:7.5[e]
BHT[d]	H	$C(CH_3)_3$	83	1:19[e]
BHT[d]	H	$CH(CH_3)_2$	79	1:4.1[e]

[a]Reference 18.
[b]*syn:anti* ratio.
[fc]BHT, 2,6-di-*tert*-butyl-4-methylphenyl.
[d]Reference 19.
[e]The ratio was determined by ^{13}C NMR.
[f]Enolate was generated with lithium tetramethylpiperidide in the absence of HMPA.
[g]Enolate was generated with lithium tetramethylpiperidide.

Fluoroalkyl ketone enolates and enol ethers have also been use in condensation reactions with ketones [22]. Interestingly, these materials fail to undergo Darzens-type side reactions (equation 18).

When ethyl trifluoroacetylacetate is treated with an allylic alkoxide, transesterification is followed by ester enolate Claisen rearrangement in a process that on decarboxylation yields stereospecifically the trifluoromethyl ketone product [*23*] (equation 19).

Control of the stereochemistry of the product has proved problematic when fluoroacetates or fluoroacetamides are used in this process. However, introduction of a fluoroalkyl substituent allows nearly complete recovery of the very high diastereoselectivities reported in the reactions of hydrocarbon amides [*9*] (equation 20) (Table 8).

[*22*] $RC(=O)CHFR^1$ $\xrightarrow{CH_3ONa(\text{or } CH_3CH_2ONa)}$ $R-C(OH)(CH_2F)-CR^1FCOR$ **18**

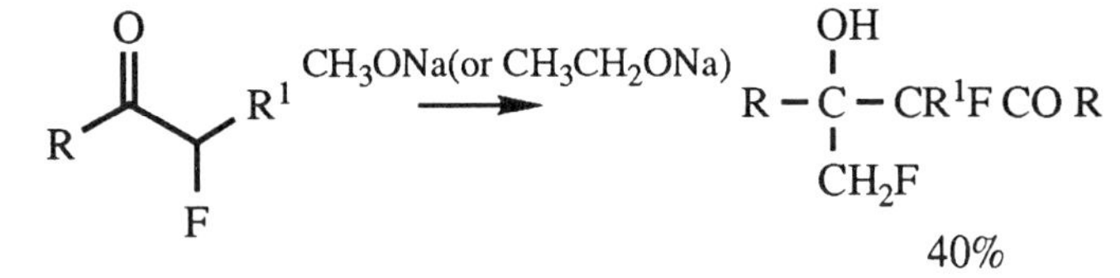

40%

[*23*] $F_3CC(=O)CH_2C(=O)OCH_2CH_3$ $\xrightarrow[150\ ^\circ C]{R^1R^2C(ONa)CH=CH_2}$ $F_3CC(=O)CH_2C(=O)OC(R^1)(R^2)CH=CH_2$ **19**

$\longrightarrow$ $F_3CC(=O)CH=C(O^-)OC(R^1)(R^2)CH=CH_2$ $\longrightarrow$ $F_3CC(=O)CH(C(=O)O^-)CH_2CH=CR^1R^2$

$\xrightarrow{-CO_2}$ $F_3CC(=O)CH_2CH_2CH=CR^1R^2$

80%

[*9*] $CF_3CFHC(=O)N(C_2H_5)_2$ $\xrightarrow[2.\ RCHO]{1.\ Bu_2BOTf\ /\ EtN(i\text{-}Pr)_2}$ **20**

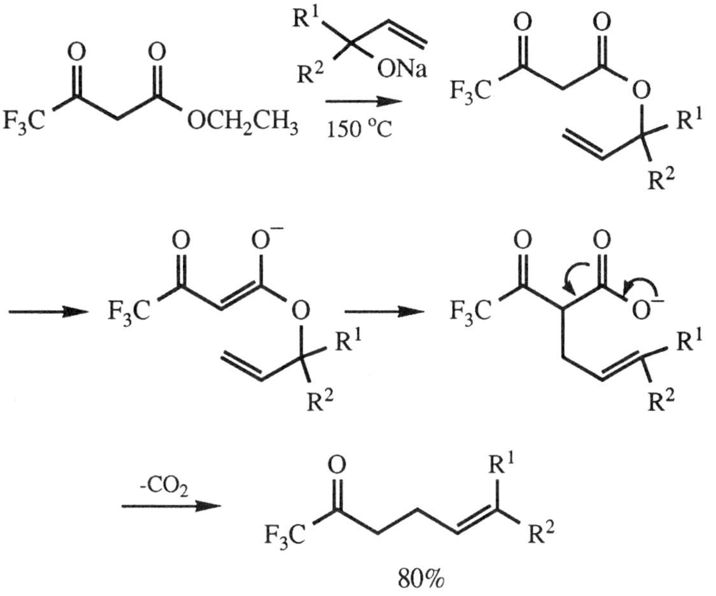

$RCH(OH)CF(CF_3)C(=O)N(C_2H_5)_2$ + $RCH(OH)CF(CF_3)C(=O)N(C_2H_5)_2$

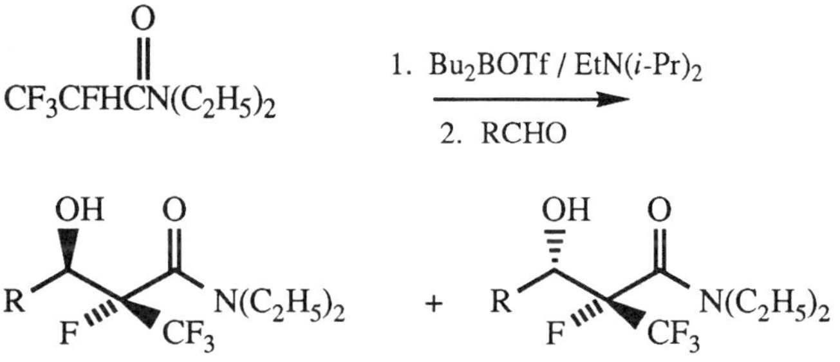

selectivity : 9 : 1
yield : 74- 88%

Table 8. Aldol Reaction of the Amide Boron Enolates $CF_3CF{=}C[OB(C_4H_9)_2]N(C_2H_5)_2$ with Aldehydes [9]

Aldehyde	*Yield* [a] *(%)*	*Isomer ratio* [b] *threo:erythro*
$CH_3CH_2CH_2CHO$	88	94:6
$(CH_3)_2CHCHO$	86	92:8
$(CH_3)_3CCHO$	74	86.14
(*E*)-$CH_3CH{=}CHCHO$	85	94:6
(*E*)-$CH_3CH{=}C(CH_3)CHO$	82	92:8
(*E*)-$C_6H_5CH{=}CHCHO$	86	100:0
C_6H_5CHO	84	93:7
p-$CH_3C_6H_4CHO$	78	100:0
p-$CH_3OC_6H_4CHO$	82	94:6
p-ClC_6H_4CHO	76	100:0

[a] Yields are of pure isolated products.
[b] The ratio was determined by ^{19}F NMR.

Fluorinated esters have synthetic utility in *Claisen condensations* [*24, 25*] (equation 21) and *Dieckmann cyclizations* [26].

[*24*] **21**

$$CH_2FCO_2C_2H_5 + (CO_2CH_2CH_3)_2 \xrightarrow{\text{EtONa}} C_2H_5O\text{–}CO\text{–}CHF\text{–}CO_2C_2H_5 \quad 90\text{–}95\%$$

Ketene acetals prepared from fluorinated esters by trimethylsilylation undergo Lewis acid-promoted aldol condensations giving satisfactory yields but low diastereoselectivity [*27*] (equation 22).

Fluorinated esters may also act as electrophiles in reactions with nonfluorinated ***ketones*** [*28*] (equation 23) or ***malononitrile*** [*29*] (equation 24). Unfortunately, the yields of β-diketones may be modest, but those of β-keto nitriles are excellent (Table 9).

A fluorinated keto ester reacts as an electrophile with hydrides, giving a hydroxy ester in a highly stereoselective reduction [30] (equation 25).

Wittig Reagents and Fluorinated Carbonyl Compounds.

Fluoroolefins may be prepared by the reaction of Wittig reagents and other phosphorus-containing ylides with fluorinated carbonyl compounds. (A discussion of the fluorinated Wittig reagents or other fluorinated phosphorus reagents with nonfluorinated carbonyl compounds is on page 581.) Triphenylphosphoranes, derived from alkyltriphenyl phosphonium salts, react with 1,1,1-trifluoroacetone [*31*] or other ***trifluoromethyl ketones*** [*32, 33*] (equation 26) (Table 10).

[27] $CF_3CH_2CO_2CH_3$ $\xrightarrow[CH_2Cl_2,\ 20\text{-}25\ ^\circ C]{TMSOTf\ /\ NEt_3}$ $CF_3CH{=}C(OSi(CH_3)_3)(OCH_3)$ **22**

$\xrightarrow{(CH_3O)_2CR^1R^2}$ $R^1R^2C(OCH_3){-}CH(CF_3){-}C(O)OCH_3$

74-92%

[28] $RC(O)CH_3$ + $R_fCO_2CH_3$ $\xrightarrow{EtONa}$ $RC(O)CH_2C(O)R_f$ **23**

36-55%

[29] $KCH(CN)_2$ + $R_fCO_2CH_2CH_3$ $\longrightarrow$ $R_f(KO)C{=}C(CN)_2$ **24**

98-100%

$CH_3CH_2OC(O){-}CF(CH_3){-}C(O)R$ $\xrightarrow{AlCl_3\ /\ Ph_3SiH}$ **25**

[30] $CH_3CH_2OC(O){-}CF(CH_3){-}CH(OH)R$ + $CH_3CH_2OC(O){-}CF(CH_3){-}CH(OH)R$

selectivity : > 35 : 1
yield : 45-74%

[33] $RC(O)CF_3$ + $Ph_3P^+CH_2CH_2CH_2C(CH_3){=}Y$ $\xrightarrow[-75\ ^\circ C\ \ RT,\ 3\ h]{BuLi}$ $R(CF_3)C{=}CHCH_2CH_2C(CH_3){=}Y$ **26**

94 - 96%

Y= OCH_2CH_2O, =CHR   Z : E 9 : 1

By analogy, *Peterson olefination* of trifluoromethyl ketones proceeds in 66% yield with an *E* /*Z* ratio of 1:1 [*34*] (equation 27).

Predictably, fluoroketones undergo olefination reactions with more reactive **arsonium ylides** [*35*] (equation 28).

References are listed on pages 643–645.

Table 9. Fluorinated Esters as Electrophiles

Nucleophile	*Electrophile*	*Product*	*Yield (%)*	*Ref.*
$KCH(CN)_2$	$CF_3(CF_2)_2CO_2CH_3$	$(CN)_2C{=}C(OK)(CF_2)_2CF_3$	91	29
$KCH(CN)_2$	$CF_3CF_2CO_2CH_2CH_3$	$(CN)_2C{=}C(OK)CF_2CF_3$	99	29
$KCH(CN)_2$	$CF_3CO_2CH_3$	$(CN)_2C{=}C(OK)CF_3$	100	29
$CH_3C(ONa){=}CH_2$	$CF_3OCF_2CO_2CH_3$	$CF_3OCF_2C(O)CH_2C(O)CH_3$	40	28
$C_2H_5C(ONa){=}CH_2$	$CF_3OCF_2CO_2CH_3$	$CF_3OCF_2C(O)CH_2C(O)CH_2CH_3$	53	28
$(C_2H_5)_2CHC(ONa){=}CH_2$	$CF_3OCF_2CO_2CH_3$	$CF_3OCF_2C(O)CH_2C(O)CH(CH_2CH_3)_2$	36	28
$(CH_3)_3CC(ONa){=}CH_2$	$CF_3OCF_2CO_2CH_3$	$CF_3OCF_2C(O)CH_2C(O)C(CH_3)_3$	41	28
$(CH_3)_3CC(ONa){=}CH_2$	$CF_3CF_2OCF_2CO_2CH_3$	$CF_3CF_2OCF_2C(O)CH_2C(O)C(CH_3)_3$	51	28
$(CH_3)_3CC(ONa){=}CH_2$	$F_2C{-}CF_2$ $F_2C{-}O{-}CF{-}CF_2CO_2CH_3$	$F_2C{-}CF_2$ $F_2C{-}O{-}CF{-}CF_2C(O)CH_2C(O)C(CH_3)_3$	47	28
$(C_2H_5)_2CHC(ONa){=}CH_2$	$CF_3CF_2CF_2CO_2CH_3$	$C_3F_7CO_2CH_2CO_2CH(C_2H_5)_2$	55	28

Table 10. Wittig Reactions with Trifluoromethyl Ketones

R in $RCOCF_3$	*Wittig Reagent*	*Yield (%)*	*Z/E*	*Ref.*
CH_3	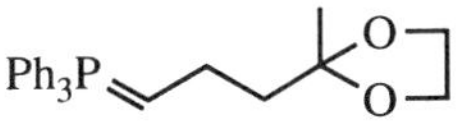 Ph_3P (structure: ylide chain with dioxolane ketal)	95	90/10	33
CH_3	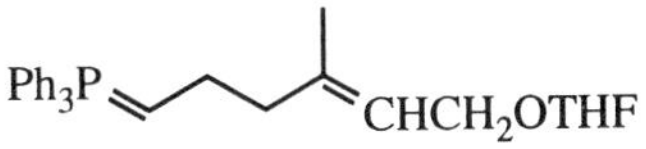 Ph_3P (structure) $CHCH_2OTHF$	96	95/5	33
$(CH_3)_2C=CHCH_2CH_2$	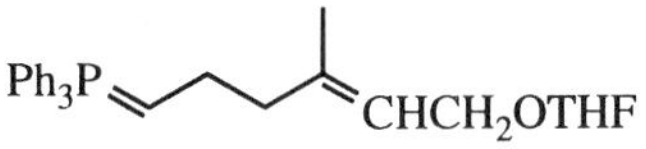 Ph_3P (structure) $CHCH_2OTHF$	94	89/11	33
CH_3	$Ph_3P=CHCOCH_3$	58	0/100	31
CH_3	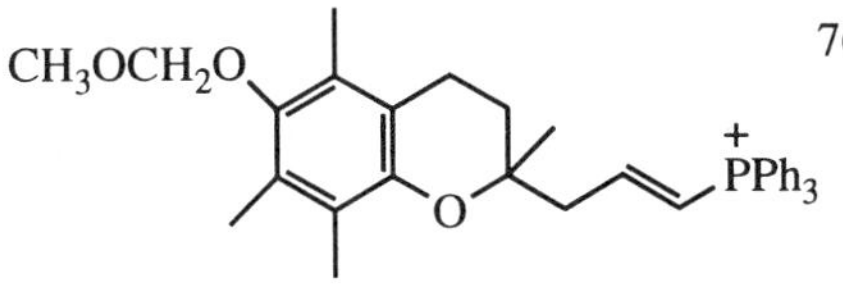 CH_3OCH_2O (chroman structure) $\overset{+}{P}Ph_3$	76	86/14	32
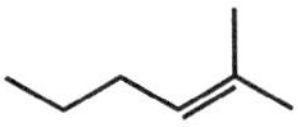 (structure)	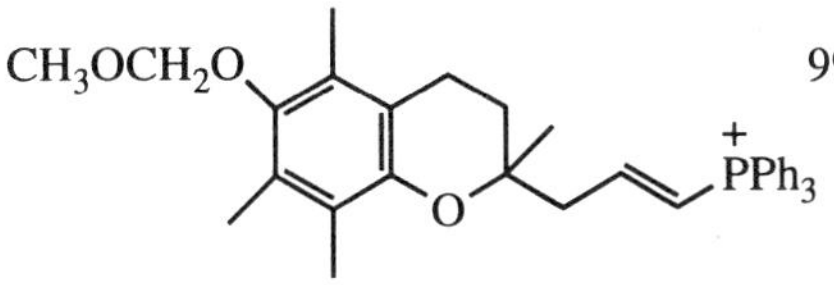 CH_3OCH_2O (chroman structure) $\overset{+}{P}Ph_3$	99	84/16	32
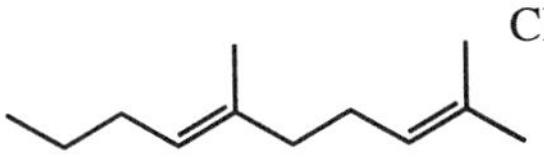 (structure)	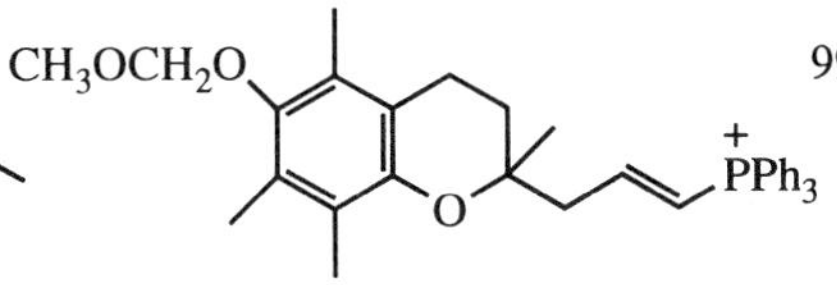 CH_3OCH_2O (chroman structure) $\overset{+}{P}Ph_3$	99	84/16	32

CF_3 O **27**

[*34*] $TMSCH_2CO_2Et$ / LiCA / THF, -78 - -23 °C → CF_3, CH_3CH_2O, O + CF_3 O, OCH_2CH_3

LiCA = Lithium dicyclohexylamide

66%
Z : E 1 : 1

$Ph_3P{=}CHCH{=}CHCO_2C_2H_5$ + $CH_3C(O)CF_3$ **28**

[*35*] Et_2O, 0 °C 4 h → $(CH_3)(CF_3)C{=}CHCH{=}CHCO_2C_2H_5$

86%

The more reactive fluoroketones also react with reagents prepared by the action of carbon tetrachloride on a trialkylphosphine to form a vinyl phosphine oxide [*36*] (equation 29).

[*36*] $R_2PCH_2R^1$ $\xrightarrow{CCl_4}$ $R_2P(Cl){=}CHR^1$ **29**

R= $C(CH_3)_3$

R^1= H, CH_3, C_3H_7, C_6H_5

$CF_3C(O)C_6H_4OCH_3$ → $R_2P(O)\text{-}C(R^1){=}C(CF_3)C_6H_4OCH_3$

70%

Horner–Emmons reagents react with ***trifluoromethyl ketones*** to form trifluoromethylated olefins; however, the double bond can isomerize out of conjugation with the carboxylic acid group with the product olefin that bears a γ-proton [*37*] (equation 30).

$$C_4H_9C(=O)CF_3 + (C_2H_5O)_2P(O)CHCO_2C_2H_5 \xrightarrow[3\ h]{} \qquad \mathbf{30}$$

[37] $$(C_4H_9)(CF_3)C{=}CHCO_2C_2H_5 \quad + \quad (C_2H_5CO_2CCH_2)(CF_3)C{=}CHCO_2C_2H_5$$

10% E : Z 9 : 1   90% E : Z 9 : 1

N-(Methoxycarbonyl)triphenylphosphine imide reacts with methyl trifluoropyruvate to form methyl *N*-methoxycarbonyl-2-imino-3,3,3-trifluoropropionate in 95% yield. This convenient building block easily adds nucleophiles such as Grignard reagents without competing side reactions at the ester group to form *trifluoromethylated amino acids* [*38*] (equation 31).

$$CF_3C(=O)C(=O)OCH_3 + Ph_3P{=}NCO_2CH_3 \xrightarrow[1.5\ h]{60\ ^{\circ}C} CF_3C(=NCO_2CH_3)C(=O)OCH_3 \qquad \mathbf{31}$$

95%

[*38*] $$\xrightarrow[R=\ C_6H_5,\ CH_3]{R\ Mg\ X} CF_3C(R)(NHCO_2CH_3)CO_2CH_3$$

53 - 80%

Fluorinated ylides have also been added to ***fluoroalkyl ketones;*** however, this olefination procedure may be accompanied by considerable side reactions resulting from deprotonation of the fluoroalkyl ketones. In most cases, the yields are not better than 30% [*39*].

Fluorinated aldehydes, where the fluorine is located α to the carbonyl [*40*] or more remotely, undergo olefination reactions cleanly [*41, 42, 43*] (equation 32) (Table 11).

Fluorinated nitriles may be subjected to Wittig olefination to form fluorinated β-keto esters in good to excellent yields [*44, 45, 46*] (equation 33).

The reactions of ***fluorinated esters and amides*** to form, respectively, enol ethers [*47*] and enamines [*48*] give high yields and are interesting synthetic transformations (equations 34 and 35) (Table 12).

Table 11. α- and β-Fluorinated α,β-Unsaturated Aldehydes in Olefination Reactions [*43*]

Ylide	*Aldehyde*	*Yield (%)*	*Z/E*
$Ph_3P{=}CHOCH_3$	$CH_3CH{=}CFCHO$	55	43/57
	$(CH_3)_2C{=}CFCHO$	57	43/57
$Ph_3P{=}CHCl$	$CH_3CH{=}CFCHO$	59	36/62
	$(CH_3)_2C{=}CFCHO$	52	50/50
$Ph_3P{=}C(CH_3)_2$	CH3CH=CFCHO	61	—
	$(CH_3)_2{=}CFCHO$	65	—
$Ph_3P{=}CHC_3H_7$	$CH_3CH{=}CFCHO$	70	87/13
	$(CH_3)_2C{=}CFCHO$	73	90/10
$Ph_3P{=}CHC_6H_5$	$CH_3CH{=}CFCHO$	95	67/33
$Ph_3P{=}CHCO_2CH_3$	$CH_3CH{=}CFCHO$	93	4/96
	$(CH_3)_2C{=}CFCHO$	84	1/99
$Ph_3P{=}C(CH_3)CO_2CH_3$	$CH_3CH{=}CFCHO$	91	1/99
	$(CH_3)_2C{=}CFCHO$	83	2/98
$[(CH_3)_2CHO]_2P(O)\bar{C}(CO_2C(CH_3)_3)(CH_2Sn(CH_3)_3)$	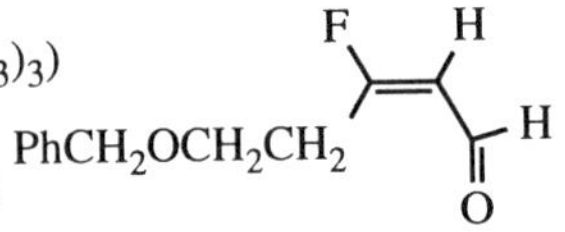 $PhCH_2OCH_2CH_2CF{=}CHCHO$	75	3/2

[*43*] 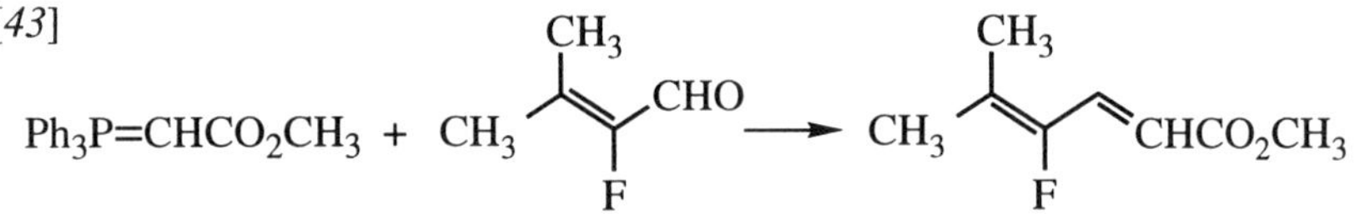 **32**

$$Ph_3P{=}CHCO_2CH_3 + (CH_3)_2C{=}CFCHO \longrightarrow (CH_3)_2C{=}CFCH{=}CHCO_2CH_3$$

Z / E 1 / 99

84%

Enamine Condensation

Fluorine-containing ***Michael addition*** acceptors have been used as synthons, a portion of a molecule recognizably related to a simpler molecule, for the introduction of fluorine into the organic molecules. Their reactions with **enamines and ketones** lead to a condensation–cyclization process.

The trifluorinated α,β-ethylenic ketone shown in equation 36 was prepared by two different methods and behaves like a Michael addition acceptor in reactions with nucleophilic species [*49*]. Its reaction with an enamine gives an annelation product [*49*] (equation 36) (Table 13). This sequence constitutes a method to build up a ring bearing a trifluoromethyl group.

$R_fCN + Ph_3P{=}CHCO_2C_2H_5 \longrightarrow$ Ph_3P—$CHCO_2C_2H_5$ / N=C—R_f **33**

[*46*] $Ph_3P{=}N{-}C(R_f){=}CHCO_2C_2H_5 \xrightarrow{H_3O^+} R_f{-}CO{-}CHCO_2C_2H_5$ 90%

[*47*] $CF_3CO_2Si(CH_3)_3 + Ph_3P{=}CHR \longrightarrow R\,CH{=}C(OSi(CH_3)_3)(CF_3)$ 17 - 78% **34**

[*48*] $R^1R^2N{-}CO{-}CF_3 + Ph_3P{=}CHR^3 \xrightarrow[\text{reflux 24 h}]{C_6H_6} R^1R^2N{-}C(CF_3){=}CHR^3$ 19 - 76% **35**

[*49*] CF_3CHO → ($(C_6H_5)_3P{=}CHCOCH_3$, 54%; $CH_3COCH_2COCH_3$, 33%) → $CF_3CH{=}CHCOCH_3$ **36**

$CF_3CH{=}CHCOCH_3$ + 1-morpholinocyclohexene → 4-(trifluoromethyl)octahydronaphthalen-2-one 57%

Table 12. Fluorinated Esters and Amides in the Wittig Reaction to Form Enol Ethers [*47*] and Enamines [*48*]

Ylide	*Ester or Amide*	*Yield (%)*	*Z/E*
$Ph_3P{=}CHPh$	$CF_3CO_2Si(CH_3)_3$	48	—
$Ph_3P{=}CH$–(cyclohexyl)	$CF_3CO_2Si(CH_3)_3$	46	—
$Ph_3P{=}CH(CH_2)_5CH_3$	$CF_3CO_2Si(CH_3)_3$	70	—
$Ph_3P{=}CHCH_2Ph$	$CF_3CO_2Si(CH_3)_3$	50, 51	—
$Ph_3P{=}CHCH_2CH_2$–(1-cyclohexenyl)	$CF_3CO_2Si(CH_3)_3$	58	—
$Ph_3P{=}CHCH_2CH_2C(=CH_2)C_6H_5$			
$Ph_3P{=}CHCH_2CH_2C(=CH_2)Ph$	$CF_3CO_2Si(CH_3)_3$	30	—
$Ph_3P{=}CH(CH_2)_2CH{=}C(CH_3)Ph$	$CF_3CO_2Si(CH_3)_3$	17	—
$Ph_3P{=}CHCH{=}CHPh$	$CF_3CO_2Si(CH_3)_3$	40	—
$Ph_3P{=}CH(CH_2)_5CH_3$	$CF_3C(O)$–N(morpholine)	42[a] 76[b]	43/57 46/54
$Ph_3P{=}CHPh$	$CF_3C(O)$–N(morpholine)	62[a] 60[b]	97/3 95/5
$Ph_3P{=}CHCH_2CH_2Ph$	$CF_3C(O)$–N(morpholine)	66[a] 57[b]	39/61 45/55

Table 12—*Continued*

Ylide	*Ester or Amide*	*Yield (%)*	*Z/E*
$Ph_3P{=}CHCH_2C_6H_{11}$	$CF_3C(O)N(CH_2)_5$	21[a] 36[b]	62/38 72/28
$Ph_3P{=}CHCH_2CH_2Ph$	$CF_3C(O)N(CH_2CH_2)_2O$	46[a] 38[b]	61/39 66/34
$Ph_3P{=}CH(CH_2)_5CH_3$	$CF_3C(O)N(CH_3)CH_2C_6H_5$	53[a] 38[b]	39/61 47/53
$Ph_3P{=}CHCH_2C_6H_{11}$	$CF_3C(O)N(CH_2CH_2)_2O$	55[a] 50[b]	43/57 62/38
$Ph_3P{=}CHCH_2CH_2Ph$	$CF_3C(O)N(CH_3)CH_2Ph$	37[a] 37[b]	47/53 49/51
$Ph_3P{=}CHCH_2Ph$	$CF_3C(O)N(CH_2Ph)CH_2Ph$	23[a] 19[b]	26/74 52/48

[a]Ylide generation in THF.
[b]Ylide generation in benzene.

References are listed on pages 643–645.

Similarly, ***fluorinated ketones*** are prepared and react with **enamines** [*50*]. This reaction involves the intermediacy of an α,β-ethylenic ketone and leads to annelation–aromatization products [*50*] (Table 13) (equation 37).

[50] $R_fCF_2I + CH_2{=}C(OCH_3){-}CH_3 \xrightarrow[\Delta]{\text{pentane}} R_fCF_2CH_2COCH_3$ **37**

$R_f = CF_3$ 40%
$R_f = CF_3(CF_2)_2$ 45%
$R_f = CF_3(CF_2)_4$ 25%

[51] $CF_3CF_2CH_2C(=O)CH_3$ + 1-(1-cyclopentenyl)pyrrolidine $\xrightarrow[100\ °C]{(C_2H_5)_3N}$ (CF₃, pyrrolidinyl-substituted indane)

The synthesis and condensation reactions of ***2,2,2-trifluoroethyl vinyl ketone*** with **enamines and ketones** were investigated [*51*] (equation 38) (Table 13). These reactions lead to the formation of 2-trifluoromethylcyclohexenones.

[51] (vinyl ketone with CF_3) + (enamine, CH_3, CH_3, N-pyrrolidinyl) $\xrightarrow[2.\ HCl\text{-}H_2O]{1.\ pentane}$ (cyclohexenone with CH_3, CH_3, CF_3) **38**

76%

1,1,1-Trifluoro-2-diazo-3-nitropropane reacts with enamines [*52*]. This reaction goes through the intermediacy of a cyclopropane derivative, which hydrolyzes to the corresponding ketone (last entry of Table 13).

Equilibrium constants for reactions of various nucleophiles with α,α,α-trifluoroacetophenone to give tetrahedral adducts were determined [*53*] (equation 39). In the equilibria, all nucleophiles were found to be less reactive with trifluoroacetophenone than with aldehydes [*53*] (equation 39).

The cyanohydrin of methyl perfluoroheptyl ketone was synthesized by a two-step process: addition of sodium bisulfite and subsequent treatment with sodium cyanide. When the ketone was reacted with sodium cyanide, cyclic addition products were obtained, instead of the product of cyanohydrin formation. This result was attributed to the solubility characteristic of a long perfluoroalkyl group, which makes the compound less soluble in water and polar organic solvents [*54*] (equation 40) (Table 14).

Table 13. Reactions of Enamines and Ketones

Reactants	Reaction Conditions	Products	Yield (%)	Ref.
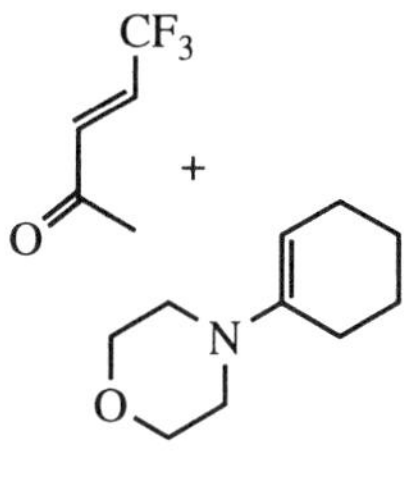 $CF_3CH{=}CHCOCH_3$ + 1-morpholinocyclohexene	1. C_5H_{12} 2. HCl–H_2O	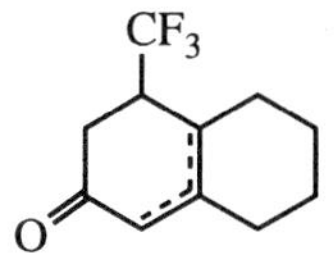 4-CF_3-octahydronaphthalen-2-one	57	49
$R_fCF_2CH_2COCH_3$ 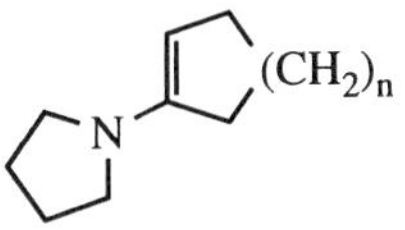 + 1-pyrrolidinocycloalkene $(CH_2)_n$	$(C_2H_5)_3N$	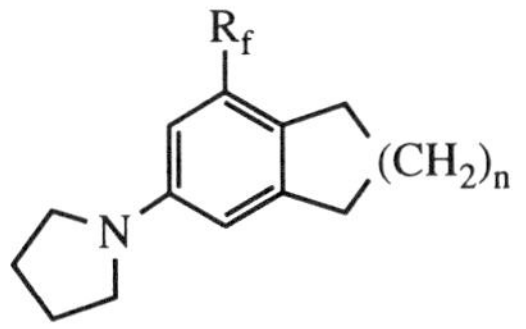 pyrrolidinyl-R_f-benzo-fused $(CH_2)_n$ ring		
		n = 1 $R_f = CF_3$	45	50
		n = 2 $R_f = CH_3$	40	50
		n = 1 $R_f = CF_3(CF_2)_2$	40	50
		n = 2 $R_f = CF_3(CF_2)_2$	33	50
		n = 1 $R_f = CF_3(CF_2)_4$	38	50
		n = 2 $R_f = CF_3(CF_2)_4$	40	50
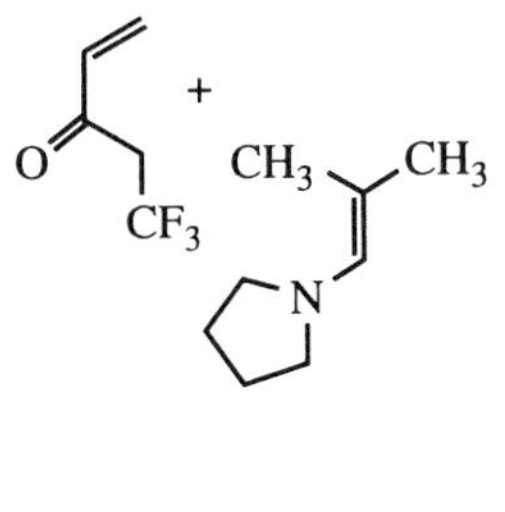 $CH_2{=}CHCOCH_2CF_3$ + $(CH_3)_2C{=}CH$–N-pyrrolidine	1. C_5H_{12}, 20 °C 2. HCl–H_2O	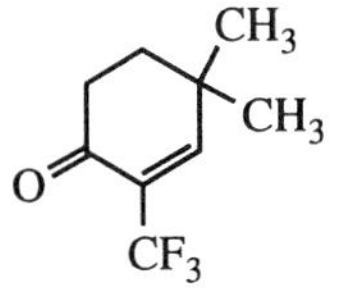 4,4-(CH_3)$_2$-2-CF_3-cyclohex-2-enone	76	51

Continued on next page.

Table 13—*Continued*

Reactants	*Reaction Conditions*	*Products*	*Yield (%)*	*Ref.*
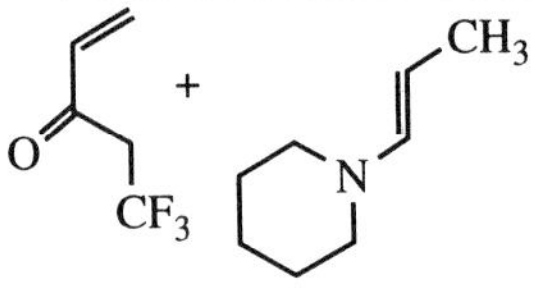	1. C_5H_{12}, 20 °C 2. HCl–H_2O, 0.5 h	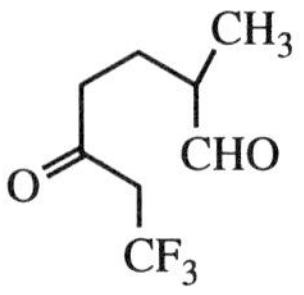	80	51
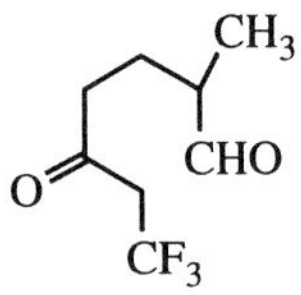	1. HCl–H_2O, 10 h	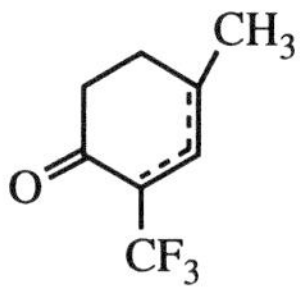	70	51
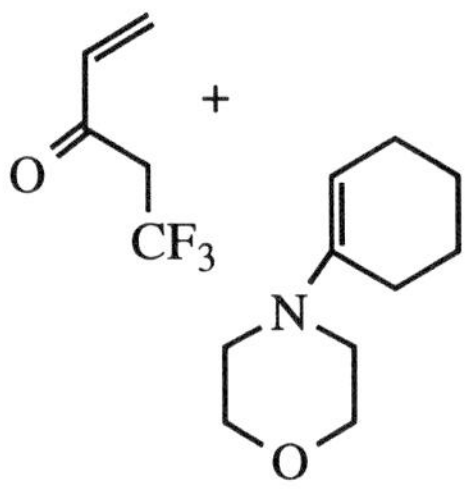	1. C_5H_{12}, Δ 2. HCl, Δ	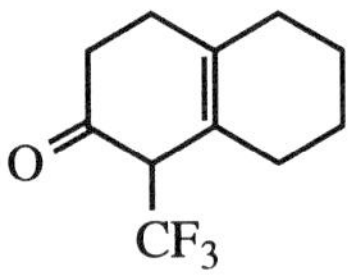	46	51
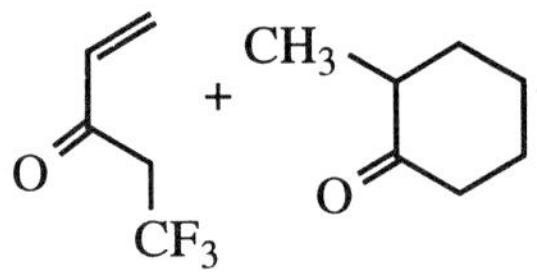	C_5H_{12}, H_2SO_4 Δ, 36 h	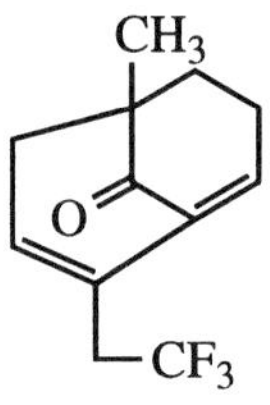	71	51

Table 13—*Continued*

Reactants	*Reaction Conditions*	*Products*	*Yield (%)*	*Ref.*
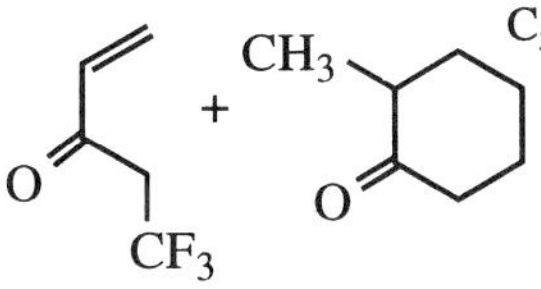 O, CF_3 + CH_3, O	C_5H_{12}, H_2SO_4 Δ, 7h	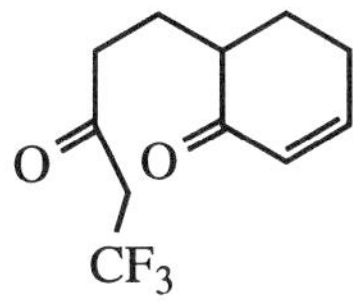 O, O, CF_3	51	51
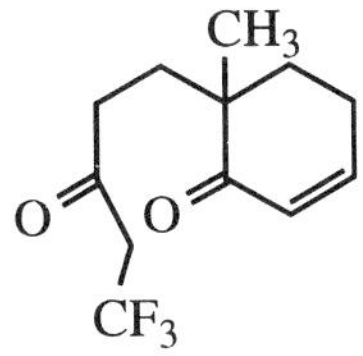 CH_3, O, O, CF_3	C_5H_{12}, 4-TsOH	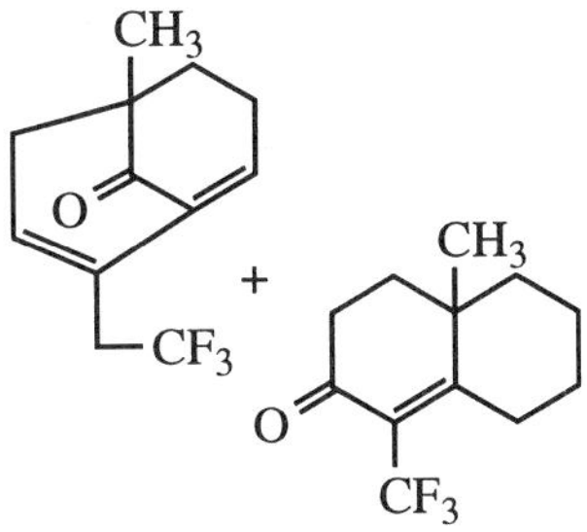 CH_3, O, CF_3 + CH_3, O, CF_3 2 : 1	45	51
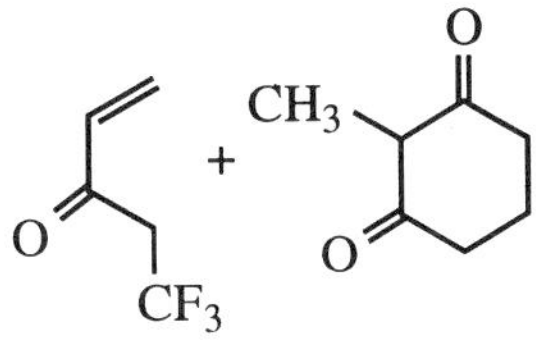 O, CF_3 + O, CH_3, O	C_5H_5OH, $(C_2H_5)_3N$, 20 °C	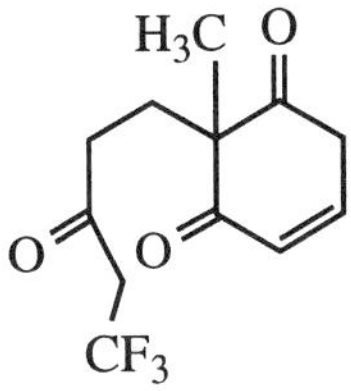 H_3C, O, O, O, CF_3	89	51

Continued on next page.

Table 13—*Continued*

Reactants	*Reaction Conditions*	*Products*	*Yield (%)*	*Ref.*
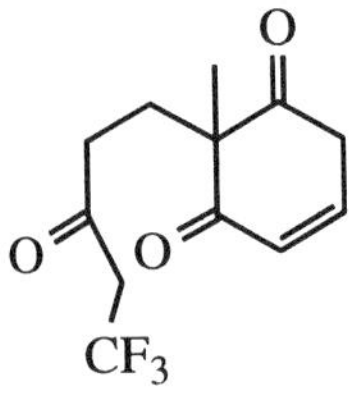 O, O, CF_3	HCl, Δ	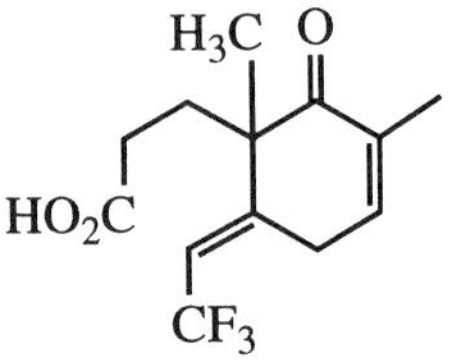 H_3C, O, HO_2C, CF_3		
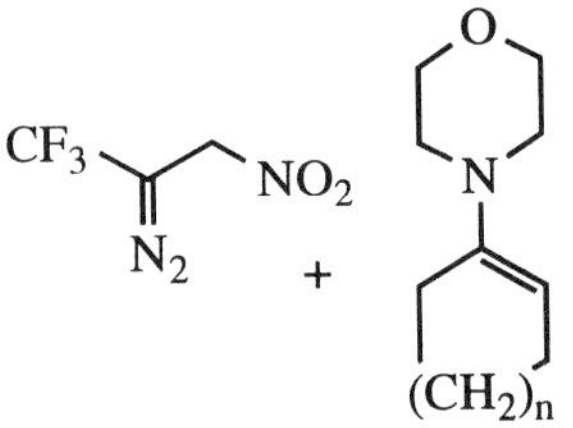 CF_3, NO_2, N_2 + O, N, $(CH_2)_n$	1. C_5H_{12}, Δ 2. HCl, Δ	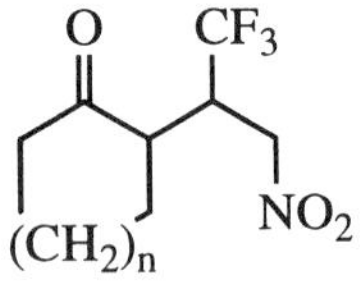 O, CF_3, $(CH_2)_n$, NO_2		
		n = 0	69	52
		n = 1	86	52

[53] $C_6H_5C(O)CF_3$ + NuH ⇌ $C_6H_5C(OH)(Nu)CF_3$ NuH = H_2O_2, HCN, HSO^-_3, $C_4H_9NH_2$, $CH_2CH_2NH_2$ (with OCH_3), NH_2OH, NH_2NH_2 **39**

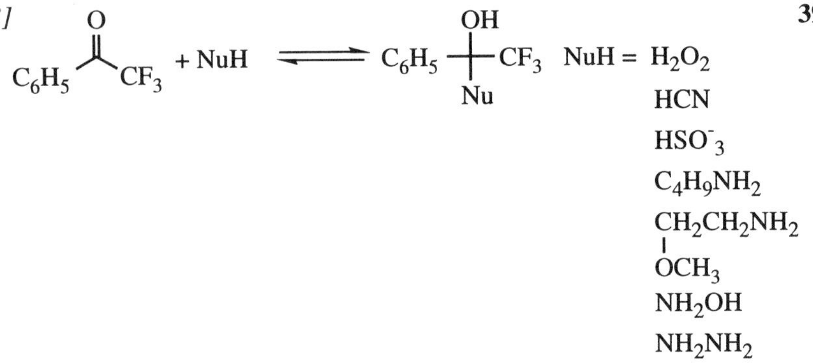

[54] $C_7F_{15}C(O)CH_3$ —NaCN, H_2O - dioxane→ O, C_7F_{15}, CH_3, NH, O, CH_3, C_7F_{15} **40**

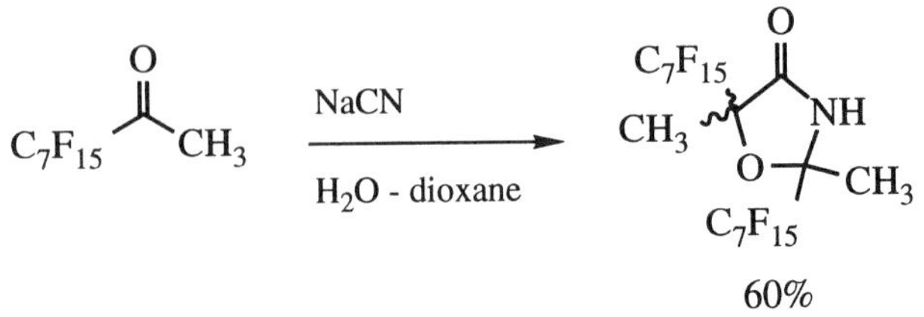

60%

References are listed on pages 643–645.

Table 14. Reactions of Fluoroketones with Nucleophiles

Reactant	Reaction Conditions	Products		Yield (%)	Ref.
$C_7H_{15}COCH_3$	1. $NaHSO_3$ 2. $NaCN$[a]	$C_7F_{15}-C(OH)(CN)-CH_3$		57	54
		2,5-bis(C_7F_{15})-2,5-dimethyl-oxazolidin-4-one (O, NH ring)	2,5-bis(C_7F_{15})-2,5-dimethyl-1,3-dioxolan-4-one		
	H_2O	45	–		
	H_2O–dioxane	60	–		
	THF	–	47		
	DME	–	60		
	CH_3CN	2	83		
	DMSO	25	25		
2,5-bis(C_7F_{15})-2,5-dimethyl-oxazolidin-4-one	50% H_2SO_4, Δ	2,5-bis(C_7F_{15})-2,5-dimethyl-1,3-dioxolan-4-one		100	54

[a]Ten equivalents of NaCN.

References for Pages 615–643

1. Ishihara, T.; Shinjo, H.; Inoue, Y.; Ando, T. *J. Fluorine Chem.* **1983,** *22,* 1.
2. Mead, D.; Loh, R.; Asato, A. E.; Liu, R. S. H. *Tetrahedron Lett.* **1985,** *26,* 2873.
3. Gazit, A.; Rappoport, Z. *J. Chem. Soc., Perkin Trans. 1* **1984,** 2863.
4. Balicki, R.; Nantka-Namirski, P. *Acta Pol. Pharm.* **1974,** *31,* 261.
5. Fujii, S.; Maki, Y.; Kimoto, H.; Cohen, L. A. *J. Fluorine Chem.* **1986,** *32,* 329.
6. Yamazaki, T.; Yamamoto, T.; Kitazume, T. *J. Org. Chem.* **1989,** *54,* 83.
7. Welch, J. T.; Eswarakrishnan, S. *J. Chem. Soc., Chem. Commun.* **1985,** 186.
8. Kuroboshi, M.; Okada, Y.; Ishihara, T.; Ando, T. *Tetrahedron Lett.* **1987,** *28,* 3501.
9. Ishihara, T.; Kuroboshi, M.; Yamaguchi, K. *Chem. Lett.* **1990,** 211.
10. Dyachenko, V. I.; Kolomiets, A. F.; Fokin, A. V. *Izv. Akad. Nauk SSSR,* **1987,** 1436; *Chem. Abstr.* **1989,** *11,* 21417.

11. Kondo, A.; Iwatsuki, S. *J. Fluorine Chem.* **1984,** *26,* 59.
12. Elkik, E.; Dahan, R.; Parlier, A. *Bull. Soc. Chim. Fr.* **1981,** 353.
13. Hanzawa, Y.; Yamada, A.; Kobayashi, Y. *Tetrahedron Lett.* **1985,** *26,* 2881.
14. Kuroboshi, M.; Ishihara, T. *Tetrahedron Lett.* **1987,** *28,* 6481.
15. Ogoshi, H.; Homma, M.; Yokota, K.; Toi, H.; Aoyama, Y. *Tetrahedron Lett.* **1983,** *24,* 929.
16. Jones, R. A.; Rustidge, D. C.; Cushman, S. M. *Synth. Commun.* **1984,** *14,* 575.
17. Lombardino, J. G. *J. Heterocycl. Chem.* **1973,** *10,* 697.
18. Welch, J. T.; Eswarakrishnan, S. *J. Org. Chem.* **1985,** *50,* 5403.
19. Welch, J. T.; Seper, K. W. *Tetrahedron Lett.* **1984,** *25,* 5247.
20. Welch, J. T.; Seper, K.; Eswarakrishman, S.; Samertino, J. *J. Org. Chem.* **1984,** *49,* 4720.
21. Welch, J. T.; Plummer, J. S. *Synth. Commun.* **1989,** *19,* 1081.
22. Elkik, E.; Imbeaux-Oudotte, M. *Tetrahedron Lett.* **1978,** 3793.
23. Camps, F.; Coll, J.; Messeguer, A.; Roca, A. *Tetrahedron Lett.* **1976,** *10,* 791.
24. Elkik, E.; Dahan, R.; Parlier, A. *Bull. Soc. Chim. Fr.* **1979,** 65.
25. Pashkevich, K. I.; Latypov, R. R.; Filyakova, V. I. *Izv. Akad. Nauk SSSR* **1986,** *11,* 2576; *Chem. Abstr.* **1987,** *107,* 197493v.
26. Elkik, E.; Parlier, A.; Dahan, R. *C. R. Hebd. Seances Acad. Sci., Ser.* **1977,** *284,* 141.
27. Yokozawa, T.; Nakai, T.; Ishikawa, N. *Tetrahedron Lett.* **1984,** *25,* 3987.
28. Shivanyuk, A. F.; Kudryavtseva, L. S.; Lozinskii, M. O.; Heplyuev, V. M.; Fialkov, Y. A.; Bratolyubova, A. G. *Ukr. Khim. Zh.* **1981,** *47,* 1078; *Chem. Abstr.* **1982,** *96,* 51425g.
29. Middleton, W. J.; Bingham, E. M. *J. Fluorine Chem.* **1982,** *20,* 397.
30. Kitazume, T.; Ikeya, T. *Chem. Expres.* **1989,** *4,* 81.
31. Sepiol, J. *J. Fluorine Chem.* **1984,** *25,* 363.
32. Koyama, M.; Tamura, M.; Ando, A.; Nagai, T.; Miki, T.; Kumadaki, I. *Chem. Pharm. Bull.* **1988,** *36,* 2950.
33. Camps, F.; Sanchez, F. J.; Messeguer, A. *Synthesis* **1988,** 823.
34. Hanzawa, Y.; Kawagoe, K.; Kobayashi, N.; Oshima, T.; Kobayashi, Y. *Tetrahedron Lett.* **1985,** *26,* 2877.
35. Huang, Y.; Shen, Y.; Zheng, J.; Zhang, S. *Synthesis* **1985,** 57.
36. Kolodyazhnyi, O. I. *Tetrahedron Lett.* **1985,** *26,* 439.
37. Trabelsi, H.; Bertaina, B.; Cambon, A. *Can. J. Chem.* **1985,** *63,* 426.
38. Soloshonok, V. A.; Gerus, I. I.; Yagupol'skii, Y. *Zh. Org. Khim.* **1986,** *22,* 1335; *Chem Abstr.* **1987,** 106, 195861w.
39. Burton, D. J.; Zawistowski, E. A. *J. Fluorine Chem.* **1972,** *1,* 347.
40. Alcazar, A.; Camps, F.; Coll, J.; Fabrias, G.; Guerrero, A. *Synth. Commun.* **1985,** *15,* 819.
41. Hosoda, A.; Taguchi, T.; Kobayashi, Y. *Tetrahedron Lett.* **1987,** *28,* 65.

42. Berglund, R. A.; Fuchs, P. L. *Synth. Commun.* **1989,** *19,* 1965.
43. Kondo, K.; Cottens, S.; Schlosser, M. *Chem. Lett.* **1984,** 2149.
44. Ding, W.; Wei, J.; Pu, J. *Youji Huaxu* **1987,** 59; *Chem. Abstr.* **1987,** *107,* 198493g.
45. Trabelsi, H.; Bollens, E.; Rouvier, E.; Cambon, A. *J. Fluorine Chem.* **1986,** *34,* 265.
46. Trabelsi, H.; Rouvier, E.; Cambon, A. *J. Fluorine Chem.* **1986,** *31,* 351.
47. Begue, J. P.; Mesureur, D. *J. Fluorine Chem.* **1988,** *39,* 271.
48. Begue, J. P.; Mesureur, D. *Synthesis* **1989,** *4,* 309.
49. Molines, H.; Wakselman, C. *J. Fluorine Chem.* **1980,** *16,* 97.
50. Molines, H.; Tordeux, M.; Wakselman, C. *Bull. Soc. Chim. Fr.* **1982,** 367.
51. Molines, H.; Wakselman, C. *J. Chem. Soc., Perkin Trans 1* **1980.**
52. Aizikovich, A. Y.; Bazyl, I. T. *Zh. Org. Khim.* **1987,** *23,* 1330.
53. Ritchie, C. D. *J. Am. Chem. Soc.* **1984,** *106,* 7187.
54. Kondo, A.; Iwatsuki, S. *J. Fluorine Chem.* **1984,** *26,* 59.

Reactions of Fluoromagnesium and Fluorolithium Compounds

by C. Tamborski

Since their initial synthesis, perfluoroorganometallic compounds of magnesium and lithium have been important intermediates (synthons) in the preparation of numerous fluorinated compounds. There are three main synthesis procedures for creating these carbon–metal bonds: (1) *direct reaction between a perfluoroalkyl or perfluoroaryl halide and a metal as magnesium or lithium,* (2) *metal–halogen interchange* (equation 1), and (3) *metalation reaction,* which denotes the replacement of an acidic hydrogen by a metal through an acid–base reaction [*1, 2, 3*] (equation 2).

[*1,2*] **1**

$$C_3F_7I + RMgX \longrightarrow RI + C_3F_7MgX$$

[*1,3*] **2**

$$CF_3C{\equiv}CH + C_2H_5MgBr \longrightarrow C_2H_6 + CF_3C{\equiv}CMgBr$$

In addition, there is a cleavage reaction whereby a perfluoroorganic group is cleaved from a metal by a base, for example, phenyllithium [*4*], ethylmagnesium bromide [*5*], or a fluoride ion [*6*] (equations 3–5).

[*4*] **3**

$$(C_6H_5)_3SnCF{=}CF_2 + C_6H_5Li \longrightarrow (C_6H_5)_4Sn + CF_2{=}CFLi$$

[*5*] **4**

$$(C_6F_5)_3SnBr + 2\ C_2H_5MgBr \longrightarrow (C_6F_5)_2Sn(C_2H_5)_2 + C_6F_5MgBr$$

[*6*] **5**

$$[(CF_3)_2CF]_2Hg + 2KF \longrightarrow 2\ (CF_3)_2CF^- + 2\ K^+ + HgF_2$$

0065–7719/95/0187–0646$08.72/1

In the early 1950s investigations into the formation of a *perfluoroalkylmagnesium halide* began [*1*] through the reaction of a perfluoroalkyl iodide and an activated magnesium at low temperature. Low yields of the perfluoroalkylmagnesium compounds were generally obtained. An important improved procedure was later used [*2*] in the synthesis of perfluoropropylmagnesium bromide through the metal– halogen exchange reaction (equation 1). Shortly afterwards, *trifluorovinylmagnesium iodide* was prepared through the direct reaction between iodotrifluoroethylene and magnesium in diethyl ether at low temperature [*7*]. The synthesis of bromopentafluorobenzene made it possible to prepare *perfluorophenylmagnesium bromide* through its reaction with magnesium in diethyl ether [*8, 9*]. The synthesis of other perfluoroorganometallic compounds soon followed, for example, *4-perfluoropyridylmagnesium bromide* [*10*], *2-perfluorobiphenylmagnesium bromide* [*11*], *1- and 2-perfluoronaphthylmagnesium chloride* [*12*], *and 2,3,4,5-tetrafluorophenylmagnesium bromide* [*13*].

The first perfluoroorganolithium compound, perfluoropropyllithium, was reported in 1954 [*14*] and was synthesized through a metal–halogen exchange reaction (equation 1) at –74 °C in diethyl ether. Because perfluorolithium compounds are in general less thermally stable than their magnesium analogues, much lower temperatures had to be used during their preparation and subsequent reaction. The successful synthesis of perfluoropropyllithium initiated the studies of other perfluoroorganolithium compounds such as perfluoroisopropyllithium [*15*], *1,4-dilithioperfluorobutane* [*16*], *pentafluorophenyllithium* [*17*], *dilithiotetrafluorobenzene* [*13*], *tetrafluorophenyllithium* [*13*], *trifluorovinyllithium* [*18*], *undecafluorobicyclo[2.2.1]heptyllithium (norboranelithium)* [*19*], *3-lithiotetrafluoropyridine* [*20*], *4,4′-dilithioperfluorodiphenyl ether* [*21*], and various mono and dilithio derivatives of biphenyl [*22, 23*]. These perfluoroorganometallic compounds, as well as other modifications, have laid the foundation for their use as synthons (intermediate compounds used in the synthesis of other compounds) in the preparation of numerous fluorine-containing compounds. These early efforts are well-documented in texts [*1*] and reviews [*24, 25*].

Fluoroaromatic Magnesium Compounds

After a vigorous initial activity, whereby new fluorinated organometallic compounds were first synthesized, much of the research effort has decreased. Some activity has been concerned with the reactions of perfluoroaromatic magnesium compounds

Heptafluoro-1-naphthylmagnesium bromide shows unusual thermal stability even after refluxing in diethyl ether for 7 h. On carbonation, a 50% yield of the acid is obtained [*26*] (equation 6).

No products from the elimination of magnesium fluorobromide forming a perfluoronaphthyne are noted. Under similar circumstances the pentafluorophenylmagnesium bromide forms a benzyne. In addition, the carbonation of the perfluoronaphthyl Grignard reagent in diethyl ether, to give reasonable yields of the acid, is unusual compared with the perfluorophenyl Grignard reagent, which gives only trace quantities of the perfluorobenzoic acid [*9*].

[*26*] **6**

$$\text{C}_{10}\text{F}_7\text{Br} + \text{Mg} \xrightarrow{\text{Et}_2\text{O}} \text{C}_{10}\text{F}_7\text{MgBr} \xrightarrow{\text{7h at } 36^\circ\text{C}} \xrightarrow[\text{2. H}^+]{\text{1. CO}_2} \text{C}_{10}\text{F}_7\text{CO}_2\text{H} \quad 50\%$$

A method for generating a **perfluoroarylmagnesium compound** is the cleavage of a pentafluorophenyl–metal bond by a nucleophile such as ethylmagnesium bromide. As an example, *tetrakis(pentafluorophenyl)tin* on reaction with ethylmagnesium bromide gives a series of products, one of which may result from pentafluorophenylmagnesium bromide [*27*] (equation 7).

Another similar example is shown by the reaction between tris(pentafluorophenyl)tin bromide and ethylmagnesium bromide (equation 8).

[*27*] **7**

$$(C_6F_5)_4Sn + C_2H_5MgBr \longrightarrow \underset{8\%}{(C_6F_5)_4Sn} + \underset{3\%}{(C_6F_5)_3SnC_2H_5}$$

$$+ \underset{29\%}{(C_6F_5)_2Sn(C_2H_5)_2} + \underset{59\%}{C_6F_5Sn(C_2H_5)_3} + \underset{1\%}{(C_2H_5)_4Sn} + \underset{36\%}{C_6F_5H}$$

[*27*] **8**

$$2(C_6F_5)_3SnBr + 2C_2H_5MgBr \longrightarrow (C_6F_5)_2Sn(C_2H_5)_2 + (C_6F_5)_4Sn$$

[*28*] **9**

$$(C_6F_5)_2Si(CH_3)H + 2RMgBr \xrightarrow[0\ ^\circ C]{THF} R_2Si(CH_3)H + 2C_6F_5MgBr$$

$R = CH_3, C_2H_5, C_6H_5$

Cleavage of the pentafluorophenyl–silicon bond by a variety of Grignard reagents has been reported [*28*]. The cleavage reaction however is solvent-dependent, requiring tetrahydrofuran instead of diethyl ether, in which no cleavage is observed (equation 9).

The presence of the cleavage product, pentafluorophenylmagnesium bromide, is shown by derivatization with trimethylchlorosilane to produce the *pentafluorophenyltrimethylsilanes* in 43–72% yield. Although the cleavage method yields a pentafluorophenylmagnesium bromide, this process has limited applicability. More convenient methods of generating the Grignard reagent, which produce higher yields and less by-products, are available.

The reaction between **perfluoroarylmagnesium halides** and ***esters of dicarboxylic acids*** gives, besides the expected keto esters, secondary alcohols as reduction products [*29, 30, 31*] (equation 10). Such a reduction is enhanced by higher temperature. The hydrogen necessary for reduction comes from the solvent, diethyl ether, which is dehydrogenated to ethyl vinyl ether, which has been identified as a by-product in a similar reaction of perfluoroalkyllithium compound [*32*].

[*29, 30, 31*] **10**

$$Ar_fMgX + C_2H_5OCORCOOC_2H_5 \xrightarrow{Et_2O} C_2H_5OC(OMgX)(Ar_f)RCOOC_2H_5 \longrightarrow$$

$$\xrightarrow[-20^o \text{ to } -30^oC]{H^+} Ar_fCORCOOC_2H_5$$

$$C_2H_5OC(OMgX)(Ar_f)RCOOC_2H_5 \xrightarrow{\text{reflux}} Ar_fCH(OH)RCOOC_2H_5$$

$Ar_f = C_6F_5$, p-ClC_6F_4, m-$CF_3C_6F_4$, m-$CF_3C_6H_4$

$R = (CH_2)_n$, $C(CH_3)_2$, CH_2-CH_2 (cyclopropane-1,1-diyl, >C<)

Fluoroaromatic Lithium Compounds

Research on perfluoroaromatic lithium compounds, like that of the magnesium compounds, has decreased since their initial discoveries. Most of the recent effort is concerned with reactions of the perfluoroaryllithium compounds with various substrates.

Reactions of **perfluorophenyllithium** with ***dimethyl oxalate*** give a mixture of products whose composition depends upon experimental conditions [*33*] (equation 11).

[*33*] **11**

$$2C_6F_5Li + CH_3OCOCOOCH_3 \xrightarrow[2.\ H^+ \text{ at } 20^oC]{1.\ -78^oC} C_6F_5COCOOCH_3 \ (\text{II})$$

$$2C_6F_5Li + CH_3OCOCOOCH_3 \xrightarrow{-78^oC,\ b} C_6F_5COCOC_6F_5 \ (\text{III})$$

$$C_6F_5COCOOCH_3 \ (\text{II}) \xrightarrow{a} (C_6F_5)_2C(OH)COOCH_3 \ (\text{I})$$

$$C_6F_5COCOC_6F_5 \ (\text{III}) \xrightarrow[CH_3O^- \text{ rearrangement}]{20^oC} (C_6F_5)_2C(OH)COOCH_3 \ (\text{I})$$

The benzilic ester (I) could arise by either of two routes: (1) attack by two molecules of pentafluorophenyllithium at the same carbon atom in the dimethyloxalate (route a) or (2) initial formation of decafluorobenzil (III) followed by a benzilic ester rearrangement during a warm-up period (route b). A comparable experiment, except treatment of the reaction mixture with hydrogen chloride at –40 °C to prevent rearrangement of the benzil (III), gives compounds I, II and III and an unexpected product, tris(pentafluorophenyl)methanol (IV). The formation of IV is surprising, and an unusual and as yet unproven elimination process must take place. The variations in yields of products with temperature at which hydrogen chloride is added are shown in Table 1.

Table 1. Effect of Temperature of Decomposition of the Product of Reaction of Pentafluorophenyllithium with Dimethyl Oxalate on Product Distribution [*33*]

	Yield of Product in Equation 11 (%)			
Temp. (°C)	*II*	*III*	*I*	*IV*
–78	47	43	—	—
–40	33	55	7	4
+20	—	—	79	11

Among other fluoroaromatic lithium compounds, **2,2′-dilithioperfluorobiphenyl** was investigated. No secondary alcohols as mentioned previously in similar reactions are noted in any examples [*33*].

Various ***chlorofluoroacetones*** on reaction with **pentafluorophenyllithium** give pentafluorophenylchlorofluoro tertiary alcohols [*34*], which on treatment with alcoholic potassium hydroxide produce epoxides (equation 12).

[*34*] **12**

$$C_6F_5Li + CF_2XCOCFYCl \longrightarrow CF_2XC(C_6F_5)(OH)CFYCl \xrightarrow[\text{alc.}]{KOH} CF_2XC(C_6F_5)\text{-}CFY \text{ (epoxide, -O-)}$$

X = Y = Cl	78%	81%
X = Cl, Y = F	80%	79%
X = Y = F	68%	79%

In earlier studies [*13*], it was shown that *pentafluorobenzene* has an acidic aryl hydrogen capable of metalation by organolithium or organomagnesium compounds to yield perfluorophenylorganometallic compounds.

Benzylic carbon–hydrogen bonds in compounds such as methylpentafluorobenzene, fluoromethylpentafluorobenzene, and difluoromethylpentafluorobenzene are not capable of metalation by butyllithium. Instead nucleophilic substitution of the *para* fluorines occurs in each example [*35*] (equation 13).

[*35*] **13**

$$C_6F_5X \ + \ C_4H_9Li \xrightarrow[\text{-76}^{\circ}\text{C}]{\text{THF}} p\text{-}C_4H_9C_6F_4X$$

$$X = CF_2H \ 70\%; \ CFH_2 \ 12.6\%; \ CH_3 \ 72\%$$

Heptafluoro-2-naphthyllithium prepared by metalation reaction can thermally decompose to a hexafluoro-1,2-naphthalyne by elimination of lithium fluoride [*36, 37*]. In this organolithium compound, fluorine elimination can occur from either position 1 or 3; however, no evidence for fluorine elimination from position 3 is observed.

The 1,2-naphthalyne can be trapped by reaction with excess furan to give the Diels–Alder adduct. In the absence of furan, 1,2-naphthalyne can react with heptafluoro-2-naphthyllithium to yield a mixture of monohydrotridecafluorobinaphthyls (equation 14).

[*36, 37*] **14**

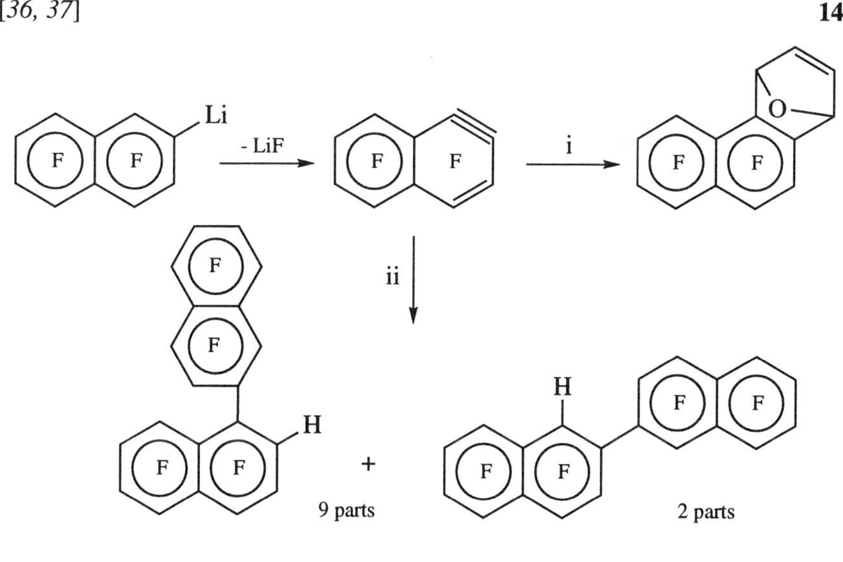

i: excess furan ii: 1. $C_{10}F_7Li$; 2. H_2O

References are listed on pages 666–669.

[*38*] **15**

BuLi → i →

$i = CO_2$ $X = CO_2H$ 61%

$i = Br_2$ $X = Br$ 76%

2-Perfluoronaphthyllithium reacts with electrophilic compounds to yield substituted perfluoronaphthalenes [*38*] (equation 15).

If an excess of butyllithium (2.5:1 molar ratio) is used during metalation, a mixture of butylated naphthyllithium compounds is formed. Reactions of this mixture with electrophiles give a mixture of 6- and 7-butyl-substituted hexafluoronaphthalene derivatives in respective ratios of 4:1 [*38*] (equation 16).

[*38*] **16**

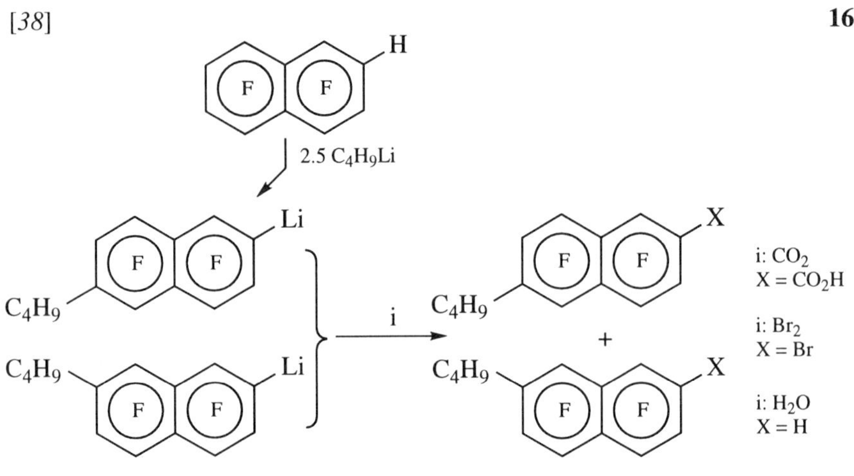

Fluoroaliphatic Magnesium and Lithium Compounds

The synthesis of new perfluoroaliphatic compounds of magnesium and lithium has continued, although at a reduced effort. Perfluoroorganometallics offer a variety of reactions with numerous substrates, which contributed to the development of the chemistry of fluorine-containing compounds.

In most instances, *perfluoroaliphatic magnesium and lithium compounds exhibit less thermal stability than perfluoroaromatic compounds.* This instability imposes some stringent restrictions on their reactions. A prior knowledge of their stability at various temperatures is fundamental to their subsequent use as organometallic intermediates. Thus it is important to select the proper reaction temperature, at which the rate of reaction with a substrate is much greater than the rate of thermal decomposition. Most of the reactions can be carried out at –78 °C, a temperature at which the organometallics have a reasonable life time. To overcome this problem, in situ synthesis of the perfluoroorganometallics [*39*], by the metal–halogen exchange reaction, may be carried out. Typically, the perfluoroalkyl iodide or bromide and the substrate electrophile, which will react with the organometallic, are placed in the reaction vessel at about –78 °C under an atmosphere of dry nitrogen. The alkyllithium or alkylmagnesium halide is then slowly added. In most instances the exchange reaction is rapid and the newly formed perfluoroorganometallic reacts with the substrate. The success of the in situ reaction depends on the difference between the rates of reaction of the perfluoroalkyl halide and the substrate E (electrophile) and the alkyllithium or alkylmagnesium compound being added (equation 17).

[*39*] **17**

$$R_fX + RM \xrightarrow{r_1} RX + R_fM$$

$$E + RM \xrightarrow{r_2} REM$$

where: R_f = perfluoroalkyl; X = Br or I
R = alkyl or phenyl; M = Li or MgX
E = esters, ketones, aldehydes, etc.

If r_1, the rate of reaction 1, is extremely high, then the secondary reaction of the substrate E with the hydrocarbon organometallic does not take place. The temperature at which the reaction mixture is subsequently terminated by hydrolysis is also an important factor.

Unlike perfluoroaliphatic organometaliics, *perfluoroolefinic and perfluoroacetylenic Grignard reagents have a greater thermal stability* and can be prepared and used as intermediates at temperatures around 0 °C [*40*].

Fluoroaliphatic Magnesium Compounds

The metal–halogen exchange reaction is useful in the synthesis of numerous perfluoroalkylmagnesium halides, some of which are shown in Table 2.

Although the metal–halogen exchange reaction is the preferred method of synthesis, the conventional Grignard synthesis through the reaction of a perfluoroorgano halide and magnesium occasionally is still used [*49, 50*].

Perfluoroalkylmagnesium compounds undergo many of the reactions of their hydrocarbon analogues. *Care must be exercised in controlling the reaction temperature* because of their thermal sensitivity.

Table 2. Perfluoroalkylmagnesium Halides (R_fMgX) Prepared by Metal–Halogen Exchange

R_f	Ref.
$CF_3(CF_2)_n$, $n = 6, 8, 10$	41, 42, 43, 44
$(CF_3)_2CFO(CF_2)_n$, n = 2	44, 45
CF_3CXY, X = F, Cl, Br; Y = Cl, Br	46
$C_2H_5OCO(CF_2)_4OCF_2CF_2$	47
$C_6F_{13}CF{=}CF$	48
$-(CF_2)_2O(CF_2)_5O(CF_2)_2$	45

Perfluoroalkylether magnesium halides react with a variety of substrates [*44, 45*] (equation 18).

18

$(CF_3)_2CFO(CF_2)_nMgX$ →

- H_2O → $(CF_3)_2CFO(CF_2)_nH$ [*44*]; n = 6 69%; n = 8 63%; n = 10 84%
- 1. $(CF_3)_2C{=}O$; 2. H_2O → $(CF_3)_2CFO(CF_2)_nC(CF_3)_2OH$ [*45*]; n = 2 49%
- 1. CO_2; 2. H_2O; 3. C_2H_5OH → $(CF_3)_2CFO(CF_2)_nC(O)OC_2H_5$ [*45*]; n = 2 86%
- $(CH_3)_3SiCl$ → $(CF_3)_2CFO(CF_2)_nSi(CH_3)_3$ [*44*]; n = 2 65%; n = 6 60%; n = 8 62%
- $(CH_3)_2SiHCl$ → $(CF_3)_2CFO(CF_2)_nSi(CH_3)_2H$ [*44*]; n = 4 68%; n = 6 69%

An unusual perfluoroalkylether magnesium bromide containing a functional ester group in the omega position is synthesized [*47*]. At low temperatures (e.g., –78 °C), the Grignard reagent reacts normally with a variety of esters to yield keto esters in yields of 1 to 91% depending on the electrophilic nature of the ester (Table 3). At a higher temperature (e.g., –20 °C), the Grignard reagent containing an ester

Table 3. Reaction of $C_2H_5O_2C(CF_2)_2O(CF_2)_2MgBr$ with Carboxylic Esters, $RCO_2C_2H_5$[a]

		GC Area (%) Yield		
$RCO_2C_2H_5$	*Time (h)*	*Unreacted Starting* Material[a,b]	*Intermolecular Reaction Product*[c,e]	*Keto Ester Product*[d,e]
$C_2H_5O_2C$	0.5	15	Trace	81
	2	12	1	84
	22	8	1	91
H	0.5	42	Trace	55
	2	26	1	70
	20	10	1	86
	48	10	1	86
C_2H_5O	0.5	75	Trace	23
	2	46	1	50
	24	13	2	80
	48	13	2	80
CH_3	0.5	84	Trace	13
	2	64	1	32
	20	14	2	80
	48	14	2	80
C_6H_5	0.5	96	Trace	2
	2	85	2	11
	20	47	4	44
	48	45[e]	5[e]	46[e]
CF_3	0.5	97	Trace	Trace
	2	92	3	Trace
	22	87	7	1
	48	84	9	1

[a]Reactions were carried out at –75 °C in diethyl ether; an aliquot sample was removed after the indicated time, hydrolyzed with 2N HCl, and analyzed by gas chromatography (GC).
[b]$C_2H_5O_2C(CF_2)_4O(CF_2)_2H$.
[c]$C_2H_5O[O{=}C(CF_2)_4O(CF_2)_2]_2H$.
[d]$C_2H_5O_2C(CF_2)_4O(CF_2)_2C(O)R$.
[e]Products characterized by GC/MS only.

group undergoes an intermolecular reaction to give oligomers (R_fH) as the main decomposition products. No intramolecular products from cyclization are observed (equation 19).

Novel polyfluoroethyl Grignard reagents containing fluorine, chlorine, and bromine are prepared through the metal–halogen exchange reaction [*46*] (equation 20).

[*47*] **19**

$$C_2H_5O\overset{O}{\overset{\|}{C}}(CF_2)_4OCF_2CF_2I \xrightarrow[-78^oC]{C_2H_5MgBr} C_2H_5I + \underset{\text{I}}{C_2H_5O\overset{O}{\overset{\|}{C}}(CF_2)_4OCF_2CF_2MgBr(I)}$$

$$\text{I} \xrightarrow[\text{reaction}]{\text{intermolecular}} C_2H_5O\overset{O}{\overset{\|}{C}}\left[(CF_2)_4OCF_2CF_2\underset{C_2H_5O}{\overset{BrMgO}{C}}\right]_{n-1}(CF_2)_4OCF_2CF_2MgBr \quad (R_fMgBr)$$

$$R_fMgBr \underset{-C_2H_5I}{\overset{+C_2H_5I}{\rightleftharpoons}} \longrightarrow \left[\begin{array}{c} Br \\ R_f \end{array} Mg \cdots \begin{array}{c} CH_2CH_3 \\ I \end{array}\right]$$

$R_fMgBr \xrightarrow{H^+} R_fH$ (76%)

$\left[\ \right] \underset{+C_2H_5MgBr}{\overset{-C_2H_5MgBr}{\rightleftharpoons}} R_fI$ (10%)

$\left[\ \right] \underset{+C_2H_5MgI}{\overset{-C_2H_5MgI}{\rightleftharpoons}} R_fBr$ (10%)

[*46*] **20**

$$\underset{\text{I-VI}}{CF_3CXYZ} + RMgHal \longrightarrow \underset{\text{Ia - VIa}}{CF_3CXYMgHal} + R\text{-}Z$$

$$\underset{\text{Ia - VIa}}{CF_3CXYMgHal} + \underset{CH_3}{\overset{CH_3}{C{=}O}} \longrightarrow \underset{\text{Ib - VIb}}{CF_3CXY\underset{CH_3}{\overset{CH_3}{C}}\text{-}OH}$$

References are listed on pages 666–669.

The various Grignard reagents react with a variety of ***aldehydes and ketones*** to yield primary alcohols (yield, 27.6%), secondary alcohols (yield, 13.1–71.1%), and tertiary alcohols (yield, 24.9–81.7%). Products of the reactions with acetone are given in Table 4.

Table 4. Reaction of $CF_3CXYMgHal$ with Acetone To Give Tertiary Alcohols

Compound in Equation 20	*X*	*Y*	*Z*	*Yield of Tertiary Alcohol (%)*
I, Ia, Ib	F	Cl	Br	34.3
II, IIa, IIb	F	Br	Br	81.7
III, IIIa, IIIb	Cl	Cl	Cl	4.7
IV, IVa, IVb	Cl	Cl	Br	19.7
V, Va, Vb	Cl	Br	Br	19.0
VI, VIa, VIb	Br	Br	Br	20.3

Perfluoroalkylmagnesium compounds generally exhibit the same type of reactions toward electrophilic reagents as their hydrocarbon analogues. Some electrophiles, however, behave differently. The *thiocyanate group* is a striking example of an electrophile that reacts very differently with a perfluoroalkylmagnesium halide and its hydrocarbon analogue [*51*] (equation 21).

21

$$R_fMgBr + RSCN \xrightarrow[-40^oC]{Et_2O} BrMgCN + R_fSR$$

$$\not\longrightarrow BrMgSR + R_fCN$$

$R_f = C_3F_7$, $R = C_6H_5CH_2$ 50% [*51*]
$R_f = C_3F_7$, $R = p\text{-}ClC_6H_4$ 45% [*51*]
$R_f = C_4F_9$, $R = C_6H_5CHClCH_2$ 22% [*52*]

Hydrocarbon Grignard reagents give the alkyl cyanides, whereas perfluorinated Grignard reagents produce the sulfides. This reaction is useful for the preparation of alkyl perfluoroalkyl sulfides to the exclusion of di- and polysulfides, which are produced by other methods.

Higher **perfluorovinylic bromides** other than trifluorovinyl bromide and pentafluoropropenyl bromide [*53*] have not been synthesized until recently [*42, 54,*

[*42*] **22**

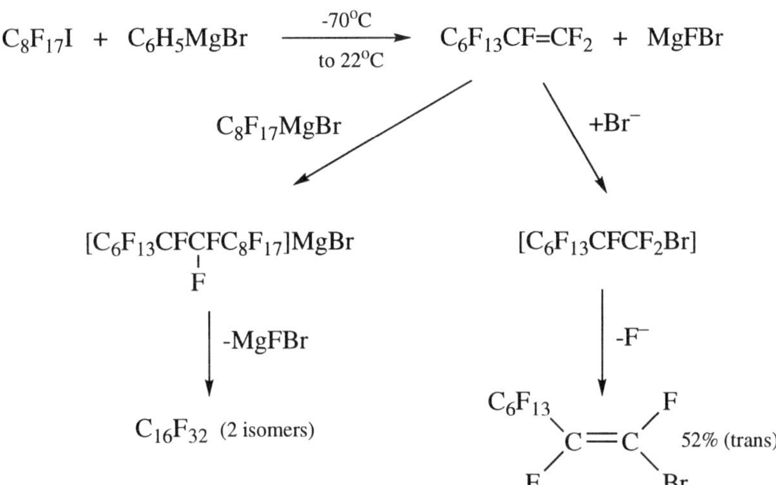

55]. Perfluorooctenyl bromide [*42*] was first prepared by the thermal decomposition of perfluorooctylmagnesium bromide (equation 22).

An extension of this reaction provides a number of other perfluorovinylic halides [*54*]. The type of reaction products from the thermal decomposition reaction and the type of hydrocarbon Grignard reagent used in the exchange reaction are solvent-dependent. When an excess of phenylmagnesium bromide is used, a variety of phenylated products are formed depending on the excess amount used [*48*] (equation 23).

[*48*] **23**

$C_6F_{13}CF{=}CFBr \xrightarrow{C_6H_5MgBr} C_6F_{13}CF{=}CFMgBr$

$C_6F_{13}CF{=}CFMgBr \xrightarrow{H_2O} C_6F_{13}CF{=}CFH$ (80%)

$C_6F_{13}CF{=}CFMgBr \xrightarrow{1.\ -MgBrF;\ 2.\ +C_6H_5MgBr} C_6F_5C(F){=}C(MgBr)C_6H_5 \xrightarrow{-MgBrF} C_6F_{13}C{\equiv}CC_6H_5$ (70%)

trans-Perfluorooctenyl bromide can be converted into the Grignard reagent by reaction with magnesium [*49*] or by a metal–halogen exchange reaction [*56*] and undergoes many typical reactions of the Grignard reagent (equation 24).

24

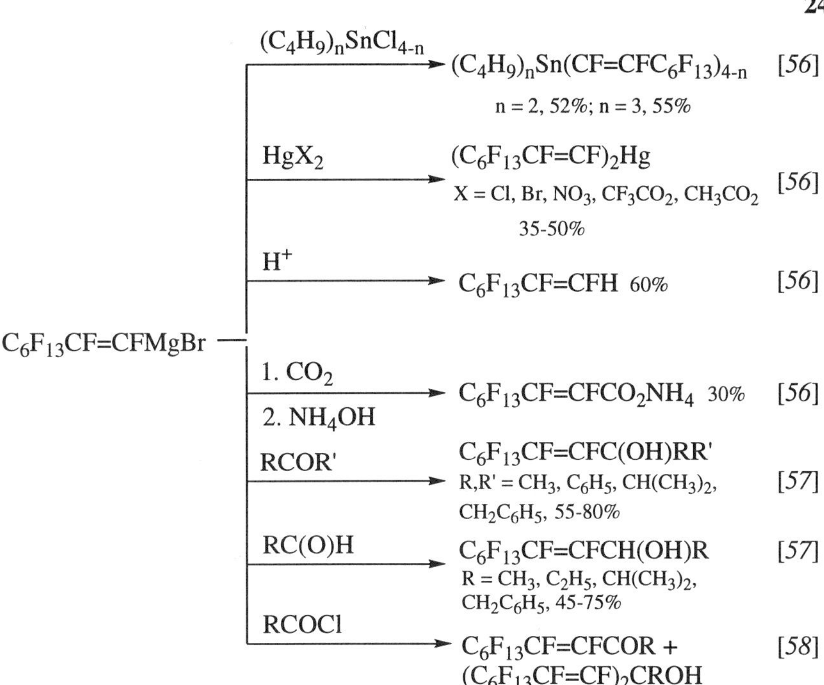

Perfluoroaliphatic Lithium Compounds

Perfluoroaliphatic lithium compounds [*1*] are prepared primarily by the metal–hydrogen exchange reaction. Fluorocycloaliphatic compounds containing an acidic hydrogen (e.g., 1*H*-undecafluorobicyclo[2.2.1]heptane) undergo a metalation reaction to give the perfluorolithium compound. The perfluorovinyllithium compounds, however, are prepared by a metal–halogen exchange or by a metalation reaction. The practical use of these synthons depends on the experimental conditions and the preparative method used. Various reaction solvents (e.g., diethyl ether, tetrahydrofuran, hexane, or combinations of these solvents) have been used. Other important factors to consider are *thermal stability of the perfluoroaliphatic lithium compound,* the type of hydrocarbon lithium reagent used in the metal–halogen exchange reaction, the reaction temperature, and the type of substrate to be reacted with the lithium compound. An example of changing experimental conditions to provide improved yields of a trifluorovinyllithium compound is demonstrated by the reaction between chlorotrifluoroethylene and butyllithium [*59*] (equation 25).

The formation of the organolithium compound is diminished by the competing addition–elimination reaction, which forms the butylated chlorodifluoroethylene.

To avoid this competing reaction, the metal–halogen exchange is performed in diethyl ether with either secondary or tertiary butyllithium at –60 °C to give the **trifluorovinyllithium compound** in near quantitative yield [*59*].

[*59*] **25**

$$CF_2{=}CFCl \xrightarrow[-80^\circ C]{C_4H_9Li,\ Et_2O} CF_2{=}CFLi + \underset{50\%}{C_4H_9CF{=}CFCl}$$

Much of the recent effort in the study of perfluoroaliphatic lithium compounds is concerned with vinyl or substituted fluorovinyl compounds. Modifications and extensions of the earlier research on the synthesis of trifluorovinyllithium provide many new fluorovinyllithium intermediates that react with numerous electrophiles to give novel and interesting fluoroolefinic compounds.

Early attempts to metalate 1,1-difluoroethylene with *n*-butyllithium in tetrahydrofuran or diethyl ether were unsuccessful. Another example whereby the course of the reaction may be altered by substitution of *sec*-butyllithium in place of *n*-butyllithium is the metalation of 1,1-difluoroethylene [*60*] (equation 26).

The **difluorovinyllithium compound** undergoes typical reactions with various ***aldehydes, ketones,*** and ***carbon dioxide*** (equation 27).

[*60*] **26**

$$CF_2{=}CH_2 + \text{sec. } C_4H_9Li \xrightarrow[-100^\circ C]{THF,\ Et_2O} \underset{100\%}{CF_2{=}CHLi}$$

27

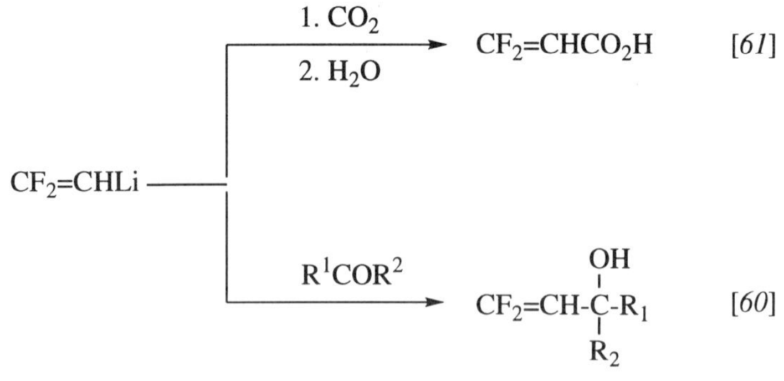

$$CF_2{=}CHLi \xrightarrow[2.\ H_2O]{1.\ CO_2} CF_2{=}CHCO_2H \quad [61]$$

$$CF_2{=}CHLi \xrightarrow{R^1COR^2} CF_2{=}CH{-}C(OH)(R_2){-}R_1 \quad [60]$$

$R_1 = C_3H_7$, $R_2 = H$ 60%
$R_1 = C_6H_5$, $R_2 = H$ 90%
$R_1R_2 = (CH_2)_5$ 85%

From 1,2-difluorodichloroethylene and an excess of butyllithium, only one monolithio derivative is formed [*62*] (equation 28).

[*62*] **28**

$$\underset{(cis/trans\ 1{:}1)}{FClC{=}CClF} \xrightarrow[Et_2O,\ \text{-}78\%]{C_4H_9Li} FClC{=}CFLi \xrightarrow[2.\ H_2O]{1.\ CO_2} \underset{cis\ 14.0\%,\ trans\ 17.6\%}{FClC{=}CFCO_2H}$$

However, 1,2-difluoro-1-chloro-2-ethoxyethylene yields two organolithium compounds (equation 29).

[*62*] **29**

$$\underset{(cis/trans\ 1{:}1)}{C_2H_5OCF{=}CFCl} \xrightarrow[Et_2O,\ \text{-}78^{o}C]{C_4H_9Li} C_2H_5OCF{=}CFLi + ClCF{=}CFLi$$

$$\downarrow 1.\ CO_2\ |\ 2.\ H_2O$$

$$\underset{34\%\ \text{total yield}\ (cis/trans\ \text{mixture})}{C_2H_5OCF{=}CFCO_2H + ClCF{=}CFCO_2H}$$

With a difluoroethylene containing hydrogen and chlorine, where both groups can be replaced by lithium, a mixture of two organolithium compounds is formed in a 2:1 ratio, indicating a *more facile replacement of chlorine* [*63*] (equation 30).

[*63*] **30**

$$\underset{(cis/trans\ 1{:}1)}{HFC{=}CFCl} \xrightarrow[THF,\ Et_2O,\ \text{-}115^{o}C]{C_4H_9Li} HFC{=}CFLi + ClFC{=}CFLi$$

$$\downarrow 1.\ CO_2\ |\ 2.\ CH_3OH$$

$$\underset{(\text{mole ratio } 2{:}1)}{HFC{=}CFCO_2CH_3 + ClFC{=}CFCO_2CH_3}$$

Fluorovinyllithium compounds, as well as perfluoroalkyllithium and perfluoroalkyletherlithium compounds [*1*], are important intermediates in the synthesis of a variety of fluorine-containing compounds. Although experimental details must be carefully adhered to, these synthons provide a convenient route to many fluoro compounds.

References are listed on pages 666–669.

Pentafluoroethanesulfonic acid (pentflic acid) is prepared by an improved procedure over previously reported methods through the use of pentafluoroethyllithium [*64*] (equation 31).

[*64*] **31**

$$C_2F_5I \xrightarrow[Et_2O,\ -78^\circ C]{CH_3Li} C_2F_5Li \xrightarrow{SO_2} C_2F_5SO_2Li \xrightarrow[100^\circ C]{AcOH,\ H_2O_2} \underset{96\%}{C_2F_5SO_3Li} \xrightarrow{H_2SO_4} \underset{85\%}{C_2F_5SO_3H}$$

This simplified procedure gives the sulfonic acid in very good yield, however the same procedure is not applicable to the synthesis of triflic acid, which would require the synthesis of trifluoromethyllithium. This procedure also is not applicable to the preparation of the sulfonic acids requiring the use of perfluoropropyl or isopropyllithium intermediates.

An unusual and different use of an **aryldifluorovinyllithium** compound is the synthesis of acetylenic compounds containing no fluorine [*65*] (equation 32).

[*65*] **32**

$$\underset{(Z + E)}{ArCF{=}CFCl} \xrightarrow[THF,\ Et_2O,\ -110^\circ C]{s\text{-}C_4H_9Li} ArCF{=}CFLi \xrightarrow{\Delta} ArC{\equiv}CF \xrightarrow[-LiF]{RLi} ArC{\equiv}CR$$

$$ArCF{=}CFLi \xrightarrow{H_2O} \underset{(Z + E)}{ArCF{=}CFH}$$

Ar = C_6H_5; R = s-C_4H_9 77%, R = t-C_4H_9 78%
Ar = p-$CH_3OC_6H_4$; R = s-C_4H_9 75%, R = C_6H_5 75%
Ar = thienyl; R = $s$$C_4H_9$ 83%, R = t-C_4H_9 85%

Trifluorovinyllithium on reaction with ***trimethylchlorosilane*** yields a trimethylsilyltrifluoroethylene, which itself is a new and useful synthon for the preparation of difluorovinylic compounds [*66*] (equation 33).

[*66*] **33**

$$CF_2{=}CFLi + (CH_3)_3SiCl \xrightarrow[-110^\circ C]{THF,\ Et_2O} CF_2{=}CFSi(CH_3)_3$$

The silylated compound can be reacted further with other organolithium compounds to give substituted silylated difluorovinyl compounds (equation 34).

[*66*] **34**

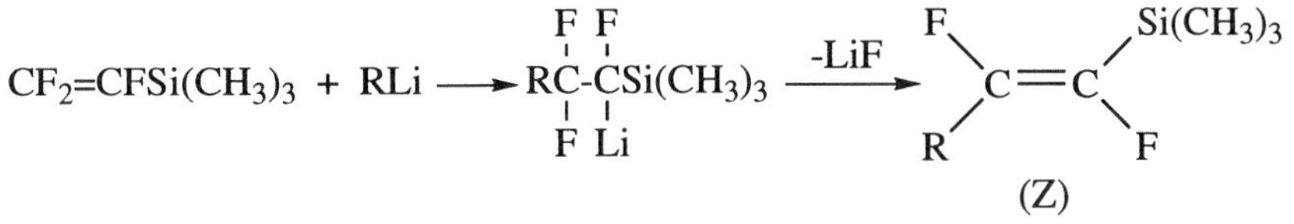

$$CF_2{=}CFSi(CH_3)_3 + RLi \longrightarrow RCF_2\text{-}CF(Li)Si(CH_3)_3 \xrightarrow{-LiF} (F)(R)C{=}C(Si(CH_3)_3)(F)\ (Z)$$

R = C_4H_9 80%, C_7H_{15} 85%,
s-C_4H_9 78%, *t*-C_4H_9 74%,
C_6H_5 50%, $(CH_3)_2C{=}C(CH_2)_2$ 79%
$(CH_3)_2C{=}CH$ 72%

The trimethylsilyl compound can be easily desilylated to yield a carbanion intermediate capable of further reactions [*66, 67*] (equation 35).

[*67*] **35**

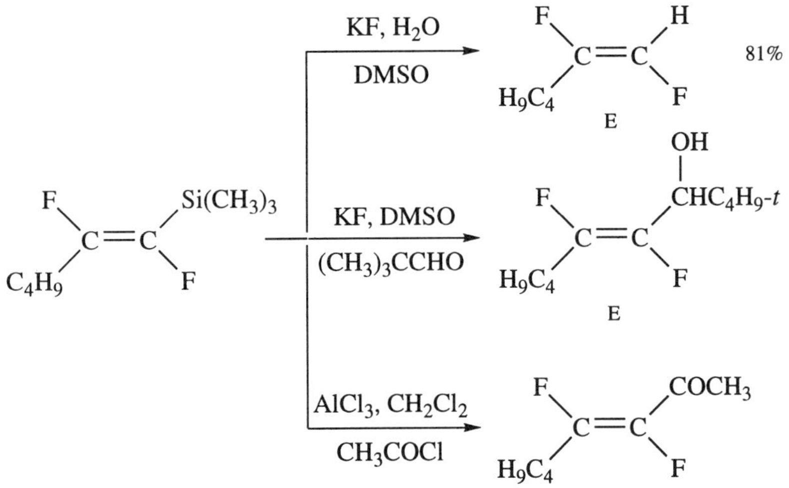

$$(F)(C_4H_9)C{=}C(Si(CH_3)_3)(F) \xrightarrow[DMSO]{KF,\ H_2O} (F)(H_9C_4)C{=}C(H)(F)\ \text{E}\quad 81\%$$

$$(F)(C_4H_9)C{=}C(Si(CH_3)_3)(F) \xrightarrow[(CH_3)_3CCHO]{KF,\ DMSO} (F)(H_9C_4)C{=}C(CH(OH)C_4H_9\text{-}t)(F)\ \text{E}$$

$$(F)(C_4H_9)C{=}C(Si(CH_3)_3)(F) \xrightarrow[CH_3COCl]{AlCl_3,\ CH_2Cl_2} (F)(H_9C_4)C{=}C(COCH_3)(F)$$

E = 90%, Z = 2%

Since the discovery of the first **perfluoroalkyllithium** and -magnesium compounds, their reactions with various ***carbonyl compounds*** [*68*] have been actively studied. These reactions could lead to functionally substituted fluoro compounds, as shown in equation 36.

The mechanism of the reaction between a hydrocarbon organolithium compound and a carbonyl compound is well-documented [*69*] and has been suggested earlier for fluorine-containing compounds [*32, 68, 70*]. More recent studies on the mechanism have resulted in a better understanding of this reaction [*71*] (equation 37).

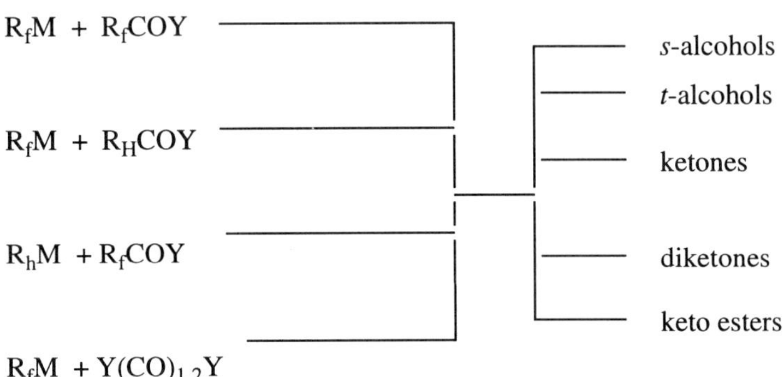

R_f and R = aliphatic, olefinic, acetylenic, aromatic
Y = H, R, OR, RCO_2, CO_2Li, CO_2NR_2, F, Cl

37

$$R_fCOOR \xrightarrow{C_6H_5Li} \left[R_f{-}\overset{\overset{\displaystyle \bar{O}\ \overset{+}{Li}}{|}}{\underset{\underset{\displaystyle C_6H_5}{|}}{C}}{-}OR \right] \xrightarrow{A} R_fLi + C_6H_5COOR$$

B: $R_fCOC_6H_5 + R_f\underset{\underset{\displaystyle OH}{|}}{C}HC_6H_5$

C: $R_fCOC_6H_5 + ROLi$

Various combinations of R_f and R (equation 36) have been studied [*39, 72, 73, 74, 75*], and it appears that the stability of the lithium salt of the hemiketal is the major factor in determining the reaction products formed via paths A, B, or C in equation 37. Other important factors that affect the course of the reaction are (1) thermal stability of the perfluoroalkyllithium compounds, (2) reaction temperature, (3) mode of addition of the reactants, (4) steric hindrance, (5) nature of the Y group (in equation 36), and (6) temperature at which the reaction is terminated by acid hydrolysis.

References are listed on pages 666–669.

Difunctional electrophiles such as diethyl carbonate and diethyl oxalate also react with perfluorooctyllithium and perfluoroalkyletherlithium compounds to yield ketones, diketones, and keto esters [*39, 74*] (equations 38–41).

[*39*] **38**

$$2.3\ C_8F_{17}I + C_2H_5OCOOC_2H_5 \xrightarrow[-78^oC]{2.3\ CH_3Li} C_8F_{17}COC_8F_{17} + C_8F_{17}CO_2C_2H_5$$

55% 16%

[*39*] **39**

$$C_8F_{17}I + R_fCOOC_2H_5 \xrightarrow[-78^oC]{CH_3Li} C_8F_{17}COR_f$$

$R_f = C_8F_{17}$ 98%
$R_f = CF(CF_3)_2OCF_2CF_2$ 96%

[*39*] **40**

$$C_8F_{17}I + C_2H_5OCOCOOC_2H_5 \xrightarrow[-78^oC]{CH_3Li} C_8F_{17}COCOOC_2H_5\ \ 94\%$$

[*39*] **41**

$$2CF(CF_3)_2OCF_2CF_2Li + C_2H_5OCOCOOC_2H_5 \xrightarrow[Et_2O]{-78^oC} CF(CF_3)_2OCF_2CF_2COCOCF_2CF_2OCF(CF_3)_2$$

Asymmetric perfluorodiketones may be prepared by a stepwise reaction [*39*] (equation 42).

[*39*] **42**

$$C_8F_{17}I + C_2H_5OCOCOOC_2H_5 \xrightarrow[-78^oC]{CH_3Li} C_8F_{17}C(OLi)(OC_2H_5)COOC_2H_5 \xrightarrow[-78^oC]{CF(CF_3)_2OF_2CF_2Li}$$

$$C_8F_{17}C(OLi)(OC_2H_5)\text{-}C(OLi)(OC_2H_5)CF_2CF_2OC(CF_3)_2F \xrightarrow{H^+} C_8F_{17}COCOCF_2CF_2OCF(CF_3)_2$$

References are listed on pages 666–669.

These studies show that the perfluoroalkyletherlithium compound shown in equation 42 has greater thermal stability than the perfluorooctyllithium, which must be prepared in situ. This in situ process has been very useful in the synthesis of other shorter chain perfluoroalkyllithium compounds. **Pentafluoroethyllithium** can be conveniently prepared in this manner and reacts with numerous ***esters*** to yield tertiary alcohols [*76, 77*] in excellent yields.

Perfluoroacetylenic lithium compounds on reaction with electrophiles yield perfluoroacetylenic functional derivatives [*78, 79*] (equation 43).

[*76, 77*] **43**

$$C_6F_{13}C{\equiv}CH \xrightarrow[\text{Et}_2\text{O, hexane}\ -45^{\circ}\text{C}]{C_4H_9Li} C_6F_{13}C{\equiv}CLi \xrightarrow{RCOY} C_6F_{13}C{\equiv}C\text{-}C(R)(R)COH$$

R = H, Y = H	75%
R = CH_3, Y = H	85%
R = C_2H_5, Y = C_2H_5	92%

Although the early reactions of perfluoroaliphatic and aromatic lithium compounds with various carbonyl compounds gave low yields and mixtures of products, the current understanding of experimental conditions makes this reaction an attractive means of preparing a variety of fluorine-containing compounds.

Reference for Pages 646–666

1. Hudlický, M. *Chemistry of Organic Fluorine Compounds;* Ellis Horwood: Chichester, U.K., 1976; pp 358–405.
2. Pierce, O. R.; Meiners, A. F.; MCBee, E. T. *J. Am. Chem. Soc.* **1953,** *75,* 2516.
3. Henne, A. L.; Nager, M. *J. Am. Chem. Soc.* **1952,** *74,* 650.
4. Seyferth, D.; Wada, T.; Raab, G. *Tetrahedron Lett.* **1960,** *22,* 20.
5. Weidenbruch, M.; Wessel, N. *Chem. Ber.* **1972,** *105,* 173.
6. Dyatkin, B. L.; Sterlin, S. R.; Martynov, B. I.; Knunyants, I. L. *Tetrahedron Lett.* **1970,** 1387.
7. Knunyants, I. L.; Sterlin, R. D.; Yatsenko, R. D.; Pinkina, L. N. *Izv. Akad. Nauk SSSR, Ser. Khim.* **1958,** 1345; *Chem. Abstr.* **1959,** *53,* 6987g.
8. Pummer, W. J.; Wall, L. A. *J. Res. Nat. Bur. Std. A* **1959,** *63,* 167.
9. Nield, E.; Stephens, R.; Tatlow, J. C. *J. Chem. Soc.* **1959,** 166.

10. Chambers, R. D.; Hutchinson, J.; Musgrave, W. K. R. *J. Chem. Soc.* **1965,** 5040.
11. Fenton, D. E.; Park, A. J.; Shaw, D.; Massey, A. G. *Tetrahedron Lett.* **1964,** 949.
12. Vorozhtsov, N. N.; Barkhash, V. A.; Ivanova, N. G.; Anichkina, S. A.; Andreevskaya, O. I. *Dokl. Akad. Nauk SSSR* **1964,** *159,* 125; *Chem. Abstr.* **1965,** *62,* 4045a.
13. Harper, R. J.; Soloski, E. J.; Tamborski, C. *J. Org. Chem.* **1964,** *29,* 2385.
14. Pierce, O. R.; McBee, E. T.; Judd, G. F. *J. Am. Chem. Soc.* **1954,** *76,* 474.
15. Chambers, R. D.; Musgrave, W. K. R.; Savory, J. *J. Chem. Soc.* **1962,** 1993.
16. Johncock, P. *J. Organomet. Chem.* **1966,** *6,* 433.
17. Coe, P. L.; Stephens, R.; Tatlow, J. C. *J. Chem. Soc.* **1962,** 3227.
18. Drakesmith, F. G.; Richardson, R. D.; Stewart, O. J.; Tarrant, P. *J. Org. Chem.* **1968,** *33,* 280, 286.
19. Campbell, S. F.; Stephens, R.; Tatlow, J. C. *Tetrahedron* **1965,** *21,* 2997.
20. Chambers, R. D.; Drakesmith, F. G.; Musgrave, W. K. R. *J. Chem. Soc.* **1965,** 5045
21. DePasquale, R. J.; Tamborski, C. *J. Organometal. Chem.* **1968,** *13,* 273.
22. Fenton, D. E.; Park, A. J.; Shaw, D.; Massey, D. *J. Organomet. Chem.* **1964,** *2,* 437.
23. Fenton, D. E.; Park, A. J.; Shaw, D.; Massey, D. *Tetrahedron Lett.* **1964,** 949.
24. Treichel, P. M.; Stone, F. G. A. *Adv. Organometal. Chem.* **1964,** *1,* 143.
25. Cohen, S. C.; Massey, A. G. *Adv. Flourine Chem.* **1970,** *6,* 83.
26. Osina, O. I.; Steingarts, V. D. *Zh. Org. Khim.* **1974,** *10,* 329; *Chem. Abstr.* **1974,** *80,* 120611y.
27. Bochkarev, M. N.; Vyazankin, N. S.; Maiorova, L. P.; Razuvaev, G. A. *Zh. Obshch. Khim.* **1978,** *48,* 2706; *Chem. Abstr.* **1979,** *90,* 152310x.
28. Sethi, S.; Howells, R. D.; Gilman, H. *J. Organometal. Chem.* ***1979,*** *69,* 377.
29. Shadrina, L. P.; Dormidontov, Y. P. *Zh. Org. Khim.* **1981,** *17,* 1408; *Chem. Abstr.* **1981,** *95,* 168741c.
30. Dormidontov, Y. P.; Shadrina, L. P.; Belogai, V. D,; Krokhalev, A. M.; Yakhlakova, O. M. *Zh. Org. Khim.* **1988,** *24,* 107; *Chem. Abstr.* **1988,** *109,* 149005g.
31. Shadrina, L. P.; Kazakov, A. F.; Dormidontov, Y. P.; Pirogova, V. V. *Zh. Org. Khim.* **1989,** *25,* 2517; *Chem. Abstr.* **1990,** *112,* 216362y.
32. McBee, E. T.; Roberts, C. W.; Curtis, S. G. *J. Am. Chem. Soc.* **1955,** *77,* 6387.
33. Chambers, R. D.; Clark, M.; Spring, D. J. *J. Chem. Soc.* **1972,** 2464.
34. Bekker, R. A.; Asratyan, G. V.; Dyatkin, B. L. *Zh. Org. Khim.* **1973,** *9,* 1635; *Chem. Abstr.* **1973,** *79,* 126014g.
35. Coe, P. L.; Oldfield, D.; Tatlow, J. C. *J. Fluorine Chem.* **1985,** *29,* 341.

36. Burdon, J.; Gill, H. S.; Parsons, I. W.; Tatlow, J. C. *J. Chem. Soc., Chem. Commun.* **1979,** 1147.
37. Burdon, J.; Gill, H. S.; Parsons, I. W.; Tatlow, J. C. *J. Fluorine Chem.* **1984,** *24,* 263.
38. Burdon, J.; Gill, H. S.; Parsons, I. W. *J. Chem. Soc.* **1980,** 2494.
39. Chen, L. S.; Chen, G. J.; Tamborski, C. *J. Fluorine Chem.* **1984,** *26,* 341.
40. Park, J. D.; Seffl, R. J.; Lacher, J. R. *J. Am. Chem. Soc.* **1956,** *75,* 59.
41. Denson, D. D.; Smith, C. F.; Tamborski, C. *J. Fluorine Chem.* **1973/74,** *3,* 247
42. Smith, C. F.; Soloski, E. J.; Tamborski, C. *J. Fluorine Chem.* **1974,** *4,* 35.
43. Howells, R. D.; Gilman, H. *J. Fluorine Chem.* **1975,** *5,* 99
44. Dua, S. S.; Howells, R. D.; Gilman, H. *J. Fluorine Chem.* **1974,** *4,* 409.
45. Denson, D. D.; Moore, G. J.; Sun, K. K.; Tamborski, C. *J. Fluorine Chem.* **1977,** *10,* 75.
46. Hemer, I.; Posta, A.; Dedek, V. *J. Fluorine Chem.* **1984,** *26,* 467.
47. Chen, L. S.; Chen, G. J.; Ryan, M. T.; Tamborski, C. *J. Fluorine Chem.* **1987,** *34,* 299
48. Moreau, P.; Albadri, R.; Commeyras, A. *Nouv. J. Chim.* **1977,** *1,* 497.
49. Hardwick, F.; Pedler, A. E.; Stephens, R.; Tatlow, J. C. *J. Fluorine Chem.* **1974,** *4,* 9.
50. Moreau, P.; Dalverny, G.; Commeyras, A. *J. Chem. Soc., Chem. Commun.* **1976,** 174.
51. Nguyen, T.; Rubinstein, M.; Wakselman, C. *J. Org. Chem.* **1981,** *46,* 1938.
52. Thoai, N.; Rubinstein, M.; Wakselman, C. *J. Fluorine Chem.* **1982,** *21,* 437.
53. Barlow, M. G. *J. Chem. Soc., Chem. Commun.* **1966,** 703.
54. Howells, R. D.; Gilman, H. *J. Fluorine Chem.* **1974,** *4,* 247, 409.
55. Thoai, N. *J. Fluorine Chem.* **1975,** *5,* 115.
56. Redwane, N.; Moreau, P.; Commeyras, A. *J. Fluorine Chem.* **1982,** *20,* 699.
57. Moreau, P.; Albadri, R.; Redwane, N.; Commeyras, A. *J. Fluorine Chem.* **1980,** *15,* 103.
58. Moreau, P.; Redwane, N.; Commeyras, A. *Bull. Soc. Chim. Fr.* **1984,** *11,* 117.
59. Gillet, J. P.; Sauvetre, R.; Normant, J. F. *Synthesis* **1986,** 355.
60. Sauvetre, R.; Normant, J. F. *Tetrahedron Lett.* **1981,** 957.
61. Gillet, J. P.; Sauvetre, R.; Normant, J. F. *Synthesis* **1982,** 297.
62. Kremlev, M. M.; Moklyachuk, L. I.; Fialkov, Y. A.; Yagupol'skii, L. M. *Zh. Org. Khim.* **1983,** 919; *Chem. Abstr.* **1983,** *99,* 104740b.
63. Yagupol'skii, L. M.; Cherednichenko, P. G.; Kremlev, M. M. *Zh. Org. Khim.* **1987,** 279; *Chem. Abstr.* **1987,** *107,* 236795s.
64. Olah, G.; Weber, T.; Bellow, D. R.; Farooq, O. *Synthesis* **1989,** 463.
65. Martin, S.; Sauvetre, R.; Normant, J. F. *Tetrahedron Lett.* **1982,** *23,* 4329.

66. Martin, S.; Sauvetre, R.; Normant, J. F. *Tetrahedron Lett.* **1983,** *24,* 5615.
67. Martin, S.; Sauvetre, R.; Normant, J. F. *J. Organometal. Chem.* **1984,** *264,* 155.
68. McBee, E. T.; Pierce, O. R.; Higgins, J. F. *J. Am. Chem. Soc.* **1952,** *74,* 1736.
69. Wakefield, B. J. *Organolithium Compounds;* Pergamon Press: New York, 1976, p. 136.
70. Haszeldine, R. N. *J. Chem. Soc.* **1953,** 1748.
71. Chen, L. S.; Chen, G. J.; Tamborski, C. *J. Fluorine Chem.* **1981,** *18,* 117.
72. Gopal, H.; Soloski, E. J.; Tamborski, C. *J. Fluorine Chem.* **1978,** *12,* 111.
73. Gopal, H.; Tamborski, C. *J. Fluorine Chem.* **1979,** *13,* 337.
74. Chen, L. S.; Tamborski, C. *J. Fluorine Chem.* **1981/82,** *19,* 43.
75. Chen, L. S.; Tamborski, C. *J. Fluorine Chem.* **1984,** *26,* 267.
76. Gassman, P. G.; O'Reilly, N. J. *Tetrahedron Lett.* **1985,** *26,* 5243.
77. Gassman, P. G.; O'Reilly, N. J. *J. Org. Chem.* **1987,** *52,* 2481.
78. Chauvin, A.; Greiner, J.; Pastor, R.; Cambon, E. *J. Fluorine Chem.* **1984,** *25,* 259.
79. Moreau, P.; Naji, N.; Commeyras, A. *J. Fluorine Chem.* **1985,** *30,* 315.

Other Organometallic Syntheses

by Donald J. Burton and Zhen-Yu Yang

Organocalcium Compounds

The reaction of perfluoroalkyl iodides with *calcium amalgam* at –20 to 40 °C has been used to prepare *perfluoroalkyl carbinols* in 30–70% yields [*1, 2*]. Aromatic aldehydes (equation 1) and ketones (equation 2) undergo this reaction; aliphatic aldehydes give high boiling mixtures. The results are interpreted in terms of the intermediate formation of an organocalcium species.

[2] **1**

$$C_6H_5CHO + Ca(Hg) + C_2F_5I \xrightarrow[\text{20 h, }-40\text{ °C}]{\text{THF}} C_6H_5CH(OH)C_2F_5 \quad 69\%$$

[2] **2**

$$C_2H_5COC_2H_5 + Ca(Hg) + C_6F_{13}I \xrightarrow[-40\text{ °C, 8 h}]{\text{THF}} (C_2H_5)_2C(OH)C_6F_{13} \quad 66\%$$

The insertion of calcium atoms into vinyl and aryl carbon–fluorine bonds has been reported. Only the resulting organometallic from the aryl derivatives appears to live long enough to be trapped by water [*3*].

Organotin Compounds

Trialkylperfluoroalkyltin derivatives can be prepared via metathesis of trimethyltin trifluoroacetate with **bis(perfluoroalkyl)cadmium** reagents [*4*] (equation 3).

[*4*] **3**

$$(CH_3)_3SnC_2F_5 \; 64\% \xleftarrow[\text{70 °C, diglyme}]{(C_2F_5)_2Cd\cdot D} (CH_3)_3SnOCOCF_3 \xrightarrow[\text{70 °C, }CH_3CN]{(i\text{-}C_3F_7)_2Cd\cdot D} (CH_3)_3SnC_3F_7\text{-}i \; 75\%$$

0065–7719/95/0187–0670$15.02/1

However, bis(trifluoromethyl)cadmium gives CF_3NO with trimethyltin nitrate [*5*] (equation 4).

The ligand-exchange reaction of tin halides with **bis(trifluoromethyl)mercury** has been used to prepare trifluoromethyltin halides [*6, 7*] (equation 5).

Bis(trifluoromethyl)cadmium reagent undergoes a related ligand-exchange process [*8*] (equation 6).

Tetrakis(trifluoromethyl)tin can be prepared via metal vapor [*9*] or fluorination techniques [*10*] and via reaction with trifluoromethyl radicals generated by radio-frequency discharge of CF_3CF_3 [*11*] (equation 7).

[*5*] **4**

$$(CF_3)_2Cd{\bullet}D + (CH_3)_3SnONO_2 \xrightarrow[RT]{CH_3CN} \underset{60\%}{CF_3NO}$$

[*7*] **5**

$$(CF_3)_2Hg + SnBr_4 \xrightarrow[42\ h]{112\ °C} \underset{46\%}{CF_3SnBr_3} + \underset{3\%}{(CF_3)_2SnBr_2}$$

[*8*] **6**

$$(CF_3)_2Cd{\bullet}glyme + SnBr_4 \xrightarrow[RT,\ 2\ h]{i} \underset{66\%}{(CF_3)_4Sn} + \underset{5\%}{(CF_3)_3SnBr} + \underset{10\%}{CF_3SnBr_3}$$

i : $Br(CH_2)_6Br$

[*11*] **7**

$$CF_3CF_3 + SnI_4 \xrightarrow[discharge]{radio\text{-}frequency} \underset{90\%}{(CF_3)_4Sn}$$

Perfluoroalkyltin halides can be prepared via oxidative addition of perfluoroalkyl iodides to tin(II) halides in dimethylformamide (DMF) [*12*]. The perfluoroalkyltin(IV) dihalide could not be isolated, but in DMF solution, the tin(IV) compound did react with aldehydes and ketones in the presence of pyridine [*12*] (equation 8). Typical perfluoroalkylcarbinols prepared by this method are shown in Table 1 [*12*].

[*12*] **8**

$$R_fI + SnCl_2 \xrightarrow[RT]{DMF} [R_fSnCl_2I] \xrightarrow[2.\ RCOR'\ \ 3.\ H^+]{1.\ Pyridine} \underset{18\text{-}82\%}{R_fRC(OH)R'}$$

References are listed on pages 718–728.

Table 1. Perfluoroalkylation of Carbonyl Compounds by Perfluoroalkyltin(IV) Halides [12]

Carbonyl Compound	R_FI	*Product*	*Yield (%)*
C_6H_5CHO	iso-C_3F_7I	$C_6H_5CH(OH)$(iso-C_3F_7)	86
C_6H_5CHO	CF_3I	$C_6H_5CH(OH)CF_3$	18
p-$CH_3C_6H_4CHO$	iso-C_3F_7I	*p*-$CH_3C_6H_4CH(OH)$(iso-C_3F_7)	82
C_4H_9CHO	C_3F_7I	$C_4H_9CH(OH)C_3F_7$	68
$C_6H_5COCH_3$	iso-C_3F_7I	$C_6H_5C(OH)(CH_3)$(iso-C_3F_7)	37
$C_2H_5COCH_3$	iso-C_3F_7I	$C_2H_5C(OH)(CH_3)$(iso-C_3F_7)	46

Trimethyl(trifluoromethyl)tin can also be prepared via in situ formation and capture of trifluoromethide by trimethyltin chloride [*13, 14*] (equation 9). This tin analogue has been used as a precursor for difluorocarbene either by thermal decomposition or by reaction with sodium iodide in 1,2-dimethoxyethane. This carbene generation procedure has been used to study difluorocarbene selectivity with steroidal olefins [*15*] (equation 10).

[*13*] **9**

$$Ph_3P + CF_2Br_2 \longrightarrow [Ph_3\overset{+}{P}CF_2Br]Br^- \xrightarrow[\substack{(CH_3)_3SnCl \\ RT \\ diglyme}]{KF} (CH_3)_3SnCF_3 \quad 37\text{–}40\%$$

[*15*] **10**

Me₃SnCF₃ / NaI

250 p. 65% 1 p.

Perfluorovinyl organotin compounds can be prepared from **perfluorovinyl Grignard reagents** and organotin halides [*16*] (equation 11). Yields are slightly improved under *Barbier* conditions [*16*].

The *trialkyl(trifluorovinyl)stannanes* were used in the Pd(0)-catalyzed coupling reaction of aryl halides [*17*] (equation 12). The product yield increased with the solvent type in the order hexamethylphosphorus triamide (HMPT) DMF > dimethyl sulfoxide (DMSO) > tetrahydrofuran (THF) > C_6H_6 > $C_2H_4Cl_2$.

[16] **11**

$$C_6F_{13}CF{=}CFMgBr \xrightarrow[5\,^{\circ}C,\ 15\ h]{(C_4H_9)_3SnCl,\ ether} C_6F_{13}CF{=}CFSn(C_4H_9)_3\ \ 55\%$$

$$C_6F_{13}CF{=}CFMgBr \xrightarrow[(C_4H_9)_2SnCl_2]{ether} 40\%\ C_6F_{13}CF{=}CFSnCl(C_4H_9)_2 + 10\%\ (C_6F_{13}CF{=}CF)_2Sn(C_4H_9)_2$$

[17] **12**

$$F_2C{=}CFSn(C_4H_9)_3 + \text{I-C}_6\text{H}_4\text{-R} \xrightarrow{[P(C_6H_5)_3]_4Pd} CF_2{=}CF\text{-C}_6\text{H}_4\text{-R}$$

$R = H, NO_2, I$ $\leq 87\%$

[18] **13**

$$\text{3-(chloromethyl)cephem } (R'NH,\ CO_2R,\ CH_2Cl) \xrightarrow[THF,\ 65^{\circ}C,\ 72\ h]{F_2C{=}CFSnBu_3,\ Pd^{\circ}} \text{3-}(CH_2CF{=}CF_2)\text{cephem}\ \ 65\%$$

In related work, 3-chloromethylcephems were coupled with tributyl(trifluorovinyl)stannane catalyzed by tri(2-furyl)phosphine palladium(0) [*18, 19*] (equation 13).

Trifluorovinyltin derivatives have been reacted with SO_2; however, only low yields of insertion into the tin–carbon bond of the trifluorovinyl group were detected [*20*].

Perfluoroalkylacetylenic tin compounds have been synthesized via the reaction of perfluoroalkylacetylenic Grignard reagents with tin halide derivatives [*21*] (equation 14). The stability of several of these perfluoroalkylacetylenic tin derivatives has been studied [*22*].

[21] **14**

$$(CH_3)_3SnCl + R_f\,C{\equiv}CMgI \longrightarrow R_f\,C{\equiv}CSn(CH_3)_3$$

$R_f = CF_3$ 64%; $R_f = C_2F_5$ 20%; $R_f = i\text{-}C_3F_7$ 77%

Bis(pentafluorophenyl)cadmium complexes have been used to prepare trimethyl(pentafluorophenyl)tin [*4*] (equation 15).

[*4*] **15**

$$(CH_3)_3SnOCOCF_3 + (C_6F_5)_2Cd{\bullet}D \xrightarrow[70\ °C]{CH_3CN} (CH_3)_3SnC_6F_5 \quad 60\%$$

Organolead Compounds

Ligand exchange of **bis(perfluoroalkyl)cadmium** or **bis(perfluoroaryl)cadmium complexes** with trimethyllead(trifluoroacetate) gives the corresponding *trimethyl(perfluoroalkyl)lead* and *trimethyl(perfluoroaryl)lead* compounds in good yields [*4*] (equation 16).

An exchange reaction between **bis(trifluoromethyl)mercury** and tetramethyllead gives *trimethyl(trifluoromethyl)plumbane* [*23*] (equation 17). This plumbane can also be prepared via the reaction of tetramethyllead with trifluoromethyl radicals produced in a radio-frequency discharge of C_2F_6 [*24*].

The reaction of trichlorotrifluoroethane with lead in the presence of carbonyl substrates gives the 1-chloro-2,2-difluoroethenyl carbinols, presumably via a fluorinated organolead intermediate [*25*] (equation 18).

[*4*] **16**

$$(CH_3)_3PbOCOCF_3 + (R_f)_2Cd{\bullet}D \xrightarrow[70\ °C]{CH_3CN} R_fPb(CH_3)_3$$

$R_f = CF_3$ 85%; C_2F_5 91%; i-C_3F_7 70%; C_6F_5 65%

[*23*] **17**

$$(CH_3)_4Pb + (CF_3)_2Hg \xrightarrow[2\ weeks]{70\ °C} CF_3Pb(CH_3)_3 \quad 93\%$$

[*25*] **18**

$$CF_3CCl_3 + C_6H_5CHO \xrightarrow{Pb} C_6H_5CH(OH)CCl{=}CF_2$$

Organozinc Compounds

The preparation of perfluoroalkylzinc compounds has been achieved earlier in ethereal solvents [*26*]. However, solvent effects play a significant role in the course of this reaction. When a mixture of acetic anhydride and methylene chloride is used, coupled and cross-coupled products can be formed [*27, 28*] (equations 19 and 20). However, the cross-coupling reaction often gives mixtures, a fact that seriously restricts the synthetic applicability of this reaction [*27, 28, 29*].

[27] 19

$$CF_3CF_2CF_2CF_2I + Zn \xrightarrow[CH_2Cl_2]{Ac_2O} CF_3(CF_2)_6CF_3$$

[27] 20

$$CF_2ClCFClCF_2CFClI + CF_2ClCFClI + Zn \xrightarrow[CH_2Cl_2]{Ac_2O} \underset{46\%}{CF_2ClCFClCF_2CFClCFClCF_2Cl} + \text{others products}$$

The importance of solvent effects in the preparation of perfluoroalkyzinc reagents is further illustrated in the reaction of ***perfluoroalkyl iodides*** with **zinc–copper couple.** In DMSO, DMF, and HMPA, the main products are the *fluoroolefins.* The formation of the fluoroolefin is facilitated when the reaction is carried out in the presence of potassium thiocyanate [*30*] (equation 21).

When the reactions in DMF or DMSO are carried out in the presence of methylene bromide, an *insertion product* is also formed [*31*] (equation 22).

The reaction of CF_3I and C_6F_5I with *dialkylzinc* in the presence of a Lewis base, such as diglyme or pyridine, quantitatively gives the corresponding fluorinated organozinc complexes [*32*] (equation 23). When R_f is C_2F_5 or iso-C_3F_7, the pure zinc complexes are not isolated.

[*30*] 21

$$C_4F_9I \xrightarrow[\text{DMSO, 80 °C, KSCN}]{\text{Zn-Cu}} \underset{40\%}{CF_3CF_2CF{=}CF_2} + \underset{47\%}{\text{E- and Z-}CF_3CF{=}CFCF_3}$$

[*31*] 22

$$3\ CH_2Br_2 + C_6F_{13}I \xrightarrow[\text{DMSO, 20 °C}]{\text{Zn-Cu}} \underset{50\%}{C_6F_{13}CH_2CH_2I} + \underset{20\%}{C_{12}F_{26}} + \underset{30\%}{C_6F_{13}H}$$

[*32*] 23

$$2\ R_fI + R_2Zn \xrightarrow{\text{Lewis base}} (R_f)_2Zn + 2\ RI$$

$$R = CH_3, C_2H_5 \qquad R_f = CF_3, C_6F_5$$

Trifluoromethylzinc compounds can be prepared via the direct reaction of dihalodifluoromethane with *zinc powder* in DMF [*33*] (equation 24). In this reaction, the DMF functions both as solvent and reactant. Mechanistic experiments support a difluorocarbene reaction intermediate. Indeed, a mixture of zinc and difluorodibromomethane in THF has been used for the synthesis of *gem*-difluo-ro-cyclopropane derivatives [*34*] (equation 25).

Trifluoromethylzinc complexes can also be prepared by reaction of bromotrifluoromethane with zinc in pyridine [*35*] (equation 26).

[*33*] **24**

$$CF_2XY + Zn \xrightarrow{DMF} CF_3ZnX + (CF_3)_2Zn$$

X = Br,Cl 80-85%

Y = Br,Cl

[*34*] **25**

$$C_6H_5C(CH_3)=CH_2 + CF_2Br_2 + Zn \xrightarrow[RT]{THF}$$ 1,1-difluoro-2-methyl-2-phenylcyclopropane (C_6H_5, CH_3, F, F)

71%

[*35*] **26**

$$3\ CF_3Br + 3\ Zn \xrightarrow[\text{3-4 atm; 60 °C; 3 h}]{pyridine} CF_3ZnBr + (CF_3)_2Zn + ZnBr_2$$

[*36*] **27**

$$Hg(CF_3)_2 + Zn(CH_3)_2 \xrightarrow[\text{RT; 8 h}]{Pyridine} Zn(CF_3)_2{\cdot}2\,Py + (CH_3)_2Hg$$

82%

Ligand-exchange reactions can be used to prepare *perfluoroalkylzinc compounds.* Solvated trifluoromethylzinc compounds can be synthesized via the reaction of dialkylzincs with bis(trifluoromethyl)mercury [*36*] (equation 27). A similar exchange process with bis(trifluoromethyl)cadmium and dimethylzinc gives a mixture of trifluoromethylcadmium and -zinc compounds [*37*].

The *perfluoroorganozinc compounds* have limited application in organic synthesis because of their lack of reactivity. However, heptafluoro-1-methylethylzinc iodide reacts with acyl fluorides or anhydrides of carboxylic acids in the presence of pyridine to give the corresponding ketones [*38*] (equations 28 and 29). In the presence of zinc fluoride, acyl chlorides can be used as substrates [*38*].

[*38*] **28**

$$C_6H_5COF + (CF_3)_2CFZnI \xrightarrow[\text{THF, RT}]{\text{Pyridine}} C_6H_5COCF(CF_3)_2 \quad 100\%$$

[*38*] **29**

$$\text{succinic anhydride} + (CF_3)_2CFZnI \xrightarrow[\text{THF, RT}]{\text{Pyridine}} (CF_3)_2CFCOCH_2CH_2COOH$$

Under sonication conditions, the reaction of perfluoroalkyl bromides or iodides with zinc can be used to effect a variety of functionalization reactions [*39, 40, 41, 42*] (equation 30). Interestingly, the ultrasound-promoted asymmetric induction with the perfluoroalkyl group on the asymmetric carbon was achieved by the reaction of perfluoroalkyl halides with optically active enamines in the presence of zinc powder and a catalytic amount of dichlorobis(π-cyclopentadienyl)titanium [*42*] (equation 31).

[*42*] **30**

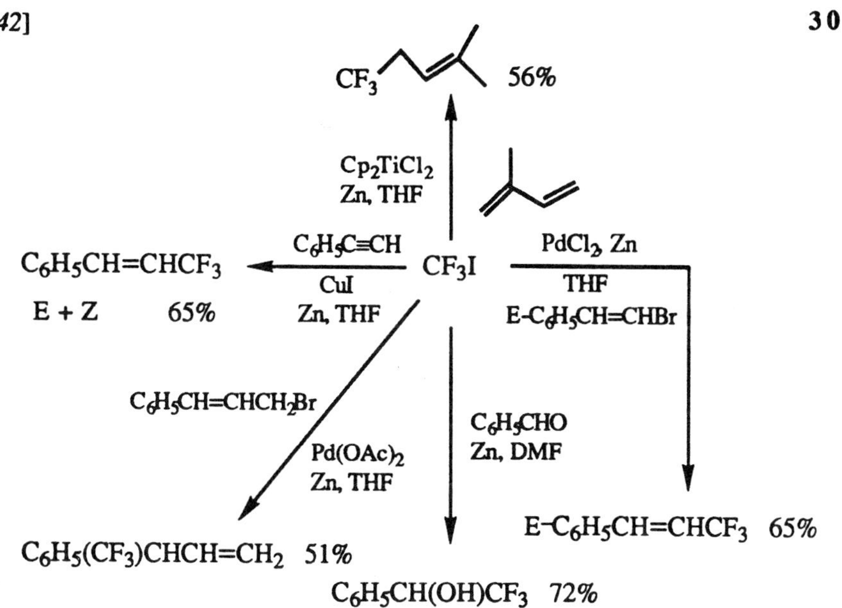

In related work, palladium, nickel, or methyl viologen (MV^{2+}) were used to catalyze the conversion of **perfluoroalkyl iodides** to α-*perfluoroalkyl carbinols* in the presence of zinc [*43, 44*] (equations 32 and 33). Only aldehydes react under these conditions.

Under *Barbier* conditions, trifluoromethyl bromide reacts with electrophiles, such as ***aldehydes, α-keto esters, activated esters,*** and ***anhydrides*** in the presence of pyridine to give trifluoromethylated compounds [*35, 45, 46*] (equations 34–37).

[*42*] **31**

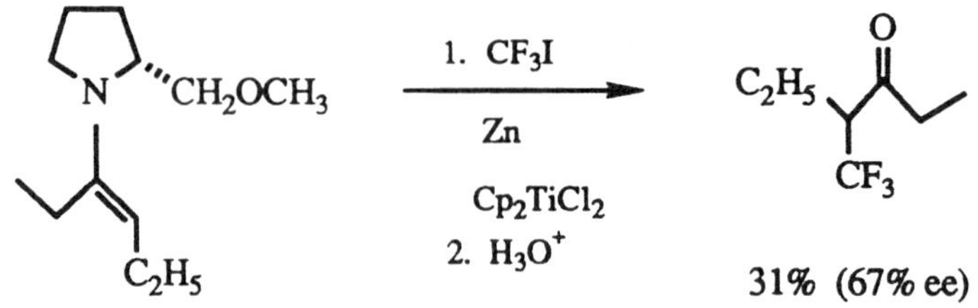

(2S)-2-(methoxymethyl)pyrrolidine enamine (CH_2OCH_3; C_2H_5) → 1. CF_3I, Zn, Cp_2TiCl_2; 2. H_3O^+ → $C_2H_5CH(CF_3)COC_2H_5$

31% (67% ee)

[*44*] **32**

$$C_6F_{13}I + C_6H_5CHO \xrightarrow[\text{RT, 13 h}]{\text{Zn, MV}^{2+}\text{, CH}_3\text{CN}} C_6H_5CH(OH)C_6F_{13}$$

65%

[*43*] **33**

$$C_6H_5CH{=}CHCHO + C_2F_5I \xrightarrow[\text{Zn, DMF, RT}]{[(C_6H_5)_3P]_2NiCl_2} C_6H_5CH{=}CHCH(OH)C_2F_5$$

42%

[*45*] **34**

$$p\text{-}FC_6H_4CHO + CF_3Br \xrightarrow[\text{2-4 bar, RT}]{\text{1. Pyridine}} p\text{-}FC_6H_4CH(OH)CF_3$$

2. H_3O^+

60%

[*46*] **35**

$$CH_3COCO_2C_2H_5 + CF_3Br \xrightarrow[\text{2-4 bar, RT}]{\text{1. Pyridine}} CH_3C(OH)(CF_3)CO_2C_2H_5$$

2. H_3O^+

35%

[*46*] **36**

$$CF_3CO_2C_2H_5 + CF_3Br \xrightarrow[\text{RT}]{\text{Pyridine}} CF_3COCF_3$$

54%

[*46*] **37**

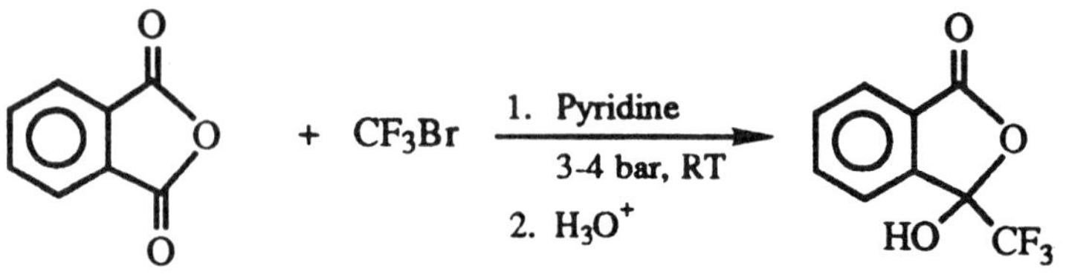

Phthalic anhydride + CF_3Br → 1. Pyridine, 3-4 bar, RT; 2. H_3O^+ → 3-hydroxy-3-(CF_3)phthalide (HO, CF_3)

61%

References are listed on pages 719–729.

Barbier conditions have been used to carry out a silicon-induced addition of perfluoroalkyl iodides to DMF to give stable hemiaminal intermediates, which give the perfluorinated aldehydes upon heating with sulfuric acid [*47*] (equation 38).

[*47*] **38**

$$CF_3(CF_2)_4CF_2I + (CH_3)_2NCHO + TDSCl \xrightarrow{Zn,\ RT} CF_3(CF_2)_4CF_2CH(OTDS)N(CH_3)_2 \ (80\%) \xrightarrow[heat]{H_2SO_4} CF_3(CF_2)_4CF_2CHO \ (75\%)$$

$TDSCl = t\text{-}C_4H_9(CH_3)_2SiCl$

The aldehydes CF_3CCl_2CHO (85%) and $CFCl_2CHO$ (76%) also were prepared from CF_3CCl_3 and $CFCl_3$ via the above procedure [*47*].

In the presence of zinc, bromotrifluoromethane and perfluoroalkyl iodides react with *thiocyanates and isocyanates* to give the corresponding **perfluoroalkyl sulfides** and substituted **perfluoroalkylacetamides,** respectively [*48*] (equation 39).

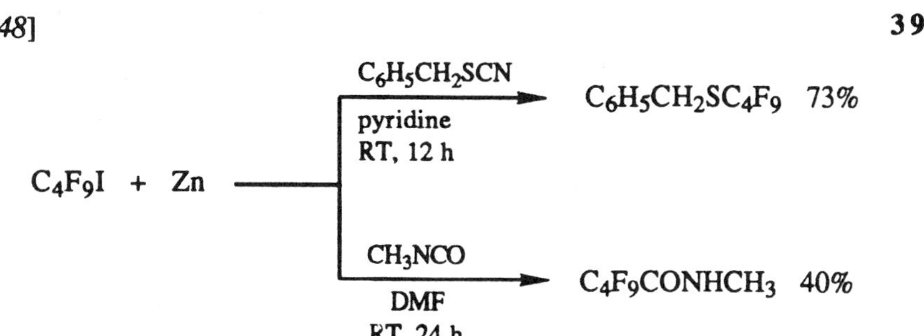

Perfluoroalkyl iodides react with carbon tetrachloride or bromotrichloromethane to give 1,1,1-trichloroperfluoroalkanes [*49*] (equation 40). Lower yields are obtained with C_4F_9I (33%) and $C_8F_{17}I$ (30%).

Trifluoromethanesulfinate can be prepared from the reaction of trifluoromethyl bromide with sulfur dioxide and zinc [*50*] (equation 41). Similar insertion occurs when perfluoroalkyl iodides are used as precursors (equations 41 and 42).

Oxidation of the trifluoromethanesulfinate gives triflic acid [*50*] (equation 43). Chlorination of the perfluoroalkylsulfinates gives the corresponding perfluoroalkyl sulfonyl chlorides [*51*] (equation 44).

[*49*] **40**

$$C_6F_{13}I + CCl_4 + Zn \xrightarrow[\text{reflux}]{CH_2Cl_2} C_6F_{13}CCl_3 \quad 51\%$$

[*50*] $$2\,CF_3Br + 2\,Zn + 2\,SO_2 \xrightarrow[\text{3 atm}]{\text{DMF}} (CF_3SO_2)_2Zn$$ **41**

[*51*] $$R_fI + SO_2 \xrightarrow[\text{DMSO}]{\text{Zn-Cu}} (R_fSO_2)_2Zn$$ **42**

[*50*] $$(CF_3SO_2)_2Zn \xrightarrow[\text{2. } H_2O_2 \text{ 3. } H_2SO_4]{\text{1. NaOH}} CF_3SO_3H{\cdot}H_2O \quad 44\%$$ **43**

[*51*] $$(R_fSO_2)_2Zn \xrightarrow[CH_3OH]{Cl_2} R_fSO_2Cl$$ **44**

$R_f = C_4F_9$ 55%; C_6F_{13} 80%; C_8F_{17} 75%

Perfluoroalkyl iodides **can be directly carboxylated with zinc** and *carbon dioxide* under ultrasonic conditions [*39*] (equation 45) or by the reaction of perfluoroalkyl iodides with carbon dioxide with a zinc–copper couple in DMSO [*51*] (equation 46). Alkylation of the intermediate carboxylate gives the corresponding ester [*52*].

In dissociating solvents, **perfluoroalkyl iodides** react with a **zinc–copper couple** to give *perfluoroalkylzinc compounds,* R_fZnI, which react with ***alkyl carbonates or pyrocarbonates*** to give *perfluorocarboxylic acids* or *esters,* respectively [*53*] (equations 47 and 48).

[*39*] $$R_fI + CO_2 \xrightarrow[\text{50-60 °C, 1 h; 2. HCl}]{\text{1. DMF ultrasound/Zn}} R_fCO_2H$$ **45**

$R_f = i\text{-}C_3F_7$ 48%; C_4F_9 61%;
C_6F_{13} 77%; C_8F_{17} 72%

[*51*] $$R_fI + CO_2 \xrightarrow[\text{RT, 3 h; 2. HCl}]{\text{1. Zn-Cu, DMSO}} R_fCO_2H$$ **46**

$R_f = C_4F_9$ 42%; C_6F_{13} 63%; C_8F_{17} 47%

[53] **47**

$$R_fI + (C_2H_5O)_2CO \xrightarrow[2.\ H_3O^+]{1.\ Zn\text{-}Cu} R_fCO_2C_2H_5$$

$R_f = C_4F_9$ 70%; C_6F_{13} 60%; C_8F_{17} 50%

[53] $$R_fI + C_2H_5OCOOCOOC_2H_5 \xrightarrow[2.\ H_3O^+]{1.\ Zn\text{-}Cu} R_fCO_2C_2H_5$$ **48**

$R_f = C_4F_9$ 65%; C_6F_{13} 60%; C_8F_{17} 50%

Under similar conditions, perfluoroalkyl iodides react with ***alkyl phosphates*** to give fluorinated phosphine oxides, phosphinates, and phosphines [*54*] (equation 49). The product formed depends upon the stoichiometry and type of iodide used. When sodium alkyl trithiocarbonates are used as substrates, *perfluoroalkyl trithiocarbonates* are formed [*55*].

[54] $$2\,R_fI + PO(OR)_3 \xrightarrow[2.\ H_3O^+]{1.\ Zn\text{-}Cu} (R_f)_2PO(OR)$$ **49**

[57] $$CF_3CCl_3 + Zn \xrightarrow{DMF} CF_3CCl_2ZnCl(DMF)_2$$ **50**

In addition to the perfluoroalkylzinc compounds, the zinc **reagent formed from 1,1,1-trifluorotrichloroethane** has received considerable attention. This zinc compound was first reported as a stable ether complex [*56*]. Later, the DMF complex was isolated and the structure was determined by X-ray diffraction and shown to be monomeric [*57*] (equation 50). This zinc reagent undergoes a variety of functionalization reactions, and some typical examples are illustrated in Table 2 [*47, 58, 59, 60, 61*]. The alcohol products (Table 2) can be converted to $ArCF{=}CXCF_3$ (X = Cl, F) by further reaction with diethylaminosulfur trifluoride (DAST) and 1,8-diazabicyclo[5.4.0]undec-7-ene (DBU) [*60*].

Sonication conditions were used with several CF_3CXYZ (X = F, Cl, Br; Y = F, Cl, Br, H; Z = Cl, Br) compounds and zinc and carbon dioxide to give CF_3CXYCO_2H [*62*] in moderate yields. Hydrogenolysis of the C–Cl and C–Br bonds in $CF_3CFClCO_2H$ and $CF_3CFBrCO_2H$ afforded CF_3CFHCO_2H [*62*].

When acetic anhydride is used in the CF_3CCl_3 and zinc reaction with aldehydes, the initial addition product undergoes an elimination reaction to give 2-chloro-1,1,1-trifluoro-2-alkenes exclusively [*60, 63*] (equation 51).

Some typical examples of this useful transformation are shown in Table 3 [*63*]. These olefinic products can be transformed into 1-aryl-3,3,3-trifluoropropynes via further reaction with sodium *tert*-butoxide [*64*].

Table 2. Functionalization Reactions of CF_3CCl_2ZnCl

Carbonyl Compound	*Product*	*Yield (%)*	*Ref.*
C_6H_5CHO	$C_6H_5CH(OH)CCl_2CF_3$	86	59
 CHO	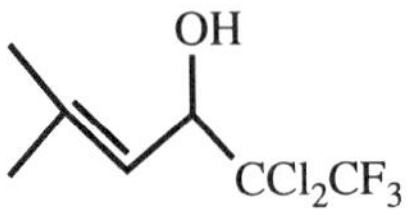 OH CCl$_2$CF$_3$	22	59
p-ClC_6H_4CHO	p-$ClC_6H_4CH(OH)CCl_2CF_3$	87	59
$C_6H_5COCO_2CH_3$	$C_6H_5C(OH)(CCl_2CF_3)CO_2CH_3$	55	60
HCHO	$CF_3CCl_2CH_2OH$	87	58
DMF–*tert*-Butyldi-methylchlorosilane	CF_3CCl_2CHO	50	47
CO_2CuCl	CF_3CCl_2COOH	18	61

[*63*] **51**

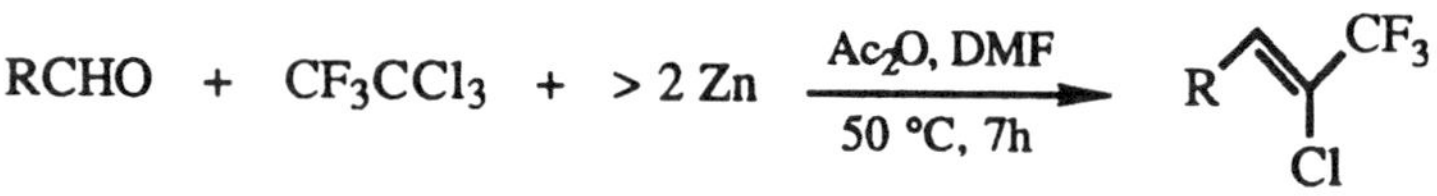

$$RCHO + CF_3CCl_3 + >2\,Zn \xrightarrow[50\ °C,\ 7h]{Ac_2O,\ DMF} R\text{–}CH{=}C(Cl)CF_3$$

Table 3. Preparation of 2-Chloro-1,1,1-trifluoro-2-alkenes from Aldehydes, Zinc, and Acetic Anhydride [*63*]

Aldehyde	*Product*	*Z:E*	*Yield (%)*
C_6H_5CHO	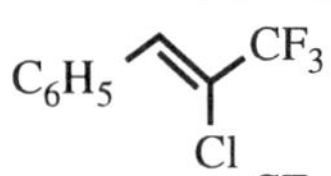 C_6H_5 CF$_3$ Cl	86:14	75
$C_6H_{11}CHO$	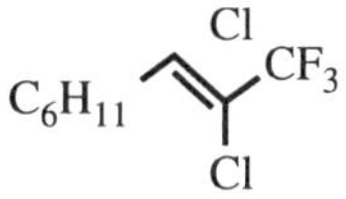 C_6H_{11} CF$_3$ Cl	85:15	50
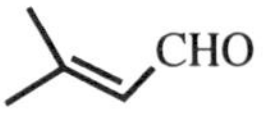 CHO	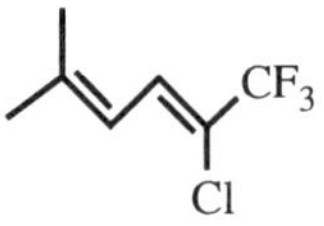 CF$_3$ Cl	88:12	53
$C_6H_5CH{=}CHCHO$	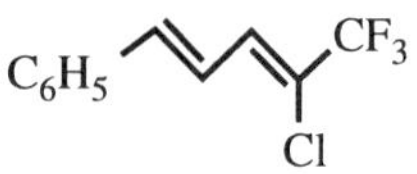 C_6H_5 CF$_3$ Cl	89:11	72

This reaction has received especial attention for the *stereocontrolled synthesis* of polyfluorinated artificial pyrethroids [*65, 66*] (equation 52).

Another useful modification of this reagent is the reaction of CF_3CCl_3 with zinc and DMF in the presence of $AlCl_3$ [*60, 63*] (equation 53). The alcohol product can be treated subsequently with DAST, thionyl chloride, or phosphorus chlorides to afford the allyl substitution product regio- and stereoselectively [*66*] (equation 54).

[*65*] **52**

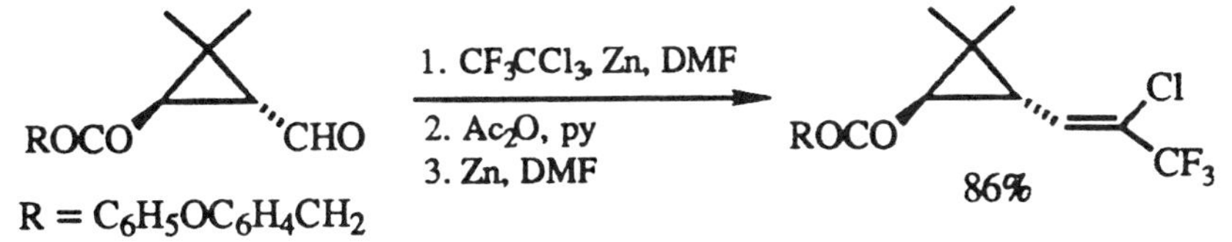

ROCO ... CHO; R = $C_6H_5OC_6H_4CH_2$ → (1. CF_3CCl_3, Zn, DMF; 2. Ac_2O, py; 3. Zn, DMF) → ROCO ... =C(Cl)CF_3 86%

[*63*] **53**

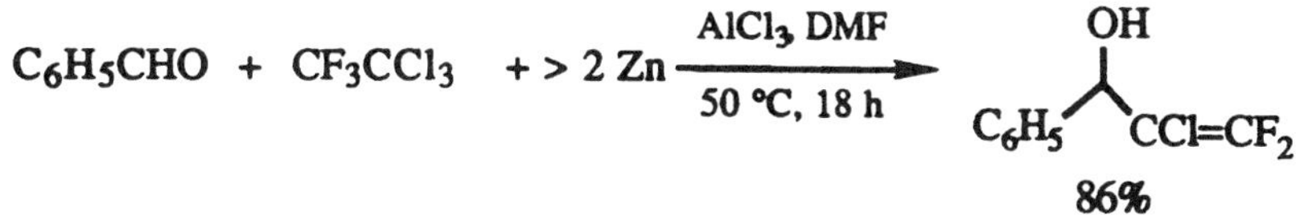

C_6H_5CHO + CF_3CCl_3 + > 2 Zn → ($AlCl_3$, DMF; 50 °C, 18 h) → $C_6H_5CH(OH)CCl{=}CF_2$ 86%

[*66*] **54**

$C_6H_5CH(OH)CCl{=}CF_2$ + $SOCl_2$ → (Et_2O; 50 °C, 12 h) → $C_6H_5CH{=}C(Cl)CF_2Cl$ 79%

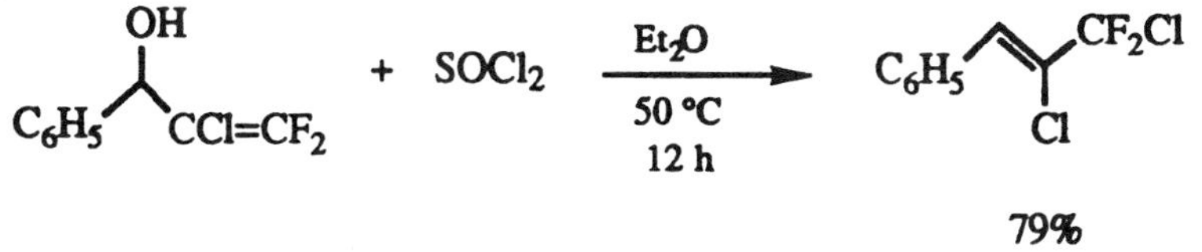

In similar work, $CF_3CCl_2CO_2CH_3$ yields methyl α-trifluoromethyl-α,β-unsaturated carboxylates when reacted with a zinc–copper couple, aldehydes, and acetic anhydride [*67*] (equation 55). This methodology gives (*Z*)-α-*fluoro*-α-β-*unsaturated carboxylates* from the reaction of **carbonyl compounds** with $CFCl_2CO_2CH_3$ and zinc and acetic anhydride [*68*].

Functionalized organofluorozinc reagents have received considerable attention as synthetic intermediates for the preparation of functionalized organofluorine

[*67*] **55**

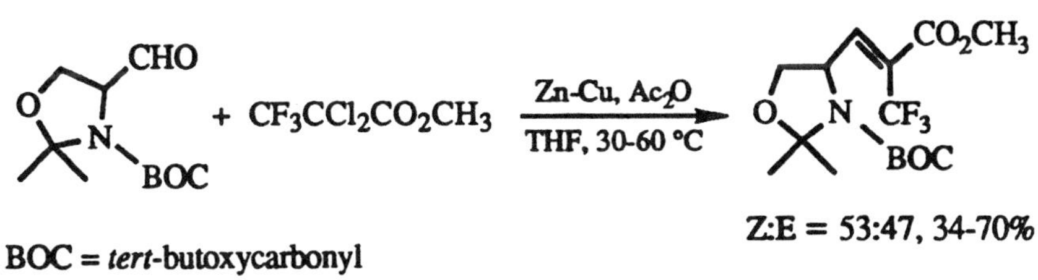

(oxazolidine)-CHO, N-BOC + $CF_3CCl_2CO_2CH_3$ → (Zn-Cu, Ac_2O; THF, 30-60 °C) → (oxazolidine)-CH=C(CF_3)CO_2CH_3, N-BOC

Z:E = 53:47, 34-70%

BOC = *tert*-butoxycarbonyl

compounds. For example, bromodifluoroacetates add to aldehydes and ketones in the presence of zinc (*Reformatsky reaction*) to give *β-hydroxy-α,α-difluoroesters* [*69, 70*] (equation 56). The intermediate zinc reagent was shown by ^{19}F and ^{13}C NMR analysis to be carbon metallated [*71*]. Kobayashi prepared this reagent from iododifluoroacetate and zinc in THF, dioxane, DMF, dimethoxyethane (DME), and CH_3CN; best results were obtained in CH_3CN [*72*]. In diethyl ether, only the succinate (coupled product) was formed [*72*].

[*70*] **56**

$$C_6H_5CHO + BrCF_2CO_2C_2H_5 + Zn \xrightarrow[\text{reflux}]{\text{THF}} C_6H_5CH(OH)CF_2CO_2C_2H_5 \quad 57\%$$

The main application of the *α,α-difluoro Reformatsky reagent* has been for the incorporation of the CF_2CO group, because this functionality often leads to enhanced biological activity of the resultant product. Consequently, the biological activity of difluorothromboxane A_2 [*69, 73*], 1-(2-deoxy-2,2-difluororibofuranosyl)pyrimidine nucleosides [*74*], 14,14-difluoro-4-demethoxydannorubicin [*75*], 2,2-difluoroarachidonic acid [*76*], dipeptidyl difluoromethyl ketones [*77, 78, 79*], and dipeptide isostere precursors [*80, 81, 82, 83, 84, 85*] have been modified by incorporation of this unit via this organozinc compound. When imines are used as substrates with this zinc reagent, lactams are formed in good yields [*86*] (equation 57). Recently, ethyl chlorodifluoroacetate (a cheaper precursor) has been used in DMF under Barbier conditions to prepare β-hydroxy-α,α-difluoroesters [*87*].

[*86*] **57**

NBn + ICF_2CO_2Me

Bn = $C_6H_5CH_2$

Zn | 0 °C

F F BnN O + F F BnN O

57.6% 13.4%

Ethyl chlorodifluoroacetate has been used in a silicon-induced *Reformatsky–Claisen reaction* of allyl chlorodifluoroacetates in the presence of zinc as a route to *2,2-difluoro unsaturated acids* [*88*] (equation 58). When this methodology is applied to chlorodifluoropropargylic esters, the corresponding *allenic esters* are formed [*88*].

[88] 58

CF_2ClCOO–$CH_2CH=CHCH_3$ $\xrightarrow[(CH_3)_3SiCl,\ 100\ °C]{Zn,\ CH_3CN}$ $CH_2=CH$–$CH(CH_3)$–CF_2COOH

47%

Another useful application of these Reformatsky reagents is their conversion to **difluoroketene silyl acetals** and subsequent reaction of these ketene silyl acetals with electrophiles [*86, 89, 90*] (equation 59).

A similar **allyl** [*91*] or **propargyl** [*92*] **Reformatsky reagent** has been used to prepare fluorinated *homoallylic* or *homopropargylic* alcohols, respectively [*91, 92*] (equations 60 and 61).

A related functionalized organozinc reagent that has found use in the synthesis of biologically interesting compounds is the stable dialkoxyphosphinyldifluoromethylzinc reagent [*93*], which is readily prepared by direct reaction of bromo- or iododifluoromethyl-

[90] 59

$IZnCF_2CO_2CH_3$ $\xrightarrow[CH_3CN]{CH_3SiCl}$ $F_2C=C(OSi(CH_3)_3)(OCH_3)$

[89] $F_2C=C(OSi(CH_3)_3)(OCH_3)$ + 2-cyclohexen-1-one ⟶ 3-($CF_2CO_2CH_3$)cyclohexanone 95%

↓ C_6H_5CHO

[90] $C_6H_5CH(OH)CF_2CO_2CH_3$ 81%

[91] 60

$H_2C=CHCF_2Br + C_6H_5CHO \xrightarrow[THF,\ RT]{Zn} C_6H_5CH(OH)CF_2CH=CH_2$

67%

[92] 61

$C_6H_5C{\equiv}CCF_2Br + C_6H_5CHO \xrightarrow[THF,\ RT]{Zn} C_6H_5C{\equiv}CCF_2CH(OH)C_6H_5$

78%

References are listed on pages 718–728.

phosphonates and zinc metal (equation 62). This zinc reagent is not as reactive as the corresponding lithium reagent but does react with reactive electrophiles or with less reactive electrophiles *catalyzed by cuprous bromide* [*94, 95, 96, 97, 98, 99*]. Some typical reactions of this reagent are summarized in equations 62–68.

[*93*] **62**

$$(RO)_2P(O)CF_2Br + Zn \xrightarrow[RT]{MG} (RO)_2P(O)CF_2ZnBr$$

$R = C_2H_5$, $i\text{-}C_3H_7$, C_4H_9 : MG = monoglyme

[*93*] **63**

$$(C_4H_9O)_2P(O)CF_2ZnBr + CH_3COCl \xrightarrow[RT,\ 20\text{-}24\ h]{TG} (C_4H_9O)_2P(O)CF_2COCH_3 \quad 77\%$$

TG = triglyme

[*94,95*] **64**

$$(C_2H_5O)_2P(O)CF_2ZnBr + ClCO_2C_2H_5 \xrightarrow[CH_3CN,\ RT]{CuBr,\ MG} (C_2H_5O)_2P(O)CF_2CO_2C_2H_5 \quad 50\%$$

MG = monoglyme

[*96*] **65**

$$(C_2H_5O)_2P(O)CF_2ZnBr \xrightarrow[CuBr,\ MG,\ RT,\ 12\ h]{H_2C{=}CHCH_2Br} (C_2H_5O)_2P(O)CF_2CH_2CH{=}CH_2 \quad 47\%$$

[*96*] **66**

$$(C_2H_5O)_2P(O)CF_2ZnBr + H_2C{=}CHCF_2Br \xrightarrow{CuBr\ |\ MG,RT} (C_2H_5O)_2P(O)CF_2CH_2CH{=}CF_2 \quad 55\%$$

Vinyl fluorinated zinc reagents can be prepared by two different methods: (1) *capture of the corresponding vinyllithium reagent* at low temperatures with a zinc salt and (2) *direct insertion of zinc into a carbon–halogen bond.*

The first method involves generation of an unstable vinyllithium derivative at low temperatures from the corresponding vinyl halide or 1-hydroalkene. Addition of zinc chloride and warming to room temperature gives the stable vinyl zinc reagent [*100, 101, 102, 103*] (equations 69–72).

These zinc reagents are useful precursors for stereocontrolled palladium-catalyzed cross-coupling reactions, as illustrated in equations 73–80 [*100, 101, 102, 103, 104, 105, 106, 107, 108*]. This methodology has been used to prepare new fluorinated analogues of codlemone [*109*].

[*97*] **67**

$(C_2H_5O)_2P(O)CF_2ZnBr$ + N-(2-bromoethyl)phthalimide → (1. CuBr, MG, RT, 10 h; 2. H_2O) → N-[2-($CF_2P(O)(OH)_2$)ethyl]phthalimide 40%

[*98*] **68**

$(C_2H_5O)_2P(O)CF_2ZnBr + C_6H_5HgCl \xrightarrow[RT,\ 12\ d]{MG} (C_2H_5O)_2P(O)CF_2HgC_6H_5$ 61%

[*100*] **69**

$F_2C{=}CFCl \xrightarrow[-110\ °C]{BuLi,\ THF} [F_2C{=}CFLi] \xrightarrow[warm]{ZnCl_2} F_2C{=}CFZnCl$

[*101*] $E{-}C_4H_9CF{=}CFH \xrightarrow[-30\ °C]{BuLi,\ THF} [E{-}C_4H_9CF{=}CFLi]$ **70**

$\xrightarrow{ZnCl_2,\ warm}$ $[E{-}C_4H_9CF{=}CFZnCl]$

[*102*] **71**

$Z{-}CHF{=}CFSi(C_2H_5)_3 \xrightarrow[THF,\ -100\ °C]{BuLi} Z{-}LiCF{=}CFSi(C_2H_5)_3 \xrightarrow[warm]{ZnCl_2} E{-}ZnCF{=}CFSi(C_2H_5)_3$

[*103*] **72**

$CF_3CH{=}CF_2 \xrightarrow[THF,\ -78°C]{t\text{-}BuLi} [CF_3CLi{=}CF_2] \xrightarrow[warm]{ZnI_2} CF_3C(ZnI){=}CF_2$

[*100*] **73**

$F_2C{=}CFZnCl$ + 2-iodopyridine $\xrightarrow[THF,\ -5\ °C]{Pd[P(C_6H_5)_3]_4}$ 2-($CF{=}CF_2$)pyridine 64%

[*100*] **74**

$$Z\text{-}s\text{-}C_4H_9CF{=}CFZnCl + \text{(2-iodothiophene)} \xrightarrow[\text{THF, }-5\ ^\circ C]{Pd[P(C_6H_5)_3]_4} \text{(2-thienyl)}\text{-}CF{=}CFC_4H_9\text{-}s(Z)$$

50%

[*104*] **75**

$$F_2C{=}CFZnCl + C_6H_{13}CH{=}CHI \xrightarrow[\text{THF, 20 °C, 1 h}]{Pd[P(C_6H_5)_3]_4} C_6H_{13}CH{=}CH\text{-}CF{=}CF_2$$

53%

[*104*] **76**

$$C_4H_9(F)C{=}C(F)ZnCl + C_6H_{13}CH{=}CHI \xrightarrow[\text{THF, 20 °C, 15 h}]{Pd[P(C_6H_5)_3]_4} C_6H_{13}CH{=}CH\text{-}CF{=}CF(C_4H_9)$$

52%

[*104*] **77**

$$ClZn(F)C{=}C(F)Si(C_2H_5)_3 + C_5H_{11}(F)C{=}C(F)I \xrightarrow[\text{THF, 20 °C, 2 h}]{Pd[P(C_6H_5)_3]_4} C_5H_{11}CF{=}CF\text{-}CF{=}CFSi(C_2H_5)_3$$

80%

[*105*] **78**

$$Z\text{-}C_7H_{15}CF{=}CFZnBr + C_4H_9C{\equiv}CI \xrightarrow[\text{THF}]{Pd[P(C_6H_5)_3]_4} E\text{-}C_7H_{15}CF{=}CFC{\equiv}CC_4H_9$$

90%

[*100*] **79**

$$Z\text{-}C_6H_5CF{=}CFZnCl + CH_3COCl \xrightarrow[\text{THF, }-5\ ^\circ C]{Pd[P(C_6H_5)_3]_4} E\text{-}C_6H_5CF{=}CFCOCH_3$$

63%

The second method involves the direct *reaction of bromo- or iodofluoroolefins with zinc powder* [*110*] (equation 81) in solvents such as DMF, N,N-dimethylacetamide (DMAC), triglyme, tetraglyme, THF, and CH_3CN. In all cases, the stereochemistry is retained. Table 4 shows some typical examples of this methodology. Similar results are obtained with 1-chloro-2-iodotetrafluorocyclobutene [*111*] and 1-iodo-2-chlorohexafluorocyclopentene [*112*].

[*100*] **80**

$$Z\text{-}C_6H_5CF{=}CFZnCl + ClCO_2C_2H_5 \xrightarrow[\text{THF},\ -5\,^\circ C]{Pd[P(C_6H_5)_3]_4} E\text{-}C_6H_5CF{=}CFCO_2C_2H_5\ \ 57\%$$

[*110*] **81**

$$R_fCF{=}CFX + Zn \xrightarrow[\text{RT to } 60\,^\circ C]{DMF} R_fCF{=}CFZnX$$

X = Br, I

Table 4. Preparation of Fluorinated Vinylzinc Reagents from (*E*)-Vinyl Halides and Zinc Metal [*110*]

Vinyl Halide	*Solvent* [a]	*Zinc Reagent*	*Yield (%)*
$F_2C{=}CFI$	DMF	$F_2C{=}CFZnI$	79
$F_2C{=}CFI$	DMAC	$F_2C{=}CFZnI$	97
$F_2C{=}CFBr$	DMF	$F_2C{=}CFZnBr$	72
(*Z*)-$CF_3CF{=}CFI$	THF	(*Z*)-$CF_3CF{=}CFZnI$	98
(*Z*)-$CF_3CF{=}CFI$	DMF	(*Z*)-$CF_3CF{=}CFZnI$	100
(*E*)-$CF_3CF{=}CFI$	TG	(*E*)-$CF_3CF{=}CFZnI$	100
(*Z*)-$CF_3CF_2CF{=}CFI$	TG	(*Z*)-$CF_3CF_2CF{=}CFZnI$	90
(*Z*)-$CF_3(CF_2)_4CF{=}CFBr$	DMF	(*Z*)-$CF_3(CF_2)_4CF{=}CFZnBr$	77
(*E*)-$CF_3C(C_6H_5){=}CFI$	THF	(*E*)-$CF_3C(C_6H_5){=}CFZnI$	78
$CF_3CF{=}C(C_6H_5)CF{=}CFBr$	DMF	$CF_3CF{=}C(C_6H_5)CF{=}CFZnBr$	71
(*E*)-$CF_3CH{=}CFI$	TG	(*E*)-$CF_3CH{=}CFZnI$	89
$F_2C{=}CBr_2$	DMF	$F_2C{=}CBrZnBr$	97

[a] DMAC, *N,N*-dimethylacetamide; TG, triglyme.

The 2-pentafluoropropenylzinc reagent has been prepared via a novel one-pot dehalogenation–insertion reaction [*103*] (equation 82).

As noted, these zinc reagents find extensive application in the preparation of *fluorinated styrenes* [*113, 114*], *aryl-substituted fluorinated propenes* [*114*], *fluorinated dienes* [*115, 116*], and *trifluorovinyl ketones* [*117*], as illustrated in equations 83–88.

[*103*] $CF_3CBr_2CF_3 \xrightarrow[\text{RT}]{\text{Zn, DMF}} CF_3C(ZnBr){=}CF_2$ 90% **82**

[*114*] **83**

$m\text{-}NO_2C_6H_4I + F_2C{=}CFZnX \xrightarrow[\text{TG, 60-80 °C}]{Pd[P(C_6H_5)_3]_4} m\text{-}NO_2C_6H_4CF{=}CF_2$ 81%

TG = Triglyme

[*114*] **84**

$E\text{-}CF_3CF{=}CFZnX + C_6H_5I \xrightarrow[\text{TG, 60 °C}]{Pd[P(C_6H_5)_3]_4} Z\text{-}C_6H_5CF{=}CFCF_3$ 82% Z:E 97:3

[*114*] **85**

$Z\text{-}CF_3CF{=}CFZnX + C_6H_5I \xrightarrow[\text{TG, 60 °C}]{Pd[P(C_6H_5)_3]_4} E\text{-}C_6H_5CF{=}CFCF_3$ 80% 100% E

[*116*] **86**

$E\text{-}CF_3CF{=}CFZnI + F_2C{=}CFI \xrightarrow[\text{TG, 75 °C, 3 h}]{Pd[P(C_6H_5)_3]_4} Z\text{-}CF_3CF{=}CFCF{=}CF_2$ 50%

[*116*] **87**

$Z\text{-}CF_3CF{=}CFZnI + F_2C{=}CFI \xrightarrow[\text{TG, 40 °C, 30 h}]{Pd[P(C_6H_5)_3]_4} E\text{-}CF_3CF{=}CFCF{=}CF_2$ 80%

[*117*] **88**

$F_2C{=}CFZnX + CH_3CH_2COCl \xrightarrow[\text{TG, RT}]{\text{CuBr}} F_2C{=}CFCOCH_2CH_3$ 83%

Fluorinated acetylenic zinc reagents have received only limited attention. The trifluoromethyl analogue can be generated via the reaction of zinc with 1,1,2-trichloro-3,3,3-trifluoropropene in DMF [*118*]. Alternatively, perfluoroalkynes can be metalated with butyllithium, and the lithium reagent reacted with zinc chloride to give the acetylenic zinc reagent [*119, 120*] (equations 89–91). This approach gives improved yields when the zinc reagent is used in coupling processes. The higher acetylenic homologues can be prepared from the 1,1,1,2,2-pentachloroperfluoroalkane and zinc. The requisite pentachloro precursors can be obtained from the corresponding perfluoroalkylethylenes, which are commercially available [*121*] (equation 92). Acid hydrolysis of the zinc reagent gives the corresponding perfluoroalkylacetylenes [*121*] (equation 93).

[*119*] **89**

$$CF_3C{\equiv}CH \xrightarrow[2.\ ZnCl_2]{1.\ BuLi} [CF_3C{\equiv}CZnCl] \xrightarrow[Pd[P(C_6H_5)_3]_4]{C_6H_5I} \underset{85\%}{C_6H_5C{\equiv}CCF_3}$$

[*120*] **90**

$$C_4F_9C{\equiv}CH \xrightarrow[THF\ \ 2.\ ZnCl_2]{1.\ BuLi} [C_4F_9C{\equiv}CZnCl]$$

$$\xrightarrow[1\text{-iodonaphthalene, THF}]{Pd[P(C_6H_5)_3]_4} \text{1-}(C{\equiv}CC_4F_9)\text{naphthalene} \quad 80\%$$

[*120*] **91**

$$C_6F_{13}C{\equiv}CH \xrightarrow[2.\ ZnCl_2]{1.\ BuLi} C_6F_{13}C{\equiv}CZnCl$$

$$\xrightarrow[\text{1-iodocyclohexene, THF}]{Pd[P(C_6H_5)_3]_4} \text{1-}(C{\equiv}CC_6F_{13})\text{cyclohexene} \quad 72\%$$

[*121*] **92**

$$R_fCH{=}CH_2 \xrightarrow[h\nu]{3\ Cl_2} R_fCCl_2CCl_3 \xrightarrow[DMF]{3\ Zn} [R_fC{\equiv}CZnCl]$$

$R_f = C_3F_7$ 66%; C_4F_9 79%; C_6F_{13} 79%; C_8F_{17} 89%; $C_{10}F_{21}$ 94%

[*121*] **93**

$$R_fC{\equiv}CZnCl + HCl \longrightarrow R_fC{\equiv}CH$$

$R_f = C_3F_7$ 77%; C_4F_9 72%; C_6F_{13} 64%; C_8F_{17} 62%; $C_{10}F_{21}$ 43%

The reaction of ***iodopentafluorobenzene*** with **dialkylzinc** in the presence of Lewis bases quantitatively gives the *pentafluorophenylzinc complexes* [*32*]. The pentafluorophenylzinc complexes can be prepared more easily via the direct reaction of bromopentafluorobenzene with zinc powder in DMF [*122*] (equations 94 and 95).

[*32*] $$2\ C_6F_5I + (CH_3)_2Zn \xrightarrow[0\ °C]{DG} (C_6F_5)_2Zn$$ **94**

[*122*] $$3\ C_6F_5Br + 3\ Zn \xrightarrow[RT]{DMF} C_6F_5ZnBr + (C_6F_5)_2Zn \quad {\sim}100\%$$ **95**

Organocadmium Compounds

The first *perfluoroalkylcadmium compounds* were prepared via metathesis between **bis(trifluoromethyl)mercury** and **dialkylcadmiums** [*123*]. An equilibrium mixture of all possible cadmium and mercury compounds was formed, the equilibrium could be shifted toward the bis(trifluoromethyl)cadmium by using excess bis(trifluoromethyl)mercury. When the reaction was carried out in glyme, a stable bis(trifluoromethyl)cadmium·glyme complex was isolated [*124, 125, 126*] (equation 96).

Other analogues were synthesized and isolated in essentially quantitative yield via direct reaction of ***primary*** and ***secondary perfluoroalkyl iodides*** with **dialkylcadmiums** in basic Lewis solvents [*127, 128*] (equation 97).

An alternative method, which avoids the prior preparation of the toxic dialkylcadmiums, is direct reaction of the ***perfluoroalkyl iodide*** with **cadmium metal** [*129*] (equation 98).

The difluoromethyl analogue can be prepared similarly [*130*] (equation 99).

[*124*] **96**

$$(CF_3)_2Hg + (CH_3)_2Cd \xrightarrow{\text{glyme}} (CF_3)_2Cd{\cdot}\text{glyme} + (CH_3)_2Hg\uparrow$$

[*127*] **97**

$$2\,R_fI + (CH_3)_2Cd \xrightarrow[\text{RT}]{CH_3CN} (R_f)_2Cd + 2\,CH_3I$$

$R_f = CF_3, C_2F_5, C_3F_7, C_4F_9, CF_3CH_2$

[*129*] **98**

$$R_fI + Cd \xrightarrow[\text{RT}]{\text{DMF}} R_fCdI + (R_f)_2Cd$$

R_f = CF_3 65%; C_2F_5 68%; C_3F_7 93%; C_4F_9 43%; C_6F_{13} 46%; C_7F_{15} 25%; C_8F_{17} 47%

[*130*] **99**

$$CHF_2X + Cd \xrightarrow[\text{RT to 50 °C}]{\text{DMF}} CHF_2CdX + (CHF_2)_2Cd$$

X = I, Br

X = I 91%

X = Br 65-75%

Solvent plays a significant role in these reactions. In contrast to the formation of organocadmium compounds in the direct reaction of perfluoroalkyl iodides and cadmium powder in DMF [*129*], dimerization of the perfluoroalkyl halide was observed in acetonitrile [*131*] (equation 100).

References are listed on pages 719–729.

[*131*]

$$C_6F_{13}I + Cd \xrightarrow[80\,°C,\ 2\ h]{CH_3CN} C_{12}F_{26} \quad 74\%$$ **100**

The **trifluoromethylcadmium reagent** can also be prepared in situ via the reaction of ***dihalodifluoromethanes*** with *cadmium metal* in DMF [*33*] (equation 101).

Several studies have investigated the preparation and properties of **trifluoromethyl and pentafluorophenyl cadmates** [*132, 133*].

Dialkylcadmium reagents are often useful alternatives to the more reactive Grignard reagents in the preparation of ketones from acyl halides. However, bis(trifluoromethyl)cadmium·glyme is decomposed by acyl halides and does not give trifluoromethyl ketones [*8, 124*]. Nevertheless, this reaction can be used as a *low-temperature source of difluorocarbene* [*8, 124*] (equation 102).

The **bis(trifluoromethyl)cadmium** complex can be used for the synthesis of CF_3NO [*5*] (equation 103). With higher homologues of the perfluoroalkylcadmium complex ($R_f = C_2F_5$ and iso-C_3F_7), a metathesis reaction occurs to give $(CH_3)_3SnR_f$ compounds [*4*].

The **difluoromethylcadmium** reagent reacts with ***allyl halides,*** and products of both α -and γ-attack, with preference for attack at the less hindered position, are detected [*130*] (equation 104).

[*33*] **101**

$$2\,Cd + 2\,CF_2XY \xrightarrow{DMF} CF_3CdX + CdY_2 + CO + [Me_2N{=}CFH]^+X^- \quad 80\text{–}95\%$$

X = Br, Cl
Y = Br, Cl

[*124*] **102**

$$CH_3COBr + (CF_3)_2Cd{\cdot}glyme \xrightarrow[-25\,°C]{\text{(tetramethylethylene)}} CH_3COF\ (95\%) + \text{1,1-difluoro-2,2,3,3-tetramethylcyclopropane}\ (53\%)$$

[*5*] **103**

$$(CF_3)_2Cd{\cdot}glyme + (CH_3)_3Sn(ONO_2) \xrightarrow[RT]{CH_3CN} CF_3NO \quad 60\%$$

[*130*] **104**

$$HCF_2CdX + H_2C{=}CHCHClCH_3 \xrightarrow[RT]{DMF} H_2C{=}CHCH(CH_3)CF_2H \quad 7$$

$$+\ CF_2HCH_2CH{=}CHCH_3 \quad 93$$

The bis(trifluoromethyl)cadmium reagent readily exchanges with other halo-organometallic complexes to give trifluoromethyl-substituted compounds, such as $(CF_3)AuP(CH_3)_3$ [*134, 135*], $(CF_3)_3AuP(CH_3)_3$ [*134, 135*], $(CF_3)_4Sn$ [*124*], $(CF_3)_4Ge$ [*124*], $(\eta^5\text{-}Cp)Co(CO)CF_3$ [*136*], $(CF_3)_4Te$ [*137*], $CF_3AgP(CH_3)_3$ [*135*], $(CF_3)_3Sb$ [*138*], $(CF_3)_3P$ [*124*], and CF_3Cu [*135*].

The functionalized cadmium compound **dialkoxyphosphinyldifluoromethylcadmium** is readily prepared by direct reaction of bromodifluoromethanephosphonates with cadmium metal [*139*] (equation 105). This cadmium reagent shows versatile chemical reactivity and reacts with a wide variety of electrophiles, as illustrated in equations 106–109 [*139, 140, 141, 142*].

[*139*] **105**

$$(C_2H_5O)_2POCF_2Br + Cd \xrightarrow[\text{RT, 2 h}]{\text{DMF}} (C_2H_5O)_2POCF_2CdBr + [(C_2H_5O)_2POCF_2]_2Cd \quad 89\%$$

[*139*] **106**

$$(C_2H_5O)_2POCF_2CdX + H_2C{=}CHCH_2Br \xrightarrow[\text{RT}]{\text{DMF}} (C_2H_5O)_2POCF_2CH_2CH{=}CH_2 \quad 31\text{-}62\%$$

[*139*] **107**

$$(C_2H_5O)_2POCF_2CdX + C_3H_7COCl \xrightarrow[\text{RT}]{\text{dioxane}} (C_2H_5O)_2POCF_2COC_3H_7 \quad 41\%$$

[*142*] **108**

$$(C_2H_5O)_2POCF_2CdX \xrightarrow[\text{2. } H_2O_2 \text{ 3. } H^+]{\text{1. } SO_2} (HO)_2POCF_2SO_3H$$

[*139*] **109**

$$(C_2H_5O)_2POCF_2CdX + C_6H_5CHO \xrightarrow[\text{TG}]{\text{NaI}} C_6H_5CH{=}CF_2 \quad 73\%$$

Only a few reports of unsaturated fluorinated cadmium compounds have appeared. The direct reaction of perfluoroalkenyl iodides or bromides with cadmium powder in DMF stereospecifically gives the **vinylcadmium compounds** [*143*] (equation 110). Table 5 shows several typical vinylcadmiums prepared by this method.

The perfluorovinylcadmium compounds react with electrophiles as expected [*144*] (equation 111) and readily undergo metathesis with cuprous halides [*145*] (equation 112).

[*143*] **110**

$$R_f\,CF{=}CFX + Cd \xrightarrow[\text{RT to 60 °C}]{\text{DMF}} R_f\,CF{=}CFCdX + (R_f\,CF{=}CF)_2Cd$$

X = Br, I

Table 5. Preparation of Vinylcadmium Reagents from Perfluorovinyl Halides and Cadmium Metal [*143*]

Vinyl Halide	*Cadmium Reagent*	*Yield (%)*
$F_2C{=}CFI$	$F_2C{=}CFCdI$	99
(*E*)-$CF_3CF{=}CFI$	(*E*)-$CF_3CF{=}CFCdI$	92
(*Z*)-$CF_3CF{=}CFI$	(*Z*)-$CF_3CF{=}CFCdI$	96
$CF_3CF{=}CICF_3$	$CF_3CF{=}C(CF_3)CdI$	91
$CF_3CF{=}C(C_6H_5)CF{=}CFBr$	$CF_3CF{=}C(C_6H_5)CF{=}CFCdBr$	61

[*144*] **111**

$$Z\text{-}CF_3CF{=}CFCdX \xrightarrow{CH_2{=}CHCH_2Br} E\text{-}CF_3CF{=}CFCH_2CH{=}CH_2 \quad 82\%$$

$$Z\text{-}CF_3CF{=}CFCdX \xrightarrow{(C_2H_5O)_2PCl} Z\text{-}CF_3CF{=}CFP(OC_2H_5)_2 \quad 100\%$$

$$Z\text{-}CF_3CF{=}CFCdX \xrightarrow{(C_6H_5)_2PCl} Z\text{-}CF_3CF{=}CFP(C_6H_5)_2 \quad 57\%$$

[*145*] **112**

$$F_2C{=}CFCdX + CuX \xrightarrow[\text{RT}]{\text{DMF}} F_2C{=}CFCu$$

X = Br, I, Cl, CN 90-100%

[*146*] **113**

$$F_2C{=}CFCF_2I + Cd \xrightarrow[\text{0 °C, 1 h}]{\text{DMF}} F_2C{=}CFCF_2CdI + (F_2C{=}CFCF_2)_2Cd$$

71%

The **perfluoroallylcadmium reagent** can be prepared similarly by direct reaction of perfluoroallyl iodide with cadmium metal [*146*] (equation 113). Reaction of this cadmium reagent with allyl bromide gives $F_2C{=}CFCF_2CH_2CH{=}CH_2$ in 78% yield [*146*].

The **perfluoroalkylacetylenic cadmium compounds** can be prepared via the reaction of the 1-iodoperfluoroalkylacetylene with cadmium powder [*147*] (equation 114).

[*147*] $R_fC{\equiv}CI + Cd \xrightarrow[RT]{DMF} RC{\equiv}CCdI + (R_fC{\equiv}C)_2Cd$ **114**

$R_f = CF_3$ 100%; C_4F_9 100%

The **perfluoroarylcadmium compounds** are synthesized via reaction of iodo- or bromopentafluorobenzene with cadmium metal [*129, 148*] (equation 115). The *ortho-* and *para-*dibromotetrafluorobenzenes can give either the mono- or biscadmium reagent depending on the reaction conditions [*149*] (equation 116). The *meta-*dibromotetrafluorobenzene gives only the monocadmium compound [*149*].

[*129*] $C_6F_5Br + Cd \xrightarrow[RT]{DMF} C_6F_5CdBr$ 90% **115**

[*149*] **116**

(F)-ring $C_6F_4(CdX)_2$ $\xleftarrow[\text{DMF, 110 °C}]{Cd}$ (F)-ring $C_6F_4Br_2$ $\xrightarrow[\text{DMF, RT to 60 °C}]{Cd}$ (F)-ring $C_6F_4(CdX)Br$

Organomercury Compounds

The photochemical or thermal reaction between ***perfluoroalkyl iodides*** and **mercury–cadmium amalgams** has been used for the synthesis of *perfluoroalkylmercury compounds* [*150*]. Functionalized analogues have been prepared similarly via this route [*151, 152*] (equation 117), and the preparation of bis(trifluoromethyl)mercury has been described [*153*].

An alternative route to the reaction of **mercuric fluoride** with *fluoroolefins* in liquid hydrogen fluoride [*154*] was developed during the early and middle 1970s. This improved method involved the reaction of fluoroolefins and mercury salts in the presence of alkali metal fluorides in aprotic solvents [*155, 156*] (equation 118).

[*152*] **117**

$$CF_3CFBrCF_2CO_2C_2H_5 + Hg \xrightarrow{heat} CF_3CF(CF_2CO_2C_2H_5)HgBr$$

50-60%

[*156*] **118**

$$2\ CF_3CF{=}CF_2 + 2\ KF + HgCl_2 \xrightarrow{DMF} [(CF_3)_2CF)]_2Hg$$

65%

Phenylmercury halides give the corresponding phenylmercury derivative, $C_6H_5HgC(CF_3)_3$ (61%) [*156*], and perfluorocyclobutene gives the corresponding cyclobutyl derivative [*156*]. Mechanistically, the reaction could be interpreted as formation of the fluorocarbanion via nucleophilic addition of fluoride ion to the fluoroolefin followed by capture of the intermediate fluorocarbanion by the mercury salt [*156*]. The regiochemistry of the reaction is consistent with this mechanism [*156*] (equation 119).

Mixed perfluoroorganomercury compounds can be prepared by a variation of this methodology [*157*] (equation 120).

[*156*] **119**

$$CF_3CF{=}CF_2 + F^- \rightleftharpoons (CF_3)_2CF^- \xrightarrow{HgX_2} [(CF_3)_2CF]_2Hg$$

[*157*] **120**

$$(CF_3)_2C{=}CF_2 + KF + CF_3HgOCOCF_3 \xrightarrow{THF} CF_3HgC(CF_3)_3$$

Similarly, $CF_3HgCBr(CF_3)_2$ was prepared in 61% yield from $CF_3CBr{=}CF_2$ [*157*], and $[CF_3CF{=}C(CF_3)]_2Hg$ was obtained in 66% yield from hexafluoro-2-butyne [*157*].

Other fluorocarbanion precursors, such as $(CF_3)_2CHCN$ [*158*] and 1-hydroperfluorobridgehead alkanes, give the corresponding mercury compound on reaction with base and mercury salts [*159*] (equation 121).The reaction of perfluoroalkyl Grignard reagents with mercury salts gives similar results [*159, 160*]. A variety of phenyl-substituted halofluoroalkyl mercurials have been synthesized by carbanion capture with phenyl mercuric halide [*161, 162, 163, 164, 165*] (equations 122 and 123), and this route has been the subject of a recent review [*166*].

[*159*] **121**

$$2\ \text{H–}(\text{F bicycle})\text{–H} + 2\ BuLi + HgCl_2 \longrightarrow (\text{H–}(\text{F bicycle})\text{–})_2Hg + 2\ BuH + 2\ LiCl$$

[*165*] **122**

$$C_6H_5HgCl + CF_3CHFBr + CH_3ONa \xrightarrow[THF,\ -35°C]{CH_3OH} \underset{65\%}{C_6H_5HgCFBrCF_3}$$

[*162*] **123**

$$C_6H_5HgCl + CH_3ONa + 2\ CHFBr_2 \xrightarrow[-25\ °C]{THF} \underset{50\text{-}55\%}{C_6H_5HgCFBr_2}$$

These precursors have received considerable attention as nonbasic precursors to halofluorocarbenes [*167*] (equation 124), and several reviews describe the scope of this mode of carbene generation [*166, 167, 168*].

[*162*] **124**

$$Z\text{-}CH_3CH{=}CHCH_3 + C_6H_5HgCFBr_2 \xrightarrow[\text{RT, 4 d}]{\text{benzene}} \text{1-bromo-1-fluoro-2,3-dimethylcyclopropane } (99\%)$$

Another major route to fluorinated organomercury compounds is *thermal or photochemical decarboxylation of fluorine-containing mercury carboxylates* [*153, 169, 170, 171, 172*], as shown for example in equation 125 [*153, 169*]. Via similar methodology, $C_6H_5HgCF_3$ (60–75%) [*171*], $(CF_3)_2Hg$ (92%) [*169*], $(O_2NCFCl)_2Hg$ (58%) [*172*], and $[(CF_3)_3C]_2Hg$ (80%) [*157*] were synthesized, and several of these mercurials have been used as fluorocarbene precursors [*166*].

[*153,169*] $$Hg(O_2CCF_3)_2 \xrightarrow[K_2CO_3]{\text{heat}} (CF_3)_2Hg \quad 55\text{-}90\%$$ **125**

The reaction of trifluoromethyl radicals, generated in a radio-frequency discharge process, with elemental mercury [*173*], mercury halides [*174*], dimethylmercury [*24*], or HgO [*175*] has been used for the preparation of CF_3HgX and $(CF_3)_2Hg$. Direct fluorination of dimethylmercury with elemental fluorine gives $(CF_3)_2Hg$ [*176*].

Trifluoromethylmercury compounds have been used to prepare other trifluoromethyl compounds [*124*] (equation 126). Related methodology gives $CF_3(CH_3)_2SnOCOCF_3$ [*177*], $(CF_3)_2Te$ (92%) [*178*], $(CF_3)_2Se$ (67%) [*178*], $(CF_3)_2Se_2$ (57%) [*178*], $(CF_3)_3Sb$ (63%) [*178*], $(CF_3)_3As$ (75%) [*178*], $(CF_3)_2AsI$ (54%) [*178*], $(CF_3)_3P$ (55%) [*178*], $(CF_3)PI_2$ (37%) [*178*], $(CF_3)_4Ge$ (22%) [*179*], CF_3GeI_3 (90%) [*179*], CF_3SnBr_3 (54%) [*6*], $(CF_3)_2SnBr_2$ (19%) [*6*], and $CF_3HgSi(C_2H_5)_3$ (67%) [*180*]. $(C_2H_5O)_2P(O)CF_2HgC_6H_5$ was prepared via an exchange reaction of $(C_2H_5O)_2P(O)CF_2ZnBr$ with C_6H_5HgCl [*98*].

[*124*] $$(CF_3)_2Hg + (CH_3)_2Cd \xrightarrow{\text{glyme}} (CF_3)_2Cd{\cdot}\text{glyme} \quad 57\%$$ **126**

Perfluorovinylmercury compounds can be prepared via the reaction of perfluorovinyl Grignard reagents with mercury salts [*16*] (equation 127); either mono- or bismercurials can be obtained. The use of alkyl or aryl mercury salts gives the mixed bismercurials, (*Z*)-$C_6F_{13}CF{=}CFHgC_2H_5$ (50%) [*16*] and (*Z*)-

$C_6F_{13}CF{=}CFHgC_6H_5$ (30%) [*16*]. Vinyllithium reagents behave similarly [*181*] (equation 128).

A decarboxylation route provides an alternative entry to perfluorovinylmercury compounds [*182*] (equation 129).

[*16*] **127**

$$\text{Z-}C_6F_{13}CF{=}CFMgBr + HgX_2 \xrightarrow[0\,°C,\ 5\,h]{\text{ether}} \text{Z-}C_6F_{13}CF{=}CFHgX \quad 35\text{-}65\%$$

$X = Cl, Br, OCOCF_3, NO_3$

[*181*] $$(CF_3)_2C{=}CFLi + HgCl_2 \longrightarrow (CF_3)_2C{=}CFHgCl$$ **128**

[*182*] **129**

$$(CF_3)_2C{=}C{=}O \xrightarrow[\text{2. } K_2CO_3 \text{ 3. heat}]{\text{1. } HgF_2} [CF_2{=}C(CF_3)]_2Hg\ (55\%) + CF_2{=}C(CF_3)HgCH(CF_3)_2\ (35\%)$$

Organocopper Compounds

The seminal work by McLoughlin and Thrower [*183*] stimulated extensive work in the preparation and reactions of fluorinated copper reagents, and this class of fluorinated organometallic compounds has received more attention than any other class. In the early work, DMSO was generally used as the solvent, although DMS, DMF, HMPA, and pyridine also have been used. The coordinating ability of the solvent plays a significant role in the mechanistic direction of the reaction. In solvents of low donor number (DN < 19), such as hexane, benzene, acetic anhydride, acetonitrile, and dioxane, perfluoroalkyl radicals are produced and can be trapped by olefins [*184*]. In solvents of high donor number (DN > 31), such as DMF, DMSO, pyridine, and HMPA, the perfluoroalkylcopper reagent is formed and can be successfully trapped by aryl iodides [*184*]. Consequently, solvents with a high donor number should be used for generation and capture of R_fCu [*184*] (equation 130). Although aryl iodides give higher yields than aryl bromides, for cost effectiveness, aryl bromides can often be utilized [*185, 186*] (equations 131 and 132).

[*184*] $$R_fI + ArI \xrightarrow[\text{DN} > 31,\ 110\text{-}120\,°C]{\text{solvent}} R_fAr$$ **130**

Typical perfluoroalkylated aromatic derivatives prepared via this type of reaction are illustrated in Table 6. This methodology has been used for the preparation 5-(perfluoroalkyl)pyrimidines [*194*], heptafluoroisopropylated aromatics

[*185*] **131**

$$C_6F_{13}I + Cu \xrightarrow[125\text{-}130\ °C]{DMSO} [C_6F_{13}Cu] \xrightarrow{p\text{-}BrC_6H_4CO_2C_2H_5} p\text{-}C_6F_{13}C_6H_4CO_2C_2H_5 \quad 93\%$$

[*186*] **132**

$$C_6F_{13}I + \text{3-bromopyridine} + 2\,Cu \xrightarrow[125\text{-}130\ °C,\ 4\,h]{DMSO} \text{3-}C_6F_{13}\text{-pyridine} \quad 82\%$$

[*195*], 5-(perfluoroalkyl)uracils [*196*], and perfluoroalkylated pyrimidine nucleosides [*197*].

In some coupling reactions of perfluorocopper reagents with haloheterocyclic compounds, nonregiospecific coupling processes were observed. For example, 3-bromobenzofuran gave six products when reacted with trifluoromethyl iodide and copper metal [*198*] (equation 133); the yields of the various products depend on the solvent used. The formation of 2-substituted benzofurans was explained by the addition of [CF_3^-] to the delocalized 2,3-double bond followed by migration of hydride and elimination of bromide ion [*198*]. The pentafluoroethyl compounds were proposed to arise from decomposition of CF_3Cu to C_2F_5Cu. Similar behavior was observed when 3-iodothiophene [*199*] or 3-bromofuran [*186*] was reacted with perfluoroalkyl iodides and copper (equation 134). A carbenoid mechanism was invoked to explain the rearranged product in the thiophene case [*199*].

Trifluoromethylation is the most important perfluoroalkylation reaction. The initial work was carried out with ***trifluoromethyl iodide*** and ***copper metal*** in the

[*198*] **133**

$$\text{3-bromobenzofuran} \xrightarrow[\text{solvent},\ 140°C]{CF_3I\,/\,Cu} \text{3-}CF_3\text{-benzofuran}\ (9\%) + \text{2,3-bis}(CF_3)\text{-benzofuran}\ (45\%) + \text{2-}CF_3\text{-benzofuran}\ (27\%)$$

$$+\ \text{3-}CF_2CF_3\text{-benzofuran}\ (9\%) + \text{2-}CF_2CF_3\text{-benzofuran}\ (9\%) + \text{benzofuran}\ (\text{trace})$$

presence of the ***aryl iodide*** [*192*, 193, 200] (Table 6). Subsequent improvements on this procedure involved using a solution of a trifluoromethylcopper complex, which was prepared by shaking ***trifluoromethyl iodide*** and **copper powder** in HMPA and filtering off the excess copper [*194, 201*]. ***Aryl halides, benzyl halides, allyl halides, vinyl halides,*** and halogenated nucleoside derivatives were trifluoromethylated via this procedure [*201*] (equation 135).

[*186*] **134**

3-Bromofuran + $C_6F_{13}I$ + Cu $\xrightarrow[125\ ^\circ C]{DMSO}$ 3-C_6F_{13}-furan (55%) + 2-C_6F_{13}-furan (35%)

Table 6. Perfluoroalkyl Aromatic Compounds Prepared via the Reaction of Perfluoroalkylcopper Reagents with Iodo- or Bromoaromatic Compounds

Aryl Halide	R_FI	*Product*	*Yield (%)*	*Reference*
1-I-3,5-bis[$C(CF_3)_2OH$]benzene ($HO(CF_3)_2C$, $C(CF_3)_2OH$, I)	C_3F_7I	1-C_3F_7-3,5-bis[$C(CF_3)_2OH$]benzene ($HO(CF_3)_2C$, $C(CF_3)_2OH$, C_3F_7)	51	187, 188
4-iodophthalonitrile (I, CN, CN)	$C_7F_{15}I$	4-C_7F_{15}-phthalonitrile (C_7F_{15}, CN, CN)	56	189
4-iodobiphenyl	CF_3I	4-R_F-biphenyl	92	190
	$C_6F_{13}I$		92	190
	C_3F_7I		90	190
iodobenzene	$CH_3O_2C(CF_2)_3I$	$C_6H_5(CF_2)_3CO_2CH_3$	72	191

Continued on next page.

Table 6—*Continued*

Aryl Halide	R_FI	*Product*	*Yield (%)*	*Reference*
Br N Br	$C_6F_{13}I$	C_6F_{13} N C_6F_{13}	89	186
Br N N	$C_6F_{13}I$	C_6F_{13} N N	71	186
Br O CO_2H	$C_6F_{13}I$	C_6F_{13} O CO_2H	51	186
I	CF_3I	CF_3	86	192
I CH_3 CH_3 CH_3 CH_3	CF_3I	CF_3 CH_3 CH_3 CH_3 CH_3	65	193
OCH_3 N CH_3O N Br	CF_3I	OCH_3 N CH_3O N CF_3	31	194
I	$Cl(CF_2)_4I$	$(CF_2)_4Cl$	86	184

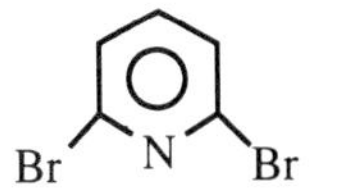

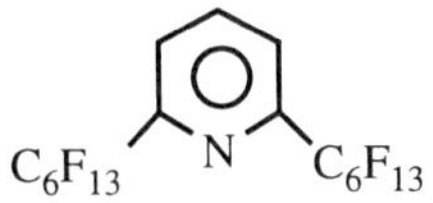

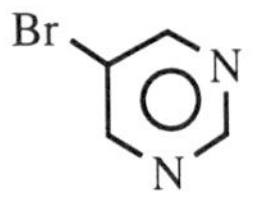

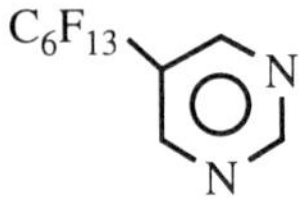

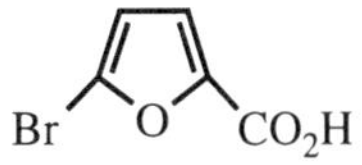

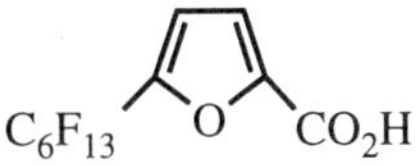

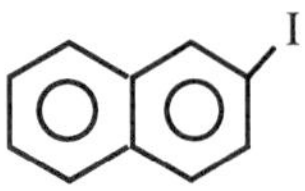

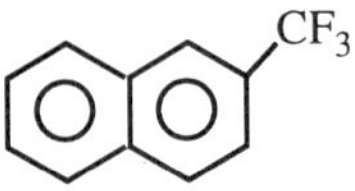

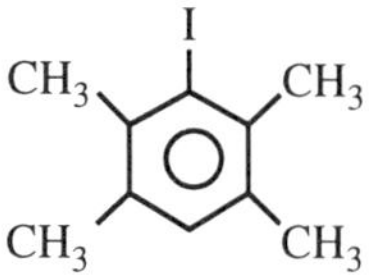

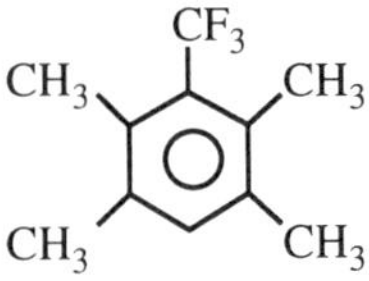

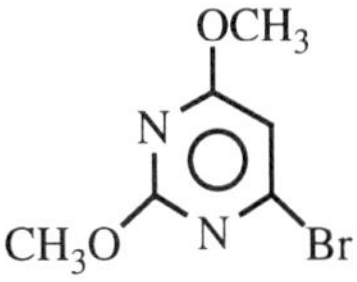

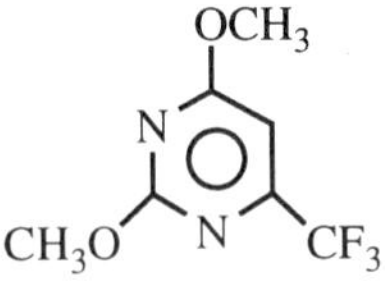

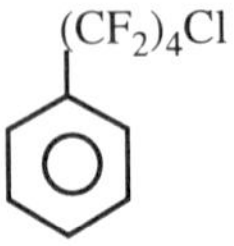

Cheaper sources of trifluoromethyl groups have been the goal of several groups. The use of *sodium trifluoroacetate and copper (I) iodide* in dipolar aprotic solvents gave regiospecific trifluoromethylation of aromatic halides [*202*] (equation 136).

[*201*] **135**

4-*tert*-butyl-1-(dibromomethylene)cyclohexane + CF_3Cu → (HMPA, RT, 20 h) 4-*tert*-butyl-1-[bis(trifluoromethyl)methylene]cyclohexane, 81%

[*202*] **136**

1-iodonaphthalene + CF_3CO_2Na + CuI → (NMP, 140-160 °C) 1-(trifluoromethyl)naphthalene, 78%

Electron-withdrawing groups and electron-releasing groups on the ring worked equally well under these conditions [*202*]. Later work with sodium trifluoroacetate and copper(I) iodide in HMPA at 150–180 °C with polymethyliodobenzenes gave similar results [*203*]. No trifluoromethylation occurred without the Cu(I) salt [*203*]. A crude Hammett study with substituted iodobenzenes indicated $[CF_3CuI^-]$ rather than $[CF_3CuI\cdot]$ as the reactive intermediate [*204*]. Although extension of this perfluoroalkylation reaction to higher perfluoroalkanoates was initially unsuccessful [*203*], subsequent studies with sodium pentafluoropropionate gave good results [*205*] (equation 137).

[*205*] **137**

4-iodonitrobenzene + $CF_3CF_2CO_2Na$ → (CuI, DMF, toluene, 145-150 °C, 2-4 h) 4-(CF_2CF_3)nitrobenzene, 58%

Aryl bromides were also perfluoroethylated under these conditions [*205*]. The key to improved yields was the azeotropic removal of water from the sodium perfluoroalkylcarboxylate [*205*]. Partial success was achieved with sodium heptafluorobutyrate [*204*]. Related work with halonaphthalene and anthracenes has been reported [*206, 207*]. The main limitation of this sodium perfluoroalkylcarboxylate methodology is the need for 2 to 4 equivalents of the salt to achieve reasonable yields.

A **trifluoromethylcopper solution** can be prepared by the reaction of **bis(trifluoromethyl)mercury** with **copper powder** in *N*-methylpyrrolidone (NMP) at 140 °C [*208*] (equation 138) or by the reaction of *N*-trifluoromethyl-*N*-nitrosotrifluoromethane sulfonamide with activated copper in dipolar aprotic solvents [*209*]. This trifluoromethylcopper solution can be used to trifluoromethylate aromatic [*209*], benzylic [*209*], and heterocyclic halides [*209*].

[*208*] **138**

$$(CF_3)_2Hg + Cu \xrightarrow[140\ °C]{NMP} [CF_3Cu] \xrightarrow[150\ °C]{p\text{-}O_2NC_6H_4CH_2Br} p\text{-}O_2NC_6H_4CH_2CF_3 \quad 58\%$$

Dihalodifluoromethanes react readily in DMF with activated **cadmium or zinc powders** to give a stable solution of **trifluoromethylcadmium or -zinc,** respectively [*33*] (equation 139).

Metathesis of the trifluoromethylcadmium or trifluoromethylzinc reagent with copper (I) salts gives solutions of **trifluoromethylcopper** [*210*] (equation 140). Depending on the stoichiometry and copper salt used, either $CF_3Cu{\cdot}L$ or $(CF_3)_2Cu^-$ can be produced [*211*].

Oxidation of $(CF_3)_2Cu^-$ with O_2, Br_2, or I_2 gives the stable copper(III) compound, $(CF_3)_4Cu^-$. The X-ray structure of an analogue of this copper(III) compound has been reported [*211*]. In the absence of HMPA, the solution of trifluoromethylcopper slowly decomposes at room temperature to give pentafluoroethylcopper [*210*] (equation 141).

At higher temperatures, an oligomeric mixture of perfluoroalkylcopper reagents is obtained [*212*] (equation 142).

[*33*]
$$2\ CF_2XY + M \xrightarrow[RT]{DMF} [CF_3MX] + (CF_3)_2M \qquad 80\text{-}95\% \tag{139}$$

X = Br, Cl M = Zn, Cd
Y = Br, Cl

[*210*]
$$[CF_3CdX] + [(CF_3)_2Cd] \xrightarrow[-50\ to\ RT]{3\ CuY} 3\ [CF_3Cu] \qquad 90\text{-}100\% \tag{140}$$

X = Br, Cl
Y = I, Br, Cl, CN

[*210*]
$$CF_3Cu \xrightarrow[RT]{DMF} CF_3CF_2Cu \tag{141}$$

[*212*]
$$[CF_3Cu] \xrightarrow[90\text{-}100\ °C]{DMF} CF_3(CF_2)_nCu \tag{142}$$

n = 1 to 14

A similar distribution of copper reagents can be obtained via the direct reaction of copper metal with dibromodifluoromethane or bromochlorodifluoromethane in DMF at 85–95 °C [*212*]. The oligomerization can be supressed via the addition of alkali metal fluorides to the reaction mixture [*212*]. When HMPA is added to the trifluoromethylcopper solution, decomposition is slowed, and this solution can be used to trifluoromethylate aromatic iodides [*210*] (equation 143).

[*210*] **143**

$$CF_2Br_2 \xrightarrow[\text{2. CuX; 3. HMPA}]{\text{1. Cd, DMF}} [CF_3Cu] \xrightarrow{m\text{-}IC_6H_4NO_2} m\text{-}CF_3C_6H_4NO_2 \quad 72\%$$

Clark has used this approach in N,N-dimethylacetamide (DMAC) to trifluoromethylate activated aryl chlorides possessing *ortho* groups capable of interacting with the metal [*213, 214*]. With less activated substrates, formation of higher-perfluoroalkyl-substituted products occurred; this problem could be partially circumvented by the addition of charcoal to the CF_2Br_2–DMAC–Cu reagent [*215*] (equation 144). Similar trifluoromethylation of activated aryl chlorides was achieved with the trifluorocopper reagent generated from CF_3SO_2Cl; however full details of this method have not yet been reported [*216*].

[*215*] **144**

$$CF_2Br_2 + 4\text{-}Cl\text{-}3\text{-}NO_2C_6H_3CF_3 + Cu \xrightarrow[\text{charcoal, 100 °C, 2 h}]{\text{DMAC}} 1,4\text{-}(CF_3)_2\text{-}2\text{-}NO_2C_6H_3 \quad 98\%$$

Trifluoromethylation of aryl, alkenyl, and alkyl halides can be accomplished by heating **methyl fluorosulfonyldifluoroacetate** and the ***appropriate halide*** precursor with **copper(I) iodide** at 60–80 °C in DMF [*217*] (equation 145). Similar trifluoromethylations of aryl, benzyl, and vinyl halides can be carried out with fluorosulfonyldifluoromethyl iodide and copper metal in DMF at 60–80 °C [*218*] (equation 146).

[*217*] **145**

$$FO_2SCF_2CO_2CH_3 + C_6H_5Br \xrightarrow[\text{80 °C, 3 h}]{\text{CuI DMF}} C_6H_5CF_3 \quad 61\%$$

[*218*] **146**

$$FO_2SCF_2I + p\text{-}ClC_6H_4I \xrightarrow[\text{80 °C, 6 h}]{\text{Cu, DMF}} p\text{-}CLC_6H_4CF_3 \quad 79\text{–}84\%$$

Trifluoromethylation of aryl iodides was carried out by the fluoride ion-induced cross-coupling reaction of aromatic iodides with trifluoromethyltrialkylsilanes in the presence of copper(I) salts [*219*] (equation 147). Some pentafluoro- ethyl derivative was also formed. This methodology was extended to pentafluoroethyl- and heptafluoropropyltriethylsilanes [*219*].

[*219*] **147**

$$\text{I-C}_6\text{H}_4\text{-NO}_2 \text{ (para)} + CF_3Si(C_2H_5)_3 \xrightarrow[\text{CuI, 80 °C}]{\text{DMF-NMP, KF}} \text{CF}_3\text{-C}_6\text{H}_4\text{-NO}_2 \text{ (para)} \quad 99\%$$

The **perfluoroalkylcopper reagents** have found extensive application as perfluoroalkylation reagents. Typical examples of perfluoroalkylation of aryl halides have been illustrated in table 6. These perfluoroalkylcopper reagents readily couple with vinyl iodides and vinyl bromides [*220, 221, 222*]. Typical examples of mono- or bisvinylic coupling are shown in equations 148 and 149. When 1-bromo-1-perfluoroalkylethylenes were reacted with perfluoroalkylcopper compounds, the expected *gem*-disubstituted olefin was not formed, but the *vic*-disubstituted product was obtained [*223*] (equation 150).

[*220*] **148**

$$C_7F_{15}I + Cu + E\text{-}ICH{=}CHI \text{ (excess)} \xrightarrow[\text{120 °C, 56 h}]{\text{DMF}} E\text{-}C_7F_{15}CH{=}CHI \quad 55\%$$

[*222*] **149**

$$C_6F_{13}I + Cu + C_6H_5CH{=}CHBr \xrightarrow[\text{110-120 °C, 16 h}]{\text{DMF}} C_6F_{13}CH{=}CHC_6H_5 \quad 70\text{-}90\%$$

[*223*] **150**

$$R_f^1CBr{=}CH_2 + R_f^2I + Cu \xrightarrow[\text{120 °C}]{\text{DMF}} R_f^1CH{=}CHR_f^2 \quad 60\text{-}70\%$$

Perfluorovinyl iodides readily undergo stereospecific coupling with the trifluoromethylcopper solution [*224*] (equation 151) prepared from dibromodifluoromethane [*210*]. With longer-chain perfluoroalkylcopper reagents, the coupling reaction is accompanied by some exchange processes [*225*] (equation 152).

[*224*] **151**

$$\text{Z-}C_3F_7CF{=}CFI + CF_3Cu \xrightarrow[-30\ °C]{DMF} \text{E-}C_3F_7CF{=}CFCF_3 \quad 52\%$$

[*225*] **152**

$$\text{Z-}CF_3CF{=}CFI \xrightarrow{C_3F_7Cu} \text{E-}CF_3CF{=}CFC_3F_7 + \text{Z,Z-}CF_3CF{=}CFCF{=}CFCF_3$$

76 : 24

Perfluoroalkylcopper reagents couple with ***1-iodoacetylenic derivatives*** to give the expected perfluoroalkyl-substituted acetylenes [*226*] (equation 153). The coupling reaction is complicated by formation of the diyne from competitive exchange processes.

Allylation is a facile process and will occur with any perfluoroalkylcopper reagents [*201, 224*]. Typical examples are shown below in equations 154 and 155.

[*226*] $$C_8F_{17}Cu + IC{\equiv}CC_6H_5 \xrightarrow{DMSO} C_8F_{17}C{\equiv}CC_6H_5 \quad 48\%$$ **153**

[*201*] **154**

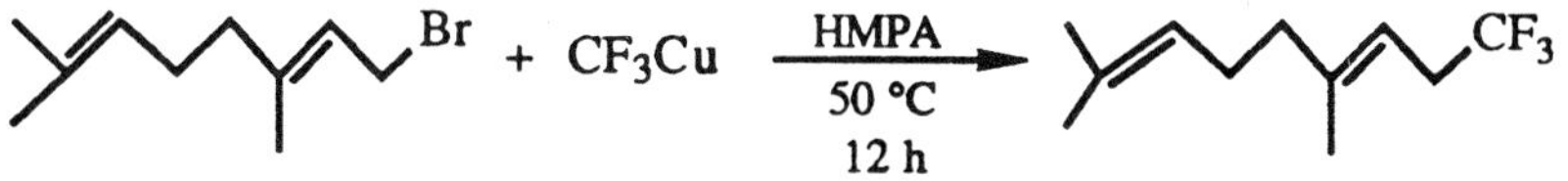

$$(CH_3)_2C{=}CHCH_2CH_2C(CH_3){=}CHCH_2Br + CF_3Cu \xrightarrow[50\ °C,\ 12\ h]{HMPA} (CH_3)_2C{=}CHCH_2CH_2C(CH_3){=}CHCH_2CF_3 \quad 50\%$$

[*224*] **155**

$$CF_3Cu + CH_2{=}CHCH(Cl)CH_3 \xrightarrow[0\ °C]{DMF} CF_3CH_2CH{=}CHCH_3$$

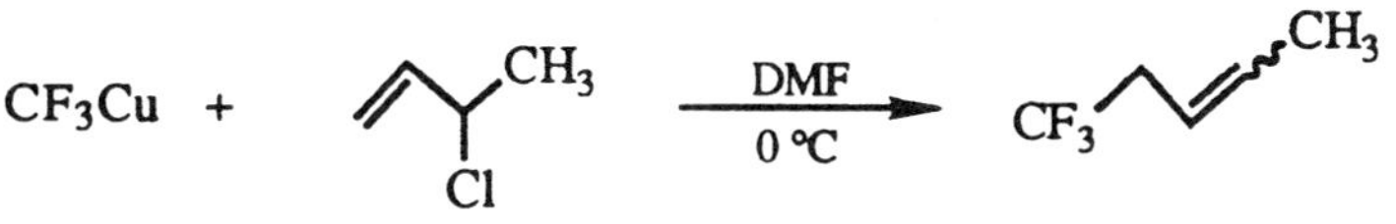

When perfluoroheptylcopper reacted with propargyl bromide, a violent reaction occurred, and less than 10% of the expected allene was obtained [*226*]. However, when propargyl chlorides or tosylates were used as substrates, the expected allenes were obtained in good yields [*227*] (equation 156).

[*227*] **156**

$$CF_3CF_2CF_2Cu + HC{\equiv}CCH(CH_3)OTs \xrightarrow[0\ °C]{DMSO} CF_3CF_2CF_2CH{=}C{=}CHCH_3 \quad 67\%$$

Functionalized allenes, such as $CF_3C(CO_2C_2H_5){=}C{=}C(CH_3)_2$ (57%) and $CF_3C[Si(CH_3)_3]{=}C{=}C(CH_3)_2$ (52%), could be prepared similarly [*227*]. Perfluoroalkylenedicopper reagents react with propargyl bromide to give perfluorodiallenes in good yields and high regioselectivity [*228*]. Difluoromethylcopper reacts similarly with propargyl chlorides and tosylates to give good yields of the corresponding difluoromethylallenes [*229*].

Perfluoroalkylcopper reagents react with ***thiocyanates*** to give ***perfluoroalkyl-substituted sulfides*** in low yields [*230*] and with benzoylformyl chloride to give the α-diketone in 49% yield [*231*]. However, other α-ketoacyl halides were prepared in less than 5% yield [*231*].

Perfluoroalkylation can be accomplished via direct reaction of ***perfluoroalkyl halides*** and **copper** with ***aromatic substrates*** [*232, 233, 234, 235, 236*]. Thus, perfluoroalkyl iodides or bromides react with functionalized benzenes in DMSO in the presence of copper bronze to give the corresponding perfluoroalkylated products directly in moderate to good yields [*233*] (equation 157). Mixtures of *ortho, meta,* and *para* isomers are obtained [*232, 233*]. The use of acetic anhydride as solvent gives similar results [*234, 235*]. Similarly, the direct reaction of perfluoroalkyl iodides and pyrroles with copper metal regiospecifically gives the 2-perfluoroalkylpyrroles [*236*] (equation 158).

[*233*] **157**

$$C_6H_5NHAc + C_8F_{17}I + Cu \xrightarrow[\text{100 °C, 16 h}]{\text{DMSO}} C_8F_{17}C_6H_4NHAc\ (58\%) + C_8F_{17}H\ (29\%)$$

[*236*] **158**

$$\text{pyrrole} + Cl(CF_2)_8I + Cu \xrightarrow[\text{90 °C, 10 h}]{\text{DMF}} \text{2-}(CF_2)_8Cl\text{-pyrrole}$$

Perfluoroalkylation of olefins and acetylenes occurs in low yields when **perfluoroalkylcopper reagents** are heated with ***alkenes or acetylenes*** in DMSO [*226, 237, 238, 239*]. Uridine derivatives react similarly [*240*].

Perfluoroalkylation of perfluoroalkylethylenes and addition of perfluoroalkyl iodides to olefins or acetylenes are catalyzed by copper metal [*238, 239*]. Similar copper-catalyzed addition of iododifluoroacetates to olefins has been observed [*241*].

Only few examples of functionalized copper reagents have been prepared. The perfluoroallylcopper reagent was prepared via metathesis of the corresponding perfluoroallylcadmium compound [*146*] (equation 159). Reaction of the allyl copper reagent with allyl bromide gave 1,1,2,3,3-pentafluoro-1,5-hexadiene in 86% yield [*146*].

Similar methodology has been used to prepare the dialkoxyphosphinyldifluoromethylcopper compound [*242*] (equation 160). Functionalization with allyl bromide, methyl iodide, and iodobenzene occurs readily, as well as stereospecific *syn* addition to hexafluoro-2-butyne [*243*].

The **methoxycarbonyldifluoromethylcopper reagent** is prepared from methyl iododifluoroacetate and copper in DMF, DMSO, or HMPA [*244*] (equation 161).

The copper species formed depends on the solvent, and three different species were detected by ^{19}F NMR, although the structure of each species was not elucidated [*245*]. This copper reagent undergoes a variety of coupling reactions with aryl, alkenyl, allyl, and acetylenic halides [*244, 245*] (equation 162).

[*146*] **159**

$$F_2C{=}CFCF_2I + Cd \xrightarrow[2.\ CuBr,\ -35\ °C]{1.\ DMF,\ 0\ °C} [F_2C{=}CFCF_2Cu]\quad 61\%$$

[*242*] **160**

$$(C_2H_5O)_2P(O)CF_2Br \xrightarrow[2.\ CuBr,\ DMF]{1.\ Cd,\ DMF} [(C_2H_5O)_2P(O)CF_2Cu]\quad 90\text{–}100\%$$

[*244*] $ICF_2CO_2CH_3 + Cu \longrightarrow [CuCF_2CO_2CH_3]$ **161**

[*245*] **162**

$$CuCF_2CO_2CH_3 \xrightarrow[HMPA]{Br\text{-}CH_2CH{=}CH\text{-}CO_2C_2H_5} CH_3O_2CCF_2\text{-}CH_2CH{=}CH\text{-}CO_2C_2H_5\quad 62\%$$

$$CuCF_2CO_2CH_3 \xrightarrow[HMPA]{Br\text{-}C{\equiv}C\text{-}CH_2CH_2\text{-}OTHP} CH_3O_2CCF_2\text{-}C{\equiv}C\text{-}CH_2CH_2\text{-}OTHP\quad 68\%$$

THP = tetrahydropyran

Reductive coupling reaction of fluorinated vinyl iodides or bromides has been used as a route to fluorinated dienes [*246, 247, 248, 249, 250*]. Generally, the vinyl iodide is heated with copper metal in DMSO or DMF; no intermediate perfluorovinylcopper reagent is detected. Typical examples are shown in equations 163–165 [*246, 247, 249*]. The X-ray crystal structure of perfluorotetracyclobutacyclooctatetraene, prepared via coupling of tetrafluoro-1,2-diiodocyclobutene with copper, is planar

[*248*]. Copper-promoted coupling of $R_f1CH{=}CIR_f2$ compounds gives the 1,2,3,4-tetrakis(perfluoroalkyl)-1,3-butadienes in 62–77% yields [*250*].

[*246*] **163**

$$\text{1-chloro-2-iodo-3,3,4,4-tetrafluorocyclobutene} + Cu \xrightarrow{180\,°} \text{2,2'-dichloro-octafluoro-bi(cyclobutenyl)} \quad 58\%$$

[*247*] **164**

$$I\text{–}C{=}C\text{–}C{=}C\text{–}I \ (\text{each } C{=}C \text{ bridged by } (CF_2)_3) \xrightarrow[\text{140 °C, 24 h}]{\text{Cu, DMF}} \text{tetrakis(perfluorocyclopenteno)cyclooctatetraene} \quad 91\%$$

[*249*] **165**

$$(CF_3)_2C{=}CICF_3 \xrightarrow{Cu} (CF_3)_2C{=}C(CF_3)C(CF_3){=}C(CF_3)_2$$

(assigned structure) 77%

The first example of a pregenerated **vinylcopper reagent** was reported by Miller, who reacted (*E*)-$CF_3CF{=}C(CF_3)Ag$ with **copper bronze** [*251*]. Later, (*Z*)-$CF_3CF{=}CFLi$ was captured with copper(I) trifluoroacetate to give (*E*)-$CF_3CF{=}CFCu$ [*252*]. In the silver precursor, the regiochemistry is dictated by the *addition of silver(I) fluoride to the perfluoroalkyne.* In the vinyllithium-exchange route, the low-temperature preparation of the unstable vinyllithium compound limits scale-up processes. Thus, both of these methods present serious problems. An alternative strategy that permits regio- and stereochemical control is the exchange reaction of the stable **vinylzinc** or **vinylcadmium reagents** with **copper(I) halides** [*145*] (equation 166).

[*145*] **166**

$$R_fCF{=}CFX + M \xrightarrow[RT]{DMF} R_fCF{=}CFMX \xrightarrow[RT]{CuX} R_fCF{=}CFCu$$

M = Cd, Zn 68-99%

Because the vinylzinc and vinylcadmium reagents can be prepared directly from the vinyl halides (I, Br) with zinc or cadmium metal, this route avoids cross-coupling processes and provides a one-pot in situ preparation of perfluorovinylcopper compounds. Table 7 shows examples of this method of preparation of vinylcopper reagents from the indicated cadmium or zinc reagent [*145*].

Table 7. Preparation of Fluorinated Vinylcopper Reagents via Exchange of Cu(I) Halides with Fluorine-Containing Vinylzinc or Cadmium Compounds [*145*]

Vinyl Halide	*Metal*	$R_fCF{=}CFCu$	*Yield (%)*
$F_2C{=}CFI$	Cd	$F_2C{=}CFCu$	99
$F_2C{=}CFBr$	Zn	$F_2C{=}CFCu$	72
(*E*)-$CF_3CF{=}CFI$	Cd	(*E*)-$CF_3CF{=}CFCu$	83
(*Z*)-$CF_3CF{=}CFI$	Zn	(*Z*)-$CF_3CF{=}CFCu$	76
(*Z*)-$CF_3(CF_2)_4CF{=}CFI$	Cd	(*Z*)-$CF_3(CF_2)_4CF{=}CFCu$	87
$CF_3C(C_6H_5){=}CFBr$	Zn	$CF_3C(C_6H_5){=}CFCu$	84

These copper reagents readily couple with a variety of electrophiles. For example, (*Z*)-$CF_3CF{=}CFCu$ couples with acetyl chloride (77%), iodobenzene (56%), benzyl bromide (56%), (*E*)-$CF_3C(C_6H_5){=}CFI$ (54%), allyl bromide (94%) and methyl iodide (87%). 2-Chlorohexafluorocyclopentenylcopper and 2-chlorotetrafluorocyclobutenylcopper also are prepared via exchange of the corresponding zinc reagent with copper(I) bromide [*111, 112, 253*]; they react with similar electrophiles as reported for (*Z*)-$CF_3CF{=}CFCu$ [*145*].

In contrast to the stability exhibited by the copper reagents containing an α-fluorine, the corresponding α-bromocopper reagent is not stable and decomposes to give an excellent yield of the triene (cumulene) product [*254*] (equation 167). The (*E*)- and (*Z*)-cumulenes were separated by chromatography, and the structures were assigned by X-ray analysis [*254*].

[*254*] **167**

$$CF_3(C_6H_5)C{=}CBr_2 \xrightarrow[\text{DMF}]{\text{Zn}} CF_3(C_6H_5)C{=}C(Br)ZnX$$

$$\downarrow$$

$$CF_3C(C_6H_5){=}C{=}C{=}C(C_6H_5)CF_3 \quad 72\%$$

Perfluoroacetylenic copper compounds also can be prepared via metathesis of the corresponding **zinc** or **cadmium reagents** with **copper(I) bromide.** The zinc or cadmium reagents are formed from the corresponding pentachloroperfluoroalkane or 1-iodoperfluoroalkyne [*121, 147*] (equations 168 and 169).

The perfluoroacetylenic copper compounds undergo coupling reactions with ***aryl iodides*** and provide a useful synthetic route to the *perfluoroalkyl aryl alkynes* [*147, 255*] (equation 170). Coupling of these copper reagents with the 1-iodoperfluoroalkynes gives the perfluorodiynes [*147, 255*] (equation 171).

[*121*] **168**

$$C_6F_{13}CCl_2CCl_3 \xrightarrow[\text{DMF}]{\text{3 Zn}} [C_6F_{13}C{\equiv}CZnCl] \xrightarrow{\text{CuBr}} [C_6F_{13}C{\equiv}CCu]$$

64-69% 100%

[*147*] **169**

$$C_4F_9C{\equiv}CI + Zn \xrightarrow{\text{DMF}} [C_4F_9C{\equiv}CZnI] \xrightarrow{\text{CuBr}} [C_4F_9C{\equiv}CCu]$$

69-72% 100%

[*147*] **170**

$$C_4F_9C{\equiv}CCu + m\text{-}O_2NC_6H_4I \xrightarrow[\text{130–150 °C}]{\text{DMF}} C_4F_9C{\equiv}CC_6H_4NO_2\text{-}m$$

76%

[*147*] $$R_fC{\equiv}CZnX \xrightarrow[\text{2) } R_fC{\equiv}CI]{\text{1) CuBr, DMF}} R_fC{\equiv}C\text{-}C{\equiv}CR_f$$ **171**

$R_f = C_4F_9$ 56% ; C_6F_{13} 67%; C_8F_{17} 70%

[*147*] **172**

$$R_f^1C{\equiv}CCu + (R_f^2)FC{=}CFI \longrightarrow (R_f^2)FC{=}CF(C{\equiv}CR_f^1)$$

$R_f^1 = C_4F_9$, $R_f^2 = CF_3$ 48%; $R_f^1 = C_4F_9$, $R_f^2 = C_5F_{11}$ 48%

[*147*] **173**

$$C_4F_9C{\equiv}CCu + (CF_3)FC{=}CFI \longrightarrow (CF_3)FC{=}CF(C{\equiv}CC_4F_9) \quad 40\%$$

The reaction of these acetylenic copper compounds with perfluorovinyl iodides stereospecifically gives the perfluoroeneynes [*147, 255*] (equations 172 and 173).

The hydrocarbon vinyl iodides behave similarly. The perfluoroacetylenic copper reagents react readily with allyl halides, and preferred attack is at the least hindered position [*147, 255*] (equation 174).

[*147*] **174**

$$CF_3C{\equiv}CCu + CH_2ClCH{=}CHCH_3 \longrightarrow CF_3C{\equiv}CCH_2CH{=}CHCH_3 \quad 90\%$$

$$+ \; CF_3C{\equiv}CCH(CH_3)CH{=}CH_2 \quad 10\%$$

Copper(I) halide-catalyzed coupling reactions of perfluoro Grignard reagents with allyl and propargyl halides have been reported [*256*]. The acetylenic copper compound may be an intermediate in these reactions.

Perfluoroarylcopper compounds are usually prepared from the corresponding ***perfluoroaryllithium, -magnesium,*** or ***-cadmium reagents*** and ***copper(I) halides*** [*257*] (equation 175).

The arylcopper reagents couple with 1-iodoarylacetylenes to give the unsymmetrical diarylacetylenes [*258*] (equation 176). Reaction with tetrabromoethyl- ene gives bis(pentafluorophenyl)acetylene in 66% yield [*258*] (equation 177). Pentafluorophenyl copper couples with (bromoethynyl)triethylsilane to give $C_6F_5C{\equiv}CSi(C_2H_5)_3$ in 85% yield [*259*].

[*257*] **175**

$$C_6F_5Br + Mg \xrightarrow[\text{reflux}]{\text{THF}} [C_6F_5MgBr] \xrightarrow[0\,°C]{\text{CuI}} [C_6F_5Cu]$$

[*258*] $$C_6F_5Cu + IC{\equiv}CC_6H_5 \xrightarrow{\text{ether}} C_6F_5C{\equiv}CC_6H_5 \quad 82\%$$ **176**

[*258*] $$2\,C_6F_5Cu + C_2Br_4 \xrightarrow{\text{THF}} C_6F_5C{\equiv}CC_6F_5 \quad 52\%$$ **177**

[*260*] **178**

$$X\text{-}C_6F_4Cu + F_2C{=}CFI \longrightarrow X\text{-}C_6F_4CF{=}CF_2 \quad 45\text{–}55\%$$

X = F, H, Br

Coupling of **perfluorophenylcopper compounds** with ***iodotrifluoroethylene*** gives the *trifluorovinylpolyfluoroaryl compounds* in moderate yields [*260*] (equation 178). The perfluoropyridyl-, perchloropyridyl-, and perchlorophenylcopper compounds behave similarly[*260*].

The **perfluoroarylcopper reagents** react with ***perfluoroalkylated acid fluorides*** to give *ketones* in excellent yields. Solvent, type of organometallic reagent, and type of acid fluoride (primary vs. secondary) influence the product yield [*261*] (equation 179). Similarly, $C_6H_5CH{=}CHCOC_6F_5$ was prepared in 52% yield via coupling of the corresponding acid chloride with pentafluorophenylcopper [*262*].

Polyfluorinated arylcopper reagents, prepared from the corresponding polyfluorinated arylcadmium compounds, add in a *syn* manner to hexafluoro-2-butyne to give the vinyl copper reagent, which reacts readily with electrophiles, such as acyl halides, methyl iodide, halogen, aryl iodides, and vinyl iodides [*122*] (equation 180).

[*261*] **179**

Cu-(F)-X + R_fOR_fCOF ⟶ COR_fOR_f-(F)-X

X = F, Br

R_fOR_f = $C_3F_7O[CF(CF_3)CF_2]_nCF(CF_3)$
$C_2F_5O[CF_2CF_2O]_nCF_2$
$CF_3O(CF_2O)_nCF_2$

[*122*] **180**

Br-(F)-X + Cd —DMF, RT→ CdX-(F)-X (50-98%) —CuBr, RT→ Cu-(F)-X (100%)

X = F, H, Br, OCH_3, CF_3

Cu-(F)-X —$CF_3C{\equiv}CCF_3$→ [$XC_6F_4(CF_3)C{=}C(Cu)CF_3$] (50-70%) —$E^+$→ $XC_6F_4(CF_3)C{=}C(E)CF_3$ (> 90%)

The corresponding (tetrafluorophenyl)biscopper has been prepared via two methods [*149, 263*] (equations 181 and 182). These biscopper reagents undergo allylation [*149, 263*], halogenation [*149*], and acylation [*149*] as expected.

The **perfluoroalkylthio- and pentafluorophenylthiocopper compounds** have been used for the introduction of the $-SR_f$ and $-SAr_f$ group into molecules [*264, 265, 266, 267*] for increased biological activity. Trifluoromethylthiocopper has been prepared via metathesis of $AgSCF_3$ with copper(I) halides [*264*] and the reaction of bis(trifluoromethyl)disulfide or bis(trifluoromethylthio)mercury with copper metal [*265*]. The thiocopper compounds undergo coupling with aryl iodides and bromides [*264, 267*] and heterocyclic iodides [*264, 265*]. Multiple trifluoromethylthio groups have been introduced via coupling reactions with di-, tri-, tetra-, penta-, and hexaiodoaromatics [*264, 265*]. Oxidation gives the corresponding sulfones [*264*] (equation 183).

An alumina-supported trifluoromethylthiocopper reagent gave improved yields of trifluoromethyl aryl sulfides in coupling reactions with this reagent [*268*] (equation 184).

[*263*] **181**

H, H; 1. BuLi, 2. CuI; Cu, Cu

[*149*] **182**

CdX, CdX + 2 CuBr; DMF, RT; Cu, Cu 100%

X = Br, —(F)—CdX

[*264*] **183**

CF_3S, SCF_3, I; CF_3SCu; CF_3S, SCF_3, SCF_3 60%; CrO_3; CF_3SO_2, SO_2CF_3, SO_2CF_3 80%

[*268*] **184**

I, NO_2 + CF_3SCu•alumina; 150 °C, t-$C_4H_9C_6H_5$; SCF_3, NO_2 100%

The analogous **trifluoromethylseleno and pentafluorophenylseleno copper compounds** are prepared via reaction of the corresponding diselenide with **copper metal** [*265, 269*]. Coupling with aryl iodides gives the arylselenium derivative [*265*] (equation 185).

[265] **185**

$$C_6F_5SeSeC_6F_5 \xrightarrow[RT]{2\,Cu} C_6F_5SeCu \xrightarrow{C_6I_6} C_6(SeC_6F_5)_6 \quad 75\%$$

Organosilver Compounds

Fluorinated olefins, such as chlorotrifluoroethylene, hexafluoropropene, perfluoroisobutylene, and hexafluorocyclobutene, react with **silver trifluoroacetate** in the presence of alkali metal fluorides to give *perfluoroalkylsilver compounds* [*270*] (equation 186).

The regiochemistry is determined by the regiochemistry of the fluoride ion addition reaction, that is, via the most stable perfluorocarbanion intermediate. Von Werner used a similar reaction to prepare silver compounds from perfluoro-2-methyl-2-butene and perfluoro-2-methyl-2-pentene [*271*]. Silver(I) fluoride adds to bis(trifluoromethyl)ketene in DMF without fluoride ion catalysis [*270*]. The analogous trifluorovinylsulfurpentafluoride reacts similarly to give the isolable pentafluorosulfur derivative [*272*] (equation 187).

[270] **186**

$$CF_3CF{=}CF_2 + CsF + AgO_2CCF_3 \xrightarrow{DMF} (CF_3)_2CFAg$$

[272]

$$SF_5CF{=}CF_2 + AgF \xrightarrow{CH_3CN} SF_5CF(CF_3)Ag{\cdot}CH_3CN$$

187

Codeposition of silver vapor with perfluoroalkyl iodides at –196 °C provides an alternative route to nonsolvated primary perfluoroalkylsilvers [*273*]. *Phosphine complexes of trifluoromethylsilver* are formed from the reaction of trimethylphosphine, silver acetate, and bis(trifluoromethyl)cadmium·glyme [*135*]. The perfluoroalkylsilver compounds react with halogens [*270*], carbon dioxide [*274*], allyl halides [*270, 274*], mineral acids and water [*275*], and nitrosyl chloride [*276*] to give the expected products. Oxidation with dioxygen gives ketones [*270*] or acyl halides [*270*]. Sulfur reacts via insertion of sulfur into the carbon–silver bond [*270*] (equation 188).

The perfluoroalkylsilver complexes exist in a dynamic equilibrium in solution with solvated silver ion and anionic perfluoroalkylsilver complexes such as $Ag[CF(CF_3)_2]^-$ [*277*]. The trifluoromethylated silver complex, $Ag(CF_3)_4^-$, is prepared via reaction of bis(trifluoromethyl)cadmium with silver nitrate in acetonitrile [*278*].

Trifluorovinyl- and (*E*)-pentafluoropropenylsilver compounds can be prepared via the exchange reaction of the corresponding **cadmium** compound with **silver trifluoroacetate** [*144*] (equation 189). The 2-pentafluoropropenyl silver compound can be synthesized via a similar exchange reaction of silver trifluoroacetate with 2-pentafluoropropenyllithium [*279*].

[*270*] **188**

$$(CF_3)_2C{=}CF_2 + AgO_2CCF_3 \xrightarrow[DMF]{KF} (CF_3)_3CAg \xrightarrow{i,\ ii} (CF_3)_3CSCH_2CH{=}CH_2 \quad 46\%$$

i : S_8
ii : $CH_2{=}CHCH_2I$

[*144*] $$CF_2{=}CFCdX + AgO_2CCF_3 \xrightarrow{DMF} F_2C{=}CFAg \quad 90\%$$ **189**

Miscellaneous Organofluorine Organometallics

Trifluoromethyl radicals, generated from hexafluoroethane in a radio-frequency discharge, react with thallium vapor to give 10–20% yields of nonsolvated $(CF_3)_2Tl$ [*280*], which readily forms triethylphosphine complexes. **Trifluoromethylindium, -gallium, and -aluminum compounds** are similarly formed [*280*]. However, the low stability of these compounds prevented their isolation [*280*]. The **perfluorophenylthallium**–dioxane complex, $(C_6F_5)_3Tl{\cdot}$dioxane, was prepared in 50% yield via reaction of $(C_6F_5)_2TlBr$ with copper in refluxing dioxane [*281*]. The aryl analogues also were prepared by reaction of thallic triflate with polyfluorobenzenes [*282*]. These polyfluorothallic compounds react with iodide ion to give the corresponding polyfluoroiodobenzenes [*282*] (equation 190).

[*282*] **190**

$$1,2,4,5\text{-}C_6H F_4 \xrightarrow{Tl(O_3SCF_3)_3} 2,3,5,6\text{-}F_4C_6H\text{-}Tl(O_3SCF_3)_2\ (60\%) \xrightarrow[H_2O]{NaI} 2,3,5,6\text{-}F_4C_6H\text{-}I\ (70\%)$$

References are listed on pages 718–728.

The radio-frequency discharge method can be used to prepare $(CF_3)_3Bi$ [*283*], $(CF_3)_2Te$ [*283*], and $(CF_3)_4Te_2$ [*283*] compounds. **Trifluoromethylgermanium** derivatives have also been prepared by direct fluorination [*284*] and exchange reactions of germanium tetraiodide with bis(trifluoromethyl)mercury [*179, 285*] or bis(trifluoromethyl)cadmium, respectively [*8, 124*]. Irradiation of a mixture of dimethyltellurium and perfluoroalkyl iodides gives $(R_f)_2Te$ and R_fTeCH_3 [*285*]. Exchange of $(CF_3)_2TeCl_2$ with bis(trifluoromethyl)cadmium in acetonitrile gives $(CF_3)_4Te$ [*137*]. **Bis(trifluoromethyl)tellurium** undergoes photochemical or thermal addition reactions with alkenes [*286*] and substitution reactions with aromatics [*287*] (equation 191).

Perfluoroalkyl- and perfluoroaryltitanium compounds were prepared in situ via reaction of the corresponding Grignard reagents with chlorotris(diethylamido)titanium [*288*]. Reaction of the titanium compounds with aldehydes resulted in fluoroalkylative amination [*288*] (equation 192).

Perfluoroalkyl or -aryl halides undergo oxidative addition with metal vapors to form nonsolvated fluorinated organometallic halides, and this topic has been the subject of a review [*289*]. Pentafluorophenyl halides react with Rieke **nickel, cobalt, and iron** to give bispentafluorophenylmetal compounds, which can be isolated in good yields as liquid complexes [*290*]. Rieke nickel can also be used to promote the reaction of pentafluorophenyl halides with acid halides [*291*] (equation 193).

[*287*] **191**

$$(CF_3)_2Te + p\text{-}CH_3C_6H_4CH_3 \xrightarrow{h\nu} 2\text{-}CF_3\text{-}1,4\text{-}(CH_3)_2C_6H_3 \quad 37\%$$

[*288*] **192**

$$R_fTi[N(C_2H_5)_2]_3 + RCHO \longrightarrow R_fCH[N(C_2H_5)_2]R \quad 52\text{-}84\%$$

[*291*] **193**

$$C_6F_5I + CH_3(CH_2)_6COCl \xrightarrow{Ni} C_6F_5CO(CH_2)_6CH_3 \quad 59\%$$

References for Pages 670–718

1. Santini, G.; Le Blanc, M.; Riess, J. G. *J. Chem. Soc., Chem. Commun.* **1975,** 678.
2. Santini, G.; Le Blanc, M.; Riess, J. G. *J. Organomet. Chem.* **1977,** *140,* 1.

3. Klabunde, K. J.; Low, J. Y. F.; Key, M. S. *J. Fluorine Chem.* **1972/73,** *2,* 207.
4. Lange, H.; Naumann, D. *J. Fluorine Chem.* **1985,** *27,* 309.
5. Lange, H.; Naumann, D. *J. Fluorine Chem.* **1984,** *26,* 93.
6. Lagow, R. J.; Eujen, R.; Gerchman, L. L.; Morrison, J. A. *J. Am. Chem. Soc.* **1978,** *100,* 1722.
7. Krause, L. J.; Morrison, J. A. *Inorg. Chem.* **1980,** *19,* 604.
8. Krause, L. J.; Morrison, J. A. *J. Am. Chem. Soc.* **1981,** *103,* 2995.
9. Juhlke, T. J.; Braun, D. W.; Bierschenk, T. R.; Lagow, R. J. *J. Am. Chem. Soc.* **1979,** *101,* 3229.
10. Lagow, R. J.; Liu, E. K. S. *Inorg. Chem.* **1978,** *17,* 618.
11. Jacob, R. A.; Lagow, R. J. *J. Chem. Soc., Chem. Commun.* **1973,** 84.
12. Kitazume, T.; Ishikawa, N. *Chem. Lett.* **1981,** 1337.
13. Kesling, H. S. Ph.D. Thesis, University of Iowa, 1975.
14. Dailey, W. P.; Ralli, P.; Wasserman, D.; Lemal, D. M. *J. Org. Chem.* **1989,** *54,* 5516.
15. Moss, R. A.; Smudin, D. J. *Tetrahedron Lett.* **1974,** 1829.
16. Redwane, N.; Moreau, P.; Commeyras, A. *J. Fluorine Chem.* **1982,** *20,* 699.
17. Sorokina, R. S.; Rybakova, L. F.; Kalinovskii, I. O.; Chernoplekova, V. A.; Beletskaya, I. P. *Zh. Org. Khim.* **1982,** *8,* 2458 (Engl. Transl. 2180).
18. Farina, V.; Baker, S. R.; Benigni, D. A.; Sapino, C., Jr. *Tetrahedron Lett.* **1988,** *29,* 5739.
19. Farina, V.; Baker, S. R.; Benigni, D. A.; Hauck, S. I.; Sapino, C., Jr. *J. Org. Chem.* **1990,** *55,* 5833.
20. Koola, J. D.; Kunze, U. *J. Organomet. Chem.* **1974,** *77,* 325.
21. Cullen, W. R.; Waldman, M. C. *J. Fluorine Chem.* **1971/72,** *1,* 41.
22. Zavgorodnii, V. S.; Polozov, B. V.; Kondrat'ev, Y. V.; Bogoradovskii, E. T.; Petrov, A. A. *Zh. Obshch. Khim.* **1977,** *47,* 2074; *Chem. Abstr.* **1978,** *88,* 6083a.
23. Eujen, R.; Lagow, R. J. *J. Chem. Soc., Dalton Trans.* **1978,** 541.
24. Guerra, M. A.; Armstrong, R. L.; Bailey, W. E., Jr.; Lagow, R. J. *J. Organomet. Chem.* **1983,** *254,* 53.
25. Shigeru, T.; Hideo, T.; Shiro, Y. *Chem. Abstr.* **1989,** *110,* 94119k, 94120d.
26. Miller, W. T., Jr.; Bergman, E.; Fainberg, A. H. *J. Am. Chem. Soc.* **1957,** *79,* 4159.
27. Keller, T. M.; Tarrant, P. *J. Fluorine Chem.* **1975,** *6,* 105.
28. Keller, T. M.; Tarrant, P. *J. Fluorine Chem.* **1975,** *6,* 297.
29. Eapen, K. C. *J. Fluorine Chem.* **1987,** *35,* 421.
30. Blancou, H.; Moreau, P.; Commeyras, A. *Tetrahedron* **1977,** *33,* 2061.
31. Blancou, H.; Commeyras, A. *J. Fluorine Chem.* **1982,** *20,* 255.
32. Lange, H.; Naumann, D. *J. Fluorine Chem.* **1984,** *26,* 435.
33. Burton, D. J.; Wiemers, D. M. *J. Am. Chem. Soc.* **1985,** *107,* 5014.

34. Dolbier, W. R., Jr.; Wojtowicz, H.; Burkholder, C. R. *J. Org. Chem.* **1990,** *55,* 5420.
35. Francese, C.; Tordeux, M.; Wakselman, C. *Tetrahedron Lett.* **1988,** *29,* 1029.
36. Liu, E. K. S. *Inorg. Chem.* **1980,** *19,* 266.
37. Liu, E. K. S.; Asprey, L. B. *J. Organomet. Chem.* **1979,** *169,* 249.
38. Sekiya, A.; Ishikawa, N. *Chem. Lett.* **1977,** 81.
39. Ishikawa, N.; Takahashi, M.; Sato, T.; Kitazume, T. *J. Fluorine Chem.* **1983,** *22,* 585.
40. Kitazume, T.; Ishikawa, N. *Chem. Lett.* **1982,** 1453.
41. Kitazume, T.; Ishikawa, N. *Chem. Lett.* **1982,** 137.
42. Kitazume, T.; Ishikawa, N. *J. Am. Chem. Soc.* **1985,** *107,* 5186.
43. O'Reilly, N. J.; Maruta, M.; Ishikawa, N. *Chem. Lett.* **1984,** 517.
44. Kitazume, T.; Ikeya, T. *J. Org. Chem.* **1988,** *53,* 2349.
45. Francese, C.; Tordeux, M.; Wakselman, C. *J. Chem. Soc., Chem. Commun.* **1987,** 642.
46. Tordeux, M.; Francese, C.; Wakselman, C. *J. Chem. Soc., Perkin Trans. 1* **1990,** 1951.
47. Lang, R. W. *Helv. Chim. Acta* **1988,** *71,* 369.
48. Tordeux, M.; Francese, C.; Wakselman, C. *J. Fluorine Chem.* **1989,** *43,* 27.
49. Grondin, J.; Blancou, H.; Commeyras, A. *J. Fluorine Chem.* **1989,** *45,* 349.
50. Wakselman, C.; Tordeux, M. *Bull. Soc. Chim. Fr.* **1986,** *6,* 868.
51. Blancou, H.; Moreau, P.; Commeyras, A. *J. Chem. Soc., Chem. Commun.* **1976,** 885.
52. Shuyama, H.; Ogawa, T.; Takahashi, M.; Hamada, M.; Yoshimitsu, M.; Oyama, K. Eur. Pat. 271 212, 1988; *Chem. Abstr.* **1988,** *109,* 210556f.
53. Benefice, S.; Blancou, H.; Commeyras, A. *Tetrahedron* **1984,** *40,* 1541.
54. Benefice-Malouet, S.; Blancou, H.; Commeyras, A. *J. Fluorine Chem.* **1985,** *30,* 171.
55. Blancou, H.; Commeyras, A. *J. Fluorine Chem.* **1982,** *20,* 267.
56. Posta, A.; Paleta, O. *Collect. Czech. Chem. Commun.* **1972,** *37,* 3946.
57. Bellus, D.; Klingert, B.; Lang, R. W.; Rihs, G. *J. Organomet. Chem.* **1988,** *339,* 17.
58. Lang, R. W. *Helv. Chim. Acta* **1986,** *69,* 881.
59. Fujita, M.; Morita, T.; Hiyama, T. *Tetrahedron Lett.* **1986,** *27,* 2135.
60. Fujita, M.; Hiyama, T. *Bull. Chem. Soc. Jpn.* **1987,** *60,* 4377.
61. Meussduerffer, J. N.; Niederpruem, H. Ger. Offen. 1 900 *758,* 1970; *Chem. Abstr.* **1970,** *73,* 87470g.
62. Hemer, I.; Havlicek, J.; Dedek, V. *J. Fluorine Chem.* **1986,** *34,* 241.
63. Fujita, M.; Hiyama, T. *Tetrahedron Lett.* **1986,** *27,* 3655.
64. Hiyama, T.; Sato, K.; Fujita, M. *Bull. Chem. Soc. Jpn.* **1989,** *62,* 1352.

65. Fujita, M.; Hiyama, T.; Kondo, K. *Tetrahedron Lett.* **1986,** *27,* 2139.
66. Fujita, M.; Kondo, K.; Hiyama, T. *Bull. Chem. Soc. Jpn.* **1987,** *60,* 4385.
67. Allmendinger, T.; Lang, R. W. *Tetrahedron Lett.* **1991,** *32,* 339.
68. Ishihara, T.; Kurobashi, M. *Chem. Lett.* **1987,** 1971.
69. Fried, J.; Hallinan, E. A.; Szwedo, M. J., Jr. *J. Am. Chem. Soc.* **1984,** *106,* 3871.
70. Fried, J.; Hallinan, E. A. *Tetrahedron Lett.* **1984,** *25,* 2301.
71. Burton, D. J.; Easdon, J. C. *J. Fluorine Chem.* **1988,** *38,* 125.
72. Kitagawa, O.; Taguchi, T.; Kobayashi, Y. *Tetrahedron Lett.* **1988,** *29,* 1803.
73. Fried, J.; John, V.; Szwedo, M. J., Jr.; Chen, C.-K.; O'Yang, C.; Morinelli, T. A.; Okwu, A. K.; Halushka, P. V. *J. Am. Chem. Soc.* **1989,** *111,* 4510.
74. Hertel, L. W.; Kroin, J. S.; Misner, J. W.; Tustin, J. M. *J. Org. Chem.* **1988,** *53,* 2406.
75. Matsuda, F.; Matsumoto, T.; Ohsaki, M.; Terashima, S. *Tetrahedron Lett.* **1989,** *30,* 4259.
76. Morikawa, T.; Nishiwaki, T.; Nakamura, K.; Kobayashi, Y. *Chem. Pharm. Bull.* **1989,** *37,* 813.
77. Gelb, M. H.; Svaren, J. P.; Abeles, R. H. *Biochemistry* **1985,** *24,* 1813.
78. Imperiali, B.; Abeles, R. H. *Biochemistry* **1987,** *26,* 4474.
79. Thaisrivongs, S.; Pals, D. T.; Kati, W. M.; Turner, S. R.; Thomasco, L. M. *J. Med. Chem.* **1985,** *28,* 1553.
80. Whitten, J. P.; Barney, C. L.; Huber, E. W.; Bey, P.; McCarthy, J. R. *Tetrahedron Lett.* **1989,** *30,* 3649.
81. Schirlin, D.; Baltzer, S.; Altenburger, J. M. *Tetrahedron Lett.* **1988,** *29,* 3687.
82. Sham, H. L.; Rempel, C. A.; Stein, H.; Cohen, J. *J. Chem. Soc., Chem. Commun.* **1990,** 904.
83. Damon, D. B.; Hoover, D. J. *J. Am. Chem. Soc.* **1990,** *112,* 6439.
84. Sham, H. L.; Wideburg, N. E.; Spanton, S. G.; Kohlbrenner, W. E.; Betebenner, D. A.; Kempf, D. J.; Norbeck, D. W.; Plattner, J. J.; Erickson, J. W. *J. Chem. Soc., Chem. Commun.* **1991,** 110.
85. Takahashi, L. H.; Radhakrishnan, R.; Rosenfeld, R. E., Jr.; Meyer, E. F., Jr.; Trainor, D. A. *J. Am. Chem. Soc.* **1989,** *111,* 3368.
86. Taguchi, T.; Kitagawa, O.; Suda, Y.; Ohkawa, S.; Hashimoto, A.; Iitaka, Y.; Kobayashi, Y. *Tetrahedron Lett.* **1988,** *29,* 5291.
87. Lang, R. W.; Schaub, B. *Tetrahedron Lett.* **1988,** *29,* 2943.
88. Greuter, H.; Lang, R. W.; Romann, A. J. *Tetrahedron Lett.* **1988,** *29,* 3291.
89. Kitagawa, O.; Hashimoto, A.; Kobayashi, Y.; Taguchi, T. *Chem. Lett.* **1990,** 1307.
90. Kitagawa, O.; Taguchi, T.; Kobayashi, Y. *Tetrahedron Lett.* **1988,** *29,* 1803.
91. Yang, Z. Y.; Burton, D. J. *J. Org. Chem.* **1991,** *56,* 1037.

92. Hanzawa, Y.; Inazawa, K.; Kon, A.; Aoki, H. *Tetrahedron Lett.* **1987,** *28,* 659.
93. Burton, D. J.; Ishihara, T.; Maruta, M. *Chem. Lett.* **1982,** 755.
94. Burton, D. J.; Sprague, L. G.; Pietrzyk, D. J.; Edelmuth, S. H. *J. Org. Chem.* **1984,** *49,* 3438.
95. Burton, D. J.; Sprague, L. G. *J. Org. Chem.* **1988,** *53,* 1523.
96. Burton, D. J.; Sprague, L. G. *J. Org. Chem.* **1989,** *54,* 613.
97. Chambers, R. D.; O'Hagan, D.; Lamont, R. B.; Jain, S. C. *J. Chem. Soc., Chem. Commun.* **1990,** 1053.
98. Sprague, L. G.; Burton, D. J.; Guneratne, R. D.; Bennett, W. E. *J. Fluorine Chem.* **1990,** *49,* 75.
99. Lindell, S. D.; Turner, R. M. *Tetrahedron Lett.* **1990,** *31,* 5381.
100. Gillet, J. P.; Sauvetre, R.; Normant, J. F. *Tetrahedron Lett.* **1985,** *26,* 3999.
101. Gillet, J. P.; Sauvetre, R.; Normant, J. F. *Synthesis* **1986,** 538.
102. Martinet, P.; Sauvetre, R.; Normant, J. F. *J. Organomet. Chem.* **1989,** *367,* 1.
103. Morken, P. A.; Lu, H.; Nakamura, A.; Burton, D. J. *Tetrahedron Lett.* **1991,** *32,* 4271.
104. Tellier, F.; Sauvetre, R.; Normant, J. F. *J. Organomet. Chem.* **1985,** *292,* 19.
105. Normant, J. F. *J. Organomet. Chem.* **1990,** *400,* 19.
106. Tellier, F.; Sauvetre, R.; Normant, J. F. *J. Organomet. Chem.* **1986,** *303,* 309.
107. Tellier, F.; Sauvetre, R.; Normant, J. F. *J. Organomet. Chem.* **1987,** *328,* 1.
108. Tellier, F.; Sauvetre, R.; Normant, J. F. *J. Organomet. Chem.* **1987,** *331,* 281.
109. Tellier, F.; Sauvetre, R.; Normant, J. F. *J. Organomet. Chem.* **1989,** *364,* 17.
110. Hansen, S. W.; Spawn, T. D.; Burton, D. J. *J. Fluorine Chem.* **1987,** *35,* 415.
111. Shin, S. K.; Choi, S. K. *J. Fluorine Chem.* **1989,** *43,* 439.
112. Choi, S. K.; Jeong, Y. T. *J. Chem. Soc., Chem. Commun.* **1988,** 1478.
113. Heinze, P. L.; Burton, D. J. *J. Fluorine Chem.* **1986,** *31,* 115.
114. Heinze, P. L.; Burton, D. J. *J. Org. Chem.* **1988,** *53,* 2714.
115. Dolbier, W. R., Jr.; Koroniak, H.; Burton, D. J.; Heinze, P. L. *Tetrahedron Lett.* **1986,** *37,* 4387.
116. Dolbier, W. R., Jr.; Koroniak, H.; Burton, D. J.; Heinze, P. L.; Bailey, A. R.; Shaw, G. S.; Hansen, S. W. *J. Am. Chem. Soc.* **1987,** *109,* 219.
117. Spawn, T. D.; Burton, D. J. *Bull. Soc. Chim. Fr.* **1986,** 876.
118. Finnigan, W. G.; Norris, W. P. *J. Org. Chem.* **1963,** *28,* 1139.
119. Bunch, J. E.; Bumgardner, C. L. *J. Fluorine Chem.* **1987,** *36,* 313.

120. Yoneda, N.; Matsuoka, S.; Miyaura, N. *Bull. Chem. Soc. Jpn.* **1990,** *63,* 2124.
121. Burton, D. J.; Spawn, T. D. *J. Fluorine Chem.* **1988,** *38,* 119.
122. MacNeil, Kathryn J. Ph.D. Thesis, University of Iowa, 1991.
123. Dyatkin, B. L.; Martynov, B. I.; Knunyants, I. L.; Sterlin, S. R.; Fedorov, L. A.; Stumbrevichute, Z. A. *Tetrahedron Lett.* **1971,** *18,* 1345.
124. Krause, L. J.; Morrison, J. A. *J. Chem. Soc., Chem. Commun.* **1980,** 671.
125. Ontiveros, C. D.; Morrison, J. A. *Inorg. Synth.* **1986,** *24,* 55.
126. Krause, L. J.; Morrison, J. A. *J. Am. Chem. Soc.* **1981,** *103,* 2995.
127. Lange, H.; Naumann, D. *J. Fluorine Chem.* **1984,** *26,* 1.
128. Lange, H.; Naumann, D. *J. Fluorine Chem.* **1988,** *41,* 185.
129. Heinze, P. L.; Burton, D. J. *J. Fluorine Chem.* **1985,** *29,* 359.
130. Hartgraves, G. A.; Burton, D. J. *J. Fluorine Chem.* **1988,** *39,* 425.
131. Chen, G. J.; Tamborski, C. *J. Fluorine Chem.* **1987,** *36,* 123.
132. Naumann, D.; Tyrva, W. *J. Organomet. Chem.* **1989,** *368,* 131.
133. Osman, A.; Tuck, D. G. *J. Organomet. Chem.* **1979,** *169,* 255.
134. Sanner, R. D.; Satcher, J. H.; Droege, M.W. *Organometallics* **1989,** *8,* 1498.
135. Nair, H. K.; Morrison, J. A. *J. Organomet. Chem.* **1989,** *376,* 149.
136. Ontiveros, C. D.; Morrison, J. A. *Organometallics* **1986,** *5,* 1446.
137. Naumann, D.; Wilkes, B. *J. Fluorine Chem.* **1985,** *27,* 115.
138. Naumann, D.; Tyrra, W.; Leifeld, F. *J. Organomet. Chem.* **1987,** *333,* 193.
139. Burton, D. J.; Takei, R.; Shin-Ya, S. *J. Fluorine Chem.* **1981,** *18,* 197.
140. Chambers, R. D.; Jaouhari, R.; O'Hagan, D. *J. Fluorine Chem.* **1989,** *44,* 275.
141. Chambers, R. D.; Jaouhari, R.; O'Hagan, D. *Tetrahedron* **1989,** *45,* 5101.
142. Burton, D. J.; Modak, A. S.; Guneratne, R. D.; Su, D.; Cen, W.; Kirchmeier, R. L.; Shreeve, J. M. *J. Am. Chem. Soc.* **1989,** *111,* 1773.
143. Burton, D. J.; Hansen, S. W. *J. Fluorine Chem.* **1986,** *31,* 461.
144. Hansen, S. W. Ph.D. Thesis, University of Iowa, 1984.
145. Burton, D. J.; Hansen, S. W. *J. Am. Chem. Soc.* **1986,** *108,* 4229.
146. Burton, D. J.; Tarumi, Y.; Heinze, P. L. *J. Fluorine Chem.* **1990,** *50,* 257.
147. Spawn, T. D. Ph.D. Thesis, University of Iowa, 1987.
148. Evans, D. F.; Phillips, R. E. *J. Chem. Soc., Dalton Trans.* **1973,** 978.
149. Burton, D. J.; Yang, Z. Y.; MacNeil, K. J. *J. Fluorine Chem.* **1991,** *52,* 251.
150. Banus, J.; Emeleus, H. J.; Haszeldine, R. N. *J. Chem. Soc.* **1950,** 3041.
151. Fields, R.; Haszeldine, R. N.; Hubbard, A. F. *J. Chem. Soc., Perkin Trans. 1* **1972,** 847.
152. Kim, Y. K.; Pierce, O. R. *J. Organomet. Chem.* **1969,** *19,* P11.
153. Eugen, R. *Inorg. Synth.* **1986,** *24,* 52.
154. Aldrich, P. E.; Howard, E. G.; Linn, W. J.; Middleton, W. J.; Sharkey, W. H. *J. Org. Chem.* **1963,** *28,* 184.

155. Dyatkin, B. L.; Sterlin, S. R.; Martynov, B. I., Knunyants, I. L. *Tetrahedron Lett.* **1970,** *17,* 1387.
156. Dyatkin, B. L.; Sterlin, S. R.; Martynov, B. I., Mysov, E. I.; Knunyants, I. L. *Tetrahedron* **1971,** *27,* 2843.
157. Martynov, B. I.; Sterlin, S. R.; Dyatkin, B. L. *Izv. Akad. Nauk SSSR* **1974**, 1642 (Engl. Transl. 1564).
158. Aktaev, N. P.; Butin, K. P.; Sokol'skii, G. A.; Knunyants, I. L. *Izv. Akad. Nauk SSSR* **1974,** 636; *Bull. Acad. Sci. USSR* 1**974,** *23,* 600; Knunyants, I. L.; Aktaev, N. P.; Semenov, V. P.; Sokol'skii, G. A. *Dokl. Akad. Nauk SSSR (Eng. Transl.)* **1974,** *219,* 771.
159. Hardwick, F.; Pedler, A. E.; Stephens, R.; Tatlow, J. C. *J. Fluorine Chem.* **1974,** *4,* 9.
160. Zissis, J. P.; Moreau, P.; Commeyras, A. *J. Fluorine Chem.* **1981/82,** *19,* 71.
161, Seyferth, D.; Murphy, G. J. *J. Organomet. Chem.* **1973,** *49,* 117.
162. Seyferth, D.; Hopper, S. P. *J. Organomet. Chem.* **1973,** *51,* 77.
163. Seyferth, D.; Woodruff, R. A. *J. Fluorine Chem.* **1972/73,** *2,* 214.
164. Seyferth, D.; Murphy, G. J. *J. Organomet. Chem.* **1973,** *52,* C1.
165. Seyferth, D.; Murphy, G. J.; Woodruff, R. A. *J. Organomet. Chem.* **1975,** *92,* 7.
166. Larock, R. C. *Organomercury Compounds in Organic Synthesis;* Springer-Verlag: Berlin, 1985.
167. Seyferth, D. *Acc. Chem. Res.* **1972,** *5,* 65.
168. Seyferth, D. In *Carbenes,* Vol. 2, Moss, R. A.; Jones, M., Jr., Eds.; Wiley: New York, 1975; pp 101-158.
169. Knunyants, I. L.; Komissarov, Y. F.; Dyatkin, B. L.; Lantseva, L. T. *Izv. Akad. Nauk SSSR Ser. Khim.* **1973,** 943 (Engl. Transl. 912).
170. Kagramanov, N. D.; Mal'tsev, A. K.; Nefedov, O. M. *Izv. Akad. Nauk SSSR, Ser. Khim.* **1977,** 1835 (Engl. Transl. *Bull. Acad. Sci. USSR* **1977,** *26,* 1697).
171. Seyferth, D.; Hopper, S. P.; Murphy, G. J. *J. Organomet. Chem.* **1972,** *46,* 201.
172. Martynov, I. V.; Brel, V. K.; Postnova, L. V.; Martynov, B. I. *Izv. Akad. Nauk SSSR, Ser. Khim.* **1984,** 2833 (Engl. Transl. *Bull. Acad. Sci. USSR* **1985,** *33,* 2597).
173. Eujen, R.; Lagow, R. D. *Inorg. Chem.* **1975,** *14,* 3128.
174. Lagow, R. J.; Gerchman, L. L.; Jacob, R. A.; Morrison, J. A. *J. Am. Chem. Soc.* **1975,** *97,* 518.
175. Schmeisser, V. M.; Walter, R.; Naumann, D. *Z. Anorg. Allg. Chem.* **1980,** *464,* 233.
176. Lagow, R. J.; Liu, E. K. S. *J. Am. Chem. Soc.* **1976,** *98,* 8270.
177. Petrosyan, V. S.; Permin, A. B.; Sacharov, S. G.; Reutov, O. A. *J. Organomet. Chem.* **1974,** *65,* C7.

178. Ganja, E. A.; Ontiveros, C. D.; Morrison, J. A. *Inorg. Chem.* **1988,** *27,* 4535.
179. Morrison, J. A.; Gerchman, L. L.; Eujen, R.; Lagow, R. J. *J. Fluorine Chem.* **1977,** *10,* 333.
180. Petrov, B. I.; Kruglaya, O. A.; Kalinia, G. S.; Vyazankin, N. S.; Martynov, B. I.; Sterlin, S. R.; Dyatkin, B. L. *Izv. Akad. Nauk SSSR, Ser. Khim.* **1973,** 189 (Engl. Transl. 196).
181. Postovoi, S. A.; Zeifman, Y. V.; Knunyants, I. L. *Izv. Akad. Nauk SSSR* **1982,** 2826 (Engl. Transl. 2498).
182. Dyatkin. B. L.; Zhuravkova, L. G.; Martynov, B. I.; Mysov, E. I.; Sterlin, S. R.; Knunyants, I. L. *J. Organomet. Chem.* **1971,** *31,* C15.
183. McLoughlin, V. C. R.; Thrower, J. *Tetrahedron* **1969,** *25,* 5921.
184. Chen, Q.-Y.; Yang, Z.-Y.; He, Y.-B. *J. Fluorine Chem.* **1987,** *37,* 171.
185. Chen, G. J.; Tamborski, C. *J. Fluorine Chem.* **1989,** *43,* 207.
186. Chen, G. J.; Tamborski, C. *J. Fluorine Chem.* **1990,** *46,* 137.
187. Griffith, J. R.; O'Rear, J. G. *Synthesis* **1974,** 493.
188. Sepiol, J.; Soulen, R. L. *J. Fluorine Chem.* **1984,** *24,* 61.
189. Keller, T. M.; Griffith, J. R. *J. Fluorine Chem.* **1978,** *12,* 73.
190. Fialkov, Y. A.; Shelyazhenko, S. V.; Yagupol'skii, L. M. *Zh. Org. Khim.* **1983,** *19,* 1048 (Engl. Transl. 933).
191. Ankudinov, A. K.; Ryazanova, R. M. *Zh. Org. Khim.* **1972,** *8,* 212 (Engl. Transl. 217).
192. Hosokawa, K.; Inukai, K. *Nippon Kagaku Kaishi* **1976,** 1791; *Chem. Abstr.* **1977,** *86,* 139662m.
193. Lukmanov, V. G.; Alekseeva, L. A.; Yagupol'skii, L. M. *Zh. Org. Khim.* **1974,** *10,* 2000 (Engl. Trans. 2019).
194. Kobayashi, Y.; Yamamoto, K.; Asai, T.; Nakano, M.; Kumadaki, I. *J. Chem. Soc., Perkin Trans. 1* **1980,** 2755.
195. Ishikawa, N.; Ochial, M. *Nippon Kagaku Kaishi* **1973,** 2351; *Chem. Abstr.* **1974,** *80,* 59599g.
196. Cech, D.; Wohlfeil, R.; Etzold, G. *Nucleic Acids Res., Spec. Publ.* **1975,** S5.
197. Kobayashi, Y.; Kumadaki, I.; Yamamoto, K. *J. Chem. Soc., Chem. Commun.* **1977,** 536.
198. Kobayashi, Y.; Kumadaki, I. *J. Chem. Soc., Perkin Trans. 1* **1980,** 661.
199. Leroy, J.; Rubinstein, M.; Wakselman, C. *J. Fluorine Chem.* **1985,** *27,* 291.
200. Kobayashi, Y.; Kumadaki, I. *Tetrahedron Lett.* **1969,** 4095.
201. Kobayashi, Y.; Kumadaki, I.; Yamamoto, K. *Tetrahedron Lett.* **1979,** 4071.
202. Matsui, K.; Tobita, E.; Ando, M.; Kondo, K. *Chem. Lett.* **1981,** 1719.
203. Suzuki, H.; Yoshida, Y.; Osuka, A. *Chem. Lett.* **1983,** 135.
204. Carr, G. E.; Chambers, R. D.; Holmes, T. F. *J. Chem. Soc., Perkin Trans. 1* **1988,** 921.

205. Freskos, J. N. *Synth. Commun.* **1988,** *18,* 965.
206. Lin, R. W. L.; Davidson, R. I. Eur. Pat. 307 519, 1989; *Chem. Abstr.* **1989,** *111,* 133807a.
207. Mintas, M.; Gusten, H.; Williard, P. G. *J. Photochem. Photobiol., A: Chem.* **1989,** *48,* 341.
208. Kondratenko, N. V.; Vechirko, E. P.; Yagupol'skii, L. M. *Synthesis* **1980,** 932.
209. Umemoto, T.; Ando, A. *Bull. Chem. Soc. Jpn.* **1986,** *59,* 447.
210. Wiemers, D. M.; Burton, D. J. *J. Am. Chem. Soc.* **1986,** *108,* 832.
211. Willert-Porada, M. A.; Burton, D. J.; Baenziger, N. C. *J. Chem. Soc., Chem. Commun.* **1989,** 1633.
212. Easdon, J. C. Ph.D. Thesis, University of Iowa, 1987.
213. Clark, J. H.; McClinton, M. A.; Blade, R. J. *J. Chem. Soc., Chem. Commun.* **1988,** 638.
214. Clark, J. H.; Denness, J. E.; McClinton, M. A.; Wynd, A. J. *J. Fluorine Chem.* **1990,** *50,* 411.
215. Clark, J. H.; McClinton, M. A.; Jones, C. W.; Landon, P.; Bishop, D.; Blade, R. J. *Tetrahedron Lett.* **1989,** *30,* 2133.
216. Heaton, C. A.; Powell, R. L. *J. Fluorine Chem.* **1989,** *45,* 86.
217. Chen, Q.-Y.; Wu, S.-W. *J. Chem. Soc., Chem. Commun.* **1989,** 705.
218. Chen, Q.-Y.; Wu, S.-W. *J. Chem. Soc., Chem. Commun.* **1989,** 2385.
219. Urata, H.; Fuchikami, T. *Tetrahedron Lett.* **1991,** *32,* 91.
220. Coe, P. L.; Milner, N. E.; Smith, J. A. *J. Chem. Soc., Perkin Trans. 1* **1975,** 654.
221. Burdon, J.; Coe, P. L.; Marsh, C. R.; Tatlow, J. C. *J. Chem. Soc., Perkin Trans. 1* **1972,** 639.
222. Pedler, A. E.; Smith, R. C.; Tatlow, J. C. *J. Fluorine Chem.* **1971/72,** *1,* 337.
223. Santini, G.; Le Blanc, M.; Riess, J. S. *Tetrahedron* **1973,** *29,* 2411.
224. Wiemers, D. M. Ph.D. Thesis, University of Iowa, 1987.
225. Heinze, P. L. Ph.D. Thesis, University of Iowa, 1986.
226. Coe, P. L.; Milner, N. E. *J. Organomet. Chem.* **1974,** *70,* 147.
227. Burton, D. J.; Hartgraves, G. A.; Hsu, J. *Tetrahedron Lett.* **1990,** *31,* 3699.
228. Hung, M.-H. *Tetrahedron Lett.* **1990,** *31,* 3703.
229. Hartgraves, G. A.; Burton, D. J. 3rd Chemical Congress of North America, Toronto, Canada, June 1988, Abstract FLUO #30.
230. Nguyen, T.; Rubinstein, M.; Wakselman, C. *J. Org. Chem.* **1981,** *46,* 1938.
231. Hudlický, M. *J. Fluorine Chem.* **1981,** *18,* 383.
232. Coe, P. L.; Milner, N. E. *J. Fluorine Chem.* **1972/73,** *2,* 167.
233. Fuchikami, T.; Ojima, I. *J. Fluorine Chem.* **1983,** *22,* 541.
234. Chen, Q.; Chen, Y.; Huang, W. *Huaxue Xuebao* **1984,** *42,* 906; *Chem. Abstr.* **1985,** *102,* 5317t.

235. Chen, Q.; Yang, Z. *Huaxue Xuebao* **1985,** *43,* 1073; *Chem. Abstr.* **1986,** *105,* 171929a.
236. Chen, Q.-Y.; Qiu, Z. *Youji Huaxue* **1987,** *1,* 44; *Chem. Abstr.* **1987,** *107,* 134149s.
237. Coe, P. L.; Milner, N. E. *J. Organomet. Chem.* **1972,** *39,* 395.
238. Chen, Q.-Y.; Yang, Z.-Y. *J. Fluorine Chem.* **1985,** *28,* 399.
239. Le Blanc, M.; Santini, G.; Guion, J.; Riess, J. G. *Tetrahedron* **1973,** *29,* 3195.
240. Cech, D.; Herrmann, R.; Staske, R.; Langen, P.; Preussel, B. *J. Prakt. Chem.* **1979,** *321,* 488.
241. Yang, Z. Y.; Burton, D. J. *J. Fluorine Chem.* **1989,** *44,* 435.
242. Maruta, M., University of Iowa, unpublished results.
243. Guneratne, R., University of Iowa, unpublished results.
244. Taguchi, T.; Kitagawa, O.; Morikawa, T.; Nishiwaki, T.; Uehara, H.; Endo, H.; Kobayashi, Y. *Tetrahedron Lett.* **1986,** *27,* 6103.
245. Kitagawa, O.; Taguchi, T.; Kobayashi, Y. *Chem. Lett.* **1989,** 389.
246. Soulen, R. L.; Choi, S. K.; Park, J. D. *J. Fluorine Chem.* **1973/74,** *3,* 141.
247. Wu, A. W.; Choi, S. K.; Park, J. D. *J. Fluorine Chem.* **1979,** *13,* 379.
248. Einstein, F. W. B.; Willis, A. C. *J. Chem. Soc., Chem. Commun.* **1981,** 526.
249. Evans, H. H.; Fields, R.; Haszeldine, R. N.; Illingworth, M. *J. Chem. Soc., Perkin Trans. 1* **1973,** 649.
250. Jeanneaux, F.; Santini, G.; Le Blanc, M.; Cambon, A.; Riess, J. G. *Tetrahedron* **1974,** *30,* 4197.
251. Miller, W. T., Jr., 9th International Symposium on Fluorine Chemistry, Avignon, France, 1979, Abstr. # 0 27.
252. Bailey, A. R., University of Iowa, unpublished results.
253. Jeong, Y. T.; Choi, S. K. *Bull. Korea Chem. Soc.* **1989,** *10,* 619.
254. Morken, P. A.; Baenziger, N. C.; Burton, D. J.; Bachand, P. C.; Davis, C. R.; Pedersen, S. D.; Hansen, S.W. *J. Chem. Soc., Chem. Commun.* **1991,** 566.
255. Burton, D. J.; Spawn, T. D. 10th ACS Winter Fluorine Conference, St. Petersburg Beach, Florida, January 1991, Abstr. #16.
256. Turbanova, E. S.; Razumnaya, S. N.; Petrov, A. A. *Zh. Org. Khim.* 1975, *11,* 2219 (Engl. Transl. 2250).
257. Cairncross, A.; Sheppard, W. A.; Wonchoba, E. *Org. Synth.* **1980,** *59,* 122.
258. Gastinger, R. G.; Tokas, E. F.; Rausch, M. D. *J. Org. Chem.* **1978,** *43,* 159.
259. Waugh, F.; Walton, D. R. M. *J. Organomet. Chem.* **1972,** *39,* 275.
260. Soloski, E. J.; Ward, W. E.; Tamborski, C. *J. Fluorine Chem.* **1972/73,** *2,* 361.
261. Gopal, H.; Tamborski, C. *J. Fluorine Chem.* **1979,** *13,* 337.
262. Dua, S. S.; Homrajani, C. *Curr. Sci.* **1981,** *50,* 1067.
263. Brooke, G. M.; Mawson, S. D. *J. Fluorine Chem.* **1990,** *50,* 101.

264. Yagupol'skii, L. M.; Kondratenko, N. V.; Sambur, V. P. *Synthesis* **1975,** 721.
265. Kondratenko, N. V.; Kolomeytsev, A. A.; Popov, V. I.; Yagupol'skii, L. M. *Synthesis* **1985,** 667.
266. Remy, D. C.; Rittle, K. E.; Hunt, C. A.; Freedman, M. B. *J. Org. Chem.* **1976,** *41,* 1644.
267. Johnston, L. J.; Peach, M. E. *J. Fluorine Chem.* **1978,** *12,* 41.
268. Clark, J. H.; Jones, C. W.; Kybett, A. P.; McClinton, M. A. *J. Fluorine Chem.* **1990,** *48,* 249.
269. Yagupol'skii, L. M. *J. Fluorine Chem.* **1987,** *36,* 1.
270. Dyatkin, B. L.; Martynov, B. I.; Martynova, L. G.; Kizim, N. G.; Sterlin, S. R.; Stumbrevichute, Z. A.; Fedorov, L. A. *J. Organomet. Chem.* **1973,** *57,* 423.
271. Probst, A.; Raab, K.; Ulm, K.; Werner, K. von *J. Fluorine Chem.* **1987,** *37,* 223.
272. Noftle, R. E.; Fox, W. B. *J. Fluorine Chem.* **1977,** *9,* 219.
273. Klabunde, K. J. *J. Fluorine Chem.* **1976,** *7,* 95.
274. Dubot, G.; Mansuy, D.; Lecolier, S.; Normant, J. F. *J. Organomet. Chem.* **1972,** *42,* C105.
275. Miller, W. T., Jr.; Burnard, R. J. *J. Am. Chem. Soc.* **1968,** *90,* 7367.
276. Banks, R. E.; Dickinson, N.; Morrissey, A. P.; Richards, A. *J. Fluorine Chem.* **1984,** *26,* 87.
277. Burch, R. R.; Calabrese, J. C. *J. Am. Chem. Soc.* **1986,** *108,* 5359.
278. Dukat, W.; Naumann, D. *Rev. Chim. Min.* **1986,** *23,* 589.
279. Lu, H. Ph.D. Thesis, University of Iowa, 1991.
280. Bierschenk, T. R.; Juhlke, T. J.; Bailey, W. I., Jr.; Lagow, R. J. *J. Organomet. Chem.* **1984,** *277,* 1.
281. Deacon, G. B.; Vince, D. G. *J. Fluorine Chem.* **1975,** *5,* 87.
282. Deacon, G. B.; Tunaley, D. *Aust. J. Chem.* **1979,** *32.* 737.
283. Juhlke, T. J.; Braun, R. W.; Bierschenk, T. R.; Lagow, R. J. *J. Am. Chem. Soc.* **1979,** *101,* 3229.
284. Liu, E. K. S.; Lagow, R. J. *J. Organomet. Chem.* **1978,** *145,* 167.
285. Lagow, R. J.; Eujen, R.; Gerchman, L. L.; Morrison, J. A. *J. Am. Chem. Soc.* **1978,** *100,* 1722.
286. Denniston, M. L.; Martin, D. R. *J. Inorg. Nucl. Chem.* **1975,** *37,* 1871.
287. Naumann, D.; Kischkewitz, J. *J. Fluorine Chem.* **1990,** *47,* 283.
288. Chen, Q. Y.; Wu, J. P. *J. Chem. Res. (S)* **1990,** 268.
289. Klabunde, K. J. *Acc. Chem. Res.* **1975,** *8,* 393.
290. Kavaliunas, A. V.; Rieke, R. D. *J. Am. Chem. Soc.* **1980,** *102,* 5944.
291. Inaba, S.; Rieke, R. D. *J. Org. Chem.* **1985,** *50,* 1373.

Additions:

Linear Additions across Double Bonds

Additions Forming Carbon–Oxygen Bonds

by L. G. Sprague

Because of their electron-deficient nature, fluoroolefins are often nucleophilically attacked by alcohols and alkoxides. Ethers are commonly produced by these addition and addition–elimination reactions. The wide availability of alcohols and fluoroolefins has established the generality of the nucleophilic addition reactions. The mechanism of the addition reaction is generally believed to proceed by attack at a vinylic carbon to produce an intermediate fluorocarbanion as the rate-determining slow step. The intermediate carbanion may react with a proton source to yield the saturated addition product. Alternatively, the intermediate carbanion may, by elimination of β-halogen, lead to an unsaturated ether, often an enol or vinylic ether. These addition and addition–elimination reactions have been previously reviewed [*1*, *2*]. The intermediate carbanions resulting from nucleophilic attack on fluoroolefins have also been trapped in situ with carbon dioxide, carbonates, and esters of fluorinated acids [*3*, *4*, *5*] (equations 1 and 2).

[*5*] 1

$$CF_2{=}CF_2 \xrightarrow[\text{THF, } 40^{\circ}\text{C, 6 h}]{CH_3SCF_2CF_2CO_2CH_3,\ CH_3ONa} CH_3SCF_2CF_2COCF_2CF_2OCH_3 \quad 53\%$$

[*3*] 2

$$CF_2{=}CF_2 \xrightarrow[40^{\circ}\text{C, 4 h}]{(CH_3O)_2CO,\ CH_3ONa} (CH_3OCF_2CF_2)_2CO \ 75\% \quad + \quad CH_3OCF_2CF_2CO_2CH_3 \ 17\%$$

The nucleophilic reaction of bromotrifluoroethene with alkoxides yields not only the expected addition and addition–elimination products but also a product from a bromophilic reaction of the carbanion intermediate [*6*] (equation 3). Similar are the reactions of sodium phenoxide with perfluorovinyl ethers in the presence of hexachloroethane or selected vicinal dibromoperfluoroalkanes. The intermediate carbanion is trapped in high yield by these sources of Cl^+ or Br^+, which suggests a

0065–7719/95/0187–0729$09.44/1

relative resistance toward β-elimination of fluoride ion [*7*]. The conformation of the carbanion has been suggested as the cause of slow elimination [*4*].

[*6*] 3

$$CF_2{=}CFBr \xrightarrow[\text{THF, RT, 1 h}]{(CH_3)_3COK} \begin{array}{l} (CH_3)_3COCF{=}CFBr\ \ 51\% \\ (CH_3)_3COCF_2CFBr_2\ \ 30\% \\ (CH_3)_3COCF_2CFHBr\ \ 4\% \end{array}$$

Base-catalyzed reaction of alcohol favors addition, whereas increasing amounts of alkoxide favor addition–elimination reactions. Perfluoro-2-methyl-2-pentene and methanol form the saturated ether, whereas two equivalents of sodium methoxide form the vinylic ether [*8*] (equation 4).

[*8*] 4

$$(CF_3)_2CHCF(OCH_3)C_2F_5\ (>80\%) \xleftarrow[-10^{o}C]{CH_3OH,\ KOH} (CF_3)_2C{=}CFC_2F_5 \xrightarrow[0^{o}\text{ to }-10^{o}C]{2\,CH_3OH,\ CH_3OH} CH_3OCF{=}C(CF_3)CF(OCH_3)C_2F_5\ (60-70\%)$$

Miller et al. [*9*] hypothesized rules on the regioselectivity of addition from the study of the base-catalyzed addition of alcohols to chlorotrifluoroethylene. *Attack occurs at the vinylic carbon with most fluorines.* Thus, isomers of dichlorohexafluorobutene react with methanol and phenol to give the corresponding saturated and vinylic ethers. The nucleophiles exclusively attack position 3 of 1,1-dichloro-1,2,3,4,4,4-hexafluoro-2-butene and position 1 of 4,4-dichloro-1,1,2,3,3,4-hexafluoro-1-butene [*10*]. In 1,1-dichloro-2,3,3,4,4,4-hexafluoro-1-butene, attack on position 2 is favored [*11*] (equation 5). Terminal fluoroolefins are almost invariably attacked at the difluoromethylene group, as illustrated by the reaction of sodium methoxide with perfluoro-1-heptene in methanol [*12*] (equation 6).

[*11*] 5

$$CF_3CF_2CF{=}CCl_2 \xrightarrow[10^{o}C,\ 12\,h]{CH_3ONa,\ CH_3OH} \begin{array}{l} CF_3CF_2CF{=}CCl_2\ \ 14\% \\ CF_3CF_2CF(OCH_3)CCl_2H\ \ 68\% \\ CF_3CF_2CF{=}CCl(OCH_3)\ \ 9\% \end{array}$$

The orientation of nucleophilic attack on a fluoroolefin is influenced also by the presence on the fluoroolefin of electron-withdrawing functional groups that stabilize the intermediate carbanion. Thus nucleophilic attack tends to be oriented to the β carbon. Methanol and "acidic" nucleophiles such as acetic acid and trifluoroacetic acid add to fluoroolefins with two geminal electron-withdrawing groups such as perfluoromethacrylyl fluoride or its methyl ester to yield 3,3-difluoro-3-methoxy-2-trifluoromethylpropanoic acid [*13*] and the respective methyl 3-acetoxy-3,3-difluoro-2-trifluoromethylpropanoates [*14*]. Highly substituted fluoroolefins react slowly with nucleophiles because of steric hindrance [*15*].

[*12*] 6

$$CF_2{=}CFCF_2C_4F_9 \xrightarrow[50^{\circ}C,\ 1\ h]{CH_3ONa,\ CH_3OH} CH_3OCF_2CFHCF_2C_4F_9\ (45\%) + CH_3OCF_2CF{=}CFC_4F_9\ (40\%,\ E{:}Z=1{:}7) + CH_3OCF{=}CFCF_2C_4F_9\ (15\%,\ E{:}Z=1{:}3)$$

The addition reactions of allylic and acetylenic alcohols produce compounds resulting from rearrangements [*16, 17*] (equation 7).

[*16*] 7

$$2\ CF_2{=}C(CF_3)_2 \xrightarrow[(C_2H_5)_3N,\ DMF]{HOCH_2C{\equiv}CCH_2OH} CCH_2OCF_2CH(CF_3)_2 {\equiv} CCH_2OCF_2CH(CF_3)_2\ (41\%)$$

+ (4-vinylidene-3,3-bis(trifluoromethyl)dihydrofuran-2-one: ring O, C=O, $C(CF_3)_2$, $C{=}C{=}CH_2$) 25%

The *addition of nucleophiles* to ***cyclic fluoroolefins*** has been reviewed by Park et al. [*18*]. The reaction with alcohols proceeds by addition-elimination to yield the *cyclic vinylic ether,* as illustrated by the reaction of 1,2-dichloro-3,3-difluorocyclopropene. Further reaction results in cyclopropane ring opening at the bond opposite the difluoromethylene carbon to give preferentially the methyl and ortho esters of (*Z*)-3-chloro-2-fluoroacrylic acid and a small amount of dimethyl malonate [*19*] (equation 8).

A polyfluorinated β,γ-unsaturated ketone is formed in situ from tributylamine and 3,4-bis(trifluoromethyl)-3-(pentafluoroethyl)-5,5,6,6,6-pentafluoro-2-hexanone. The enol form of the unsaturated ketone cyclizes via an intermolecular addition–elimination reaction that involves exclusive attack by oxygen rather than by carbon. This reaction demonstrates the "hardness" of a F–C= site toward

[*19*] 8

CH_3O OCH_3

CH_3ONa, CH_3OH / $0^{o}C$, 2 h

$2\,CH_3ONa, CH_3OH$ / $0^{o}C$, 3 h

F F

Cl OCH_3

F F

63 %

Z-$CHCl{=}CFCO_2CH_3$ 31 %

Z-$CHCl{=}CFC(OCH_3)_3$ 11 %

$CH_3O_2CCH_2CO_2CH_3$ 0.4 %

[*20*] 9

$CF_3CF_2CH(CF_3)C(CF_3)(C_2F_5)COCH_3$ $\xrightarrow[\text{tetraglyme}]{(C_4H_9)_3N}$

CF_3 CF_3 CF_2CF_3 CF_3CF_2 F ^{-}O $\xrightarrow{-F^{-}}$ CF_3 CF_3 CF_2CF_3 CF_3 O

83 %

$\xrightarrow[\text{Pt}]{600-610^{o}C}$ CF_3 CF_3 CF_3 O $CH_2CF_2CF_3$

74 %

nucleophilic attack. Pyrolysis converts the initial product to the isomeric furan derivative [*20*] (equation 9).

Pyridine-1-oxide and some of its ring-substituted derivatives act as nucleophiles and generally give the corresponding 2-(1,2,2,2-tetrafluoro)pyridines upon reaction with ***hexafluoropropene*** [*21, 22*]. The reaction involves a novel rearrangement, as illustrated in the proposed mechanism (equation 10 and Table 1). Initial attack of the fluoroolefin by the oxygen produces an intermediate 1,5 dipole, followed by collapse and extrusion of carbonyl fluoride.

Inorganic and perfluoroorganic hypohalites add across olefinic bonds (equation 11 and Table 1). The reactions of hypofluorites have been included in a review on their chemistry [*23*]. In reactions of perfluoorganic hypofluorites, considerable fluorination of the C=C bond may be observed. With hypochlorites such as

[*21*] 10

$$CF_2{=}CFCF_3 + \text{pyridine N-oxide} \xrightarrow[16.5\,h]{62^{\circ}C} \text{[N-}OCF_2CF^{-}CF_3\text{, 2-H pyridinium]} \longrightarrow \text{[bicyclic N–O–}CF_2\text{–}CF(CF_3)\text{ adduct]} \longrightarrow \text{2-}CFHCF_3\text{-pyridine} + COF_2$$

44%

Table 1. Examples of Addition of Hypohalites to Fluoroolefins

	Yield (%) of Reaction with Hypohalite[a]			
Fluoroolefin	*CF_3OF*	*CF_3OCl*	*FSO_3I*	*CF_3CO_2Cl*
CFH=CFH	[*24*]	88 [*24*]		65 [*25*]
$CF_2{=}CFCF_3$	34 [*24*]	66 [*24*]	98 [*26*]	84 [*27, 28*]
$CF_2{=}CCl_2$	[*24*]	90 [*24, 29*]		81 [*25*]
$CF_2{=}CFCl$			75 [*26*]	65 [*25*]
$CF_2{=}CF_2$			61 [*26*]	54 [*25*]

[a]Numbers in brackets are references.

[*24*] 11

$$CH_2{=}CF_2 \xrightarrow[-160^{\circ}\text{ to }22^{\circ}C,\ 20\,h]{CF_3OF,\ CFCl_3/CF_2Cl_2} CF_3OCH_2CF_3\ 74\% + CF_3OCF_2CH_2F\ 2\% + CF_3CH_2F\ \text{trace}$$

CF_3OCl, chlorofluorination is less common [*24*]. Some ClF addition may result from ClF impurity present.

The irradiation of **bis(nonafluoro-*tert*-butyl)peroxide** in octafluorocyclopentene or decafluorocyclohexene yields the corresponding vicinal *bis(nonafluoro-tert-butyl) diethers* in 55 and 63%, respectively [*30*]. Negligible amounts of dimeric or oligomeric products are obtained. Conversely, the similar reaction with bis(trifluoromethyl)peroxide yields primarily 2,2′-bis(trifluoromethoxy)dicycloalkanes (equation 12). The dissimilar behavior is believed to be due to the reduced stability of the trifluoromethyloxy radical and thus its reduced population, which favors oligomerization [*30*].

[*30*] **12**

CF_3OOCF_3, -20°C; 3 h, hν (2500W) — 77 %

Bis(trifluoromethyl)trioxide, CF_3OOOCF_3, adds to only one equivalent of ***octafluorocyclopentene*** to give *cis* and *trans* isomers of a mixed trifluoromethyl ether–trifluoromethylperoxide (equation 13). A radical mechanism is involved [*31*], unlike in the electrophilic reactions of CF_3OOCl [*32*].

[*31*] **13**

CF_3OOOCF_3; 67°C, 8 h — 42% cis; 38% trans

The known chemistry of the **bis(trifluoromethyl)nitroxyl** [*33, 34*] includes additions to both olefins [*35, 36, 37*] and fluoroolefins [*38*]. The analogous *N*-trifluoromethylsulfamate oxide, $K^+ [CF_3(SO_3)NO]^-$, similarly yields the vicinal 2:1 addition products with ***chlorotrifluoroethene*** or ***tetrafluoroethene*** [*39*]. The nitroso derivative, $(CF_3)_2NONO$, reacts with a limited set of halogenated olefins to add the $(CF_3)_2NO$- and -NO moieties [*40*].

References for Pages 729–734

1. Chambers, R. D.; Mobbs, R. H. *Advances in Fluorine Chemistry;* Butterworths: Washington, D.C., **1965;** Vol. 4, p 50.
2. Lovelace, A. M.; Rausch, D. A.; Postelnek, W. *Aliphatic Fluorine Compounds*; ACS Monograph Series; Reinhold: New York, 1958; p. 155.
3. Krespan, C. G. U.S. Pat. 4 304 927, **1981**; *Chem. Abstr.* **1982,** *96,* 103669m.
4. Krespan, C. G.; Van-Catledge, F. A.; Smart, B. E. *J. Am. Chem. Soc.* **1984,** *106,* 5544.
5. Krespan, C. G.; Smart, B. E. *J. Org. Chem.* **1986,** *51,* 320.
6. Jiang, X.; Ji, G.; Shi, Y. *J. Fluorine Chem.* **1987,** *37,* 405.
7. Feiring, A. E.; Wonchoba, E. R. *J. Org. Chem.* **1992,** *57,* 7014.
8. England, D. C. *J. Org. Chem.* **1981,** *46,* 147.
9. Miller, W. T., Jr.; Fager, E. W.; Griswold, P. H. *J. Am. Chem. Soc.* **1948,** *70,* 431.
10. Hu, C.; Xu, Z. *Youji Huaxue* **1989,** *9,* 26; *Chem. Abstr.* **1989,** *111,* 56998j.
11. Hu, C.; Xu, Z. *J. Fluorine Chem.* **1989,** *42,* 69.

12. Gross, U.; Storek, W. *J. Fluorine Chem.* **1984,** *26,* 457.
13. Krespan, C. G. U.S. Pat. 3 962 325, 1976; *Chem. Abstr.* **1976,** *85,* 62685r.
14. Utebaev, U.; Rokhlin, E. M.; Lur'e, E. P. *Izv. Akad. Nauk SSSR* **1976,** 142; *Chem. Abstr.* **1977,** *87,* 38799b.
15. Evan, H. H.; Fields, R.; Haszeldine, R. N.; Illington, M. *J. Chem. Soc., Perkin Trans. 1* **1973,** 649.
16. Andreev, V. G.; Sorochkin, Y. I.; Kolomiets, A. F.; Sokol'skii, G. A. *Zh. Vses. Khim. O-va.* **1979,** *24,* 663; *Chem. Abstr.* **1980,** *92,* 180642s.
17. Andreev, V. G.; Kolomiets, A. F.; Sokol'skii, G. A. *Dokl. Akad. Nauk SSSR* **1980,** *250,* 1386; *Chem. Abstr.* **1980,** *93,* 45853s.
18. Park, J. D.; McMurtry, R. J.; Adams, J. H. *Fluorine Chem. Rev.* **1968,** *2,* 55.
19. Soulen, R. L.; Paul, D. W. *J. Fluorine Chem.* **1977,** *10,* 261.
20. Bartlett, S.; Chambers, R. D.; Kely, N. M. *Tetrahedron Lett.* **1980,** *21,* 1891.
21. Mailey, E. A.; Ocone, L. R. *J. Org. Chem.* **1968,** *33,* 3343.
22. Banks, R. E.; Haszeldine, R. N.; Robinson, J. M. *J. Chem. Soc., Perkin Trans. 1*, **1976,** 1226.
23. Mukhametshin, F. M. *New Fluorinating Agents in Organic Synthesis;* Springer-Verlag: Berlin, 1989; p. 69.
24. Johri, K. K.; DesMarteau, D. D. *J. Org. Chem.* **1983,** *48,* 242.
25. Tari, I.; DesMarteau, D. D. *J. Org. Chem.* **1980,** *45,* 1214.
26. Schack, C. J.; Christe, K. O. *J. Fluorine Chem.* **1982,** *20,* 283.
27. Schack, C. J.; Christe, K. O. *J. Fluorine Chem.* **1978,** *12,* 325.
28. Christe, K. O.; Schack, C. J. U.S. Pat. 4 216 338, 1980; *Chem. Abstr.* **1980,** *93,* 204090j.
29. Dicelio, L.; Schumacher, H. J. *J. Photochem.* **1979,** *11,* 1.
30. Toy, M. S.; Stringham, R. S. *J. Fluorine Chem.* **1976,** *7,* 375.
31. Hohorst, F. A.; Paukstelis, J. V.; DesMarteau, D. D. *J. Org. Chem.* **1974,** *39,* 1298.
32. Walker, N.; DesMarteau, D. D. *J. Am. Chem. Soc.* **1975,** *97,* 13.
33. Banks, R. E. *Fluorocarbon and Related Chemistry,* **1976,** *3,* 239.
34. Banks, R. E. *Fluorocarbon and Related Chemistry*, **1974,** *2,* 223 .
35. Banks, R. E.; Birchall, J. M.; Brown, A. K.; Haszeldine, R. N.; Moss, F. *J. Chem. Soc., Perkin Trans. 1* **1975,** 2033.
36. Fernandes, T. R.; Haszeldine, R. N.; Tipping, A. E. *J. Chem. Soc., Dalton Trans.* **1978,** 1024.
37. Banks, R. E.; Birchall, J. M.; Haszeldine, R. N.; Nona, S. N. *J. Fluorine Chem.* **1980,** *16,* 391.
38. Banks, R. E.; Haszeldine, R. N.; Stevenson, M. J. *J. Chem. Soc.* **1966,** 901.
39. Banks, R. E.; Dickinson, N. *J. Fluorine Chem.* **1981,** *18,* 299.
40. Ang, H. G.; So, K. K. *J. Fluorine Chem.* **1985,** *27,* 125.

Additions Forming Carbon-Sulfur Bonds

by K. B. Baucom

Hexafluoropropene and **sulfur** react at 425 °C in the vapor phase to give hexafluorothioacetone and its dimer [*1*]. In dimethylformamide, it reacts with potassium fluoride and sulfur to give hexafluorothioacetone dimer, which further reacts with hexafluoropropene to give the *E* and *Z* isomers of perfluoro-2,4,6-tris(trifluoromethyl)-5-thia-3-heptene [2] (equation 1).

[*1*] **1**

$CF_2{=}CFCF_3$ —(S, C; 425 °C)→ $(CF_3)_2C{=}S$ (74%) + hexafluorothioacetone dimer (26%)

$CF_2{=}CFCF_3$ —(KF, S, DMF)→ [2] hexafluorothioacetone dimer (>80%) → $S[CF(CF_3)_2]_2$

[2] dimer —(KF, $CF_2{=}CFCF_3$)→ $(CF_3)_2CF(F)C{=}C(CF_3)SCF(CF_3)_2$ (75%)

Perfluoroisobutylene and **potassium sulfide** give 2,4-bis[hexafluoroisopropylidene]-1,3-dithietane and a small amount of the bis(perfluoro-*tert*-butyl)trisulfide. The trisulfide is the result of a reaction of the nascent perfluoro-*tert*-butyl carbanion with sulfur. The same products in different yields are obtained with sulfur and cesium fluoride [*3*] (equation 2).

3,4-Dichlorotetrafluoro-2,5-dihydrothiophene is converted to the perfluoro-*p*-dithiin, probably by an addition–elimination process [*3*] (equation 3).

Perfluoro-2-butyne reacts with sulfur in carbon disulfide to give 1,2-bis(trifluoromethyl)vinylene trithiocarbonate ***A*** and a subsequent product ***B*** [*3*] (equation 4).

References are listed on page 741.

[3] **2**

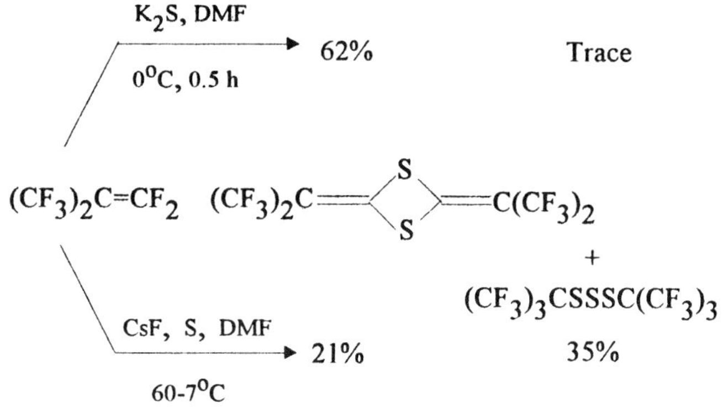

$(CF_3)_2C{=}CF_2$ —(K_2S, DMF; 0°C, 0.5 h)→ $(CF_3)_2C{=}C\langle S_2\rangle C{=}C(CF_3)_2$ 62%; Trace

$(CF_3)_2C{=}CF_2$ —(CsF, S, DMF; 60-7°C)→ $(CF_3)_2C{=}C\langle S_2\rangle C{=}C(CF_3)_2$ 21% + $(CF_3)_3CSSSC(CF_3)_3$ 35%

[3] **3**

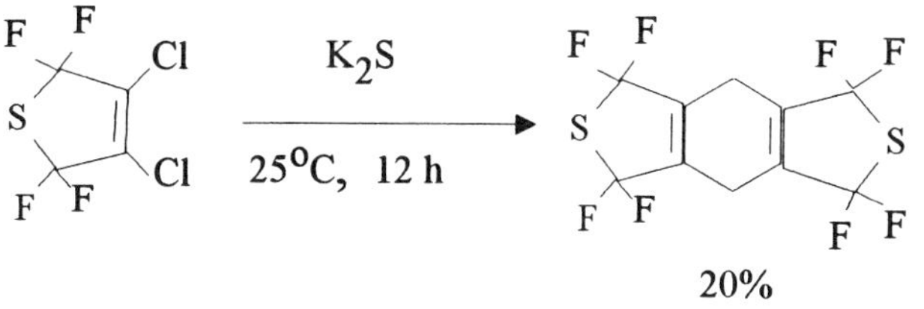

2,2,5,5-tetrafluoro-3,4-dichloro-2,5-dihydrothiophene —(K_2S; 25°C, 12 h)→ tetrafluoro-bis(thieno) fused product, 20%

[3] **4**

$CF_3C{\equiv}CCF_3$ —(S, CS_2; 200 °C)→ 4,5-bis(CF_3)-1,3-dithiole-2-thione, *A* 11% + tetrakis(CF_3)-dithiolylidene product, *B* 33%

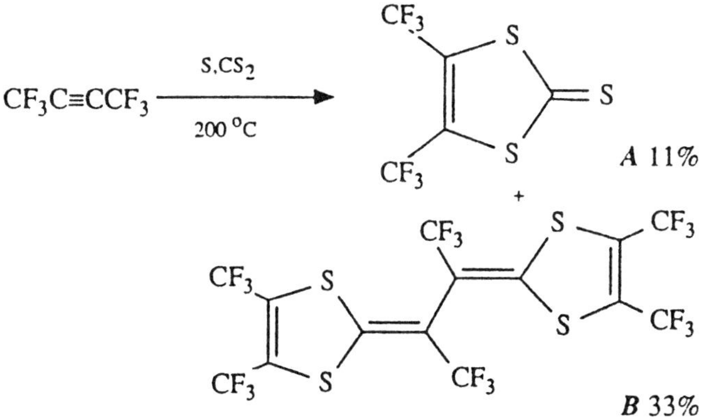

[4,5] **5**

$$(CF_3)_2C{=}CF_2 \xrightarrow[\text{Diglyme, 0°C}]{Et_2NC(S)SK \cdot 3H_2O} (CF_3)_2C{=}C\langle S_2\rangle C{=}C(CF_3)_2$$

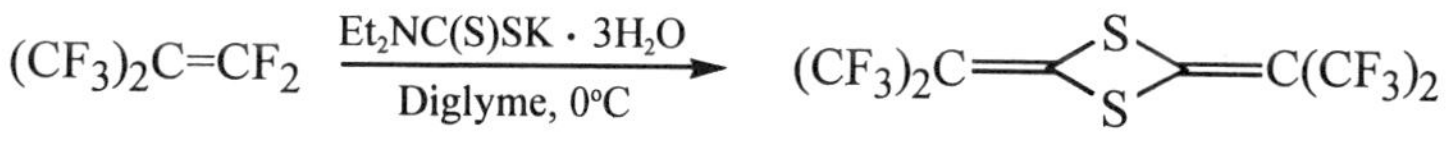

Potassium-*N,N*-diethyldithiocarbamate in diglyme reacts with ***perfluoroisobutylene*** to give 2,4-bis(hexafluoroisopropylidene)-1,3-dithietane as a crystalline solid [*4, 5*] (equation 5).

Under similar conditions ***hexafluoropropene*** gives *S*-(1,1,2,3,3,3-hexafluoropropyl)-*N,N*-dimethyldithiocarbamate in 63% isolated yield. Under slightly different conditions, the (*Z*)- and (*E*)-*S*-(1,2,3,3,3-pentafluoropropenyl)-*N,N*-dimethyldithiocarbamates are also formed [*6*] (equation 6).

[*6*] **6**

$$CF_2{=}CFCF_3 \xrightarrow[\text{DMA, 20 °C}]{(CH_3)_2NC(S)SNa} \underset{29\%}{CF_3CHFCF_2SC(S)N(CH_3)_2} + \underset{Z\ 58\%,\ E\ 5\%}{CF_3CF{=}CFSC(S)N(CH_3)_2}$$

Morpholinosulfur trifluoride and ***tetrafluoroethylene*** in the presence of potassium fluoride and 18-crown-6 give morpholinopentafluoroethylsulfur difluoride, which is subsequently oxidized sequentially to pentafluoroethanesulfinic acid and pentafluoroethanesulfonic acid [*7*] (equation 7).

[*7*] **7**

$$F_3S\text{-}N(\text{morpholine}) + CF_2{=}CF_2 \xrightarrow[\text{18-Crown-6}]{KF} CF_3CF_2CF_2S\text{-}N(\text{morpholine})\ 77.4\%$$

$$\xrightarrow{1.\ H_2O;\ 2.\ H_2SO_4} \underset{83.8\%}{CF_3CF_2SO_2H} \xrightarrow[H_2O_2]{CF_3CO_2H} \underset{75\%}{CF_3CF_2SO_3H}$$

Pentafluorosulfur chloride adds under free-radical conditions to ***fluorinated olefins,*** to acetylene, to ketene, to methyl trifluorovinyl ether [*1*] and to perfluoroallene [*8*].

Pentafluorosulfur chloride and **pentafluorosulfur bromide** add to ***vinylsilanes*** to give stable adducts in good yields, the bromide being much more reactive than the chloride. The regiochemistry has been established [*9*] (equations 8 and 9).

Pentafluorosulfur bromide adds to various olefins to give predominantly the adducts with the pentafluorosulfur radical as the initiating species [*10, 11*] (Table 1).

Telomers of fluoroolefins with pentafluorosulfur chloride and bromide are reported [*12*]. The telomers of tetrafluoroethylene with pentafluorosulfur end groups are formed by the reaction of S_2F_{10} with 1,2-diiodotetrafluoroethane and tetrafluoroethylene [*13*] (equation 10).

References are listed on page 741.

[9] **8**

$$CH_2{=}CHSi(CH_3)_3 \xrightarrow[180^oC,\ 22\ h]{SF_5Cl} SF_5CH_2CHClSi(CH_3)_3\quad >85\%$$

$$CH_2{=}CHSi(CH_3)_3 \xrightarrow[25^oC,\ 1\ h]{SF_5Br} SF_5CH_2CHBrSi(CH_3)_3\quad 72\%$$

[9] **9**

$$SF_5Br + CH_2{=}CHSiCl_3 \xrightarrow[3\ h]{0\ ^oC} SF_5CH_2CHBrSiCl_3 \quad 93\%$$

Table 1. Additions of Pentafluorosulfur Bromide to Fluoroolefins

Olefin	*Time*	*Temp. (°C)*	*Product*	*Yield (%)*	*Ref.*
$CH_2{=}CFCl$	6 days	RT[a]	$SF_5CH_2CFClBr$	51	10
$CH_2{=}CHCF_3$	18 days	RT	$SF_5CH_2CHBrCF_3$	39	10
$CHF{=}CHCl$	4.5 days	80	$SF_5CHFCHClBr$	49.5	10
$CHF{=}CFCl$	5 days	80	$SF_5CHFCFClBr$	70	10
$CF_2{=}CHCl$	14 days	100	$SF_5CF_2CHClBr$	73	10
$CF_2{=}CFCl$	12 h	RT	$SF_5CF_2CFClBr$	60	11
$CHF{=}CF_2$	12 h	RT	SF_5CHFCF_2Br	46	11
$CH_2{=}CF_2$	12 h	RT	$SF_5CH_2CF_2Br$	70	11
$CH_2{=}CHF$	12 h	RT	SF_5CH_2CHFBr	70	11

[a] RT, room temperature.

[13] **10**

$$ICF_2CF_2I \xrightarrow[150\ ^oC,\ 4\ h]{S_2F_{10}} SF_5CF_2CF_2I \xrightarrow[150\ ^oC]{CF_2{=}CF_2} SF_5(CF_2CF_2)_{1\text{-}9}I$$

Trifluorothiolacetic acid adds to a series of ***olefins*** under ultraviolet irradiation. The addition appears to start by a $CF_3COS\cdot$radical attack, giving the more stable radical intermediate [*14*] (Table 2).

References are listed on page 741.

Table 2. Addition of Trifluorothiolacetic Acid to Alkenes under Ultraviolet Irradiation [*14*]

Olefin	*Adduct*	*Yield (%)*
$CH_2{=}CH_2$	$CF_3COSCH_2CH_3$	49
$CH_2{=}CHF$	$CF_3COSCH_2CH_2F$	31
$CH_2{=}CF_2$	$CF_3COSCH_2CF_2H$	29
$CHF{=}CF_2$	$CF_3COSCHFCF_2H$	66

trans*-Chlorotetrafluoro(trifluoromethyl)sulfur** where chlorine is *trans* to trifluoromethyl is prepared by addition of chlorine to trifluoromethylsulfur trifluoride in the presence of cesium fluoride [*15*] (equation 11). It reacts with ***olefins and ***acetylenes*** under ultraviolet irradiation in Pyrex glass to give predominantly the adduct, with the trifluoromethylsulfur tetrafluoride group retaining its *trans*-configuration in all cases [*15*] (Table 3).

[*15*] $$CF_3SF_3 \xrightarrow{Cl_2,\ CsF} trans\text{-}CF_3SF_4Cl$$ **11**

Table 3. Addition of *trans*-Chlorotetrafluoro(trifluoromethyl)sulfur to Alkenes [*15*]

Alkene or Alkyne	*Adduct*[a]	*Yield (%)*
$CH_2{=}CH_2$	RCH_2CH_2Cl	52
$CH_2{=}CHCH_3$	$RCH_2CHClCH_3$	59
$CF_2{=}CF_2$	RCF_2CF_2Cl[b]	85
$CF_2{=}CFCF_3$	$RCF(CF_3)CF_2Cl$	44
	$RCF_2CFClCF_3$	1
$CH_2{=}CF_2$	RCH_2CF_2Cl	57
$CHF{=}CH_2$	$RCHFCH_2Cl$	47
$CF_2{=}CFCl$	RCF_2CFCl_2	23
	$R(CF_2CFCl)_2Cl$	15
$CH{\equiv}CH$	$RCH{=}CHCl$	62

[a] R is $CF_3SF_4^-$.

[b] Higher telomers are obtained at higher tetrafluoroethylene loadings.

References are listed on page 741.

Bis(trifluoromethyl)aminosulfenyl chloride reacts with ***olefins*** in the dark at –78 °C to give stereospecifically *anti*-adducts [*16*] (equation 12).

[*16*] **12**

$(CF_3)_2NSCl$

$SN(CF_3)_2$ / Cl — 89% (threo)

$SN(CF_3)_2$ / Cl — 92% (erythro)

Cl / $SN(CF_3)_2$ — 89%

References for Pages 736–741

1. Hudlický, M. *Chemistry of Organic Fluorine Compounds;* Ellis Horwood: Chichester, U.K., 1976; p 410.
2. England, D. C. *J. Org. Chem.* 1981, *46,* 153.
3. Krespan, C. G. England, D. C. *J. Org. Chem.* **1968,** *33,* 1850.
4. England, D. C. U.S. Pat. 3 694 460, 1972; *Chem. Abstr.* **1973,** *78,* 16161z.
5. Sterlin, S. R.; Zhurarkova, L. G.; Dyatkin, B. L.; Knunyants, I. L. *Izv. Akad. Nauk SSSR,* **1971,** 2517; *Chem. Abstr.* **1972,** *76* ,126829q
6. Kitazume, T.; Sasaki, S.; Ishikawa, N. *J. Fluorine Chem.* **1978,** *12,* 193.
7. Radchenko, O. A.; Il'chenko, A. Y.; Yagupol'skii, L. M. *Zh. Org. Khim.* **1991,** *17,* 500; *Chem. Abstr.* **1981,** *95,* 132234a
8. Witucki, E. F. *J. Fluorine Chem.* **1982,** *20,* 803.
9. Berry, A. D.; Fox, W. B. *J. Fluorine Chem.* **1975,** *6,* 175.
10. Mir, Q. C.; Debuhr, R.; Hang, C.; White, H. F.; Gard, G. L. *J. Fluorine Chem.* **1980,** *16,* 373.
11. Steward, J.; Kegley, L.; White, J. F.; Gard, G. L. *J. Org. Chem.* **1969,** *34,* 760.
12. Case, J. R.; Ray, N. H.; Roberts, H. L. *J. Chem. Soc.* **1961,** 2066, 2070.
13. Hutchinson, J. *J. Fluorine Chem.* **1973/74,** *3,* 429.
14. Weeks, P.; Gard, G. L. *J. Fluorine Chem.* **1971/72,** *1,* 295.
15. Darragh, J. I.; Haran, G.; Sharp, D. W. A. *J. Chem Soc., Dalton Trans.* **1973,** *21,* 2289.
16. Servia, C. F.; Tipping, A. E. *J. Fluorine Chem.* **1981,** *19,* 91.

Additions Forming Carbon–Nitrogen Bonds

by L. G. Sprague

The nucleophilic attack of nitrogen bases leads to a variety of products as the result of addition or addition–elimination reactions. The regioselectivity resembles that of attack by alcohols and alkoxides: an intermediate carbanion is believed to be involved. In the absence of protic reagents, the fluorocarbanion generated by the addition of sodium azide to polyfluorinated olefins can be captured by carbon dioxide or esters of fluorinated acids [*1, 2, 3*] (equation 1).

[*3*] 1

$$CF_2{=}CF_2 \xrightarrow[\text{DMSO, }50^{\circ}\text{C, 4 h}]{\text{1. NaN}_3\text{, CO}_2\;\; \text{2. (CH}_3\text{O)}_2\text{SO}_2} N_3CF_2CF_2CO_2CH_3 \;\; 91\%$$

Anhydrous ammonia adds to ***fluoroolefins*** to produce **nitriles**. This phenomenon is used to characterize chemically the terminal difluoromethylene olefin that is claimed to be in equilibrium with the internal isomer [*4*] (equation 2). Thus, isomerization to the terminally unsaturated isomer prior to attack by ammonia yields the cyanoenamine.

[*4*] 2

$$(CF_3)_2CFS(CF_3)C{=}CFCF(CF_3)_2 \rightleftarrows (CF_3)_2CFSC({=}CF_2)CF_2CF(CF_3)_2$$

$$\xrightarrow[25^{\circ}\text{C}]{\text{NH}_3\text{, THF}} (CF_3)_2CFS(NC)C{=}C(NH_2)CF(CF_3)_2$$

The addition of **primary amines** to ***fluoroolefins*** under anhydrous conditions yields *imines.* The hexafluoropropene dimer, perfluoro-2-methyl-2-pentene, and *tert*-butylamine react to yield a mixture of two compounds in a 9:4 ratio [*4*] (equation 3) rather than just the major ketenimine–imine, as previously reported [*5*]. It is claimed that this result is possible by means of isomerization to the terminally unsaturated difluoromethylene isomer prior to nucleophilic attack.

Secondary amines add to ***fluoroolefins*** under anhydrous conditions to give *fluorinated tertiary amines* in good yields. If the fluoroolefin is added to the amine without cooling the reaction mixture, or if an excess of the secondary amine is used, there is a tendency toward *dehydrofluorination of the tertiary amine*. The products of the reaction of secondary amines with a terminal perfluoroolefin are *easily hydrolyzed to the substituted amide.* Under hydrolytic conditions, amides are prepared in a one-step synthesis in good yield. The ease with which hydrolysis occurs is exploited to prepare fluorination reagents such as *Ishikawa's reagent* from diethylamine and hexafluoropropene [*6*] or the *Yarovenko reagent* from dialkylamine and chlorotrifluoroethylene [*7*]. *These reagents are useful for conversion of carboxylic acids to acyl fluorides [8] or of alcohols to alkyl fluorides.*

[*4*] 3

$$(CF_3)_2C{=}CFC_2F_5 \rightleftarrows CF_2{=}C(CF_3)CF_2C_2F_5$$

$$\xrightarrow[\text{ether, -50}^{\circ}\text{C}]{(CH_3)_3CNH_2} (CH_3)_3CN{=}C{=}C(CF_3)C(C_2F_5){=}NC(CH_3)_3 \quad 44\%$$

$$(CH_3)_3CN{=}C{=}C(CF_3)CF_2C_2F_5 \quad 19\%$$

Primary and secondary amines, when added to ***fluorine-containing alkynes,*** yield secondary or tertiary fluoroalkylenamines. For example, 1-hydroperfluoro-1-decyne reacts with secondary amines in a variety of solvents at 5 °C to give *N,N*-dialkyl-1,2-dihydroperfluoro-1-decenamines as products. The configuration of the double bond is probably *E* [*9*].

Fluoride ion produced from the nucleophilic addition–elimination reactions of fluoroolefins can catalyze *isomerizations and rearrangements*. The reaction of perfluoro-3-methyl-1-butene with dimethylamine gives as products 1-*N,N*-dimethylamino-1,1,2,2,4,4,4-heptafluoro-3-trifluoromethylbutane, *N,N*-dimethyl-2,2,4,4,4-pentafluoro-3-trifluoromethylbutyramide, and approximately 3% of an unidentified olefin [*10*]. The butylamide results from hydrolysis of the observed tertiary amine, and thus they share a common intermediate, 1-*N,N*-dimethylamino-1,1,2,4,4,4-hexafluoro-3-trifluoromethyl-2-butene, the product from the initial addition–elimination reaction (equation 4). The expected product from simple addition was not found.

[*10*] 4

$$(CF_3)_2CFCF{=}CF_2 \xrightarrow[\text{RT, 10 min}]{(CH_3)_2NH,\ \text{-HF}} \left[(CF_3)_2C{=}CFCF_2N(CH_3)_2\right] \xrightarrow{HF}$$

$$(CF_3)_2CHCF_2CF_2N(CH_3)_2 \quad 55\%$$

$$(CF_3)_2CHCF_2CON(CH_3)_2 \quad 10\%$$

References are listed on pages 745–746.

Examples of the reactions of ***fluoroolefins*** and **trialkylamines** are less common. ***Hexafluorocyclobutene*** and triethyl- or tributylamine give the corresponding trialkylammonium 2,2,3,3,4,4-hexafluorocyclobutane ylide [*11*], not the isomeric vinylammonium fluoride as originally claimed [*12*] (equation 5). The observed ylide is stable; its formation can be explained by the rearrangement of the initially formed vinylammonium fluoride via addition of fluoride to the vinylic β-carbon. Further reaction with boron trifluoride gives the vinylammonium tetrafluoride salt [*11*]. Tetrakis(trifluoromethyl)-2,5-difluorofuran and triethylamine react in an. addition–elimination manner to yield an intermediate allylic tetraalkylammonium salt, which disproportionates via a novel defluorination reaction to tetrakis(trifluoromethyl)furan [*13*] (equation 6). The defluorination of the 2,5-difluorofuran is also achieved by pyrolysis over iron.

[*11*] 5

F $\xrightarrow[\text{RT, 5 days}]{N(C_2H_5)_3}$ F $^{-}$ $\overset{+}{N}(C_2H_5)_3$ 80% $\xrightarrow[\text{THF, RT}]{BF_3{\bullet}O(C_2H_5)_2}$ F $\overset{+}{N}(C_2H_5)_3$ BF_4^- quantitative

[*13*] 6

CF_3, CF_3, F, F, O, CF_3, CF_3 $\xrightarrow{N(CH_2CH_3)_3}$ CF_3, CF_3, $N^+(CH_2CH_3)_3$, F, CF_3, O, CF_3

$\xrightarrow{-\ N(CH_2CH_3)_3^+\ F^-}$ CF_3, CF_3, CF_3, O, CF_3 41%

***N*-Halogenated compounds** such as *N*-chlorotrifluoroacetamide, *N*-chloroimidosulfuryl fluoride, and *N,N*-dichlorotrifluoromethylamine *add across C=C bonds to form saturated amides [14], imidosulfuryl fluorides [15], and amines [16],* respectively. Allylic halogenation also occurs with the use of *N*-bromo- or *N*-chloroperfluoroamides. The primary amine *N,N*-dichlorotrifluoromethylamine selectively affords 1:1 or 2:1 adducts with either tetrafluoroethylene or chlorotrifluoroethylene [*16*] (equation 7). The reaction mechanism is believed to involve thermal free radicals, with control achieved principally by reaction temperature. The 1:1 adduct is formed even in the presence of a large excess of olefin.

[*16*] 7

$$CF_2{=}CFCl + CF_3NCl_2 \xrightarrow[12\text{-}16\ h]{65\text{-}70^{\circ}C} CF_3N(Cl)CF_2CFCl_2\ \ 80\%$$

$$CF_2{=}CFCl + CF_3NCl_2 \xrightarrow[12\text{-}14\ h]{95\text{-}100^{\circ}C} CF_3N(CF_2CFCl_2)_2\ \ 60\%$$

The reactions of ***N*-bromobis(trifluoromethyl)amine** with olefins occur through two distinct modes of addition. From photochemically or thermally generated radicals it yields 1:1 adducts with electron-rich and electron-deficient. olefins [*10, 17, 18, 19, 20, 21*]. As the chain-propagating species, the $(CF_3)_2N\cdot$ radical attacks the C=C bond to form an intermediate organic radical that precedes the product. Yields are generally high; however, the reaction of electron-deficient perfluoro-3-methyl-1-butene affords a complex mixture of products which results from scrambling of the $(CF_3)_2N\cdot$ and $Br\cdot$ radicals [*10*] (equation 8). Under ionic conditions (liquid phase, in the dark at low temperatures), the reaction with electron-rich olefins occurs through electrophilic addition [*17, 18, 19, 22, 23, 24*]. From the results, the amine is interpreted to be polarized with a pseudopositive bromine. Cyclic bromonium ions are proposed as intermediates in some addition reactions, with exclusive attack of the bis(trifluoromethyl)amide anion from the *anti* direction [*17, 22, 23*].

[*10*] 8

$$(CF_3)_2CFCF{=}CF_2 \xrightarrow[h\nu,\ 70\ h]{(CF_3)_2NBr} (CF_3)_2NN(CF_3)_2\ \ 36\%$$

$$(CF_3)_2CFCFBrCF_2Br\ \ 18\%$$

$$(CF_3)_2CFCFBrCF_2N(CF_3)_2\ \ 20\%$$

$$CF_3N{=}CF_2\ \ 5\%$$

References for Pages 742–745

1. Krespan, C. G.; Van-Catledge, F. A.; Smart, B. E. *J. Am. Chem. Soc.* **1984,** *106,* 5544.
2. Krespan, C. G.; Smart, B. E. *J. Org. Chem.* **1986,** *51,* 320.
3. Krespan C. G. U.S. Pat. 4 474 700, **1984;** *Chem. Abstr.* **1985,** *102,* 167672g.
4. England, D. C.; Piecara, J. C. *J. Fluorine Chem.* **1981,** *17,* 265

5. Flowers, W. T.; Haszeldine, R. N.; Owen, C. R.; Thomas, A. *J. Chem. Soc., Chem. Commun.* **1974,** 134.
6. Takaoko, A.; Iwakiri, H.; Ishikawa, N. *Bull. Chem. Soc. Jpn.* **1979,** *52,* 3377.
7. Yarovenko, N. N.; Raksha, M. A. *Zh. Obshch. Khim.* **1959,** *29,* 2159; *Chem. Abstr.* **1960,** *54,* 9724h.
8. Cox, D. G.; Sprague, L. G.; Burton, D. J. *J. Fluorine Chem.* **1983,** *23,* 383.
9. Le Blanc, M.; Santini, G.; Riess, J. G. *Tetrahedron Lett.* **1975,** 4151.
10. Haszeldine, R. N.; Mir, I. D.; Tipping, A. E. *J. Chem. Soc., Perkin Trans. 1* **1979,** 565.
11. Burton, D. J.; Howells, R. D.; Vander Valk, P. D. *J. Am. Chem. Soc.* **1977,** *99,* 4830.
12. Pruett, R. L.; Bahner, C. T.; Smith, H. A. *J. Am. Chem. Soc.* **1952,** *74,* 1633.
13. Chambers, R. D.; Lindley, A. A.; Philpot, P. D.; Fielding, H. C.; Hutchinson, J.; Whittaker, G. *J. Chem. Soc., Chem. Commun.* **1978,** 431.
14. Lessard, J.; Mondon, M.; Touchard, D. *Can. J. Chem.* **1981,** *59,* 431.
15. Kluver, H.; Glemser, O. *Chem. Ber.* **1977,** *110,* 1597.
16. Sarwar, G.; Kirchmeier, R. L.; Shreeve, J. M. *Inorg. Chem.* **1989,** *28,* 2187.
17. Barlow, M. G.; Fleming, G. L.; Haszeldine, R. N.; Tipping, A. E. *J. Chem. Soc. (C)* **1971,** 2744.
18. Coy, D. H.; Fleming, G. L.; Haszeldine, R. N.; Newlands, M. J.; Tipping, A. E. *J. Chem. Soc., Perkin Trans. 1* **1972,** 1880.
19. Coy, D. H.; Haszeldine, R. N.; Newlands, M. J.; Tipping, A. E. *J. Chem. Soc., Perkin Trans. 1* **1973,** 1062.
20. Coy, D. H.; Haszeldine, R. N.; Newlands, M. J.; Tipping, A. E. *J. Chem. Soc., Perkin Trans. 1* **1973,** 1066.
21. Fleming, G. L.; Haszeldine, R. N.; McAllister, J. R.; Tipping, A. E. *J. Chem. Soc., Perkin Trans. 1* **1975,** 1633.
22. Haszeldine, R. N.; Mir, I. D.; Tipping A. E. *J. Chem. Soc., Perkin Trans. 1* **1976,** 556.
23. Haszeldine, R. N.; Obo, W. G.; Sparkes, G. R.; Tipping, A. E. *J. Chem. Soc., Perkin Trans. 1* **1980,** 372.
24. Hart, T. W.; Haszeldine, R. N.; Tipping, A. E. *J. Chem. Soc., Perkin Trans. 1* **1980,** 1544.

Additions Forming Carbon–Carbon Bonds

by S. F. Sellers

The addition of halogenated aliphatics to carbon–carbon double bonds is the most useful type of carbon–carbon bond-forming synthetic method for highly halogenated substrates. Numerous synthetic procedures have been developed for these types of reactions, particularly for the addition of perfluoroalkyl iodides to alkenes using thermal or photolytic initiators of free radical reactions such as organic peroxides and azo compounds [*1*].

These types of reactions have been used extensively to prepare *fluorotelomers*. For example, Zonyl products, manufactured by E. I. du Pont de Nemours & Co., Inc., are based on materials prepared from **perfluoroethyl iodide** and ***tetrafluoroethylene*** [*2*] (equation 1).

[*2*]
$$C_2F_5I + n\ CF_2{=}CF_2 \xrightarrow{150\text{-}200^{\circ}C} C_2F_5(CF_2CF_2)_nI \qquad \mathbf{1}$$

Similarly, difunctional materials can be made from ***tetrafluoroethylene*** and **1,2-diiodotetrafluoroethane** in a thermal process [*3*] (equation 2).

[*3*]
$$ICF_2CF_2I + n\ CF_2{=}CF_2 \xrightarrow{200\text{ - }220^{\circ}C} I(CF_2CF_2)_{n+1}I \qquad \mathbf{2}$$

Much recent work on these types of reactions involves the development of methods for the addition reactions to be carried out under milder conditions.

The reaction of **perfluoroalkyl iodides** with electron donor nucleophiles such as **sodium arene** and **alkane sulfinates** in aprotic solvents results in radical addition to alkenes initiated by an electron-transfer process. The additions can be carried out at room temperature, with high yields obtained for strained olefins [*4*] (equations 3–5).

[*4*] **3**

$$\xrightarrow[\substack{p\text{-}MeC_6H_4SO_2Na \\ DMF,\ 25^{\circ}C}]{C_8F_{17}I}$$

C_8F_{17}

100%

I

[*4*] **4**

$$CH_2{=}CHC_6H_{13} + C_8F_{17}I \xrightarrow[DMF,\ 25^{\circ}C]{p\text{-}CH_3C_6H_4SO_2Na} C_8F_{17}CH_2CHIC_6H_{13}$$

62%

References are listed on pages 751–752.

[*4*] **5**

$$CH_2{=}CH_2 + C_8F_{17}I \xrightarrow[\text{DMF, } 25^\circ C]{p\text{-}CH_3C_6H_4SO_2Na} C_8F_{17}CH_2CH_2I \ (17\%) + C_8F_{17}CH_2CH_2SO_2C_6H_4CH_3 \ (33\%)$$

[*5*] **6**

$$C_8F_{17}I + CH_2{=}CH_2 \xrightarrow{\text{Ru(C), } 120^\circ C, \text{ 12 h}} C_8F_{17}CH_2CH_2I \ (93\%)$$

$$C_8F_{17}I + CH_2{=}CH_2 \xrightarrow{\text{Pt(C), } 100^\circ C, \text{ 12 h}} C_8F_{17}CH_2CH_2I \ (94\%)$$

[*6*] **7**

$$\text{cyclohexene} \xrightarrow[\text{CuO, } Ac_2O, \ 100^\circ C, \text{ 6 h}]{R_FI} \text{1-}R_F\text{-2-iodocyclohexane } (59\%) + R_F\text{-cyclohexane } (11\%)$$

[*7*] **8**

$$C_6H_{13}CH{=}CH_2 \xrightarrow[\text{Sn-Ac OAg, Me OH}]{C_6F_{13}I} C_6F_{13}CH_2CHIC_6H_{13} \quad 71\%$$

[*9*] **9**

$$(CH_3)_2CHCH{=}CH_2 \xrightarrow[\text{Mg, DMF, } 80^\circ C, \text{ 10 h}]{Cl(CF_2)_4I} Cl(CF_2)_4CH_2CHICH(CH_3)_2 \quad 65\%$$

The addition of perfluoroalkyl iodides to alkenes is also *effected catalytically by transition metals* such as Ru, Pt, Ni, W, and Mo [*5*]; Cu [*6*]; Sn [*7*]; Fe [*8*]; and Mg [*9*] (equations 6–9).

The activity of *homogeneous catalysts* also has been demonstrated. *Wilkinson's catalyst, tris(triphenylphosphine) rhodium chloride,* induces perfluoroalkyl iodides to add to olefins at 80 °C [*10*] (equation 10). *Tetrakis(triphenylphosphine)–palladium* promotes the addition to both alkenes and alkynes in hexane [*11*].

Amines and amine salts induce addition at 120–140 °C [*12*] (equation 11).

The scope of haloaliphatic additions to alkenes has been extended to the preparation of 1,2-bis(perfluoroalkyl)iodoethanes by the addition of **perfluoroalkyl iodides** to ***fluoroalkyl alkenes*** [*13*] (equation 12) or to alkadienes [*14*] (equation 13).

References are listed on pages 751–752.

[*10*] **10**

$$Cl(CF_2)_4I + C_5H_{11}CH{=}CH_2 \xrightarrow[C_6H_6, 80^\circ C,\ 1.5\ h]{(Ph_3P)_3RhCl} Cl(CF_2)_4CH_2CHIC_5H_{11}\quad 75\%$$

[*12*] **11**

$$CH_2{=}CHC_5H_{11} + C_4F_9I \xrightarrow{Bu_4NI,\ 135^\circ C,\ 6\ h} C_4F_9CH_2CHIC_5H_{11}\quad 82\%$$

$$CH_2{=}CHC_5H_{11} + C_4F_9I \xrightarrow{Et_3N,\ 135^\circ C,\ 3\ h} C_4F_9CH_2CHIC_5H_{11}\quad 65\%$$

[*13*] **12**

$$CH_2{=}CHC_4F_9 \xrightarrow[195^\circ C,\ 4\ h]{C_4F_9I} C_4F_9CHICH_2C_4F_9\quad 84\%$$

[*14*] **13**

$$CH_2{=}CH(CH_2)_2CH{=}CH_2 \xrightarrow[\text{Azobisisobutyronitrile},\ 70^\circ C,\ 6\ h]{2\ C_4F_9I} C_4F_9CH_2CHI(CH_2)_2CHICH_2C_4F_9\ \ 80\% + C_4F_9CH_2CHI(CH_2)_2CH{=}CH_2$$

[*15*] **14**

$$C_3F_7I + CH_2{=}CHOCOCH_3 \longrightarrow C_3F_7CH_2CHIOCOCH_3$$

[*16*] **15**

$$ICF_2CO_2CH_3 + CH_2{=}CH(CH_2)_3CH_3 \xrightarrow{Cu,\ 60^\circ C} CH_3(CH_2)_3CHICH_2CF_2CO_2CH_3\quad 75\%$$

The functionality of either the haloaliphatic compound or the alkene can sometimes be maintained in this type of radical addition [*15, 16*] (equations 14 and 15).

In a series of papers, Tedder and co-workers reported the factors determining the reactivity of perfluorinated radicals with various fluoroethylenes. Relative Arrhenius parameters for trifluoromethyl radicals [*17*] and pentafluoroethyl radicals [*18*] were determined, with higher selectivity demonstrated for the higher homologue. *Selectivity of addition* to unsymmetrical olefins was found also to increase with greater radical branching [*19*].

Fluorinated alkenes *can insert into C–H bonds* at elevated temperatures, a relatively unusual example of simultaneous C–C and C–H bond formation [*20, 21*] (equations 16 and 17).

[*20*] **16**

$$CF_3CF{=}CF_2 + CHCl_3 \xrightarrow{280^{\circ}C,\ 116\ h} CCl_3CF_2CHFCF_3$$

[*21*] **17**

$$(CH_3)_3CH + CF_2{=}CFH \xrightarrow[84\ atm]{280^{\circ}C} HCF_2CFHC(CH_3)_3\quad 33\%$$

Access to a variety of unsaturated perfluorinated telomers is provided by the *homotelomerization of perfluorinated alkenes,* generally initiated and *catalyzed by fluoride ion* [22].

Tetrafluoroethylene gives highly branched, internally unsaturated alkenes on contact with cesium fluoride in aprotic solvents [*23, 24*] (equation 18).

[*24*] **18**

$$CF_2{=}CF_2 \xrightarrow[DMF,\ 80^{\circ}C]{CsF} \begin{array}{l} CF_3CF{=}CFCF_3 + C_2F_5C(CF_3){=}CFCF_3 \\ +C_2F_5C(CF_3){=}C(CF_3)C_2F_5 \\ +(C_2F_5)_2C(CF_3)C(CF_3){=}CFCF_3 \end{array}$$

Similarly, ***hexafluoropropylene*** undergoes fluoride ion-induced homotelomerization to give a series of dimers and trimers. These telomerizations can be induced by other nucleophiles, such as amines. Indeed, the selectivity of the process can be changed significantly by varying reagents and reaction conditions [*25, 26*] (equations 19 and 20).

Perfluorocycloalkenes also undergo telomerization [*27, 28*] (equations 21 and 22).

[*25*] **19**

$$CF_3CF{=}CF_2 \xrightarrow[40^{\circ}C]{Et_3N} CF_3CF{=}CFCF(CF_3)_2\quad 96\%$$

[*26*] **20**

$$CF_3CF{=}CF_2 \xrightarrow[40^{\circ}C]{(CF_3CHFCF_2OCH_2CH_2)_3N} \begin{array}{c} CF_3CF{=}C[CF(CF_3)_2]_2 \\ 30\% \\ (CF_3)_2CF(C_2F_5)C{=}C(CF_3)_2 \\ 66\% \end{array}$$

References are listed on pages 751–752.

[*14*] 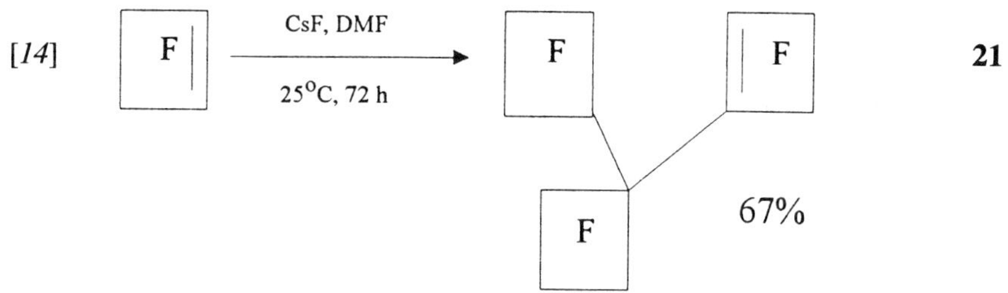

21

[*15*] 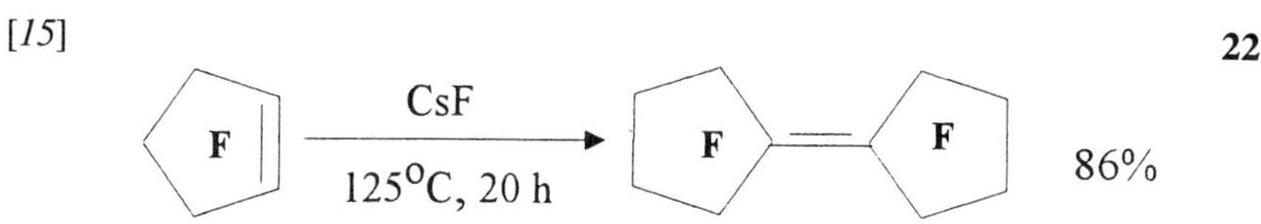

22

References for Pages 747–751

1. Hudlický, M. *Chemistry of Organic Fluorine Compounds,* 2nd ed.; Ellis Horwood: Chichester, U.K., 1976; pp 421–431.
2. Haszeldine, R. N. *J. Chem. Soc.* **1953,** 3761.
3. Bedford, C. D.; Baum, K. *J. Org. Chem.* **1980,** *45,* 347.
4. Feiring, A. E. *J. Org. Chem.* **1985,** *50,* 3269.
5. Werner, K. von *J. Fluorine Chem.* **1985,** *28,* 229.
6. Chen, Q.-Y.; Yang, Z.-Y. *J. Fluorine Chem.* **1985,** *28,* 399.
7. Kuroboshi, M.; Ishihara, T. *J. Fluorine Chem.* **1988,** *39,* 299.
8. Chen, Q.-Y.; He, Y.-B.; Yang, Z.-Y. *J. Fluorine Chem.* **1986,** *34,* 255.
9. Chen, Q.-Y.; Qin, Z.-M.; Yang, Z.-Y. *J. Fluorine Chem.* **1987,** *36,* 149.
10. Chen, Q.-Y.; Yang, Z.-Y. *J. Fluorine Chem.* **1988,** *39,* 217.
11. Ishara, T.; Kuroboshi, M.; Okada, Y. *Chem. Lett.* **1986,** 1895.
12. Brace, N. O. *J. Org. Chem.* **1979,** *44,* 212.
13. Jeanneaux, F.; Le Blanc, M.; Cambon, A.; Guion, J. *J. Fluorine Chem.* **1974,** *4,* 261.
14. Brace, N. O. *J. Org. Chem.* **1973,** *38,* 3167.
15. Rakhimov, A. I.; Kaluga, V. I. *Zh. Org. Khim.* **1980,** *16,* 223; *Chem.Abstr.* *92,* 214861e.
16. Yang, Z. Y.; Burton, D. J. *J. Fluorine Chem.* **1989,** *45*, 435.
17. Cape, J. N.; Greig, A. C.; Tedder, J. M.; Walton, J. C. *J. Chem. Soc., Faraday Trans. 1* **1975,** *71,* 592.
18. El-Soueni, A.; Tedder, J. M.; Walton, J. C. *J. Fluorine Chem.* **1978,** *11,* 407.
19. El-Soueni, A.; Tedder, J. M.; Vertommen, L. L.T.; Walton, J. C. *Colloq. Int. CNRS* **1977,** *278,* 411.
20. Haszeldine, R. N.; Rowland, R.; Tipping. A. E.; Tyrell, G. *J. Fluorine Chem.* **1982,** *21,* 253.

21. Modarai, B. *J. Org. Chem.* **1976,** *41,* 1980.
22. Young, J. A. *Fluorine Chemistry Reviews,* Vol. 1; Tarrant, P., Ed.; Marcel Dekker: New York, 1967; pp 359–395.
23. Graham, D. P. *J. Org. Chem.* **1966,** *31,* 955.
24. Fielding, H. C; Rudge, A. J. Br. Pat. 1 082 127, 1977.
25. Dresdner, R. D.; Tlumac, F. N.; Young, J. A. *J. Org. Chem.* **1965,** *30,* 3524.
26. Martini, T.; Halasz, von S. P. *Tetrahedron Lett.* **1974,** 2129.
27. Chambers, R. D.; Taylor, G.; Powell, R. L. *J. Chem. Soc., Perkin Trans. 1* **1980,** 426.
28. Chambers, R. D.; Gribble, M. Y.; Marper, E. *J. Chem. Soc., Perkin Trans. 1* **1973,** 1710.

Additions Forming Bonds between Carbon and Silicon, Tin, and Phosphorus

by R. A. DuBoisson

Hydrosilylations of fluorine-containing alkenes are free radical reactions initiated by UV light or organic peroxides. The direction of addition is the same as with fluorinated alkyl halides. However, the reaction between **hydrosilanes** and ***fluorine-containing olefins*** catalyzed by platinum group metal complexes may result in bidirectional addition and/or formation of a vinylic silane, the latter by dehydrogenative silylation[*1*]. The natures of both the silane and the catalyst affect the outcome of the reaction[*1*]. A random selection of some typical new reactions of **silanes** are shown in Table 1 [*1, 2, 3, 4*].

Stannanes also add across double bonds of ***fluorinated olefins*** in a free radical reaction. Trimethylstannane undergoes stereospecific addition to hexafluorocyclobutene to afford *trans*-1,2,3,3,4,4-hexafluoro-1-(trimethylstannyl)cyclobutane [*5*] (equation 1)

The addition of **phosphines** to ***fluoroolefins*** is of free radical nature [*6, 7, 8*] (equations 2–5).

[*5*] $\begin{array}{c}CF{=}CF\\ |\quad\;|\\ CF_2{-}CF_2\end{array} + SnH(CH_3)_3 \xrightarrow{20\ ^\circ C} \begin{array}{c}HCF{-}CFSn(CH_3)_3\\ |\quad\;|\\ CF_2{-}CF_2\end{array}$ 94% **1**

[*6*] $(CH_3)_2PH + CH_2{=}CHCF_3 \xrightarrow{h\nu} (CH_3)_2PCH_2CH_2CF_3$ 78% **2**

[*7*] $(CH_3)_2PH + CH_2{=}CF_2 \xrightarrow{h\nu} (CH_3)_2PCF_2CHF_2$ 100% **3**

[*6*] **4**

$P(CF_3)_3 + CH_2{=}CHF \xrightarrow{h\nu} (CF_3)_2PCH_2CHFCF_3$ (3%) $+ (CF_3)_2PCHFCH_2CF_3$ (20%)

[*8*] **5**

$(CH_3)_2PP(CH_3)_2 + CF_2{=}CF_2 \xrightarrow{240\text{-}280\ ^\circ C} (CH_3)_2PCF_2CF_2P(CH_3)_2$

In contrast to the reaction of hexafluoropropene with phosphine, dimethylphosphine affords not the 1:1 adduct but a mixture of pentafluoropropenylphosphines [*9*] (equation 6).

Dialkylphosphinous acids react with ***perfluoroalkenes*** under free radical conditions to form carbon–phosphorus bonds [*10*] (equation 7).

References are listed on pages 755–756.

Table 1. Addition of Silanes to Fluorinated Olefins

Fluoroolefin	*Silane*	*Conditions*	*Main Product*	*Yield (%)*	*Ref.*
$CF_2=CF_2$	$SiHF_3$		$CHF_2CF_2SiF_3$		2
$CH_2=CHCF_3$	$SiH(C_2H_5)_3$	$Ru_3(CO)_{12}$, 70 °C, 6 h	$CF_3CH=CHSi(C_2H_5)_3$	78	1
$CH_2=CHCF_3$	$SiH(C_2H_5)_3$	$RhCl(PPh_3)_3$, 70 °C, 6 h	$CF_3CH_2CH_2Si(C_2H_5)_3$ +	67	1
			$CF_3CH=CHSi(C_2H_5)_3$	9	
$CH_2=CHCF_3$	$SiH(OC_2H_5)_3$	$Ru_3(CO)_{12}$, 120 °C, 24 h	$CF_3CH_2CH_2Si(OC_2H_5)_3$	52	1
$CH_2=CHCF_3$	$SiH(OC_2H_5)_3$	RhCl(PPh3)$_3$, 120 °C, 24 h	$CF_3CH_2CH_2Si(OC_2H_5)_3$	85	1
$CH_2=CHCF_3$	$(CH_3)_2SiHCl$	$PdCl_2(PhCN)_2$/ $2PPh_3$, 100 °C, 14 h	$CF_3CH(CH_3)Si(CH_3)_2Cl$	88	1
$CF_2=CFCF_3$	SiH_4	$h\nu$, Hg	$CF_3CHFCF_2SiH_3$ +	51	3
			$CHF_2CF(CF_3)SiH_3$	34	
$CHF=CClCF_3$	$SiHCl_3$	$h\nu$, 300 h	$CF_3CHClCHFSiCl_3$ +	65	4
			$CF_3CH_2CHFSiCl_3$	29	
$CH_2=CHC_2F_5$	CH_3SiHCl_2	$PdCl_2(PhCN)_2$/ $2PPh_3$ 100 °C, 14 h	$C_2F_5CH(CH_3)SiCl_2CH_3$	38	1
CF=CF (ring with) CF_2—CF_2	$SiHCl_3$	$h\nu$	HCF—$CFSiCl_3$ (ring with) CF_2—CF_2	80	3
$CH_2=CHC_6F_5$	$SiH(C_2H_5)_3$	$Ru_3(CO)_{12}$, 70 °C, 18 h	$C_6F_5CH=CHSi(C_2H_5)_3$	92	1
$CH_2=CHC_6F_5$	$SiHCl_3$	$RhCl(PPh_3)_3$, 120 °C, 24 h	$C_6F_5CH_2CH_2SiCl_3$	84	1

[9] **6**

$$\underset{(Z\ 92\%;\ E\ 5\%)}{CF_3CF=CFP(CH_3)_2} \xleftarrow[\text{500 h}]{\text{dark, 50 °C}} \underset{+(CH_3)_2PH}{CF_3CF=CF_2} \xrightarrow{h\nu,\ 50\text{ °C}} \underset{(Z\ 57\%;\ E\ 26\%)}{CF_3CF=CFP(CH_3)_2}$$

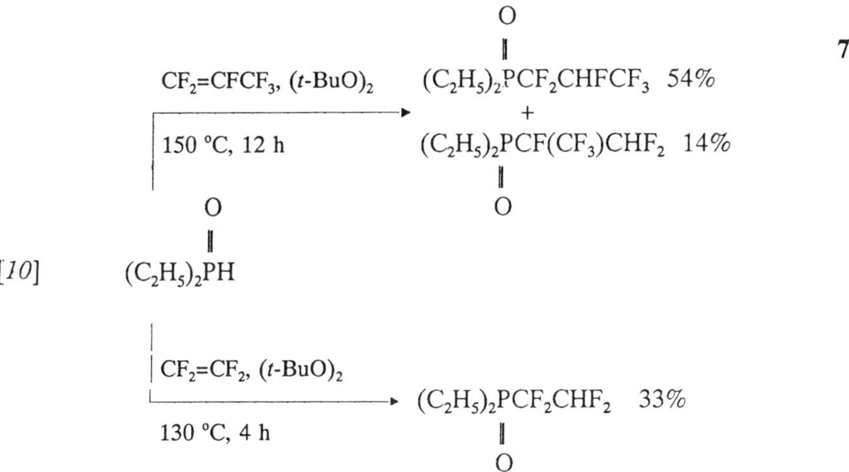

A special case is addition of **trifluoromethyl-** and **bis(trifluoromethyl)phosphorus** radicals to alkenes, referred to as *insertion of olefins into phosphorus–carbon bond* [*6*] (equations 8 and 9).

[*6*]

$$P(CF_3)_3 + CH_2{=}CH_2 \xrightarrow{h\nu} (CF_3)_2PCH_2CH_2CF_3 \quad 86\% \qquad \mathbf{8}$$

$$P(CF_3)_3 + CH_2{=}CHF \xrightarrow{h\nu} (CF_3)_2PCHFCH_2CF_3 \ 19\% + (CF_3)_2PCH_2CHFCF_3 \ 4\% \qquad \mathbf{9}$$

References for Pages 753–755

1. Ojima, I.; Fuchikami, T.; Yatabe, M. *J. Organomet. Chem.* **1984,** *260,* 335.
2. Haszeldine, R. N.; Robinson, P. J.; Simmons, R. F. *J. Chem. Soc.* **1964,** 1890.
3. Attridge, C. J.; Barlow, M. G.; Bevan, W. I.; Cooper, D.; Cross, G. W.; Haszeldine, R. N.; Middleton, J.; Newlands, M. J.; Tipping, A. E. *J. Chem. Soc., Dalton Trans.* **1976,** 694.
4. Haszeldine, R. N., Pool, R.; Tipping, A. E. *J. Chem. Soc., Dalton Trans.* **1979,** 2292.
5. Cullen, W. R.; Styan, G. E. *J. Organomet. Chem.* **1966,** *6,* 633.
6. Cooper, P.; Fields, R.; Haszeldine, R. N.; Mitchell, G. H.; Nona, S. N. *J. Fluorine Chem.* **1982,** *21,* 317.
7. Brandon, R.; Haszeldine, R. N.; Robinson, P. J. *J. Chem. Soc., Perkin Trans. 2* **1973,** 1295.

8. Brandon, R.; Haszeldine, R. N.; Robinson, P. J. *J. Chem. Soc., Perkin Trans. 2* **1973,** 1301.
9. Cooper, P.; Fields, R.; Haszeldine, R. N. *J. Chem Soc., Perkin Trans. 1* **1975,** 702.
10. Haszeldine, R. N.; Hobson, D. L.; Taylor, D. R. *J. Fluorine Chem.* **1976**, *8,* 115.

Linear Additions to Acetylenic Compounds

by R. L. Soulen

Examples of linear addition reactions that form C–O, C–S, C–N, C–P, C–C, and C–Si bonds are reviewed. Only a few of the growing number of linear additions that form a carbon–transition metal bond are included.

Additions Forming Carbon–Oxygen, Carbon–Sulfur, Carbon–Nitrogen, and Carbon–Phosphorus Bonds

Conjugated terminal fluoroenynes add **water** to the double bond at –5 °C to form α-fluoro allenic acid fluorides. At 20 °C, the allenic acid fluorides and concentrated sulfuric acid give γ- and δ-lactones (equation 1). In the presence of concentrated sulfuric acid, conjugated nonterminal fluoroenynes add water to the triple bond and then to the double bond to form α-fluoro-β-diketones and exocyclic α-fluoroenones [*1*] (equation 2).

[*1*] **1**

$$CF_2{=}CF{-}C{\equiv}CC_6H_{13} \xrightarrow[-5°C,\ 7\ h]{H_2SO_4} \left[\overset{+}{C}F_2{-}CF{=}C{=}CHC_6H_{13}\right] \longrightarrow$$

$$\xrightarrow[-HF]{H_2O} FCOCF{=}C{=}CHC_6H_{13} \xrightarrow[20°C]{H_2SO_4} \left[FCOCF{=}CH{-}\overset{OH}{\overset{|}{C}}HC_6H_{13}\right] \longrightarrow$$

$$\longrightarrow \text{FC=CH–CHC}_6\text{H}_{13}\text{–O–OC (ring) } 56\% \;+\; \text{FC=CH–CH}_2\text{–CHC}_5\text{H}_{11}\text{–O–OC (ring) } 29\%$$

Water adds to the *triple bond* of perfluorobutylalkynols in refluxing 98% formic acid to give the perfluorobutyl keto alcohol. Methanolic potassium hydroxide or prolonged reflux in formic acid converts the keto alcohol to 2,2-dimethyl-5-perfluoropropyl-2,2-dimethylfuran-3(2*H*)-one [2] (equation 3).

Hydrolysis of alkyl perfluoroalkylacetylenic esters with aqueous sodium hydroxide gives (*Z*)-β-alkoxy-β-perfluoroalkylacrylic acids [*3*] (equation 4).

0065–7719/95/0187–0757$08.00/1

[1] 2

$$C_4H_9CF{=}CFC{\equiv}CC_4H_9 \xrightarrow[20°C,\ 1h]{H_2SO_4} \left[C_4H_9CF(OH)CHFC(OH){=}CHC_4H_9\right] \xrightarrow{-HF} \left[C_4H_9COCF{=}C(OH)CH_2C_4H_9\right] \longrightarrow C_4H_9COCHFCOCH_2C_4H_9\ (12\%) + \text{2-methyl-1-(}C_4H_9COCF{=}\text{)cyclopentane}\ (70\%)$$

[2] 3

$$C_4F_9C{\equiv}CC(CH_3)_2OH \xrightarrow[reflux, 8\ h]{HCOOH} C_4F_9CH_2COC(CH_3)_2OH$$

$$C_4F_9CH_2COC(CH_3)_2OH \xrightarrow[reflux, 60\ h]{HCOOH} C_4F_9CH_2C(OH)(CH_3)COCH_3\ (15\%) + \text{4-}C_3F_7\text{-2,2-dimethyl-3(2H)-furanone}\ (85\%)$$

$$C_4F_9CH_2COC(CH_3)_2OH \xrightarrow[KOH, 1\ h]{CH_3OH} \text{4-}C_3F_7\text{-2,2-dimethyl-3(2H)-furanone}\ (100\%)$$

[3] 4

$$CF_3C{\equiv}CCOOR \xrightarrow{\text{aqueous 10\% NaOH}} (F_3C)(RO)C{=}C(H)(COOH)\ \ 94\%$$

$R = CH_3, C_2H_5$

At room temperature, base-catalyzed addition of **thiophenol** to ***phenyl trifluoromethyl acetylene*** gives exclusively (*Z*)-3-phenyl-3-phenylthio-1,1,1-trifluoropropene. At 150 °C, the predominant isomer is the *E* isomer (64%), with 23% of the *Z* isomer and 13% of the other two regioisomers [*4*] (equation 5).

Similarly, **6-mercaptopurine** in alkaline medium adds across the triple bond in ***3,3,3-trifluoropropyne*** to form the carbon–sulfur bond to the carbon more remote from the trifluoromethyl group. In trifluoromethyl-*tert*-butyldiacetylene, the bond is formed to the carbon adjacent to the trifluoromethyl group [*5*] (equation 6).

Binucleophiles give regiospecific addition to ***trifluoromethylacetylene*** with and without subsequent cyclization. Initial attack by the softer nucleophile occurs preferentially at the carbon β to the trifluoromethyl group [*6*] (equation 7).

[*4*] **5**

$F_3CC{\equiv}CC_6H_5 \xrightarrow[\text{KOtBu}]{\text{PhSH}}$ (F$_3$C)(H)C=C(SC$_6$H$_5$)(C$_6$H$_5$) + (F$_3$C)(H)C=C(C$_6$H$_5$)(SC$_6$H$_5$)

RT	92%	0%
150°C	23%	64%

[*5*] **6**

$CF_3C{\equiv}CH$ + RSNa (60°C, 3.5 h) → (H)(F$_3$C)C=C(H)(SR) 71%

$CF_3C{\equiv}CC{\equiv}CC(CH_3)_3$ + RSNa (20°C, 3.5 h) → (F$_3$C)(RS)C=C(H)(C≡CC(CH$_3$)$_3$) 81%

R = purin-6-yl (H–N, N, N, N ring)

[*6*] **7**

$CF_3C{\equiv}CH$ + H_2NNH_2, C_2H_5OH (-70°C, 2 h) → $(CF_3CH_2CH{=}N)_2$ 83%

$CF_3C{\equiv}CH$ + $H_2NCH_2CH_2NH_2$, THF (-70°C, 3 day) → 2-(CHCH$_2$CF$_3$)-imidazolidine (N, N ring) 55%

$CF_3C{\equiv}CH$ + $HSCH_2CH_2OH$, KOH (-70°C, 12 h) → $CF_3CH{=}CH{-}S{-}CH_2CH_2OH$ (HOCH$_2$CH$_2$) E, Z 72%

$CF_3C{\equiv}CH$ + $HSCH_2CH_2SH$, KOH (-70°C, 12 h) → $CF_3CH{=}CHSCH_2{-}CH_2SCH{=}CHCF_3$ Z, Z 71%

Examples of photochemical methods of addition to acetylene derivatives are the addition of **methyl disulfide** to ***hexafluoro-2-butyne*** [*7*] (equation 8), of **trifluoromethanethiol** to ***methyl propiolate*** [*8*] (equation 9), of **methanethiol** to ***trifluoromethylacetylene*** and ***hexafluoro-2-butyne*** [*9*] (equation 8), and of **trimethylsilane** to ***tetrafluoropropyne*** [*10*].

Pentafluorosulfur bromide adds to ***pentafluorosulfuracetylene*** to yield a 1:1 adduct that, on dehydrobromination, gives bis[1,2-(pentafluorosulfur)]acetylene [*11*] (equation 10).

At room temperature, *trans*-chlorotetrafluorotrifluoromethylsulfur and acetylene give a 62% yield of *trans*-1-chloro-2-tetrafluorotrifluoromethylsulfurethylene [*12*]. Hexafluoro-2-butyne and FSO_2OCl or FSO_2OBr react to give a mixture of 1:1 and 1:2 adducts [*13*] (equation 11).

[7] $CF_3C{\equiv}CCF_3 + CH_3SSCH_3 \xrightarrow[100°C,\ 1\ h]{UV} CH_3S(CF_3)C{=}C(CF_3)SCH_3$ **8**

95% cis/ trans = 50:50

[8] $HC{\equiv}CCOOCH_3 + CF_3SH \xrightarrow[46h]{UV} CF_3SCH{=}CHCOOCH_3$ **9**

57% cis/ trans = 77.5:22.5

[11] **10**

$SF_5Br + HC{\equiv}CSF_5 \xrightarrow[3\ h]{105°C} (F_5S)(Br)C{=}C(H)(SF_5) \xrightarrow[RT]{KOH} F_5SC{\equiv}CSF_5$

75% 64%

[13] **11**

$CF_3C{\equiv}CCF_3 + XY \xrightarrow[-30°C]{F\ 113} (X)(F_3C)C{=}C(CF_3)(Y) + CF_3C(X)(X)-C(Y)(Y)CF_3 + CF_3C(X)(Y)-C(X)(Y)CF_3$

$Y = OSO_2F$				
	X = Cl	38.8%	25.4%	17%
	X = Br	34.9%	21.2%	0%

Alkynyl trifluoromethyl sulfones add **dimethylformamide** and **dimethylsulfoxide** to give products in good yields. The consistent stereospecificity of the reaction (*syn* addition product only) suggests that an unstable [2+2] cycloaddition intermediate is initially formed [*14*] (equation 12).

[14] **12**

$C_4H_9C{\equiv}CSO_2CF_3 \xrightarrow[RT,\ 1\ day]{DMF} (C_4H_9)((CH_3)_2N)C{=}C(SO_2CF_3)(CHO)$ 50%

$C_4H_9C{\equiv}CSO_2CF_3 \xrightarrow[RT,\ 1day]{DMSO} (C_4H_9)(O^-)C{=}C(SO_2CF_3)(S^+(CH_3)_2)$ 90%

Perfluoroalkylacetylene and **primary amines** [*15*] and **secondary amines** [*16*] react in ether solution to give intermediates that, on acidic hydrolysis, yield *enaminoketones* and a smaller amount of the unsaturated fluoro aldehydes (equation 13).

Other examples of addition reactions that form carbon–nitrogen bonds are the reactions of ***hexafluoro-2-butyne*** with various **heterocycles** [*17*], **aromatic**

nitrones [*18*], **heterocyclic *N*-oxides** [*19*], and **1,2-dihydro-1,3,5-triazines** [*20*] (equation 14).

Gas-phase UV photolysis of **tetrafluorodiphosphine** in the presence of ***hexa-fluoro-2-butyne*** gives equal quantities of the products of *syn* and *anti* addition to the triple bond [*21*] (equation 15).

13

$R_FCF_2C{\equiv}CH$, $R_F = C_5F_{11}, C_7F_{15}$

Left: RNH_2, Et_2O, 35-40°C → $R_FCF{=}CHCH{=}NR$

Right: R_2NH, Et_2O, 0- 50°C, 3-36 h → $R_FCF_2CH{=}CHNR_2$

30% HCl, 35°C, 1 h (each)

$R_FCOCH{=}CHNHR$ + $R_FCF{=}CHCHO$ + $R_FCOCH{=}CHNR_2$

35-65% <10% <10% 15-46%

[*15*] [*16*]

14

[*20*]

2-R-4-NH-6-CHC_6H_5-dihydrotriazine + $CF_3C{\equiv}CCF_3$ →(benzene, 80-90°C, 30 h) 2,6-di-R-pyrimidine with 4,5-di-CF_3 + 2,4-di-R-6-C_6H_5-1,3,5-triazine

$R = \underline{p}\ ClC_6H_5$ 60% 7%

15

[*21*]

$$RC{\equiv}CR' + F_2PPF_2 \xrightarrow[RT,\ 2\text{-}4\ h]{UV} RC(PF_2){=}C(PF_2)R$$

R	R'	%
CF_3	H	25
CF_3	CH_3	60
CF_3	CF_3	65

Additions Forming Carbon–Carbon and Carbon–Silicon Bonds

Electrochemically generated **trifluoromethyl radicals** add to ***1-hexyne*** to give a 1:4 mixture of (*E*)- and (*Z*)-1,1,1-trifluoro-2-heptene [*22*]. Kinetic data on the addition of photochemically generated trifluoromethyl radicals to acetylene and substituted acetylenes were reported [*23*]. Alcohols and aldehydes add to hexa-fluoro-2-butyne in the presence of peroxide and γ-ray initiation [*24*] (equations 16 and 17).

[24] **16**

$$CF_3C{\equiv}CCF_3 + CH_3CHO \longrightarrow (CF_3)HC{=}C(COCH_3)(CF_3) + CF_3CH(COCH_3)CH(COCH_3)CF_3$$

benzoyl peroxide	18%	46%
γ-ray	30%	0%

[24] **17**

$$CF_3C{\equiv}CCF_3 + CH_3OH \xrightarrow[\text{RT}]{\gamma\text{-ray}} (CF_3)(CH_3O)C{=}C(H)(CF_3) + (CF_3)(HOCH_2)C{=}C(H)(CF_3)$$

64% 12%

At reflux, **tetrahydrofuran** slowly adds to terminal perfluoroalkylethylenes, ***perfluoroalkylacetylenes,*** and ethyl perfluoroalkylpropynoates [25] (equation 18).

By contrast, the ionic addition of **enamines** to ***hexafluoro-2-butyne*** is exothermic and gives dieneamines that, on acidic hydrolysis, yield fluoroalkenyl ketones [26] (equation 19).

1-Diethylaminopropyne and *trans*-octafluoro-2-butene give a single product from the *addition of a vinylic fluorine across the triple bond* [27] (equation 20).

[25] **18**

$$\text{THF} + C_6F_{13}C{\equiv}CH \xrightarrow[\text{18 days}]{\text{reflux}} \text{2-(}CH{=}CHC_6F_{13}\text{)-THF}$$

56%

[26] **19**

$$\text{1-}N(CH_3)_2\text{-cyclohexene} + CF_3C{\equiv}CCF_3 \xrightarrow[\text{F 113}]{CHCl_3} \text{2-}(CF_3C{=}CHCF_3)\text{-1-}(CH_3)_2N\text{-cyclohexene} \xrightarrow[50^\circ C,\ 4\ h]{H_3O^+} \text{2-}(CF_3C{=}CHCF_3)\text{-cyclohexanone}$$

not isolated 50% *trans* isomer

[27] **20**

$$CH_3C{\equiv}CN(C_2H_5)_2 + (CF_3)FC{=}C(F)(CF_3) \xrightarrow[\text{RT, 2 h}]{\text{F 113, dioxane}} (CF_3)FC{=}C(CF_3)\text{-}C(CH_3){=}C(F)N(C_2H_5)_2$$

References are listed on pages 765–766.

Examples of **perfluoroalkyl iodide** addition to the *triple bond* include free radical addition of perfluoropropyl iodide to 1-heptyne [*28*] (equation 21), thermal and free radical-initiated addition of iodoperfluoroalkanesulfonyl fluorides to acetylene [*29*] (equation 22), thermal addition of perfluoropropyl iodide to hexafluoro-2-butyne [*30*] (equation 23), and palladium-catalyzed addition of perfluorobutyl iodide to phenylacetylene [*31*] (equation 24). The *E* isomers predominate in these reactions. Photochemical addition of trifluoromethyl iodide to vinylacetylene gives predominantly the 1:4 adduct by addition to the double bond [*32*]. Platinum-catalyzed addition of perfluorooctyl iodide to 1-hexyne in the presence of potassium carbonate, carbon monoxide, and ethanol gives ethyl β-perfluorooctyl-α-butylpropenoate [*33*] (equation 25).

[*28*] **21**

$$C_3F_7I + HC{\equiv}C(CH_2)_4CH_3 \xrightarrow[70°C,\ 20\ h]{AIBN} C_3F_7CH{=}CI(CH_2)_4CH_3$$

91%, Z / E = 1:19

[*29*] **22**

$$FO_2S(CF_2)_2O(CF_2)_2I + HC{\equiv}CH \xrightarrow[130°C,\ 36\ h]{Bz_2O_2} FO_2S(CF_2)_2O(CF_2)_2CH{=}CHI$$

78%, Z , E mixture

[*30*] **23**

$$C_3F_7I + CF_3C{\equiv}CCF_3 \xrightarrow[7\ days]{220°C} CF_3CI{=}C(CF_3)C_3F_7$$

68%, Z / E = 3:4

[*31*] **24**

$$R_fI + HC{\equiv}CR \xrightarrow[\substack{hexane \\ 60\text{-}67°C,\ 24\ h}]{Pd\ catalyst} R_fCH{=}CIR$$

57-76% E>Z

$R_f = C_3F_7, C_4F_9, C_6F_{13}$ $R = C_6H_5, C_6H_{13}, C_4H_9, (CH_3)_3Si$

[*33*] **25**

$$C_8F_{17}I + C_4H_9C{\equiv}CH + CO \xrightarrow[C_2H_5OH,\ K_2CO_3]{Pt\ catalyst} C_8F_{17}CH{=}C(C_4H_9)COOC_2H_5$$

Bz_2O_2 = benzoyl peroxide AIBN = 2,2'-azobis-2-(methylpropionitrile)
PdP = $Pd(PPh_3)_4$ PtP = $PtCl_2(PPh_3)_2$

Electrochemical addition of perfluorobutyl iodide to 2-methyl-3-butyn-2-ol followed by basic dehydroiodination and thermal cleavage gives perfluorobutylacetylene in an overall yield of 83% [*34*] (equation 26).

Addition of **1,8-diiodoperfluorooctane** to ***bis(trimethylsilyl)acetylene*** followed by potassium fluoride desilylation gives 3,3,4,4,5,5,6,6,7,7,8,8,9,9,10,10-hexadecafluoro-1,1,1-dodecadiyne [*35*] (equation 27).

Cesium fluoride-promoted *telomerization* of equimolar amounts of perfluoro-2-butyne and 2-bromoperfluoro-2-butene gives a mixture of hexadienes or, with excess of perfluoro-2-butyne, linear telomers [*36*] (equation 28). *Cyclotrimerization* of perfluoro-2-butyne to hexakis(trifluoromethyl)benzene is catalyzed by acetylacetonato(1,5-cyclooctadiene)rhodium(I) [*37*] (equation 29), tetrakis(trifluorophosphine)nickel(0) [*38*], and π-bis(benzene)chromium [*39*].

[*34*] **26**

$$C_4F_9I + HC{\equiv}C(CH_3)_2COH \xrightarrow{i} C_4F_9CH{=}CIC(CH_3)_2OH \xrightarrow{ii} C_4F_9CH{\equiv}CC(CH_3)_2OH \xrightarrow{iii} 83\%\ C_4F_9C{\equiv}CH$$

i = carbon fiber electrode, 0.7 A, RT, 2 h
ii = KOH, methanol, RT
iii = KOH pellets, 80°-120°C

[*35*] **27**

$$I(CF_2)_8I + (CH_3)SiC{\equiv}CSi(CH_3)_3 \xrightarrow{i} \underset{86\%}{(CH_3)_3SiC{\equiv}C(CF_2)_8C{\equiv}CSi(CH_3)_3} \xrightarrow{ii} HC{\equiv}C(CF_2)_8C{\equiv}CH$$

i = trace I_2, 200°C, 57 h
ii = KF dihydrate, methanol, RT, 20 h

[*36*] **28**

$$CF_3C{\equiv}CCF_3 + CF_3CF{=}CBrCF_3 \xrightarrow[30°C,\ 2\ h]{CsF,\ CH_3CN} CF_3CF{=}C(CF_3)C(CF_3){=}CBrCF_3$$

89% EE>> EZ or ZZ

[37] **29**

$$CF_3C{\equiv}CCF_3 \xrightarrow[benzene,\ RT, 17\ h]{Rh(acac)(cod)} [C_6(CF_3)_6]\text{—}Rh\text{—}[CF_3C{=}CCF_3][acac]$$

56%

Thermal insertion reactions of dihydridobis(η^5- cyclopentadienyl)molybdenum with trifluoromethylacetylene give exclusively *syn* addition. At –40 °C, hexafluoro-2-butyne inserts to the Mo–H bond to give *anti* addition [*40*] equation 30).

[*40*] **30**

$$\underset{50\%}{CF_3C(R_2MoH){=}CH_2} \xleftarrow[toluene,\ -70°C]{CF_3C{\equiv}CH} R_2MoH_2 \xrightarrow[CH_2Cl_2,\ -40°C]{CF_3C{\equiv}CCF_3} (F_3C)(R_2MoH)C{=}C(H)(CF_3)$$

$R = C_2H_5$

References are listed on pages 765–766.

References for Pages 757–764

1. Tellier, F.; Sauvetre, R.; Normant, J. F. *J. Organomet. Chem.* **1987,** *328,* 1.
2. Gomez, L.; Calas, P.; Commeyras, A. *J. Chem. Soc., Chem. Commun.* **1985,** 1493.
3. Huang, Y.-Z.; Shen, Y.-C.; Xin, Y.-K.; Wang, Q.-W.; Wu., W.-C. *Yu Chi Hua Hsueh* **1981,** *3,* 193; *Chem. Abstr.* **1981,** *95,* 168205f.
4. Bumgardner, C. L.; Bunch, J. E.; Whangbo, M. H. *Tetrahedron Lett.* **1986,** *27,* 1883.
5. Turbanova, E. S.; Orlova, N. A.; Ratsino, E. V.; Sokolov, L. B. *Zh. Org. Khim.* **1981,** *17,* 736 (Engl. Transl.); *Chem. Abstr.* **1981,** *95,* 150598b.
6. Stepanova, N. P.; Lebedev, V. B.; Orlova, N. A.; Turbanova, E. S.; Petrov, A. A. *Zh. Org. Khim.* **1988,** *24,* 6929 (Engl. Transl.); *Chem. Abstr.* **1989,** *110,* 7617e.
7. Toy, M. S.; Stringham, R. S. *Ind. Eng. Chem., Prod. Res. Dev.* **1983,** *22,* 8.
8. Harris, J. F., Jr. *J. Org. Chem.* **1972,** *37,* 1340.
9. Sharp, D. W. A.; Miguel, H. T. *Isr. J. Chem.* **1978,** *17,* 144; *Chem. Abstr.* **1978,** *89,* 107970f.
10. Haszeldine, R. N.; Pool, C. R.; Tipping, A. E. *J. Chem. Soc., Perkin Trans. 1* **1974,** 2293.
11. Berry, A. D.; DeMarco, R. A.; Fox, W. B. *J. Am. Chem. Soc.* **1979,** *101,* 737.
12. Darragh, J. I.; Haran, G.; Sharp, D. W. A. *J. Chem. Soc., Dalton Trans.* **1973,** 2289.
13. Fokin, A. V.; Studnev, Y. N.; Rapkin, A. I.; Krotovitch, I. N.; Verenikin, O. V. *Izv. Akad. Nauk SSSR* **1985,** *5,* 1094; *Chem. Abstr.* **1985,** *103,* 214819t.
14. Hanack, M.; Wilhelm, B. *Angew. Chem.* **1989,** *101,* 1083; *Angew. Chem. Int. Ed. Engl.* **1989,** *28,* 1057.
15. Gallucci, J.; Le Blanc, M.; Riess, J. G. *J. Chem. Res. (S)* **1978,** *11,* 430.
16. Le Blanc, M.; Santini, G.; Gallucci, J.; Riess, J. G. *Tetrahedron* **1977,** *33,* 1453.
17. Davidson, J. L.; Murray, I. E. P.; Preston, P. N. *J. Chem. Res. Synop.* **1981,** *5,* 126.
18. Kobayashi, Y.; Kumadaki, I.; Yoshida, T. *Heterocycles* **1977,** *8,* 387.
19. Kobayashi, Y.; Kumadaki, I.; Fujino, S. *Heterocycles* **1977,** *7,* 871.
20. Kapran, N. A.; Lukamanov, V. G.; Yagupol'skii, L. M.; Cherkasov, V. M. *Khim. Geterotsikl. Soedin.* **1977,** *1,* 122; *Chem. Abstr.* **1977,** *86,* 171389p.
21. Morse, J. G.; Mielcarek, J. J. *J. Fluorine Chem.* **1988,** *40,* 41.
22. Brooks, C. J.; Coe, P. L.; Pedler, A. E.; Tatlow, J. C. *J. Chem. Soc., Perkin Trans. 1* **1978,** 202.

23. El-Soueni, A. E.; Tedder, J. M.; Walton, J. C. *J. Chem. Soc., Faraday Trans. 1* **1981,** *77,* 89.
24. Chambers, R. D. ; Jones, C. G. P.; Silvester, M. J. *J. Fluorine Chem.* **1986,** *32,* 309.
25. Chauvin, A.; Greiner, J.; Pastor, R.; Cambon, A. *J. Fluorine Chem.* **1985,** *27,* 385.
26. Blazejewski, J. C.; Cantacuzene, D. *Tetrahedron Lett.* **1973,** *43,* 4241.
27. Blazejewski, J. C.; Cantacuzene, D.; Wakselman, C. *Tetrahedron Lett.* **1974,** *24,* 2055.
28. Brace, N. O.; Van Elswyk, J. E. *J. Org. Chem.* **1976,** *41,* 766.
29. Chen, L. F.; Mohtasham, J.; Gard, G. L. *J. Fluorine Chem.* **1989,** *43,* 329.
30. Fields, R.; Haszeldine, R. N.; Kumadaki, I. *J. Chem. Soc. Perkin Trans. 1* **1982,** 2221.
31. Ishihara, T.; Kuroboshi, M.; Okada, Y. *Chem. Lett.* **1986,** *11,* 1895.
32. El-Soueni, A.; Tedder, J. M.; Walton, J. C. *J. Fluorine Chem.* **1981,** *17,* 51.
33. Urata, H.; Yugari, H.; Fuchikami, T. *Chem. Lett.* **1987,** *5,* 833.
34. Calas, P.; Moreau, P.; Commeyras, A. *J. Chem Soc., Chem. Commun.* **1982,** 433.
35. Baum, K.; Bedford, C. D.; Hunadi, R. J. *J. Org. Chem.* **1982,** *47,* 2251.
36. Miller, W. T.; Hummel, R. J.; Pelosi, L. F. *J. Am. Chem. Soc.* **1973,** *95,* 6850.
37. Barlex, D. M.; Jarvis, A. C.; Kemmitt, R. D. W.; Kimura, B. Y. *J. Chem. Soc., Dalton Trans.* **1972,** 2549.
38. Davidson, J. L.; Sharp, D. W. A. *J. Fluorine Chem.* **1976,** *7,* 145.
39. Huang, Y.; Li, J.; Zhou, J.; Chen, J.; Zhu, Z. *Sci. Sin. Ser. B (Engl. Ed.)* **1983,** *26,* 1249.
40. Nakamura, A.; Otuska, S. J. *Mol. Catal.* **1975/76,** *1,* 285.

Cycloadditions Forming Three- and Four-Membered Rings

by B. E. Smart

Cycloadditions Forming Three-Membered Rings

The addition of **fluorinated carbenes** or **carbenoids** to *alkenes* and *alkynes* is the most popular, one-step method for making *fluorinated cyclopropanes* and *cyclopropenes*. *α-Fluorocarbenes* are particularly well behaved, because they all have *singlet ground states* [*1, 2*] and therefore usually *add stereospecifically* to alkenes and *do not insert* into C–H bonds competitively with addition. Moreover, quantitative competition studies of carbene additions to alkenes near room temperature show that α-fluorocarbenes are more selective than other α-halocarbenes, with difluorocarbene being the most selective electrophilic carbene known [*3, 4*]. The relative selectivities, however, can be quite temperature dependent [*5, 6*]. The numerous preparations and cycloadditions of fluorocarbenes have been reviewed thoroughly [*7, 8, 9, 10*].

Among the many methods of generating **difluorocarbene,** the treatment of *bromodifluoromethylphosphonium bromides* with potassium or cesium fluoride is particularly useful at room temperature or below [*11, 12, 13*]. The sodium iodide-promoted decomposition of *phenyl(trifluoromethyl)mercury* is very effective at moderate temperatures [*8, 14*]. *Hexafluoropropylene oxide* [*15*] and *chlorodifluoroacetate* salts [*7*] are excellent higher temperature sources of difluorocarbene.

The phosphonium salt method works best with nucleophilic olefins [*11, 12, 16, 17, 18, 19*] (Table 1 and equations 1–3) and has been used in mechanistically important studies of difluorocarbene additions to norbornadienes [*20, 21, 22, 23*] that provided the first example of a concerted homo-1,4-addition (equation 4). A recent modification uses catalytic 1,4,7,10,13,16-hexaoxacyclooctadecane (18-crown-6) to shorten reaction times and increase yields with less nucleophilic olefins [*12*] (Table 1). Neither procedure, however, compares with the use of **phenyl(trifluoromethyl)mercury** or **(trifluoromethyl)trimethyltin reagents** [*8, 14*] for efficient reactions with less nucleophilic olefins (equations 3 and 5) and cyclic dienes [*24, 25*] (equations 6 and 7).

The use of dibromodifluoromethane with zinc [*26*] or lead metal [*27*] is a simple, economical method for generating *difluorocarbene* under mild conditions, but the reaction is limited to highly nucleophilic olefins (equation 8).

0065–7719/95/0187–0767$09.80/1

[*17*] $(C_6H_5)_3P^+CF_2Br\ Br^-$ / CsF, triglyme, 26 °C → F_2 38% + F_2 22% **1**

[*19*] $(C_6H_5)_3P^+CF_2Br\ Br^-$ / KF, triglyme, RT → F_2 92% **2**

[*19*] $(C_6H_5)_3P^+CF_2Br\ Br^-$ / KF, triglyme, RT → 8%

[*14*] $C_6H_5HgCF_3$ / NaI, C_6H_6, 80 °C → 83% F_2 **3**

[*20*] **4**

$(C_6H_5)_3P^+CF_2Br\ Br^-$ / KF, triglyme, 25 °C (54-62%) → F_2 34% + F_2 55-62% + F_2 4-11%

[*11*] $[(CH_3)_2N]_3P^+CF_2Br\ Br^-$ / CsF, triglyme, 85 °C → 21%

[*12*] 2 $(C_6H_5)_3P^+CF_2Br\ Br^-$ 2 KF / 18-crown-6, glyme, RT → 37%

[*8*] $C_6H_5HgCF_3$, NaI / C_6H_6, 80 °C → 83%

[*8*] $(CH_3)_3SnCF_3$, NaI / glyme, 80-85 °C → 89%

F_2 **5**

References are listed on pages 791–796.

Table 1. *gem*-Difluorocyclopropanes from Bromo(difluoromethyl)-triphenylphosphonium Bromide

$R^1R^4C{=}CR^2R^3$ + n $(C_6H_5)_3P^+CF_2Br\ Br^-$ + n MF ⟶ R^1 R^2 R^4 R^3 F F

R^1	R^2	R^3	R^4	*n*	*M*	*Conditions*[a]	*Yield (%)*	*Ref.*
CH_3	CH_3	CH_3	CH_3	1	Cs	A	79 [b]	11
CH_3	CH_3	CH_3	H	1	Cs	A	67	11
C_4H_9	CH_3	CH_3	H	2	K	B	89	12
CH_3	H	H	CH_3	0.25	Cs	A	66 [b]	11
CH_3	CH_3	H	H	0.25	Cs	A	6 [b]	11
C_3H_7	CH_3	H	H	2	K	B	43	12
CH_3	H	CH_3	H	0.25	Cs	A	12 [b]	11
C_3H_7	H	CH_3	H	2	K	B	42	12
C_4H_9O	H	H	CH_3	1	Cs	A	60	12
C_4H_9O	H	H	CH_3	2	K	B	82	12
CH_2OTHP [c]	CH_3	CH_3	H	2	K	B	90	18
CH_2OTHP [c]	CH_3	C_6H_5	H	2	K	B	76	18
$CH_2{=}CH$	H	H	H	1	K [d]	A	55	16

[a] A, triglyme, room temperature, 24 h; B, glyme, 0.1 equiv., 18-crown-6, 25 °C.
[b] Gas–liquid chromatography yield.
[c] THP, 2*H*-tetrahydropyran.
[d] 4KF.

[*24*] cyclohexa-1,3-diene $\xrightarrow[\text{Bu}_4\text{NI, 18-crown-6, 85 °C}]{C_6H_5HgCF_3\text{, NaI}}$ F_2 56% **6**

[*25*] cyclohexa-1,4-diene $\xrightarrow[C_6H_6\text{, 85 °C}]{C_6H_5HgCF_3\text{, NaI}}$ F_2 + F_2 F_2 **7**

34% 45%

[26] [27]

$R^1(R^4)C{=}C(R^2)R^3$ —(A: CF_2Br_2, Zn, Cat. I_2, THF, RT; B: Pb, cat. Bu_4NBr, $ClCH_2CH_2Cl$, 50-70 °C)→ 1,1-difluoro-2-R^1-2-R^4-3-R^2-3-R^3-cyclopropane (F_2) **8**

R^1	R^2	R^3	R^4	A	B
CH_3	CH_3	CH_3	CH_3	96%	90%
C_6H_5	H	H	CH_3	71%	55%
C_6H_5	H	H	H	15%	17%
CH_3	CH_3	H	H	7%	–

The reaction of chlorodifluoromethane with alkoxide ions generated in low concentration from halide ions and epoxides [*28, 29*] is an interesting, higher temperature method that gives good to excellent yields of *gem*-difluorocyclopropanes from just moderately nucleophilic olefins (equation 9).

[*29*]

$R^1(R^4)C{=}C(R^2)R^3$ + $CHClF_2$ —(epichlorohydrin CH_2—$CHCH_2Cl$ (O), cat. Bu_4NCl, 120–140 °C)→ 1,1-difluoro-2-R^1-2-R^4-3-R^2-3-R^3-cyclopropane (F_2) **9**

R^1	R^2	R^3	R^4	
CH_3	CH_3	H	H	72%
CH_3	H	CH_3	H	60%
C_6H_5	H	H	H	95%
CH_2=CH	H	H	H	61%

Hexafluoropropylene oxide (HFPO), which decomposes reversibly to difluorocarbene and trifluoroacetyl fluoride with a half-life of about 6 h at 165 °C [*30*], is a versatile reagent. Its pyrolysis with olefins is normally carried out at 180–200 °C, and yields are usually good with either electron-rich or electron-poor olefins [*31, 32, 33, 34, 35, 36, 37*] (Table 2). The high reaction temperatures allow the cyclopropanation of very electron poor double bonds [*38*] (equation 10) but can result in rearranged products [*39, 40, 41*] (equations 11–13).

High-temperature carbene additions to ***allenes*** are especially prone to give rearranged methylenecyclopropanes [*42, 43, 44*] (equations 14 and 15), and there-

Table 2. *gem*-Difluorocyclopropanes from Alkenes and Hexafluoropropylene Oxide

$$R^1R^4C{=}CR^2R^3 + CF_3CF(O)CF_2 \longrightarrow \text{1,1-difluoro-2-}R^1\text{-2-}R^4\text{-3-}R^2\text{-3-}R^3\text{-cyclopropane} + CF_3COF$$

R¹	*R²*	*R³*	*R⁴*	*Temp. (°C)*	*Time (h)*	*Yield (%)*	*Ref.*
Cl	Cl	H	F	200	8	50	31
F	Cl	H	F	180	6	57	32
F	F	Cl	H	180	6	75	33
F	Cl	F	H	180	6	80	33
F	Cl	F	Cl	185	8	85	33
F	F	H	F	185	6	65	32
CH_2Cl	H	H	Cl	190	4	62	34
CH_2Cl	F	F	Cl	185	10	94	35
CF_2Br	H	H	Br	185	8	67	35
CF_3O	F	F	F	210	8	60	33
CO_2CH_3	F	F	F	225	7	17	36
C_5H_{11}	H	H	H	200	3	88	37

[*38*] Perfluoro-3-methylindene (F-ring, CF_3, F, F_2) $\xrightarrow[\text{225-230 °C}]{\text{HFPO}}$ difluorocyclopropane adduct (CF_3, F_2, F, F_2) 83% **10**

HFPO = hexafluoropropylene oxide

fore, other routes to fluorinated methylenecyclopropanes have been developed [*34, 35, 45, 46, 47*] (equations 16–19).

gem-Difluorospiropentanes are conveniently prepared from methylenecyclopropanes and hexafluoropropylene oxide [*35*] (equation 20).

The pyrolysis of **sodium chlorodifluoroacetate** is still a widely used, classical method for generating difluorocarbene, especially with enol and allyl acetates [*48, 49, 50, 51*] (equation 21). A convenient alternative that avoids the hygroscopic salt uses methyl chlorodifluoroacetate with 2 equivalents of a lithium chloride–hexamethylphosphoric triamide complex at 75–80 °C in triglyme [*52*]. Yields are excellent with electron-rich olefins but are less satisfactory with moderately nucleophilic alkenes (4–5% yields for 2-butenes).

[*39*] **11**

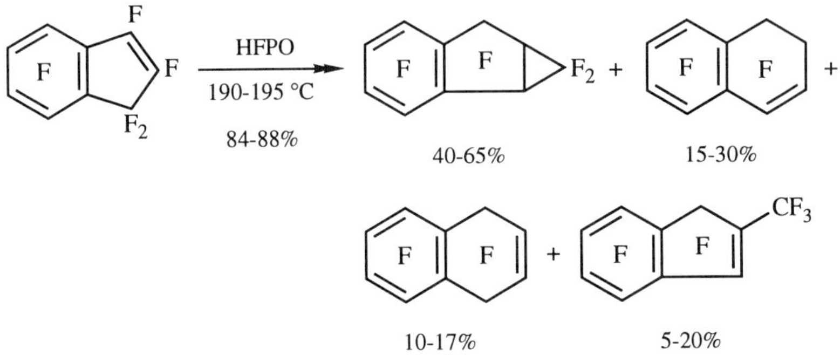

HFPO, 190-195 °C, 84-88%

40-65% + 15-30% + 10-17% + 5-20%

[*40*] **12**

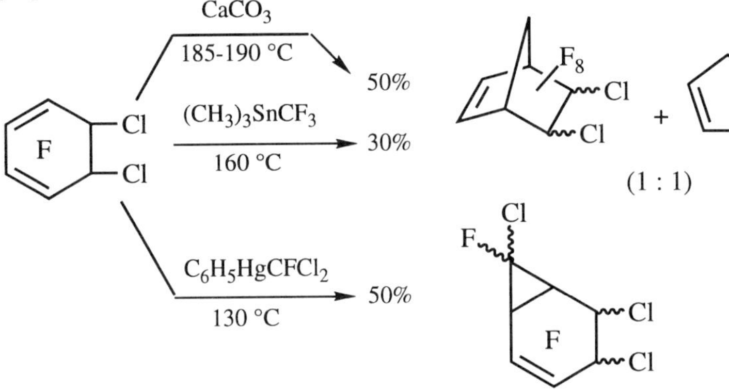

2 HFPO, $CaCO_3$, 185-190 °C → 50%

$(CH_3)_3SnCF_3$, 160 °C → 30%

(1 : 1)

$C_6H_5HgCFCl_2$, 130 °C → 50%

[*41*] $CF_2{=}C(CF_3)C(O)C_2F_5$ → HFPO, 225 °C → 46% **13**

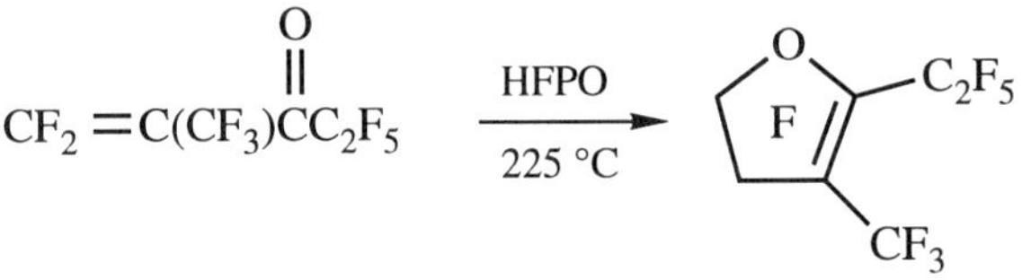

[*42*] **14**

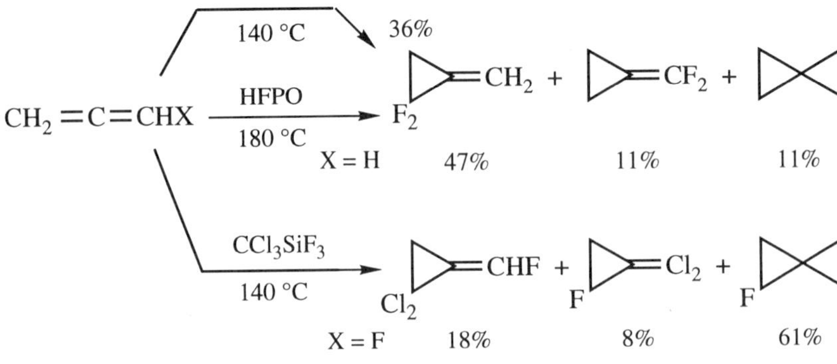

$CH_2{=}C{=}CHX$

$(CH_3)_3SnCF_3$, 140 °C → 36%

HFPO, 180 °C, X = H → 47% + 11% + 11% [*43*]

CCl_3SiF_3, 140 °C, X = F → 18% + 8% + 61% [*43*]

[*44*] $CF_2=C=C(R)(CF_3)$ —(HFPO, 175 °C)→ 2-R-2-CF_3-3,3-difluoro-1-($=CF_2$)cyclopropane; R = CF_3, 52%; C_3F_7, 70% **15**

—(CCl_3SiF_3, 140 °C)→ 2-R-2-CF_3-3,3-difluoro-1-($=CCl_2$)cyclopropane; R = CF_3, 89%; C_3F_7, 84%

[*35, 45*] Zn, dioxane, 100 °C: X = F, 73%

X_2, F_2 cyclopropane with Cl and CX_2Cl → X_2, F_2 cyclopropane $=CX_2$ **16**

[*34*] Zn, $ZnCl_2$, DMSO, 155 °C: X = H, 79%

[*35*] F_2 cyclopropane with Br and CF_2Br —(Zn, dioxane, 80-85 °C)→ F_2 cyclopropane $=CF_2$ 63% **17**

[*46*] F_2, F_2 cyclopropane with $CH_2SeC_6H_5$ and H —(1. O_3, -78 °C; 2. 100-150 °C)→ F_2, F_2 cyclopropane $=CH_2$ 54% **18**

[*47*] pyrazoline (N=N) with $=CHF$ —(CCl_4, 140 °C)→ F-cyclopropane $=CH_2$ (90.5%) + cyclopropane $=CHF$ (9.5%) **19**

References are listed on pages 791–796.

[*35*] 20

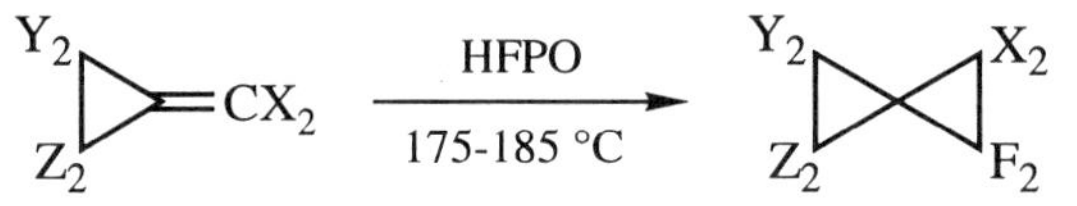

X	Y	Z	
H	H	H	34%
F	F	H	63%
F	F	F	40%

[*48*] 21

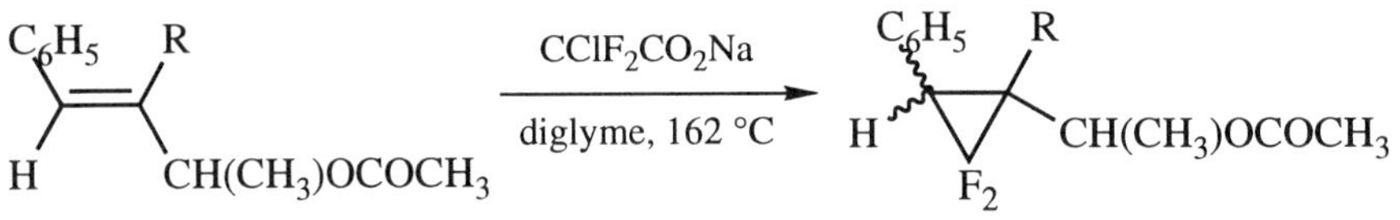

R = H, 88%; CH_3, 64%

To generate **chlorofluoro**- and **bromofluorocarbene,** *phenyl(dichlorofluoromethyl)*- and *phenyl(dibromofluoromethyl)mercury* are vastly superior to other reagents [*8, 14*]. High yields (>80%) of fluorohalocyclopropanes are typical, and even normally unreactive olefins can be used (equation 22).

[*8,14*] 22

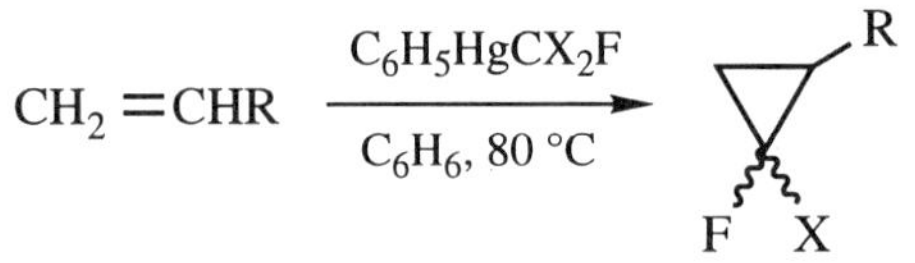

X = (Br, Cl) R = $CH_2Si(CH_3)_3$, 51%, 95%; $OCOCH_3$, 95%, 86%; CN, 24%, 40%

Phase-transfer systems for generating fluorohalocarbenes from the corresponding haloforms are simple and effective with nucleophilic olefins [*53, 54, 55, 56, 57, 58, 59*] (Table 3 and equations 23 and 24), and the process was extended recently to difluorocarbene [*60*]. Chlorofluorocarbene also can be generated from fluorotrichloromethane and titanium(0) produced in situ ($TiCl_4$ + $LiAlH_4$). Yields of *gem*-chlorofluorocyclopropanes range from excellent (85% for α-methylstyrene) to poor (12% for 1-hexene) [*61*].

Modest yields of monofluorocyclopropanes are obtained by UV irradiation of diiodofluoromethane with olefins [*62*], but in general, there are *no practical methods for preparing fluorocarbene* [*7*]. *Monofluorocyclopropanes* therefore are usually made by reduction of *gem-bromofluoro-* or *gem-chlorofluorocyclopropanes* [*63*].

Substituted fluorocarbenes have been generated by several methods, but organometallic transfer reagents again are most useful synthetically [*64, 65, 66, 67, 68*] (equations 25–27). **Diazirines** cleanly produce carbenes thermally [*69, 71*] or photochemically [*70*] (equation 28) but are more useful for mechanistic studies [*3*, *4*] than for synthesis. Because of the low selectivity of cycloadditions of bis(trifluoromethyl)carbene [*72*, *73*], alternative cyclopropanation routes to bis(trifluoromethyl)cyclopropanes were investigated but with only limited success [*74*, *75*] (equation 29). There have been few reports on **unsaturated fluorocarbenes.** ***Difluorovinylidene*** ($CF_2{=}C$) is formed by mercury-sensitized photolysis [*76*] or multiphoton irradiation [*77*] of trifluoroethylene and by photolysis of difluoropropadienone [*78*], but these methods have little preparative value.

Table 3. *gem*-Fluorohalocyclopropanes from Phase-Transfer Cycloadditions

$R^1R^4C{=}CR^2R^3 + CHX_2F \longrightarrow$ 1-F-1-X-2-R^1-2-R^4-3-R^2-3-R^3-cyclopropane

Conditions[a]	*X*	*R^1*	*R^2*	*R^3*	*R^4*	*Yield (%)*	*Ref.*
60% KOH,	Cl	CH_3	CH_3	H	H	70	53
TMBAC,		$AcO(CH_2)_3$	CH_3	H	H	40	
RT		$HOCH_2$	CH_3	H	H	24	
50% KOH,	Cl	CH_3	CH_3O	H	CH_3	75	54
18-crown-6,	Cl	CH_3	H	C_2H_5O	H	85	
0 °C							
58% KOH,	Cl	C_2H_5O	H	H	C_6H_5O	76	55
CH_2Cl_2, CCl_4							
TEBAC, 40 °C							
50% NaOH,	Br	*t*-C_4H_9	H	H	H	57	56
CH_2Cl_2		C_6H_5O	H	H	H	71	
TEBAC, 40 °C		C_6H_5	H	H	H	84	
50% NaOH,	I	C_6H_5	H	H	H	60	57
CH_2Cl_2		C_6H_5	H	H	CH_3	17	
TEBAC, 20 °C							

[a] TMBAC, trimethylbenzylammonium chloride; RT, room temperature; TEBAC, triethylbenzylammonium chloride.

References are listed on pages 791–796.

[*58*] $\xrightarrow[\text{TOMAC, ether, -2 °C}]{CHCl_2F\text{, 50\% NaOH}}$ 52% **23**

TOMAC = $(C_8H_{17})_3NCH_3$ Cl

[*59*] **24**

$\xrightarrow[CH_2Cl_2\text{, TEBAC, 0 °C-RT}]{CHBr_2F\text{, 50\% NaOH}}$ Br F

90%

TEBAC = $C_6H_5CH_2N(C_2H_5)_3$ Cl

[*64*] $\xrightarrow[C_6H_6\text{, 158 °C}]{C_6H_5HgCBrFCF_3}$ 87% CF_3 F

[*65*] $\xrightarrow[\text{150 °C}]{CHF_2CF_2SiF_3}$ 57% F CHF_2 **25**

[*66*] $\xrightarrow[C_6H_6\text{, 144 °C}]{C_6H_5HgCClFCO_2C_2H_5}$ 69% F $CO_2C_2H_5$

[*65*] $CX_2{=}CX_2$ $\xrightarrow[\text{150 °C}]{CHF_2CF_2SiF_3}$ X_2 X_2 F CHF_2 **26**

X = H, 82%,
X = F, 74%

[*67*] $\xrightarrow[C_6H_6\text{, 135 °C}]{C_6H_5HgCBrClCF_3}$ 27% Cl CF_3

CH_3 CH_3 C=C H H **27**

[*68*] $\xrightarrow[\text{175 °C}]{ClCF_2CF_2SiF_3}$ 63% F CF_2Cl

[*69*] 170 °C, $R = CF_3$, 31%

[*70*] (R)(F)-diazirine + tetramethylethylene —hν, $R = C_6H_5$→ 49% **28** (R, F cyclopropane)

[*71*] 115 °C, $R = C_6H_5O$, 39%

[*74*] $(CF_3)_2C{=}C(X)(Y)$ —$(CH_3)_2S^+\bar{C}HCO_2C_2H_5$, DMF, RT→ cyclopropane (X, $CO_2C_2H_5$, Y, H, CF_3, CF_3)

$X = H,\ Y = CO_2C_2H_5$, 37%
$X = H,\ Y = CH_2OCH_2C_6H_5$, 47% **29**

[*75*] —$(CH_3)_2S^+\bar{C}HCO_2C_2H_5$, CH_3CN, RT→ $(CF_3)_2C{=}CFC^-(CO_2C_2H_3)(^+S(CH_3)_2)$

$X = Y = F$ 71%

Various carbene-transfer reactions can be used with both electron-rich and electron-poor ***alkynes*** to make *fluorinated cyclopropenes* [*9, 13, 79, 80, 81, 82*] (Table 4). Haloacetylenes are too thermally unstable for most cycloaddition conditions, and simple fluorinated cyclopropenes are made by other methods [*32, 45, 83, 84*] (equations 30–32).

Cycloadditions Forming Four-Membered Rings

Fluorinated cyclobutanes and cyclobutenes are relatively easy to prepare because of the propensity of many *gem*-difluoroolefins to thermally cyclodimerize and cycloadd to alkenes and alkynes. Even with dienes, fluoroolefins commonly prefer to form cyclobutane rather than six-membered-ring Diels–Alder adducts. Tetrafluoroethylene, chlorotrifluoroethylene, and 1,1-dichloro-2,2-difluoroethylene are especially reactive in this context. Most evidence favors a stepwise diradical or, less often, a dipolar mechanism for [2+2] cycloadditions of fluoroalkenes [*85, 86*], although arguments for a symmetry-allowed, concerted $[2_s+2_a]$ process persist [*87*]. The scope, characteristic features, and mechanistic studies of fluoroolefin

Table 4. Fluorinated Cyclopropenes from Cycloadditions to Alkenes

$R^1C{\equiv}CR^2$ + [CXY] ⟶ (cyclopropene with R^1, R^2 on the double bond and X, Y on the saturated carbon)

Conditions	*X, Y*	*R¹*	*R²*	*Yield, (%)*	*Ref.*
$(CH_3)_3SnCF_3$, 143–145 °C	F, F	CF_3	H	15	79
	F, F	C_2F_5	H	23	
	F, F	$(CH_3)_3Sn$	CF_3	39	
	F, F	$(CH_3)_3Ge$	CF_3	84	
CCl_3SiF_3, 140 °C	Cl, Cl	CF_3	CF_3	93	80
3-chloro-3-(trifluoromethyl)diazirine (CF_3, Cl, N=N), 200 °C	Cl, CF_3	CF_3	CF_3	56	81
$C_6H_5HgCBr_3$, C_6H_6, 80 °C	Br, Br	CF_3S	CF_3S	90	82
$3(C_6H_5)_3P^+CF_2Br\ Br^-$, 3KF 18-crown-6, glyme, 25 °C	F, F	C_6H_5	H	79	13
		C_6H_5	CH_3	66	
		C_3H_7	CH_3	46	

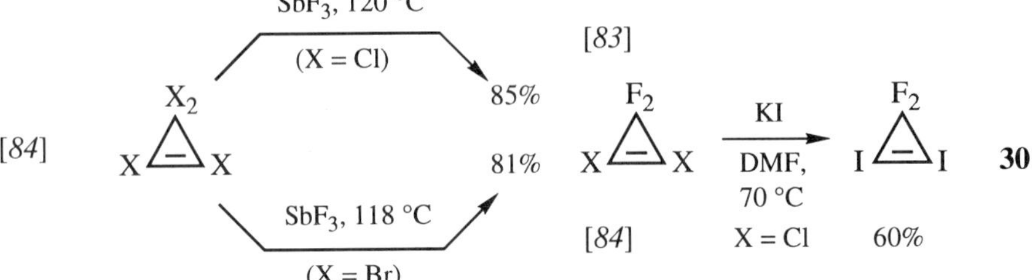

[*32*] F_2 FCl△FCl —Zn / C_2H_5OH, 55 °C→ F_2 F△F 85% **31**

[*45*] F_2, F_2 ▷$=CF_2$ —cat. $ZnBr_2$ / diglyme, 100 °C→ F_2 F△CF_3 100% **32**

cycloadditions have been reviewed extensively [*85, 88*], and only more recent developments will be featured here.

A significant, recent theoretical study of the reaction between tetrafluoroethylene and 1,3-butadiene [*89*] substantiates the hypothesis that the low π-bond energy in tetrafluoroethylene, which is due mainly to the preference of a CF_2 radical center for pyramidal geometry [*90*], is largely responsible for its kinetic reactivity in biradical cycloadditions. *Geminal fluorine substitution (CF_2=) is normally required for high reactivity, but it is neither necessary nor sufficient for four-membered-ring formation.* For instance, nonfluorinated methylenecyclopropanes with strained, weak π-bonds can undergo thermal [2+2] cycloadditions [*34, 91*], and 1,2-dichloro-1,2-difluoroethylene, although much less reactive than 1,1-dichloro-2,2-difluoroethylene, does form cyclobutane products with alkenes and dienes [*92, 93*]. 1,1-Difluoroethylene, whose π-bond strength is comparable with that of ethylene, and perfluoromethylenecyclopropane [*94*] are examples of *gem*-difluoroalkenes that uncharacteristically do not cyclodimerize.

The reaction rates and product yields of [2+2] cycloadditions are expectedly enhanced by electronic factors that favor radical formation. Olefins with geminal capto–dative substituents are especially efficient partners (equations 33 and 34) because of the synergistic effect of the electron acceptor (capto) with the electron donor (dative) substituents on radical stability [*95*].

[*95*] CF_2=CFCl —CH_2=CHCN / 160 °C, 8 h→ 65%; —CH_2=CRCN / 120 °C, 8 h→ F_2 cyclobutane with F, Cl, CN, R **33**

R = Cl, 72%; C_6H_5S, 81%; *t*-C_4H_9S, 90%

[*95*] CF_2=$CRSC_6H_5$ —$(CH_3)_3C$(NC)C=CH_2 / 120 °C, 10 h→ F_2 cyclobutane with SC_6H_5, R, CN, $C(CH_3)_3$ **34**

R = H, 47%; F, 57%; Cl, 88%

Olefins with strained, relatively weak π-bonds form cyclobutanes under rather mild conditions [*96*] (equations 35 and 36). By contrast, *cis*-1-methylcyclooctene hardly reacts with 1,1-dicholo-2,2-difluoroethylene after 15 days at 150 °C, and norbornene gives only a 9% yield of cycloadduct after 3 days at 120 °C [*96*].

[*96*] $(CH_2)_m$ / $(CH_2)_n$ cycloalkene $\xrightarrow[80\ °C,\ 84\ h]{CF_2{=}CCl_2}$ cyclobutane adduct (F_2, Cl_2) + cyclobutane adduct (Cl_2, F_2) **35**

m, n = 1	30.7%	19.3%
m = 2, n = 0	58.5%	19.5%

[*96*] CH_3-substituted cycloalkene $\xrightarrow[80\ °C,\ 84\ h]{CF_2{=}CCl_2}$ bicyclic cyclobutane (CH_3, Cl_2, F_2, H) 24% **36**

The behavior of strained, ***fluorinated methylenecyclopropanes*** depends upon the position and level of fluorination [*34*]. 1-(Difluoromethylene)cyclopropane is much like tetrafluoroethylene in its preference for [2+2] cycloaddition (equation 37), but its 2,2-difluoro isomer favors [4+2] cycloadditions (equation 38). Perfluoromethylenecyclopropane is an exceptionally reactive dienophile but does not undergo [2+2] cycloadditions, possibly because of steric reasons [*34, 45*].

Cycloadditions involving most possible combinations of simple fluoroalkenes and alkenes or alkynes have been tried [*85*], but kinetic activation enthalpies ($\Delta H^{\ddagger}$) for only the dimerizations of tetrafluoroethylene (22.6–23.5 kcal/mol), chlorotrifluoroethylene (23.6 kcal/mol), and perfluoropropene (31.6 kcal/mol) and the cycloaddition between chlorotrifluoroethylene and perfluoropropene (25.5 kcal/mol) have been determined accurately [*97, 98*]. Some cycloadditions involving more functionalized alkenes are listed in Table 5 [*99, 100, 101, 102, 103*].

With a suitable combination of electron-deficient fluoroalkene and electron-rich addend, cycloaddition can proceed by a *dipolar mechanism involving zwitterion intermediates.* Like its isomer, 1,2-bis(trifluoromethyl)-1,2-dicyanoethylene [*85*], 1,1-bis(trifluoromethyl)-2,2-dicyanoethylene forms cyclobutanes by an ionic mechanism [*104, 105, 106*] (equations 39 and 40).

Ynamines also react with many ***fluorinated alkenes*** by an ionic mechanism to give fluorocyclobutenes, accompanied by varying amounts of diene depending on the fluoroalkene [*107, 108*] (equation 41).

Fluorinated 1,3-dienes can show a strong preference for biradical [2+2] additions over concerted [4+2] additions, and perfluoro-1,3-butadiene is notoriously unreactive in Diels–Alder cycloadditions [*109*]. Like dimerization [*85*], other diene cycloadditions can give a secondary tricyclic product that arises by an unusual criss-cross cycloaddition [*100, 110*] (equation 42). More substituted dienes and enynes smoothly dimerize to give cyclobutane primary products (equations 43–45) but apparently no secondary tricyclic product [*99, 111*].

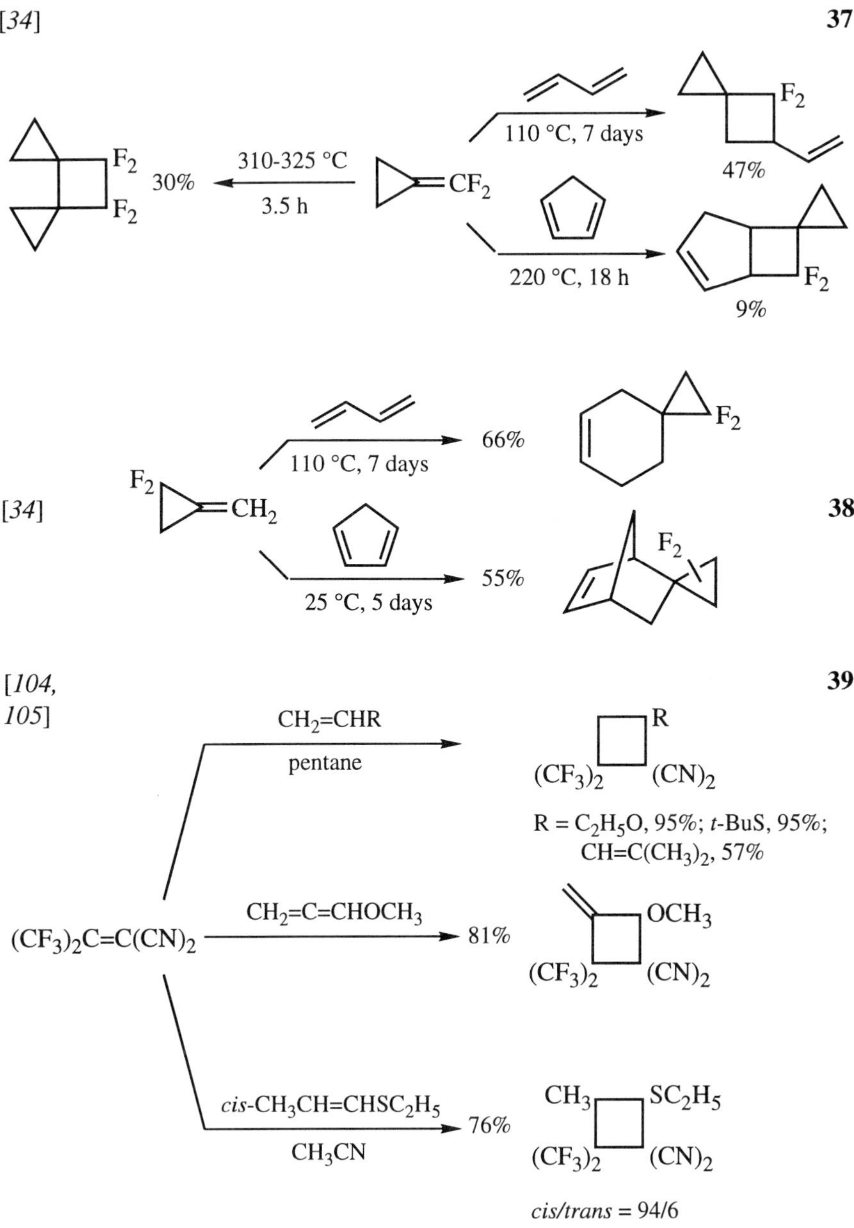

References are listed on pages 791–796.

Table 5. [2+2] Cycloadditions of Fluorinated Olefins

Reactants	*Product (Yield, %)*	*Conditions*	*Ref.*
$2CF_2=CR^1R^2$	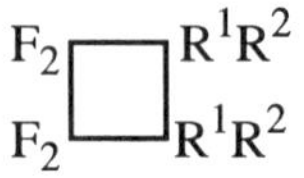 cyclobutane: F_2, F_2, R^1R^2, R^1R^2		99
$R^1 = F, R^2 = C_6H_5$	90 (E/Z = 56/44)	90 °C, 3 h	
$R^1 = F, R^2 = C_4H_3S$ [a]	91 (E/Z = 48/52)	100 °C, 1 h	
$R^1 = F, R^2 = C_5H_4N$ [b]	25 (E/Z = 40/60)	80 °C, 48 h	
$R^1 = H, R^2 = 4\text{-}CH_3C_6H_4$	87 (E/Z = 50/50)	180 °C, 40 h	
$R^1 = H, R^2 = 4\text{-}CH_3OC_6H_4$	80 (E/Z = 56/44	140 °C, 10 h	
$CF_2=CF_2$ $CH_2=CHR$	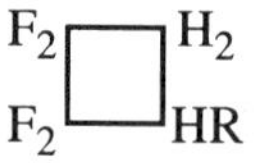 cyclobutane: F_2, F_2, H_2, HR		
$R = Si(CH_3)_3$	86	180 °C, 8 h	100
$R = SiCl_3$	84		
$R = Si(OCH_3)_3$	75		
$R = Si(CH_3)_3$	17	400–410 °C, flow system	101
$R = Si(CH=CH_2)_3$	75		
$R = CH_2Si(CH_3)_3$	25		
$CF_2=CF_2$ $CF_2=CFCO_2CH_3$	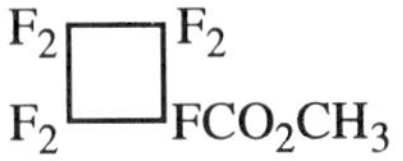 cyclobutane: F_2, F_2, F_2, FCO_2CH_3 14	200 °C, 1 h 225 °C, 7 h	36
$CF_2=CF_2$ $CH_2=C(CH_2O_2CCH_3)(CO_2C_2H_5)$	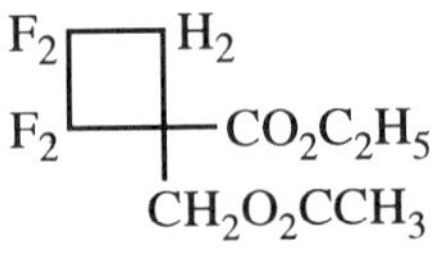 cyclobutane: F_2, F_2, H_2, $C(CO_2C_2H_5)(CH_2O_2CCH_3)$ 67	C_6H_6, 270 °C, 18 h	102

Table 5—*Continued*

Reactants	*Product (Yield, %)*	*Conditions*	*Ref.*
CF_2=CF_2 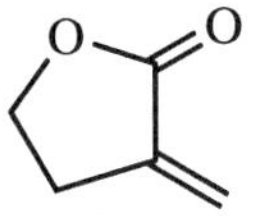 (structure: 3-methylene-γ-butyrolactone)	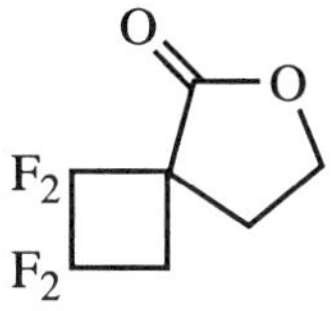 (structure: spiro lactone with F_2, F_2 cyclobutane)	C_6H_6, 150 °C, 18 h	102
CF_2=$CFSF_5$ CH_2=CHCH=CH_2	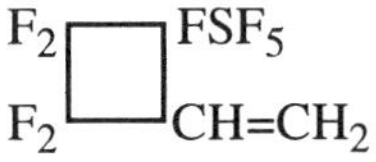 F_2, F_2 cyclobutane with FSF_5, CH=CH_2 70 (*E/Z* = 50/50)	190 °C, 22 h	103

[a] 2-Thienyl.
[b] 2-Pyridyl.

[*106*] **40**

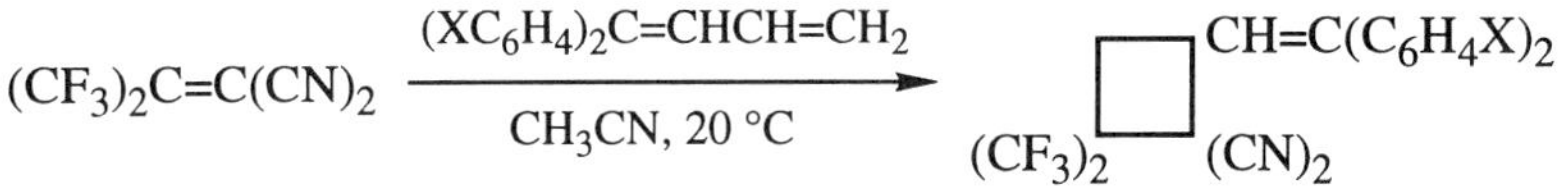

$(CF_3)_2C{=}C(CN)_2$ —[$(XC_6H_4)_2C{=}CHCH{=}CH_2$; CH_3CN, 20 °C]→ cyclobutane with $(CF_3)_2$, $(CN)_2$, $CH{=}C(C_6H_4X)_2$

X = H, 63%; 4-CH_3, 96%; 3-CF_3, 85%

41

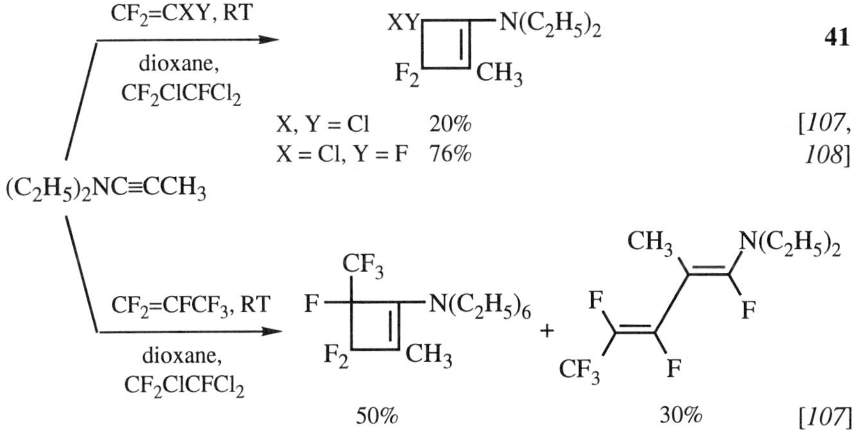

$(C_2H_5)_2NC{\equiv}CCH_3$ —[$CF_2{=}CXY$, RT; dioxane, $CF_2ClCFCl_2$]→ cyclobutene with XY, F_2, $N(C_2H_5)_2$, CH_3

X, Y = Cl 20%
X = Cl, Y = F 76%

[*107*, *108*]

$(C_2H_5)_2NC{\equiv}CCH_3$ —[$CF_2{=}CFCF_3$, RT; dioxane, $CF_2ClCFCl_2$]→ cyclobutene with F, CF_3, F_2, $N(C_2H_5)_6$, CH_3 (50%) + diene with CH_3, $N(C_2H_5)_2$, F, F, CF_3, F (30%) [*107*]

References are listed on pages 791–796.

[*110*] $CF_2{=}CFCF{=}CF_2$ + $CH_2{=}CHCH{=}CH_2$ —175 °C→ 23% **42**

F, F_2, F_2, F

[*100*] F_2, F_2 (cyclobutane)–$SiCl_3$ —600 °C, $-SiCl_3F$→ [$CF_2{=}CFCH{=}CH_2$] → 67%

[*99*] *E*–$CF_2{=}CFCH{=}CHC_6H_{13}$ —100 °C, 2 h→ [F]($CH{=}CHC_6H_{13}$)($CH{=}CHC_6H_{13}$) **43**

89% (*E/Z* = 58/42)

[*99*] *E*–$CF_2{=}CXCF{=}CFC_7H_{15}$ —150 °C, 24 h→ [F](X)(X)($CF{=}CFC_7H_{15}$)($CF{=}CFC_7H_{15}$) **44**

X = F 90% (*E/Z* = 54/46)
X = H 85% (*E/Z* = 97/3)

[*111*] 2 $CF_2{=}CFC{\equiv}CC_7H_{15}$ —120 °C, 8 h→ [F]($C{\equiv}CC_7H_{15}$)($C{\equiv}CC_7H_{15}$) **45**

80% (*E/Z* = 3/2)

Perfluoropolyenes also can rearrange to four-membered-ring products upon fluoride ion or Lewis acid catalysis [*112, 113, 114*] (equations 46 and 47). These intramolecular "cycloadditions" are multistep processes involving carbanion or carbocation intermediates.

Fluoroalkyl acetylenes are powerful enophiles and Diels–Alder dienophiles but also can give good yields of cyclobutenes in their cycloadditions [*115, 116, 117*] (equations 48 and 49).

Fluorinated allenes are especially reactive in cycloadditions because of their highly strained double bonds [*118, 119*]. 1,1-Difluoro- and 1-fluoroallene readily undergo both [2+2] and [4+2] cycloadditions [*118, 124*] (equations 50–52). Extensive studies of stereochemistry and regioselectivity show that cyclobutane formation involves initial C(2) bond formation to give 2-ethylallyl diradical intermediates [*120, 122, 123*], whereas [4+2] cycloadducts arise from concerted addition exclusively to the nonfluorinated C(2)–C(3) double bond of the allene, which is consistent with frontier molecular orbital control [*118, 124*].

[*112*] $(CF_3)_2C{=}C(CF{=}C(CF_3)_2)_2$ —cat. CsF, CH_3CN, RT→ $(CF_3)_2$, F_2 cyclobutane with two $=C(CF_3)_2$ groups 95% **46**

[*113*] $CF_3CF{=}CFCF{=}CFR$ —cat. SbF_5, 90 °C, R = F→ cyclobutane with F, CF_3, CF_3, F and two $=CFCF_3$ groups 80-90%

47

[*114*] $CF_3CF{=}CFCF{=}CFR$ —cat. SbF_5, 100 °C, R = C_2F_5→ F-cyclobutene with CF_3, C_2F_5 (61.6%) + F-cyclopentene with CF_3, CF_3 (18.4%)

[*115*] $CF_3C{\equiv}CCF_3$ + isobutene —145 °C→ CF_3, CF_3, H diene product 80%

48

[*116*] $CF_3C{\equiv}CCF_3$ + dimethylallene —80 °C, 48 h→ cyclobutene with CF_3, CF_3 (73%) + cyclobutene with CF_3, CF_3 (24%)

Fluoroalkyl allenes and higher cumulenes also can take part in [2+2] cycloadditions via biradical pathways [*125, 126*] (equations 53 and 54).

Nonfluorinated allenes also readily react with fluoroalkenes to give diverse fluorinated alkylidenecyclobutanes [*127, 128, 129, 130*] (equations 55 and 56), except for tetramethylallene, which rearranges to 2,4-dimethyl-1,3-pentadiene under the reaction conditions prior to cycloaddition (equation 57). Systematic studies of 1,1-dichloro-2,2-difluoroethylene additions to alkyl-substituted allenes establish a two-step, diradical process for alkylidenecyclobutane formation [*131, 132, 133*].

[*117*] **49**

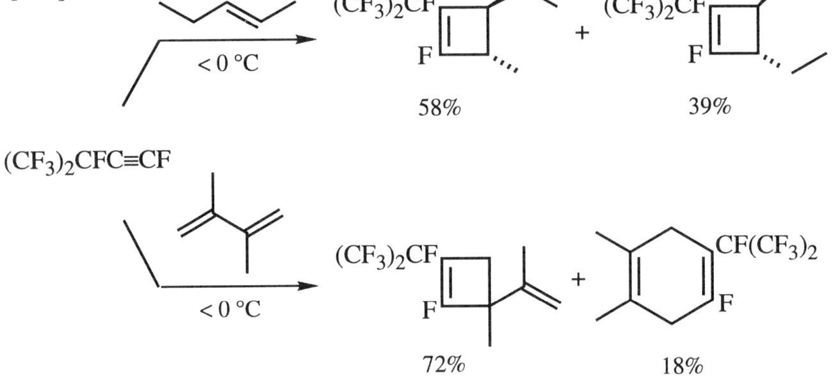

$(CF_3)_2CFC{\equiv}CF$

< 0 °C → $(CF_3)_2CF$, F cyclobutene products: 58% + 39%

< 0 °C → $(CF_3)_2CF$, F cyclobutene 72% + $CF(CF_3)_2$, F cyclohexadiene 18%

[*120, 121*] **50**

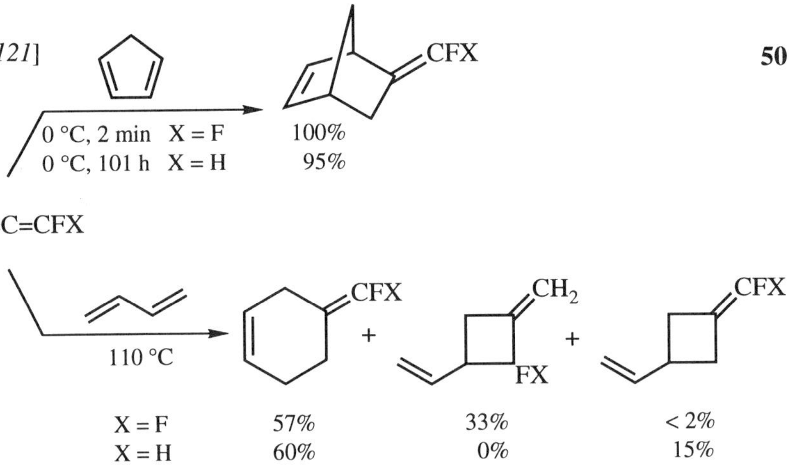

0 °C, 2 min X = F 100%
0 °C, 101 h X = H 95%

CFX

$CH_2{=}C{=}CFX$

110 °C → CFX + CH_2, FX + CFX

X = F	57%	33%	< 2%
X = H	60%	0%	15%

[*120*] **51**

$CH_2{=}CHCN$

100 °C, 12 h X = F → CH_2, NC, FX 46.7% + CFX, NC 17.3%
135 °C, 24 h X = H → 14% + 31%

$CH_2{=}C{=}CFX$

$CF_2{=}CCl_2$

130 °C, 8 h X = H → F_2, Cl_2, CH_2, FH 20% + F_2, Cl_2, CFH 20%

References are listed on pages 791–796.

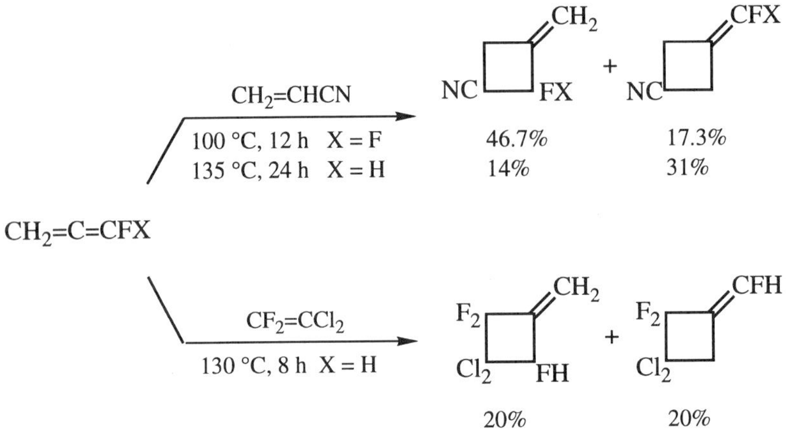

[*122*] **52**

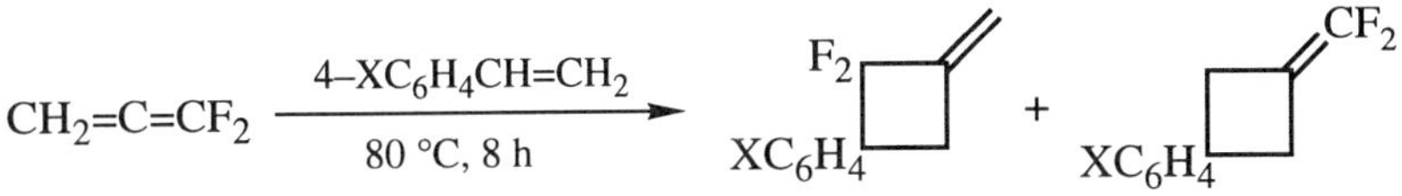

$CH_2{=}C{=}CF_2$ —($4\text{–}XC_6H_4CH{=}CH_2$; 80 °C, 8 h)→ [cyclobutane: F_2, XC_6H_4, =CH₂] + [cyclobutane: XC_6H_4, =CF_2]

X		
X = H	49.8%	10.2%
= OCH_3	49.3%	8.7%
= NO_2	62.5%	11.5%

[*125*] **53**

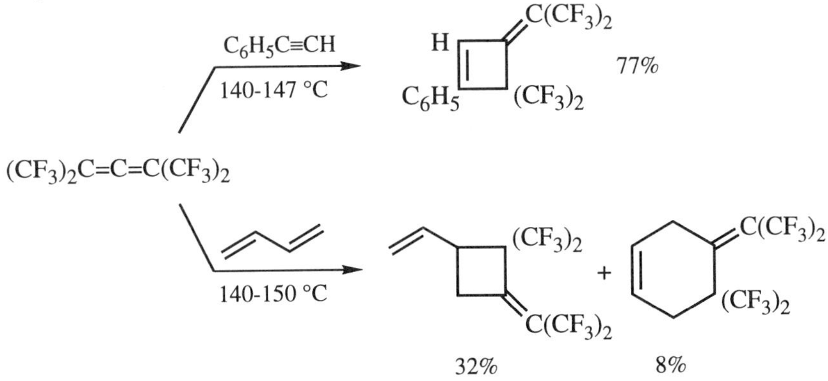

$(CF_3)_2C{=}C{=}C(CF_3)_2$ —($C_6H_5C{\equiv}CH$; 140-147 °C)→ [cyclobutene: H, C_6H_5, =$C(CF_3)_2$, $(CF_3)_2$] 77%

$(CF_3)_2C{=}C{=}C(CF_3)_2$ —(butadiene; 140-150 °C)→ [vinylcyclobutane: $(CF_3)_2$, =$C(CF_3)_2$] 32% + [cyclohexene: =$C(CF_3)_2$, $(CF_3)_2$] 8%

[*126*] $(C_6H_5)_2C{=}C{=}C{=}C(CF_3)_2$ —($CH_2{=}CR_2$; 100-110 °C)→ [cyclobutane: $(C_6H_5)_2C{=}$, =$C(CF_3)_2$, R_2] **54**

R	
R = CH_3O,	63%
$(CH_3)_2N$	54%

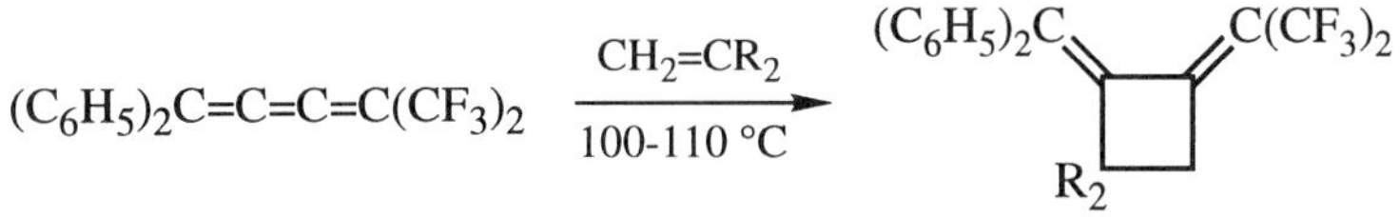

[*127, 128*] $CH_2{=}C{=}CH_2$ —($CF_2{=}CXY$; 150 °C)→ [cyclobutane: =CH_2, F_2, XY] + [cyclobutane: =CH_2, F_2, XY] **55**

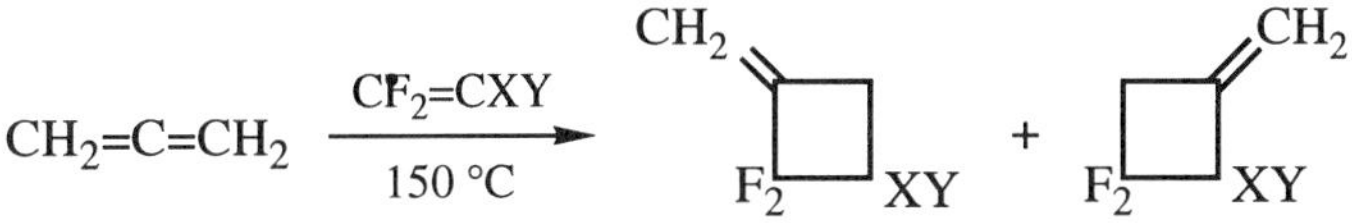

X, Y = F	30%	
X, Y = Cl	85%	4%
X = Cl, Y = F	74%	13%

[*129, 130*] $CH_2{=}C{=}C(CH_3)_2$ —($CF_2{=}CXY$; 150 °C)→ [cyclobutane: =$C(CH_3)_2$, F_2, XY] + [cyclobutane: =CH_2, $(CH_3)_2$, F_2, XY] **56**

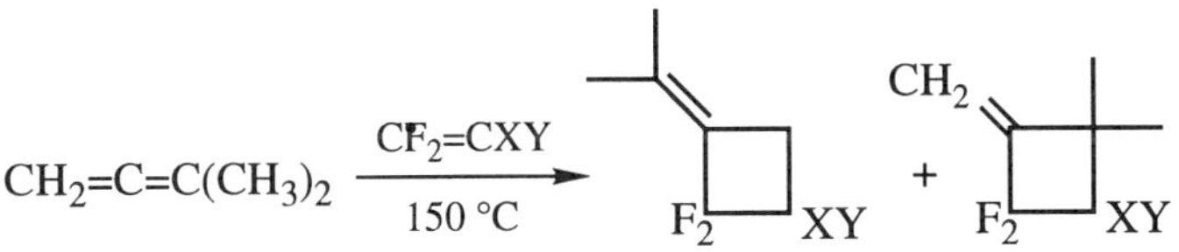

X, Y = F	70%	19%
X, Y = Cl	65%	16%
X = Cl, Y = F	39%	22%

[*127, 130*] **57**

$CF_2{=}CXY$, 120-150 °C

X, Y = F	65%
X, Y = Cl	93%
X = Cl, Y = F	80%

Both the carbon–carbon and carbon–oxygen double bonds of fluoroketenes can take part in [2+2] cycloadditions, but with cyclopentadiene, only cyclobutanones are produced via concerted $[2_a+2_s]$ additions [*134*] (equation 58). Cycloadditions involving the carbon–oxygen double bonds to form oxetanes are discussed on page 855. ***Difluoroketene*** is very short lived and difficult to intercept but has been trapped successfully by very electron rich addends to give 2,2-difluorocyclobutanones in moderate yields [*135*] (equation 59).

Thermally unfavorable [2+2] *cycloadditions often can be promoted photochemically* [*136, 137, 138*]; some examples for fluorinated derivatives are given in equations 60–62.

[*134*] **58**

RCHFC(O)Cl → ($(C_2H_5)_3N$, ether, -78 °C to 4 °C)

R = F		0%
R = H	3%	29%
R = CH_3	23%	–
R = CF_3	10%	–
R = C_6H_5	38%	–

[*135*] $ClCF_2COCl$ **59**

$OSi(CH_3)_3$ / $OSi(CH_3)_3$; Zn, ethyl acetate, 40 °C → 34%; OTMS, O, F_2, OTMS

$(CH_2)_n$, $OSi(CH_3)_3$ / $OSi(CH_3)_3$; Zn, ether → $(CH_2)_n$, OTMS, O, F_2, OTMS

n = 1, 47%; 2, 55%

[*136*] **60**

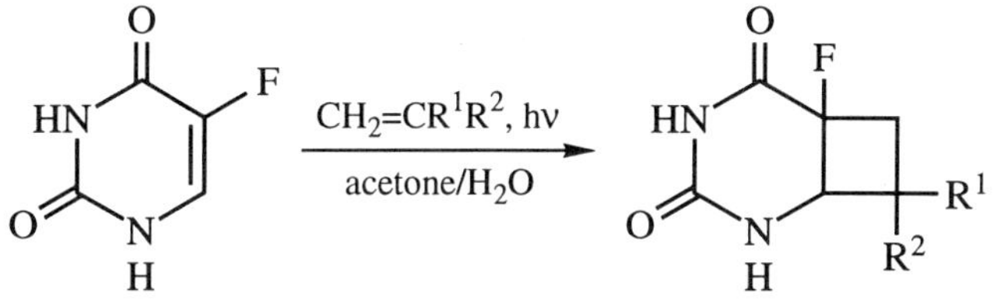

$CH_2{=}CR^1R^2$, hν; acetone/H_2O

$R^1, R^2 = CH_3$ 90%
$R^1 = OCOCH_3, R^2 = CH_3$ 57%

[*137*] **61**

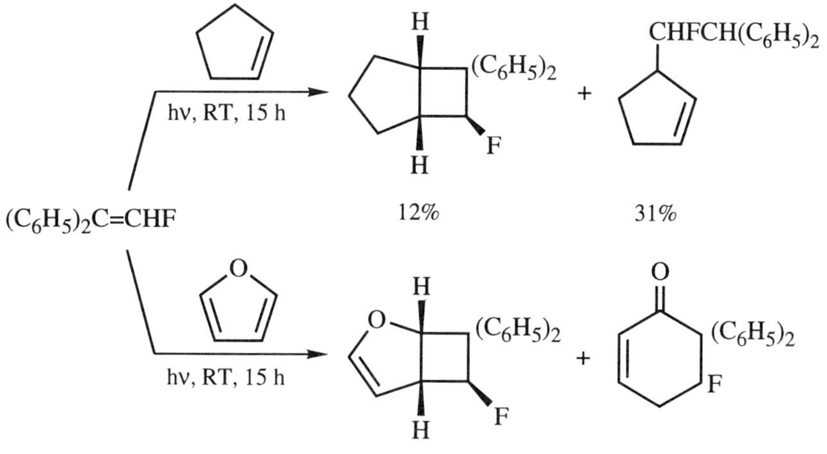

$(C_6H_5)_2C{=}CHF$

hν, RT, 15 h

$(C_6H_5)_2$ F 12% + $CHFCH(C_6H_5)_2$ 31%

hν, RT, 15 h

$(C_6H_5)_2$ F 23% + $(C_6H_5)_2$ F 20%

[*138*] **62**

$(CF_2)_n$ X X

, hν; cat. $C_6H_5COCH_3$, 66 h

X X $(CF_2)_n$

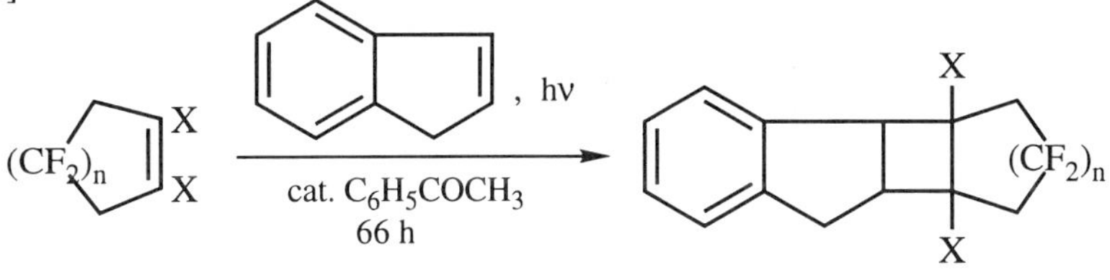

n = 0, X = Cl 88%
n = 0, X = F 40%
n = 1, X = Cl 85%

The *photochemical cycloadditions* of ***alkenes*** and ***alkynes*** with aromatic compounds have received by far the most attention. Yields of [2+2] cycloadducts can be good, but reaction times are often long and secondary rearrangement products are common [*139, 140, 141, 142, 143, 144, 145, 146*] (equations 63–65). The pioneering mechanistic and synthetic work on aromatic photocycloadditions has been reviewed [*147*].

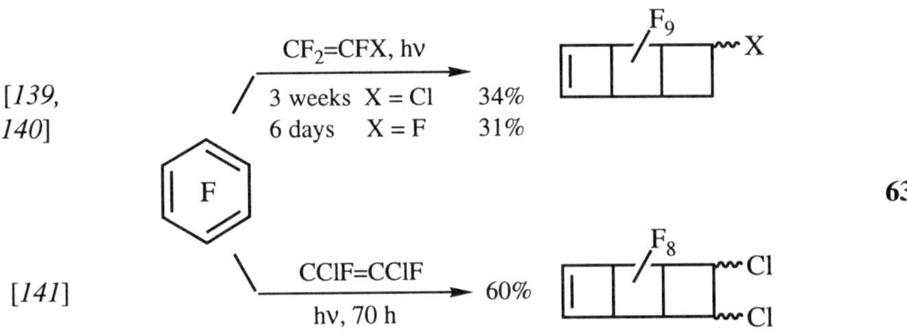

63

[*142*] **64**

[*143*]

, hν

c-C_6H_{12}, 12 h

F H F H F 59%

\+ F F F H F F F F H 26%

F

\+ 15% F

$C_6H_5C{\equiv}CR$, hν

c-C_6H_{12}, 60 h

F C_6H_5 R

R = CH_3, 65%; C_3H_7, 60%; t-C_4H_9S, 86%

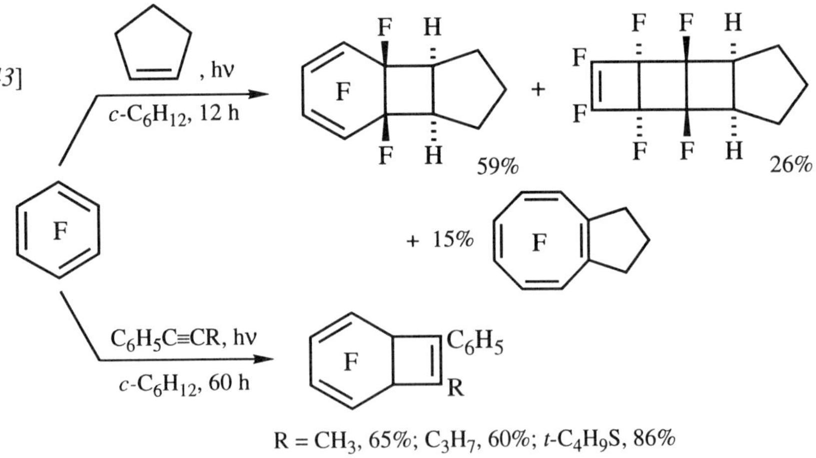

[*144*] **65**

[*145*]

$CH_2{=}CH_2$, hν, 900 h → 37% + 21%

cyclopentene, hν, C_6H_{12}, 3 h → 40%

t-$C_4H_9C{\equiv}CC_6H_5$, hν, C_6H_{12}, 20 h → 14% + 12% + 34%

[*146*]

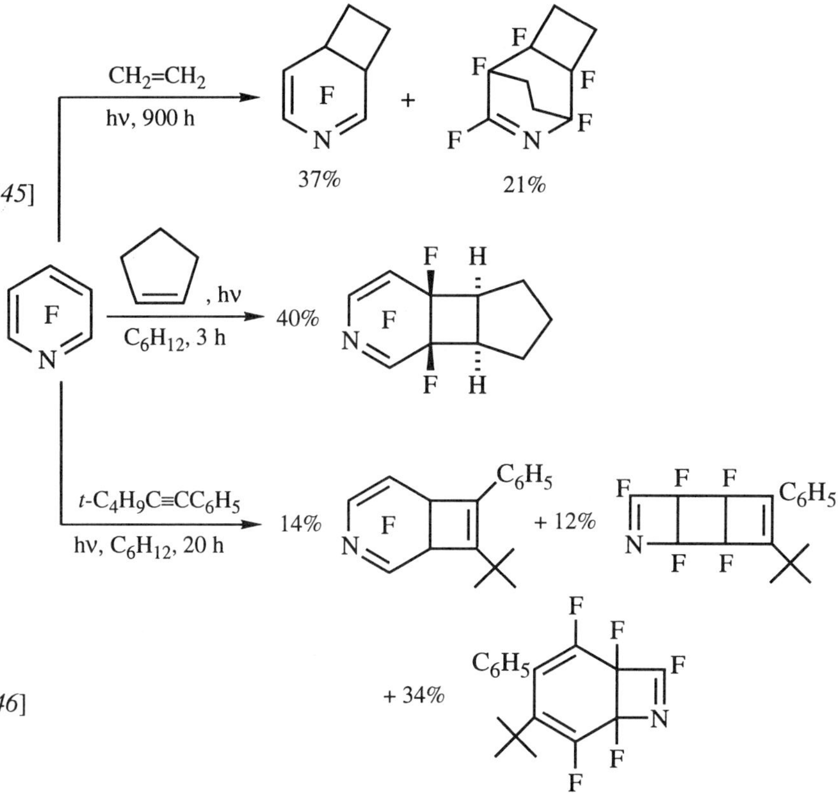

References for pages 767–791

1. Dixon, D. A. *J. Phys. Chem.* **1986,** *90,* 54.
2. Carter, E. A.; Goddard, W. A., III *J. Chem. Phys.* **1988,** *88,* 1752.
3. Moss, R. A. *Acc. Chem. Res.* **1980,** *13,* 58.
4. Moss, R. A. *Acc. Chem. Res.* **1989,** *22,* 15.
5. Giese, B.; Lee, W.-B.; Meister, J. *Liebigs Ann. Chem.* **1980,** 725.
6. Houk, K. N.; Rondan, N. G.; Mareda, J. *Tetrahedron* **1985,** *41,* 1555.
7. Burton, D. J.; Hahnfeld, J. L. *Fluorine Chem. Rev.* **1977**, *8,* 119.
8. Seyferth, D. In *Carbenes,* Vol. II; Moss, R. A.; Jones, M., Jr., Eds.; John Wiley: New York, 1975; pp 101–158.
9. Hudlický, M. *Chemistry of Organic Fluorine Compounds,* 2nd ed.; Ellis Horwood: Chichester, U.K., 1976; pp 443–450.
10. *Houben-Weyl Methoden der Organishen Chemie,* Regitz, M.; Giese, B., Eds.; Verlag Georg Thieme: Stuttgart, 1989; Vol. E19, part B.
11. Burton, D. J.; Naae, D. G. *J. Am. Chem. Soc.* **1973,** *95,* 8467.

12. Bessard, Y.; Müller, U.; Schlosser, M. *Tetrahedron* **1990,** *46,* 5213.
13. Bessard, Y.; Schlosser, M. *Tetrahedron* **1991,** *47,* 7323.
14. Seyferth, D. *J. Organometal. Chem.* **1973,** *62,* 33.
15. Millauer, H.; Schwertfeger, W.; Siegemund, G. *Angew. Chem. Int. Ed. Engl.* **1985,** *24,* 161.
16. Dolbier, W. R., Jr.; Sellers, S. F. *J. Am. Chem. Soc.* **1982,** *104,* 2494.
17. Dolbier, W. R., Jr.; Sellers, S. F. *J. Org. Chem.* **1982,** *47,* 1.
18. Schlosser, M.; Bessard, Y. *Tetrahedron* **1990,** *46,* 5222.
19. Greenberg, A.; Yang, N. *J. Org. Chem.* **1990,** *55,* 372.
20. Jefford, C. W.; Mareda, J.; Gehret, J. C. E.; Kabengele, nT.; Graham, W. D.; Burger, U. *J. Am. Chem. Soc.* **1976,** *98,* 2585.
21. Misslitz, U.; Jones, M., Jr.; De Meijere, A. *Tetrahedron Lett.* **1986,** *26,* 5403.
22. Jefford, C. W.; Huy, P. T. *Tetrahedron Lett.* **1980,** *21,* 755.
23. Houk, K. N.; Rondan, N. G.; Paddon-Row, M. N.; Jefford, C. W.; Juy, P. T.; Burrow, P. D.; Jordan, K. D. *J. Am. Chem. Soc.* **1983,** *105,* 5563.
24. Dolbier, W. R., Jr.; Odaniec, M.; Gomulka, E.; Jaskolski, M.; Koroniak, H. *Tetrahedron* **1984,** *40,* 3945.
25. Sellers, S. F.; Dolbier, W. R., Jr.; Koroniak, H.; Al-Fekri, D. M. *J. Org. Chem.* **1984,** *49,* 1033.
26. Dolbier, W. R., Jr.; Wojtowicz, H.; Burkholder, C. R. *J. Org. Chem.* **1990,** *55,* 5420.
27. Fritz, H. P.; Kornrumpf, W. *Z. Naturforsch.* **1981,** *31B,* 1375.
28. Kimpenhaus, W.; Buddrus, J. *Chem. Ber.* **1976,** *109,* 2370.
29. Kamel, M.; Kimpenhaus, W.; Buddrus, J. *Chem. Ber.* **1976,** *109,* 2351.
30. Mahler, W.; Resnick, P. R. *J. Fluorine Chem.* **1973,** *3,* 451.
31. Craig, M. C.; McPhail, R. A.; Spiegel, D. A. *J. Phys. Chem.* **1978,** *82,* 1056.
32. Sargent, P. B.; Krespan, C. G. *J. Am. Chem. Soc.* **1969,** *91,* 415.
33. Sargeant, P. B. *J. Org. Chem.* **1970,** *35,* 678.
34. Dolbier, W. R., Jr.; Seabury, M.; Daly, D.; Smart, B. E. *J. Org. Chem.* **1986,** *51,* 974.
35. Dolbier, W. R., Jr.; Sellers, S. F.; Al-Sader, B. H.; Fielder, T. H., Jr.; Elsheimer, S.; Smart, B. E. *Isr. J. Chem.* **1981,** *21,* 176.
36. England, D. C.; Solomon, L.; Krespan, C. G. *J. Fluorine Chem.* **1973,** *3,* 63.
37. Moore, E. P. U.S. Pat. 3 338 978, 1967; *Chem. Abstr.* **1968,** *68,* 114045.
38. Chuikov, I. P.; Karpov, V. M.; Platonov, V. E.; Yakobson, G. G. *Izv. Akad. Nauk SSSR* **1988,** 1839 (Engl. Transl. 1645).
39. Karpov, V. M.; Platonov, V. E.; Yakobson, G. G. *Izv. Akad. Nauk SSSR* **1976,** 2295 (Engl. Transl. 2141).
40. Dailey, W. P.; Ralli, P.; Wasserman, D.; Lemal, D. M. *J. Org. Chem.* **1989,** *54,* 5516.

41. England, D. C. *J. Org. Chem.* **1981,** *46,* 147.
42. Birchall, J. M.; Fields, R.; Haszeldine, R. N.; McLean, R. J. *J. Fluorine Chem.* **1980,** *15,* 487.
43. Bunegar, M. J.; Fields, R.; Haszeldine, R. N. *J. Fluorine Chem.* **1980,** *15,* 497.
44. Bosbury, P. W. L.; Fields, R.; Haszeldine, R. N.; Lomax, G. R. *J. Chem. Soc., Perkin Trans. 1* **1982,** 2203.
45. Smart, B. E. *J. Am. Chem. Soc.* **1974,** *96,* 927.
46. Dolbier, W. R., Jr.; Sellers, S. F.; Al-Sader, B. H.; Smart, B. E. *J. Am. Chem. Soc.* **1980,** *102,* 5398.
47. Dolbier, W. R., Jr.; Burkholder, C. R. *J. Am. Chem. Soc.* **1984,** *106,* 2139.
48. Kobayashi, Y.; Taguchi, T.; Morikawa, T.; Takase, T.; Takanashi, H. *J. Org. Chem.* **1982,** *47,* 3232.
49. Crabbé, P.; Luche, J.-L.; Damiano, J.-C.; Luche, M.-J.; Cruz, A. *J. Org. Chem.* **1979,** *44,* 2929.
50. Kobayashi, Y.; Taguchi, T.; Mamada, M.; Shimizu, H.; Murohashi, H. *Chem. Pharm. Bull.* **1979,** *27,* 3123.
51. Taguchi, T.; Takigawa, T.; Tawara, Y.; Morikawa, T.; Kobayashi, Y. *Tetrahedron Lett.* **1984,** *25,* 5689.
52. Wheaton, G. H.; Burton, D. J. *J. Fluorine Chem.* **1977,** *9,* 25.
53. Schlosser, M.; Chan, L. V. *Helv. Chim. Acta* **1975,** *58,* 25.
54. Bessière, Y.; Savary, D. N.-H.; Schlosser, M. *Helv. Chim. Acta* **1977,** *60,* 1739.
55. Nguyen, T.; Molines, H.; Wakselman, C. *Synth. Commun.* **1985,** *15,* 925.
56. Müller, C.; Stier, F.; Weyerstahl, P. *Chem. Ber.* **1977,** *110,* 124.
57. Weyerstahl, P.; Mathias, R.; Blume, G. *Tetrahedron Lett.* **1973,** 611.
58. Christl, M.; Freitag, G.; Brüntrup, G. *Chem. Ber.* **1978,** *111,* 2307.
59. Anke, L.; Reinhard, D.; Weyerstahl, P. *Liebigs Ann. Chem.* **1981,** 591.
60. Balcerzak, P.; Fedorynski, M.; Jonczyk, A. *J. Chem. Soc., Chem. Commun.* **1991,** 826.
61. Dolbier, W. R., Jr.; Burkholder, C. R. *J. Org. Chem.* **1990,** *55,* 589.
62. Hahnfeld, J. L.; Burton, D. J. *Tetrahedron Lett.* **1975,** 1819.
63. Boche, G.; Walborsky, H. M. In *Cyclopropane Derived Reactive Intermediates;* Patai, S.; Rappoport, Z., Eds.; John Wiley: Chichester, 1990; Chapter 1.
64. Seyferth, D.; Murphy, G. J. *J. Organometal. Chem.* **1973,** 52, C1.
65. Haszeldine, R. N.; Rowland, R.; Speight, J. G.; Tipping, A. E. *J. Chem. Soc., Perkin Trans. 1* **1980,** 314.
66. Seyferth, D.; Woodruff, R. A. *J. Org. Chem.* **1973,** *36,* 4031.
67. Seyferth, D.; Mueller, D. C. *J. Am. Chem. Soc.* **1971,** *93,* 3714.
68. Haszeldine, R. N.; Pool, C. R.; Tipping, A. E.; Watt, O'B. R. *J. Chem. Soc., Perkin Trans. 1* **1976,** 513.
69. Dailey, W. P. *Tetrahedron Lett.* **1987,** *28,* 5801.

70. Moss, R. A.; Lawrynowicz, W. *J. Org. Chem.* **1984,** *49,* 3828.
71. Moss, R. A.; Kmiecik-Lawrynowicz, G.; Krogh-Jespersen, K. *J. Org. Chem.* **1986,** *51,* 2168.
72. Chambers, R. D. *Fluorine in Organic Chemistry;* John Wiley: New York, 1973; pp 133–134.
73. Krespan, C. G.; Middleton, W. J. *Fluorine Chem. Rev.* **1971,** *5,* 57.
74. Taguchi, T.; Hosoda, A.; Torisawa, Y.; Shimazaki, A.; Kobayashi, Y.; Tsushima, K. *Chem. Pharm. Bull.* **1985,** *33,* 4085.
75. Zeifman, Y. V.; Lantseva, L. T. *Izv. Akad. Nauk SSSR* **1983,** 2149 (Engl. Transl. 1941).
76. Norstrum, R. J.; Gunning, H. E.; Strausz, O. P. *J. Am. Chem. Soc.* **1976,** *98,* 1454.
77. Stachnik, R. A.; Pimentel, G. C. *J. Phys. Chem.* **1984,** *88,* 2205.
78. Brahms, J. C.; Dailey, W. P. *J. Am. Chem. Soc.* **1990,** *112,* 4046.
79. Cullen, W. R.; Waldman, M. C. *J. Fluorine Chem.* **1971**, *1,* 151.
80. Birchall, J. M.; Burger, K.; Haszeldine, R. N.; Nona, S. N. *J. Chem. Soc., Perkin Trans. 1* **1981**, 2080.
81. Grayston, M. W.; Lemal, D. M. *J. Am. Chem. Soc.* **1976,** *98,* 1278.
82. Haas, A.; Krächter, H.-U. *Chem. Ber.* **1988**, *121,* 1833.
83. Glück, C.; Poignée, V.; Schwager, H. *Synthesis* **1987,** 260.
84. Sepiol, J.; Soulen, R. L. *J. Org. Chem.* **1975,** *40,* 3791.
85. Hudlický, M. *Chemistry of Organic Fluorine Compounds,* 2nd ed.; Ellis Horwood: Chichester, U.K., 1976; pp 450–463.
86. Bartlett, P. D. *Quart. Rev.* **1970**, *24,* 473.
87. Roberts, D. W. *Tetrahedron* **1985**, *41,* 5529.
88. Smart, B. E. In *The Chemistry of Functional Groups,* Supplement D; Patai, S.; Rappoport, Z., Eds.; John Wiley: New York, 1983; Part 2, Chapter 14, pp 638–642.
89. Getty, S. J.; Borden, W. T. *J. Am. Chem. Soc.* **1991**, *113,* 4334.
90. Wang, S. Y.; Borden, W. T. *J. Am. Chem. Soc.* **1989,** *111,* 7282.
91. De Meijere, A.; Wenck, H.; Sayed-Mahdavi, F.; Viehe, H. G.; Gallez, V.; Erden, I. *Tetrahedron* **1986,** *42,* 1291.
92. Bartlett, P. D.; Mallet, J. J.-B. *J. Am. Chem. Soc.* **1976,** *98,* 143.
93. Wheland, R.; Bartlett, P. D. *J. Am. Chem. Soc.* **1973,** *95,* 4003.
94. Smart, B. E. *J. Am. Chem. Soc.* **1974,** *96,* 929.
95. DeCock, C.; Piettre, S.; Lahousse, F.; Janousek, Z.; Merényi, R.; Viehe, H. G. *Tetrahedron* **1985,** *41,* 4183.
96. Becker, K. B.; Hohermuth, M. K. *Helv. Chim. Acta* **1982,** *65,* 229.
97. Atkinson, B.; Tsiamis, C. *Int. J. Chem. Kinetics* **1979,** *11,* 1029.
98. Atkinson, B.; Tsiamis, C. *Int. J. Chem. Kinetics* **1979,** *11,* 585.
99. Tellier, F.; Sauvêtre, R.; Normant, J. F.; Dromzee, Y.; Jeannin, Y. *J. Organometal. Chem.* **1987**, *331,* 281.

100. England, D. C.; Weigert, F. J.; Calabrese, J. C. *J. Org. Chem.* **1984,** *49,* 4816.
101. Sheludyakov, V. D.; Zhun′, V. I.; Loginov, S. V.; Shcherbinin, V. V.; Turkel′taub, G. N. *Zh. Obshch. Khim.* **1984,** *54,* 2108 (Engl. Transl. 1983).
102. Johnson, W. M. P.; Holan, G. *Aust. J. Chem.* **1981,** *34,* 2355.
103. Banks, R. E.; Barlow, M. G.; Haszeldine, R. N. *J. Chem. Soc., Perkin Trans. 1* **1974,** 1266.
104. Huisgen, R.; Brückner, R. *Tetrahedron Lett.* **1990,** *31,* 2553.
105. Brückner, R.; Huisgen, R. *Tetrahedron Lett.* **1990,** *31,* 2557.
106. Drexler, J.; Lindermayer, R.; Hassan, M. A.; Sauer, J. *Tetrahedron Lett.* **1985,** *26,* 2555, 2559.
107. Blazejewski, J. C.; Cantacuzène, D.; Wakselman, C. *Tetrahedron Lett.* **1974,** 2055.
108. Bellus, D.; Martin, P.; Sauter, H.; Winkler, T. *Helv. Chim. Acta* **1980,** *63,* 1130.
109. Perry, D. R. *A. Fluorine Chem. Rev.* **1967,** *1,* 253.
110. Weigert, F. J. *J. Fluorine Chem.* **1991,** *52,* 125.
111. Tellier, F.; Sauvêtre, R.; Normant, J. F. *J. Organometal. Chem.* **1987,** *331,* 328.
112. Ter-Gabrielyan, E. G.; Gambraryan, M. P. *Izv. Akad. Nauk SSSR* **1986,** 1341 (Engl. Transl. 1217).
113. Petrov, V. A.; German, L. S.; Belen'kii, G. G. *Izv. Akad. Nauk SSSR* **1989,** 391 (Engl. Transl. 339).
114. Petrov, V. A.; Belen'kii, G. G.; German, L. S. *Izv. Akad. Nauk SSSR* **1990,** 920 (Engl. Transl. 826).
115. Chia, H.-A.; Kirk, B. E.; Taylor, D. R. *J. Chem. Soc., Perkin Trans. 1* **1974,** 1209.
116. Kirk, B. E.; Taylor, D. R. *J. Chem. Soc., Perkin Trans. 1* **1974,** 1844.
117. Chambers, R. D.; Sheppard, T.; Tamura, M.; Bryce, M. R. *J. Chem. Soc., Chem. Commun.* **1989,** 1657.
118. Dixon, D. A.; Smart, B. E. *J. Phys. Chem.* **1989,** *93,* 7772.
119. Smart, B. E. In *Molecular Structure and Energetics,* Liebman, J.; Greenberg, A., Eds.; VCH Publishers: New York, 1986, Vol. 3, pp 170–173.
120. Dolbier, W. R., Jr.; Burkholder, C. R. *J. Org. Chem.* **1984,** *90,* 2381.
121. Dolbier, W. R., Jr.; Piedrahita, C. A.; Houk, K. N.; Strosier, R. W.; Gandour, R. W. *Tetrahedron Lett.* **1978,** 2231.
122. Dolbier, W. R., Jr.; Wicks, G. E. *J. Am. Chem. Soc.* **1985,** *107,* 3626.
123. Dolbier, W. R., Jr.; Seabury, M. *J. Am. Chem. Soc.* **1987,** *109,* 4393.
124. Dolbier, W. R., Jr. *Acc. Chem. Res.* **1991,** *24,* 63.
125. Mirzabekyants, N. S.; Bargamova, M. D.; Cheburkov, Y. A.; Knunyants, I. L. *Izv. Akad. Nauk SSSR* **1976,** 1099 (Engl. Transl. 1068).
126. Gotthardt, H.; Jung, R. *Tetrahedron Lett.* **1984,** *25,* 4217.
127. Taylor, D. D.; Warburton, M. R.; Wright, D. B. *J. Chem. Soc. C* **1971,** 385.

128. Taylor, D. D.; Warburton, M. R.; Wright, D. B. *J. Chem. Soc., Perkin Trans. 1* **1972,** 1365.
129. Taylor, D. D.; Wright, D. B. *J. Chem. Soc. C* **1971,** 391.
130. Taylor, D. D.; Wright, D. B. *J. Chem. Soc., Perkin Trans. 1* **1973,** 445.
131. Pasto, D. J.; Warren, S. E.; Weyenberg, T. *J. Org. Chem.* **1986,** *51,* 2106.
132. Pasto, D. J.; Yang, S. H. *J. Org. Chem.* **1986,** *51,* 1676.
133. Pasto, D. J.; Sugi, K. D.; Malandra, J. L. *J. Org. Chem.* **1991,** *56,* 3781.
134. Dolbier, W. R., Jr.; Lee, S. K.; Phanstiel, O. *Tetrahedron* **1991,** *47,* 2065.
135. Habibi, M. H.; Saidi, K.; Sams, L. C. *J. Fluorine Chem.* **1987,** *37,* 177.
136. Wexler, A.; Balchunis, R. J.; Swenton, J. S. *J. Org. Chem.* **1984,** *49,* 2732.
137. Sket, B.; Zupan, M. *Tetrahedron Lett.* **1978,** 2607.
138. Kimoto, H.; Muramatsu, H.; Inukai, K. *J. Fluorine Chem.* **1977,** *9,* 417.
139. Barefoot, A. C., III; Sanders, W. D.; Buzby, J. M.; Grayston, M. W.; Lemal, D. M. *J. Org. Chem.* **1980**, *45,* 4292.
140. Rahman, M. M.; Secor, B. A.; Morgan, K. M.; Shafer, P. R.; Lemal, D. M. *J. Am. Chem. Soc.* **1990,** *112,* 5986.
141. Lemal, D. M.; Buzby, J. M.; Barefoot, A. C., III; Grayston, M. W.; Laganis, E. D. *J. Org. Chem.* **1980,** *45,* 3118.
142. Sket, B.; Zupan, M. *J. Chem. Soc., Perkin Trans. 1* **1977,** 365.
143. Sket, B.; Zupancic, N.; Zupan, M. *Tetrahedron* **1984,** *40,* 3795.
144. Barlow, M. G.; Brown, D. E.; Haszeldine, R. N. *J. Chem. Soc., Perkin Trans. 1* **1978,** 363.
145. Sket, B.; Zupancic, N.; Zupan, M. *J. Org. Chem.* **1982,** *47,* 4462.
146. Sket, B.; Zupan, M. *Tetrahedron* **1984,** *45,* 1755.
147. Zupan, M.; Sket, B. *Isr. J. Chem.* **1978,** *17,* 92.

Cycloadditions Forming Five- and Six-Membered Rings

by W. R. Dolbier

Cycloadditions Forming Five-Membered Rings

Although cyclizations of fluorinated substrates to form five-membered rings were only rarely encountered in the literature prior to 1972, there has been an explosion of activity in this area recently, particularly with regard to 1,3-dipolar cycloadditions.

1,3-Dipolar cycloadditions. Interest in 1,3-dipolar cycloadditions increased dramatically during the past 20 years, largely because of the pioneering studies of Huisgen [*1, 2*]. The versatility of this class of pericyclic reactions in the synthesis of five-membered-ring heterocyclic compounds is comparable with that of the Diels–Alder reaction in the synthesis of six-membered-ring carbocyclic systems (equation 1).

1

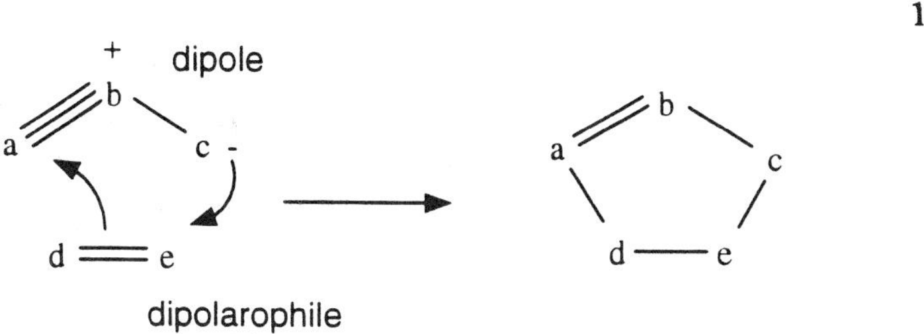

As in the Diels–Alder reaction, there are two equally important substrate components (addends) that are required to make a successful 1,3-dipolar cycloaddition. The *1,3-dipole* is generally the *electron-rich* component. The *dipolarophile,* like the dienophile in the Diels–Alder reaction, depends upon its *electron deficiency* for its reactivity. Thus, in general, the reactivities of a particular combination of addends depend upon the relative energies of the lowest unoccupied molecular orbital (LUMO) of the dipolarophile (the lower the better) and of the highest occupied molecular orbital (HOMO) of the 1,3-dipole (the higher the better) [*3*].

LUMO and HOMO energies of carbon–carbon double bonds are not affected dramatically by *direct* (i.e., vinylic) fluorine substitution [*4*], but *allylic* fluorine substitution leads to a significant lowering of both HOMO and LUMO energies. Such considerations lead to the prediction that little change in the reactivity of 1,3-dipoles or dipolarophiles will result from direct fluorine substitution, whereas

0065–7719/95/0187–0797$12.14/1

a definite enhancement of dipolarophilicity and diminution of dipole reactivity should result from perfluoroalkyl substitution.

Application of this synthetically important ring-forming methodology to the synthesis of fluorinated heterocyclics usually involves the reactions of conventional, non-fluorine-containing 1,3-dipoles with fluorinated dipolarophiles. Interest in the preparation and use of fluorinated dipoles, with the goal of diversifying the available methodology for the incorporation of perfluoroalkyl groups into heterocyclic rings, has increased, however. Results from these two areas will be discussed, beginning with a presentation of recent results in the use of fluorinated dipolarophiles.

Reactions of fluorinated dipolarophiles. Electron-deficient unsaturated species generally make better dipolarophiles; therefore, ***fluorinated alkenes*** become better dipolarophiles when vinylic fluorines are replaced by perfluoroalkyl groups. For example, perfluoro-2-butene is unreactive with *diazomethane,* but more highly substituted perfluoroalkenes, such as perfluoro-2-methyl-2-pentene, undergo cycloadditions in high yields [5] (equation 2). Note the regiospecificity that is observed in this reaction.

[5] **2**

$CH_2{=}N_2$ (Et$_2$O, RT):
- $CF_3CF{=}CFCF_3$ → No Reaction
- $(CF_3)_2C{=}CFC_2F_5$ → pyrazoline (H–N, N=CH; CF_3, CF_3; C_2F_5, F) 94%
- $CF_3(C_2F_5)C{=}C(C_2F_5)CF_3$ → pyrazoline (H–N, N=CH; CF_3, C_2F_5; CF_3, C_2F_5) 93%

An unusual seven-membered-ring-forming rearrangement was observed in the addition of diazomethane to a perfluorinated cyclobutene [5] (equation 3).

[5] **3**

Cyclobutene (R_f, R_f, F_2, F_2; $R_f = C_4F_7$) + $CH_2{=}N_2$ → bicyclic pyrazoline (H, H; N=N; R_f, R_f; F_2, F_2) → seven-membered ring (H, N–H, N; R_f, R_f; F_2, F_2) 85%

Perfluoroisobutylene undergoes cycloadditions with *azides* only at elevated temperatures; the reaction can lead to subsequent loss of nitrogen [*6*] (equation 4).

In another high-temperature reaction, chlorotrifluoroethylene undergoes cycloaddition with the ***azomethine ylide*** generated from the thermal electrocyclic ring opening of an aziridine, a reaction that contributes to a good overall synthesis of 3,4-difluoropyrroles [*7*] (equation 5).

[*6*] **4**

$CF_2{=}C(CF_3)_2$ → ($C_6H_5CH_2N_3$, 190 °C, 16 h) 1-benzyl (PhCH₂) triazoline with CF_3, CF_3, F_2, N=N — 87%

$CF_2{=}C(CF_3)_2$ → ($C_6H_5N_3$, 170 °C, 18 h, - N_2) N-Ph aziridine with CF_3, CF_3, F_2 — 60%

[*7*] **5**

N-$C(CH_3)_3$ aziridine with CO_2CH_3, H ⇄ (200 °C / benzene, 2 h) $CH_2{=}N^+(C(CH_3)_3)$–$CH^-(H)CO_2CH_3$

↓ $CF_2{=}CFCl$

N-$C(CH_3)_3$ pyrrolidine with CO_2CH_3, H, FCl, F_2 → (1. CH_3O-; 2. heat,, - CO_2) N-$C(CH_3)_3$ 3,4-difluoropyrrole (F, F) 25%

In contrast, perfluoroalkyl-substituted, unsaturated azides undergo biscycloadditions under mild conditions [*8*] (equation 6).

The high reactivity of monoperfluoroalkyl-substituted alkenes was demonstrated prior to 1972 via the facile reaction of diazomethane with 3,3,3-trifluoropropene [*9*]. More recently, such dipolarophiles were found to be similarly reactive

[8] 6

$R_fCH{=}CHCH_2N_3$ $\xrightarrow[\text{quant}]{\text{RT, 6 mo}}$ [bis-triazoline product with two R_f groups]

$R_f = C_2F_5, C_6F_{13}$

[10] 7

$R_fCH{=}CH_2$ + $C_6H_5CH{=}N^+(O^-)C_6H_5$ $\xrightarrow[\text{reflux}]{\text{toluene}}$ [5-R_f isoxazolidine (Ph, H, N–O, Ph)] + [4-R_f isoxazolidine (Ph, H, N–O, Ph)]

$R_f = C_4F_9$	32.5%	17.5%
$R_f = C_6F_{13}$	36%	4%

with *C*-phenyl-*N*-phenyl ***nitrones*** [10] (equation 7). Similar reactions of *N*-methyl nitrone were also reported [10].

Cycloadditions of diazomethane with fluorinated cyclobutenes provide insight into those factors that govern the reactivity and regioselectivity of such reactions. Although 3,3,4,4-tetrafluorocyclobutene undergoes reactions at ambient temperature in 5 min [11, 12], complete reaction with the less reactive perfluorocyclobutene requires 14 days [13] (equation 8). Note also the regioselectivity observed in the reaction of diazomethane with 3,3-difluorocyclobutene [14] (equation 9).

Numerous examples demonstrate that perfluoroalkylated ***alkynes*** are quite reactive dipolarophiles. Terminal alkynes, such as 3,3,3-trifluoropropyne, exhibit

8

$CH_2{=}N_2$ + [F_2, F_2 cyclobutene with X, X] $\xrightarrow{Et_2O,\ RT}$ [bicyclic pyrazoline with F_2, F_2, X, X, N=N]

[11, 12]	X = H	5 min	71%
[13]	X = F	14 d	55%

regioselectivity in their reactions with *p*-methoxyphenyl azide [*15*], phenyl nitrile oxide [*16*], and diazomethane [*17*] (equations 10 and 11).

[*14*] 9

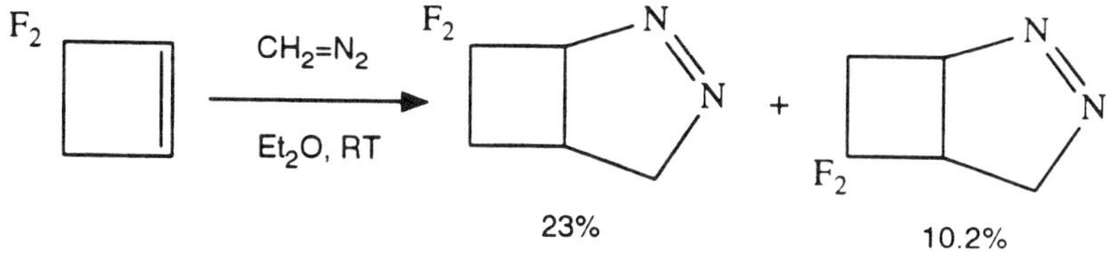

F_2 $CH_2{=}N_2$ / Et_2O, RT F_2 N N + N N F_2

23% 10.2%

[*15*] 10

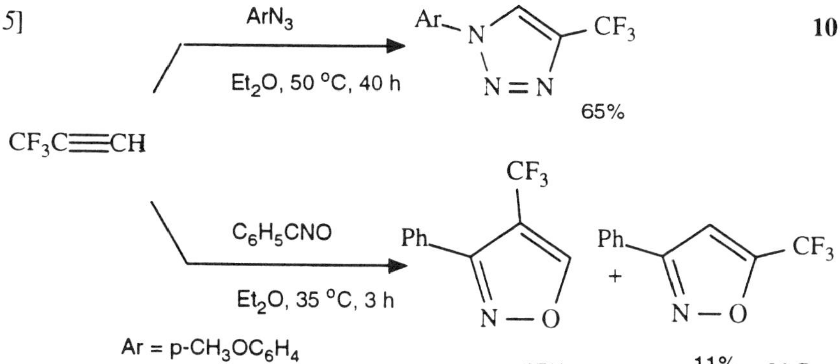

ArN$_3$ / Et_2O, 50 °C, 40 h Ar N CF_3 N=N 65%

$CF_3C{\equiv}CH$

C_6H_5CNO / Et_2O, 35 °C, 3 h CF_3 Ph N—O + Ph CF_3 N—O

Ar = p-$CH_3OC_6H_4$

67% 11% [*16*]

[*17*] 11

$C_4F_9C{\equiv}CH$ $CH_2{=}N_2$ / Et_2O, RT C_4F_9 N—N H + C_4F_9 N—N H

37.8% 22.2%

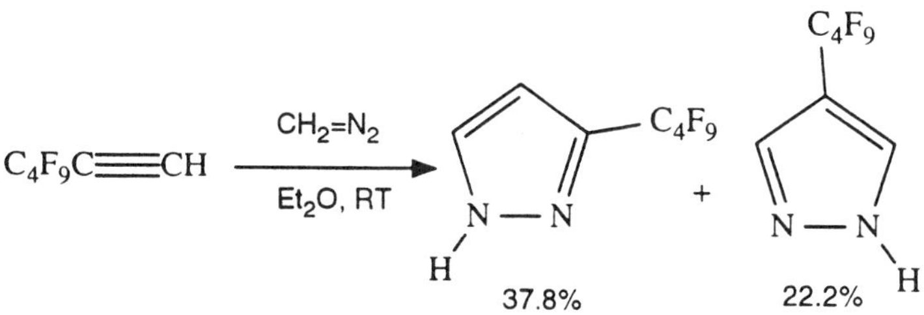

The additions of phenyl azide and **phenylnitrile oxide** to pentafluorophenylacetylene are also regiospecific [*15, 18*] (equation 12). Interestingly, in the latter reaction, *phenylacetylene* gives regiochemistry that is opposite to that observed for pentafluorophenylacetylene [*18*].

Perfluoro-2-butyne is, as expected, even more reactive in its reaction with phenyl **azide** [*15*]. Its reaction with *N*-phenylsydnone is also reported [*19*] (equation 13).

Lastly, perfluoropropyne undergoes cycloaddition with **diphenyldiazomethane** to give a single product, although the regiochemistry is undetermined

[*15*] 12

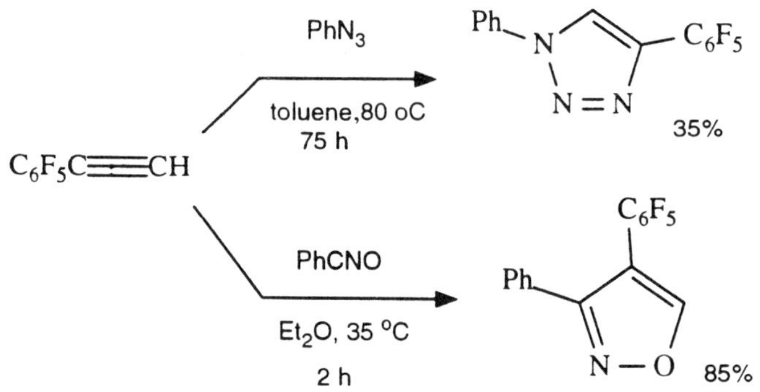

$C_6F_5C{\equiv}CH$ → (PhN_3, toluene, 80 oC, 75 h) → 1-Ph-4-C_6F_5-1,2,3-triazole, 35%

[*18*]

$C_6F_5C{\equiv}CH$ → (PhCNO, Et_2O, 35 °C, 2 h) → 3-Ph-4-C_6F_5-isoxazole, 85%

[*15*] 13

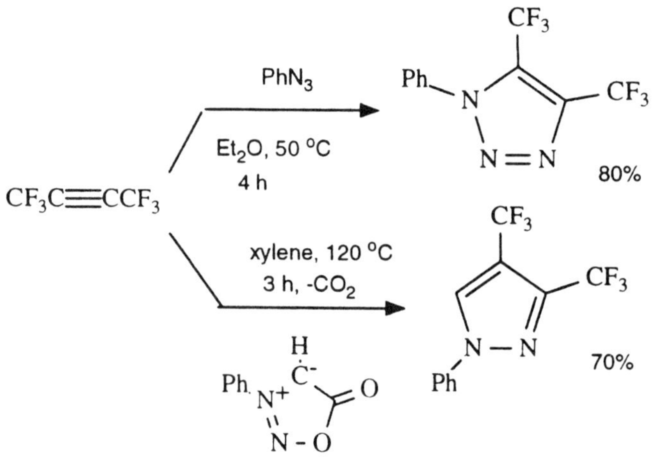

$CF_3C{\equiv}CCF_3$ → (PhN_3, Et_2O, 50 °C, 4 h) → 1-Ph-4,5-bis(CF_3)-1,2,3-triazole, 80%

[*19*]

$CF_3C{\equiv}CCF_3$ → (xylene, 120 °C, 3 h, -CO_2; N-phenylsydnone) → 1-Ph-3,4-bis(CF_3)pyrazole, 70%

[*20*] 14

$CF_3C{\equiv}CF$ → ($Ph_2C{=}N_2$, Et_2O, RT) → 3,3-diphenyl-3H-pyrazole with CF_3(F) and F(CF_3), 35%

[*19*]

$CF_3C{\equiv}CF$ → (toluene, 120 °C, 3 h; N-phenylsydnone) → 1-Ph-pyrazole with F(CF_3) and CF_3(F)

4-F 42.2%

4-CF_3 20.8%

References are listed on pages 835–839.

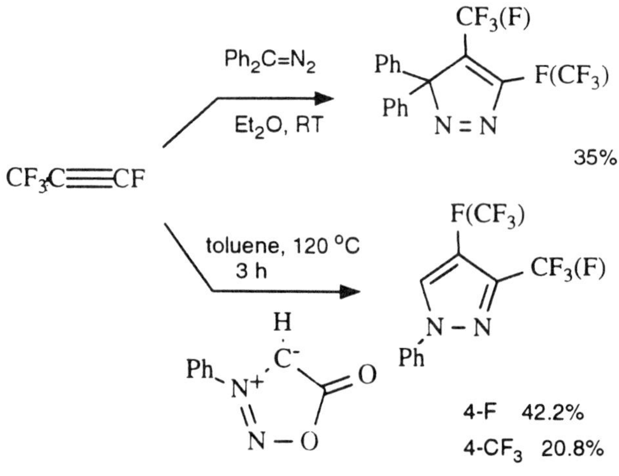

[*20*]. However, the regiochemistry of its addition to ***N*-phenylsydnone** is known [*19*] (equation 14).

The 1,3-dipolar cycloadditions of ***fluorinated allenes*** provide a rich and varied chemistry. Allenes, such as 1,1-difluoroallene and fluoroallene, that have fluorine substitution on only one of their two cumulated double bonds are very reactive toward 1,3-dipoles. Such activation derives from the electron-attracting inductive and hyperconjugative effects of the allylic fluorine substituent(s) that give rise to a considerable lowering of the energy of the LUMO of the C(2)–C(3) π bond [*21*].

For example, the most reactive of the allenes, 1,1-difluoroallene, reacts at room temperature with **nitrones, nitrile oxides,** and **diazoalkanes** to give cycloadducts in high yield [*22, 23, 24, 25*] (equation 15).

15

[*22*] $CF_2{=}C{=}CH_2$ + PhCH=N⁺(O⁻)CH₃ —($CHCl_3$, RT, 3 h)→ isoxazolidine with exocyclic CF_2, N–CH_3, Ph/H, 95%

[*23*] $CF_2{=}C{=}CH_2$ —(PhCNO, CCl_4, RT, 1 h)→ 42% + 33% (isomeric CF_2-methylene isoxazolines with Ph)

[*24*] $CF_2{=}C{=}CH_2$ —($CH_2{=}N_2$, Et_2O, RT, 5 min)→ 4-difluoromethylene-1-pyrazoline, 95%

The regiospecificity observed above for **diazomethane** addition is lost when other diazoalkanes are used, for example, dimethyldiazomethane's reaction with 1,1-difluoroallene [*24*] (equation 16).

16

[*24*] $CF_2{=}C{=}CH_2$ —($(CH_3)_2C{=}N_2$, Et_2O, 0 °C, 5 min)→ 3,3-dimethyl-4-difluoromethylene-1-pyrazoline (60%) + 5,5-dimethyl-4-difluoromethylene-1-pyrazoline (39%)

On the other hand, alkyl substitution on the 1,1-difluoroallene does not affect the regiospecificity of diazomethane addition [*24*] (equation 17).

Fluoroallene is somewhat less reactive than 1,1-difluoroallene, requiring 1 h for complete reaction with diazomethane compared with the almost instantaneous reaction of difluoroallene [*23, 25*]. Note the high stereoselectivity in this reaction (equation 18)!

Such *syn* stereoselectivity also was observed for a number of additions of **nitrones** to fluoroallene, such as its reaction with *N*-phenyl-*C*-phenyl nitrone [*25, 26, 27*] (equation 19).

[*24*] **17**

$CF_2{=}C{=}C(CH_3)_2 \xrightarrow[Et_2O,\ RT,\ 2\ h]{CH_2{=}N_2}$ 4-difluoromethylene-3,3-dimethyl-1-pyrazoline (CH$_3$, CH$_3$, CF$_2$, N=N) 90%

[*23*] **18**

$CHF{=}C{=}CH_2 \xrightarrow[Et_2O,\ RT,\ 1\ h]{CD_2{=}N_2}$ (D, D, F, H, N=N) 47.5% + (D, D, H, F, N=N) 6.5%

[*26*] **19**

$CHF{=}C{=}CH_2 \xrightarrow[CHCl_3,\ RT]{Ph(H)C{=}N^+(O^-)Ph}$ (Ph, H, F, H, N–O, Ph) 81% + (Ph, H, H, F, N–O, Ph) 14%

It is likely that the *syn* selectivity exhibited in cycloadditions of fluoroallene is due to electrostatic interactions [*23, 25*]. As in the case of difluoroallene, the reactions of fluoroallene with diazoalkanes and nitrile oxides are facile, but such reactions, other than that shown in equation 18, are neither regio- nor stereospecific [*23, 25*]. Indeed, the addition of phenylnitrile oxide to fluoroallene occurs with preferential *anti* addition for both regioisomeric products (equation 20).

Perfluoroallene is also quite reactive in its additions to nitrones [*20*], diazoalkanes [*20*], and **sydnones** [*19*]. With sydnones, the isolated product derives from a fluoride ion rearrangement of the primary adduct (equation 21).

Fluorinated α,β-***unsaturated carbonyl compounds*** also are reactive dipolarophiles. Because of the highly activating carbonyl substituent, these 1,3-dipolar cycloadditions are rapid and regiospecific. Good examples are the additions of

[23] **20**

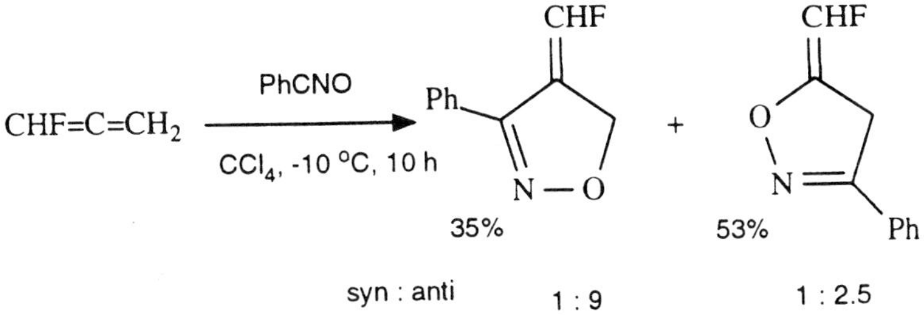

$CHF{=}C{=}CH_2$ —(PhCNO; CCl_4, -10 °C, 10 h)→ 4-(CHF=)-3-Ph-isoxazoline (35%) + 5-(CHF=)-3-Ph-isoxazoline (53%)

syn : anti 1 : 9 1 : 2.5

[20] **21**

$CF_2{=}C{=}CF_2$ + Ph(H)C=N⁺(O⁻)CH_3 —(benzene, RT)→ 71%

[20] $CF_2{=}C{=}CF_2$ —($PhCH{=}N_2$; benzene, RT)→ 78%

[19] $CF_2{=}C{=}CF_2$ + Ph-sydnone —(xylene, 120 °C, 2 h)→ 63%

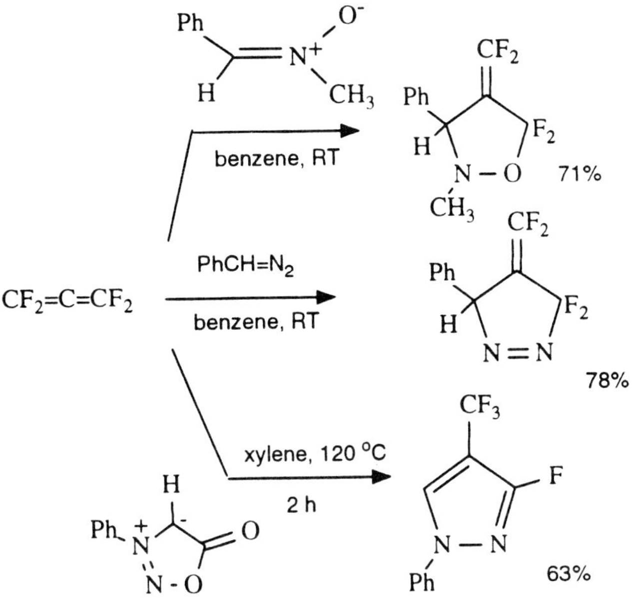

[28] **22**

CF_3(H)C=C(H)C(O)CH_3 —($CH_2{=}N_2$; Et_2O, RT, 1 min)→ 91%

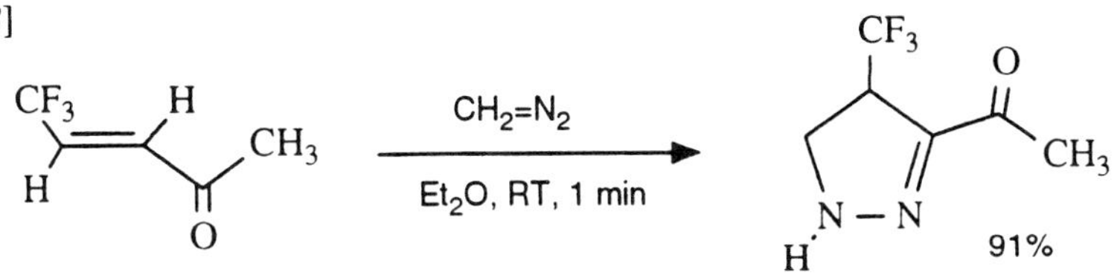

diazomethane to trifluoromethyl-substituted ketones and esters [*28, 29*]. Note the opposite *regiochemistry* observed in these two reactions (equations 22 and 23).

The latter regiochemistry is observed also for the additions of phenylazide to the same ester [*29*] and addition of diazomethane to a similar ketone [*29*] (equation 24). Such results are a good indication that the geminal trifluoromethyl groups, rather than the ester function, are dominating the regiochemical determination.

[*29*] **23**

$CH_2{=}N_2$, Et_2O, -20 °C → 100%

[*29*]

PhN$_3$, Et_2O, 60 °C, 15 d → 95%

24

[*29*]

$CH_2{=}N_2$, Et_2O, -20 °C → 100%

25

[*29*]

$CH_2{=}N_2$, Et_2O, -20 °C → 70% + 30%

[*17*] **26**

$C_4F_9C{\equiv}CCO_2C_2H_5$ → ($CH_2{=}N_2$, Et_2O, RT) 75%

In contrast, 30% of the alternative regioisomer is obtained when diazomethane adds to the analogous ***nitrile*** [*29*] (equation 25).

A regiospecific addition takes place when diazomethane reacts with an ***acetylenic ester*** [*17*] (equation 26).

α,β-Unsaturated esters and ketones with one β-perfluoroalkyl substituent are activated with respect to the analogous hydrocarbon substrates but add with the same regiochemistry as the hydrocarbon analogues, that is, with the carbonyl function directing the addition. In the case of β,β-bisperfluoroalkyl-substituted substrates, however, the orientation of addition is directed by the perfluoroalkyl substituents.

Some 1,3-dipolar cycloadditions to hetero π bond systems have been reported, including a couple of examples of additions of **azides** to the activated ***nitrile function*** of trifluoroacetonitrile [*30, 31*] (equation 27).

Bis(trifluoromethyl)oximes also are highly reactive dipolarophiles, as evidenced by the facile regiospecific addition of diazomethane [*32*] (equation 28).

27

CF_3CN → ($C_5H_{11}N_3$; 25 h, 110 °C) or ($(CH_3)_3SiN_3$; 14 d, 20 °C) → 1-R-5-CF_3-tetrazole

[*30*] R = C_5H_{11} 79%

[*31*] R = $(CH_3)_3Si$ 50%

28

[*32*]

$(CF_3)_2C{=}NOCF(C_6H_5)_2$ → ($CH_2{=}N_2$; Et_2O, RT) → 5,5-bis(CF_3)-4-$OCF(Ph)_2$-Δ²-1,2,4-... (N=N ring) 99%

Reactions of fluorinated dipoles. In recent years, much effort has been devoted to the preparation of trifluoromethyl-substituted 1,3-dipoles with the goal of using them to introduce trifluoromethyl groups into five-membered-ring heterocycles. **Fluorinated diazoalkanes** were the first such 1,3-dipoles to be prepared and used in synthesis. A number of reports of cycloadditions of mono- and bis(trifluoromethyl)diazomethane appeared prior to 1972 [*9*]. Other types of fluorine-substituted 1,3-dipoles were virtually unknown until only recently. However, largely because of the efforts of Tanaka's group, a broad knowledge of the chemistry of trifluoromethyl-substituted nitrile oxides, nitrile imines, nitrile ylides, and nitrones has been accumulated recently.

Trifluoromethyldiazomethane behaves as a typical diazoalkane in its additions to carbon–carbon multiple bonds [*9*]. For example, its reactions with ethylene and acetylene are slow; 3,3,3-trifluoropropene reacts 20 times faster than ethylene, and methyl propiolate react 2–3 times faster than 3,3,3-trifluoropropyne. Although trifluoromethydiazomethane appears to react faster with electron-deficient olefins, it is less reactive than diazomethane, and its range of reactivity is considerably smaller [*9, 33*] (equation 29).

The reaction of 2 equivalents of trifluoromethyldiazomethane with trifluoroacetonitrile was reported [*34*] (equation 30).

[*33*] $CF_3CH{=}N_2$ → ($CH_2{=}CRX$, neat, RT) **29**

91% R = H, X = CH_2Cl, 2 d
98% R = CH_3, X = CO_2CH_3, 15 d

[*33*] ($CH_2{=}CHX$, neat, RT, 8 h)

98% X = CO_2CH_3
87% X = CN

[*33*] ($HC{\equiv}CX$, neat, RT)

97% X = CH_2Br, 17 d
100% X = C_6H_5, 4 w
81% X = CH_3, 8 w

[*34*] $CF_3CH{=}N_2$ → (CF_3CN) 84% **30**

The regio- and stereoselectivities of cycloadditions of **trifluoroacetonitrile oxide,** which is generated in situ by treatment of the trifluoroacetohydroxamyl bromide etherate with triethylamine in toluene (equation 31), have been determined in a series of studies by Tanaka [*35, 36, 37, 38*]. The highly reactive nitrile oxide reacts regioselectively with a variety of activated terminal **alkenes** and **alkynes** (equations 32 and 33).

[35] **31**

$$CF_3(Br)C{=}NOH \xrightarrow[\text{toluene}]{Et_3N} CF_3C{\equiv}N^+{-}O^-$$

[35] **32**

$$CF_3C{\equiv}N^+{-}O^- \xrightarrow[\text{toluene, RT}]{CH_2{=}CHR} \text{3-}CF_3\text{-5-R-4,5-dihydroisoxazole}$$

R	Yield
R = C_6H_5	84%
R = $CH_2OC_6H_5$	80%
R = CH_2OH	60%
R = CO_2CH_3	82%

[35] **33**

$$CF_3C{\equiv}N^+{-}O^- \xrightarrow[\text{toluene, RT}]{RC{\equiv}CH} \text{3-}CF_3\text{-5-R-isoxazole}$$

R	Yield
R = C_6H_5	93%
R = CH_2OH	38%

On the other hand, its cycloadditions with 1,2-disubstituted alkenes under similar conditions produce stereospecifically a mixture of regioisomeric products [*35*] (equation 34). In contrast, its reaction with the unsymmetrical alkyne 1-phenylpropyne leads to a single product [*35*] (equation 35).

Reactions of trifluoroacetonitrile oxide with conjugated dienes also lead to regiospecific additions [*36*] (equation 36). Its addition to the strain-activated double

[35] **34**

$$CF_3C{\equiv}N^+{-}O^- \xrightarrow[\text{toluene, RT}]{CH_3CH{=}CHR} \text{3-}CF_3\text{-4-}CH_3\text{-5-R-isoxazoline} + \text{3-}CF_3\text{-4-R-5-}CH_3\text{-isoxazoline}$$

R		
R = C_6H_5	56%	9.2%
R = CO_2CH_3	16.8%	22.2%

[35] **35**

$$CF_3C{\equiv}N^+{-}O^- + CH_3C{\equiv}CC_6H_5 \xrightarrow{\text{toluene, RT}} \text{3-}CF_3\text{-4-}C_6H_5\text{-5-}CH_3\text{-isoxazole} \quad 23\%$$

bonds of norbornene, as has been observed for other 1,3-dipoles, is stereospecific and takes place exclusively *exo* [*37*]. This specificity is lost in reactions with norbornadiene, which gives a mixture of *exo* and *endo* products (equation 37).

[*36*] **36**

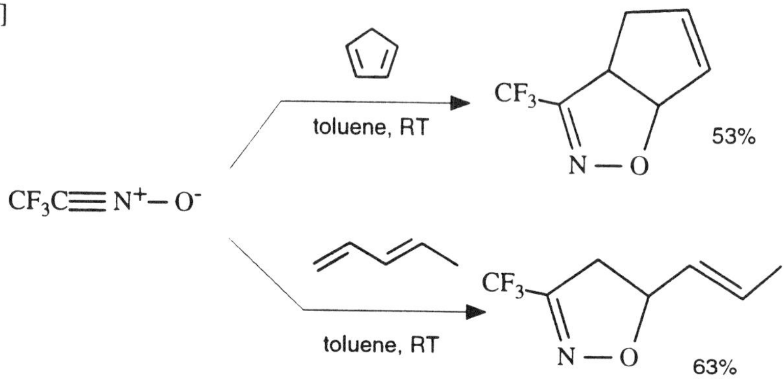

$CF_3C{\equiv}N^+{-}O^-$ — cyclopentadiene, toluene, RT → bicyclic isoxazoline (CF_3, N—O), 53%

$CF_3C{\equiv}N^+{-}O^-$ — 1,3-pentadiene, toluene, RT → 3-CF_3-5-propenyl isoxazoline (N—O), 63%

[*37*] **37**

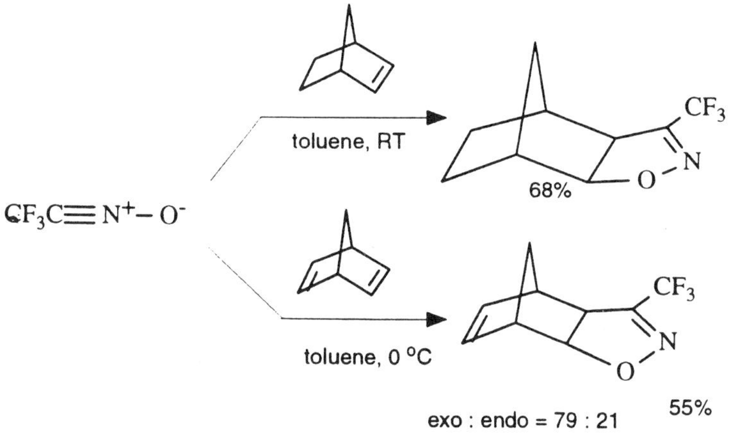

$CF_3C{\equiv}N^+{-}O^-$ — norbornene, toluene, RT → CF_3 (N, O) adduct, 68%

norbornadiene, toluene, 0 °C → CF_3 (N, O) adduct, exo : endo = 79 : 21, 55%

[*38*] **38**

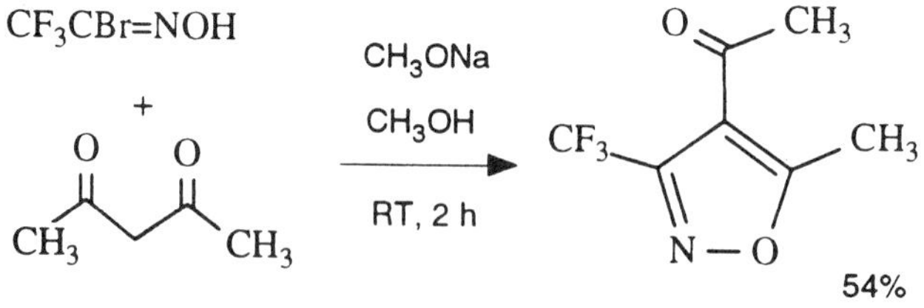

$CF_3CBr{=}NOH$ + $CH_3COCH_2COCH_3$ (O, O, CH_3, CH_3) — CH_3ONa, CH_3OH, RT, 2 h → CF_3 isoxazole with $COCH_3$ (O, CH_3) and CH_3 (N—O), 54%

Trifluoroacetonitrile oxide also reacts with stabilized enolate ions, such as that derived from 2,4-pentanedione, to give good yields of 1,3-dipolar adducts [38] (equation 38).

The analogous **trifluoroacetonitrile phenylimine,** generated from *N*-phenyl-trifluoroacetohydrazonyl bromide (equation 39), exhibits similar reactivity and regioselectivity in its reactions with activated alkenes and alkynes [*39*] (equations 40 and 41).

[*39*] **39**

$$CF_3CBr{=}NNHC_6H_5 \xrightarrow{Et_3N,\ toluene} CF_3C{\equiv}N^+{-}N^-{-}C_6H_5$$

[*39*] **40**

$$CF_3C{\equiv}N^+{-}N^-{-}C_6H_5 \xrightarrow[toluene,\ 80\text{-}90\ ^\circ C]{CH_2{=}CHR}$$ 3-CF_3-5-R-1-Ph-pyrazoline (N — N, Ph)

R = C_6H_5, 1 h 97%

R = CO_2CH_3, 5 h 85%

R = OC_4H_9, 12 h 89%

[*39*] **41**

$$CF_3C{\equiv}N^+{-}N^-{-}C_6H_5 \xrightarrow[toluene,\ RT,\ 2\ h]{HC{\equiv}CR}$$ 3-CF_3-5-R-1-Ph-pyrazole + 3-CF_3-4-R-1-Ph-pyrazole

R = C_6H_5	45%	
R = CO_2CH_3	63%	17%

In a manner analogous to classic nitrile imines, the additions of trifluoromethylacetonitrile phenylimine occur regiospecifically with activated terminal alkenes but less selectively with alkynes [*39*]. The nitrile imine reacts with both dimethyl fumarate and dimethyl maleate in moderate yields to give exclusively the *trans* product, presumably via epimerization of the labile H at position 4 [*40*] (equation 42). The nitrile imine exhibits *exo* selectivities in its reactions with norbornene and norbornadiene, which are similar to those seen for the nitrile oxide [*37*], and even greater reactivity with enolates than that of the nitrile oxide [*38, 41*]. Reactions of trifluoroacetonitrile phenyl imine with isocyanates, isothiocyanates, and carbodiimides are also reported [*42*].

Treatment of *N*-benzyltrifluoroacetimidoyl chloride with triethylamine in toluene at room temperature leads to in situ generation of **trifluoroacetonitrile benzylide** [*43*] (equation 43), which reacts with methyl acrylate to form cycloadducts [*43*] (equation 44). Although the kinetic ratio of products favors the *cis* adduct (3:1), thermodynamic equilibration leads to an excess of the *trans* isomer (7:1).

[*40*] **42**

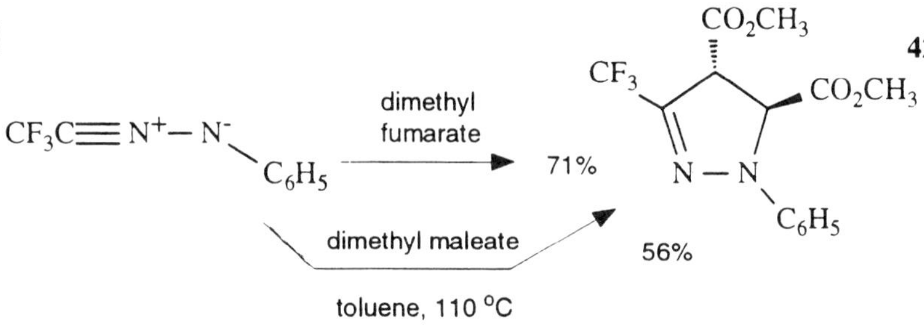

$CF_3C\equiv N^+-N^-C_6H_5$ —(dimethyl fumarate, 71%; dimethyl maleate, 56%; toluene, 110 °C)→ pyrazoline with CF_3, CO_2CH_3, CO_2CH_3, $N-N$, C_6H_5

[*43*] **43**

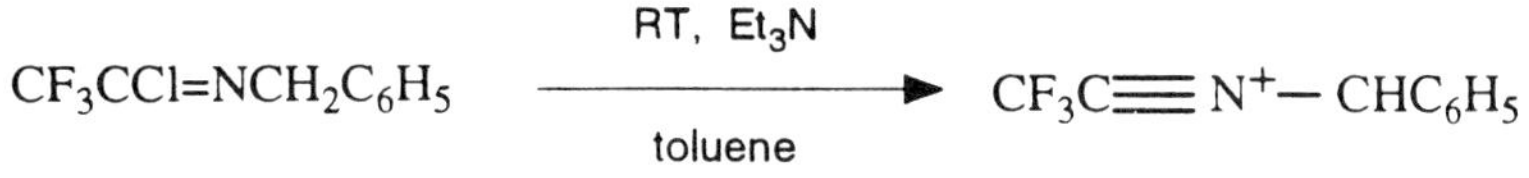

$$CF_3CCl{=}NCH_2C_6H_5 \xrightarrow[\text{toluene}]{\text{RT, }Et_3N} CF_3C\equiv N^+-CHC_6H_5$$

[*43*] **44**

$$CF_3C\equiv N^+-CHC_6H_5 \xrightarrow[\text{toluene, RT, 7 d}]{CH_2{=}CHCO_2CH_3}$$ CF_3, CO_2CH_3, N, C_6H_5 34%

cis : trans = 3/1

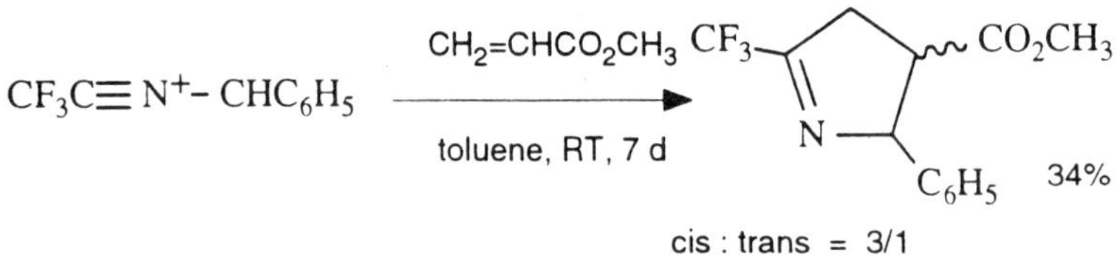

A novel pyrolytic method of generating nitrile ylides in situ was reported by Burger [*44*] (equation 45). Such nitrile ylides react with various dipolarophiles: alkynes [*44*] (equation 46), nitriles [*45*] (equation 47), dimethyl azodicarboxylate [*45*], aldehydes [*45*], and nitroso compounds [*46*].

[*44*] **45**

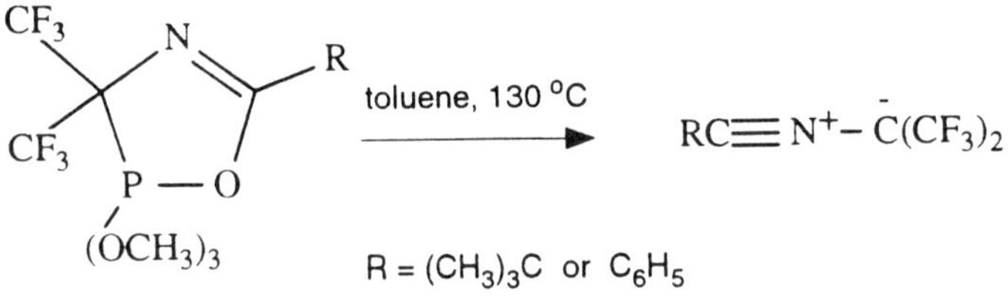

CF_3, CF_3, N, R, P — O, $(OCH_3)_3$ —(toluene, 130 °C)→ $RC\equiv N^+-\bar{C}(CF_3)_2$

R = $(CH_3)_3C$ or C_6H_5

[*44*] **46**

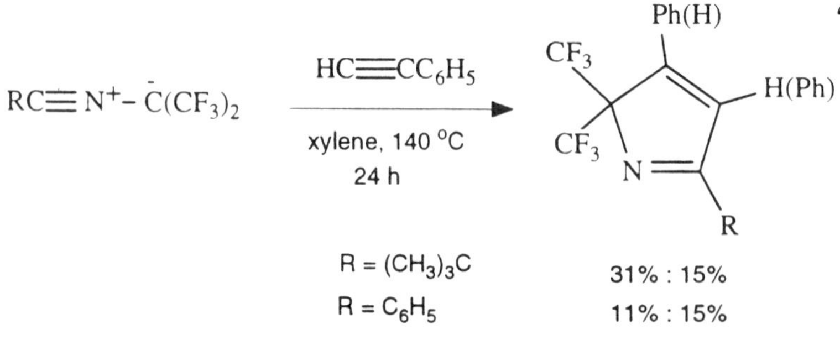

$$RC\equiv N^+-\bar{C}(CF_3)_2 \xrightarrow[\text{xylene, 140 °C, 24 h}]{HC\equiv CC_6H_5}$$ CF_3, CF_3, Ph(H), H(Ph), N, R

R = $(CH_3)_3C$ 31% : 15%

R = C_6H_5 11% : 15%

[*45*] **47**

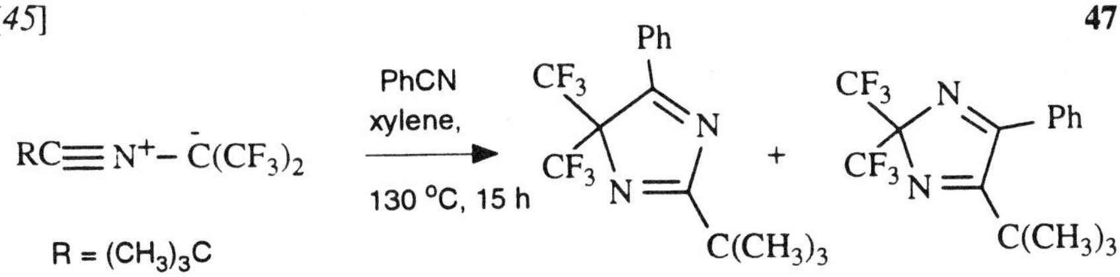

$$RC\equiv N^{+}-\bar{C}(CF_3)_2 \xrightarrow[130\ ^\circ C,\ 15\ h]{PhCN,\ xylene,}$$

R = $(CH_3)_3C$

Ph, CF_3, CF_3, N, N, $C(CH_3)_3$ + CF_3, CF_3, N, N, Ph, $C(CH_3)_3$

Recently, Burger devised an improved method of carrying out mild, regiospecific cyclizations that involve an intermediate that acts as a synthon for a **nitrile ylide** of HCN [*47*] (equation 48). With this methodology, cycloadditions with activated alkenes, alkynes, and azo compounds were carried out [*47*] (equation 49). All such reported reactions were regiospecific and had the same orientational preference.

[*47*] **48**

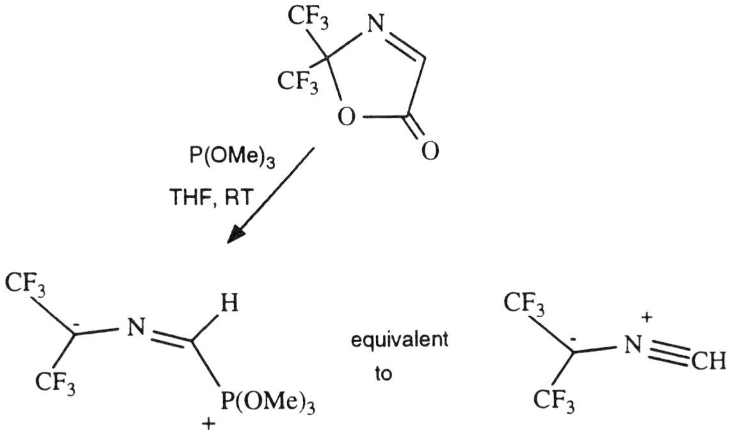

CF_3, CF_3, N, O, O

$P(OMe)_3$
THF, RT

CF_3, CF_3, N, H, $P(OMe)_3$ + equivalent to CF_3, CF_3, $N^{+}\equiv CH$

[*47*] **49**

CF_3, CF_3, N, O, O

$P(OMe)_3$
THF, RT

$HC\equiv CCO_2CH_3$ $CH_2{=}CHCN$

CF_3, CF_3, N, CO_2CH_3 85% CF_3, CF_3, N, CN 85%

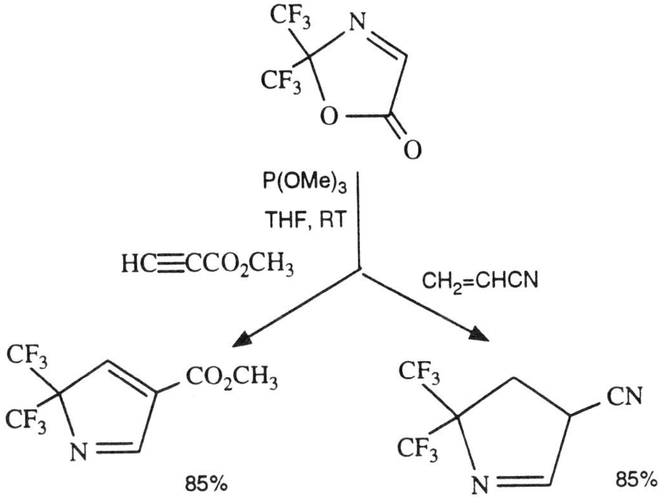

References are listed on pages 835–839.

Nitrones are among the most highly studied and useful reagents for the synthesis of five-membered-ring heterocycles. The first fluorinated nitrone, *N*-methyl-*C*-(trifluoromethyl)nitrone, was prepared recently and used to introduce trifluoromethyl groups into such heterocycles.

Nitrone hydrate is converted into nitrone by boiling in benzene with azeotropic removal of water [*48*] (equation 50). This in situ formation of nitrone is carried out in the presence of various alkenes and alkynes, which undergo cycloaddition with the nitrone [*48, 49*] (equations 51 and 52).

[*48*] **50**

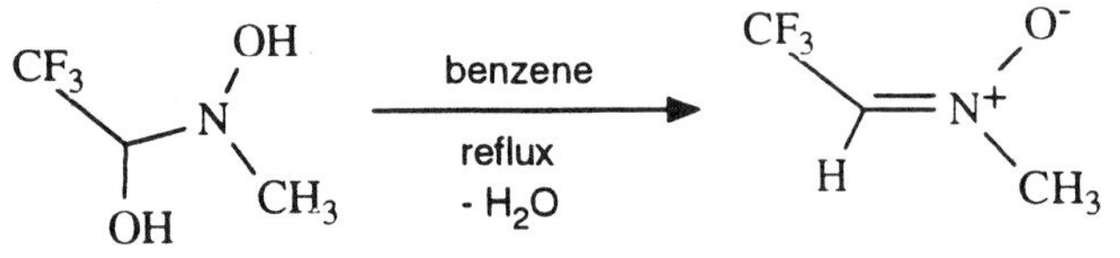

$$CF_3CH(OH)N(OH)CH_3 \xrightarrow[\text{reflux, } -H_2O]{\text{benzene}} CF_3CH{=}N^+(O^-)CH_3$$

[*49*] **51**

$$CF_3CH{=}N^+(O^-)CH_3 \xrightarrow[\text{benzene, reflux}]{CH_2{=}CHR} \text{3-}CF_3\text{-5-R-2-}CH_3\text{-isoxazolidine}$$

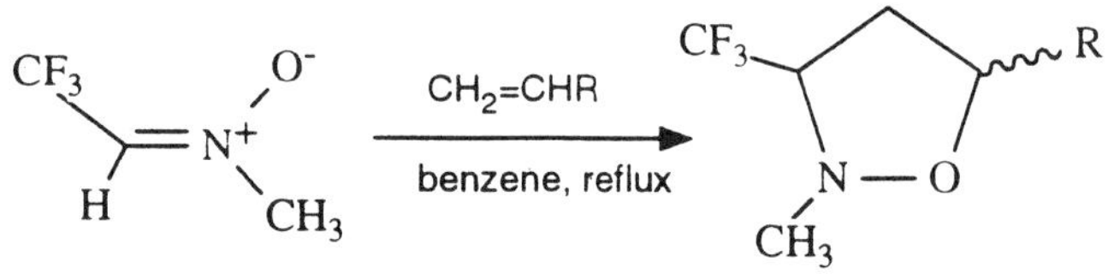

R	trans	cis
C_6H_{13}	41%	17%
C_6H_5	59%	12%
CO_2CH_3	60%	10%

[*48*] **52**

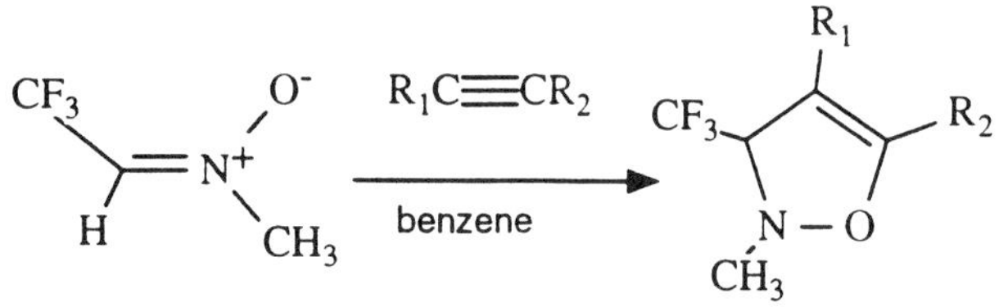

$$CF_3CH{=}N^+(O^-)CH_3 \xrightarrow[\text{benzene}]{R_1C{\equiv}CR_2} \text{3-}CF_3\text{-4-}R_1\text{-5-}R_2\text{-2-}CH_3\text{-4-isoxazoline}$$

R_1	R_2	Conditions	Yield (%)
H	C_6H_{13}	60 °C, 4 h	26
H	C_6H_5	80 °C, 4 h	91
H	CO_2CH_3	50 °C, 14 h	52
	C_6H_5	80 °C, 20 h	37
CO_2CH_3	CO_2CH_3	50 °C, 4 h	36

[*4 + 1*] *Cycloadditions.* 1,3-Dipolar cycloadditions certainly have been the most important means of forming fluorinated five-membered rings. Fluorinated substances, however, have also been used in other classic five-membered-ring-forming cyclizations. For example, **isocyanides** have been used as single-atom addends in formal [4+1] cycloadditions with ***heterodienes.*** Fluorinated α,β-unsaturated esters [*50*], amides [*50*], and ketones [*51*], as well as 1,3-diazadienes [*52*], react with alkyl isocyanides to give heterocyclic adducts (equations 53–55).

[*50*] **53**

$CF_2{=}C(CF_3)CO_2CH_3$ → (RN≡C:, hexane, -70 °C; R = c-C_6H_{13}) 60%

[*51*] **54**

→ (RN≡C:, Et_2O, RT, 3 d; R = c-C_6H_{11}) 54%

[*52*] **55**

→ (RN≡C:, Et_2O, 60 °C, 30 h; R = Bu) 92%

The use of *radical cyclizations to make five-membered rings* has become a very important tool for synthetic chemists. Although there has been a virtual explosion of reports in the literature regarding the cyclization of 5-hexenyl radicals to cyclopentyl carbinyl radicals in all types of hydrocarbon systems [*53*], the use of this cyclization for the synthesis of fluorine-containing cyclopentanes has been largely ignored.

The first example of a cyclization of fluorine-containing 5-hexenyl radicals was the study of the radical-initiated cyclodimerization reaction of 3,3,4,4-tetrafluoro-4-iodo-1-butene. In this reaction, the intermediate free radical adds either to more of the butene or to an added unsaturated species [*54, 55*] (equation 56). Electron-deficient alkenes are not as effective trapping agents as electron-rich alkenes and alkynes [*55*].

[*54, 55*] **56**

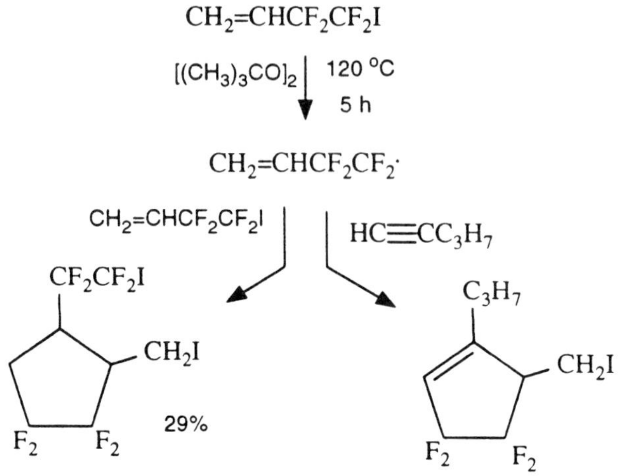

$CH_2{=}CHCF_2CF_2I$

$[(CH_3)_3CO]_2$, 120 °C, 5 h

$CH_2{=}CHCF_2CF_2^{\cdot}$

$CH_2{=}CHCF_2CF_2I$ | $HC{\equiv}CC_3H_7$

CF_2CF_2I, CH_2I, F_2, F_2 29% | C_3H_7, CH_2I, F_2, F_2

A brief paper by Kobayashi indicated some of the potential of this cyclization for making fluorine compounds [*56*] (equation 57). Watanabe et al. provided examples of incorporation of a trifluoromethyl group into a five-membered ring [*57*] (equation 58). Nevertheless, the full potential of this methodology in preparing fluorine-containing materials has not been realized.

[*56*] **57**

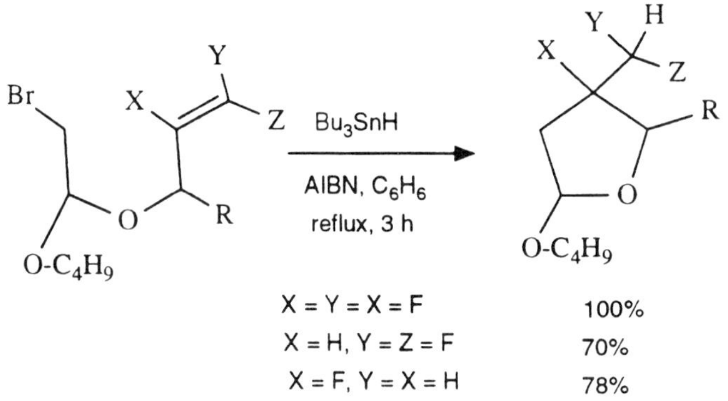

Br, X, Y, Z, O, R, O-C_4H_9 → (Bu_3SnH; AIBN, C_6H_6, reflux, 3 h) → X, Y, H, Z, R, O, O-C_4H_9

X = Y = X = F 100%

X = H, Y = Z = F 70%

X = F, Y = X = H 78%

[*57*] **58**

CF_3, Cl, Cl, C_2H_5, O → (Bu_3SnH; AIBN, C_6H_6, 65 °C, 4 h) → CH_3, CF_3, Cl, O, C_2H_5 76%

References are listed on pages 835–839.

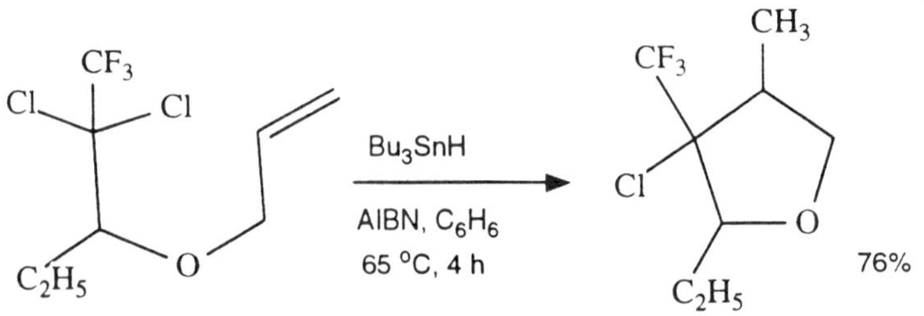

Another reaction that has potential is the intramolecular *ene* reaction, which preferentially results in five-membered rings. Because fluorinated ketones have a propensity for undergoing intermolecular *ene* reactions [*58*], this reaction has tremendous synthetic possibilities. Only one example of an *ene* reaction of an unsaturated trifluoromethyl ketone has been reported, but it indicates the potential of the methodology [*59*] (equation 59).

Lastly, a termolecular cyclization of unknown mechanism provides a useful synthesis of tetrakis(trifluoromethyl)thiophene [*60*] (equation 60).

[*59*] **59**

C_6H_5, CH_3, CF_3, O; $EtAlCl_3$, CH_2Cl_2, - 15 °C, 7 h → H, C_6H_5, OH, CF_3 100%

[*60*] **60**

$CF_3C{\equiv}CCF_3$ + S; sulfolane, 110 °C → 61% CF_3, S, CF_3, CF_3, CF_3

Cycloadditions Forming Six-Membered Rings

In the [2+4] pericyclic cycloaddition reaction known as the **Diels–Alder reaction,** fluorine-containing compounds have been widely used as dienes, dienophiles, or both. Much of the fundamental work, including many comprehensive and systematic studies, was done before 1972, and Hudlicky provides an excellent summary of this work [*9*]. Additional sources for early work in this area are reviews in *Organic Reactions* [*61*] and *Fluorine Chemistry Reviews* [*62*].

To understand *the effect of fluorine on the reactivity of dienophiles and dienes* in the Diels–Alder reaction, one must recognize that, in general, as in 1,3-dipolar cycloaddition, electron-deficient dienophiles and electron-rich dienes are most reactive in this cycloaddition. With direct fluorine substitution not affecting significantly the HOMO or LUMO energies [*4*], one would predict that the reactivities of dienes or dienophiles should not be affected drastically by *vinylic* fluorine substitution. On the other hand, with *allylic* fluorine substitution lowering the energies of both HOMO and LUMO, one would expect an enhancement of dienophilic reactivity (and diminution of diene reactivity) when a π system has perfluoroalkyl substituents.

Fluorinated dienophiles. Although ethylene reacts with butadiene to give a 99.98% yield of a Diels–Alder adduct [*63*], tetrafluoroethylene and 1,1-dichloro-2,2-difluoroethylene prefer to react with 1,3-butadiene via a [2+2] pathway to form almost *exclusively cyclobutane adducts* [*61, 64*] (equation 61). This obvious difference in the behavior of hydrocarbon ethylenes and fluorocarbon ethylenes is believed to result not from a lack of reactivity of the latter species toward [2+4] cycloadditions but rather from the fact that the rate of *nonconcerted* cyclobutane formation is greatly enhanced [*65*].

[*61*] **61**

$CF_2{=}CCl_2$ + $CH_2{=}CHCH{=}CH_2$ —heat→ [vinylcyclobutane, F_2, Cl_2] + [cyclohexene, F_2, Cl_2]

>99 : 1

As part of his elegant and comprehensive examination of the competition between [2+2] and [2+4] reactions of **fluorinated ethylenes,** Bartlett found that trifluoroethylene's [2+4] reaction with butadiene competed somewhat with its [2+2] cycloaddition [*66*] (equation 62).

[*66*] **62**

$CF_2{=}CFH$ + $CH_2{=}CHCH{=}CH_2$ —215 °C, 24 h→ [vinylcyclobutane, F_2, FH] 60% + [vinylcyclobutane, FH, F_2] 27% + [cyclohexene, F_2, FH] 13%

In a definitive study of butadiene's reaction with 1,1-dichloro-2,2-difluoroethylene, Bartlett concluded that [2+4] adducts of acyclic dienes with fluorinated ethylenes are formed through a mixture of *concerted and nonconcerted, diradical pathways* [*67*]. The degree of observed [2+4] cycloaddition of fluorinated ethylenes is related to the relative amounts of *transoid* and *cisoid* conformers of the diene, with very considerable (i.e., 30%) Diels–Alder adduct being observed in competition with [2+2] reaction, for example, in the reaction of 1,1-dichloro-2,2-difluoroethylene with cyclopentadiene [*9, 68*].

In contrast to the relative lack of Diels–Alder reactivity exhibited by fluorinated ethylenes, **ethylenes substituted with perfluoroalkyl groups show greatly**

enhanced dienophilic reactivity. Numerous examples were given in Hudlicky's book [*9*], and a couple of nice stereochemical and regiochemical studies of the Diels–Alder chemistry of 3,3,3-trifluoropropene have been reported [*69, 70*] (equation 63). No [2+2] products are observed in these reactions.

[*69*] **63**

$CH_2{=}CHCF_3$ → (cyclopentadiene, 135 °C, 84 h) → bicyclic adduct with CF_3; endo : exo = 56% : 14%

[*70*]

$CH_2{=}CHCF_3$ + 2-(trimethylsilylmethyl)butadiene (CH_2SiMe_3) → (150 °C, 72 h) → cyclohexene with CH_2SiMe_3, CF_3(H), H(CF_3); 26% (12%)

For dienophiles with vinylic fluorine, such as perfluoropropene, the [2+4] reactivity is diminished so that [2+2] cycloaddition dominates [*71*] (equation 64).

Knunyants showed that such perfluoroalkenes, under forcing conditions, undergo Diels–Alder reactions with cyclic dienes such as cyclohexadiene [*72*] (equation 65) or furan [*73*] (equation 66).

[*71*] **64**

$CF_2{=}CFCF_3$ + $CH_2{=}CHCH{=}CH_2$ → (180 °C, 70 h) → cyclobutane (F_2, F, CF_3, vinyl) 87% + cyclohexene (F_2, CF_3, F) 4%

Alkynes substituted with one or two trifluoromethyl groups are also highly reactive dienophiles [*9*]. Indeed, hexafluoro-2-butyne is used increasingly as a definitive acetylenic dienophile in "difficult" Diels–Alder reactions. It was used, for example, to prepare novel "inside–outside" bicycloalkanes via its reaction with *cis,trans*-1,3-undecadiene [*74*] (equation 67) and to do a tandem Diels–Alder reaction with a 1,1-bis(pyrrole)methane [*75*] (equation 68). Indeed, its reactions with pyrrole derivatives and furan have been used in the syntheses of 3,4-bis(trifluoromethyl)pyrrole [*76, 77*] (equation 69) and 1,4-bis(trifluoromethyl)benzene-2,3-oxide [*78*] (equation 70), respectively.

Although one trifluoromethyl group activates an alkene as a dienophile, two or more can actually inhibit dienophilic reactivity, as evidenced by the fact that

1,1-dichlorohexafluoroisobutylene [*65*] and perfluoro-3,4-dimethylhex-3-ene [*79*] are unreactive with butadiene at temperatures up to 250 °C. Such lack of reactivity might well derive from the inhibiting steric effect of the bulky perfluoroalkyl groups.

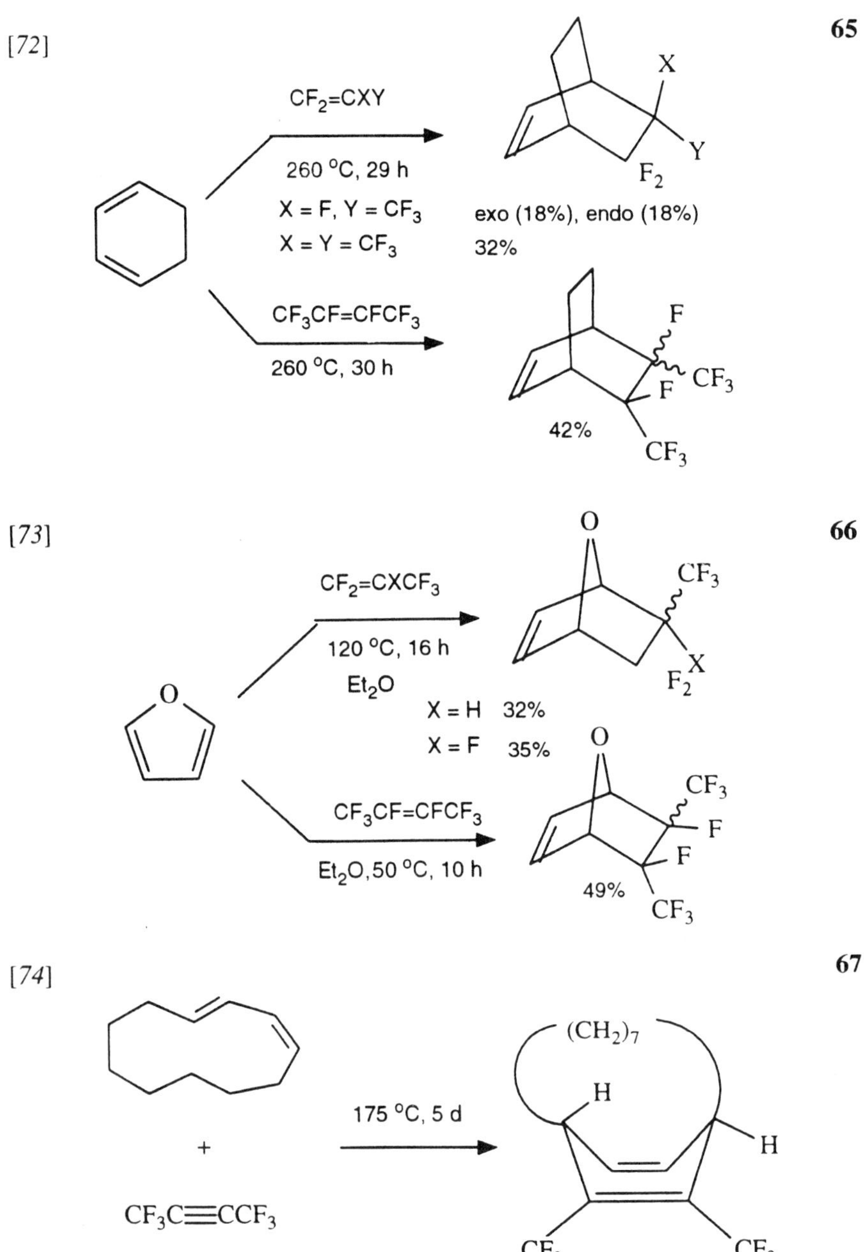

[*75*] **68**

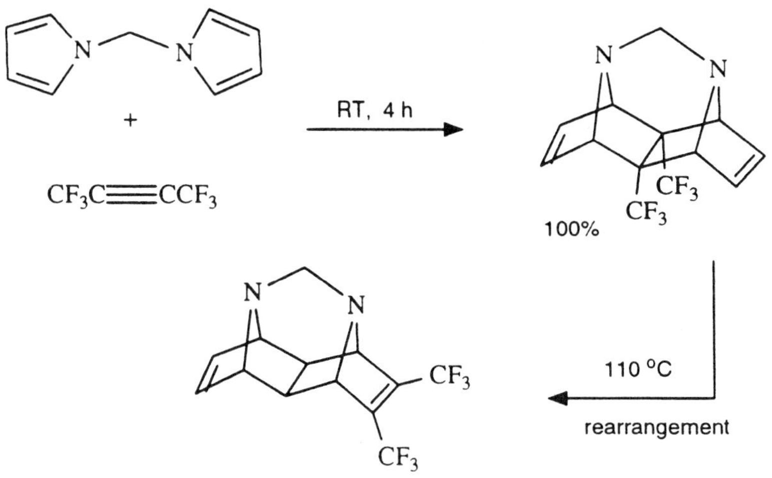

RT, 4 h

$CF_3C{\equiv}CCF_3$

100%

110 °C

rearrangement

[*76, 77*] **69**

$N-CO_2C(CH_3)_3$

100 °C, 15 h

$CF_3C{\equiv}CCF_3$

96%

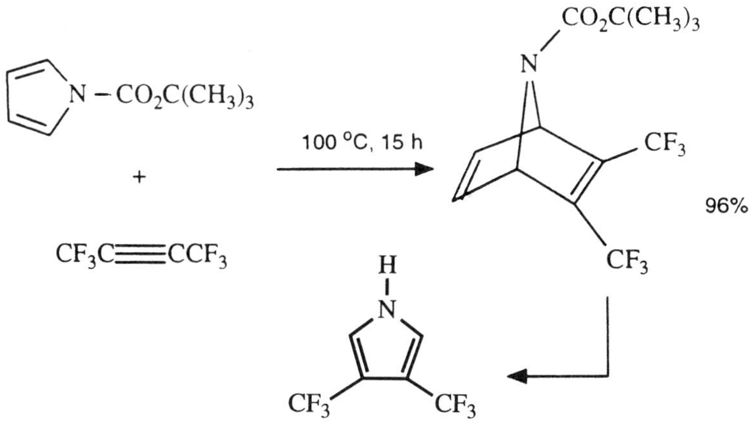

On the other hand, Chambers found that the related perfluorobiscyclobutylidene undergoes Diels–Alder reactions with butadiene, cyclopentadiene, and furan [*79, 80*] (equation 71). Such enhanced reactivity may be due to strain in the substrate.

Similarly, partially fluorinated and perfluorinated methylenecyclopropanes [*81, 82*], cyclopropenes [*83, 84, 85*], cyclobutenes [*73, 86*], and bicyclic alkenes [*87, 88, 89, 90*] apparently derive dienophilic reactivity from relief of their ground-state strain during reaction. Thus 2,2-difluoromethylenecyclopropane and perfluoromethylenecyclopropane undergo exclusive [2+4] cycloadditions [*81, 82*] (equations 72 and 73), whereas (difluoromethylene)cyclopropane undergoes only [2+2] cycloadditions [*81*].

[*78*] **70**

0 °C, 100 h
81%
hυ
120 °C
24 h
85%
80%

[*79,80*] **71**

CH_2=CHCH=CH_2
80 °C
72%
80 °C, 16 h
70%

Fluorinated cyclopropenes have long been known to be reactive dienophiles [*9*], and the past 20 years have provided us with a few more nonexceptional examples [*83, 84, 85*].

An interesting probe of reactivity was presented by Burton in his study of cycloadditions of 1,2-disubstituted-3,3,4,4-tetrafluorocyclobutenes and 1,2-disubstituted-3,3,4,4,5,5-hexafluorocyclopentenes with butadiene, 2-methylbutadiene, and 2,3-dimethylbutadiene [*86*]. On the basis of the extent of their conversions to adducts, the relative reactivities of the cyclobutenes and of the cyclopentenes are as shown in equation 74. A typical reaction is shown in equation 75.

Fluorocyclobutenes are generally more reactive than fluorocyclopentenes, and only 1-chloro-3,3,4,4-tetrafluorocyclobutene is reactive enough to give significant conversion with butadiene. The most reactive of the dienes is 2,3-dimethylbutadiene, a result consistent with the percentage of *cisoid* conformers being a critical factor in determining diene reactivity.

Strained pertrifluoromethyl-substituted valence tautomers of aromatic systems, such as tetrakis(trifiuoromethyl)Dewar thiophene [*87*] and hexakis(trifluoromethyl)benzvalene [*88, 89*], undergo Diels–Alder reactions with various cyclic and acyclic dienes (equations 76 and 77).

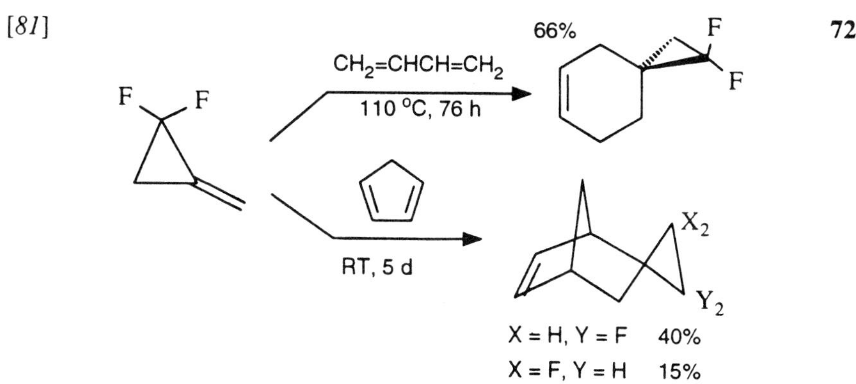

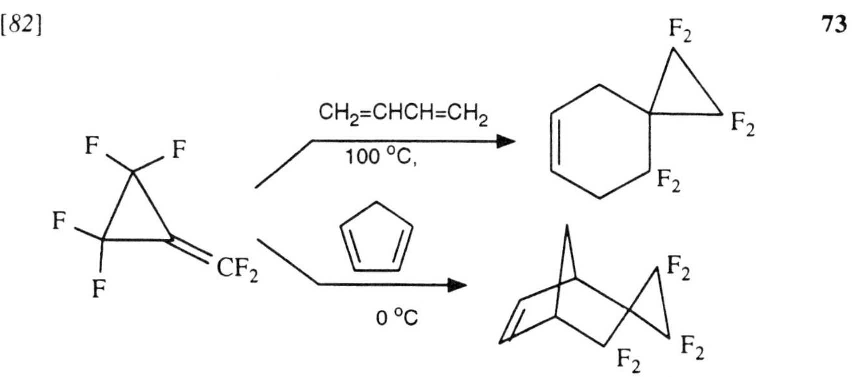

[*86*] **74**

F_2 F_2 X Y F_2 F_2 F_2 X Y

X = Y = H > X = H, Y = Cl > X = H, Y = F >

X = Cl, Y = F > X = Y = F

[*86*] **75**

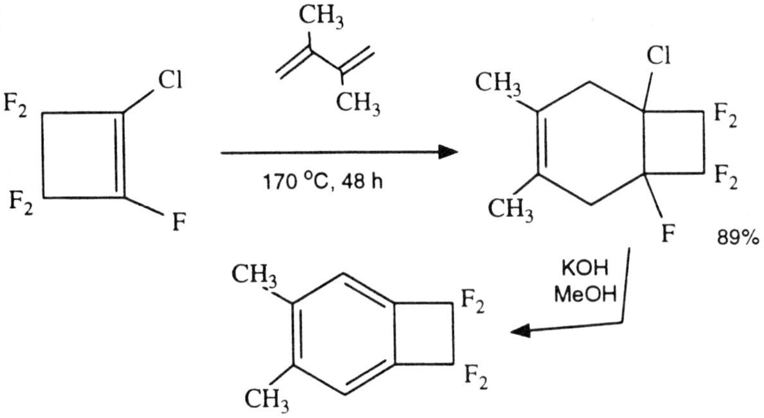

170 °C, 48 h

89%

KOH MeOH

[*87*] **76**

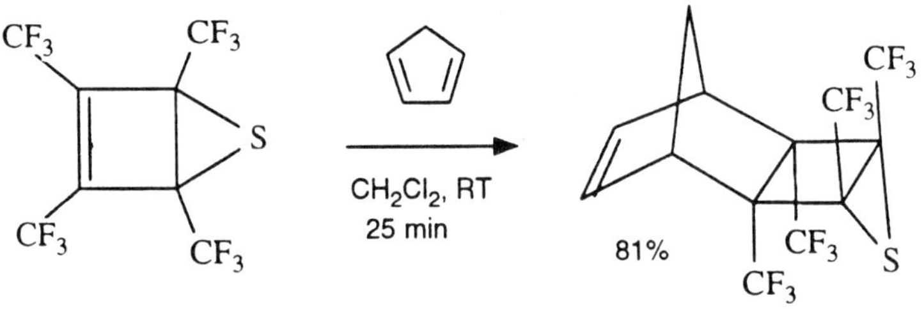

CH_2Cl_2, RT 25 min

81%

[*88*] **77**

CH_2=CHCH=CH_2

90 °C, 4 h

78%

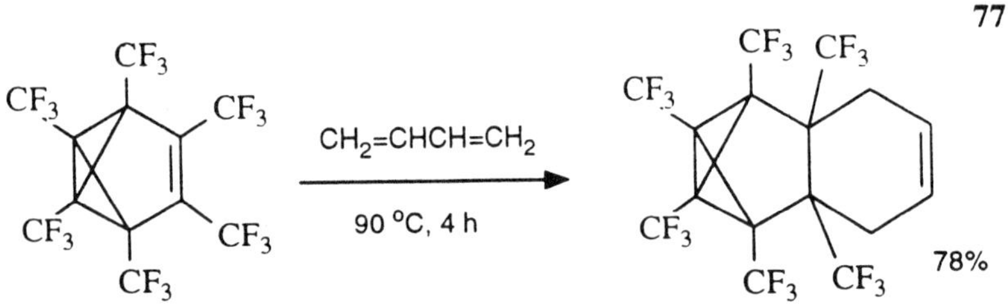

Allene itself is not a good dienophile; its reaction with cyclopentadiene requires temperatures of >200 °C and gives a 49% yield [*90*]. **Fluoroallene and 1,1-difluoroallene** are much more reactive dienophiles, the latter reacting instantly and quantitatively at –20 °C [*91, 92*], and the former taking 4 days to react quantitatively with cyclopentadiene at 0 °C [*25, 27, 93*] (equation 78).

Because the fluorine substituents both inductively and hyperconjugatively withdraw electron density from the C(2)–C(3) π bond, the LUMO is located there, and Diels–Alder reactions take place exclusively with this bond [*25*]. 1,1-Difluoroallene and fluoroallene react readily with a large selection of cyclic and acyclic dienes, and acyclic dienes, [2+2] cycloadditions compete with the Diels–Alder processes. As shown in the example in equation 79, a significantly different regiochemistry is observed for the [2+4] cycloaddition compared with the [2+2]

[*91*] **78**

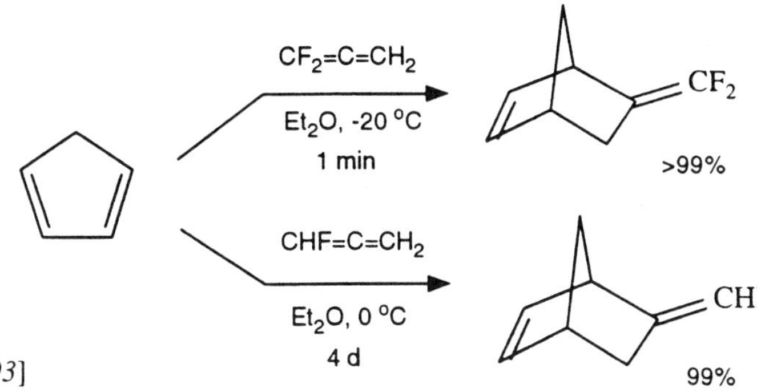

[*25, 93*]

[*91*] **79**

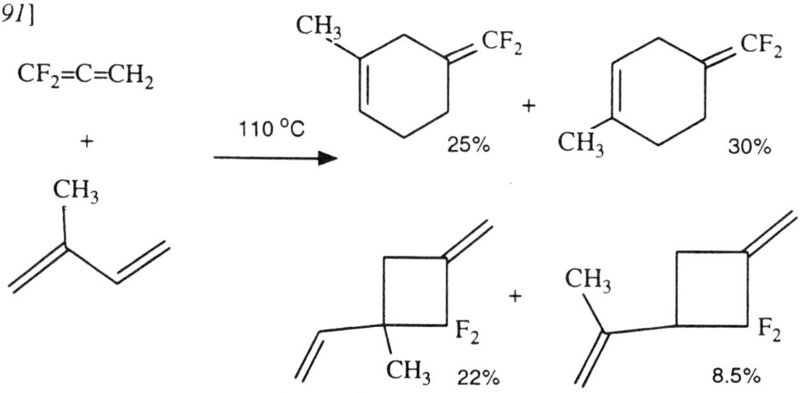

process, a result that speaks strongly against a common mechanism for the two reactions [*91*].

Fluoroallene also undergoes reaction exclusively at the C(2)–C(3) π bond and exhibits a slight *syn* selectivity in its Diels–Alder reactions [*25, 26, 93*] (equation 80), much less than that observed in its 1,3-dipolar cycloadditions.

Allenes carrying trifluoromethyl groups, such as 1,1-dichloro-3,3-bis(trifluoromethyl)allene, undergo facile, room-temperature Diels–Alder reactions with cyclopentadiene (87%) and furan (95%) [*94*].

Substituting vinylic hydrogen in α,β-**unsaturated carbonyl compounds** with vinylic fluorine does not affect their dienophilic character negatively. Indeed, 3,3-difluoroacrylic acid is more reactive toward furan than its nonfluorinated counterpart [*95*] (equation 81). Consistent with this observation is the fact that tetrafluorobenzoquinone forms only a bis-Diels–Alder adduct in 68% yield in its reaction with cyclopentadiene at room temperature [*96, 97*].

Various trifluoromethyl-containing α,β-**unsaturated acids, esters, ketones, and nitriles** have been used as dienophiles. Details regarding regiochemistry and stereochemistry have been reported [*28, 98, 99*] (equations 82–84).

[25,93] **80**

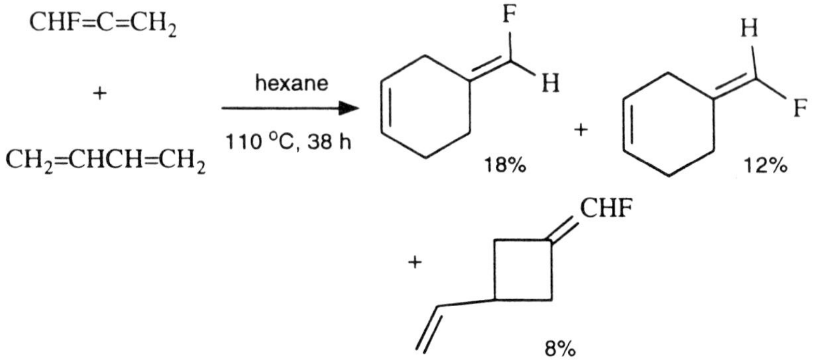

$CHF{=}C{=}CH_2$ + $CH_2{=}CHCH{=}CH_2$ → (hexane, 110 °C, 38 h) 18% + 12% + 8%

CHF

F H H F

[95] **81**

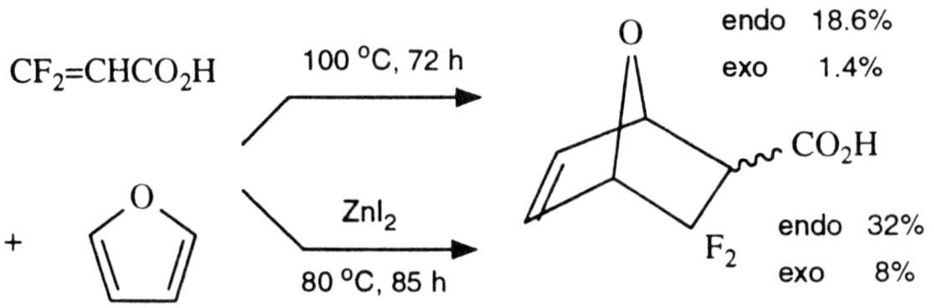

$CF_2{=}CHCO_2H$ + furan → (100 °C, 72 h) endo 18.6%, exo 1.4%; (ZnI_2, 80 °C, 85 h) endo 32%, exo 8%

O CO_2H F_2

[98] **82**

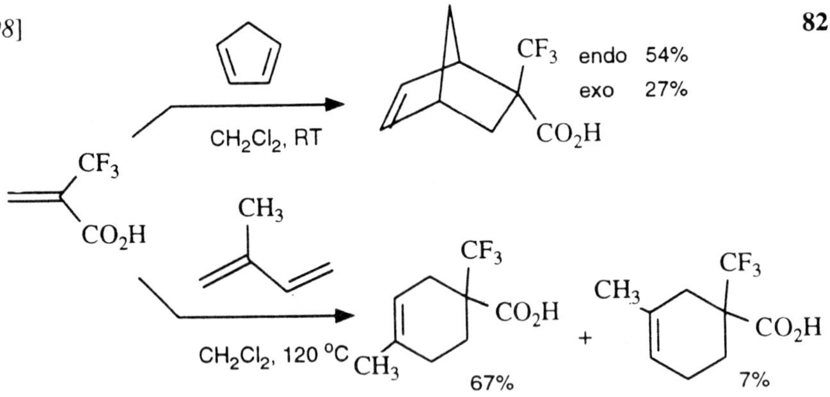

CF_3 CO_2H + cyclopentadiene → (CH_2Cl_2, RT) endo 54%, exo 27%

CF_3 CO_2H

+ isoprene (CH_3) → (CH_2Cl_2, 120 °C) 67% + 7%

CF_3 CO_2H CH_3 ; CH_3 CF_3 CO_2H

[99] **83**

CF_3 CO_2H + furan → (RT, 13 d) endo 74%, exo 18%

O CO_2H CF_3

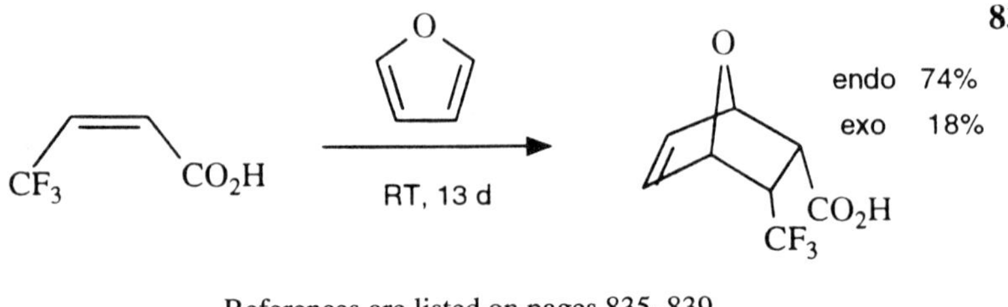

[*28*] **84**

$CF_3CH{=}CHCOCH_3$ + cyclopentadiene, 80 °C, 1 h → 5-acetyl-6-(trifluoromethyl)bicyclo[2.2.1]hept-2-ene: endo 62%, exo 18%

[*100*] **85**

$RC{\equiv}CCO_2H$ + isoprene, THF, 45 °C, 72 h → 2-R-4-methylcyclohexa-1,4-dienecarboxylic acid + 2-R-5-methylcyclohexa-1,4-dienecarboxylic acid

R		
CF_3	70%	30%
CF_2H	71%	29%
$H(CF_2)_3$	68%	32%

Fluoroalkyl-substituted **acetylenic acids** are more reactive dienophiles than dimethyl acetylenedicarboxylate [*100*] (equation 85).

2,2-Bis(trifluoromethyl)-1,1-dicyanoethylene is a very reactive dienophile. It undergoes facile and high-yield [2+4] cycloadditions with 1,3-dienes, cyclopentadiene, and anthracene [*101*] (equation 86). It is reactive enough in a Diels–Alder reaction with styrene [*102*] (equation 86).

Tetrafluorobenzyne, generally generated by the treatment of pentafluorobenzene with butyllithium at –78 °C in ether in the presence of the substrate diene, is a versatile dienophile [*9, 103, 104*]. In an interesting study of the use of substituted benzynes to synthesize isoindoles, tetrafluorobenzyne, 4-fluorobenzyne, and 4-(trifluoromethyl)benzyne were shown to react in moderate yields with *N*-(trimethylsilyl)pyrroles, with the adducts being easily converted to the respective fluorinated isoindoles [*105*] (equation 87).

[*101*] **86**

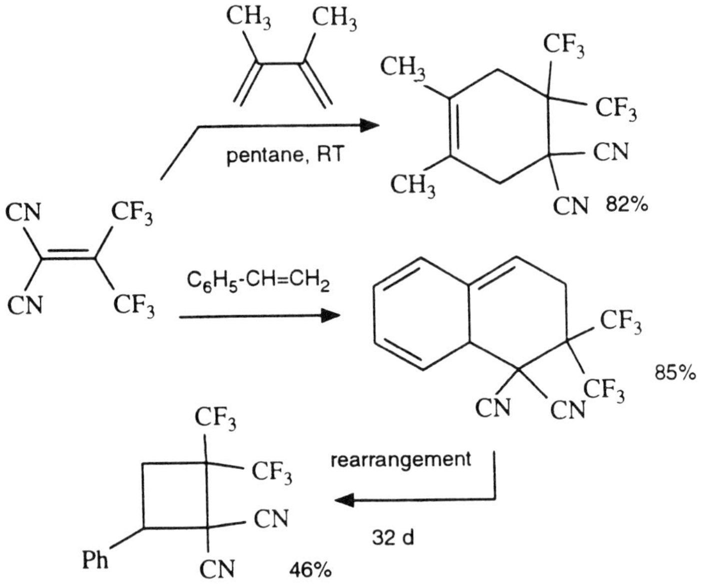

pentane, RT

82%

C_6H_5-CH=CH_2

85%

rearrangement

32 d

[*102*]

46%

[*105*] **87**

BuLi, -78 °C

Et_2O, pentane

35%

77%

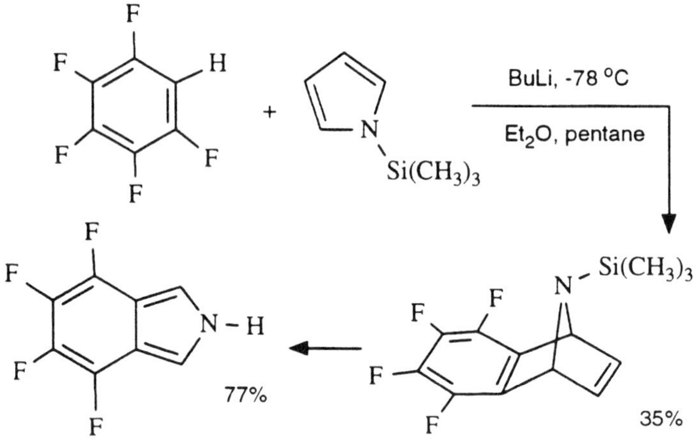

Interestingly, 3-fluorobenzyne has shown a surprising regioselectivity in its additions to 2-substituted furans [*106*] (equation 88).

Fluorinated heterodienophiles and heterodienes. Diels–Alder reactions in which the dienophiles have perfluoroalkyl-substituted multiple bonds between carbon and a heteroatom are quite common. Reported earlier were reactions of perfluoroketones, thiones, ketimines, thioesters, nitroso compounds, and nitriles [*9*]. Examples of α-**fluoroimines** [*107*], ω-**hydroperfluorothioaldehydes** [*108*], **perfluorosulfines** [*109, 110*], and **selenocarbonyldifluoride** [*111*] (equations 89–92) have been reported recently.

[*106*] 88

[*107*] 89

[*108*] 90

Polyfluoro-2-acyliminopropanes exhibit an interesting dichotomy of reactivity: they react as dienophiles with cyclopentadiene but as heterodienes with acrolein [*112*] (equation 93).

Fluorine-substituted heterodienes are particularly prone to *inverse-electron-demand Diels–Alder reactions* with electron-rich dienophiles, as can be seen from the examples in equations 94–97 [*113, 114, 115, 116, 117*].

[*110*] **91**

100 °C

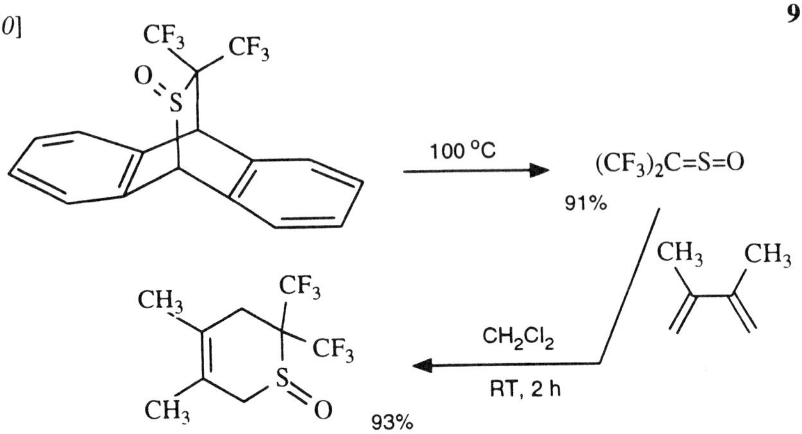

$(CF_3)_2C{=}S{=}O$

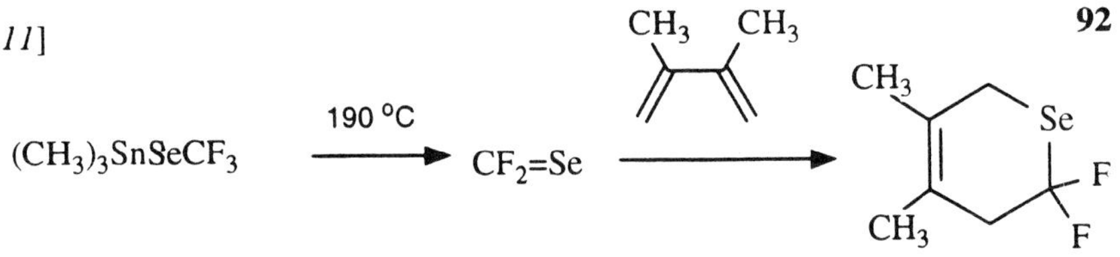

91%

CH_3 CH_3

CH_2Cl_2

RT, 2 h

CH_3 CF_3 CF_3 CH_3 S O 93%

[*111*] **92**

$(CH_3)_3SnSeCF_3$ → 190 °C → $CF_2{=}Se$ → CH_3 CH_3 Se F F

[*112*] **93**

20 °C, 3 d

91%

CF_3 CF_3 N CH_3 O

O CF_3 CH_3 N CF_3

O H

CH_3 O H N O CF_3 CF_3 70%

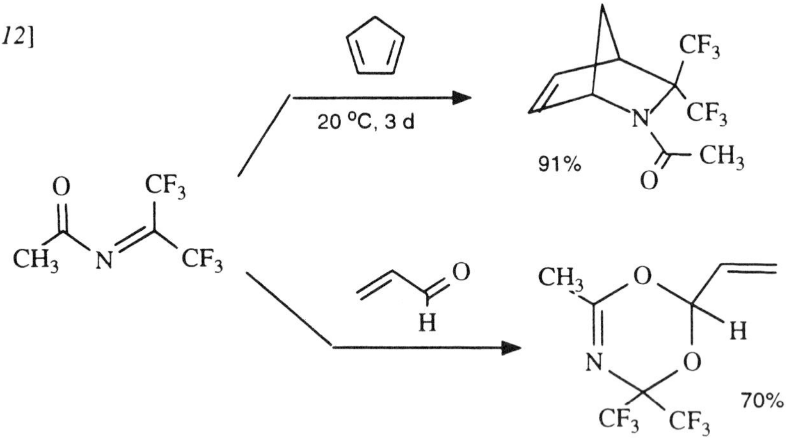

The use of trifluoromethyl-substituted **α-pyrones** in Diels–Alder reactions with all types of dienophiles provides an interesting route to trifluoromethylbenzenes [*118*] (equation 98).

Hydrocarbon **oxadiazoles** do not readily undergo Diels–Alder reactions, but 2,5-bis(trifluoromethyl)-1,3,4-oxadiazole reacts with a number of strained or elec-

[113] **94**

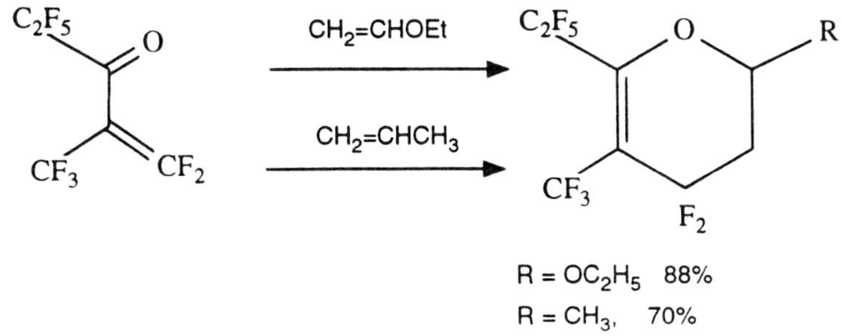

$C_2F_5C(O)C(CF_3)=CF_2$ → (CH_2=CHOEt or CH_2=$CHCH_3$) → 2-C_2F_5-3-CF_3-4,4-F_2-6-R-dihydropyran

R = OC_2H_5 88%

R = CH_3, 70%

[114] **95**

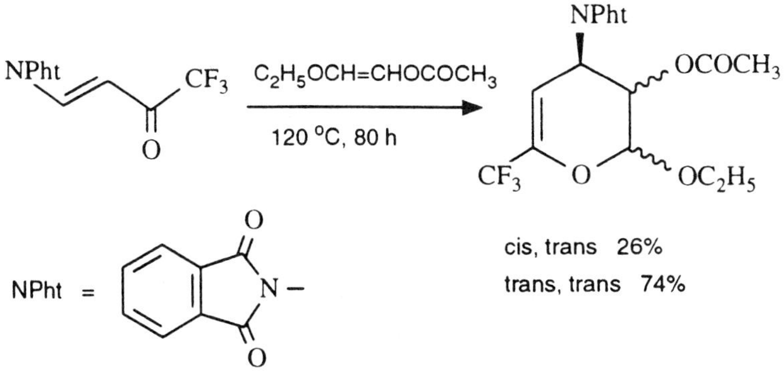

NPht–CH=CH–C(O)–CF_3 → C_2H_5OCH=$CHOCOCH_3$, 120 °C, 80 h

NPht = (phthalimido)

cis, trans 26%

trans, trans 74%

[115] **96**

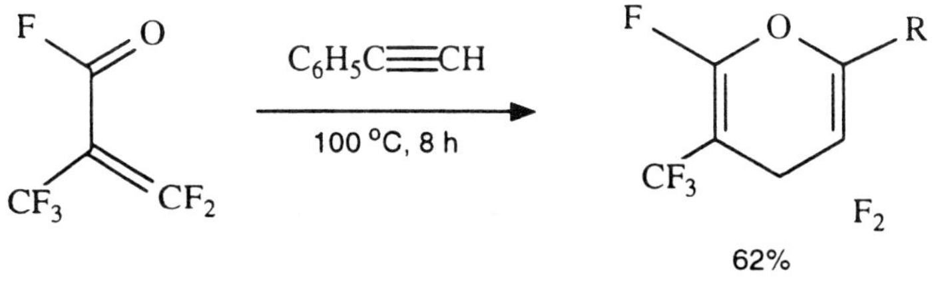

FC(O)C(CF_3)=CF_2 → $C_6H_5C{\equiv}CH$, 100 °C, 8 h

62%

[117] **97**

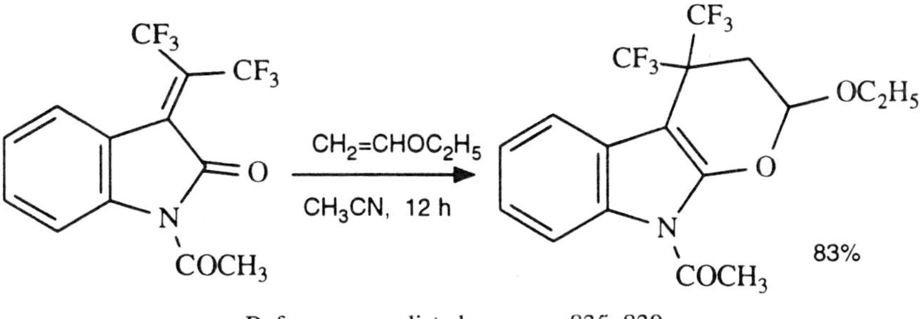

CH_2=$CHOC_2H_5$, CH_3CN, 12 h

83%

[*118*] 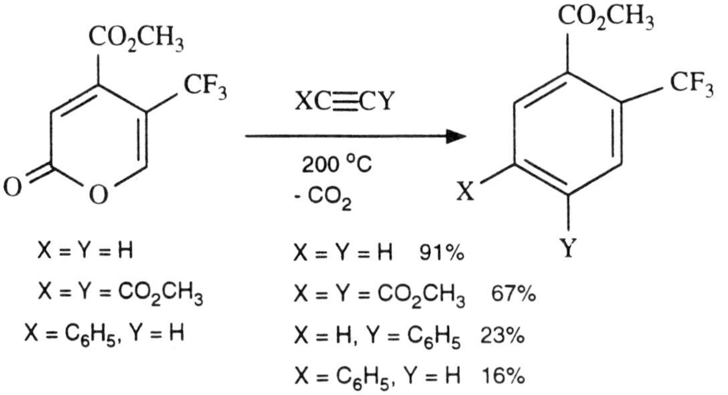 **98**

$XC{\equiv}CY$, 200 °C, $-CO_2$

$X = Y = H$
$X = Y = CO_2CH_3$
$X = C_6H_5$, $Y = H$

$X = Y = H$ 91%
$X = Y = CO_2CH_3$ 67%
$X = H$, $Y = C_6H_5$ 23%
$X = C_6H_5$, $Y = H$ 16%

[*119*] 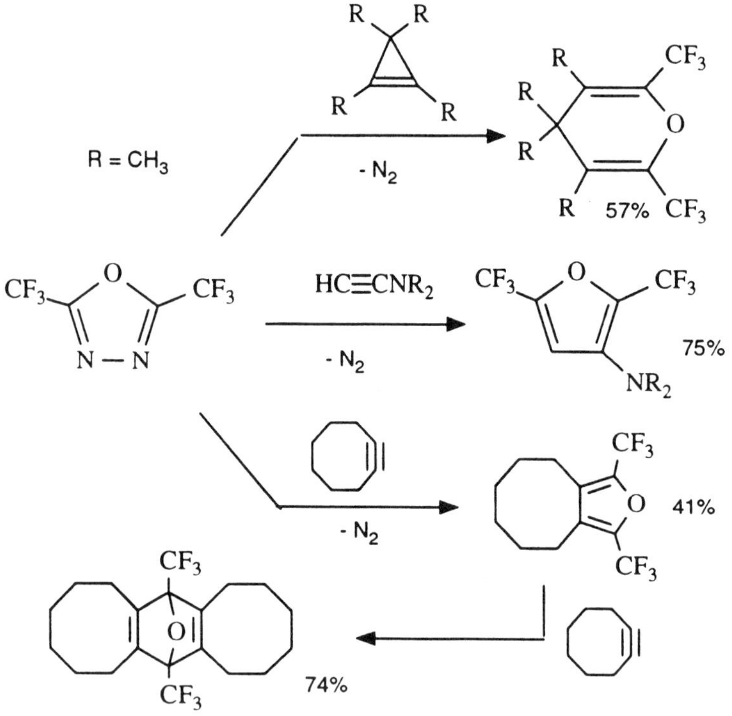 **99**

$R = CH_3$

$-N_2$ 57%

$HC{\equiv}CNR_2$, $-N_2$ 75%

$-N_2$ 41%

74%

tron-rich dienophiles to give interesting overall transformations after spontaneous elimination of nitrogen [*119*] (equation 99).

3,6-Bis(trifluoromethyl)-1,2,4,5-tetrazine, which is synthesized from 2,5-bis(trifluoromethyl)-1,3,4-oxadiazole by addition of hydrazine followed by mild oxidation [*120*] (equation 100), is a reactive and very interesting Diels–Alder diene [*121*]. Its normal reaction is to add the dienophile and then to extrude N_2 to form a diazene, as exemplified by its reaction with propyne [*120*] (equation 101).

As the last reaction of equation 99 suggests, trifluoromethyl-substituted furans are themselves reactive Diels–Alder dienes [*122*] (equation 102).

[*120*] **100**

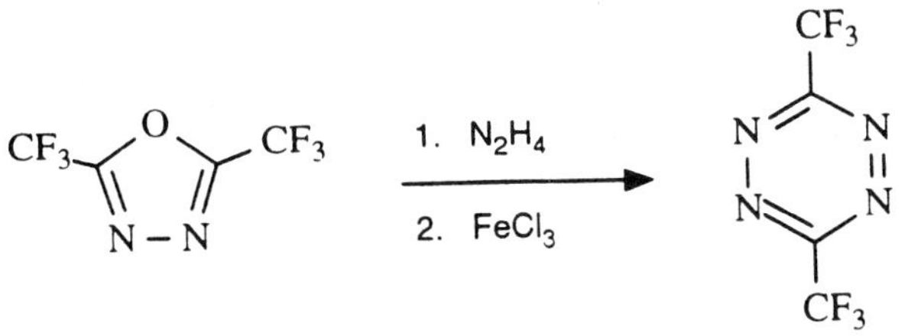

CF_3 CF_3 O N – N

1. N_2H_4

2. $FeCl_3$

CF_3 N N N N CF_3

[*120*] **101**

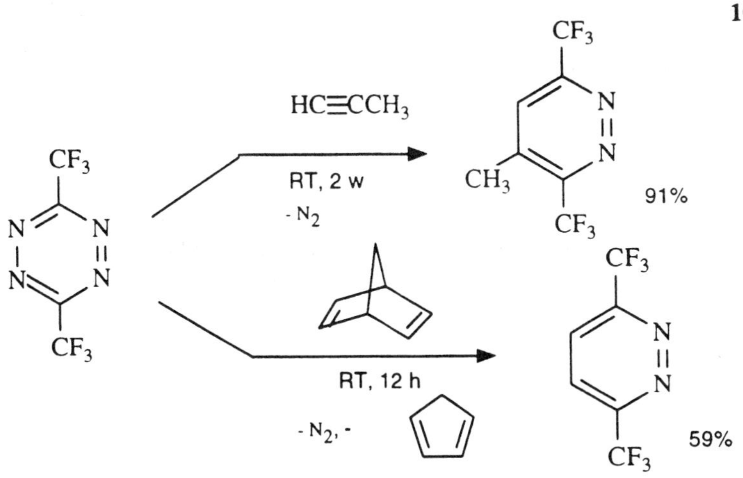

$HC{\equiv}CCH_3$

RT, 2 w

- N_2

91%

RT, 12 h

- N_2, -

59%

[*122*] **102**

$HC{\equiv}CCO_2C_2H_5$

150 °C

$CO_2C_2H_5$

$HC{\equiv}CCO_2C_2H_5$

OC_2H_5

$CO_2C_2H_5$

H_3O^+

$CO_2C_2H_5$

52%

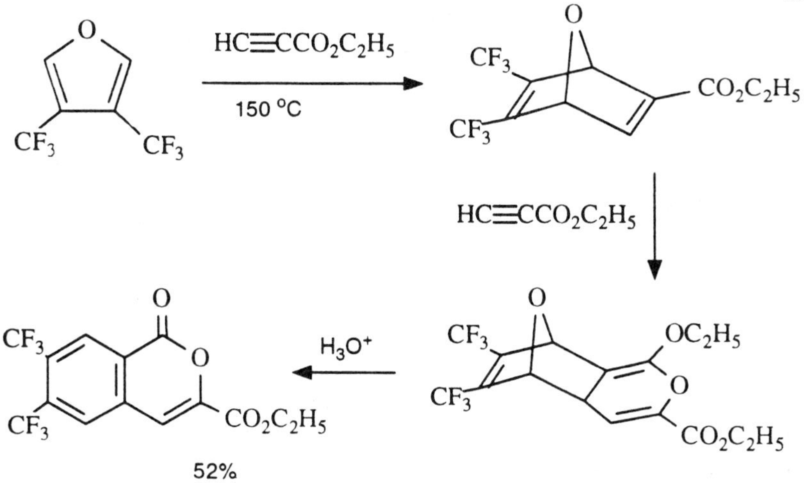

Although ***hexafluoro-1,3-butadiene*** is better known for its [2+2] reactions, its Diels–Alder reactions, particularly with electron-deficient alkenes such as acrylonitrile and perfluoropropene, are not unknown [*9*]. The first report of a Diels–Alder reaction is with an acetylenic dienophile. Although the major product of its reaction with phenylacetylene is its [2+2] adduct, a 3.5% yield of products of a Diels–Alder reaction is also observed [*123*] (equation 103).

[123] **103**

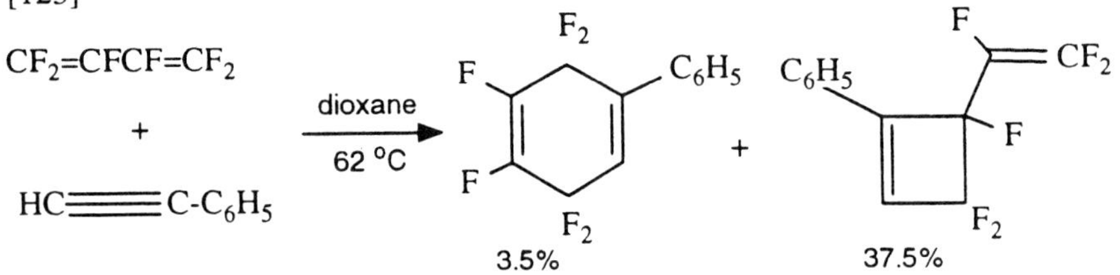

$CF_2{=}CFCF{=}CF_2$ + $HC{\equiv}C\text{-}C_6H_5$ —(dioxane, 62 °C)→ [F, F, F_2, F_2, C_6H_5 cyclohexadiene] 3.5% + [C_6H_5, F, CF_2, F, F_2 cyclobutene] 37.5%

In a related *photoreaction,* hexafluorobutadiene forms the [2+4] adduct with *p*-(*N,N*-dimethylamino)styrene in 11% yield, whereas no such adduct could be detected for the same reaction with *p*-methoxystyrene [*124*] (equation 104). The analogous thermal reaction at 120 °C leads only to [2+2] adducts.

[*124*] **104**

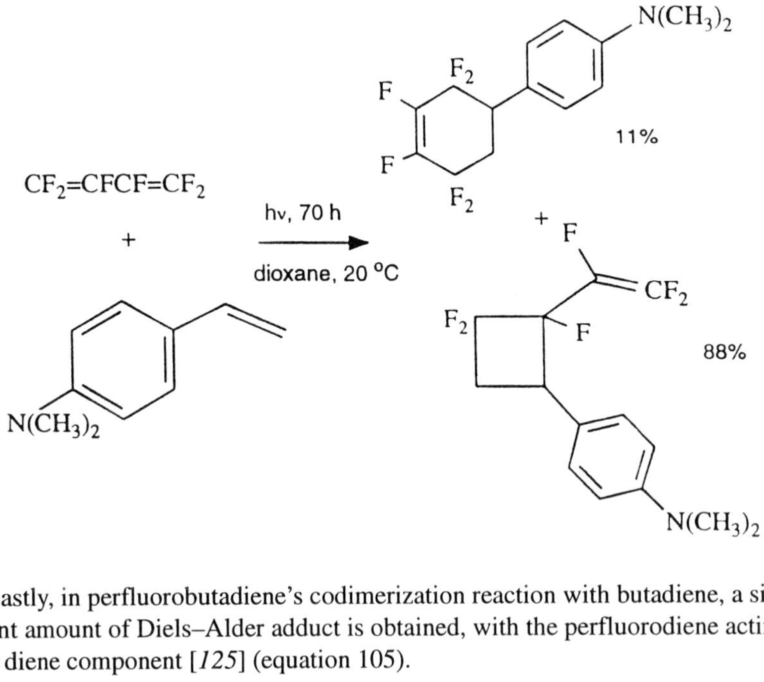

$CF_2{=}CFCF{=}CF_2$ + p-$N(CH_3)_2$-styrene —(hν, 70 h; dioxane, 20 °C)→ [F, F, F_2, F_2, $N(CH_3)_2$ cyclohexene adduct] 11% + [F, CF_2, F, F_2, $N(CH_3)_2$ cyclobutane adduct] 88%

Lastly, in perfluorobutadiene's codimerization reaction with butadiene, a significant amount of Diels–Alder adduct is obtained, with the perfluorodiene acting as the diene component [*125*] (equation 105).

[*125*] **105**

$CF_2{=}CFCF{=}CF_2$ + $CH_2{=}CHCH{=}CH_2$ $\xrightarrow[8\ h]{175\ ^\circ C}$

30% + 39%

8% + 23%

References for Pages 797–835

1. Huisgen, R. *Angew. Chem. Int. Ed. Engl.* **1963,** *5,* 633.
2. Huisgen, R. *Angew. Chem. Int. Ed. Engl.* **1963,** *5,* 565.
3. Hook, K. N.; Sims, J.; Watts, C. R.; Luskus, L. J. *J. Am. Chem. Soc.* **1973,** *95,* 7301.
4. Brundle, C. R.; Robin, M. B.; Kuebler, N. A.; Basch, H. *J. Am. Chem. Soc.* **1972,** *94,* 1451.
5. Bryce, M. R.; Chambers, R. D.; Taylor, 6. *J. Chem. Soc., Chem. Commun.* **1983,** 5.
6. Zeifman, Y. V.; Lantseva, L. T. *Izv. Akad. Nauk SSSR* **1986,** *35,* 248 (Engl. Transl. 231).
7. Leroy, J.; Rubinstein, M.; Wakselman, C. *J. Fluorine Chem.* **1984,** *25,* 255.
8. Manhart, E.; von Werner, K. *Synthesis* **1978,** 705.
9. Hudlický, M. *Chemistry of Organic Fluorine Compounds,* 2nd ed.; Ellis Horwood: Chichester, U.K., 1976.
10. Fayn, J.; Cambon, A. *J. Fluorine Chem.* **1988,** *40,* 63.
11. Dolbier, W. R., Jr.; Al-Fekri, D. M. *Tetrahedron Lett.* **1983,** *24,* 4047.
12. Dolbier, W. R., Jr.; Al-Fekri, D. M. *Tetrahedron* **1987,** *43,* 39.
13. Bryce, M. R.; Chambers, R. D.; Taylor, G. *J. Chem. Soc. Perkin Trans. 1* **1984,** 509.
14. Dolbier, W. R., Jr.; Al-Fekri, D. M. *J. Org. Chem.* **1987,** *52,* 1872.
15. Stepanova, N. P.; Orlova, N. A.; Galishev, V. A.; Turbanova, E. S.; Petrov, A. A. *Zh. Org. Khim.* **1985,** *51,* 979 (Engl. Transl. 889).
16. Turbanova, E. S.; Stepanova, N. P.; Lebedev, V. B.; Galishev, V. A.; Petrov, A. A. *Zh. Org. Khim.* **1983,** *19,* 221 (Engl. Transl. 204).

17. Froissard, J.; Greiner, J.; Pastor, R.; Cambon, A. *J. Fluorine Chem.* **1984,** *26,* 47.
18. Turbanova, E. S.; Stepanova, N. P.; Sakharov, V. N.; Galishev, V. A.; Petrov, A. A. *Zh. Org. Khim.* **1983,** *19,* 223 (Engl. Transl. 205).
19. Blackwell, G. B.; Haszeldine, R. N.; Taylor, D. R. *J. Chem. Soc., Perkin Trans. 1* **1982,** 2207.
20. Blackwell, G. B.; Haszeldine, R. N.; Taylor, D. R. *J. Chem. Soc., Perkin Trans. 1* **1983,** 1.
21. Domelsmith, L. N.; Houk, K. N.; Dolbier, W. R., Jr.; Piedrahita, C. A. *J. Am. Chem. Soc.* **1978,** *100,* 6908.
22. Dolbier, W. R., Jr.; Burkholder, C. R. *Isr. J. Chem.* **1985,** *26,* 115.
23. Dolbier, W. R., Jr.; Purvis, G. D., III; Seabury, M.; Wicks, G. E.; Burkholder, C. B. *Tetrahedron* **1990,** *24,* 7991.
24. Dolbier, W. R., Jr.; Burkholder, C. R.; Winchester, W. R. *J. Org. Chem.* **1984,** *49,* 1518.
25. Dolbier, W. R., Jr. *Acc. Chem. Res.* **1991,** *24,* 63.
26. Dolbier, W. R., Jr.; Wicks, G. E.; Burkholder, C. R.; Palenik, G. J.; Gawron, M. *J. Am. Chem. Soc.* **1985,** *107,* 7183.
27. Dolbier, W. R., Jr.; Wicks, G. E.; Burkholder, C. R. *J. Org. Chem.* **1987,** *52,* 2196.
28. Ogoshi, H.; Mizushima, H.; Toi, H.; Aoyama, Y. *J. Org. Chem.* **1986,** *51,* 2366.
29. Saunier, Y. M.; Danion-Gougot, R.; Danion, D.; Carrie, R. *Tetrahedron* **1976,** *32,* 1995.
30. Melnikov, A. A.; Gokolova, M. M.; Pervozvanskaya, M. A.; Melnikov, V. V. *Zh. Org. Khim.* **1979,** *15,* 1861 (Engl. Transl. 1677).
31. Lazukina, L. A.; Kukhar, V. P. *Zh. Org. Khim.* **1979,** *15,* 2216 (Engl. Transl. 2009).
32. Banks, R. E.; Richards, A. *J. Chem. Soc., Chem. Commun.* **1985,** 205.
33. Fields, R.; Tomlinson, J. P. *J. Fluorine Chem.* **1979,** *13,* 147.
34. Fields, R.; Tomlinson, J. P. *J. Fluorine Chem.* **1979,** *13,* 19.
35. Tanaka, K.; Masuda, H.; Mitsuhashi, K. *Bull. Chem. Soc. Jpn.* **1984,** *57,* 2184.
36. Tanaka, K.; Masuda, H.; Mitsuhashi, K. *Bull. Chem. Soc. Jpn.* **1985,** *58,* 2061.
37. Tanaka, K.; Masuda, H.; Mitsuhashi, K. *Bull. Chem. Soc. Jpn.* **1986,** *59,* 3901.
38. Tanaka, K.; Kishida, M.; Maeno, S.; Mitsuhashi, K. *Bull. Chem. Soc. Jpn.* **1986,** *59,* 2631.
39. Tanaka, K.; Maeno, S.; Mitsuhashi, K. *Chem. Lett.* **1982,** 543.
40. Tanaka, K.; Maeno, S.; Mitsuhashi, K. *J. Heterocyclic Chem.* **1985,** *22,* 565.

41. Tanaka, K.; Suzuki, T.; Maeno, S.; Mitsuhashi, K. *J. Heterocyclic Chem.* **1986,** *23,* 1535.
42. Tanaka, K.; Honda, O.; Minoguchi, K.; Mitsuhashi, K. *J. Heterocyclic Chem.* **1987,** *24,* 1391.
43. Tanaka, K.; Daikaku, H.; Mitsuhashi, K. *Chem. Lett.* **1983,** 1463.
44. Burger, K.; Roth, W. D.; Neumayr, K. *Chem. Ber.* **1976,** *109,* 1984.
45. Burger, K.; Einhellig, K. *Chem. Ber.* **1973,** *106,* 3421.
46. Burger, K.; Einhellig, K.; Roth, W. D.; Daltrozzo, E. *Chem. Ber.* **1977,** *110,* 605.
47. Burger, K.; Neuhauser, H.; Eggersdorfer, M. *Synthesis* **1987,** 924.
48. Tanaka, K.; Ohsuga, M.; Sugimoto, Y.; Okafiyi, Y.; Mitsuhashi, K. *J. Fluorine Chem.* **1988,** *39,* 39.
49. Tanaka, K.; Sugimoto, Y.; Okafuji, Y.; Tachikawa, M.; Mitsuhashi, K. *J. Heterocyclic Chem.* **1989,** *26,* 381.
50. Avetisyan, E. A.; Gambaryan, N. P. *Izv. Akad. Nauk SSSR* **1975,** *24,* 1898 (Engl. Transl. 1784).
51. Simonyan, L. A.; Avetisyan, E. A.; Safronova, Z. V.; Gambaryan, N. P. *Izv. Akad. Nauk SSSR* **1977,** *26,* 2061 (Engl. Transl. 1906).
52. Burger, K.; Wassmuth, U.; Penninger, S. *J. Fluorine Chem.* **1982,** *20,* 813.
53. Giese, B. *Radicals in Organic Synthesis: Formation of Carbon–Carbon Bonds;* Pergamon Press: New York, 1986.
54. Piccardi, P.; Modena, M. *J. Chem. Soc., Chem. Commun.* **1971,** 1041.
55. Piccardi, P.; Massardo, P.; Modena, M.; Santoro, E. *J. Chem. Soc., Perkin Trans. 1* **1974,** 1848.
56. Morikawa, T.; Nishiwaki, T.; Iitaka, Y.; Kobayashi, Y. *Tetrahedron Lett.* **1987,** *28,* 671.
57. Watanabe, Y.; Yokozawa, T.; Takata, T.; Endo, T. *J. Fluorine Chem.* **1988,** *39,* 431.
58. England, D. C. *J. Am. Chem. Soc.* **1961,** *83,* 2205.
59. Aubert, C.; Begue, J. P. *Tetrahedron Lett.* **1988,** *29,* 1011.
60. Chambers, R. D.; Jones, C. G. P.; Silvester, J. J.; Speight, D. B. *J. Fluorine Chem.* **1984,** *25,* 47.
61. Roberts, J. D.; Sharts, C. M. *Org. Reactions* **1962,** *12,* 1.
62. Perry, D. R. *A. Fluorine Chem. Rev.,* **1967,** *1,* 253.
63. Bartlett, P. D.; Schueller, K. E. *J. Am. Chem. Soc.* **1968,** *90,* 6077.
64. Swenton, J. S.; Bartlett, P. D. *J. Am. Chem. Soc.* **1968,** *90,* 2056.
65. Bartlett, P. D.; Wheland, R. C. *J. Am. Chem. Soc.* **1972,** *94,* 2145.
66. Bartlett, P. D.; Jacobson, B. M.; Walker, L. E. *J. Am. Chem. Soc.* **1973,** *95,* 146.
67. Bartlett, P. D.; Mallet, J. J.-B. *J. Am. Chem. Soc.* **1976,** *98,* 143.
68. Turro, N. J.; Bartlett, P. D. *J. Org. Chem.* **1965,** *30,* 1849.
69. Gaede, B.; Balthazar, T. M. *J. Org. Chem.* **1983,** *48,* 276.
70. Ojima, I.; Yatabe, M.; Fuchikami, T. *J. Org. Chem.* **1982,** *47,* 2051.

71. Banks, R. E.; Barlow, M. G. *J. Chem. Soc., Perkin Trans. 1* **1974,** 1266.
72. Albekov, V. A.; Benda, A. F.; Gontar, A. F.; Solilskii, G. A.; Knunyants, I. L. *Izv. Akad. Nauk SSSR* **1988,** *37,* 885 (Engl. Transl. 785).
73. Albekov, V. A.; Benda, A. F.; Gontar, A. F.; Sokolskii, G. A.; Knunyants, I. L. *Izv. Akad. Nauk SSSR* **1988,** *37,* 897 (Engl. Transl. 777).
74. Gassman, P. G.; Korn, S. R.; Bailey, T. F.; Johnson, T. H.; Finer, J.; Clardy, J. *Tetrahedron Lett.* **1979,** 3401.
75. Visnick, M.; Battiste, M. A. *J. Chem. Soc., Chem. Commun.* **1985,** 1621.
76. Leroy, J.; Cantacuzene, D.; Wakselman, C. *Synthesis* **1982,** *4,* 313.
77. Kaesler, R. W.; Le Goff, E. *J. Org. Chem.* **1982,** *47,* 4779.
78. Prinzbach, H.; Bingmann, H.; Markiert, J.; Gischer, G.; Knowthe, L.; Eberbach, W.; Brokatzky-Geiger, J. *Chem. Ber.* **1986,** *119,* 589.
79. Chambers, R. D.; Kirk, J. R.; Taylor, G. *J. Fluorine Chem.* **1983,** *22,* 393.
80. Bayliff, A. E.; Bryce, M. R.; Chambers, R. D.; Kirk, J. R.; Taylor, G. *J. Chem. Soc., Perkin Trans. 1* **1985,** 1191.
81. Dolbier, W. R., Jr.; Seabury, M.; Daly, D.; Smart, B. E. *J. Org. Chem.* **1986,** *51,* 974.
82. Smart, B. E. *J. Am. Chem. Soc.* **1974,** *96,* 929.
83. Birchall, J. M.; Burger, K.; Haszeldine, R. N.; Nona, S. N. *J. Chem. Soc., Perkin Trans. 1* **1981,** 2080.
84. Muller, P.; Etienne, R.; Pfyffer, J.; Pineda, N.; Schipoff, M. *Tetrahedron Lett.* **1978,** 3151.
85. Kobayashi, Y.; Yoshida, T.; Hanzawa, Y.; Kumakaki, I. *Tetrahedron Lett.* **1980,** *21,* 4601.
86. Burton, D. J.; Link, B. A. *J. Fluorine Chem.* **1983,** *22,* 397.
87. Kobayashi, Y.; Kumadaki, I.; Ohsawa, A.; Sekine, Y.; Ando, A. *J. Chem. Soc., Perkin Trans. 1* **1977,** 2355.
88. Kobayashi, Y.; Honda, M.; Hanzawa, I.; Kumadaki, I.; Ohsawa, A. *J. Chem. Soc., Perkin Trans. 1* **1979,** 1743.
89. Kobayashi, Y.; Kumakaki, I.; Ohsawa, A.; Hanzawa, Y.; Honda, M.; Miyashita, W. *Tetrahedron Lett.* **1977,** 1795.
90. Carboni, R. A.; Lindsey, R. V. *J. Am. Chem. Soc.* **1959,** *81,* 4342.
91. Dolbier, W. R., Jr.; Piedrahita, C. A.; Houk, K. N.; Strozier, R. W.; Gandour, R. W. *Tetrahedron Lett.* **1978,** 2231.
92. Dolbier, W. R., Jr.; Burkholder, C. R. *J. Org. Chem* **1984,** *49,* 2381.
93. Dolbier, W. R., Jr.; Burkholder, C. R. *Tetrahedron Lett.* **1980,** 785.
94. Bosbury, P. W. L.; Fields, R.; Haszeldine, R. N. *J. Chem. Soc., Perkin Trans. 1* **1982,** 2203.
95. Leroy, J.; Molines, H.; Wakselman, C. *J. Org. Chem.* **1987,** *52,* 2900.
96. Wilson, R. M. *J. Org. Chem.* **1983,** *48,* 707.
97. Hudlický, M.; Bell, H. M. *J. Fluorine Chem.* **1975,** *5,* 189.
98. Hanzawa, Y.; Suzuki, M.; Kobayashi, Y. *Tetrahedron Lett.* **1989,** *30,* 571.

99. Leroy, J.; Fischer, N.; Wakselman, C. *J. Chem. Soc., Perkin Trans. 1* **1990,** 1281.
100. Kuwabara, M.; Fukunishi, K.; Nomura, M.; Yamanaka, H. *J. Fluorine Chem.* **1988,** *41,* 227.
101. Middleton, W. J.; Bingham, E. M. *J. Fluorine Chem.* **1982,** *20,* 397.
102. Bruckner, R.; Huisgen, R.; Schmid, J. *Tetrahedron Lett.* **1990,** *31,* 7129.
103. Hankinson, B.; Heaney, H.; Price, A. P.; Sharma, R. P. *J. Chem. Soc., Perkin Trans. 1* **1973,** 2569.
104. Buxton, P. C.; Hales, N. J.; Hankinson, B.; Heaney, H.; Ley, S. V.; Sharma, R. P. *J. Chem. Soc., Perkin Trans. 1* **1974,** 2681.
105. Anderson, P. S.; Christy, M. E.; Englehardt, E. L.; Lundell, G. F.; Ponticello, G. S. *J. Heterocyclic Chem.* **1977,** *14,* 213.
106. Gribble, G. W.; Keavy, D. J.; Branz, S. E.; Kelly, W. J.; Pals, M. A. *Tetrahedron Lett.* **1988,** *29,* 6227.
107. Albekov, V. A.; Benda, A. F.; Gontar, A. T.; Sokolskii, G. A.; Knunyants, I. L. *Izv. Akad. Nauk SSSR* **1986,** *35,* 1437 (Engl. Transl. 1305).
108. Shermolovich, Y. G.; Slyusarenko, E. I.; Markovskii, L. N. *Zh. Org. Khim.* **1988,** *54,* 1931 (Engl. Transl. 1741).
109. Schwab, M.; Sundermeyer, W. *Chem. Ber.* **1986,** *119,* 2458.
110. Elsaesser, A.; Sundermeyer, W. *Chem. Ber.* **1985,** *118,* 4553.
111. Grobe, J.; Van, D. L.; Welzel, J. *J. Organomet. Chem.* **1988,** *340,* 153.
112. Kryukov, L. N.; Krykova, L. Y. *Zh. Org. Khim.* **1982,** *18,* 1873 (Engl. Transl. 1638).
113. England, D. C. *J. Org. Chem.* **1981,** *46,* 147.
114. Tietze, L. F.; Hartfiel, U. *Tetrahedron Lett.* **1990,** *31,* 1697
115. England, D. C.; Donald, E. A.; Weigert, F. J. *J. Org. Chem.* **1981,** *46,* 144.
116. England. D. C.; Krespan, C. G. *J. Fluorine Chem.* **1973,** *3,* 91.
117. Righetti, P. P.; Gambra, A.; Tacconi, G.; Desimoni, G. *Tetrahedron* **1981,** *37,* 1779.
118. Martin, P.; Streith, J.; Rihs, G.; Winkler, T.; Bellus, D. *Tetrahedron Lett.* **1985,** *26,* 3947.
119. Thalhammer, F.; Wallfahrer, U.; Sauer, J. *Tetrahedron Lett.* **1988,** *29,* 3231.
120. Barlow, M. G.; Haszeldine, R. N.; Pickett, J. A. *J. Chem. Soc., Perkin Trans. 1* **1978,** 378.
121. Hoferichter, R.; Seitz, G.; Wassermuth, H. *Chem. Ber.* **1989,** *122,* 711.
122. Abubaker, A. B.; Booth, B. L.; Tipping, A. E. *J. Fluorine Chem.* **1990,** *47,* 353.
123. Kazmina, N. B.; Kurbakova, A. P.; Leitas, L. A.; Kvasov, B. A.; Mysov, E. J. *Zh. Org. Khim.* **1986,** *55,* 1668 (Engl. Transl. 1500).
124. Kazmina, N. B.; Mysov, E. I.; Antipin, M. Y.; Akhmedov, A. I.; Struchkov Y. T. *Izv. Akad. Nauk SSSR* **1981,** *30,* 842 (Engl. Transl. 623).
125. Weigert, F. J. *J. Fluorine Chem.* **1991,** *52,* 125.

Addition Reactions across Polyfluoroalkyl- and Perfluoro-alkyl-Substituted CO and CN Multiple Bond Systems

by K. Burger, N. Sewald, C. Schierlinger, and E. Brunner

Linear Nucleophilic Additions to CO and CN Multiple Bonds

Fluorine-containing carbonyl compounds and imines can be divided into several subgroups according to their reactivity toward nucleophiles. Acyl fluorides, imide fluorides, and certain *N*-fluoroimines predominantly or exclusively undergo substitution reactions with nucleophiles. This type of reaction has been discussed in detail [*1*]. The reactions of hexafluoroacetone with inorganic and organometallic compounds have been reviewed recently [*2*]. Mono- and difluoroalkyl-substituted carbonyl compounds often show a reactivity toward nucleophiles that is analogous to that of other α-halogenoketones. Therefore, their chemistry is not included.

This review concentrates on compounds containing perfluoroalkyl groups adjacent to CO or CN multiple bonds. Because of their low-lying lowest unoccupied molecular orbital (LUMO), they exhibit a significant increase in reactivity toward nucleophiles compared with their fluorine-free analogues [*3, 4, 5*].

Perfluorinated carbonyl compounds, especially hexafluoroacetone, are highly electron-deficient species and react vigorously with a wide variety of HX nucleophiles. The reaction of these ketones and of most polyfluorinated imines toward nucleophiles can be generalized by the scheme shown in equation 1.

$$F_3C(R^1)C{=}Y + HX \longrightarrow F_3C(R^1)C(YH)(X) \qquad F_3C(R^1)C{=}Y + X^{\ominus} \longrightarrow F_3C(R^1)C(Y^{\ominus})(X) \qquad (1)$$

$X = OH, OR^3, SH, SR^3, NH_2, NHR^3, NR^3R^4, NHNH_2, NHOH, N_3$
$Y = O, NR^2$

With water, these ketones form *stable monohydrates* or higher nonstoichiometric hydrates [*2, 6, 7, 8, 9, 10*]; dehydration of the geminal diols is achieved by treatment with hot concentrated sulfuric acid or phosphorus pentoxide [*4*].

0065–7719/95/0187–0840$13.04/1

Nucleophiles like alcohols [*2, 8*], hydrogen sulfide [*3*], thiols [*2, 10*], ammonia, amines, hydrazines, hydroxylamines [*3, 11, 12, 13, 14, 15*], azides [*2*], other pseudohalides [*2*], phosphonates [*2, 16, 17, 18, 19, 20*], and phosphanes [*2, 19*] add rapidly across the CO or CN double bond to yield *stable adducts*. The phosphonate adducts undergo a subsequent alcohol→ester rearrangement [*19, 20*] (equation 2).

[*19*] **2**

$$(CF_3)_2C{=}O + (RO)_2P(O)H \xrightarrow{RT} (CF_3)_2C(OH)P(O)(OR)_2 + (CF_3)_2C(H)OP(O)(OR)_2$$

base, 100% (first product → second product)

R	%	%
CH_3	6	94
C_2H_5	12	88
C_4H_9	95	5

Hexafluoroacetone azine accepts nucleophiles (ROH, RSH, R_2NH) in positions 1 and 2 to yield hydrazones [*21*]. **Phosphites** give open-chain products via a skeletal rearrangement [22]. Radical addition reactions are also reported [*23*].

Treatment of trifluoropyruvates with tosylhydrazine and phosphorus oxychloride–pyridine yields trifluoromethyl-substituted diazo compounds [*24*] (equation 3).

[*24*] **3**

$$F_3C{-}C(=O){-}CO_2C_2H_5 \xrightarrow[\text{2. pyridine, } POCl_3]{\text{1. TosNHNH}_2\text{, } CH_2Cl_2} F_3C{-}C(=N_2){-}CO_2C_2H_5$$

85%

When ***hexafluoroacetone*** reacts with **amides, urethanes** [*25*], **thioamides** [*26*], **amidines** [*27*], **sulfonamides** [*28, 29*], **sulfinamides** [*30*], and ***O,O*-dialkyl-amidophosphates** [*31*], the corresponding semiamidals are formed in nearly quantitative yield. The thermal stability of these adducts toward the retro reaction increases with the nucleophilicity of the amino compound [*5*]. Many polyfluorinated carbonyl compounds react likewise [*32, 33*]. On treatment of ureas [*34*], thioureas [*34*], thioamides [*26*], and *C,N*-diarylamidines [*27, 35*] first with hexa- fluoroacetone and then with dehydrating agents, heterodienes are obtained (equation 4).

From semiamidals, the corresponding trifluoromethyl-substituted *N-acyl imines* [*25, 28*], *1,3-diazabutadienes* [*27*], *N-sulfonyl imines* [*5, 29*], *N-sulfinyl imines* [*30*], and *N-phosphoryl imines* [*31*] can be obtained in high yields on reaction with powerful dehydrating reagents like $POCl_3$–pyridine or trifluoroacetic anhydride–pyridine [*2, 5*].

Halogen-free *N*-acyl aldimines and *N*-acyl ketimines tautomerize readily to give enamides [*36*]. In contrast, perfluorinated *N*-acyl imines are stable compounds. These electron-deficient imines not only exhibit high thermal stability but also show unique properties both as electrophiles and as strongly polarized hetero-1,3-dienes.

4

i: $(CF_3CO)_2O$, C_5H_5N; ii: $POCl_3$, C_5H_5N: iii: $SOCl_2$, C_5H_5N

X	Y			
C	O	77-79%	22-81%	[*25*]
C	S		65-88%	[*26*]
C	NR^2	100%	80-90%	[*27*]
S	O	68-74%	37-42%	[*30*]
P(OEt)	O	85-98%	89%	[*31*]

R^1 = aryl, alkyl R^2 = aryl, *t*-C_4H_9

The reaction of trifluoromethyl-substituted *N*-acyl imines toward nucleophiles in many aspects parallels that of the parent polyfluoro ketones. Heteronucleophiles and carbon nucleophiles, such as enamines [*37, 38*], enol ethers [*38, 39, 40*], hydrogen cyanide [*34*], trimethylsilylcarbonitrile [*2, 41*], alkynes [*42*], electron-rich heterocycles [*43*], 1,3-dicarbonyl compounds [*44*], organolithium compounds [*45, 46, 47, 48*], and Grignard compounds [*49, 50*], readily undergo hydroxyalkylation with hexafluoroacetone and amidoalkylation with acyl imines derived from hexafluoroacetone.

The reaction of trifluoropyruvates or their acyl imines with carbon nucleophiles offers a convenient route to α-t*rifluoromethyl-substituted* α-hydroxy or α-*amino acids*. Reduction of the keto or imino function yields 3,3,3-trifluorolactates or 3,3,3-trifluoroalanine esters [*32*].

A wide variety of nucleophiles, such as 1-alkylpyrroles, furans, thiophenls [51], phenols [*52*], anilines [*53, 54*], indoles [*55*], CH-acidic compounds [*56, 57*], as well as organolithium [*58*], Grignard [*57, 59*], organocadmium, and organozinc compounds [*50*], undergo C-hydroxyalkylation with trifluoropyruvates to yield derivatives of α-*trifluoromethyl* α-hydroxy acids.

Similarly, 1-alkylpyrroles, indoles, furans, thiophenes [60], α-picoline [61], enols, malonates [*16*], and organometallic compounds [*50, 62*] react with acyl imines of trifluoropyruvates to give derivatives of α-*trifluoromethyl* α-amino acids.

A surprising feature of the reactions of hexafluoroacetone, trifluoropyruvates, and their acyl imines is the C-hydroxyalkylation or C-amidoalkylation of activated aromatic hydrocarbons or heterocycles even in the presence of unprotected amino or hydroxyl functions directly attached to the aromatic core. Normally, aromatic amines first react reversibly to give N-alkylated products that rearrange thermally to yield C-alkylated products. With aromatic heterocycles, the reaction usually takes place at the site of the maximum π electron density [*53*] (equation 5).

[*53*] **5**

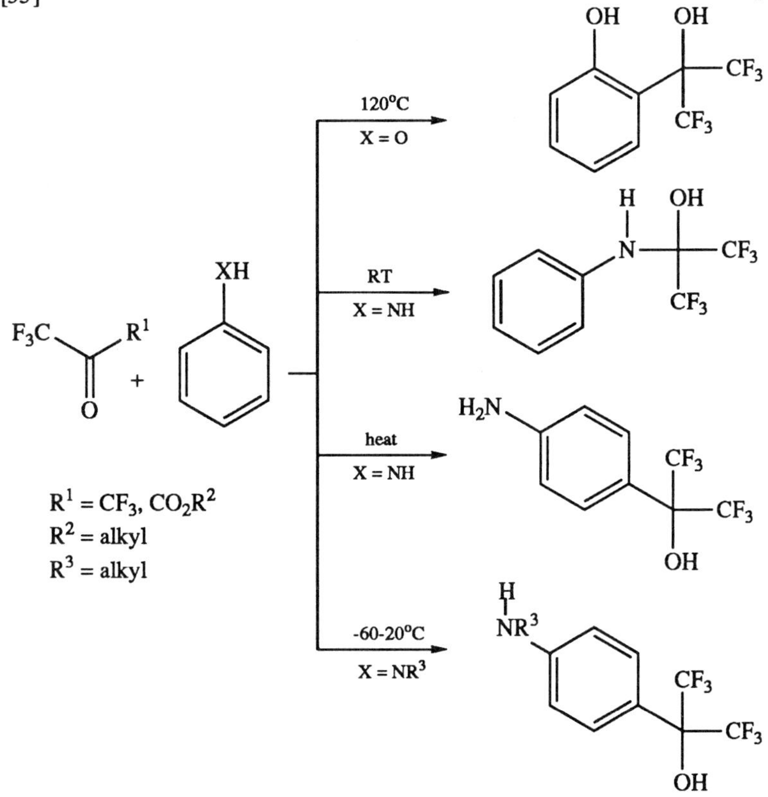

F_3C–CO–R^1 + C_6H_5XH →
- 120°C, X = O → 2-$(CF_3)_2C(OH)$-phenol
- RT, X = NH → C_6H_5NH–$C(OH)(CF_3)_2$
- heat, X = NH → 4-H_2N–C_6H_4–$C(OH)(CF_3)_2$
- -60-20°C, X = NR^3 → 4-HNR^3–C_6H_4–$C(OH)(CF_3)_2$

$R^1 = CF_3, CO_2R^2$
R^2 = alkyl
R^3 = alkyl

Perfluorinated nitriles are highly electrophilic compounds because of the adjacent electron-accepting substituent. Therefore, the addition of **amines, alcohols,** and **mercaptans** forming *amidines, imidates,* and *thioimidates* is well-documented [*63*].

The thioimidates obtained from perfluorinated nitriles and methylmercaptan can be transformed into thiohydroxyimidates on treatment with hydroxylamine [*64*]. CH-acidic compounds, such as methylketones [*65, 66*], nitroalkanes [*67*], and enamines [*68*], add to perfluorinated nitriles in the same way (equation 6).

[*64*] **6**

$$R_FC{\equiv}N \xrightarrow[\text{sealed}]{CH_3SH,\ K_2CO_3\ (cat.)} R_F{-}C({=}NH){-}SCH_3 \xrightarrow[\text{125°C, 15 h, sealed}]{NH_2OH{\cdot}HCl} R_F{-}C({=}N{-}OH){-}SCH_3$$

$R_F = CF_3$ 60%
$R_F = C_2F_5$ 41%
$R_F = C_3F_7$ 42%

Hypochlorites [*69*], **chlorofluorosulfates** [*70*], **chloroamines** [*71*], and **sulfenyl chlorides** [*72*] react with perfluorinated nitriles according to their polarity to form chloroiminoethers, chloroiminofluorosulfates, α-chlorohydrazones, and *N*-sulfenyltrifluoro acetimide chlorides, respectively.

Perfluorinated isonitriles are of special interest as ligands in metallorganic chemistry [*73*]. Generally, *N*-perfluoroalkylated imines result from addition reactions. *N*-Trifluoromethylhalomethaneimines are obtained on reaction with HX (X = F, Cl, Br) [*74*]. Via this route, the remarkably stable and distillable *N*-trifluoromethylformamide is obtained from trifluoromethylisonitrile and trifluoroacetic acid [*75*] (equation 7).

[*75*] **7**

$$F_3C{-}N{\equiv}C \xrightarrow[\text{RT},\ -(CF_3CO)_2O]{CF_3CO_2H} F_3C{-}NH{-}C({=}O){-}H$$

46%

Fluorinated alkylisocyanates and ***isothiocyanates*** add **alcohols, mercaptans,** and **amines** to yield stable 1:1 adducts [*76*]. When the reaction is performed with **carboxylic acids,** the anhydrides first formed decompose to give ***N*-fluoroalkylated amides** [*77*].

***N*-Chlorocarbamates** can be obtained on addition of hypochlorites [*69*]. The reaction with α,α-difluorotrimethylamine, a versatile new source of the unhydrated fluoride ion, leads to *N*-trifluoromethyl amidines [*78*] (equation 8).

Cyclocondensation Reactions

Cyclocondensation reactions starting from two components are possible only when both of them have two reactive centers. An initial electrophilic–nucleophilic interaction yielding a linear product is followed by a second electrophilic–nucleophilic interaction in the final cyclization step [*79*]. The cyclocondensation step is controlled by a series of rules (*Baldwin rules*) [*80*]. There are various types of such interactions (equation 9).

[*78*] **8**

$$F_3C{-}N{=}C{=}O \xrightarrow[\text{sealed tube}]{[Me_2N^{\oplus}{=}CHF]F^{\ominus}\ \ 50^{\circ}C,\ 6\ h} F_3C{-}N{=}CH\text{-}N(CH_3)_2$$

91%

9

n- - - -e / n- - - -e; n- - - -e / e- - - -n; n / n ... e,e; e / e ... n,n

The distance between the two reactive centers in each component is given by numbering the skeleton atoms; for example, 1,3 [n,n] represents a 1,3-dinucleophilic compound. For further details *see* reference 79.

Cyclocondensation reactions with perfluoroalkyl-substituted CO and CN multiple bond systems can be divided into several subgroups, according to the charge pattern of both reactants. On the basis of this simple concept, hetero-1,3-dienes should undergo two types of condensation reactions, classified by the number of skeleton atoms of the diene being incorporated into the ring system (equation 10).

Cyclocondensation Reactions with [n,e] Substrates

Four-Membered Heterocycles

When **aminophosphanes,** representing a 1,2 [n,e] substrate, react with two equivalents of ***hexafluoroacetone,*** four-membered *oxazaphosphetanes* are formed [*81*] (equation 11).

Five-Membered Heterocycles

Cyanoformamidines having both nucleophilic and electrophilic capacity react with hexafluoroacetone to give five-membered heterocycles [*82*] (equation 12).

Hexafluoroacetone and certain perfluorinated or partially fluorinated ketones, aldehydes, and imines react with **α-functionalized carboxylic acids (e.g.,** α-amino, α-*N*-methylamino [*83, 84*], α-hydroxy [*85*], and α-mercapto [*86*] acids) to give *five-membered heterocyclic systems* (equation 13).

The hexafluoroacetone derivatives are highly volatile compounds. They can therefore be used for gas-chromatographic analysis of mixtures of α-amino and α-hydroxy acids [*84, 85*]. Furthermore, they show potential for synthesis of various natural and nonnatural α-amino, α-hydroxy, and α-mercapto acids, as well as for synthesis of small peptides and depsipeptides [*87*]. Multifunctional carboxylic acids *can be selectively protected at the* α *position.* When this strategy is applied to aspartic acid, a simple, preparative, and regioselective carboxyl group derivatization is possible [*88*]. Efficient syntheses for L- and D-dihydroorotic acid [*89*] have been developed (equation 14).

10

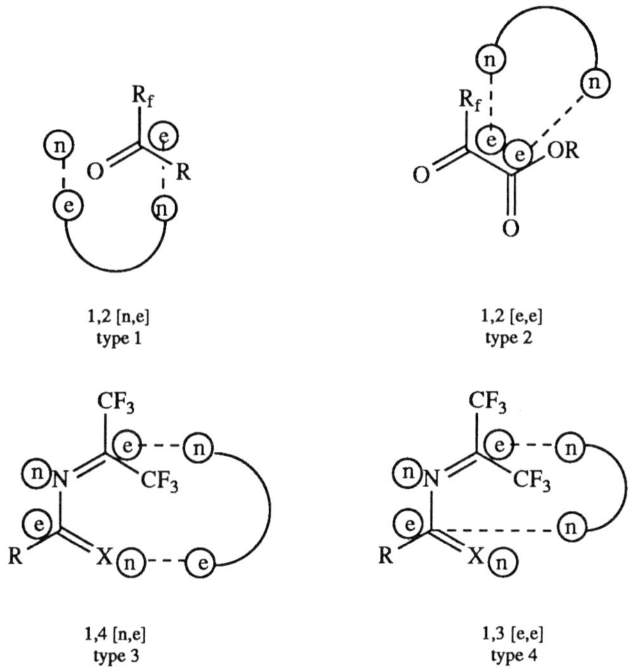

1,2 [n,e]
type 1

1,2 [e,e]
type 2

1,4 [n,e]
type 3

1,3 [e,e]
type 4

[*81*] **11**

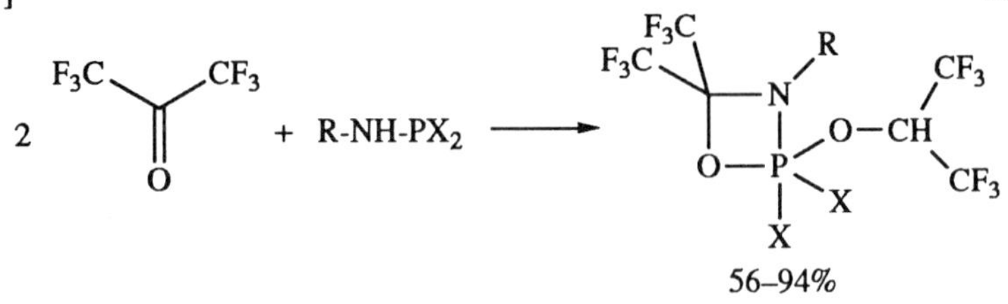

$$2\ (F_3C)_2C{=}O + R\text{-}NH\text{-}PX_2 \longrightarrow$$

56–94%

$X = OCH(CF_3)_2$, F $R = H, CH_3, t\text{-}C_4H_9$

[*82*] **12**

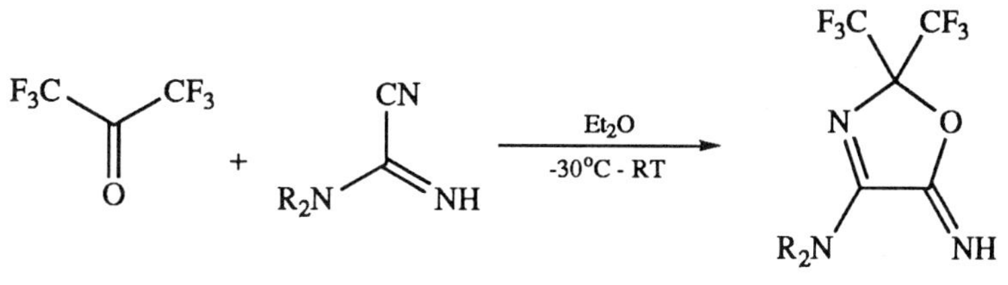

$R = CH_3, C_2H_5, CH(CH_3)_3$

Et_2O, -30°C - RT

75 - 84%

[*83*, *84*, *85*, *86*] **13**

$R{-}CH(XH){-}CO_2H$ + $(F_3C)_2C{=}O$ → DMSO, RT

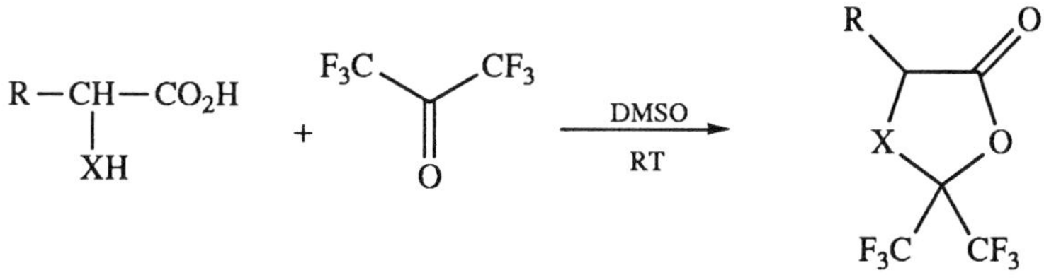

$X = NH, NCH_3, O, S$

51 - 99%

[*89*] **14**

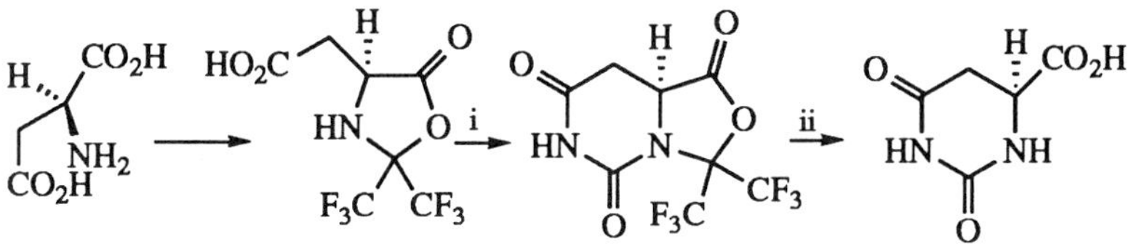

74% 98%

i: $ClSO_2N{=}C{=}O$, toluene, RT to 110°C

ii. 2N HCl, RT, 12 h

The efficiency of this method was demonstrated by the elegant two-step *synthesis of aspartame* [*87*]. Protection of the α-amino group and activation of the α-carboxylic group are accomplished in only one step. Deprotection of the amino functionality occurs during aminolysis, such as with methyl phenylalaninate (H-Phe-OMe in equation 15).

The 2,2-bis(trifluoromethyl)-4-methyl-2*H*-5-oxazolone, readily available from 2,2-bis(trifluoromethyl)-1,3-oxazolidin-5-one, is a synthetic equivalent of activated pyruvate [*90*] (equation 16).

Novel fluorinated 2,2-bis(trifluoromethyl)dioxolanes containing alkyne groups [*91*] have been synthesized from hexafluoroacetone and either 1,4-dibromo-2-butene or propargylic alcohol [*92*].

[*87*] **15**

HO$_2$C–CH$_2$–(oxazolidinone, HN, O, C(CF$_3$)$_2$, C=O) $\xrightarrow[\text{Et}_2\text{O, RT}]{\text{H-Phe-OMe}}$ HO$_2$C–CH$_2$–CH(NH$_2$)–C(=O)–NH–CH(CH$_2$C$_6$H$_5$)–CO$_2$CH$_3$

aspartame, 72%

[*90*] **16**

HN–CH$_2$–C(=O)–O–C(CF$_3$)$_2$ (ring) $\xrightarrow[\text{reflux}]{\text{SO}_2\text{Cl}_2\text{, AIBN}}$ N=CH–C(=O)–O–C(CF$_3$)$_2$ (ring), 72% $\xrightarrow[0^\circ\text{C - RT}]{\text{CH}_2\text{N}_2}$ N=C(CH$_3$)–C(=O)–O–C(CF$_3$)$_2$ (ring), 71%

$\xrightarrow[\text{RT}]{\text{RCH(NH}_2\text{)CO}_2\text{CH}_3}$ H$_3$C–C(=O)–C(=O)–NH–CH(R)–CO$_2$CH$_3$

58 - 72%

R = H, CH$_3$, CH(CH$_3$)$_2$, CH$_2$C$_6$H$_5$

Six-Membered Heterocycles

Trifluoromethyl-substituted dioxazine or pyrimidine derivatives are obtained on reaction of ***hexafluoroacetone*** with **cyanoguanidines** [*93*] or **cyanoacetamides** [*94*] (equation 17).

Bis(trifluoromethyl)-substituted pyrimidines are also available from ***trifluoroacetonitrile*** on reaction with **enamines** and **ynamines** [*68*]. With dimethylaminocrotonates, a cyclocondensation takes place to give *2-pyridones*. 5-Cyano-6-trifluoromethyluracil is available via a similar route [*95*] (equation 18).

Reaction type 3 (equation 10), where the complete hetero-1,3-diene skeleton is incorporated into the newly formed ring system, occurs with compounds having both a nucleophilic center and an electrophilic center. If these two functionalities are in positions 1 and 2, various types of six-membered ring systems become accessible.

4,4-Bis(trifluoromethyl)-1,3-diaza-1,3-butadienes require only room temperature to react with **acetyl cyanide** to yield *1,4,5,6-tetrahydropyrimidin-6-ones* [*96*]. Likewise, certain open-chain 1,3-diketones (acetylacetone and acetoacetates) and the heterodiene form six-membered ring systems [*97*] (equation 19).

[*94*] **17**

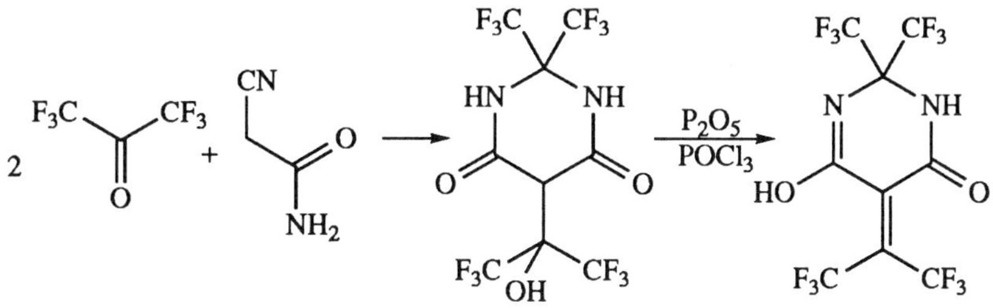

P_2O_5 / $POCl_3$

100% 78%

[*95*] **18**

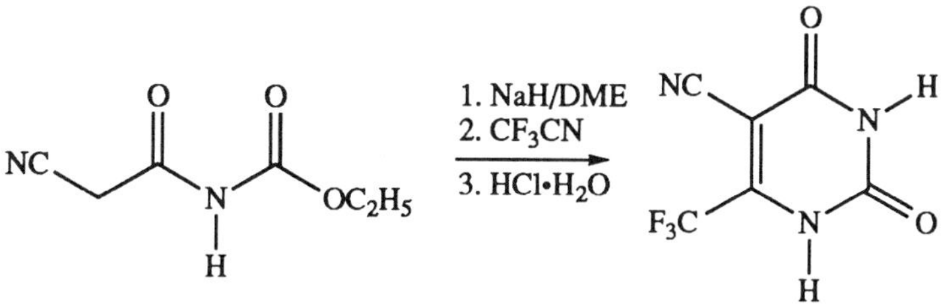

1. NaH/DME
2. CF_3CN
3. $HCl{\cdot}H_2O$

75%

[*96*] **19**

RT, 1h

-HCN

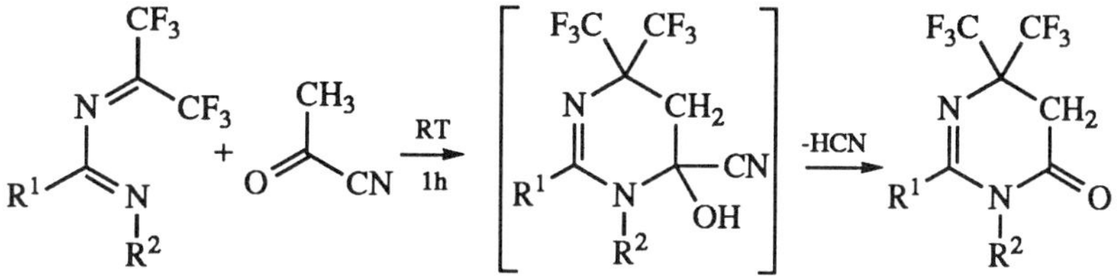

R^1 = aryl
R^2 = aryl

76-81%

4,4-Bis(trifluoromethyl)-4,5-dihydrooxazin-6-ones [*28*] and their O-acetylated derivatives [*96*] are formed on treatment of ***acyl imines*** with acetyl chloride–triethylamine at room temperature. The reaction was interpreted as a cycloaddition reaction involving a *ketene* [*28*]. However, the periselectivity and regiochemistry of this reaction are not in agreement with results obtained from the reaction of substituted ketenes with acyl imines where only the C=O double bond is involved in the cycloadditon process [*38, 96, 98*]. The only reasonable conclusion is that a ketene is not an intermediate in this reaction. Periselectivity as well as regioselectivity can be explained by a two-step mechanism involving a carbanionic attack at C-4 followed by a 6-exo-trig ring closure [*80*] (equation 20).

[*96*] **20**

On the basis of these findings, the reaction of acyl imines with methanesulfonyl chloride–triethylamine is not expected to proceed via a *sulfene intermediate* as previously proposed [*99*]. Again, a carbanion intermediate accounts nicely for the experimental facts. The electrophilicity of the hetero-1,3-diene is extremely high, therefore the carbanion, formed on reaction of triethylamine with methanesulfonyl chloride, should undergo nucleophilic attack at C-4 of the hetero-1,3-diene faster than sulfene formation by chloride elimination.

Cyclocondensation Reactions with [n,n] Substrates

Three-Membered Heterocycles

When hexafluoroacetone-*O*-tosyloxime reacts with primary amines acting as 1,1 [n,n] compounds, *3,3-bis(trifluoromethyl)diaziridines* are obtained [*100*]; with *O*-alkylhydroxylamines, the corresponding diazirine is formed [*101*] (equation 21).

21

$$(CF_3)_2C{=}N{-}O{-}SO_2{-}C_6H_4{-}CH_3 \xrightarrow[\text{DME, NEt}_3]{+R^1NH_2} \text{3,3-bis(trifluoromethyl)diaziridine (HN—NR)}\ \text{30-85\%}\ [100]$$

$$\xrightarrow{+R^2ONH_2} \text{3,3-bis(trifluoromethyl)diazirine (N=N)}\ [101]$$

$R^1 = CH_2C{\equiv}CH, CH_2CO_2CH_3, CH_2CO_2C_2H_5, CH_2CH_2OCOCH_3$
$CH_2CH_2N(CH_3)_2, CH(CH_3)CO_2CH_3, CH(CH_3)CO_2C_2H_5$

R^2 = alkyl

Five-Membered Heterocycles

Three skeleton atoms of a hetero-1,3-diene are incorporated into a five-membered ring system on reaction with 1,2 [n,n] compounds (e.g., hydrazines; reaction type 4, equation 10). On further heating in the presence of azobisisobutyronitrile (AIBN), the bis(trifluoromethyl)-substituted 1,2,4-triazolines are transformed into 5-trifluoromethyl-1,2,4-triazoles in high yield [*102*] (equation 22).

Six-Membered Heterocycles

Polyfluorinated α-diketones react with **1,2-diamino compounds,** such as *ortho*-phenylenediamine, to give *2,3-substituted quinoxalines* [*103*]. Furthermore, the carboxyl function of trifluoropyruvates offers an additional electrophilic center. Cyclic products are obtained with binucleophiles [*13, 104*]. With aliphatic or aromatic 1,2-diamines, six-membered heterocycles are formed. Anilines and phenols undergo C-alkylation with trifluoropyruvates in the *ortho* position followed by ring closure to form γ-lactams and γ-lactones [*11, 13, 52, 53, 54*] (equation 23).

Semicarbazide hydrochloride [*105*] and amidrazones [*106*] react with ***trifluoropyruvates*** to give six-membered heterocycles. A variety of trifluoromethyl substituted heterocyclic systems is available, starting from the hydrate of ***trifluoropyruvic acid,*** a versatile 1,2-bielectrophilic building block (reaction type 2, equation 10) [*107*] (equation 24).

Six-membered rings also result from the reaction of 1,3-diazabutadienes with 1,3 [n,n] compounds such as amidines and certain cyclic 1,3-diketones [*97*].

[102] **22**

- R^2NH_2, RT, 2-4 h; 30–78%

AIBN, 140°C, 4-8 h; - CHF_3; 46-60%

R^1 = aryl
R^2 = aryl
R^3 = H, CH_3, CH_2CH_2OH, CH_2CH_2CN

23

80% [13]

[13]

60%

benzene, 120°C

26-77% [54]

R = H, 4-CH_3, 4-CH_3O, 4-C_6H_5, 2-Cl, 3-Cl, 4-Cl, 4-$CO_2C_2H_5$

Cycloaddition Reactions

Perfluoroalkyl groups adjacent to multiple bond systems lower the frontier molecular orbitals (FMOs). Therefore, cycloaddition reactions preferentially occur with electron-rich multiple-bond systems. The preference of bis(trifluoromethyl)-substituted hetero-1,3-dienes for polar reactions makes them excellent model compounds for developing new types of diene reactions deviating from the well documented Diels–Alder scheme (pathway 1). A systematic study of the reactions of diene (1=2–3=4)–dienophile (5≡6) combinations reveals new synthetic possibilities that have not yet been fully exploited as tools for preparative organic chemistry (equation 25).

Three-Membered Heterocycles

Three-membered heterocyclic compounds are available when one-atom fragments like ***carbenes*** [*108*] or phosphorus compounds [*109*] add to C=O or C=N bonds (equation 26).

The ***addition of diazo compounds*** generally leads to three-membered rings, although in special cases, linear adducts with an intact diazo group [*110*] or 1,3,4-oxadiazol-3-ines [*111*] can be isolated. Most diazo compounds are unstable and yield oxirans and aziridines [*112, 113, 114*]. *Aziridines* are obtained exclusively on reaction of certain polyfluorinated acyl imines with diazomethane [*115*].

Four-Membered Heterocycles

Four-membered heterocycles are easily formed via [2+2] cycloaddition reactions [*63*]. These cycloaddition reactions normally represent multistep processes with dipolar or biradical intermediates. The fact that heterocumulenes, like isocyanates, react with electron-deficient C=X systems is well-known [*116*]. Via this route, β-lactones are formed on addition of ketene derivatives to hexafluoroacetone [*117, 118*]. The presence of a trifluoromethyl group adjacent to the C=N bond in quinoxalines, 1,4-benzoxazin-2-ones, 1,2,4-triazin-5-ones, and 1,2,4-triazin-3,5-diones accelerates [2+2] photocycloaddition processes with ketenes and allenes [*106*] to yield the corresponding azetidine derivatives. Starting from olefins, *fluorinated oxetanes* are formed thermally and photochemically [*119, 120*]. The reaction of 5*H*-1,2-azaphospholes with fluorinated ketones leads to [2+2] cycloadducts [*121*] (equation 27).

With linear ***iminophosphoranes*** and ***alkylidenephosphoranes,*** polyfluorinated ketones readily undergo *aza-Wittig* and *Wittig reactions,* respectively [*3*]. In these cases, the [2+2] cycloadducts [*49, 50, 122*] are usually not isolated; instead, they decompose with formation of polyfluorinated imines and olefins, respectively. Analogously, treatment of polyfluorinated ketones with *N*-sulfinyl imines [*5, 123*] yields the corresponding imines via four-membered intermediates; the reaction occurs with loss of sulfur dioxide (equation 28).

[*105*] **24**

F_3C–CO–CO_2H + H_2N–CO–$NHNH_2$ • HCl → CF_3–C(=N–NH–$CONH_2$)–CO_2H (73%) —$SOCl_2$→ 6-(trifluoromethyl)-1,2,4-triazine-3,5-dione (82%)

[*106*]

F_3C–CO–CO_2R^1 + R^2–C(=NH)–$NHNH_2$ —EtOH, reflux→ 3-R^2-6-CF_3-1,2,4-triazin-5-one (57-75%)

R^1 = alkyl R^2 = aryl

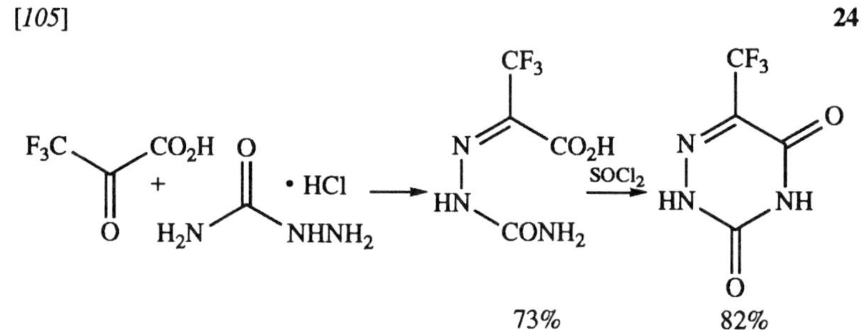

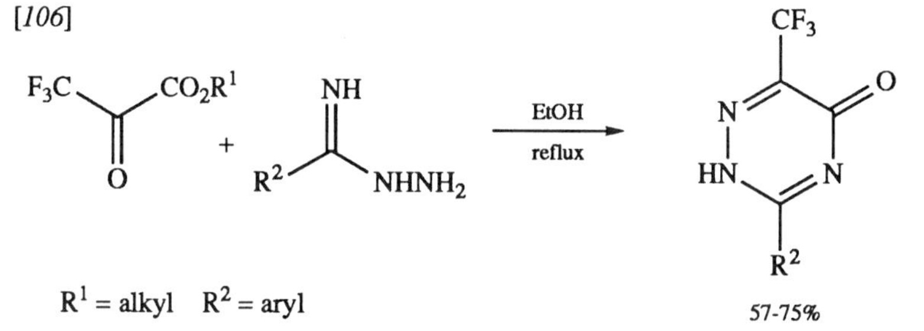

25

pathway 1 (5≡6), pathway 2 (5≡6), pathway 3 (5≡6), pathway 4 (5=6–7=8)

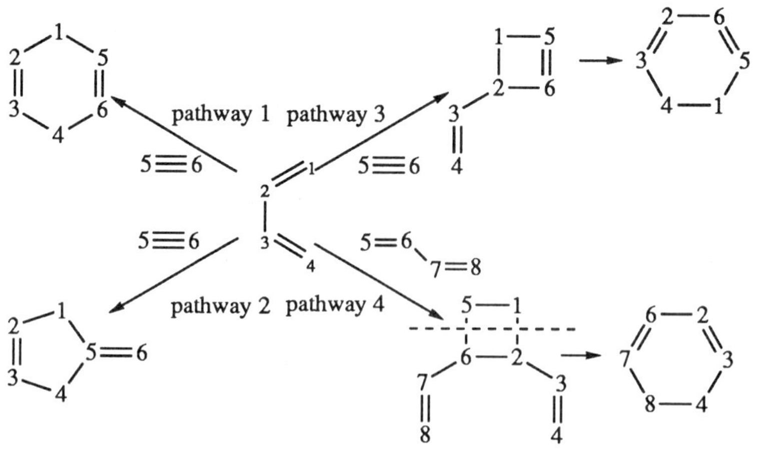

[*108*] **26**

$XCF_2C(=O)CFX_2$ —($PhHgCCl_2Br$)→ XCF_2, X_2CF oxirane with CCl_2 33-74%

X = Cl, F

27

$(F_3C)_2C=O$ —($Ph_2C=C=O$; KF, CH_3CN or DMF)→ β-lactone with C_6H_5, C_6H_5, F_3C, F_3C 59% [*117*]

$(F_3C)_2C=O$ —($Me_3Si\text{-}CH=C=O$; 0°C, 15 d)→ β-lactone with $(CH_3)_3Si$, F_3C, F_3C 61% [*118*]

F_3C-triazinone (N–H, C_6H_5) —(2 $CH_2=C=O$; hν)→ bicyclic product with F_3C, C_6H_5, $N\text{-}C(=O)CH_3$ 61% [*106*]

Azaphosphole (C_6H_5, C_6H_5, CO_2CH_3, X, X, Y) —($F_3C(R)C=O$)→ F_3C, R oxazaphosphetidine product (C_6H_5, C_6H_5, CO_2CH_3, X, X, Y) [*121*]

46-85%

X = CH_3, C_6H_5
Y = H, CO_2CH_3
R = CF_3, CH_3, C_6H_5

[123] **28**

$$\begin{array}{c}F_3C \\ CH_3O_2C\end{array}\!\!>\!C{=}O + RN{=}S{=}O \xrightarrow{80^\circ C} \left[CH_3O_2C{-}C(CF_3){-}O{-}S(=O){-}N(R) \right] \longrightarrow \begin{array}{c}F_3C \\ CH_3O_2C\end{array}\!\!>\!C{=}N{\sim}R$$

72-88%

$R = C_6H_5, C_6H_5CO$

Five-Membered Heterocycles

[4+1] Cycloaddition Reactions

Highly reactive dipolar bis(trifluoromethyl)-substituted heterodienes are excellent traps for "one-skeleton-atom species", even when these are short lived. The hetero-1,3-dienes add electron-rich and electron-poor carbenes [*124, 125, 126, 127, 128, 129, 130*], carbene complexes [*131*], carbene analogues [$SnCl_2$, $Sn(C_5H_5)_2$, $GeCl_2$, GeI_2] [*132, 133*], isonitriles [*128*], etc., to give [4+1] cycloadducts (equation 29).

[128] **29**

$$(CF_3)_2C{=}N{-}C(R^1){=}N{-}R^3 + C{\equiv}N{-}R^2 \xrightarrow[24\text{-}36\ h]{\text{toluene, }50\text{-}70^\circ C} \text{4,4-bis(CF}_3\text{)-2-R}^1\text{-1-R}^3\text{-5-(=N-R}^2\text{)-imidazoline}$$

77-93%

$R^1 = C_6H_5$
$R^2 = t\text{-}C_4H_9, n\text{-}C_4H_9, \text{aryl}, CH_2C_6H_5, c\text{-}C_6H_{11}$
$R^3 = \text{aryl}$

Isonitriles and ***some nitriles*** can be viewed formally as having a nucleophilic center and an electrophilic center at the same skeleton atom. Trimethylsilylcyanide and cyanoformates add to hetero-1,3-dienes in a stepwise process to give five-membered ring systems with the same structure as the isonitrile adducts [*41*].

Five-membered ring systems are also formed on transfer of single-skeletona-tom fragments, usually in a stepwise process: CR_2 from diazo alkanes [*28*], NH from azoimide (hydrazoic acid) [*134*], O from peroxy acids [*135*], S from phosphorus pentasulfide [*26*] or S_8 [*136*], Se from phosphorus pentaselenide [*137*] or Se_8 [*136*], and Te from antimony telluride [*137*]. Especially when oxygen is in the terminal position of the hetero-1,3-diene, it is often replaced by chalcogenes during ring formation (equation 30).

30

R^1 = aryl R^2 = aryl

X	reagent	product Z	product Y	reaction conditions	yield %	ref.
S, NR^2	HN_3	S, NR^2	NH	100-240 °C	17–45	[*134*]
NR^2	RCO_3H	NR^2	O	RT	35–64	[*135*]
O	P_2S_5	S	S	140 °C	42–70	[*26*]
S	S_8	S	S	100-110 °C	83–95	[*136*]
S	P_2Se_5	S	Se	110-120 °C	71–89	[*137*]
O	P_2Se_5	Se	Se	130 °C	17–21	[*137*]
S	Se_8	S	Se	100-110 °C	79–88	[*136*]
S	Sb_2Te_3	S	Te	80-90 °C	14–22	[*137*]

The chalcogene heterocycles have been used as stable precursors for ***sulfur-*** and ***selenium-containing hetero-1,3-dienes*** in cycloaddition reactions. 3*H*-1,2,4-Thiaselenazoles are a convenient source of 4,4-bis(trifluoromethyl)-1-thia-3-aza-buta-1,3-dienes, and 3*H*-diselenazoles are a convenient source of 4,4-bis(trifluoromethyl)-1-selena-3-azabuta-1,3-dienes as well as bis(trifluoro-methyl)-substituted *nitrile ylides* [*137*].

Of all known [4+1] cycloadducts, the tin heterocycles exhibit the most interesting preparative potential [*133*]. On heating they are transformed into 5-fluoro-4-trifluoromethyl-1,3-azoles [*132, 133*]. The fluorine atom at C-5 can be replaced by various nucleophiles. By this route, the 4-trifluoromethyl-1,3-azole moiety can be introduced into many compounds of biological interest (equation 31).

[*132*] **31**

CF_3 N CF_3 R X — $SnCl_2$, xylene, RT → [F_3C CF_3 N $SnCl_2$ R X] — 80-120°C, 4-24 h → CF_3 N F R X (22-68%) — Nuc^-, 25-50°C → CF_3 N Nuc R X (49-89%)

R = aryl
X = O, NR, S
Nuc = H, O-alkyl, O-aryl
S-alkyl, S-aryl,
$N(alkyl)_2$

A wide variety of ***α-trifluoromethyl α-amino acids*** are readily available from the reaction of 5-fluoro-4-trifluoromethyl-1,3-azoles with allylic alcohols [*138, 139*]. α-Trifluoromethyl-substituted α-amino acids show antibacterial and antihypertensive activity. Some are highly specific enzyme inhibitors (suicide inhibitors) and may be important as bioregulators [*140*]. Furthermore, they are interesting candidates for peptide modification.

Bis(trifluoromethyl)-substituted phosphoranes are obtained on reaction of bis(trifluoromethyl)-substituted heterodienes and P(III) species [*141, 142, 143*]. In the case of sulfur- and selenium-containing phosphoranes, the corresponding heterodienes are generated in situ from heterocyclic precursors [*137*] (equation 32).

4,4-Bis(trifluoromethyl)-substituted hetero-1,3-dienes and **acetylenes** react to give open-chain trifluoromethyl-substituted *N*-propargylic amides, 4*H*-1,3-oxazines, and 2-oxazolines [*42, 144*]. The formation of 2-oxazolines is an example of pathway 2 (equation 25), where only *one carbon atom of the acetylene moiety is incorporated* into the new ring system. The selectivity of this reaction can be controlled efficiently in favor of the five-membered ring system by altering the reaction conditions. In the presence of 4-dimethylaminopyridine, the five-mem-

[137] **32**

(CH$_3$)$_2$N-substituted 2,2-bis(CF$_3$)-1,3,4-diselenazoline + 2-methoxy-1,3,2-dioxaphospholane (POCH$_3$) $\xrightarrow[\text{-30°C to RT}]{\text{hexane}}$ spiro phosphorane product (F$_3$C, CF$_3$, N, OCH$_3$, P–O, Se, O, (CH$_3$)$_2$N)

69%

bered ring system becomes the main or exclusive reaction product. The value of 4-dimethylaminopyridine in the manipulation of periselectivity and regiochemistry in polar cycloaddition processes was recognized only recently [*42*] (equation 33).

An unusual formation of five-membered heterocyclic systems where *only the nitrile carbon atom becomes part of the new ring skeleton* occurs in the reaction of trifluoroacetonitrile with glycol and thioglycol [*145*] (equation 34).

[*42*] **33**

R–C(=O)–N=C(CF$_3$)$_2$ + HC≡C-C$_6$H$_5$ $\xrightarrow[\text{80°C, 18 h}]{\text{DMAP, toluene}}$ 2-R-4,4-bis(CF$_3$)-5-benzylidene-oxazoline (F$_3$C, CF$_3$, N, O, R, C$_6$H$_5$)

80%

R = C$_6$H$_5$, C(CH$_3$)$_3$; DMAP = 4-dimethylaminopyridine

[*145*] **34**

F$_3$C—C≡N + HX–CH$_2$CH$_2$–YH $\xrightarrow{\text{NaF, RT}}$ 2-CF$_3$-2-NH$_2$ ring (X, Y)

X,Y = O	85%
X = O; Y = S	98%
X,Y = S	30%

[3 + 2] Cycloaddition reactions.

Perfluorinated and partially fluorinated substituents directly bonded to hetero multiple bond systems lower the energies of FMOs. Consequently, they are highly reactive in HOMO (1,3-dipole)–LUMO (dipolarophile)-controlled [3+2] cycloaddition reactions. In some instances, the fluorine-containing substituents also stabilize the newly formed ring system. The possibility of multistep processes has to be taken into account.

Polyfluoroalkyl- and perfluoroalkyl-substituted CO and CN multiple bonds as dipolarophiles. ***Diazo alkanes*** are well known to react with **carbonyl compounds,** usually under very mild conditions, to give *oxiranes* and *ketones.* The reaction has been interpreted as a nucleophilic attack of the diazo alkane on the carbonyl group to yield diazonium betaines or 1,2,3-oxadiazol-2-ines as reaction intermediates, which generally are too unstable to be isolated. Aromatic diazo compounds react readily with partially fluorinated and perfluorinated ketones to give 1,3,4-oxadiazol-3-ines in high yield. At 25 °C and above, the aryloxadiazolines lose nitrogen to give epoxides [*111*].

Halogenated and halogenoalkyl-substituted **imines** react with ***diazo alkanes*** under very mild conditions and preferentially yield aziridines [*5, 146, 147*]. Diazonium betaines have been considered as intermediates of these reactions [*148, 149*]. When diazomethane reacts with certain imines of hexafluoroacetone [*5, 146, 150*], 1,1-bis(trifluoromethyl)-2-azabuta-1,3-dienes [*147*], or hexafluoroacetone azine [*21*], stable [3+2] cycloadducts are obtained. The latter two hetero-1,3-dienes can add two molecules of diazomethane (equation 35).

[*147*] **35**

CH_2N_2, RT, 2-3 d

$R^1 = CH_3$	30%	32%
$R^1 = CH(CH_3)_2$	42%	51%

The *N*-benzenesulfonyl imines of hexafluoroacetone readily react with ***nitrile oxides*** to give [3+2] adducts, apparently in a multistep reaction [*151*] (equation 36).

Although only a few examples of [3+2] cycloaddition reactions of this type have been described so far, most 1,3-dipoles should react in this way with predictable regiochemistry [*5, 146, 150, 151*].

Five-membered ring systems can be obtained from ***hetero-1,3-dienes*** on reaction with **oxiranes** and **thiiranes.** To avoid competition from a possible 1,4-addition, the nucleophilic attack of the terminal heteroatom of the diene has to be sterically or electronically hindered by incorporation of the heteroatom into electron-poor heteroaromatic ring systems or by introducing electron-withdrawing substituents at C-2 [*152*]. *N*-Benzenesulfonyl imines, which react as imines, react with oxiranes and thiiranes to give exclusively oxazolidines and thiazolidines, respectively [*153, 154*].

[*151*] **36**

$C_6H_5SO_2N{=}C(CF_3)_2 + {}^{\ominus}O{-}N^{\oplus}{\equiv}C{-}CHF{-}CF_3 \longrightarrow$ [2,2-bis(trifluoromethyl)-3-$C_6H_5{-}SO_2$-4-(CHF–CF_3)-oxadiazoline: ring F$_3$C–C(CF$_3$)–O–N=C(CHF–CF$_3$)–N(SO$_2$C$_6$H$_5$)]

Perfluoroalkyl-substituted nitriles react readily with various 1,3-dipoles, such as ***azomethine ylides*** [*155, 156, 157*], ***azomethine imines*** [*158*], ***diazo alkanes*** [*159*], ***azides*** [*160, 161, 162*], and ***nitrile ylides*** [*163*] to give stable five-membered ring systems (equation 37).

Trifluoromethyl-substituted 1,3-dipoles. Trifluoromethyl-substituted 1,3-dipoles are especially valuable building blocks for the synthesis of a wide variety of trifluoromethyl-substituted five-membered ring systems. The building-block strategy of introducing trifluoromethyl groups into ring systems very often is superior to selective introduction of the fluorinated substituents in a final step of the reaction sequence. Incorporation of trifluoromethyl groups into a 1,3-dipolar species usually increases reactivity.

Trifluoromethyl-substituted 1,3-dipoles of the propargyl–allenyl type and trifluoromethyl-substituted nitrilium betaines. Trifluoromethyl- [*164, 165*] and bis(trifluoromethyl)-substituted [*166, 167*] ***nitrile ylides*** have been generated by different methods and trapped with various dipolarophiles to yield [3+2] [*168*] and [3+1] cycloadducts [*169*], respectively.

Bis(trifluoromethyl)-substituted nitrile ylides undergo *dimerization* reactions in the absence of trapping reagents [*143, 168, 170*] (equation 38).

When 5-*tert*-butyl-2,2,2-trimethoxy-3,3-bis(trifluoromethyl)-2,3-dihydro-1,4,2-oxazaphosphole is pyrolyzed at 700–860 °C and the cycloreversion products are condensed at –196 °C, the nitrile ylide formed can be identified by infrared spectroscopy (equation 39) [*171*].

Likewise, trifluoromethyl-substituted ***nitrile imines*** [*172*] and ***nitrile oxides*** [*173, 174, 175*] have been used to synthesize trifluoromethyl-substituted five-membered ring systems of the *pyrazole, isoxazole, isoxazoline,* and *1,2,4-oxadiazole* type (equation 40). In the absence of trapping reagents, trifluoroacetonitrile oxide dimerizes to give a trifluoromethyl-substituted furoxan or a *1,4-dioxa-2,5-diazine,* depending on the provenance of the nitrile oxide [*173*].

[*128*] **37**

CF_3CN; CH_2Cl_2, $-H_2$ [*155*]

$CHCO_2C_2H_5$ → $CO_2C_2H_5$, CF_3

62%

CF_3CN; sealed, RT, 12 w [*159*]

$Si(CH_3)_3$ ... H, CF_3

93%

$R—N_3$ → R_FCN; sealed, 130-150°C [*159*]

R, CF_3

22-96%

R = *n*-C_8H_17, C_6H_5, $C_3F_7CH_2$
R_F = CF_3, C_3F_7

Trifluoromethyl-substituted diazonium betaines [176]. Synthetic routes to trifluoromethyl-substituted diazo alkanes, such as 2,2,2-trifluorodiazoethane [*177, 178, 179*] and alkyl 3,3,3-trifluoro-2-diazopropionates [*24*], have been developed. Rhodium-catalyzed decomposition of 3,3,3-trifluoro-2-diazopropionates offers a simple preparative route to *highly reactive carbene complexes,* which have an enormous synthetic potential [*24*]. [3+2] Cycloaddition reactions were observed on reaction with nitriles to give 5-alkoxy-4-trifluoromethyloxazoles [*180*] (equation 41).

Trifluoromethyl-substituted 1,3-dipoles of the allyl type. Trifluoromethyl-substituted *azomethine imines* are readily available on reaction of ***hexafluo-***

[*166,167*] **38**

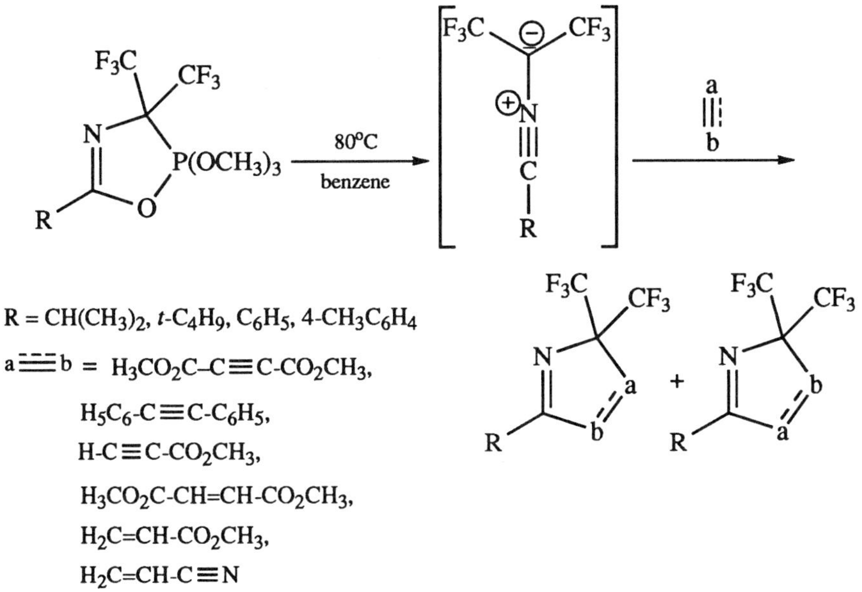

$80^{\circ}C$, benzene

$R = CH(CH_3)_2$, t-C_4H_9, C_6H_5, 4-$CH_3C_6H_4$

a$\equiv$b = H_3CO_2C–$C\equiv C$-CO_2CH_3,

H_5C_6-$C\equiv C$-C_6H_5,

H-$C\equiv C$-CO_2CH_3,

H_3CO_2C-CH=CH-CO_2CH_3,

H_2C=CH-CO_2CH_3,

H_2C=CH-$C\equiv N$

[*171*] **39**

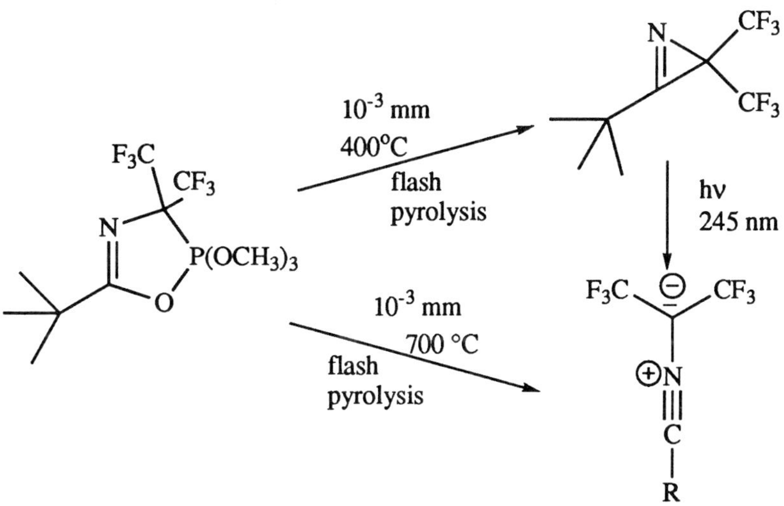

10^{-3} mm, $400^{\circ}C$, flash pyrolysis

hν 245 nm

10^{-3} mm, 700 °C, flash pyrolysis

roacetone azine with **olefins** and **acetylenes** [*181, 182*]. They are therefore the most thoroughly investigated trifluoromethyl-substituted 1,3-dipoles [*183*].

Trifluoromethyl-substituted ***nitrones*** have been prepared [*184, 185, 186*] and used for [3+2] cycloaddition reactions [*185, 186*] (equation 42).

[*172*] **40**

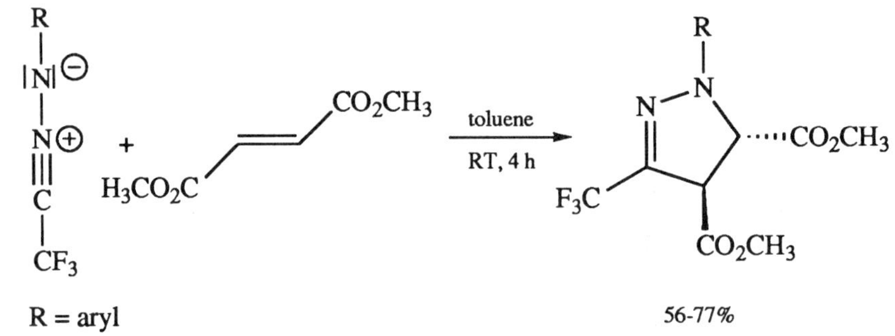

R = aryl

toluene, RT, 4 h

56-77%

[*173*]

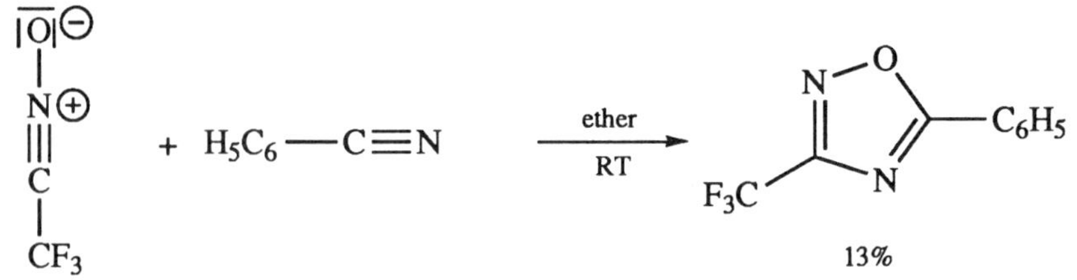

ether, RT

13%

[*24*] **41**

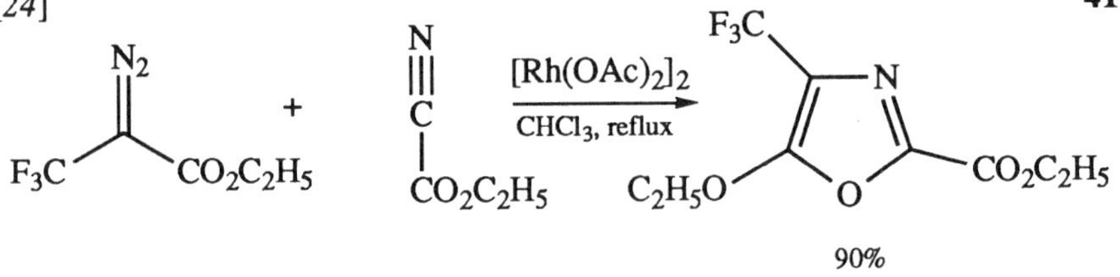

$[Rh(OAc)_2]_2$, $CHCl_3$, reflux

90%

[*185*] **42**

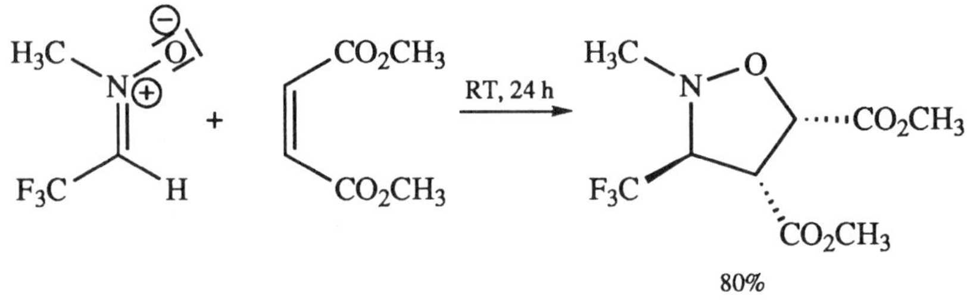

RT, 24 h

80%

Trifluoromethyl-substituted ***azimines*** are surprisingly stable compounds. They are accessible by 1,3-dipole metathesis from trifluoromethyl-substituted azomethine imines and certain nitroso compounds [*187, 188*]. On photolysis, an electrocyclic ring closure first gives the *triaziridines,* which are stable at room temperature. On heating above 80–100 °C, a valence tautomerization takes place and azimines are formed [*189*] (equation 43).

[*188, 189*] **43**

$R^3N{=}O$, $-(CF_3)_2CO$; hν

$R^1 = CH_3$ $R^2 = CH_3, C_6H_5, C(CH_3){=}CH_2$

$R^3 = CF_3$	68-95%	79-85%
$R^3 = C_6F_5$	78-93%	
$R^3 = C_6H_5$	43%	

Criss-Cross Cycloaddition Reactions

The first example of a cycloaddition reaction of a multiple bond to a diene was reported in 1917. Surprisingly, it was found that benzal azine adds to 2 equivalents of several unsaturated systems, when offered in excess, to yield bicyclic compounds. This reaction was named *"criss-cross" cycloaddition* [*190*]. Exploitation of the preparative potential of criss-cross cycloaddition began only in the early 1970s, when hexafluoroacetone azine became available on a larger scale [*191, 192*]. The study of this reaction proved to be an impetus for the development of azine chemistry [*183, 193*].

Hexafluoroacetone azine reacts with 2 equivalents of terminal **olefins** [*194*] and **acetylenes** [*182*] to give *1,5-diazabicyclo[3.3.0]octanes* and *1,5-diazabicyclo[3.3.0]octa-2,6-dienes,* respectively (equation 44).

The above cycloaddition process consists of two separate [3+2] cycloaddition steps and represents a 1,3–2,4 addition of a multiple bond system to a hetero-1,3-diene [*181*]. The structure of the azomethine imine intermediate has been proved unequivocally by X-ray analysis [*195*]. Ethylene [*194*]; acetylene [*182*]; many alkyl- and aryl- as well as geminal dialkyl- and diaryl-substituted alkenes [*196, 197, 198, 199*], dienes [*200*], and alkynes [*182, 201*]; certain cyclic alkenes [*198, 199,*

[*182, 198*] **44**

$H_2C{=}C(CH_3)_2$, RT — 85%

$H_2C{=}C(CH_3)_2$, 90-100°C, 12 h — 91%

$HC{\equiv}CR$, RT

$HC{\equiv}CR$, RT

R = H 83%

R = C_6H_5 46%

202]; vinyl ethers [*203*]; alkoxyacetylenes and ynamines [*204*]; acrylates [*205*]; and propiolates [*201*] react according to this scheme. Under appropriate reaction conditions the azomethine imines can be isolated.

A simple preparative route to the previously unknown 1*H*-3-pyrazolines via azomethine imines was developed. ***Olefins*** of the type $R^1CH{=}CHR^2$ react with ***hexafluoroacetone azine*** to give *azomethine imines,* which undergo a sequence of prototropic shifts to form 1*H*-3-pyrazolines [*196, 202*]. On heating, the latter are transformed into *3-trifluoromethylpyrazoles* [*196, 206*] and, on treatment with bases, into *1,2,5,6-tetrahydropyrimidines* [*206*] (equation 45).

[*202, 206*] **45**

CF_3 CF_3 N N F_3C CF_3 + $R^1CH{=}CHR^2$

R^1, R^2 = alkyl, aryl

sealed

30-90°C

3-20 d

H R^2 CF_3 H R^1 $N^{\oplus}$ CF_3 $N^{\ominus}$ F_3C CF_3

R^2 CF_3 N CF_3 H R^1 NH F_3C CF_3

32-93%

base, heat

R^2 N CF_3 CF_3 NH R^1 H F_3C CF_3

AIBN, heat

H R^2 N CF_3 CF_3 NH R^1 F_3C CF_3

79-92%

R^2 CF_3 N CF_3 H R^1 N F_3C

18-76%

References are listed on pages 877–887.

An unusual *1,4-migration of a trifluoromethyl group* was observed when azomethine imines were synthesized from hexafluoroacetone azine and alkoxyacetylenes. The rearrangement, which occurs at temperatures as low as 0 °C, results in the formation of *N*-(perfluoro-*tert*-butyl)pyrazoles [*207*] (equation 46).

[*207*] **46**

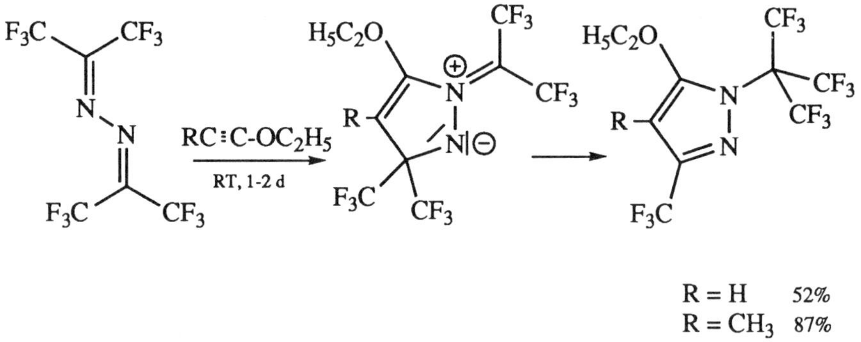

RC≡C-OC$_2$H$_5$, RT, 1-2 d

R = H 52%
R = CH_3 87%

The ***azomethine imines*** exhibit the typical cycloaddition behavior expected of 1,3-dipolar species [*183*]. Numerous [3+2] cycloaddition reactions have been performed [*201, 204*]. **Tetracyanoethylene** adds to azomethine imines across the nitrile function instead of the C=C double bond. This reaction is a rare example of this type of periselectivity [*208*] (equation 47).

[*208*] **47**

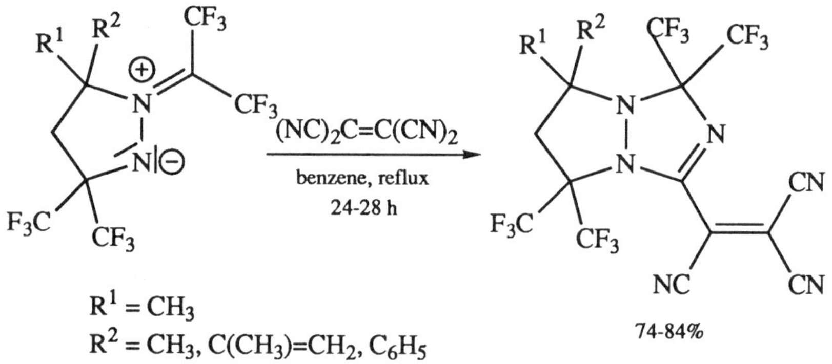

$(NC)_2C=C(CN)_2$, benzene, reflux, 24-28 h

$R^1 = CH_3$
$R^2 = CH_3$, $C(CH_3)=CH_2$, C_6H_5

74-84%

Certain 1,5-diazabicyclo[3.3.0]oct-2-enes can be transformed unexpectedly into 4*H*-5,5-dihydro-1,2-diazepines on heating [*209*]. 1,5-Dipoles formed on heating of 1,5-diazabicyclo[3.3.0]oct-2-enes [*210*] can be trapped with olefins to give [3+2] cycloadducts. At elevated temperatures, they undergo a [3+2] cycloreversion. This reaction sequence offers a simple route to dienes with interesting substitution patterns, for example, 1,1-bis(trifluoromethyl)-1,3-butadiene [*211*]. The [3+2] cycloadducts that arise from the reaction of the 1,5-dipoles with acetylenes undergo an electrocyclic ring opening on further heating. When this reaction sequence is repeated, manifold extended 1,x-dipoles become available [*212*].

Because the criss-cross cycloaddition reaction is a sequence of two [3+2] cycloaddition steps, the reaction with α,ω-diolefins offers a new entry into macromolecular chemistry. New types of polymers with interesting structures and properties can be synthesized [*213, 214, 215*] (equation 48).

[*213, 214, 215*] **48**

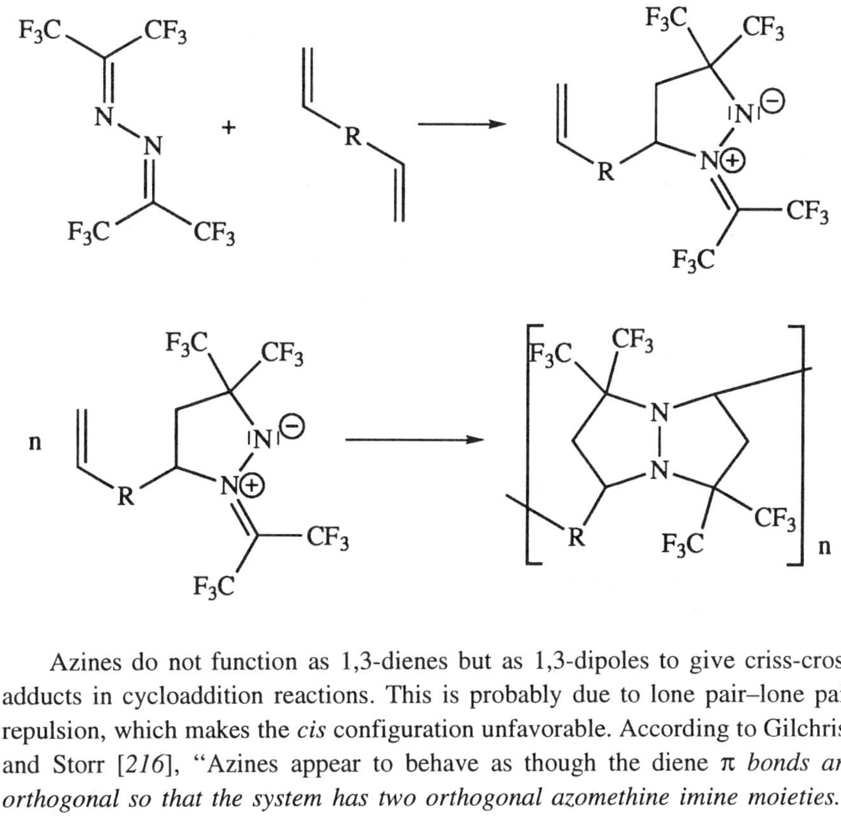

Azines do not function as 1,3-dienes but as 1,3-dipoles to give criss-cross adducts in cycloaddition reactions. This is probably due to lone pair–lone pair repulsion, which makes the *cis* configuration unfavorable. According to Gilchrist and Storr [*216*], "Azines appear to behave as though the diene π *bonds are orthogonal so that the system has two orthogonal azomethine imine moieties." Consequently, 1,4-cycloaddition reactions are only known when the azine moiety is incorporated into a ring system [106, 217, 218, 219*]. A number of 1,2-cycloaddition reaction is described. They occur only when a possible 1,x-dipolar intermediate is perfectly stabilized. But even in many of these cases, a [3+2] cycloadduct is the precursor of the [2+2] adduct.

The Diels–Alder adduct isolated from the reaction with 2,3-dimethylbuta-1,3-diene at elevated temperatures [*200*] is in fact the product of a two-step reaction: a 1,3-cycloaddition followed by a 3,2-sigmatropic rearrangement [*199*] (equation 49).

[*199, 200*] **49**

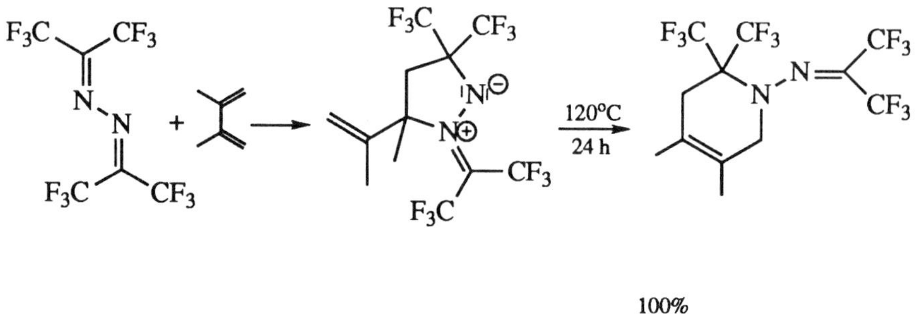

100%

Six-Membered Heterocycles

Polyfluoroalkyl- and Perfluoroalkyl-Substituted CO and CN Multiple Bonds as Dienophiles in [4+2] Cycloaddition Reactions

In general, the reactions of 1,3-dienes with aliphatic and aromatic aldehydes and ketones proceed poorly unless highly reactive dienes and/or Lewis acid catalysts are used [*220*]. These reactions can be promoted also by high-pressure techniques [*221*]. Because most [4+2] cycloadditions of dienes with C=O dienophiles are HOMO (diene)–LUMO (dienophile)-controlled processes, electron-deficient aldehydes, dienes, dienophiles, and ketones, especially perfluorinated and partially fluorinated ones, have been used successfully for the synthesis of six-membered rings [*221, 222, 223, 224*]. Certain bis(trifluoromethyl)-substituted hetero-1,3-dienes add hexafluoroacetone to give the [4+2] cycloadducts in high yields [*225*]. The [4+2] cycloaddition of pentafluoronitroacetone to 2,3-dimethyl-1,3-butadiene is followed by a [3+2] cycloaddition of the nitro group to the newly formed C=C double bond of the dihydropyrane ring, and a caged product is obtained [*226, 227*] (equation 50).

[*226, 227*] **50**

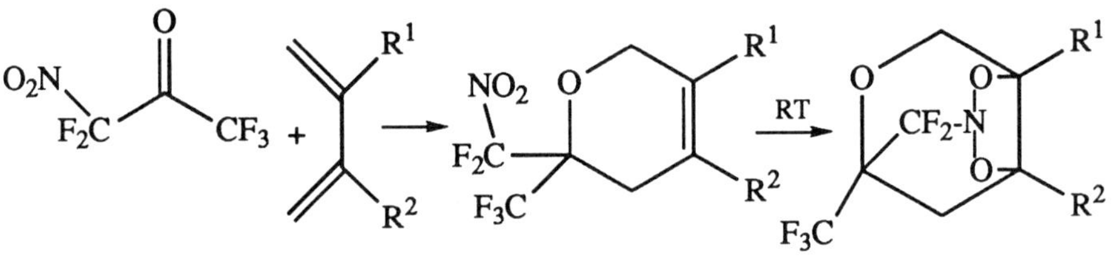

R^1 = H, CH_3 R^2 = H, CH_3

The dienophilic character of imines parallels that of carbonyl compounds. Consequently, *electron-deficient imines are the most reactive dienophiles* of this class, particularly those having *C*-perfluoroalkyl [*5, 146, 150, 228*], *N*-acyl [*126, 127*], or *N*-sulfonyl groups [*148, 229, 230*].

One of the features of *Diels–Alder reactions* with most alkyl and aryl nitriles that has made them rather unattractive as dienophiles is the requirement of very high reaction temperatures. Again, only when electron-withdrawing substituents are directly bonded to the nitrile function do [4+2] cycloaddition reactions occur at reasonably low temperatures [*148, 231, 232*]. A high-yield [4+2] cycloaddition was observed on reaction of 4,4-bis(trifluoromethyl)-1-thia-3-aza-1,3-butadienes with trifluoroacetonitrile at 150 °C [*225*].

Azetes, which are stabilized by *tert*-butyl groups, react with **trifluoroacetonitrile** to give *pyrimidines* [*233*] (equation 51).

[*233*] **51**

CF_3CN, CH_2Cl_2, 20 °C

CF_3

91%

[4+2] Cycloaddition reactions where bis(trifluoromethyl)-substituted hetero-1,3-dienes act as dienophiles have been described for open-chain and cyclic dienes [*115, 126, 127*]. The balance of the diene–dienophile activity of bis(trifluoromethyl)-substituted hetero-1,3-dienes can be influenced strongly by the substituents bonded to the imino nitrogen atom. For instance, *N*-(arylsulfonyl) derivatives of trifluoroacetaldimine and hexafluoroacetone imine do not act as dienes but exhibit only the dienophile reactivity of electron-deficient imines [*5, 229, 234, 235, 236, 237*] (equation 52).

[4+2] Cycloaddition Reactions with Trifluoromethyl-Substituted Hetero-1,3-dienes

Bis(trifluoromethyl)-substituted heterodienes are electron-deficient species. They therefore react preferentially with electron-rich multiple bond systems to give [4+2] cycloadducts (*Diels–Alder reaction with inverse electron demand*) [*238*].

On the other hand, these are compounds with marked 1,4-dipolar character, having electron lone pairs at the terminal heteroatom and an electrophilic center at C-4. Consequently, they can react with polarized multiple bond systems, even when these are extremely electron-poor [*225*].

Bis(trifluoromethyl)-substituted heterodienes, being the most reactive heterodienes known to date, must be regarded as having unique synthetic potential.

[4+2] Cycloaddition reactions of bis(trifluoromethyl)-substituted heterodienes have been performed with open-chain [*28*] and cyclic olefins [*26*], cyclobutadienes [*239*], enol ethers [*3, 26, 28, 38, 240*], enamines [*26, 38, 100*], acetylenes [*42, 100, 144*], alkoxyacetylenes [*100*], ynamines [*26, 144*], aldehydes and ketones [*98, 130, 225, 241, 242, 243*], imines [*244*], nitriles [*26, 100, 130, 225, 241, 242, 245, 246, 247, 248*], cyanamides [*245, 249*], cyanoguanidines [*250*], sulfoxides [*251*], ketenes [*28, 38, 98, 100, 225, 252*], isocyanates, and isothiocyanates [*38*].

[*229*] **52**

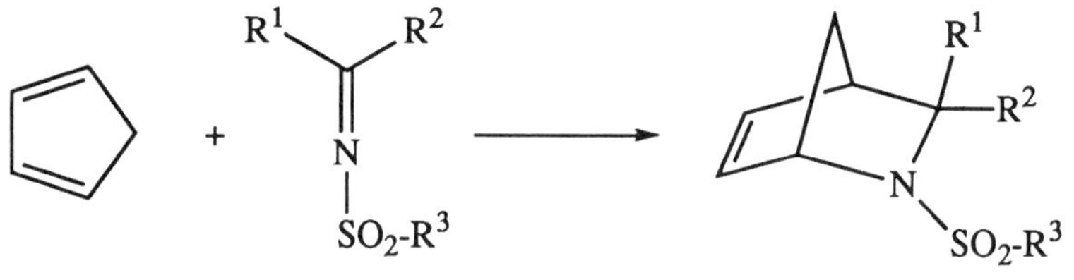

$R^1 = CF_3, CF_2Cl$
$R^2 = CF_3, CF_2Cl, CF_2NO_2$
$R^3 = CH_3, C_6H_5$

40-77%

The reactions of bis(trifluoromethyl)-substituted hetero-1,3-dienes are predominantly LUMO-controlled processes [*238*]. With polar or highly polarizable dienophiles, the tendency to undergo stepwise cycloaddition reactions is considerable. Notably these ***hetero-1,3-dienes*** react with α,β-*unsaturated hetero multiple bond systems* across the hetero multiple bond exclusively [*243, 246, 248*] (equation 53).

[*248*] **53**

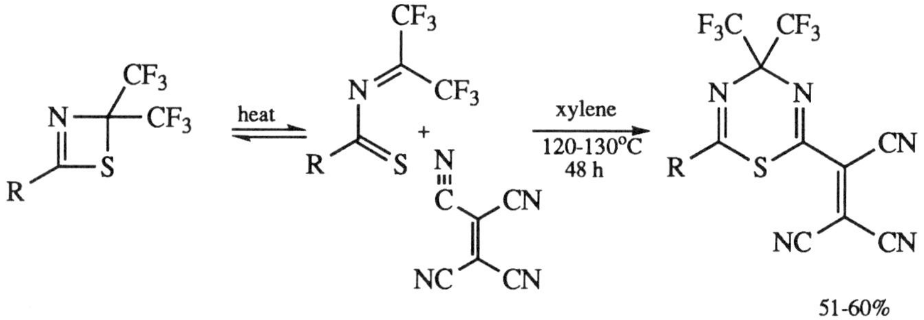

heat

xylene
120-130°C
48 h

51-60%

R = aryl

The observed regiochemistry can be explained readily by a charge-controlled reaction. The highly electrophilic center of the hetero-1,3-diene at C-4, bearing the two trifluoromethyl groups, is attacked by the center of highest electron density of the dienophile to give an optimally stabilized 1,x-dipolar intermediate. We are convinced that the rare examples of [4+2] cycloaddition reactions exhibiting this type of periselectivity and regiochemistry [*253, 254, 255, 256, 257*] belong to a common mechanistic type.

In contrast, when α,β-unsaturated multiple bond systems act as dienophiles in concerted [4+2] cycloaddition reactions, they react across the C=C double bond. Periselectivity as well as regiochemistry are explained on the basis of the size of the orbital coefficients and the resonance integrals [*258*].

4,4-Bis(trifluoromethyl)-1,3-diaza-1,3-butadienes react with cyanamides, acetonitrile, and benzonitrile to give [4+2] cycloadducts. In contrast, with electron-poor nitriles (trifluoroacetonitrile, trichloroacetonitrile, and 4-chlorophenylazonitrile), six-membered ring systems are formed with reorganization of the hetero-1,3-diene skeleton (pathway 3, equation 25). The reaction sequence consists of three separate steps: [2+2] cycloaddition, electrocyclic ring opening, and electrocyclic ring closure [*245*] (equation 54).

[*245*] **54**

CF_3, N, CF_3, C_6H_5, N, C_9H_{11}

$F_3C\text{-}C{\equiv}N$, 80°C, 3 d, sealed tube, toluene

F_3C, CF_3, N, N, C_9H_{11}, F_3C, N, C_6H_5

80%

C_9H_{11} = 2,4,6-trimethylphenyl

The reaction of ***4,4-bis(trifluoromethyl)-1,3-diaza-1,3-butadienes*** with certain α,β-**unsaturated ketones** yields *pyrimidine d*erivatives. A two-step mechanism, metathesis–electrocyclic ring closure and metathesis–intramolecular *ene reaction,* is a plausible explanation for the experimental results (pathway 4, equation 25) [*259*].

This new reaction type should be transferable to nonfluorinated hetero-1,3-dienes that are capable of stepwise cycloaddition reactions.

Certain trifluoromethyl-substituted 1,2,4,5-tetrazines [*260, 261*] and 1,2,4-triazines [*106*] can be used as cyclic hetero-1,3-dienes and provide efficient preparative routes to partially fluorinated heterocycles (equations 55 and 56).

[*260*] **55**

$R = H, CH_3, OCH_3, N(CH_3)_2, SCH_3$

i: 12-24 h, sealed
benzene or toluene, 110-140°C

$[O_2]$

40-87%

[*106*] **56**

$CHCl_3$
reflux
$-N_2$

n = 1 61%
n = 2 31%

3-Trifluoromethyl-1-oxa-2-aza-1,3-butadienes can be prepared in situ by 1,4-HBr elimination from the corresponding α-bromo oxime on treatment with base, and provide an interesting and versatile four-skeleton-atom building block [*262*] (equation 57).

Seven-Membered Heterocycles

Oxiranes exhibit 1,3 [e,n] capacity. Therefore, seven-membered ring systems can be synthesized on reaction with ***hetero-1,3-dienes.*** The reaction is catalyzed by 4-dimethylaminopyridine. On catalysis with boron trifluoride, the regiochemistry is reversed [*263*] (equation 58).

Formation of five-membered ring systems (1,2-addition) can compete with formation of the seven-membered heterocycles (1,4-addition). If the first step of the reaction sequence, namely the nucleophilic attack of the terminal heteroatom of the diene, is hindered by steric or electronic effects, the five-membered ring product is formed exclusively.

[*262*] **57**

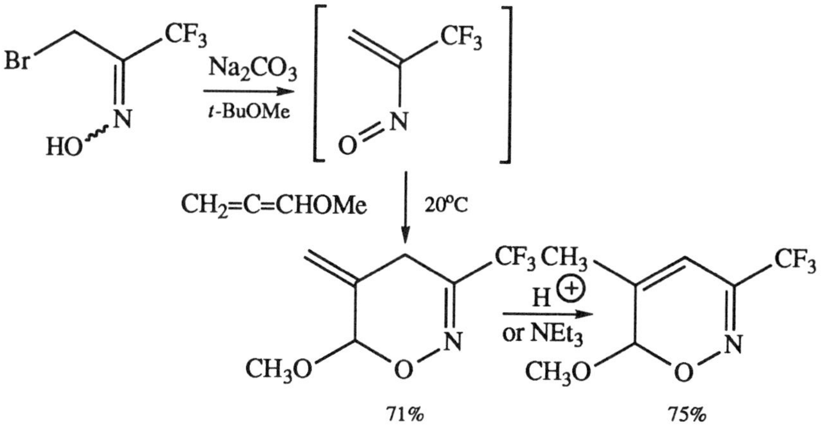

Br, CF_3, HO, N, Na_2CO_3, *t*-BuOMe, CF_3, O=N, CH_2=C=CHOMe, 20°C, CF_3, N, CH_3O, O, 71%, $H^{\oplus}$ or NEt_3, CH_3, CF_3, N, CH_3O, O, 75%

[*263*] **58**

CF_3, N, CF_3, R^1, X, +, O, Cl, DMAP, 100 °C, toluene, 2-4 h, F_3C, CF_3, N, O, R^1, X, 37-69%, CH_2Cl, BF_3, 100 °C, toluene, 5-8 h, F_3C, CF_3, N, O, R^1, X, CH_2Cl

X = O, S, NR^2
R^2 = aryl, *t*-C_4H_9
DMAP = 4-dimethylaminopyridine

X = N-(2,6-dimethylphenyl); 50%

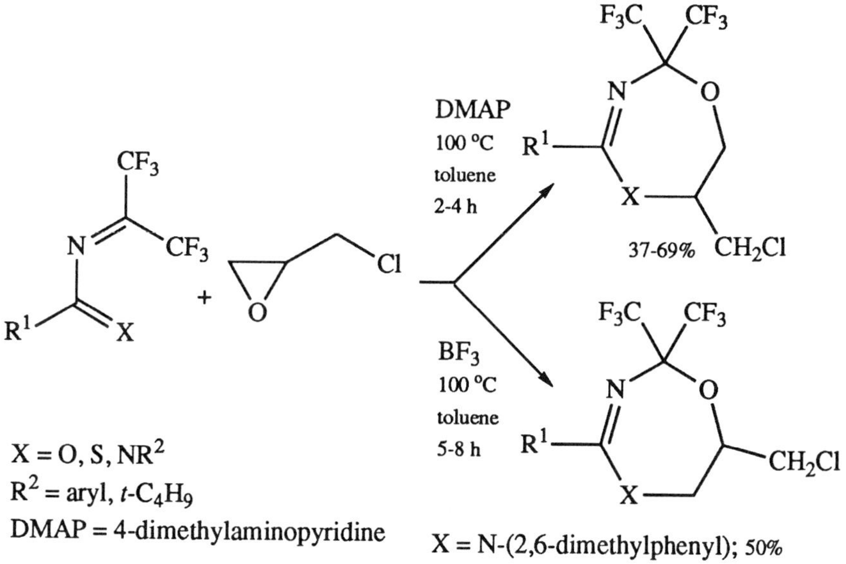

Miscellaneous

Polyfluorinated ketones readily undergo *ene reactions* [*114, 264, 265*]; some reactions can be catalyzed efficiently by Lewis acids [*266*] (equation 59).

Ketones and imines with adjacent perfluoroalkyl moieties show sharply reduced abilities to interact with electrophiles. Hexafluoroacetone is not protonated even in superacidic media.

[*114*] **59**

$F_3C(CH_3O_2C)C{=}O$ + cycloalkene $HC{=}CH$–CH_2–$(CH_2)_n$ $\xrightarrow[\text{9-16 h, sealed}]{160\text{-}200^\circ C}$ F_3C–C(OH)(CO_2CH_3)–CH(cycloalkenyl, $HC{=}CH$, $(CH_2)_n$)

n = 2,3 38-44%

Chlorination and bromination of perfluoroketimines take place only in the presence of alkali fluorides as catalyst [*267*]. Chlorine fluoride adds readily across the CO or CN bond of polyfluorinated ketones and imines [*268, 269*].

Generally, the addition of chlorine or bromine to trifluoroacetonitrile leads to a mixture of partially halogenated imines, amines, and azo alkanes [*270, 271*]. In special cases, such as the HgF_2-mediated addition of bromine, *N,N*-dihaloperfluoro-2-alkylamines can be obtained in good yields [*272*].

On reaction of methyl trifluoropyruvate with phosphorus pentachloride, 1,1,2,2-tetrachloro-3,3,3-trifluoro-1-methoxypropane is obtained [*273*]. Hexafluoroacetone hydrazone is oxidized to bis(trifluoromethyl)diazomethane upon treatment with phosphorus pentachloride [*274*].

Treatment of hexafluoroacetone with a P(III) species results either in formation of five-membered ring systems via reductive CC coupling of two molecules of hexafluoroacetone [*275, 276, 277, 278, 279, 280, 281*] (equation 60) or in reductive fluoride elimination [*282*] (equation 61).

[*275-281*] **60**

F_3C–C(=O)–CF_3 + $PR^1R^2R^3$ ⟶ 1,3,2-dioxaphospholane: $(F_3C)_2C$–$C(CF_3)_2$, O, O, $P(R^1)(R^2)(R^3)$

28-82%

$R^1 = CH_3, F, t\text{-}C_4H_9$

$R^2 = CH_3, F, t\text{-}C_4H_9$

$R^3 = OSi(CH_3)_3, F, Cl, N_3, OCH_3, OCH(CF_3)_2, N(CH_3)_2, N(C_2H_5)_2, N(allyl)_2$

[282] **61**

$F_3C-C(=O)-CF_3$ + $P(OR)_3$ → $F_2C{=}C(CF_3)-O-P(=O)(OR)_2$ $\xrightarrow[H_2SO_4]{conc.}$ $F_2C{=}C(CF_3)OH$

R = CH_3, $CH(CH_3)_2$

72-90%

Reference for Pages 840–877

1. Knunyants, I. L.; Gontar, A. F. In *Soviet Scientific Reviews, Chemistry Reviews;* Harwood Acad. Pub.: London, 1983, Vol. 5.
2. Witt, M.; Dhathathreyan, K. S.; Roesky, H. W. *Adv. Inorg. Chem. Radiochem.* **1986,** *30,* 223.
3. Gambaryan, N. P.; Rokhlin, E. M.; Zeifman, Y. V.; Ching-Yun, C.; Knunyants, I. L. *Angew. Chem.* **1966,** *78,* 1008; *Angew. Chem. Int. Ed. Engl.* **1966,** *5,* 947.
4. Saloutin, V. I.; Pashkevich, K. I.; Postovskii, I. Y. *Usp. Khim.* **1982,** *51,* 1287 (Engl. Transl. 736).
5. Fokin, A. V.; Kolomiets, A. F.; Vasil'ev, N. V. *Usp. Khim.* **1984,** *53,* 398 (Engl. Transl. 238).
6. McDonald, R. S.; Teo, K. C.; Stewart, R. *J. Chem. Soc., Perkin Trans. 2* **1983,** 297.
7. Chen, L. S.; Fratini, A. V.; Tamborski, C. *J. Fluorine Chem.* **1986,** *31,* 381.
8. Fomin, A. N.; Saloutin, V. I.; Pashkevich, K. I.; Bazhenov, D. V.; Grishin, Y. K.; Ustynyuk, Y. A. *Izv. Akad. Nauk SSSR* **1983,** 2626 (Engl. Transl. 2361).
9. Pashkevich, K. I.; Fomin, A. N.; Saloutin, V. I.; Bazhenov, D. V.; Grishin, Y. K. *Izv. Akad. Nauk SSSR* **1982,** 1359; (Engl. Transl. 1210).
10. Skryabina, Z. E.; Saloutin, V. I.; Pashkevich, K. I.; Bazhenov, D. V.; Grishin, Y. K.; Ustynyuk, Y. A. *Izv. Akad. Nauk SSSR* **1986,** 2046 (Engl. Transl. 1862).
11. Chkanikov, N. D.; Sviridov, V. D.; Zelenin, A. E.; Galakhov, M. V.; Kolomiets, A. F.; Fokin, A. V. *Izv. Akad. Nauk SSSR* **1990,** 383 (Engl. Transl. 323).
12. Sviridov, V. D.; Chkanikov, N. D.; Shapiro, A. B.; Klimova, N. V.; Pyatin, B. M.; Kolomiets, A. F.; Fokin, A. V. *Izv. Akad. Nauk SSSR* **1989,** 2348 (Engl. Transl. 2161).
13. Saloutin, V. I.; Piterskikh, I. A.; Pashkevich, K. I.; Kodess, M. I. *Izv. Akad. Nauk SSSR* **1983,** 2568 (Engl. Transl. 2312).

14. Saloutina, L. V.; Zapevalov, A. Y.; Kolenko, I. P. *Zh. Org. Khim.* **1986,** *22,* 2250 (Engl. Transl. 2019).
15. Saloutina, L.V.; Zapevalov, A.Y.; Kodess, M. I.; Kolenko, I. P.; German, L. S. *Izv. Akad. Nauk SSSR* **1982,** 2615 (Engl. Transl. 2310).
16. Osipov, S. N.; Sokolov, V. B.; Kolomiets, A. F.; Martynov, I. V.; Fokin, A. V. *Izv. Akad. Nauk SSSR* **1987,** 1185 (Engl. Transl. 1098).
17. Il'in, G. F.; Kolomiets, A. F.; Fokin, A. V. *Zh. Obshch. Khim.* **1981,** *51,* 2143 (Engl. Transl. 1845).
18. Kryukov, L. N.; Kryukova, L. Y.; Kolomiets, A. F. *Zh. Obshch. Khim.* **1982,** *52,* 2133 (Engl. Transl. 1899).
19. Janzen, A. F.; Pollitt, R. *Can. J. Chem.* **1970,** *48,* 1987.
20. Heine, J.; Röschenthaler, G. V. *Chem. Ztg.* **1989,** *113,* 186.
21. Burger, K.; Tremmel, S.; Schickaneder, H. *J. Fluorine Chem.* **1976,** *7,* 471.
22. Burger, K.; Thenn, W.; Fehn, J. *Chem. Ber.* **1974,** *107,* 1526.
23. Burger, K.; Zettl, C. *Chem. Ztg.* **1980,** *104,* 71.
24. Shi, G.; Xu, Y. *J. Chem. Soc., Chem. Commun.* **1989,** 607.
25. Steglich, W.; Burger, K.; Dürr, M.; Burgis, E. *Chem. Ber.* **1974,** 107, 1488.
26. Burger, K.; Ottlinger, R.; Albanbauer, J. *Chem. Ber.* **1977,** *110,* 2114.
27. Burger, K.; Penninger, S. *Synthesis* **1978,** 524.
28. Zeifman, Y. V.; Gambaryan, N. P.; Simonyan, L. A.; Minasyan, R. B.; Knunyants, I. L. *Zh. Obshch. Khim.* **1967,** *37,* 2476 (Engl. Transl. 2355).
29. Zeifman, Y. V.; Knunyants, I. L. *Dokl. Akad. Nauk SSSR* **1967,** *173,* 354 (Engl. Transl. 249).
30. Burger, K.; Albanbauer, J.; Käfig, F.; Penninger, S. *Liebigs Ann. Chem.* **1977,** 624.
31. Kryukov, L. N.; Vitovskii, V. S.; Kryukova, L. Y.; Isaev, V. L.; Sterlin, R. N.; Knunyants, I. L. *Zh. Vses. Khim. O-va. im. D. I. Mendeleeva* **1976,** *21,* 63; *Chem. Abstr.* **1977,** *87,* 84447h.
32. Burger, K.; Hoess, E.; Gaa, K.; Sewald, N.; Schierlinger, C. *Z. Naturforsch. B: Chem. Sci.* **1991,** *46,* 361.
33. Weygand, F.; Steglich, W.; Lengyel, I.; Fraunberger, F.; Maierhofer, A.; Oettmeier, W. *Chem. Ber.* **1966,** *99,* 1944.
34. Sotnikov, N. V.; Sokol'skii, G. A.; Knunyants, I. L. *Izv. Akad. Nauk SSSR* **1977,** 2168 (Engl. Transl. 2009).
35. Middleton, W. J.; Krespan, C. G. *J. Org. Chem.* **1965,** *30,* 1398.
36. Malassa, I.; Matthies, D. *Chem. Ztg.* **1987,** *111,* 181 and 253.
37. Sviridov, V. D.; Chkanikov, N. D.; Galakhov, M. V.; Kolomiets, A. F.; Fokin, A. V. *Izv. Akad. Nauk SSSR* **1989,** 1652 (Engl. Transl. 1515).
38. Burger, K.; Wamuth, U.; Forster, B.; Penninger, S. *Z. Naturforsch. B: Chem. Sci.* **1984,** *39,* 1442.
39. Ishihara, T.; Shinjo, H.; Inone, Y.; Ando, T. *J. Fluorine Chem.* **1983,** *22,* 1.

40. Kryukova, L. Y.; Kryukov, L. N.; Sokol'skii, G. A. *Zh. Org. Khim.* **1987,** *23,* 230; (Engl. Transl. 205).
41. Burger, K.; Huber, E.; Kahl, T.; Partscht, H.; Ganzer, M. *Synthesis* **1988,** 44.
42. Burger, K.; Sewald, N.; Huber, E.; Ottlinger, R. *Z. Naturforsch. B: Chem. Sci.* **1989,** *44,* 1298.
43. Nikishin, G. I.; Glukhovtsev, V. G.; Karakhanova, N. K.; Il'in, Y. V. *Izv. Akad. Nauk SSSR* **1981,** 479; *Chem. Abstr.* **1981,** *95,* 80599b.
44. Il'in, G. F.; Kolomiets, A. F.; Fokin, A. V.; Sokol'skii, G. A. *Zh. Org. Khim.* **1980,** *16,* 1096; *Chem. Abstr.* **1980,** *93,* 167813z.
45. Boere, R. T.; Willis, C. J. *Can. J. Chem.* **1985,** *63,* 3530.
46. Sun, K. K.; Tamborski, C.; Eapen, K. C. *J. Fluorine Chem.* **1981,** *17,* 457.
47. Moreau, P.; Naji, N.; Commeyras, A. *J. Fluorine Chem.* **1985,** *30,* 315.
48. Seyferth, D.; Murphy, G. J.; Woodruff, R. A. *J. Am. Chem. Soc.* **1974,** *96,* 5011.
49. Zeifman, Y. V. *Izv. Akad. Nauk SSSR* **1990,** 202 (Engl. Transl. 186).
50. Soloshonok, V. A.; Gerus, I. I.; Yagupol'skii, Y. L.; Kukhar, V. P. *Zh. Org. Khim.* **1987,** *23,* 2308 (Engl. Transl. 2034).
51. Golubev, A. S.; Kolomiets, A. F.; Fokin, A. V. *Izv. Akad. Nauk SSSR* **1989,** 2369 (Engl. Transl. 2180).
52. Dyachenko, V. I.; Kolomiets, A. F.; Fokin, A. V. *Izv. Akad. Nauk SSSR* **1987,** 2511 (Engl. Transl. 2332).
53. Fokin, A. V. *Actual. Chim.* **1987,** 163; Chkanikov, N. D. *J. Fluorine Chem.* **1991,** *54,* 119.
54. Zelenin, A. E.; Chkanikov, N. D.; Kolomiets, A. F.; Fokin, A. V. *Izv. Akad. Nauk SSSR* **1986,** 2081 (Engl. Transl. 1895).
55. Zelenin, A. E.; Chkanikov, N. D.; Ivanchenko, Y.; Tkachev, V. D.; Rusakova, V. A.; Kolomiets, A. F.; Fokin, A. V. *Khim. Geterotsikl. Soedin.* **1987,** 1200 (Engl. Transl. 959).
56. Golubev, A. S.; Galachov, M. V.; Kolomiets, A. F.; Fokin, A. V. *Izv. Akad. Nauk SSSR* **1989,** 2127 (Engl. Transl. 1959).
57. Soloshonok, V. A.; Yagupol'skii, Y. L.; Kukhar, V. P. *Zh. Org. Khim.* **1989,** *25,* 2523 (Engl. Transl. 2263).
58. Soloshonok, V. A.; Yagupol'skii, Y. L.; Kukhar, V. P. *Zh. Org. Khim.* **1988,** *24,* 1638 (Engl. Transl. 1478).
59. Sewald, N.; Burger, K. *Z. Naturforsch. B: Chem. Sci.* **1990,** *45,* 871.
60. Osipov, S. N.; Chkanikov, N. D.; Shklyaev, Y. V.; Kolomiets, A. F.; Fokin, A. V. *Izv. Akad. Nauk SSSR* **1989,** 2131 (Engl. Transl. 1962).
61. Osipov, S. N.; Chkanikov, N. D.; Kolomiets, A. F.; Fokin, A. V. *Izv. Akad. Nauk SSSR* **1989,** 213 (Engl. Transl. 201).
62. Burger, K.; Sewald, N. *Synthesis* **1990,** 871.
63. Hudlický, M. *Chemistry of Organic Fluorine Compounds;* Ellis Horwood: Chichester, U.K., 1976.

64. Barnette, W. E. *J. Fluorine Chem.* **1984,** *26,* 161.
65. Aizikovich, A. Y.; Pashkevich, K. I.; Gorshkov, V. V.; Rudaya, M. N.; Postovskii, I. Y. *Zh. Obsh. Khim.* **1980,** *50,* 1866 (Engl. Transl. 1523).
66. Lee, L. F.; Sing, Y. L. *J. Org. Chem.* **1990,** *55,* 380.
67. Aizikovich, A. Y.; Sheptun, Y. N. *Zh. Org. Khim.* **1985,** *21,* 2212 (Engl. Transl. 2025).
68. Burger, K.; Wamuth, U.; Hein, F.; Rottegger, S. *Liebigs Ann. Chem.* **1984,** 991.
69. Fokin, A. V.; Studnev, Y. N.; Rapkin, A. J.; Pasevina, K. J. *Izv. Akad. Nauk SSSR* **1980,** 2623; *Chem. Abstr.* **1981,** *94,* 191645n.
70. Fokin, A. V.; Studnev, Y. N.; Rapkin, A. J.; Krotovich, I. N.; Verenikin, O. V. *Izv. Akad. Nauk SSSR* **1981,** 2370 (Engl. Transl. 1953).
71. Sarwar, G.; Kirchmeier, R. L.; Shreeve, J. M. *Inorg. Chem.* **1989,** *28,* 3345.
72. Höfs, H.-U.; Mews, R.; Noltemeyer, M.; Sheldrick, G. M.; Schmidt, M.; Henkel, G.; Krebs, B. *Z. Naturforsch. B: Chem. Sci.* **1983,** *38,* 454.
73. Lentz, D.; Kroll, J. *Chem. Ber.* **1987,** *120,* 303.
74. Lentz, D.; Oberhammer, H. *Inorg. Chem.* **1985,** *24,* 4665.
75. Lentz, D.; Brüdgam, J.; Hartl, H. *Angew. Chem.* **1987,** *99,* 951; *Angew. Chem. Int. Ed. Engl.* **1987,** *26,* 921.
76. Lutz, W.; Sundermeyer, W. *Chem. Ber.* **1979,** *112,* 2158.
77. Til'kunova, N. A.; Gontar', A. F.; Sizov, Y. A.; Bykhovskaya, E. G.; Knunyants, I. L. *Izv. Akad. Nauk SSSR* **1977,** 2381 (Engl. Transl. 2214).
78. Knunyants. I. L.; Delyagina, N. I.; Igumnov, S. M. *Izv. Akad. Nauk SSSR* **1981,** 860 (Engl. Transl. 639).
79. Gilchrist, T. L. *Heterocyclic Chemistry;* Pitman: London, 1985; p 55.
80. Baldwin, J. *J. Chem. Soc., Chem. Commun.* **1976,** 734.
81. Francke, R.; Dakternieks, D.; Gable, R. W.; Hoskins, B. F.; Roeschenthaler, G. V. *Chem. Ber.* **1985,** *118,* 922.
82. Gruetzmacher, H.; Roesky, H. W. *Chem. Ber.* **1986,** *119,* 2127.
83. Simmons, H. E.; Wiley, D. W. *J. Am. Chem. Soc.* **1960,** *82,* 2288.
84. Weygand, F.; Burger, K.; Engelhardt, K. *Chem. Ber.* **1966,** *99,* 1461.
85. Weygand, F.; Burger, K. *Chem. Ber.* **1966,** *99,* 2880.
86. Gold, M.; Burger, K. *J. Fluorine Chem.* **1987,** *35,* 87.
87. Burger, K.; Rudolph, M. *Chem. Ztg.* **1990,** *114,* 249.
88. Burger, K.; Gold, M.; Neuhauser, H.; Rudolph, M. *Chem. Ztg.* **1991,** *115,* 77.
89. Burger, K.; Neuhauser, H.; Rudolph, M. *Chem. Ztg.* **1990,** *114,* 251.
90. Burger, K.; Eggersdorfer, M. *Liebigs Ann. Chem.* **1979,** 1547.
91. Resnick, P. R. *Polym. Prepr.,* **1990,** *31,* 312.
92. Hung, M. H. *J. Fluorine Chem.* **1991,** *52,* 159.
93. Davydov, A. V. *Zh. Obshch. Khim.* **1981,** *51,* 2362 (Engl. Transl. 2037).

94. Igumnov, S. M.; Sotnikov, N. V.; Sokol'skii, G. A.; Knunyants, I. L. *Zh. Vses. Khim. O-va. im. D. I. Mendeleeva* **1981,** *26,* 97; *Chem. Abstr.* **1981,** *94,* 192260v.
95. Sing, Y. L.; Lee, L. F. *J. Org. Chem.* **1985,** *50,* 4642.
96. Burger, K.; Huber, E.; Sewald, N.; Partscht, H. *Chem. Ztg.* **1986,** *110,* 83.
97. Burger, K.; Partscht, H.; Huber, E. *Chem. Ztg.* **1985,** *109,* 185.
98. Burger, K.; Ottlinger, R. *J. Fluorine Chem.* **1978,** *11,* 29.
99. Gambaryan, N. P.; Zeifman, Y. V. *Izv. Akad. Nauk SSSR* **1969,** 2059 (Engl. Transl. 1915).
100. Kostyanovsky, R. G.; Shustov, G. V.; Zaichenko, N. L. *Tetrahedron* **1982,** *38,* 949.
101. Shustov, G. V.; Tavakalyan, N. B.; Pleshkova, A. P.; Kostyanovsky, R. G. *Khim. Geterotsikl. Soedin.* **1981,** 810 (Engl. Transl. 600).
102. Burger, K.; Kahl, T. *Chem. Ztg.* **1988,** *112,* 109.
103. Hergenrother, P. M.; Hudlický, M. *J. Fluorine Chem.* **1978,** *12,* 439.
104. Chkanikov, N. D.; Vershinin, V. L.; Galakhov, M. V.; Kolomiets, A. F.; Fokin, A. V. *Izv. Akad. Nauk SSSR* **1989,** 126 (Engl. Transl. 113).
105. Dipple, A.; Heidelberger, C. *J. Med. Chem.* **1966,** *9,* 715.
106. Katagiri, N.; Watanabe, H.; Kaneko, C. *Chem. Pharm. Bull.* **1988,** *36,* 3354.
107. Mustafa, M. E.; Takaoka, A.; Ishikawa, N. *Bull. Soc. Chim. Fr.* **1986,** 944.
108. Seyferth, D.; Tronich, W.; Smith, W. E.; Hopper, S. P. *J. Organomet. Chem.* **1974,** *67,* 341.
109. Niecke, E.; Gudat, D.; Schoeller, W. W.; Rademacher, P. *J. Chem. Soc., Chem. Commun.* **1985,** 1050.
110. Burger, K.; Schierlinger, C.; Neuhauser, H. *Chem. Ztg.* **1989,** *113,* 247.
111. Shimizu, N.; Bartlett, P. D. *J. Am. Chem. Soc.* **1979,** *100,* 4260.
112. Griffith, J. R.; O'Rear, J. G. *Ind. Eng. Chem., Prod. Res. Dev.* **1974,** *13,* 148.
113. Coe, P. L.; Holton, A. G. *J. Fluorine Chem.* **1977,** *10,* 553.
114. Golubev, A. S.; Kolomiets, A. F.; Fokin A. V. *Izv. Akad. Nauk SSSR* **1988,** 127 (Engl. Transl. 117).
115. Osipov, S. N.; Kolomiets, A. F.; Fokin, A. V. *Izv. Akad. Nauk SSSR* **1988,** 132 (Engl. Transl. 122).
116. Shozda, R. J. *J. Org. Chem.* **1967,** *32,* 2960.
117. Zubovics, Z.; Ishikawa, N. *J. Fluorine Chem.* **1976,** *8,* 43.
118. Zaitseva, G. S. *Zh. Obshch. Khim.* **1983,** *53,* 2068 (Engl. Transl. 1867).
119. Albert, R. M.; Butler G. B. *J. Org. Chem.* **1977,** *42,* 674.
120. Barlow, M. G.; Coles, B.; Haszeldine, R. N. *J. Fluorine Chem.* **1980,** *15,* 381.
121. Sheldrick, W. S.; Schomburg, D.; Schmidpeter, A.; Criegern, V. T. *Chem. Ber.* **1980,** *113,* 55.
122. Kolodyazhnyi, O. I. *Zh. Obshch. Khim.* **1984,** *54,* 966 (Engl. Transl. 861).

123. Golubev, A. S.; Kolomiets, A. F.; Fokin, A. V. *Izv. Akad. Nauk SSSR* **1990,** 2461; *Chem. Abstr.* **1991,** *114,* 101095n.
124. Hoffmann, R. W.; Steinbach, K.; Lilienblum W. *Chem. Ber.* **1976,** *109,* 1759.
125. Burger, K.; Ottlinger, R. *Chem. Ztg.* **1977,** *101,* 402.
126. Kryukov, L. N.; Kryukova, L. Y.; Kolomiets, A. F. *Zh. Org. Khim.* **1981,** *17,* 2629 (Engl. Transl. 2347).
127. Kryukov, L. N.; Kryukova, L. Y.; Kolomiets, A. F. *Zh. Org. Khim.* **1982,** *18,* 1873 (Engl. Transl. 1638).
128. Burger, K.; Wassmuth, U.; Penninger, S. *J. Fluorine Chem.* **1982,** *20,* 813.
129. Gambaryan, N. P.; Kaitmazova, G. S.; Kagramanova, E. M.; Simonyan, L.A.; Safronova, Z. V. *Izv. Akad. Nauk SSSR* **1984,** 1102 (Engl. Transl. 1012).
130. Safronova, Z. V.; Simonyan, L. A.; Zeifman, Y. V.; Gambaryan, N. P. *Izv. Akad. Nauk SSSR* **1979,** 1826 (Engl. Transl. 1688).
131. Fischer, E. O.; Weiss, K.; Burger, K. *Chem. Ber.* **1973,** *106,* 1581.
132. Burger, K.; Geith, K.; Hübl, D. *Synthesis* **1988,** *189,* 194, 199.
133. Burger, K.; Geith, K.; Sewald, N. *J. Fluorine Chem.* **1990,** *46,* 105.
134. Burger, K.; Kahl, T. *J. Fluorine Chem.* **1987,** *36,* 329.
135. Burger, K.; Kahl, T. *J. Fluorine Chem.* **1987,** *37,* 53.
136. Burger, K.; Hübl, D.; Huber, E. *Chem. Ztg.* **1986,** *110,* 87.
137. Burger, K.; Ottlinger, R.; Goth, H.; Firl, J. *Chem. Ber.* **1980,** *113,* 2699.
138. Burger, K.; Geith, K.; Gaa, K. *Angew. Chem.* **1988,** *100,* 860; *Angew. Chem. Int. Ed. Engl.* **1988,** *27,* 848.
139. Burger, K.; Gaa, K.; Geith, K.; Schierlinger, C. *Synthesis* **1989,** 850.
140. Welch, J. T. *Tetrahedron* **1987,** *43,* 3123.
141. Kryukov, L. N.; Kryukova, L. Y.; Isaev, V. L.; Sterlin, R. N.; Knunyants, I. L. *Zh. Vses. Khim. O-va. im. D.I. Mendeleeva* **1977,** *22,* 228; *Chem. Abstr.* **1977,** *87,* 38820b
142. Burger, K.; Penninger, S.; Tremmel, S. *Z. Naturforsch. B: Chem. Sci.* **1980,** *35,* 749.
143. Burger, K. In *Organophosphorus Reagents in Organic Synthesis;* Cadogan, J. I. G., Ed.; Academic: London, 1979; p 492 and literature cited therein.
144. Burger, K.; Wassmuth, U.; Huber, E.; Neugebauer, D.; Riede, J.; Ackermann, K. *Chem. Ztg.* **1983,** *107,* 271.
145. Glemser, O.; Shreeve, J. M. *Inorg. Chem.* **1979,** *18,* 2319.
146. Zeifman, Y. V.; Gambaryan, N. P.; Knunyants, I. L. *Izv. Akad. Nauk SSSR* **1965,** 1472 (Engl. Transl. 1431).
147. Burger, K.; Fehn, J.; Gieren, A. *Liebigs Ann. Chem.* **1972,** *757,* 9.
148. Logothetis, A. L. *J. Org. Chem.* **1964,** *29,* 3049.
149. Dyatkin, B. L.; Makarov, K. N.; Knunyants, I. L. *Tetrahedron* **1971,** *27,* 51.

150. Knunyants, I. L.; Zeifman, Y. V. *Izv. Akad. Nauk. SSSR* **1967,** 711 (Engl. Transl. 695).
151. Vasil'ev, N. V.; Kolomiets, A. F.; Sokol'skii, G. A. *Zh. Vses. Khim. O-va. im. D. I. Mendeleeva* **1980,** *25,* 703; *Chem. Abstr.* **1981,** *94,* 174991e.
152. Kryukov, L. N.; Kryukova, L. Y.; Kolomiets, A. F. *Zh. Vses. Khim. O-va. im. D.I. Mendeleeva* **1986,** *31,* 112; *Chem. Abstr.* **1987,** *106,* 32905x.
153. Fokin, A. V.; Kolomiets, A. F.; Il'in, G. F.; Fedyushina, T. L. *Izv. Akad. Nauk SSSR* **1982,** 1872 (Engl. Transl. 1663).
154. Il'in, G. F.; Kolomiets, A. F.; Sokol'skii, G. A. *Zh. Org. Khim.* **1979,** *15,* 2216 (Engl. Transl. 2008).
155. Kobayashi, Y.; Kumadaki, I.; Kobayashi, E. *Heterocycles* **1981,** *15,* 1223.
156. Banks, R. E.; Thomson, J. *J. Fluorine Chem.* **1983,** *22,* 589.
157. Banks, R. E.; Mohialdin, S. N. *J. Fluorine Chem.* **1986,** *34,* 275.
158. Banks, R. E.; Hitchen, S. M. *J. Fluorine Chem.* **1980,** *15,* 179.
159. Crossman, J. M.; Haszeldine, R. N.; Tipping, A. E. *J. Chem. Soc., Dalton Trans.* **1973,** 483.
160. Lazukina, L. A.; Kukhar', V. P. *Zh. Org. Khim.* **1979,** *15,* 2216 (Engl. Transl. 2009).
161. Mel'nikov, A. A.; Sokolova, M. M.; Pervozvanskaya, M. A.; Mel'nikov, V. V. *Zh. Org. Khim.* **1979,** *15,* 1861 (Engl. Transl. 1677).
162. Carpenter, W. R. *J. Org. Chem.* **1962,** *27,* 2085.
163. Stegmann, W.; Gilgen, P.; Heimgartner, H.; Schmid, H. *Helv. Chim. Acta* **1976,** *59,* 1018.
164. Steglich, W.; Gruber, P.; Heininger, H. U.; Kneidl, F. *Chem. Ber.* **1971,** *104,* 3816.
165. Gruber, P.; Müller, L.; Steglich, W. *Chem. Ber.* **1973,** *106,* 2863.
166. Burger, K.; Fehn, J. *Chem. Ber.* **1972,** *105,* 3814.
167. Burger, K.; Albanbauer, J.; Manz, F. *Chem. Ber.* **1974,** *107,* 1823.
168. Hansen, H. J.; Heimgartner, H. In *1,3-Dipolar Cycloaddition Chemistry;* Padwa, A., Ed.; Wiley: New York, 1984; p 177 and literature cited therein.
169. Burger, K.; Fehn, J.; Müller, E. *Chem. Ber.* **1973,** *106,* 1.
170. Burger, K.; Goth, H.; Einhellig, K.; Gieren, A. *Z. Naturforsch. B: Chem. Sci.* **1981,** *36,* 345.
171. Wentrup, C.; Fischer, S.; Berstermann, H. M.; Kuzaj, M.; Lüerssen, H.; Burger, K. *Angew. Chem.* **1986,** *98,* 99; *Angew. Chem. Int. Ed. Engl.* **1986,** *25,* 85.
172. Tanaka, K.; Maeno, S.; Mitsuhashi, K. *J. Heterocycl. Chem.* **1985,** *22,* 565.
173. Middleton, W. J. *J. Org. Chem.* **1984,** *49,* 919.
174. Tanaka, K.; Masuda, H.; Mitsuhashi, K. *Bull. Chem. Soc. Jpn.* **1986,** *59,* 3901.
175. Truskanova, T. D.; Vasil'ev, N. V.; Gontar, A. F.; Kolomiets, A. F.; Sokol'skii, G. A. *Khim. Geterotsikl. Soedin.* **1989,** 972 (Engl. Transl. 815).

176. Regitz, M.; Heydt, H. In *1,3-Dipolar Cycloaddition Chemistry;* Padwa, A., Ed.; Wiley: New York, 1984; p 393 and literature cited therein.
177. Atherton, J. H.; Field, R. *J. Chem. Soc. C* **1968,** 1507.
178. Niecke, E.; Flick, W. *J. Organomet. Chem.* **1976,** *104,* C 23.
179. Laganis, E. D.; Lemal, D. M. *J. Am. Chem. Soc.* **1980,** *102,* 6633.
180. Shi, G.; Xu, Y. *J. Org. Chem.* **1990,** *55,* 3383.
181. Burger, K.; Thenn, W.; Gieren, A. *Angew. Chem.* **1974,** *86,* 481; *Angew. Chem. Int. Ed. Engl.* **1974,** *13,* 474.
182. Burger, K.; Schickaneder, H.; Thenn, W. *Tetrahedron Lett.* **1975,** 1125.
183. Grashey, R. In *1,3-Dipolar Cycloaddition Chemistry;* Padwa, A.; Ed.; Wiley: New York, 1984; p 733 and literature cited therein.
184. Banks, R. E.; Haszeldine, R. N.; Jackson, P. E. *J. Fluorine Chem.* **1978,** *12,* 153.
185. Tanaka, K.; Sugimoto, Y.; Okafuji, Y.; Tachikawa, M.; Mitsuhashi, K. *J. Heterocycl. Chem.* **1989,** *26,* 381.
186. Tanaka, K.; Ohsuga, M.; Sugimoto, Y.; Okafuji, Y.; Mitsuhashi, K. *J. Fluorine Chem.* **1988,** *39,* 39.
187. Bell, D.; Tipping, A. E. *J. Fluorine Chem.* **1978,** *11,* 567.
188. Burger, K.; Dengler, O.; Gieren, A.; Lamm, V. *Chem. Ztg.* **1982,** *106,* 408.
189. Kaupp, G.; Dengler, O.; Burger, K.; Rottegger, S. *Angew. Chem.* **1985,** *97,* 329; *Angew. Chem. Int. Ed. Engl.* **1985,** *24,* 341.
190. Bailey, J. R.; Moore, N. H. *J. Am. Chem. Soc.* **1917,** *39,* 279.
191. Burger, K.; Fehn, J.; Thenn, W. *Angew. Chem.* **1973,** *85,* 541; *Angew. Chem. Int. Ed. Engl.* **1973,** *12,* 502.
192. Vershinin, V. L. Vasil'ev, N. V.; Kolomiets, A. F.; Sokol'skii, G. A. *Zh. Org. Khim.* **1984,** *20,* 1806 (Engl. Transl. 1646).
193. Wagner-Jauregg, T. *Synthesis* **1976,** 349.
194. Forshaw, T. P.; Tipping, A. E. *J. Chem. Soc. C* **1971,** 2404.
195. Gieren, A.; Narayanan, P.; Burger, K.; Thenn, W. *Angew. Chem.* **1974,** *86,* 482; *Angew. Chem. Int. Ed. Engl.* **1974,** *13,* 475.
196. Armstrong, S. E.; Tipping, A. E. *J. Chem. Soc., Perkin Trans. 1* **1975,** 538
197. Armstrong, S. E.; Tipping, A. E. *J. Chem. Soc., Perkin Trans. 1* **1975,** 1902.
198. Burger, K.; Thenn, W.; Rauh, R.; Schickaneder, H. *Chem. Ber.* **1975,** *108,* 1460.
199. Burger, K.; Dengler, O.; Hübl, D. *J. Fluorine Chem.* **1982,** *19,* 589.
200. Armstrong, S. E.; Tipping, A. E. *J. Chem. Soc., Perkin Trans. 1* **1975,** 1411.
201. Burger, K.; Hein, F.; Zettl, C.; Schickaneder, H. *Chem. Ber.* **1979,** *112,* 2609.
202. Burger, K.; Schickaneder, H.; Hein, F.; Elguero, J. *Tetrahedron* **1979,** *35,* 389.
203. Burger, K.; Hein, F. *Liebigs Ann. Chem.* **1982,** 853.

204. Burger, K.; Hein, F. *Liebigs Ann. Chem.* **1979,** 133.
205. Burger, K.; Schickaneder, H.; Hein, F.; Gieren, A.; Lamm, V.; Engelhardt, H. *Liebigs Ann. Chem.* **1982,** 845.
206. Burger, K.; Hein, F.; Dengler, O. *J. Fluorine Chem.* **1982,** *19,* 437.
207. Hein, F.; Burger, K.; Firl, J. *J. Chem. Soc., Chem. Commun.* **1979,** 792.
208. Burger, K.; Schickaneder, H.; Pinzel, M. *Liebigs Ann. Chem.* **1976,** 30.
209. Burger, K.; Schickaneder, H.; Zettl, C. *Liebigs Ann. Chem.* **1982,** 1749.
210. Burger, K.; Schickaneder, H.; Zettl, C. *Synthesis* **1976,** 803.
211. Burger, K.; Schickaneder, H.; Zettl, C. *Angew. Chem.* **1977,** *89,* 60; *Angew. Chem. Int. Ed. Engl.* **1977,** *16,* 54.
212. Burger, K.; Schickaneder, H.; Zettl, C. *Angew. Chem.* **1977,** *89,* 61; *Angew. Chem. Int. Ed. Engl.* **1977,** *16,* 55.
213. Nuyken, O.; Maier, G.; Burger, K. *Makromol. Chem.* **1988,** *189,* 2245.
214. Nuyken, O.; Maier, G.; Burger, K.; i Albet, A. S. *Makromol. Chem.* **1989,** *190,* 1953.
215. Nuyken, O.; Maier, G.; Burger, K. *Makromol. Chem.* **1990,** *191,* 2455.
216. Gilchrist, T. L.; Storr, R. C. *Organic Reactions and Orbital Symmetry;* Cambridge University Press: Cambridge, U.K., 1979; p 138.
217. Barlow, M. G.; Haszeldine, R. N.; Pickett, J. A. *J. Chem. Soc., Perkin Trans. 1* **1978,** 378.
218. Carboni, R. A.; Lindsey, R. V. *J. Am. Chem. Soc.* **1959,** *81,* 4342.
219. Seitz, G.; Mohr, R. *Arch. Pharm.* **1986,** *319,* 690.
220. Boger, D. L.; Weinreb, S. N. *Hetero Diels–Alder Methodology in Organic Synthesis;* Academic Press: San Diego, 1987; p 94 and literature cited therein.
221. Jurczak, I.; Chmielewski, M.; Filipek, S. *Synthesis* **1979,** 41.
222. England, D. C. *J. Am. Chem. Soc.* **1961,** *83,* 2205.
223. Linn, W. I. *J. Org. Chem.* **1964,** *29,* 3111.
224. Taylor, D. R.; Wright, D. B. *J. Chem. Soc., Perkin Trans. 1* **1973,** 956.
225. Burger, K.; Huber, E.; Schöntag, W.; Partscht, H. *Chem. Ztg.* **1986,** *110,* 79.
226. Simonyan, L. A.; Gambaryan, N. P.; Petrovskii, P. V.; Knunyants, I. L. *Izv. Akad. Nauk SSSR* **1968,** 370 (Engl. Transl. 357).
227. Espenbetov, A. I.; Yanovskii, A. I.; Struchkov, Y. I.; Simonyan, L. A.; Gambaryan, N. P. *Izv. Akad. Nauk SSSR* **1982,** 607 (Engl. Transl. 536).
228. Simonyan, L. A.; Zeifman, Y. V.; Gambaryan, N. P. *Izv. Akad. Nauk SSSR* **1968,** 1916 (Engl. Transl. 1830).
229. Il'in, G. F.; Kolomiets, A. F. *Zh. Vses. Khim. O-va. im. D.I. Mendeleeva* **1980,** *25,* 705; *Chem. Abstr.* **1981,** *94,* 139594f.
230. Roesky, H. W.; Giere, H. H. *Z. Anorg. Allg. Chem.* **1970,** *378,* 177.
231. Janz, G. J. In *1,4-Cycloaddition Reactions;* Hamer, J.; Ed.; Academic Press: New York, 1967; p 98 and literature cited therein.

232. Feast, W. J.; Hughes, R. R.; Musgrave, W. K. R. *J. Fluorine Chem.* **1977,** *9,* 271.
233. Hees, U.; Ledermann, M.; Regitz, M. *Synlett* **1990,** 401.
234. Kresze, G.; Albrecht, R. *Chem. Ber.* **1960,** *97,* 490.
235. Kresze, G.; Wagner, U. *Liebigs Ann. Chem.* **1972,** *762,* 106.
236. Krow, G. R.; Rodebaugh, R.; Markowski, J.; Ramey, K. C. *Tetrahedron Lett.* **1973,** 1899.
237. Krow, G. R.; Pynn, G. R.; Rodebaugh, R.; Markowski, J. *Tetrahedron* **1974,** *30,* 2977.
238. Sauer, J.; Sustmann, R. *Angew. Chem.* **1980,** *92,* 773; *Angew. Chem. Int. Ed. Engl.* **1980,** *19,* 779.
239. Michels, G.; Hermesdorf, M.; Schneider, J.; Regitz, M. *Chem. Ber.* **1988,** *121,* 1775.
240. Sinitsa, A. D.; Drach, B. S.; Kisilenko, A. A. *Zh. Org. Khim.* **1973,** *9,* 685 (Engl. Transl. 706).
241. Kryukov, L. N.; Kryukova, L. Y.; Kurykin, M. A.; Sterlin, R. N.; Knunyants, I. L. *Zh. Vses. Khim. O-va. im. D.I. Mendeleeva* **1979,** *24,* 463; *Chem. Abstr.* **1979,** *91,* 193269e.
242. Kryukov, L. N.; Kryukova, L. Y.; Kolomiets, A. F. *Zh. Org. Khim.* **1982,** *18,* 1837 (Engl. Transl. 1638).
243. Burger, K.; Schöntag, W.; Wassmuth, U. *Z. Naturforsch. B: Chem. Sci.* **1982,** *37,* 1669.
244. Burger, K.; Kahl, T. *J. Fluorine Chem.* **1989,** *42,* 51.
245. Burger, K.; Wassmuth, U.; Partscht, H.; Gieren, A.; Hübner, T.; Kaerlein, T. *Chem. Ztg.* **1984,** *108,* 205.
246. Burger, K.; Schöntag, W.; Wassmuth, U. *J. Fluorine Chem.* **1983,** *22,* 99.
247. Burger, K.; Goth. H. *Angew. Chem.* **1980,** *92,* 836; *Angew. Chem. Int. Ed. Engl.* **1980,** *19,* 810.
248. Burger, K.; Goth, H.; Schöntag, W.; Firl, J. *Tetrahedron* **1982,** *38,* 287.
249. Burger, K.; Simmerl, R. *Liebigs Ann. Chem.* **1984,** 982.
250. Aksinenko, A. Y.; Sokolov, V. B.; Korenchenko, O. V.; Chekhlov, A. N.; Fokin, E. A.; Martynov, I. V. *Izv. Akad. Nauk SSSR* **1989,** 2815 (Engl. Transl. 2580).
251. Kryukov, L. N.; Kryukova, L. Y.; Kolomiets, A. F.; Sokol'skii, G. A. *Zh. Org. Khim.* 1980, *16,* 463; *Chem. Abstr.* **1980,** *92,* 198368s.
252. Zeifman, Y. V.; Gambaryan, N. P.; Minasyan, R. B. *Izv. Akad. Nauk SSSR* **1965,** 1910 (Engl. Transl. 1881).
253. Goerdeler, J.; Schenk, H. *Chem. Ber.* **1965,** *98,* 2954.
254. Jäger, G. *Chem. Ber.* **1972,** *105,* 137.
255. Gillard, M.; T'Kint, C.; Sonveaux, E.; Ghosez, L. *J. Am. Chem. Soc.* **1979,** *101,* 5837.
256. Larson, E. R.; Danishefsky, S. *Tetrahedron Lett.* **1982,** 1975.
257. Burger, K.; Partscht, H. *Chem. Ztg.* **1982,** *106,* 303.

258. Fleming, I. *Grenzorbitale und Reaktionen organischer Verbindungen;* VCH Publishers: Weinheim, 1988.
259. Burger, K.; Partscht, H.; Wassmuth, U.; Gieren, A.; Betz, H.; Weber, G.; Hübner, T. *Chem. Ztg.* **1984,** *108,* 213.
260. Seitz, G.; Hoferichter, R.; Mohr, R. *Angew. Chem.* **1987,** *99,* 345; *Angew. Chem. Int. Ed. Engl.* **1987,** *26,* 332.
261. Seitz, G.; Mohr, R. *Chem. Ztg.* **1987,** *111,* 81.
262. Reissig, H. U.; Zimmer, R. *Angew. Chem.* **1988,** *100,* 1576; *Angew. Chem. Int. Ed. Engl.* **1988,** *27,* 1518.
263. Burger, K.; Maier, G.; Kahl, T. *Chem. Ztg.* **1988,** *112,* 111.
264. Kobayashi, Y.; Nagai, T.; Kumadaki, I. *Chem. Pharm. Bull.* **1984,** *32,* 5031.
265. Nagai, T.; Miki, T.; Kumadaki, I. *Chem. Pharm. Bull.* **1986,** *34,* 4782.
266. Nagai, T.; Kumadaki, I.; Miki, T.; Kobayashi, Y.; Tomizawa, G. *Chem. Pharm. Bull.* **1986,** *34,* 1546.
267. Ruff, J. K. *J. Org. Chem.* **1967,** *32,* 1675.
268. Fokin, A. V.; Uzun, A. T.; Stolyarov, V. P. *Usp. Khim.* **1977,** *46,* 1995 (Engl. Transl. 1057).
269. Peterman, K. E.; Shreeve, J. M. *Inorg. Chem.* **1974,** *13,* 2705.
270. Chang, S. C.; DesMarteau, D. D. *Inorg. Chem.* **1983,** *22,* 805.
271. Geisel, M.; Waterfeld, A.; Mews, R. *Chem. Ber.* **1985,** *118,* 4459.
272. Waterfeld, A.; Mews, R. *J. Chem. Soc., Chem. Commun.* **1982,** 839.
273. Pashkevich, K. I.; Saloutin, V. I.; Bobrov, M. B. *J. Fluorine Chem.* **1988,** *41,* 421.
274. Weigert, F. J. *J. Fluorine Chem.* **1972,** *1,* 445.
275. Volkholz, M.; Stelzer, O.; Schmutzler, R. *Chem. Ber.* **1978,** *111,* 890.
276. Elkmeier, H. B.; Hodges, K. C.; Stelzer, O.; Schmutzler, R. *Chem. Ber.* **1978,** *111,* 2077
277. Dakternieks, D.; Röschenthaler, G.-V.; Sauerbrey, K.; Schmutzler, R. *Chem. Ber.* **1979,** *112,* 2380.
278. Janzen, A. F.; Lemire, A. E.; Marat, R. K.; Queen, A. *Can. J. Chem.* **1983,** *61,* 2264.
279. Prishchenko, A. A.; Livantsov, M. V.; Moshnikov, S. A.; Lutsenko, I. F. *Zh. Obshch. Khim.* **1987,** *57,* 1910 (Engl. Transl. 1708).
280. Heine, J.; Röschenthaler, G. V. *Z. Naturforsch. B: Chem. Sci.* **1988,** *43,* 196.
281. Weferling, N.; Schmutzler, R. *Chem. Ber.* **1989,** *122,* 1465.
282. Bekker, R. A.; Melikyan, G. G.; Luz'e, E. P.; Dyatkin, B. L.; Knunyants, I. L. *Dokl. Akad. Nauk SSSR* **1974,** *217,* 1320 (Engl. Transl. 572).

Eliminations

by M. Hudlický and A. E. Pavlath

The high electronegativity of fluorine plays an important part in the elimination reactions of organic fluorine compounds, frequently resulting in unexpected results. In addition to the elimination of halogen, hydrogen halides, water and carbon dioxide, even larger molecules might be removed.

Dehydrohalogenation

Hydrogen iodide is easily eliminated by strong bases from perfluoroalkylethyl iodides to give terminal alkenes. With perfluoroalkylpropyl iodides, however, replacement of iodine by nucleophiles predominates over the elimination reaction [*1*] (equation 1).

In the presence of dicyclohexyl-18-crown-6 ether, **potassium fluoride** converts fluorinated vinylic iodides to acetylenes [2] (equation 2).

Crown ether 18-crown-6 plays a role in the generation of *phenylfluorocarbene* by ***1,1-dehydrobromination*** of α-bromo-α-fluorotoluene with potassium *tert*-butoxide [*3*] (equation 3).

Double dehydrobromination of *cis*- and *trans*-1-(bromodifluoromethyl)-2-bromocyclohexane with **potassium hydroxide** gives, as the final product, 1-cyclohexene-1-carboxylic acid as a result of the hydrolysis of the intermediate, 1-cyclohexenecarbonyl fluoride [*4*] (equation 4).

[*1*] **1**

$$R_f(CH_2)_nCH_2CH_2I \xrightarrow[\text{or MeONa, MeOH, 30°C}]{\text{NaOH, EtOH, 50°C}} \begin{array}{lcc} & n=0 & n=1 \\ R_f(CH_2)_nCH{=}CH_2 & 99\% & \\ R_f(CH_2)_nCH_2CH_2OCH_3 & & 85\% \end{array}$$

[*2*] **2**

$$CF_3CH{=}CICH_3 \xrightarrow[\text{dioxane, reflux 3-4 h}]{\text{KF, dicyclohexyl-18-crown-6}} CF_3C{\equiv}CCH_3 \quad 45\%$$

[*3*] **3**

$$C_6H_5CHBrF \xrightarrow[\text{18-Crown-6}]{(CH_3)_3COK} [C_6H_5\ddot{C}F]$$

0065–7719/95/0187–0888$08.90/1

[4] (2-bromo-1-(bromodifluoromethyl)cyclohexane; 27% *cis*, 73% *trans*) $\xrightarrow[\text{RT} \rightarrow 200° \text{ distn.}]{\text{KOH}}$ 1-(CF_2Br)cyclohexene (39%) + (CF_2=)cyclohexene (9%) + 1-(COF)cyclohexene (17%) 4

[3] $$CH_3SCHCl_2 \xrightarrow[\text{18-Crown 6}]{(CH_3)_3COK} [CH_3S\ddot{C}Cl]$$ 5

1,1-Elimination of **hydrogen chloride** from 1,1-dichloromethyl sulfide with **potassium *tert*-butoxide** affords *chloromethylthiocarbene* [3] (equation 5).

Dehydrochlorination of ***bis(trifluoromethylthio)acetyl chloride*** with **calcium oxide** gives bis(trifluoromethylthio)ketene [5] (equation 6). Elimination of hydrogen chloride or hydrogen bromide by means of **tetrabutylammonium** or **potassium fluoride** from ***vinylic chlorides*** or ***bromides*** leads to acetylenes or allenes [6] (equation 7). Addition of dicyclohexyl-18-crown-6 ether raises the yields of potassium fluoride-promoted elimination of hydrogen bromide from (*Z*)-β-bromo-*p*-nitrostyrene in acetonitrile from 0 to 53–71%. In dimethyl formamide, yields increase from 28–35% to 58–68%.

[5] $$(CF_3S)_2CHCOCl \xrightarrow[140°,\ 12\ h]{Na_2SO_4,\ CaO} (CF_3S)_2C{=}C{=}O \quad 46\%$$ 6

[6] 7

$$O_2N\text{-}C_6H_4\text{-}CH{=}CRX \xrightarrow[\text{B } Bu_4NF,\ MeCN,\ 25°C]{\text{A } KF,\ DMSO} O_2N\text{-}C_6H_4\text{-}C{\equiv}CR$$

Z	R = H, X = Cl;	A : 120°, 4 h	50%
Z	R =H, X = Br;	A : 80°, 30 min	77%
Z	R = CH_3, X = Cl;	A : 100°, 45 min	85%
Z	R = CH_3, X = Br;	A : 100°, 10 min	94%
Z	R = H, X = Cl;	B : 9 h	97%
Z	R = H, X = Br;	B : 2 h	92%
Z	R = CH_3, X = Cl;	B : 8 h	60%
Z	R = CH_3, X = Br;	B : 2 h	87%

$$O_2N\text{-}C_6H_4\text{-}CH{=}C{=}CHR$$

E	R = CH_3, X = Cl;	A : 100°, 35 min	70%
E	R = CH_3, X = Br;	A : 100°, 20 min	93%
E	R = CH_3, X = Cl;	B : 6 h	48%
E	R = CH_3, X = Br;	B : 4 h	45%

Dehydrochlorination of trifluoroacetohydroxamyl chloride yields trifluoroacetonitrile oxide that dimerizes depending on the dehydrochlorinating agents either to bis(trifluoromethyl)-1,1,2,5-dioxadiazine or to 3,4-bis(trifluoromethyl)furoxan [*7*] (equation 8).

[*7*] $CF_3C(Cl){=}NOH$ —(CsF, diglyme or Et_3N, Et_2O / NaOH/asbestos)→ $CF_3C{\equiv}N{\rightarrow}O$ → 3,6-bis(trifluoromethyl)-1,4,2,5-dioxadiazine (34%) + 3,4-bis(trifluoromethyl)furoxan (30%) **8**

1,3-Elimination of hydrogen chloride from 2*H*-perfluoro-2-methylpropane-1-sulfenyl chloride by the **triethylamine–boron trifluoride** complex results in cyclization to perfluoro-1,1-dimethylthiirane (perfluoroisobutylene sulfide) [*8, 9*] (equation 9).

$$[9]\quad (CF_3)_2CHCF_2SCl \xrightarrow[\text{50°C, 3 h, sealed}]{Et_3N.BF_3} (CF_3)_2C\overset{S}{—}CF_2 \quad 60\% \qquad \mathbf{9}$$

As expected, elimination of **hydrogen fluoride** in most cases is considerably more difficult. In bimolecular reactions, the rates are about 2–3 orders of magnitude lower than those of hydrogen chloride and about 4–5 orders of magnitude lower than those of hydrogen bromide [*10*].

For compounds in which hydrogen is sufficiently "acidified" by enough fluorines in β positions, the ElcB mechanism may be operating [*11*].

Monofluoroalkanes and vicinal difluoroalkanes are dehydrofluorinated if strong enough bases are applied [*10, 12*]. In 5-fluorononane and fluorocyclododecane, elimination by means of **sodium methoxide** in methanol gives *cis*- and *trans*-alkenes in respective yields of 8 and 21% and in ratios of 1:2.2–2.4; however, the bulky **lithium diisopropyl amide** in tetrahydrofuran produces *trans*-isomers almost exclusively. The strength of the base does not have much effect on the rate of elimination, but the lithium cation causes considerable acceleration [*10*] (equation 10).

6,7-Difluorododecane is unaffected by sodium methoxide in methanol, but its treatment with **potassium *tert*-butoxide** in tetrahydrofuran eliminates hydrogen fluoride stereospecifically: *meso* and DL compounds give, respectively, (*E*)- and (*Z*)-6-fluoro-6-dodecene [*12*] (equation 11).

Geminal difluorides usually require a strong base for conversion to vinylic fluorides, but some **aluminum oxides** such as neutral γ-alumina or Woelm alumi-

[*10*] **10**

$RCH_2CHFCH_2R \longrightarrow$ (H)(R)C=C(H)(CH_2R) + (H)(R)C=C(CH_2R)(H)

R=C_3H_7	MeONa, MeOH 125°C, 48 h	2.3%	5.7%
	Me_3COK, DMSO 50°C, 24 h	5.5%	23.5%
	LDA, Et_2O 40°C, 48 h	4.4%	69.6%
R,R=-$(CH_2)_9$-	MeONa, MeOH 125°C, 48 h	6.5%	14.5%
	LDA, Et_2O 40°C, 48 h	1.6%	80.4%

[*12*] **11**

$C_5H_{11}CHFCHFC_5H_{11} \xrightarrow[75°C,\ 16\ h]{Me_3COK,\ THF} C_5H_{11}CH{=}CFC_5H_{11}F$

meso	*E* 66%
DL	*Z* 56%[a]

[a] 4% of $C_5H_{11}C{\equiv}CC_5H_{11}$ is formed.

[*14*] 1,1-difluorocycloalkane $(CH_2)_n$ $\xrightarrow{Al_2O_3\ Woelm}$ 1-fluorocycloalkene $(CH_2)_n$ **12**

n	Conditions	Yield
n = 2	70°, 15 h	66%
3	70°, 15 h	63%
4	hexane, reflux, 25 h	20%
9-*trans*	hexane, RT, overnight	46%

[*14*] 1,1-difluoro-4-(OCH_3)cyclohexane $\xrightarrow[reflux\ 16\ h]{alumina,\ hexane}$ 1-fluoro-4-(OCH_3)cyclohexene 43% **13**

num oxide dehydrofluorinate readily under unexpectedly mild conditions. Just stirring the geminal difluorides with a large excess of the alumina at room or slightly elevated temperature affords excellent yields [*13, 14*] (equations 12 and 13).

Dehydrofluorination of compounds in which a single hydrogen is flanked by heavily fluorinated groups is achieved by aqueous or alcoholic **alkali hydroxides** and proceeds probably by the ElcB mechanism. Thus 1-(2*H*-hexafluoropropyl)adamantane [*15*] and 1,3-bis(2*H*-hexafluoropropyl)adamantane [*16*] are dehydrofluorinated in respective yields of 75 and 81% (equation 14).

[*15*] CF_2CHFCF_3 —NaOH, 82-85°→ CF=$CFCF_3$ *cis* 6% *trans* 69% **14**

Polyfluorocyclohexanes are dehydrofluorinated by **potassium hydroxide** to polyfluorocyclohexenes and polyfluorocyclohexadienes (where applicable). The reaction is sometimes very complex [*17*] (equation 15).

Dehydrofluorination of ***dihydrododecafluorocycloheptanes*** gives a mixture of monohydroundecafluorocycloheptenes and decafluorocycloheptadienes [*18*] (equation 16).

[*17*] H, F —KOH, H_2O, reflux 30 min→ F 82% **15**

[*18*] H, F, H —KOH, H_2O, reflux, 8 h→ F, H 4.2% + F, H 10.7% **16**

↓ F 24.2% ↓ F 5.7%

Similar dehydrofluorination occurs with ***polyfluorinated heterocycles.*** 2,2-Bis(trifluoromethyl)-3,4-difluorooxetane gives 3-fluoro-4,4-bis(trifluoromethyl)-2-oxete [*19*] (equation 17), and heptafluoro-*p*-dioxane yields hexafluoro-*p*-dioxene [*20*] (equation 18).

[19] $(CF_3)_2C$—O / CHF—CHF $\xrightarrow[\text{reflux, 8 h}]{\text{KOH}}$ $(CF_3)_2C$—O / FC=CH 59% **17**

[20] CF_2–O–CHF / CF_2–O–CF_2 (ring) $\xrightarrow[150\text{-}160°]{\text{KOH}}$ CF_2–O–CF / CF_2–O–CF (ring, CF=CF) 52.5% **18**

With highly fluorinated compounds, even relatively weak bases cause elimination of hydrogen fluoride. Such reagents are especially desirable in dehydrofluorinations of polyfluoro compounds containing functionalities that would suffer from stronger bases [*21*] (equation 19).

A peculiar dehydrofluorination occurs when trifluoromethyl dihydropyridine derivatives are treated with organic bases. A double-bond shift and a hydrogen migration convert one trifluoromethyl group to a difluoromethyl group and aromatize the ring [22] (equation 20).

Elimination of hydrogen fluoride from ***vinylic fluorides*** yields allenes [*23*] or acetylenes [*24*] (equations 21 and 22).

[*21*] $CF_3CH(CF_3)CO_2CH_3$ $\xrightarrow[\text{reflux 2 h}]{Et_3N.BF_3}$ $CF_2{=}C(CF_3)CO_2CH_3$ 80% **19**

[22] **20**

4-CF_3-4-H-3,5-bis(CO_2R)-2,6-bis(CF_3)-1,4-dihydropyridine (N–H) $\xrightarrow[-H^+]{\text{DBUa,b}}$ carbanion (4-CF_3, 3,5-CO_2R, 2,6-CF_3, N–H) $\xrightarrow{-F^-}$ 4-CF_3-3,5-bis(CO_2R)-2-CF_3-6-(=CF_2) (N–H) $\longrightarrow$ 4-CF_3-3,5-bis(CO_2R)-2-CF_3-6-CHF_2-pyridine

$R = C_6H_5$ [a]THF, reflux 18 h 67%

$R = CF_3$ [b]diglyme, 110-120°, 45 min 12%

[c]DBU is 1,8-diazabicyclo[5.4.0]undecane

[23] **21**

$$(CF_3)_2CHCF{=}C(CO_2CH_3)_2 \xrightarrow[\text{-20°C, 5 days}]{Et_3N.BF_3,\ Et_2O} (CF_3)_2C{=}C{=}C(CO_2CH_3)_2 \ \ 38\%$$

\+

5% [2,2,3-trifluoro-4-(trifluoromethyl)-6-methoxy-2H-pyran-5-carboxylic acid methyl ester: ring O, F_2C, CF_3, F, OCH_3, CO_2CH_3]

[24] $$CF_2{=}CH_2 \xrightarrow[\text{-110° to -80°C}]{\text{s-BuLi}} CF{\equiv}CH \quad >90\%$$ **22**

Dehydrofluorination of *N,O*-bis(trifluoromethyl)hydroxylamine with potassium fluoride gives an excellent yield of a perfluorinated oxime ether [*25*] with a carbon–nitrogen double bond (equation 23).

[25] $$CF_3ONHCF_3 \xrightarrow[\text{RT, sealed, 14 h}]{KF} CF_3ON{=}CF_2$$ **23**

Simultaneous *elimination of chloride ion and carbon dioxide* occurs during heating of methyl chlorodifluoroacetate with lithium chloride in hexamethylphosphoric triamide (HMPA). The *difluorocarbene* generated in this way is trapped by electron-rich alkenes to form 1,1-difluorocyclopropanes [*26*] (equation 24).

[26] **24**

$$CClF_2CO_2CH_3 \xrightarrow[\text{80°, 24 h}]{\text{LiCl, HMPA}} [\,:CF_2\,] \xrightarrow{Me_2C{=}CMe_2} (CH_3)_2C{-}C(CH_3)_2 \text{ (bridged by } CF_2) \quad 86\%$$

Some chlorofluoro compounds lose hydrogen fluoride in preference to hydrogen chloride. A ***geminal chlorofluorocyclopropane*** suffers dehydrofluorination with the concomitant ring cleavage to give a chlorinated diene [*27*] (equation 25).

[27] 1-chloro-1-fluoro-2-ethylcyclopropane (Cl, F, C_2H_5) $\xrightarrow[\text{250-300°C}]{\gamma\text{-}Al_2O_3}$ **25**

$$CH_3CCl{=}CH{-}CH{=}CH_2 \quad 38\%$$
$$\uparrow$$
$$CH_3CH{=}CH{-}CCl{=}CH_2 \quad 19\%$$

The last example represents a fairly rare ***elimination of hydrogen fluoride in preference to hydrogen chloride,*** a reaction that deserves a more detailed discussion.

A comparison of bond dissociation energies of carbon–halogen bonds shows that the carbon–fluorine bond is much stronger than the carbon–chlorine, carbon–bromine, and carbon–iodine bonds: 108–116, 83.5, 70, and 56 kcal/mol, respectively [28]. This makes the E1 mechanism, in which the rate-determining step is breaking of the carbon–halogen bond, highly unlikely.

By the same token, in base-promoted E2 dehydrohalogenations, the rate of elimination of the halogen as a halide ion is expected to be I > Br > Cl >> F. This "element effect" has indeed been documented in many instances, a few examples of which are listed in Table 1.

Table 1. Relative Rates of Base-Promoted Dehydrohalogenations of Halogenated Compounds

Halogen Compound	*Base, Solvent*	*Temp (°C)*	*X = F*	*X = Cl*	*X = Br*	*X = I*	*Ref.*
$PhCHBrCF_2X$	EtONa, EtOH	25	1	400,000	30,000,000		29
$PhCH_2CH_2X$	EtONa, EtOH	30	1	68	4,100	26,000	30, 31
BuCHXMe	MeONa, MeOH	100	1	714	27,214	107,143	*32*

The ratios of the rates of elimination of hydrogen halides depend evidently on the structure of the halogenated compound and span many orders of magnitude. The rate of *dehydrofluorination is much slower than that of dehydrochlorination, dehydrobromination, and dehydroiodination.* However, there are exceptions.

In the reaction of ***dimethyl and diethyl α-bromo-α′-fluorosuccinates*** with **potassium phthalimide,** elimination of hydrogen fluoride predominates over that of hydrogen bromide in a ratio of 3:2 [*33*]. When the above-mentioned esters are refluxed with sodium azide in methanol, dialkyl bromofumarates and dialkyl azidofumarates are formed mainly, indicating preferential elimination of hydrogen fluoride [*34*]. When dimethyl and diethyl α-bromo-α′-fluorosuccinates are heated with **aqueous methanolic potassium acetate,** hydrogen fluoride is eliminated in preference to hydrogen bromide at a rate higher by 1–2 orders of magnitude [*34*]. The rates of elimination of hydrogen fluoride from *erythro-* and *threo-*α-bromo-α′-fluoro esters are only slightly different from each other, in contrast to the dehydrobromination of diethyl DL- and meso-α,α′-dibromosuccinates, where the rates differ considerably. The elimination of hydrogen bromide from the DL-dibromo ester is faster than the dehydrofluorination of both bromofluoro esters, and that, in turn, is faster than the dehydrobromination of the meso-dibromo ester [*34*].

The *stereochemistry* of the dehydrofluorination of the diethyl α,α′-dibromosuccinates is in accord with the expectation: the *meso* compound gives a 100% yield of the *cis*-bromo ester (diethyl bromomaleate), and the DL isomer gives a 95% yield of the *trans* ester and a 5% yield of the *cis* ester (diethyl bromofumarate and bromomaleate, respectively). In the case of diethyl α-bromo-α′-fluorosuccinates, the *threo* isomer affords a 100% yield of diethyl bromofumarate, but the *erythro* isomer gives only a 30% yield of diethyl bromomaleate and a 70% yield of diethyl bromofumarate. Diethyl bromofumarate is not formed by rearrangement of diethyl bromomaleate, because the latter does not rearrange under the conditions used [*34*].

Results similar to those with dimethyl and diethyl α-bromo-α′-fluorosuccinates are obtained when the free acids ***erythro***- and ***threo-α-bromo-α′-fluorosuccinic acids*** are treated with **aqueous alkalies** [*35*]. Under these conditions, hydrogen fluoride is eliminated exclusively in preference to hydrogen bromide to give predominantly or exclusively bromofumaric acid from both diastereomers. The reaction is very fast and follows second-order kinetics [*35*] (Table 2).

Table 2. Dehydrofluorination of *erythro*- and *threo*-α-Bromo-α′-fluorosuccinic Acids by Sodium Deuteroxide in Deuterium Oxide [*35*]

Acid	*Temp. (°C)*	*Half-life (min)*	K_2 *(mol* $L^{-1}s^{-1}$*)*
erythro-$HO_2CCHBrCHFCO_2H$	30	9.25	0.00405
erythro-$HO_2CCHBrCHFCO_2H$	45	3.0	0.01133
threo-$HO_2CCHBrCHFCO_2H$	30	18.75	0.00307
threo-$HO_2CCHBrCHFCO_2H$	45	4.8	0.01010

It is interesting that the rates of elimination of hydrogen fluoride from all fluorinated succinic acids span hardly 1 order of magnitude, whereas the rates of preferential dehydrofluorination of both α-bromo-α′-fluorosuccinic acids are higher by almost 2 orders of magnitude [*35, 36, 37, 38*] (Table 3).

All the facts seem to indicate that the elimination of hydrogen fluoride from vicinal fluorohalo compounds proceeds by a *carbanion or carbanionlike mechanism* in which the *acidity of hydrogen* plays a decisive role. The acidity of hydrogen is enhanced by the number of fluorine atoms in the β position, be it by inductive effect [*39, 40*] or by hyperconjugation [*41*]. But even stronger is the effect of *α-halogen, which increases the rate of elimination by almost 2 orders of magnitude* [*42*] (Table 4).

From the thorough studies of the mechanism of elimination of hydrogen halides from vicinal fluorohalo compounds, it follows that the result of elimination depends on the stability of the carbanionic species in which the negative charge is on the carbon β to fluorine and α to chlorine or bromine [*43, 44, 45, 46, 47, 48, 49, 50, 51, 52, 53, 54*].

Preferential elimination of hydrogen fluoride from vicinal halofluoro compounds occurs also in the ***cyclohexane series*** [*55, 56, 57*], ***acenaphthene series*** [*58*], and ***benzodihydrofuran*** series [*59,60*]. Here, the strength of the base and the stereochemistry play important roles.

Table 3. Relative Rates of Dehydrofluorination of α-Bromo-α′-fluorosuccinic Acids Compared with Fluorinated Succinic Acids by Using 0.05 M Disodium Salts of the Acids in 0.20 M Solutions of NaOD in D_2O at 6 °C [35, 36, 37]

Acid	K_2 *(mol L^{-1} s^{-1})*	*Relative Rates*
$HO_2CCF_2CHFCO_2H$	0.00013	1
DL-$HO_2CCHFCHFCO_2H$	0.00016	1.23
meso-$HO_2CCHFCHFCO_2H$	0.00031	2.38
$HO_2CCHFCH_2CO_2H$	0.00043	3.31
$HO_2CCF_2CH_2CO_2H$	0.00115	8.85
threo-$HO_2CCHBrCHFCO_2H$	0.033[a]	254
erythro-$HO_2CCHBrCHFCO_2H$	0.037[a]	285

[a]Estimate made by extrapolation.

Table 4. Rates of Dehydrofluorination of 1-Phenyl-2,2,2-trifluoroethane and 1-Chloro-1-phenyl-2,2,2-trifluoroethane [42]

Compound	*Base, Solvent*	*Temp. (°C)*	K_2 *(mol $L^{-1}s^{-1}$)*	*Relative Rate*
$PhCH_2CF_3$	EtONa, EtOH	50	0.000000345	1
$PhCHClCF_3$	EtONa, EtOH	50	0.0000256	85.3

cis-1-Bromo-2-fluorocyclohexane in which the most acidic hydrogen to be eliminated is antiperiplanar to both bromine and fluorine hydrogen bromide is eliminated exclusively by the usual *anti* elimination [55, 56] (equation 26).

[55,56] Br F (cyclohexane) —Me_3COK, Me_3COH or $NaNH_2$, Me_3COH, THF→ F (cyclohexene) **26**

The preferential *syn* elimination of hydrogen fluoride from ***trans-1-bromo-2-fluorocyclohexane*** to give 1-bromocyclohexene is achieved only when a strong base such as **sodamide** is used [55, 56]. **Potassium *tert*-butoxide** causes elimination of hydrogen bromide to form 3-fluorocyclohexene [56] (equation 27).

Preferential elimination of hydrogen fluoride occurs also in highly fluorinated ***1H,2-chlorodecafluorocyclohexanes*** and in ***1H,1,2-dichlorononafluorocyclohexanes*** [57]. It takes place predominantly but not exclusively only in those isomers where hydrogen and fluorine are in antiperiplanar conformation. When hydrogen and fluorine are in synplanar positions, only hydrogen chloride is eliminated on treatment with **18 N potassium hydroxide** at room temperature for 14 h [57] (equation 28).

[*56*] 27

[*55,56*]

Me_3COK, Me_3COH

$NaNH_2$, THF

[*57*] 28

All bonds not shown are to fluorine.

In 1-chloro-2-fluoroacenaphthene [*58*] and in 2,3-dihalo-2,3-dihydrobenzofuran [*59, 60*], potassium *tert*-butoxide eliminates hydrogen fluoride in preference to hydrogen chloride. *trans*-2-Chloro-3-fluoro-2,3-dihydrobenzofuran loses hydrogen fluoride quantitatively on treatment with sodamide in *tert*-butyl alcohol [*60*] (equation 29).

[*60*] 29

$NaNH_2$, Me_3COH
THF, RT, 1 min

97%

Dehalogenation

Although fluorocarbons are considered very stable compounds, they can be defluorinated to unsaturated derivatives under certain mild conditions. Hexadecafluorobicyclo[4.4.0]dec-1(6)-ene reacts with activated zinc powder at 80–100 °C to yield partially and fully aromatized products [*61*]. The final product composition depends on the solvent. Dioxane, acetonitrile, and dimethylformamide, in this order, effect increasing unsaturation (equation 30).

[61] **30**

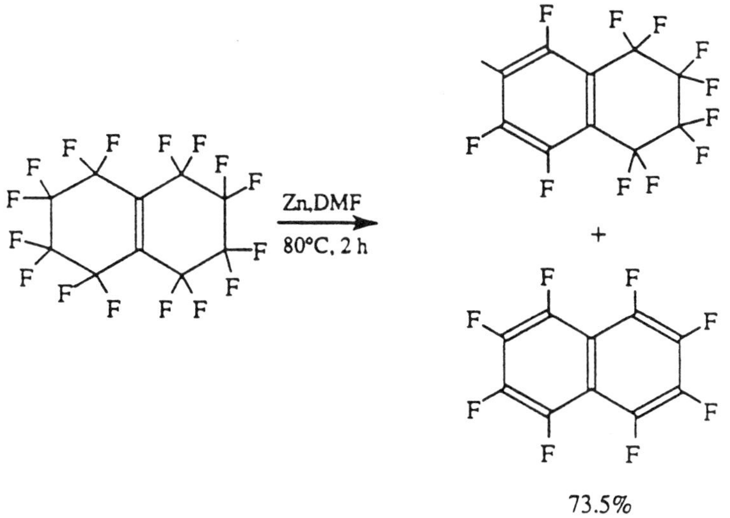

Zn,DMF
80°C, 2 h

+

73.5%

Sodium amalgam converts perfluoro(3,4-dimethyl-3-hexene) to perfluoro-(3,4-dimethyl-2,4-hexadiene at room temperature in 70% yield [*62*] (equation 31).

Defluorination occurs even with **sodium fluoride** at 530 °C when tetrafluorothiolene is converted to 2,5-difluorothiophene [*63*]. Dehydrofluorination would be expected at such high temperature, but defluorination is favored. The product composition also excludes a disproportionation reaction mechanism (equation 32).

[62] 31

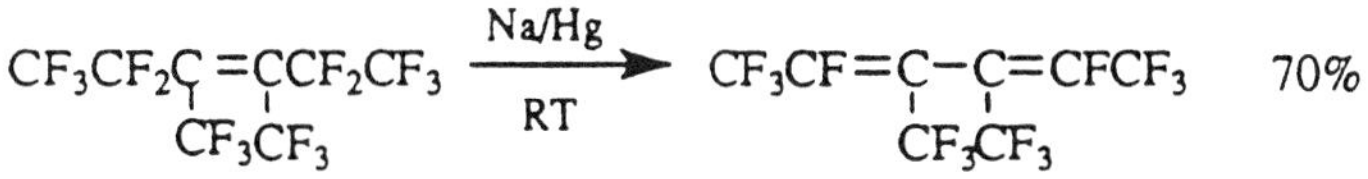

$$CF_3CF_2C(CF_3)=C(CF_3)CF_2CF_3 \xrightarrow[RT]{Na/Hg} CF_3CF=C(CF_3)-C(CF_3)=CFCF_3 \quad 70\%$$

[63] 32

NaF
530°C

32.7%

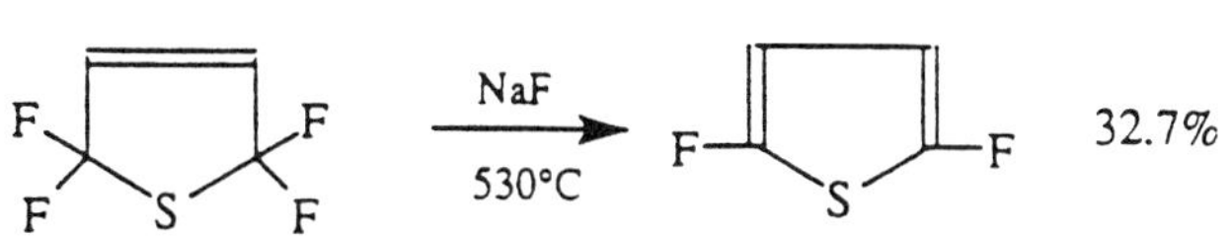

When other halogens are also present, *dehalogenation* reactions may result in a mixture of products resulting from the elimination of fluorine and mainly of the other halogen [*64*]. Chlorofluoro compounds may be preferentially dechlorinated without the loss of fluorine by using **zinc** in dimethylformamide solution [*65*] (equation 33).

[65] **33**

$$R_FCCl_2CCl_3 \xrightarrow[90\text{-}100\ ^\circ C]{Zn, DMF} [\,R_FC{\equiv}CZnCl\,] \xrightarrow{HCl} R_FC{\equiv}CH$$

$R_F = C_3F_7$	77%
C_4F_9	72%
C_6F_{13}	64%
C_8F_{17}	62%

In a similar reaction, chloro(trifluoromethyl)sulfine is formed by the dechlorination of 1,1-dichloro-2,2,2-trifluoroethanesulfinyl chloride with **copper** at 200 °C in the gas phase [*66*] (equation 34).

Mercury can be used to dehalogenate acyl halides to ketenes when a halogen is present next to the COCl group [*67*] (equation 35).

[66] **34**

$$CF_3CCl_2SOCl \xrightarrow[200^\circ C \text{ in gas phase}]{Cu} CF_3CCl{=}SO \quad 62\%$$

[67] **35**

$$SF_5C(CH_3)BrCOCl \xrightarrow[RT]{Hg} SF_5C(CH_3){=}CO \quad 94\%$$

However, when the chlorine atoms are more dispersed through a polychlorofluorocarbon, both *chlorine and fluorine* may be removed by **zinc.** In these cases, ***triphenylphosphine*** in dioxane can be used to prepare dechlorinated products in high purity and good yield [*68*](equations 36 and 37).

[68] **36**

$$CF_2ClCFClCF_2CFClCF_2COOEt \xrightarrow[55^\circ C,\ dioxane]{(C_6H_5)_3P} CF_2{=}CFCF_2CFClCF_2COOEt \quad 60\%$$

[68] **37**

$$CF_2ClCFClCF_2CFClCF_2CF{=}CF_2 \xrightarrow[55^\circ C,\ dioxane]{(C_6H_5)_3P} CF_2{=}CFCF_2CFClCF_2CF{=}CF_2 \quad 59\%$$

A halogen atom may be removed together with a fluorine from a vicinal carbon atom. Monofluoroacetylene can be obtained by the conversion of 1,1-difluoro-2-bromoethylene to its Grignard derivative followed by β-elimination at 58–60 °C [*69*] (equation 38).

[69] **38**

$$CF_2{=}CHBr \xrightarrow[\text{THF, 60°C}]{\text{Mg}} CF_2{=}CHMgBr \longrightarrow CF{\equiv}CH \quad 41.5\%$$

By using freshly precipitated **copper** powder at 200 °C, perfluroallenes can be prepared by the *elimination of bromine fluoride or iodine fluoride* from the corresponding starting materials [*70*]. The reaction is carried out at low pressure, 1–2 mm of Hg, and the products quickly dimerize into cyclobutane derivatives (equation 39).

[70] **39**

$$(CF_3)_2C{=}CBrCF_3 \xrightarrow[\text{200°C, in gas phase}]{\text{Cu}} (CF_3)_2C{=}C{=}CF_2 \quad 97\%$$

In a similar reaction, iodine fluoride also can be removed from perfluorobutyl-, perfluorohexyl-, and perfluorooctyl iodide by using **zinc–copper** couple in dimethyl sulfoxide or dimethylformamide [*71*].

The product is a mixture; however, the composition may be altered by the nucleophile used in the reaction. For example, the reaction of perfluorobutyl iodide in the presence of sodium acetate or sodium bisulfite results mostly in perfluoro-2-butene and nonafluorobutane, whereas with potassium thiocyanate, perfluoro-1-butene is the dominant product, together with perfluoro-2-butene and a small amount of nonafluorobutane (equation 40).

With 1,2,2-trifluoro-1,2-dichloroethyl iodide, **zinc** reacts exothermically, but iodine is eliminated intermolecularly; the reaction yields 1,2,3,4-tetrachloroperfluorobutane [*72*] (equation 41).

Elimination of any combination of fluorine, chlorine, bromine, or iodine is also possible when the halogen atoms to be removed are on the same carbon atom. The

[71] **40**

$$R_FCF_2CF_2I \xrightarrow[\text{DMSO or DMF}]{\text{Zn/Cu}} R_FCF{=}CF_2 \quad 62\text{-}70\%$$

[72] **41**

$$CF_2ClCFClI \xrightarrow[\text{AcOEt, } CH_2Cl_2\text{, exothermic}]{\text{Zn}} CF_2ClCFClCFClCF_2Cl \quad 66\%$$

reaction products are various carbenes. Active **titanium** powder removes chlorine from fluorotrichloromethane at 0 °C resulting in chlorofluorocarbene [*73*], whereas difluorodibromomethane and **lead** give difluorocarbene at 40 °C almost quantitatively [*74*]. The carbenes thus formed react in situ with olefins (equations 42 and 43).

With fluorodiiodomethane, *photolysis* gives fluorocarbene, but a small quantity of the intermediate monoradical can be trapped [*75*] (equation 44).

[73] **42**

$$CFCl_3 \xrightarrow[\text{THF, 0°C}]{\text{Ti}} CFCl: \quad 77\%$$

[74] **43**

$$CF_2Br_2 \xrightarrow[\text{CH}_2\text{Cl}_2\text{, 40°C}]{\text{Pb}} CF_2: \quad 90\%$$

Elimination of Nitrogen

Various fluorinated cyclic compounds containing -N=N- bonding can be decomposed by the *elimination of nitrogen.* The photolysis of phenylfluorodiazirine results in the formation of the intermediate phenylfluorocarbene, which can react instantaneously with olefins [*76*] (equation 45).

The six-membered ring in polyfluoropyridazines is much more stable, and pyrolysis at 680–725 °C is needed to eliminate nitrogen, resulting in various fluorinated acetylene compounds in 80–90% yield [*77*](equation 46).

On the other hand, as expected, five- and six-membered rings containing three vicinal nitrogen atoms can be decomposed easily. By using either photolysis or mild

[75] **44**

$$CHFI_2 \xrightarrow{h\nu} \underset{10\%}{CFHI\cdot} \longrightarrow \underset{45\%}{CFH:}$$

[76] **45**

$$C_6H_5CFN_2 \xrightarrow[\text{25°C}]{h\nu} C_6H_5CF: \quad 42\text{-}76\%$$

[77] **46**

$$\text{tetrakis}(C_6F_5)\text{pyridazine} \xrightarrow{\text{680-725°C}} \underset{82\text{-}90\%}{C_6F_5C{\equiv}CC_6F_5}$$

thermolysis, 1-alkoxy-2,2-bis(trifluoromethyl)aziridines can be prepared from the corresponding triazolines in 43–87% yield [*78*] (equation 47).

Similarly, low-temperature photolysis of 4,5,6-fluorosubstituted 1,2,3-triazines results in the elimination of nitrogen, but the product composition depends on the substituents. When the substituents are fluorine atoms, the intermediate product is a four-membered, nitrogen-containing ring that quickly dimerizes. When all the substituents are perfluoroalkyl groups, the pyrolysis results in a mixture perfluoroalkyl acetylenes and perfluoroalkyl cyanides [*79*] (equations 48 and 49).

[*78*] **47**

$$\text{4,4-bis}(CF_3)\text{-5-OR-triazoline} \xrightarrow[CCl_4,\ 6\ h]{h\nu} \text{1-OR-2,2-bis}(CF_3)\text{aziridine} \quad 44\text{-}87\%$$

[*79*] **48**

$$\text{4,5,6-trifluoro-1,2,3-triazine} \xrightarrow[RT]{h\nu} \text{trifluoroazete}$$

[*79*] **49**

$$\text{4,5,6-tris}(R_F)\text{-1,2,3-triazine} \xrightarrow[RT]{h\nu} R_FC{\equiv}CR_F + R_FCN$$

Dehydration

The *elimination of water* from a fluorinated compound generally follows a reaction path similar to that of its nonfluorinated counterpart, although the presence of the highly electronegative fluorine atoms may have unexpected effects. Various monofluoro alcohols can be *dehydrated via their tosyl esters* at 75 °C by using potassium *tert*-butoxide [*80*] (equation 50).

The dehydration of highly fluorinated alcohols generally requires **phosphorus pentoxide** or **sulfuric acid,** but hexafluoroisobutyl alcohol is easily converted to hexafluoroisobutylene by **potassium hydroxide** with or without solvent at 20–

[*80*] **50**

$$CH_3CH(OTs)CHFCHCH_3 \xrightarrow[75°C]{t\text{-BuOK}} CH_3CF{=}CHCH_3\ \ 16.4\% + CH_3CHFCH{=}CH_2\ \ 69.6\%$$

50 °C. With chlorinated methanes and ethanes as solvents, the yields vary between 45 and 80% [*81*] (equation 51).

Trifluoromethyl homoallyl alcohols also dehydrate easily with **phosphorus oxychloride–pyridine** complex, but it is very difficult to remove water from their saturated analogues by the same method [*82*] (equation 52).

The preparation of fluorinated acid anhydrides can be done by simple methods. Phosphorus pentoxide removes a molecular equivalent of water from trifluoroacetic acid and trifluoromethanesulfonic acid and forms the mixed anhydride [*83*] (equation 53).

When a hydrogen atom is present next to the carboxyl group, an *intramolecular dehydration* may occur, resulting in a ketene. Two highly fluorinated, sulfur-containing ketenes were prepared this way [*67, 84*] (equations 54 and 55).

[81] **51**

$$(CF_3)_2CHCH_2OH \xrightarrow{20°C} (CF_3)_2C{=}CH_2$$

KOH	57.9%
NaOH	26.4%

[82] **52**

$$RCH{=}CHCH_2C(OH)(CF_3)_2 \xrightarrow[\text{pyridine, } 110°C,\ 65\ h]{POCl_3} RCH{=}CHCH{=}C(CF_3)_2$$

$R = C_6H_5$, 81%

$R = C_6H_5S$, 90%

[83] **53**

$$CF_3COOH + CF_3SO_3H \xrightarrow[-20°C]{P_2O_5} CF_3COOSO_2CF_3 \quad 75\%$$

[84] **54**

$$(CF_3S)_2CHCOOH \xrightarrow[100°C]{P_2O_5} (CF_3S)_2C{=}C{=}O \quad 65\%$$

[67] **55**

$$SF_5CHRCOOH \xrightarrow[140\text{-}160°C]{P_2O_5} SF_5CR{=}C{=}O \quad 60\text{-}70\%$$

Certain fluorine compounds may be used to dehydrate nonfluorinated alcohols through their complexes or derivatives. With alcohols, **1,1,1-trichlorotrifluoroacetone** forms intermediate hemiketals that easily decompose to olefins in the presence of a catalytic amount of *p*-toluenesulfonic acid [*85*]. Similarly, the

References are listed on pages 908–912.

decomposition of enolic trifluorosulfonates results in allenic [*86*], acetylenic [*87*], and diacetylenic [*88*] compounds (equation 56). In certain cases, compounds with high optical purities can be obtained through asymmetric elimination of **trifluomethanesulfonic acid** by chiral *N,N*-dimethyl-1-phenylethylamine [*89*].

[87]

$$(CH_3)_3C\underset{\displaystyle O_3SCF_3}{\underset{|}{C}}{=}CH_2 \xrightarrow[60°C]{\text{Pyridine}} (CH_3)_3CC{\equiv}CH \quad 90\% \qquad \mathbf{56}$$

Elimination of Carbon Oxides

Frequently, the decomposition of fluorine compounds results in a product with fewer carbon atoms through loss of carbon monoxide, carbon dioxide, or even an organic molecule. Perfluorinated α-diketones undergo *decarbonylation* with **cesium fluoride** in diglyme [*90*] (equation 57).

[90]

$$CF_3COCOR_F \xrightarrow[\text{in diglyme, 110-115°C}]{CsF} CF_3COR_F \quad 76\text{-}83\% \qquad \mathbf{57}$$

The *elimination of carbon monoxide* from nonfluorinated acyl fluorides, however, does not result in a fluorine compound. Although it was claimed earlier that benzoyl fluoride can be converted into fluorobenzene by using tris(triphenylphosphine)rhodium chloride, recent studies proved that the product is benzene and not fluorobenzene [*91*].

The *photolytic* reaction of a perfluoro anhydride in the vapor phase at 50 °C results in the elimination of not only carbon monoxide but also of carbon dioxide [*92*]. The tetrafluorocyclobutadiene formed is not stable and dimerizes easily (equation 58).

[92]

(tetrafluorobicyclic anhydride, F, F, F, F, O, O, O) $\xrightarrow[50°C]{h\nu}$ (tetrafluorocyclobutadiene, F, F, F, F) 41% **58**

The *decarboxylation* reactions of fluorinated carboxylic acids are similar to those of their nonfluorinated counterparts, but predictably many exceptions exist. The oxidation of the potassium salts of perfluoro acids with potassium persulfate leads to decarboxylation and coupling [*93*] (equation 59).

[93]

$$R_FCOOK \xrightarrow[80°C]{} R_F{-}R_F \quad 90\text{-}99\% \qquad \mathbf{59}$$

The *pyrolysis of perfluoro carboxylic salts* can result both in mono- and bimolecular products. At 210–220 °C, silver salts give mostly the coupled products; at 160–165 °C in *N*-methylpyrrolidinone, the corresponding copper salts also give the simple decarboxylated compounds in nearly equal amounts. The decomposition of the copper salts in the presence of iodobenzene at 105–125 °C results in a phenyl derivative, in addition to the olefin and coupled product [*94*] (equations 60–62).

[94]

$$R_FCF{=}CFCOOAg \xrightarrow{210\text{–}220°C} (R_FCF{=}CF)_2 \quad 86\text{-}87\% \qquad \mathbf{60}$$

[94]

$$(R_FCF{=}CFCOO)_2Cu \xrightarrow[\text{N-methyl-pyrrolidinone}]{160\text{–}165°C} R_FCF{=}CFH \quad 25\text{-}42\% \qquad \mathbf{61}$$

[94]

$$R_FCF{=}CFCOOAg \xrightarrow[C_6H_5I]{105\text{-}125°C} R_FCF{=}CFC_6H_5 \quad 35\% \qquad \mathbf{62}$$

[95]

$$[(CF_3)_2CFCF{=}CFCOO]_2Hg \xrightarrow[220\text{--}240°C]{K_2CO_3} [(CF_3)_2CFCF{=}CFCF]_2Hg \qquad \mathbf{63}$$

On the other hand, however, the mercuric salt of a similar perfluoro acid gives a bis(perfluoroalkenyl)mercuric compound [*95*] (equation 63).

Carbon dioxide may be eliminated even from the ester of a fluorinated acid by using lithium chloride–hexamethylphosphoric triamide complex at reflux temperature. The intermediate carbene is formed in 78–90% yield [*96, 97*] (equation 64).

[96,97]

$$CF_2ClCOOCH_3 \xrightarrow[\text{75-80°C, triglyme}]{LiCl.2HMPA} CF_2{:} + CO_2 + CH_3Cl \quad 78\text{-}90\% \qquad \mathbf{64}$$

Other Eliminations

The high-temperature pyrolysis of sulfonyl fluoride results in the *elimination of sulfur dioxide*, although secondary reactions also occur, depending on the residence time. With perfluorooctanesulfonyl fluoride, long residence times result in perfluoro(C_8–C_{16}) compounds, and shorter residence times lead to perfluorohexadecane [*98*] (equation 65).

[98] 65

$$C_8F_{17}SO_2F \xrightarrow[460\text{-}550^\circ C]{} C_nF_{2n+2} \quad n=4\text{-}16 \quad 44.5\text{-}60.9\%$$

In some cases, however, especially with shorter perfluoroalkanesulfonyl fluorides, sulfur dioxide may be removed by **antimony pentafluoride** at temperatures between 20 and 90 °C depending on the substituents [*99*] (equation 66).

[99] 66

$$(R_F)_2CFSO_2F \xrightarrow[20\text{-}90^\circ C]{SbF_5} (R_F)_2CF_2$$

The elimination of sulfur dioxide is apparently more difficult than that of either carbon monoxide or carbon dioxide. When fluorosulfonyldifluoroacetyl halide (chloride, bromide, or iodide) is photolyzed, *carbon monoxide* is quantitatively eliminated to give halodifluoromethanesulfonyl fluoride with increasing ease from chloride to bromide to iodide [*95, 100*] (equation 67).

[95,100] 67

$$FSO_2CF_2COX \xrightarrow{h\nu} \underset{75\text{-}90\%}{FSO_2CF_2X} \quad X = Cl, Br \text{ or } I$$

Similarly, fluorosulfonyldifluoroacetic acid can be decarboxylated easily to give difluoromethanesulfonic acid by using catalytic amounts of sodium chloride, but the reaction can proceed readily to the next step, resulting in loss of sulfur dioxide and formation of difluorocarbene [*101, 102*] (equations 68 and 69).

[101] 68

$$FSO_2CF_2COOH \xrightarrow{H_2O} FSO_2CF_2H$$

[102] 69

$$FSO_2CF_2COO^- \xrightarrow[Na^+ \text{ or } Li^+]{MeCN} CF_2: + CO_2 + SO_2 + F^-$$

Various highly fluorinated **alkyltrifluorosilanes** can be pyrolyzed at 140–200 °C with the loss of silicon tetrafluoride without affecting the carbon chain. The resulting carbenes react with olefins in situ [*103, 104*] (equations 70–72)

[103] 70

$$FCH_2CF_2SiF_3 \xrightarrow{140^\circ C} FCH_2CF: \quad 79\%$$

[103] 71

$$FCCl_2CF_2SiF_3 \xrightarrow{140^\circ C} FCCl_2CF:$$

[104] 72

$$CF_3CF_2SiF_3 \xrightarrow{200^\circ C} CF_3CF:$$

References are listed on pages 908–912.

When the alkyltrifluorosilane is pyrolyzed, a more complex mixture is obtained, but trichloromethyltrifluorosilane gives dichlorocarbene under similar conditions [*105*] (equation 73).

[105] $$CCl_3SiF_3 \xrightarrow{140°C} CCl_2: \quad >69\%$$ **73**

Elimination reactions of fluorine compounds are not limited to the removal of simple molecules. Frequently, large molecules or combination of smaller ones are formed as by-products, especially in pyrolytic reactions. For example, perhalogenated acid chlorides lose not only *carbon monoxide but also chlorine fluoride* [*106, 107*] (equations 74 and 75).

[106] **74**

$$CF_3OCF_2CF_2COCl \xrightarrow[Na_2CO_3,\ 4h]{230\text{-}280°C} CF_3OCF{=}CF_2$$

[107] **75**

$$H(CF_2)_4CF_2CF_2COCl \xrightarrow[Na_2CO_3,\ 4h]{230\text{-}280°C} H(CF_2)_4CF{=}CF_2 \quad 71\%$$

The reactions of some fluorinated ethers may result in the *elimination of alkyl fluorides.* In the case of 2-methoxyperfluoro-2-butene, treatment with antimony pentafluoride gives perfluoro-3-buten-2-one and methylfluoride [*107*]. By reacting 2-chloro-1,1,2-trifluorodiethyl ether with boron trifluoride etherate or with aluminum chloride, chlorofluoroacetyl fluoride can be obtained with the elimination of ethyl fluoride [*108*] (equations 76 and 77).

[107] **76**

$$C_2F_5CF{=}C(OCH_3)CF_3 \xrightarrow[RT]{SbF_5} \underset{84\%}{CF_3CF{=}CFCOCF_3} + CH_3F$$

[108] **77**

$$CClFHCF_2OC_2H_5 \xrightarrow{BF_3 \cdot Et_2O} \underset{45.4\%}{CClFHCOF} + \underset{43.8\%}{C_2H_5F}$$

References for Pages 888–908

1. Brace, N. O.; Marshall, L. W.; Pinson, C. J.; van Wingerden, G. *J. Org. Chem.* **1984,** *49,* 2361.
2. Mielcarek, J. J.; Morse, J. G.; Morse, K. W. *J. Fluorine Chem.* **1978,** *12,* 321.
3. Moss, R. A.; Joyce, M. A.; Pilkiewicz, F. G. *Tetrahedron Lett.* **1975,** 2425.
4. Elsheimer, S.; Michael, M.; Landavaso, A.; Slattery, D. K.; Weeks, J. *J. Org. Chem.* **1988,** *53,* 6151.
5. Haas, A.; Lieb, M.; Praas, H. W. *J. Fluorine Chem.* **1989,** *44,* 329.

6. Naso, F.; Ronzini, L. *J. Chem. Soc., Perkin Trans. 1* **1974,** 340.
7. Middleton, W. J. *J. Org. Chem.* **1984,** *49,* 919.
8. Bekker, R. A.; Popkova, V. Y.; Knunyants, I. L. *Izv. Akad. Nauk SSSR* **1980,** 1692; *Chem. Abstr.* **1980,** *93,* 186054u.
9. Bekker, R. A.; Rozov, L. A.; Popkova, V. Y. *Izv. Akad. Nauk SSSR* **1983,** 2575 (Engl. Transl. 2317); *Chem. Abstr.* **1984,** *101,* 23242y.
10. Matsubara, S.; Matsuda, H.; Hamatani, T.; Schlosser, M. *Tetrahedron* **1988,** 2855.
11. Leffek, K. T.; Schroeder, G. *Can. J. Chem.* **1982,** *60,* 3077.
12. Matsuda, H.; Hamatani, T.; Matsubara, S.; Schlosser, M. *Tetrahedron* **1988,** 2865.
13. Boswell, G. A., Jr. U.S. Pat. 3 413 321, 1968; *Chem. Abstr.* **1969,** *70,* 58140g.
14. Strobach, D. R.; Boswell, G. A., Jr. *J. Org. Chem.* **1971,** *36,* 818.
15. Podkhalyuzin, A. T.; Nazarova, M. P.; Yanketevich, A. Z. *Zh. Org. Khim.* **1976,** *12,* 910; *Chem. Abstr.* **1976,** *85,* 20677j.
16. Podkhalyuzin, A. T.; Nazarova, M. P. *Dokl. Akad. Nauk SSSR* **1976,** *229,* 105; *Chem. Abstr.* **1976,** *85,* 142719v.
17. Coe, P. L.; Mott, A. W.; Tatlow, J. C. *J. Fluorine Chem.* **1982,** *20,* 167.
18. Khalil, A. E. M. M.; Stephens, R.; Tatlow, J. C. *J. Fluorine Chem.* **1983,** *22,* 31.
19. Barlow, M. G.; Coles, B.; Haszeldine, R. N. *J. Fluorine Chem.* **1980,** *15,* 387.
20. Coe, P. L.; Dodman, P.; Tatlow, J. C. *J. Fluorine Chem.* **1975,** *6,* 115.
21. Knunyants, I. L.; Zeifman, Y. V.; Lushnikova, T. V.; Rokhlin, E. M.; Abduganiev, E. G.; Utebaev, U. *J. Fluorine Chem.* **1975,** *6,* 227.
22. Lee, L. F.; Stikes, G. L.; Molyneaux, J. M.; Sing, Y. L.; Chupp, J. P.; Woodard, S. S. *J. Org. Chem.* **1990,** *55,* 2872.
23. Rozov., L. A.; Zeifman, Y. V.; Gambaryan, N. P.; Cheburkov, Y. A.; Knunyants, I. L. *Izv. Akad. Nauk SSSR* **1976,** 2750 (Engl. Transl. 2560); *Chem. Abstr.* **1977,** *86,* 120902t.
24. Sauvetre, R.; Normant, J. F. *Tetrahedron Lett.* **1982,** *23,* 4325.
25. Lam, W. Y.; DesMatreau, D. D. *J. Fluorine Chem.* **1982,** *18,* 441.
26. Wheaton, G. A.; Burton, D. J. *J. Fluorine Chem.* **1976,** *8,* 97; **1977,** *9,* 25.
27. Volchkov, N. V.; Lipkind, M. B.; Zabolotskich, A. V.; Nefedov, O. M. *Izv. Akad. Nauk SSSR* **1990,** 228; *Chem. Abstr.* **1990,** *113,* 23040d.
28. Gordon, A. J.; Ford, R. A. *The Chemist's Companion;* Wiley Interscience: New York, 1972; pp 107–112.
29. Koch, H. F.; Dahlberg, D. B.; McEnfee, M. F.; Klecha, C. J. *J. Am. Chem. Soc.* **1976,** *98,* 1060.
30. DePuy, C. H.; Froemsdorf, D. H. *J. Am. Chem. Soc.* **1957,** *79,* 3710.
31. DePuy, C. H.; Bishop, C. A. *J. Am. Chem. Soc.* **1960,** *82,* 2535.
32. Bartsch, R. A.; Bunnett, J. F. *J. Am. Chem. Soc.* **1968,** *90,* 408.

33. Bose, A. K.; Das, K. G.; Funke, P. T. *J. Org. Chem.* **1964,** *29,* 1202.
34. Hudlický, M. *J. Fluorine Chem.* **1972,** *2,* 1.
35. Hudlický, M. *J. Fluorine Chem.* **1984,** *25,* 353.
36. Hudlický, M.; Hall, J. A. *J. Fluorine Chem.* **1983,** *22,* 73.
37. Hudlický, M.; Glass, E. T. *J. Fluorine Chem.* **1983,** *23,* 15.
38. Hudlický, M. *Collect. Czech. Chem. Commun.* **1991,** *56,* 1680.
39. Streitwieser, A., Jr.; Holtz D. *J. Am. Chem. Soc.* **1967,** *89,* 692.
40. Streitwieser, A., Jr.; Marchand, A. P.; Pudjaatmaka, A. H. *J. Am. Chem. Soc.* **1967,** *89,* 693.
41. Schleyer, R. P.; Kos, A. Y. *Tetrahedron* **1983,** *39,* 1141.
42. Koch, H. F.; Tumas, W.; Knoll, R. *J. Am. Chem. Soc.* **1981,** *103,* 5423.
43. Hine, J.; Burske, N. W.; Hine, M.; Langford, P. B. *J. Am. Chem. Soc.* **1957,** *79,* 1406.
44. Hine, J.; Wiesboeck, R.; Ghirardelli, R. G. *J. Am. Chem. Soc.* **1961,** *83,* 1219.
45. Hine, J.; Wiesboeck, R.; Ramsay, O. B. *J. Am. Chem. Soc.* **1961,** *83,* 1222.
46. Hine, J.; Langford, P. B. *J. Org. Chem.* **1962,** *27,* 4149.
47. Koch, H. F.; Dahlberg, D. B.; Toczko, A. G.; Solsky, R. L. *J. Am. Chem. Soc.* **1973,** *95,* 2029.
48. Streitwieser, A., Jr.; Koch, H. F. *J. Am. Chem. Soc.* **1964,** *86,* 408.
49. Koch, H. F.; Dahlberg, D. B.; Lodder, G.; Root, K. S.; Touchette, N. A.; Solsky, R. L.; Zuck, R. M.; Wagner, L. J.; Koch, N. H.; Kuzemko, M. A. *J. Am. Chem. Soc.* **1983,** *105,* 2394.
50. Koch, H. F.; Koch, J. G.; Donovan, B. D.; Toczko, A. G.; Kielbania, A. J., Jr. *J. Am. Chem. Soc.* **1981,** *103,* 5417.
51. Koch, H. F.; Koch, J. G.; Koch, N. H.; Koch, A. S. *J. Am. Chem. Soc.* **1983,** *105,* 2388.
52. Koch, H. F.; Koch, A. S. *J. Am. Chem. Soc.* **1984,** *106,* 4536.
53. Koch, H. F. *Acc. Chem. Res.* **1984,** *17,* 137.
54. Koch, H. F. In *Comprehensive Carbanion Chemistry;* Buneel, E.; Durst, T., Eds.; Elsevier, Amsterdam: 1987; pp 321-360.
55. Lee, Y. G.; Bartsch, R. A. *J. Am. Chem. Soc.* **1979,** *101,* 228.
56. Hudlicky, M. *J. Fluorine Chem.* **1986,** *32,* 441.
57. Campbell, S. F.; Lancashire, F.; Stephens, R.; Tatlow, J. C. *Tetrahedron* **1967,** *23,* 4435.
58. Baciocchi, E.; Ruzziconi, R.; Sebastiani, G. V. *J. Org. Chem.* **1982,** *47,* 3237.
59. Baciocchi, E.; Ruzziconi, R.; Sebastiani, G. V. *J. Am. Chem. Soc.* **1983,** *105,* 6114.
60. Bartsch, R. A.; Cho, B. R.; Pugia, M. J. *J. Org. Chem.* **1987,** *52,* 5494.
61. Hu, C.; Long, F.; Xu, Z. *J. Fluorine Chem.* **1990,** *48,* 29.
62. Briscoe, M. W.; Chambers, R. D.; Mullins, S. J.; Nakamura, T.; Drakesmith, F. G. *J. Chem. Soc., Chem. Commun.* **1990,** 1127

63. Burdon, J.; Parsons, I. W. *J. Fluorine Chem.* **1979,** *13,* 159.
64. Roh, Z.; Vachta, J.; Trnka, B.; Kubik, J.; Siler, J. Czech. Pat. CS 203 738, 1983; *Chem. Abstr.* **1984,** *100,* 85242
65. Burton, D. J.; Spawn, T. D. *J. Fluorine Chem.* **1988,** *38,* 119.
66. Fritz, H.; Sundermeyer, W. *Chem. Ber.* **1989,** *122,* 1757.
67. Bittner, J.; Seppelt, K. *Chem. Ber.* **1990,** *123,* 2187.
68. Johncock, P. *Synthesis* **1977,** 551.
69. Smirnov, K. M.; Tomilov, A. P. *Zh. Vses. Khim. Obshchest.* **1974,** *19,* 350; *Chem. Abstr.* **1975,** *82,* 63076.
70. Bosbury, P. W. L.; Fields, R.; Haszeldine, R. N.; Moran, D. *J. Chem. Soc., Perkin Trans. 1* **1976,** 1173.
71. Blancou, H.; Moreau, P.; Commeyras, A. *Tetrahedron* **1977,** *33,* 2061.
72. Keller, T. M.; Tarrant, P. *J. Fluorine Chem.* **1975,** *6,* 105.
73. Dolbier, W. R., Jr.; Burkholder, C. R. *Tetrahedron Lett.* **1988,** *29,* 6749.
74. Fritz, H. P.; Kornrumpf, W. Z. *Naturforsch. B. Anorg. Chem., Org. Chem.* **1981,** *36B,* 1375.
75. Hahnfeld, J. L.; Burton, D. J. *Tetrahedron Lett.* **1975,** 1819.
76. Moss, R. A.; Lawrynowicz, W. *J. Org. Chem.* **1984,** *49,* 3828.
77. Chambers, R. D.; Clark, M.; MacBride, J. A. H.; Musgrave, W. K. R.; Srivastava, K. C. *J. Chem. Soc., Perkin Trans. 1* **1974,** 125.
78. Kostyanovskii, R. G.; Prosyanik, A. V.; Mishchenko, A. I.; Zaichenko, N. L.; Chervin, I. I.; Markov, V. I. *Izv. Akad. Nauk SSSR* **1980,** 882; *Chem. Abstr.* **1980,** *93,* 71405.
79. Chambers, R. D.; Shepherd, T.; Tamura, M.; Hoare, P. *J. Chem. Soc., Perkin Trans. 1* **1990,** 983
80. Baklouiti, A.; Chaabouni, M. M. *J. Fluorine Chem.* **1981,** *19,* 181.
81. Misaki, S.; Takamatsu, S. *J. Fluorine Chem.* **1984,** *24,* 531.
82. Nagai, T.; Hama, M.; Yoshioka, M.; Yuda, M.; Yoshida, N.; Ando, A.; Koyama, M.; Miki, T.; Kumadaki, I. *Chem. Pharm. Bull.* **1989,** *37,* 177.
83. Taylor, S. L.; Forbus, R. T., Jr.; Martin, J. C. *Org. Synth.* **1986,** *64,* 217.
84. Haas, A.; Lieb, M.; Praas, H. W. *J. Fluorine Chem.* **1989,** *44,* 320.
85. Abdel-Baky, S.; Moussa, A. *Synth. Commun.* **1988,** *18,* 1795.
86. Stang, P. J.; Hargrove, R. J. *J. Org. Chem.* **1975,** *40,* 657.
87. Hargrove, R. J.; Stang, P. J. *J. Org. Chem.* **1974,** *39,* 581.
88. Stang, P. J.; Dixit, V. *Synthesis* **1985,** 962.
89. Kashihara, H.; Suemune, H.; Kawahara, T.; Sakai, K. *Tetrahedron Lett.* **1987,** *28,* 6489.
90. Kurykin, M. A.; German, L. S. *Izv. Akad. Nauk SSSR* **1988,** 2649; *Chem. Abstr.* **1989,** *110,* 212054.
91. Ehrenkaufer, R. E.; MacGregor, R. R.; Wolf, A. P. *J. Org. Chem.* **1982,** *47,* 2489.
92. Gerace, M. J.; Lemal, D. M.; Ertl, H. *J. Am. Chem. Soc.* **1975,** *97,* 5584.

93. Serguchev, Y. A.; Davydova, V. G. *Zh. Org. Khim.* **1982,** *18,* 2610; *Chem. Abstr.* **1983,** *98,* 178676.
94. Cherstkov, V. F.; Galakhov, M. V.; Mysov, E. I.; Sterlin, S. R.; German, L. S. *Izv. Akad. Nauk SSSR* **1989,** 1336; *Chem. Abstr.* **1990,** *112,* 76319.
95. Cherstkov, V. F.; Galakhov, M. V.; Sterlin, S. R.; German, L. S.; Knunyants, I. L. *Izv. Akad. Nauk SSSR* **1983,** 1208; *Chem. Abstr.* **1983,** *99,* 140071
96. Wheaton, G. A.; Burton, D. J. *J. Fluorine Chem.* **1976,** *8,* 97.
97. Wheaton, G. A.; Burton, D. J. *J. Fluorine Chem.* **1977,** *9,* 25.
98. Napoli, M.; Fraccaro, C.; Scipioni, A.; Armelli, R.; Pianca, M. *J. Fluorine Chem.* **1984,** *24,* 377.
99. Ermolov, A. F.; Eleev, A. F.; Kutepov, A. P.; Sokol'skii, G. A. *Zh. Org. Khim.* **1981,** *17,* 2239; *Chem. Abstr.* **1982,** *96,* 68269.
100. Volkov, N. D.; Nazaretyan, V. P.; Yagupol'skii, L. M. *Synthesis* **1979,** 972; *Zh. Org. Khim.* **1977,** *13,* 1788; *Chem. Abstr.* **1977,** *87,* 167508.
101. Chen, Q,; Wu, S. *J. Org. Chem.* **1989,** *54,* 3023.
102. Chen, Q,; Wu, S. J. *J. Fluorine Chem.* **1990,** *47,* 509.
103. Bevan, W. I.; Haszeldine, R. N. *J. Chem. Soc., Dalton Trans.* **1974,** 2509.
104. Sharp, K. G.; Schwager, I. *Inorg. Chem.* **1976,** *15,* 1697.
105. Birchall, J. M.; Gilmore, G. N.; Haszeldine, R. N. *J. Chem. Soc., Perkin Trans. 1* **1974,** 2530.
106. Baranova, L. A.; Ryazanova, R. M.; Tumanova, A. V.; Sokolov, S. V. *Zh. Org. Khim.* **1972,** *8,* 2305; *Chem. Abstr.* **1973,** *79,* 4952
107. England, D. C.; Piecara, J. S. *J. Fluorine Chem.* **1985,** *28,* 417.
108. Hudlický, M. *J. Fluorine Chem.* **1985,** *29,* 349.

Molecular Rearrangements and Pyrolysis

by H. Koroniak

This chapter includes migration of halogen atoms, catalyst-assisted rearrangements, and thermal and photochemical rearrangements.

Migration of Halogen Atoms

Isomerization of halogenated fluoroparafins associated with *migration of halogen atoms is catalyzed by some Lewis acids* such as aluminum chloride, aluminum bromide, ferric chloride, titanium tetrachloride, and antimony pentachloride [*1*]. In these isomerizations, products tend to have geminal, rather than vicinal, fluorine atoms. The reaction has great preparative value and is complementary to the halogen-exchange reaction, which usually does not lead to trifluoromethyl derivatives. High yields of this rearrangement are sometimes decreased by intermolecular disproportionation. On the other hand, haloalkanes containing fluorine can be prepared successfully by thermal disproportionation of chlorofluoroalkanes in the presence of magnesium, aluminum, or chromium oxides pretreated with hydrogen fluoride [*2, 3, 4*]. Modification of this method involves use of catalysts obtained in situ from aluminum chloride or aluminum bromide, complexed with 1,1-dichloro- or 1,1-dibromo-1,2,2,2-trifluoroethane [*5*].

An efficient catalyst for thermal isomerizations of halofluorocarbons [*6, 7, 8, 9*] is prepared by treatment of alumina with dichlorodifluoromethane at 200–300 °C [*9*] or aluminum chloride with chlorofluorocarbons in the presence of metals [*10*] or palladium on alumina [*11*]. These catalysts are far more efficient than aluminum halides themselves (equations 1 and 2).

[12] $$CHFClCClF_2 \xrightarrow[100^\circ C]{AlCl_3} CF_3CHCl_2 \quad 100\% \qquad \textbf{1}$$

[13] $$CF_3CFClCF_2Cl \xrightarrow[400^\circ C,\ flow]{AlCl_3} CF_3CCl_2CF_3 \quad 55\% \qquad \textbf{2}$$

Isomerization of fluoroolefins by a shift of a double bond is catalyzed by halide ions [*1*]. The presence of crown ether makes this reaction more efficient [*14*]. Prolonged reaction time favors the rearranged product with an internal double bond (equations 3–5). Isomerization of perfluoro-1-pentene with cesium fluoride yields perfluoro-2-pentenes in a *Z:E* ratio of 1:6 [*15*]. Antimony pentafluoride also causes isomerization of olefins leading to more substituted products [*16*].

0065–7719/95/0187–0913$08.90/1

[17] $(CF_2CF{=}CF_2)_2 \xrightarrow[20^\circ C,\ 80h]{CsF} (CF_3CF{=}CF)_2$ 100% **3**

[14] $(CF_3)_2CFCF{=}CFCF_3 \xrightarrow[18\text{-crown-6},\ 40^\circ C,\ 3h]{KF,\ CH_3CN} (CF_3)_2C{=}CFC_2F_5$ 96% **4**

[18] $R_FCF_2CF{=}CF_2 \xrightarrow[diglyme,\ 40^\circ C]{CsF} R_FCF{=}CF_3$ 93-96% **5**

$R_F{=}C_3F_7,\ C_6F_{13}$

Isomerization of olefins can be accomplished also in the gas phase in a flow reactor with alumina or nickel oxide as catalyst. Thus perfluoro-2-butene can be prepared by passing perfluoro-1-butene over these catalysts at 150–400 °C [*19*]. Similar results may be obtained in an autoclave with the same catalysts. Perfluoro-1-butene isomerizes almost quantitatively at 250 °C after 100 h to perfluoro-2-butene [*20*].

Lewis acids promote migration of fluorine in halofluoroalkenes to yield isomers, which can be transformed easily into perfluorinated alkynes [*21, 22*] (equation 6).

[21,22] **6**

$CBrF_2CBr{=}CF_2 \xrightarrow[\text{or } AlBr_3]{AlCl_3} CF_3CBr{=}CFBr \xrightarrow[dioxane]{Zn} CF_3C{\equiv}CF$

>95% 43%

2,3-Dibromotetrafluoropropene passed over copper yields tetrafluoroallene (39%) and perfluoropropyne (11%) [*23*]. Similarly, perfluorocyclobutene in a nickel flow reactor in contact with cesium fluoride at temperatures over 450 °C yields perfluorobutyne (40%) [*24*]. At 590–600 °C, the yield is greater than 90% [*25*].

Rearrangement of fluorine with concomitant ring opening takes place in fluorinated epoxides. Hexafluoroacetone can be prepared easily from perfluoropropylene oxide by isomerization with a fluorinated catalyst like alumina pretreated with hydrogen fluoride [*26, 27, 28*]. In ring-opening reactions of epoxides, the distribution of products, ketone versus acyl fluoride, depends on the catalyst [*29*] (equation 7). When cesium, potassium, or silver fluoride are used as catalysts, dimeric products also are formed [*29*].

[29] **7**

$F_2C(O)CF(CF_2)_4H \xrightarrow{100^\circ C} H(CF_2)_4COCF_3 + H(CF_2)_5COF$

Catalyst	$H(CF_2)_4COCF_3$	$H(CF_2)_5COF$
SbF_5	96%	0%
$SbCl_5$	86%	0%
$AlCl_3$	56%	7%

Catalyst-Assisted Rearrangements

In most cases, the cleavage of a carbon–carbon bond causes rearrangements of the carbon skeleton. Ring contraction, ring expansion, and alkyl group migration are observed under different conditions. These transformations proceed in most cases in the presence of catalysts at elevated temperatures. Examples where only temperature causes rearrangements will be discussed in the next section.

Cesium fluoride in dimethylformamide catalyzes the *isomerization of fluorinated cyclobutenes,* perfluorobipyrimidines, and their oligomers to products with expanded rings [*30, 31, 32*]. The product distribution in cobalt trifluoride fluorination depends strongly on the temperature of the reaction [*33*]. Fluorinated 1-dimethylamino-5,6,7,8-tetrafluoro-1,4-dihydro-1,4-ethenonaphthalene rearranges in protic media to a biphenyl derivative [*34*] (equation 8).

[34] **8**

EtOH, reflux; $N(CH_3)_2$; 85%

Studies of solvolysis of similar polyfluorinated polycyclic aromatic systems, such as 2,3-(tetrafluorobenzo)bicyclo[2.2.2]octadienes and related compounds, proved the ionic mechanism of this rearrangement [*35, 36, 37*] (equation 9). Possible nonclassical carbonium ion involvement has been discussed [*38, 39, 40, 41*].

[35] **9**

AcOAg, 20°C; OAc; *exo, endo* 87% + OAc 8%

Strong acids or superacid systems generate stable fluorinated carbocations [*40, 42*]. Treatment of tetrafluorobenzbarrelene with arenesulfonyl chlorides in nitromethane–lithium perchlorate yields a crystalline salt with a rearranged benzobarrelene skeleton [*43*]. Ionization of polycyclic adducts of difluorocarbene and derivatives of bornadiene with antimony pentafluoride in fluorosulfonyl chloride yields stable cations [*44, 45*].

In most cases, fluorinated substituents, when not directly placed at the reaction center, do not influence much the rearrangement routes [*1*]. *Favorskii rearrangement* of a substrate bearing a trifluoromethyl group proceeds as expected [*46*] (equation 10).

[46] **10**

$$\text{3-}F_3C\text{-}C_6H_4\text{-}CHBr\text{-}COBr \xrightarrow[-20^\circ C \text{ to } 0^\circ C]{MeONa,\ MeOH} \text{3-}F_3C\text{-}C_6H_4\text{-}CH{=}CH\text{-}COOCH_3$$

79% Z:E = 99:1

The *Curtius degradation* of acyl azides prepared either by treatment of acyl halides with sodium azide or trimethylsilyl azide [*47*] or by treatment of acyl hydrazides with nitrous acid [*1*] yields primarily alkyl isocyanates, which can be isolated when the reaction is carried out in aprotic solvents. If alcohols are used as solvents, urethanes are formed. Hydrolysis of the isocyanates and the urethanes yields primary amines.

In the *Schmidt reaction* of fluorinated dicarboxylic acids, the appropriate amides can be obtained in fairly good yield [*48*]. Complications arise from possible cyclization if the fluorine atom is in the δ-position relative to the newly formed amino group [*1*]. Fluorinated aromatic ethers, upon heating in dimethylformamide, undergo Smiles rearrangement to give diarylamines [*49, 50*] (equation 11).

11

[49]

$$\text{2-}CH_3CONH\text{-}C_6H_4\text{-}O\text{-}C_5F_4N \xrightarrow[70\text{-}80^\circ C,\ 1h]{DMF,\ K_2CO_3} \text{2-}HO\text{-}C_6H_4\text{-}N(COCH_3)\text{-}C_5F_4N$$

51%

Benzidine rearrangement of fluorinated hydrazobenzenes proceeds with a high yield in the presence of a strong acid [*51*] (equation 12). Ring contraction occurs when 2-bromoheptafluoronaphthalene is treated with antimony pentafluoride [*52*] (equation 13).

[51]

$$HC_6F_4\text{-}NH\text{-}NH\text{-}C_6F_4H \xrightarrow[-20^\circ C \text{ to } +20^\circ C]{SbF_5\ /\ HF} H_2N\text{-}C_6F_4\text{-}C_6F_4\text{-}NH_2$$ **12**

[52] **13**

$$\text{octafluoro-2-bromonaphthalene (heptafluoro-2-bromonaphthalene)} \xrightarrow[\text{up to } 300^{\circ}\text{C}]{SbF_5} \text{perfluoro-1-methylindane, } 20\text{-}25\%$$

Treatment of perfluorinated alkenes with oxidizing agents like sodium hypochlorite gives stable epoxides [*53, 54*], which can be cleaved by the fluoride anion to give alkenes, ketones, and acid fluorides [*53, 54, 55*]. Thus perfluoropropionyl fluoride can be prepared by passing hexafluoropropylene oxide through potassium or cesium fluoride deposited on carbon [*56*]. Treatment of perfluoropropene oxide with anhydrous hydrogen fluoride in an autoclave at 100 °C for 24 h yields hexafluoroacetone almost quantitatively [*57*]. Antimony pentafluoride can be used also to open an epoxide ring [*58*]. Perfluorinated unsaturated acyl fluorides can be prepared by the reaction of fluorosulfonates with cesium fluoride [*59*] (equation 14).

[59] **14**

$$R_FCF{=}CFCF_2OSO_2F \xrightarrow[50\text{-}80^{\circ}C]{CsF} R_FCF{=}CFCOF \; (78\text{-}83\%) \xrightarrow[\text{reflux}]{MeOH} R_FCF{=}CFCOOCH_3 \; (100\%)$$

$$R_F = C_3F_7,\ (CF_3)_2CF$$

Stereoselective synthesis of trifluoromethylated olefins [*60, 61, 62*], dienes [*63*], and fluorinated hydroxyalkenoic acids and their esters can be accomplished by *Claisen rearrangement* or, in general, 3,3-*sigmatropic rearrangement* [*64, 65, 66*]. Pyrolysis of allyl pentafluorophenyl ether gives 4-allylpentafluoro-2,5-cyclohexadienone [*1*]. Fluorinated aromatics with an unsaturated side chain isomerize similarly to benzofuran or its analogues [*67, 68, 69, 70*] (equation 15). 3-Fluorosalicylaldehyde is synthesized in high yield from 2-fluorophenol in three steps via the Claisen rearrangement of allyl phenyl ether and subsequent oxidation of the product [*71*].

[70] **15**

$$C_6F_5OCH_2C{\equiv}CH \xrightarrow[\text{gas phase, } 370^{\circ}C]{SiO_2} \text{4,5,6,7-tetrafluoro-2-(fluoromethyl)benzofuran } (CH_2F),\ 8\%$$

An alkyl group can also migrate from oxygen to nitrogen or phosphorus [*1, 72*] (*Michaelis–Arbuzov rearrangement*). With this methodology, tetrafluoropyridine phosphonates and phosphinates can be obtained [*73, 74*]. **Chlorine fluoride** can react with fluoroalkyl sulfites. The reaction leads to fluorosulfates via the Arbuzov rearrangement [*75*] (equation 16).

Migration from nitrogen to carbon is observed also in *aza–Cope rearrangement* [*76*]. Ring expansion occurs in thermal *rearrangement of aziridine* derivatives [*77*] (equation 17).

Migration of the perfluoromethylthio group occurs in perfluoroalkyl acetylenic thiol esters and substituted *N*-methylpyrroles [*78, 79*] (equation 18).

[*75*]

$$(RO)_2S{=}O \xrightarrow[-196^{o}C\text{ to RT}]{ClF,\ 14h} ROSO_2F \quad 45\text{-}100\% \qquad (16)$$

$$R = CH_2CF_3,\ CH(CF_3)_2$$

[*77*]

$$C_6H_5NHCON\text{(aziridine, }CF_3, C_6H_5) \xrightarrow[CH_3COC_2H_5\text{, reflux}]{NaI} \text{oxazoline}(H_5C_6, CF_3, C_6H_5NH) \quad 48\% \qquad (17)$$

[*79*]

$$R_FC{\equiv}CCOSCH_3 \xrightarrow[2.\,HCl]{1.\,NaOH/H_2O} (R_F)(CH_3S)C{=}CHCOOH \quad 92\text{-}97\% \qquad (18)$$

Thermal Rearrangements and Pyrolysis

Polyfluoroparafins, fluorocarbons, and other perfluoro derivatives show remarkable heat stability. They are usually stable at temperatures below 300 °C. Thermal decomposition at 500–800 °C, however, causes all possible splits in the molecules and produces complex mixtures that are difficult to separate. For preparative purposes, only pyrolyses that do not yield complicated mixtures of products are of interest [*1*]. The pyrolytic reactions of polyfluoro and perfluoro derivatives, when carried out at 500–1100 °C, represent the most useful route to preparative generation of perfluoroolefins on the laboratory scale [*1*].

The yields of thermal rearrangements of some **perfluorinated olefins** are very low. The fact that perfluorocyclobutene yields perfluoro-1,3-butadiene at 650 °C only in a 12% yield [*1*] is due to the higher thermodynamic stability of perfluorocyclobutene compared with the-open chain product [*124*].

Pyrolysis studies have often been applied to chlorotrifluoroethylene and tetrafluoroethylene. Pyrolysis of chlorotrifluoroethylene at 560–590 °C yields 70–83% of a mixture containing both linear and cyclic dimers of chlorotrifluoroethylene and chlorofluoropropenes. During flow pyrolysis at 750–800 °C, chlorotrifluoroethylene yields a complex mixture of chlorofluorinated benzenes, toluenes, and xylenes [*81*]. *p*-Difluorobenzene at 1096 °C in a flow reactor (contact time, 5 s) yields a mixture of *o*-, *m*-, and *p*-difluorobenzenes (*o*:*m*:*p* ratio, 4:18:78); *o*- and *m*-isomers behave similarly [*82*].

In general, *hexafluoropropene* and branched *perfluoroalkyl halides* can be prepared in high yield by thermal fragmentation of ***chlorofluoroalkanes*** [*83, 84*] or halogenated cycloalkanes [*85*]. *Chlorofluorocarbenes* also can be generated from chlorotrifluoroethylene in the presence of oxygen in a discharge flow reactor [*86*].

Pyrolysis of tetrafluoroethylene has been studied in detail as a function of temperature and pressure. The reaction gives a different distribution of several products where perfluorocyclobutene, isomeric perfluorobutenes, and perfluoropropenes prevail [*1*].

Thermal decomposition is a major route to smaller perfluorinated molecules. Tetrafluoroethylene pyrolyzed at 1100–1300 °C with carbon dioxide gives a mixture of tetrafluoromethane (19.9%), hexafluoroethane (61.3%), and carbonyl fluoride (18.6%) [*87*].

A new thermal preparation of fluorinated species is *copyrolysis*. Copyrolysis of fluorinated compounds like perfluorobenzene, fluorinated aromatic anhydrides, and fluorinated heteroaromatics with tetrafluoroethylene or other fluorinated olefins is a useful method of preparing fluorinated olefins [*88, 89*], functionalized fluoroaromatics [*90, 91, 92, 93, 94, 95*], fluorinated benzocycloalkanes [*80, 96, 97, 98, 99, 100*], fluorinated heterocycles [*80, 93, 101, 102, 103*], and fluorinated polycyclic compounds [*104*] (equations 19 and 20).

[101] 19

620-650°C

19% 7%

[104] 20

650-720°C

40-45%

A mixture of nitromethane and hexafluorobenzene, when thermolyzed at 550 °C, yields pentafluorotoluene and pentafluorophenol as major products. The formation of nitrosyl and nitryl fluorides is probably a driving force in this transformation [*105*]. A potential general preparative route to various perfluorovinyl amines is pyrolytic decarboxylation of potassium salts of perfluoro-2-(dialkyamino)propionic acid in the presence of sodium or potassium carbonate [*106*]. During thermolysis of fluorinated aromatic azides in the presence of an excess of 1,3,5-trimethyl- or 1,3,5-trimethoxybenzene, diaryl amines are formed, probably via nitrene insertion into a carbon–hydrogen bond [*107*]. Perfluorinated nitroso alkanes, in the absence of air, undergo complete thermal decomposition at 80 °C to yield complex mixtures of imines, nitroso compounds, and amines [*108*].

cis-1,2-Dimethyl-3,3,4,4-tetrafluorocyclobutane isomerizes reversibly to its *trans*-isomer via a free radical cleavage of the C(1)–C(2) bond. The competitive reaction is irreversible formation of the more stable 1,1-difluoropropene [*109*]. Pyrolytic dehalogenation of 3,4-dichloro-1,2,3,4-tetrafluorocyclobutene was unsuccessful, whereas pyrolysis of the *trans*-3,4-diiodo analogue yielded perfluorocyclooctatetraene and other products [*110*] (equation 21).

[110] **21**

250°C; 58%

580°C; 31% + 24% + ICF=CFCF=CFI *EE + EZ + ZZ isomers* 37%

Bis(trifluoromethyl)disulfide loses sulfur when heated at 425–435 °C and gives perfluorodimethyl sulfide in 88% yield [*111*]. Similarly, thermolysis of perfluorotetramethyldithietan gives a high yield of perfluoro-2,3-dimethyl-2-butene [*112*] (equation 22).

[112] **22**

$(F_3C)_2C(S)_2C(CF_3)_2 \xrightarrow{325^\circ C} (F_3C)_2C{=}C(CF_3)_2$ 96%

Perfluorotetramethylthiadiphosphanorbornadiene and bis(trifluoromethyl)-thiadiphosphole can be prepared by thermolysis of an adduct of methanol and hexakis(trifluoromethyl)-1,4-diphosphabarrelene with sulfur [*113*] (equation 23).

Pyrolysis of the adduct of hexafluorinated Dewar benzene and phenyl azide results in ring expansion giving azepine, which photochemically yields an intramolecular 2+2 adduct, a good dienophile for the Diels–Alder reaction [*114, 115*] (equation 24). Thermolysis of fluorinated derivatives of 1,5-diazabicyclo-

[3.3.0]oct-2-ene leads to diazepine derivatives with extrusion of hexafluoroisobutylene [*116*] (equation 25).

[113] 23

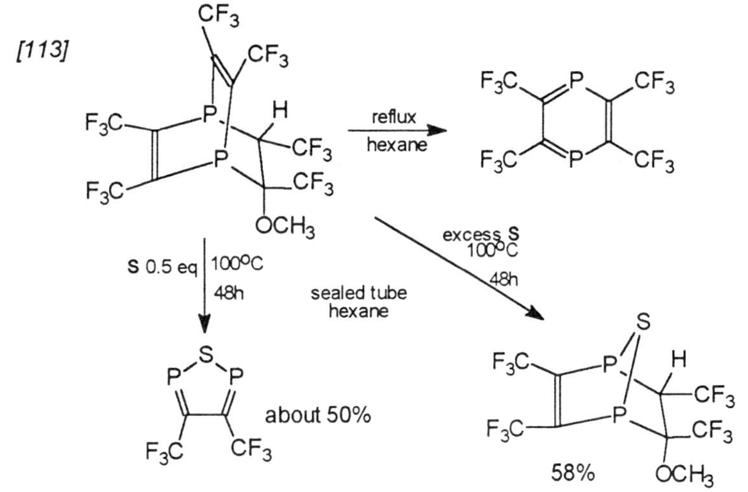

[114,115] 24

[116] 25

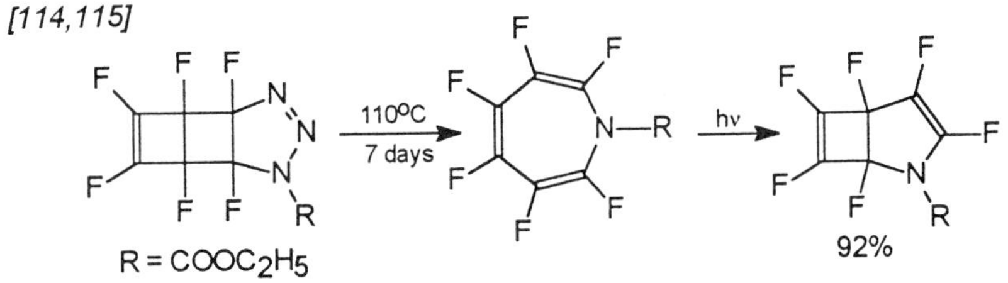

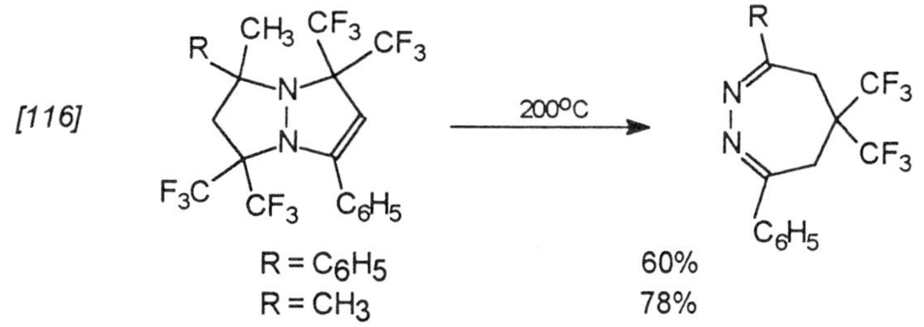

Fluorination and skeletal transformation of fluorinated cycloalkanes occurs in the reaction with antimony pentafluoride at high temperature [*117*]. In the case of perfluorinated benzocyclobutanes, an unexpected alicyclic ring cleavage has been observed. Perfluorinated alkyl benzocyclobutanes, when treated with antimony pentafluoride, can be converted to perfluorinated styrenes and then transformed to perfluorinated indans [*118, 119*].

Photolysis sometimes gives the same products as pyrolysis. An example is photolysis of perfluorooxazetidines [*1*]. Frequently, however, the results of photolysis and pyrolysis are different. Thermal rearrangement of fluorinated pyridazines at 300 °C yields a mixture of fluorinated pyrimidines and pyrazines [*120, 121, 122, 123*].

Rearrangement studies give an interesting insight into the specific effect of fluorine on the thermodynamic stability and rearrangement kinetics of fluorinated cyclopropanes. Fluorine decreases the thermodynamic stability of the cyclopropyl ring, in contrast with the generally observed effect of fluorine increasing the stability of molecules to which it is introduced [*124*].

Fluorine specifically weakens the carbon–carbon bond opposite the carbon atom bearing fluorine by about 4–5 kcal/mol per fluorine atom [*125*]. It has been shown experimentally that isomerization of *cis*-1,1-difluoro-2,3-dimethylcyclopropane to the *trans*-isomer has an activation energy (E_a)of 49.7 kcal/mol [*126*], which is about 10 kcal/mol lower than that of the parent hydrocarbon [*127*] (equation 26).

[126,127] **26**

	E_a (kcal/mol)	log A
R = H	59.4	15.25
R = F	49.8	14.7

The strength of the carbon–carbon bond adjacent to a difluoromethylene group is hardly affected by the presence of fluorine on the cyclopropyl ring. When heated, 2,2-difluoromethylenecyclopropane undergoes methylene–cyclopropane rearrangement [*128*]. Under kinetic control, 2,2-difluoro-1-methylenecyclopropane and (difluoromethylene)cyclopropane are formed in a 2:1 ratio, although the latter is slightly more stable [*129*] (equation 27).

[128,129] **27**

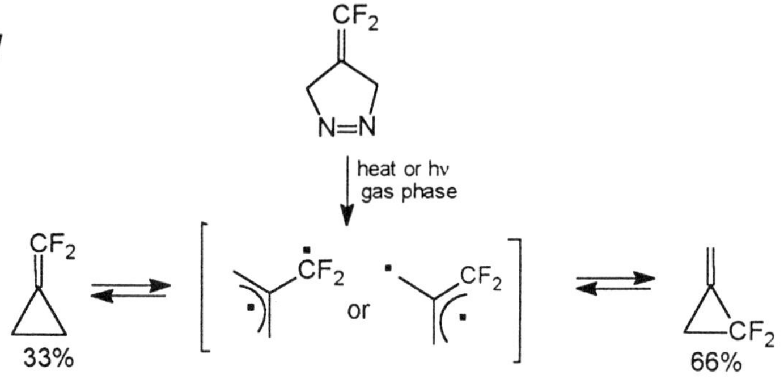

heat or hν
gas phase

33% 66%

1,1-Difluorospiro[2.2]pentane rearranges to 2,2-difluoromethylene cyclobutane [*130*] (equation 28).

2,2-Difluoro-1-vinylcyclopropane undergoes a free radical 1,3-rearrangement with ring expansion yielding difluorocyclopentene derivatives [*131, 132*] (equation 29). Similar, but more complex rearrangement occurs with 1,1-difluoro-4-methylenespiro[3.2]hexane (equation 30).

[130] 28

heat gas phase
heat gas phase
CF_2 >85%
$=CF_2$

[131,132] 29

200°C gas phase
R=H 96% 4 %
R=CH_3 97% 3%
conversion > 90%

[129,131] 30

200°C gas phase
5%
95%

In some cases, when difluorocyclopropyl derivatives have the appropriate geometry, they rearrange thermally in a concerted manner, and the energy required is lower than that required by radical formation [*133, 134*] (equations 31–33).

The common side reaction in most thermal studies of fluorine-substituted cyclopropanes is *difluorocarbene extrusion*. Increasing the number of fluorine substituents on the cyclopropane ring significantly increases the rate of difluorocarbene extrusion [*135, 136, 137*].

Electrocyclic ring opening of cyclobutenes is a general method to obtain selectively appropriate dienes. This reaction follows orbital symmetry rules, and the structure of the product depends on the geometry of the starting material. Thermal ring opening of *trans*-perfluoro-3,4-dimethylcyclobutene gives (*Z,Z*)- and (*E,E*)-perfluoro-2,4-hexadienes. The reaction is kinetically controlled, and at low temperatures, only the (*Z,Z*)-isomer is formed [*138, 139, 140*]. The selectivity of this reaction has been explained theoretically and depends on the electron-donating or electron-withdrawing properties of a substituent at position 3 or 4 of cyclobutene [*141*]. On the basis of the stability of the transition state, the electron-donating substituent at position 3 or 4 preferably rotates outward [*141*].

At elevated temperatures perfluoro-3,4-dimethyl (or -diethyl) cyclobutenes equilibrate with isomeric perfluorobutadienes [*138, 139*] via nonstereoselective ring opening or *cis–trans* isomerization. The distribution of the products of the rearrangement of perfluoro-3,4-dimethylcyclobutene at 355 °C and of perfluoro-3,4-diethylcyclobutene at 300 °C at equilibrium is shown in equation 34 and depends on reaction temperature [*138, 139*].

[133] **31**

185°C / retro Diels-Alder; 185°C slow

31% 69% traces

[134] **32**

53-87°C, 1,5 hydrogen shift

95%

[134] **33**

200°C radical

65% + 30%

[138,139] **34**

R	cis	trans	EE	EZ	ZZ
CF_3	5%	29%	14%	35%	17%
C_2F_5	0%	8%	27%	47%	18%

Results of studies of the electrocyclic ring opening of 3-fluoro-, 3,3-difluoro-, and 3-trifluoromethylcyclobutene are consistent with the theoretical predictions of the effect of fluorine on this reaction [*142*]. Surprisingly, fluorinated analogues of hexatriene–cyclohexadiene systems undergo complex rearrangements mainly via free radical mechanisms and not by electrocyclic ring opening as expected [143] (equation 35).

Photochemical Rearrangements

In 1966, a photochemical rearrangement by ultraviolet (UV) irradiation of hexafluorobenzene to hexafluorobicyclo[2.2.0]hexa-2,5-diene was achieved. Since then, many reactions analogous to the valence tautomerism of benzene and bicyclo[2.2.0]hexadiene (i.e., Dewar benzene), as well as of fluorinated benzvalene and prismane, have been described [*1*]. Studies on the synthesis and properties of analogues of Dewar benzene have been reported [*144, 145*].

UV irradiation of a mixture of hexafluorobenzene in the presence of oxygen gives Dewar benzene oxide also as a minor product, which undergoes thermal transformation to hexafluorocyclohexa-2,4-dienone [*146*] (equation 36).

When irradiated, fluorinated isomers of Dewar benzene yield prismane derivatives that rearrange thermally to benzene. Photolysis of hexakis(trifluoromethyl)benzvalene ozonide gives tetrakis(trifluoromethyl)cyclobutadiene and its dimer [*147*].

[143] **35**

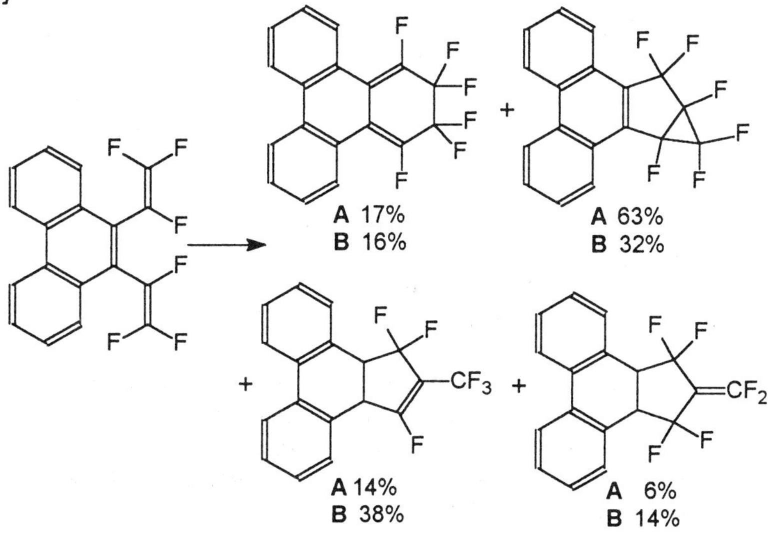

A 17%
B 16%

A 63%
B 32%

A 14%
B 38%

A 6%
B 14%

A reaction time 4.7h at 184.5°C; conversion 24%
B reaction time 16.5h at 184.5°C; conversion 60%

[146] **36**

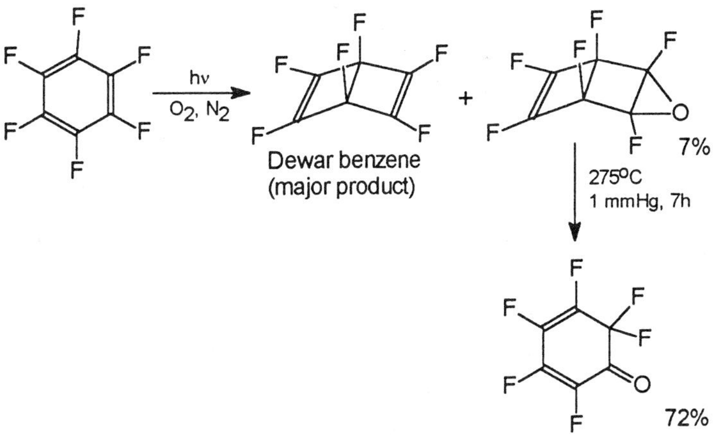

hν
O_2, N_2

Dewar benzene
(major product)

7%

275°C
1 mmHg, 7h

72%

UV irradiation of perfluoro *o*-, *m*-, and *p*-xylenes in the gas phase gives a mixture of all possible Dewar benzene isomers. Prismane valence-bond isomers are proposed to be intermediates [*148*].

Pyridines and pyridazines bearing bulky perfluorinated groups behave similarly [*149, 150, 151*], although their photochemical rearrangement is not always clearly documented [*152*]. Azaprismanes, however, have been isolated and fully characterized [*153*] (equation 37).

[153] **37**

UV irradiation of hexafluorobenzene with indene or cycloalkenes gives high yields of 2+2 adducts, which undergo further intramolecular cycloaddition to form hexafluoropolycycloalkanes [*154*] (equation 38). Photolysis of fluorinated derivatives of vinylbenzenes afford benzocyclobutenes, whereas allyl benzenes yield Dewar benzene-type products [*155*].

[154] **38**

Dewar pyrrole [*156*] and Dewar thiophene stabilized by the presence of fluorinated substituents have been successfully isolated, and their chemical properties have been studied [*157, 158, 159, 160, 161*]. The olefinic bond in these species is a good component in Diels–Alder and 1,3-dipolar cycloadditions [*157, 160, 161*]. Tetrakis(trifluoromethyl)thiophene, when irradiated, yields hexafluoro-2-butyne and 1,2,3,4-tetrakis(trifluoromethyl)-5-thiabicyclo[2.1.0]pent-2-ene (Dewar thiophene) [*162*] (equation 39). Dewar thiophene rearranges to thiophene at 160 °C [*163*]. The sulfur atom in Dewar thiophene can be removed easily to yield derivatives of cyclobutene [*157*].

[162,163] **39**

F_3C, CF_3, F_3C, CF_3, S — hν, λ>214 nm or λ=254 nm/Hg → ; ← 160°C, 5h — major product

Photochemical isomerization of partially fluorinated benzenes also leads to relatively stable Dewar isomers [*164, 165*]; however, at elevated temperatures, they rearrange to benzene derivatives [*166*].

Photolysis of 2,3,5,6-tetrakis(trifluoromethyl)-1,4-diphoshabenzene gives 1,3,4,6-tetrakis(trifluoromethyl)-2,5-diphosphatricyclo[3.1.0.0^{2,6}]hex-3-ene, an analogue of benzvalene containing phosphorus atoms in the ring system [*167, 168*] (equation 40).

[167,168] **40**

heat or hν → ; hν, 72h perfluoropentane → 41%

Fluorinated bicyclo[2.2.0]hex-5-en-2-ones prepared from Dewar benzene derivatives serve as synthons of cyclobutenes [*169*]. Photochemical cleavage of fluorinated azacyclohexadiene in the gas phase also gives fluorinated cyclobutene [*170*] (equation 41).

syn- And *anti*-octakis(trifluoromethyl)tricyclo[4.2.0.0^{2,5}]hexa-3,7-diene, when irradiated in a fluorocarbon solvent, yield perfluorooctamethylcubane and perfluorooctamethylcuneane, respectively [*171*] (equation 42).

Perfluoroolefins isomerize photochemically to yield less substituted olefins [*172*]. Photolysis of polyfluorotriarylnitrones leads to polyfluorotriaryloxaziridines [*173*] (equation 43).

[170] **41**

$CF(CF_3)_2$ ... λ=254 nm, gas phase, 72h → ... NCF_3, $(CF_3)_2FC$ ... 87%

[171] **42**

hν, 36h → 20% + 15%; 300°C 16h; 300°C 1h → 100%

syn or anti

R=CF_3

[173] **43**

hν → >60%

R, R_1, R_2 = C_6H_5, C_6F_5

Irradiation of cyclic ketones having perfluoroalkyl groups causes cleavage of a ring to yield acyclic products [*174*] (equation 44). Similarly, perfluorinated ketones undergo decarbonylation when irradiated [*175*]. Gas-phase photolysis of perfluorodiazoketones, in the presence of a trapping agent, yields fluorinated furan as a major product [*176*] (equation 45).

[174] **44**

hν, C_6H_6, 10% CH_3OH → $RCH{=}CH(CH_2)_3CH(OCH_3)_2$ 29%-74%

R = CF_3, C_2F_5, C_6F_{13}

[176] 45

$$(F_3C)C(=O)C(F_3C)=N^+=N^- \xrightarrow[F_3CC\equiv CCF_3]{h\nu} \text{tetrakis(trifluoromethyl)furan (major product)}$$

Photocyclization and subsequent dehydrofluorination yield specifically fluorinated arenes [*177*] (equation 46).

Perfluorotropilidene isomers have been obtained during UV irradiation of perfluoronorbornadiene and subsequent thermal rearrangement of the primarily obtained products [*178*].

Irradiation at 20 °C of *perfluoro ethers* containing a carbonyl group causes almost quantitative decarbonylation [*179*]. Aziridine undergoes photochemical addition with methyl trifluoroacetate [*180*] (equation 47).

Trifluoromethyl or pentafluorophenyl *phosphonates* can be produced photochemically by irradiation of triethylphosphites with fluorinated iodides [*181*] (equation 48).

[177] 46

$$\text{2-}(CF_2{=}CH)\text{biphenyl} \xrightarrow[\text{cyclohexane}]{h\nu,\ 24h} \text{9,9-difluoro-9,10-dihydrophenanthrene} \xrightarrow{KOH} \text{9-fluorophenanthrene}\ \ 60\%$$

[180] 47

$$\text{2-}H_5C_6\text{-3,3-dimethyl-azirine} \xrightarrow[CF_3COOCH_3]{h\nu} \text{4-}H_5C_6\text{-5-}CF_3\text{-2,2-dimethyl-oxazoline}\ \ 77\%$$

[181] 48

$$(C_2H_5O)_3P \xrightarrow{h\nu,\ RI} (C_2H_5)_2P(R)=O$$

$R = CF_3, C_6F_5$ 32%-51%

References are listed on pages 930–937.

References for pages 913–929

1. Hudlický, M. *Chemistry of Organic Fluorine Compounds;* Ellis Horwood: Chichester, U.K., 1976.
2. Asahi Glass Co. Ltd. Jpn. Pat. 57 197 232, 1982; *Chem. Abstr.* **1983,** *98,* 215170q.
3. Asahi Glass Co. Ltd. Jpn. Pat. 57 197 233, 1982; *Chem. Abstr.* **1983,** *98,* 215171r.
4. Okazaki, S.; Eriguchi, H. *Chem. Lett.* **1980,** 891.
5. Gozzo, F.; Troiani, N.; Piccardi, P. Eur. Pat. Appl. EP 79 481, 1983; *Chem. Abstr.* **1983,** *99,* 130303p.
6. Morikawa, S.; Samejima, S.; Yoshitake, M.; Tatematsu, S.; Tanuma, T. Jpn. Pat. 02 115 135, 1990; *Chem. Abstr.* **1990,** *113,* 114636u.
7. Morikawa, S.; Samejima, S.; Yoshitake, M.; Tatematsu, S.; Tanuma, T. Jpn. Pat. 02 108 639, 1990; *Chem. Abstr.* **1990,** *113,* 97029d.
8. Morikawa, S.; Yoshitake, M.; Tatematsu, S. Jpn. Pat. 02 40 332, 1990; *Chem. Abstr.* **1990,** *113,* 23097c.
9. Morikawa, S.; Yoshitake, M.; Tatematsu, S.; Yoneda, S.; Yanase, G. Jpn. Pat. 01 258 630, 1989; *Chem. Abstr.* **1990,** *112,* 178054q.
10. Zawalski, R. C. U.S. Pat. 4 925 993, 1989; *Chem. Abstr.* **1990,** *113,* 151806v
11. Gervasutti, C. Eur. Pat. Appl. EP 363 925, 1990; *Chem. Abstr.* **1990,** *113,* 131552f.
12. Sonoyama, H.; Osaka, Y. Jpn. Pat. 78 121 710, 1978; *Chem. Abstr.* **1979,** *90,* 54456e.
13. Daikin Kogyo Co. Jpn. Pat. 60 78 925, 1985; *Chem. Abstr.* **1985,** *103,* 122978r.
14. Ozawa, M.; Komatsu, T.; Matsuoka, K. Ger. Pat. 2 706 603, 1977; *Chem. Abstr.* **1977,** *87,* 183969m.
15. Kurykin, M. A.; German, L. S. *Izv. Akad. Nauk SSSR* 1**981,** 2646; *Chem. Abstr.* **1982,** *96,* 68253x.
16. Filyakova, T. I.; Belen'kii, G. G.; Lure, E. P.; Zapevalov, A. Y.; Kolenko, I. P.; German, L. S. *Izv. Akad. Nauk SSSR* **1979,** 681; *Chem. Abstr.* **1979,** *91,* 38844.
17. Daikin Kogyo Co. Jpn. Pat. 82 85 329, 1982; *Chem. Abstr.* **1982,** *97,* 144327n.
18. Filyakova, T. I.; Kodess, M. I.; Peschanskii, N. V.; Zapevalov, A. Y.; Kolenko, I. P. *Zh. Org. Khim.* **1987,** *23,* 1858; *Chem. Abstr.* **1988,** *109,* 37476e
19. Daikin Kogyo Co. Jpn. Pat. 80 130 926, 1980; *Chem. Abstr.* **1981,** *94,* 102806a
20. Daikin Kogyo Co. Jpn. Pat. 80 133 321, 1980; *Chem. Abstr.* **1981,** *94,* 83591h.

21. Haszeldine, R. N.; Banks, R. E.; Taylor, D. R. U.S. Pat. 3 709 948, 1973; *Chem. Abstr.* **1973,** *78,* 71391z.
22. Haszeldine, R. N.; Banks, R. E.; Taylor, D. R. Br. Pat. 1 285 335, 1972; *Chem. Abstr.* **1972,** *77,* 151443n.
23. Banks, R. E.; Davies, W. D.; Haszeldine, R. N.; Taylor, D. R. *J. Fluorine Chem.* **1977,** *10,* 487.
24. Chambers, R. D.; Taylor, G. Br. Pat. 2 044 768, 1980; *Chem. Abstr.* **1981,** *94,* 175815n.
25. Chambers, R. D.; Jones, C. G. P.; Taylor, G.; Powell, R. L. *J. Fluorine Chem.* **1981,** *18,* 407.
26. Asahi Glass Co. Ltd. Jpn. Pat. 58 62 130, 1983; *Chem. Abstr.* **1984,** *100,* 5861z.
27. Asahi Glass Co. Ltd. Jpn. Pat. 58 62 131, 1983; *Chem. Abstr.* **1984,** *100,* 5862a.
28. Osaka, Y.; Higashizuka, T. Jpn. Pat. 78 25 512, 1978; *Chem. Abstr.* **1978,** *89,* 59660f.
29. Zapevalov, A. Y.; Kolenko, I. P.; Plashkin, V. S. *Zh. Org. Khim* **1975,** *11,* 1622; *Chem. Abstr.* **1975,** *83,* 192954x.
30. Chambers, R. D.; Kirk, J. R.; Taylor, G.; Powell, R. L. *J. Chem. Soc., Perkin Trans. 1* **1982,** 673.
31. Chambers, R. D.; Taylor, G.; Powell, R. L. *J. Chem. Soc., Chem. Commun.* **1979,** 1062.
32. Barnes, R. N.; Chambers, R. D.; Hewitt, C. D. *J. Fluorine Chem.* **1986,** *34,* 59.
33. Coe, P. L.; Sellers, S. F.; Tatlow, J. C.; Fielding, H. C.; Whittaker, G. *J. Fluorine Chem.* **1981,** *18,* 417.
34. Heaney, H.; Ley, S. V. *J. Chem. Soc., Perkin Trans. 1* **1974,** 2698.
35. Lobanova, T. P.; Barkhash, V. A. *Zh. Org. Khim.* **1973,** *9,* 2281; *Chem. Abstr.* **1974,** *80,* 36906r.
36. Slyn'ko, N. M.; Mironova, M. V.; Barkhash, V. A. *Zh. Org. Khim.* **1976,** *12,* 1907; *Chem. Abstr.* **1977,** *86,* 71247x.
37. Spivak, A. Y.; Lobanova, T. P.; Chertok, V. S.; Podgornaya, M. L.; Barkhash, V. A. *Zh. Org. Khim.* **1976,** *12,* 1210; *Chem. Abstr.* **1976,** *85,* 93415g.
38. Povolotskaya, N. N.; Kollegova, M. I.; Rumyantseva, A. G.; Spivak, A. Y.; Barkhash, V. A. *Zh. Org. Khim.* **1972,** *8,* 1037; *Chem. Abstr.* **1972,** *77,* 61665g.
39. Rumyantseva, A. G.; Petrov, A. K.; Kollegova, M. I.; Barkhash, V. A. *Zh. Org. Khim.* **1972,** *8,* 1030; *Chem. Abstr.* **1972,** *77,* 61666h.
40. Kamshii, G. T.; Mamatyuk, V. I.; Spivak, A. Y.; Barkhash, V. A. *Zh. Org. Khim.* **1979,** *15,* 1221; *Chem. Abstr.* **1979,** *91,* 192523w.
41. Kamshii, G.T.; Mamatyuk, V. I.; Slyn'ko, N. M.; Barkhash, V. A. *Zh. Org. Khim.* **1976,** *12,* 2546; *Chem. Abstr.* **1977,** *86,* 188884u.

42. Nisnevich, G. A.; Vyalkov, A. I.; Kamshii, G. T.; Mamatyuk, V. I.; Barkhash, V. A. *Zh. Org. Khim.* **1983,** *19,* 2081; *Chem. Abstr.* **1984,** *100,* 138351e.

43. Bodrikov, I. V.; Chumakov, L. V.; Novikova, T. I.; Nisnevich, G. A.; Mamatyuk, V. I.; Gatilov, Y. V.; Bagryanskaya, I. Y; Barkhash, V. A. *Zh. Org. Khim.* **1986,** *22,* 316; *Chem. Abstr.* **1986,** *105,* 152331h.

44. Hart, H.; Stein, D. L. *Tetrahedron Lett.* **1982,** *23,* 3435.

45. Nisnevich, G. A.; Mamatyuk, V. I.; Barkhash, V. A. *Zh. Org. Khim.* **1983,** *19,* 110; *Chem. Abstr.* **1983,** *99,* 37779f.

46. Engler, T. A.; Falter, W. *Tetrahedron Lett.* **1986,** *27,* 4119.

47. Lutz, W.; Sundermayer, W. *Chem. Ber.* **1979,** *112,* 2158.

48. Yagupol'skii, L. M.; Alekseenko, A. N.; Il'chenko, A. Y. *Zh. Org. Khim.* **1977,** *13,* 2621; *Chem. Abstr.* **1978,** *88,* 89067q.

49. Kolchina, E. F., Gerasimova, T. N., *Izv. Akad. Nauk SSSR* **1990,** 850; *Chem. Abstr.* **1990,** *113,* 58281k.

50. Kolchina, E. F.; Kargapolova, I. Y.; Gerasimova, T. N. *Izv. Akad. Nauk SSSR* **1986,** 1855; *Chem. Abstr.* **1987,** *107,* 39719u.

51. Andreevskaya, O. I.; Furin, G. G.; Yakobson, G. G. *Zh. Org. Khim.* **1977,** *13,* 1684; *Chem. Abstr.* **1977,** *87,* 167639y.

52. Pozdnyakovich, Y. V.; Bardin, V. V.; Shtark, A. A.; Shteingarts, V. D. *Zh. Org. Khim.* **1979,** *15,* 656; *Chem. Abstr.* **1979,** *91,* 20153j.

53. Bryce, M. R.; Chambers, R. D.; Kirk, J. R. *J. Chem. Soc., Perkin Trans. 1* **1984,** 1391.

54. Zapevalov, A. Y.; Filyakova, T. I.; Kolenko, I. P.; Peschanskii, N. V.; Kodess, M. I. *Zh. Org. Khim.* **1984,** *20,* 2267; *Chem. Abstr.* **1985,** *102,* 149004m.

55. Cuzzato, P.; Castellan, A.; Pasquale, A. Eur. Pat. 260 713, 1988; *Chem. Abstr.* **1988,** *109,* 230301x.

56. Daikin Kogyo Co. Jpn. Pat. 58 38 231, 1983; *Chem. Abstr.* **1983,** *99,* 5225f.

57. Rammelt, P. P.; Siegemund, G. Eur. Pat. 54 227, 1982; *Chem. Abstr.* **1982,** *97,* 181732m.

58. Zapevalov, A. Y.; Filyakova, T. I.; Kolenko, I. P.; Kodess, M. I. *Zh. Org. Khim.* **1986,** *22,* 93; *Chem. Abstr.* **1987,** *106,* 4771g.

59. Cherstkov, V. F.; Sterlin, S. R.; German, L. S.; Knunyants, I. L. *Izv. Akad. Nauk SSSR* **1983,** 1872; *Chem. Abstr.* **1984,** *100,* 5812j.

60. Hanzawa, Y.; Kawagoe, K.; Kimura, N.; Kobayashi, Y. *Chem. Pharm. Bull.* **1986,** *34,* 3953.

61. Normant, J. F.; Reboul, O.; Sauvetre, R.; Deshayes, H.; Masure, D.; Villieras, J. *Bull. Soc. Chim. Fr.* **1974,** 2072.

62. Yamazaki, T.; Ishikawa, N. *Bull. Soc. Chim. Fr.* **1986,** 937.

63. Hanzawa, Y.; Kawagoe, K.; Yamada, A.; Kobayashi, Y. *Tetrahedron Lett.* **1985,** *26,* 219.

64. Kubota, T.; Kondoh, Y.; Suda, Y.; Tanaka, A.; Katoh, S.; Ohyama, T.; Tanaka, T. *Chem. Express* **1987,** *2,* 619.
65. Malherbe, R.; Rist, G.; Bellus, D. *J. Org. Chem.* **1983,** *48,* 860.
66. Taguchi, T.; Morikawa, T.; Kitagawa, O.; Mishima, T.; Kobayashi, Y. *Chem. Pharm. Bull.* **1985,** *33,* 5137.
67. Brooke, G. M. *J. Chem. Soc., Perkin Trans. 1* **1982,** 107.
68. Brooke, G. M.; Cooperwaite, J. R. *J. Chem. Soc., Perkin Trans. 1* **1985,** 2643.
69. Brooke, G. M.; Cooperwaite, J. R. *J. Chem. Soc., Perkin Trans. 1* **1985,** 2637.
70. Brooke, G. M.; Wallis, D. I. *J. Chem. Soc., Perkin Trans. 1* **1981,** 1417.
71. Martan, M.; Engel, D. J. U.S. Pat. 4 124 643, 1978; *Chem. Abstr.* **1979,** *90,* 87041u.
72. Dittrich, R.; Haegele, G. *Phosphorus Sulphur* **1981,** *101,* 127.
73. Boenigk, W.; Haegele, G. *Chem. Ber.* **1983,** *116,* 2418.
74. Boenigk, W.; Fischer, U.; Haegele, G. *Phosphorus Sulphur* **1983,** *16,* 263.
75. Kumar, R. C.; Kinkead, S. A.; Shreeve, J. M. *Inorg. Chem.* **1984,** *23,* 3112.
76. Welch, J. T.; DeCorte, B.; DeKimpe, N. *J. Org. Chem.* **1990,** *55,* 4981.
77. Quinze, K.; Laurent, A.; Mison, P. *J. Fluorine Chem.* **1989,** *44,* 233.
78. Gerstenberger, M. R. C.; Haas, A.; Liebig, F. *J. Fluorine Chem.* **1982,** *19,* 461.
79. Shen, Y.; Zheng, J.; Huang, Y. *J. Fluorine Chem.* **1987,** *36,* 471.
80. Dvornikova, K. V.; Platonov, V. E.; Yakobson, G. G. *J. Fluorine Chem.* **1978,** *11,* 1.
81. Ivanova, S. M.; Boikov, Y. A.; Barabanov, V. G.; V'yunov, K. A.; Ginak, A. I. *Zh. Org. Khim.* **1982,** *18,* 2463; *Chem. Abstr.* **1983,** *98,* 125500t.
82. Scott, L. T.; Highsmith, J. R. *Tetrahedron Lett.* **1980,** *21,* 4703.
83. Freudenreich, R.; Mielke, I.; Rettenbeck, K.; Schoettle, T. Eur. Pat. Appl. EP 337 127, 1988; *Chem. Abstr.* **1990,** *112,* 157661v.
84. Tonelli, C.; Tortelli, V. Eur. Pat. 361 282, 1990; *Chem. Abstr.* **1990,** *113,* 114633r.
85. German, L. S.; Grigor'ev, A. S.; Kolbanovskii, Y. A.; Chepik, S. D. *Izv. Akad. Nauk SSSR* **1990,** 371; *Chem. Abstr.* **1990,** *113,* 39874q.
86. Meunier, H.; Purdy, J. R.; Thrush, B. A. *J. Chem. Soc., Faraday Trans. 2* **1980,** *76,* 1304.
87. Couture, M. J.; Hayashi, D. U.S. Pat. 4 365 102, 1982; *Chem. Abstr.* **1983,** *98,* 142942f.
88. Birchall, J. M.; Fields, R.; Haszeldine, R. N.; McLean R. J. *J. Fluorine Chem.* **1980,** *15,* 487.
89. Davies,T.; Haszeldine, R. N.; Rowland, R.; Tipping, A. E. *J. Chem. Soc., Perkin Trans. 1* **1983,** 109.

90. Ivanova, E. P.; Karpov, V. M.; Platonov, V. E.; Tataurov, G. P.; Yakobson, G. G.; Yakhlakova, O. M. *Izv. Akad. Nauk SSSR* **1972,** 733; *Chem. Abstr.* **1972,** *77,* 75040m.
91. Platonov, V. E.; Furin, G. G.; Malyuta, N. G.; Yakobson, G. G. *Zh. Org. Khim.* **1972,** *8,* 430; *Chem. Abstr.* **1972,** *76,* 140278.
92. Platonov, V. E.; Gatilova, V. P.; Dvornikova, K. V.; Yakobson, G. G. *Izv. Akad. Nauk SSSR* **1974,** 1668; *Chem. Abstr.* **1974,** *81,* 104876d.
93. Platonov, V. E., Maksimov, A. M., Yakobson, G. G. *Izv. Akad. Nauk SSSR* **1977,** 2387; *Chem. Abstr.* **1978,** *88,* 74250v
94. Savchenko, T. I.; Petrova, T. D.; Platonov, V. E.; Yakobson, G. G. *Zh. Org. Khim.* **1979,** *15,* 1025; *Chem. Abstr.* **1979,** *91,* 91270f.
95. Yakobson, G. G.; Platonov, V. E.; Furin, G. G.; Malyuta, N. G.; Ermolenko, N. V. *Izv. Akad. Nauk SSSR* **1971,** 2615; *Chem. Abstr.* **1972,** *76,* 126741e.
96. Malyuta, N. G.; Platonov, V. E.; Furin, G. G.; Yakobson, G. G. *Tetrahedron* **1975,** *31,* 1201.
97. Platonov, V. E.; Senchenko, T. V.; Malyuta, N. G.; Yakobson, G. G. *Izv. Akad. Nauk SSSR* **1973,** 2827; *Chem. Abstr.* **1974,** *80,* 108225t.
98. Platonov, V. E.; Senchenko, T. V.; Yakobson, G. G. *Zh. Org. Khim.* **1976,** *12,* 816; *Chem. Abstr.* **1976,** *85,* 20910e.
99. Senchenko, T. V.; Platonov, V. E.; Yakobson, G. G. *Izv. Sib. Otd. Akad. Nauk SSSR* **1978,** 129; *Chem. Abstr.* **1978,** *89,* 42855a.
100. Dvornikova, K. V.; Platonov, V. E.; Yakobson, G. G. *Zh. Org. Khim.* **1975,** *11,* 2383; *Chem. Abstr.* **1976,** *84,* 59002g.
101. Maksimov, A. M.; Platonov, V. E.; Yakobson, G. G. *Izv. Akad. Nauk SSSR* **1986,** 138; *Chem. Abstr.* **1987,** *106,* 4789u.
102. Maksimov, A. M.; Platonov, V. E.; Yakobson, G. G.; Deryagina, E. N.; Voronkov, M. G. *Zh. Org. Khim.* **1979,** *15,* 1839; *Chem. Abstr.* **1980,** *92,* 58524q.
103. Savchenko, T. I.; Petrova, T. D.; Platonov, V. E.; Yakobson, G. G. *Zh. Org. Khim.* **1979,** *15,* 1018; *Chem. Abstr.* **1979,** *91,* 91269n.
104. Platonov, V. E.; Senchenko, T. V.; Yakobson, G. G. *Izv. Akad. Nauk SSSR* **1976,** 2843; *Chem. Abstr.* **1977,** *86,* 121019x.
105. Fields, E. K.; Meyerson, S. *J. Org. Chem.* **1972,** *37,* 751.
106. Abe, T.; Hayashi, E. *Chem. Lett.* **1988,** 1887.
107. Banks, R. E.; Madany, I. M. *J. Fluorine Chem.* **1985,** *30,* 211.
108. Banks, R. E.; Flowers, W. T.; Haszeldine, R. N. *J. Fluorine Chem.* **1979,** *13,* 267.
109. Dolbier, W. R., Jr.; Daly, D. T.; Koroniak, H. *Tetrahedron* **1986,** *42,* 3763.
110. Barlow, M. G.; Crawley, M. W.; Haszeldine, R. N. *J. Chem. Soc., Perkin Trans. 1* **1980,** 122.
111. Lawles, E. W.; Harman, L. D. *J. Inorg. Nucl. Chem.* **1969,** *31,* 1541.

112. Bell, A. N.; Fields, R.; Haszeldine, R. N.; Moran, D. *J. Chem. Soc., Perkin Trans. 1* **1980,** 487.
113. Kobayashi, Y.; Fujino, S.; Kumadaki, I. *J. Am. Chem. Soc.* **1981,** *103,* 2465.
114. Barlow, M. G.; Culshaw, S.; Haszeldine, R. N.; Morton W. D. *J. Chem. Soc., Perkin Trans. 1* **1982,** 2105.
115. Barlow, M. G.; Harrison, G. M.; Haszeldine, R. N.; Morton, W. D.; Shaw-Luckman, P.; Ward, M. D. *J. Chem. Soc., Perkin Trans. 1* **1982,** 2101.
116. Burger, K.; Schickaneder, H.; Prox, A. *Tetrahedron Lett.* **1976,** 4255.
117. Karpov, V. M.; Mezhenkova, T. V.; Platonov, V. E.; Yakobson, G. G. *Bull. Soc. Chim. Fr.* **1986,** 980.
118. Karpov, V. M.; Mezhenkova, T. V.; Platonov, V. E.; Yakobson, G. G. *Izv. Akad. Nauk SSSR* **1990,** 1114; *Chem. Abstr.* **1990,** *113,* 131690z.
119. Karpov, V. M.; Mezhenkova, T. V.; Platonov, V. E.; Yakobson, G. G. *Izv. Akad. Nauk SSSR* **1987,** 1918; *Chem. Abstr.* **1988,** *108,* 221384d.
120. Chambers, R. D.; Clark, M.; Maslakiewicz, J. R.; Musgrave, W. K. R. *Tetrahedron Lett.* **1973,** 2405.
121. Chambers, R. D.; Musgrave, W. K. R.; Sargent, C. R. *J. Chem. Soc., Perkin Trans. 1* **1981,** 1071.
122. Chambers, R. D.; Sargent, C. R. *J. Chem. Soc., Chem. Commun.* **1979,** 446.
123. Chambers, R. D.; Sargent, C. R.; Clark, M. *J. Chem. Soc., Chem. Commun.* **1979,** 445.
124. Smart, B. E. *The Chemistry of Functional Groups,* Supplement D; Patai, S.; Rappoport, R. Z., Eds.; John Wiley: New York, 1983; pp 603–655.
125. O'Neal, H. E.; Benson, S. W. *J. Phys. Chem.* **1968,** *72,* 1866.
126. Dolbier, W. R., Jr.; Enoch, H. O. *J. Am. Chem. Soc.* **1977,** *99,* 4532.
127. Flowers, M. D.; Frey, H. M. *Proc. Roy. Chem. Soc. (London, Ser. A)* **1960,** *257,* 121.
128. Dolbier, W. R., Jr.; Fielder, T. H. *J. Am. Chem. Soc.* **1978,** *100,* 5577.
129. Dolbier, W. R., Jr. *Acc. Chem. Res.* **1981,** *14,* 195.
130. Dolbier, W. R., Jr.; Al-Sader, B. H.; Sellers, S. F.; Elsheimer, S. *J. Am. Chem. Soc.* **1981,** *103,* 715.
131. Dolbier, W. R., Jr.; Al-Sader, B. H.; Sellers, S. F.; Koroniak, H. *J. Am. Chem. Soc.* **1981,** *103,* 2138.
132. Dolbier, W. R., Jr.; Sellers, S. F. *J. Am. Chem. Soc.* **1982,** *104,* 2494.
133. Sellers, S. F.; Dolbier, W. R., Jr.; Koroniak, H.; Al-Fekri, D. M. *J. Org. Chem.* **1984,** *49,* 1033.
134. Dolbier, W. R., Jr.; Sellers, S. F. *J. Org. Chem.* **1982,** *47,* 1.
135. Herbert, F. P.; Kerr, J. A.; Trotmann-Dickenson, A. F. *J. Chem. Soc.* **1965,** 5710.
136. Atkinson, B.; McKeegan, D. *Chem. Commun.* **1966,** 189.

137. Querro, E. D.; Ferrero, J. C.; Staricco, E. H. *Int. J. Chem. Kinet.* **1977,** *9,* 339.
138. Dolbier, W. R., Jr.; Koroniak, H.; Burton, D. J.; Heinze, P. L.; Bailey, A. R.; Shaw, G. S.; Hansen, S. W. *J. Am. Chem. Soc.* **1987,** *109,* 219.
139. Dolbier, W. R., Jr.; Koroniak, H.; Burton, D. J.; Bailey, A. R.; Shaw, G. S.; Hansen, S. W. *J. Am. Chem. Soc.* **1984,** *106,* 1871.
140. Dolbier, W. R., Jr.; Koroniak, H.; Burton, D. J.; Heinze, P. L. *Tetrahedron Lett.* **1986,** *27,* 4387.
141. Rondan, N. G.; Houk, K. N. *J. Am. Chem. Soc.* **1985,** *107,* 2099.
142. Dolbier, W. R., Jr.; Gray, T. A.; Keaffaber, J. J.; Celewicz, L.; Koroniak, H. *J. Am. Chem. Soc.* **1990,** *112,* 363.
143. Dolbier, W. R., Jr.; Palmer, K.; Koroniak, H.; Zhang, H.-Q.; Goedkin, V. L. *J. Am. Chem. Soc.* **1991,** *113,* 1059.
144. Dabbagh, A. M. M.; Flowers, W. T.; Haszeldine, R. N.; Robinson P. J. *J. Chem. Soc., Perkin Trans. 2* **1979,** 1407.
145. Dabbagh, A. M. M.; Flowers, W. T.; Haszeldine, R. N.; Robinson, P. J. *J. Chem. Soc., Chem. Commun.* **1975,** 323.
146. Barlow, M. G.; Haszeldine, R. N.; Peck, C. J. *J. Chem. Soc., Chem. Commun.* **1980,** 158.
147. Masamune, S.; Machiguchi, T.; Aratani, M. *J. Am. Chem. Soc.* **1977,** *99,* 3524.
148. Barlow, M. G.; Haszeldine, R. N.; Kershaw, M. J. *J. Chem. Soc., Perkin Trans. 1* **1975,** 2005.
149. Chambers, R. D.; MacBride, J. A. H.; Maslakiewicz, J. R.; Srivastava, K. C. *J. Chem. Soc., Perkin Trans. 1* **1975,** 396.
150. Chambers, R. D.; Maslakiewicz, J. R.; Srivastava, K. C. *J. Chem. Soc., Perkin Trans. 1* **1975,** 1130.
151. Chambers, R. D.; Clark, M.; Maslakiewicz, J. R.; Musgrave, W. K. R.; Urben, P. G. *J. Chem. Soc., Perkin Trans. 1* **1974,** 1513.
152. Chambers, R. D.; Maslakiewicz, J. R. *J. Chem. Soc., Chem. Commun.* **1976,** 1005.
153. Hees, U.; Vogelbacher, U. J.; Michels, G.; Regitz, M. *Tetrahedron* **1989,** *45,* 3115.
154. Sket, B.; Zupancic, N.; Zupan, M. *J. Chem. Soc., Perkin Trans. 1* **1987,** 981.
155. Brovko, V. V.; Sokolenko, V. A.; Yacobson, G. G. *Zh. Org. Khim.* **1974,** *10,* 2385; *Chem. Abstr.* **1975,** *82,* 111807f.
156. Kobayashi, Y.; Ando, A.; Kawada, K.; Kumadaki, I. *J. Org. Chem.* **1980,** *45,* 2968.
157. Kobayashi, Y.; Kumadaki, I.; Ohsawa, A.; Sekine, Y.; Ando, A. *Heterocycles* **1977,** *6,* 1587.
158. Kobayashi, Y.; Ando, A.; Kawada, K.; Ohsawa, A.; Kumadaki, I. *J. Org. Chem.* **1980,** *45,* 2962.

159. Kobayashi, Y.; Kawada, K.; Ando, A.; Kumadaki, I. *Tetrahedron Lett.* **1984,** *25,* 1917.
160. Kobayashi, Y.; Kumadaki, I.; Ohsawa, A.; Sekine, Y.; Ando, A. *Symp. Heterocycl.* **1977,** 89; *Chem. Abstr.* **1978,** *89,* 163321g.
161. Kobayashi, Y.; Kumadaki, I.; Ohsawa, A.; Sekine, Y.; Mochizuki, H. *Chem. Pharm. Bull.* **1975,** *23,* 2773.
162. Wiebe, H. A.; Braslawsky, S.; Heicklen, J. *Can. J. Chem.* **1972,** *50,* 2721.
163. Kobayashi, Y.; Kumadaki, I.; Ohsawa, A.; Sekine, Y. *Tetrahedron Lett.* **1975,** 1639.
164. Ratajczak, E.; Szuba, B.; Price, D. J. *Photochem.* **1980,** *13,* 233.
165. Kobayashi, Y.; Ohsawa, A.; Baba, M.; Sato, T.; Kumadaki, I. *Chem. Pharm. Bull.* **1976,** *24,* 2219.
166. Kobayashi, Y.; Ohsawa, A. *Chem. Pharm. Bull.* **1976,** *24,* 2225.
167. Kobayashi, Y.; Fujino, S.; Hamana, H.; Hanzawa, Y.; Morita, S.; Kumadaki, I. *J. Org. Chem.* **1980,** *45,* 4683.
168. Kobayashi, Y.; Fujino, S.; Hamana, H.; Kumadaki, I.; Hanzawa, Y. *J. Am. Chem. Soc.* **1977,** *99,* 8511.
169. Soelch, R. R.; McNierney, E.; Tannenbaum, G. A.; Lemal, D. M. *J. Org. Chem.* **1989,** *54,* 5502.
170. Barnes, R. N.; Chambers, R. D.; Hercliffe, R. D.; Middleton, R. *J. Chem. Soc., Perkin Trans. 1* **1981,** 3289.
171. Pelosi, L. F.; Miller, W. T. *J. Am. Chem. Soc.* **1976,** *98,* 4311.
172. Bell, A. N.; Fields, R.; Haszeldine, R. N.; Kumadaki, I. *J. Chem. Soc., Chem. Commun.* **1975,** 866.
173. Petrenko, N. I.; Shelkovnikov, V. V.; Eroshkin, V. I.; Gerasimova, T. N. *J. Fluorine Chem.* **1987,** *36,* 99.
174. Semisch, C.; Margaretha, P. *J. Fluorine Chem.* **1986,** *30,* 471.
175. Glazkov, A. A.; Ignatenko, A. V.; Slavinskii, N. V.; Krukovskii, S. P.; Ponomarenko, V. A. *Izv. Akad. Nauk SSSR* **1978,** 702; *Chem. Abstr.* **1979,** *91,* 30408k.
176. Mahaffy, P. G.; Visser, D.; Torres, M.; Bourdelande, J. L.; Strausz, O. P. *J. Org. Chem.* **1987,** *52,* 2680.
177. Lapouyade, R.; Hanafi, N.; Morand, J. P. *Angew. Chem.* **1982,** *94,* 795.
178. Dailey, W. P.; Lemal, D. M. *J. Am. Chem. Soc.* **1984,** *106,* 1169.
179. Martini, T. *Tetrahedron Lett.* **1976,** 1865.
180. Uebelhart, P.; Gilgen, P.; Schmid, H. *Org. Photochem. Synth.* **1976,** *2,* 72.
181. Burton, D. J.; Flynn, R. M. *Synthesis* **1979,** 615.

Chapter 5

Fluorinated Compounds as Reagents

Fluorinated Compounds as Reagents

by Peter J. Stang and Viktor V. Zhdankin

In the past 20 years, the application of fluorinated reagents in chemistry experienced tremendous growth. Some well-known derivatives of hydrofluoric acid and trifluoroacetic acid found new applications, and many new classes of synthetically useful fluorinated organic and inorganic compounds have been synthesized.

This chapter focuses on the use of fluorinated reagents in organic synthesis and does not cover such large areas as fluorinated organic solvents and fluorinated complexing agents. Fluorinated reagents usually have some advantages and special features in comparison with the corresponding nonfluorinated analogues. Usually, fluorinated compounds possess higher thermal stability and greater resistance to the action of strong oxidizers. The extremely high electron-withdrawing properties of the fluorine atom cause such unusual and advantageous features as superacidity of perfluoroalkanesulfonic acids, high activity of inorganic fluorides as Lewis acids, and strong electrophilic properties of many fluorine-substituted reagents.

Several comprehensive reviews of some classes of fluorinated reagents will be cited in the corresponding subdivisions of the present chapter.

Hydrofluoric Acid

Anhydrous hydrofluoric acid (hydrogen fluoride) is a relatively strong acid that is inert to reduction or oxidation and is a good solvent for many organic compounds. Its low boiling point facilitates product recovery and recycling. Because of these properties, hydrofluoric acid is widely used in organic synthesis as a reaction medium, combining the properties of an acidic reagent and a solvent.

Stability toward reduction makes hydrogen fluoride a good medium for different *hydrogenation processes* [*1*, *2*]. It is a useful solvent for the hydrogenation of benzene in the presence of Lewis acids [*1*]. Anhydrous hydrofluoric acid has pronounced catalytic effect on the hydrogenations of various aromatic compounds, aliphatic ketones, acids, esters, and anhydrides in the presence of platinum dioxide [2] (equations 1–3).

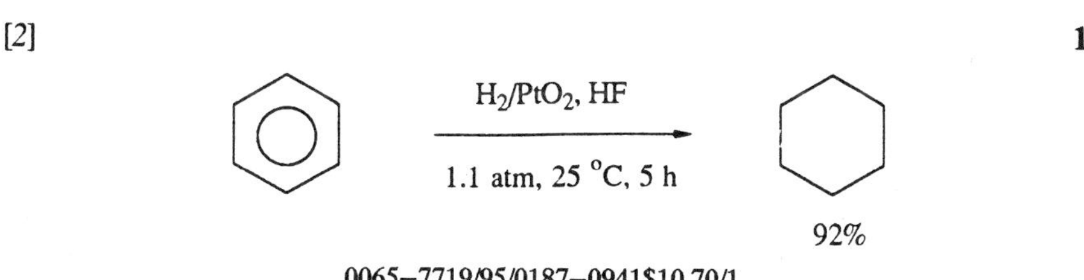

0065–7719/95/0187–0941$10.70/1

[2] **2**

$$CH_3(CH_2)_{10}COOH \xrightarrow[340\text{ atm, }25\ ^\circ C,\ 20\text{ h}]{H_2/PtO_2,\ HF} \underset{21\%}{CH_3(CH_2)_{10}CH_3} + \underset{42\%}{[CH_3(CH_2)_{11}]_2O}$$

[2] **3**

$$[CH_3(CH_2)_{10}CO]_2O \xrightarrow[340\text{ atm, }25\ ^\circ C,\ 20\text{ h}]{H_2/PtO_2,\ HF} \underset{77\%}{[CH_3(CH_2)_{11}]_2O}$$

Compared with reactions in other strong acids used in hydrogenations, such as hydrochloric, sulfuric, and trifluoroacetic acid, reactions in hydrofluoric acid are faster and proceed more selectively under milder conditions [*2*].

A nitro group in aromatic nitro compounds can be hydrogenated in hydrofluoric acid under mild conditions [*3, 4*]. This reduction is usually accompanied by some side processes. Catalytic hydrogenation of nitrobenzene in hydrogen fluoride gives *p*-fluoroaniline in high yield and aniline as the principal by-product [*3*] (equation 4). When the hydrogenation of an aromatic nitro compound is carried out in the presence of the second aromatic component, a condensation leading to diphenylamines occurs [*4*] (equation 5).

[*3*] **4**

$$C_6H_5NO_2 \xrightarrow[3.8\text{ atm, }50\ ^\circ C]{H_2/PtO_2,\ HF} \underset{61\%}{p\text{-}FC_6H_4NH_2} + \underset{12.6\%}{C_6H_5NH_2}$$

[*4*] **5**

$$X\text{-}C_6H_4NO_2 + C_6H_5Y \xrightarrow[13\text{-}20\text{ atm, }50\ ^\circ C]{H_2/PtO_2,\ HF} \underset{30\text{-}82\%}{X\text{-}C_6H_4\text{-}NH\text{-}C_6H_4\text{-}Y}$$

$X = OCH_3, OC_2H_5, CH_3, C_2H_5, C_6H_5$
$Y = OH, OCH_3, NH_2, NHC_6H_5$

Commercially available hydrogen fluoride usually is not suitable for catalytic hydrogenation because of its sulfur dioxide content. An oxidative treatment with manganese dioxide and distillation are needed for the preparation of hydrogenation-grade hydrogen fluoride [*3, 4*].

Anhydrous hydrogen fluoride is a superior reagent for various cyclizations. It converts aryl-substituted diethyl arylmalonates into tetralones in good yields [*5*] (equation 6).

References are listed on pages 971–975.

In contrast to other acids, anhydrous hydrogen fluoride does not cause hydrolysis and decarboxylation of the malonic acid residues in these reactions [*5*]. It is a good *reagent for the cyclization* of α-benzamidoacetophenones to 2,5-diphenyloxazoles [*6*] (equation 7). The same reaction with concentrated sulfuric acid gives cyclic product with only a 12% yield [*6*].

[*5*] 3-CH_3O-C_6H_4-$C(CO_2C_2H_5)_2$-$CH_2CH_2CO_2C_2H_5$ $\xrightarrow[\text{RT, 18h}]{\text{HF (excess)}}$ 6-methoxy-1,1-bis($CO_2C_2H_5$)-4-oxo-tetralin, 95% **6**

[*6*] $C_6H_5C(O)$-NH-CH(R)-$C(O)C_6H_5$ $\xrightarrow[\text{RT}]{\text{HF}}$ 4-R-2,5-di(C_6H_5)oxazole **7**

R = H	91%
R = CH_3	95%

Anhydrous hydrogen fluoride is widely used for final *removal of various protecting groups* used in peptide chemistry. Even better results are achieved with pyridinium polyhydrogen fluoride, **Olah's reagent** [*7*], which is prepared by dissolving anhydrous hydrogen fluoride in pyridine to form a solution containing up to 70% hydrogen fluoride (w/w). This solution is stable; it contains free hydrogen fluoride and acts as a reservoir for hydrogen fluoride in a convenient liquid medium. This reagent in the presence of anisole removes most of the currently employed protective groups from the amino acid derivatives with almost quantitative regeneration of the amino acids and is especially useful in the Merrifield synthesis of peptides [*8*].

Aqueous hydrofluoric acid dissolved in acetonitrile is a good catalyst for *intramolecular Diels–Alder reactions* [*9*]. This reagent promotes highly stereoselective cyclizations of different triene esters (equation 8). The use of other acids, such as hydrochloric, acetic, and trifluoroacetic acid, results in complete polymerization of the starting trienes [*9*] (equation 8).

Aqueous hydrofluoric acid catalyzes such reactions as *nitration and oxidation.* Thus anthraquinone can be nitrated easily with nitric acid in 80–98% aqueous hydrofluoric acid [*10*].

Fluorides

The fluorides of main group elements are widely applied in organic chemistry as catalysts for a wide variety of reactions. Ionic fluorides of group 1 and group 2 metals and of tetraalkylammonium serve as a source of relatively basic fluoride anion, catalyzing different condensations. **"Naked" fluoride anion** from tetraethylammonium fluoride or potassium fluoride–18-crown-6 in aprotic solvents acts as a powerful *base catalyzing Michael-type addition* reactions of acetonitrile or nitromethane to an activated double bond [*11, 12*] (equations 9 and 10).

[*9*] **8**

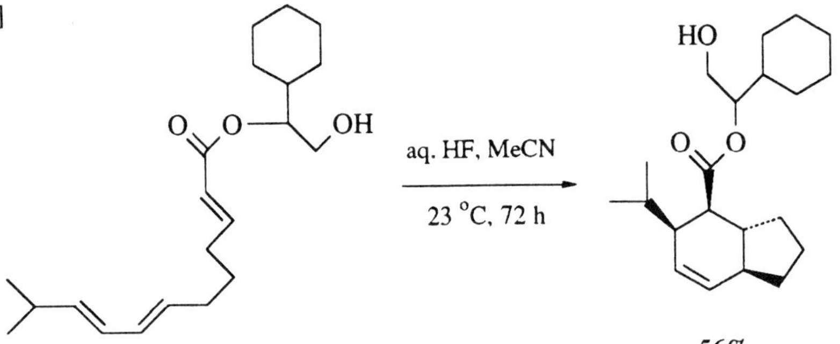

OH ; O ; O

aq. HF, MeCN → 23 °C, 72 h

HO ; O ; O

56%

[*11*] **9**

$$CH_3CN + (C_6H_5)_2C{=}CHNO_2 \xrightarrow[\text{MeCN, RT, 2 h}]{(C_2H_5)_4N^+F^-} (C_6H_5)_2C(CH_2CN)CH_2NO_2$$

45%

[*12*] **10**

$$CH_3NO_2 + C_6H_5CH{=}CHCOC_6H_5 \xrightarrow[\text{MeCN, 81 °C, 1.5 h}]{\text{KF/18-Crown-6}} C_6H_5CH(CH_2NO_2)CH_2COC_6H_5$$

94%

The fluoride anion has a pronounced catalytic effect on the aldol reaction between enol silyl ethers and carbonyl compounds [*13*]. This reaction proceeds at low temperature under the influence of catalytic amounts (5–10 mol %) of **tetrabutylammonium fluoride,** giving the aldol silyl ethers in high yields (equation 11).

This condensation finds considerable generality: enol silyl ethers of a variety of ketones and both aromatic and aliphatic aldehydes are usable. For enol silyl ethers of substituted cyclohexanones the reaction is regio- and stereospecific [*13*].

Covalent fluorides of group 3 and group 5 elements (boron, tin, phosphorus, antimony, etc.) are widely used in organic synthesis as strong Lewis acids. **Boron trifluoride etherate** is one of the most common reagents used to catalyze many organic reactions. A representative example is its recent application as a *catalyst in the cycloadditions* of 2-aza-1,3-dienes with different dienophiles [*14*]. Boron trifluoride etherate and other fluorinated Lewis acids are effective activators of the electrophilic properties of polyvalent iodine compounds (such as iodosobenzene and iodonium ylides) in reactions forming new carbon–carbon bonds [*15*].

[*13*] **11**

$$\text{1-(trimethylsilyloxy)cyclohexene} + C_6H_5CHO \xrightarrow[\text{THF, -78 °C, 3.5 h}]{(C_4H_9)_4N^+F^-} \text{2-[(trimethylsilyloxy)(phenyl)methyl]cyclohexanone}$$

80%

The use of **antimony pentafluoride** as a component of superacids for the generation of carbocations from various organic compounds was reviewed recently [*16*].

Fluorides of boron, phosphorus, and antimony react with hydrogen fluoride, producing complex **fluoroboric, fluorophosphoric, and fluoroantimonic acid,** which are used in organic chemistry as catalysts and as valuable reagents for the preparation of various organic and inorganic salts. Fluoroboric acid in ether solution catalyzes the coupling of silyl enol ethers by iodosobenzene [*17*]. Fluoroantimonic acid in excess hydrogen fluoride is a powerful *catalyst for the carboxylation* of alcohols with carbon monoxide [*18*]. Iodobenzene diacetate reacts with aqueous fluoroboric, fluorophosphoric, and fluoroantimonic acid to give stable, highly electrophilic reagents (equation 12), which can be applied for the coupling of enol silyl ethers or the functionalization of alkenes and alkynes [*19*].

[*19*] **12**

$$C_6H_5I(OCOCH_3)_2 \xrightarrow[\text{RT, 1 h}]{\text{HX, CHCl}_3\text{,}} C_6H_5{-}\overset{+}{I}{-}O{-}\overset{+}{I}{-}C_6H_5 \;\; X^- \;\; X^-$$

$X = BF_4, PF_6, SbF_6,$ 60-80%

Many ionic fluoroborates are currently used in organic synthesis. **Nitronium fluoroborate** is one of the most powerful nitrating reagents [*20*]. **Triphenylmethyl fluoroborate (trityl fluoroborate)** is used to introduce the trityl protective group to primary hydroxyl groups [*21*]. **Alkenyl and alkynyl iodonium fluoroborates** are effective alkenylating or alkynylating reagents toward various enolate anions, organocuprates, and other nucleophiles (for a review *see* reference 15).

Trifluoroacetic Acid and Its Derivatives

Trifluoroacetic acid is a useful medium for many organic reactions. It is a strong carboxylic acid that combines the properties of a good solvent for most organic substrates with low nucleophilicity and high stability toward strong oxidizers and mild reducing reagents.

Because of its high polarity and low nucleophilicity, a trifluoroacetic acid medium is usually used for the investigation of such carbocationic processes as *solvolysis, protonation of alkenes, skeletal rearrangements, and hydride shifts* [*22–24*]. It also has been used for several synthetically useful reactions, such as electrophilic aromatic substitution [*25*], reductions [*26, 27*], and oxidations [*28*]. Trifluoroacetic acid is a good medium for the nitration of aromatic compounds. *Nitration of benzene or toluene* with sodium nitrate in trifluoroacetic acid is almost quantitative after 4 h at room temperature [*25*]. Under these conditions, toluene gives the usual mixture of mononitrotoluenes in an *o:m:p* ratio of 61.6:2.6:35.8.

A trifluoroacetic acid medium can be used for the *reduction of acids, ketones, and alcohols with sodium borohydride* [*26*] or triethylsilane [*27*]. Diarylketones are smoothly reduced by sodium borohydride in trifluoroacetic acid to diarylmethanes (equation 13).

[*26*] **13**

$$R^1C(=O)R^2 \xrightarrow[\text{RT, 15-30 h}]{NaBH_4,\ CF_3COOH} R^1CH_2R^2$$

R^1, R^2 = aryl 73-94%

The reduction is general for a variety of substituted benzophenones. Such substituents as CH_3, OH, OCH_3, F, Br, $N(CH_3)_2$, NO_2, COOH, $COOCH_3$, $NHCOC_6H_5$, and CN survive the reaction conditions and do not alter the course of the reduction. Diarylmethanols are reduced to diarylmethanes under the same conditions and probably are the intermediates in the reduction of ketones [*26*].

Triethylsilane also can be used as a *reducing agent in trifluoroacetic acid* medium [*27*]. This reagent is used for the reduction of benzoic acid and some other carboxylic acids under mild conditions (equation 14). Some acids (phthalic, succinic, and 4-nitrobenzoic) are not reduced under these conditions [*27*].

[*27*] **14**

$$RCOOH \xrightarrow[50\ ^\circ C]{(C_2H_5)_3SiH,\ CF_3COOH} RCH_3$$

40-97%

R = CH_3, 4-$CH_3OC_6H_4$, 4-$CH_3C_6H_4$, ferrocenyl, etc.

Trifluoroacetic acid is a useful medium for a number of oxidation reactions. It is highly resistant to strong oxidants, even to permanganates and chromates. For instance, various alkanes, cycloalkanes, and arenes can be oxidized degradatively by potassium permanganate in trifluoroacetic acid under mild conditions [*28*].

A very common oxidizing reagent is **peroxytrifluoroacetic acid,** which is usually generated in situ from trifluoroacetic acid [*29, 30, 31*] or trifluoroacetic anhydride [*32, 33, 34*] and hydrogen peroxide. Peroxytrifluoroacetic acid is one of the *most efficient epoxidizing reagents* [*35*]. It can be used to prepare epoxides

References are listed on pages 971–975.

from hindered and deactivated alkenes, which are resistant to the usual oxidizing agents (*m*-chloroperoxybenzoic, peroxybenzoic, and peroxyacetic acid) [*32, 33, 34*]. Hindered cyclohexenes [*33*] and cyclopentenes [*34*] are epoxidized with peroxytrifluoroacetic acid under mild conditions (equations 15 and 16). Such epoxidation is used in the synthesis of the valuable natural product cemepoxide and its analogues [*33*].

[*33*] **15**

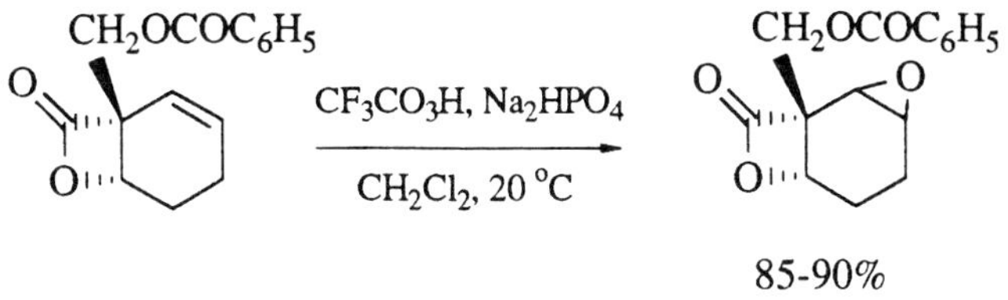

$CH_2OCOC_6H_5$ → CF_3CO_3H, Na_2HPO_4 / CH_2Cl_2, 20 °C → $CH_2OCOC_6H_5$

85-90%

[*34*] **16**

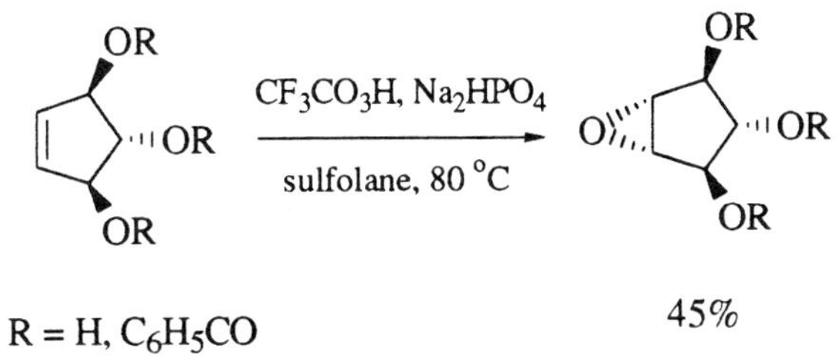

OR, OR, OR → CF_3CO_3H, Na_2HPO_4 / sulfolane, 80 °C → OR, OR, OR

R = H, C_6H_5CO

45%

Peroxytrifluoroacetic acid is used for numerous oxidations of saturated hydrocarbons and aromatic compounds. It *oxidizes alkanes, alkanols, and carboxylic acids* with formation of hydroxylation products [*29*]. Oxidation of cyclohexane with peroxytrifluoroacetic acid proceeds at room temperature and leads to cyclohexyl trifluoroacetate in 75% yield; 1-octanol under similar conditions gives a mixture of isomeric octanediols in 59% yield, and palmitic acid gives a mixture of hydroxypalmitic acids in 70% yield [*29*].

Peroxytrifluoroacetic acid in the presence of sulfuric acid can be employed for the selective one-step oxidation of the saturated side chain of cholestenone [*31*] (equation 17).

[*31*] **17**

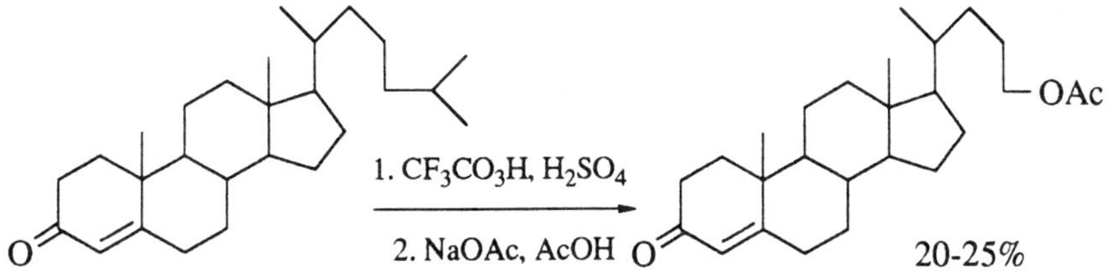

1. CF_3CO_3H, H_2SO_4 / 2. NaOAc, AcOH → OAc

20-25%

An excess of *peroxytrifluoroacetic acid oxidizes aromatic rings* to complete destruction [*30*]. If an aliphatic side chain is present, an aliphatic carboxylic acid is produced. The reactions of aromatic compounds with peroxytrifluoroacetic acid are exothermic and complete in 1 h. Oxidation of toluene gives acetic acid in 95% yield; Similar oxidation of *n*-propylbenzene yields butyric acid (82%) and oxidation of acenaphthene gives succinic acid (80%). Under these conditions pyridine and quinoline produce the corresponding *N*-oxides in quantitative yields [*30*].

Trifluoroacetic (hexafluoroacetic) anhydride is a valuable reagent for organic synthesis. Like trifluoroacetic acid, it can be used as a medium for a number of organic reactions. Various aromatic compounds can be nitrated in trifluoroacetic anhydride with inorganic nitrates at room temperature [*36*]. **Ammonium nitrate in trifluoroacetic anhydride** nitrates benzene, toluene, aniline, chlorobenzene, methoxybenzene, and polymers with aromatic groups; mononitration products are formed in yields of 90–100%. However, phenols are oxidized under these conditions to quinonoid products [*36*]. Ammonium nitrate in trifluoroacetic anhydride is also a useful reagent for the *nitration of enol acetates* leading to α-nitroketones in almost quantitative yields [*37*]. Highly electrophilic trifluoroacetyl nitrate generated in situ from the nitrate salt and trifluoroacetic anhydride is considered to be the reactive intermediate in these nitrations [*36, 37*].

Tetrabutylammonium iodide in trifluoroacetic anhydride is an *effective reducing reagent* [*38*]. This system can be used for direct reduction of arenesulfonic acids to the corresponding thiols or disulfides in moderate yields under mild conditions (equation 18). Alkanesulfonic acids are reduced by this system to disulfides with 30–57% yields [*38*].

[*38*] **18**

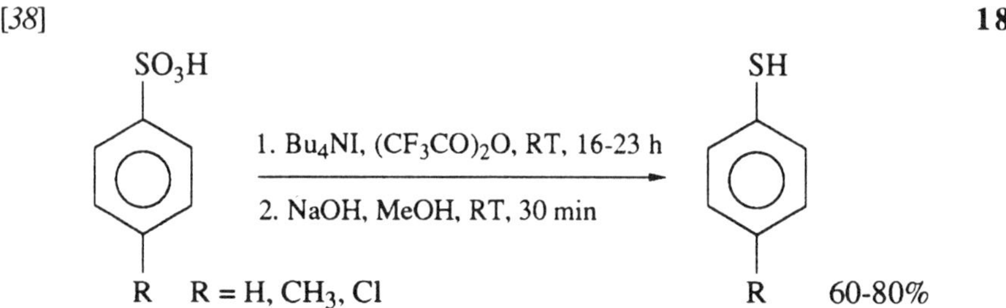

Sodium iodide in trifluoroacetic anhydride reacts with epoxides to form the corresponding alkenes in high yields [*39*]. The reduction is stereospecific and generates olefins of the same geometry as the starting epoxides [*39*].

Dimethyl sulfoxide reacts with trifluoroacetic anhydride at low temperature to give a complex that is an efficient reagent for the *oxidation of alcohols to carbonyl compounds* [*40, 41*]. This reagent can be used to oxidize primary and secondary aliphatic alcohols; cycloalkyl alcohols; and allylic, homoallylic, benzylic, acetylenic, and steroidal alcohols (equation 19).

Yields of carbonyl compounds can be increased by using highly hindered amines (such as diisopropylethylamine) in the second step of the reaction [*41*]. The system dimethyl sulfoxide–trifluoroacetic anhydride is widely used in the synthesis of natural products. It is used for the oxidation of hydroxyl moieties in the key steps in the synthesis of L-ascorbic acid (vitamin C) [*42*] and in a highly stereoselective synthesis of (–)-(*R*)-mevalonolactone [*43*].

[*40,41*] **19**

$$R^1CH(OH)R^2 \xrightarrow[\text{2. } (C_2H_5)_3N,\ -50\ ^\circ C \text{ to RT}]{\text{1. } (CH_3)_2SO/(CF_3CO)_2O,\ CH_2Cl_2,\ <-50\ ^\circ C,\ 30 \text{ min}} R^1COR^2 \quad 50\text{-}98\%$$

R^1, R^2 = H, Alk, Ar, cycloalkyl, etc.

Trifluoroacetic anhydride in a mixture with sulfuric acid is an efficient reagent for the *sulfonylation of aromatic compounds* [*44*]. The reaction of benzene with this system in nitromethane at room temperature gives diphenyl sulfone in 61% yield. Alkyl and alkoxy benzenes under similar conditions form the corresponding diaryl sulfones in almost quantitative yield, whereas yields of sulfones from deactivated arenes such as chlorobenzene are substantially lower [*44*]. The same reagent (trifluoroacetic anhydride–sulfuric acid) reacts with adamantane and its derivatives with formation of isomeric adamantanols, adamantanones, and cyclic sultones [*45*].

Trifluoroacetic anhydride is an efficient *dehydrating reagent* [*46, 47*]. In the presence of pyridine, it smoothly dehydrates amides and aldoximes to the corresponding nitriles [*46*] and adducts of CH–acids and 1,2,3-indantrione [*47*] (equation 20).

[*47*] **20**

2-hydroxy-2-(CHXY)-1,3-indandione $\xrightarrow[\text{dioxane, } 0\text{ - }20\ ^\circ C,\ 2 \text{ h}]{(CF_3CO)_2O,\ C_5H_5N}$ 2-(XYC=)-1,3-indandione 93-97%

X, Y = $COOC_2H_5$, $COOCH_3$, $COCH_3$, $COC(CH_3)_3$, COC_6H_5, CN, $CONH_2$, CH_3, H

Trifluoroacetic anhydride is a good reagent for various *cyclizations*. In the presence of a catalytic amount of phosphoric acid, it is used for the macrocyclization of ω-(2-thienyl)alkanoic acids [*48*] (equation 21).

This cyclization procedure was conveniently applied to the preparation of the key intermediate of a five-step synthesis of (±)-muscone [*48*].

[*48*] **21**

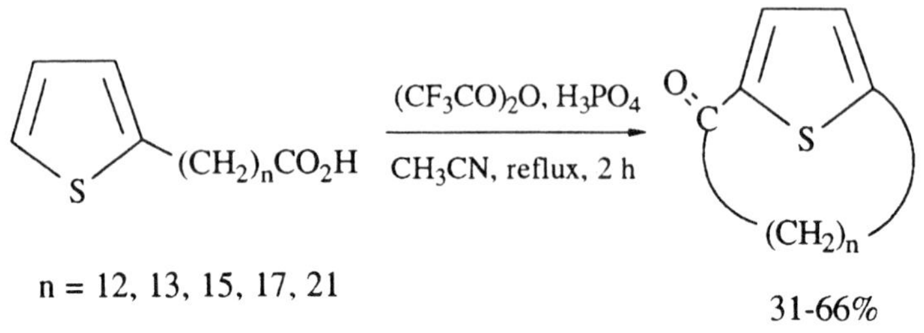

$(CF_3CO)_2O$, H_3PO_4; CH_3CN, reflux, 2 h

(thiophene)$(CH_2)_nCO_2H$ → (cyclic ketone)$(CH_2)_n$

n = 12, 13, 15, 17, 21

31-66%

Salts of Trifluoroacetic Acid

Trifluoroacetates of silver, mercury(II), thallium(III), lead(IV), and iodine(III) are synthetically valuable reagents that combine the properties of strong electrophiles, oxidizers, and Lewis acids. Furthermore, trifluoroacetate anions are stable to oxidation, are weak nucleophiles, and usually do not cause any contamination of the reaction mixture.

Silver trifluoroacetate is a suitable catalyst for various *cationic rearrangements* involving multiple carbon–carbon bonds [*49, 50*]. In the presence of silver trifluoroacetate, 2-propynyl acetates rearrange to the butadienyl acetates to give dienes that are useful in Diels–Alder reactions [*49*] (equation 22).

[*49*] **22**

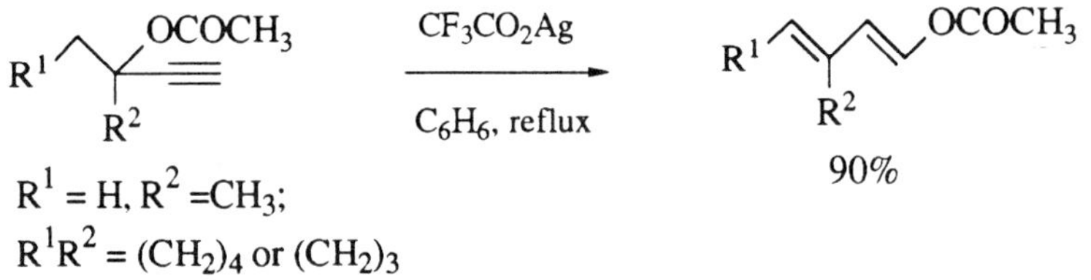

OCOCH$_3$, R^1, R^2 → CF_3CO_2Ag; C_6H_6, reflux → R^1, OCOCH$_3$, R^2

R^1 = H, R^2 =CH_3;

R^1R^2 = $(CH_2)_4$ or $(CH_2)_3$

90%

Silver trifluoroacetate is used in a one-step synthesis of bicyclo[3.2.2]nona-6,8-diene-3-one from 2-methoxyallyl bromide and benzene [*50*] (equation 23).

Analogous reactions of toluene, *p*-xylene, and mesitylene yield the corresponding substituted bicyclo[3.2.2]nona-6,8-diene-3-ones [*50*].

[*50*] **23**

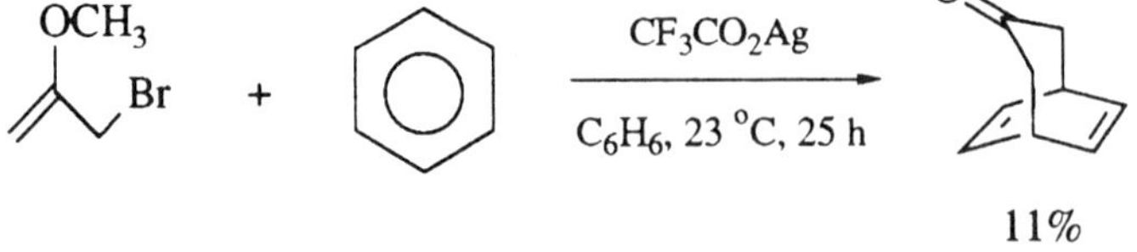

OCH_3, Br + (benzene) → CF_3CO_2Ag; C_6H_6, 23 °C, 25 h → O

11%

A useful oxidizing reagent is **silver(III) tristrifluoroacetate** which can be generated from "silver peroxide" (AgO) and a trifluoroacetic acid–trifluoroacetic anhydride mixture [*51*]. This reagent readily oxidizes alicyclic and bicyclic hydrocarbons such as norbornane, adamantane, cyclohexane, and cyclooctane to the corresponding trifluoroacetates [*51*].

Mercury(II) trifluoroacetate is a good electrophile that is highly reactive toward carbon–carbon double bonds [*52, 53, 54*]. When reacting with olefins in nucleophilic solvents, it usually gives exclusively mercurated solvoadducts, but never products of skeletal rearrangement. *Solvomercuration–demercuration of alkenes* with mercury(II) trifluoroacetate is a remarkably effective procedure for the preparation of esters and alcohols with Markovnikov's regiochemistry [*52, 53*] (equation 24).

[*53*] **24**

$RCH{=}CH_2 \xrightarrow[\text{THF, } H_2O\text{, RT}]{(CF_3CO_2)_2Hg} RCH(OH)CH_2HgOCOCF_3 \xrightarrow{\text{NaBH}_4\text{, NaOH, RT}} RCH(OH)CH_3$ 70-97%

R = alkyl, cycloalkyl, aryl, etc.

The mercuration–demercuration sequence was applied to the stereospecific synthesis of 2-deoxy-α-hexopyranoside derivative via a mercury(II) trifluoroacetate-promoted cyclization [*54*] (equation 25).

[*54*] **25**

1. $(CF_3CO_2)_2Hg$, THF, -78 °C, 3h
2. $NaBH_4$, NaOH, RT, 3h

70%

Bn = $C_6H_5CH_2$

The use of the other mercury(II) salts (acetate and bromide) in this cyclization led to a mixture of anomeric 2-deoxyhexopyranosides with low yields [*54*].

Thallium(III) trifluoroacetate is a *versatile oxidant* for organic compounds [*55, 56, 57*]. It reacts with alkenes at room temperature to form oxiranes, ketones, and 1,2-diols [*55*]. Usually these oxidations are accompanied by cyclizations and rearrangements. The reaction of thallium(III) trifluoroacetate with substituted cinnamic acids results in instantaneous oxidative dimerization leading to bislactone lignans, which belong to a naturally occurring family of compounds [*56*] (equation 26).

Oxidation of 4*H*-pyran-4-thiones with thallium(III) trifluoroacetate was used in the one-pot synthesis of 1,6-dioxa-6*a*-thiapentalenes, a hypervalent heterocyclic system [*57*] (equation 27).

[56] **26**

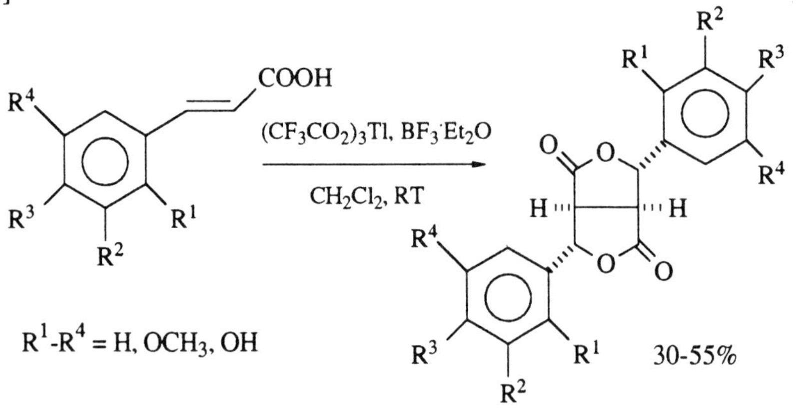

$$\xrightarrow[\text{CH}_2\text{Cl}_2,\ \text{RT}]{(\text{CF}_3\text{CO}_2)_3\text{Tl},\ \text{BF}_3\cdot\text{Et}_2\text{O}}$$

R^1-R^4 = H, OCH_3, OH

30-55%

[57] **27**

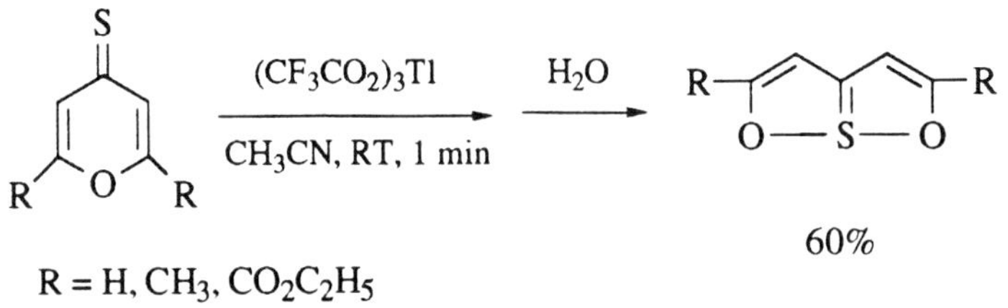

$$\xrightarrow[\text{CH}_3\text{CN, RT, 1 min}]{(\text{CF}_3\text{CO}_2)_3\text{Tl}} \xrightarrow{\text{H}_2\text{O}}$$

R = H, CH_3, $CO_2C_2H_5$

60%

Lead(IV) trifluoroacetate is a strong electrophilic and oxidizing reagent. It is a valuable reagent for the *hydroxylation of aromatic compounds* [*58, 59*]. Lead(IV) trifluoroacetate also reacts with silylated benzenes with the exclusive formation of the corresponding trifluoroacetate esters [*59*] (equation 28).

[59] **28**

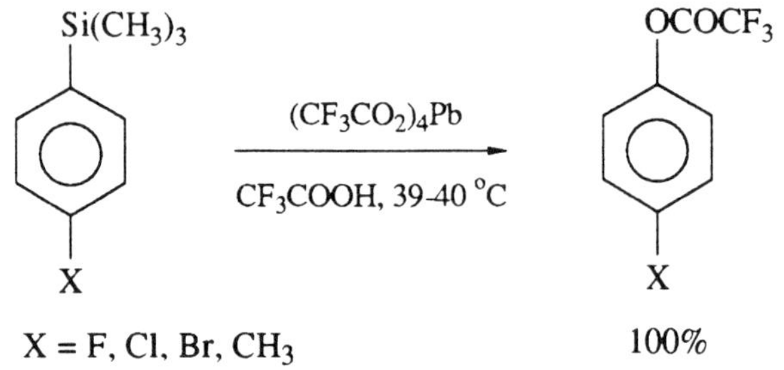

$Si(CH_3)_3$ $$\xrightarrow[\text{CF}_3\text{COOH, 39-40 °C}]{(\text{CF}_3\text{CO}_2)_4\text{Pb}}$$ $OCOCF_3$

X = F, Cl, Br, CH_3

100%

Iodine(III) trifluoroacetates (iodine tristrifluoroacetate and iodosobenzene bis-trifluoroacetate) resemble lead(IV), thallium(III), and mercury(II) reagents in their reactions but do not share the undesirable high toxicity typical for the heavy metals. **Iodine tristrifluoroacetate** is a very *powerful oxidant* that can introduce the trifluoroacetoxy group even into alkanes [*60, 61*]. Because branched alkanes react faster than normal alkanes, this reaction can be used to remove the former from a mixture with the latter [*60*]. Smilarly, iodine tristrifluoroacetate reacts with cyclic ethers, introducing one or two trifluoroacetoxy groups in the position vicinal to the ether oxygen [*61*]. The reaction with alkenes leads mainly to the formation of 1,2-bistrifluoroacetoxyalkanes [*62*].

Iodosobenzene bistrifluoroacetate is a versatile *mild oxidant* that has been used to oxidize a broad range of organic compounds, such as alkenes, alkynes, carbonyl compounds, and alcohols. Its application in organic synthesis has been summarized in several recent reviews devoted to polyvalent iodine compounds [*63, 64, 65*].

Fluorinated Sulfonic Acids

The most important fluorinated sulfonic acids used as reagents in organic chemistry are fluorosulfonic acid, trifluoromethanesulfonic (triflic) acid, nonafluorobutanesulfonic acid, and perfluororesinsulfonic acid (Nafion-H). These acids have several advantages over other acid systems that make them very useful reagents. First, perfluorinated sulfonic acids are the strongest known Brønsted acids, much stronger than hydrochloric, nitric, sulfuric, and even perchloric acid [*66, 67, 68*]. Second, perfluorinated sulfonic acids are nonoxidizing and are stable to hydrolysis or the action of strong nucleophiles, and they possess superior thermal stability and resistance to oxidation. Several reviews already exist on the synthetic application of fluorosulfonic acid [*70*], triflic acid [*66, 67, 68, 69*], and Nafion-H [*71*].

Among the fluorinated sulfonic acids, **trifluoromethanesulfonic acid (triflic acid)** is the most valuable and widely used in organic synthesis, because it is commercially available and stronger and more chemically stable than fluorosulfonic acid. During the past two decades, triflic acid has experienced a tremendous growth in its use as an acidic catalyst for different kinds of carbocationic reactions. Its ability to protonate alkenes and even alkanes make triflic acid a superior *reagent for oligomerization and polymerization* [*67*], *Friedel–Crafts alkylations and acylations* [*72, 73, 74, 75, 76*], *cationic rearrangements* of saturated polycyclic compounds [*77, 78, 79, 80, 81*], etc. The alkylation and transalkylation of alkylbenzenes [*72, 73*] or phenols [*74*] in triflic acid proceed much more rapidly and cleanly than in other strongly acidic media. Triflic acid is a much better catalyst than aluminum chloride in the Friedel–Crafts acylations of a variety of aromatic substrates and acid halides [*75*]. Intramolecular acylation of arylalkanoic acids is a highly efficient and mild one-pot procedure for the preparation of the corresponding cyclic ketones [*76*] (equation 29).

This cyclization also gives good results in the case of derivatives of naphthalene, anthracene, phenanthrene, and other aromatic substrates [*76*].

In general, fluorinated sulfonic acids can be used as catalysts for various *cationic cyclizations*. Typical examples are the triflic acid catalysis in the double cyclization of *N,N*-dibenzylpropynylamine [*82*] (equation 30) and the **fluorosulfonic acid**-catalyzed condensation of phenylacetaldehyde [*83*] (equation 31).

[*76*] **29**

$(CH_2)_n$, R, HOOC

1. $SOCl_2$, C_6H_6, reflux, 1 h
2. CF_3SO_3H, CH_2Cl_2, -78 to 20 °C, 15 h

$(CH_2)_n$, R, O

n = 1,2; R = H, CH_3 80-95%

[*82*] **30**

CH_3, N

CF_3SO_3H (catal.)
0 to 20 °C, 15 h

CH_3, N

96%

[*83*] **31**

CH_2CHO

FSO_3H
CCl_4, 0 °C, 1 h

O

14%

Triflic acid is strong enough to protonate polycyclic saturated hydrocarbons [*77, 78, 79*], and even *n*-butane [*80, 81*], and to initiate *skeletal rearrangements.* Acidic treatment of homoadamantane [*77*] (equation 32), 2-homoprotoadamantane [*78*] (equation 33), or *cis*-2,3-trimethylenebicyclo[3.3.0]octane [*79*] (equation 34) causes their rearrangement to isomeric hydrocarbons.

Analogously, triflic acid isomerizes *n*-butane into 2-methylpropane [*80, 81*]. Interestingly, perfluorobutanesulfonic acid, which is of similar strength as triflic acid, does not catalyze the isomerization of *n*-butane [*80*].

The reaction of triflic acid with 2-hydroxy-2-adamantanecarboxylic acid results in decarbonylation to adamantanone [*84*]. However, in the presence of carbon monoxide, the same reactants give 4,5-homoadamantanedione as the product of a pinacolone-type rearrangement (equation 35).

Carbonylation of 1-adamantyl triflate in the presence of triflic acid also gives a derivative of homoadamantane as the result of a similar rearrangement with ring expansion [*85*] (equation 36).

Triflic acid is a useful reagent for the *removal of protecting groups* (deblocking) in synthetic proteins [*67*]. At the same time, it is an excellent catalyst for the protection of a variety of phenols in the form of their *tert*-butyl ethers [*86*] (equation 37).

[77] **32**

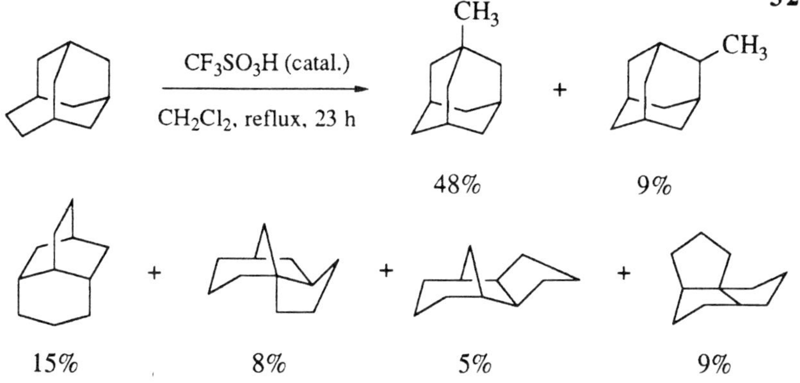

CF_3SO_3H (catal.)

CH_2Cl_2, reflux, 23 h

CH_3 CH_3

48% 9%

15% 8% 5% 9%

[78] **33**

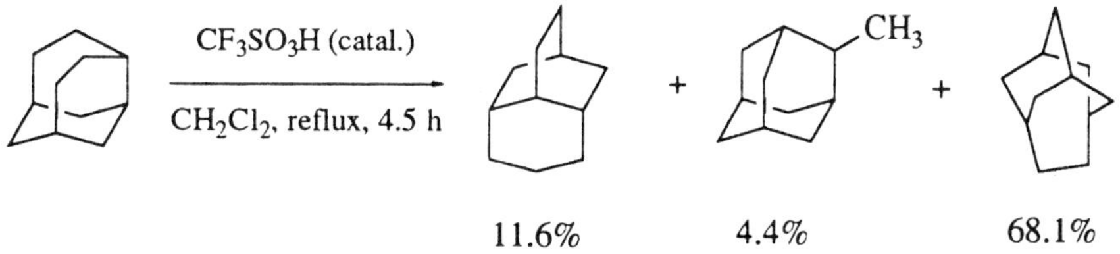

CF_3SO_3H (catal.)

CH_2Cl_2, reflux, 4.5 h

CH_3

11.6% 4.4% 68.1%

[79] **34**

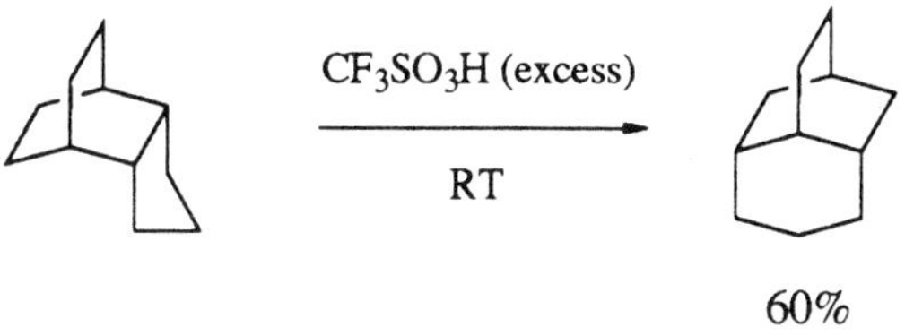

CF_3SO_3H (excess)

RT

60%

[84] **35**

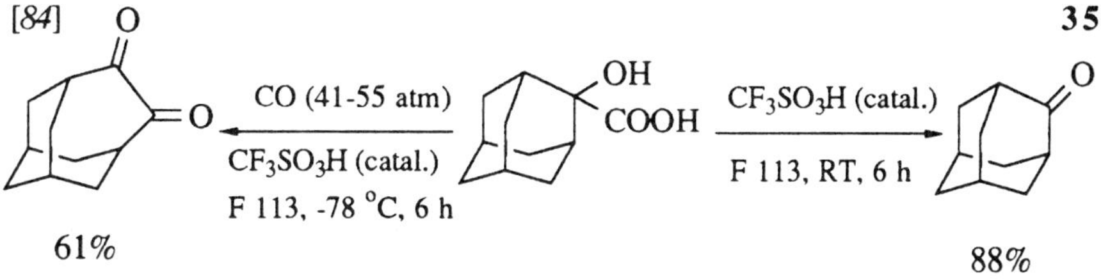

O O OH COOH O

CO (41-55 atm)

CF_3SO_3H (catal.)

F 113, -78 °C, 6 h

CF_3SO_3H (catal.)

F 113, RT, 6 h

61% 88%

[85] **36**

OSO_2CF_3

CO (1 atm)

CF_3SO_3H (catal.)

CCl_4, 30 °C, 20 h

O O-C OH

70%

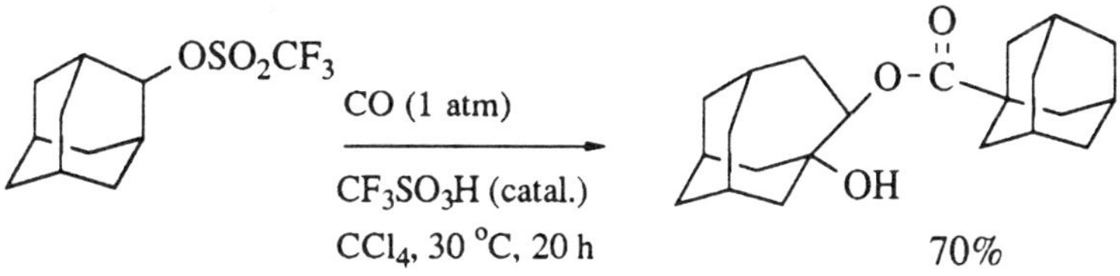

References are listed on pages 971–975.

[*86*] **37**

$$\text{R}^1\text{–R}^5\text{-substituted phenol (OH)} + \text{isobutylene} \xrightarrow[\text{CH}_2\text{Cl}_2,\ -78\ ^\circ\text{C, 4 h}]{\text{CF}_3\text{SO}_3\text{H (catal.)}} \text{R}^1\text{–R}^5\text{-substituted aryl–OC(CH}_3)_3$$

R = H, OCH_3, Br, $CH_2CH{=}CH_2$, OCH_2O 70-99%

The quantitative deprotection of phenol ethers can be achieved by the action of a catalytic amount of triflic acid in trifluoroethanol at –5 °C [*86*].

Triflic acid is an excellent catalyst for the *nitration of aromatic compounds* [*87*]. In a mixture with nitric acid, it forms the highly electrophilic nitronium triflate, which can be isolated as a white crystalline solid. Nitronium triflate is a powerful nitrating reagent in inert organic solvents and in triflic acid or sulfuric acid. It nitrates benzene, toluene, chlorobenzene, nitrobenzene, *m*-xylene, and benzotrifluoride quantitatively in the temperature range of –110 to 30 °C with exceptionally high positional selectivity [*87*].

In the past few years, **perfluororesinsulfonic acid (Nafion-H)** has found growing application in organic chemistry [*71*]. The types of Nafion-H most frequently used in organic synthesis are perfluorinated polymers with the structures $\sim(F_2C{-}CF_2)_x{-}(CF_2{-}CF)_y{-}OCF_2{-}CF{-}OCF_2{-}CF_2{-}SO_3H$ or $\sim(F_2C{-}CF_2)_x{-}(CF_2{-}CF)_y{-}OCF_2{-}CF{-}OCF(SO_3H){-}CF_3$ with an *x/y* ratio ranging from 2 to 50. Nafion-H resin has a strongly acidic character comparable with triflic acid and can be prepared easily from the commercially available potassium salts. It is generally used as a *catalyst for the same reactions catalyzed by triflic acid* but has an advantage of being easily recovered by simple filtration of the reaction mixture. A comprehensive account of the application of Nafion–H in organic synthesis, including catalysis in Friedel–Crafts alkylations and acylations, transalkylation, nitration, sulfonation, esterification, rearrangements, and condensations, was presented in a recent review [*71*].

Trifluoromethanesulfonic Anhydride and Chloride

Trifluoromethanesulfonic (triflic) anhydride is commercially available or can be prepared easily by the reaction of triflic acid with phosphorus pentoxide [*66*]. This moderately hygroscopic colorless liquid is a useful reagent for the preparation of various organic derivatives of triflic acid. A large variety of **organic ionic triflates** can be prepared from triflic anhydride. A recent example is the preparation of unusual oxo-bridged *dicationic salts of different types* [*88, 89, 90, 91, 92, 93*] (equations 38–44). Stabilized dication ether salts of the Hückel aromatic system and some other systems (equations 38 and 39) can be prepared in one step by the reaction of triflic anhydride with certain activated ketones such as cyclopropenone, tropone, and pyridone [*88*].

[*88*] **38**

$$\text{R}_2\text{C}_3{=}\text{O} \xrightarrow[\text{CHCl}_3,\ \text{RT, 30 min}]{(\text{CF}_3\text{SO}_2)_2\text{O}} [\text{R}_2\text{C}_3^{+}{-}\text{O}{-}^{+}\text{C}_3\text{R}_2] \cdot 2\ {}^{-}\text{OTf}$$

82%

R = C_3H_7 or C_6H_5; Tf = CF_3SO_2

[*88*] **39**

$$\text{C}_7\text{H}_6{=}\text{O} \xrightarrow[\text{CCl}_4,\ 0\ ^\circ\text{C}]{(\text{CF}_3\text{SO}_2)_2\text{O}} [\text{C}_7\text{H}_6^{+}{-}\text{O}{-}^{+}\text{C}_7\text{H}_6] \cdot 2\ {}^{-}\text{OTf}$$

68%

Like the reaction of ketones, the interaction of triflic anhydride with thiourea and substituted thioureas also gives dicationic triflates (equations 40 and 41); however, two sulfur atoms form the bridge in this case. This result indicates that triflic anhydride is acting as an oxidizing reagent toward thiourea [*89*].

[*89*] **40**

$$2\ (\text{R}_2\text{N})_2\text{C{=}S} \xrightarrow[\text{CH}_2\text{Cl}_2,\ \text{RT, 24 h}]{(\text{CF}_3\text{SO}_2)_2\text{O}} [(\text{R}_2\text{N})_2\text{C-S-S-C}(\text{NR}_2)_2]^{2+} \cdot 2\ {}^{-}\text{OTf}$$

R = H, CH_3 62-80%

[*89*] **41**

$$2\ (\text{C}_6\text{H}_5\text{N})_2\text{C}_3\text{H}_2{=}\text{S} \xrightarrow[\text{CH}_2\text{Cl}_2,\ \text{RT, 10 min}]{(\text{CF}_3\text{SO}_2)_2\text{O}} [(\text{C}_6\text{H}_5\text{N})_2\text{C}_3\text{H}_2^{+}{-}\text{S}{-}\text{S}{-}^{+}\text{C}_3\text{H}_2(\text{NC}_6\text{H}_5)_2] \cdot 2\ {}^{-}\text{OTf}$$

78%

A variety of *oxo-bridged dicationic triflates* with the positive charges localized on the heteroatoms can be prepared from triflic anhydride and the corresponding substrates [*90, 91, 92, 93*]. Treatment of hexamethylphosphoric triamide with triflic anhydride yields the diphosphonium triflate salt [*90*] (equation 42).

Cyclic dication ether triflate salts can be prepared by the reaction of diamides with triflic anhydride [*91*] (equation 43).

Oxo-bridged diiodonium triflate (*Zefirov's reagent*), a useful reagent for the synthesis of triflate esters and iodonium salts, can be prepared by the treatment of iodosobenzene with triflic anhydride [*92*] or by the reaction of iodobenzene diacetate with triflic acid [*93*] (equation 44).

[*90*] **42**

$$[(CH_3)_2N]_3P{=}O \xrightarrow[\text{RT, 10 min}]{(CF_3SO_2)_2O} \{[(CH_3)_2N]_3P\text{-}O\text{-}P[N(CH_3)_2]_3\}^{2+} \cdot 2\,^{-}OTf$$

66.5%

[*91*] **43**

$$(CH_3)_2N\text{-}\overset{O}{\overset{\|}{C}}\text{-}(CH_2)_n\text{-}\overset{O}{\overset{\|}{C}}\text{-}N(CH_3)_2 \xrightarrow[\text{CH}_2\text{Cl}_2,\ 0\,^{\circ}\text{C, 30 min}]{(CF_3SO_2)_2O} \left[(CH_3)_2NC \overset{O}{\frown} CN(CH_3)_2 \ \ (CH_2)_n\right]^{2+}$$

n = 2,3,4

• 2 $^{-}$OTf 66-68%

[*92,93*] **44**

$$C_6H_5IO \xrightarrow[\text{CH}_2\text{Cl}_2,\ \text{RT, 20 min}]{(CF_3SO_2)_2O} C_6H_5\text{—}\overset{+}{I}\text{—}O\text{—}\overset{+}{I}\text{—}C_6H_5 \xleftarrow[\text{CHCl}_3,\ \text{RT, 2 h}]{CF_3SO_3H} C_6H_5I(O_2CCH_3)_2$$

• 2 $^{-}$OTf

90-93%

Triflic anhydride is a useful reagent for the preparation of *covalent triflate esters from alcohols, ketones, and other organic substrates* [*66*]. In many cases, very reactive triflates can be generated in situ and subjected to subsequent transformation without isolation [*94, 95, 96, 97*]. Typical examples are cyclization of amides into dihydroisoquinolines (equation 45) and synthesis of *N*-hydroxy-α–amino acid derivatives (equation 46) via the intermediate covalent triflates.

[*94*] **45**

R^3, R^2, R^4, R^4, NH, =O, R^1 $\xrightarrow[\text{ether, RT}]{(CF_3SO_2)_2O}$ R^3, R^2, R^4, R^4, N, OTf, R^1

60-70 °C

R^3, R^2, R^4, R^4, N, R^1 90-94%

R^1-R^4 = H, OCH_3, C_6H_5, $CH_3OC_6H_4$

[95] **46**

$$\mathrm{R^2O\text{-}C(=O)\text{-}CH(R^1)(OH)} \xrightarrow[\mathrm{CH_2Cl_2,\ -78\ to\ 0\ ^\circ C}]{\mathrm{(CF_3SO_2)_2O}} \mathrm{R^2O\text{-}C(=O)\text{-}CH(R^1)(OTf)} \xrightarrow[\mathrm{CH_2Cl_2,\ RT,\ 25\ min}]{\mathrm{PhCH_2ONH_2}} \mathrm{C_6H_5CH_2O\text{-}NH\text{-}CH(R^1)\text{-}C(=O)OR^2}\ \ 78\text{-}89\%$$

$R^1 = CH_3, CH_2C_6H_5, C_6H_5$, etc.
$R^2 = CH_3, C_2H_5$

The intermediate formation of covalent triflates is assumed also in the reaction of alcohols with triflic anhydride in the presence of nitriles to give the corresponding amides [*96*] (equation 47).

[96] **47**

$$\mathrm{R^2\text{-}C(R^1)(R^3)\text{-}OH} + \mathrm{R^4\text{-}CN} \xrightarrow[\mathrm{2.\ NaHCO_3,\ H_2O}]{\mathrm{1.\ (CF_3SO_2)_2O,\ CH_2Cl_2,\ -20\ to\ 20\ ^\circ C,\ 2\text{-}5\ h}} \mathrm{R^2\text{-}C(R^1)(R^3)\text{-}NHCOR^4}\ \ 50\text{-}98\%$$

R^1 - R^4 = H, CH_3, n-C_4H_9, t-C_4H_9, C_6H_5, 1-adamantyl

In contrast to the normal Ritter reaction, the reaction shown in equation 47 gives good results starting from primary and secondary alcohols. Other advantages are the mild reaction conditions and easy workup [*96*].

A *new synthesis of acetylenes* using the reaction of triflic anhydride with acyl ylides (equation 48) was developed recently [*97*]. Vinyl triflates, generated in situ, are proposed to be the key intermediates.

[97] **48**

$$\mathrm{Ph_3\overset{+}{P}\text{-}\overset{-}{C}R^1\text{-}C(=O)\text{-}R^2} \xrightarrow[\mathrm{0\ to\ 20\ ^\circ C,\ 2\ h}]{\mathrm{(CF_3SO_2)_2O,\ C_6H_6}} \left[\mathrm{Ph_3\overset{+}{P}(R^1)C{=}C(R^2)OTf\ \ ^{-}OTf}\right] \xrightarrow[\mathrm{-20\ to\ 5\ ^\circ C,\ 17\ h}]{\mathrm{2\%\ Na/Hg,\ THF}} \mathrm{R^1\text{-}C{\equiv}C\text{-}R^2}\ \ 83\text{-}93\%$$

$R^1 = CH_3, C_4H_9$
$R^2 = C_6H_5, C_4H_9, C_5H_{11}$, etc.

References are listed on pages 971–975.

The **chloride of triflic acid (trifluoromethanesulfonyl chloride)** is an effective sulfonylating agent. Like triflic anhydride, it usually reacts with alcohols and other nucleophiles with the formation of the corresponding derivatives of triflic acid [*69*]. However, in some reactions, it acts as a *chlorinating reagent* [*98*]. The reactions of trifluoromethanesulfonyl chloride with 1,3-dicarbonyl compounds or some carboxylic esters in the presence of a base result in the formation of chlorinated products in high yields (equation 49).

[*98*] **49**

$$\text{R-C(=O)-CH}_2\text{-C(=O)-R}^1 \xrightarrow[\text{CH}_2\text{Cl}_2,\ 20\text{ to }25\ ^\circ\text{C},\ 1\text{ h}]{\text{CF}_3\text{SO}_2\text{Cl}/(\text{C}_2\text{H}_5)_3\text{N}} \text{R-C(=O)-CCl}_2\text{-C(=O)-R}^1$$

96-100%

$R,R^1 = CH_3, OCH_3, OC_2H_5$

When applied to 5- or 6-hydroxycarboxylic acids, this reaction gives *cyclization* products in good yield [*99*] (equations 50 and 51).

[*99*] **50**

$C_2H_5O_2C$, $CO_2C_2H_5$, OH; CF_3SO_2Cl/DBU; CH_2Cl_2, 22 °C, 3 h; $C_2H_5CO_2$, $CO_2C_2H_5$, O

DBU = 1,5-diazabicyclo[5.4.0]undec-5-ene 100%

[*100*] **51**

H, H, N_3, OH, N, O, CO_2CH_3, C_6H_5; $CF_3SO_2Cl/(C_2H_5)_3N$; CH_2Cl_2, 20 to 25 °C, 1 h; H, H, N_3, O, N, O, C_6H_5, CO_2CH_3

69%

The use of this cyclization in the synthesis of isooxapenams demonstrates its usefulness in forming oxygen heterocycles [*100*] (equation 51).

Trimethylsilyl Trifluoromethanesulfonate

Trimethylsilyl trifluoromethanesulfonate (trimethylsilyl triflate) is the most synthetically useful representative of the family of trialkylsilyl perfluoroalkanesulfonates (for a review, *see* reference 101). This reagent is commercially available or can be prepared easily by the reaction of chlorotrimethylsilane and triflic acid [*101*]. It has wide application in organic synthesis as an *excellent silylating reagent* and as a strong Lewis acid [*see* the chapter on silylation, page 615]. Silylated intermediates can be generated from trimethylsilyl triflate in situ and involved in subsequent transformations without isolation. An example of such a reaction is an intramolecular aldol-type condensation of 1,4-dicarbonyl compounds initiated by trimethylsilyl triflate. This reaction was applied in the total synthesis of naturally occurring tricyclic sesquiterpene capnellanes [*102*] (equation 52).

[*102*] **52**

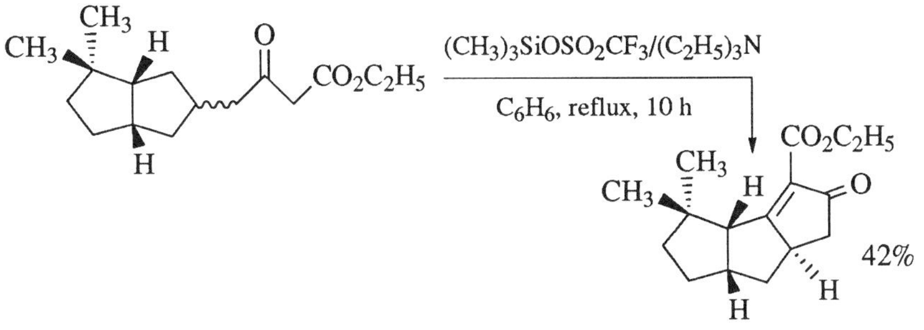
CH3
CH3
H
O
CO2C2H5
H
(CH3)3SiOSO2CF3/(C2H5)3N
C6H6, reflux, 10 h
CO2C2H5
CH3
CH3
H
O
42%
H
H

For some condensations with silylated substrates as starting compounds, trimethylsilyl triflate can be used as a catalyst [*103, 104, 105*]. A typical example of such a reaction is the *aldol-type condensation of silyl enol ethers and acetals* catalyzed by 1–5 mol% of trimethylsilyl triflate [*103*] (equation 53).

[*103*] **53**

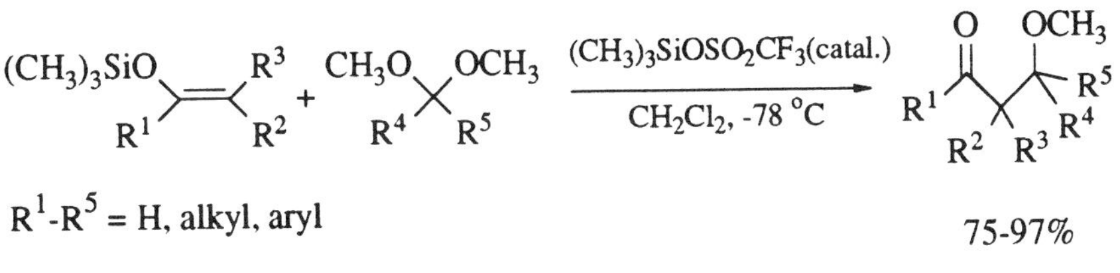
(CH3)3SiO
R3
CH3O
OCH3
+
R1
R2
R4
R5
(CH3)3SiOSO2CF3(catal.)
CH2Cl2, -78 °C
O
OCH3
R1
R5
R4
R2
R3
R1-R5 = H, alkyl, aryl
75-97%

Similarly, trimethylsilyl triflate can be used as a catalyst for the alkylation of 2-methoxy-1,3-oxazolidines [*104*] or 1-acetoxyadamantane [*105*] with allylsilane and for the reduction of acetals to ethers with trialkylsilanes [*106*].

Esters of Fluorosulfonic and Perfluoroalkanesulfonic Acids

A vast variety of perfluoroalkanesulfonic esters are known, and their application in synthetic organic chemistry is increasing rapidly (for a comprehensive review, *see* reference 66). These compounds are readily available by the reaction of organic halides with silver perfluoroalkanesulfonate or by the reaction of an alcohol with the corresponding perfluoroalkanesulfonic anhydride or chloride. Alkyl perfluoroalkanesulfonates are powerful alkylating reagents because of the excellent leaving-group properties of the perfluoroalkanesulfonic group [*66*]. The most widely used are commercially available **methyl triflate** and **methyl fluorosulfate (magic methyl)** (for a review, *see* reference 107). Both of these compounds are commonly used methylating reagents for a variety of organic substrates [*66*].

Another triflate ester that recently has found growing application in organic synthesis is commercially available **trimethylsilylmethyl trifluoromethanesulfonate.** This *powerful alkylating reagent* can be used for the synthesis of various methylides by an alkylation–desilylation sequence. A representative example is the generation and subsequent trapping by 1,3-dipolar cycloaddition of indolium methanides from the corresponding indole derivatives and trimethylsilylmethyl trifluoromethanesulfonate [*108*] (equation 54).

[*108*] **54**

$CO_2C_2H_5$ $(CH_3)_3SiCH_2OSO_2CF_3$ CH_2Cl_2, 0 °C, 2.5 h $CO_2C_2H_5$ ^{-}OTf $(CH_3)_3Si$ CsF

$CO_2C_2H_5$ CO_2CH_3 $CH_3O_2CC{\equiv}CCO_2CH_3$ $CO_2C_2H_5$ CO_2CH_3 57% CH_2^{-}

An interesting class of covalent triflates are *vinyl* and *aryl* or *heteroaryl* triflates. Vinyl triflates are used for the direct solvolytic generation of vinyl cations and for the generation of unsaturated carbenes via the α-elimination process [*66*]. A triflate ester of 2-hydroxypyridine can be used as a catalyst for the acylation of aromatic compounds with carboxylic acids [*109*] (equation 55).

[*109*] **55**

ArH + RCOOH —(2-pyridyl OTf; dioxane, RT, 2 h)→ ArCOR

ArH = dimethoxybenzenes, anisol, xylenes, fluorene, diphenyl ether, etc. 50-99%

R = alkyl, aryl

References are listed on pages 971–975.

Aryl perfluoroalkylsulfonates, readily available by the reaction of the corresponding phenols and acid anhydrides or chlorides, are used as reagents in organometallic coupling reactions (for a recent review, *see* reference 69).

Amides and Azide of Trifluoromethanesulfonic Acid

Amides of trifluoromethanesulfonic acid (triflamides) can be prepared by the reaction of the corresponding amines with triflic anhydride. The most applicable in organic synthesis are *N*-phenyltriflamides, which can be used as mild and selective triflating reagents [*110, 111*].

N-Phenyltriflamide is a useful reagent for the *mild oxidation of alkyl halides to carbonyl compounds* through a multistep one-pot procedure [*112*] (equation 56).

[*112*] **56**

$$R^1CHBrCOR^2 \xrightarrow[\text{acetone, reflux, 4-5 days}]{CF_3SO_2NHC_6H_5/K_2CO_3/KI} R^1COCOR^2 \quad 62\text{-}79\%$$

$R^1 = CH_3, C_3H_7, C_6H_{13}$

$R^2 = C_8H_{17}, C_6H_5, OC_2H_5$

This reaction gives good results for a variety of activated and unactivated alkyl halides [*112*]. Oxidation of 2-bromoketones with *N*-phenyltriflamide was used in a one-pot synthesis of pyrazines by the sequence of reactions shown in equation 57 [*113*]. The procedure was successfully applied to the synthesis of deoxyaspergillic acid [*114*].

[*113*] **57**

$$R^1COCHBrR^2 + CF_3SO_2NHC_6H_5 \xrightarrow[\text{3. } O_2/CH_3OH/CH_3ONa]{\text{1. } K_2CO_3/DMF \quad \text{2. } H_2NCH_2CH_2NH_2} \text{2-}R^1\text{-3-}R^2\text{-pyrazine} \quad 64\text{-}70\%$$

$R^1 = C_6H_5, i\text{-}C_3H_7, C_4H_9, C_8H_{13},$

$R^2 = CH_3$

Another nitrogen-containing triflate derivative, **trifluoromethanesulfonyl azide**, is prepared by the reaction of triflic anhydride with aqueous sodium azide [*115*] and is used as an efficient reagent for the synthesis of alkylazides from alkylamines (equation 58).

To reduce the danger of an explosion in this reaction, trifluoromethanesulfonyl azide can be generated in methylene chloride in situ and used for subsequent transformations without isolation [*115*].

[*115*] **58**

$$RNH_2 \xrightarrow[CH_2Cl_2,\ 0\ ^\circ C,\ 2\ h]{CF_3SO_2N_3} RN_3$$

$R = C_6H_{13}$ or $(CH_3)_2CCH_2C(CH_3)$ — 66-77%

Salts of Trifluoromethanesulfonic Acid

A large variety of salts of triflic acid formed both from metals and nonmetals are known. Many of these salts are versatile reagents for organic synthesis because of such properties of the triflate anion as very low nucleophilicity and low coordinating ability. However, despite low nucleophilicity, the triflate anion can combine with carbocationic intermediates under appropriate conditions to form triflate esters [*116, 117, 118*].

Triflates of aluminum, gallium and boron, which are readily available by the reaction of the corresponding chlorides with triflic acid, are effective *Friedel–Crafts catalysts* for alkylation and acylation of aromatic compounds [*119, 120*]. Thus alkylation of toluene with various alkyl halides in the presence of these catalysts proceeds rapidly at room temperature in methylene chloride or nitromethane. Favorable properties of the triflates in comparison with the corresponding fluorides or chlorides are considerably decreased volatility and higher catalytic activity [*120*].

Triflates of titanium and tin are effective catalysts for various condensations of carbonyl compounds [*121, 122, 123, 124, 125*]. *Claisen and Dieckmann-type condensations* between ester functions proceed under mild conditions in the presence of **dichlorobis(trifluoromethanesulfonyloxy)titanium(IV)** and a tertiary amine (equations 59 and 60). These highly regio- and stereoselective condensations were used successfully in the synthesis of carbohydrates [*122*].

[*121*] **59**

$$C_6H_5CH_2CH_2CO_2CH_3 + C_6H_5CO_2CH_3 \xrightarrow[CH_2Cl_2,\ 0\ to\ 20\ ^\circ C,\ 10\ h]{Cl_2Ti(OTf)_2/R_3N} C_6H_5CH_2CH(COC_6H_5)CO_2CH_3$$

71%

[*121*] **60**

$$\text{diethyl 3-methylhexanedioate } (CO_2C_2H_5,\ CO_2C_2H_5) \xrightarrow[CH_2Cl_2,\ 0\ to\ 20\ ^\circ C,\ 10\ h]{Cl_2Ti(OTf)_2/R_3N} \text{ethyl 3-methyl-2-oxocyclopentanecarboxylate } (CO_2C_2H_5)$$

56%

Stannous triflate is an efficient catalyst for aldol-type condensations [*123, 124, 125*]. Under conditions of kinetic control, it provides excellent diastereoselectivity in various cross-aldol reactions (equation 61).

[*123-125*] **61**

R^1-C(=O)-CH_2R^2 —[$Sn(OTf)_2/R_3N$; CH_2Cl_2, -78 °C, 0-5 h]→ —[R^3CHO; 2-5 h]→ R^1C(=O)CH(R^2)CH(OH)R^3

R^1-R^3 = alkyl, aryl, heteroaryl 41-79%

Magnesium triflate and zinc triflate are outstanding catalysts for the *introduction of the thioketal group for the protection of the ketone function* [*126*]. The reaction of a variety of ketones with ethane 1,2-dithiol in the presence of these triflates proceeds under mild conditions to form the corresponding thioketals in high yield (equation 62).

[*126*] **62**

OSi(*t* -Bu)Ph_2 —[$HS(CH_2)_2SH$, $Zn(OTf)_2$; CH_2Cl_2, 23-42 °C, 5.5 h]→ OSi(*t* -Bu)Ph_2 >85%

Triflate salts of copper and silver also are versatile reagents in organic synthesis [*127, 128, 129, 130, 131*]. **Copper(II) triflate** is an efficient dehydrating agent for a variety of alcohols [*127*]. Catalytic amounts of this salt in decalin or heptane at room temperature or with moderate heating cause rapid *dehydration of tertiary, secondary, and primary alcohols and diols* to form the corresponding alkenes with Zaitsev orientation and *E* stereochemistry in yields from 30 to 92% [*127*]. Copper(II) triflate is a good reagent for the oxidative coupling of ketone enolates with the formation of 1,4-diketones (equation 63) and for a number of catalytic transformations, such as stereoselective reduction of acetylenic sulfones by alkyl hydrosilanes [*128*] (equation 64).

Silver(I) triflate is widely applied to the preparation of various derivatives of triflic acid, both covalent esters [*66*] and ionic salts. For example, it can be used for the in situ generation of iodine(I) triflate, a very effective iodinating reagent for aromatic and heteroaromatic compounds [*130*] (equations 65 and 66).

Silver(I) triflate and copper(I) triflate can be applied as catalysts. A representative example is the preparation of alkynyl tosylates by the catalytic decomposition of alkynyl iodonium salts in the presence of these salts [*131*] (equation 67).

[*128*] **63**

$$2\,R\text{-}C(OLi){=}CR^1R^2 \xrightarrow[-78\text{ to }25\,^{\circ}C,\ 1\text{ h}]{Cu(OTf)_2,\ THF} R\text{-}CO\text{-}CR^1R^2\text{-}CR^1R^2\text{-}CO\text{-}R$$

R-R^2 = alkyl, aryl — 63-83%

[*129*] **64**

$$R\text{-}C{\equiv}C\text{-}SO_2C_6H_5 + HSi(C_2H_5)_2CH_3 \xrightarrow[15\,^{\circ}C,\ 6\text{ h}]{Cu(OTf)_2,\ i\text{-}PrOH} R\text{-}CH{=}CH\text{-}SO_2C_6H_5$$

R = CH_3, C_6H_5, C_5H_{11}, cyclohexyl, etc. — 55%

[*130*] **65**

$$RC_6H_5 \xrightarrow[0\text{ to }150\,^{\circ}C]{AgOTf/I_2} RC_6H_4I$$

R = H, OCH_3, CF_3 — 45-100%

[*130*] **66**

$$\text{thiophene} \xrightarrow[CH_2Cl_2,\ 0\,^{\circ}C]{AgOTf/I_2} \text{2-iodothiophene}$$

70%

[*131*] **67**

$$R\text{-}C{\equiv}C\text{-}\overset{+}{I}C_6H_5\ \ p\text{-}CH_3C_6H_4SO_3^- \xrightarrow[25\,^{\circ}C,\ 2\text{-}20\text{ h}]{M^{+\,-}OTf,\ CH_3CN} R\text{-}C{\equiv}C\text{-}OSO_2C_6H_4CH_3$$

R = alkyl, phenyl; M = Ag, Cu — 37-88%

Ionic triflate derivatives of nonmetallic elements such as selenium, sulfur, phosphorus, and iodine form an important class of reagents for organic chemistry. Highly electrophilic **phenylselenyl triflate** can be used in the cyclization of 5- and 6-hydroxyalkenes, affording the corresponding tetrahydrofurans and pyrans [*132*] (equation 68).

[*132*] **68**

$$CH_2{=}C(R^1)\text{-}CHR^2\text{-}(CH_2)_n\text{-}C(OH)R^3R^4 \xrightarrow[CH_2Cl_2,\ -78\text{ to }20\,^{\circ}C]{PhSeOTf} C_6H_5SeCH_2\text{-}\underset{\text{ring}}{C(R^1)\text{-}CHR^2\text{-}(CH_2)_n\text{-}CR^3R^4\text{-}O}$$

R^1-R^4 = H, CH_3; n =1,2 — 63-80%

Dimethyl sulfoxide (DMSO) and triphenylphosphine oxide react with triflic anhydride to form hygroscopic and moderately stable salts [*133, 134*]. The salt derivative of DMSO is an efficient and selective oxidative reagent for alcohols. It oxidizes various alcohols to the corresponding carbonyl derivatives in methylene chloride at room temperature; the best yields (up to 97%) are obtained from the reaction of benzylic alcohols [*133*]. The second salt, **triphenylphosphine ditriflate,** acts as a "general oxygen activator" [*134*], transforming the hydroxyl in alcohols or acids into a better leaving group with subsequent elimination or nucleophilic substitution (equation 69).

[*134*] $$\underset{R = \text{Alkyl}}{RCOOH} \xrightarrow[CH_2Cl_2,\ RT]{1.\ Ph_3\overset{+}{P}\text{-}OTf\ ^{-}OTf} \xrightarrow[RT,\ 12\ h]{2.\ PhNH_2} \underset{80\%}{RCONHPh}$$ **69**

In recent years, triflate derivatives of trivalent iodine have found more and more applications as versatile reagents in organic synthesis [*135, 136, 137, 138, 139, 140, 141, 142, 143, 144, 145*]. **μ-Oxo-bis[(trifluoromethanesulfonyloxy)(phenyl)iodine],** which is readily available from iodosobenzene and triflic anhydride (equation 44), is especially valuable for the preparation of alkynyl iodonium triflates from silyl- or stannylacetylenes [*136*] (equation 70).

[*136*] **70**

$$R{-}{\equiv}{-}Si(CH_3)_3 \xrightarrow[CH_2Cl_2,\ 0\,^{\circ}C,\ 30\ min]{Ph{-}I(OTf){-}O{-}I(OTf){-}Ph} R{-}{\equiv}{-}\overset{+}{I}C_6H_5\ ^{-}OTf$$

45-88%

$R = H, (CH_3)_3Si, t\text{-}C_4H_9, C_6H_{13}, C_6H_5$

Another useful reagent for the preparation of alkynyl iodonium triflates is **[cyano(trifluoromethylsulfonyloxy)(phenyl)]iodine** [*137, 138, 139, 140*] prepared from iodosobenzene, trimethylsilyl triflate, and trimethylsilyl cyanide (equation 71). This reagent reacts with various stannylacetylenes under very mild conditions to form the corresponding alkynyl iodonium salts in high yields [*139*] (equation 72).

In contrast to the previous method (equation 70), reaction 72 made possible the preparation of iodonium triflates from functionalized acetylenes bearing an electron-withdrawing group such as tosyl, cyano, or carbonyl [*138*]. Of special interest is the application of this method to the synthesis of the bisiodonium acetylenic salt [*139, 140*] (equation 73).

Alkynyl iodonium triflates prepared by the above reactions (equations 70, 72, and 73) have become valuable reagents in organic chemistry, serving as premier progenitors for a variety of *functionalized acetylenes* and other useful compounds [*138, 139, 140*]. Examples of some chemical transformations of **bis[phenyl(trifluoromethylsulfonyloxy)iodo]acetylene,** including nucleophilic substitution of the iodonium moiety and cycloaddition reactions, are given in equations 74 and 75.

[*137*] $$C_6H_5IO \xrightarrow[CH_2Cl_2,\ -30\ to\ 0\ ^\circ C,\ 15\ min]{(CH_3)_3SiOTf,\ (CH_3)_3SiCN} C_6H_5\overset{+}{I}CN\ {}^-OTf \quad 89\%$$ **71**

[*138*] **72**

$$X{-}C{\equiv}C{-}Sn(C_4H_9)_3 \xrightarrow[CH_2Cl_2,\ -40\ ^\circ C,\ 15\ min]{C_6H_5\overset{+}{I}CN\ {}^-OTf} X{-}C{\equiv}C{-}\overset{+}{I}C_6H_5\ {}^-OTf \quad 36\text{-}77\%$$

X = CN, *p*-$CH_3C_6H_4SO_2$, C_6H_5CO, etc.

[*139-140*] **73**

$$(C_4H_9)_3Sn{-}C{\equiv}C{-}Sn(C_4H_9)_3 \xrightarrow[CH_2Cl_2,\ -40\ ^\circ C]{C_6H_5\overset{+}{I}CN\ {}^-OTf} C_6H_5\overset{+}{I}{-}C{\equiv}C{-}\overset{+}{I}C_6H_5 \cdot 2\ {}^-OTf \quad 81\%$$

[*140*] **74**

$$C_6H_5\overset{+}{I}{-}C{\equiv}C{-}\overset{+}{I}C_6H_5 \cdot 2\ {}^-OTf \xrightarrow[CH_3CN,\ -35\ ^\circ C]{C_6H_5SNa} C_6H_5S{-}C{\equiv}C{-}SC_6H_5 \quad 66\%$$

$$\xrightarrow[CH_3CN,\ -35\ ^\circ C]{C_6H_5OLi} C_6H_5O{-}C{\equiv}C{-}OC_6H_5 \quad 57\%$$

$$\xrightarrow[CCl_4,\ 20\ ^\circ C]{(C_6H_5)_3P} (C_6H_5)_3\overset{+}{P}{-}C{\equiv}C{-}\overset{+}{P}(C_6H_5)_3 \cdot 2\ {}^-OTf \quad 59\%$$

[*140*] **75**

$$C_6H_5\overset{+}{I}{-}C{\equiv}C{-}\overset{+}{I}C_6H_5 \cdot 2\ {}^-OTf \xrightarrow[CH_3CN,\ -35\ ^\circ C]{\text{cyclopentadiene}} \text{2,3-bis}(\overset{+}{I}C_6H_5)\text{norbornadiene} \cdot 2\ {}^-OTf \quad 69\%$$

Another type of iodonium salts, **perfluoroalkylphenyliodonium triflates** (FITS reagents), has undergone considerable development recently [*141, 142, 143, 144, 145*]. These compounds, available from perfluoroalkyliodides (equation 76), are very effective *electrophilic perfluoroalkylating agents.* They react with carbanions, aromatic compounds, alkenes, alkynes, silyl enol ethers, and other nucleophiles under mild conditions to introduce the perfluoroalkyl moiety into organic substrates (equation 77) (*see* the section on alkylation, page 446).

[*141*] $$C_nF_{2n+1}I \xrightarrow[\text{2. } CF_3SO_3H,\ C_6H_6]{\text{1. } CF_3CO_3H} C_nF_{2n+1}\overset{+}{I}C_6H_5 \cdot CF_3SO_3^- \qquad \textbf{76}$$

n = 2-10 70-90%

[*142*] $$R^-M^+ \xrightarrow[\text{THF, -110 to 20 °C, 0.3-1 h}]{C_nF_{2n+1}\overset{+}{I}C_6H_5 \cdot CF_3SO_3^-} C_nF_{2n+1}R \qquad \textbf{77}$$

32-82%

R = alkyl, vinyl, alkynyl, etc.
M = Li, MgCl; n = 2,3,8

An interesting development of this research is the preparation of polymer-supported FITS reagent from bis(trifluoroacetoxy)iodoperfluoroalkanes and Nafion-H [*145*]. FITS–Nafion reacts with organic substrates that react to usual FITS reagents, but the products of the perfluoroalkylation reaction can be separated easily from the insoluble resin by filtration [*145*].

Other Fluorinated Organic Reagents

The introduction of fluorine atoms usually modifies the properties of an organic compound toward some useful directions such as higher reactivity, thermal stability, and resistance to oxidation and reduction. This strategy allows the creation of new valuable organic reagents. A large variety of fluorinated organic reagents have been introduced in pure and applied chemistry in recent years. **Fluorinated solvents, complexones, and crown ethers** possess some unique properties that justify their application in industry, analytical chemistry, and other areas. **Optically active trifluoromethyl-substituted carbinols** $R(CF_3)CHOH$ (R, cyclohexyl, phenyl, 1-naphthyl, or 9-anthryl) [*146*] and *acids* $R^1R^2(CF_3)CCO_2H$ (R^1, aryl; R^2, OCH_3) [*147*] are used as *chiral solvating agents for asymmetric transformations* and as resolving reagents. Commercially available (–)-(*S*)- or (+)-(*R*)-$Ph(CF_3)C(OCH_3)CO_2H$ is an efficient resolving agent for the resolution of polycyclic arene oxides and diols [*147*].

Pentafluorophenyl derivatives have received wide application in biochemistry as highly selective and efficient *protective reagents* (for a review, *see* reference 148).

Fluorinated organic reagents are widely used in synthetic organic chemistry. Following are several examples.

Pentafluorophenyl acetate is a highly *selective acetylating reagent,* useful for acetylations at hydroxyl and amino groups under mild conditions. When applied to the acetylation of amino alcohols, it gives selective formation of *N*-acetyl derivatives at room temperature and *N,O*-diacetylated products under moderate heating in the presence of triethylamine [*149*].

[Bis(2,2,2-trifluoroethyl)]phosphite can be used to prepare esters of phosphorous acid by a transesterification reaction with alcohols and phenols [*150*] (equation 78).

[*150*] **78**

$$(CF_3CH_2O)_2P(=O)H \xrightarrow[2.\ NH_3,\ H_2O]{1.\ ROH,\ MeCN,\ 70\ ^\circ C,\ 4\ h} (RO)(^-O)P(=O)H\ \ NH_4^+$$

$R = C_6H_5, C_6H_5CH_2$, etc. 60-95%

A variety of fluorinated oxidizing agents are used in organic synthesis. **Bis(*p*-trifluoromethylphenylsulfonyl) peroxide** is a highly efficient reagent for the *oxidative deamination of amines to the corresponding carbonyl compounds.* This oxidation proceeds under very mild conditions (–78 °C), and the yields of products are generally higher than yields from other procedures [*151*] (equation 79).

[*151*] **79**

$$RCH_2NHR^1 \xrightarrow[2.\ HCl,\ H_2O,\ 150\ ^\circ C]{1.\ (CF_3C_6H_4SO_2O)_2,\ KOH,\ C_2H_5OCOCH_3,\ -78\ ^\circ C,\ 6\text{-}7\ h} RCHO$$

R,R^1 = alkyl, aryl 20-90%

An efficient *oxidizing reagent* for oxyfunctionalization of saturated hydrocarbons, **methyl(trifluoromethyl)dioxirane,** can be prepared from aqueous potassium peroxomonosulfate and 1,1,1-trifluoropropanone [*152*]. The reaction of this reagent with adamantane gives the corresponding tris(hydroxy)adamantane in high yield [*153*] (equation 80).

[*153*] **80**

$$\text{adamantane} \xrightarrow[CF_3COCH_3,\ CH_2Cl_2,\ -20\ ^\circ C,\ 2\ h]{(CF_3)(CH_3)C(O_2)\ \text{(excess)}} \text{1,3,5-tris(hydroxy)adamantane}$$

80%

References are listed on pages 971–975.

References for pages 941–970

1. Wristers, J. *J. Am. Chem. Soc.* **1975,** *97,* 4312.
2. Feiring, A. E. *J. Org. Chem.* **1977,** *42,* 3255.
3. Fidler, D. A.; Logan, J. S.; Boudakian, M. M. *J. Org. Chem.* **1961,** *26,* 4014.
4. Weinmayr, V. *J. Am. Chem. Soc.* **1955,** *77,* 1762.
5. Askam, V.; Qazi, T. U. *J. Chem. Soc., Perkin Trans. 1* **1977,** 1263.
6. Daub, G. H.; Ackerman, M. E.; Hayes, F. N. *J. Org. Chem.* **1973,** *38,* 828.
7. Olah, G. A.; Welch, J.; Vankar, Y. D.; Nojima, M.; Kerekes, I.; Olah, J. A. *J. Org. Chem.* **1979,** *44,* 3872.
8. Matsuura, S.; Niu, C.; Cohen, J. S. *J. Chem. Soc., Chem. Commun.* **1976,** 451.
9. Roush, W. R.; Gillis, H. R.; Essenfeld, A. P. *J. Org. Chem.* **1984,** *49,* 4674.
10. Comninellis, C.; Javet, P.; Plattner, E. *Tetrahedron Lett.* **1975,** 1429.
11. Hoz, S.; Albeck, M.; Rappoport, Z. *Synthesis* **1975,** 162.
12. Belsky, I. *J. Chem. Soc., Chem. Commun.* **1977,** 237.
13. Noyori, R.; Yokoyama, K.; Sakata, J.; Kuwajima, I.; Nakamura, E.; Shimizu, M. *J. Am. Chem. Soc.* **1977,** *99,* 1265.
14. Barluenga, J.; Joglar, J.; Gonzales, F. J.; Fustero, S. *Synlett* **1990,** 129.
15. Moriarty, R. M.; Vaid, R. K. *Synthesis* **1990,** 431.
16. Olah, G. A.; Surya Prakash, G. K.; Sommer, J. *Superacids;* Wiley: New York, 1985.
17. Zhdankin, V. V.; Mullikin, M.; Tykwinski, R.; Berglund, B.; Caple, R.; Zefirov, N. S.; Koz'min, A. S. *J. Org. Chem.* **1989,** *54,* 2605.
18. Takahashi, Y.; Tomita, N.; Yoneda, N.; Suzuki, A. *Chem. Lett.* **1975,** 997.
19. Zhdankin, V. V.; Tykwinski, R.; Berglund, B.; Mullikin, M.; Caple, R.; Zefirov, N. S.; Koz'min, A. S. *J. Org. Chem.* **1989,** *54,* 2609.
20. Zlotin, S. G.; Krayushkin, M. M.; Sevost'yanova, V. V.; Novikov, S. S. *Izv. Akad. Nauk SSSR* **1977,** 2286 (Engl. Transl. 2121).
21. Hannessian, S.; Staub, A. P. A. *Tetrahedron Lett.* **1973,** 3555.
22. Allen, A. D.; Tidwell, T. T. *J. Am. Chem. Soc.* **1982,** *104,* 3145.
23. Kirmse, W.; Brandt, S. *Chem. Ber.* **1984,** *117,* 2524.
24. Fichtner, M. W.; Haley, N. F. *J. Org. Chem.* **1981,** *46,* 3141.
25. Spitzer, U. A.; Stewart, R. *J. Org. Chem.* **1974,** *39,* 3936.
26. Gribble, G. W.; Kelly, W. J.; Emery, S. E. *Synthesis* **1978,** 763.
27. Kursanov, D. N.; Parnes, Z. N.; Kolomnikova, G. D.; Kalinkin, M. I. *Izv. Akad. Nauk SSSR* **1978,** 2413 (Engl. Transl. 2147).
28. Stewart, R.; Spitzer, U. A. *Can. J. Chem.* **1978,** *56,* 1273.
29. Deno, N. C.; Messer, L. A. *J. Chem. Soc., Chem. Commun.* **1976,** 105.
30. Liotta, R.; Hoff, W. S. *J. Org. Chem.* **1980,** *45,* 2887.
31. Manley, R. P.; Curry, K. W.; Deno, N. C.; Meyer, M. D. *J. Org. Chem.* **1980,** *45,* 4385.
32. Holbert, G. W.; Ganem, B. *J. Chem. Soc., Chem. Commun.* **1978,** 248.

33. Holbert, G. W.; Ganem, B. *J. Am. Chem. Soc.* **1978,** *100,* 352.
34. Sinnott, M. L.; Widdows, D. *J. Chem. Soc., Perkin Trans. 1* **1981,** 401.
35. Swern, D. *Organic Peroxides;* Wiley: New York, 1973; Vol. 2, p 403.
36. Crivello, J. V. *J. Org. Chem.* **1981,** *46,* 3056.
37. Dampawan, P.; Zajac, W. W. *Synthesis* **1983,** 545.
38. Numata, T.; Awano, H.; Oae, S. *Tetrahedron Lett.* **1980,** *21,* 1235.
39. Sonnet, P. E. *J. Org. Chem.* **1978,** *43,* 1841.
40. Omura, K.; Sharma, A. K.; Swern, D. *J. Org. Chem.* **1976,** *41,* 957.
41. Huang, S. L.; Omura, K.; Swern, D. *Synthesis* **1978,** 297.
42. Crawford, T. C.; Breitenbach, R. *J. Chem. Soc., Chem. Commun.* **1979,** 388.
43. Eliel, E. L.; Soai, K. *Tetrahedron Lett.* **1981,** *22,* 2859.
44. Tyobeka, T. E.; Hancock, R. A.; Weigel, H. *J. Chem. Soc., Chem. Commun.* **1980,** 114.
45. Kovalev, V. V.; Fedorova, O. A.; Shokova, E. A. *Zh. Org. Khim.* **1987,** *23,* 1882 (Engl. Transl. 1672).
46. Carotti, A.; Campagna, F.; Ballini, R. *Synthesis* **1979,** 56.
47. Carotti, A.; Campagna, F.; Casini, G.; Ferappi, M.; Giardina, D. *Gazz. Chim. Ital.* **1979,** *109,* 329.
48. Catoni, G.; Galli, C.; Mandolini, L. *J. Org. Chem.* **1980,** *45,* 1906.
49. Cookson, R. C.; Cramp, M. C.; Parsons, P. J. *J. Chem. Soc., Chem. Commun.* **1980,** 197.
50. Hill, A. E.; Hoffmann, H. M. R. *J. Am. Chem. Soc.* **1974,** *96,* 4597.
51. Crivello, J. V. *Synth. Commun.* **1976,** *6,* 543.
52. Brown, H. C.; Kurek, J. T.; Rei, M. H.; Thompson, K. L. *J. Org. Chem.* **1985,** *50,* 1171.
53. Brown, H. C.; Geoghegan, P. J.; Kurek, J. T. *J. Org. Chem.* **1981,** *46,* 3810.
54. Suzuki, K.; Mukaiyama, T. *Chem. Lett.* **1982,** 683.
55. Lethbridge, A.; Norman, R. O. C.; Thomas, C. B. *J. Chem. Soc., Perkin Trans. 1* **1973,** 2763.
56. Taylor, E. C.; Andrade, J. G.; Steliou, K.; Jagdmann, G. E.; McKillop, A. *J. Org. Chem.* **1981,** *46,* 3078.
57. Reid, D. H.; Webster, R. G. *J. Chem. Soc., Chem. Commun.* **1972,** 1283.
58. Campbell, J. R.; Kalman, J. R.; Pinhey, J. T.; Sternhell, S. *Tetrahedron Lett.* **1972,** 1763.
59. Kalman, J. R.; Pinhey, J. T.; Sternhell, S. *Tetrahedron Lett.* **1972,** 5369.
60. Buddrus, J. *Chem. Ber.* **1982,** *115,* 2377.
61. Buddrus, J.; Plettenberg, H. *Angew. Chem. Int. Ed. Engl.* **1976,** *15,* 436.
62. Buddrus, J.; Plettenberg, H. *Chem. Ber.* **1980,** *113,* 1494.
63. Varvoglis, A. *Chem. Soc. Rev.* **1981,** *10,* 377.
64. Varvoglis, A. *Synthesis* **1984,** 709.
65. Merkushev, E. B. *Russian Chem. Rev.* **1987,** *56,* 825.
66. Stang, P. J.; Hanack, M.; Subramanian, L. R. *Synthesis* **1982,** 85.

67. Stang, P. J.; White, M. R. *Aldrichimica Acta* **1983,** *16,* 15.
68. Howells, R. D.; Mc Cown, J. D. *Chem. Rev.* **1977,** *77,* 69.
69. Huang, W.-Y.; Chen, Q.-Y. In *The Chemistry of Sulfonic Acids, Esters and Their Derivatives;* Patai, S.; Rappoport, Z., Eds.; Wiley: Chichester, U.K., 1991, Chapter 21, p 903.
70. Gillespie, R. J. *Acc. Chem. Res.* **1968,** *1,* 202.
71. Olah, G. A.; Iyer, P. S.; Prakash, G. K. S. *Synthesis* **1986,** 513.
72. Bakoss, H. J.; Roberts, R. M. G.; Sadri, A. R. *J. Org. Chem.* **1982,** *47,* 4053.
73. Roberts, R. M. G. *J. Org. Chem.* **1982,** *47,* 4050.
74. Rajadhyaksha, R. A.; Chaudhary, D. D. *Ind. Eng. Chem. Res.* **1987,** *26,* 1276.
75. Butler, I. R.; Morley, J. O. *J. Chem. Res., Synop.* **1980,** 358.
76. Hulin, B.; Koreeda, M. *J. Org. Chem.* **1984,** *49,* 207.
77. Takaishi, N.; Inamoto, Y.; Fujikura, Y.; Aigami, K.; Golicnic, B.; Mlinaric-Majerski, K.; Majerski, Z.; Osawa, E.; Schleyer, P. V. R. *Chem. Lett.* **1976,** 763.
78. Takaishi, N.; Inamoto, Y.; Aigami, K.; Fujikura, Y.; Osawa, E.; Kawanisi, M.; Katsushima, T. *J. Org. Chem.* **1977,** *42,* 2041.
79. Inamoto, Y.; Aigami, K.; Takaishi, N.; Fujikura, Y.; Tsuchihashi, K.; Ikeda, H. *J. Org. Chem.* **1977,** *42,* 3833.
80. Choukroun, H.; Germain, A.; Brunel, D.; Commeyras, A. *Nouv. J. Chim.* **1981,** *5,* 39.
81. Choukroun, H.; Germain, A.; Brunel, D.; Commeyras, A. *Nouv. J. Chim.* **1983,** *7,* 83.
82. Takayama, H.; Suzuki, T.; Nomoto, T. *Chem. Lett.* **1978,** 865.
83. Kagan, J.; Agdeppa, D. A.; Chang, A. I.; Chen, S. A.; Harmata, M. A.; Melnick, B. *J. Org. Chem.* **1981,** *46,* 2916.
84. Olah, G. A.; Wu, A. *J. Org. Chem.* **1991,** *56,* 2531.
85. Takeuchi, K.; Akiyama, F.; Miyazaki, T.; Kitagawa, I.; Okamoto, K. *Tetrahedron* **1987,** *43,* 701.
86. Holcombe, J. L.; Livinghouse, T. *J. Org. Chem.* **1986,** *51,* 111.
87. Coon, C. L.; Blucher, W. G.; Hill, M. E. *J. Org. Chem.* **1973,** *38,* 4243.
88. Stang, P. J.; Maas, G.; Smith, D. L.; McCloskey, J. A. *J. Am. Chem. Soc.* **1981,** *103,* 4836.
89. Maas, G.; Stang, P. J. *J. Org. Chem.* **1981,** *46,* 1606.
90. Aaberg, A. A.; Gramstad, T.; Husebye, S. *Acta Chem. Scand.* **1980,** *A34,* 717.
91. Gramstad, T.; Husebye, S.; Saebo, J. *Acta Chem. Scand.* **1985,** *B39,* 505.
92. Hembre, R. T.; Scott, C. P.; Norton, J. R. *J. Org. Chem.* **1987,** *52,* 3650.
93. Zefirov, N. S.; Zhdankin, V. V.; Dan'kov, Y. V.; Koz'min, A. S. *Zh. Org. Khim.* **1984,** *20,* 446 (Engl. Transl. 401).
94. Nagubandy, S.; Fodor, G. *Heterocycles* **1981,** *15,* 165.

95. Feenstra, R. W.; Stokkingreef, E. H. M.; Nivard, R. J. F.; Ottenheijm, H. C. J. *Tetrahedron Lett.* **1987,** *28,* 1215.
96. Garcia, M. A.; Martinez, A. R.; Teso, V. E.; Garcia, F. A.; Hanack, M.; Subramanian, L. R. *Tetrahedron Lett.* **1989,** *30,* 851.
97. Bestmann, H. J.; Kumar, K.; Schaper, W. *Angew. Chem. Int. Ed. Engl.* **1983,** *23,* 167.
98. Hakimelahi, G. H.; Just, G. *Tetrahedron Lett.* **1979,** 3643.
99. Hakimelahi, G. H.; Just, G. T*etrahedron Lett.* **1979,** 3645.
100. Hakimelahi, G. H.; Just, G. *Can. J. Chem.* **1981,** *59,* 941.
101. Emde, H.; Domsch, D.; Feger, H.; Frick, U.; Goetz, A.; Hergott, H. H.; Hofmann, K.; Kober, W.; Kraegeloh, K.; Oesterle, T.; Steppan, W.; West, W.; Simchen, G. *Synthesis* **1982,** 1.
102. Shibasaki, M.; Mase, T.; Ikegami, S. *J. Am. Chem. Soc.* **1986,** *108,* 2090.
103. Murata, S.; Suzuki, M.; Noyori, R. *J. Am. Chem. Soc.* **1980,** *102,* 3248.
104. Conde-Frieboes, K.; Hoppe, D. *Synlett* **1990,** 99.
105. Sasaki, T.; Nakanishi, A.; Oluro, M. *J. Org. Chem.* **1982,** *47,* 3219.
106. Tsunoda, T.; Suzuki, M.; Noyori, R. *Tetrahedron Lett.* **1979,** 4679.
107. Alder, R. W. *Chem. Ind. (London)* **1973,** 983.
108. Fishwick, C. W. G.; Jones, A. D.; Hitchell, M. B.; Egglestone, D. S.; Baures, P. W. *Chem. Lett.* **1990,** 359.
109. Keumi, T.;Yoshimura, K.; Shimada, M.; Kitajima, H. *Bull. Chem. Soc. Jpn.* **1988,** *61,* 455.
110. Hendrickson, J. B.; Bergeron, A. *Tetrahedron Lett.* **1973,** 4607.
111. Tilley, J. W.; Zawoiski, S. *J. Org. Chem.* **1988,** *53,* 386.
112. Bergeron, R. J.; Hoffman, P. G. *J. Org. Chem.* **1979,** *44,* 1835.
113. Bergeron, R. J.; Hoffman, P. G. *J. Org. Chem.* **1980,** *45,* 161.
114. Bergeron, R. J.; Hoffman, P. G. *J. Org. Chem.* **1980,** *45,* 163.
115. Cavender, C. J.; Shiner, V. J. *J. Org. Chem.* **1972,** *37,* 4679.
116. Zefirov, N. S.; Koz'min, A. S.; Sorokin, V. D.; Zhdankin, V. V. *Zh. Org. Khim.* **1982,** *18,* 1768 (Engl. Transl. 1546).
117. Zefirov, N. S.; Zhdankin, V. V.; Sorokin, V. D.; Dan'kov, Y. V.; Kirin, V. N.; Koz'min, A. S. *Zh. Org. Khim.* **1982,** *18,* 2608 (Engl. Transl. 2301).
118. Zefirov, N. S.; Zhdankin, V. V.; Makhon'kova, G. V.; Dan'kov, Y. V.; Koz'min, A. S. *J. Org. Chem.* **1985,** *50,* 1872.
119. Olah, G. A.; Laali, K.; Farooq, O. *J. Org. Chem.* **1984,** *49,* 4591.
120. Olah, G. A.; Farooq, O.; Farnia, S. M. F.; Olah, J. A. *J. Am. Chem. Soc.* **1988,** *110,* 2560.
121. Tanabe, Y.; Mukaiyama, T. *Chem. Lett.* **1984,** 1867.
122. Stevens, R. W.; Mukaiyama, T. *Chem. Lett.* **1983,** 595.
123. Mukaiyama, T.; Stevens, R. W.; Iwasawa, N. *Chem. Lett.* **1982,** 353.
124. Mukaiyama, T.; Iwasawa, N. *Chem. Lett.* **1984,** 753.
125. Mukaiyama, T.; Iwasawa, N.; Stevens, R. W.; Haga, T. *Tetrahedron* **1984,** *40,* 353.

126. Corey, E. J.; Shimoji, K. *Tetrahedron Lett.* **1983,** *24,* 4607.
127. Laali, K.; Cerzina, R. J.; Flajnik, C. M.; Geric, C. M.; Dombroski, A. M. *Helv. Chim. Acta* **1987,** *70,* 607.
128. Kobayashi, Y.; Taguchi, T.; Tokuno, E. *Tetrahedron Lett.* **1977,** 3741.
129. Ryu, I.; Kusumoto, N.; Ogawa, A.; Kambe, N.; Sonoda, N. *Organometallics* **1989,** *8,* 2279.
130. Kobayashi, Y.; Kumadaki, I.; Tsutomu, Y. *J. Chem. Res. (S)* **1977,** 215.
131. Stang, P. J.; Surber, B. W.; Chen, Z. C.; Roberts, K. A.; Anderson, A. G. *J. Am. Chem. Soc.* **1987,** *109,* 228.
132. Murata, S.; Suzuki, T. *Tetrahedron Lett.* **1987,** *28,* 4297.
133. Hendrickson, J. B.; Schwartzman, S. M. *Tetrahedron Lett.* **1975,** 273.
134. Hendrickson, J. B.; Schwartzman, S. M. *Tetrahedron Lett.* **1975,** 277.
135. Zefirov, N. S.; Zhdankin, V. V.; Koz'min, A. S. *Izv. Akad. Nauk SSSR, Ser. Khim.* **1983,** 1682 (Engl. Transl. 1530).
136. Stang, P. J.; Arif, A. M.; Crittell, C. M. *Angew. Chem. Int. Ed. Engl.* **1990,** *29,* 287.
137. Zhdankin, V. V.; Crittell, C. M.; Stang, P. J.; Zefirov, N. S. *Tetrahedron Lett.* **1990,** *31,* 4821.
138. Stang, P. J.; Williamson, B. L.; Zhdankin, V. V. *J. Am. Chem. Soc.* **1991,** *113,* 5870.
139. Stang, P. J.; Zhdankin, V. V. *J. Am. Chem. Soc.* **1990,** *112,* 6437.
140. Stang, P. J.; Zhdankin, V. V. *J. Am. Chem. Soc.* **1991,** *113,* 4571.
141. Umemoto, T.; Kuriu, Y.; Shuyama, H.; Miyano, O.; Nakayama, S. *J. Fluorine Chem.* **1986,** *31,* 37.
142. Umemoto, T.; Kuriu, Y. *Tetrahedron Lett.* **1981,** *22,* 5197.
143. Umemoto, T.; Kuriu, Y.; Nakayama, S. *Tetrahedron Lett.* **1982,** *23,* 1169.
144. Umemoto, T.; Kuriu, Y.; Nakayama, S.; Miyano, O. *Tetrahedron Lett.* **1982,** *23,* 1471.
145. Umemoto, T. *Tetrahedron Lett.* **1984,** *25,* 81.
146. Bacciarelli, M.; Forni, A.; Moretti, I.; Torre, G. *J. Org. Chem.* **1983,** *48,* 2640.
147. Balani, S.; Boyd, D. R.; Cassidy, E. S.; Greene, R. M. E.; McCombe, K. M.; Sharma, N. D.; Jennings, W. B. *Tetrahedron Lett.* **1981,** *22,* 3277.
148. Jarman, M. *J. Fluorine Chem.* **1989,** *42,* 3.
149. Kisfaludy, L.; Mohacsi, T.; Low, M.; Drexler, F. *J. Org. Chem.* **1979,** *44,* 654.
150. Gibbs, D. E.; Larsen, C. *Synthesis* **1984,** 410.
151. Hoffman, R. V.; Kumar, A. *J. Org. Chem.* **1984,** *49,* 4011.
152. Mello, R.; Fiorentito, M.; Fusco, C.; Curci, R. *J. Am. Chem. Soc.* **1989,** *111,* 6749.
153. Mello, R.; Cassidei, L.; Fiorentito, M.; Fusco, C.; Curci, R. *Tetrahedron Lett.* **1990,** *31,* 3067.

Chapter 6

Properties of Fluorinated Compounds

Physical and Physicochemical Properties

by B. E. Smart

Fluorinated organic compounds often may seem abnormal in comparison with hydrocarbon or other halocarbon compounds, but their behavior usually is quite intelligible and predictable when the effects of fluorination on molecular properties are understood. This chapter discusses these characteristic effects.

The fundamental properties of the fluorine atom underlie many of its characteristic substituent effects. Its high ionization potential and low polarizability (Table 1) imply very weak inter- and intramolecular interactions in fluorocarbons. Its extreme electronegativity ensures strong inductive electron withdrawal and polarized $^{\delta+}$C–Fd$^{\delta}$ bonding, and consequently, C–F bonds are more ionic and stronger than other C–X bonds. The C–F bond dipole also can impart significant polar character to partially fluorinated materials, whose physical properties therefore can differ from those of either their hydrocarbon or fluorocarbon analogues. In sum, the fluorine atom's special electronic substituent effects in organic molecules arise from its unique combination of properties: (1) high electronegativity; (2) relatively small size; (3) three tightly bound, nonbonding electron pairs; and (4) excellent 2s and 2p orbital overlap with the corresponding orbitals on carbon.

Physical Properties

The physical properties of saturated fluorocarbons and their analogous hydrocarbons differ in many respects [*4, 5*]. Saturated fluorocarbons have the lowest dielectric constants, surface tensions, and refractive indexes of any liquids at room

Table 1. Atomic Physical Properties[a]

Atom	*IP (kcal/mol)*	*EA (kcal/mol)*	α_v (Å^3)	r_v, Å	χ_p
H	313.6	17.7	0.667	1.20	2.20
F	401.8	79.5	0.557	1.47	3.98
Cl	299.0	83.3	2.18	1.75	3.16
Br	272.4	72.6	3.05	1.85	2.96
I	241.2	70.6	4.7	1.98	2.66

[a]IP, ionization potential [*1*]; EA, electron affinity [*1*]; a_v, atom polarizability [*2*]; r_v, van der Waals radius [*3*]; χ_p, Pauling electronegativity.

0065–7719/95/0187–0979$10.16/1

temperature. They have significantly greater densities, viscosities, critical temperatures, and compressibilities but much lower cohesive energies and internal pressures than the corresponding hydrocarbons. Some specific comparative data for C_6F_{14} and C_6H_{14} are given in Table 2.

Table 2. Comparative Physical Properties of *n*-Hexanes[a]

n-Hexane	*bp (°C)*	T_c *(°C)*	d^{25} *(g/cm³)*	η^{25} *(cP)*	*Ref.*
C_6F_{14}	57	174	1.672	0.66	6
$F(CF_2)_3(CH_2)_3H$	64	200	1.265	0.48	7
$C_6H_{13}F$	91.5	246	0.794	0.46	7
C_6H_{14}	69	235	0.655	0.29	6
n-Hexane	γ^{25} *(dyne/cm)*	β *(10⁻⁶ atm⁻¹)*	n_D^{25}	ε	*Ref.*
C_6F_{14}	11.4	254	1.252	1.69	6
$F(CF_2)_3(CH_2)_3H$	14.3	198	1.290	5.99	7
$C_6H_{13}F$	19.8	—	1.372	5.63	7
C_6H_{14}	17.9	150	1.372	1.89	6

[a] T_c, critical temperature; d^{25}, density; η^{25}, viscosity; γ^{25}, surface tension; β, compressibility at 1 atm; n_D^{25}, refractive index;ε, dielectric constant.

These unusual properties of fluorocarbons reflect their nonpolar character, low polarizability, and overall relatively weak intermolecular attractions. Saturated perfluoro-*tert*-amines and -dialkyl ethers also closely resemble fluorocarbons rather than typical amines or ethers in their physical properties [*4, 8*].

The properties of hydrofluorocarbons often fall between those of their corresponding fluorocarbons and hydrocarbons, but there can be significant differences. The dielectric constants of $C_6H_{13}F$ and $CF_3(CF_2)_2(CH_2)_2CH_3$ and the surface tension and boiling point of $C_6H_{13}F$ are notably higher than those of either C_6F_{14} or C_6H_{14}. These differences reflect the more polar character of the hydrofluorocarbons owing to their net C–F and C–C bond dipoles.

There are several compilations and reviews of fluorocarbon physical properties [*4, 5, 6, 9, 10, 11, 12*], and only the boiling points, surface energies and activities, and solvent properties are discussed in this section to illustrate the characteristic fluorine substituent effects.

Boiling Points

The boiling points of homologous, linear saturated fluorocarbons and hydrocarbons closely track each other, and the differences between the boiling points of fluorocarbons and hydrocarbons with the same number of carbon atoms is surprisingly small considering the much higher molecular weight of the fluorocarbon [*9*]. The boiling points of linear and cyclic saturated fluorocarbons and perfluoro-1-

alkenes can be predicted accurately by simple, empirical equations [*13*]. Butane and perfluorobutane boil only 0.5 °C apart but differ in molecular weight by 180, whereas carbon tetrafluoride and hexane have almost the same molecular weights, and yet their boiling points differ by nearly 200 °C! Perfluoroethers and amines [bp of $(C_4F_9)_3N$, 174–178 °C vs. bp of $(C_4H_9)_3N$, 216.5 °C] likewise boil at much lower temperatures than their nonfluorinated analogues despite their greater molecular weights. Moreover, in stark contrast to their nonfluorinated counterparts, the difference in boiling points of saturated fluorocarbon isomers and perfluoroether isomers is practically negligible (Table 3). These boiling point trends clearly reflect the extremely low intermolecular interactions in perfluorinated compounds [*4*].

Table 3. Boiling Points of Perfluorinated and Nonfluorinated Isomers

Compound	*Structure and bp (°C)*			*Ref.*
C_5F_{12}	 29.3	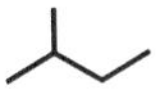 30.1	 29.5	10
C_5H_{12}	36.1	27.9	9.5	
$C_{10}F_{18}$	 142.5		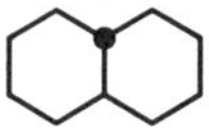 141.0	15
$C_{10}H_{18}$	193.3		185.3	
C_8F_{16}	 102.3		 101.8	10
C_8H_{16}	129.7		120.1	
$C_6F_{14}O$	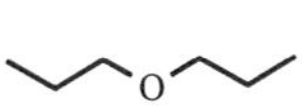 56		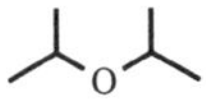 54	16
$C_6H_{14}O$	90		69	
$C_8F_{18}O$	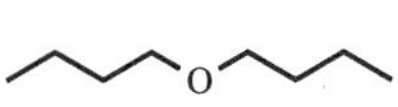 111		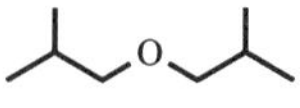 98	16
$C_8H_{18}O$	142		122	

References are listed on pages 1003–1010.

Saturated hydrofluorocarbons exhibit quite different boiling point patterns. For the series of methanes and ethanes in Table 4, the boiling points parallel dipole moments: the highest boiling member of the series has the highest dipole moment. The correlation between dipole moment (μ) and boiling point is not general, however. For instance, CH_3CHF_2 (μ, 2.27 D; bp, –24.7 °C) has nearly the same dipole moment as CH_3CF_3 but boils about 23 °C higher. This fact suggests the added importance of attractive intermolecular $^{\delta+}C–F^{\delta-}$, $^{\delta-}C–H^{\delta+}$ bond dipole interactions [*14*] or $CH_3CF_2H \cdots FCHFCF_3$ hydrogen bonding [9]. Boiling point patterns for hydrofluoro-, hydrochlorofluoro-, and chlorofluoroethane isomers that reveal a complex interplay of polar and polarizability effects have been established [*14*].

For hydrofluoroethylenes, the relationship between boiling point and dipole moment also is somewhat irregular (Table 5), but for *E* and *Z* isomers, the isomer with the higher dipole moment normally boils higher, as indicated by the following isomeric pairs: (*Z*)-CHF=CHF (μ, 2.42; bp, –25 °C) vs. (*E*)-CHF=CHF (μ, 0; bp, –51 °C) and (*Z*)-CF_3CH=$CHCF_3$ (bp, 33 °C) vs. (*E*)-CF_3CH=$CHCF_3$ (bp, 9 °C) [*10*].

Surface Tension and Activity

Saturated fluorocarbons have the lowest surface tensions of any organic liquids and can completely wet almost any surface. Perfluorinated amines and ethers also

Table 4. Boiling Points and Dipole Moments of Methanes and Ethanes [*17, 18*]

Compound	*bp (°C)*	*μ (D)*	*Compound*	*bp (°C)*	*μ (D)*
Methanes			Ethanes		
CH_4	–161	0.0	CH_3CH_3	–88.6	0.0
CH_3F	–78	1.85	CH_3CF_3	–47.5	2.32
CH_2F_2	–52	1.97	CF_3CF_3	–78.2	0.0
CHF_3	–82	1.65			
CF_4	–128	0.0			

Table 5. Boiling Points and Dipole Moments of Ethylenes [*17, 18*]

Ethylene	*bp (°C)*	*μ (D)*
CH_2=CH_2	–102	0.0
CH_2=CHF	–72	1.43
CH_2=CF_2	–83	1.38
CHF=CF_2	–51	1.40
CF_2=CF_2	–76	0.0

have low surface tensions (15–16 dyne/cm) [*8*]. The surface tensions of hydrofluorocarbons are always higher than their fluorocarbon counterparts, but they can be greater, smaller, or equal to those of their hydrocarbon analogues depending upon fluorine content (Table 6).

Table 6. Liquid [*11*] and Polymer [*20*] Surface Tensions

Liquid	Υ *(dyne/cm)*	*Polymer*	Υ_c *(dyne/cm)*
C_5F_{12}	9.4	$–CF_2CF(CF_3)–$	16.2
C_5H_{12}	15.2	$–CF_2CF_2–$	18.5
C_6F_{14}	11.4	$–CH_2CF_2–$	25
$C_6F_{13}H$	12.6[a]	$–CH_2CHF–$	28
$C_5H_{11}CF_3$	17.9[a]	$–CH_2CH_2–$	31
$C_6H_{13}F$	19.8[a]	$–CF_2CClF–$	31
C_6H_{14}	17.9		
cyclo-$C_6F_{11}CF_3$	15.4		
cyclo-$C_6H_{11}CH_3$	23.3		

[a]Reference 7.

The effects of fluorination on solid-surface free energies parallel the liquid trends. Perfluorinated polymers have the lowest critical surface tensions, which directly relate to their antistick properties [*19*], but substitution of fluorine by hydrogen or by the more polarizable chlorine atom markedly raises their surface free energy.

The lowest wettable surface known ($\gamma_c = 6$ dyne/cm) is a monolayer of perfluorolauric acid on platinum, whose surface is made up of closely packed CF_3 groups [*20*]. Fluorinated graphite, $(C_2F)_n$ and $(CF)_n$, also have surface tensions approaching 6 dyne/cm [*21*]. Perfluorinated materials, however, are not required for low surface energies; only the outermost surface groups must be perfluorinated [*20, 22*].

Fluorocarbons with a hydrophilic functional group are very active surfactants [*23*]. Less than 1% of ionic or nonionic surfactants with perfluoroalkyl groups can reduce the surface tension of water from 72 to 15–20 dyne/cm, compared with 25–35dyne/cm for typical hydrocarbon surfactants [*24*]. Perfluoroether surfactants are about as active as their perfluoroalkyl counterparts of similar chain length [*25, 26*], but fluorosurfactants with more polar alkyl end groups are considerably less active than their perfluoroalkyl analogues (Table 7).

Fluorosurfactants have a much greater tendency to form micelles than do hydrocarbon ones. The effect of each CF_2 group on the critical micelle concentration (cmc) is equivalent to that of 1.5–1.8 CH_2 groups [*24, 29, 30*]. Lightly fluorinated surfactants behave differently. The cmcs of $CF_3(CH_2)_nCO_2Na$ are *higher* than their hydrocarbon analogues [*27*], which indicates that the polar CF_3CH_2 group *increases* hydrophilicity (p. 987).

Table 7. Properties of Fluorinated Surfactants

Sufactant	γ_{cmc}[a]	*cmc (mM)*[b]	*Ref.*
$C_nF_{2n+1}CO_2Li$			
$n = 6$	27.8	98 (30 °C)	24
$n = 8$	24.6	10.6 (30 °C)	24
$n = 10$	20.5	0.39 (60 °C)	24
$C_7F_{15}CO_2H$	15.2	2.8 (30 °C)	24
$H(CF_2)_7CO_2H$	21.8	—	20
$C_8F_{17}CO_2NH_4$	14.8	6.7 (30 °C)	24
$CF_3[OCF_2CF(CF_3)]_2OCF_2CO_2NH_4$	17.5	7.0 (25 °C)	25
$CF_3(CH_2)_8CO_2Na$	—	167 (25 °C)	27
$C_8F_{17}CH_2CON[(C_2H_4O)_3CH_3]_2$	19	0.012 (25 °C)	28

[a]Aqueous surface tension above cmc at 25 °C.
[b]Values were determined at temperatures in parentheses.

The size and shape of micelles also are affected by fluorination. Whereas hydrocarbon surfactants usually form spherical micelles, linear fluorocarbon surfactants tend to produce larger, rodlike species [*31, 32*]. This is attributed to two inherent characteristics of the $(CF_2)_n$ chain: (1) it adopts a helical rather than a linear zigzag conformation [*33, 34, 35, 36*], and (2) it is much stiffer than the $(CH_2)_n$ chain [*35, 37, 38*]. The relatively stiff, helical $(CF_2)_n$ chains thus prefer cylindrical to spherical packing.

Solubility

The relative polarities of various solvents according to a new empirical solvatochromatic polarity scale [*39*] are listed in Table 8. Fluorinated compounds represent both extremes in polarity.

Saturated fluorocarbons are the least polar organic compounds and are poor solvents for all materials except those with very low cohesive energies, such as gases and other fluorocarbons [*12, 41*]. They are more like Ar and Kr in their solvent–solute interactions [*6*]. Saturated fluorocarbons are practically insoluble in water [*9, 42*] and only slightly soluble in hydrocarbons, but they can dissolve some low-molecular-weight hydrocarbons [*9, 12*]. Heats of solution for fluorocarbons and hydrocarbons markedly differ [*12, 43*], and their mixtures exhibit grossly nonideal behavior that often results in phase separation at room temperature [*44, 45*]. The relatively high solubility of oxygen and other gases in fluorocarbons compared with hydrocarbons or water [*41*] results from the large cavities (free volume) in fluorocarbons that can accommodate the gases [*46, 47*], and not from any special attractive interactions [*6*]. The gas solubility directly correlates with the compressibility of the fluorocarbon [*48*].

Table 8. Solvent Spectral Polarity (P_s) Index [39]

Solvent	P_s	*Solvent*	P_s
C_6F_{14}	0.00	C_6H_{14}	2.56
cyclo-$C_6F_{11}CF_3$	0.46	*cyclo*-$C_6H_{11}CH_3$	3.34
C_8F_{18}	0.55	C_8H_{18}	2.86
$(C_4F_9)_3N$	0.68	$(C_4H_9)_3N$	3.93
CF_3CCl_3[a,b]	3.22	CH_3CCl_3	7.03
CCl_3F	3.72	CCl_4	4.64
C_6F_6	4.53	C_6H_6	6.95
C_6H_5F	7.52	C_6H_5Cl	8.30
1,2-$C_6H_4F_2$	7.86	1,2-$C_6H_4Cl_2$	8.94
CHClFCHClF[a]	8.19	CH_2ClCH_2Cl	8.61
CF_3CH_2OH	10.20	CH_3CH_2OH	8.05
$(CF_3)_2CHOH$	11.08	$(CH_3)_2CHOH$	7.85

[a]Reference 40.
[b]For CCl_2FCClF_2, $P_s = 3.22$ [39].

Fluorocarbon solvent polarity (P_s) increases when fluorine is substituted by chlorine or hydrogen, and hydrofluorocarbons can be more polar than their hydrocarbon analogues. Notably, $P_s(C_6H_5F) = 7.52$ and P_s (1,2-$C_6H_4F_2$) = 7.86 vs. $P_s(C_6H_6) = 6.95$, whereas both $C_8F_{17}H$ ($P_s = 4.33$) and $C_4F_9CH_2CH_3$ ($P_s = 4.01$) are considerably more polar than their hydrocarbon counterparts [*40*].

The extraordinarily high polarity of the fluorinated alcohols reflects their strong hydrogen-bonding capability [*54*]. The P_s values of 10.2 for CF_3CH_2OH and 11.08 for $(CF_3)_2CHOH$ compare with 10.6 and 12.1 for 50% formic acid and water, respectively [*39*].

Physicochemical Properties

The effects of fluorination on lipophilicity and acidity are discussed in this section. Several useful substituent constants are compiled in Table 9.

Lipophilicity

Several Hansch–Leo π substituent parameters derived from octanol–water partition coefficients (P), $\pi_X = \log P\,(C_6H_5X) - \log P\,(C_6H_6)$, are listed in Table 9. Fluorination clearly increases lipophilicity, and its effect can be particularly large for heteroatom substituents.

The generalization that fluorination always increases lipophilicity is incorrect, however. It holds for aromatic fluorination and fluorination adjacent to most but

Table 9. Hydrophobic [*49*] and Electronic Substituent Constants [*50*]

Substituent	π^a	$\sigma_m b$	$\sigma_p b$	F^c	R^c
F	0.14	0.34	0.06	0.45	–0.39
Cl	0.71	0.37	0.23	0.42	–0.19
OH	–0.67	0.12	–0.37	0.33	–0.70
NO_2	–0.27	0.72	0.78	0.65	0.13
CH_3	0.56	–0.07	–0.17	0.01	–0.18
CF_3	0.88	0.43	0.54	0.38	0.16
CH_3CH_2	1.02	–0.07	–0.15	0.00	–0.15
CF_3CH_2	—	0.12	0.09	0.15	–0.06
CF_3CF_2	1.68	0.47	0.52	0.44	0.08
OCH_3	–0.02	0.12	–0.27	0.29	–0.56
OCF_3	1.04	0.38	0.35	0.39	–0.04
SCH_3	0.61	0.15	0.00	0.23	–0.23
SCF_3	1.44	0.40	0.50	0.36	0.14
$C(O)CH_3$	–0.55	0.39	0.50	0.33	0.17
$C(O)CF_3$	0.02	0.63	0.80	0.54	0.26
SO_2F	0.05	0.80	0.91	0.72	0.19
SO_2CH_3	–1.63	0.60	0.72	0.53	0.19
SO_2CF_3	0.55	0.83	0.96	0.74	0.22
$NHSO_2CH_3$	–1.18	0.23	0.03	0.28	–0.25
$NHSO_2CF_3$	0.92	0.44	0.39	0.45	–0.06
C_6H_5	1.96	0.06	–0.01	0.12	–0.13
C_6F_5	—	0.26	0.27	0.27	0.00

[a]Hansch–Leo hydrophobic substituent parameter.
[b]Hammett sigma constants.
[c]Modified Swain–Lupton constants.

not all atoms or groups with π electrons, whereas monofluorination or trifluoromethylation of saturated alkanes *decreases* lipophilicity.

Fluorination of carbon adjacent to heteroatoms invariably increases lipophilicity, as does fluorination of double bonds (Table 10), but α-fluorination of aliphatic carbonyl groups is an exception. α-Fluoroketones or aldehydes that form stable

Table 10. Octanol–Water Partition Coefficients

Compound	*log P*	*Ref.*	*Compound*	*log P*	*Ref.*
$CH_2{=}CH_2$	1.13	51	$CH_3(CH_2)_nOH$		
$CH_2{=}CF_2$	1.24	51	$n = 1$	–0.32	52
$CF_2{=}CF_2$	2.00	51	$n = 2$	0.34	52
CH_4	1.09	49	$n = 3$	0.88	52
CH_3F	0.51	49	$n = 4$	1.40	52
CH_2F_2	0.20	49	$n = 5$	1.64	52
CHF_3	0.64	49	$CF_3(CH_2)_nOH$		
CF_4	1.18	49	$n = 1$	0.36	52
CH_3CH_3	1.81	49	$n = 2$	0.39	52
CH_3CHF_2	0.75	49	$n = 3$	0.90	52
CF_3CF_3	2.00	49	$n = 4$	1.15	52
C_5H_{12}	3.11	49	$n = 5$	1.36	52
$C_5H_{11}F$	2.33	49			

hydrates and hemiketals or acetals [*53*], as well as α-fluorocarboxylic acids that are strong hydrogen-bond donors [*54*], have greater heats of hydration than their nonfluorinated analogues. Their apparent lipophilicities consequently depend on the choice of partitioning solvents used in the measurement (Table 11).

Table 11. Solvent Effects on Solvent–Water Partition Coefficients (*P*) [*49*]

Compound	*log P* ($C_9H_{19}OH$–H_2O)	*log P* (C_6H_{14}–H_2O)
$CH_3COCH_2COCH_3$	0.26	0.02
$CF_3COCH_2COCH_3$	0.59	–0.50
Compound	*log P* ($C_2H_5OC_2H_5$–H_2O)	*log P* (C_6H_6–H_2O)
CH_3CO_2H	–0.36	–1.74
CF_3CO_2H	–0.27	–1.89

Consistent with their polar character discussed in the previous section, hydrofluorocarbons are often less lipophilic than their hydrocarbon counterparts. The relative lipophilicities of CH_nF_{4-n} exactly parallel their order of dipole moments, with CH_2F_2 being the most hydrophilic compound in the series. The comparative data for CH_3CHF_2, $C_5H_{11}F$, and $CF_3(CH_2)_nOH$ (n = 4, 5) in Table 10 also show that fluorination decreases lipophilicity. In general, monofluorination and trifluoromethylation appear to decrease lipophilicity when the site of fluorination is removed from any heteroatom or C=C double bond by three or more C–C bonds.

Acidity

The strong electron-withdrawing effect of fluorine and fluorinated substituents ensures that the solution acidities of acids, alcohols, and amides are always increased by fluorination [*9*]. Increases in acidity by several pK_a units are common (Table 12). Especially pronounced are the effects of CF_3SO_2 and FSO_2, which are among the most electron-withdrawing groups known [*55, 56*]. For instance, $(FSO_2)_3CH$ is about as acidic as sulfuric acid [*57*], and $CF_3SO_2CH_3$ is 10^{13} times more acidic than $CH_3SO_2CH_3$ in dimethyl sulfoxide [*58*]. The strongest of all acids are the anhydrous Brønsted superacids [*59*]: CF_3SO_3H ($H_0 = -13.8$) and FSO_3H ($H_0 = -15.1$). Anhydrous FSO_3H and HF ($H_0 = -15.1$) are the strongest pure acids known, and HF–0.5% SbF_5 ($H_0 = -21$) is the most acidic material so far measured [*60*]. (H_0 is the Hammett acidity function.)

Table 12. Acid Strengths in Water at 25 °C [*61*]

Acid	*pK_a*	*Acid*	*pK_a*
CH_3SO_3H	–1.9	C_6H_5OH	10.0
CF_3SO_3H	–5.1	C_6F_5OH	5.5[a]
CH_3CO_2H	4.8	$(CH_3)_3COH$	19.0
CF_3CO_2H	–0.2	$(CF_3)_3COH$	5.4
O, NH, O	9.6	O, F_2, NH, F_2, O	2.1[b]

[a]Reference 62.
[b]Reference 63.

Fluorination increases C–H acidity, with certain special exceptions, but the effects of α- and β-fluorination differ. For C–H acids whose conjugate anions are nonplanar, α-fluorination always increases acidity, but less so than other halogens (Table 13). The acidities of CHX_3, CF_3CHX_2, and $(CF_3)_2CHX$ [*64*] increase in the order F < Cl < Br. For planar conjugate anions, however, α-fluorination can *decrease* acidity, but the effect is often unpredictable. Examples are given in Table 14.

The increasing order of acidity, $CF_3H < CF_3CF_2H < (CF_3)_2CFH < (CF_3)_3CH$, and the fact that pentakis(trifluoromethyl)cyclopentadiene (p$K_a \leq -2$ [*69*]) is about 10^{18} times more acidic than cyclopentadiene, whereas pentafluorocyclopentadiene is only slightly more acidic [*70*], illustrate that a trifluoromethyl group has a much

Table 13. Equilibrium C–H Acidities [*65, 66*]

Compound	pK_{CsCHA}	*Compound*	pK_{CsCHA}
CHF_3	30.5	$(CF_3)_2CFH$	25.2
$CHCl_3$	24.4	$(CF_3)_3CH$	~21[a]
$CHBr_3$	22.7	C_6H_6	43
CF_3CHF_2	28.2	$C_6H_4F_2$ (1, 2)	34.5[b]
CF_3CHCl_2	24.4	$C_6H_2F_4$ (1, 2, 4, 5)	26
CF_3CHBr_2	23.7	C_6HF_5	25.9

[a]Estimated but likely too high; *see* reference 67.
[b]For C–H_3, C–H_6.

Table 14. Equilibrium Acidities of Esters, Diketones, and Nitromethanes [*64*]

Compound	pK_a	*Compound*	pK_a
$CH_3CO_2CH_3$	24[a]	CH_3NO_2	10.21
$CH_2FCO_2C_2H_5$	21[a]	CH_2FNO_2	9.5
$CHF_2CO_2C_2H_5$	25[a]	CHF_2NO_2	12.40
$NO_2CH_2CO_2C_2H_5$	5.82	$CH_2(NO_2)_2$	3.63
$NO_2CHFCO_2C_2H_5$	6.28	$CHF(NO_2)_2$	7.70
$C_6H_5COCH_2COC_6H_5$	10.7[b]		
$C_6H_5COCHFCOC_6H_5$	8.5[b]		

[a]Polarographic values.
[b]Reference 68.

larger, uniformly acidifying effect than fluorine [*64*]. The electronic effects governing the different influences of α- and β-fluorination on anion stability and C–H acidity are well understood (p. 996).

The basicities of amines, ethers, and carbonyl compounds are invariably decreased by fluorination. 2,2,2-Trifluoroethylamine (pK_b = 3.3 [*61*]) and $C_6F_5NH_2$ (pK_b = -0.36 [*62*]) are about 10^5 times less basic than $CH_3CH_2NH_2$ and $C_6H_5NH_2$, respectively, and $(CF_3)_2CHNH_2$ (pK_b = 1.22 [*71*]) is over 10^9 times less basic than *i*-$C_3H_7NH_2$. The relative gas-phase acidities in Table 15 illustrate the large effect of fluorination. Perfluoro-*tert*-amines $(R_f)_3N$ and ethers R_fOR_f have no basic character in solution [*8, 74*], and CF_3COCF_3 is not protonated by superacids [*72*].

Chemical Reactivity

The effects of fluorination on bond strengths, reactive intermediates, and steric interactions that directly relate to chemical reactivity are summarized in this section.

Table 15. Gas-Phase Proton Affinities [73]

Compound	$-\Delta H^0$ *(kcal/mol)*	*Compound*	$-\Delta H^0$ *(kcal/mol)*
$CF_3N(CH_3)_2$	193.8	$CF_3CH_2OC_2H_5$	186.4
$(CH_3)_3N$	225.1	$(C_2H_5)_2O$	200.2
$C_5F_5N^a$	177.5	$(CF_3)_2CO$	161.4
$C_5H_5N^b$	213.7	$(CH_3)_2CO$	196.7

[a]Pentafluoropyridine.
[b]Pyridine.

Bond Strengths and Reactivity

Extensive reviews of the effects of fluorination on structure and bonding are available [*75, 76, 77*], and only the characteristic trends in bond strengths will be covered here. The bond energies cited are average values corrected for the revised heats of formation of alkyl radicals [*78*], but their precision is seldom better than ±2 kcal/mol for the fluoro compounds.

Fluorine forms the strongest single bond to carbon of any element. On average, the C–F bond is 25 kcal/mol stronger than the C–Cl bond in monohaloalkanes (Table 16). Moreover, fluorination characteristically affects other C–X bond strengths. *α-Fluorination always markedly increases the strengths of C–F and C–O bonds,* but not C–H, C–Cl, or C–Br bonds. By contrast, β-fluorination significantly increases C–H bond strengths but has little effect on C–F bonds. Remarkably, the C–H bond in $(CF_3)_3CH$ is estimated to be 15 kcal/mol stronger than the $(CH_3)_3C$–H bond [*80*].

The strong C–F bond and poor leaving-group ability of the fluoride ion [*81, 82, 83*] make alkyl fluorides much less reactive than other alkyl halides toward nucleophilic displacement. Monofluoroalkanes are typically 10^2–10^6 times less reactive than their corresponding alkyl chlorides in S_N1 or S_N2 displacements [*82, 83*], although their hydrolysis can be acid-catalyzed [*82, 84*], whereas alkyl CF_3 and CF_2H groups are inert to fluoride ion displacement under normal reaction conditions. Nucleophilic attack on carbon in polyfluorohaloalkanes also is highly disfavored, but attack on halogen by one- or two-electron transfer processes is possible [*85*] (*see* p. 446 *et seq.*) Equations 1–3 formally are simple displacements, but the first and second involve a radical-anion mechanism ($S_{RN}1$) [*86, 87*], whereas the third is a difluorocarbene-mediated process [*88*].

Even the chemically robust perfluoroalkanes can undergo electron-transfer reactions (equation 4) because of their relatively high electron affinities [*89*]. Strong reducing agents like alkali metals [*90*] or sodium naphthalide [*91*] are normally required for reaction, but perfluoroalkanes with low-energy, *tert*-C–F σ* anti-

[*86*] $$C_6F_{13}I \xrightarrow[C_6H_6]{O_2N\bar{C}(CH_3)_2\ (C_4H_9)_4N^+} C_6F_{13}C(CH_3)_2NO_2 \quad 71\%$$ **1**

Table 16. Bond Dissociation Energies (D^0) of Methanes, Ethanes, and Ethers [75, 76, 78]

	D^0 (C–X) (kcal/mol)			
Compound	*H*	*F*	*Cl*	*Br*
CH_3X	104.3	108.3	82.9	69.6
CH_2X_2	104.3	119.5	81.0	64
CHX_3	104.3	127.5	77.7	62
CX_4	104.3	130.5	72.9	56.2
	D^0 (C–X) (kcal/mol)			
Compound	*H*	*F*	*Cl*	*Br*
CH_3–X	104.3	108.3	82.9	69.6
CF_3–X	106.7	130.5	87.1	70.6
	D^0 (C–X) (kcal/mol)			
Compound	*CH_3*	*CF_3*	*OCH_3*	*OCF_3*
CH_3–X	88.8	101.2	83.2	—
CF_3–X	101.2	98.7	—	105.2
	D^0 (C–X) (kcal/mol)			
Compound	*H*	*F*	*Cl*	*Br*
CH_3CH_2–X	100.1	107.9	83.7	69.5
CF_3CH_2–X	106.7	109.4	83.2[a]	—
CF_3CF_2–X	102.7	126.8	82.7	69.5

[a]Theoretical estimate [79].

[87] 2

$$CF_2Br_2 \xrightarrow[\text{DMF, -40 °C}]{\text{2 Cl–}C_6H_4\text{–}S^-Na^+} \text{Cl–}C_6H_4\text{–}SCF_2S\text{–}C_6H_4\text{–Cl}$$

35%

bonding orbitals can react by electron transfer with certain nucleophiles under exceptionally mild conditions [92, 93] (equation 5).

Fluorination usually strengthens C–C bonds, and the effect of partial fluorination can be especially pronounced. The C–C bonds in poly(tetrafluoroethylene) are about 8 kcal/mol stronger than those in poly(ethylene), and CF_3CF_3 has a 10 kcal/mol stronger C–C bond than CH_3CH_3, but CF_3CH_3 has an even stronger, polar

[88] 3

$$CF_2Br_2 \xrightarrow[\text{diglyme, 25 °C}]{(C_6H_5)_3P} [(C_6H_5)_3\overset{+}{P}Br]\ [CF_2Br^-]$$

$$[(C_6H_5)_3\overset{+}{P}Br]\ [CF_2Br^-] \rightleftharpoons [(C_6H_5)_3PBr_2] + [CF_2]$$

$$[CF_2] + (C_6H_5)_3P \rightleftharpoons [(C_6H_5)_3\overset{+}{P}CF_2^-]$$

$$[(C_6H_5)_3\overset{+}{P}CF_2^-] \xrightarrow{(C_6H_5)_3PBr_2} (C_6H_5)_3\overset{+}{P}CF_2Br\ Br^-\ \ 100\% + (C_6H_5)_3P$$

$$—CF_2CF_2— \xrightarrow{e^-} [—CF_2CF_2—]^{\dot{-}} \xrightarrow{-F^-} [—CF_2\dot{C}F_2—]$$ 4

$$\xrightarrow{e^-} [—CF_2\bar{C}F—] \xrightarrow{-F^-} —CF{=}CF— \xrightarrow{e^-} \text{etc.}$$

[92] 5

Perfluorodecalin $\xrightarrow[\text{DMF, 60-70 °C, 10 days}]{C_6H_5S^-Na^+}$ octakis(phenylthio)naphthalene

F, F_2, F_2, F_2, F_2, F_2, F_2, F_2, F_2, F; C_6H_5, S, SC_6H_5, SC_6H_5, C_6H_5, S, S, C_6H_5, SC_6H_5, SC_6H_5, C_6H_5, S

C–C bond. Fluorination increases the bond strengths in cycloalkanes, including cyclobutanes [*75, 94*], but by contrast, it *decreases* C–C bond strengths and increases ring strain in cyclopropanes and other three-membered ring compounds [*75, 94, 95*].

The high thermal stabilities of most saturated fluorocarbons and perfluorinated ethers reflect their very strong C–F, C–O, and C–C bonds. Their stabilities are limited only by their C–C bond strengths, which decrease moderately with increasing chain length but significantly with branching [*96*] (p. 1001). Carbon tetrafluoride measurably undergoes C–F homolysis only above 2000 °C, whereas CF_3CF_3 and $CF_3CF_2CF_3$ decompose around 1000 °C [*96*], but tetrafluoroethylene homopolymer decomposes rapidly above 500 °C, and its branched copolymers with perfluoroalkenes are substantially less stable [*97*]. Perfluoroethers are commonly even more thermally stable than fluorocarbons [*8*]. For example, poly(tetrafluoroethylene oxide) decomposes about 10 times slower than poly(tetrafluoroethylene) at 585 °C [*97*].

Perfluorocyclopropanes and -epoxides, however, are unusually thermolabile. Compared with octafluorocyclobutane, which only slowly decomposes at 500 °C [*98*], hexafluorocyclopropane smoothly extrudes difluorocarbene at about 170 °C [*99*], as do tetrafluoroethylene oxide [*100*] and hexafluoropropylene oxide [*101*].

Saturated hydrofluorocarbons ordinarily are less stable than their fluorocarbon analogues even though they may have stronger C–C bonds, but they decompose primarily by HF elimination. Pentafluoroethane, for example, exclusively loses HF at 925 °C, and C–C bond scission becomes significant only at 1125 °C [*102*]. Cyclopropanes again are special exceptions. The activation energy for thermal C–C bond cleavage in 1,1-difluorocyclopropane (56.4 kcal/mol) is 17.8 kcal/mol *greater* than that for hexafluorocyclopropane [*95*]. The elimination of HF from hydrofluorocarbons, of course, is markedly accelerated by bases (*see* p. 888).

The C–F bonds in 1-fluoroalkenes and fluorobenzenes also are very strong (Table 17), but alkene π bond strengths vary with the level of fluorination (Table 18). Both $CHF{=}CF_2$ and $CF_2{=}CF_2$ have significantly weaker π bonds than $CH_2{=}CH_2$, $CH_2{=}CHF$, and $CH_2{=}CF_2$, consistent with other data indicating that tri- and tetrafluorination thermodynamically destabilize double bonds [*75*]. The low π bond energy in $CF_2{=}CF_2$ underlies its propensity to undergo thermal biradical [2+2] cycloadditions [*103*] (*see* p. 767).

Table 17. Bond Dissociation Energies (D^0) of Ethylenes and Benzenes [*78*]

	D^0 (C–X) (kcal/mol)		
Compound	*H*	*F*	*Cl*
C_2H_3–X	111.0	125.2	95.2
C_6H_5–X	103.0	115.5	83.4

Table 18. π-Bond Dissociation Energies [*75*]

Compound	D_π^0 *(kcal/mol)*	*Compound*	D_π^0 *(kcal/mol)*
$CH_2{=}CH_2$	63–64[a]	$CHF{=}CF_2$	55.4[b]
$CH_2{=}CHF$	60.3[b]	$CF_2{=}CF_2$	52.3
$CH_2{=}CF_2$	62.8		

[a]Reference 103.
[b]Approximate theoretical value [*104*].

Fluorination likewise significantly destabilizes the multiple bonds in allenes and acetylenes [*105*]. Fluoro- and difluoroacetylene are dangerously explosive, and hexafluoro-2-butyne is very susceptible to both concerted and biradical addition reactions [*106, 107*] (*see* pp. 757 and 767).

Perfluoroalkyl groups thermodynamically destabilize double bonds and small rings, but they can *kinetically* stabilize highly strained molecules [*75*]. This remarkable "perfluoroalkyl effect" has made possible the isolation of structures that are uncommon in hydrocarbon chemistry, especially valence-bond isomers of aromatics and heteroaromatics such as ***1***, ***2***, and ***3*** [*108*].

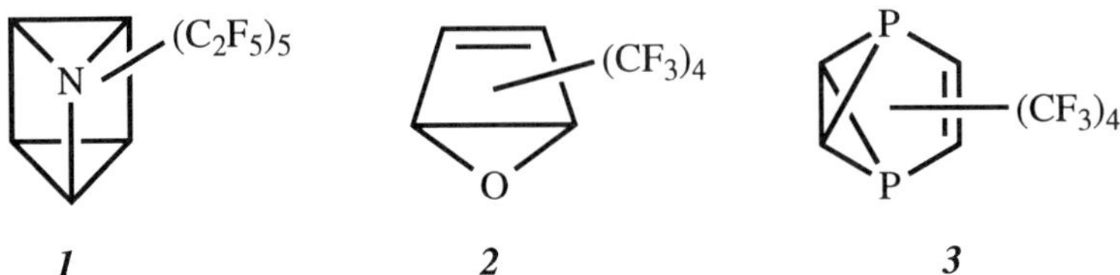

Carbocations

α-Fluorine stabilizes carbocations by resonance (***4***), which dominates its destabilizing inductive effect in fully developed ions. The gas-phase ion stabilities increase in the order $^+CH_3 < {}^+CF_3 < {}^+CH_2F < {}^+CHF_2$ and $^+CH_2CH_3 \ll {}^+CF_2CH_3 \cong {}^+CHFCH_3$ [*109, 110*], and the stability of α-haloalkyl carbocations in solution increases in the order Br < Cl < F [*111*]. Numerous long-lived fluorinated cations have been generated in solution, including the $(CH_3)_2CF^+$ and CH_3CF_2+ ions, but no fluoromethyl or simple 1-fluoroalkyl cations have been detected under non-exchanging conditions [*112*]. The trifluoromethyl cation is long-lived in both the gas phase and solid state [*113*] but not in solution [*112, 114*].

β-Fluorine or fluorine further removed from the cation center always inductively destabilizes carbocations [*115, 116*]. No simple β-fluoroalkyl cations have been observed in either the gas phase or solution, and unlike the cases of the other halogens, there is no evidence for formation of alkyl fluoronium ions (***5***) in solution [*117, 118*], although $(CH_3)_2F^+$ is long-lived in the gas phase [*119*]. The only β-fluorinated cations observed in solution are those that benefit from additional conjugative stabilization, such as α-trifluoromethylbenzyl cations [*112*] and perfluorinated allyl [*120*], cyclopropenium [*112*], and tropylium [*121*] ions.

$$>\overset{+}{C}-\ddot{\underset{..}{F}}: \longleftrightarrow >C=\ddot{\underset{..}{F}}^{+}$$

4

$\overset{+}{F}$, CH_2, CH_2, $(CH_2)_n$

5

The effects of α- vs. β-fluorination imply that *fluoroolefins normally react regioselectively with electrophiles to minimize the number of fluorines* β *to the electron-deficient carbon in the transition state.* Diverse types of electrophilic additions follow this rule (equations 6–8, for example), although there are exceptions, especially for ionic additions of halomethanes to fluoroolefins [*124*].

[*112*] $$CH_2{=}CF_2 \xrightarrow{FSO_3H} CH_3CF_2OSO_2F$$ **6**

[*122*] $$C_{10}H_{21}CH{=}CF_2 \xrightarrow[AlCl_3]{C_2H_5COCl} C_{10}H_{21}CH(CF_2Cl)COC_2H_5$$ **7**

[*123*] $$CF_2{=}CFX \xrightarrow{SO_3} \underset{O\text{---}SO_2}{CF_2\text{-}CFX}$$ **8**

$(X = H, Cl, CF_3, C_4H_9)$

Polyfluoroalkenes are relatively unreactive toward electrophiles, particularly when perfluoroalkyl groups are present. Neither tetrafluoroethylene nor hexafluoropropene is protonated by superacids, and 3,3,3-trifluoropropene is not protonated by fluorosulfonic acid but instead ionizes to the 1,1-difluoroallyl cation [*112*]. Hydrofluoro- and halofluoroethylenes can react with a variety of electrophiles, but their reactivity decreases with increasing fluorine content [*125*]. Quantitative relative rate data for hydrolysis and solvolysis reactions reveal how large the β-fluorine deactivating effects can be. For example, α-trifluoromethylstyrene undergoes acid-catalyzed hydration 10^{10} times slower than α-methylstyrene, and $C_6H_5C(CF_3)_2OTs$ is some 10^{18} times less reactive than cumyl tosylate toward solvolysis [*115, 116*].

Polyfluoroaromatics show similar reactivity patterns, wherein the ease of electrophilic substitution generally decreases with increasing fluorine content. Pentafluorobenzene undergoes substitution only under forcing conditions, and perfluoroaromatics completely resist substitution, which would require displacement of F^+, but instead give addition products when they react with electrophiles [*111, 112, 126*].

The effect of monofluorination on alkene or aromatic reactivity toward electrophiles is more difficult to predict. Although α-fluorine stabilizes a carbocation relative to hydrogen, its opposing inductive effect makes olefins and aromatics more electron deficient. *Fluorine therefore is activating only for electrophilic reactions with very late transition states* where its resonance stabilization is maximized. The faster rate of addition of trifluoroacetic acid and sulfuric acid to 2-fluoropropene vs. propene is an example [*115, 116*], but cases of such enhanced fluoroalkene reactivity in solution are quite rare [*127*]. By contrast, there are many examples where the *ortho–para*-directing fluorine substituent is also activating in electrophilic aromatic substitutions [*128*].

Carbanions

The effects of fluorination on carbanion stability are largely deduced from C–H acidity data (p. 988) [*64*]. α-Halogens stabilize carbanions in the order Br > Cl > F, which is opposite the inductive electron-withdrawing order and reflects the predominant role of I_π electron repulsion, **6** [129]. This destabilizing interaction increases in the order Br < Cl < F and is maximized in planar carbanions. α-Fluorocarbanions thus highly favor pyramidal structures. The pyramidal trifluoromethyl anion is calculated to have an inversion barrier of almost 120 kcal/mol compared with under 2 kcal/mol for the methyl anion [130], and polyfluoroallyl anions are predicted to favor unusual pyramidal over classical planar, delocalized structures [*131, 132*]. For pyramidal anions, α-fluorine is always stabilizing relative to hydrogen, albeit less so than other halogens, whereas in planar anions it can be destabilizing (p. 988).

β-Fluorination always strongly stabilizes carbanions both by induction and by negative (anionic) hyperconjugation, *7*. The latter "no-bond resonance" has been controversial, but its importance is now well established both theoretically [*133, 134*] and experimentally [*67*]. The X-ray crystal structures of salts ***8*** [*135*] and ***9*** [*136*] provide cogent evidence for negative hyperconjugation.

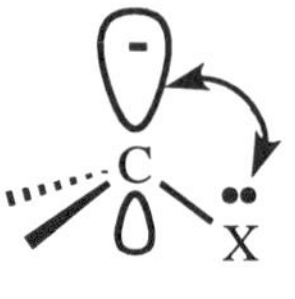

6 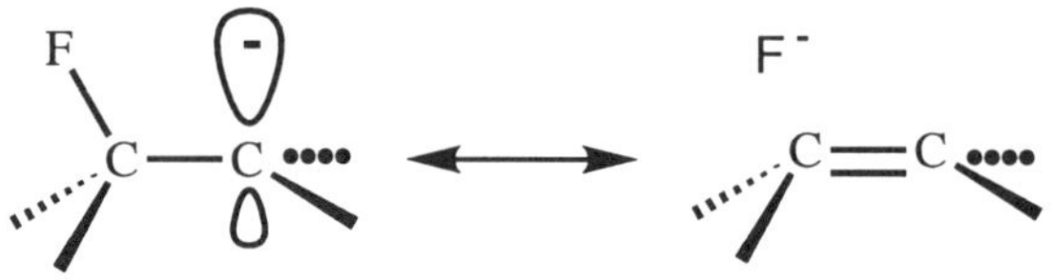 **7**

$CF_3O^- TAS^+$

8

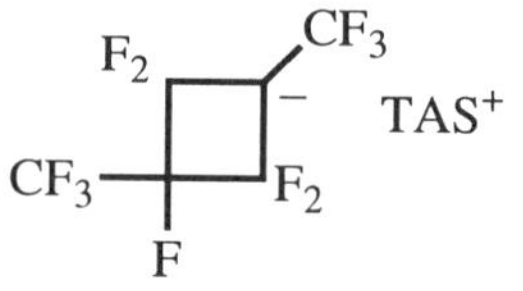

($TAS^+ \equiv [(CH_3)_2N]_3S^+$)

9

Several long-lived perfluorocarbanions have been observed in solution [*137, 138*], and many stable salts such as ***9, 10*** [*137, 139*], ***11*** [*137*], ***12*** [*137*], and ***13*** [*140*] have been isolated. Notably, no long-lived α-fluorocarbanion has been detected, which is consistent with the order of anion stability, $(R_f)_3CF^- >> (R_f)_2CF^- >> R_fCF_2^-$, deduced from C–H acidities.

Unsaturated fluorocarbons are much more reactive toward nucleophiles than their hydrocarbon counterparts owing to fluorine's ability to both stabilize carbanions and inductively increase the electrophilicity of multiple bonds and aromatic rings. Nucleophilic attack dominates the chemistry of unsaturated fluorocarbons, and the role of fluoride ion in fluorocarbon chemistry is analogous to that of the proton in hydrocarbon chemistry [*129*]. Like the related electrophilic reactions for hydrocarbons, there are fluoride-promoted isomerizations and dimerizations (equation 9), oligomerizations (equation 10), additions (equation 11), and anionic Friedel–Crafts alkylations (equation 12) that all proceed via carbanionic intermediates [*129, 141*].

$R_f\bar{C}(CF_3)_2$ TAS$^+$

10 ($R_f = CF_3, C_3F_7, SO_2F$)

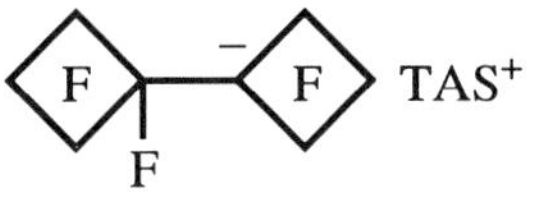

F F TAS$^+$ F

11

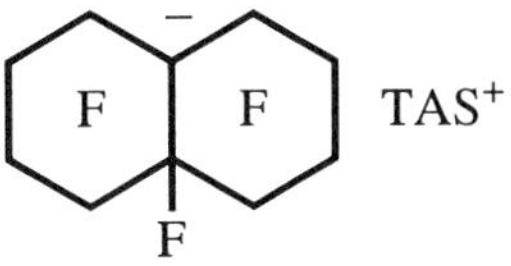

F F TAS$^+$ F

12

X CF$_3$ CF$_3$ CF$_3$ CF$_3$ TAS$^+$

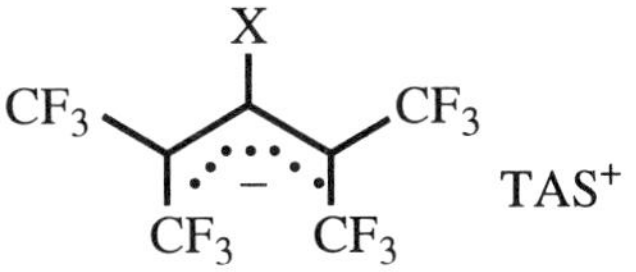

13 (X = F, CN)

$$2\,CF_3CF{=}CF_2 \xrightarrow{F^-} (CF_3)_2CFCF{=}CFCF_3 \rightleftharpoons (CF_3)_2C{=}CFC_2F_5 \qquad \mathbf{9}$$

$$4\,CF_2{=}CF_2 \xrightarrow{F^-} C_2F_5(CF_3)C{=}C(CF_3)C_2F_5 \qquad \mathbf{10}$$

$$CF_3CF{=}CF_2 \xrightarrow[F^-]{HCONH_2} CF_3CFHCF_3 \qquad \mathbf{11}$$

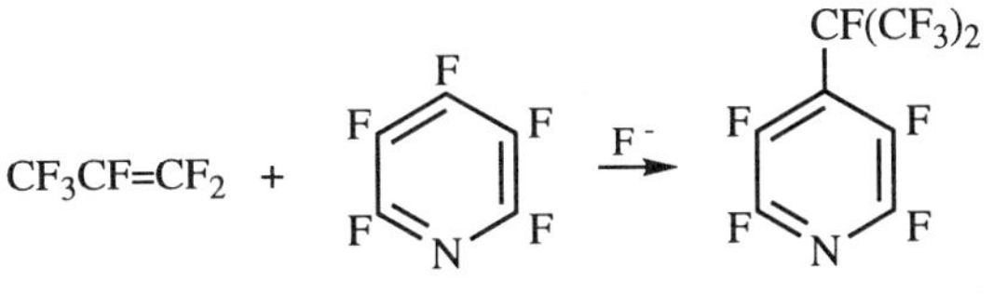

$$CF_3CF{=}CF_2 + C_5F_5N \xrightarrow{F^-} 4\text{-}(CF(CF_3)_2)C_5F_4N \qquad \mathbf{12}$$

The reactivity and orientation of nucleophilic attack on fluoroolefins are determined by the stability of the possible carbanion intermediates. *Fluoroolefins react regioselectively with nucleophiles so as to maximize the number of fluorines* β *to the electron-rich carbon in the transition state.* The reactivities increase in the order $CF_2{=}CF_2 < CF_2{=}CFCF_3 << CF_2{=}C(CF_3)_2$ and $CF_2{=}CF_2 < CF_2{=}CFCl < CF_2{=}CFBr$, and nucleophiles attack exclusively at the $CF_2{=}$ end of these olefins [*129, 141*]. The regiochemistry of nucleophilic attack normally is predictable, but the product distribution arising from addition and addition–elimination pathways depends upon the olefin, nucleophile, and reaction conditions (equations 13–15). The various factors that control product distributions have been reviewed [*142, 143, 144*].

[*81*] $C_6H_5C(CF_3)=CF_2 \xrightarrow[C_2H_5OH,\ -78\ °C]{NaOC_2H_5} C_6H_5C(CF_3)=CFOC_2H_5$ (85%) + $C_6H_5CH(CF_3)CF_2OC_2H_5$ (15%) **13**

[*142*] $CF_2=CFCF_3 \xrightarrow[CH_3OH,\ 60\ °C]{KOH} CH_3OCF_2CFHCF_3$ (83%) + $CH_3OCF=CFCF_3$ (17%) **14**

[*144*] [F] $\xrightarrow[CH_3OH,\ 60\ °C]{KOH}$ [F]–OCH_3, F (52%) + [F]–OCH_3, OCH_3 (12%) **15**

Elimination of hydrogen halides from polyfluoroalkanes by bases also usually involves carbanion intermediates (E1cB mechanism) [*81*], and orientation is therefore governed by relative C–H acidities and leaving-group mobility. Some examples are shown in equations 16–18 [*145*].

$$CF_3C\mathbf{H}_2CHBrCH_3 \xrightarrow[C_2H_5OH]{KOH} CF_3CH=CHCH_3 \quad \mathbf{16}$$

$$C\mathbf{H}Cl_2CCl_2CHF_2 \xrightarrow[C_2H_5OH]{KOH} CCl_2=CClCHF_2 \quad \mathbf{17}$$

$$CCl_3CH_2CF_2Cl \xrightarrow[H_2O]{KOH} CCl_2=CHCF_2Cl \quad \mathbf{18}$$

The effects of fluorination on nucleophilic aromatic substitution are more complex, and extensive surveys of regiochemical and kinetic data are available [*146, 147*]. Interestingly, for C_6F_5X derivatives, nucleophiles attack *para* to X for most substituents, even though the rates may vary by several orders of magnitude. For example, the relative reactivities of C_6F_5X toward $NaOCH_3$ (CH_3OH at 60 °C) are X = CH_3 (0.6) < H(1.0) << CF_3 (5×10^3), and toward $NaOC_6H_5$ (dimethylacetamide at 106 °C) are X = F (0.9) < H (1.0) < Cl (32) << CF_3 (2.4×10^4) [*146, 147*]. These reactivities reflect I_π repulsion in ***14*** (X = F) and the combined inductive and resonance stabilization by the trifluoromethyl group in ***14*** (X = CF_3).

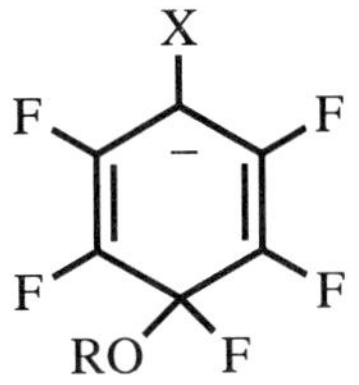

X

F F

F F

RO F

14

Free Radicals

α-Fluororadicals prefer to be pyramidal to minimize I_π repulsion, and fluoromethyl radicals deviate from planarity in the order $\cdot CH_2F < \cdot CHF_2 < \cdot CF_3$ with a computed inversion barrier of about 25 kcal/mol for $\cdot CF_3$ [*148*]. The pyramidal character of α-fluororadicals is the source of the low π-bond energy in tetrafluoroethylene [*103*] and low rotational barriers in polyfluoroallyl radicals [*149*]. β-Fluorination, by contrast, has little effect on radical geometry, and bridging by fluorine is negligible [*150*]. Fluorine atom migration is rare, but there is good evidence for the $CH_2FCF_2\cdot \rightarrow CF_3CH2\cdot$ rearrangement [*151*]. In contrast to hydrocarbon radicals, however, perfluoroalkyl radicals never disproportionate (2 $R_fCF_2CF_2\cdot \rightarrow R_fCF{=}CF_2 + R_fCF_2CF_3$) under normal reaction conditions. Because this disproportionation is eliminated as a terminating step in the free-radical polymerization of tetrafluoroethylene, extraordinarily high molecular weights can be achieved.

The effect of fluorination on radical stability is somewhat ambiguous and varies with the electronic character of the radical center [*148, 152, 153*] but generally is small. Fluorination, however, can significantly affect radical abstraction and addition reactions, primarily due to polar, steric, or bond strength factors [*154, 155*]. Fluorine inductively deactivates C–H abstraction by *electrophilic* radicals, and the effect can be quite large. For example, the chlorine radical abstracts hydrogen about 2.2×10^4 times faster from ethane than from 1,1,1-trifluoroethane [*156*]. This reactivity reflects both the stronger C–H bond in the latter (Table 16) and the relatively unfavorable polarization in the transition state (***15***). Fluorine can be activating for abstraction by *nucleophilic* radicals, however. Methyl radical reacts faster with trifluoromethane than methane despite the stronger C–H bond in trifluoromethane (Table 16), whereas the opposite is true for hydrogen abstraction by the electrophilic bromine radical [*154*]. This difference can be attributed to the favorable polarization in ***16***. Empirical correlations of the rates of hydrogen abstraction from hydrofluorohaloalkanes have been developed, especially for hydroxyl radical reactions that provide a mechanism for their removal from the atmosphere [*157, 158*].

$$\overset{\delta^+}{[CF_3CH_2} \cdots H \cdots \overset{\delta^-}{Cl]} \qquad \overset{\delta^-}{[CF_3} \cdots H \cdots \overset{\delta^+}{CH_3]}$$

15 ***16***

The rates and orientation of free radical additions to fluoroalkenes depend upon the nature of the attacking radical and the alkene, but polar effects again are important. For instance, methyl radical adds 9.5 times faster to tetrafluoroethylene than to ethylene at 164 °C, but the trifluoromethyl radical adds 10 times faster to ethylene [*155*]. The more favorable polar transition states combine the nucleophilic radical with the electron-deficient olefin ***17*** and vice versa (***18***). These polar effects account for the tendency of perfluoroalkenes and alkenes to produce highly regular, alternating copolymers (*see* Chapter starting on page 1101).

The complex directive effects in radical additions have been reviewed, and rules for predicting rates and regiochemistry have been developed [*154, 155*].

$$[\overset{\delta^+}{CH_3} \cdots \overset{\delta^-}{CF_2} \overset{\bullet\bullet}{—} {}^{\bullet}CF_2] \qquad [\overset{\delta^-}{CF_3} \cdots \overset{\delta^+}{CH_2} \overset{\bullet\bullet}{—} {}^{\bullet}CH_2]$$

17 ***18***

Carbenes

The structure, stability, and reactivity of carbenes are affected by fluorination. All α-fluorocarbenes are ground-state singlets, principally because of resonance stabilization, ***19*** (analogous to ***4***), and difluorocarbene has a singlet–triplet separation of about 44 kcal/mol [*159, 160*]. Difluorocarbene is relatively unreactive compared with CH_2, and the reactivities and electrophilicities of carbenes decrease in the series $CH_2 > CBr_2 > CCl_2 > CFCl > CF_2$ and $CH_2 > CHF > CF_2$ [*161*]. α-Fluorocarbenes are less prone to C–H insertion and generally add stereospecifically to olefins.

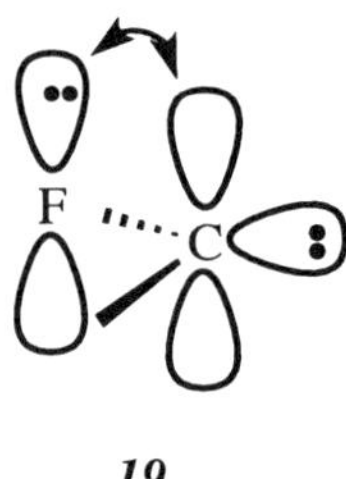

19

β-Fluorocarbenes like bis(trifluoromethyl)carbene and trifluoromethylcarbene by contrast are highly electrophilic, ground-state triplet species that display little selectivity in their reactions [*159, 162*].

The generation and cycloaddition reactions of fluorocarbenes are covered in more detail on p. 767 *et seq*.

Steric Effects

Fluorine and oxygen, *not* hydrogen, are almost identical in size, judging by their van der Waals radii, $r_v(F) = 1.47$ Å and $r_v(O) = 1.52$ Å [*3, 163*]. The modified Taft steric parameters, E_s^0(F) = –0.46 and E_s^0 (OH) = –0.55 [*49*], and the similar biphenyl rotation barriers for the fluorine and hydroxyl derivatives (Table 19) further indicate that the fluorine and hydroxyl groups are essentially "chemical isosteres".

The largest known fluorine steric effect is the k_H/k_F rate ratio of 10^{11} at 25 °C for the *meta* ring flip in ***20*** [*166*]. Other dynamic processes that reveal bona fide fluorine steric effects are the barriers to *i*-C_3H_7 group rotation in ***21*** (ΔG_{298}‡ = 6.9 kcal/mol [*167*]) and to inversion in ***22*** (X = F) (Table 19), which are each about 5 kcal/mol higher than in the unsubstituted systems.

The course of each of the free radical reactions shown in equations 19–21, where fluorine substitution alters the normal *trans* stereochemistry of addition, is ascribed to *endo*-fluorine steric effects.

Table 19. Inversion and Rotation Barriers

Compound	*ΔG_{340}‡ (kcal/mol)*	*Ref.*
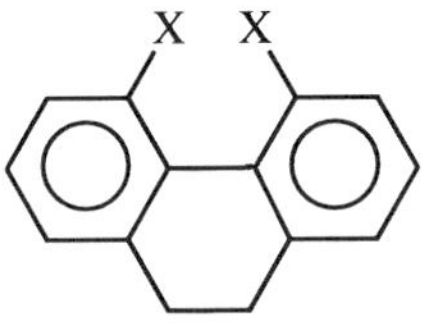 **22**		164
X = H	< 6	
X = F	11.1	
X = CH_3	17.8	
X = CF_3	>29	
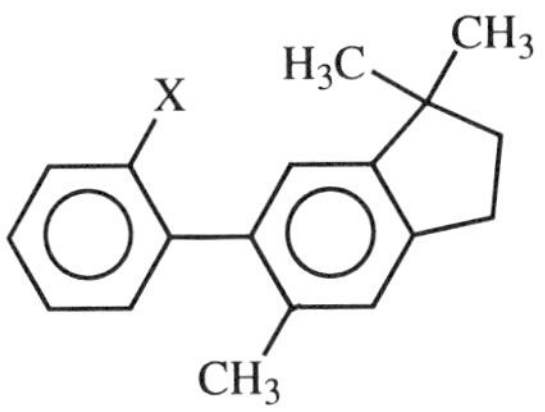		165
X = OH	16.1	
X = F	14.2	
X = CH_3	19.3	
X = $CH(CH_3)_2$	22.2	
X = CF_3	22.0	

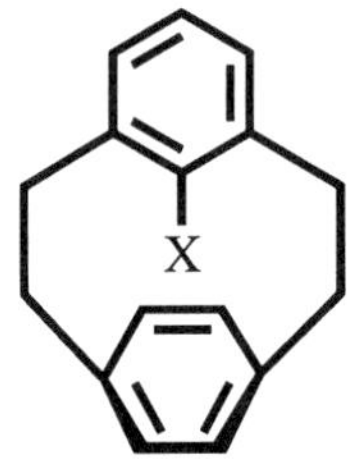

X

20

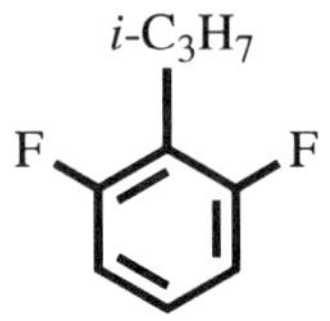

i-C_3H_7

F F

21

19

[168]

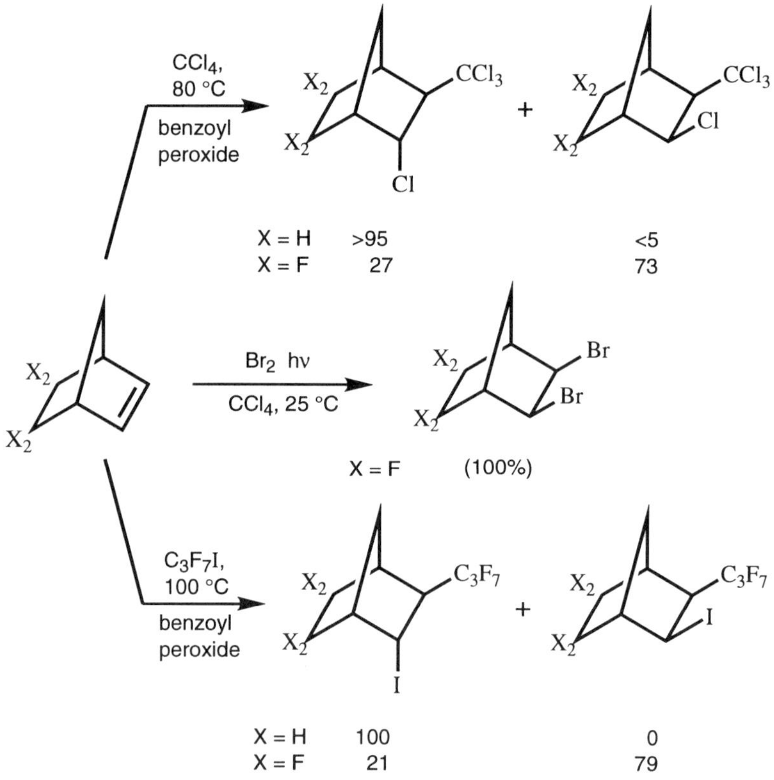

CCl_4, 80 °C

benzoyl peroxide

X_2 X_2 CCl_3 Cl

+

X_2 X_2 CCl_3 Cl

X = H	>95	<5
X = F	27	73

[169]

X_2 X_2

Br_2 hν

CCl_4, 25 °C

X_2 X_2 Br Br

X = F (100%)

C_3F_7I, 100 °C

benzoyl peroxide

X_2 X_2 C_3F_7 I

+

X_2 X_2 C_3F_7 I

[168]

X = H	100	0
X = F	21	79

[170]

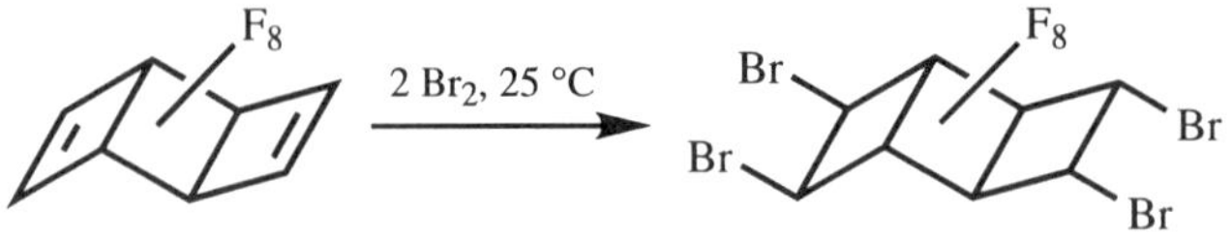

F_8

2 Br_2, 25 °C

Br Br F_8 Br Br

20

References are listed on pages 1003–1010.

[171] Br, Br, F_6 — Cl_2, hν / $CFCl_3$, 0-25 °C → Br, Br, F_6, Cl, Cl **21**

The trifluoromethyl group is considerably larger than the methyl group by any measure of steric size, and the data in Table 20 indicate that *trifluoromethyl is sterically at least as large as isopropyl.* There are no steric parameters reported for other perfluoroalkyl groups, but C=N rotation barriers in $(CF_3)_2C{=}NR$ (13.1 kcal/mol for t-C_4H_9; 13.8 kcal/mol for i-C_3F_7 [*172*]) suggest that t-C_4H_9 and i-C_3F_7 are comparable in size. The remarkable thermolability of highly branched fluorocarbons, however, clearly implies that perfluoroalkyl groups are all much larger than their alkyl counterparts. The activation energy for C–C bond homolysis in $(i\text{-}C_3F_7)_2C(CF_3)CF_2CF_3$ (36 kcal/mol [*173*]) is 15 kcal/mol *lower* than that for $(t\text{-}C_4H_9)(i\text{-}C_3H_7)C(CH_3)\,CH_2CH_3$ [*174*]!

Table 20. Alkyl Group Steric Parameters[a]

Alkyl Group	$E_s{}^0$	*A*	ν	r_v
CH_3	–1.24	1.70	0.52	1.80
$(CH_3)_2CH$	–1.71	2.20	0.76	2.2
$(CH_3)_2CHCH_2$	–2.17	—	0.98	—
CF_3	–2.40	2.4–2.5	0.91	2.2

[a] E_s0, Modified Taft steric parameter [E_s0 (H) = 0] [*49*]; *A*, cyclohexane ΔG^0 (axial–equatorial value) [*175*]; ν, Charton steric parameter [*176*]; r_v, effective van der Waals radius.

References for pages 979–1003

1. Sen, K. D.; Jorgensen, C. K. *Electronegativity;* Springer-Verlag: New York, 1987.
2. Nagel, J. K. *J. Am. Chem. Soc.* **1990,** *112,* 4740.
3. Bondi, A. *J. Phys. Chem.* **1964,** *68,* 441.
4. Reed, T. M., III In *Fluorine Chemistry;* Simons, J. H., Ed.; Academic Press: New York, 1964; Vol. V, Chapter 2.
5. Bryce, H. G. In *Fluorine Chemistry;* Simons, J. H., Ed.; Academic Press: New York, 1964; Vol. V, Chapter 4.
6. Maciejewski, A. *J. Photochem. Photobio., A: Chemistry* **1990,** *51,* 87.
7. Mahler, W., Du Pont, personal communication.
8. Slinn, D. S. L.; Green, S. W. In *Preparation, Properties, and Industrial Applications of Organofluorine Compounds;* Banks, R. E., Ed.; Ellis Horwood: Chichester, U.K., 1982; Chapter 2.

9. Hudlicky, M. *Chemistry of Organofluorine Compounds,* 2nd ed.; Ellis Horwood: Chichester; U. K., 1976; Chapter 7.
10. Lovelace, A. M.; Rausch, D. A.; Postelnek, W. *Aliphatic Fluorine Compounds;* Reinhold: New York, 1958.
11. *Lange's Handbook of Chemistry;* 13th ed.; Dean, J. A., Ed.; McGraw–Hill: NewYork, 1985.
12. Barton, A. M. *F. Handbook of Solubility Parameters and Other Cohesive Parameters;* CRC Press: Boca Raton, 1983.
13. Aten, A. H. W., Jr. *J. Fluorine Chem.* **1976,** *8,* 934.
14. Woolf, A. A. *J. Fluorine Chem.* **1990,** *50,* 89.
15. Smith, J. K.; Patrick, C. R. *Proc. Chem. Soc.* **1961,** 138.
16. Persico, D. F.; Huang, H.-N.; Lagow, R. J.; Clark, L. C., Jr. *J. Org. Chem.* **1985,** *50,* 5156.
17. Nelson, R. D., Jr.; Lide, D. R., Jr.; Maryott, A. A. *Selected Values of Electric Dipole Moments in the Gas Phase;* NSRD-NBS 10, Government Printing Office: Washington, DC, 1967.
18. McClellan, A. L. *Tables of Experimental Dipole Moments;* W. H. Freeman: San Francisco, 1963; Vol. I.
19. Wu, S. Polymer Interface and Adhesion; Marcel Dekker: New York, 1982.
20. Pittman, A. G. In *Fluoropolymers;* Wall, L. A., Ed.; John Wiley: New York, 1972; Vol. XXV, Chapter 13.
21. Watanabe, N.; Nakajima, T.; Touhara, H. *Graphite Fluorides;* Elsevier: Oxford, 1988.
22. Arduengo, A. J., III; Moran, J. R.; Rodriguez-Parada, J.; Ward, M. D. *J. Am. Chem. Soc.* **1990,** *112,* 6153.
23. Fielding, H. C., Jr. In *Organofluorine Chemicals and Their Industrial Applications;* Banks, R. E., Ed.; Ellis Horwood: Chichester, U. K., 1979; Chapter 11.
24. Kuneida, H.; Shinoda, K. *J. Phys. Chem.* **1976,** *80,* 2468.
25. Caporiccio, G.; Burzio, F.; Carniselli, G.; Biancardi, V. *J. Colloid Interface Sci.* **1984,** *98,* 202.
26. Martini, G.; Ottaviani, M. F.; Ristori, S.; Lenti, D.; Sanguineti, A. *Colloids Surf.* **1990,** *45,* 177.
27. Muller, N.; Birkhahn, R. H. *J. Phys. Chem.* **1967,** *71,* 957.
28. Selve, C.; Ravey, J.-C.; Stebe, M.-J.; El Moudjahid, C.; Moumni, E. M.; Delpuech, J.-J. *Tetrahedron* **1991,** *47,* 4.
29. Mathis, G.; Leempoel, P.; Ravey, J.-C.; Selve, C.; Delpuech, J. J. *J. Am. Chem. Soc.* **1984,** *106,* 6162.
30. Matos, L.; Ravey, J.-C.; Serratrice, G. *J. Colloid Interface Sci.* **1989,** *128,* 341.
31. Burkitt, S. J.; Ottewill, R. H.; Hayter, J. B.; Ingram, B. T. *Colloid Polym. Sci.* **1987,** *265,* 619.

32. Hoffman, H.; Klaus, J.; Reizlein, K.; Ulbricht, W.; Ibel, K. *Colloid Polym. Sci.* **1982,** *260,* 435.
33. Dixon, D. A.; Van-Catledge, F. A. *Int. J. Supercomputer Appl.* **1988,** *2,* 62.
34. Schneider, J.; Erdelen, C.; Ringsdorf, H.; Rabolt, J. F. *Macromolecules* **1989,** *22,* 3475.
35. Zhang, W. P.; Dorset, D. L. *Macromolecules* **1990,** *23,* 4322.
36. Wolf, S. G.; Deutsch, M.; Landau, E. M.; Lahav, M.; Leiserowitz, L.; Kjaer, K.; Als-Nielsen, J. *Science* **1988,** *242,* 1286.
37. Eaton, D. F.; Smart, B. E. *J. Am. Chem. Soc.* **1990,** *112,* 2821.
38. Naselli, C.; Swalen, J. D.; Rabolt, J. F. *J. Chem. Phys.* **1989,** *90,* 3855.
39. Freed, B. K.; Biesecker, J.; Middleton, W. J. *J. Fluorine Chem.* **1990,** *48,* 63.
40. Krespan, C. G.; Smart, B. E., Du Pont, unpublished results.
41. Wilhelm, E.; Battino, R. *Chem. Rev.* **1973,** *73,* 11.
42. Kabalnov, A. S.; Makarov, K. N.; Shcherbakova, O. V. *J. Fluorine Chem.* **1990,** *5,* 271.
43. Fusch, R.; Chambers, E. J.; Stephenson, W. K. *Can. J. Chem.* **1987,** *65,* 2624.
44. Rowlinson, J. S.; Swinton, F. L. *Liquids and Liquid Mixtures,* 3rd ed.; Butterworths: London, 1982.
45. Carlfors, J.; Stibbs, P. *J. Phys. Chem.* **1984,** *88,* 4410.
46. Afzal, J.; Ashlock, S. R.; Fung, B. M.; O'Rear, E. A. *J. Am. Chem. Soc.* **1986,** *90,* 3019.
47. Hamza, M. H. A.; Serratrice, G.; Stébé, M.-J.; Delpuech, J. J. *J. Am. Chem. Soc.* **1981,** *103,* 3733.
48. Serratrice, G.; Delpuech, J. J.; Diguet, R. *Nouv. J. Chem.* **1982,** *6,* 489.
49. Hansch, C.; Leo, A. *Substituent Constants for Correlation Analysis in Chemistry and Biology;* John Wiley: New York, 1979.
50. Hansch, C.; Leo, A.; Taft, R. W. *Chem. Rev.* **1991,** *91,* 165.
51. Sangster, J. *J. Chem. Phys. Ref. Data* **1989,** *18,* 1111.
52. Muller, N. *J. Pharm. Sci.* **1986,** *75,* 987.
53. Feiring, A. E.; Smart, B. E. In *Ullmann's Encyclopedia of Industrial Chemistry*; VCH Publishers: New York, 1988; Vol. A11, pp 367–370.
54. Abraham, M. H.; Grellier, P. L.; Prior, D. V.; Duce, P. P.; Morris, J. J.; Taylor, P. J. *J. Chem. Soc., Perkin Trans. 2* **1989,** 699.
55. Yagupol'skii, L. M.; Il'chenko, A. Y.; Kondratenko, N. V. *Russ. Chem. Rev.* **1974,** *43,* 32.
56. Yagupol'skii, L. M. *J. Fluorine Chem.* **1987,** *36,* 1.
57. Kluter, G.; Pritzkow, H.; Seppelt, K. *Angew. Chem. Int. Ed. Engl.* **1980,** *19,* 942.
58. Bordwell, F. G. *Acc. Chem. Res.* **1988,** *21,* 456.
59. Olah, G. A.; Prakash, G. K. S.; Sommer, J. *Superacids;* John Wiley: New York, 1985.

60. Gillespie, R. J.; Liang, J. *J. Am. Chem. Soc.* **1988,** *110,* 6053.
61. Stewart, R. *The Proton: Applications to Organic Chemistry;* Academic Press: Orlando, 1985.
62. Filler, R. *Fluorine Chem. Rev.* **1977,** *8,* 1.
63. Hine, J.; Hahn, S.; Hwang, J. *J. Org. Chem.* **1988,** *53,* 884.
64. Reutov, O. A.; Beletskaya, I. P.; Butin, K. P. *CH–Acids;* Pergamon: Oxford, 1978.
65. Streitwieser, A., Jr.; Holtz, D.; Ziegler, G. R.; Stoffer, J. O.; Brokaw, M. L.; Guibe, F. *J. Am. Chem. Soc.* **1976,** *98,* 5229.
66. Streitwieser, A., Jr.; Scannon, P. J.; Heimeyer, H. H. *J. Am. Chem. Soc.* **1972,** *94,* 7936.
67. Sligh, J. H.; Stephens, R.; Tatlow, J. C. *J. Fluorine Chem.* **1980,** *15,* 411.
68. Purrington, S. T.; Bumgardner, C. L.; Lazaridis, N. V.; Singh, P. *J. Org. Chem.* **1987,** *52,* 4307.
69. Laganis, E. D.; Lemal, D. M. *J. Am. Chem. Soc.* **1980,** *102,* 6633.
70. Paprott, G.; Seppelt, K. *J. Am. Chem. Soc.* **1984,** *106,* 406.
71. Roberts, R. D.; Ferran, H. E., Jr.; Gula, M. J.; Spenser, T. A. *J. Am. Chem. Soc.* **1980,** *102,* 7054.
72. Olah, G. A.; Pittman, C. U., Jr. *J. Am. Chem. Soc.* **1966,** *88,* 3310.
73. Lias, S. G.; Liebman, J. F.; Levin, R. D. *J. Chem. Phys. Ref. Data* **1984,** *13,* 695.
74. Banks, R. E. *Fluorocarbons and Their Derivatives;* Macdonald: London, 1970; pp 131, 161–162.
75. Smart, B. E. In *Molecular Structure and Energetics;* Liebman, J. F.; Greenberg, A., Eds.; VCH Publishers: Deerfield Beach, FL, 1986; Vol. III, Chapter 4.
76. McMillen, D. F.; Golden, D. M. *Annu. Rev. Phys. Chem.* **1982,** *33,* 493.
77. Yokozeki, A.; Bauer, S. H. *Top. Curr. Chem.* **1975,** *53,* 71.
78. Griller, D.; Kanabus-Kaminska, J. M.; Maccol, A. *J. Mol. Struc. (Theochem)* **1988,** *163,* 125.
79. Dixon, D. A.; Smart, B. E., Du Pont, unpublished results.
80. Evans, B. S.; Weeks, I.; Whittle, E. *J. Chem. Soc., Faraday Trans. 1* **1983,** *79,* 147.
81. Koch, H. F. In *Comprehensive Carbanion Chemistry;* Buncel, E.; Durst, T., Eds.; Elsevier: Amsterdam, 1987; Part C, Chapter 6.
82. Hudlicky, M. *Chemistry of Organofluorine Compounds,* 2nd Ed.; Ellis Horwood: Chichester; U. K., 1976; Chapter 5.
83. Chambers, R. D. *Fluorine in Organic Chemistry;* John Wiley: New York, 1973; Chapter 5.
84. Namavari, M.; Satayamurthy, N.; Phelps, M. E.; Barrio, J. R. *Tetrahedron Lett.* **1990,** *31,* 4973.
85. Wakselman, C.; Kaziz, C. In *Fluorine, The First Hundred Years;* Banks,

R. E.; Sharp, D. W. A.; Tatlow, J. C., Eds.; Elsevier: New York, 1986; Chapter 12.

86. Feiring, A. E. *J. Org. Chem.* **1983,** *48,* 347.
87. Rico, I.; Cantacuzene, D.; Wakselman, C. *J. Org. Chem.* **1983,** *48,* 1979.
88. Burton, D. J. *J. Fluorine Chem.* **1983,** *23,* 339.
89. Kebarle, P.; Chowdhury, S. *Chem. Rev.* **1987,** *87,* 513.
90. Hudlický, M. *Chemistry of Organofluorine Compounds;* 2nd ed.; Ellis Horwood: Chichester, U.K., 1976; pp 563–566.
91. Benning, R. C.; McCarthy, T. J. *Macromolecules* **1990,** *23,* 2648.
92. MacNicol, D. D.; Robertson, C. D. *Nature (London)* **1988,** *332,* 59.
93. Cooper, D. L.; Allan, N. L.; Powell, R. L. *J. Fluorine Chem.* **1990,** *49,* 421.
94. Liebman, J. F.; Dolbier, W. R., Jr.; Greenberg, A. *J. Phys. Chem.* **1986,** *90,* 394.
95. Dolbier, W. R., Jr. *Acc. Chem. Res.* **1981,** *14,* 195.
96. Banks, R. E. *Fluorocarbons and Their Derivatives;* Macdonald: London, 1970; pp 17–19.
97. Wall, L. A. In *Fluoropolymers;* Wall, L. A., Ed.; John Wiley: New York, 1972; Vol. XXV, Chapter 12.
98. Butler, J. N. *J. Am. Chem. Soc.* **1962,** *84,* 1393.
99. Birchall, M. J.; Fields, R.; Haszeldine, R. N.; McClean, R. J. *J. Fluorine Chem.* **1980,** *15,* 487.
100. Tarrant, P.; Allison, C. G.; Barthold, K. P.; Stump, E. C., Jr. *Fluorine Chem. Rev.* **1971,** *5,* 77.
101. Millauer, H.; Schwertfeger, W.; Siegemund, G. *Angew. Chem. Int. Ed. Engl.* **1985,** *24,* 161.
102. Aviyente, V.; Inel, Y. *Can. J. Chem.* **1990,** *68,* 1332.
103. Wang, S. Y.; Borden, W. T. *J. Am. Chem. Soc.* **1989,** *111,* 7282.
104. Dixon, D. A.; Fukunaga, T.; Smart, B. E. *J. Am. Chem. Soc.* **1986,** *108,* 1585.
105. Dixon, D. A.; Smart, B. E. *J. Phys. Chem.* **1989,** *93,* 7772.
106. Hudlický, M. *Chemistry of Organofluorine Compounds,* 2nd ed.; Ellis Horwood: Chichester, U.K., 1976; pp 465–481.
107. Bruce, M. I.; Cullen, W. R. *Fluorine Chem. Rev.* **1969,** *4,* 790.
108. Kobayashi, Y.; Kumadaki, I. *Top. Curr. Chem.* **1984,** *123,* 103.
109. Blint, R. J.; McMahon, T. B.; Beauchamp, J. L. *J. Am. Chem. Soc.* **1974,** *96,* 1269.
110. Williams, A. D.; Le Breton, P. R.; Beauchamp, J. L. *J. Am. Chem. Soc.* **1976,** *98,* 2705.
111. Olah, G. A.; Mo, Y. K. In *Carbonium Ions;* Olah, G. A.; Schleyer, P. v. R., Eds.; John Wiley: New York, 1976; Vol. v, Chapter 36.
112. Olah, G. A.; Mo, Y. K. *Adv. Fluorine Chem.* **1973,** *7,* 69.

113. Prochaska, F. T.; Andrews, L. *J. Am. Chem. Soc.* **1978,** *100,* 2102.
114. Olah, G. A.; Heiliger, L.; Prakash, G. K. S. *J. Am. Chem. Soc.* **1989,** *111,* 8020.
115. Allen, A. D.; Tidwell, T. T. In *Advances in Carbocation Chemistry;* Creary, X., Ed.; JAI Press: Greenwich, 1989; Chapter 1.
116. Tidwell, T. T. *Angew. Chem. Int. Ed. Engl.* **1984,** *23,* 20.
117. Koser, G. F. In *The Chemistry of Functional Groups, Supplement D;* Patai, S.; Rappoport, Z., Eds.; John Wiley: Chichester, U. K., 1983; Part 2, Chapter 25.
118. Olah, G. A.; Prakash, G. K. S.; Krishnamurthy, V. V. *J. Org. Chem.* **1983,** *48,* 5116.
119. Cacace, F. *Acc. Chem. Res.* **1988,** *21,* 215.
120. Galakhov, M. V.; Petrov, V. A.; Bakhmutov, V. I.; Belen'kii, G. G.; Kvasov, B. A.; German, L. S.; Fedin, E. I. *Bull. Acad. Sci. USSR* **1985,** *34,* 279.
121. Dailey, W. P.; Lemal, D. M. *J. Am. Chem. Soc.* **1984,** *106,* 1169.
122. Suda, M. *Tetrahedron Lett.* **1980,** *21,* 255.
123. Knunyants, I. L.; Sokolski, G. A. *Angew. Chem. Int. Ed. Engl.* **1972,** *11,* 583.
124. Paleta, O. *Fluorine Chem. Rev.* **1977,** *8,* 39.
125. Belen'kii, G. G.; German, L. S. In *Soviet Scientific Reviews, Section B, Chemistry Reviews;* Vol'pin, M. E., Ed., Ellis Horwood: London, 1984; Vol. 5, pp 183–218.
126. Yakobson, G. G.; Furin, G. G. In *Soviet Scientific Reviews, Section B, Chemistry Reviews;* Vol'pin, M. E., Ed., Ellis Horwood: London, 1984; Vol. 5, pp 255–296.
127. Johnson, W. J.; Daub, G. W.; Lyle, T. A.; Niwa, M. *J. Am. Chem. Soc.* **1980,** *102,* 7802.
128. Taylor, R. *Electrophilic Aromatic Substitution;* John Wiley: Chichester, 1990.
129. Chambers, R. D.; Bryce, M. R. In *Comprehensive Carbanion Chemistry;* Buncel, E.; Durst, D., Eds., Elsevier: Amsterdam, 1987; Part C, Chapter 5.
130. Marynick, D. S. *J. Mol. Struct.* **1982,** *87,* 161.
131. Dixon, D. A.; Fukunaga, T.; Smart, B. E. *J. Phys. Org. Chem.* **1988,** *1,* 153.
132. Hammons, J. H.; Hrovat, D. A.; Borden, W. T. *J. Phys. Org. Chem.* **1990,** *3,* 635.
133. Dixon, D. A.; Fukunaga, T.; Smart, B. E. *J. Am. Chem. Soc.* **1986,** *108,* 4027.
134. Reed, A. E.; Schleyer, P. v. R. *J. Am. Chem. Soc.* **1990,** *112,* 1434.
135. Farnham, W. B.; Smart, B. E.; Middleton, W. J.; Calabrese, J. C.; Dixon, D. A. *J. Am. Chem. Soc.* **1985,** *107,* 4565.

136. Farnham, W. B.; Dixon, D. A.; Calabrese, J. C. *J. Am. Chem. Soc.* **1988,** *110,* 2607.
137. Smart, B. E.; Middleton, W. J.; Farnham, W. B. *J. Am. Chem. Soc.* **1986,** *108,* 4905.
138. Bayliff, A. E.; Chambers, R. D. *J. Chem. Soc., Perkin Trans. 1* **1988,** 201.
139. Smart, B. E.; Middleton, W. J. *J. Am. Chem. Soc.* **1987,** *109,* 4982.
140. Farnham, W. B.; Middleton, W. J.; Fultz, W. C.; Smart, B. E. *J. Am. Chem. Soc.* **1986,** *108,* 3125.
141. Smart, B. E. In *The Chemistry of Functional Groups, Supplement D;* Patai, S.; Rappoport, Z., Eds.; John Wiley: Chichester, U. K., 1983; Part 1, Chapter 14.
142. Chambers, R. D. *Fluorine in Organic Chemistry;* John Wiley: New York, 1973; Chapter 7.
143. Rappoport, Z. *Adv. Phys. Org. Chem.* **1969,** *1,* 7.
144. Park, J. D.; McMurtry, R. J.; Adams, J. H. *Fluorine Chem. Rev.* **1968,** *2,* 55.
145. Chambers, R. D. *Fluorine in Organic Chemistry;* John Wiley: New York, 1973; Chapter 6.
146. Rodionov, P. P.; Furin, G. G. *J. Fluorine Chem.* **1990,** *47,* 361.
147. Kobrina, L. S. *Fluorine Chem. Rev.* **1974,** *7,* 1.
148. Pasto, D. J.; Krasnansky, R.; Zercher, C. *J. Org. Chem.* **1987,** *52,* 3062.
149. Hammons, J. H.; Coolidge, M. B.; Borden, W. T. *J. Phys. Chem.* **1990,** *94,* 5468.
150. Kaplan, L. *Bridged Free Radicals;* Marcel Dekker: New York, 1972.
151. Kotaka, M.; Kahida, T.; Sato, S. *Z. Naturforsch., B. Chem. Sci.* **1990,** *45,* 721.
152. Creary, X.; Sky, A. F.; Mehrsheikh-Mohammadi, M. E. *Tetrahedron Lett.* **1988,** *29,* 6839.
153. Dust, J. M.; Arnold, D. R. *J. Am. Chem. Soc.* **1983,** *105,* 1221.
154. Tedder, J. M. *Angew. Chem. Int. Ed. Engl.* **1982,** *21,* 401.
155. Tedder, J. M.; Walton, J. C. *Adv. Phys. Org. Chem.* **1978,** *16,* 51.
156. Tschuikow-Roux, E.; Yano, T.; Niedzielski, J. *J. Chem. Phys.* **1985,** *82,* 65.
157. Atkinson, R. *Chem. Rev.* **1985,** *85,* 69.
158. Liu, R.; Huie, R. E.; Kurylo, M. J. *J. Phys. Chem.* **1990,** *94,* 3247.
159. Dixon, D. A. *J. Phys. Chem.* **1986,** *90,* 54.
160. Carter, E. A.; Goddard, W. A., III *J. Chem. Phys.* **1988,** *88,* 1752.
161. Moss, R. A. *Acc. Chem. Res.* **1990,** *13,* 58.
162. Chambers, R. D. *Fluorine in Organic Chemistry;* John Wiley: New York, 1973; pp 129–131, 133–134.
163. Williams, D. E.; Houpt, D. J. *Acta Crystallogr.* **1986,** *B42,* 286.
164. Cosmo, R.; Sternhell, S. *Aust. J. Chem.* **1987,** *40,* 35.

165. Bott, G.; Field, L. D.; Sternhell, S. *J. Am. Chem. Soc.* **1980,** *102,* 5618.
166. Sherrod, S. A.; daCosta, R. L.; Barnes, R. A.; Boekelheide, V. *J. Am. Chem. Soc.* **1974,** *96,* 1565.
167. Schaefer, T.; Veregin, R. P.; Laatikainen, R.; Sebastian, R.; Marat, K.; Chareton, J. L. *Can. J. Chem.* **1982,** *60,* 2611.
168. Smart, B. E. *J. Org. Chem.* **1973,** *38,* 2027.
169. Smart, B. E. *J. Org. Chem.* **1973,** *38,* 2035.
170. Barefoot, A. C., III; Sanders, W. D.; Buzby, J. M.; Grayston, M. W.; Lemal, D. M. *J. Org. Chem.* **1980,** *45,* 4292.
171. Dailey, W. P.; Correa, R. A.; Harrison, E., III; Lemal, D. M. *J. Org. Chem.* **1989,** *54,* 5512.
172. Dawson, W. H.; Hunter, D. H.; Willis, C. J. *J. Chem. Soc., Chem. Commun.,* **1980,** 874.
173. Fernandez, R. E. *Diss. Abstr. Int.* **1988,** *B48,* 3569.
174. Hellmann, S.; Beckhaus, H.-D.; Rüchardt, C. *Chem. Ber.* **1983,** *116,* 2238.
175. Hirsch, J. *Top. Stereochem.* **1967,** *1,* 199.
176. Gallo, R. *Prog. Phys. Org. Chem.* **1983,** *14,* 115.

Biological Properties

by R. Filler and K. Kirk

The biochemical aspects of compounds containing the carbon–fluorine bond have developed rapidly since the elegant studies in the early 1950s by Sir Rudolph Peters, who elucidated the mechanism of the toxic action of fluoroacetate by invoking the concept of "lethal synthesis". In the late 1950s, Duschinsky, Heidelberger, and colleagues conducted pioneering studies on the tumor-inhibitory effects of nucleotides of fluorinated pyrimidines, such as 5-fluorouracil (5-FU) [*1, 2*]. These studies spurred intense interest in the application of fluorinated compounds as antimetabolites and mechanistic probes.

With the publication of several books on this subject [*3–9*] during the past 20 years, the pace of research on the biochemistry of the C–F bond has quickened, and many new investigators have joined the field. This section presents an overview so that the reader can sense the flavor and excitement of research in this area and observe the directions of some of the significant investigations. The most thorough and up-to-date coverage will be found in reference 9, from which much of the subsequent discussion has been drawn.

Fluorine-Containing Carboxylic Acids

Peters originally ascribed the acute toxicity of **fluoroacetate** to the *inhibition of the enzyme aconitase* by fluorocitrate, formed during the Krebs (citric acid) cycle (*10*). Subsequent studies have demonstrated, however, that a mitochondrial enzyme, responsible for the formation of a citrate–glutathione thioester, is irreversibly inhibited by fluorocitrate. This finding strongly suggests that the toxic effects of fluorocitrate are related to inhibition of mitochondrial citrate transport [*11, 12*]. The substitution of a C–F bond for a C–H bond in fluoroacetic, 3-fluoropropionic, and 3-fluoropyruvic acid provides a prochiral mechanistic probe for the study of enzymatic reactions associated with the synthesis and metabolism of biologically important carboxylic acids [*13*]. The use of 3-fluoropyruvate is shown in equation 1 [*14*].

[*11*] **1**

B:
pro-S
H H
F CO_2^-
O
Fluoropyruvate

Trans-carboxylase
(Retention at C-3)

^-O_2C H
F CO_2^-
O
3-Fluorooxalacetate

NADH
L-Malate Dehydrogenase

$-O_2C$ H
F CO_2^-
HO H
2,3-Fluoromalate

0065–7719/95/0187–1011$08.00/1

Fluorine-Containing Ketones

An adjacent trifluoromethyl group sharply increases the electrophilic character of the carbonyl carbon. Compounds that readily form hydrates and hemiacetals show a time-dependent reversible inhibition of the enzyme acetylcholinesterase (equation 2), in which the tight complex makes inhibition only partially reversible [*15*]. In comparison with a nonfluorinated analogue, several **aliphatic ketones flanked by CF_3 and CF_2 groups**, are exceptionally *potent reversible inhibitors* of acetylcholinesterase, as documented by comparison of inhibition constants K_i, shown in equation 3 [*16*].

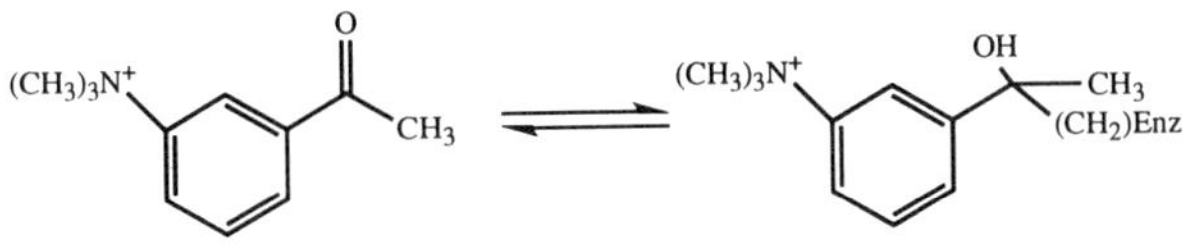

2

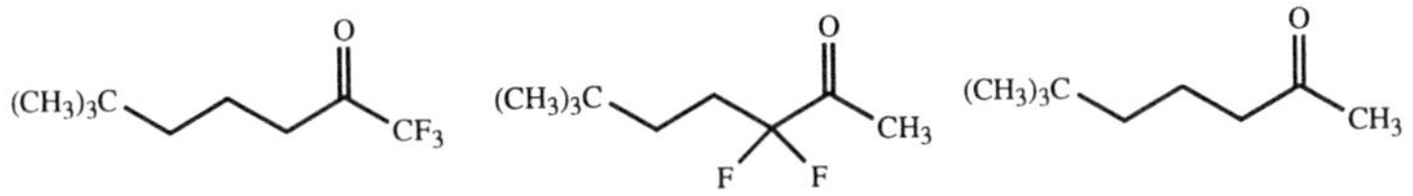

3

K_i (nM) 16 1.6 310,000

Effects of Fluorine on the Biological Properties of Steroids and Terpenoids

9-α-Fluoro-11-β-hydroxysteroids, such as *dexamethasone* (*1*) markedly *enhance glucocorticoid (liver glycogen deposition) and anti-inflammatory behavior relative* to nonfluorinated analogues, but with increased mineralocorticoid (retention of Na^+, Cl^-, and HCO_3^-) activity [*17, 18*]. Similar behavior is observed with 6-α-fluorosteroids, such as *fluprednisolone* (**2**). The greater biological activity of fluorosteroids has been attributed, in part, to an increased effectiveness of the

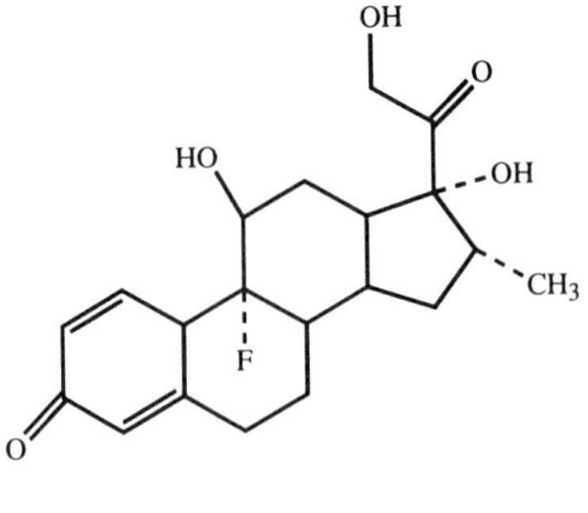

1 dexamethasone

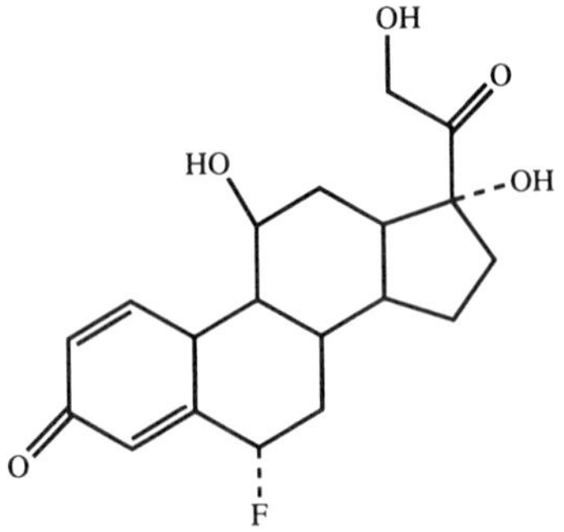

2 fluprednisolone

References are listed on pages 1018–1020.

11-β-OH group to serve as a hydrogen bond donor toward a substrate, because the neighboring 9-fluoro substituent increases the acidity of the 11-β-OH. Other data suggest that additional factors are probably involved, such as distortion of the A ring and reduced rate of oxidation of 11-β-OH to C=O. A number of fluorine-containing nonsteroidal anti-inflammatory drugs have replaced the powerful steroids during the past two decades.

^{18}F-Fluorinated steroids were investigated as positron-emitting probes for androgen and progesterone receptors [*19*]. Compound ***3*** is a promising *positron emission tomography* (PET) imaging agent for the progesterone receptor [*20*].

Among the **D vitamins**, multiple fluorine substituents in the side chain of 25-hydroxy-D_3 (***4***) markedly increases bone *resorptive activity* [*21, 22*]. *The enhanced activity may be due to blockade of degradation caused by the presence of fluorine in specific positions.*

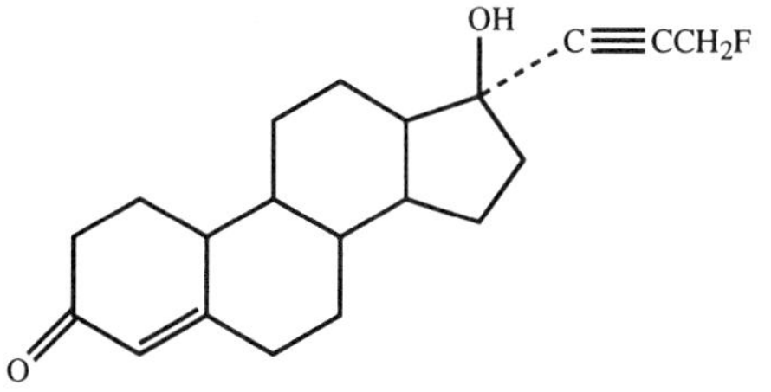

3 17α-(3-fluoro-1-propynyl)-nortesterone

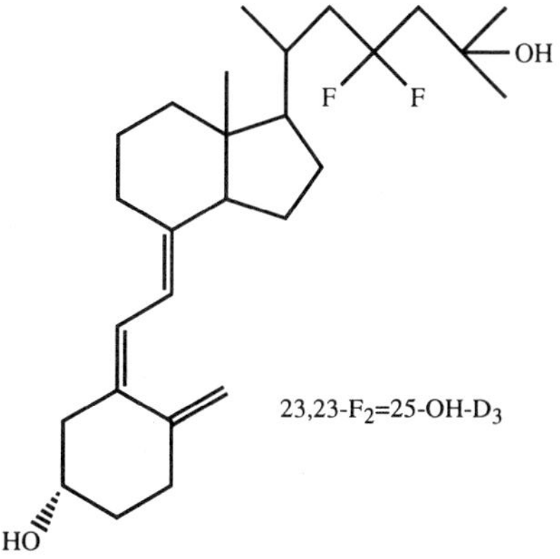

23,23-F_2=25-OH-D_3

4

Fluorinated Pyrimidines and Their Nucleosides

A landmark scientific breakthrough resulted from the synthesis and metabolic studies of **5-fluorouracil** (*5*) (fl^5 ura or 5-FU), **5-fluoroorotic acid** (***6***) (fl^5 oro), and **5-fluorocytosine** (*7*) (fl^5 cyt) by Heidelberger and co-workers [2]. Several excellent recent reviews are available on this important field [*9, 23, 24, 25, 26*].

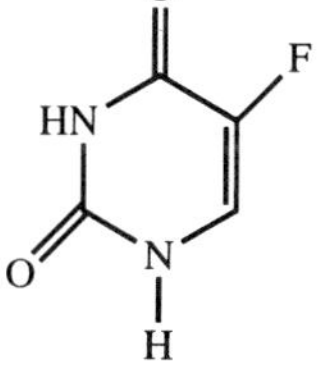

fl^5ura

5

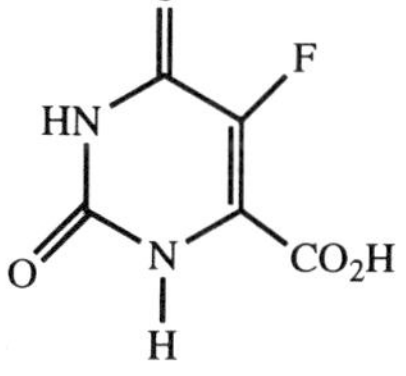

fl^5ora

6

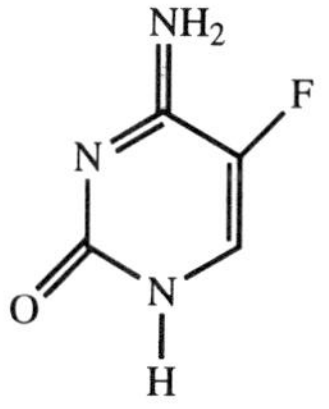

fl^5cyt

7

References are listed on pages 1018–1020.

A required step for the in vivo **incorporation of uracil into DNA** is the thymidylate synthase-catalyzed methylation (using formate) of deoxyuridine monophosphate (dUMP) at C-5 to produce thymidine monophosphate (dTMP). The introduction of fluorine in place of hydrogen at C-5 to form (fl^5 dUMP) alters the chemical reactivity at that site. Thus, the essential C-5 methylation of uracil is prevented because fluorine, a "deceptor" group, is unreactive to formate. Because (fl^5 ura) and its anabolites are concentrated in cancer cells, this enzymatic blockade inhibits tumor growth by causing "thymineless death" of malignant cells, another example of lethal synthesis. Additional toxicity arises by biosynthesis of the unnatural deoxyuridine triphosphate (fl^5 dUTP) and its incorporation into RNA [2]. 5-FU and its derivatives continue to serve as important agents in the treatment of solid tumors, such as those of the breast, ovary, and gastrointestinal tract, but the major impact of these pivotal studies has extended far beyond the original objectives to include, more recently, *new antiviral agents*.

Fluorine-Containing Carbohydrates

The replacement of a C–OH bond by a C–F bond has facilitated studies on the *transport and metabolism of sugars* and on the participation of carbohydrates in biosynthetic pathways. The F and OH groups are roughly of comparable size, and there are numerous instances when a C–F bond mimics a C–OH bond. Recently, **fluorodeoxy sugars** were used as probes for the study of hydrogen bonding in the glycogen phosphorylase–glucose complex [*27*]. The α and β anomers of 1-fluorohexoses serve as excellent substrates for glycosidases, enzymes that catalyze the hydrolytic cleavage of simple glycosides and oligosaccharides.

2-Deoxy-2-fluoroglucosides (*8*) are mechanism-based *glucosidase inhibitors*. Fluorine at C–2 slows the rate of the acetal C–OR (R = 2,4-dinitrophenyl) bond cleavage in ***8*** by destabilizing the proposed oxocarbonium ion-like transition state for glucosidase-catalyzed hydrolyses [*28*].

Fluorosugars, such as 3-deoxy-3-fluoroglucose, are useful probes of the mechanism and specificity of glucose transport systems [*29*], whereas 4-deoxy-4-fluoromannose inhibits the glycosylation of a viral protein [*30*]. New synthetic methodologies have increased the availability of **^{18}F-labeled sugars**, whose use as PET scanning agents is expanding. An area of intense activity is the study of nucleosides containing fluorine in the sugar moiety as *antiviral agents*. Of special interest are those that *exhibit anti-human immunosuppressive virus* (HIV) activity. Fluorinated 2′,3′-dideoxynucleosides, such as F-ddC (***9***), show promise as agents for the treatment of *acquired immune deficiency syndrome* (AIDS) [*31*].

Fluorinated Amino Acids

The leucine-reversible toxicity of δ,δ,δ-**trifluoroleucine (*10*) to microorganisms, such as *Escherichia coli*, has been attributed to incorporation of *10*** into protein [*32*] and to the action of ***10*** as a false feedback *inhibitor of* α*-isopropylmalate*

References are listed on pages 1018–1020.

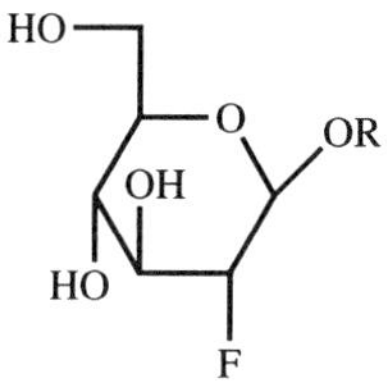

8

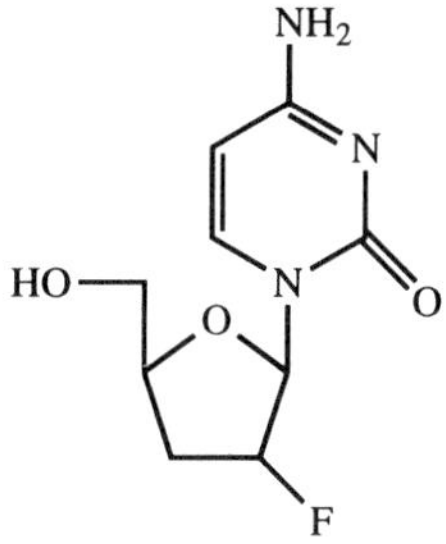

9

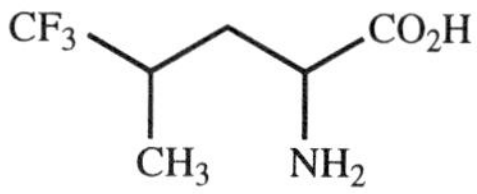

Trifluoroleucine

10

synthase, an enzyme critical to the biosynthesis of leucine [*33*]. Analogues of 5-fluorotryptophan were used to explore the mechanism of interaction of Trp with tryptophan synthase by ^{19}F NMR spectroscopy [*34*]. A pteridine-linked phenylalanine hydroxylase catalyzes the rapid conversion of 4-fluorophenylalanine (4-FPhe) to tyrosine [*35*]. Despite this facile enzymatic cleavage, 4-FPhe does not undergo the NIH shift to the 3-halo-tyrosines, which is observed with 4-XPhe (X = Cl, Br) [*36*]. 3-Fluorotyrosine, but not the 2-isomer, is extremely toxic, because it is metabolized by tyrosine metabolic enzymes to fluoroacetate and, ultimately, to fluorocitrate [*37*]. Ring-fluorinated analogues of dihydroxyphenylalanine **(DOPA)** ***(11, 12,*** **and** ***13*****)** precursors of the corresponding dopamines, were investigated in detail [*38, 39, 40*]. The ^{18}F species are of special interest as PET scanning agents to quantitate *dopaminergic activity*, for example, as a diagnostic tool for Parkinson's disease.

Both ***threo-*** **(*14*)** and ***erythro*-4-fluoro-DL-glutamic acid (*15*)** are noncompetitive *inhibitors of glutamine synthase,* an enzyme that catalyzes the synthesis of glutamine from L-glutamic acid and ammonia. This inhibition may explain the

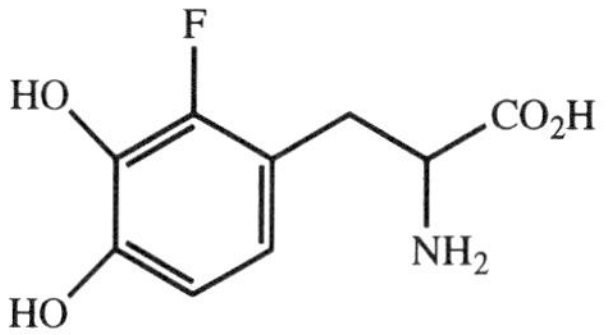

11 2-F-DOPA

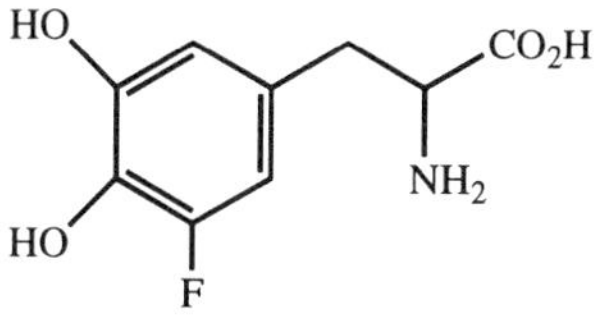

12 5-F-DOPA

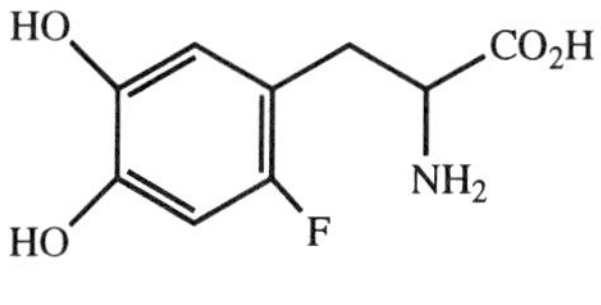

13 6-F-DOPA

References are listed on pages 1018–1020.

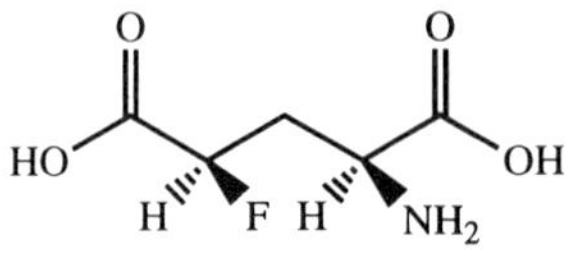

L-*erythro*-4-F-Glu
14

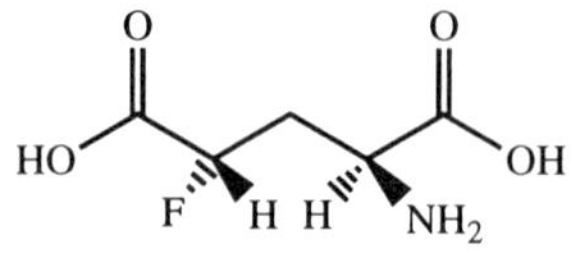

L-*threo*-4-F-Glu
15

antitumor and antiviral activities of ***14*** and ***15*** [*41*]. *McGuire and Coward* [*42*] found that DL-*threo*-4-fluoroglutamate was an effective concentration-dependent inhibitor of polyglutamylation of tetrahydrofolate and the anticancer drug methotrexate (MTX) (***16a***), whereas the *erythro* isomer was only weakly inhibitory. The *threo* isomer behaved as an alternative substrate and was incorporated only slightly less effectively than L-glutamate [*42*]. DL-*erythro*- and DL-*threo*-4-fluoromethotrexate (FMTX) ***(16b),*** are biosynthesized from the corresponding 4-fluoroglutamates. FMTX is a poorly glutamylated mimic of MTX [*43, 44*]. Each isomer of FMTX was equivalent to MTX in the ability to inhibit *dihydrofolate reductase* activity and bind to intracellular *dihydrofolate reductase* when intracellular drug concentration was limiting. The growth inhibitory effects of *erythro*-FMTX against a human leukemia and H-35 rat hepatoma cell lines were almost the same as those of MTX. *threo*-FMTX was slightly less potent [*44*]. A recently synthesized fluoroglutamate-containing peptide is a good substrate for folylpolyglutamate synthetase [*45*].

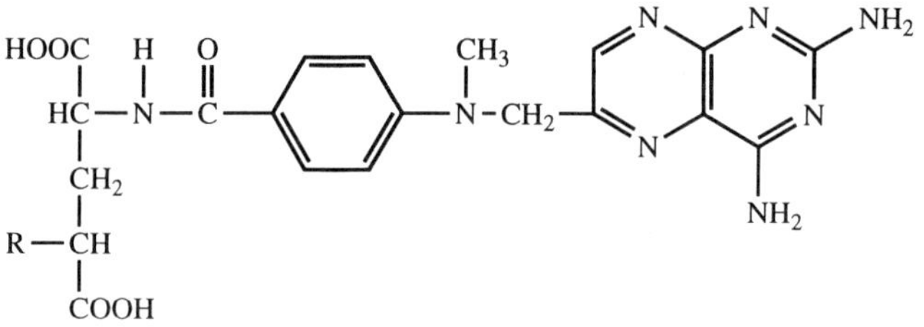

16a R = H
16b R = F

β-F-D-alanine exhibits in vivo antibacterial activity [*46*], due to its inactivation of alanine racemase and related enzymes by serving as a suicide substrate. Detailed studies led to an initial mechanism [*47*] and a revised mechanism [*48*] for the inhibition. A number of aliphatic monofluoro γ-amino acids *inhibit the activity of γ-aminobutyrate aminotransferase* (GABA-T). By blocking GABA-T, these compounds are effective anticonvulsants, because they permit an increase in the levels of the important neurotransmitter **γ-aminobutyric acid** (GABA) in the brain [*49*].

Fluorinated Neuroactive Amines

Fluorine has played a remarkable role in the development of selective, yet potent, *central nervous system (CNS) agents*. These drugs include antidepressants, neuroleptics (antipsychotics), sedative–hypnotics, and anxiolytics. In this section, we review briefly a few examples in which fluorine-containing amines affect neuronal function. **Ring-fluorinated catecholamines**, such as 2-, 5-, and 6-fluoronorepinephrine (FNE) (***17a, 17b,*** and ***17c,*** respectively), exhibit dramatic fluorine-induced adrenergic selectivities [*50*]. Thus, 2-norepinephrine was comparable with norepinephrine (NE) as a β-adrenergic agonist but was devoid of α-adrenergic activity, whereas 6-FNE mimicked NE as an α-adrenergic agonist. Preliminary studies of fluoro analogues (***18a*** and ***18b***) of serotonin [5-hydroxytryptamine (5HT)] indicate minimal differences between the fluoro compounds and 5-HT in interaction with 5-HT receptors, despite significant differences in the acidities of the phenolic function [*51*].

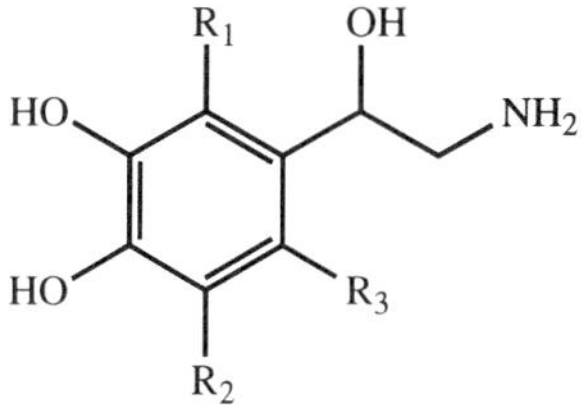

17 $R_1 = F, R_2 = R_3 = H$
17 $R_2 = F, R_1 = R_3 = H$
17 $R_3 = F, R_1 = R_2 = H$

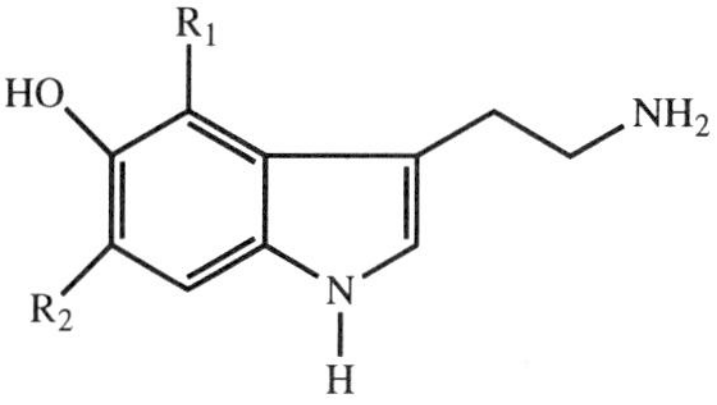

18 $R_1 = H, R_2 = F$
18 $R_1 = R_2 = F$

Monoamine oxidase (MAO) inactivates serotonergic and catecholaminergic neurotransmitters. MAO (A and B) inhibitors exhibit mood-elevating properties. **5-Fluoro-α-methyltryptamine** (***19***) is an important MAO A-*selective inhibitor*. In the treatment of certain depressive illnesses, 4-fluorotranylcypromine (***20b***) is 10 times more potent than the parent tranylcypromine (TCP; ***20a***). The enhanced in vivo activity may be due to increased lipophilicity of ***20b*** and/or to blockade of metabolic *para-hydroxylation* [52].

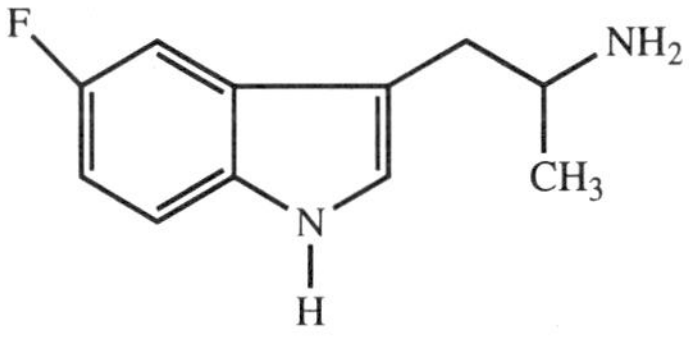

19

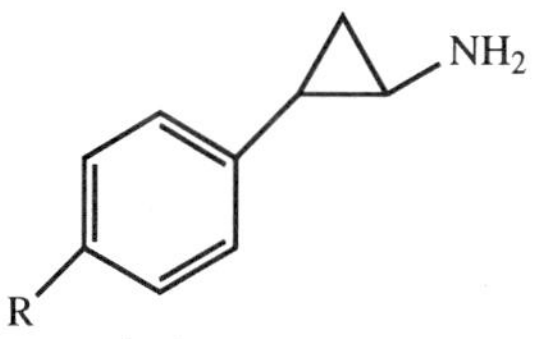

20a R = H
20b R = F

References are listed on pages 1018–1020.

References for Pages 1011–1018

1. Duschinsky, R.; Pleven, E.; Heidelberger, C. *J. Am. Chem. Soc.* **1957,** *79,* 4559.
2. Douglas, K. T. *Med. Res. Rev.* **1987,** *7,* 441.
3. *Carbon–Fluorine Compounds: Chemistry, Biochemistry, and Biological Activities, A Ciba Foundation Symposium;* Elsevier: Amsterdam, 1972.
4. *Biochemistry Involving Carbon–Fluorine Bonds;* Filler, R., Ed., ACS Symposium Series 28; American Chemical Society: Washington, DC, 1976.
5. *Biomedicinal Aspects of Fluorine Chemistry;* Filler, R.; Kobayashi, Y., Eds., Kodansha Ltd.: Tokyo, and Elsevier Biomedical: Amsterdam, 1982.
6. *Fluorinated Carbohydrates: Chemical and Biochemical Aspects;* Taylor, N. F., Ed., ACS Symposium Series 374; American Chemical Society: Washington, DC, 1988.
7. *Selective Fluorination in Organic and Bioorganic Chemistry;* Welch, J. T., Ed., ACS Symposium Series 456, American Chemical Society: Washington, DC, 1991.
8. Welch, J. T.; Eswarakrishnan, S.; *Fluorine in Bioorganic Chemistry*; Wiley: New York, 1991.
9. Kirk, K. L.; *Biochemistry of the Halogens, Vol. II. Biochemistry of Halogenated Organic Compounds;* Plenum: New York, 1991.
10. Peters, R. A.; Wakelin, R. W.; Buffa, P.; Thomas, L. C. *Proc. Roy. Soc. (London), B* **1953,** *140,* 497.
11. Kun, E.; Kirsten, E; Sharma, M. L. *Proc. Natl. Acad. Sci. U.S.A.* **1977,** *74,* 4942.
12. Kirsten, E.; Sharma, M. L.; Kun, E. *Mol. Pharmacol.* **1978,** *14,* 172.
13. Walsh, C. *Adv. Enzymol.* **1983,** *55,* 197.
14. Goldstein, J. A.; Cheung, Y.-F.; Marletta, M. A.; Walsh, C. *Biochemistry* **1978,** *17,* 5567.
15. Brodbeck, U.; Schweikert, K.; Gentinetta, R.; Rottenberg, M. *Biochim. Biophys. Acta* **1979,** *567,* 357.
16. Gelb, M. H.; Svaren, J. P.; Abeles, R. H. *Biochemistry* **1985,** *24,* 1813.
17. Fried, J. *Cancer,* **1957,** *10,* 752.
18. Fried, J.; Borman, A. *Vit. Horm.* **1958,** *16,* 303.
19. Brandes, S. J.; Katzenellenbogen, J. A. *Nucl. Med. Biol.* **1988,** *15,* 53.
20. Brandes, S. J.; Katzenellenbogen, J. A. *Mol. Pharmacol.* **1987,** *32,* 391.
21. Stern, P. H.; Tanaka, Y.; DeLuca, H. F.; Ikekawa, N.; Kobayashi, Y. *Mol. Pharmacol.* **1981,** *20,* 460.
22. Stern, P. H.; Mavreas, T.; Tanaka, Y.; DeLuca, H. F.; Ikekawa, N.; Kobayashi, Y. *J. Pharmacol. Exp. Ther.* **1984,** *229,* 9.
23. Douglas, K. T. *Med. Res. Rev.* **1987,** *7,* 441.
24. Heidelberger, C.; Danneberg, P.; Moran, R. G. *Adv. Enzymol.* **1983,** *54,* 58.

25. Myers, C. E. *Pharmacol. Rev.* **1981,** *33,* 1.
26. Santi, D. V.; Pogolotti, A. L. Jr.; Newman, E. M.; Wataya, T. In *Biomedicinal Aspects of Fluorine Chemistry;* Filler, R.; Kobayashi, Y., Eds.; Kodansha Ltd.: Tokyo, and Elsevier Biomedical: Amsterdam, 1982, pp 123-142.
27. Withers, S. G.; Street, I. P.; Percival, M. D. In *Fluorinated Carbohydrates: Chemical and Biological Aspects;* Taylor, N. F., Ed.; ACS Symposium Series 374; American Chemical Society: Washington, DC, 1988, pp 59-77.
28. Withers, S. G.; Street, I. P.; Bird, P.; Dolphin, D. H. *J. Am. Chem. Soc.* **1987,** *109,* 7530.
29. Barnett, J. E. G. In *Carbon–Fluorine Compounds: Chemistry, Biochemistry, and Biological Activities, A Ciba Foundation Symposium*; Elsevier: Amsterdam, 1972; pp 95–115.
30. Grier, T. J.; Rasmussen, J. R. *J. Biol. Chem.* **1984,** *259,* 1027.
31. Okabe, M.; Sun, R.-C.; Zenchoff, G. B. *J. Org. Chem.* **1991,** *56,* 4392 and references therein.
32. Rennert, O. M.; Anker, H. S. *Biochemistry* **1963,** *2,* 471.
33. Calvo, J. M.; Freundlich, M.; Umbarger, H. E. *J. Bacteriol.* **1969,** *97,* 1272.
34. Miles, W. W.; Phillips, R. S.; Yeh, H. J. C.; Cohen, L. A. *Biochemistry* **1986,** *25,* 4240.
35. Kaufman, S. *Biochim. Biophys. Acta* **1961,** *51,* 619.
36. Guroff, G.; Daly, J. W.; Jerina, D. M.; Renson, J.; Witkop, B.; Udenfriend, S. *Science* **1967,** *157,* 1524.
37. Weissman, A.; Koe, B. J. *J. Pharmacol. Exp. Ther.* **1967,** *155,* 135.
38. Firnau, G.; Sood, S.; Pantel, R.; Garnett, E. S. *Mol. Pharmacol.* **1981,** *19,* 130.
39. Garnett, E. S.; Firnau, G.; Nahmias, C. *Nature* **1983,** *305,* 137.
40. Creveling, C. R.; Kirk, K. L. *Biochem. Biophys. Rev. Commun.* **1985,** *130,* 1123.
41. Firsova, N. A.; Selivanova, K. M.; Alekseeva, L. V.; Evstigneeva, Z. G. *Biokhimiya* **1986,** *51,* 850.
42. McGuire, J. J.; Coward, J. K. *J. Biol. Chem.* **1985,** *260,* 6747.
43. McGuire, J. J.; Haile, W. H.; Coward, J. K. *Biochem. Pharmacol.* **1989,** *38,* 4321.
44. McGuire, J. J.; Graber, M.; Licato, N.; Vincenz, C.; Coward, J. K.; Nimec, Z.; Galivan, J. *Cancer Res.* **1989,** *49,* 4517.
45. Licato, J. J.; Coward, J. K.; Nimec, Z.; Galivan, J.; Bolanowska, W. E.; McGuire, J. J. *J. Med. Chem.* **1990,** *33,* 1022.
46. Kollonitsch, J.; Barash, L.; Kahan, F. M.; Kropp, H. *Nature* **1973,** *243,* 346.
47. Wang, E.; Walsh, C. *Biochemistry* **1978,** *17,* 1313.
48. Likos, J. J.; Ueno, H.; Feldhaus, R. W.; Metzler, D. E. *Biochemistry* **1982,** *21,* 4377.

49. Silverman, R. B.; George, C. *Biochemistry* **1988,** *27,* 3285 and references therein.
50. Kirk, K. L.; Creveling, C. R. *Med. Res. Rev.* **1984,** *4,* 189.
51. Hollingsworth, E.; Daly, J. W.; Kirk, K. L., unpublished results.
52. Coutts, R. T.; Rao, T. S.; Micetich, R. G.; Hall, T. W. E. *Cell Mol. Neurobiol.* **1987,** *7,* 271.

Chapter 7

Analysis of Organic Fluorine Compounds

Destructive Analytical Methods

by Dayal T. Meshri

Since the 1950s the exponential growth of organic fluorine chemistry has been attributed to newly discovered applications in medicine, biochemistry, agriculture, electronics, plastics, and other industries. These advances have produced thousands of new fluorine-containing gaseous, liquid, and solid compounds, and consequently have compelled analytical chemists to develop new techniques for their characterization and identification. This need is now greater than ever before. The characterization of inorganic or organic fluorides may be divided into two major classes: nondestructive and destructive procedures.

The modern electronic industry has played a very important role in the development of instrumentation based on physical–analytical methods. As a result, a rapid boom in the fields of infrared, nuclear magnetic resonance (NMR), Raman, and mass spectroscopy and vapor-phase (or gas–liquid) chromatography has been observed. Instruments for these methods have become indispensable tools in the analytical treatment of fluorinated mixtures, complexes, and compounds. The detailed applications of the instrumentation are covered later in this chapter.

Destructive analytical procedures, which were mostly developed at the beginning of this century for inorganic compounds, have been modified, revised, and established for the determination of fluorine in organofluorine compounds.

Of the many destructive analytical techniques, only the most widely used methods, especially those based on the personal experience of the author during the past 28 years will be discussed fully. More background information on the various techniques for fluorine analysis can be found in the literature [*1, 2, 3, 4, 5*].

Conversion to Fluoride Ions

All covalently bound fluorine atoms, such as those in C–F, N–F, P–F, and S–F groups, must first be converted into fluoride ions. Although this conversion could be achieved via many routes, the author's preferred method for organofluorine compounds has been fusion with sodium or potassium metals. This technique has been applied to fluorinated gases and low- and high-boiling liquids and solids. For inorganic salts, double salts, and complex anions, the use of a fusion mixture (2 parts of K_2CO_3, 1 part of Na_2CO_3, and 1 part of Na_2O_2 by weight) is preferred. After fusion, Willard–Winter distillation [*6, 7, 8*], null point potentiometry [*9, 10*], and ion-selective electrode methods [*11*] are used for quantitative analysis.

0065–7719/95/0187–1023$08.00/1

Fusion with Sodium Metal

Caution: All alkali metals react violently upon contact with water. Read all Material Safety Data Sheets (MSDS) very carefully prior to handling of alkali metals and handle these metals only under the direct supervision of trained and qualified personnel.

A piece of sodium metal stored under kerosene in a metal container is removed from a jar and blotted with dry napkin or filter paper. With a sharp knife, the layer of oxides is removed until a shiny surface appears. The removed layer is then destroyed carefully by adding very small pieces (not larger than 0.5 cm) to precooled 200 mL of methanol or ethanol.

Caution: Addition of 0.5 cm pieces of sodium metal to methanol or ethanol must be done in a chemical hood and behind a safety shield. Addition should be slow to minimize evaporation loss of methanol or ethanol. No flames or burner should be permitted in the area. Disposal of sodium metal must be carried out in someone's presence.

A small piece (about 2 g) of the shiny metal is then cut into small pieces that can be easily introduced into a narrow-mouth, thick-walled fusion tube by using a glass rod or a stainless steel spatula. The tube is then connected to a vacuum line by Tygon tubing or a flexible Cajon connection and evacuated under dynamic vacuum for 10 min; the stopcock of the vacuum line is closed, and the whole assembly is weighed on an analytical balance. The total weight of the assembly is recorded up to the fourth or fifth decimal. The fusion assembly is again connected to a vacuum line and cooled in liquid nitrogen, and a small sample of gaseous or low-boiling liquid (approximately 0.1–0.2 g) is distilled into the fusion tube. Every effort should be made to cool only the lower part of the fusion tube so that the sample condenses at the bottom and the narrow portion and the neck of the fusion tube stays at ambient temperature. After the transfer, the valve of the vacuum line is opened, and any noncondensable materials are removed from the fusion assembly. The valve of the fusion assembly is then closed, and the assembly is kept behind a safety shield and allowed to reach room temperature at a calculated pressure of less than 1.0 atm.

If after 1 h there is no gas leak or a violent reaction, the assembly is weighed again. The difference between the second weight and the first weight is the weight of the compound to be analyzed. The bottom of the fusion assembly is again cooled in liquid nitrogen, and the narrow neck is kept at room temperature. The assembly is clamped far below the narrow neck and carefully sealed at the neck while the sample is at liquid-nitrogen temperature. The sealed tube is removed from the liquid nitrogen bath and kept behind a protective shield in the hood for 30 min. If the assembly is sealed properly and no gas has leaked at room temperature, the sample is immersed in a 50–60 °C warm water bath for 20 min. The outside surface of the fusion assembly is wiped, the assembly is clamped, and its bottom is gradually heated with an open flame. Within a few minutes, the sodium metal melts, and the organic fluorine compound starts reacting. The inner shining surface of the fusion tube turns darker and darker as fluorine is being removed and carbon is deposited on the suface of the fusion tube. Flaming is continued for 20–25 min until all material has reacted; excess sodium metal is sublimed to form a shiny surface. The fusion tube is cooled; placed in a glass or stainless steel beaker; and broken by putting a small scratch with a file at the neck of the glass assembly, wetting the scratch, and touching with hot wire or glass rod. Excess sodium metal is destroyed with about 50 mL of methanol or ethanol until the reaction is complete. The solution is diluted with 100 mL of water and filtered through a Gooch crucible. The residue is washed four times with 20-mL portions of distilled water. The filtrate is transferred into a 250-mL volumetric flask, and distilled water is added to the mark. The solution is now ready for quantitative analysis by Willard–Winter distillation, null point potentiometry, or ion-selective electrode method. Perchloric acid should not be used as the distillation medium in the Willard–Winter method, because the presence of ethoxide or methoxide in the stock solution could create safety problems. Alkali metal fusions and peroxide fusions are also described in other references [*12, 13, 14, 15*]. Peroxide fusion is chosen to avoid the formation of cyanide or sulfide if nitrogen or sulfur are present in the compound.

Oxygen-Flask Combustion

Another destructive method for the determination of organic fluorine is oxygen-flask combustion [*3, 16, 17*].

In the oxygen-flask combustion technique, a weighed sample (1.5–5.0 mg) is wrapped in an ash-free filter paper or a methyl cellulose capsule that is fixed to a platinum wire or gauze attached to a ground-glass stopper. An Erlenmeyer flask containing 20 mL of distilled water is filled with oxygen. The sample is ignited, and the ground-glass stopper is inserted quickly and air-tightly. The sample is allowed to burn in the oxygen atmosphere. The flask is swirled occasionally until the combustion gases are completely absorbed. This process takes almost 40–45 min. The ground-glass joints and the walls of the flask are rinsed thoroughly with distilled water. The sample is ready for analysis by Willard–Winter distillation, null point potentiometry, or ion-selective electrode method.

Tube Combustion

The third destructive technique is tube combustion [*18*]. A sample is heated in the presence of moist air or oxygen, and the combustion products are passed over platinized silica in a silica tube at 1100 °C. This method is very simple and effective for decomposing organofluorine compounds whether they are volatile gases, high-boiling liquids, or solids. Hydrogen fluoride formed in the tube combustion process is swept away and absorbed in water. This method has an advantage over alkali fusion: it produces a low-ionic-strength hydrogen fluoride solution that could be used directly in the ion-selective electrode method. Tube combustion is not suitable for samples containing boron, because fluoroborate ions generated by combustion interfere with the determination of fluoride ions. Thus for boron-containing samples, peroxide or alkali fusion method must be used.

Quantitative Determination of Fluoride Ion

There are five major methods for the determination of fluoride ions in aqueous solution: (1) volumetric, (2) gravimetric, (3) null point potentiometry, (4) fluoride ion-selective electrode (also known as specific ion electrode), and (5) ion chromatography.

Gravimetric and volumetric analyses are well described in the literature [*6, 19, 20, 21, 22, 23, 24, 25, 26*]; only the remaining three methods will be discussed.

Null Point Potentiometry

This method involves very simple and inexpensive equipment that could be set up in any laboratory [*9, 10*]. The equipment consists of a 250-mL beaker (used as an external half-cell), two platinum foil electrodes, a glass tube with asbestos fiber sealed in the bottom (used as an internal half-cell), a microburet, a stirrer, and a portable potentiometer. The asbestos fiber may be substituted with a membrane. This method has been used to determine the fluoride ion concentration in many binary and complex fluorides and has been applied to unbuffered solutions from Willard–Winter distillation, to ion-exchange eluant, and to pyrohydrolysis distillates obtained from oxygen-flask or tube combustions. The solution concentrations range from 0.1 to 5×10^{-4} M. This method is based on complexing by fluoride ions of one of the oxidation states of the redox couple, and the potential difference measured is that between the two half-cells. Initially, each cell contains the same ratio of cerium(IV) and cerium(III) ions.

As a result, the electromotive force (EMF) of the cell is zero. In the presence of fluoride ions, cerium(IV) forms a complex with fluoride ions that lowers the cerium(IV)–cerium(III) redox potential. The inner half-cell is smaller, and so only 5 mL of cerium(IV)–cerium (III) solution is added. To the external half-cell, 50 mL of the solution is added, but the EMF of the cell is still zero. When 10 mL of the unknown fluoride solution is added to the inner half-cell, 100 mL of distilled water is added to the external half-cell. The solution in the external half-cell is mixed thoroughly by turning on the stirrer, and 0.5 M sodium fluoride solution is added from the microburet until the null point is reached. The quantity of known fluoride in the titrant will be 10 times the quantity of the unknown fluoride sample, and so the microburet readings must be corrected prior to actual calculations.

The redox solution for 0.095 M fluoride ion solution is prepared by dissolving 12.7 g of cerium(IV) ammonium sulfate in 200 mL of distilled water and 14 mL of 18 M sulfuric acid. Cerium(III) sulfate (2.8 g) is added, and the solution is diluted to 1 L.

Fluoride Ion-Selective Electrode

The availability of ion-selective electrodes has revolutionized the field of analytical chemistry. The fluoride ion-selective electrode has greatly assisted in the task of determining fluoride ion concentration, particularly after samples have been destroyed by oxygen-flask or tube combustion methods. Even samples destroyed by alkali metal or peroxide fusion can be analyzed after neutralization. A major advantage of this method is that it allows fluoride ions to be determined quickly, accurately, and economically.

There are two types of fluoride ion-selective electrodes available [*27*]: Orion model 96-09-00, a combination fluoride electrode, and model 94-09-00, which requires a reference electrode. The author prefers to use Orion model 94-09-00 because it has a longer operational life and is less expensive. When an electrode fails, the reference electrode is usually less expensive to replace. The Fisher Accumet pH meter, model 825 MP, automatically computes and corrects the electrode slope. It gives a direct reading for pH, electrode potential, and concentration in parts per million. The fluoride ion-specific electrode can be used for direct measurement [*28, 29*] or for potentiometric titration with Th^{4+} or La^{3+} nitrate solutions, with the electrode as an end point indicator.

Direct analysis with the fluoride ion-selective electrode requires addition of total ionic strength adjustor buffer solution (TISAB) to the standard and to unknown samples. Some advantages of this addition are that it provides a constant background ion strength; ties up interfering cations such as aluminum or iron, which form a complex with fluoride ions; and maintains the pH between 5.0 and 5.5. According to the manufacturer's claim, reproducibility of direct-electrode measurement is ±2.0%, and the accuracy for fluoride ion measurement is ±0.2% [*27*].

Ion Chromatography

Ion chromatography is the most modern technique for fluoride ion determination [*13, 30*]. This method has several advantages over the fluoride ion-selective electrode. It is very sensitive (to part-per-billion levels of F ions), requires a small sample size (50-µL injection), is selective and simple, and can simultaneously analyze several anions. When used with the proper eluant, it can analyze Br, Cl, F, BF_4, PF_6, and PO_3F^{2-}, and $PO_4{}^{3-}$ ions simultaneously. This technique is similar to high-pressure liquid chromatography. A solution of unknown sample is injected into the stream of a suitable eluant and passed through a separation column containing a selected anion-exchange resin. Detection is by the continuous measurement of the conductivity of the eluant, which increases as the anions are separated. A major disadvantage of this method is the initial cost of the equipment.

References for Pages 1023–1027

1. *Fluorine Chemistry;* Simons, J. H., Ed.; Academic Press: New York, 1954; Vol. 3, Chapter 3.
2. Mazor, L. *Analytical Chemistry of Organic Halogen Compounds;* Pergamon Press: New York, 1975; Chapter 4.

3. Hudlický, M. *Chemistry of Organic Fluorine Compounds;* Ellis Horwood: Chichester, U.K., 1976; Chapter 8, pp 558–580.
4. Turnbull, S. G.; Benning, A. F.; Feldmann, G. W.; Linch, A. L.; McHarness, R. C.; Richards, M. K. *Ind. Eng. Chem. Anal. Ed.* **1946,** *39,* 286.
5. Newman, E. J.; Fennell, R. W.; Dixon, E. J., et. al. *Analyst* **1971,** *96,* 384.
6. Willard, H. H.; Winter, O. B. *Ind. Eng. Chem. Anal. Ed.* **1933,** *5,* 7.
7. Willard, H. H.; Toribara, T. Y.; Holland, L. N. *Anal. Chem.* **1947,** *19,* 343.
8. Willard, H. H.; Horton, C. A. *Anal. Chem.* **1950,** *22,* 1190.
9. O'Donnel, T. A.; Stewart, D. F. *Anal. Chem.* **1961,** *33,* 1337.
10. O'Donnel, T. A.; Stewart, D. F. *Anal. Chem.* **1962,** *34,* 1347.
11. Orion Research Inc. *Instruction Manual;* Orion: Boston, 1988.
12. Belcher, R.; Caldas, E. F.; Clark, S. J.; Macdonald, A. *Mikrochim. Acta* **1953,** *3/4,* 283.
13. Wang, C. Y.; Tarter, J. G. *Anal. Chem.* **1983,** *55,* 1775.
14. Eger, C.; Yarden, A. *Anal. Chem.* **1956,** *28,* 512.
15. Rittner, R. C.; Ma, T. S. *Mikrochim. Acta* **1972,** 404.
16. Schoniger, W. *Mikrochim. Acta* **1956,** 869.
17. Davies, G. J.; Leonard, M. A. *Analyst* **1985**, *110,* 1205.
18. Kakabadse, G. J. ; Monohin, B.; Bather, J.; Weller, E. C.; Woodbridge, P. *Nature (London)* **1971,** *229,* 626.
19. Huckaby, W. B.; Welch, E. T.; Metler, A. V. *Anal. Chem.* **1947,** *19,* 154.
20. Gilkey, W. K.; Rohs, H. L.; Hansen, H. V. *Ind. Eng. Chem. Anal. Ed.* **1936,** *8,* 150.
21. Richter, F. *Z. Anal. Chem.* **1942,** *124,* 161.
22. Churchill, H. V. *Ind. Eng. Chem. Anal. Ed.* **1945,** *17,* 720.
23. Singer, L.; Armstrong, W. D. *Anal. Chem.* **1959,** *31,* 105.
24. Singer, L.; Armstrong, W. D.; Vogel, J. J. *J. Lab. Clin. Med.* **1959,** *74,* 354.
25. Wade, M. A.; Yamamura, S. S. *Anal. Chem.* **1965,** *37,* 1276.
26. Cady, G. H. *Anal. Chem.* **1976**, *48,* 655.
27. Orion Research Inc. *Instruction Manual, Fluoride Electrodes;* Orion: Cambridge, 1990.
28. Terry, M.; Kasler, F. *Microkhim. Acta* **1971,** 569.
29. Alberto, E. *Analyst* **1988,** *113,* 1299–1303.
30. Fernando, S.; Antonio, C.; Vincente, H. *Afinidad* **1990,** *47,* 127–130.

Modern Methods of Separation, Identification, and Structure Determination

by Mark L. Robin

Very little in the way of advances has occurred since 1971 in the applications of ultraviolet or infrared spectroscopy to the analysis of fluorinated organic compounds. Therefore, only gas–liquid chromatography, liquid chromatography, mass spectrometry, and electron scattering for chemical analysis (ESCA) are discussed. The application of nuclear magnetic resonance (NMR) spectroscopy to the analysis of fluorinated organic compounds is the subject of another section of this chapter.

Gas–Liquid Chromatography

Gas–liquid chromatography has been applied successfully to the separation and identification of numerous fluorinated organic compounds and is an indispensable tool for the synthetic fluorine chemist. Details regarding the separation of specific mixtures may be found in the experimental sections of journal articles describing the syntheses of fluorinated organic compounds. In general, separations of low-molecular-weight fluorinated organic compounds may be accomplished on relatively nonpolar poly(organosiloxane) phases such as the methylsilicone phases OV–1, OV–101, SP–2100, SE–30, or CP–SIL 5 [*1, 2, 3, 4*]. In general, these stationary phases may be applied to the analysis of fluorinated organic compounds boiling above approximately –50 °C; for highly volatile fluorinated organic compounds, columns using Porapak Q (styrene–divinylbenzene) or other porous polymer beads are useful. Less volatile, higher molecular weight fluorinated organic compounds may often be conveniently separated on poly(organosiloxane) phases of intermediate polarity, such as the 50:50 methyl–phenylsilicone phases OV–17 or SP–2250 [*5, 6*]. Many fully halogenated compounds have been separated on alumina–potassium chloride columns, but because they decompose on the column, hydrogen-containing compounds cannot be analyzed in this fashion [*7*].

The high thermal and chemical stability of fluorocarbons, combined with their very weak intermolecular interactions, makes them ideal stationary phases for the separation of a wide variety of organic compounds, including both hydrocarbons and fluorine-containing molecules. Fluorinated stationary phases include perfluoroalkanes, fluorocarbon surfactants, poly(chlorotrifluoroethylene), poly(perfluoroalkyl) ethers, and other functionalized perfluoro compounds. The applications of fluorinated compounds as stationary phases in gas–liquid chroma-

0065–7719/95/0187–1029$08.00/1

tography was reviewed recently [*8, 9*]. These highly fluorinated stationary phases have been used successfully in the separation of perfluoroalkanes and perfluoroalkenes [*10, 11, 12, 13, 14*]; perfluorinated aromatics [*13, 14*]; chlorofluorocarbons [*10, 11, 13, 15, 16*]; and various compounds including alkanes, alkenes, acids, aldehydes, alcohols, esters, ketones, amines, and haloalkanes [*10, 11, 17, 18, 19*].

Derivatization is frequently employed in gas–liquid chromatography to improve the thermal stability and increase the volatility and chromatographic performance of polar molecules. Fluorinated organic compounds are used to derivatize a large variety of organic molecules. The introduction of a fluorine-containing unit not only increases the volatility of a compound but also enhances detection via the electron-capture detector because of the high electron affinity of such fluorinated units [*4*]. Derivatives of alcohols, phenols, and amines are prepared readily by reaction with the fluorinated anhydrides. Trifluoroacetylation has been used in the separation of sulfate esters [*20*]; tetroses and aldopentoses [*21*]; aminophenols [*22*]; hexoses [*23*]; amines [*24*]; carbohydrates and their methyl glycosides [*25*]; and a variety of hydroxyl-containing compounds, including cholesterol, testosterone, 1-naphthol, octadecanol, and 10-nonadecanol [*26*]. The use of volatile *N*(*O,S*)-trifluoroacetyl amino acid alkyl esters is particularly useful and allows the separation of all 20 protein amino acids [*6, 27*]. The gas chromatographic analysis of pharmaceutically important phenol alkylamines has been accomplished via trimethylsilylation of the phenol function, followed by trifluoroacetylation of the amine function with *N*-methylbis(trifluoroacetamide) [*5, 28*]. Besides trifluoroacetylation, pentafluoropropionylation and heptafluorobutyrylation with the appropriate fluorinated anhydrides have been applied to amino acids [*29, 30*], sulfate esters [*20*], hydroxyl-containing compounds [*26*], amines [*31*], fatty acids [*32*], and histidine derivatives [*33*]. The heptafluorobutyryl group can also be introduced into organic molecules by using 1-(heptafluorobutyryl)imidazole, which has been used in the determination of ethanolamines and isopropanolamines in air at the part-per-billion level [*34*].

Pentafluorobenzyl bromide has been used in the derivatization of mercaptans [*35*] and phenols [*36*], in the analysis of prostaglandins [*37*], and in quantitative GC–MS [*38*]. 1,3-Dichlorotetrafluoroacetone is used for the derivatization of amino acids to the corresponding cyclic oxazolidinones and allows the rapid analysis of all 20 protein amino acids [*6*]. Pentafluorophenyldialkylchlorosilane derivatives have facilitated the gas chromatographic analysis of a wide range of functionally substituted organic compounds, including steroids, alcohols, phenols, amines, carboxylic acids, and chlorohydrins [*4*].

The resolution of optically active compounds by gas chromatography with chiral phases is a well-established procedure, and the separation of *N*-perfluoroacetylated amino acid ester enantiomers in 1967 was the first successful application of enantioselective gas–liquid chromatography [*39*]. Amino acids have been resolved as their *N*-trifluoroacetyl esters on chiral diamide phases such as *N*-lauroyl-L-valine *tert*-butylamide or *N*-docosanoyl-L-valine *tert*-butylamide [*40, 41,*

References are listed on pages 1033–1036.

42, 43], and epoxy alcohols have been resolved as their trifluoroacetyl esters on hexakis(2,3,6-tri-*O*-alkyl)cyclodextrin phases [*44*]. Enantiomers of various substituted glutamic acids have been resolved as their *N*-trifluoroacetyl diisopropyl ester derivatives on an XE-60-(*S*)-valine-(*S*)-phenylethylamide column [*39*]. Enantiomers of fluorine-containing organic molecules have also been separated on achiral columns. The use of the *N,O*-heptafluorobutyryl-*O*-isobutyl esters of a series of serines provided resolution of the *erythro* and *threo* forms on an SE–30 column [*3*], and the diastereomers of 4-fluoroglutamic acid have been resolved as their *N*-trifluoroacetyl methyl esters on an OV–225 column or as the methyl 3-fluoro-2-pyrrolidone-5-carboxylate derivatives, formed via the thermal cyclization of 4-fluoroglutamic acid dimethyl ester hydrochloride, on an OV–17 or OV–330 column [*45*]. Fluorinated stationary phases have been used to resolve optical isomers of 1,6-dioxaspiro[4.4]nonanes via complexation gas chromatography on nickel(II) bis[6-(heptafluorobutanoyl)-(5*S*)-carvonate] as a chiral stationary phase [*46*], 3-methylcyclopentene on optically active dicarbonylrhodium(I) 3-(trifluoroacetyl)-(1*R*)-camphorate in squalene [*47*], and secondary methyl ethers on nickel(II) bis[3-(heptafluorobutyryl)-(1*R*)-camphorate] and nickel(II) bis[3-(heptafluorobutyryl)-(1*R*,2*S*)-pinan-4-onate] [*48*].

Liquid Chromatography

In 1980 the first heptadecafluorodecyl (HDFD)-bonded phase for reversed-phase liquid chromatography was used to separate a number of fluorinated and nonfluorinated compounds, including alcohols, phenols, arenes, esters, cyanides, and various herbicides [*49*]. The HDFD phase offers no improvements for the separation of nonfluorinated compounds but is highly selective toward fluorinated solutes. Benzene and fluorobenzene are easily separated on the HDFD column but cannot be separated on a decyl column. In general, as the degree of fluorination of solutes increases, their relative retentions on the HDFD surface increases more than they do on the corresponding hydrocarbon surface, a fact indicating the presence of fluorine–fluorine interactions [*50*]. Perfluorinated alkyl phases have also been used in the liquid chromatographic separation of proteins [*51*] and in the separation of solutes containing various functional groups, including aldehydes, ketones, alcohols, nitro compounds, cyanides, ethers, and carboxylic acids [*52*]. The applications of fluorinated compounds as phases in liquid chromatography were reviewed [*8*]. Perfluoro organic compounds are useful as stationary phases in liquid chromatography for the separation of both hydrocarbon and fluorine-containing compounds, but their potential has not been explored fully.

Mass Spectrometry

The mass spectrometry of fluorinated organic compounds was reviewed in 1961 [*53*], and this review should be consulted for information on the applications of mass spectrometry in organofluorine chemistry up to that time. With the advent of instrumentation allowing for gas chromatographic separation followed by mass spectrometry of the separated components, gas chromatography–mass spectrometry (GC–MS) has become an indispensable method for the separation and identification of fluorinated organic compounds in mixtures. Mass spectrometry is useful for the determination of the molecular weight and structure of fluorinated organic compounds, although in the case of conventional electron-impact (EI) mass spectrometry, the parent peak is often missing or very small. For example, the electron-impact ionization of polyfluorinated and perfluoro compounds at an electron energy of 70 eV often yields no parent ions; instead, only abundant low-mass fragments, such as $[CF_3]^+$ (*m/z* 69), $[C_3F_5]^+$ (*m/z* 131), and $[C_4F_7]^+$ (*m/z* 181), that are not always characteristic of the structure of the molecule under investigation are formed [*54*]. Chemical-ionization (CI) mass spectrometry is also used to analyze fluorinated organic compounds and, in general, results in less fragmentation compared with EI mass spectrometry. Because of their large electron attachment cross sections, polyfluorinated and perfluoro compounds can be identified by negative-ion chemical ionization (NICI) mass spectrometry [*55*]. For small perfluoro compounds such as CF_4 and C_2F_6, dissociative electron attachment is the only negative-ion-forming process, but for perfluoro compounds of higher molecular weight, electron attachment can occur to yield stable parent negative ions [*56*]. Intense negative parent ions are often formed in the NICI mass spectra of polyfluorinated and perfluoro compounds [*57, 58, 59*], allowing for molecular weight determination of those species, for which the parent peak may be missing or very weak in the conventional EI mass spectrum. For perfluorinated alkanes, alkenes, and aromatics, the molecular ion $[M]^-$ is in general the most abundant ion observed in the NICI mass spectrum [*60*]. For higher molecular weight fluorocarbons, the NICI mass spectrum may yield *only* molecular ion information and almost no structural information [*61*]. To obtain structural information in such cases, collision-activated dissociation (CAD) mass spectrometry, which produces more intense high-mass fragments, may be employed. Applications of CAD mass spectrometry for the elucidation of structural information have included the investigation of perfluoro-1,3-dimethyladamantane [*61*] and of highly fluorinated alkanes, alcohols, and ethers [*62*].

Electron Scattering for Chemical Analysis (ESCA)

Practical applications of ESCA were developed only recently. Fluorine, being the most electronegative element, induces the largest chemical shifts in the carbon 1s (C1s) binding energies, and hence, fluorinated compounds exhibit spectral maxima in regions where other atoms do not [*63*]. Because of these large chemical shifts in C1s binding energies and because fluoropolymers are often difficult to study by conventional methods, fluoropolymers were some of the first polymeric systems investigated by ESCA [*64*]. The surface structures of a number of fluoropolymers have been investigated, including poly(vinylfluoride), poly(vinylidenefluoride), poly(trifluoroethylene), poly(tetrafluoroethylene), and poly(chlorotri-

References are listed on pages 1033–1036.

fluoroethylene) [*65, 66*]. ESCA also has been applied to the determination of the surface structure of fluorographites [*67*] and plasma polymers of perfluorobenzene, perfluorocyclohexa-1,4-diene, perfluorocyclohexa-1,3-diene, perfluorocyclohexene, perfluorocyclohexane, *cis*- and *trans*-1,2-difluoroethylene, and 1,1-difluoroethylene [*68, 69*]. Clark and Brennan recently investigated the low-energy electron beam polymerization of hexafluorobenzene via ESCA, and the C1s ESCA spectra show that CF_3-, CF_2-, CF-, and C-type functional groups are present in the polymer formed, indicating that some rearrangement reactions have taken place [*70*]. Because ESCA is surface sensitive, it is ideally suited for the analysis of surface coatings and has been applied to the study of the surface fluorination of polymers, for example, high-density polyethylene [*71, 72, 73*].

ESCA has been used to determine the molecular structure of the fluoride ion-induced trimerization product of perfluorocyclobutene [*74*] and the products of the sodium borohydride reduction of perfluoroindene [*75*]. ESCA is also used to analyze and optimize gas-phase reactions, such as the bromination of trifluoromethane to produce bromotrifluoromethane, a valuable fire suppression agent [*76*]. The ionization energies for several hundred fluorine-containing compounds are summarized in a recent review [*77*].

References for Pages 1029–1033

1. Bonvell, S. I.; Monheimer, R. H. *J. Chromatogr. Sci.* **1980,** *18,* 18.
2. Gilbert, J.; Startin, J. R. *J. Chromatogr.* **1980**, *189,* 86.
3. Klein, R. J. *Chromatogr.* **1979**, *170,* 468.
4. Poole, C. F.; Sye, W. F.; Singhawangcha, S.; Hsu, F.; Zlatkis, A. *J. Chromatogr.* **1980,** *199,* 123.
5. Donike, M. *J. Chromatogr.* **1975,** *103,* 91.
6. Husek, P. *J. Chromatogr.* **1982,** *234,* 381.
7. Noy, T. *J. Chromatogr.* **1987,** *393,* 343.
8. Varughese, P.; Gangoda, M. E.; Gilpin, R. K. *J. Chromatogr. Sci.* **1988,** *26,* 401.
9. Pomaville, R. M.; Poole, C. F. *Anal. Chim. Acta* **1987,** *200,* 151.
10. Dhanesar, S. C.; Poole, C. F. *J. Chromatogr.* **1983,** *267,* 388.
11. Dhanesar, S. C.; Poole, C. F. *Anal. Chem.* **1983,** *55,* 1462.
12. Muller, U.; Dietrich, P.; Prescher, D. *J. Chromatogr.* **1983,** *259,* 243.
13. Pscheidl, H.; Oberdorfer, E.; Moller, E.; Haberland, D. *J. Chromatogr.* **1986,** *365,* 383.
14. Vernon, F.; Edwards, G. T. *J. Chromatogr.* **1975,** *110,* 73.
15. Baiulescu, G. E.; Ilie, V. A. *Stationary Phases in Gas Chromatography;* Pergamon: Oxford, 1975; p 195.
16. Glajch, J. L.; Schindel, W. G. *LC-GC Mag.* **1986,** *4,* 574.
17. Blaser, W. W.; Kracht, W. R. *J. Chromatogr. Sci.* **1978,** *16,* 111.
18. Dhanesar, S. C.; Poole, C. F. *Anal. Chem.* **1983,** *55,* 2148.

19. Neu, H. J.; Heeg, F. J. *J. High Resolut. Chromatogr. Chromatogr. Commun.* **1980,** *3,* 537.
20. Murray, S.; Baillie, T. A. *Biomed. Mass Spectrom.* **1979,** *6,* 82.
21. Decker, P.; Schweer, H. *J. Chromatogr.* **1982,** *236,* 369.
22. Coutts, R. T.; Hargesheimer, E. E.; Pasutto, F. M.; Baker, G. B. *J. Chromatogr. Sci.* **1981,** *19,* 151.
23. Decker, P.; Schweer, H. *J. Chromatogr.* **1982,** *243,* 372.
24. Charles, R.; Gil-Av, E. *J. Chromatogr.* **1980,** *195,* 317.
25. Koenig, W. A.; Benecke, I.; Bretting, H. *Angew. Chem.* **1981,** *93,* 688.
26. Schwartz, D. P.; Allen, C. *J. Chromatogr.* **1981,** *208,* 55.
27. Michael, G. *J. Chromatogr.* **1980,** *196,* 160.
28. Christophersen, A. S.; Hovland, E.; Rasmussen, K. E. *J. Chromatogr.* **1982,** *234,* 107.
29. Abdalla, S.; Bayer, E.; Frank, H. *Chromatographia,* **1987,** *20,* 83.
30. Kuesters, E.; Allgaier, H.; Jung, G.; Bayer, E. *Chromatographia,* **1984,** *18,* 287.
31. Tiljander, A.; Skarping, G.; Dalene, M. J. *J. Chromatogr.* **1989,** *479,* 145.
32. Larsson, L.; Sonesson, A.; Jiminez, J. *Eur. J. Clin. Microbiol.* **1987,** *6,* 729.
33. Rogoskin, V. A.; Krylov, A.; Khlebnikova, N. S. *J. Chromatogr.* **1987,** *423,* 33.
34. Langvardt, P. W.; Melcher, R. G. *Anal. Chem.* **1980,** *52,* 669.
35. Wu, H. L.; Funazo, K.; Tanaka, M.; Shono, T. *Anal. Lett.* **1981,** *14,* 1625.
36. *Fed. Reg.* **1979,** *44,* 69484 (EPA Method 604).
37. Rosenfeld, J. M.; Mureika-Russel, M.; Love, M. *J. Chromatogr.* **1989,** *489,* 263.
38. Leis, H. J.; Gleispacj, H.; Malle, E. *Rapid Commun. Mass Spectrom.* **1988,** *2,* 263.
39. Maurs, M.; Ducrocq, C.; Righini-Tapie, A.; Azerad, R. *J. Chromatogr.* **1985,** *325,* 444.
40. Chang, S. C.; Charles, R.; Gil-Av, E. *J. Chromatogr.* **1982,** *238,* 29.
41. Chang, S. C.; Charles, R.; Gil-Av, E. *J. Chromatogr.* **1982,** *235,* 87.
42. Charles, R.; Beitler, U.; Feibush, B.; Gil-Av, E. *J. Chromatogr.* **1975,** *112,* 121.
43. Iwase, H. *Chem. Pharm. Bull.* **1975,** *23,* 1608.
44. Koenig, W. A.; Lutz, S.; Wenz, G.; Goergen, G.; Neumann, C.; Gaebler, A.; Boland, W. *Angew. Chem.* **1989,** *101,* 180.
45. Tolman, V.; Vlasakova, V.; Zivny, K. *J. Chromatogr.* **1984,** *315,* 421.
46. Schurig, V. *Naturwissenschaften* **1987,** *74,* 190.
47. Schurig, V. *Chromatographia* **1980,** *13,* 263.
48. Halterman, R. L.; Rousch, W. R.; Hoong, L. K. *J. Org. Chem.* **1987,** *52,* 1152.
49. Berendsen, G. E.; Pikaart, K. A.; deGalan, L.; Olieman, C. *Anal. Chem.* **1980,** *52,* 1990.

50. Billiet, H. A. H.; Schoenmakers, P. J.; deGalan, L. *J. Chromatogr.* **1981,** *218,* 443.
51. Xindu, G.; Carr, P. W. *J. Chromatogr.* **1983,** *269,* 96.
52. Sadek, P. C.; Carr, P. W. *J. Chromatogr.* **1984,** *288,* 25.
53. Majer, J. R. In *Advances in Fluorine Chemistry;* Stacey, M.; Tatlow, J. C.; Sharpe, A. G., Eds.; Butterworths: London, 1961; Vol. 2, p 55.
54. Huang, S. *Org. Mass Spectrom.* **1989,** *24,* 1065.
55. Christodoulides, A. A.; Christophorou, L. G.; Pai, R. Y.; Tung, C. M. *J. Chem. Phys.* **1974,** *70,* 1156.
56. Hunter, S. R.; Christophorou, L. G. *J. Chem. Phys.* **1984,** *80,* 6150.
57. Gregor, I. K.; Guilhaus, M. *J. Fluorine Chem.* **1983,** *23,* 549.
58. Scherer, K. V.; Yamanouchi, K.; Ono, T. *J. Fluorine Chem.* **1990,** *50,* 47.
59. Waddell, K. A.; Blair, I. A.; Wellby, J. *Biomed. Mass Spectrom.* **1983,** *10,* 83.
60. Huang, S. K.; Despot, K. A.; Sarkahian, A.; Bierl, T. W. *Biomed. Environ. Mass Spectrom.* **1990,** *19,* 202.
61. Huang, S.; Klein, D. H.; Adcock, J. L. *Rapid Commun. Mass Spectrom.* **1988,** *2,* 204.
62. Huang, S.; Tuinman, A. *Org. Mass Spectrom.* **1990,** *25,* 225.
63. Dilks, A. In *Electron Spectroscopy: Theory, Techniques and Applications;* Brundle, C. R.; Baker, A. D., Eds.; Academic: London, 1981; Vol. 4, Chapter 5.
64. Clark, D. T.; Kilcast, D.; Feast, W. J.; Musgrave, W. K. R. *J. Polym. Sci., Polym. Chem.* **1973,** *11,* 389.
65. Clark, D. T.; Thomas, H. R. *J. Polym. Sci., Polym. Chem.* **1978,** *16,* 791.
66. Clark, D. T.; Feast, W. J. *J. Macromol. Sci.–Rev. Macromol. Chem.* **1975,** *C12,* 191.
67. Clark, D. T.; Peeling, J. *J. Polym. Sci,, Polym. Chem.* **1976,** *14,* 2941.
68. Clark, D. T.; Dilks, A.; Shuttleworth, D. In *Polymer Surfaces;* Clark, D. T.; Feast, W. J., Eds.; Wiley: London, 1978.
69. Clark, D. T.; Shuttleworth, D. *J. Polym. Sci., Polym. Chem.* **1979,** *17,* 1317.
70. Clark, D. T.; Brennan, W. J. *J. Fluorine. Chem.* **1988,** *40,* 419.
71. Clark, D. T.; Feast, W. J.; Musgrave, W. K. R.; Ritchie, I. *J. Polym. Sci., Polym. Chem.* **1975,** *13,* 857.
72. Clark, D. T.; Feast, W. J.; Musgrave, W. K. R.; Ritchie, I. In *Advances In Polymer Friction and Wear;* Lee, L. H., Ed.; Plenum Press: New York, 1975, Vol. 5A.
73. Yagi, T.; Pavlath, A. E.; Pittmann, A. G. *J. Appl. Polym. Sci.* **1982,** *27,* 4019.
74. Clark, D. T.; Chambers, R. D.; Adams, D. B. *J. Chem. Soc. Perkin Trans. 1* **1975,** *7,* 647.

75. Adams, D. B.; Clark, D. T.; Feast, W. J.; Kilcast, D.; Musgrave, W. K. R.; Preston, W. E. *Nature (London), Phys. Sci.* **1972,** *239,* 47.
76. Bock, H.; Solouki, B.; Aygen, S.; Hirabayashi, T.; Mohmand, S.; Rosmus, P.; Wittmann, J. *J. Mol. Struct.* **1980,** *60,* 31.
77. Akopyan, N. E.; Golovin, A. V.; Rodin, A. A.; Sergeev, Y. L. *Russ. Chem. Rev.* **1988,** *57,* 745 (Engl. Transl.); *Usp. Khim.* **1988,** *57,* 1297.

Nuclear Magnetic Resonance Spectroscopy of Organofluorine Compounds

by T. Stephen Everett

Over the past 40 years fluorine nuclear magnetic resonance (^{19}F-NMR) spectroscopy has become the most prominent instrumental method for structure elucidation of organofluorine compounds. Consequently the amount of spectral data published has grown almost exponentially. Unfortunately NMR data for fluorinated compounds are not as well, or as easily, organized as proton data. To facilitate retrieval of fluorine NMR information and comparison of data, acquisition parameters should be clearly defined. Guidelines for publication of NMR data have been established by the International Union for Pure and Applied Chemistry (IUPAC) [*1*]. The following rules for acquisition and reporting of NMR data should be strictly observed:

1. Use fluorotrichloromethane (CCl_3F) as the standard reference.
2. Use a standard sign convention of (+) for signals downfield from (left of) CCl_3F and (–) for signals upfield from (right of) CCl_3F. Therefore the vast majority of C–F signals are negative.
3. Give a clear indication of solvent, concentration, and temperature. These parameters have a much greater effect on chemical shifts and coupling constants for fluorine than for protons.
4. Give an exact description of instrumentation—magnetic field strength, Continuous Wave (CW) or Fourier Transformed (FT), pulse sequence, decoupling, etc.
5. Use standardized plotting of spectral presentations, comparable to 0–10 ppm used for proton spectra. The range from +50 to –250 ppm covers most organofluorine signals but is too wide to show finer splitting and would require individual signals to be expanded.

Extensive secondary literature coverage of fluorine NMR data is available for articles published through 1981. Massive amounts of fluorine data were compiled in periodic reviews in *Annual Reports on NMR Spectroscopy* [*2, 3, 4, 5, 6, 7, 8, 9*] and in two volumes of *Progress in NMR Spectroscopy* [*10, 11*]. Books written by Dungan and Van Wazer [*12*] and Mooney [*13*], both published in 1970, are still widely referenced today. The majority of ^{19}F-NMR data compiled were from highly fluorinated and perfluorinated compounds.

0065–7719/95/0187–1037$13.40/1

In the past 10 years, with increased availability and utility of NMR instrumentation, the volume of NMR information on fluorinated compounds has greatly expanded, especially because of the following developments:

1. FT instruments of higher magnetic field strengths, providing greater spectral resolution and sensitivity
2. Extensive use beyond the field of chemistry, into biochemistry, biology, and medical science, including in vivo spectroscopy and imaging of biological samples
3. New techniques (multiple pulse methods) in structure elucidation [*14*]
4. Acquisition of newly accessible heteronuclei (such as ^{15}N and ^{17}O) [*15*]

Over 20 journals or series with an emphasis on NMR spectroscopy are now available:

Advances in Magnetic Resonance
Applied Magnetic Resonance
Applied Spectroscopy
Annual Reports on NMR Spectroscopy
Biological Magnetic Resonance
Bulletin of Magnetic Resonance
Canadian Journal of Applied Spectroscopy
Journal of Applied Spectroscopy
Journal of Magnetic Resonance
Journal of Molecular Spectroscopy
Magnetic Resonance in Biology
Magnetic Resonance in Chemistry (formerly *Org. Magn. Reson.*)
Magnetic Resonance Imaging
Magnetic Resonance in Medicine
Magnetic Resonance Review
NMR–Basic Principles and Progress
NMR in Biomedicine
Nuclear Magnetic Resonance
Organic Magnetic Resonance (now *Magn. Reson. Chem.*)
Progress in Nuclear Magnetic Resonance Spectroscopy
Spectrochimica Acta, Part A: Molecular Spectroscopy
Spectroscopy Letters
Spectroscopy: An International Journal

In contrast to this expansion in the primary sources of data, secondary ^{19}F-NMR literature coverage has declined. Fields [*16*], Jameson [*17*], and Bovey[*18*] have contributed noteworthy fluorine NMR chapters in recently published books. Jameson's review [*17*] presents tables and figures vividly displaying the magnitude

References are listed on pages 1078–1086.

of fluorine chemical shifts and coupling constants for both organic and inorganic fluorides. Japan Halon, in conjunction with Varian Instruments Ltd., has published proton and fluorine spectra of the company's fluorinated products [*19*]. Three computerized databases have been created and should gain widespread use. Preston Scientific Ltd.'s database covering more than 3000 fluorinated samples was initiated in 1986 [*20*]. Weigert has compiled databases for fluorocarbons [*21*], C_xF_y, and chlorofluorinated ethanes and ethylenes [*22*], $C_2H_xCl_yF_z$.

The primary function of this section is to organize data to facilitate NMR structure elucidation of organofluorine compounds. Selectively fluorinated aliphatics are emphasized, whereas fluorinated aromatics are covered in less detail. Inorganic nitrogen, phosphorus, silicon, and sulfur fluorides are not included, although compounds containing these and other heteroatoms attached to CF_3 are the focus of multinuclear data presented later (*see* Table 16).

The following types of studies will not be presented individually but may have contributed supporting data to coverage by compound type: conformational analyses [*23, 24, 25, 26, 27*], fluoropolymers [*28, 29, 30, 31, 32*], solid-state NMR [*33*], and solvent effects [*34, 35, 36, 37*]. Many excellent articles with in-depth NMR interpretation of one specific compound or of a small, structurally related group of compounds can be found in the chemical literature. A few of these, not incorporated elsewhere in this section are referenced here: carbonyl fluorides [*38, 39, 40*], fluoropropanes [*41, 42, 43*], fluorocyclopropanes [*44, 45, 46*], fluorobutanes [*47*], perfluorocyclobutanone [*48*], fluorohexanes [*49*], and vinyl fluorides [*50, 51, 52, 53, 54*].

Basic data coverage in this section commences with fluorinated methanes, ethanes, ethylenes, acrylates, aliphatic monofluorides (primary, secondary, and tertiary), and geminal difluorides (primary and secondary). The compilation continues with six categories of trifluoromethylated compounds and finishes with aromatic fluorides. ^{19}F-NMR applications extend greatly beyond simple structure determination. Six of these special topics are briefly surveyed, including analytical NMR methods and biochemical NMR studies.

Tabulated data are presented with fluorine chemical shifts in parts per million (ppm) from fluorotrichloromethane (FC11), negative values upfield (to the right). Coupling constants (*J*) are recorded in hertz, with number of intervening bonds indicated by the superscript to the left and the coupling nuclei indicated by the subscript to the right ($^2J_{HF}$ = proton–fluorine geminal coupling, $^3J_{FF}$ = fluorine–fluorine vicinal coupling, etc.). Coupling patterns are designated as follows: (s) = singlet, (d) = doublet, (t) = triplet, (q) = quartet, and (m) = multiplet, to the right of the chemical shift. Distinguishable splitting patterns of five or more peaks are written out: quint, sext, sept, oct, etc. Broadened signals are indicated by a "b", for example, (bs) for a broad singlet . In most cases fluorine signals are reported proton coupled, proton signals are reported fluorine coupled, and carbon signals are reported fluorine coupled and proton decoupled. Solvents, instrumentation, and other parameters used in data acquisition vary considerably for the literature values reported.

Basic NMR Structure Elucidation of Organofluorine Compounds

Fluorine NMR data can be collected readily on most spectrometers, requiring only minor adjustments to instrumentation used to run proton samples. The fluorine-19 nucleus is easily detected (relative abundance, 100%; spin, 1/2) and generates a wealth of spectral information to assist in structure elucidation. To take full advantage of all the spectral evidence available, ^{1}H, ^{13}C, and ^{19}F chemical shifts and coupling constants should be acquired and correlated.

The majority of ^{19}F-NMR signals, with fluorine bound to carbon, are detected *upfield from CCl_3F*. Acyl fluorides are the only derivatives that give fluorine signals downfield from the reference. Chemical shifts for organofluorides fall in a 300-ppm range, shown in the correlation diagram in Figure 1. Expected trends in deshielding of the fluorine nucleus are clearly evident for $CF_3 > CF_2 > CF$ and 3° > 2° > 1°. Similar correlation diagrams can be found in other references [*16, 17, 55*].

The NMR spectra of fluorinated molecules are often quite fascinating because of the *distinctive geminal (^{2}J) and vicinal (^{3}J) couplings* of fluorine to proton and fluorine to fluorine. In the tables that follow, chemical shifts, peak multiplicity, and magnitudes of spin–spin coupling interactions are provided. Splitting patterns readily reveal those carbon and hydrogen atoms that are within three bonds of fluorine. $^{2}J_{HF}$ values of 45–55 Hz are common (but can range up to 85 Hz), and these signals tend to be prominent in the proton-NMR spectrum because of the magnitude of splitting. $^{3}J_{HF}$ couplings are smaller, with values of 7–30 Hz being common. In carbon-13 NMR, J_{CF} values of hundreds of hertz for a one-bond coupling quickly fall off to tens of hertz for a two-bond coupling constant. Appreciable carbon–fluorine couplings of greater than 1 Hz may be seen for nuclei separated by four or five bonds.

Not all consequences of fluorine incorporation in a molecule aid in structure identification. *Long-range coupling* and conformational preferences can compli-

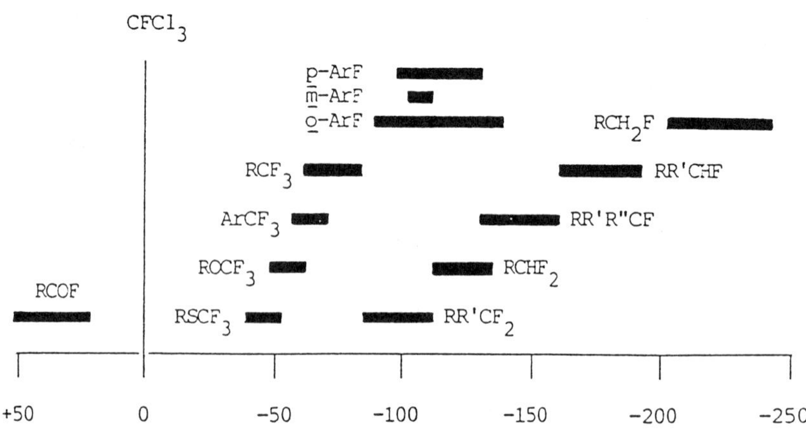

Figure 1. Correlation diagram for ^{19}F-NMR chemical shifts.

References are listed on pages 1078–1086.

cate the NMR interpretation of fluorinated compounds. Fluorinated aromatics show complex splitting patterns due to proton–fluorine coupling throughout the ring system and to benzylic substituents. Solvent, temperature, and concentration of an NMR sample can alter fluorine chemical shifts and coupling constants by several parts per million or hertz.

Fluorohalomethanes

One-carbon fluorinated compounds are used as synthetic reagents and in many theoretical inquiries of structure and spectroscopy. Twenty fluorohalomethanes are listed in Table 1 [*56, 57, 58, 59, 60, 61, 62*]. Fluorocarbons 11B1, 22B1, and 31B1 and 12 of 15 iodofluoromethanes are not listed because of a lack of NMR data. Most of the $^{1}J_{CF}$ values are taken from the older literature [*57, 59*], obtained from

Table 1. NMR Data for Fluoromethanes

Fluorocarbon Number	*Compound*	*^{19}F-NMR Signal*	$^{2}J_{HF}$	$^{1}J_{CF}$	*Ref.*
11	CCl_3F	0.0 (s)	—	337	56
11B2	CBr_2ClF	+5.4 (s)	—		56
11B3	CBr_3F	+6.9 (s)	—	372	56, 57
12	CCl_2F_2	–6.8 (s)	—	324	56, 57
12B1	$CBrClF_2$	+0.1 (s)	—	342	56
12B2	CBr_2F_2	+6.5 (s)	—	357	56, 57
12I2	CF_2I_2	+18.6 (s)	—	378	58
13	$CClF_3$	–27.9 (s)	—	299	56, 59
13B1	$CBrF_3$	–17.9 (s)	—	324	56, 59
13I1	CF_3I	–5.0 (s)	—	345	56, 57
14	CF_4	–61.7 (s)	—	259	56, 57
21	$CHCl_2F$	–80.8 (d)	54	294	56, 57
21B1	CHBrClF	–80.4 (d)	55	304	60
	CDBrClF	–82.2 (t)	—[a]	303	60
21B2	$CHBr_2F$	–84.5 (d)	50		56
22	$CHClF_2$	–71.8 (d)	63		56
23	CHF_3	–78.3 (d)	80	274	56, 57
31	CH_2ClF	–168.7 (t)	49		61
31I1	CH_2FI	–191.3 (t)	49		62
32	CH_2F_2	–143.4 (t)	50	235	57, 59
41	CH_3F	–271.9 (q)	46	157	57

[a] $^{2}J_{DF}$ = 8 Hz.

the detection of ^{13}C satellites in the fluorine spectra. Proton and carbon chemical shifts would supplement this table, but presently are unavailable or scattered throughout the chemical literature.

Fluoroethanes, Fluoroethylenes, and Fluoroacrylates

A complete database of ^{19}F-NMR information on 66 fluorinated ethanes and ethylenes of the formula $C_2H_xCl_yF_z$ is available on floppy disk through Project Seraphim [*22*]. Table 2 presents only the nine nonchlorinated fluoroethanes, and Table 3 presents the six nonchlorinated fluoroethylenes from this more extensive list.

1,2-Difluoroethane is a relatively simple compound with a complex NMR signal. The preferred gauche conformation of the molecule generates magnetic nonequivalence and an AA′A″A‴XX′ spin system (Figure 2). This "simple" fluorinated compound has been the subject of many theoretical studies [*63, 64, 65*]. By comparison, its structural isomer, 1,1-difluoroethane, gives a clear doublet of quartets in the 19F-NMR spectrum (Figure 3).

Fluorine spectra of two ***fluoroethylenes,*** both from the Japan Halon compilation [*19*], are shown in Figures 4 and 5. The splitting pattern of fluoroethene (doublet of doublets of doublets) is clarified by using a branching display above the peaks, from which coupling constants can be measured easily. The AA′XX′ spectrum of 1,1-difluoroethene is also shown.

Table 2. NMR Data for Fluoroethanes [*22*]

Fluorocarbon Number	*Compound*		*^{19}F-NMR Signal*	$^2J_{HF}$	$^3J_{HF}$	$^3J_{FF}$
116	CF_3CF_3		–89.0 (s)	—	—	—
125	CF_3CHF_2		–90.2 (td)	—	3	3
		(CF_3)				
			–141.9 (dq)	53	—	3
		(CF_2)				
134	CHF_2CHF_2		–140.2 (dd)	52	5	—
134a	CF_3CH_2F•		–78.5 (dt)	—	8	16
		(CF_3)				
		(CF)	–241.0 (tq)	46	—	16
143	CHF_2CH_2F		–129.9 (ddt)	55	13	18
		(CF_2)				
		(CF)	–238.9 (ttd)	46	6	18
143a	CH_3CF_3		–64.5 (q)	—	13	—
152	CH_2FCH_2F		–225.9 (tt)	45	17	—
152a	CH_3CHF_2		–110.0 (dq)	57	21	—
161	CH_3CH_2F		–211.5 (tq)	49	27	—

Table 3. NMR Data for Fluoroethylenes [*22*]

Fluoro-carbon Number	*Compound*	*^{19}F-NMR Signal*	Coupling Constant					
			$^2J_{FF}$	$^2J_{HF}$	$^3J_{FF}$ *trans*	$^3J_{FF}$ *cis*	$^3J_{HF}$ *trans*	$^3J_{HF}$ *cis*
1114	$CF_2{=}CF_2$	−134.0 (s)	—	—	—	—	—	—
1123	F_a H C = C F_b F_c	(F_a) −125.7 (ddd)	87	—	119	—	—	4
		(F_b) −99.7 (ddd)	87	—	—	33	13	—
		(F_c) −205.0 (ddd)	—	71	119	33	—	—
1132	CHF=CHF							
	cis	−165.0 (ddd)	—	73	—	19	20	—
	trans	−186.3 (ddd)	—	74	125	—	—	4
1132a	$CH_2{=}CF_2$	−81.3 (dd)		—	—	—	34	1
1141	$CH_2{=}CHF$	−113.0 (ddd)	—	85	—	—	52	20

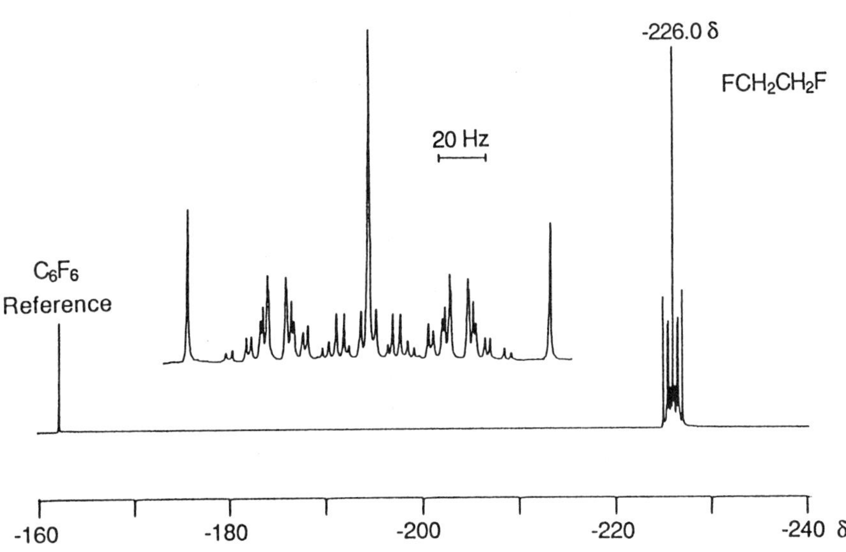

Figure 2. 75-MHz ^{19}F-NMR spectrum of 1,2-difluoroethane, an AA′A″A‴XX′ spin system.

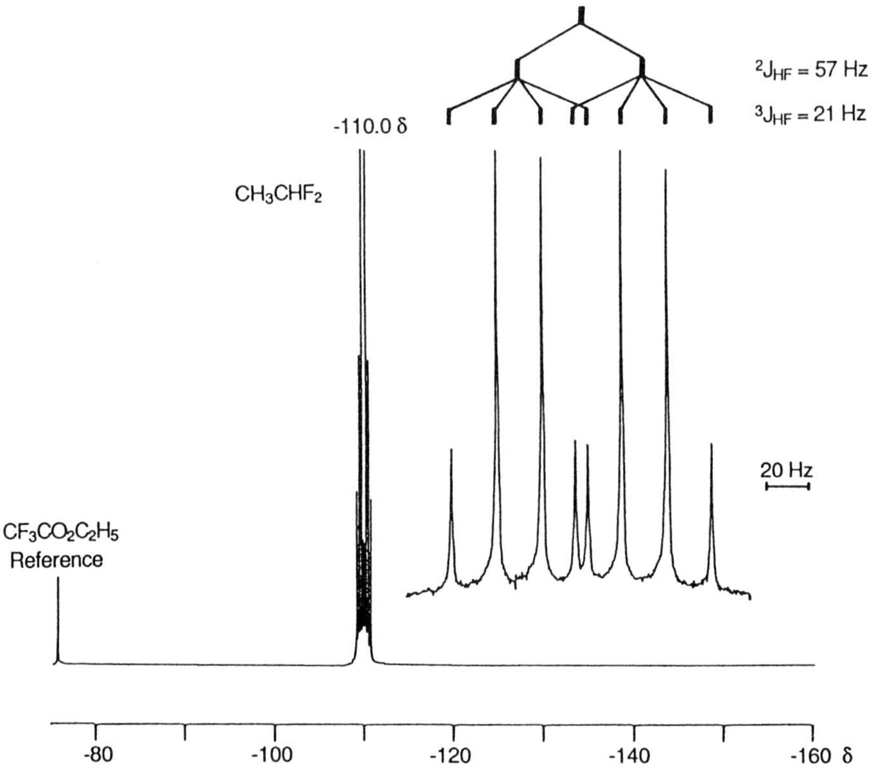

$^2J_{HF}$ = 57 Hz
$^3J_{HF}$ = 21 Hz
-110.0 δ
CH_3CHF_2
20 Hz
$CF_3CO_2C_2H_5$
Reference
-80 -100 -120 -140 -160 δ

Figure 3. 75-MHz ^{19}F-NMR spectrum of 1,1-difluoroethane.

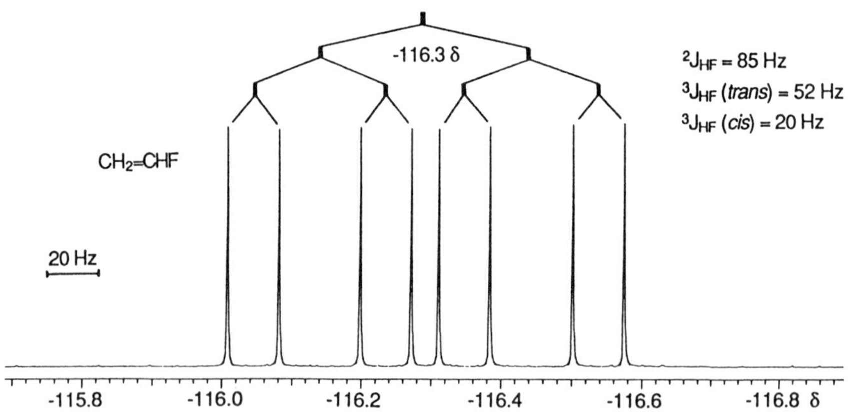

Figure 4. 282-MHz ^{19}F-NMR spectrum of fluoroethene. (Reproduced with permission of Japan Halon/Varian.)

References are listed on pages 1078–1086.

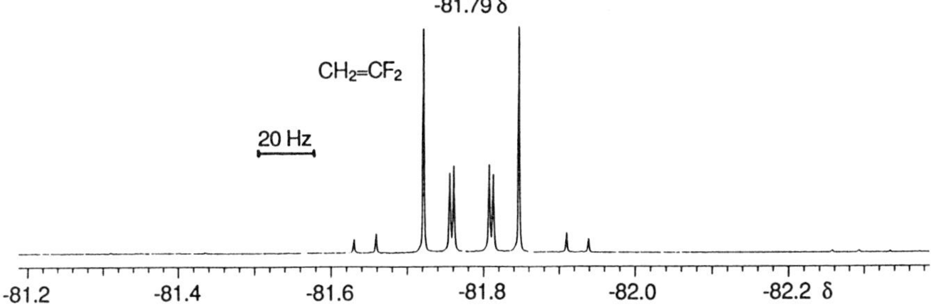

Figure 5. 282-MHz ^{19}F-NMR spectrum of 1,1-difluoroethene, an AA′XX′ spin system. (Reproduced with permission of Japan Halon/Varian.)

Four ***fluorinated acrylates*** are presented in Table 4 [*66, 67, 68, 69*] to supplement the vinyl fluoride data. Perfluoroacrylic acid [*70*] and methyl perfluoroacrylate [*69, 70*] are both reported in the literature. The *cis* and *trans* isomers of fluoroethylenes are readily distinguished by the magnitude of their coupling constants, with ^3J *trans* typically more than 3 times the magnitude of ^3J *cis*.

Table 4. NMR Data for Fluoroacrylates

		Coupling Constant						
Compound	^{19}*F-NMR Signal*	$^2J_{FF}$	$^2J_{HF}$	$^3J_{FF}$ *trans*	$^3J_{FF}$ *cis*	$^3J_{HF}$ *trans*	$^3J_{HF}$ *cis*	*Ref.*
trans $CHF{=}CHCO_2H$	−106.0 (dd)	—	80	—	—	—	14	66
$CH_2{=}CFCO_2C_2H_5$	−117.6 (dd)	—	—	—	—	44	12	67
$(F_a)(F_b)C{=}C(H)(CO_2C_2H_5)$								68
	(F_a) −70.0 (dd)	16	—	—	—	—	3	
	(F_b) −66.0 (dd)	16	—	—	—	22	—	
$(F_a)(F_b)C{=}C(F_c)(CO_2CH_3)$	(F_a) −89.4 (dd)	26	—	—	35	—	—	69
	(F_b) −99.7 (dd)	26	—	111	—	—	—	
	(F_b) 187.1 (dd)	—	—	111	35	—	—	

Aliphatic Monofluorides

Monofluorinated aliphatic compounds are presented in Table 5 [*19, 56, 71, 72, 74, 75, 77*], Table 6 [*56, 71, 72, 77, 78, 81, 83*], and Table 7 [*56, 71, 84*], divided by the degree of substitution at the C–F position. Many of these compounds are

Table 5. NMR Data for Primary Monofluorides

Compound RCH_2F	^{19}F*-NMR Signal*	$^2J_{HF}$	$^3J_{HF}$	*Ref.*
CH_3CH_2F	–211.5 (tq)	48	27	71
$CH_3CH_2CH_2F$	–218.6 (tt)	48	24	71
$CH_3(CH_2)_nF$[a]	–218.0 (tt)	48	24	56
$(CH_3)_2CHCH_2F$	–220.3 (td)	48	17	71
$(CH_3)_3CCH_2F$	–222.8 (t)		—	71
$C_6H_5CH_2F$	–206.3 (t)	49	—	71
$C_6H_5CH_2CH_2F$	–216.0 (tt)	47	22	71
$C_6H_5CH_2CH_2CH_2F$	–220.3 (tt)	47	25	72
FCH_2CO_2Na	–218.9 (t)	48	—	56
$FCH_2CO_2C_2H_5$	–230.2 (t)	47	—	56
$FCH_2COC_6H_5$	–226.1 (t)	48	—	74
FCH_2COCH_3	–225.7 (tq)[b]	48	—	56
FCH_2CONH_2	–222.8 (td)[b]	48	—	56
$FCH_2CH_2NH_3Cl$	–226.8 (tt)	46	24	56
FCH_2CH_2OH	–226.6 (tt)	48	30	56
$FCH_2CH_2OCH_3$	–223.1 (tt)	49	31	71
FCH_2CH_2SH	–212.8 (tt)	50	21	75
FCH_2CN	–232.0 (t)	46	—	56
$CH_2{=}CHCH_2F$	–216.0 (t)	48	—	71
$HC{\equiv}CCH_2F$	–217.5 (td)[c]	48	—	71
$BrCH_2CH_2F$	–212.4 (tt)	47	20	19
$BrCH_2CH_2CH_2F$	–228.8 (m)	45		77
$C_6H_5CH(CH_2F)CO_2H$	–217.0 (td)	46	14	56

[a] $n > 2$.
[b] $^4J_{HF} = 4$.
[c] $^4J_{HF} = 6$.

readily available through newer methodologies of selective fluorination. Almost half have been reported by Weigert [*71*], and most can be synthesized from the corresponding alcohols by using diethylaminosulfur trifluoride (DAST). Fluorodecarboxylation of carboxylic acids using xenon difluoride is one of several alternative routes.

Alipathic monofluorides have a well-defined chemical shift range of –206 to –232 ppm for *primary monofluorides,* –165 to –195 ppm for *secondary monofluorides* (excluding cyclopropyl fluoride), and –128 to –170 ppm for *tertiary*

Table 6. NMR Data for Secondary Monofluorides

Compound RR′CHF	*^{19}F-NMR Signal*	*$^{2}J_{HF}$*	*$^{3}J_{HF}$*	*Ref.*
$(CH_3)_2CHF$	–165.2 (d sept)	73	31	71, 77
$CH_3CH_2CHFCH_3$	–173.2			71
CH_2=$CHCHFCH_3$	–171.6			71
$CH_3CH_2CH_2CHFCH_3$	–172.8			71
$(CH_3CH_2)_2CHF$	–172.9	46		77
cyclo-C_3H_5F	–212.6 (ddt)	65	22/10 *cis/trans*	78
cyclo-$C_6H_{11}F$	(ax) –186.0 (eq) –165.5			71
cyclo-$C_8H_{15}F$	–158.2			81
1-Fluoroindane	–159.9			71
$C_6H_5CHFCH_3$	–166.8 (dq)	48	24	71
$C_6H_5CH_2CHFCH_3$	–170.8 (dddq)	48	24/24/19[a]	72
$C_6H_5CHFCH_2CH_3$	–176.0 (ddd)	48	24/20[a]	72
$CHF(CO_2C_2H_5)_2$	–195.3 (d)	48	—	56
$CH_3CH_2CHFCO_2C_2H_5$	–193.7 (dt)	49	25	56
$NH_2CH_2CHFCO_2H$	–190.2 (ddd)	50	26/20[a]	56
$C_6H_5CHFCO_2H$	–181.3 (d)	47	—	56
$C_6H_5CHFCO_2C_2H_5$	–174.0 (d)	47	—	83
$C_6H_5CHFCH_2OH$	–180.0 (ddd)	47	23/19[a]	83

[a]Multiple values arise from 2 or more nonequivalent, vicinal hydrogens.

monofluorides (excluding one bicyclic compound) in the examples provided. Geminal coupling constants are typically 45–50 Hz, whereas vicinal J_{HF} values show greater variation, with 20 Hz or more common. Figures 6 and 7 present spectra of two primary monofluorides, fluoroacetamide and fluoroacetone, highlighted to show $^{4}J_{HF}$ coupling across the carbonyl. Only one of the N–H protons is coupled to fluorine because of restricted rotation around the amide C–N bond [*73*].

Fluorocyclohexane has been studied by ^{19}F-NMR spectroscopy to gain insights into inversion of the six-membered ring, not only in solution [*71*] but also in the gaseous [*79*] and solid [*80*] states. Similarly, fluorocyclooctane [*81*] and 1-fluorocyclooctene [*82*] conformations have been studied by variable-temperature fluorine NMR. Numerous bicyclic systems with a fluorine substituent at the bridgehead position have been reported [*84*]. These rigid systems are advantageous for the generation of substituent chemical shift (SCS) data (discussed later).

Table 7. NMR Data for Tertiary Monofluorides

Compound RR′R″CF	*^{19}F-NMR Signal*	*$^{3}J_{HF}$*	*Ref.*
$(CH_3)_3CF$	–130.8 (m)		71
$CH_3CH_2CF(CH_3)_2$	–139.2 (m)		71
$(CH_3CH_2)_2CFCH_3$	–149.1 (m)		71
$(CH_3CH_2)_3CF$	–156.2 (m)		71
$C_6H_5CF(CH_3)_2$	–137.9 (sept)	22	56
$(C_6H_5)_2CFCH_3$	–135.0 (q)	23	71
$(C_6H_5)_3CF$	–126.7 (s)	—	71
1-Fluoro-1-methyl-cyclohexane	–152.3 (m)		71
1-Fluorobicyclo-[2.2.1]heptane	–182.0		84
1-Fluorobicyclo-[2.2.2]octane	–147.6		84
1-Fluoroadamantane	–127.8		84
Fluoxymesterone[a]	–169.8 (ddm)	10/30 *cis/trans*	56
$HC{\equiv}CCF(CH_3)_2$	–129.3 (m)		71
$CH_3CF(CO_2C_2H_5)_2$	–157.8 (q)	22	56
$C_6H_5CF(CO_2C_2H_5)_2$	–161.0 (s)	—	56

[a]

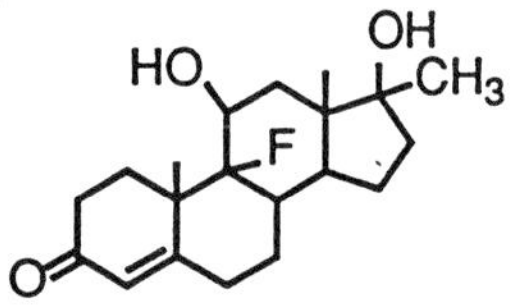

Geminal Difluorides

Geminal difluorinated aliphatics are presented in Table 8 [*22, 56, 71, 75, 76, 77*] and Table 9 [*71, 76, 87, 88, 89*], divided by degree of substitution at the F–C–F position. The fewer examples in this category reflects the limited synthetic accessibility to these compounds. Like the monofluorinated aliphatics, many of these difluorides have been reported by Weigert [*71*] and can be synthesized from the corresponding aldehyde or ketone by using DAST.

Difluoromethyl moieties have a well-defined chemical shift range from –110 to –129 ppm and couplings of approximately 57 Hz ($^{2}J_{HF}$) and 17 Hz ($^{3}J_{HF}$). Fluorine NMR signals for difluoromethylene moieties range from –85 to –111 ppm (excluding difluorocyclopropyls), with vicinal J_{HF} values similar to the primary cases. More complicated splitting patterns arise from *diastereotopic fluorines* when a chiral center is present in the geminal difluoride. Diastereotopic fluorines may differ in chemical

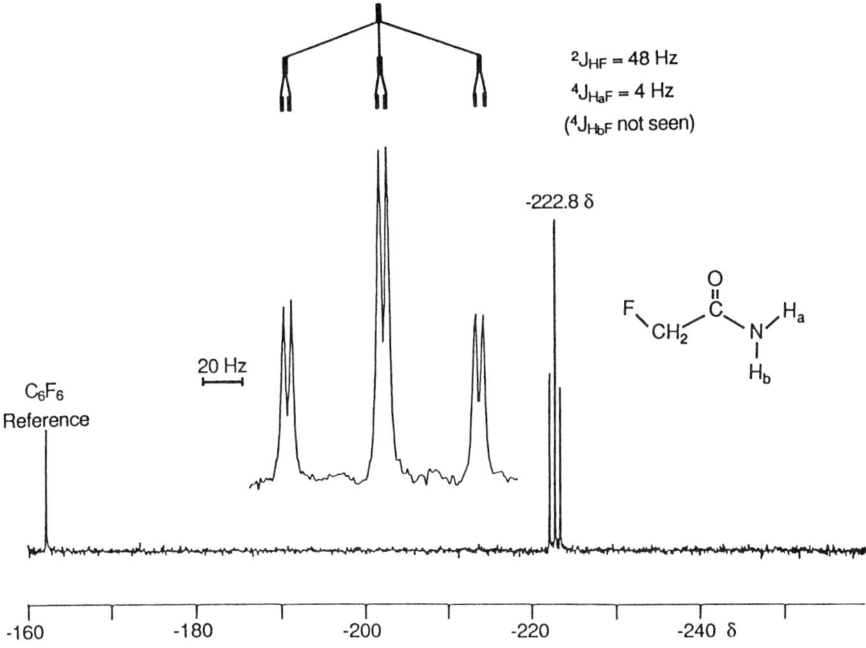

Figure 6. 75-MHz ^{19}F-NMR spectrum of fluoroacetamide showing $^4J_{HF}$ coupling of 4 Hz.

shifts by several parts per million, with $^2J_{FF}$ values over 400 Hz. Figure 8 presents one such diastereotopic arrangement. Two compounds of biological interest with diastereotopic fluorines merit brief mention. α-Difluoromethylornithine (*1*) [*85*] shows two nonequivalent fluorine signals (two doublets of doublets) at –128.1 and –133.2 ppm ($^2J_{FF}$ = 276 Hz; $^2J_{HF}$ = 55 Hz). An unsymmetrical 3′–5′ nucleic acid dimer has been reported [*86*] containing a difluoromethylphosphonate linkage with diastereotopic fluorines due to the chiral phosphorus atom (*2*). Its fluorine NMR consists of two ddd's with coupling constants of $^2J_{FF}$ = 462 Hz, $^2J_{PF}$ = 93 Hz, and $^2J_{HF}$ = 48 Hz.

1 H_2N ... CO_2H H_2N CHF_2

2 O P RO OR' CHF_2

Comparison of Mono-, Di-, and Trifluoromethyl Moieties

Table 10 [*22, 56, 77, 90, 91, 92, 93, 94*] compares NMR signals of monofluoromethyl, difluoromethyl, and trifluoromethyl moieties attached to six different atoms (C, Si, N, P, O, and S). An approximate halving of the magnitude of the

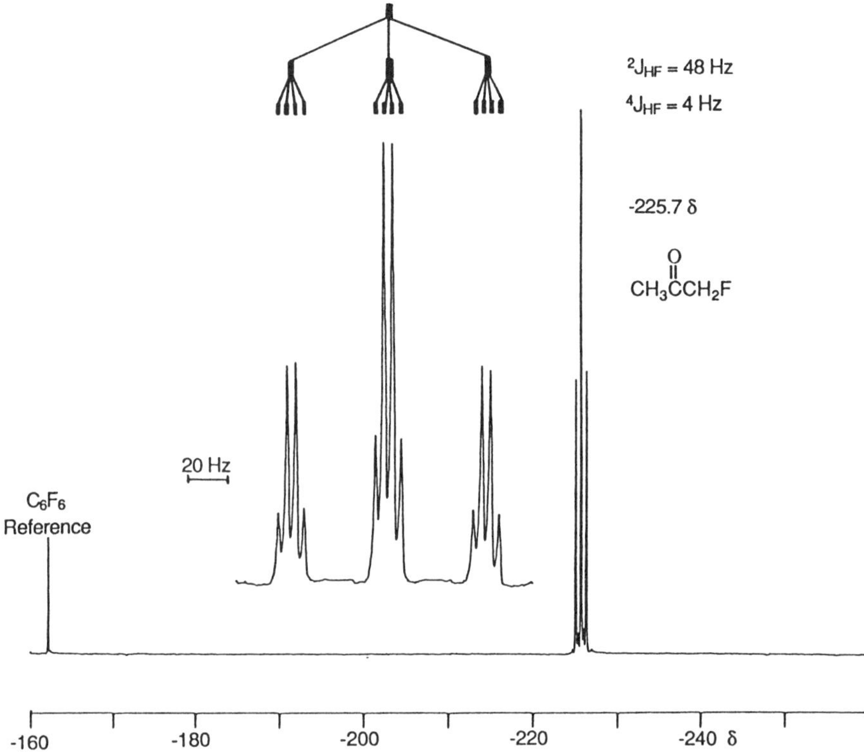

Figure 7. 75-MHz ^{19}F-NMR spectrum of fluoroacetone showing $^4J_{HF}$ coupling of 4 Hz.

Table 8. NMR Data for Primary Geminal Difluorides

Compound $RCHF_2$	*^{19}F-NMR Signal*	*$^2J_{HF}$*	*$^3J_{HF}$*	*Ref.*
CH_3CHF_2	–110.0 (dq)	57	21	22
$CH_3CH_2CHF_2$	–120.0 (dt)	57	17	71
$CH_3CH_2CH_2CHF_2$	–116.8 (dt)	57	17	71
$(CH_3)_2CHCHF_2$	–126.7 (ddm)[a]	57	15	71
$(CH_3)_3CCHF_2$	–128.6 (dm)[a]	57	—	71
$C_6H_5CHF_2$	–111.2 (d)	57	—	71
$C_6H_5CH_2CHF_2$	–115.1 (dt)	56	17	77
CHF_2CH_2SH	–116.9	55	15	75
CHF_2CN	–119.8 (d)	52	—	76
CHF_2CO_2H	–128.2 (d)	53	—	56

[a] $^4J_{HF}$ = 1.

Table 9. NMR Data for Secondary Geminal Difluorides

Compound RR′CF$_2$	*^{19}F-NMR Signal*	*$^{3}J_{HF}$*	*Ref.*
$(CH_3)_2CF_2$	–84.5 (sept)	18	71
$CH_3CH_2CF_2CH_3$	–93.3 (qt)	18/15	71
$(CH_3CH_2)_2CF_2$	–102.4 (quint)	21	71
$(CH_3)_3CCF_2CH_3$	–102.2 (q)	20	71
$C_6H_5CF_2CH_3$	–87.7 (q)	18	71
$C_6H_5CH_2CF_2CH_3$	–89.6		71
$C_6H_5CF_2CH_2CH_3$	–97.9		71
$(C_6H_5)_2CF_2$	–88.5 (s)	—	71
CH_3CF_2CN	–85.4	18	76
$CF_2(CO_2C_2H_5)_2$	–109.0 (s)	—	87
1,1-Difluoro-2,2,3,3-tetramethyl-cyclopropane	–148.9 (m)	—	88
1,1-Difluoro-cyclohexane	(ax) –104 (eq) –88		71
2,2-Difluoro-cyclohexanone	–111		87
7,7-difluoro-norcarane	–129.2 (dt)[a] –159.0 (d)[a]	*cis* 14 *trans* 0	88
1,1-Difluoro-cyclooctane	–89.0 (quint)	15	89

chemical shift occurs as one proceeds from CH_2F to CHF_2 to CF_3. Many novel fluoromethylated silanes have been synthesized in the past two years [*90, 96, 97, 98, 99, 100*]. Fluoromethyl moieties attached to nitrogen are specifically represented by the substituted pyrazole ***3*** . Herbicidal activities of this mono-, di-, and trifluoromethyl series of compounds were studied [*91*].

3 R = CH_2F, CHF_2, CF_3

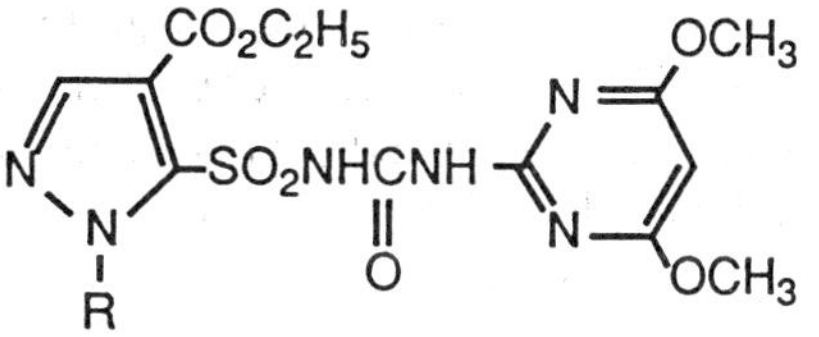

Burton [*92*] published extensive NMR information for ***fluorinated quaternary phosphonium salts*** that are used as Wittig reagents. Clear data–structure correlations allow NMR information on other compounds to be predicted. The *trifluoromethyl analogue* is a recent addition to this series [*93*]. The fluorine NMR of

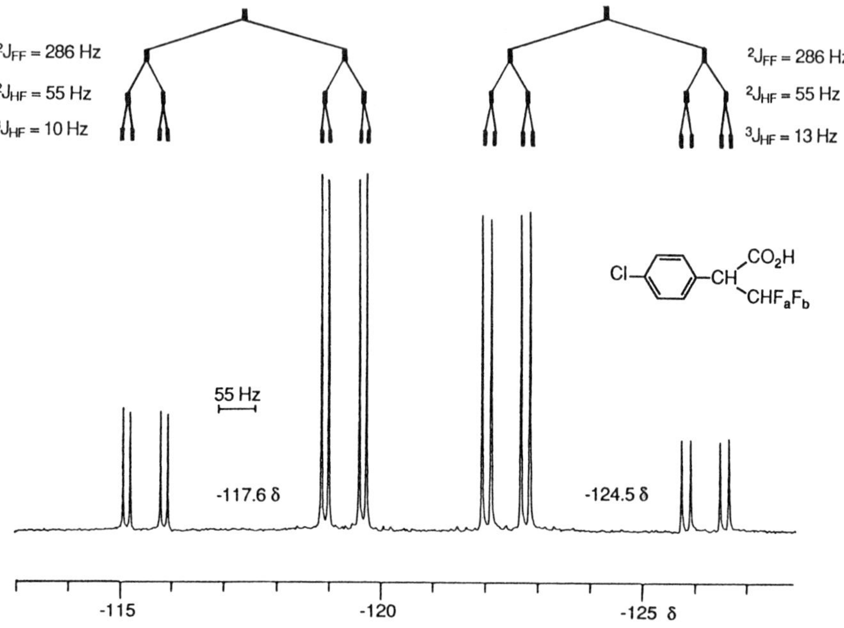

Figure 8. 75-MHz ^{19}F-NMR spectrum of 3,3-difluoro-2-(4-chlorophenyl)-propanoic acid showing diastereotopic geminal fluorine signals.

fluoromethyltriphenylphosphonium iodide appears in Figure 9 and clearly shows the magnitude of geminal phosphorus–fluorine coupling.

Comparison of Fluoromethyl Moieties Attached to Carbon and Silicon

Multinuclear NMR data for homologous series of ***fluoromethylated malonates*** [72] and ***trimethylsilanes*** [97] are compiled in Table 11. In both series, fluoromethyl attachment is to a quaternary site. These compounds are readily synthesized using fluorohalomethanes to incorporate the final fluoromethyl moiety. All the malonates, except diethyl methyltrifluoromethylmalonate (***4***) [*93*], are isopropyl-substituted diethyl esters [*72*]. The silane data, with the exception of trimethyltrifluoromethylsilane (***5***) [*95*], are from reference 97. Chemical shift data are very comparable, with the malonates having higher proton and fluorine chemical shifts but slightly lower carbon values. The magnitudes of $^1J_{CF}$ and $^2J_{HF}$ coupling are similar for both sets of compounds.

4 $CH_3-C(CO_2C_2H_5)_2-CF_3$ **5** $(CH_3)_3Si-CF_3$

Table 10. NMR Data for Variously Attached Mono-, Di- and Trifluoromethyl Moieties

Compound	*^{1}H-NMR Signal*	*^{19}F-NMR Signal*	*$^2J_{HF}$*	*Ref.*
CH_3CH_2F		−211.5 (tq)[a]	49	22
CH_3CHF_2		−110.0 (dq)[b]	57	22
CH_3CF_3	—	−64.5 (q)[c]	—	22
H_3SiCH_2F	4.61 (dq)	−265.0 (tq)[d]	47	90
H_3SiCHF_2	6.14 (tq)	−128.9 (dq)[e]	46	90
H_3SiCF_3	—	−55.2 (q)[f]	—	90
R_2NCH_2F[g]	6.54 (d)	−166.1 (t)	50	91
R_2NCHF_2[g]	8.23 (t)	−95.3 (d)	57	91
R_2NCF_3[g]	—	−51.6 (s)	—	91
$[(C_6H_5)_3PCH_2F]Br$		−242.9 (dt)[h]	44	92
$[(C_6H_5)_3PCHF_2]Br$		−126.2 (dd)[i]	47	92
$[(C_6H_5)_3PCF_3]OTf$	—	−58.3 (d)[j]	—	93
$C_6H_5OCH_2F$	5.10 (d)	−148.7 (t)	54	77
$C_6H_5OCHF_2$	6.30 (t)	−76.0 (d)	78	94
$C_6H_5OCF_3$	—	−58.3 (s)	—	56
$C_6H_5SCHF_2$	6.80 (t)	−90.0 (d)	60	94
$C_6H_5SCF_3$	—	−43.2 (s)	—	56

[a] $^3J_{HF} = 27$.
[b] $^3J_{HF} = 21$.
[c] $^3J_{HF} = 13$.
[d] $^3J_{HF} = 15$.
[e] $^3J_{HF} = 11$.
[f] $^3J_{HF} = 8$.

[g] R_2N =

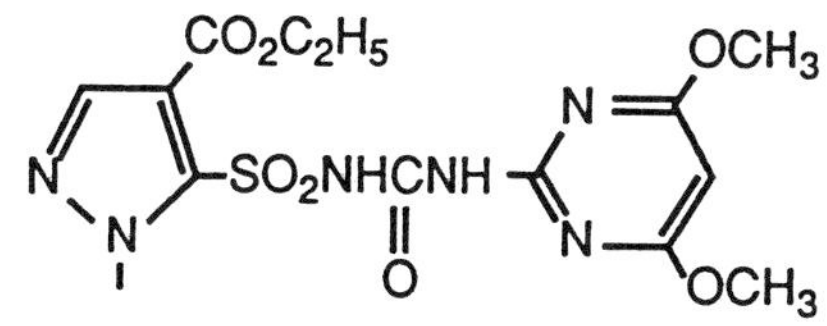

[h] $^2J_{PF} = 58$.
[i] $^2J_{PF} = 80$.
[j] $^2J_{PF} = 90$.

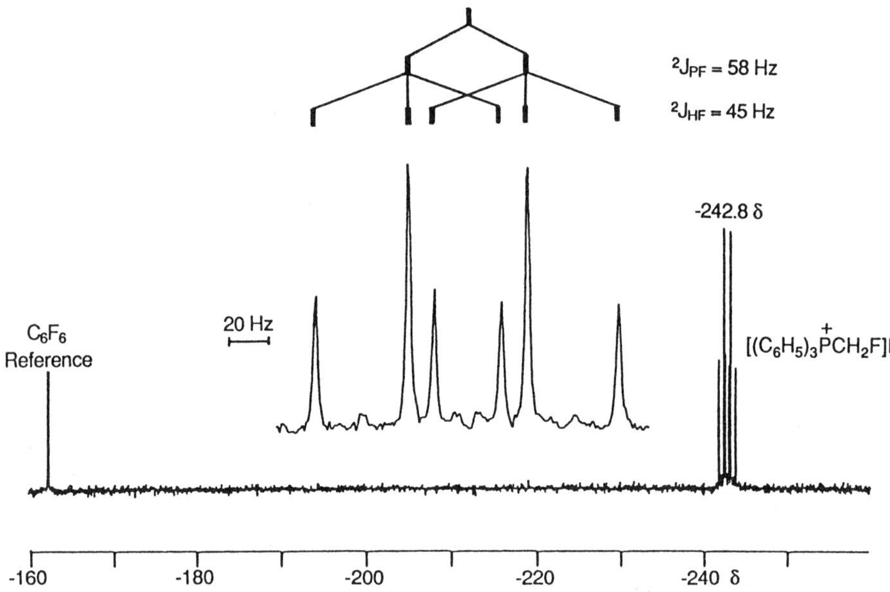

Figure 9. 75-MHz ^{19}F-NMR spectrum of fluoromethyltriphenylphosphonium iodide.

Table 11. Multinuclear Spectral Data for Fluoromethylated Malonates [72] and Fluoromethylated Trimethylsilanes [97]

Compound	*^{1}H-NMR Signal*	*^{13}C-NMR Signal^a*	*^{19}F-NMR Signal*	*^{29}Si-NMR Signal^a*	*$^1J_{CF}$*	*$^2J_{HF}$*	*$^2J_{SiF}$*
Mal-CH_2F	4.8 (d)	83 (d)	–227 (t)	—	175	47	—
Mal-CHBrF	6.9 (d)	93 (d)	–142 (d)	—	262	47	—
Mal-CHClF	6.7 (d)	101 (d)	–141 (d)	—	252	48	—
Mal-CHF_2	6.3 (t)	116 (t)	–125 (d)	—	248	54	—
Mal-CF_3	—		–71 (s)	—		—	—
Mal-$CBrF_2$	—	120 (t)	–44 (s)	—	316	—	—
TMS-CH_2F	4.4 (d)		–277 (t)	–2 (d)		47	27
TMS-CHClF	6.1 (d)	102 (d)	–170 (d)	5 (d)	254	47	27
TMS-CHF_2	5.7 (t)	122 (t)	–140 (d)	0 (t)	254	47	29
TMS-CF_3	—		–67 (s)	4 (q)		—	38
TMS-$CClF_2$	—	135 (t)	–63 (s)	10 (t)	327	—	32
TMS-$CBrF_2$	—	132 (t)	–58 (s)	12 (t)	338	—	29

[a]Fluorine coupled/proton decoupled.

Trifluoromethyl Compounds

Table 12 [*19, 56, 101, 102, 103*], Table 13 [*56, 73, 104*], and Table 14 [*56, 105, 106*] cover six common CF_3-containing subunits of general interest and utility: trifluoroethanes, trifluoroacetates, trifluoroacetamides, trifluoromethanesulfonates, trifluoromethyl ketones, and trifluoromethyl carbinols. The majority of these data were acquired under uniform conditions (10% solutions in $CDCl_3$ at room temperature [*56*]) to provide for a valid comparison of ^{19}F-NMR values within a narrow range of approximately 10 ppm. Most of these compounds are common,

Table 12. NMR Data for 1,1,1-Trifluoroethanes

Compound CF_3CH_2R	*^{1}H-NMR Signal*	*^{19}F-NMR Signal*	*$^3J_{HF}$*	*Ref.*
CF_3CH_2F	4.61 (dq)	–77.8 (dt)	8	19
CF_3CH_2Cl	3.83 (q)	–71.9 (t)	8	56
CF_3CH_2Br	3.66 (q)	–69.3 (t)	9	56
CF_3CH_2I	3.55 (q)	–65.8 (t)	10	56
CF_3CH_3	1.88 (q)	–61.1 (q)	13	56
$CF_3CH_2CF_3$		–63.6	10	101
CF_3CH_2CN		–66.8	9	101
$CF_3CH_2NH_2$		–76.2 (t)	9	56
$(CF_3CH_2)_3N$	3.46 (q)	–71.5 (t)	9	102
$CF_3CH_2NHNH_2$		–71.0 (t)		56
CF_3CH_2R[a]		–69.6 (t)	10	56
CF_3CH_2OH	3.93 (q)	–77.8 (t)	9	56
$CF_3CH_2O_2CCF_3$		–74.4 (t)	8	56
$CF_3CH_2O_2CCH{=}CH_2$	4.53 (q)	–74.2 (t)	8	19
CF_3CH_2OTs[b]	4.35 (q)	–74.4 (t)	8	56
CF_3CH_2OTf[c]		–74.7 (t)	7	56
$CF_3CH_2OCH_2CF_3$		–75.2 (t)	8	56
$CF_3CH_2OCH{=}CH_2$	4.05 (q)	–74.6 (t)	8	56
$CF_3CH_2OC_6H_5$		–74.6 (t)		103
$(CF_3CH_2O)_3P$		–75.9 (td)[d]	8	56

[a]R = –N⟨piperazine ring⟩N–CH_3

[b]Ts = *p*-toluenesulfonyl (Tosyl).

[c]Tf = trifluoromethanesulfonyl (Triflyl).

[d] $^4J_{PF}$ = 4.

Table 13. NMR Data for Trifluoroacetyl and Trifluoromethanesulfonyl Derivatives [56]

Trifluoroacetyl Derivative	*^{19}F-NMR Signal*	*Trifluoromethanesulfonyl (Triflyl) Derivative*	*^{19}F-NMR Signal*
CF_3COCl	–75.7 (s)	CF_3SO_2Cl	–75.5 (s)
CF_3CO_2H	–76.2 (s)	CF_3SO_3H	–80.0 (s)
$CF_3CO_2CH_3$	–75.6 (s)	$CF_3SO_3CH_3$	–74.8 (s)
$CF_3CO_2C_2H_5$	–75.8 (s)	$CF_3SO_3C_2H_5$	–75.6 (s)
$CF_3CO_2CH_2CF_3$	–75.2 (s)	$CF_3SO_3CH_2CF_3$	–74.5 (s) –74.7 (t)[a]
$CF_3CO_2C_6H_5$	–75.3 (s)	$CF_3SO_3CF_3$[b]	–75.3 (s) –52.9 (s)[c]
$(CF_3CO)_2O$	–75.5 (s)	$(CF_3SO_2)_2O$	–72.2 (s)
$CF_3CO_2Si(CH_3)_3$	–76.7 (s)	$CF_3SO_3Si(CH_3)_3$	–77.5 (s)
$(CF_3CO_2)_2IC_6H_5$	–74.0 (s)	$(CF_3SO_3)_2Si(t\text{-}C_4H_9)_2$	–75.9 (s)
$CF_3COSC_2H_5$	–76.1 (s)	$(CF_3SO_2)_2NC_6H_5$	–71.1 (s)
CF_3CONH_2	–76.8 (s)[d]		
$CF_3CONHCH_3$	–76.5 (s)		
$CF_3CONHCH_2CO_2H$	–76.4 (s)		
$(CF_3CO)_2NCH_3$	–71.3 (s)		
1-(CF_3CO)-imidazole	–71.3 (s)		

[a] $^3J_{HF}$ = 7 Hz for CF_3CH_2.
[b] Reference 104.
[c] Fluorine chemical shift for CF_3O.
[d] Reference 73 reports a doublet, $^4J_{HF}$ = 2 Hz, for this sample in acetone.

Table 14. NMR Data for Trifluoromethyl Ketones and Corresponding Trifluoromethyl Carbinols

Trifluoromethyl Ketone	*^{19}F-NMR Signal for CF_3COR*	*^{19}F-NMR Signal for $CF_3CH(OH)R$*	*$^3J_{HF}$*	*Ref.*
CF_3COCH_3	–80.5 (s)	–82.4 (d)	8	56, 105
CF_3COCF_3	–84.1 (s)	–75.1 (d)	6	105
$CF_3COCH_2COCH_3$	–77.3 (s)			56
$CF_3COCH_2COCF_3$	–77.0 (s)			56
$CF_3COCH_2CO_2C_2H_5$	–75.1 (s)	–80.3 (d)	6	56
$CF_3COC_6H_5$	–71.9 (s)	–77.8 (d)	7	56, 106
9-(CF_3CO)-anthracene	–76.5 (s)	–74.7 (d)	8	56,106
3-(CF_3CO)-camphor	–70.7 (s)			56
3-(CF_3CO)-indole	–73.0 (s)	–78.3 (d)	6	56
$CF_3CO_2C_2H_5$	–75.8 (s)	–84.4 (d)	4	56

References are listed on pages 1078–1086.

commercially available reagents with nondescriptive singlets or doublets in fluorine NMR.

The ^{19}F-NMR signals of most ***1,1,1-trifluoroethanes*** (CF_3CH_2R) appear as triplets in the region from –66 to –78 ppm. Proton–fluorine three-bond coupling is typically 8–10 Hz. Data for 20 compounds in this category are given in Table 12.

The ^{19}F-NMR spectra of ***trifluoroacetates, trifluoroacetamides, and trifluoromethanesulfonates*** show sharp singlets with chemical shifts from -71 to -80 ppm (Table 13). Although the chemical shift differences are small, these reagents are used as fluorine-tagging agents in the analysis of mixtures by ^{19}F-NMR spectroscopy or in sample derivatization for enhanced detection by liquid chromatography, gas chromatography, or gas chromatography–mass spectrometry (*see Analytical Methods*).

Trifluoromethyl ketones and related ***trifluoromethyl carbinols*** are presented for comparison in Table 14. Biochemical interest in trifluoromethyl ketones as enzyme inhibitors is based on their ability to form relatively stable hydrates and thereby act as transition state analogues of substrates of hydrolytic enzymes. The final entry in this table gives the chemical shift for the commercially available trifluoroacetaldehyde ethyl hemiacetal and shows a pronounced upfield shift (from –75.8 to –84.4 ppm) for the carbonyl-to-carbinol transformation. A –8.6-ppm shift (–84.1 to –92.7 ppm) is also seen for the conversion of hexafluoroacetone to its hydrate (Table 15).

Table 15 [*8, 19, 56, 59, 93, 107, 108, 109, 110, 111, 112, 113, 114*] contains NMR data for a variety of trifluoromethyl compounds that do not fit into the more common categories already covered. Most carbon-attached trifluoromethyl (CF_3–C) ^{19}F-NMR signals fall in the 25-ppm range from –60 to –85 ppm, a notable exception being trifluoromethylacetylenes, with signals at approximately –50 ppm.

Trifluoromethylated compounds in which the *CF_3 is directly bonded to an atom other than carbon* are considered in Table 16 [*57, 59, 95, 109, 115, 116, 117, 118, 119, 120, 121, 122, 123, 124, 125*]. These heteroatom-containing compounds are of interest to both organic and inorganic chemists and have been studied by NMR for numerous reasons:

1. Basic consideration of stability–reactivity. Some of these act as trifluoromethylating agents: $Te(CF_3)_2$, $(CH_3)_3SiCF_3$.
2. Novel structure and bonding, such as $Xe(CF_3)_2$, with the first reported Xe–C bond. Although this compound is unstable, its structure is supported by ample NMR evidence [*119*].
3. Study of CF_3– as a pseudohalogen in ligand-exchange studies.
4. Characterization of organometallic agents through multinuclear analysis by correlating fluorine chemical shifts and fluorine–heteroatom coupling constants.

Trends in these NMR data clearly stand out as one proceeds down the periodic table, such as for the oxygen family: fluorine signals are shifted downfield and $^1J_{CF}$

Table 15. NMR Data for Miscellaneous Trifluoromethyl Compounds

Compound	^{19}*F-NMR Signal*	$^{3}J_{HF}$	*Ref.*
$CF_3CHBrCl$	–76.4 (d)	5	56
$CF_3CH_2CH_2Br$	–66.3 (t)	10	19
$CF_3CH_2CH_2I$	–67.6 (t)	10	19
$CF_3CH_2CH_2OH$	–64.8 (t)	11	19
$CF_3CH_2CO_2H$	–61.0 (t)	10	107
$CF_3CH_2CH_2CO_2C_2H_5$	–67.6 (t)	10	56
$CF_3CH{=}CH_2$	–66.9 (d)	4	19
$CF_3C{\equiv}CH$	–52.1 (d)[a]	—	19
$CF_3C{\equiv}CCF_3$	–53.5 (s)	—	56
$CF_3C{\equiv}CC_6H_5$	–49.7 (s)	—	93
$(CF_3)_3CH$	–64.2 (d)	7	56
$(CF_3)_3CCH_3$	–68.2 (s)	—	108
$(CF_3)_4C$	–62.3 (s)	—	109
CF_3OH	–54.5		110
CF_3OCF_3	–62.0 (s)	—	59
$(CF_3O)_4C$	–59.0 (s)	—	111
CF_3NH_2	–48.9 (t)	10	112
$H_2NCH(CF_3)CO_2H$	–70.9 (d)	8	56
$(CH_3)_2CHCH(CF_3)CO_2H$	–65.4 (d)	8	56
$(CH_3)_3CCH(CF_3)CO_2H$	–61.8 (d)	9	56
$CF_3C(OH)_2CF_3$	–92.7		8
$CF_3C(OH)(CH_3)CN$	–82.5 (s)	—	56
$CF_3C(OCH_3)(C_6H_5)CN$	–78.1 (s)	—	56
[structure: 2-(CF_3)cyclohexanone; labels O, CF_3]	–69.2 (d)	9	93

values increase. Some values are also given in Table 16 for fluorine–heteroatom two-bond couplings. In a similar manner (but not covered here), fluoride, perfluoroacyl, *p*-fluorophenyl, pentafluorophenyl, and trifluoro-β-diketonate ligands have been used in fluorine NMR studies of heteroatomic structures (for example, in the spectroscopy of fluoroalkyl and fluoroaryl derivatives of transition metals [7]).

Table 15—Continued

Compound	*^{19}F-NMR Signal*	*$^{3}J_{HF}$*	*Ref.*
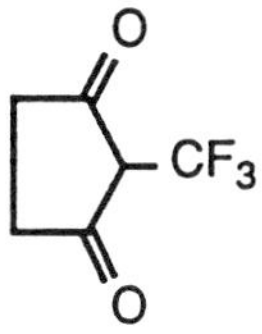	–61.0 (m)		113
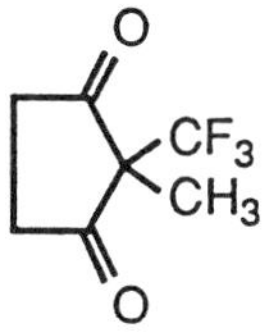	–69.8 (s)	—	93
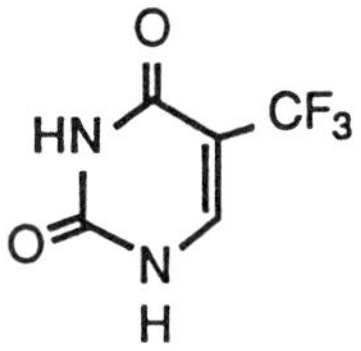	–61.7 (s)	—	19
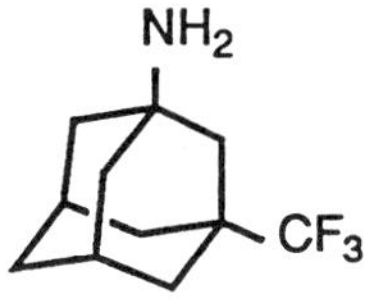	–82.5 (s)	—	114

[a] $^{4}J_{HF}$ = 3 Hz.

The final group of trifluoromethylated compounds to be considered is ***benzotrifluorides***. The $ArCF_3$ unit has a narrow range of ^{19}F-NMR chemical shifts, because its electronic environment remains largely unaffected by other substituents on the aromatic ring. Data in Table 17 [*56*] clearly show the lack of influence *meta* and *para* substituents have on the aromatic CF_3 singlet: *meta*-substituted benzotrifluorides, –63.4 ± 0.3 ppm; *para*-substituted benzotrifluorides, –63 ± 1 ppm. Substituents *ortho* to CF_3 lead to downfield shifts of the fluorine signal (compared with unsubstituted benzotrifluoride) of up to 7 ppm. A J_{HF} coupling of 2 Hz is seen in *o*-trifluoromethylbenzaldehyde and *o*-trifluoromethylcinnamic acid, whereas a J_{FF} coupling of 12.5 Hz is seen in *o*-fluorobenzotrifluoride. Several NMR references on benzotrifluorides are available [*126, 127, 128, 129*].

Table 16. Fluorine–Heteroatom NMR Interactions in the Trifluoromethyl Moiety

Atom	*Compound*	*^{19}F-NMR Signal*	*Het-NMR Signal*	*$^{1}J_{CF}$*	*$^{2}J_{Het/F}$*	*Ref.*
^{11}B	$B(CF_3)_3 \cdot NH(C_2H_5)_2$	–62	–11		30	115
^{13}C	$C(CF_3)_4$	–62	—			109
^{29}Si	$(CH_3)_2Si(CF_3)_2$	–64	–2		44	95
Ge	$Ge(CF_3)_4$	–46	—		—	116
$^{117/119}Sn$	$Sn(CF_3)_4$	–42	—		531 (^{119}Sn) 507 (^{117}Sn)	116
^{15}N	$N(CF_3)_3$	–59	—	269		59
^{31}P	$P(CF_3)_3$	–52	–2	316	85	117
As	$As(CF_3)_3$	–48	—	340	—	117
Sb	$Sb(CF_3)_3$	–41	—	359	—	117
Bi	$Bi(CF_3)_3$	–30	—		—	118
^{17}O	$O(CF_3)_2$	–62	—	265	—	59
^{33}S	$S(CF_3)_2$	–39	—	309	—	57
^{77}Se	$Se(CF_3)_2$	–33	718	331	—	57
^{125}Te	$Te(CF_3)_2$	–22	1368	352	32	117
^{129}Xe	$Xe(CF_3)_2$	–10	—		1940	119
Mo	$Mo(CO)_3CpCF_3$[a]	+12	—	367	—	121
^{183}W	$W(CO)_3CpCF_3$[a]	+9	–3177	352	42	121
Mn	$Mn(CO)_5CF_3$	+10	—		—	122
Fe	$Fe(CO)_4(CF_3)_2$	+3	—	364	—	123
Co	$Co(CO)_4(CF_3)_2$	+11	—		—	122
^{195}Pt	$PtBipy(CF_3)_2$[b]	–24	—		741	124
Cu	$[CuCF_3]$	–29	—		—	125
Zn	$Zn(CF_3)_2$	–43	—		—	120
$^{111/113}Cd$	$Cd(CF_3)_2$	–36	410 (^{113}Cd)		228 (^{111}Cd) 392 (^{113}Cd)	120
^{199}Hg	$Hg(CF_3)_2$	–34	—	356	1264	57

[a] Cp = cyclopentadienyl.
[b] Bipy = bipyridyl.

Table 17. NMR Data for Benzotrifluorides [*56*]

R in $RC_6H_4CF_3$	*^{19}F-NMR Signal* ortho	meta	para
H	–63.2	–63.2	–63.2
Br	–63.2	–63.4	–63.3
Cl	–63.1	–63.4	–63.1
I	–63.3	–63.4	—
NO_2	–60.5	–63.4	–63.6
F	–61.9 (d)	–63.4	–62.5
OH	–61.4	–63.3	–62.0
NH_2	–63.2	–63.4	–61.7
COCl	–63.1	–63.5	–63.0
CN	–62.4	–63.7	–64.0
NCO	–63.1	–63.5	–63.0
CF_3	—	–63.6	–63.8
CHO	–56.1 (d)	–63.4	–63.7
CO_2H	–59.9	–63.4	–63.8
CH_2Br	—	–63.2	–63.2
CH_2OH	–60.5	–63.2	–63.0
CH_2NH_2	–60.1	–63.1	–62.9
CH_2CN	–61.3	–63.3	–63.3
$COCH_3$	–58.7	–63.3	–63.6
CH_2CO_2H	–60.4	–63.2	–63.1
$CH{=}CHCO_2H$	–59.4 (d)	–63.4	–63.5
COC_6H_5	–58.5	–63.2	–63.5

Aromatic Fluorides

NMR data on the most common fluorinated aromatics are presented in three categories. Monofluorophenyls and pentafluorophenyls have found the greatest application in synthesis and spectroscopy. Data for polyfluorinated benzenes are also presented.

Table 18 [*56*] presents ^{19}F chemical shift data for 82 commercially available compounds, ***monofluorinated benzenes*** with 28 different substituents positioned *ortho, meta,* and *para.* The *ortho*-substituted fluorobenzenes have the widest chemical shift range, from –94 to –142 ppm, whereas *meta*-substituted compounds are tightly grouped in a 5-ppm range from –110 to –115 ppm. Most of these signals are unresolved multiplets at magnetic field strengths below 100 MHz. The signals

Table 18. NMR Data for Monosubstituted Fluorobenzenes [*56*]

	Fluorine Chemical Shift			*Substituent Chemical Shift (SCS)*		
Substituent	*ortho*	*meta*	*para*	*ortho*	*meta*	*para*
F	–138.9	–110.5	–120.1	–25.4	+3.0	–6.6
Cl	–116.0	–111.4	–116.5	–2.5	+2.1	–3.0
Br	–107.7	–111.1	–115.8	+5.8	+2.4	–2.3
I	–94.0	–111.0	–114.7	+19.5	+2.5	–1.2
NO_2	–118.3	–109.6	–102.7	–4.8	+3.9	+10.8
OH	–141.6	–112.2	–124.6	–28.1	+1.3	–11.1
NH_2	–135.9	–113.7	–127.3	–22.4	–0.2	–13.8
COCl	–108.7	–111.3	–101.4	+4.8	+2.2	+12.1
CN	–107.0	–110.2	–103.0	+6.5	+3.3	+10.5
NCO	–122.8	–111.7	–116.4	–9.3	+1.8	–2.9
CF_3	–115.0	–111.3	–108.1	–1.5	+2.2	+5.4
CHO	–122.4	–111.9	–103.0	–8.9	+1.6	+10.5
CO_2H	–108.6	–112.5	–104.6	+4.9	+1.0	+8.9
CH_2Br	–117.5	–113.0	–113.5	–4.0	+0.5	0
CH_2Cl	–118.3	–113.0	–113.8	–4.8	+0.5	–0.3
$CONH_2$	–113.4	–112.2	–107.9	+0.1	+1.3	+5.6
CH_3	–118.2	–114.8	–119.2	–4.7	–1.3	–5.7
OCH_3	–136.1	–112.3	–124.9	–22.6	+1.2	–11.4
CH_2OH	–120.3	–113.6	–115.5	–6.8	–0.1	–2.0
CH_2NH_2	–120.5	–113.7	–116.7	–7.0	–0.2	–3.2
CH_2CN	–117.8	–112.2	–114.5	–4.3	+1.3	–1.0
$CH{=}CH_2$	—	–114.1	–114.9	—	–0.6	–1.4
$COCH_3$	–110.3	–112.5	–106.0	+3.2	+1.0	+7.5
CH_2CO_2H	–117.5	–113.5	–115.8	–4.0	0	–2.3
$NHCOCH_3$	–131.5	–112.2	–118.6	–18.0	+1.3	–5.1
CO_2Et	–110.2	–113.1	–106.6	+3.3	+0.4	+6.9
C_6H_5	–118.6	–113.2	–116.4	–5.1	+0.3	–2.9
COC_6H_5	–111.7	—	–106.6	+1.8	—	+6.9

of *para*-substituted fluorobenzenes appear in the range from –101 to –127 ppm, usually as clear triplets of triplets. Compilations by Dungan and Van Wazer [*12*] and Mooney [*13*] include hundreds of fluorinated aromatics and still serve as the most convenient sources of these NMR values.

Another important component of Table 18 is substituent chemical shift (SCS) data derived for each of the 82 entries. The SCS is simply the difference in ^{19}F-NMR chemical shifts of the substituted compounds and that of unsubstituted fluorobenzene (–113.5 ppm in $CDCl_3$). These values numerically represent the influence a substituent has on the shielding or deshielding of the fluorine nucleus and depend upon substituent position (*o, m,* or *p*). Fluorine chemical shifts can be predicted for polysubstituted fluorobenzene systems simply by adding the SCS value of each substituent to a base value of –113.5 ppm.

Fifolt [*130*] reported this ***chemical shift additivity*** method for fluorobenzenes in two deuterated solvents: d_6-acetone and d_6-dimethyl sulfoxide (DMSO). Close correlations between experimental and calculated fluorine chemical shifts were seen for 50 compounds. Data presented in Table 18 result from measurements in deuterochloroform as the solvent [*56*]. Fluorine chemical shifts calculated by this additivity method can be used to predict approximate values for any substituted benzene with one or more fluorines and any combination of the substituents, to differentiate structural isomers of multisubstituted fluorobenzenes [fluoronitrotoluenes (***6, 7,*** and ***8***) in example 1, Table 19], and to assign chemical shifts of multiple fluorines in the same compound [2,5-difluoroaniline (***9***) in example 2, Table 19]. Calculated chemical shifts can be in error by more than 5 ppm (upfield) in some highly fluorinated systems, especially when one fluorine is *ortho* to two other fluorines. Still, the calculated values can be informative even in these cases [2,3,4,6-tetrafluorobromobenzene (***10***) in example 3, Table 19].

Table 20 [*56*] lists fluorine chemical shifts for the 12 fluorobenzenes in the series $C_6H_{6-x}F_x$. Complete proton, carbon, and fluorine spectral analyses of this series have been published [*131, 132, 133, 134*]. Approximate coupling constants are $^3J_{HF}$ = 10 Hz, $^4J_{HF}$ = 5 Hz, $^5J_{HF}$ = 1 Hz, $^3J_{FF}$ = 20 Hz, $^4J_{FF}$ = 5 Hz, and $^5J_{FF}$ = 15 Hz.

Considerable interest in the NMR spectroscopy of ***pentafluorophenyl*** compounds has been shown [*135, 136, 137*]. Table 21 [*56*] gives data for more than 30 of these compounds. In most cases, the chemical shift increases upfield in the series *ortho, para,* and *meta,* with approximate respective values of –140, –150 and –160 ppm. At magnetic field strengths below 100 MHz, only the triplet or tripletlike splitting pattern for the *para*-fluorine signal can be readily interpreted. Fluorine–fluorine coupling throughout the ring and coupling with benzylic substituents make complete analyses difficult, but several cases have been reported, such as pentafluorothiophenol [*138*] and (pentafluorophenyl)sulfur fluorides [*139*] among others.

NMR studies of ***fluoronaphthalenes*** have included monofluorinated [*140, 141*] as well as polyfluorinated [*142, 143, 144, 145*] compounds. ***Fluorinated***

Table 19. Examples of Chemical Shift Additivity Method for Fluorinated Benzenes [*56*]

Example 1: Fluoronitrotoluene Isomers

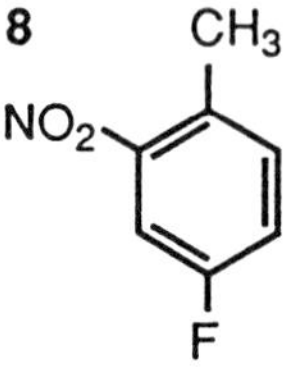

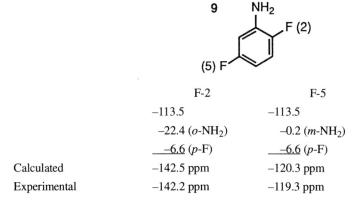

	6	7	8
	–113.5	–113.5	–113.5
	–4.7 (o-CH_3)	–4.8 (o-NO_2)	+3.9 (m-NO_2)
	+10.8 (p-NO_2)	–5.7 (p-CH_3)	–5.7 (p-CH_3)
Calculated	–107.4 ppm	–124.0 ppm	–115.3 ppm
Experimental	–106.7 ppm	–123.4 ppm	–114.2 ppm

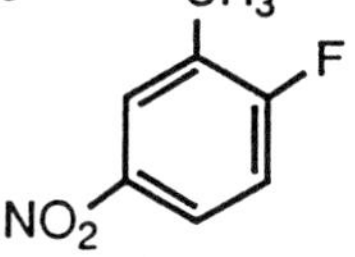

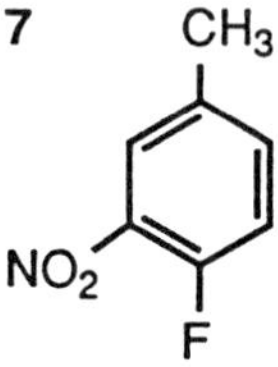

Example 2: 2,5-Difluoroaniline

9

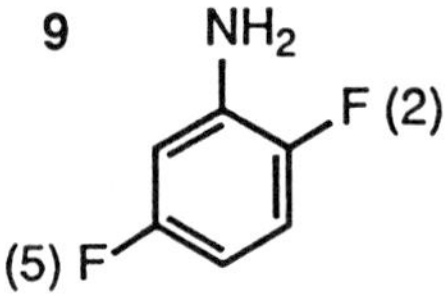

	F-2	F-5
	–113.5	–113.5
	–22.4 (o-NH_2)	–0.2 (m-NH_2)
	–6.6 (p-F)	–6.6 (p-F)
Calculated	–142.5 ppm	–120.3 ppm
Experimental	–142.2 ppm	–119.3 ppm

Example 3: 2,3,4,6-Tetrafluorobromobenzene

10

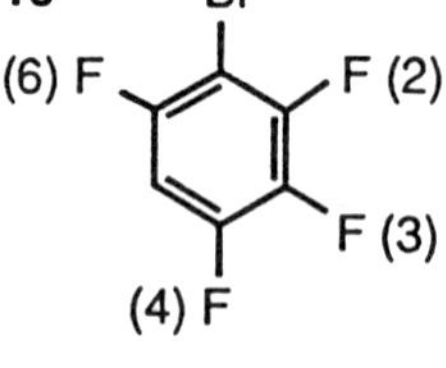

	F-2	F-3	F-4	F-5
Calculated	–127.1 ppm	–168.5 ppm	–135.2 ppm	–108.3 ppm
Experimental	–124.8 ppm	–162.3 ppm	–132.7 ppm	–109.5 ppm

Table 20. NMR Data for Fluorinated Benzenes [*56*]

Compound $C_6H_{6-x}F_x$	*Fluorine Position(s)*	*^{19}F-NMR Signal*
Fluorobenzene		–113.5 (ttd)
1,2-Difluorobenzene		–138.9 (m)
1,3-Difluorobenzene		–110.5 (m)
1,4-Difluorobenzene		–120.1 (tt)
1,2,3-Trifluorobenzene	1,3	–135.4
	2	–161.8
1,2,4-Trifluorobenzene	1	–143.8
	2	–133.8
	4	–116.0
1,3,5-Trifluorobenzene		–108.0
1,2,3,4-Tetrafluorobenzene	1,4	–139.9
	2,3	–156.4
1,2,3,5-Tetrafluorobenzene	1,3	–132.1
	2	–166.3
	5	–114.2
1,2,4,5-Tetrafluorobenzene		–139.7 (t)
Pentafluorobenzene	2,6	–138.9
	3,5	–162.3
	4	–153.8 (t)
Hexafluorobenzene		–162.2 (s)

pyridine is possibly the most analyzed heteroaromatic system, with various fluoropyridines studied by ^{15}N-NMR [*146, 147*] as well as ^{19}F-NMR [*148, 149, 150, 151, 152*] spectroscopy. Selectively fluorinated ***polycyclic aromatic hydrocarbons*** have been the subjects of long-range and through-space coupling studies by Sardella [*153, 154, 155*] (discussed later). Data for many fluorinated heteroaromatic systems including inter-ring coupling constants of quinolines have been reported [*156, 157, 158*].

Beyond Basic Structure Elucidation

Fluoride Ion

^{19}F-NMR chemical shifts reported for the fluoride ion vary considerably [*159, 160, 161, 162, 163*]. The electronic environment associated with a fluoride ion is much more exposed to the medium than a covalent C–F moiety. Chemical shifts

Table 21. NMR Data for Pentafluorophenyl Compounds [56]

	^{19}F-NMR Signal for Fluorine at Position(s)		
R in RC_6F_5	*2,6*	*3,5*	*4*
F	–162.2	–162.2	–162.2
Br	–132.6	–160.5	–154.5
Cl	–140.6	–161.2	–155.9
SO_2Cl	–134.0	–157.1	–140.0
I	–119.4	–159.7	–152.4
NO_2	–145.7	–158.2	–146.5
H	–138.9	–162.3	–153.8
OH	–164.0	–164.2	–168.7
SH	–137.1	–161.8	–158.4
NH_2	–162.7	–165.3	–173.6
$NHNH_2$	–158.6	–164.0	–167.6
COCl	–137.8	–159.2	–145.3
CN	–131.5	–158.1	–142.2
CF_3	–139.8	–159.8	–147.1
CHO	–144.6	–160.7	–143.8
CO_2H	–136.7	–160.4	–146.6
CH_2Br	–142.7	–161.8	–153.4
$CONH_2$	–140.2	–160.4	–150.2
CH_3	–143.8	–164.1	–159.4
OCH_3	–158.2	–164.5	–164.2
CH_2OH	–145.0	–162.5	–154.6
$CH{=}CH_2$	–143.9	–163.6	–156.6
$COCH_3$	–141.3	–160.7	–149.7
CO_2CH_3	–138.7	–161.0	–149.2
CH_2CO_2H	–142.7	–162.6	–155.2
$CH(OH)CH_3$	–145.1	–162.5	–156.1
$CH{=}CHCO_2H$	–139.5	–161.8	–150.6
$CH_2CO_2CH_3$	–142.8	–162.9	–155.8
C_6H_5	–137.9	–160.8	–150.3
SC_6F_5	–132.2	–160.2	–150.1
COC_6H_5	–140.3	–160.3	–150.9
$P(C_6F_5)_2$	–130.3	–159.8	–148.0

for fluoride ion are greatly influenced by solvent, concentration, and counterion, because the nature of the anion can range from that of a "naked fluoride" to a highly solvated species. Chemical shift values are listed in Table 22 [*162*].

Fluorinated Carbocations and Carbanions

^{19}F-NMR spectroscopy is ideal for detecting charged fluorinated intermediates and has been applied to the study of increasingly stable carbocation and carbanion species. Olah [*164, 165*] has generated stable fluorocarbocations in SbF_5/SO_2ClF at low temperatures. The relatively long-lived perfluoro-*tert*-butyl anion has been prepared as both the cesium and tris(dimethylamino)sulfonium (TAS) salts by several groups [*166, 167, 168*]. Chemical shifts of fluorinated carbocations and carbanions are listed in Table 23.

Table 22. Chemical Shifts of Fluoride Ion

Fluoride Source	*Solvent*	*Chemical Shift (ppm)*
$(CH_3)_4NF$	Acetonitrile	–73.2 (s)
$(CH_3)_4NF$	Methylene chloride	–97.0 (s)
$(CH_3)_4NF$	Ethanol	–136.7 (s)
NaF	Water	–119.2 (bs)
HF	Acetonitrile	–184.0 (s)

Table 23. Chemical Shifts of Fluorinated Carbanions and Carbocations

Compound	*^{19}F-NMR Chemical Shift (ppm)*	*Ref.*
$(CF_3)_3C^-Cs^+$	–46	167
$(CF_3)_3C^-TAS^{+\,a}$	–45	168
$(C_6H_5)_2CF^+SbF_6^-$	+11	164
$C_6H_5CF_2^+SbF_6^-$	+15	165

^{a} TAS, tris(dimethylamino)sulfonium.

Long-Range and Through-Space Coupling

The fluorine nucleus is subject to coupling interactions transmitted through many bonds when positioned within a conjugated system. These long-range couplings over five or more bonds can be detected as J_{HF}, J_{CF}, or J_{FF} splittings in proton, carbon, or fluorine NMR (a $^{14}J_{FF}$ has been reported [*169*]).

Long-range coupling information can provide insights into conformational preferences [*170*] and relative transmissive abilities of intervening atoms or moieties [*169*]. Long-range coupling data for fluoroanilines [*171*], fluoroanisoles [*172, 173, 174*], benzyl fluoride [*175*], bis(fluorophenyl)methane [*176*], bis(fluorophenyl) ether [*177*], polycyclic aromatic hydrocarbons [*153, 154, 155*], and heteroaromatic systems [*156, 157, 158*] have been reported.

Through-space coupling, an overlap interaction of nonbonded orbitals, has been studied extensively [*178, 179, 180, 181, 182*]. Intramolecular pairs of fluorine atoms crowded against each other exhibit large J_{FF} coupling constants, as in the case of substituted 1,8-difluoronaphthalenes (***11;*** J_{FF} = 59–75 Hz) [*178*]. Even larger coupling constants of approximately 170 Hz are seen in 4,5-difluorophenanthrene (***12***) [*179*]. Direct through-space coupling has also been measured for ^{15}N–^{19}F interactions, such as in naphthalenone oxime (***13;*** J_{NF} = 22.4 Hz) [*181*]. Most recently, indirect F–F coupling through an intervening phenyl group has been shown for 1,5,8-trifluoro-9,10-diphenylanthracene (***14;*** J_{FF} = 6.4 Hz) [*182*], in further support of the orbital overlap theory of through-space interactions.

11 **12**

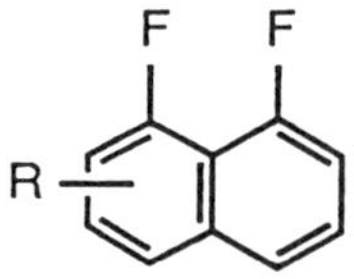

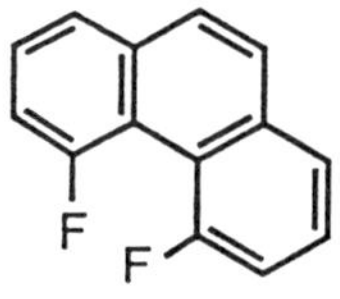

13 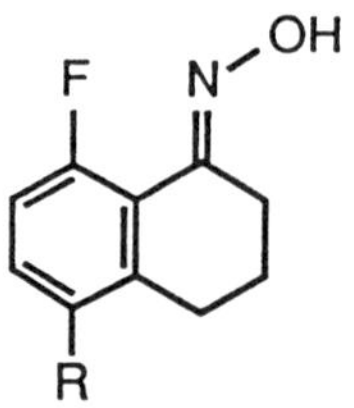 **14**

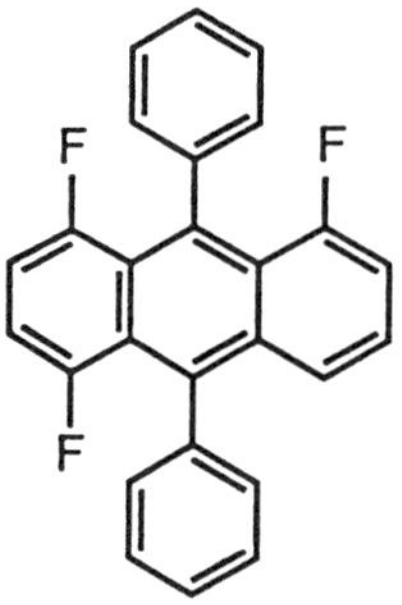

Substituent Chemical Shifts

Insights into the transmission of polar-substituent effects in saturated systems have been advanced in numerous papers [*183, 184, 185*]. Rigid model systems are used to eliminate conformational uncertainties, and ^{13}C- and/or ^{19}F-NMR spectroscopy is used to monitor chemical shift differences resulting from changes in substituents. Fluorinated bicyclo[2.2.1]heptane (***15***) [*183*], bicyclo[2.2.2]octane (***16***) [*184*], and more recently adamantane (***17***) [*185*] often serve as the model ring systems. Appropriately positioned fluoro or *p*-fluorophenyl moieties are influenced by through-bond and through-space sigma-electron delocalization.

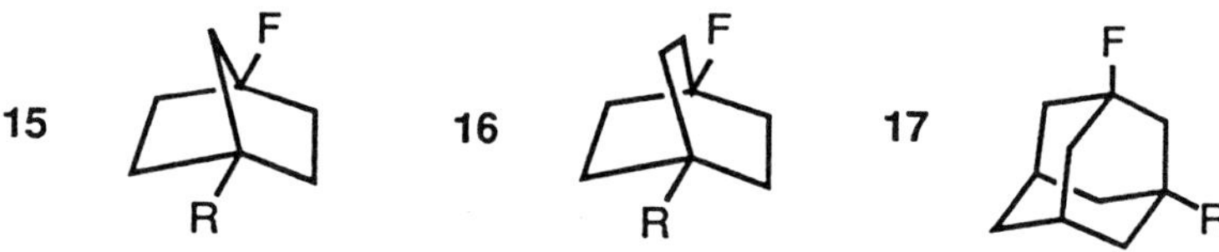

Analytical Methods

Analytical ***fluorine-tagging reagents*** provide for qualitative and quantitative identification of components in a mixture through the ^{19}F-NMR detection of chemical shift differences and peak integrations of the fluorinated derivatives. Tagging reagents [*103, 188, 189, 190*] are highly reactive compounds that convert samples to fluorinated derivatives quickly and quantitatively. Easy detection is then possible, because the fluorine nucleus has a high NMR sensitivity free of background interference and a wide chemical shift range to resolve fluorine signals arising from subtle differences in molecular structure.

Tagging reagents react with nucleophilic nitrogen-, oxygen-, and sulfur-containing functional groups to form fluorinated (usually acyl) derivatives. Included in this category are trifluoroacetic anhydride [*186*], pentafluoropropionic anhydride [*187*], trifluoroacetyl chloride [*188*], *p*-fluorobenzoyl chloride [*189, 190*], trifluoromethanesulfonyl chloride [*191*], hexafluoroacetone [*192*], and 2,2,2-trifluorodiazoethane [*103*]. Compound types tagged include alcohols, thiols, primary and secondary amines, phenols, sterols, and amino acids.

The reliability of these analytical methods may be questionable when chemical shift differences of derivatives are of the same magnitude as variations encountered from solvent, concentration, and temperature influences. Reported fluorine chemical shift ranges for trifluoroacetylated alcohols (1 ppm), *p*-fluorobenzoylated sterols (1 ppm), and *p*-fluorobenzoylated amino acids (0.5 ppm) are quite narrow, and correct interpretation of the fluorine NMR spectra of these derivatized mixtures requires strict adherence to standardized sampling procedure and NMR parameters.

^{19}F-NMR spectroscopy has been used to assess optical purity when ***chiral fluorinated derivatizing agents*** are used in the resolution of enantiomers. Diastereomeric derivatives give separate fluorine signals that can be identified and quantified for each optical isomer, although these chemical shift differences are usually small (tenths of a ppm). Chiral derivatizing agents include α-methoxy-α-trifluoromethylphenylacetic acid (***18;*** MTPA or Mosher's acid) [*193, 194, 195*] and related derivatives [*196, 197*], trifluoromethylarylcarbinols including 2,2,2-trifluoro-1-(9-anthryl)ethanol (***19;*** Pirkel's reagent) [*198, 199, 200*], derivatives of α-fluoroacetic acid [*201, 202*] such as 2-fluoro-2-phenylacetic acid (***20***) [*203*], and 1-amino-2-fluoro-2-phenylethane (***21***) [*204*]. Compounds resolved by this methodology include alcohols, thiols, amines, carboxylic acids, and amino acids. Similarly, chiral lanthanide shift reagents, derived from chiral fluoroalkyl β-diketones such as 3-(trifluoroacetyl)camphor (***22***) [*205, 206*], have been added to racemic mixtures to generate diasteromeric adducts with differentiable NMR

References are listed on pages 1078–1086.

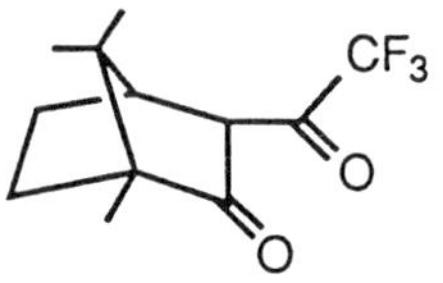

spectra. The preparation and properties of these and other chiral fluoroorganic compounds have been recently reviewed [*207*].

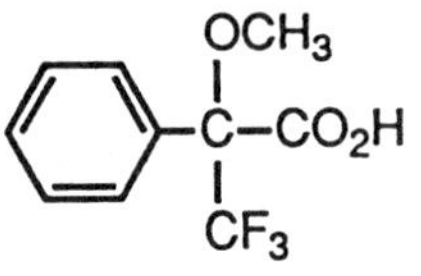

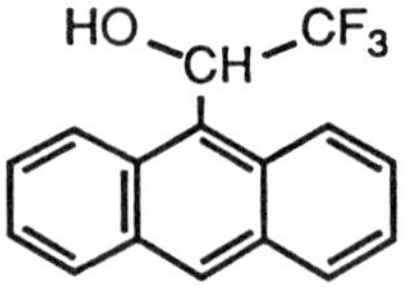

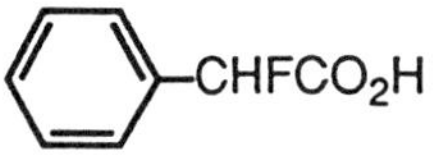

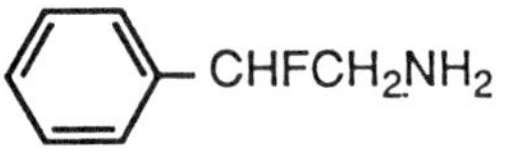

Biochemically Related Studies

Incorporation of fluorine into a biological substrate opens a spectral window for viewing biomolecular structure and dynamics in solution. With minimal background interference, fluorine NMR can provide clear spectral information for fluorine-containing macromolecules, in contrast to an indecipherable mass of signals from proton or carbon NMR. Whether the fluorinated unit is termed a probe, tag, marker, or reporter group, its function is the same: to act as a beacon of spectral information.

Several problems in basic methodologies have persisted over the 25 years since ^{19}F-NMR spectroscopy was first applied to a biochemical question. Most limiting is that of NMR sensitivity. High substrate concentrations, relative to the naturally occurring biological levels, are required for NMR detection. Although most NMR studies use millimolar and sometimes submillimolar concentrations, many biomolecules exist at micromolar or lower levels.

A restricted number of applicable "fluorine markers" may also be viewed as a limitation. The trifluoroacetyl moiety has been the most commonly used group, used in studies of amino acids, steroids, and carbohydrates. Enhanced ^{19}F-NMR detection should be possible with specifically designed agents of optimal biological and spectral properties.

Two excellent overviews of fluorine NMR in biochemistry have been published [*208, 209*]. The diverse studies can be organized by compound type or application. Applications include *detection of organofluorine compounds in human plasma* [*210*] and *magnetic resonance imaging (MRI)* of fluorine compounds (discussed below). ^{19}F-NMR spectroscopy has been used to follow the metabolic fates of fluorinated compounds such as fluoroanilines [*211, 212*], trifluoromethylanilines [*213*], fluoroacetates [*214*], chain-fluorinated polyamines [*85*], fluorodeoxyglucose [*215*], and 5′-deoxy-5-fluorouridine [*216*].

The following biochemically related ^{19}F-NMR studies of structure or function have been published: nucleic acid components (mainly 5-fluorouracil) [*86,*

216, 217], carbohydrates [*215, 218, 219, 220, 221*], amino acids and proteins [*222, 223, 224*], membrane lipids [*225, 226*], steroids (mainly corticosteroids) [*227, 228, 229, 230*], synthetic blood substitutes [*231, 232*], and both structure elucidation [*233, 234, 235, 236, 237*] and in vivo spectroscopy [*238, 239, 240*] of anesthetics. Figure 10 shows the fluorine spectrum of the anesthetic sevoflurane.

Fluorine MRI is a relatively new venture in the rapidly expanding area of noninvasive diagnostic imaging. Both proton and fluorine nuclei can be detected on standard instrumentation and generate no ionizing radiation. The primary advantage of fluorine over proton imaging is its negligible biological background. Fluorinated compounds have been used to follow dynamic processes such as blood oxygenation [*241*] and metabolism [*242, 243*].

Dozens of compounds have been used in in vivo fluorine NMR and MRI studies, chosen more for their commercial availability and established biochemistry than for ease of fluorine signal detection [*244*]. Among the more common of these are halothane and other fluorinated anesthetics [*245, 246*]; fluorodeoxyglucose [*242, 243*]; and perfluorinated synthetic blood substitutes, such as Fluosol [*246*], a mixture of perfluorotripropylamine and perfluorodecalin. Results have been limited by chemical shift effects (multiple signals spread over a wide spectral range) and long acquisition times.

Novel agents have been designed to optimize desirable biological and imaging properties. The perfluoro-*tert*-butyl moiety is being studied as a general fluorine

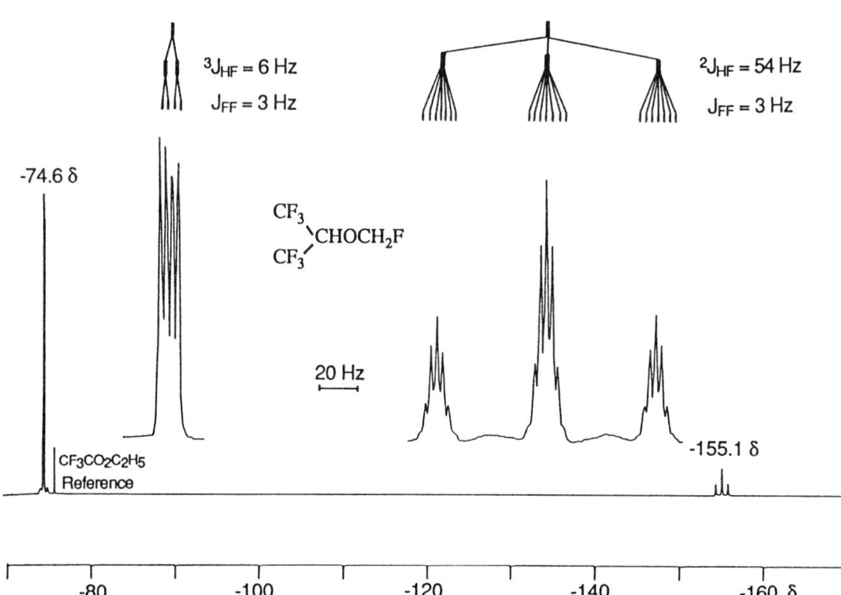

Figure 10. 75-MHz ^{19}F-NMR spectrum of sevoflurane showing long-range fluorine–fluorine coupling.

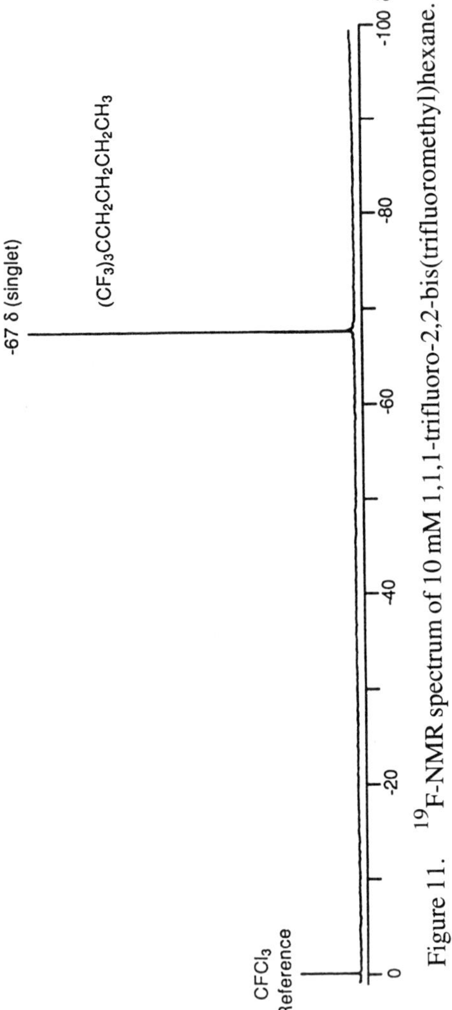

Figure 11. ^{19}F-NMR spectrum of 10 mM 1,1,1-trifluoro-2,2-bis(trifluoromethyl)hexane.

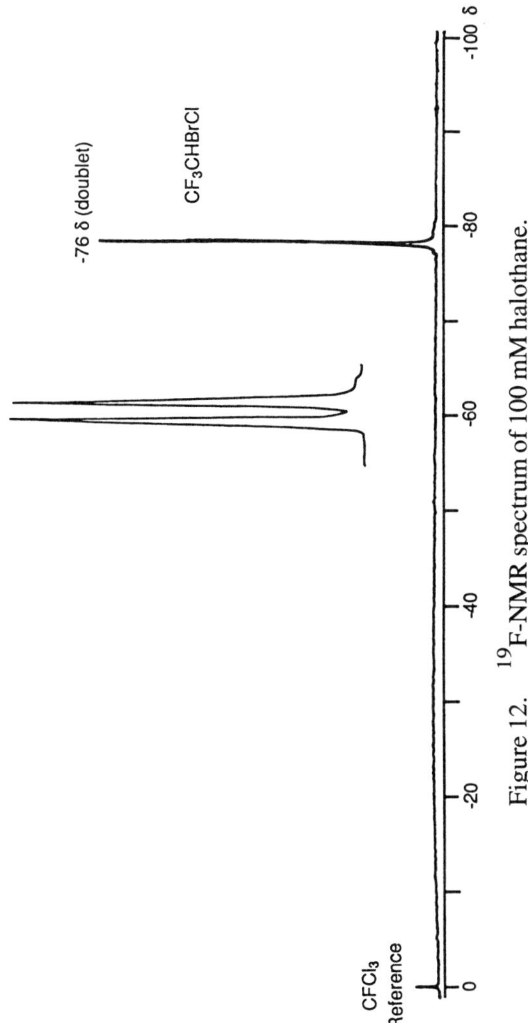

Figure 12. ^{19}F-NMR spectrum of 100 mM halothane.

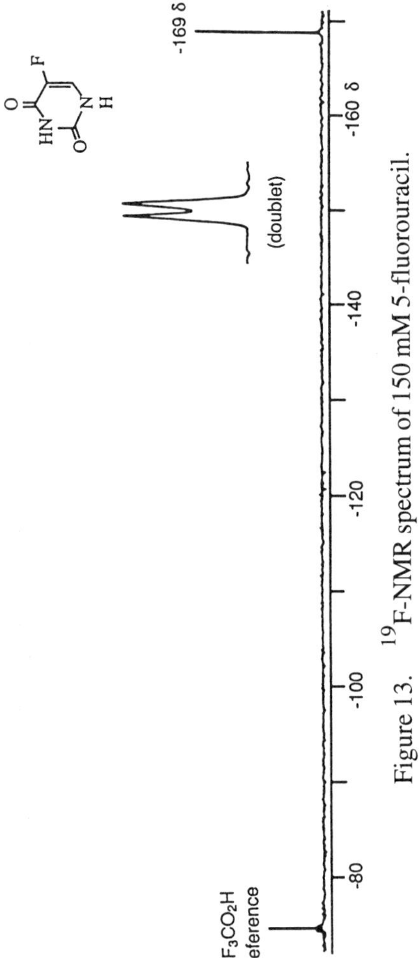

Figure 13. ^{19}F-NMR spectrum of 150 mM 5-fluorouracil.

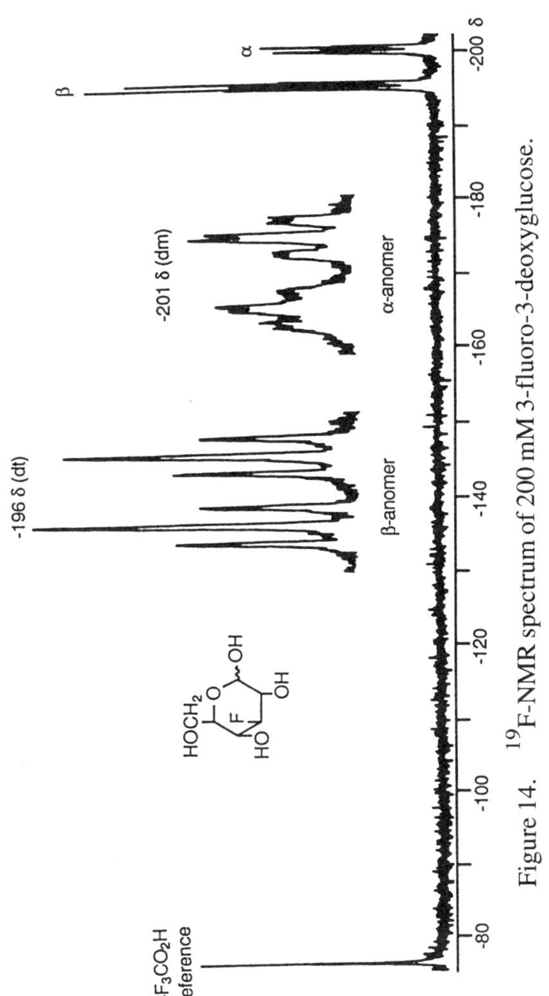

Figure 14. ^{19}F-NMR spectrum of 200 mM 3-fluoro-3-deoxyglucose.

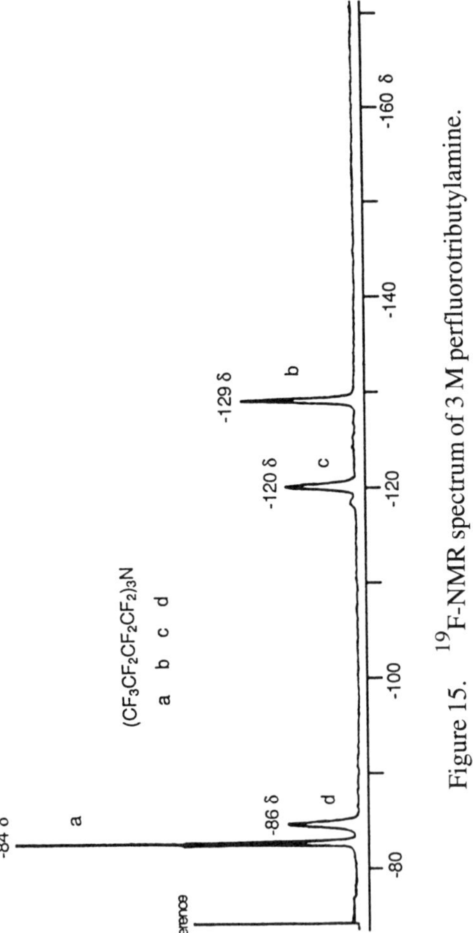

Figure 15. ^{19}F-NMR spectrum of 3 M perfluorotributylamine.

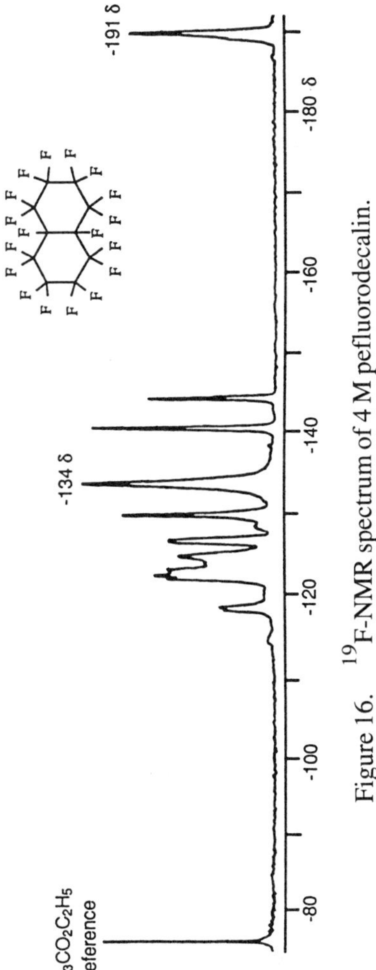

Figure 16. ^{19}F-NMR spectrum of 4 M pefluorodecalin.

NMR–MRI reporter group. Nine magnetically equivalent fluorine nuclei generate a single intense resonance for easy detection in spectroscopy or imaging. This signal is a sharp singlet not split by neighboring nuclei or spread over a wide frequency range. These spectral properties ensure a maximum signal-to-noise ratio in signal acquisition. Figures 11 through 16 compare the structures and spectra of six compounds in varying concentrations for detection by NMR–MRI. Submillimolar concentrations of perfluoro-*tert*-butyl fatty acids (***23***) have been detected in vivo [*247*]. Another reporter group currently under evaluation is 3,5-bis(trifluoromethyl)phenylacetate, which has been attached to 2-oleylglycerine (***24***) for in vivo studies [*248*].

23 $(CF_3)_3C(CH_2)_nCO_2H$

24 $CH_3(CH_2)_7CH{=}CH(CH_2)_7CO_2CH(CH_2O_2CCH_2C_6H_3(CF_3)_2)_2$

References for pages 1037–1078

1. Anon. *Pure Appl. Chem.* **1976,** *45,* 217.
2. Mooney, E. F.; Winson, P. H. *Annu. Rep. NMR Spectrosc.* **1968,** *1,* 243.
3. Jones, K.; Mooney, E. F. *Annu. Rep. NMR Spectrosc.* **1970,** *3,* 261.
4. Jones, K.; Mooney, E. F. *Annu. Rep. NMR Spectrosc.* **1971,** *4,* 391.
5. Fields, R. *Annu. Rep. NMR Spectrosc.* **1972,** *5A,* 99.
6. Cavalli, L. *Annu. Rep. NMR Spectrosc.* **1976,** *6B,* 43.
7. Fields, R. *Annu. Rep. NMR Spectrosc.* **1977,** *7,* 1.
8. Wray, V. *Annu. Rep. NMR Spectrosc.* **1980,** *10B,* 1.
9. Wray, V. *Annu. Rep. NMR Spectrosc.* **1983,** *14,* 1.
10. Emsley, J. W.; Phillips, L. *Prog. Nucl. Magn. Reson. Spectrosc.* **1971,** *7,* 1.
11. Emsley, J. W.; Phillips, L.; Wray, V. *Prog. Nucl. Magn. Reson. Spectrosc.* **1976,** *10,* 83.
12. Dungan, C. H.; Van Wazer, J. R. *Compilation of Reported ^{19}F NMR Chemical Shifts: 1951 to mid-1967;* Wiley–Interscience, New York, 1970.
13. Mooney, E. F. *An Introduction to 19F NMR Spectroscopy;* Heyden & Sons, London, 1970.
14. Morris, G. A. *Magn. Reson. Chem.* **1986,** *24,* 371.
15. *NMR of Newly Accessible Nuclei;* Laszlo, P., Ed.; Academic Press: New York, 1983.
16. Fields, R. *J. Fluorine Chem.* **1986,** *33,* 287.

17. Jameson, C. A. In *Multinuclear NMR;* Mason, J., Ed., Plenum Press: New York, 1987.
18. Bovey, F. A.; Jelinski, L.; Mirau, P. A. *Nuclear Magnetic Resonance Spectroscopy,* 2nd ed.; Academic Press: New York, 1988.
19. *Proton and Fluorine Nuclear Magnetic Resonance Spectral Data;* Varian Instruments/Japan Halon: Tokyo, 1988.
20. *FNMR, a ^{19}F Spectral Database;* Preston Scientific Ltd.: Manchester, England, 1986.
21. Weigert, F. J.; Karel, K. J. *J. Fluorine Chem.* **1987,** *37,* 125.
22. Weigert, F. J. *J. Fluorine Chem.* **1990,** *46,* 375. Database available on disk through Project Seraphim, Department of Chemistry, Eastern Michigan University, Ypsilanti, MI 48197.
23. Hamman, S.; Beguin, C. G.; Charlon, C.; Luu-Duc, C. *Org. Magn. Reson.* **1983,** *21,* 361.
24. Hamman, S.; Benaissa, T.; Beguin, C. G. *Magn. Reson. Chem.* **1988,** *26,* 621.
25. Shapiro, B. L.; Thomas, W. A.; McClanahan, J. L.; Johnson, M. D., Jr. *J. Magn. Reson.* **1973,** *11,* 355.
26. Salman, S. R.; Farrant, R. D.; Lindon, J. C. *Magn. Reson. Chem.* **1990,** *28,* 645.
27. Rudaya, M. N.; Saloutin, V. I.; Fomin, A. N.; Pashkevich, K. I. *Izv. Akad. Nauk SSSR* **1987,** 1412.
28. Ikeda, K.; Ogoma, Y.; Fujii, T.; Hachimori, A.; Kondo, Y.; Hayakawa, T.; Iwatsuki, M.; Akaike, T. *Polymer* **1990,** *31,* 344.
29. Matyjaszewski, K. *Wiad. Chem.* **1979,** *33,* 463; *Chem. Abstr.* **1979,** *91,* 193652z.
30. Fink, H. P.; Geiss, D.; Walenta, E.; Purz, H. J.; Schmolke, R.; Frigge, K. *Acta Polym.* **1989,** *40,* 186.
31. Lutringer, G.; Meurer, B.; Weill, G. *Makromol. Chem.* **1989,** *190,* 2815.
32. Majumdar, R. N.; Harwood, H. J. *Appl. Polym. Anal. Charact.* **1987,** 423.
33. A single issue of *Magnetic Resonance in Chemistry* (**1990,** *28*) is devoted exclusively to the NMR of solids, although there are no specific articles on solid-state fluorine NMR.
34. Cox, R. H.; Smith, S. L. *J. Magn. Reson.* **1969,** *1,* 432.
35. Abraham, R. J.; Cooper, M. A.; Siverns, T. M.; Swinton, P. F.; Weder, H. G. *Org. Magn. Reson.* **1974,** *6,* 331.
36. Muller, N. *J. Magn. Reson.* **1977,** *28,* 203.
37. Muller, N. *J. Pharm. Sci.* **1986,** *75,* 987.
38. Schaumburg, K. *J. Magn. Reson.* **1972,** *7,* 177.
39. Hansen, P. E.; Nicolaisen, F. M.; Schaumburg, K. *J. Am. Chem. Soc.* **1986,** *108,* 625.
40. Hansen, P. E.; Berg, A.; Schaumburg, K. *Magn. Reson. Chem.* **1987,** *25,* 508.

41. DeMarco, A.; Gatti, G. *J. Magn. Reson.* **1972,** *6,* 200.
42. Dahbi, A.; Hamman, S.; Beguin, C. G. *Magn. Reson. Chem.* **1986,** *24,* 337.
43. Kvicala, J.; Paleta, O.; Dedek, V. *J. Fluorine Chem.* **1989,** *43,* 155.
44. Cavalli, L. *Org. Magn. Reson.* **1970,** *2,* 233.
45. Camps, F.; Coll, J.; Fabrias, G.; Guerrero, A. *J. Fluorine Chem.* **1985,** *29,* 261.
46. Gassen, K. R.; Baasner, B. *J. Fluorine Chem.* **1990,** *49,* 127.
47. Burdon, J.; Ezmirly, S. T.; Huckerby, T. N. *J. Fluorine Chem.* **1988,** *40,* 283.
48. Dabbit, O.; Sutcliffe, L. H. *Spectrochim. Acta* **1980,** *36A,* 379.
49. Kestner, T. A. *J. Fluorine Chem.* **1987,** *36,* 77.
50. Natiello, M. A.; Contreras, R. H.; Gavarini, H. O. *Chem. Phys.* **1985,** *98,* 279.
51. Osten, H. J.; Jameson, C. J.; Craig, N. C. *J. Chem. Phys.* **1985,** *83,* 5434.
52. Heinze, P. L.; Burton, D. J. *J. Fluorine Chem.* **1986,** *31,* 115.
53. Bosch, M. P.; Camps, F.; Fabrias, G.; Guerrero, A. *Magn. Reson. Chem.* **1987,** *25,* 347.
54. Burton, D. J.; Spawn, T. D.; Heinze, P. L.; Bailey, A. R.; Shin-Ya, S. *J. Fluorine Chem.* **1989,** *44,* 167.
55. Gordon, A. J.; Ford, R. A. *The Chemist's Companion;* Wiley: New York, 1972.
56. Everett, T. S. *Multinuclear NMR Spectral Identification of Organofluorine Compounds;* unpublished data.
57. Harris, R. K. *J. Mol. Spectrosc.* **1963,** *10,* 309.
58. Elsheimer, S.; Dolbier, W. R.; Murla, M. *J. Org. Chem.* **1984,** *49,* 205.
59. Muller, N.; Carr, D. T. *J. Phys. Chem.* **1963,** *67,* 112.
60. Doyle, T. R.; Vogl, O. *J. Am. Chem. Soc.* **1989,** *111,* 8510.
61. Sartori, P.; Habel, W. *J. Fluorine Chem.* **1980,** *16,* 265.
62. Burton, D. J.; Greenlimb, P. E. *J. Org. Chem.* **1975,** *40,* 2796.
63. Hirano, T.; Nonoyama, S.; Miyajima, T.; Kurita, Y.; Kawamura, T.; Sato, H. *J. Chem. Soc., Chem. Commun.* **1986,** 606.
64. Dixon, D. A.; Smart, B. E. *J. Phys. Chem.* **1988,** *92,* 2729.
65. Wiberg, K. B.; Murcko, M. A.; Laidig, K. E.; MacDougall, P. J. *J. Phys. Chem.* **1990,** *94,* 6956.
66. Molines, H.; Wakselman, C. *J. Fluorine Chem.* **1984,** *25,* 447.
67. Thenappan, A.; Burton, D. J. *J. Fluorine Chem.* **1990,** *48,* 153.
68. Leroy, J.; Molines, H.; Wakselman, C. *J. Org. Chem.* **1987,** *52,* 290.
69. England, D. C.; Solomon, L.; Krespan, C. G. *J. Fluorine Chem.* **1974,** *3,* 63.
70. Nguyen, T.; Wakselman, C. *Synth. Commun.* **1990,** *20,* 97.
71. Weigert, F. J. *J. Org. Chem.* **1980,** *45,* 3476.
72. Everett, T. S. *J. Chem Ed.* **1988,** *65,* 422.

73. Akiyama, H.; Yamauchi, F.; Ouchi, K. *J. Chem. Soc., Chem. Commun.* **1970,** 355.
74. Merritt, R. F.; Ruff, J. K. *J. Org. Chem.* **1965,** *30,* 328.
75. Harris, J. F., Jr.; Stacey, J. W. *J. Am. Chem. Soc.* **1963,** *85,* 749.
76. Burdon, J.; Knights, J. R.; Parsons, I. W.; Tatlow, J. C. *J. Chem. Soc., Perkin Trans. 1* **1976,** 1930.
77. Patrick, T. B.; Johri, K. K.; White, D. H. Bertrand, W. S.; Mokhtar, R.; Kilbourn, M. R.; Welch, M. J. *Can. J. Chem.* **1986,** *64,* 138.
78. Rahman, M.; McKee, M. L.; Shevlin, P. B. *J. Am. Chem. Soc.* **1986,** *108,* 6296.
79. Chu, P.-S.; True, N. S. *J. Phys. Chem.* **1985,** *89,* 5613.
80. Wasylishen, R. E. *Can. J. Chem.* **1986,** *64,* 2094.
81. Weigert, F. J.; Middleton, W. J. *J. Org. Chem.* **1980,** *45,* 3289.
82. Weigert, F. J.; Strobach, D. R. *Org. Magn. Reson.* **1970,** *2,* 303.
83. Watanabe, S.; Fujita, T.; Usui, Y. *J. Fluorine Chem.* **1986,** *31,* 247.
84. Bradshaw, T. K.; Hine, P. T.; Della, E. W. *Org. Magn. Reson.* **1980,** *16,* 26.
85. Hull, W. E.; Kunz, W.; Port, R. E.; Seiler, N. *NMR Biomed.* **1988,** *1,* 11.
86. Bergstrom, D. E.; Swartling, D. J. *Mol. Struct. Energ. (Fluorine-Containing Molecules)* **1988,** *8,* 259.
87. Patrick, T. B.; Scheibel, J. J.; Cantrell, G. L. *J. Org. Chem.* **1981,** *46,* 3917.
88. Wheaton, G. A.; Burton, D. J. *J. Fluorine Chem.* **1977,** *9,* 25.
89. Anderson, J. E.; Glazer, E. S.; Griffith, D. L.; Knorr, R.; Roberts, J. D. *J. Am. Chem. Soc.* **1969,** *91,* 1386.
90. Beckers, H.; Buerger, H. *J. Organomet. Chem.* **1990,** *385,* 207.
91. Morimoto, K.; Makino, K.; Yamamoto, S.; Sakata, G. *J. Heterocyclic Chem.* **1990,** *27,* 807.
92. Van Hamme, M. J.; Burton, D. J.; Greenlimb, P. E. *Org. Magn. Reson.* **1978,** *11,* 275.
93. Umemoto, T.; Ishihara, S. *Tetrahedron Lett.* **1990,** *31,* 3579.
94. Langlois, B. R. *J. Fluorine Chem.* **1988,** *41,* 247.
95. Ruppert, I.; Schlich, K.; Volbach, W. *Tetrahedron Lett.* **1984,** *25,* 2198.
96. Prakash, G. K. S.; Krishnamurti, R.; Olah, G. A. *J. Am. Chem. Soc.* **1989,** *111,* 393.
97. Broicher, V.; Geffken, D. *J. Organomet. Chem.* **1990,** *381,* 315.
98. Pawelke, G. *J. Fluorine Chem.* **1989,** *42,* 429.
99. Stahly, G. P.; Bell, D. R. *J. Org. Chem.* **1989,** *54,* 2873.
100. Buerger, H.; Eujen, R.; Moritz, P. *J. Organomet. Chem.* **1991,** *401,* 249.
101. Aktaev, N. P.; Ilin, G. F.; Sokolskii, G. A.; Knunyants, I. L. *Izv. Akad. Nauk SSSR* **1977,** 1112.
102. Buerger, H.; Krumm, B.; Pawelke, G. *J. Fluorine Chem.* **1989,** *44,* 147.
103. Koller, K. L.; Dorn, H. C. *Anal. Chem.* **1982,** *54,* 529.
104. Olah, G. A.; Weber, T.; Farooq, O. *J. Fluorine Chem.* **1989,** *43,* 235.

105. Olah, G. A.; Pittman, C. U., Jr. *J. Am. Chem. Soc.* **1966,** *88,* 3310.
106. Francese, C.; Tordeux, M.; Wakselman, C. *J. Chem. Soc., Chem. Commun.* **1987,** 642.
107. Wakselman, C.; Tordeux, M. *J. Fluorine Chem.* **1982,** *21,* 99.
108. Delyagina, N. I.; Pervova, E. Y.; Knunyants, I. L. *Izv. Akad. Nauk SSSR* **1972,** 376.
109. Maraschin, N. J.; Catsikis, B. D.; Davis, L. H.; Jarvinen, G.; Lagow, R. J. *J. Am. Chem. Soc.* **1975,** *97,* 513.
110. Seppelt, K. *Angew. Chem.* **1977,** *89,* 325.
111. Lin, W.-H.; Clark, W. D.; Lagow, R. J. *J. Org. Chem.* **1989,** *54,* 1990.
112. Kloeter, G.; Lutz, W.; Seppelt, K.; Sundcrmcycr, W. *Angew. Chem.* **1977,** *89,* 754.
113. Blazejewski, J. C.; Dorme, R.; Wakselman, C. *Synthesis* **1985,** 1120.
114. Feldhoff, R.; Haas, A.; Lieb, M. *J. Fluorine Chem.* **1990,** *49,* 225.
115. Brauer, D. J.; Buerger, H.; Dorrenbach, F.; Krumm, B.; Pawelke, G.; Weuter, W. *J. Organomet. Chem.* **1990,** *385,* 161.
116. Lagow, R. J.; Gerchman, L. L.; Jacob, R. A.; Morrison, J. A. *J. Am. Chem. Soc.* **1975,** *97,* 518.
117. Ganja, E. A.; Ontiveros, C. D.; Morrison, J. A. *Inorg. Chem.* **1988,** *27,* 4535.
118. Morrison, J. A.; Lagow, R. J. *Inorg. Chem.* **1977,** *16,* 1823.
119. Turbini, L. J.; Aikman, R. E.; Lagow, R. J. *J. Am. Chem. Soc.* **1979,** *101,* 5833.
120. Burton, D. J.; Wiemers, D. M. *J. Am. Chem. Soc.* **1985,** *107,* 5014.
121. Naumann, D.; Varbelow, H. G. *J. Fluorine Chem.* **1988,** *41,* 415.
122. McClellan, W. R. *J. Am. Chem. Soc.* **1961,** *83,* 1598.
123. Dukat, W.; Naumann, D. *J. Chem. Soc., Dalton Trans.* **1989,** 739.
124. Clark, H. C.; Manzer, L. E. *J. Organomet. Chem.* **1973,** *59,* 411.
125. Wiemers, D. M.; Burton, D. J. *J. Am. Chem. Soc.* **1986,** *108,* 832.
126. Moreland, C. G.; Bumgardner, C. L. *J. Magn. Reson.* **1971,** *4,* 20.
127. Bartle, K. D.; Hallas, G.; Hepworth, J. D. *Org. Magn. Reson.* **1973,** *5,* 479.
128. Schuster, I. I. M. *J. Magn. Reson.* **1975,** *17,* 104.
129. Newmark, R. A.; Hill, J. R. *Org. Magn. Reson.* **1977,** *9,* 589.
130. Fifolt, M. J.; Sojka, S. A.; Wolfe, R. A.; Hojnicki, D. S.; Bieron, J. F.; Dinan, F. J. *J. Org. Chem.* **1989,** *54,* 3019.
131. Wray, V.; Lincoln, D. N. *J. Magn. Reson.* **1975,** *18,* 374.
132. Ernst, L.; Lincoln, D. N.; Wray, V. *J. Magn. Reson.* **1976,** *21,* 115.
133. Wray, V.; Ernst, L.; Lustig, E. *J. Magn. Reson.* **1977,** *27,* 1.
134. Ernst, L.; Wray, V. *J. Magn. Reson.* **1977,** *28,* 373.
135. Bruce, M. I. *J. Chem. Soc. A* **1968,** 1459.
136. Cohen, E. A.; Bourn, A. J. R.; Manatt, S. L. *J. Magn. Reson.* **1969,** *1,* 436.
137. Pushkina, L. N.; Stepanov, A. P.; Zhukov, V. S.; Naumov, A. D. *Org. Magn. Reson.* **1972,** *4,* 607.
138. Lustig, E.; Hansen, E. A.; Ragelis, E. P. *Org. Magn. Reson.* **1969,** *1,* 295.

139. Meakin, P.; Ovenall, D. W.; Sheppard, W. A.; Jesson, J. P. *J. Am. Chem. Soc.* **1975,** *97,* 522.
140. Adcock, W.; Alste, J.; Rizvi, S. Q. A; Aurangzeb, M. *J. Am. Chem. Soc.* **1976,** *98,* 1701.
141. Cassidei, L.; Sciacovelli, O. *Spectrochim. Acta* **1985,** *41A,* 1459.
142. Matthews, R. S. *Magn. Reson. Chem.* **1982,** *18,* 226.
143. Matthews, R. S. *J. Fluorine Chem.* **1990,** *48,* 7.
144. Matthews, R. S. *J. Fluorine Chem.* **1990,** *50,* 381.
145. Cassidei, L.; Sciacovelli, O.; Forlani, L. *Spectrochim. Acta* **1982,** *38A,* 755.
146. Jakobsen, H. J.; Brey, W. S. *J. Chem. Soc., Chem. Commun.* **1979,** 478.
147. Sibi, M. P.; Lichter, R. L. *Org. Magn. Reson.* **1980,** *14,* 494.
148. Thomas, W. A.; Griffin, G. E. *Org. Magn. Reson.* **1970,** *2,* 503.
149. Emsley, J. W.; Lindon, J. C.; Salman, S. R. *J. Chem. Soc., Faraday Trans. 2* **1972,** 1343.
150. Matthews, R. S. *J. Magn. Reson.* **1978,** *30,* 537.
151. Denisov, A. Y.; Mamatyuk, V. I.; Shkurko, O. P. *Magn. Reson. Chem.* **1985,** *23,* 482.
152. Sandall, J. P. B.; Sutcliffe, L. H. *Magn. Reson. Chem.* **1990,** *28,* 268.
153. Sardella, D. J.; Boger, E. *Magn. Reson. Chem.* **1986,** *24,* 287.
154. Sardella, D. J.; Boger, E. *Magn. Reson. Chem.* **1989,** *27,* 13.
155. Hsee, L. C.; Sardella, D. J. *Magn. Reson. Chem.* **1990,** *28,* 688.
156. Matthews, R. S. *Org. Magn. Reson.* **1976,** *8,* 240.
157. Matthews, R. S. *Org. Magn. Reson.* **1976,** *8,* 628.
158. Matthews, R. S. *J. Magn. Reson.* **1978,** *29,* 65.
159. Miller, J. M.; Clark, J. H. *J. Chem. Soc., Chem. Commun.* **1982,** 1318.
160. Hudlický, M. *J. Fluorine Chem.* **1985,** *28,* 461.
161. Jache, A. W. *Mol. Struct. Energ. (Fluorine-Containing Molecules)* **1988,** *8,* 165.
162. Christe, K. O.; Wilson, W. W. *J. Fluorine Chem.* **1990,** *46,* 339.
163. Christe, K. O.; Wilson, W. W. *J. Fluorine Chem.* **1990,** *47,* 117.
164. Olah, G. A.; Stephenson, M.; Shih, J. G.; Krishnamurthy, V. V.; Prakash, G. K. S. *J. Fluorine Chem.* **1988,** *40,* 319.
165. Prakash, G. K. S.; Heiliger, L.; Olah, G. A. *J. Fluorine Chem.* **1990,** *49,* 33.
166. Dyatkin, B. L.; Delyagina, N. I.; Sterlin, S. R. *Usp. Khim.* **1976,** *45,* 1205. (Engl. Transl. *Russ. Chem. Rev.* **1976,** *45,* 607.)
167. Bayliff, A. E.; Bryce, M. R.; Chambers, R. D.; Matthews, R. S. *J. Chem. Soc., Chem. Commun.* **1985,** 1018.
168. Smart, B. E.; Middleton, W. J.; Farnham, W. B. *J. Am. Chem. Soc.* **1986,** *108,* 4905.
169. Mitchell, P. J.; Phillips, L.; Roberts, S. J.; Wray, V. *Org. Magn. Reson.* **1974,** *6,* 126.
170. Vikic-Topic, D.; Meic, Z. *J. Mol. Struct.* **1986,** *142,* 371.

171. Schaefer, T.; Penner, G. H. *Can. J.Chem.* **1985,** *63,* 2253.
172. Schaefer, T.; Laatikainen, R.; Wildman, T. A.; Peeling, J.; Penner, G. H.; Baleja, J.; Marat, K. *Can. J. Chem.* **1984,** *62,* 1592.
173. Schaefer, T.; Penner, G. H. *Can. J. Chem.* **1988,** *66,* 1635.
174. Schaefer, T.; Sebastian, R. *Can. J. Chem.* **1989,** *67,* 1027.
175. Schaefer, T.; Beaulieu, C.; Sebastian, R.; Penner, G. H. *Can. J. Chem.* **1990,** *68,* 581.
176. Schaefer, T.; Niemczura, W.; Danchura, W.; Wildman, T. A. *Can. J. Chem.* **1979,** *57,* 1881.
177. Schaefer, T.; Penner, G. H.; Takeuchi, C.; Tseki, P. *Can. J. Chem.* **1988,** *66,* 1647.
178. Mallory, F. B.; Mallory, C. W.; Fedarko, M. *J. Am. Chem. Soc.* **1974,** *96,* 3536.
179. Mallory, F. B.; Mallory, C. W.; Ricker, W. M. *J. Am. Chem. Soc.* **1975,** *97,* 4770.
180. Mallory, F. B.; Mallory, C. W.; Ricker, W. M. *J. Org. Chem.* **1985,** *50,* 457.
181. Mallory, F. B.; Mallory, C. W. *J. Am. Chem. Soc.* **1985,** *107,* 4816.
182. Mallory, F. B.; Mallory, C. W.; Baker, M. B. *J. Am. Chem. Soc.* **1990,** *112,* 2577.
183. Adcock, W.; Abeywickrema, A. N.; Kok, G. B. *J. Org. Chem.* **1984,** *49,* 1387.
184. Adcock, W.; Iyer, V. S. *J. Org. Chem.* **1985,** *50,* 1538.
185. Adcock, W.; Kok, G. B. *J. Org. Chem.* **1987,** *52,* 356.
186. Voelter, W.; Breitmaier, E.; Jung, G.; Bayer, E. *Org. Magn. Reson.* **1970,** *2,* 251.
187. Zuber, G. E.; Staiger, D. B.; Warren, R. J. *Anal. Chem.* **1983,** *55,* 64.
188. Sleevi, P.; Glass, T. E.; Dorn, H. C. *Anal. Chem.* **1979,** *51,* 1931.
189. Spratt, M. P.; Meng, Y.; Dorn, H. C. *Anal. Chem.* **1985,** *57,* 76.
190. Spratt, M. P.; Armistead, D.; Motell, E.; Dorn, H. C. *Anal. Chem.* **1985,** *57,* 359.
191. Shue, F. F.; Yen, T. F. *Anal. Chem.* **1982,** *54,* 1641.
192. Ho, F. F.-L. *Anal. Chem.* **1974,** *46,* 496.
193. Dale, J. A.; Mosher, H. S. *J. Am. Chem. Soc.* **1968,** *90,* 3732.
194. Dale, J. A.; Dull, D. L.; Mosher, H. S. *J. Org. Chem.* **1969,** *34,* 2543.
195. Kalyanam, N.; Lightner, D. A. *Tetrahedron Lett.* **1979,** 415.
196. Nabeya, A.; Endo, T. *J. Org. Chem.* **1988,** *53,* 3358.
197. You, T.; Mosher, H. S. *Youji Huaxue* **1989,** *9,* 518.
198. Pirkle, W. H.; Sikkenga, D. L.; Pavlin, M. S. *J. Org. Chem.* **1977,** *42,* 384.
199. Pirkle, W. H.; Beare, S. D. *J. Am. Chem. Soc.* **1967,** *89,* 5485.
200. Pirkle, W. H.; Simmons, K. A. *J. Org. Chem.* **1981,** *46,* 3239.
201. Takeuchi, Y.; Ogura, H.; Ishii, Y.; Koizumi, T. *J. Chem. Soc., Perkin Trans. 1* **1989,** 1721.

202. Takeuchi, Y.; Ogura, H.; Ishii, Y.; Koizumi, T. *Chem. Pharm. Bull.* **1990,** *38,* 2404.
203. Barrelle, M.; Hamman, S. *J. Chem. Res. Synop.* **1990,** 100.
204. Hamman, S. *J. Fluorine Chem.* **1989,** *45,* 377.
205. McCreary, M. D.; Lewis, D. W.; Wernick, D. L.; Whitesides, G. M. *J. Am. Chem. Soc.* **1974,** *96,* 1038.
206. Sweeting, L. M.; Crans, D. C.; Whitesides, G. M. *J. Org. Chem.* **1987,** *52,* 2273.
207. Bravo, P.; Resnati, G. *Tetrahedron: Asymmetry* **1990,** *1,* 661.
208. Gerig, J. T. *Biol. Magn. Reson.* **1978,** *1,* 139.
209. Gerig, J. T. In *Biomedicinal Aspects of Fluorine Chemistry;* Filler, R.; Kobayashi, Y., Eds.; Kodansha, Ltd.: Tokyo, 1982.
210. Guy, W. S.; Taves, D. R.; Brey, W. S., Jr., In *Biochemistry Involving Carbon–Fluorine Bonds;* Filler, R., Ed.; ACS Symposium Series 28, American Chemical Society; Washington, DC, 1976.
211. Rietjens, I. M. C. M.; Vervoort, J. *Xenobiotica* **1989,** *19,* 1297.
212. Vervoort, J.; De Jager, P. A.; Steenbergen, J.; Rietjens, I. M. C. M. *Xenobiotica* **1990,** *20,* 657.
213. Wade, K. E.; Troke, J.; Macdonald, C. M.; Wilson, I. D.; Nicholson, J. K. *Methodol. Surv. Biochem. Anal.* **1988,** *18,* 383.
214. Tecle, B.; Casida, J. E. *Chem. Res. Toxicol.* **1989,** *2,* 429.
215. Nakada, T.; Kwee, I. L.; Conboy, C. B. *J. Neurochem.* **1986,** *46,* 198.
216. Malet-Martino, M. C.; Armand, J.-P.; Lopez, A.; Bernadou, J.; Beteille, J.-P.; Bon, M.; Martino, R. *Cancer Res.* **1986,** *46,* 2105.
217. Bernadou, J.; Armand, J.-P.; Lopez, A.; Malet-Martino, M. C.; Martino, R. *Clin. Chem.* **1985,** *31,* 846.
218. Dax, K.; Glanzer, B. I.; Schulz, G.; Vyplel, H. *Carbohydr. Res.* **1987,** *162,* 13.
219. Hall, L. D.; Manville, J. F.; Bhacca, N. S. *Can. J. Chem.* **1969,** *47,* 1.
220. Hall, L. D.; Manville, J. F. *Can J. Chem.* **1969,** *47,* 19.
221. Hall, L. D.; Steiner, P. R.; Pedersen, C. *Can. J. Chem.* **1970,** *48,* 1155.
222. Sykes, B. D.; Hull, W. E. *Methods Enzymol.* **1978,** *49G,* 270.
223. Sykes, B. D.; Weiner, J. H. *Magn. Reson. Biol.* **1980,** *1,* 171.
224. Sievers, R. E.; Bayer, E.; Hunziker, P. *Nature* **1969,** *223,* 179.
225. Macdonald, P. M.; Sykes, B. D.; McElhaney, R. N. *Can. J. Biochem. Cell Biol.* **1984,** *62,* 1134.
226. Higashi, N.; Matsumoto, T.; Niwa, M. *Sci. Eng. Rev. Doshisha Univ.* **1988,** *29,* 154.
227. Grode, S. H.; Gillis, R. W. *J. Magn. Reson.* **1989,** *82,* 122.
228. Smith, L. L.; Ezell, E. L. *Magn. Reson. Chem.* **1990,** *28,* 414.
229. Hughes, D. W.; Bain, A. D.; Robinson, V. J. *Magn. Reson. Chem.* **1991,** *29,* 387.

230. Levisalles, J.; Rudler-Chauvin, M.; Martin, J. A. *Bull. Soc. Chim. Fr.* **1980,** 167.
231. Kong, C. F.; Holloway, G. M.; Parhami, P.; Fung, B. M. *J. Phys. Chem.* **1984,** *88,* 6308.
232. Lin, W.-H.; Lagow, R. J. *J. Fluorine Chem.* **1990,** *50,* 345.
233. Koehler, K. A.; Stone, E. E.; Shelton, R. A.; Jarnagin, F.; Koehler, L. S.; Fossel, E. T. *J. Magn. Reson.* **1978,** *30,* 75.
234. Balonga, P. E.; Kowalewski, V. J.; Contreras, R. H. *Spectrochim. Acta* **1986,** *42A,* 23.
235. Balonga, P. E.; Kowalewski, V. J.; Contreras, R. H. *Spectrochim. Acta* **1988,** *44A,* 819.
236. Huang, C.; Venturella, V. S.; Cholli, A. L.; Venutolo, F. M.; Silbermann, A. T.; Vernice, G. G. *J. Fluorine Chem.* **1989,** *45,* 239.
237. Cholli, A. L.; Huang, C.; Venturella, V. S.; Pennino, D. J.; Vernice, G. G. *Appl. Spectrosc.* **1989,** *43,* 24.
238. Wyrwicz, A. M.; Pszenny, M. H.; Schofield, J. C.; Tillman, P. C.; Gordon, R. E.; Martin, P. A. *Science* **1983,** *222,* 428.
239. Burt, C. T.; Moore, R. R.; Roberts, M. F. *J. Magn. Reson.* **1983,** *53,* 163.
240. Evers, A. S.; Berkowitz, B. A.; d'Avignon, D. A. *Nature,* **1987,** *328,* 157.
241. Fishman, J. E.; Joseph, P. M.; Floyd, T. F.; Mukherji, B.; Sloviter, H. A. *Magn. Reson. Imag.* **1987,** *5,* 279.
242. Nakada, T.; Kwee, I. L.; Card, P. J.; Matwiyoff, N. A.; Griffey, B. V.; Griffey, R. H. *Magn. Reson. Med.* **1988,** *6,* 307.
243. Nakada, T.; Kwee, I. L. *Magn. Reson. Imag.* **1987,** *5,* 259.
244. Holland, G. N.; Bottomley, P. A.; Hinshaw, W. S. *J. Magn. Reson.* **1977,** *28,* 133.
245. Wyrwicz, A. M.; Ryback, K. R.; Chew, W.; Hurd, R. *J. Magn. Reson.* **1986,** *69,* 572.
246. McFarland, E.; Koutcher, J. A.; Rosen, B. R.; Teicher, B.; Brady, T. J. *J. Comput. Assist. Tomogr.* **1985,** *9,* 8.
247. Rogers, W. J.; Everett, T. S.; Buchalter, M. B.; Rademakers, F. E.; Shapiro, E. P. In *Abstracts 8th Annual Meeting Soc. Magn. Reson. Med.;* 1989.
248. Weichert, J.; Chenevert, T.; Fechner, K.; Swanson, S.; Longino, M.; Counsell, R.; Glazer, G. In *Abstracts 7th Annual Meeting Soc. Magn. Reson. Med.;* 1988.

Chapter 8

Applications of Fluorinated Compounds

Refrigerants, Propellants, and Foam-Blowing Agents

by R. L. Powell

The introduction of the chlorofluorocarbon (CFC) fluids in the early 1930s marked the inception of the halofluorocarbon industry. Both in terms of tonnage produced and product value, the CFCs have dominated the organofluorocarbon industry and have provided the feedstocks for the development of other fluorocarbon products such as poly(tetrafluoroethylene).

Each compound of this type is designated by a number derived from its structure, preceded by the designation CFC (chlorofluorocarbon) for a completely halogenated molecule such as CCl_2F_2; by HCFC (hydrochlorofluorocarbon) for chlorine-containing molecules such as $CHClF_2$, which also contain hydrogen; or by HFC for molecules such as CF_3CH_2F, which contain only carbon, hydrogen, and fluorine [*1*]. In commercial literature, manufacturers use their own trade name instead of CFC or HCFC. The best known of these trade names is "Freon," the trademark of DuPont, which is often used incorrectly as a generic term to describe chlorofluorinated fluids. The designations CFC and HCFC should be used in the scientific literature, unless the source of the fluid needs to be specified by reference to the manufacturer.

Since 1974, the halofluorocarbon fluid industry has been affected profoundly by the hypothesis of Rowland and Molina [2] implicating chlorofluorocarbons and, to a lesser extent, the hydrochlorofluorocarbons in the depletion of stratospheric ozone. In the mid-1980s, the discovery of the reduction of the ozone layer over Antarctica during its spring, the so-called "ozone hole", provided the first direct observational support for the hypothesis [*3*]. Following a more precise calibration of the instruments used to observe stratospheric ozone concentrations, it was discovered in 1990 that the ozone layer between latitudes 40° and 60° north is being depleted at a rate of 0.8% per year during the winter and early spring [*4*]. These latitudes cover the world's major industrialized countries.

Refrigeration

A great expansion of refrigeration commenced with the introduction of the chlorofluorocarbon refrigerants in the 1930s and has seen the spread of refrigerators into every home in the industrialized world. Although domestic refrigerators are filled with dichlorodifluoromethane (CFC 12), chlorodifluoromethane (HCFC 22) is used in many air-conditioning units. Supermarket refrigeration is dominated by the azeotrope of HCFC 22 and chloropentafluoroethane (CFC 115), generally

0065–7719/95/0187–1089$08.00/1

known as refrigerant 502. Trichlorofluoromethane (CFC 11) is used in water chiller systems widely used in North America for air-conditioning large buildings.

Other Uses of Chlorofluorocarbons

Chlorofluorocarbons not only proved to be excellent refrigerants but also expanded into other applications.

Trichlorofluoromethane (CFC 11) is an excellent blowing agent for polyurethane insulation because of its low vapor thermal conductivity coupled with its boiling point of 24 °C, close to room temperature, and low toxicity. The chlorofluorocarbon, dissolved in the polyol component of the formulation, is evaporated by the heat of the polyol–isocyanate reaction to generate the cells of the foam. Trichlorofluoromethane-blown foams have 30–40% better insulating properties than equivalent foams blown with carbon dioxide. Such foams insulate refrigerator cabinets, replacing the rockwool fiber common until the end of the 1950s.

Chlorofluorocarbon-blown foam blocks are used to insulate the walls and roofs of some buildings, thus reducing heat losses and helping to conserve fossil fuels. In this area, polyurethane foam competes with polystyrene foam, which until recently was blown with dichlorodifluoromethane (CFC 12) but is now blown with a mixture of chlorodifluoromethane (HCFC 22) and 1-chloro-1,1-difluoroethane (HCFC 142b).

In the past, the single biggest application of chlorofluorocarbons has been as aerosol propellants, an application that grew rapidly from the 1950s to the mid-1970s. The low toxicity, nonflammability, and good solvent properties of trichlorofluoromethane (CFC 11) and dichlorodifluoromethane (CFC 12) made them especially attractive for this purpose. The first aerosol packages were used to dispense insecticides such as pyrethroids to kill flying insects, but subsequently, the range was extended to domestic and personal-care products, such as deodorants, hair sprays, shaving foam, paint lacquers, and even foodstuffs, such as cream and soft cheese. However, with the signing of the Montreal Protocol to limit the worldwide production and consumption of chlorofluorocarbons, their application as aerosol propellants has ceased essentially in all major industrial countries. Only medical aerosols used to dispense metered doses of antiasthmatic drugs, necessary to relieve life-threatening conditions, still use dichlorodifluoromethane as propellant, but even this use will be phased out in the near future as toxicity testing of nonchlorinated alternatives is completed.

The production of chlorofluoro- and hydrochlorofluorocarbons expanded rapidly after 1945, reaching about 1 million tons in 1986 despite a drop in production in the late 1970s, when the United States and several other countries banned the use of chlorofluorocarbons as aerosol propellants. In the early 1990s, production and consumption are in decline.

References are listed on pages 1097–1098.

Processes to Chlorofluorocarbons and Hydrochlorofluorocarbons

Although many syntheses of chlorofluorocarbons and hydrochlorofluorocarbons have been published, those actually used in manufacturing processes are limited. By far the most important is the original Swarts reaction [5].

In the industrial process, the chlorocarbon and liquid hydrogen fluoride feeds are pumped simultaneously into a complex liquid mixture of Sb(III) and Sb(V) chlorofluorides at temperatures in the 60–150 °C range. The products are generally more volatile than the reactants and therefore distill preferentially from the reactor vessel; thus the reactor can be operated continuously.

Table 1 lists the chlorocarbon feedstocks for the production of the major chlorofluorocarbons and hydrochlorofluorocarbons.

Table 1. Chlorocarbon Feedstocks for the Production of the Major Chlorofluorocarbons and Hydrochlorofluorocarbons

Feedstock	*Reactor Temperature (°C)*	*Product(s)*
CCl_4	100	CCl_3F, CCl_2F_2
$CHCl_3$	60	$CHClF_2$
$CCl_2{=}CCl_2$	150	CCl_2FCClF_2, $CClF_2CClF_2$
$CCl_2{=}CH_2$	80	$CClF_2CH_3$

Although antimony pentafluoride can fluorinate 1,1,2-trichloro-1,2,2-trifluoroethane to chloropentafluoroethane, this route is not used industrially because antimony pentafluoride and hydrogen fluoride are too corrosive. Both dichlorotetrafluoroethane and chloropentafluoroethane are produced by vapor-phase fluorination of tetrachloroethene with proprietary "chromia" catalysts at 300 to 500 °C (equation 1).

$$CCl_2 = CCl_2 + \xrightarrow[300\text{–}500\ ^\circ C]{Cl_2,\ HF/''chromia''} CF_3CF_2Cl + CF_2ClCF_2Cl + CF_3CFCl_2 + HCl$$

By varying the temperature and the ratio of hydrogen fluoride to tetrachloroethene, the ratio of chloropentafluoroethane to dichlorotetrafluoroethanes can be varied at will. The ratio of 1,2-dichlorotetrafluoroethane to 1,1-dichlorotetrafluoroethane depends upon the temperature of the catalyst, higher temperatures favoring the latter, with yields that can vary from 10 to 50% depending upon operating conditions.

Chlorotrifluoromethane (CFC 13) is produced by the disproportionation of CCl_2F_2 over $AlCl_3$ catalyst at 80–100 °C in near-quantitative conversion.

$$2CCl_2F_2 \xrightarrow[80\text{–}100\ ^\circ C]{AlCl_3} CF_3Cl + CCl_3F$$

References are listed on pages 1097–1098.

The Ozone Layer Problem

In 1974, Rowland and Molina [2] conjectured that the high stability of the chlorofluorocarbons, which made them so attractive for refrigeration, enables them to pass unchanged through the troposphere and thus to the stratosphere, where the intense UV radiation from the sun could split off chlorine atoms. The free chlorine atoms could catalyze the destruction of ozone molecules, which form a layer in the stratosphere between 20–30 km above the earth's surface. This ozone layer absorbs much of the energetic solar UV radiation and prevents it from reaching the biosphere surface where it could adversely affect living organisms.

Initial evidence for the effect of chlorofluorocarbons on the ozone layer was very limited, based mainly on computer models of the atmosphere. In the 1970s, there was no observational support for ozone depletion. However, refrigerant producers initiated research programs and identified a number of potential alternatives. The research programs were given renewed impetus in the mid-1980s following political pressure in the United States and the discovery of the ozone depletion over the Antarctic during its spring. This was the first observational evidence for the possible effect of chlorofluorocarbons on stratospheric ozone.

The Montreal Protocol

The concern over the effect of chlorofluorocarbons on the ozone layer culminated in the Montreal Protocol, the first international agreement to protect the global environment. Originally negotiated in 1987, the protocol now controls, throughout much of the world, the use and production of chlorofluorocarbons, halons (bromofluorocarbons), carbon tetrachloride, and methylchloroform (base solvent for degreasing formulations used in the engineering industry). The original version of the protocol provided for a 50% reduction in the production and use of the five controlled CFCs by mid 1998. In light of further evidence of continuing stratospheric ozone depletion, subsequent revisions of the protocol in 1990 (London) and 1992 (Copenhagen) introduced timetables for the complete phase-out of CFCs and halons. Originally HCFCs were considered part of the solution, but at the London conference they were deemed to be "transitional substances". Their value in hastening the move away from CFCs was recognized, but it was expected they would be phased out between 2020 to 2040. At Copenhagen a specific timetable for HCFC phase-out was established. The European Union has enacted regulations for CFCs to be phased out by the end of 1994, one year ahead of the latest requirements of the Montreal Protocol, and has proposed a more stringent timetable for HCFC phase-out. Table 2 summarizes the situation at the end of 1993. Provisions, in the form of funding and agreements to ensure access to appropriate technologies, have been made for special needs of the developing countries.

Chlorofluorocarbon Replacements

Chlorofluorocarbon alternatives must retain the attractive properties of chlorofluorocarbons, especially low toxicity, non-flammability, and good thermody-

Table 2. Percentage Reductions in Consumption of Halocarbons Required by the Regulations

	Copenhagen Amendments to Montreal Protocol					*European Community's Regulation*				
	Agreed and in Force			*Proposed*		*Enacted and in Force*			*Proposed*	
	CFCs	*Other CFCs*	*Halons*	*HCFCs*	*HBFCs*	*CFCs*	*Other CFCs*	*Halons*	*HCFCs*	*HBFCs*
Cap Level[a]				3.1%					2.5%	
Base Year	1986	1989	1986	1989	1989	1986	1986	1986	1989	1989
Year										
1993	—	−20	—	—	—	−50	−50	Freeze	—	—
1994	−75	−75	−100	—	—	−85	−85	−100	—	—
1995	—	—	—	—	—	−100	−100	—	—	Freeze
1996	−100	−100	—	Freeze	−100	—	—	—	—	−100
2000	—	—	—	—	—	—	—	—	−25	—
2004	—	—	—	−35	—	—	—	—	−60	—
2008	—	—	—	—	—	—	—	—	−80	—
2010	—	—	—	−65	—	—	—	—	—	—
2012	—	—	—	—	—	—	—	—	−95	—
2015	—	—	—	−90	—	—	—	—	−100	—
2020	—	—	—	−99.5	—	—	—	—	—	—
2030	—	—	—	−100	—	—	—	—	—	—

[a]The cap is the percentage of the "calculated" level of chlorofluorocarbons consumed in the base year plus the calculated level of hydrofluorocarbons consumed the same base year. It applies only to HCFCs. "Calculated" in the context means that the amount of each substance is adjusted by its ozone depletion potential (ODP), a measure of its potential to deplete stratospheric ozone relative to that of CFC 11.

Table 3. Non-Chlorine Containing Replacements for CFCs and HCFCs

Fluid	*Formula or Composition*	*bp (°C)*	*Flammability*[a]	*Fluid Replaced*	*Application*
32	CH_2F_2	−51	F		
125	CF_3CF_2H	−48	N		
143a	CF_3CH_3	−48	F		
32/125		−52	N	22	Air conditioning
32/125/134a	23/25/52		N	22	Air conditioning
	30/40/30		N	502	Supermarket freezers
	10/70/20		N	502	Supermarket freezers
125/143a	55/45		N	502	Supermarket freezers
125/143a/134a	45/52/4		N	502	Supermarket freezers
E125[b]	CF_3OCHF_2	−37	N	22 or 125?	Air conditioning Refrigeration
134a	CF_3CH_2F	−27	N	12	Domestic refrigerators Automobile air conditioning Medical aerosols Industrial refrigeration

152a	CHF_2CH_3	−24	F	12	Domestic refrigeration
227ea	CF_3CHFCF_3	−19	N	12	Medical aerosols Firefighting
236fa	$CF_3CH_2CF_3$	−1	N	114	Air conditioning in naval ships
236cb	$CF_3CH_2CH_2F$	−1		114	Air conditioning in naval ships
E134[b]	CHF_2OCHF_2	5	N	114	Air conditioning in naval ships
356ffa	$CF_3CH_2CH_2CF_3$	26	N	113	Foam blowing
245ca	$CHF_2CF_2CH_2F$	26	F	11	Large scale air conditioning
236fcb	$CF_3CH_2CF_2CH_3$	40	N	113	Solvent
	$CF_3CHFCHFCF_2CF_3$	47	N	113	Solvent

[a]F, flammable; N, nonflammable.
[b]A fluorinated dimethyl ether is assigned the same number as the corresponding C_2 alkane prefixed with an "E", a nomenclature recently adopted by the National Institute of Science and Technology (NIST) in the United States.

References are listed on pages 1097–1098.

namic performance, but avoid their adverse environmental impact. Only hydrofluorocarbons (HFCs) possess all of these desirable attributes. The compounds to emerge from research programs are listed in Table 3.

HFC 134a was the first of the new refrigerants to enter production when ICI Chemicals & Polymers Limited started the world's first commercial-scale plant in the United Kingdom in October 1990. Although HFC 134a is a direct substitute for CFC 12, the major domestic refrigeration and mobile air conditioning refrigerant, it has proved impossible to find single-fluid alternatives for other major refrigerants such as HCFC 22 and azeotrope R502. The generally preferred replacement for HCFC 22 is a blend of HFCs 32, 125, and 134a in the percentage weight ratio 23:25:52. A variety of R502 alternatives are currently being offered based on blends of HFCs 32, 125, and 134a or on blends of HFCs 125 and 143a, with or without HFC 134a. Small-scale commercial production of difluoromethane (HFC 32), pentafluoroethane (HFC 125), and 1,1,1-trifluoroethane (HFC 143a) has commenced. 1,1-difluoroethane (HFC 152a) has been available as small-tonnage refrigerant for many years, but its marked flammability inhibits its wide use as a CFC replacement. 1,1,1,2,3,3,3-heptafluoropropane (HFC 227a) seems likely to find application as a medical aerosol propellant and perhaps as a fire-fighting agent. Other compounds listed in are still in the developmental stage, and their commercial introduction is not yet certain. The important thermodynamic and physical properties of the HFCs have been determined [*7*].

To facilitate the phase-out of the CFCs a number of transitional materials have been introduced. 1,1-Dichloro-2,2,2-trifluoroethane (HCFC 123) is being developed as a replacement for CFC 11 as the refrigerant in centrifugal chillers, which are commonly used to air condition large buildings in the United States. 1,1-dichlorolfluoroethane (HCFC 141b) is being introduced in the United States and Japan as a blowing agent for the polyurethane foam insulation in refrigerators. The azeotrope of chlorodifluoromethane and perfluoropropane and the "near-azeotropic blends" of these two fluids with propane have been introduced as "drop-in" replacements for R502, notably in Japan. A blend of pentafluoroethane and chlorodifluoromethane has also been promoted as a transitional replacement for R502.

The special solvent properties of 1,1,2-trichloro-1,2,2-trifluoroethane (CFC 113) are related to the presence of chlorine as well as fluorine atoms. Considerable work looking into HCFCs as replacements including mixtures of 1,1-dichloro-1,1,1-trifluoroethane (HCFC 123) and 1,1-dichloro-1-fluoroethane (HCFC 141b) and halogenated propanes, notably 1,1-dichloro-2,2,3,3,3-pentafluoroethane ($CF_3CF_2CHCl_2$, HCFC 225cb) and 1,3-dichloro-1,1,2,2,3-pentafluoroethane ($ClCF_2CF_2CHClF$, HCFC 225ca) was done originally. There are a variety of alternative cleaning technologies, including a return to simple chlorinated solvents such as dichloromethane, tri- and tetra-chloroethene, which are destroyed so rapidly in the lower atmosphere that they do not deplete the stratospheric ozone layer. Aqueous surfactant and hydrocarbon solvent cleaning systems are also being developed. As transitional solutions, perfluorocarbon fluids are being allowed for the cleaning of critical electronic and mechanical units for which aqueous and other CFC 113 replacements are currently unacceptable.

Routes to Chlorofluorocarbon Replacements

In this period of intense development cloaked in industrial secrecy, not all manufacturers have declared the routes they chose for their initial production plants. Preferred long-term processes are even less clear. In some cases, the technologies that sufficed for the production of chlorofluorocarbons and hydrochlorofluorocarbons would be pushed to their limits as processes for the replacements. For example 1,1,1,2-tetrafluoroethane (HFC 134a) can be produced by the fluorination of 1-chloro-2,2,2-trifluoroethane (HCFC 133a) with antimony pentafluoride, but considerable overfluorination to 1-chloro-1,2,2,2-tetrafluoroethane (HCFC 124) and pentafluoroethane (HFC 125) is observed [*8, 9*]. In any case, the reaction mixture would be highly corrosive.

A much more selective reaction is possible by using vapor-phase fluorination over a "chromia" catalyst at 300 to 400 °C, but conversions are thermodynamically limited to 10–20% under acceptable operating conditions [*10, 11*]. Despite this disadvantage, this process has been selected by ICI and Hoechst for their first plants.

The only other route being considered seriously by industrial companies is the hydrogenation of 1,1-dichloro-2,2,2-tetrafluoroethane (CFC 114a) or 1-chloro-1,2,2,2-tetrafluoroethane (HCFC 124) over a catalyst such as 5% palladium on charcoal. In particular, DuPont [*12*], Asahi Glass [*13*], and Ausimant [*14*] have published patents and papers on this chemistry. This reaction was first reported in an ICI patent originating in 1977 [*15*].

The Future

Refrigeration technology is moving into a new phase with the introduction of hydrofluorocarbons that combine the safety of chlorofluorocarbons with a much lower environmental impact. The considerable care being taken to choose the new refrigerants under the close scrutiny of the public and regulatory bodies will ensure that the new products will benefit mankind for years to come.

Reference forPages 1089–1097

1. ASHRAE Standard, BSR/ASHRAE 34–1989R, First Public Draft, March 1991, American Society of Heating, Refrigeration, and Air Conditioning Engines.
2. Molina, M. J.; Rowland, F. S. *Nature,* **1974**, *810,* 249.
3. *Scientific Assessment of Stratospheric Ozone: 1989,* World Meterological Organization Global Ozone Research and Monitoring Project, Report No. 20, *1* and *2,* 1989.
4. Stolarski, R. S.; Bloomfield, P.; McPeters, R. D.; Herman, J. R. *Geo. Phys. Res. Lett.* **1991,** *18,* 1015.
5. Swarts, F. *Bull. Acad. Roy. Belg.* **1892,** *11,* 309, 374.
6. Acknowledgement to ICI Chemicals and Polymers Limited for allowing this section to be based on the company's information note (August 1991) on the Montreal Protocol.

7. *A Survey of Current World Wide Research on the Thermophysical Properties of Alternative Refrigerants,* McLinden, M. O.; Haynes, W. M.; Watson, J. R. T.; Watanabe, K. *National Institute of Standards and Technology,* U.S.A., *NISTIR* **1991, 3969.**
8. McCulloch, A.; Powell, R. L.; Young, B. D. Eur. Pat. EP 300 724 A, 1989; *Chem. Abstr.* **1989,** *110,* 231101q.
9. Gumprecht, W. H. U.S. Pat. 4 851 595, 1989; *Chem. Abstr.* **1990,** *112,* 20668h.
10. Darragh, J. I.; Powell, R. L.; Potter, S. E. Br. Pat. 1 589 924, 1981.
11. Wanske, W.; Siegemund, G.; Schnieder, W. Eur. Pat. EP 417 680, 1991; *Chem. Abstr.* **1991,** *114,* 184784k.
12. Kielhorn, F. F.; Manogue, W. H. WO 91/05752, 1991; *Chem. Abstr.* **1991,** *115,* 158497u.
13. Morikawa, S.; Samejuna, S.; Yositake, M.; Tatematsu, S. Eur. Pat. EP 347 830 A3, 1989; *Chem. Abstr.* **1990,** *112,* 734785t.
14. Gertvasutti, G.; Marangoni, L.; Marra, W. *J. Fluorine Chem.* **1981/1982,** *19,* 1.
15. Darragh, J. I. Belgium Pat. BE 867 285 A, 1978; *Chem. Abstr.* **1978,** *90,* 151557w.

Fire Suppression Agents

by Mark L. Robin

The unique combination of properties associated with certain fluorinated methanes and ethanes has led to their widespread use in fire suppression systems. The three halogenated fire suppression agents in general use today are bromotrifluoromethane (CF_3Br, Halon 1301), bromochlorodifluoromethane (CF_2BrCl; Halon 1211) and, in small volumes, primarily in the republics of the former Soviet Union and Eastern European nations, 1,2-dibromotetrafluoroethane ($BrCF_2CF_2Br$; Halon 2402).

These agents extinguish fire by a mechanism involving both a physical and a chemical contribution. The agents absorb heat, resulting in the slowing down of heat-producing reactions and a lowering of the flame temperature [*1, 2*]. The agents also suppress fires chemically by dissociating in the flame to produce species that interfere with the normal chain reactions responsible for flame propagation [*3, 4, 5, 6*].

These fire suppression agents are characterized by high fire suppression efficiency, low toxicity, low residue formation following extinguishment, low electrical conductivity, and long-term storage stability [*7*]. High fire suppression efficiency and high liquid densities permit the use of compact storage containers. Because the agents produce no corrosive or abrasive residues upon extinguishment, they are used to protect such areas as libraries and museums, where the use of water or solid extinguishing agents could cause secondary damage equal to or exceeding that caused by direct fire damage. Because they are electrically nonconducting, the agents can be used to protect electrical and electronic equipment. Bromotrifluoromethane and bromochlorodifluoromethane are effective fire suppression agents at concentrations safe for human exposure and, because of their low toxicity, may be used in areas that are normally occupied by humans and in areas where the egress of personnel may be undesirable or impossible. Bromotrifluoromethane and bromochlorodifluoromethane fire suppression systems are used to protect nuclear power facility control rooms, aircraft carriers, submarines, radar, military command and control centers, armored vehicles, museums, libraries, electronic control rooms, commercial aircraft, machinery spaces, engine nacelles, petroleum production units, and processing and storage areas for flammable materials [*8*].

In addition to their applications as extinguishing agents, these agents also are used as inerting agents, that is, as agents that prevent the initiation of a fire or explosion [*2*]. In these applications, the space to be protected is filled with the inerting agent to a concentration sufficient to render the contents of the space nonflammable.

0065–7719/95/0187–1099$08.00/1

During the past 20 years, the use of these highly efficient, clean, nontoxic fire suppression agents has prevented the loss of human life, and these agents currently protect billions of dollars worth of equipment. Fixed fire suppression systems using halons number in the thousands, and even more portable extinguishers are in use. Approximately 40 million pounds of halons are manufactured worldwide each year, with annual sales of $80–100 million [*9*]. Because of their unique combination of properties, they have served as near ideal fire suppression agents. However, because of their recent implication in the destruction of stratospheric ozone, the production and use of these life-saving agents is being severely restricted. As a result, intensive research efforts are currently underway in both the industrial and academic sectors to find suitable replacements for these agents [*10, 11*].

References for Pages 1099–1100

1. Ewing, C. T.; Hughes, J. T.; Carhart, H. W. *Fire and Materials* **1984,** *8,* 148.
2. Ford, C. L. In *Halogenated Fire Suppressants;* Gann, R. G., Ed.; ACS Symposium Series 16; American Chemical Society: Washington, DC, 1975; pp 1–63.
3. Biordi, J. C.; Lazzara, C. P.; Papp, J. F. In *Halogenated Fire Suppressants;* Gann, R. G., Ed.; ACS Symposium Series 16; American Chemical Society: Washington, DC, 1975; pp 256–294.
4. Biordi, J. C.; Lazzara, C. P.; Papp, J. F. *J. Phys. Chem.* **1976,** *80,* 1042.
5. Sheinson, R. S.; Penner-Jahn, J. E.; Indritz, D. *Fire Technology* **1989,** *15,* 437.
6. Westbrook, C. K. *Combust. Sci. Technol.* **1983,** *34,* 201.
7. Ford, C. L. In *NFPA Fire Protection Handbook,* 15th ed.; National Fire Protection Association: Quincy, MA, 1981; pp 18–21.
8. Pitts, W. M.; Nyden, M. R.; Gann, R. G.; Mallard, W. G.; Tsang, W. *Natl. Inst. Stand. Technol. Note 1279,* 1990.
9. Hogue, C. *Int. Environ. Rep.* **1990,** November 21, pp 493–495.
10. Robin, M. L. Presented at the Halon Alternatives Technical Working Conference, Albuquerque, New Mexico, 1991; paper 1.
11. Tapscott, R. E.; Floden, J. R. Presented at the 200th National Meeting of the American Chemical Society: Washington, DC, 1990; FLUO 29.

Organic Fluoropolymers

by W. W. Schmiegel

Although the technological basis of all fluorine-containing plastics and most elastomers continues to be the free radical polymerization of fluoroolefins, which themselves are based on the vastly greater fluorocarbon refrigerant industry, important advances have been made in the past two decades. These include primarily the production of polymers that are more resistant to degradation by heat, oxidation, bases, and solvents, as well as polymers that are more easily processable, that is, able to be converted into their final forms for use, whether by thermoplastic or thermoset processes [*1, 2, 3, 4*].

Examples of polymers with *improved resistance to thermooxidative degradation* are ***fluoroelastomers*** with increased fluorine contents and fluorocarbon plastics with end groups fluorinated by postpolymerization treatment. Although sometimes at the expense of decreased oil resistance, improved base resistance of fluoroelastomers has been achieved by replacement of ***vinylidene fluoride (VDF)*** by ethylene or propylene in copolymers of ***tetrafluoroethylene (TFE), hexafluoropropylene (HFP),*** and ***perfluoro(methyl vinyl ether) (PMVE).*** Needed processability improvements have been achieved by end group modification with chain-transfer agents during polymerization. In this way, the concentration of ionic end groups derived directly from inorganic initiators such as persulfates is significantly reduced and the ionic self-association of end groups into ionomeric aggregates is diminished. The resultant lower bulk viscosity of such polymers allows fabrication of larger and more complex finished parts because of increased flow during molding at a given temperature and pressure. Control over molecular weight distribution either during polymerization or by polymer blending has also afforded processability improvements.

A major development in ***fluoroplastics*** is the recent small-scale production of ***Teflon AF***, a noncrystalline (amorphous) fluorocarbon polymer with a high *glass transition temperature* (240 °C). This optically transparent TFE copolymer is soluble in certain fluorocarbons and has the same chemical and oxidative stability as crystalline TFE homopolymers [*5*].

The monomer compositions, trade names, and manufacturers of major types of fluoropolymers are listed in Table 1 [*1, 2, 3, 4, 6, 7, 8*].

0065–7719/95/0187–1101$08.00/1

Table 1. Major Commercial Organic Fluoropolymers [*1, 2, 3, 4, 6, 7, 8*]

Polymer Type	*Acronym*	*Monomer*	*Trademark*	*Producer*	*Country*
Plastic homopolymer	PTFE	CF_2=CF_2	Algoflon, Halon	Ausimont	Italy USA
			Fluon	ICI	UK
			Hostaflon	Hoechst	Germany
			Polyflon	Daikin	Japan
			Teflon	Du Pont	USA Netherlands
	PCTFE	CF_2=CFCl	Actar	Allied Signal	USA
			Aclon	Ausimont	Italy USA
			Daiflon	Daikin	Japan
			Kel–F	3M	USA
	PVDF	CH_2=CF_2	Dyflor	Dynamit Nobel	Germany
			Kureha KF	Kureha Chemical	Japan
			Kynar	Atochem	USA
			Neoflon VDF	Daikin	Japan
			Solef	Solvay	Belgium
	PVF	CH_2=CHF	Tedlar	Du Pont	USA
Plastic copolymer	FEP	CF_2=CF_2	Neoflon	Daikin	Japan
		CF_2=$CFCF_3$	Teflon FEP	Du Pont	USA Netherlands
	PFA	CF_2=CF_2	Hostaflon TFA	Hoechst	Germany
		CF_2=$CFOR_f$	Neoflon AP	Daikin	Japan
			Teflon	Du Pont	USA
	ETFE	CF_2=CF_2	Aflon COP	Asahi Glass	Japan
		CH_2=CH_2	Halon ET	Ausimont	Italy USA
		CF_2=$CFOR_f$	Hostaflon ET	Hoechst	Germany
		or	Neoflon EP	Daikin	Japan
		CH_2=CHR_f	Tefzel	Du Pont	USA

Table 1—Continued

Polymer Type	*Acronym*	*Monomers*	*Trademark*	*Producer*	*Country*
	ECTFE	CF_2=CFCl CH_2=CH_2	Halar	Ausimont	USA
	(None)	CF_2=CF_2	Teflon AF (amorphous)	Du Pont	USA
Elastomeric copolymer		CH_2=CF_2	Dai-el	Daikin	Japan
		CF_2=$CFCF_3$	Fluorel Tecnoflon Viton	3M Ausimont Du Pont	USA Italy USA Netherlands
		CH_2=CF_2	Dai-el	Daikin	Japan
		CF_2=CF_2 CF_2=$CFCF_3$	Fluorel Tecnoflon Viton	3M Ausimont Du Pont	Italy USA Netherlands
		CH_2=CF_2 CF_2=CF_2 CF_2=$CFOCF_3$	Viton GLT, GFLT	Du Pont	USA Netherlands
		CF_2=CF_2 CF_2=$CFOCF_3$	Kalrez	Du Pont	USA
		CF_2=CF_2 CH_2=$CHCH_3$	Aflas 150	Asahi Glass	Japan
		CF_2=CF_2 CH_2=$CHCH_3$ CH_2=CF_2	Aflas 200	Asahi Glass	Japan
		CF_2=CF_2 CH_2=CH_2 CF_2=$CFOCF_3$	VTR–6186	Du Pont	USA

Monomer Synthesis

In addition to the pyrolysis of chlorodifluoromethane [*9*], another commercially important synthesis of TFE is based on trifluoromethane [*10*] (equation 1).

[10]
$$2CHF_3 \xrightarrow{700\text{-}800°C} CF_2{=}CF_2 \qquad 1$$

References are listed on pages 1116–1118.

Besides the dechlorination of 1,2-dichloro-1,1-difluoroethane [*11, 12*], two more commercial routes to VDF have been developed [*9, 13, 14, 15, 16*] (equations 2 and 3).

[9, 13, 14, 15] $$CH_3CClF_2 \xrightarrow{700\text{-}900°C} CH_2{=}CF_2 + HCl$$ 2

[16] $$CH_3CF_3 \xrightarrow{1100\text{-}1300°C} CH_2{=}CF_2 + HF$$ 3

Little 1-*H*-pentafluoropropylene is produced anymore, because its existence is no longer justified as a less stable alternative to HFP in VDF-based elastomers, given the expiration of patents covering the basic VDF/HFP/(TFE) compositions.

Perfluoro-1,1-dimethyl dioxole is prepared from hexafluoroacetone and ethylene oxide in four proprietary steps [*5*] (equation 4).

[5] 4

$(CF_3)_2C{=}O$ + CH_2—CH_2 (epoxide, O bridging) → $(CF_3)_2C$<(O—CH_2)(O—CH_2) →

$(CF_3)_2C$<(O—CCl_2)(O—CCl_2) → $(CF_3)_2C$<(O—CClF)(O—CClF) → $(CF_3)_2C$<(O—CF)(O—CF) (with CF=CF double bond)

Polymerization

Free-radical polymerization of fluoroolefins continues to be the only method of producing high-molecular-weight polymers. Attempt at anionic polymerization, usually with fluoride initiation, lead to very low molecular weight polymers because of facile chain transfer via uncontrollable fluoride elimination from propagating carbanions and subsequent re-initiation on unreacted monomer. Cationic initiation is ineffective, because the electron-deficient double bond has too little affinity for a positive charge and the resultant fluorocarbon cation would be inductively destabilized. Thus, unlike anionic propagation centers, which readily lose fluoride in a chain-transfer process, and cationic propagation centers, which do not form, free radical propagation centers add efficiently to monomers because of the absence of electrostatic repulsion at carbon and the lack of an efficient pathway for unimolecular decomposition to form a terminal double bond. Another way to view this difference is to consider the relative energetics for double bond formation from a β-fluoro carbanion compared with a β-fluoro carbon-centered free radical.

References are listed on pages 1116–1118.

High-molecular-weight homopolymers of TFE, CTFE, VDF, and ***vinyl fluoride (VF)*** can be prepared by commercial processes, but homopolymerization of PMVE and especially HFP requires extreme conditions and is not practiced commercially.

Copolymerization of fluoroolefins has resulted in a wide variety of materials from plastics with improved processing and performance characteristics and lower crystallinity, including entirely noncrystalline polymers, which may be elastomeric at temperatures as low as –30 °C or may exhibit structural rigidity up to 240 °C.

Typically *fluoroolefin polymerizations* are conducted in aqueous emulsions with free radical initiators like ammonium persulfate or persulfate–bisulfite redox couples, chain-transfer agents to control molecular weight and end-group structure, surfactants to emulsify polymer particles, and various modifiers to influence either initiation or monomer solubility. Numerous proprietary variations of these parameters apply to actual commercial processes, and it is generally difficult to deduce from patents which product corresponds to which process.

Solution polymerization, when practiced, generally takes advantage of initiator systems that avoid the formation of ionic end groups and requires fluorinated solvents to minimize *chain transfer,* except when low-molecular-weight products are deliberately sought. Examples of solvents that efficiently promote chain transfer are chlorocarbons and aliphatic esters and ketones. The primary considerations in selecting a chain-transfer agent are its ability to donate a hydrogen or, in the case of chlorocarbons, a chlorine atom to the propagating polymer radical and its ability as a free radical to initiate efficiently further polymerization.

Properties of Fluoropolymers

The fluorine content and its distribution along the chain are the principal determinants of the properties of fluoropolymers. Generally, the higher the fluorine content, the greater the chemical and thermooxidative stability, solvent resistance, weatherability, and melting point. However, because of its in-chain ether structure, ***polyhexafluoropropylene oxide (PHFPO),*** which unfortunately has not been prepared in elastomer-grade molecular weights, is an exception to the general principle that associates a high fluorine content with low segmental mobility. Because of the random orientation of the asymmetric carbon, this anionically initiated polyether also fails to crystallize and therefore exhibits excellent low-temperature flexibility. Its limitation to low molecular weight nonetheless allows production of highly inert, wide-temperature service fluids such as *heat-transfer media and lubricants* [*17*].

Fluoroplastics

Polytetrafluoroethylene (PTFE)

Polytetrafluoroethylene is a linear polymer of the general formula $–(CF_2 CF2)_n–$. Its molecular weight (10^6 to 10^7) and its *melting point* (327 °C) are extremely high. The *usable temperature range* of PTFE extends from below

–200 °C to above 260 °C. Because of its extremely high melt viscosity (10–100 GPa·s at 380 °C), conventional thermoplastic processing techniques are inapplicable and special fabrication techniques have had to be developed.

PTFE is insoluble in all known solvents and resists attack by most chemicals, although its surface is readily degraded by alkali metal–ammonia solutions. Such solvated electron media etch PTFE to produce a brown–black layer quite unlike the original white, low-friction, nonstick surface.

The physical properties of PTFE are dominated by the extremely long and rigorously linear chain structure. Steric hindrance between fluorine substituents prevents the adoption of a planar zigzag structure and instead requires the chain to assume a helical zigzag structure that is characterized by a net 180° twist along the chain axis every 13–15 carbon atoms. The *high melting point (327 °C)* reflects the small entropy of fusion of its stiff chains. The interchain bonding energy is very low (3.2 kJ/mol), as is the surface energy (18.6 mN/m). The melting point of *as*-polymerized PTFE is 342 °C, but once melted, its original melting point cannot be recovered and is fixed at 327 °C. This phenomenon is due to the irreversible reduction of a rodlike crystallinity characteristic of 92–98% of the virgin material to values considerably lower and dependent on molecular weight and the rate of cooling from the melt. Commercial grades of PTFE are about 45–75% crystalline. The mechanical properties of PTFE are influenced by the processing history of the sample. Variables such as preform pressure, sintering temperature, and cooling rate, as well as polymer properties such as molecular weight, particle size, and particle size distribution, determine mechanical behavior. Affected properties include tensile strength, elongation, flex life, stiffness, resilience, impact strength, and permeability. Relatively unaffected properties are low-temperature flexibility, thermal stability, and coefficient of friction.

The *frictional properties of PTFE* are unique. Its unusually low static coefficient of friction decreases with increasing load and is lower than the dynamic coefficient of friction. This precludes stick-slip behavior. The low surface energy also prevents wetting by liquids other than low-surface-tension fluids like fluorocarbons.

The electrical properties of PTFE are dominated by its *extremely low dielectric constant* (2.1). This value is invariant over a broad range of temperatures (−40 to 250 °C) and frequencies (5 Hz to 10 GHz). Similarly, PTFE has an unusually *low dissipation factor,* which is also quite independent of temperature and frequency. This behavior results from the high degree of dipolar symmetry of the perfluorinated and unbranched chains. The dielectric strength, resistivity, and arc resistance are very high.

The *chemical resistance of PTFE* is almost universal. It resists attack by aqua regia, hot fuming nitric acid, hot caustic, chlorine, chlorosulfonic acid, and all solvents. Despite this broad chemical resistance, PTFE is attacked by *molten alkali metals, ammonia solutions of such metals, chlorine trifluoride, and gaseous fluorine* at elevated temperature and pressure. PTFE swells or dissolves in certain highly fluorinated oils near its melting point. Specific lists of chemicals compatible with PTFE are available [*18*].

The *thermal stability of PTFE* is exceptionally high in both nitrogen and air. Thermal decomposition rates are immeasurably slow at 440 °C but high at 540 °C.

PTFE decomposes to TFE with first-order kinetics and a 347.4-kJ/mol activation energy under vacuum pyrolysis conditions. It is extremely flame resistant and does not burn in air. Its limiting oxygen index (LOI), the minimum oxygen content of an atmosphere under ambient conditions that sustains combustion, is 96%, which means that it requires almost pure oxygen for combustion.

High-energy electron beam or gamma radiation degrades PTFE by generating radicals that lead to chain scission. The presence of oxygen accelerates the rate of scission and degradation. Mechanical properties (Table 2) decrease markedly after high dosages [*1, 2*].

Table 2. Mechanical and Electrical Properties of Commercial PTFE Resins [*1*]

Property	*Granular*	*Fine Powder*
Tensile strength, MPa	7–28	18–24
Elongation, %	100–200	300–600
Flexural modulus, MPa	350–630	280–630
Coefficient of thermal expansion, K^{-1}	1.2×10^4	
Dielectric constant (60 Hz to 2 GHz)	2.1	2.1
Dissipation factor (60 Hz to 2 GHz)	0.0003	
Volume resistivity (ohm·cm)	$>10^{18}$	$>10^{18}$
Dielectric strength (2 mm thickness, kV/mm)	23.6	23.6

Tetrafluoroethylene–Hexafluoropropylene Copolymers (FEP)

Copolymerization of TFE and HFP results in a linear, perfluorinated polymer, commonly referred to as an **FEP *(fluorinated ethylene-propylene) resin,*** of the following general structure:

$$-[(CF_2CF_2)_xCF_2\underset{\displaystyle |}{\overset{\displaystyle CF_3}{}}CF]_y-$$

Typical HFP contents are about 8.5 mol %. Crystallinity as polymerized is about 70% and drops to about 50% when molded. The steric requirements of its pendant trifluoromethyl groups greatly reduce its tendency to crystallize relative to PTFE. The incorporation of HFP also decreases its molecular weight to only about 0.01

References are listed on pages 1116–1118.

of that of PTFE. These two characteristics are primarily responsible for the 10^6-fold lower melt viscosity of FEP relative to PTFE and allow the *use of standard thermoplastic processing techniques.* Most properties of FEP resins are quite similar to those of PTFE. These include chemical resistance, nonflammability, and electrical and frictional properties. The mechanical properties of FEP from –25 to 200 °C also are similar to those of PTFE; however, FEP rapidly loses its tensile properties above 200 °C, whereas PTFE can be used to at least 260 °C. In nitrogen, FEP is about 10 times as resistant to high-energy radiation as PTFE. In air, both degrade at the same rate [*1, 2*].

Tetrafluoroethylene–Perfluoro(propyl vinyl ether) Copolymers (PFA)

Copolymerization of TFE and perfluoro(propyl vinyl ether) (PPVE) yields melt-processable, linear polymers known as ***perfluoroalkoxy (PFA) resins*** of the following general structure:

$$\begin{array}{c} \qquad\qquad\qquad OCF_2CF_2CF_3 \\ \qquad\qquad\quad | \\ -[(CF_2CF_2)_xCF_2CF]_y- \end{array}$$

These polymers were developed after FEP resins and are designed to retain the *excellent thermal stability and high-temperature mechanical properties of PTFE resins while retaining the melt processability of FEP resins.* The perfluoroalkoxy substituent introduces less steric strain in the backbone than does the less flexible trifluoromethyl substituent of FEP resin [*19*] and is also much more effective on a molar basis in interrupting crystallinity. The crystallinity of PFA just after polymerization is about 70% and drops to about 60% on cooling from the melt. The *melting point* of commercial grades (*305–310 °C*) is about 20 °C below that of PTFE; however, the thermal stabilities and temperatures of maximum use of PFA and PTFE resins are about the same. The mechanical, frictional, adhesive, and chemical properties of PFA resins are similar to those of PTFE. Besides a lower melt viscosity and higher critical shear rate compared with HFP, PFA resins also exhibit good optical properties as thin films. Energy transmission is high in the infrared, visible, and ultraviolet regions. It resistance to high-energy radiation is similar to those of PTFE and FEP [*1, 2*].

Tetrafluoroethylene–Ethylene Copolymers (ETFE)

Copolymerization of TFE and ethylene yields highly alternating, linear polymers of the general structure $-(CF_2CF_2CH_2CH_2)_n-$. These materials, known as ***ETFE resins,*** possess high tensile strength, moderate stiffness, high flex resistance, superior impact strength and cut-through resistance, and good abrasion resistance. Their electrical properties are similar to those of PTFE, FEP, and PFA resins. ETFE resins are highly resistant to gross degradation by high-energy radiation and, on exposure, undergo a net gelation reaction rather than chain scission. This tendency is exploited when deliberate cross-linking is desired to improve high-temperature mechanical properties and can be enhanced by dispersing radical traps, such as certain polyfunctional allyl compounds, into the resin. The critical shear rate of ETFE resins is about 250 times greater than that of FEP. This property ensures *superior processability* relative to FEP and confers excellent *extrusion and injection molding behavior.*

Simple copolymers of TFE and ethylene suffer from poor stress-crack resistance, which severely limits their service life in certain applications. To overcome this deficiency, commercial ETFE resins are prepared by terpolymerization and contain from about 0.1 to 10 mol % of additional monomers such as perfluoro(alkyl vinyl) ethers or perfluoroalkyl ethylenes [*20, 21, 22*]. The resistance of ETFE to chemicals and solvents is excellent. *Strong acids and bases do not affect ETFE;* however, strong oxidizing acids, sulfonic acids, and organic bases attack the polymer at elevated temperatures. Some high-boiling esters such as diisobutyl adipate dissolve ETFE resins above 200 °C. Their limiting oxygen index (31%) *precludes flammability in air* [*1, 2*].

Polychlorotrifluoroethylene (PCTFE)

Polychlorotrifluoroethylene is a linear polymer of the following general formula:

$$
\begin{array}{c}
\quad\;\, Cl \\
\quad\;\, | \\
-(CF_2CF)_n-
\end{array}
$$

PCTFE is semicrystalline and *melt-processable* and exhibits good mechanical properties, *chemical inertness, resistance to radiation,* and resistance to permeation by vapors. When quenched quickly from the melt, PCTFE is optically clear despite its substantial crystallinity (45%). When cooled slowly from the melt, it is translucent and has 65% crystallinity. Its *melting point* is ca. 210 °C, and its *useful temperature range is from –200 to 180 °C.* Although resistant to degradation by most chemicals, PCTFE is *swollen by chlorocarbons, ketones, and simple aromatic hydrocarbons.* Its limiting oxygen index is 95%. PCTFE is highly resistant to UV and gamma radiation. Its coefficient of friction is similar to that of FEP [*1, 23*].

Chlorotrifluoroethylene–Ethylene Copolymers (ECTFE)

Copolymerization of CTFE and ethylene yields linear, semicrystalline polymers known as ***ECTFE resins.*** They have a highly alternating structure:

$$
\begin{array}{c}
\;\; Cl \qquad\qquad\quad \\
\;\; | \qquad\qquad\quad \\
-(CF_2CFCH_2CH_2)_n-
\end{array}
$$

ECTFE resins have high tensile strength, moderate stiffness, high flex resistance and impact strength, and excellent cut-through resistance. They exhibit *good electrical properties, chemical and radiation resistance,* and *nonflammability.* Commercial ECTFE resins that contain 50 mol % CTFE are about 50–60% crystalline and 80–90% alternating and have *melting points of 235–245 °C.* Their *useful temperature range is from –100 to 150 °C.* ECTFE and homopolymer PCTFE have similar chemical resistance. Stability toward acids, bases, and halogens is good, but amines, esters, and ketones can attack ECTFE resins, particularly when warm. The polymer is insoluble below 120 °C and exhibits superior barrier properties with respect to water vapor and other gases. *ECTFE is nonflammable* and burns only when exposed to flame. Its limiting oxygen index is 64% [*1, 24*].

Poly(vinylidene fluoride) (PVDF)

Poly(vinylidene fluoride) is a linear, semicrystalline polymer of the general structure $-(CH_2CF_2)_n-$. About 5% of the monomer units are reversed from their normal head-to-tail orientation in the chain [*25*]. The alternating methylene and difluoromethylene groups create strong dipoles that can interact with highly polar solvents and confer solubility. Its resistance to abrasion, stress fatigue, cold flow, and radiation is excellent. Its *useful temperature range extends from –40 to 150 °C.* Copolymerization with HFP (<15%) improves failure properties like elongation at break and impact strength but decreases the tensile modulus. Compared with ETFE and ECTFE, unmodified PVDF has a higher yield strength and lower elongation at room temperature. The electrical properties of PVDF are unique. Its high dielectric constant (8–9) and its dissipation factor preclude its use as a primary insulator but allows its use in the construction of *piezoelectric and pyroelectric films* [*26*]. Such films are prepared from the phase-1 (beta-form) morphology, which is obtained on melt extrusion, when such films are oriented by stretching, metallizing on both sides, and then permanent polarization by a high-voltage field. Subsequent stretching or compression of the polarized film generates a piezoelectric voltage. Similarly, heating the film creates a pyroelectric voltage. *PVDF is resistant to most inorganic acids and bases, alcohols, halogenated hydrocarbons, and aromatics;* however, polar solvents like acetone, ethyl acetate, dimethylformamide (DMF), and (DMAC) readily dissolve the polymer. Even relatively weak bases like aliphatic amines easily attack PVDF, especially when swollen or dissolved. It is *nonflammable* in air and has a limiting oxygen index of 44% [*1, 27, 28*].

Poly(vinyl fluoride) (PVF)

Poly(vinyl fluoride) is a linear, semicrystalline polymer of the following general structure:

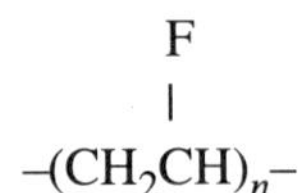

References are listed on pages 1116–1118.

As normally polymerized, PVF *melts between 185 and 210 °C* and contains 12–18% inverted monomer units. It is normally considered a *thermoplastic,* but because of its instability above its melting point, it cannot be processed by conventional thermoplastic techniques. Instead it is generally *extruded into films in a solvent-swollen (organosol) form,* and the solvent is subsequently evaporated and recovered. Such films can be oriented further to achieve specific mechanical properties. PVF films are exceptionally weather and radiation resistant considering their modest fluorine content. PVF is insoluble below 100 °C but, at higher temperatures, it dissolves in polar solvents like amides, ketones, tetramethylene sulfone, and tetramethylurea. *Resistance to acids and bases at room temperature is good* [*1, 29*].

Tetrafluoroethylene–Perfluoro(2,2-dimethyl-1,3-dioxole) Copolymers (Teflon AF)

This unique family of amorphous ***perfluoroplastics*** combines the thermal and chemical stability of semicrystalline, melt-processable perfluoroplastics with optical transparency, solubility in selected fluorocarbon solvents, and increased tensile modulus. Teflon AF amorphous fluoropolymers are random copolymers with glass transition temperatures as high as 300 °C and *upper use temperatures up to 285 °C.* The glass transition temperature increases with increasing dioxole content, and present commercial products have an *upper limit of 240 °C. Refractive index* decreases with increasing dioxole content to values as low as 1.29, which is lower than for any other fluoropolymer. The coefficient of linear thermal expansion in the glassy state ranges from 80 to 100 ppm/K. From 1 to 10 000 MHz at 23 °C, the dielectric constant ranges from 1.89 to 1.93 and the dissipation factor ranges from 7.3×10^{-5} to 35×10^{-5}.

Light transmission of a 220-nm film is 95% over the visible and near-infrared range (400–2000 nm) and decreases from 95 to 80% from 400 to 200 nm in the ultraviolet. Solubility in perfluorinated ethers such as Fluorinert FC–75 is as high as 15%. The weight loss in air at 260 °C/4 h is zero; at 380 °C/1 h, it is 0.53% [5].

Fluoroelastomers

To be considered *an elastomer,* a polymer generally *has to exhibit flexibility and recovery from elongation or compression above 0 °C.* This criterion requires sufficient rotational mobility among chain segments at and above this temperature to allow the elastomeric specimen to regain its original or nearly original dimensions when released from a substantial (e.g., 50%) deformation. The temperature at which a polymer emerges on heating from its glassy state, a state characterized by low segmental mobility but without crystallinity, and becomes segmentally mobile is known as the glass transition temperature and defines the lower end of its useful elastomeric temperature range. This value depends somewhat on the method of measurement but generally encompasses a range of up to 10 Kelvin, es-

References are listed on pages 1116–1118.

pecially for random copolymers. To prevent irreversible deformation by slow flow, these highly viscous liquids are cross-linked. The constraint imposed by the cross-links allows the uniaxial orientation upon simple extension to relax to the original higher entropy state of disordered chains at equilibrium and thereby regain the original macroscopic dimensions. Recovery from compression, equivalent to biaxial extension, can be viewed in similar terms.

All commercially produced *fluorocarbon elastomers are copolymers of one or two symmetrical olefins and an additional unsymmetrical olefin* whose function is to interrupt the structural regularity that would result from its absence in the molecule and, therefore, preclude crystallization. This is true of random as well as alternating copolymers because of the random distribution of asymmetric centers from nonstereoregular addition by the propagating radical from the unsymmetrical olefin to the next monomer. The resulting copolymer has equal numbers of randomly distributed D and L centers and resists crystallization, in proportion to the concentration of the unsymmetrical monomer in the case of nonalternating copolymers. A further contribution to the interruption of structural regularity in both random and alternating copolymers, and indeed homopolymers, comes from occasional head-to-head incorporation (monomer inversion) of an unsymmetrical monomer, which results in reduced regioselectivity. This can occur with either vinyl or vinylidene monomers. This inversion, which depends on polymerization temperature, is self-correcting in the sense that the normal direction of addition resumes after a single inversion.

The description of the physical properties of fluoroelastomers is necessarily less precise than that of fluoroplastics because of the major effect of adding curatives and fillers to achieve useful cross-linked materials of a given hardness and specific mechanical properties. Generally, two parameters are varied: increasing cross-link density increases modulus and decreases elongation, and raising filler levels increases hardness and decreases solvent swell because of the decreased volume fraction of the elastomer. In addition to these two major variables, the major determinants of vulcanizate behavior are the chemical and thermal stabilities of its cross-links. The selection of elastomer, of course, places limits on the overall resistance to fluids and chemicals and on its service temperature range.

Fluorocarbon elastomers are conveniently divided into three groups: (1) the most widely used group of copolymers of VDF and one or two perfluoroolefins, (2) a less widely used group based on TFE and simple hydrocarbon olefins, and (3) a much smaller group comprising copolymers of TFE and perfluoroolefins [*1, 7, 30*].

Vinylidene Fluoride-Based Copolymers

The most widely used fluoroelastomers are ***copolymers of VDF and HFP*** and optionally also ***TFE.*** HFP interrupts the crystallinity of otherwise crystalline PVDF. TFE increases the fluorine content for increased solvent and heat resistance without raising the glass transition temperature as much as would an equivalent amount of additional HFP. The dipolymer is random except that there are no contiguous HFP units and can therefore be represented as follows:

References are listed on pages 1116–1118.

$$-(CH_2CF_2CF_2\overset{\displaystyle CF_3}{\overset{|}{C}}F)_x(CH_2CF_2)_y$$

A VDF unit also follows HFP in the structurally much more complex terpolymers [*30*]. *Dipolymers contain 60 wt % VDF* (66% F) *and terpolymers contain 33–50 wt % VDF* (66–69.5% F) and generally less than 28 wt % TFE. Molecular weights range from 1.5×10^5 to 10^6. Their respective *continuous service temperature ranges are –18 to 210 °C and –12 to 230 °C.* Dipolymer vulcanizates retain over 50% of their tensile strength for more than 1 year at 200 °C or for 2 months at 260 °C.

Both the dipolymers and terpolymers have *excellent resistance to hydrocarbons* found in petroleum-based fuels and lubricants. The 69.5% F terpolymer resists swelling in blended fuels that contain methanol and can be used in contact with certain phosphate ester-based hydraulic fluids. Terpolymers are preferred for contact with aromatic solvents, although either type performs well in higher alcohols. VDF-based elastomers dissolve in polar aprotic solvents such as ketones, esters, amides, and certain ethers. These elastomers are therefore not suitable for contact with fluids that contain substantial amounts of these solvents because of excessive swell and consequent loss of mechanical properties.

Dehydrofluorination by primary and secondary aliphatic amines occurs at room temperature and is the basis of *diamine cross-linking,* which occurs by dehydrofluorination and subsequent nucleophilic substitution of the double bond. The locus of dehydrofluorination is a VDF unit flanked by two perfluoroolefin units. This selectively base-sensitive methylene group also undergoes elimination as the first step in phase-transfer-catalyzed cross-linking with quaternary ammonium or phosphonium salts, bisphenols, and inorganic oxides and hydroxides as HF acceptors [*31, 32*].

Resistance to aqueous alkali is far greater than to weaker soluble bases like amines; however, surface cracking eventually occurs. This process is accelerated if swelling solvents or phase-transfer cations also are present. *Resistance to aqueous acids is generally good.* Ozone resistance and weatherability are excellent, and gas permeability is low. Specific compounding affords nonflammable or self-extinguishing vulcanizates. Resistance to electron radiation is moderate and independent of composition. Failure occurs by embrittlement (excessive cross-linking) rather than by chain scission.

PMVE-containing VDF/TFE-based elastomers have significantly better low-temperature properties than corresponding HFP-containing terpolymers without sacrificing fluid resistance or thermal stability. In addition to their higher cost due to PMVE, they also require the incorporation of a cure site monomer for efficient cross-linking. Even though they react readily with base, they eliminate the elements of trifluoromethanol, which gives carbonyl fluoride, and, unlike HFP-containing polymers, give a double bond that is insufficiently reactive to undergo efficient substitution by basic bisnucleophilic cross-linking agents [*31*]. They are therefore also somewhat more resistant to embrittlement by soluble bases. These polymers are instead cross-linked by peroxides and polyfunctional allyl compounds that act on bromine- or iodine-containing sites introduced by cure site monomers or chain-transfer agents [*33, 34*].

Propylene- and Ethylene-Based Fluoroelastomers

Replacement of a vinylidene fluoride unit by an ethylene or propylene unit in a locally perfluorinated chain environment greatly reduces the acidity of the methylene hydrogens. Copolymers of TFE and propylene are therefore considerably more resistant to bases and polar solvents than VDF-based elastomers. TFE and propylene form a highly alternating structure:

$$-(CF_2CF_2CH_2\underset{\displaystyle CH_3}{\overset{|}{C}}H)_n- $$

This structure is similar to that of the copolymer TFE and ethylene, except that the random orientation of the methyl group from nonstereospecific free radical copolymerization of propylene affords a noncrystalline structure [*35*]. The relatively low fluorine content (54%) of these elastomers compared with VDF-based elastomers (66–69.5%) makes them significantly less *resistant to swelling by hydrocarbons.* Because of strict alternation, these elastomers have a relatively *high glass transition temperature (–2 °C)* and consequently limited low-temperature properties. Furthermore, they must be polymerized with a cure site monomer or receive a postpolymerization treatment to adequately activate them for vulcanization [*36*]. To counteract the limited cure response, low-temperature flexibility, and hydrocarbon resistance, these polymers have also been modified with substantial amounts (ca. 35 wt %) of VDF [*37, 38*]. This provides some improvements but inevitably decreases the resistance to bases and polar solvents.

A somewhat different approach has been chosen in designing a base-resistant fluoroelastomer based on ethylene. Because of the symmetry of ethylene compared with propylene, PMVE was chosen as the asymmetric monomer and TFE as the major fluoromonomer. This monomer pair simultaneously provides a higher fluorine content (64%), avoids formation of base-sensitive sites and strongly hydrogen-bonding methylene groups, and affords outstanding solvent resistance and improved low-temperature properties. The only known solvents for this hydrofluoroelastomer are fluorocarbons and chlorofluorocarbons. Its inertness requires incorporation of a cure site monomer, which allows efficient vulcanization by peroxides in the presence of polyfunctional allylic radical traps. The outstanding properties of this elastomer are achieved at considerably higher cost compared with propylene-based elastomers [*39*].

References are listed on pages 1116–1118.

Perfluoroelastomers

Copolymerization of TFE with sufficient PMVE (40 wt %) to interrupt PTFE crystallinity affords an elastomer that has a thermal stability and chemical resistance approaching that of PTFE. Unlike the amorphous copolymer of TFE and perfluoro-2,2-dimethyl-1,3-dioxole, this copolymer has a low enough glass transition temperature (0 °C) to be an elastomer. Its very high fluorine content (73%) renders it *insoluble in all solvents except fluorocarbons and chlorofluorocarbons.* Its inertness requires incorporation of a cure site monomer for efficient vulcanization. Weight losses in air are 6.5 and 12% at 316 °C/7 days for the dipolymer and functional terpolymer, respectively. A vulcanizate cross-linked through a nitrile-containing cure site monomer fully retains its 100% tensile modulus, tensile strength, and elongation upon exposure to air at 288 °C/18 d. At 316 °C/18 d the 100% modulus decreases by less than 20%, tensile strength decreases by less than 20%, and elongation is unchanged [*1, 40, 41, 42*].

Processing and Applications

Fluoroplastics

The processing and applications of PTFE as described in the first edition of this book are essentially unchanged. All the fluoroplastics described, with the exception of PTFE and PVF, can be processed by conventional thermoplastic techniques such as injection molding and extrusion [*1, 30*].

FEP is primarily used in electrical applications such as a *wire covering.* Its chemical inertness also makes it ideal for *lining pipes and tanks, molding laboratory ware, and constructing heat exchangers* for use in corrosive environments. Its low frictional properties are well suited for *roll and conveyor covers,* and its high light transmission and weatherability fit requirements for *solar collector windows.*

PFA has applications similar to those of FEP but which require somewhat higher thermal stability and greater retention of mechanical properties. It finds extensive use in critical chemical-handling operations in the *semiconductor industry.*

ETFE is used in many critical wiring applications that require excellent cut-through resistance, such as *wiring in aircraft frames and power wiring in mass-transport systems.* It is also used for mechanical parts that require greater hardness than can be achieved with FEP and PFA. ECTFE has similar uses, but greater care must be taken in processing to prevent decomposition and corrosion.

PCTFE is used for laboratory ware and the construction of *chemical apparatus.* Its transparency makes it well suited for *sight glasses and tubes.* Thin films are used in *waterproof specialty packaging.*

PVDF is used as a *coating for metallic architectural substrates,* such as roofing, panel siding, and windows, and for *wire insulation* in the electronics and aircraft industries. It is also used to *mold pipes.*

References are listed on pages 1116–1118.

PVF is used primarily as a weather-resistant and environmentally protective film on the surface of laminated sheets of metal, hardboard, paper, and a variety of nonfluorinated plastics. Major uses are in *architectural siding, aircraft cabin interior panels, duct lining, and pipe covering.*

Teflon AF can be extruded and injection or compression molded. It can be formed by spin casting or applied from solution by spray coating or direct application [5]. Its extremely low refractive index and its solubility make it highly useful as a *coating for the core of optical fibers.*

Fluoroelastomers

Fluoroelastomers are prepared on two-roll rubber mills and in internal mixing equipment like Banbury mixers and twin-screw extruders. Unlike other elastomers except polyurethanes and thermoplastic elastomers, no processing oils or plasticizers are used because of the inadequate stability and compatibility of available materials. Fillers are used to reinforce or simply extend the compound to a given hardness. The amount of cross-linking agent used in an elastomer that is not limited by its cure site concentration determines the degree of cross-linking or cross-link density. This parameter primarily controls its elongation at break and its modulus. The fully formulated elastomer is *vulcanized by heating* under pressure in a mold for less than *15 min at 160–210 °C* or by *extrusion* through a shaping die followed by unconfined heating as in a steam autoclave. Depending on the thermal stability of the raw polymer, fluoroelastomers are *heated for up to 1 day in an air oven at 200–280 °C to stabilize the vulcanizate properties* by driving reactions to completion and degassing volatile reaction products from the continuous elastomer phase or its interfaces with filler particles or other insoluble constitments.

Fluoroelastomers are widely used in the automotive and other transportation industries as *oil seals for rotating and reciprocating shafts, O-ring static seals, gaskets, and hoses.* Other applications include hydraulic and pneumatic seals, hose and seals for the chemical process and oil industries, expansion joints for flue ducts of power plants, pollution control equipment, and fuser rolls for duplicating machines. ***Perfluoroelastomer*** parts are used in applications that require flexibility and sealing in extremely aggressive environments that surpass the resistance limits of hydrofluoroelastomers. A variety of *O-rings, shaft seals, gaskets, and hoses* are available for service in the manufacturing and aerospace industries. Their high cost is justified when no other material is functional and when premature part failure is unacceptably costly [*1*].

References for Pages 1101–1116

1. Carlson, D. P.; Schmiegel, W. W. *Ullmann's Encyclopedia of Chemical Technology,* 5th ed.; VCH: Weinheim, Germany, 1988; Vol. A11, p 393.
2. Gangal, S. V. In *Encyclopedia of Polymer Science and Engineering;* Mark, H. F.; Bikales, N. M.; Overberger, C. G.; Menges, G., Eds., Wiley: New York, 1989; Vol. 16, p 577.

3. Kawachi, S.; Nanba, N. J. *Syn. Org. Chem. (Japan)* **1984,** *42,* 829.
4. Brady, R. F. *Chem. Br.* **1990,** 427.
5. Resnick, P. R. *Polym. Prepr.* **1990,** *31,* 312; *Chem. Abstr.,* **1991,** *114,* 144578q.
6. Ramney, M. W. *Fluorocarbon Resins;* Noyes Data Corporation: Park Ridge, NJ, 1971.
7. Arnold, R. G.; Barney, A. L.; Thompson, D. C. *Rubber Chem. Technol.* **1971,** *46,* 619.
8. Wall, L. A. *Fluoropolymers;* Wiley-Interscience: New York, 1972; Vol. XXXV.
9. Downing, F. B.; Benning, A. F.; McHarness, R. C. U.S. Pat. 2 551 573, 1945; *Chem. Abstr.* **1951,** *45,* 9072e.
10. Hauptschein, M.; Fainberg, A. H. U.S. Pat. 3 009 966, 1960; *Chem. Abstr.* **1962,** *56,* 9961c.
11. Farlow, M. W.; Muetterties, L. U.S. Pat. 2 894 966, 1955; *Chem. Abstr.* **1959,** *53,* 19881d.
12. Calfee, J. D.; Miller, C. B. U.S. Pat. 2 734 090, 1954; *Chem. Abstr.* **1956,** *50,* 9441h.
13. Russel, M. M.; Bernhart, W. S. U S. Pat. 2 774 799, 1954; *Chem Abstr.* **1957,** *51,* 955b.
14. Miller, C. B. U.S. Pat. 2 628 989, 1951; *Chem Abstr.* **1954,** *48,* 1406i.
15. Scherer, O.; Steinmetz, A.; Kuehn, H.; Wetzel, W.; Grafen, K. U.S. Pat. 3 183 277, 1961; *Chem Abstr.* **1961,** *55,* 12295e.
16. Hauptschein, M.; Fainberg, A. H. U.S. Pat. 3 188 356, 1961; *Chem. Abstr.* **1965,** *63,* 6859a.
17. Eleuterio, H. S. *J. Macromol. Sci. Chem.* **1972,** *A-6,* 1027.
18. *J. Teflon* (Du Pont) **1970,** *Jan.–Feb.,* 11.
19. *Du Pont Innovations* **1973,** *5*(1), 16.
20. Carlson, D. P. U.S. Pat. 3 624 250, 1971; *Chem. Abstr.* **1969,** *71,* 13533s.
21. Ukihashi, H.; Yamabe, M. U.S. Pat. 4 123 602, 1978; *Chem. Abstr.* **1979,** *90,* 88051j.
22. Chandrasekaran, S. U.S. Pat. 3 853 811, 1975; *Chem. Abstr.* **1975,** *82,* 99371z.
23. Chandrasekaran, S. In *Encyclopedia of Polymer Science and Engineering;* Mark, H. F.; Bikales, N. M.; Overberger, C. G.; Menges, G., Eds.; Wiley: New York, 1985; Vol. 3, p 463.
24. Miller, W. A. In *Encyclopedia of Polymer Science and Engineering;* Mark, H. F. ; Bikales, N. M.; Overberger, C. G.; Menges, G., Eds.; Wiley: New York, 1985; Vol. 3, p 480.
25. Wilson, C. W.; Santee, E. R. *J. Polym. Sci. Part C* **1965,** *8,* 97.
26. *Machine Design* **1987,** *April 16,* 133.
27. Dohany, J. E.; Humphrey, J. S. In *Encyclopedia of Polymer Science and*

Engineering; Mark, H. F.; Bikales, N. M.; Overberger, C. G.; Menges, G., Eds.; Wiley: New York, 1989; Vol. 17, p 532.

28. Lovinger, A. J. In *Developments in Cystalline Polymers;* Basset, D. C., Ed.; Elsevier: New York, 1982; Chapter 5.
29. Brasure, D.; Ebnesajjad, S. In *Encyclopedia of Polymer Science and Engineering;* Mark, H. F.; Bikales, N. M.; Overberger, C. G.; Menges, G., Eds.; Wiley: New York, 1989; Vol. 17, p 468.
30. Lynn, M. M.; Worm, A. T. In *Encyclopedia of Polymer Science and Engineering;* Mark H. F.; Bikales, N. M. ; Overberger, C. G.; Menges, G., Eds.; Wiley: New York, 1987; Vol. 7, p 257.
31. Schmiegel, W. W. *Angew. Makromol. Chem.* **1979,** *76/77,* 39.
32. Schmiegel, W. W. *Kautsch. Gummi Kunstst.* **1978,** *31,* 137.
33. Apotheker, D.; Finlay, J. B.; Krusic, P. S., Logothetis, A. L. *Rubber Chem. Technol.* **1982,** *55,* 1004.
34. Schmiegel, W. W.; Logothetis, A. L. *Polymers for Fibers and Elastomers;* Arthur J. C., Ed.; ACS Symposium Series 260; American Chemical Society: Washington, DC, 1984; p 260.
35. Brasen, W. R.; Cleaver, C. S.; French Patent 1 469 510, 1967; *Chem. Abstr.* **1968,** *68,* 13885w.
36. Kojima, G.; Wachi, H. *Rubber Chem. Technol.* **1978,** *51,* 940.
37. Harrell, J. R.; Schmiegel, W. W. U.S. Pat. 3 859 259, 1975; *Chem. Abstr.* **1975,** *82,* 112956r.
38. Kojima, G.; Wachi, H. *International Rubber Conference,* Kyoto, Japan; 1985, p 242.
39. Moore, A. L. *Elastomerics* **1986,** *118,* 9.
40. Kalb, G. H. In *Polymerization Reactions and New Polymers;* Platzer, N. A., Ed.; Advances in Chemistry Series 129; American Chemical Society: Washington, DC, 1973; p 13.
41. Barney, A. L.; Kalb, G. H.; Khan, A. A. *Rubber Chem. Technol.* **1971,** *44,* 660.
42. Breazeale, A. F. U.S. Pat. 4 281 092, 1981; *Chem. Abstr.* **1982,** *96,* 7951b.

Fluorinated Pharmaceuticals

by Arthur J. Elliott

Prior to 1975, discussions of fluorinated pharmaceuticals focused mainly on anti-inflammatory steroids, antipsychotic phenothiazines, and 5-fluorouracil and its derivatives as anticancer agents [*1, 2*]. During the past 25 years, and especially in the last decade, fluorinated drugs have been marketed for the treatment of a wide variety of diseases. This section will focus on those presently marketed drugs where the presence of fluorine has produced significant therapeutic advantages.

Anti-infectives

The introduction of fluorine into anti-infective agents has produced a number of products. The fluorinated quinolone antibacterials, as a group, are characterized by low incidence of side effects and cure rates as high as 90%. There are presently six commercial products of general structure ***1***, two of which are shown.

Norfloxacin (*1;* R = C_2H_5, R^1 = H), a typical example, exhibits broad-spectrum activity and is useful in the treatment of *upper respiratory tract and urinary infections* [*3*]. **Lomefloxacin** (*2*)**,** a very recent introduction, is a third-generation product that, given once daily, is especially useful against pathogens resistant to cephalosporins, penicillins, and aminoglycosides [*4*]. **Floxacillin** (*3*) is a stable, orally active antibacterial with improved activity over the nonfluorinated product (cloxacillin) [*5*].

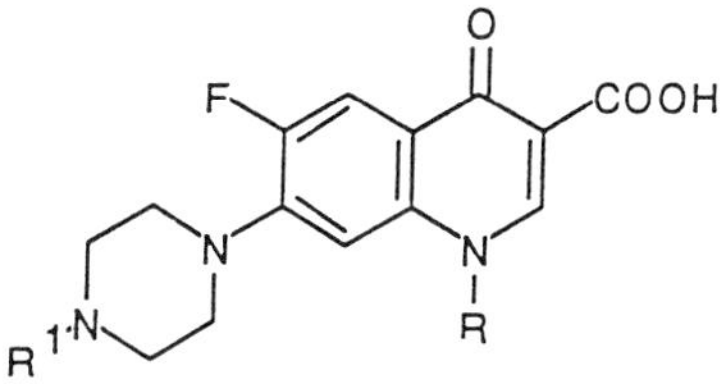

1

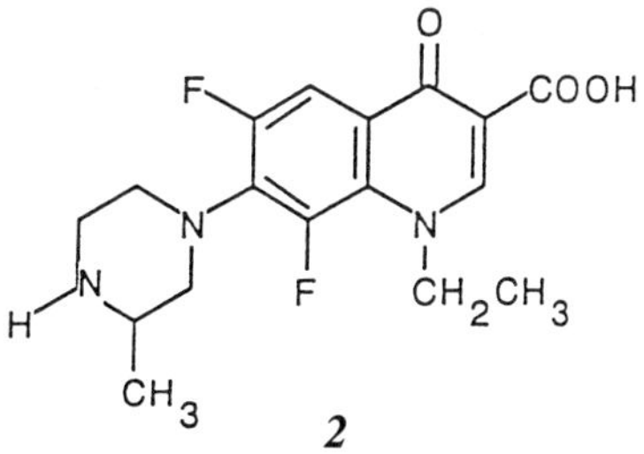

2

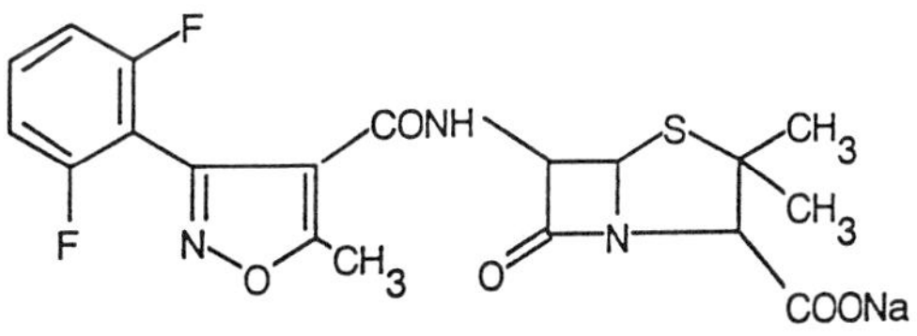

3

0065–7719/95/0187–1119$08.00/1

Fluconazole (*4*) is the first member of a new generation of *orally active antifungal agents,* highly effective in the treatment of *dermal and vaginal infections* [*6, 7*]. **5-Fluorocytosine** (*5*) is also used to treat serious systemic fungal infections [*5*].

The recently introduced antimalarial **halofantrine** (**6**) is an orally active blood schizonticide reported to be more than 95% effective in the *treatment of malaria* [*8*]. **Mefloquine hydrochloride** (*7*) continues to be useful in the prophylaxis and treatment of malaria [*9*].

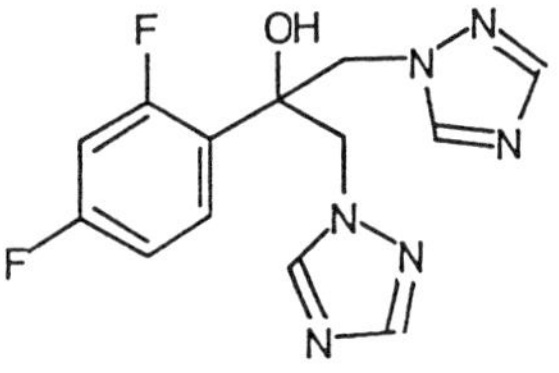

4 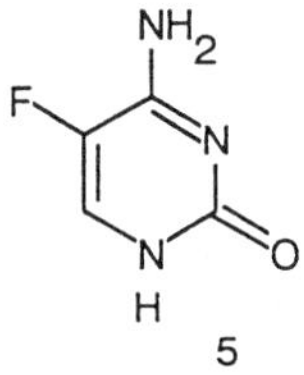 ***5***

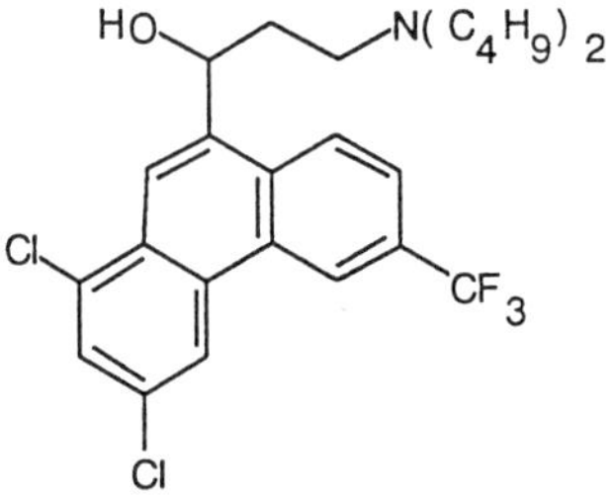

6 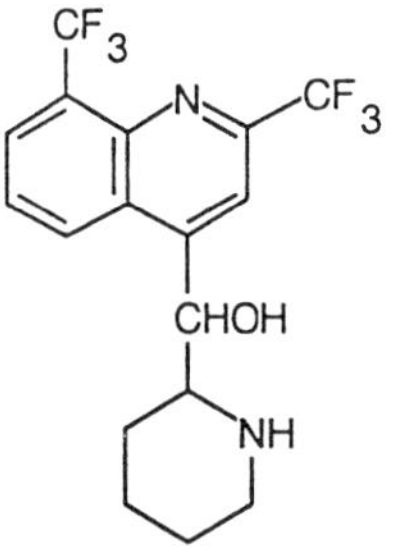 ***7***

Anticancer Agents

Interest continues in prodrugs of *5-fluorouracil* (5-FU). **Doxifluridine** (*8*) was recently introduced and appears to be more potent and less toxic than 5-FU [*10*]. **Flutamide** (*9*) and **nilutamide** (*10*) are both available for the treatment of prostatic cancer [*11, 12*].

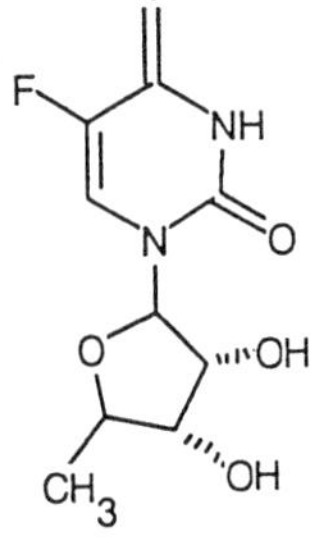

8 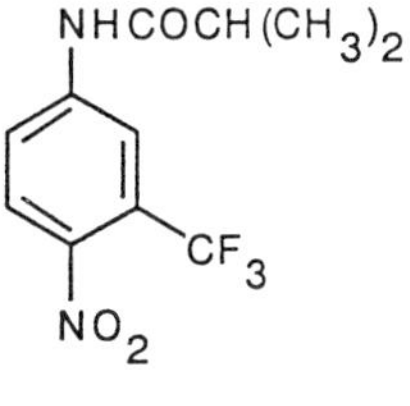 ***9*** 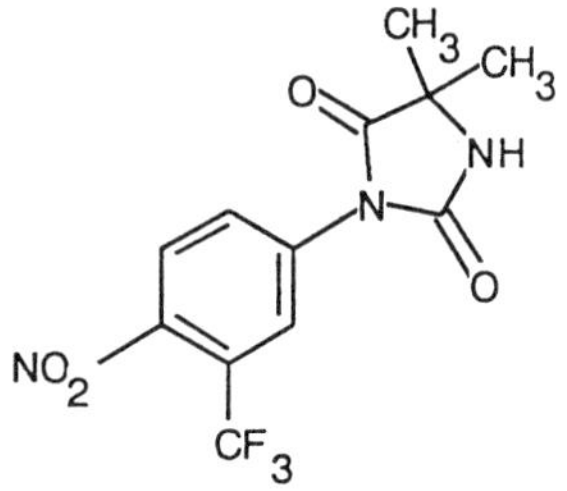 ***10***

Nonstereoidal Anti-inflammatory Agents

Fluorinated agents such as **flurbiprofen** (*11*), **flunisal**(*12*), **diflunisal** (*13*), and **sulindac** (*14*), acting as prostaglandin synthesis inhibitors, have been available for some time [*5*]. The recent introduction of **flunoxaprofen** (*15*), a *lipoxygenase inhibitor,* is notable. It reportedly produces considerably less severe gastric disturbance [*13*].

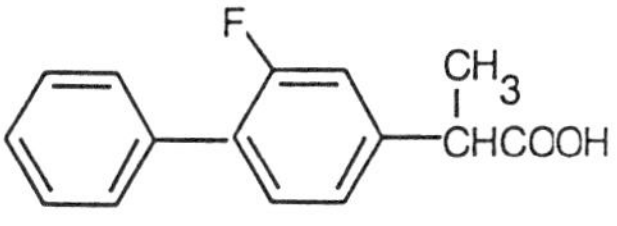

11

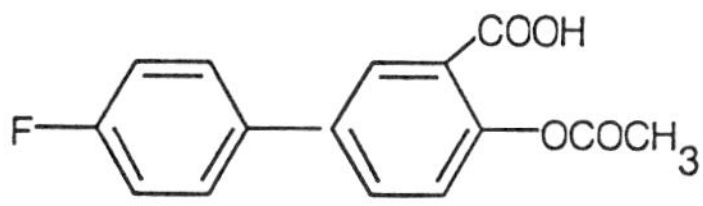

12

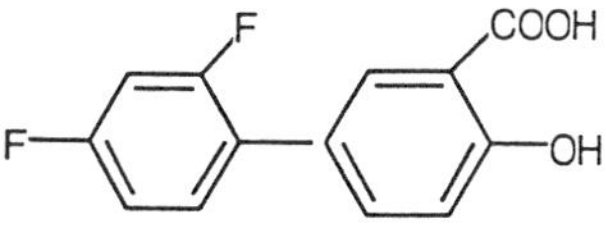

13

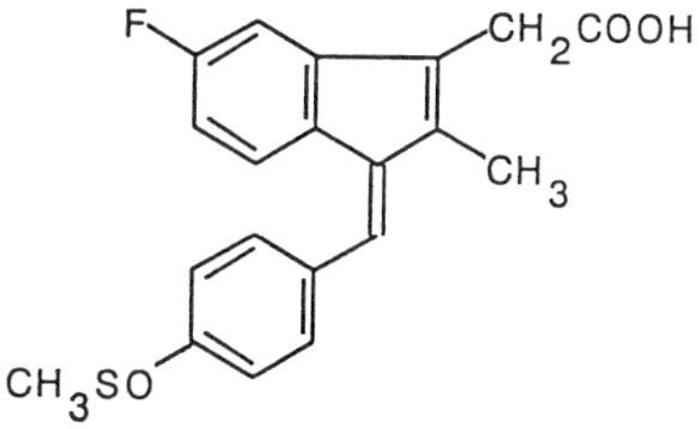

14

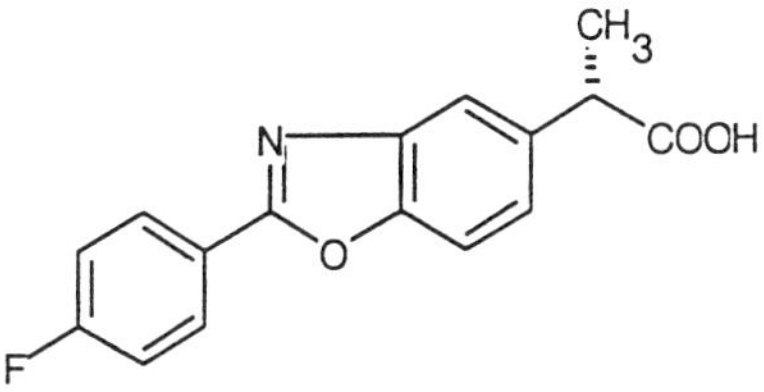

15

Central Nervous System(CNS) Agents

The role of fluorine in the development of CNS agents has been reviewed [*14*]. Fluorinated phenothiazines, typified by **fluphenazine** (*16*) are more potent than **chlorpromazine** in the management of *severely disturbed psychotic patients* [*14*].

The butyrophenones, of which **haloperidol** (*17*) is the oldest example, are typically 50 times more potent than **chlorpromazine**. **Pimozide** (*18*) is an example of the diarylbutylamines, which are potent neuroleptics [*14*].

Two recently introduced *antidepressants* are notable in that they are selective serotonin-uptake inhibitors. **Citalopram** (*19*) is reported to be as effective as **amitriptyline** in the *treatment of endogenous depression* [*15, 16*]. **Fluoxetine** (*20*) as the hydrochloride is approved for major *depressive disorders* including those with concomitant *anxiety*. Interestingly, it also appears useful in the *treatment of obesity* [*17*].

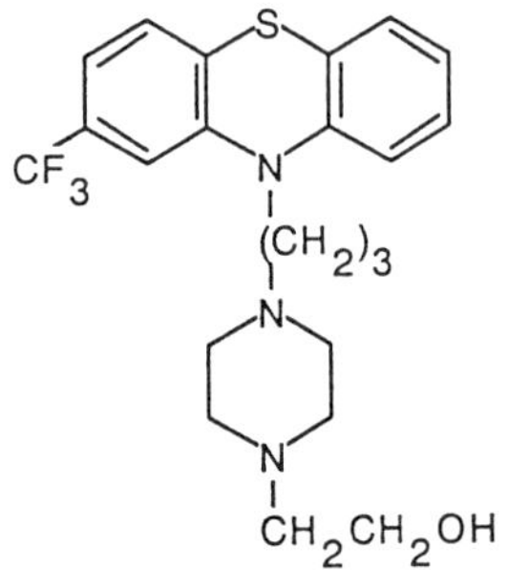

16

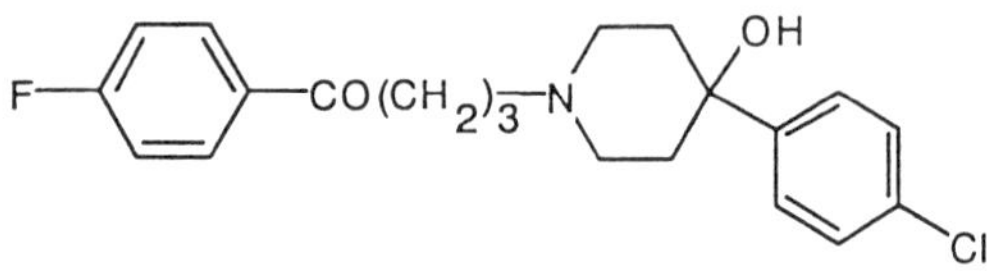

17

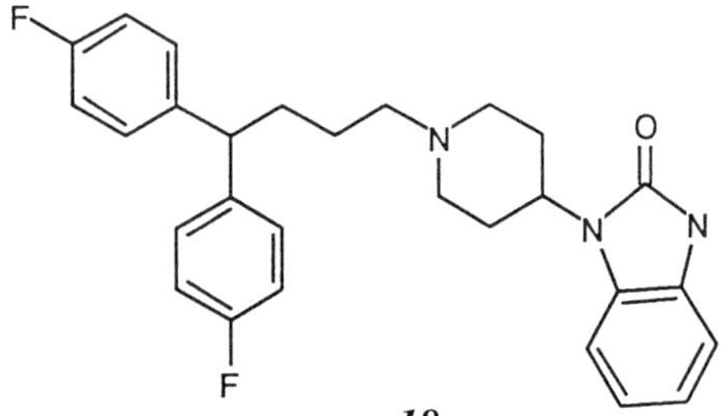

18

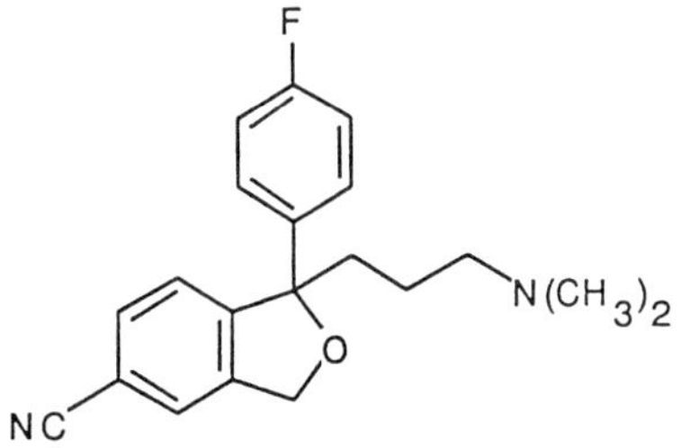

19

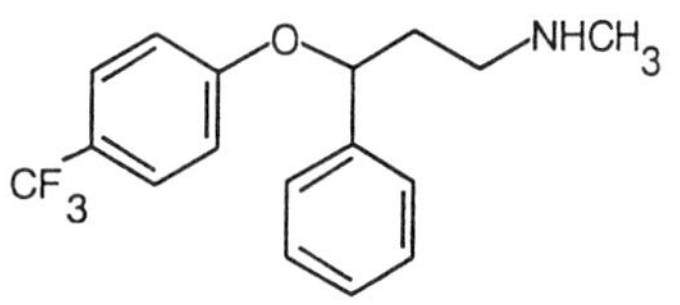

20

The most important fluorinated benzodiazepine, **flurazepam** (*21*) has found considerable use (and abuse) as a *hypnotic* [*14*]. **Flumazenil** (*22*) is a fast-acting *antidote* in the treatment of benzodiazepine intoxication and in the reversal of the CNS effects of benzodiazepines during anesthesia [*18, 19*].

Flupirtine (*23*) administered as the maleate, is a centrally active, nonaddicting *analgesic* slightly *more potent than aspirin* and is especially useful in the *management of postoperative and dental pain* [*20*].

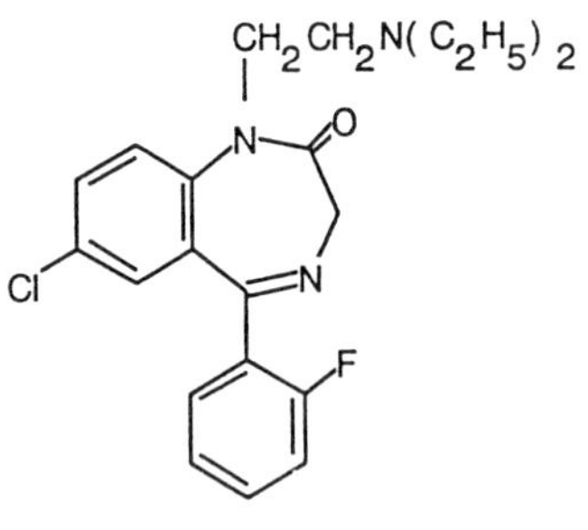

21

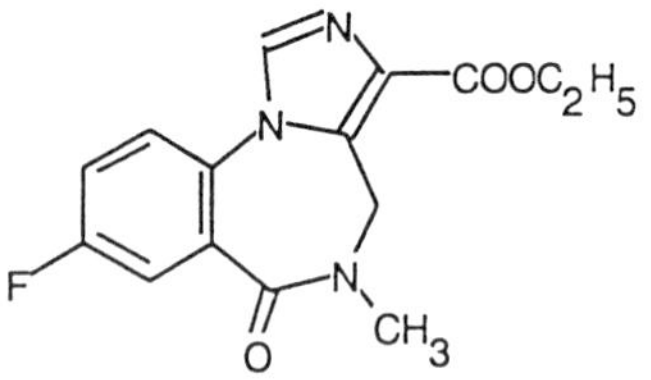

22

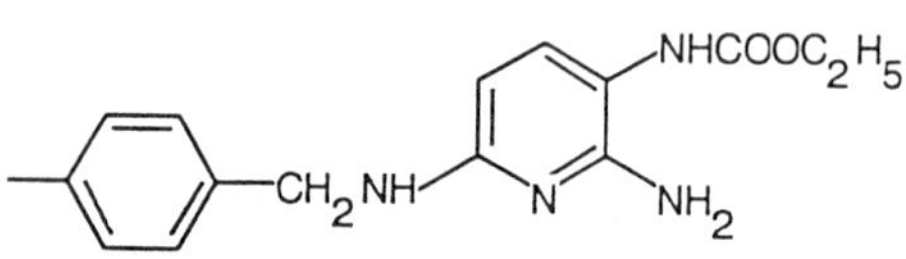

23

References are listed on pages 1124–1125.

The *anticonvulsant* **progabide** (*24*) is useful in a wide variety of *seizure disorders.* It was synthesized as a γ-aminobutyric acid (GABA) prodrug but its activity appears to reside in the parent drug and its acid metabolite, as well as the GABA liberated [*21*].

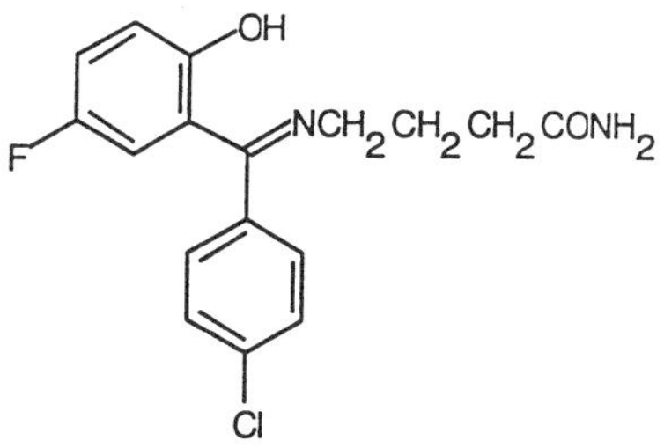

24

Miscellaneous

Ketanserin (*25*) is an *antihypertensive agent* acting as an antagonist at both the serotonin-S_2 and α_1-adrenergic receptors [*22*]. **Bendroflumethiazide** (*26*) is the most notable in a class of *diuretic–antihypertensive* agents [*5*].

Cisapride (*27*) is effective in the treatment of *reflux esophagitis and constipation,* with a novel mechanism thought to involve acetylcholine release in the gut. It is devoid of CNS and cardiac side effects [*23*].

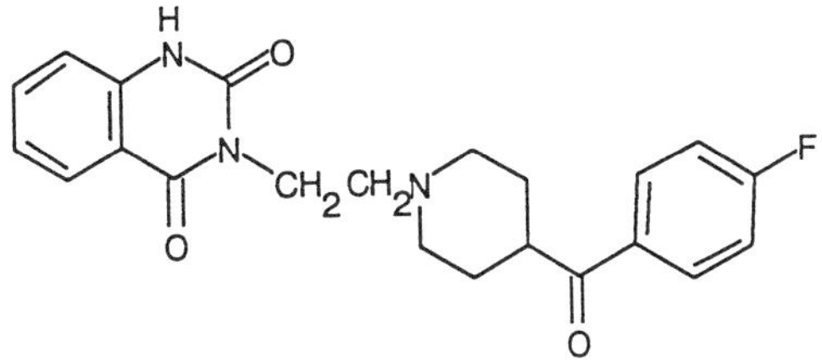

25

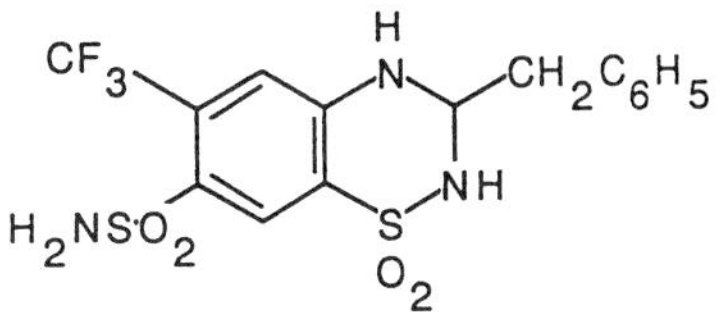

26

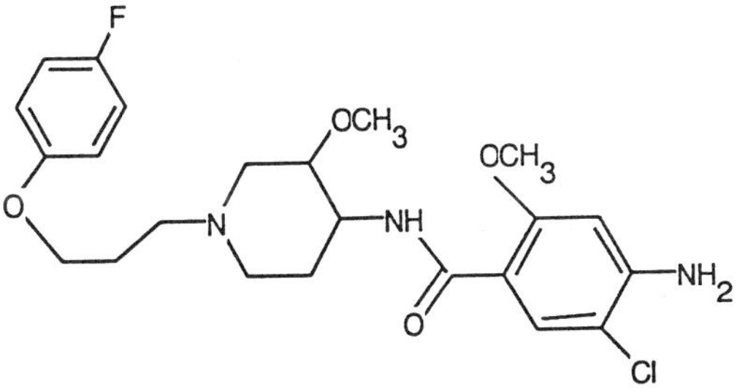

27

References are listed on pages 1124–1125.

Tolrestat (*28*) is a long-acting aldose reductase inhibitor useful in the prophylaxis of diabetic neuropathy, retinopathy, and cataracts [*24, 25*]. Lastly, the introduction of fluorine into steroids probably ushered in the modern era of fluorinated drugs. Many commercial products were introduced [*2*]. Typical of the products are the *anti-inflammatory glucocorticoids* **dexamethasone** (*29;* X = F, Y = H), **paramethasone** (*29;* X = H, Y = F), and **flumethasone** (*29;* X = H, Y = F).

28 **29**

References for Pages 1119–1124

1. Hudlický, M. *Chemistry of Organic Fluorine Compounds,* 2nd ed.; Ellis Horwood: Chichester, U.K., 1976; pp 563–566.
2. Sheppard, W. A.; Sharts, C. M. *Organic Fluorine Chemistry;* W. A. Benjamin: New York, 1969; p 454.
3. Allen, R. C. *Annu. Rep. Med. Chem.* **1987,** *22,* 117.
4. Spinorin, C. *Annu. Rev. Microbiol.* **1989,** *43,* 601.
5. Filler, R; Naqui, S. M. In *Biomedicinal Aspects of Fluorine Chemistry;* Filler, R.; Kobayashi, Y., Eds.; Elsevier: New York, 1982; p 2.
6. Isalska, B. J.; Stanbridge, T. N. *Br. Med. J.* **1988,** *297,* 178.
7. Odds, F. C.; Cheeseman, S. L.; Abbott, A. B. *J. Antimicrob. Chemother.* **1986,** *18,* 473.
8. Wirima, J.; Khoromana, C.; Molyneux, M. E.; Gilles, H. M. *Lancet* **1988,** *2,* 250.
9. Prous, J. R., *Annu. Drug Data Rep.* **1985,** *7,* 810.
10. Prous, J. R., *Annu. Drug Data Rep.* **1988,** *10,* 67.
11. Neri, R.; Kassem, N. In *Progress in Cancer Research and Therapy;* Brescianti, F. et al., Eds.; Raven Press: New York, 1984; Vol. 31, p 507.
12. Allen, R. C. *Annu. Rep. Med. Chem.* **1988,** *23,* 338.
13. Quaglia, G.; Periti, M.; Dimarzio, L.; Salvaggio, A. *Curr. Ther. Res.* **1986,** *39,* 66.
14. Elliott, A. J. In *Biomedicinal Aspects of Fluorine Chemistry;* Filler, R.; Kobayashi, Y., Eds.; Elsevier: New York, 1992, p 55.

15. Burrows, G. D.; McIntyre, I. M.; Judd, F. K.; Norman, R. R. *J. Clin. Psychiatry* **1988,** *49* (Suppl.), 18.

16. Dyck, L. E.; Boulton, A. A. *Neurochem. Res.* **1989,** *14,* 1047.

17. Benfield, P.; Heel, R. C.; Lewis, S. P. *Drugs* **1986,** *32,* 481.

18. File, S. E.; Pellow, S. *Psychopharmacol.* **1986,** *88,* 1.

19. File, S. E.; Pellow, S. *Adv. Biochem. Psychopharmacol.* **1986,** *41,* 187.

20. Anonymous *Drugs Fut.* **1983,** *8,* 773.

21. Bergmann, K. J. *Clin. Neuropharmacol.* **1985,** *8,* 13.

22. Vanhoutte, P. M. *J. Cardiovasc. Pharmacol.* **1985,** *7* (Suppl. 7), S105.

23. Van Outryve, M.; Vanderlinden, I.; Dedullen, G.; Rutgeerts, L. *Curr. Ther. Res.* **1988,** *43,* 408.

24. Nagata, M.; Robinson, W. G., Jr. *Invest. Ophthalmol. Vis. Sci.* **1987,** *28,* 1867.

25. Masson, E. A.; Boulton, A. J. M. *Drugs* **1990,** *39,* 190.

^{18}F-Labeled Radiopharmaceuticals

by D. F. Halpern

Positron emission tomography (PET) is an in vivo imaging method that uses short-lived, positron-emitting radiotracers to track biochemical processes in humans and animals. First used in scientific and clinical research, it is being used more frequently to detect disease-related biochemical changes before the anatomical changes caused by the disease can be visualized by standard medical imaging techniques. The use of a short-half-life, positron-emitting isotope like **fluorine-18** (^{18}F half-life, 109.8 min) demands a high-technology research center where ^{18}F can be generated by cyclotron irradiation of an appropriate molecule like neon-20 or oxygen-18. Only in this way can rapid synthesis of a fluorine-18 radiopharmaceutical and its subsequent administration be accomplished in a matter of hours [*1*].

The first radiotracer routinely accepted in PET research and used in almost every PET center is 2-deoxy-2-[^{18}F]fluoro-D-glucose (^{18}FDG) (Table 1). The predominant application of ^{18}FDG is the quantitation of brain glucose metabolism in normal and abnormal brain tissue. An immense synthetic effort over the past decade has provided improved syntheses of this compound. The synthesis of ^{18}FDG often uses 1,3,4,6-tetra-*O*-acetyl-2-*O*-trifluoromethanesulfonyl-βD-D-mannopyranose as the starting material in combination with a variety of ^{18}F-containing nucleophiles (equation 1). The importance of dopamine in Parkinson's disease and research directed toward understanding the role of dopamine and dopamine receptors in the brain have stimulated chemists to provide the necessary radiopharmaceuticals. Because dopamine does not cross the blood–brain barrier, the synthesis of the prodrug 6-[^{18}F]fluoro-DOPA, which crosses the blood–brain barrier and then

[*3*] **1**

CH_3CO_2, O_2CCH_3, O_2CCH_3, OTf, CH_3CO_2 $\xrightarrow[CH_3CN,\ 85°C]{K^{18}F,\ Kryptofix^{®}\ 2.2.2}$ CH_3CO_2, O_2CCH_3, O_2CCH_3, CH_3CO_2, ^{18}F

$\xrightarrow{H_3O^+,\ 120°C}$ ^{18}FDG (40-55%)

0065–7719/95/0187–1126$08.00/1

metabolizes to 6-[^{18}F]fluorodopamine, has provided scientists with a tool to study brain dopamine metabolism with PET. Radiotracers that detect the presence of receptors on tumors have allowed the identification of patients who might respond to a hormone-targeted therapy. In one example, 16α-[^{18}F]fluoroestradiol-17β uptake in human patients with breast cancer correlated with the density of estrogen receptors measured in biopsy specimens [*1*] (Table 1).

The traditional routes of *nucleophilic substitution, electrophilic substitution, or addition* can be used to rapidly incorporate fluorine-18 into a desired molecule. Kilbourn's book, *Fluorine-18 Labelling of Radiopharmaceuticals* [2], provides an

Table 1. Synthesis of 2-Deoxy-2-[^{18}F]fluoro-D-glucose

Substrate	*Method*[a]	*Synthesis Time (min)*[b]	*Radio-chemical Yield (%)*[c]	*Ref.*
1,3,4,6-Tetra-*O*-acetyl-2-*O*-trifluoro-methanesulfonyl-β-D-mannopyranose	A[d]	70	12–17	3
	B[d]	<60	40–55	4
	C	40	41 ± 15	5
	C	70	36	6
Methyl 3-*O*-benzyl-4,6-*O*-benzylidine-2-*O*-trifluoromethanesulfonyl-β-D-mannopyranoside	A	80	34–37	7
	B	60	37–43	7
1,6-Anhydro-3,4-di-*O*-benzyl-2-*O*-trifluoro-methanesulfonyl-β-D-mannopyranose	B	80	<54[e]	7

[a]A, Tetrabutylammonium [^{18}F]fluoride; B, Kryptofix 2.2.2/[^{18}F]fluoride ion; C, ion-exchange resin containing ^{18}F.
[b]After the end of bombardment (EOB).
[c]Relative to the initial amount of [^{18}F]F_2.
[d]A microcomputer-controlled automated synthesis.
[e]Not chemically pure.

[*17*] **2**

K^{18}F, Kryptofix® 2.2.2, DMSO, 120°C

NO_2 CHO → ^{18}F CHO → → → HO OH ^{18}F $CH_2CH_2NH_2$

Radiochemical yield (EOB)	68%	20%
Time (minutes from EOB)	18	105

References are listed on pages 1131–1132.

excellent review of this subject. Most of these syntheses can be described as "no carrier added," because fluorine-19 is not intentionally added during the synthesis [*1, 2*]. Several laboratories have even automated their radiolabeling procedures using a ^{18}F-containing nucleophile [*4, 8, 9*]. The two common sources of ^{18}F used in nucleophilic substitution reactions are **quaternary ammonium [^{18}F]fluorides** [*3*] and **Kryptofix 2.2.2** (4,7,13,16,21,24-hexaoxa-1,10-diazabicyclo[8.8.8]hexacosane) in combination with **potassium [^{18}F]fluoride.** Agents like ***1-bromo-3-[^{18}F]fluoropropane*** [*10*], ***[^{18}F]epifluorohydrin*** [*11*], ***[^{18}F]fluoroethyl triflate*** [*12*], and ***ethyl [^{18}F]fluoroacetate*** [*13*] have also been used as ^{18}F-containing

[*14*] **3**

$[^{18}F]F_2$, $CHCl_3$, R T

[25% Radiochem. yield (maximum yield = 50%)]

[*24*] **4**

$CH_3(CH_2)_3CH_2CH{=}CHCH_2(CH_2)_6CO_2H$

0.5% $[^{18}F]F_2$ in Ne, $FCCl_3$, -70°C, (12-16%)

$CH_3(CH_2)_3CH_2CHF{-}CH^{18}FCH_2(CH_2)_6CO_2H$

intermediates. Kryptofix 2.2.2 with [^{18}F]fluoride is often used to displace a halogen (equation 2) [*13, 14*] or a nitro group (equation 3) [*15, 16, 17, 18*]. One concern about this synthetic route is the detection of Kryptofix 2.2.2 and its elimination from the radiopharmaceutical. Needless to say, procedures for its detection [*19*] and elimination [*20*] have been developed. Table 2 provides an overview of some of

Table 2. Fluorine-18 Exchange by Nucleophilic Displacement

Radiopharmaceutical	*Synthesis Time (min)*[a]	*Radio-chemical Yield (%)*[b]	*Ref.*
2-Deoxy-2-[^{18}F]fluoroacetamido-D-glucopyranose	90	9	13
2-Deoxy-2-[^{18}F]fluoro-D-galactose	90	36–39	21
6-[^{18}F]fluoro-L-DOPA	100	10	15
	100–110	12	22
6-[^{18}F]fluorodopamine	105	20	17
16α-[^{18}F]fluoroestradiol-17β	90	22	24
21-[^{18}F]fluoro-16-α-ethyl-19-norprogesterone	40	4–30	23
[^{18}F]fluoromisonidazole	140	40	11
N-3-([^{18}F]fluoropropyl)-*N*-norbuprenorphine	100	15	10
N-3-([^{18}F]fluoropropyl)-*N*-nordiprenorphine	100	15	10
(±)-6-[^{18}F]fluoronorepinephrine	93	20	18
(+)- or (–)-6[^{18}F]fluoronorepinephrine	128	6	18
L-*p*-[^{18}F]fluorophenylalanine	120	—	16
[^{18}F]fluororaclopride	50	15	12

[a]After the end of bombardment (EOB).
[b]Relative to the initial amount of [^{18}F]F_2 (maximum possible, 50%).

Table 3. Comparison of Electrophilic Fluorination of Aromatic Amino Acids with [^{18}F]fluorine or [^{18}F]acetyl Hypofluorite[a] [25]

Amino Acid	*Agent*[b]	*Radio-chemical Yield (%)*[c]	*Isomer Distribution (%)*		
			2-	*3-*	*4-*
Phenylalanine	A	28.0	72.5	13.9	13.6
	B	43.4	81.7	11.0	7.3
Tyrosine	A	30.8	7.4	92.6	
	B	43.7	5.4	94.6	
DOPA	A	10.2	75.2	24.0	0.8
	B	15.4	88.9	9.7	1.4
O-Acetyltyrosine	A	15.9	71.2	28.8	
	B	19.7	83.2	16.8	

[a]Substrate-to-agent ratio, 1.2–1.4; 0 °C.
[b]A, [^{18}F]fluorine diluted with an unreactive gas; B, [^{18}F]acetyl hypofluorite.
[c]For comparison with [^{18}F]F_2, the radiochemical yields obtained with acetyl hypofluorite need to be divided by 2.

Table 4. Fluorine-18 Exchange by Electrophiles

Radiopharmaceutical	*Fluorinating Agent*[a]	*Synthesis Time (min)*[b]	*Radio-chemical Yield (%)*[c]	*Ref.*
2-[^{18}F]fluorophenylalanine	C	<60	30	27
3-[^{18}F]fluorotyrosine	C	<60	30	27
4-[^{18}F]-L-*m*-tyrosine	B	60	25	28
4-Borono-2-[^{18}F]fluoro-DL-phenylalanine	B	80	25–35	29
dl-erythro-9,10-[^{18}F]difluoro-palmitic acid	A	150	12–16	26
1-[^{18}F]fluoropyridone	A	35	48	30
6-[^{18}F]fluoropyridoxal	B	35–40	18	31
2-Deoxy-2-[^{18}F]fluoro-D-galactose	A	60	20	32
6-[^{18}F]fluorometaraminol	B	<60	20–42	33
6-[^{18}F]fluoro-L-DOPA	B	—	12[d]	34
3-[^{18}F]fluorodiazepam	A	40	25	14

[a]A, [^{18}F]fluorine diluted with an unreactive gas; B, [^{18}F]acetyl hypofluorite; C, [^{18}F]trifluoroacetyl hypofluorite.
[b]After the end of bombardment (EOB).
[c]Relative to the initial amount of [^{18}F]F_2 (maximum possible, 50%).
[d]Decay corrected.

the radiopharmaceuticals appearing in the literature over the past few years that have been prepared by nucleophilic substitution.

Electrophilic fluorinating agents like **[^{18}F]acetyl hypofluorite** or **[^{18}F]fluorine,** diluted with an unreactive gas like neon, are routinely used to provide ^{18}F-fluorinated radiopharmaceuticals, as illustrated in equation 4 and Table 3. The initial reaction conditions are usually optimized using [^{19}F]fluorine to ensure that a pure product can be obtained in a reasonable period. Although radiochemical yields of these processes vary widely, the ability to isolate a pure radiopharmaceutical in a short time with an acceptable radiochemical yield is a challenge because of the rapid decay of the radioisotope. The versatility of these electrophilic reagents can be shown by the variety of radiochemicals that have been prepared (Table 4). One limitation of electrophilic fluorination is its lack of selectivity. Careful choice of the appropriate reaction conditions, solvent, or route can sometimes circumvent this limitation (equation 4).

References for Pages 1126–1130

1. Fowler, J. S.; Wolf, A. P. *Annu. Rep. Med. Chem.* **1989,** *24,* 277.
2. Kilbourn, M. R. *Fluorine-18 Labelling of Radiopharmaceuticals;* National Academy Press: Washington, DC, 1990.
3. Brodack, J. W.; Dence, C. S.; Kilbourn, M. R.; Welch, M. *J. Appl. Radiat. Isot.* **1988,** *39,* 699.
4. Hamacher, K.; Blessing, G.; Nebeling, B. *Appl. Radiat. Isot.* **1990,** *41,* 49.
5. Toorongian, S. A.; Muholland, G. K.; Jewett, D. M.; Bachelor, M. A.; Kilbourn, M. R. *Nucl. Med. Biol.* **1990,** *17,* 273.
6. Kunst, E. J.; Wortmann, R.; Machulla, H. J. *J. Radioanal. Nucl. Chem.* **1989,** *132,* 85.
7. Haradahira, T.; Maida, M.; Kojima, M. J. *Labelled Compd. Radiopharm.* **1988,** *25,* 497.
8. Chaley, T.; Mattacchiere, R.; Velez, J. W.; Dahl, J. R.; Margouleff, D. *Appl. Radiat. Isot.* **1990,** *41,* 29.
9. Padgett, H. C.; Schmitt, D. G.; Luxen, A.; Bida, G. T.; Satyamurthy, N.; Barrio, J. R. *Appl. Radiat. Isot.* **1989,** *40,* 433.
10. Bai, L.; Teng, R.; Shive, C.; Wold, A. P.; Dewey, S. L.; Holland, M. J.; Simon, E. J. *Nucl. Med. Biol.* **1990,** *17,* 217.
11. Grierson, J. R.; Link, J. M.; Mathis, C. A.; Rasey, J. S.; Krohn, K. A. *J. Nucl. Med.* **1989,** *30,* 343.
12. Kiesewetter, D. O.; Brucke, T.; Finn, R. D. *Appl. Radiat. Isot.* **1989,** *40,* 455.
13. Tada, M.; Oikawa, A.; Iwata, R.; Fujiwara, T.; Kubota, K.; Matsuzawa, T.; Sugiyama, H.; Ido, T.; Ishiwata, K.; Sato, T. J. *Labelled Compd. Radiopharm.* **1989,** *27,* 1317.
14. Luxen, A.; Satyamurthy, N.; Bida, G. T.; Phelps, M. E. *J. Fluorine Chem.* **1987,** *36,* 83.
15. Lemaire, C.; Guillaume, M.; Cantineau, R.; Christiaens, L. *J. Nucl. Med.* **1990,** *31,* 1247.
16. Lemaire, C.; Guillaume, M.; Christiaens, L.; Palmer, A. J.; Cantineau, R. *Appl. Radiat. Isot.* **1987,** *38,* 1033.
17. Ding, Y. S.; Fowler, J. S.; Gattey, S. J.; Dewey, S. L.; Wolf, A. P.; Schleyer, D. J. *J. Med. Chem.* **1991,** *34,* 861.
18. Ding, Y. S.; Fowler, J. S.; Gattey, S. J.; Dewey, S. L.; Wolf, A. P. *J. Med. Chem.* **1991,** *34,* 767.
19. Chaly, T.; Dahl, J. R. *Nucl. Med. Biol.* **1989,** *16,* 385.
20. Moerlein, S. M.; Brodack, J. W.; Siegel, B. A.; Welch, M. J. *Appl. Radiat. Isot.* **1989,** *40,* 741.

21. Haradahira, T.; Maida, M.; Kai, Y.; Kojima, M. *J. Labelled Compd. Radiopharm.* **1988,** *25,* 721.

22. Ding, Y. S.; Shive, C. Y.; Fowler, J. S.; Wolf, A. P.; Plenevaux, A. *J. Fluorine Chem.* **1990,** *48,* 189.

23. Pomper, M. G.; Katzenellenbogen, J. A.; Welch, M. J.; Brodack, J. W.; Mathias, C. J. *J. Med. Chem.* **1988,** *31,* 1360.

24. Brodack, J. W.; Kilbourn, M. R.; Welch, M. J.; Katzenellenbogen, J. A. *Appl. Radiat. Isot.* **1986,** *37,* 217.

25. Coenen, H. H.; Franken, K.; Kling, P.; Stöcklin, G. *Appl. Radiat. Isot.* **1988,** *39,* 1243.

26. Schmall, B.; Finn, R. D.; Rapoport, S. I.; Noronha, J. G.; Degeorge, J. J.; Kiesewetter, D. O.; Simpson, N. R.; Larson, S. M. *Nucl. Med. Biol.* **1990,** *17,* 805.

27. Murakami, M.; Takahashi, K.; Kondo, Y.; Mizusawa, S.; Nakamichi, H.; Sasaki, H.; Hagami, E.; Ida, H.; Kanno, I. J. *Labelled Compd. Radiopharm.* **1988,** *25,* 773.

28. Perlmutter, M.; Satyamurthy, N.; Luxen, A.; Phelps, M. E.; Barrio, J. R. *Appl. Radiat. Isot.* **1990,** *41,* 801.

29. Ishiwata, K.; Ido, T.; Mejia, A. A.; Ichihashi, M.; Mishima, Y. *Appl. Radiat. Isot.* **1991,** *42,* 325.

30. Oberdorfer, F.; Hofmann, E.; Maier-Borst, W. *Appl. Radiat. Isot.* **1988,** *39,* 685.

31. Diksic, M.; Traving, B. C.; Nagahiro, S. *Nucl. Med. Biol.* **1989,** *16,* 413.

32. Oberdofer, F.; Traving, B. C.; Maier-Borst, W.; Hull, W. E. *J. Labelled Compd. Radiopharm.* **1988,** *25,* 466.

33. Mislanhar, S. G.; Gildersleeve, D. L.; Wieland, D. M.; Massin, C. C.; Mulholland, G. K.; Toorongian, S. A. *J. Med. Chem.* **1988,** *31,* 362.

34. Adam, M. J.; Jivan, S. *Appl. Radiat. Isot.* **1988,** *39,* 1203.

Fluorinated Anesthetics

by L. L. Ferstandig

The fluorinated inhalation anesthetics (*fluroxene* 2,2,2-trifluoroethyl vinyl ether), *methoxyflurane (2,2-dichloro-1,1-difluoroethyl methyl ether), and halothane* (1-bromo-1-chloro-2,2,2-trifluoroethane) were thoroughly discussed in the earlier edition of this book. Fluroxene is no longer used mostly because it is flammable at the higher concentration levels used in anesthesia. Methoxyflurane, because of its high solubility in fatty tissues, remains in human patients for long periods, all the while undergoing metabolism generating oxalic acid and fluoride ion [*1*]. Because some obese patients having long surgical procedures succumbed to fluoride poisoning, methoxyflurane is no longer used for human patients but it still finds some veterinary use.

Halothane remains the leading anesthetic in many parts of the world. However, it is believed to cause a fulminant hepatitis in rare, susceptible individuals, especially after repeated use within short intervals. It was believed, but now disputed, that this hepatitis resulted from toxic metabolites [*2*]. (Actually, the major metabolite is trifluoroacetic acid, which as a salt in body fluids, is benign.) As rare as the hepatitis cases were (1 in 20 000), they frequently resulted in malpractice suits, especially in the United States. This problem led to a search for more ideal nonflammable anesthetics that are also metabolized to a lesser extent [*3*]

Enflurane (2-chloro-1,1,2-trifluoroethyl difluoromethyl ether; Table 1 lists physical properties) became the next successful anesthetic, meeting both the criteria of nonflammability and very low metabolism. This anesthetic is manufactured by adding methanol to chlorotrifluoroethylene [*4*] followed by careful chlorination of the methyl group to the dichloromethyl derivative [*5*] (equation 1). The latter compound is then fluorinated to replace the two chlorines on the methyl group. The problem in this synthesis is overchlorination, which is minimized by deliberate underchlorination and recycling of the underchlorinated material. Enflurane is still in use today, but its major shortcoming is that it sometimes produces uncontrolled movements in patients. Even though it is metabolized to a lesser extent than halothane, it produces more fluoride ion but apparently not enough to cause renal dysfunction [*6*].

An isomer of enflurane named *isoflurane* (1-chloro-2,2,2-trifluoroethyl difluoromethyl ether) does not produce uncontrolled movements, is nonflammable, and is metabolized to an even lesser extent than enflurane [*7*]. As of this writing, isoflurane is the fastest growing anesthetic in more economically developed countries, but because of cost, it has not overtaken halothane in the rest of the world.

0065–7719/95/0187–1133$08.00/1

Table 1. Physical Properties of Fluorinated Inhalation Anesthetics

Anesthetic	*bp (°C)*	*Density (25/25)*	*MAC (%)*[a]
Fluroxene	43.1	1.135	3.40
Methoxyflurane	104.7	1.410	0.16
Halothane	50.2	1.872	0.75
Enflurane	56.5	1.520	1.68
Isoflurane	48.5	1.496	1.15
Sevoflurane	58.6	1.525	1.71
Desflurane	22.8	1.467 (15/4)	7.30

[a]MAC is the minimum concentration (v/v %) required to anesthetize 50% of the animals tested.

[5] *1*

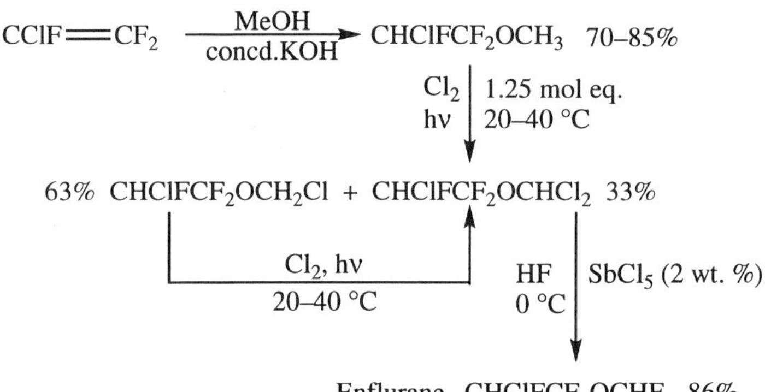

In the original route to isoflurane, the methyl ether of trifluoroethanol is made with dimethyl sulfate [*8*] followed by careful chlorination of the methyl group to make the dichloromethyl ether. This ether is fluorinated with hydrogen fluoride and an antimony catalyst and the final step is monochlorination of the α carbon of the ethyl group [*8*] (equation 2).

In a later version of the synthesis [*9*], the trifluoroethyl difluoromethyl ether is made directly from trifluoroethanol and chlorodifluoromethane (equation 2) and then chlorinated to give the final product. Again, the major problem is overchlorination, because all the hydrogens are readily replaced by chlorine. Separation of the overchlorinated by-products poses a special problem because of close boiling points. This problem can be solved by adding acetone to create a more easily separable azeotrope of acetone and isoflurane [*10*].

References are listed on pages 1136–1137.

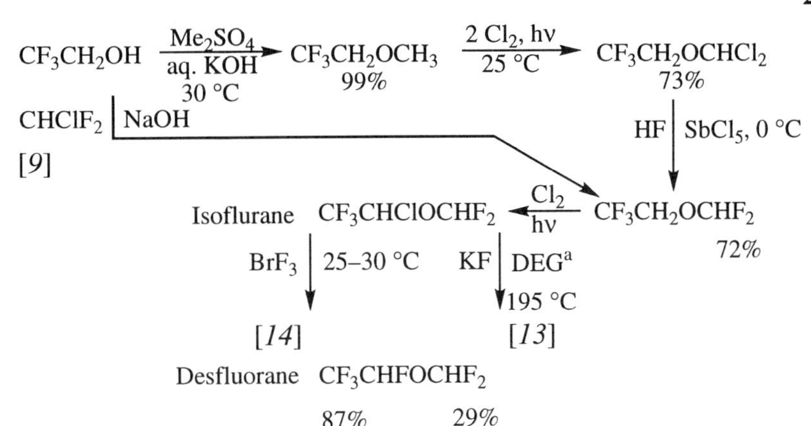

[a]DEG is diethylene glycol

Other fluorinated anesthetics are *sevoflurane,* $(CF_3)_2CHOCH_2F$, which is in use in Japan, and *desflurane,* $CF_3CHFOCHF_2$, which was approved recently for use in the United States. Sevoflurane is made by methylation of hexafluoroisopropyl alcohol followed by monochlorination of the methyl group and subsequent replacement of the chlorine by fluorine through potassium fluoride [*11*] (equation 3). Another route to sevoflurane involves the direct fluorination of the methyl ether with bromine trifluoride [*11*]. Still another interesting path uses paraformaldehyde and hydrogen fluoride with hexafluoroisopropyl alcohol [*11*] (equation 3). Sevoflurane has been tested for many years in the United States but is still not approved for use. Two major objections remain: the metabolism to produce fluoride ion and the reaction of sevoflurane [*12*] with soda lime or Baralyme used to remove both carbon dioxide and water from the gases recirculated to the patient. This reaction produces unknown toxic compounds.

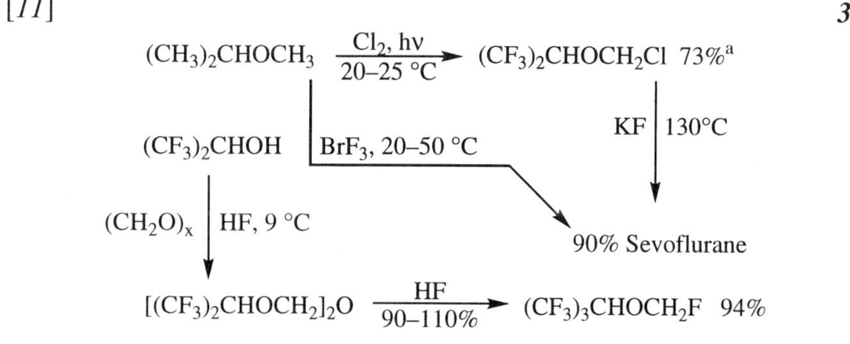

[a]Conversion is 57%.

Desflurane can be made by fluorination of isoflurane with potassium fluoride [*13*] or with bromine trifluoride [*14*] (equation 2). Although approximately six times more desflurane than isoflurane is required for anesthesia, it achieves the anesthetic state and the patient recovers from this state more rapidly. This feature seems to be of interest to some anesthesiologists. One negative property is that it irritates the breathing passageway, especially in children. Desflurane is flammable at higher levels of anesthetic use. It is the least metabolized of all the commercial anesthetics, producing trifluoroacetic acid and fluoride ion [*15*].

Recently, halothane, enflurane, and isoflurane were separated into their pure enantiomers by gas chromatography [*16*]. In fact, the pure enantiomers of halothane and enflurane [*17*] and of isoflurane [*18*] also have been synthesized. It is of some theoretical interest that one of the enantiomers of isoflurane is about twice as effective in anesthesia as the other [*19*]. Of the two major theories on the mechanism of anesthesia this selectivity seems to favor the mechanism involving anesthetic binding with the protein layer that controls the passage of ions in the pain message pathway. The other mechanism proposes that specific lipid membranes are perturbed by solution of the anesthetic and that this perturbation interferes with pain message delivery. That route suggests physical interaction of the anesthetic with the lipids, and any role favoring one enantiomer over the other seems less likely [*19*].

There has been some controversy over the effect of traces of anesthetic gases in the operating room on the health of personnel working there daily. Numerous animal studies using low levels of anestheltic gases have failed to show any effects, and several epidemiological studies show that human health is not affected by traces of anesthetic gases [*20*].

Many other classes of fluorine compounds, from hexafluorobenzene to fluorinated heterocycles, have some anesthetic properties, but all have been found wanting for one reason or another. More details on some of these can be found in reference 21. This reference also discusses the proposed metabolic pathways in some detail.

References for Pages 1133–1136

1. Mazze, R. I.; Trudell, J. R.; Cousins, M. J. *Anesthesiology* **1971,** *35,* 247.
2. Shingu, K.; Eger, E. I.; Johnson, B. H. *Anesth. Analg.* **1982,** *61,* 824.
3. Terrell, R. C. *Brit. J. of Anaest.* **1984,** *56,* 3S.
4. Park, J. D.; Vail, D. K.; Lea, K. R.; Lacher, J. R. *J. Am. Chem. Soc.* **1948,** *70,* 1550.
5. Terrell, R. C., U.S. Pat. 3 469 011, 1969; *Chem. Abstr.* **1970,** *72,* 3025.
6. Chase, R. E.; Holaday, D. A.; Fiserova-Bergerova, V.; et al. *Anesthesiology* **1971,** *35,* 262 .
7. Greenstein, L. R.; Hitt, B. A.; Mazze, R. I. *Anesthesiology,* **1975,** *42,* 420.
8. Croix, L. S.; Terrell, R. C. U.S. Pat. 3 535 388, 1970; Ger. Pat. 1 814 962, 1970; *Chem. Abstr.* **1970,** *72,* 3004.
9. Croix, L. S.; Terrell, R. C. U.S. Pat. 3 637 477, 1972; Ger. Pat. 1 814 962, 1970; *Chem. Abstr.* **1970,** *72,* 3004.

10. Croix, L. S. U.S. Pat. 3 720 587, 1973; Ger. Pat. 2 234 309, 1973; *Chem. Abstr.* **1973,** *78,* 110530.
11. Regan, B. M. U.S. Pat. 3 683 092, 1972; *Chem. Abstr.* **1972,** *77,* 156346.
12. Mazze, R. I. *Anesthesiology* **1992,** *77,* 1062.
13. Terrell, R. C. U.S. Pat. 4 762 856, 1988; *Chem. Abstr.* **1988,** *109*, 189832.
14. Eur. Pat. 341 004, 1990; *Chem. Abstr.* **1990,** *112,* 178052.
15. Sutton, T. S.; Koblin, D. D.; Gruenke, L. D.; et al. *Anesth. and Analg.* **1991,** *73,* 180.
16. Meinwold, J.; Thompson, W. R.; Pearson, D. L.; et al. *Science* **1991,** *251,* 560.
17. Pearson, D. L. Ph. D. Thesis, Cornell University, Ithaca, NY, 1990.
18. Huang, C. G.; Rozov, L. A.; Halpern, D. F.; Vernice, G. G. *J. Org. Chem.* **1993,** *58,* 7382.
19. Franks, N. P.; Lieb, W. R. *Science* **1991,** *254,* 427.
20. Ferstandig, L. L. *Anesth. Analg.* **1978,** *57,* 328.
21. Jones, W. G. H. In *Preparation, Properties, and Industrial Applications of Organofluorine Compounds;* Banks, R. E., Ed.; Ellis Horwood: Chichester, U.K., 1982; Chapter 5.

Perfluorinated Liquids in Biology and Medicine

by L. C. Clark

Most advances in medicine brought about by chemistry are related to the pharmacological activities of particular compounds. With perfluorinated liquids, we now have, for the first time, a new class of chemicals that has complete lack of activity as its main characteristic of value for biological uses.

Perfluorinated liquids, which had been commercially available since the end of World War II where they played a role in the United State's development of the atomic bomb, had found some limited uses in biological work. But in 1966, an experiment [*1*] involving a mouse breathing in such an oxygen-bubbled liquid and surviving drew attention to the potential uses of these remarkable liquids. This experiment effectively demonstrated all the main biologically valuable attributes of completely fluorinated liquids in one dramatic move. Photographs and movies of this mouse breathing perfluoroalkylfuran are widely used even now to illustrate the helpful properties of these liquids, which are still new to most people. The main medically useful properties of perfluorinated liquids are their impressive lack of reactivity and their ability to dissolve large quantities of oxygen. Some examples of recent applications of perfluorinated liquids follow.

Liquid Breathing

Before birth, our lungs are filled with amniotic fluid, which drains away rapidly after birth as the lungs fill with air. If the newborn is fully mature, a lipid coating called lung surfactant keeps the lungs from collapsing by surface tension. If the lungs are not fully mature, the infant can be kept alive only with the help of a membrane oxygenator connected to the blood circulation system, often for several weeks. Preliminary work [*2*] has shown that such infants can safely breathe perfluorocarbon liquids and maintain good oxygen levels in blood as long as liquid ventilation is continued. As soon as a fluorocarbon liquid with the ideal physicochemical properties can be synthesized, such infants can be supported until their lungs are mature enough to make the essential surfactant. There will be other uses for fluorocarbon liquids to sustain or improve pulmonary function in damaged lungs.

Eye Surgery

When the retina becomes detached from the back of the eye, as in the case of nearsightedness or injury sustained in car accidents, it can be weighted down by injecting liquid fluorocarbon. After the retina is reattached by laser surgery or by

0065–7719/95/0187–1138$08.00/1

spot-freezing, the liquid is removed [*3*]. Fluorocarbons in this case are valuable because of their high density (specific gravity of about 2.00). Because they are inert, fluorocarbon liquids will not hurt the retina itself. Their oxygen-dissolving properties may also be of help. Also because of their high density, fluorocarbon liquids are used to float unwanted material out of the eye during ophthalmic surgery. Perfluorooctane is used because it has almost the same refractive index as the gel, the vitreous, that fills the eye. Perfluorophenanthrene also is used because it is even less reactive than perfluorooctane and because it does not break up as readily into tiny droplets, which would make it more difficult to remove after the reattachment is complete and the healing has begun. Hundreds of patients have been spared a life of blindness because of this application of fluorocarbons. The future may see the complete replacement of the vitreous gel by a fluorocarbon of the right kind.

Fluorocarbons as X-Ray Contrast Liquids and Imaging Agents

Shortly after the introduction of fluorocarbons into biology, Long [*4*] recognized that adding a bromine atom to a fluorinated organic molecule would add X-ray opacity to its already valuable characteristics. Certain stable iodinated organic compounds have been used for this purpose for many years. Iodinated perfluorochemicals are not stable and tend to decompose on standing or when exposed to light. The corresponding bromine-substituted perfluorinate, however, is remarkably stable. Perfluorooctylbromide (Alliance Pharmaceutical) is finding wide use in biomedicine mainly because of its radio-opacity. The most widely used fluorocarbon X-ray contrast agent, used either as a neat liquid or as an emulsion, is perfluorooctylbromide.

Computed tomography was first applied in medical imaging using roentgen rays. Nuclear magnetic resonance (NMR), long an extremely valuable tool in fluorine chemistry, is now also used for imaging in a rapidly developing technology. The NMR properties of the fluorine atom are ideal, and there are many fluorine atoms in a typical perfluorinated liquid. Because (T1) in the NMR is affected by the paramagnetic properties of the oxygen atom, it is possible to make images of the oxygen tension distributions in a living animal after it is infused with certain fluorocarbons [*5*]. This remarkable combination of the physical chemistry of perfluorinates—fluorine for where it is and oxygen for how the oxygen needs of the organs are met—makes possible new advances in the diagnosis and treatment of many diseases.

Fluorocarbons for Making Synthetic Blood Substitutes

Shortly after the liquid-breathing experiment [*1*] was published, the same liquid was used to perfuse an isolated heart from an animal [*6*], by intermittent perfusion with typical aqueous buffer and fluorocarbon. Then Sloviter [7] showed that an emulsion of this fluorocarbon liquid, made by ultrasonication with an albumin solution, could sustain the life of the brain of a rat as well as or better than perfusing

References are listed on pages 1141–1142.

it with natural anticoagulated blood. This experiment was followed soon after by a remarkable demonstration by Geyer [*8*] that an oxygen-breathing rat could survive in apparent good condition when its blood is completely replaced with an emulsion made with perfluorotributylamine and a synthetic nonionic emulsifier, a balanced salt solution, called Ringer's solution, and a suitable oncotic agent. As the fluorocarbon was removed from the bloodstream by the liver and spleen over a period of several days, the rat regenerated enough blood cells to survive and return to normal living conditions.

Later it was discovered that perfluorotributylamine (FC43; 3M Co.) did not leave the body for many months. On the other hand, fluorocarbons with very high vapor pressures, such as FC75 (3M Co.), cause bubbles to form in the blood. Those with very low vapor pressures, such as FC43 (3M Co.), do not cause bubbles but reside in the body too long.

Experiments were started by Clark [*9*] and by Naito [*10*] to find a fluorocarbon that was volatile enough to leave the body in a reasonable time after its function as a blood substitute was completed and yet not so volatile as to cause the bubble problem. Perfluorodecalin appeared to be the answer, being transpired in a matter of days or weeks rather than years. Naito, of the Green Cross Corporation, continued the development of this kind of blood substitute by adding perfluorotripropylamine and manufacturing an emulsion for clinical use on a relatively large scale under the trademark Fluosol DA [*11*]. Because of the inherent instability of perfluorodecalin-based emulsions, it had to be shipped and stored frozen until used. Despite these inconveniences, this emulsion has considerable human use, mainly in situations where only half-liter doses were required, as in balloon catheter treatment of coronary artery occlusion, and as a means, with oxygen breathing, to enhance the effectiveness of oxygen-dependent chemotherapy and radiation therapy for cancer. It was also shown during this time that infusion of fluorocarbon emulsions in animals had a protective effect in that it decreased the amount of heart tissue stunned or injured during an experimental heart attack [*12*]. This protection is due to the fact that fluorocarbons only dissolve, but do not bind, oxygen and therefore give it up much more readily on the basis of equilibration with the tissue in need of oxygen. Further, the emulsion particles are much smaller than red cells and may be able to get through partially clogged arteries.

Almost all of the biomedical research done in the 25 years following the liquid-breathing work was conducted with commercially available fluorocarbons manufactured for various industrial uses by the electrochemical Simons process (fluorination in a hydrofluoric acid solution) or the cobalt fluoride process (fluorination with this solid in a furnace at about 200 °C). These processes tended to yield many by-products, partly because they were, to some extent, free radical reactions and partly because it was difficult to easily achieve complete fluorination. Aromatic hydrocarbons gave better products with the cobalt trifluoride [*13*] method, whereas saturated hydrocarbons yielded better products with fluorination using diluted or cooled gaseous fluorine (Lagow). Incompletely fluorinated material was either

References are listed on pages 1141–1142.

found to be or thought to be toxic when given to animals; a number of ways were found to eliminate incompletely fluorinated substances, including refluxing with sodium hydroxide solutions or washing with diethylamine [*14*]. Preparative chromatography is used in our laboratory [*15*] to separate closely boiling liquids, such as isomers of perfluorodecalin, and to eliminate small amounts of highly fluorinated small molecules that sometimes tag along and are impossible to separate by any other process.

The need for blood substitutes is very great, and fluorocarbons offer great promise as the oxygen- and carbon dioxide-transporting component. Such synthetic blood substitutes will be emulsions of fluorocarbon with an aqueous phase containing the essential sodium, potassium, calcium, chloride, and bicarbonate ions; a high-molecular-weight oncotic agent such as dextran; and an emulsifier for the fluorocarbon. Unlike donated natural human blood, which needs to be typed, crossmatched, and tested for infectious agents, synthetic blood can be used in all persons, and probably in all other animals as well, and can be sterilized by steam and stored for months without refrigeration.

The complex interplay of physicochemical and biological characteristics that regulate the all-important rate at which fluorocarbons may migrate within and finally leave the body, through the lungs and the skin, is not yet completely understood. Certainly, variables are involved, such as vapor pressure, solubility in body tissues, molecular size and shape, lipid solubility, electron configuration, and critical solution temperatures [*16, 17*].

The final steps to a synthetic blood depend completely upon good chemistry tailored to meet the exact needs of the body. Fluorocarbons, such as perfluorodecalin, recently have been found to induce hyperinflated lungs when given either intravenously as an emulsion or intratracheally as a neat liquid [*18, 19*]. But this and other physiological side effects are now understood, and research is well advanced to prevent undesirable side effects in medical applications of fluorocarbon liquids.

For a state-of-the-art view of the science of biomedical fluorocarbons, the reader is referred to an early symposium on the subject [*20*].

Nomenclature

The word "fluorocarbon" has been used here in a generic way to indicate completely fluorinated liquids, including ethers and amines, rather than in its stricter meaning of having only carbon and fluorine atoms. The definition of the word "perfluorinated" is somewhat vague, because it can include compounds that retain double bonds or protons.

References for Pages 1138–1141

1. Clark, L. C., Jr.; Gollan, F. *Science* **1966,** *152,* 1755.
2. Greenspan, J. S.; Wolfson, M. R.; Rubenstein, S. D.; Shaffer, T. H. *Lancet* **1989,** *Nov. 4,* 1095.

3. Clark, L. C., Jr. U.S. Pat. 4 490 351, 1984.
4. Long, D. M.; Liu, M.; Alrenga, P.; Szanto, P. S. *Surg. Forum.* **1971,** *22,* 207.
5. Clark, L. C., Jr.; Ackerman, J. L.; Thomas, S. R.; Millard, R. W.; Hoffmann, R. E.; Pratt, R. G.; Ragle-Cole, H.; Kinsey, R. A.; Janakiraman, R. *Adv. Exp. Med. Biol.* **1984,** *180,* 835.
6. Gollan, F.; Clark, L. C., Jr. *The Physiologist.* **1966,** *9,* 191.
7. Sloviter, H. A.; Yamada, H.; Ogoshi, S. *Fed. Proc., Fed. Am. Soc. Exp. Biol.* **1970,** *29,* 1755.
8. Geyer, R. P. *Fed. Proc., Fed. Am. Soc. Exp. Biol.* **1970,** *29,* 175.
9. Clark, L. C., Jr.; Becattini, F.; Kaplan, S. *Ala. J. Med. Sci.* **1972,** *9,* 16.
10. Naito, R. *Technical Information Serial* **1974,** *1,* September 11 (The Green Cross Corp., Osaka, Japan).
11. Tremper, K. K.; Levine, E. M.; Waxman, K. *Int. Anesth. Clin.* **1985,** *23,* 185.
12. Glogar, D. H.; Kloner, R. A.; Muller, J.; DeBoer, W. V.; Braunwald, E.; Clark, L. C., Jr. *Science* **1981,** *211,* 1439.
13. Haszeldine, R. N.; Smith, F. *J. Chem. Soc.* **1950,** 3617.
14. Clark, L. C., Jr.; Moore, R. Abstract presented at the American Chemical Society Fourth Winter Fluorine Conference, Daytona Beach, FL, January 28–February 2, 1979.
15. Hoffmann, R.E. Unpublished manuscript.
16. Moore, R. E.; Clark, L. C., Jr. *Proc. 5th Int. Symp. Perfluorochem. Blood Substitutes,* Mainz, 1981; W. Zuckschwerdt Verlag: Munich, 1982; p 50.
17. Yamanouchi, K.; Tanaka, M.; Yokoyama, K.; Awazu, S.; Kobayashi, Y. *Chem. Pharm. Bull.* **1985,** *33,* 1221.
18. Clark, L. C., Jr.; Hoffmann, R. E.; Davis, S. L. *Biomater. Artificial Cells Immobilization Biotechnol.* **1992,** *20,* 1085.
19. Hoffmann, R. E.; Bhargava, H. K.; Davis, S. L.; Clark, L. C., Jr. *Biomater. Artificial Cells Immobilization Biotechnol.* **1992,** *20,* 1073.
20. Clark, L. C., Jr. *Fed. Proc., Fed. Am. Soc. Exp. Biol.* **1970,** *29,* 1696.

Fluorinated Agrochemicals

by R. W. Lang

Little was known about fluorinated agrochemicals when the first edition [*1*] of this book was published.

In 1978, only 25 (4%) of some 600 pesticides contained fluorine [*2*]. The second edition of *The Agrochemicals Handbook* [*3*] reported an 8.5% contribution of fluorinated agrochemicals in 1988. Today, about 10% of all commercially available agrochemicals in the market contain fluorine. This ever-growing interest in selectively fluorinated molecules for biological applications is due mainly to the fact that strategically positioned fluorine substituents in a molecule may greatly modify its physicochemical properties and thus its biological activity [*4*].

In the 1960s, simple chemical structures with broad biological activity were used as preventive pesticides at dose rates of kilograms per hectare. Today, highly complex and/or specifically fluorinated compounds are used at dose rates of only grams per hectare and mainly for curative applications [*5*]. The strategic change toward integrated pest management and the enormous enhancement in chemical activity makes fluorine-containing molecules, which normally cost much more to prepare than their nonfluorinated analogues, competitive and commercially viable commodities.

Examples of modern fluorine-containing herbicides, insecticides, and fungicides are shown below. Most of these compounds are prepared starting from fluorinated bulk chemicals, such as fluorinated carbocyclic and heterocyclic compounds, benzotrifluoride, fluorinated acetic acid derivatives, and Freons. Direct fluorination is used only occasionally by producers of fine chemicals and will, at least in the near future, remain the domain of producers of bulk chemicals who have the necessary technical expertise.

Fungicides

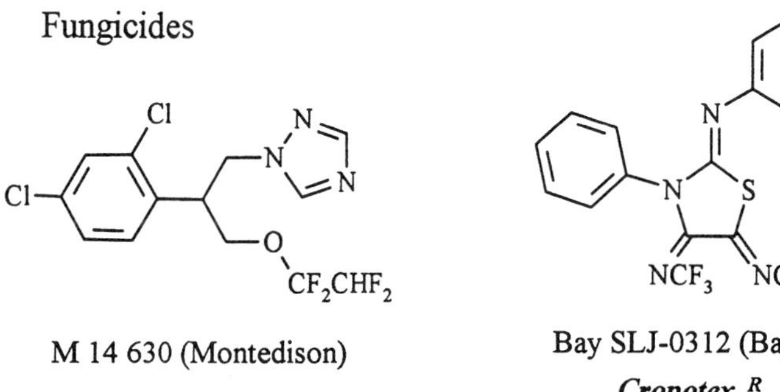

M 14 630 (Montedison)

Bay SLJ-0312 (Bayer)
Cropotex *R*

0065–7719/95/0187–1143$08.00/1

Fungicides (Continued)

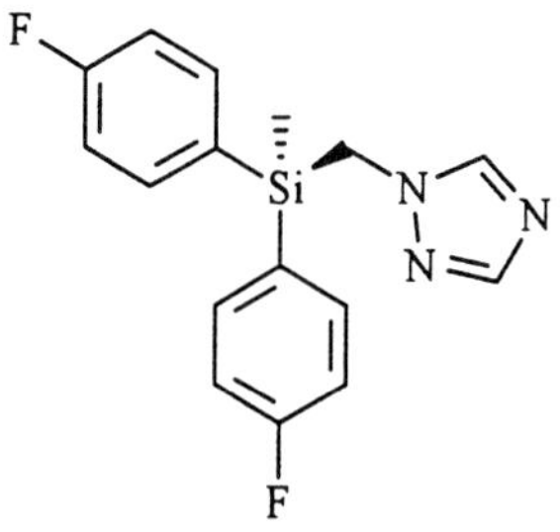

DPX H6573 (DuPont)
Punch *R*

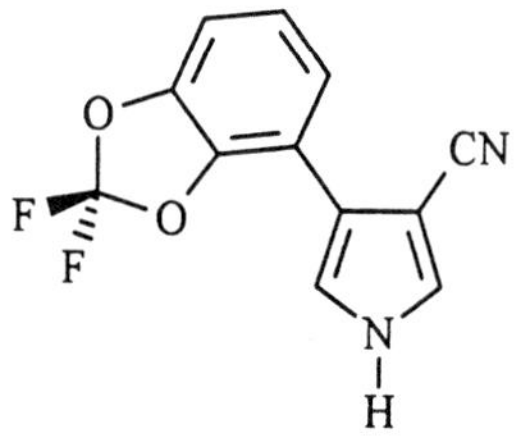

CGA 173 506 (Ciba)
Celeste *R*

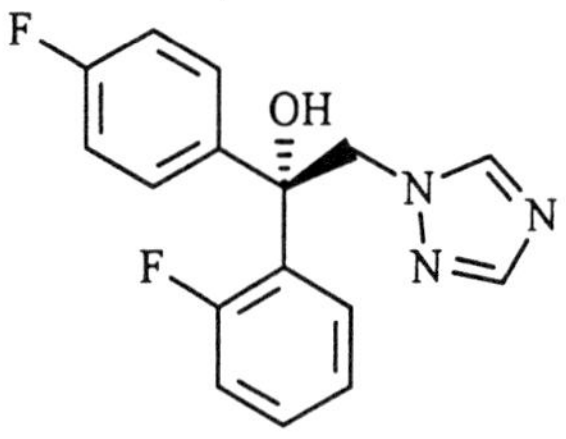

PP 450 (ICI)
Impact *R*

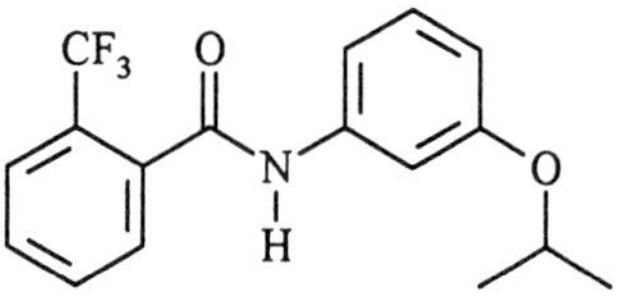

NNF-136 (Nihon Nohyaku)
Moncut *R*

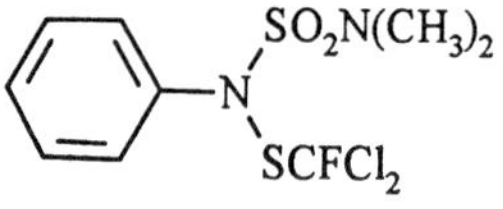

Bay 47 531 (Bayer)
Euparene *R*

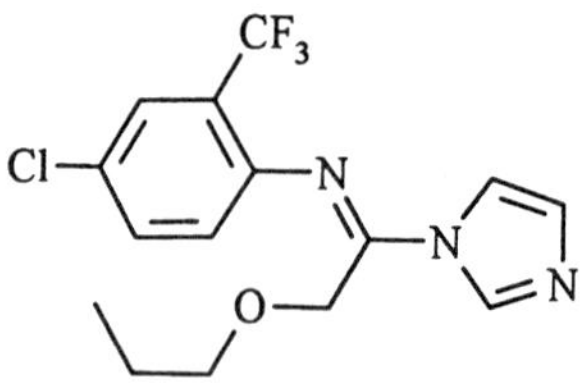

NF-144 (Nippon Soda)
Trifmine *R*

References are listed on pages 1147.

Herbicides

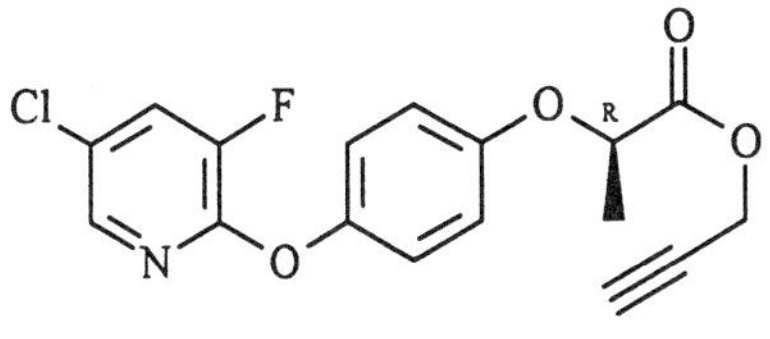

CGA 184 927 (Ciba)
Topik [R]

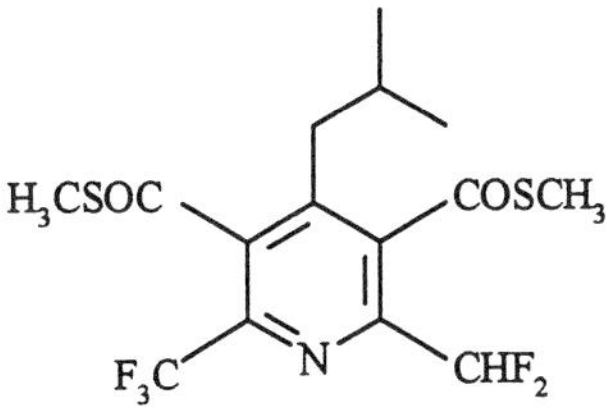

Mon 7200 (Monsanto)
Dimension [R]

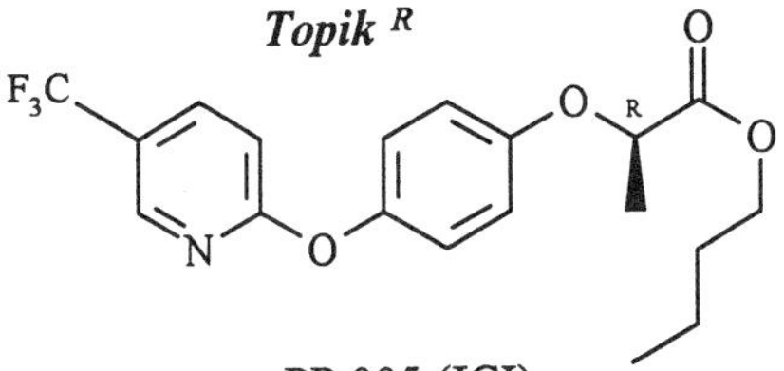

PP 005 (ICI)
Fusilade 5 [R]

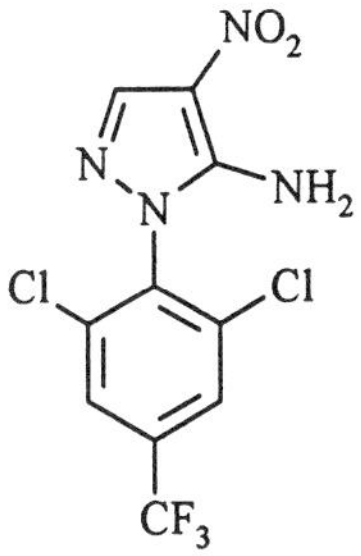

Nicpyraclofen (Bayer)

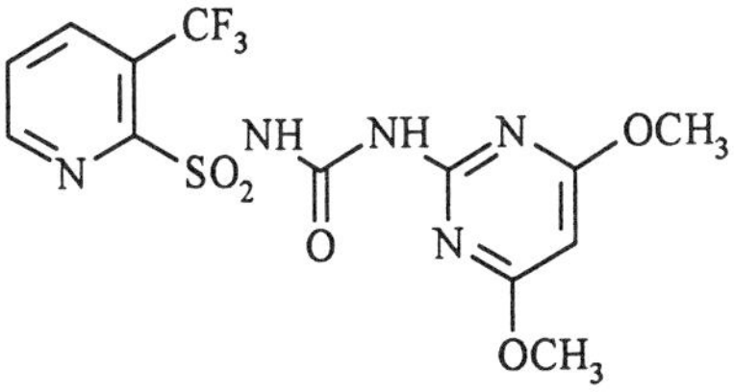

SL 160 (Ishihara)

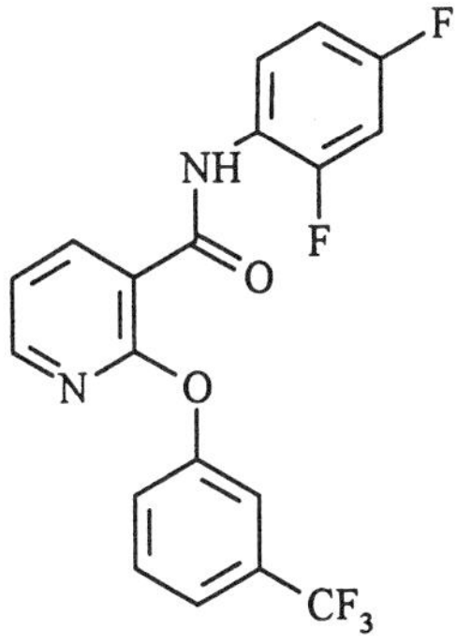

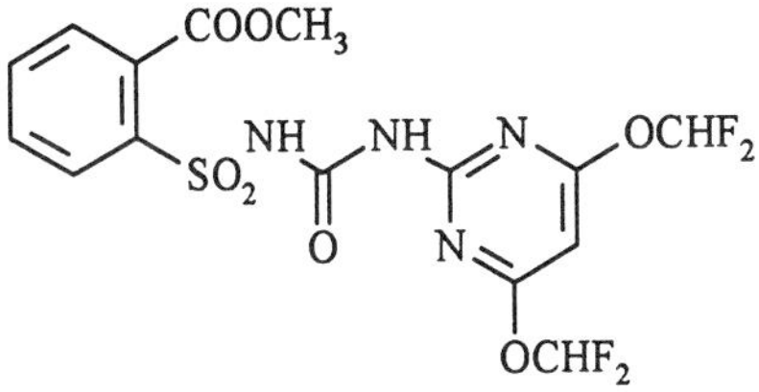

CGA 136 872 (Ciba)
Beacon [R]

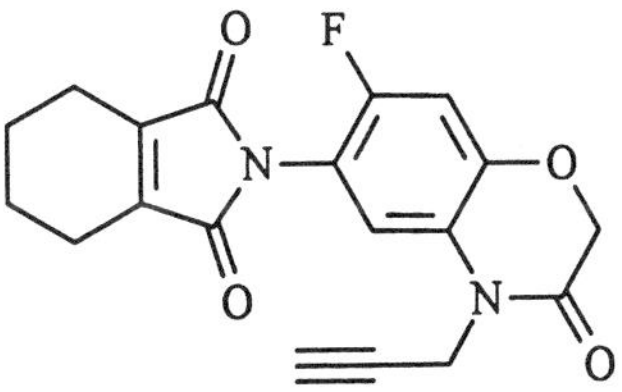

M&B 38 544
(Rhône-Poulenc)

S-53482 (Sumitomo)

References are listed on pages 1147.

Insecticides

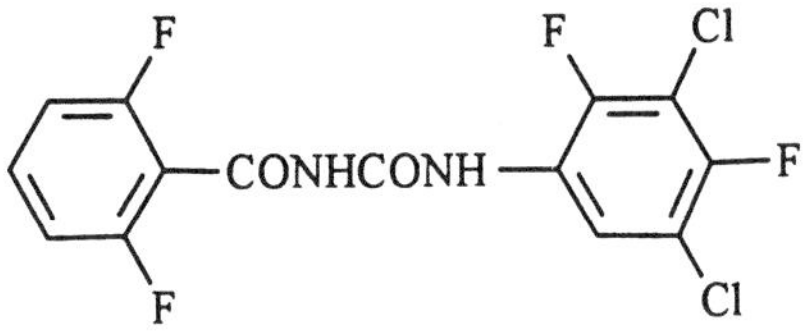

CME 134 (Celamerck)
Dart *R*

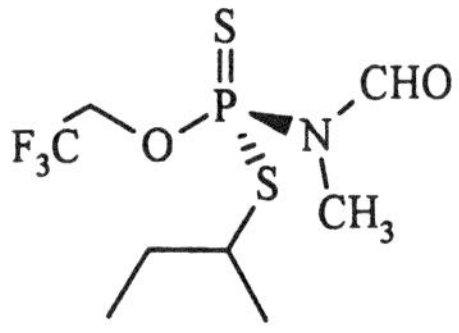

RH 2593
(Rohm&Haas)

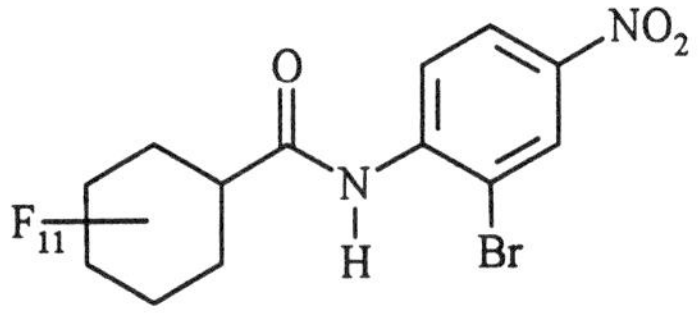

EL 499 (Eli Lilly)

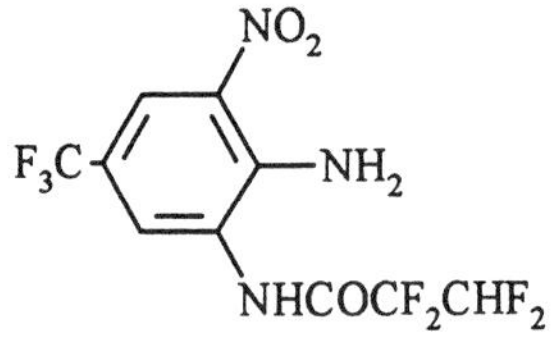

$NHCOCF_2CHF_2$

Nifluridide (Eli Lilly)

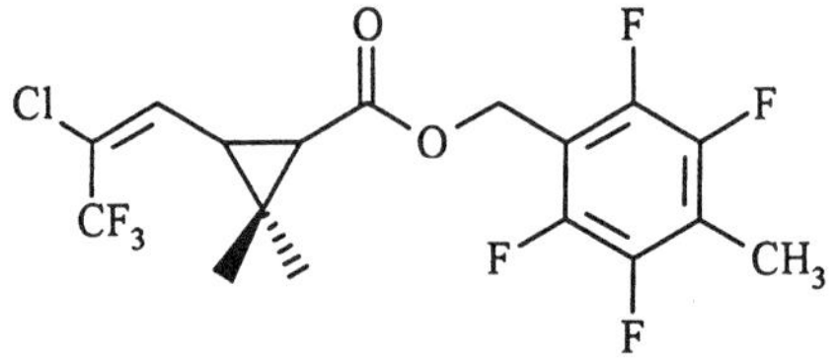

PP 993 (ICI)
Force *R*

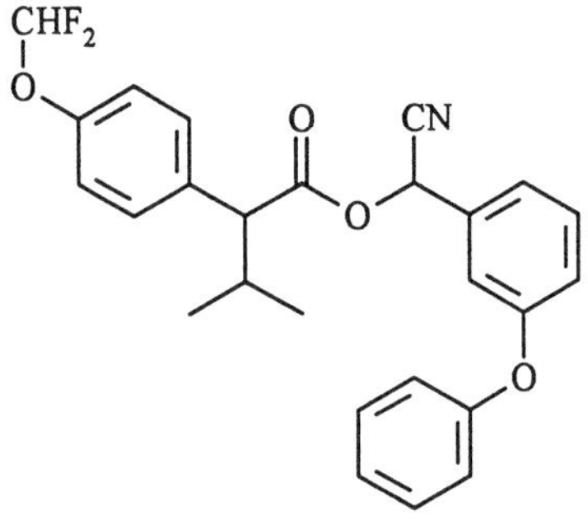

$CF_3(CF_2)_7SO_2NHC_2H_5$

GX 071 (Griffin)
Finitron *R*

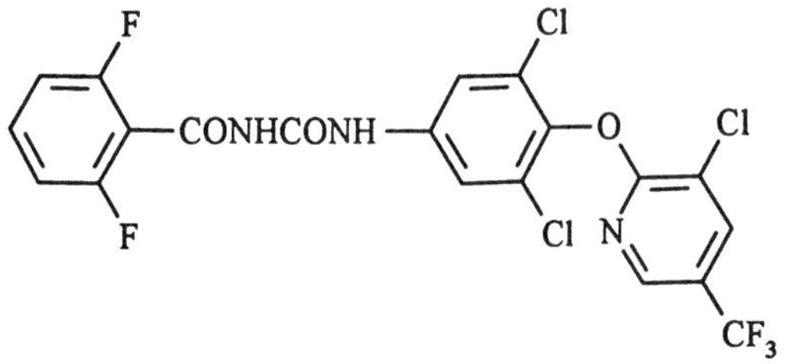

AC 222 705 (Cyanamid)
Cybolt *R*

CGA 112 913 (Ciba)
Jupiter *R*

References for Pages 1143–1146

1. Hudlicky, M. *Chemistry of Organofluorine Compounds;* Ellis Horwood: Chichester, U.K., 1976
2. *Organofluorine Compounds and Their Industrial Applications*; Banks, R. E., Ed.; Ellis Horwood: Chichester, U.K., 1979
3. *The Agrochemicals Handbook,* 2nd ed.; The Royal Society of Chemistry: London, 1988.
4. *Biomedicinal Aspects of Fluorine Chemistry*; Filler, R.; Kobayashi, Y., Eds.; Elsevier Biomedical Press: Amsterdam, Netherlands 1982.
5. Bellus, D. *Chimia* **1991,** *45*, 154.

AUTHOR INDEX

Bibliographic references appear at many places throughout the book. Therefore, this index uses three numbers to direct the user to the desired information. The first, **bold** number indicates the bibliograph page containing the full citation, the second, *italic* number gives the reference number itself, and the final, normal type number(s) lists the text page(s) on which that reference is cited. The monographs and reviews on pages 5–23 have no reference numbers, so the middle column in the index below is blank. A few references are not cited in the text. For those cases, the last column below is blank. The Russian soft sing, transliterated as ' , is ignored in alphabetization.

SUBJECT INDEX

A

B

E

G

H

I

J

K

L

M

Q

R